Schwenkhagen/Meinhold

Wörterbuch Elektrotechnik und Elektronik

Deutsch-Englisch/Englisch-Deutsch

Hans Fritz Schwenkhagen
Heinz Meinhold

Dictionary of Electrical and Electronic Engineering

German-English/English-German

Fifth revised edition

Cornelsen
GIRARDET

Hans Fritz Schwenkhagen
Heinz Meinhold

Wörterbuch Elektrotechnik und Elektronik

Deutsch-Englisch/Englisch-Deutsch

Fünfte, überarbeitete Auflage

Cornelsen
GIRARDET

CIP-Titelaufnahme der Deutschen Bibliothek

Schwenkhagen, Hans-Fritz:
Wörterbuch Elektrotechnik und Elektronik: deutsch-englisch ; englisch-deutsch / Hans Fritz Schwenkhagen; Heinz Meinhold. – 5., überarb. Aufl. – Düsseldorf: Cornelsen Girardet, 1989
ISBN 3-464-49400-4
NE: Meinhold, Heinz:; HST

Bestellnummer 494004
5. Auflage
Druck 4 3 2 / 92

Vertrieb: Cornelsen Verlagsgesellschaft, Bielefeld
Einbandgestaltung: Wenzel Schmidt, Essen
Satz: Satz-Rechen-Zentrum Hartmann & Heenemann, Berlin
Druck: Fürst & Sohn, Berlin
Bindearbeiten: Lüderitz & Bauer, Berlin

ISBN 3-464-49400-4

Vorwort

Auch in der 5., stark überarbeiteten Auflage wurde die Grundkonzeption des seit Jahren bewährten Werkes beibehalten.

Berücksichtigt wurden der Wortschatz der Elektrotechnik und Elektronik, der angrenzenden Fachgebiete sowie grundlegende Termini aus der Mathematik und den Naturwissenschaften. Das Vokabular wurde auf den neuesten Stand gebracht, veraltete sowie sachfremde Begriffe wurden herausgenommen oder ersetzt. Der Inhalt – besonders auf dem Gebiet der EDV und der Mikroelektronik – ist wesentlich erweitert und ergänzt worden.

Zur Unterscheidung der je nach Sachgebiet verschiedenartigen Bedeutung von Fachwörtern wurde die Dezimal-Klassifikation (DK) beibehalten, die sich in Wissenschaft und Technik bewährt hat. Für viele der meist aus dem anglikanischen Sprachbereich einfließenden Fachwörter der Mikroelektronik gibt es keine treffenden deutschen Bezeichnungen. Die Anwender verwenden diese Ausdrücke in der Originalsprache. Im englischen Teil des Wörterbuchs wird für solche Begriffe anstelle einer Übersetzung eine kurze Erläuterung zugefügt.

Leonberg, 1989 *Dr. Ing. Heinz Meinhold*

Inhaltsverzeichnis / Contents

Auszug aus der Dezimal-Klassifikation DK
Excerpt from Universal Decimal Classification U.D.C.

Im Interesse der Raumersparnis sind nur die DK-Zahlen aufgeführt, die im Wörterbuch Verwendung gefunden haben. Viele Bezeichnungen sind aus dem gleichen Grunde gekürzt.

For sake of brevity, the list of numbers has been cut down to the numbers referred to in the dictionary. For the same reason, many headings are abbreviated.

3	**Sozialwissenschaft**	**Social Sciences**
33	Volkswirtschaft	Economics
34	Rechtswissenschaft	Jurisprudence, Law, Legislation (incl. Patents)
35	Öffentliche Verwaltung	Public Administration
351.814.33	Luftverkehrslenkung	Air Traffic Control
5	**Naturwissenschaften**	**Natural Sciences**
51	Mathematik	Mathematics
52	Astronomie, Geodäsie	Astronomy, Geodesy
527	Navigation	Navigation
53	Physik	Physics
531	Allgemeine Mechanik	General Mechanics
532	Mechanik der Flüssigkeiten	Mechanics of Fluids
533	Mechanik gasförmiger Körper	Gases, Pneumatics
534	Mechanische Schwingungen, Akustik	Mechanical Vibrations, Acoustics
535	Optik	Optics
536	Wärmelehre	Theory of Heat
537	Elektrizitätslehre	Theory of Electricity
538	Magnetismus	Magnetism
539	Atomphysik, Molekularphysik	Molecular and Atomic Physics
54	Chemie	Chemistry
55	Geologie, Geophysik, Meteorologie	Geology, Geophysics, Meteorology
6	**Angewandte Wissenschaften**	**Applied Sciences**
61	Medizin	Medical Sciences
62	Ingenieurwesen, Industrie	Engineering, Industry
620.1	Werkstoffprüfung	Testing of Materials
621	Maschinenbau, Elektrotechnik	Mechanical Engineering, Electrical Engineering
621.0	Theorie des Maschinenbaus	Theory of Mechanical Engineering
621.1	Dampfkraftmaschinen	Steam Engines
621.2	Wasserkraft	Water Power
621.3	Elektrotechnik (Siehe detailliertes Verzeichnis)	Electrical Engineering (For details, see itemized list)

621.4	Verbrennungskraftmaschinen	Combustion-Engines
621.5	Pneumatische Energie	Pneumatic Energy
621.56	Kühlung	Refrigeration
621.6	Maschinen und Vorrichtungen zum Fördern von Flüssigkeiten	Apparatus for Conveyance of Liquids
621.7	Spanlose Metallbearbeitung	Non-cutting Shaping
621.791	Löten, Schweißen, Schneiden	Soldering, Welding, Cutting
621.8	Maschinenelemente	Machine Elements
621.876	Aufzüge, Fördermittel	Lifts, Escalators
621.9	Spanende Umformung	Cutting-Shaping
622	Bergbau	Mining
623	Wehrtechnik	Military Engineering
624	Bauingenieurwesen	Civil Engineering
625	Verkehrstechnik (Eisenbahn und Straßen)	Traffic, Transportation (Railway, Roads)
626	Wasserbau	Hydraulic Engineering
628	Gesundheitstechnik (Umweltschutz, Wasserversorgung usw).	Sanitary Technology
629	Fahrzeugtechnik	Vehicles (other than rail-borne)
629.113	Kraftfahrzeuge	Automotive Industry and Technique
629.12	Schiffe	Ships and Shipbuilding
629.13	Luftfahrzeuge	Aircraft Engineering
63	Landwirtschaft	Agriculture
64	Hauswirtschaft	Housekeeping
643	Haushaltgeräte	Household Appliances
65	Handel und Verkehr	Commerce, Traffic
651	Bürowesen	Office Organization
654.19	Rundfunk	Radio Broadcasting
655	Druckerei, Verlagswesen	Printing, Publishing
656.7	Luftverkehr	Air Traffic
66	Chemische Industrie	Chemical Industry
67	Verschiedene Industrien, z. B. Textilindustrie	Various Industries, e. g. Textile Industry
68	dgl.	Various Industries
681	Feinmechanik	Precision Mechanics
681.3	Datenverarbeitung	Data Processing
681.5	Automatisierungstechnik	Automatization
681.84	Plattenspieler, Diktiergeräte, Aufnahme- und Wiedergabegeräte	Record Player, Dictating Machines Tape Recorder, and Playback Equipment
69	Bauhandwerk	Building Industry
696	Installationshandwerk	Wiring and Plumbing

621.3	**Elektrotechnik** (Detailliertes Verzeichnis)	**Electrical Engineering** (Itemized Classification)
621.3.049.76	Mikroelektronik	Microelectronic

621.31	**Allgemeine Elektrotechnik**	**Electrical Engineering in General**
621.311	Kraftwerke	Power Stations

621.313	Elektrische Maschinen	Electric Machines
621.314	Umformung elektrischer Energie	Conversion of Electrical Energy
621.314.1	Gleichstromumformer	d.c. Converter
621.314.2	Wechselstromtransformatoren	a.c. Transformers
621.314.5	Umformung von Wechselstrom in Gleichstrom	a.c./d.c. and d.c./a.c. Conversion, Convertors
621.314.6	Gleichrichtung und Gleichrichter	Rectification and Rectifiers
621.314.63	Trockengleichrichter, Sperrschichtgleichrichter	Metal Oxide Rectifiers, Barrier Film Rectifiers
621.315	Übertragung elektrischer Energie	Transmission of Electrical Energy
621.315.1	Freileitungen	Overhead Lines
621.315.2	Kabel	Cables
621.315.6	Isolierstoffe, Isolatoren	Insulating Materials, Insulators
621.316	Verteilung und Regelung elektrischer Energie	Distribution and Control of Electrical Energy
621.316.5	Schaltgeräte, Steckverbindungen, Fassungen	Switchgear, plug-in device and sockets
621.316.7	Regelung, Regler	Regulation, Regulators
621.316.9	Schutzeinrichtungen	Protection
621.317	Messen, Steuern, Regeln	Measurements, Control Engineering
621.318	Technische Anwendungen des Magnetismus	Applied Magnetism
621.318.5	Relais	Relays
621.319	Technische Anwendungen der Elektrostatik	Applied Electrostatics
621.319.4	Kondensatoren	Capacitors
621.32	**Elektrische Lichttechnik, Lampen**	**Electric Lighting Equipment, Lamps**
621.326	Glühlampen	Incandescent Lamps
621.33	**Elektrische Zugförderung**	**Electric Traction**
621.332	Leitungen, Oberleitungen	Supply Lines, Distribution
621.35	**Elektrochemische Technik**	**Applied Electrochemistry**
621.352	Galvanische Elemente	Primary Cells
621.355	Speicherung elektrischer Energie	Storage, Accumulators
621.357	Technische Anwendung der Elektrochemie. Elektrolyse, Galvanostegie	Industrial Applications of Electrochemistry. Electro-Deposition. Electroplating

621.36	**Thermo-Elektrizität. Elektrowärme**	**Thermo-Electricity. Electric Heating**
621.37	**Elektromagnetische Schwingungen und Impulse**	**Elektromagnetic Waves and Pulses**
621.371	Wellenausbreitung im freien Raum	Free-Space Wave Propagation
621.372	Wellenausbreitung auf Leitungen	Transmission on Lines
621.372.2	Leitungstheorie	Transmission Line Theory
621.372.4	Zweipole	Two-Pole Circuit Elements
621.372.5	Vierpole	Quadripoles
621.372.8	Wellenleiter	Wave Ducts
621.373	Schwingungs- und Impulserzeuger	Wave- and Pulse-Generators
621.374	Impulsbetätigte Schaltungen, Digitalschaltungen	Pulse-operated Circuits, Digital Circuits
621.375	Verstärker	Amplifiers
621.376	Modulation, Demodulation	Modulation, Demodulation
621.377	Verzögerungsglieder, Informationsspeicher	Delay-Circuits, Memories and Stores
621.38	**Elektronik, Photoelektronik, Röntgentechnik**	**Electronic, Optronic, X-Ray-Equipment**
621.382	Halbleiterbauelemente	Semiconductor Components
621.382.2	Dioden	Diodes
621.382.3	Transistoren	Transistors
621.383	Photozellen	Photocells
621.384	Erzeugung und Anwendung aktiver Strahlen	Generation and Application of Active Radiation
621.385	Elektronenröhren	Electronic Tubes
621.386	Röntgenröhren und Zubehör	X-Ray-Tubes and Accessories
621.39	**Elektrische Nachrichtentechnik, Fernwirktechnik**	**Telecommunication, Remote Control**
621.391	Informations- und Signaltheorie	Information and Signal Theories
621.392	Theorie der Leitungen und Vierpole	Theory of Communication Lines. Quadripoles
621.394	Telegraphie	Telegraphy
621.394.1	Verfahren	Modes and Procedures

621.394.3	Vermittlungseinrichtungen	Telegraph Exchanges
621.394.5	Übertragungstechnik	Transmission Technique
621.394.6	Telegraphenapparate und Fernschreiber	Apparatus and Teleprinter
621.395	Telefonie	Telephony
621.395.1	Betrieb und Betriebsarten	Modes and Procedures
621.395.2	Nebenstellenanlagen	PBX = private branch exchange
621.395.3	Öffentliche Fernsprechanlagen, Vermittlungstechnik	Public Telephone Network. Switching Technique
621.395.4	Mehrfachtelefonie	Multiple Telephony
621.395.6	Fernsprechapparate und Zubehör	Telephone Apparatus and Accessories
621.396	Funktechnik	Radiocommunication
621.396.2	Verfahren	Operation Mode
621.396.4	Mehrfachverkehr, Übertragung auf Leitungen	Multi-Transmission, Transmission on Lines
621.396.6	Apparate und Schaltungen	Apparatus and Circuits
621.396.61	Sender	Transmitters. Oscillators
621.396.62	Empfänger	Receivers
621.396.67	Antennen, Erdungen	Aerials. Earths
621.396.69	Bauelemente	Components
621.396.9	Anwendungen: Wetterdienst, Rundfunk, Peildienst	Applications: Meteorological Service. Broadcasting. Direction Finding
621.396.93	Verkehr mit Fahrzeugen, Schiffen, Flugkörpern; Funknavigation, Funkbaken	Intercommunication with Vehicles, Vessels, Missiles; Radio-Navigation. Radio-Beacons
621.396.94	Satellitenfunk	Communication via Satellites
621.396.96	Radar, Funkmeß	Radar
621.397	Fernsehen	Television
621.398	Fernwirktechnik, Fernmeßtechnik	Remote, Control and Telemetering Systems

Teil I
Deutsch-Englisch

A

A Ampere
a-Ader A-wire, positive wire
AB = Aussetzbetrieb intermittent duty, periodic duty, intermittent service
AB-Verstärker class-AB amplifier
abarbeiten to spend; to work off
abgearbeitetes Elektron spent electron (621.385)
Abbau destruction (54); dismantling (621); mining (622)
Abbaubeleuchtung coal-face lighting
abbauen to destruct, to break down; to dismantle; to work, to mine
AB-Betrieb class AB operation
abbeizen to pickle
Abbeizbad pickling bath
Abbiegeblock former (621.313)
abbiegen to bend off; to offset, to deflect
abbiegen nach to turn to (625)
abbilden to copy; to represent (51)
Abbildung illustration, figure, picture; representation (51)
Abbildungsmagnet focusing magnet
Abbildungsmaßstab projection ratio, scale
Abbindedraht bonding wire (621.3)
abbinden to bond (together) (621.3); to set (691.5); to bind
Abbindezeit setting time (69)
abblättern to scale (off)
abblenden to dim
Abblendlicht dim light
Abblendregler dimmer, dimmer control
Abblendschalter dimmer switch
abböschen to scarp, to slope
Abbrand born-up, burning-off; burn off (621.311)
abbrandfest resistant to burning
Abbrandfestigkeit resistance to burning
abbrechen to break (off); to dismantle (621); to stop; to truncate (681.3)
abbremsen to test under brake load, to make a brake test; to brake; to slow down
abgebremstes Drehmoment stalling torque, breakdown torque
Abbremszeit stop time
Abbrenn-... sparking, arcing (621.316.5); (621.791)
Abbrennkontakte sparking contacts, arcing tips
Abbrennschweißung flash welding
Abbrenn-Stumpfschweißung arcing-butt welding, flash-butt welding (US)
Abbrennzeit flashing period
abbrennen to burn off, to burn down
abgebranntes Brennelement spent-fuel element
Abbruch breakdown, dismantling; demolition; abort (681.3)
abdämmen to dam (up)
Abdampf dead steam, waste steam, exhaust steam
abdampfen to evaporate, to volatilize (54)
Abdeck-... cover, capping
Abdecker resist (621.382)
Abdeckfrequenz blanketing frequency (621.396)
Abdeckhaube cover, hood
Abdeckleiste capping (lath)
Abdeckplatte cover plate, cover sheet
abdecken to cap, to cover; to block out; to uncover
abdichten to seal (up), to pack
Abdichtung sealing, packing
abdrehen to turn off (621.316); (621.9)
abdrosseln to choke
Abdrückvorrichtung pushing-off device, forcing-off device
A-Beamter, A-Beamtin „A“ operator (621.395)
Aberration aberration
aberregen de-energize
A-Betrieb class A operation (621.375)
Abfall fall, drop-out (621.3); decay, slope (51); waste (621), refuse (628); offcut (637), scrap (671/3); drop (621.318, 537)
Abfallenergie surplus power
Abfallmaterial waste (material)
Abfallprodukt by-product
Abfallspannung drop-out voltage, drop-away (621.318)
Abfallstrom release current, drop-out current
Abfallverzögerung delay time, release lag, drop-out time lag
Abfallzeit fall time (621.316.9), (621.318), (621.382), (621.39); release time, drop-out time, delay time, fall time; decay time (621.382)
Abfallzeitkonstante fall-time coefficient
abfallen to drop (off), to decline; to go to waste; to slope
abfallende Flanke trailing edge
abfangen to prop, to support (69); to intercept (623); to capture (539)
abfasen to chamfer
abfedern to spring
Abfederung springing (629.113)

abflachen to flatten (out), to level off; to smooth (621.3)
abgeflacht oblate
Abflachung flat
abfließen to leak off, to leak away, to flow off (621.3), (621.6)
abfließender Strom outgoing current
Abfluß drain(age), outflow; discharge
Abfrage inquiry, enquiry (621.395), (681.3); interrogation command (621.398)
Abfrageautomatik automatic scanning unit (621.398)
Abfragegerät interrogator
Abfrageimpuls interrogating pulse
Abfrageklinke answering jack
Abfragelogik interrupt logic
Abfrageplatz attendant's cabinet, operators position
Abfrage- und Schlußlampe answer lamp
Abfrageschnur answering cord
Abfragesender interrogator
Abfragestation inquiry station
Abfragestöpsel answering plug
Abfragetaste enquiry key, speaking key, talking key, answer key
Abfragetechnik polling (681.3)
Abfrageterminal inquiry terminal
Abfragen answering (621.395)
abfragen to answer; to interrogate, to query; to scan (621.398); to poll (681.3)
abfühlen to sense; to read (681.3)
abführen to discharge, to drain away (621.3); to dissipate (536)
Abgabe dissipation (536); discharge (621.3)
Abgabe von elektrischer Ladung discharge of electricity
Abgabe von Leistung power output
Abgangszelle feeder box (621.316)
Abgas waste gas, flue gas, exhaust gas
abgeben to give out, to give forth
abgegebene Leistung power output, delivered power
abgehend outgoing
abgehende Amtsleitung outgoing junction line
abgehende Speiseleitung outgoing feeder
Abgleich adjustment, balance, building out, alignment
90°-Abgleich quadrature adjustment, power-factor compensation (621.317)
Abgleich auf Tonlosigkeit silent balance
Abgleichbedingung condition of balance, equation for balance (621.317)
Abgleichfehler adjustment error, unbalance
Abgleichgenauigkeit accuracy of adjustment
Abgleichkondensator building out capacitor, trimmer
Abgleichmotor balancing motor; trimmer (621.395)
Abgleichpotentiometer balancing potentiometer
Abgleichpunkt balance point
Abgleichtoleranz accuracy of adjustment
Abgleichvorrichtung adjusting device
Abgleichwicklung adjusting winding
Abgleichwiderstand adjustable resistance, balancing resistance
Abgleichzustand balance (621.317)
abgleichbar adjustable
abgleichen to adjust, to readjust, to balance, to compensate, to equalize (621.3); to line up (621.39); to align; to build-out (621.395)
abgraten to remove the burr
abgreifen to pick off, to tap (off), to pick-up (621.317)
abgreifen an to tap off from (621.3)
abgrenzen to delimit, to mark off, to demarcate
Abgriff slider, tap, wiper (621.3); pick-up (621.317), probe (621.319)
mit Abgriff versehen tapped (621.3)
Abgriffarm des Potentiometers potentiometer arm, moving arm of the potentiometer
Abgriffpunkt tapping point
Abgriffschelle variable slider, sliding clip, tapping clip (621.3)
Abgriffspule pick-up coil (621.317)
abhacken to chop off
abhaken to unhook; to tick off, to check off (65)
abhängen to hang up, to abandon a call (621.395); to detach, to uncouple (621.33)
abhängen von to depend (up)on
abhängig dependent
Abhängig-Maximal-Zeit (AMZ) inverse time limit
abhängiges Maximal-Zeitrelais inverse time-lag relay
abhängiger Überstromzeitschutz inverse-time overcurrent protection
abhängige Variable dependent variable (51)
abhängig verzögerte Auslösung inverse time-lag tripping
abhaspeln to reel off, to uncoil
Abhilfe remedy, relief
Abhilfsmaßnahmen (pl.) corrective action
abhören to overhear, to listen in (621.395); to intercept (621.396)
Abhörsicherheit privacy (621.395/396)
abisolieren to strip, to skin
Abisolierzange wire stripper, stripping pliers

abkanten to fold, to bend edges at right angles
abkippen to dump
abklemmen to disconnect (621.3); to squeeze off
abklingen to die (away, down, out), to diminish; to decay (621.3)
abklingen auf to settle down to
Abklingcharakteristik decay characteristic
Abklingkonstante damping coefficient
Abklingresttoleranz ripple to lerance (621.382)
Abklingzeit recovery time (621.396), (621.382), delay time; ringing time (621.397); ripple time (621.382)
abkneifen to pinch off
abknipsen to nip off
abkröpfen to bend, to crank
Abkröpfung bending, crank
abkühlen to cool, to chill; (plötzlich:) to quench
  sich abkühlen to cool down
Abkühlung cooling, chilling; (plötzliche:) quenching
Abkühlzeit cooling period
abkürzen to shorten; to abridge; to abbreviate
  abgekürzte Rechnung short-cut method
  abgekürzter Versuch quick test
Abkürzung shortening; abbreviation
abladen to unload; to dump
Ablage deviation (621.396)
ab Lager off-the-shelf
ablagern to deposit (54), (55)
Ablagerung deposit, deposition (54), (55)
  Ablagerungsvorgang depositing
ablängen to sever, to end, to part, to cut to length
ablassen to discharge, to let off, to drain
Ablaßhahn discharge cock, drain cock
Ablauf drain, drainage; waste outlet
  Ablaufbahn incline
  Ablaufgrube (drainage) sump
  Ablaufhaspel pay-off reel; uncoiler
  Ablaufplan flow chart, flow diagram
  Ablaufspule uncoiling spool, unwinding spool
  Ablaufsteuerung sequence control
  Ablaufverfolgung tracing (681.3)
ablaufen to escape; to flow to waste, to run to waste; to elapse
  ablaufen lassen to run off; to drain
  ablaufende Bürstenkante brush heel (621.313)
  ablaufende Kante leaving edge, trailing edge
  ablaufende Polkante trailing pole horn, trailing tip
ablauffähiges Programm active program
abläuten to ring off (621.395)
AB-Leistung periodic rating (621.313)
ableiten to drain away, to deflect, to by-pass, to branch, to discharge to earth (621.3); to derive (51) (**aus:** from); to drop
  abgeleitete Einheit derived unit
  abgeleitete Funktion derived function
  abgeleitetes Normal substandard, secondary standard
Ableitblech deflector plate
Ableiter arrester, (surge) diverter, lightning arrester (US)
  Ableiterfunkenstrecke arrester gap
  Ableiterstrom discharge current
Ableitkapazität leakage capacitance
Ableitkondensator bypass capacitor
Ableitung derivation, differentiation (51); leak, leakance, leakage (621.3); leak conductance (621.372) **n-te Ableitung** n'th derivative
  Ableitungsbelag shunt conductance per unit length, leakance per unit length
  Ableitungsdämpfung leakance loss
  Ableitungsstrom leakage current
  Ableitungsverlust loss by leakage
  Ableitungswiderstand leak resistance, resistance to earth
Ableitweg current leakage path
Ableitwiderstand bleeder (resistor), diverter (621.316.9)
Ablenk-... deflecting, deflector
  Ablenkanode deflecting plate
  Ablenkblech deflector (plate), baffle plate, jet deflector (621.1)
  Ablenkebene deflection plane (621.397)
  Ablenkempfindlichkeit deflection sensitivity
  Ablenkfrequenz sweep-repetition rate (621.397)
  Ablenkgenerator sweep generator; sweep oscillator
  Ablenkmagnet deflection magnet, beam-bending magnet
  Ablenkmittel deflection means, deflection coils (621.397)
  Ablenkplatte deflecting plate, deflector electrode (621.385); arc buffle, splash buffle (621.314.6)
  Ablenkrolle angle pulley
  Ablenkschaltung deflection circuit, sweep circuit
  Ablenkspule scanning coil, deflection coil, deflection yoke (621.385/397)
  Ablenksteuerung deflection control (621.397)
  Ablenkverstärker sweep amplifier, deflection amplifier
ablenkbar deviable
ablenken to diffract (535); to deflect, to divert

Ablenkung diffraction (535); deflection (621.397); deviation
ablesen an to read by, to read (off) from
Ablesefehler error in reading
Ablesegenauigkeit reading accuracy
Ablesung reading
  Ablesungen machen to take readings, to take observations
  Ablesungsreihe set of readings
  Ablesungszeitraum accounting period (621.317)
ablochen to key punch (681.31)
ablöten to unsolder, to unsweat
Abluft exhaust air, outlet air
abmanteln to unsheathe; to strip (621.315.2)
Abmaß allowance, to lerance, deviation
abmessen to (take the) gauge, to measure off
Abmessung dimension, size; measurement
  Abmessung über alles overall dimension
  Abmessungs-... dimensional
  mit den Abmessungen... measuring...
abmontieren to dismantle, to knock down, to dismount, to take off
Abnahme decrease, diminution (**an:** in), decline, reduction; acceptance, inspection (65); consumption, demand (620); (621.317); decrement (681.3)
  Abnahme der Lichtausbeute efficiency depreciation
  Abnahme der Stromstärke fall of current
  Abnahmeabteilung inspection department (621)
  Abnahmebeamter (quality) inspector, acceptance official
  Abnahmebedingungen acceptance specifications, acceptance stipulations
  Abnahmeprotokoll acceptance certificate, test report
  Abnahmeprüfung acceptance test, proof test, inspection test
  Abnahmeverweigerung rejection, refusal of acceptance
abnehmen to decrease, to diminish (**an:** in), to decline, to wane; to drop, to decline, to fall off (Spannung) (621.3); to pick-up (621.317); to accept, to test for acceptance (65)
  abgenommene Energie energy consumed
  abgenommene Leistung power consumed, consumer's demand
  abnehmbar detachable, removable
Abnehmer consumer (621.316)
  Abnehmeranlage consumer's installation
  Abnehmerleitung trunk
  Abnehmerrolle trolley (621.33)
  Abnehmerseite demand side
  Abnehmersicherung consumer's fuse
  Abnehmerspule pick-up coil (621.317)
abnieten to unrivet
abnutzen to wear off, to wear down, to wear out
  sich abnutzen to rub off (621.9); to get worn out
Abnutzung wear, abrasion; deterioration
  Abnutzung ausgleichen to take up wear
  Abnutzung beim Einlaufen break-in wear
  abnutzungsfest wearable, long wearing
  Abnutzungsfestigkeit wearability
abplatten to flatten
abplatzen to snap off
abprallen to rebound, to bounce
abquetschen to pinch off; to press (67)
Abrechnungszeitraum accounting period (621.31)
abreiben to rub off, to abrade, to abrase
abreichern to deplete (539)
Abreiß-... arcing, interruption, breakaway
  Abreißbogen interruption arc
  Abreißbuchse breakaway receptacle
  Abreißfunke wipe spark, breaking spark
  Abreißkabel breakaway cable
  Abreißkontakt arcing contact (621.316.5)
  Abreißstecker breakaway plug, quick disconnect plug
  Abreißzündung high (low) tension ignition, make-and-break ignition (621.4)
abreißen to break (off), to be interrupted; s. a. **abbrechen** u. **abmontieren**
abrichten to true, to adjust (621)
Abrieb (surface) abrasion
  abriebfest resistant to wear, wear resisting, long wearing
  Abriebfestigkeit resistance to wear, wear resistivity, abrasion resistance, abrasion strengh
abriegeln to block off, to bolt
abrollen to uncoil, to roll off
Abrollhaspel pay-off reel, uncoiler
Abruf fetch (681.3)
abrufen to ring off (621.395); to fetch (681.3)
Abrufen request
  Abrufzeichen proceed-to-send-signal (621.39); start dialling signal, start pulsing signal
abrunden to round (off), to radius
Abrundung rounding; fillet (621.791)
  Abrundungsradius blending radius
abrutschen to whip sidewards, to glide off
Absatz shoulder (621.3)
  Absatz drehen to shoulder down
Absaug-... exhaust, suction

Absauganlage exhaust plant, (fog) suction plant, exhaust installation, suction installation (621.357)
Absaugelektrode drain
absaugen to exhaust, to suck off
abschaben to abrade, to shave (off)
Abschalt-... arcing, interruption, breaking
Abschaltautomatik automatic shut-down equipment
Abschaltdauer arcing time (621.316.9)
Abschalteinrichtung interrupting device, breake device
Abschaltfunke break spark, wipe spark
Abschaltklinke break jack
Abschaltkontakt disconnecting contact
Abschaltleistung breaking capacity, interrupting capacity, interruptive capacity (621.316.5); interrupting rating (621.316.9); interrupting power
Abschaltlichtbogen breaking arc
Abschaltstellung off-position
Abschaltstrom breaking current (621.316.5): interruption current
Abschaltthyristor turn-off thyristor, cut-off thyristor
Abschaltüberspannung break impulse
Abschaltvorgang (circuit-)breaking, disconnecting
Abschaltzeit breaking period, interrupting time, turn-off time, release time
abschalten to interrupt, to break, to de-energize; to turn off, to switch off, to cut off, to interrupt (621.3); (e i n e n Fehler:) to clear (621.3)
abgeschaltete Leitung cut-off feeder, dead line (621.3)
es ist abgeschaltet power is off
Abschaltung circuit-breaking, tripping, opening (of a circuit-breaker), disconnecting, disconnection, isolation
Abschaltung der Last taking off the load
Abschaltung eines Kreises circuit opening, circuit breaking
abschattieren to shade off, to gradate
Abschattung shading; obscuration; obstruction effect
Abschattungsfaktor shadow factor
Abschattungszone blind zone
abscheiden to precipitate (54); (**sich:**) to segregate; to deposit (54)
abscheren to shear (off)
abscheuern to rub off, to abrade
abschilfern to exfoliate, to flake off
Abschirm-... shield(ing), screening
Abschirmbecher can (621.35); screening can, shielding can (621.396)
Abschirmblech screening plate
Abschirmgeflecht shielding harness (621.396)
Abschirmgehäuse screening cabinet
Abschirmkäfig screened cabin, Faraday cage
Abschirmkasten screening box
Abschirmlage shield layer (621.396)
Abschirmring shading ring
Abschirmrohr screening tube, shielding tube
Abschirmwicklung shield winding
Abschirmwirkung screening effect, shielding effect
abschirmen to shield, to screen; to enclose
abgeschirmte Doppelleitung shielded pair
abgeschirmtes Gehäuse shielded enclosure
abgeschirmte Leitung screened cable, shielded line
abgeschirmte Rahmenantenne screened loop aerial, shielded loop aerial
abgeschirmter Raum shielded compartment, test cage
abgeschirmte Zuleitung screened lead
abgeschirmte Zündanlage shielded ignition wiring
Abschirmung shielding, screening
abschleifen to grind off, to abrade
sich abschleifen to wear off
abschleudern to catapult
abschließen to close, to seal; to conclude, to terminate; to lock
abgeschlossen sein mit to terminate in (621.3)
in sich abgeschlossen self-contained
abgeschlossener Hohlraum closed cavity
abgeschlossene Leitung terminated line
abgeschlossener Raum confined space
abgeschlossenes System sealed system
Abschluß conclusion; termination (621.3), seal, sealing (621.385), (621.791)
mit einem Abschluß versehen to terminate
Abschlußbedingung terminal condition (621.3)
Abschlußkabel termination cable
Abschlußkappe end cap
Abschlußkasten terminal box
Abschlußübertrager terminating transformer
Abschlußwiderstand termination, terminating impedance; (o h m s c h :) terminating resistance, terminal resistance
Abschmelzen melting; blowing, fusing, fusion (621.3)
Abschmelzdauer fusing time
Abschmelzelektrode fusible arc-welding electrode
Abschmelzschweißung flash welding

Abschmelzsicherung (safety) fuse
Abschmelzstrom fusing current, blowing current (621.316.9)
Abschmelztemperatur fusing temperature
Abschmelzvorgang fusing
Abschmelzzeit melting time; fusing time
abschmelzen to melt; to blow, to fuse (621.3); to seal (off) (621.385)
abgeschmolzene Röhre sealed valve; sealed tube (US)
abschmieren to lubricate
Abschmierfett lubricating grease
Abschmiernippel oil gun nipple
abschneiden to cut off, to clip off, to pare; to level (**mit:** with)
abschneiden mit level with
abgeschnittenes Stück paring
Abschnitt segment (51); section, element, cut, cutting; period, interval
abschnittsweise link-by-link (621.395)
in Abschnitte zerlegen to section, to sectionalize
Abschnürung pinch-off (621.382)
Abschnürspannung cut-off voltage
abschrägen to bevel (off), to slant, to slope; to chamfer
abgeschrägte Kante bevelled edge
abgeschrägte Pole bevelled poles (621.313)
abgeschrägte Verbindung scarf joint (694)
Abschrägung bevel, bevelling, slanting, slope; chamfer
abschrauben to unscrew, to unfasten, to screw off
abschrecken to quench, to chill
abschruppen to plane off (621.9)
abschuppen to flake off, to exfoliate; to scale
abschürfen to abrade, to abrase; to chafe (61), (621.9)
abschwächen to weaken; to reduce; to attenuate (621.3)
Abschwächung attenuation (621.3); reduction, weakening
Abschwächungskoeffizient attenuation coefficient
abseihen to filter, to strain, to percolate
absetzen to neck, to shoulder down (621)
sich absetzen to settle, to deposit (54)
abgesetztes Gerät remote equipment
absichern to install a fuse, to protect with fuse (621.316); to secure
Absicherung fuse rating, fusing
absichtlich geerdeter Leiter purposely identified conductor (621.315)
absinken auf to fall back to, to settle down to
absinken von ... auf ... to drop off from ... to ...
absitzen lassen to allow to settle
Absolut-... absolute
Absolutgeschwindigkeit absolute velocity
Absolutlader absolute loader (681.3)
Absolutmessung absolute measurement
Absolutwert absolute value
absolut absolute
absolute Adresse absolute address
absolutes Ampere abampere
absoluter Block absolute block (625)
absolute elektrische Einheiten absolute electrical units
absolute Empfindlichkeit absolute sensitivity
absoluter Leistungspegel absolute level
absoluter Nullpunkt (der Temperatur) absolute zero (of temperature)
absolutes Ohm abohm, true ohm
absoluter Pegel absolute level (US), actual level
absoluter permissiver Block absolute permissive block (625)
absolute (Temperatur-)Skala absolute scale
absolutes (c.g.s.-)System absolute system of units
absondern to separate, to isolate, to segregate; to secrete (61)
Absorber-Kühlschrank absorption-type refrigerator
absorbierbar absorbable
absorbieren to absorb
absorbierend absorbing, absorptive
absorbierte Dosis absorbed dose (621.384)
Absorption absorption
Absorption in der Atmosphäre atmospheric absorption (621.396)
Absorption von Schwingungen absorption of vibrations
Absorptionsbande absorption band
Absorptionsfähigkeit absorptivity
Absorptionsfläche absorption cross section (621.396); absorption area, capture area
Absorptionsfrequenzmesser grip-dip (frequency) meter
Absorptionskoeffizient absorption factor, absorptance (535)
Absorptionskreis absorption circuit
Absorptionskühlschrank absorption-type refrigerator
Absorptionsleistungsmesser absorption power meter
Absorptionslinie absorption line

Absorptionsmittel absorbent; space cloth (621.396)
Absorptionsmodulation absorption control (621.396)
Absorptionsquerschnitt absorption cross-section, effective area
Absorptionsrohr absorption modulator (621.396), losser tube (US)
Absorptionsröhre absorber valve (621.396)
Absorptionsspektrum absorption spectrum
Absorptionsverlust absorption loss
Absorptionsvermögen absorptive power, absorption capacity
Absorptionswellenmesser absorption wavemeter
abspalten to rend apart, to split (off)
sich abspalten to split off, to splinter off
Abspann-... stay, guy
Abspannanker stay, guy anchor (621.315.1)
Abspannbock (für Erdseil) dead-end pedestal (for ground wire)
Abspanndraht back stay, stay wire, guy wire (621.315.1); span wire (621.332)
Abspannisolator terminal insulator, tension insulator, shackle insulator, strain type insulator; dead-end insulator (US)
Abspannkausche dead-end thimble
Abspannklemme anchor clamp, strain clamp; dead-end clamp (US); ending fitting
Abspannmast strain structure; dead-end tower, anchor tower, terminal pole, anchor pole
Abspannöse strain ear (621.32)
Abspannpfahl dead man (621.315)
Abspannpflock guy anchor, anchor log (621.315.1)
Abspannseil guy rope, stay rope
Abspannturm supporting to wer
abspannen to step the voltage down (621.314); to stay, to guy (621.315.1); to tie (72)
Abspanner step-down transformer (621.314)
Abspanntransformator step-down transformer
Abspannung straining; dead-ending
Absperr-... isolating, stop
Absperrschieber slide valve, gate-seating action valve
Absperrventil isolating valve, check valve, stop valve, shut-off valve
absperren to isolate, to bar, to block off, to cut off; to fence off
Absperrung isolation, seclusion, stoppage
abspielen to playback (681.84)
abspritzen to spray, to cleanse, to wash
abspulen to unreel, to uncoil, to reel out, to run off
Abstand distance, space, spacing; clearance, separation; interval; gap
Abstand der Resonanzspitzen spacing of the humps
auf Abstand gewickelte Wicklung spaced winding
im Abstande x unter... distand x below...
im Abstand von... voneinander spaced... apart
in Abständen von... anordnen to space at intervals of...
mit Abstand anordnen to spread apart
mit Abstand gewickelt space wound (621.317)
mit Abstand voneinander hinlegen to space
Abstand von Mitte zu Mitte centre-to-centre distance
Abstandsbuchse spacing bush
Abstandshalter spacer, spreater; bead (621.315.2); double strain yoke (621.315.1)
Abstandshalterung durch Scheiben bead support (621.315.2)
Abstandsmesser range meter
Abstandsring spacer ring
Abstandsscheibe disc spacer
Abstandsstück distance piece, spacer, spacing bar
Abstandszünder proximity fuse (623)
abstechen to tap, to run off (621.9)
abstecken to stake (off), to set out (52); to step off (51)
abstehen to stand off (**von:** from); to grow stale
abgestandenes Wasser stale water
absteifen to support, to stay, to stiffen
abstellen to cut off (621.1); to stop (up) (621.1/4); to switch off, to turn off, to disconnect (621.3)
Abstellknopf stop button
Abstellstab shut-off rod, emergency rod; scram rod (US) (539)
Abstimm-... tuning
Abstimmanzeiger tuning indicator (621.396.62)
Abstimmanzeigeröhre cathode-ray tube, tuning indicator
Abstimmbereich tuning range, tunable range
Abstimmblindspritze tuning stub
Abstimmdiode tuning variable capacitance diode
Abstimmeinheit tuner
Abstimmeinrichtung tuning control, tuning equipment
Abstimmfehler tuning error
Abstimmfrequenz t.r.f. = tuned radio frequency
Abstimmglied tuner (621.396)

Abstimmkern tuning slug
Abstimmkern aus Eisenpulver iron-dust tuning slug
Abstimmknopf tuning control (621.396.6); tuning knob
Abstimmkondensator tuning condenser, tuning capacitor
Abstimmkreis tuning circuit
Abstimmkreisglied tuned circuit element
Abstimmpeilfehler plain tuning error (621.396.9)
Abstimmpunkt tuning point
Abstimmschärfe sharpness of tuning; tuning precision
Abstimmskala tuning scale
Abstimmspule tuning coil, tuning inductor
abstimmbar tunable
abstimmen to tune, to synchronize
abstimmen auf to resonate with (534); to tune in for (621.396.6)
abgestimmt sein auf to be unison with
abgestimmter Anodenkreis tuned anode circuit, tank circuit, plate-tuned circuit
abgestimmtes Düppelband tuned rope
abgestimmtes Filter tuned filter
abgestimmter Gitterkreis tuned grid (circuit)
abgestimmter HF-Kreis tuned r.f. circuit
abgestimmter HF-Verstärker tuned r.f. amplifier
abgestimmter Hohlraumresonator tuned resonating cavity
abgestimmter Kreis tuned circuit
abgestimmte Kreise isochronous circuits
abgestimmtes Relais tuned relay
abgestimmter Siebkreis tuned filter
abgestimmter Tonfrequenzruf harmonic selective ringing
abgestimmter Verstärker tuned amplifier, resonance amplifier
Abstimmung tuning; synchronizing, balancing, setting
Abstimmung auf tuning in for
abstoppen to bring to a stop
abstoßen to repel; to push off (621.33), (629.12)
sich abstoßen to repel, to fly apart
abstoßend repellent, repulsive
abstoßende Kraft force of repulsion
sich abstoßende Pole repelling poles
Abstoßung repulsion
Abstoßungskraft repulsive force, repelling power
abstrahlen to radiate, to transmit
abgestrahlt radiated, transmitted, outgoing (621.396)
Abstrahlung radiation
Abstrahlwinkel angel of departure, tilt
Abstreifblech stripper plate
abstreifen to strip (off)
abgestreiftes Atom stripped atom
abstufen to gradate, to graduate; to grade; to step (62)
abgestufte Isolation graded insulation, co-ordinated insulation
Abstufung gradation, graduation
abstumpfen to truncate (51); to neutralize (54); to blunt (621)
abgestumpfte Schneidkante dull cutting edge
abstützen to buttress, to strut
absuchen to scan
Abszisse abscissa (pl.-ae)
Abszissenachse axis of abscissae
Abtast-... scanning, sensing, sampling
Abtasteinrichtung scanning gear, sensing device
Abtastfleck scanning spot
Abtastfolge scanning sequence
Abtastfrequenz sampling frequency
Abtastgerät scanner, reader, sensing head
Abtastgeschwindigkeit scan speed, scanning speed, scanning rate, spot speed
Abtastintervall sampling period
Abtastkopf scanning head
Abtastregelsystem sampled-data control system
Abtastschaltung scanning circuit
Abtast- und Halteschaltung sample and hold circuit
Abtastspitze probe
Abtaststeuerung sampling control
Abtaststift follower, former pin (621.91)
Abtaststrahl scanning beam
Abtastzeile scanning line
abtasten to scan (621.397); to sense, to feel, to pick-up; to trace
Abtaster scanner, reader; sampler
Abtastung scanning (621.397), sampling
Abtastung mit Zeilensprung interlaced scanning
Abtastung ohne Zeilensprung progressive scanning
abteilen to partition, to sectionalize, to divide; s. auch **abtrennen**
Abteilung department
Abteilungsleiter head of department
abtragen to mark off, to measure off, to plot (51)
abtreiben to drive off (54)
Abtreiber still (66)

abtrennen to separate, to detach, to sever; to isolate (621.3); to disconnect (621.3), (625)
abgetrennter Teil isolated section
Abtrennung isolation, disconnection (621.3); separation
Abtrieb power take-off (621.8)
Abtriebsdrehzahl speed of driving end
Abtriebsleistung power take-off
Abtriebsmoment torque output (621.3)
Abtriebsseite indrive end
Abtriebswelle driven shaft
Abtrift drift, leeway (527)
abtropfen to drip down, to drip off
abtropfende Lösung drippings (pl.) of solution
Abtropfenlassen drainage of excess liquid
AB-Verstärker class AB amplifier
abwägen to balance, to poise
Abwälzverfahren hob method
abwandeln to modify
Abwandern der Frequenz frequency drift
Abwandern des Nullpunktes zero shift, zero drift
Abwärme waste heat
Abwärts... down..., downward...
Abwärtsgehen fall
Abwärtshub downward stroke
Abwärtskompoundierung decompounding
Abwärtsregelung reduction (of speed)
Abwärtsschweißung downhand welding
Abwärtstransformator step-down transformer
Abwärtsübersetzung step-down ratio
abwaschbar washable
Abwasser waste water
abwechselnd alternating; in turns, by turns
abwechselnd eintreten to alternate
abwechselnd verschiedenes Vorzeichen haben to alternate in polarity
abweichen to depart, to diverge, to deviate, to differ (**von:** from)
abweichend von unlike
Abweichung departure, divergence, deviation, difference; differential (51)
Abweichung in der Frequenz difference in frequency, frequency deviation
Abweichung vom Gleichgewicht unbalance (621)
Abweichung von der Mittellinie excentricity
Abweichung vom Nennwert deviation from the rated value
Abweichung von der Norm variation from standard, anomaly
Abweichung von Null null drift
Abweichung des Nullpunkts zero error
Abweichung vom Sollwert deviation from reference value
abweisen to deflect
Abweiser deflector
Abweisblech deflector plate (621.2); flashing (69)
Abweiseranode beam-forming plate
abwerfen to throw off; to lock out (621.395); to release (621.318)
abwerten to degrade, to depreciate
Abwertung degradation, depreciation
Abwesenhheitsdienst absent subscriber service (621.395.3)
abwetzen to fray
abwickeln to uncoil, to uncurl, to wind off
Abwicklung einer Fläche development of a surface
Abwurf throwing off
Abwurf auf den Bedienplatz call-forwarding to attendant
Abwurf auf Teilnehmerschaltung line lock out
Abwurfimpuls reject pulse
Abwurfleitung line for release
Abwurfwicklung release winding
abwürgen to twist off; to stall (621.4)
abzapfen to draw off (54)
Abziehbild decal (621.317); transferpicture
Abziehfeder pull-off spring
Abziehvorrichtung puller, draw-off tackle; extractor
Abziehzange stripping pliers
abziehen to subtract (51); to draw off, to drain off (54); to seal off (621.385); to hone (621.9)
Abzug fume cupboard (54); damp shaft, outlet (621); print (77); copy (65)
Abzugshaube vent hood
abzwacken to snap off
Abzweig subcircuit, side chain, stub (621.3); branch (line) (621.3); branch-T (621.3), (66); tap (621.395.4)
Abzweigdose junction box, access fitting (621.315)
Abzweigklemme branch joint, tee joint
Abzweigleitung branch conductor, branch circuit, branch line
Abzweigmuffe branch joint, T-joint, Y-joint, multiple joint
Abzweigpunkt branching point
Abzweigschalter branch switch
Abzweigstecker socket-outlet adapter
Abzweigstück pipe tee
Abzweigverstärker drop repeater
Abzweigwiderstand shunt resistance
abzweigen to branch off

abzwicken to nip off, to pinch off
achromatisch achromatic
Achromatismus achromatism
Achse axis (pl.: axes), axle
auf der Achse sitzend axle mounted
aus der Achse verschoben off the axis
Achsabstand centre distance, wheel base
Achsaufhängung axle suspension (621.33)
Achsbruch axle fracture
Achsbüchse axle box (621.33)
Achsdruck axle load, axle pressure
Achsdynamo axle generator (system) (621.33)
Achsdynamoregler axle-generator regulator
Achsdynamo-Umschalter axle-generator pole-changer
Achsenschnitt crossover
Achsenwinkel axis angle (621.8)
Achslager journal bearing, axle bearing
Achslast axle load, weight per axle
Achsmotor gearless motor, direct-drive motor (621.33)
Achsreibung journal friction
Achsrichtung, in along axis
Achsschenkel axle... journal
Achsstand axle base
Achsstummel stub axle, wheel stud
Achszähler axle counter
Achteck octagon
achteckig octagonal
Achtercharakteristik bilateral characteristic, octogonal characteristic (534)
Achterfeld eight-element dipol array (621.396)
Achterkurve, Achtermuster figure-eight pattern; figure of eight reception (621.396)
Achterschaltung superphantom circuit (621.394)
Achterschutz protection by balanced relays
Achtschlitzanode eight-segment plate
Achtschlitzmagnetron eight-segment magnetron
actio et reactio action and reaction
Adapter adapter
Adcock-Peiler Adcock direction finder
addierbar additive, summable
addieren to add, to totalize, to sum (up)
sich addieren zu to sum up to
Addierschaltung adder (circuit)
Addierwerk (accumulating) register, adder
Addition addition, summation
Additionsbefehl add instruction
Additionszeichen sign of addition
additive Mischung additive mixing
Ader vein (61); core, strand, wire, conductor (621.315)
Aderabschirmung core screening
Aderpaar pair (of wires)
Adervierer quad
Adhäsion adhesion
adhäsiv adhesive
Adiabate adiabatic line
adiabatisch adiabatic
adiabatische Expansion adiabatic expansion
adiabatische Kompression adiabatic compression
Adjunkte adjunct (51)
Admittanz admittance
Adresse address
Adreßbus address bus
Adresse von Adresse indirect address (681.31)
Adressenteil address part, address section (681.31)
Adreßregister address register
Adressierung addressing
Adressierungsart addressing mode
adsorbieren to adsorb
adsorbierte Stoffe adsorbed matter (sing.)
Adsorption adsorption
Adsorptionsvermögen adsorptive power
A/D-Wandler A/D converter
A.E. = Ångström-Einheit Å.U. = Ångstrom Unit = 10^{-8} cm
Aerodynamik aerodynamics (pl.)
aerodynamisch aerodynamic
Aerometer aerometer, hydrometer
Aeronautik aeronautics (pl.)
Aerosol aerosol
Aerostatik aerostatics (pl.)
Affinität affinity
Agens agent
Aggregat aggregate
Aggregatzustand state (of aggregation)
Ah = Amperestunde ah = ampere hour
Ah-Zähler quantity meter, coulombmeter
Ähnlichkeit similarity; analogy
Ähnlichkeitssatz law of similarity
Aiken-Code Aiken-Code
Akkord common chord, accord (78)
Akkordarbeit task work, piece work
Akku = Akkumulator (s. dort)
Akkulampe accumulator lamp
Akkumulator accumulator (621.3), (621.6), (681.3); storage cell, secondary cell (621.3)
Akkumulatorenbatterie accumulator battery, storage battery, secondary battery, secondary cell
Akkumulatorbleiplatte accumulator lead plate
Akkumulatorengefäß accumulator container
Akkumulatoren-Gitterplatte accumulator grid plate

Akkumulatorenglas containing cell
Akkumulatorladegerät accumulator charger, battery charger
Akkumulatorenlokomotive accumulator locomotive, storage-battery electric locomotive
Akkumulatorenplatte accumulator plate
Akkumulatorensäure accumulator acid
Akkumulatorentriebwagen battery rail car
Akkumulatorenwagen battery-driven vehicle
Akkumulatorenzelle accumulator cell, battery cell
aktinisch actinic
Aktion action
Aktionsradius range
Aktionsspannungen neuroelectricity (sing.), neuroelectric potentials
Aktionsstrom action current
aktiv active
aktive Amperewindung operating ampere turns
aktive chemische Masse active paste (621.355)
aktive Fehlerdämpfung active balance, return loss
aktiver Filter active filter
aktive Fläche junction surface (621.382)
aktives Gewicht weight of core-and-coil assembly (621.314.2)
aktive Leiterlänge active wires (pl.)
aktiver Niederschlag active deposit
aktives Material active material (621.35)
aktiver Pegel active level
aktiver Pegelübergang active transition
aktive Reserve equipment in parallel operation
aktiver Strahler exciter
aktiver Stromkreis (mit Energiequelle) active electric network, source
aktiver Teil (eines Reaktors) loaded region (of a reactor)
aktiver Zweipol two-terminal source (621.3)
aktivieren to activate
Aktivierung activation
Aktivierungsenergie activation energy, intensity of activation
Aktivierungsquerschnitt activation cross section
Aktivität activity
Aktivkohle activated carbon
aktualisieren to update
aktuelles Befehlsregister current instruction register
Akustik acoustics (pl.), sound work
akustisch acoustic (al)
akustische Anzeige audible indication, audible signal
akustischer Blindwiderstand acoustic reactance
akustisches Echo sound echo
akustische Größe acoustical quantity
akustischer Koppler acoustic coupler
akustisches Ohm mechanical ohm
akustische Rückkopplung acoustic feedback
akustischer Scheinwiderstand acoustic impedance
akustische Schwingung acoustic (al) vibration
akustisches Signal sound signal, audible signal
akustisches Signal auslösen to sound an alarm
akustischer Speicher acoustic memory
akustische Verkleidung acoustic lining
akustische Verzögerungsschaltung sonic delay line
akustischer Wandler acoustical electrical transducer
akustischer Widerstandsoperator acoustic impedance
akustischer Wirkwiderstand acoustic resistance
Akzeptor acceptor
Akzeptorverunreinigung acceptor impurity
Alarm alarm, warning
Alarmanzeige alarm unit, alarm display
Alarmeinrichtung alarm system, alarm device
Alarmempfänger personal alerter
Alexanderson-Generator Alexanderson alternator
Algebra algebra
algebraisch algebraic (al)
algebraisch behandeln to deal algebraically with
algebraische Funktion algebraic function
algebraische Gleichung algebraic equation
ALGOL ALGOL = algorithmic language
Algorithmus algorithm
Alkali alkali
Alkalielement caustic soda cell
Alkalimetalle alkali metals
Alkalizelle potassium cell; alkaline photocell
alkalin alkaline
alkalisch alkaline, alkalious
alkalischer Akkumulator alkaline accumulator
alkalische Batterie Edison battery, alkaline (storage)battery
alkalischer Sammler alkaline accumulator, alkaline cell, alkaline (type) battery, steel alkaline battery
Alkohol alcohol, spirit
allein weiterlaufend self-propagating
allgemein general
allgemeine Beleuchtung multidirectional illumination; general lighting; universal lighting

allgemeine Feldtheorie unified field theory
allgemeine Lösung general solution, complementary solution (51)
allgemeine Regel general rule
allmählich gradual; by degrees, by and by
allmählich abklingen to tail off (621.385)
allmählich abnehmen to die away, to die down, to die out
allotropisch allotropic(al)
Allpass all-pass network
Allrichtungsfunkfeuer omnidirectional beacon
Allstrom-... all-mains, universal mains
Allstromempfänger a.c./d.c. receiver, all-mains receiver, universal receiver
Allstrommotor universal motor
Allwellen... all wave
Allwellenempfänger all-wave receiver (621.396)
Allwellenfilter all-pass network
Allwetterlandung all-weather landing
Allzweck-... all-purpose, general-purpose, allround
Allzweckmotor general-purpose motor
Alpha alpha
alphanumerisch alphanumeric
Alphateilchen alpha particle
Alphastrahlen alpha rays
Alphastrahlung alpha radiation
Alt-..., alt old
alt werden to age
alter Fachmann old hand
Alteisen scrap
Altgrad degree (51)
Altkupfer reclaimable copper, scrab copper
Altmaterial old material
Altöl waste oil
altern to age
alternieren to alternate
alternierender Burst alternating burst
Alterung ageing, seasoning, curing, maturing (621.318.2); weakening (621.3)
durch Alterung brüchig werden to age-harden (621.315)
Alterung eines Dauermagneten weakening of a permanent magnet
Alterungsbeständigkeit resistance to ageing
Alterungshärtung age hardening
Alterungsprüfung ageing test
Alterungsriß season crack (621)
Alterungsverfahren ageing process
Altkupfer reclaimable copper, scrap copper
Alu = Aluminium alu = aluminium
Aluminium aluminium; aluminum (US)
Aluminium schmiert beim Drehen aluminium lugs when turned
Aluminiumband strip aluminium
Aluminiumbecher aluminium can (621.396.6)
Aluminiumdraht aluminium wire
Aluminium-Elektrolyt-Kondensator aluminium-electrolytic capacitor
Aluminiumfreileitung aluminium overhead line
Aluminiumguß aluminium castings (pl.), cast aluminium
Aluminiumlegierung aluminium alloy
Aluminiumknetlegierung wrought aluminium alloy
Aluminiumleiter aluminium conductor
Aluminiumlot aluminium solder
Aluminiummantel aluminium sheath, aluminium cover
Aluminiummantelkabel aluminium-covered cable, aluminium-sheathed cable
Aluminiumofen aluminium furnace
Aluminiumoxid alumina
aluminothermisches Schweißverfahren thermite welding
Amalgam amalgam
Amalgamzelle mercury cell
amalgamieren to amalgamate
Amalgamierungsbad blue dip (621.357)
A-Mast A-pole
Amateur amateur
Amateurfunkdienst amateur radio
Amateursender amateur transmitter
Ameisensäure formic acid
Amerika-Kabel transatlantic cable
Amerikanische Drahtlehre A.W.G. = American Wire Gauge
Amerikanischer Standard-Code für Informationsaustausch American Standard Code for Information Interchange = ASCII
Amin amine
Ammoniak ammonia
Ammonium ammonium
amorph amorphous
amorpher Halbleiter glassy semiconductor, ovonic
Ampere ampere
Amperemeter ammeter, ampmeter, amperemeter
Amperemeter mit Nebenwiderstand shunted ammeter
Amperemeter mit Thermoumformer thermocouple ammeter
Ampèrescher Rahmen Ampère's table

Ampèresche Schwimmerregel Ampère's rule
Ampere-Stäbe ampere-wires, ampere-conductors (621.313)
Amperestunde ah = ampere hour
Amperestunden-Kapazität ampere-hour-capacity (621.35)
Amperestundenzähler amperehour meter, quantity meter, coulombmeter
Amperewaage current balance
Amperewindungen ampere turns = amp turns = a.t.
amphoter amphoteric
amphoterer Elektrolyt amphoteric electrolyte
Amplitude amplitude (621.3); modulus (51)
Amplitudenabhängigkeit amplitude response
Amplitudenabnahme reduction in amplitude
Amplitudenanstieg amplitude increase
Amplitudenbegrenzer (amplitude) limiter, clipper, peak limiter (621.396)
Amplitudenbegrenzung chopping (621.396)
Amplitudenfrequenzgang amplitude-frequency response
Amplitudengang amplitude response
Amplitudenhub amplitude sweep, amplitude swing
Amplitudenmodulation = A.M amplitude modulation = a.m.
amplitudenmoduliert amplitude-modulated
Amplitudenregelung amplitude control
Amplitudenresonanz amplitude resonance
Amplitudenschwankung amplitude swing
Amplitudenselektor amplitude discriminator
Amplitudensieb synchron separator (621.397)
Amplitudentastung amplitude keying
Amplitudenverzerrung amplitude distortion
Amplitudenzeitwert amplitude-time value
Amplidynschaltung amplidyn circuit
Amt board, office; exchange (621.395)
Amtsanschluß public exchange connection
Amtsbatterie exchange battery
Amtsbündel interoffice trunk (621.395)
Amtsfreizeichen dial to ne
Amtsgespräch public exchange call, city call
Amtskabel interoffice trunk
Amtsleitung trunk line, (direct) exchange line, subscriber's line
Amtsnummer exchange number
Amtstrennung telephonist release
Amtsverbindung junction line, junction circuit (621.395)
Amtswähler office selector, code selector, junction selector
Amtszeichen dialling tone (621.395)
AMZ-Abschaltung inverse time lag circuit breaking
AMZ-Auslösung inverse time lag circuit breaking
AMZ-Glied inverse time lag relay
analog analogic (al) (**zu:** to); analog, analogous, analogue
analoge Anzeige analog indication
Analog-Digital-Umsetzer digitizer, analog-digital converter
Analog-Digital-Umsetzung analog-digital conversion
Analog-Frequenz-Umsetzung analog-frequency conversion
Analogmeßgerät analog measuring instrument
Analogrechner analog computer
analoger Schaltkreis linear circuit
Analysator analyser (535)
Analyse analysis (51), (54)
Analysenverfahren method of analysis (54)
analysenrein analytically pure
analysieren to analyse
analytisch analytic (al)
analytische Funktion analytical function
analytische Geometrie analytical geometry, co-ordinate geometry
Anaphorese anaphoresis
anätzen to etch (66)
anbieten to offer (621.395)
anbinden an to fasten to, to tie to; to keep pointed at (621.396.67)
anblasen to start up (66)
Anblasekühlung fan cooling, forced-air cooling, forced ventilation (621.313/314)
mit Anblasekühlung fan-cooled, fan-ventilated
anbrennen to ignite, to fire; to catch fire
anbringen to apply, to affix, to mount, to attach (**an:** to)
Anderson-Brücke Anderson's bridge
Anderthalbseitenband-Verfahren sesqui-sideband system
Änderung change, alteration; modification
Änderungsgeschwindigkeit rate of change
andrehen to screw on; to switch on (621.396), to turn on (621.316), (628); to start up (621.4)
Andrehmotor pony motor
andrücken to press on, to force on
Andrückfeder holding spring, pressure spring
Andrückplatte holding-down plate
Andrückschraube holding-down screw
aneinanderfügen to join together
aneinanderreihen to arrange in tandem
aneinandergereiht in tandem, end to end

aneinanderschließen to (inter)connect, to couple: to lock together
Anemometer anemometer, wind gauge
Aneroidbarometer aneroid barometer, diaphragm gauge
anfachen to fan, to blow; to reinforce (534)
anfahren to start up
Anfahr... starting
Anfahrdrehmoment starting torque
Anfahrverlust accelerating loss (621.33); starting losses (pl.) (621.313)
Anfahrvorgang excursion (539)
Anfahrzugkraft starting tractive force
Anfahrzugleistung starting tractive power
Anfang beginning; start, initiation, inception; source (621.392)
Anfang eines Abzweiges entry (621.3)
Anfangsbedingungen initial condition
Anfangsdruck initial pressure
Anfangsgeschwindigkeit intial velocity
Anfangsglied leading term (51)
Anfangsgründe first principles
Anfangspermeabilität intial permeability
Anfangspunkt starting point, start position; point of origin (51)
Anfangsreaktanz initial reactance; direct axis subtransient reactance (621.313)
Anfangsspannung initial voltage
Anfangswert initial value
anfeuchten to moisten, to wet, to damp
anflanschen an to flange to
Anflug touch, tinge; approach (656.7)
Anflug mit automatischer Peilung ADF approach
Anflugaufschaltung automatic approach coupling
Anflugfolge approach sequence
Anfluggrundlinie centre line, flight base line
Anflugradar approach radar
Anflugrichtungsfeuer approach light
Anflugschneise approach lane
Anflugüberwachungsradar approach control radar
Anflugweg approach path
Anflugwinkel angle of approach
Anflugwinkelfeuer angle-of-approach light
anfordern to request (621.395) (621.398)
anfressen to attack, to corrode; to pit (628.15)
angefressener Kontakt pitted contact
a-n-Funkleitbake equisignal radio range beacon (656.7)
angebbar in expressible in terms of
Angebot traffic offert (621.39); offer (33)
angelagertes Elektron trapped electron
angelochter Lochstreifen chadless tape
angreifen to corrode; to attack
Angriffspunkt point of application, working point
anhaften to adhere, to stick to
anhalten to stop, to arrest
anheben to lift, to raise; to boost (621.3) (US); to recondition (621.396); to open (621.6); to accentuate (621.376)
angehobene Trägerwelle reconditioned carrier (621.396)
anheizen to fire (up), to heat up; to start up, to stoke up (621.1)
Anheizzeit heating time; cathode-heating time (621.385); firing time (64); warming-up time
Anheizung initial heating
Anhydrid anhydride
Anilin aniline
Anion anion
anionaktiv anionic
anisotrop anisotropic
anisotrope Leitfähigkeit anisotropic conductivity
Anisotropiefeldstärke anisotropy field
Anker tie (72); keeper; armature (538), (621.3); anchor (621), (629.12); stay (621.315.1); rotating element (621.317)
Anker mit Kommutator commutator armature
Anker mit kurzgeschlossener Wicklung short-circuited rotor, squirrel-cage rotor
Anker mit Nuten grooved armature, slotted armature
Anker mit offener Wicklung open-circuit armature
Anker mit Stabwicklung barwound armature
Anker mit Wellenwicklung wavewound armature
Ankerabfall release of the armature
Ankerachse armature spindle (621.317)
Ankeranschlag armature stop (621.318)
Ankerbalken tie beam
Ankerbandage armature band
Ankerbewegung armature travel
Ankerblech armature lamination, armature stamping, armature core disc
Ankerblechpaket armature core plate (stack)
Ankerblechschnitte armature stampings, armature punchings
Ankerblechsegmet segmetal core disc
Ankerbolzen anchor bolt (621)
Ankerdraht tie wire; armature wire (621.313); guy wire (621.315.1)
Ankerdruckplatte armature-end plate
Ankereisen armature core

Ankereisenverlust armature-iron loss
Ankerendblech armature head (621.313)
Ankerfahne armature-end connection
Ankerfeld armature field
Ankergang escapement (681.11)
Ankergegenfeld opposing magnetic field of armature
Ankerhub armature travel, armature stroke
Ankerhubbügel armature pusher
Ankerhülse armature sleeve (621.313)
Ankerkern armature core
Ankerkraftfluß armature flux
Ankerkühlschlitz armature duct
Ankerkurzschlußbremsung breaking by short-circuiting the armature
Ankerleiter armature conductor
Ankernut armature slot
Ankernutwellen armature-slot ripple
Ankerquerfeld cross-induction field
Ankerrückwirkung armature reaction, armature ampere turns (621.313)
Ankerschraube foundation bolt (621)
Ankerspanner strainer (621)
Ankerspannschloß turnbuckle (621)
Ankerspannung armature voltage
Ankerspannungsregelung variable voltage speed control
Ankerspeichenstern armature spider
Ankerspule armature coil
Ankerstab armature bar, armature conductor
Ankerstern armature spider
Ankerstreufluß armature-stray flux
Ankerstrom armature current
Ankerstromkreis armature circuit
Ankerwelle armature shaft (621.313)
Ankerwellenkeil armature key
Ankerwicklung armature winding
Ankerwicklung ohne Eisenkern coreless armature
Ankerwickelmaschine armature winder
Ankerwiderstand armature resistance
ankerben to score
ankitten to cement on
anklemmen to secure under terminals; to connect
anklopfen call waiting (621.395)
anknipsen to snap on
ankommen to arrive, to come in
ankommende Amtsverbindung incoming junction line, incoming exchange line
ankommendes Gespräch incoming call (621.395)
ankommender Ruf incoming call
ankommende Welle incident wave
ankoppeln to couple
angekoppelter Widerstand coupled impedance
ankörnen to punch mark
Ankündigungssignal preliminary signal (625)
ankuppeln an to couple up with
ankurbeln to crank
Anlage plant; installation, equipment
eine Anlage außer Betrieb setzen to put a plant out of service
eine Anlage ist außer Betrieb a plant is shut down
eine Anlage läuft a plant moves
eine Anlage ist spannungslos power is off
Anlage übertage surface installation
Anlage untertage underground installation
Anlagenbildsteuerung mimic-diagram control
Anlagentechnik project engineering, system engineering
Anlaß-... starting
Anlaßbatterie starter battery
Anlaßbefehl start order
Anlaßdrossel starting reactor
Anlaßhebel starting lever
Anlaßleistung starting capacity
Anlaßschalter starting switch, switch starter, breaker starter, breaker controller
Anlaßschaltung starting connection
Anlaßschütz contactor starter
Anlaßschützensteuerung contactor-type starting control
Anlaßstoß rush of current
Anlaßstrom starting current
Anlaßstromstoß starting-current rush
Anlaßstufe step of the starter
Anlaßtransformator auto-transformer starter, starting transformer
Anlaßventil starting valve
Anlaßvorgang starting (process)
Anlaßwicklung starting winding
Anlaßwiderstand starter resistor, starting resistance, starting rheostat
Anlaßzeit starting period, starting time, run-up time
Anlaßzeitrelais time-limit start relay
Anlassen, asynchron asynchronous starting, induction starting
Anlassen mit Sterndreieckschalter star-delta starting, wye-delta starting
Anlassen mit kapazitiver Hilfsphase capacitor starting
Anlassen mit verminderter Spannung reduced-voltage starting
Anlassen mit Widerstand rheostatic starting
anlassen to start (up) (621.4); to temper (621.7)

Anlasser starter (621.3), (621.4)
Anlasser mit Drehzahlregelung starter with speed regulation
Anlasser mit Zentrifugalregler centrifugal starter
Anlasserkontakt starter contact
Anlasserkurbel starting handle (621.4)
Anlasserritzel starter pinion
Anlasserstufe starter contact
Anlauf start, starting (up); start up (621.374)
Anlauf- und Sicherheitskupplung starting and safety clutch
Anlaufbedingung starting condition
Anlauffarbe temper colour
Anlaufgebiet counter range (539)
Anlaufgegenmoment starting counter to rque
Anlaufkäfig starting winding
Anlaufkondensator starting capacitor
Anlauflänge start distance (681.3), (681.84)
Anlaufleistung starting power (621.317)
Anlaufmoment starting torque, torque at starting
Anlaufspule starting coil
Anlaufstrom starting current
Anlaufstromgebiet residual current region (621.385)
Anlaufvorgang starting (process), start-up (621.34/1)
Anlaufwert threshold (621.317); reaction value (621.316.7)
Anlaufwicklung start(ing) winding; auxiliary phase
Anlaufzeit starting period, starting time; response time (621.316.7); period rise (539); warming-up period (621.396.6)
anlaufen to start (up); to rub (621.1); to tarnish (621.7)
anläuten to ring up, to ring the ringer (621.395)
anlegen to inject, to apply, to impress (voltage) (621.3); to set to
anlegen an to apply to; to tap on to, to switch on to (621.3)
angelegte Frequenz applied frequency
angelegte Spannung applied voltage, applied potential
Anlegewandler clip-on transformer, clamp-on transformer, hook-on transformer, split-core type transformer
anleimen to glue on
Anleitung instruction, manual
anleuchten to floodlight
angeleuchtet floodlit
anliegen gegen to bear against, to bear on (72)
anliegende Bremse dragging brake, applied brake
anlöten to solder on, to sweat on
anmelden to book, to file, to place (621.395); to report, to give notice of (621.396.62); to apply for (34)
anmeldepflichtig notifiable
Annäherung approach; approximation (51)
Annäherungsmeßverfahren successive approximation methode
Annäherungsschalter proximity switch
Annäherungswert approximate value
Annährungszünder proximity fuze
Anode anode, plate (621.385)
Anodenabgriff anode-tapping point
Anodenabschirmung anode shield
Anodenabstimmkreis plate-tuned circuit
Anodenausgleichsdrossel anode-balancing coil (621.314.5)
Anodenbasisschaltung cathode follower, grounded-anode circuit; grounded-plate circuit (US)
Anodenbasisschaltung mit Gitter-Kathodeneingang bootstrap circuit
Anodenbatterie anode battery, "B" battery, plate battery (621.396)
Anodenbatterie-HF-Drossel shunt feed choke (621.396.61)
Anodenbatteriespannung plate supply (voltage)
Anodenbelastung plate load, plate dissipation
Anodenbelastungswiderstand plate load
Anodenblech plate (621.385)
Anodenbrumm plate supply ripple
Anodendunkelraum anode dark space
Anodeneffekt anode effect (621.35)
Anodenfall anode drop (621.387)
Anodenfläche plate area (621.385)
Anodengleichrichter plate power detector, anode (circuit-)detector
Anodengleichrichtung plate detection, anode-bend detection
Anodengleichspannung plate voltage
Anodengleichstrom anode current
Anodengleichstromleistung d.c. anode power input
Anodengleichstrom-Widerstand anode d.c. resistance
Anodenglimmlicht anode glow, positive glow
Anodenkappe anode valve cap, to p cap, anode cap
Anodenkennlinie anode characteristic
Anodenknick anode bend
Anodenkreis plate circuit, anode circuit

Anodenmantel anode shell
Anodenmodulation anode modulation, plate modulation, series modulation
Anodenmodulationsröhre anode modulator, plate modulator
Anodenoberfläche plate surface (621.385)
Anodensättigung anode (current) saturation
Anodenschlamm anode slime, anode mud (621.35)
Anodenschwingkreis tank circuit (oscillator), plate tank, tuned-plate circuit
Anodensegment eines Magnetrons anode segment of a magnetron
anodenseitig steuerbarer Thyristor N-gate thyristor
Anodensiebkreis anode-supply filter circuit
Anodenspannung anode voltage, plate voltage, plate supply
Anodenspannungsabfall anode voltage drop
Anodenspitzenstrom peak plate current
Anodenstrom anode current, plate current
Anodenstromkennlinie anode current characteristic (621.385)
Anodenstrommodulation anode current modulation
Anodenstromquelle plate supply
Anodenstromsättigung plate current saturation
Anodenstummel anode butt (621.365)
Anodenverlustleistung anode dissipation, plate dissipation
Anodenwiderstand anode resistance; anode resistor, plate load (resistor) (621.396)
Anodenzerstäubung anode sputtering
Anodenzündspannung anode break-down voltage
anodisch anodic
anodisch behandeln to anodize, to de-plate
anodische Beizung anode pickling
anodische Erzeugung von Schutzschichten anodic production of solids
anodisches Glimmlicht anode glow
anodische Korrosion anodic corrosion
anodische Oxidation anodic oxidation
anodische Passivierung anodic passivation
anodische Polarisation anodic polarization (621.35)
anodische Reinigung anode cleaning (621.35)
anodische Schicht anode layer
anodische Stromausbeute anode efficiency
anomal anomalous, abnormal
anomal polarisierte Wellen abnormally polarized waves
anomale Dispersion anomalous dispersion
Anomalie anomaly, abnormality
Anordnung arrangement, layout, assembly; configuration; order
Anordnungsplan layout drawing
anorganisch inorganic
anorganische Chemie inorganic chemistry
anorganisches Material dead matter
anorganische Substanz inorganic matter
anösen to eyelet
anpassen to adjust, to adapt, to fit; to match (621.396.6)
angepaßte Last optimum load, matched load, match
Anpassung adjustment; adaptation, match, matching (621.392)
Anpassungsdämpfung return loss
Anpassungsfähigkeit adaptability, flexibility, versatility
Anpassungsfaktor voltage standing wave ratio
Anpassungsglied adapter; tuner (621.396); building-out section (621.395)
Anpassungsimpedenz matching impedance
Anpassungskoeffizient return-current coefficient
Anpassungsleitung matching line, matching section (621.396)
Anpassungsmesser standing wave detector
Anpassungsstab shim rod (539)
Anpassungsstück matching bar, matching stub, phasing section
Anpassungsübertrager matching transformer
Anpassungsverhältnis standing wave ratio (621.317/396)
Anpassungswiderstand impedance match
Anpassungswirkungsgrad matching efficiency
anpeilen to take a bearing, to take bearings on
anprallen an to strike against
anregen to excite; to stimulate (534); to initiate
angeregtes Atom excited atom
Anregekontakt initiating contact
Anregung excitation; stimulation (534); activation (54)
Anregungsband excitation band (539)
Anregungsenergie excitation energy (539)
Anregungssignal forward signal (621.316.7)
Anregungssubstanz chemical agitator
Anregungssystem exciter system (621.396)
anreichern to enrich, to concentrate; to enhance (621.382)
angereicherter Reaktor enriched pile (539)
Anreicherung enrichment, concentration
Anreicherungsbetrieb enhancement-mode operation
Anreicherungsfaktor enrichment factor

Anreicherungs-IG-FET enhancement-type field-effect transistor
anreißen to set off, to lay off, to scribe (621)
Anreißnadel scriber
Anreißplatte marking-off table (621)
Anreißwinkel square
Anreißzirkel scribing compasses (pl.)
AN-Richtfunkfeuer AN Radio Range
Anruf call, calling (621.395)
Anruf beantworten to answer a call
Anrufklappe call indicator
Anrufklingel ringer
Anruflampe calling lamp
Anrufleitung ring wire
Anrufordner allotter
Anrufrelais calling relay; line relay
Anrufsuchen hunting, finding
Anrufsucher finder switch, tandem selector, line finder; transmitter finder
Anrufteilung call sharing
Anrufverteilung call distribution
Anrufvorgang calling
anrufen to call, to ring up, to ring the ringer (621.395)
anrufende Leitung calling line
anrufender Teilnehmer caller, party calling
angerufener Teilnehmer party called, called subscriber
Anrufer caller
anrußen to soot
Ansage anouncement
Ansage einer neuen Rufnummer changed-number anouncement
Ansagedienst recorded information service
Ansagegerät announcement device
Ansatz shoulder
Ansatzgriff extension handle
Ansatzschraube shoulder screw
Ansatzstück extension, joining piece
Ansatzstutzen slip fitter
ansäuern to acidify
Ansäuerung acidification
ansaugen to suck, to draw in
anschalten an to switch on to , to connect up to
Anschaltenetz access network
Anschauungsbild visual aid chart, telltale picture
Anschlag (end) stop, tappet (621); touch (78); stroke (681)
auf einen Anschlag treffen to meet a stop
bis zum Anschlag nach . . . drehen to turn all the way to . . .; to turn to the stop
Anschlagbolzen stop pin
Anschlagkraft impact force
Anschlagleiste locating edge
Anschlagpuffer snubber
Anschlagring shaft collar, stop collar
Anschlagschraube stop-screw, set screw
Anschlagstange stop bar
Anschlagstift peg, stop-beg, stop pin
Anschlagwinkel try square (621.9)
anschlagen to meet a stop; to type (681); to press a key (78)
anschlagen an to hit against, to strike
anschlagfreier Drucker non-impact printer
anschließen an to attach to; to join to, to connect with (621), (621.3)
angeschlossener Stromkreis branch circuit, associated circuit
Anschluß (branch) connection, branch circuit, (621.3); connection; terminal, joint (621), (621.3); line (621.395); port (621.3.049.76)
Anschluß je Anschlußleitung terminal per line
Anschluß je Sprechstelle terminal per station
Anschluß herstellen to make a connection
Anschluß für verminderte Leistungsentnahme reduced kva tap
Anschlußart mode of connection
Anschlußbahn branch line (625)
Anschlußbedingungen (power) supply regulations (696)
Anschlußbereich exchange area (621.395)
Anschlußbesetzton line-busy to ne
Anschlußbolzen terminal stud
Anschlußdose junction box, connection box (621.316); wall socket (621.395); conduit box (696)
Anschlußdraht terminal wire, connecting wire
Anschlußfahne tag (621.313)
Anschlußfeld connection panel, connection board
Anschlußflansch connecting flange
Anschlußkabel connection cable; subscriber's cable (621.395); consumer's cable (621.31)
Anschlußkästchen stage pocket
Anschlußkasten outlet box, conduit box, connection box, terminal box
Anschlußkennung line verification
Anschlußklemme (connecting) terminal (621.3), binding post (621.317)
Anschlußkopf terminal head
Anschlußlage position
Anschlußleiste connecting block, terminal strip
Anschlußleitung lead-in, leading in wire, power supply line, connecting wire; subscriber's line (621.395)
Anschlußlitze flexible lead, pigtail (621.313)

Anschlußnummer subscriber's number (621.395)
Anschlußöse terminal lug
Anschlußpunkt point of connection, junction point (621.3)
Anschlußschnur flex, flexible cable, flexible cord
Anschlußschraube binding post, binding screw, terminal screw, connecting screw
Anschlußsperre service denial
Anschlußstecker connecting plug
Anschlußstelle wiring point
Anschlußstrippe jumper
Anschlußstück connecting piece, connecting link
Anschlußstutzen connection stub; cable gland
anschmelzen to sweat on
anschmieden to forge on
anschmoren to freeze together (621.3)
angeschmort pitted (621.3)
Anschnittfläche cut face
Anschnittsteuerung phase-angle control
anschweißen to weld on
Anschwingstrom pre-oscillation current, starting current
Anschwingzeit build-up period
anschrauben to screw on, to bolt on
anseilen to guy
Ansetzen von Kristallen efflorescence
Ansicht view, elevation (621)
Ansicht in auseinandergenommenem Zustand exploded view
Ansicht der Einzelteile exploded view
Ansicht von hinten rear view
Ansicht im Schnitt section view
Ansicht schräg von vorn oblique front view
Ansicht von unten bottom view
Ansicht von vorn front view
Anspitzschaltung peaker
Ansprechen response, responsivity
Ansprechen des Ableiters sparkover
Ansprechband response band
Ansprechbereich responding range
Ansprechempfindlichkeit response sensitivity
Ansprechgeschwindigkeit speed of response
Ansprechgrenzen sensitivity limits
Ansprechpegel operating level
Ansprechpunkt operating point (621.318.5), set point
Ansprechschwelle threshold value, minimum response value
Ansprechspannung threshold voltage, pick-up voltage (621.318)
Ansprechstellung operating position
Ansprechstrom fusing current (621.316.9); actuating current, operating current
Ansprechvermögen responsivity
Ansprechverzug make time
Ansprechwert threshold; sensitiveness (621.316.7); pick-up value, response value (621.318)
Ansprechzeit operating time; melting time (621.316.9); response time
ansprechen (auf) to respond (to), to be responsive (to), to come into action (on)
anspringen to start (up) (621.313)
anstauchen to upset (621.791)
Anstauchung upset
anstechen to tap; to prick
anstellen to turn on (621.3); to lean on (to)
Anstellmotor screwdown motor (621.7)
Anstellwinkel angle of pitch, blade angle
ansteuern to drive, to control; to trigger
Ansteuerungsimpuls drive pulse, trigger pulse
Ansteuerungsstrom drive current
Anstieg increase, rise; built-up curve
Anstieggeschwindigkeit rate of rise; slew(ing) rate (621.375)
Anstiegszeit rise time, build-up time
Anstoß initiation, impulse
Anstoß geben to initiate, to push to start
anstoßen to start into action
anstrahlen to floodlight (621.32); to irradiate
angestrahlt floodlit
Anstrahlung irradiation, floodlighting
Anstrich paint (ing), coat (ing), finishing coat, washing (69)
Anströmgeschwindigkeit afflux velocity (539)
anstückeln to piece
Anteil des sichtbaren Lichts visibility factor
Anteil des sichtbaren Lichts an der Gesamtstrahlung luminosity factor
Antenne aerial; antenna (US)
Antenne für Mehrfachempfang spaced aerial
Antenne mit schwenkbarer Charakteristik steerable antenna
Antennenableiter aerial discharger
Antennenabstand aerial spacing
Antennen-Abstimmungsglied aerial tuning unit
Antennen-Abstimmkondensator aerial tuning condenser
Antennen-Abstimmspule aerial tuning inductance
Antennenanlage antenna assembly, antenna system
Antennenanordnung aerial array
Antennen-Anpassungsglied aerial tuning unit
Antennenanschluß aerial terminal

Antennenbild image aerial
Antennencharakteristik aerial characteristic, aerial pattern
Antennendiagramm aerial diagram, lobe pattern
Antennendraht aerial wire
Antennendrosselmodulation high power choke modulator
Antenneneffekt antenna effect
Antenneneingangswiderstand aerial feed impedance
Antennenelement aerial element
Antennenfeld aerial array, aerial bay
Antennenfuß aerial mount, aerial pedestal, aerial base, aerial support
Antennengewinn aerial gain, antenna gain
Antennengruppe group aerial, aerial array
Antennenisolator aerial insulator
Antennenkopplung aerial coupling
Antennenkopplungskreis aerial-coupling circuit
Antennenkreis aerial circuit
Antennenkreistastung back-shunt keying
Antennenleistung antenna power
Antennenlitze aerial wire
Antennennachbildung dump aerial, dummy aerial
Antennennachlaufsteuerung aerial following control
Antennenreflektor aerial reflector
Antennenrichtdiagramm antenna directivity pattern, antenna radiation pattern
Antennenschalter aerial switch; lightning switch
Antennenschutzfunkenstrecke anchor gap (621.396.67)
Antennenschwerpunkt aerial center
Antennenspeiseleitung (aerial) feeder
Antennenspule aerial coil
Antennenstrahlungswiderstand aerial radiation resistance
Antennenstrom aerial current
Antennensystem aerial system
Antennenträger aerial support
Antennentransformator jigger, antenna matching device
Antennenuntersatz aerial mount, aerial pedestal
Antennenwahlschalter antenna selector
Antennenwand billboard array of aerials, flat aerial, aerial curtain
Antennenweiche aerial filter, antenna filter, duplexer
Antennenwiderstand aerial resistance
Antennenwirkfläche effective area
Antennenwirkungsgrad aerial efficiency
Antennenzeile co-linear array; (v e r t i k a l :) stacked array
Antennenzuführungsleitung rattail, feeder
Anthygronleitung Anthygron weatherproof wire
Anthygronbleimantelleitung Anthygron served lead-sheathed wire
Antikathode anti cathode, target (621.386)
Antimon antimony, stibium
Antiparallelschaltung antiparallel coupling, back-to-back coupling
antippen to prod, to tap
Antistatiktuch antistatic cloth
Antiteilchen antiparticle
Antivalenz nonequivalent, either-or
Antivalenzschaltung exclusive OR circuit
antreiben to drive, to move; to impel, to propel, to power, to actuate
antreibend motive, driving; impulsive, propulsive
angetrieben werden von to receive motion from, to be powered by
angetriebene Welle driven shaft
Antrieb drive, driving (gear), impulse, propulsion, (prime) power; propulsion (539)
Antriebseinrichtung driving device
Antriebseinrichtung für Drehbewegung rotary actuator
Antriebseinrichtung für Schubbewegung linear actuator
Antriebsdrehzahl driving speed
Antriebshebel actuating arm, operating handle, operating lever
Antriebskraft driving force, moving force, motive power
Antriebsleistung driving power, motive power, propulsion power
Antriebsmaschine driving engine, prime mover; propelling machine (629.12)
Antriebsmechanismus driving mechanism, operating mechanism
Antriebsmittel impellent
Antriebsmoment driving to rque
Antriebsrad driving wheel
Antriebsräder driving gear (sing.)
Antriebsriemen drive belt
Antriebsritzel driver pinion
Antriebsrolle capstan
Antriebsscheibe driving disc, driving pulley
Antriebsseil driving rope
Antriebsseite coupling end, drive end, drive side
Antriebsseite eines Motors back of a motor

Antriebsvorrichtung propelling equipment
Antriebswelle driving shaft, spindle (621); propeller shaft (629)
Antwort response (621.316.7)
Antworteinrichtung transponder
Antworteinrichtung mit Frequenzumsetzung cross-band transponder
Antwortzeit response time (681.3)
anvisieren to sight
anwachsen to build up, to increase, to augment, to grow
Anwahl selection (621.316.7)
Anwahlbefehl selective command (621.398)
Anwahlsteuerung selective control
anwärmen to warm (up)
Anwärmofen preheater
Anwärmzeit warming-up period (621.1), (621.385)
Anweisung instruction, direction, regulation; instruction, statement (681.3)
Anweisungen beachten to adhere to directions
anwenden to employ, to apply, to make use of
anwenden auf to apply to
angewandte Physik applied physics
Anwender user
Anwender-Programm application program, user program
Anwender-programmierbare Logik field programmable logic array
Anwender-Software (standard-)user software
Anwenderstecker application connector
Anwendungsbereich range of application, scope of application
anwerfen to start up (621.4); to crank (621.4)
Anwurf-... starting
Anwurfkurbel starting handle, starting crank (621.4)
Anwurfmoment starting torque
Anwurfmotor pony motor, starting motor (621.4), (629.113)
Anwurfvorrichtung starting device
Anwurfzeit initiation time
anzapfen to tap, to bleed (621.3)
angezapfte Leitung tapped line
Anzapf-... tap..., tapping..., tapped... (621.3); bled..., bleeder... (621.1)
Anzapfstelle tapping point
Anzapftransformator tapped voltage transformer
Anzapfumschalter tap-changing gear
Anzapfung bleeding, tap, tapping; branch connection
Anzapfung an tapping off
Anzapfungsumschaltung tap changing
Anzeichen symptom (**für:** of)
anzeichnen to scribe, to lay off, to mark (621)
Anzeige indication, reading, display
Anzeige mit Festzeichenunterdrückung moving target indication (621.396.96)
Anzeigebereich indicating range
Anzeigeeinrichtung indicating device, indicator, read-out device
Anzeigefehler indication error, error of pointing
Anzeigegerät indicator, display (unit)
Anzeigeinstrument indicating instrument, dial instrument
Anzeigelampe indicating lamp; pilot bulb (621.3)
Anzeigeregister display register
Anzeigeröhre indicator tube, read out tube
Anzeigetafel indicator (light-)panel, annunciator board
Anzeigeträgheit inertia of needle
Anzeigewahrscheinlichkeit slip-scan ratio (621.396.96)
anzeigen to indicate, to record
anzeigendes Instrument direct reading instrument, indicating instrument
Anzeiger indicator
anziehen to attract, to pull towards oneself (53), to tighten, to screw up (621); to operate (621.318.5)
sich anziehende Pole attracting poles
Anziehung attraction, pull
Anziehungskraft attractive power, force of attraction, mutual attraction; nuclear attraction (539)
Anziehungskraft der Erde gravitation
Anziehungszentrum centre of attraction
Anzugserregung energizing power
Anzugsmoment starting torque
Anzugsspannung pull-in voltage
Anzugsstrom starting current
Anzugsverzögerung operate lag
Anzugswicklung operate winding
anzünden to light; to fire, to ignite
aperiodisch aperiodic, dead-beat, frequency-inselective
aperiodisch gedämpft aperiodic, critically damped; dead-beat (621.317)
aperiodisch gedämpftes Galvanometer dead-beat galvanometer
aperiodisch gedämpftes Instrument aperiodic instrument
aperiodische Antenne aperiodic aerial, untuned aerial
aperiodische Dämpfung aperiodic damping

aperiodischer Kreis aperiodic circuit (621.396)
im aperiodischen Grenzzustand critically damped
Apertur angular aperture (535); aperture (621.396.67)
Aperturfehler aperture distortion (621.397)
Apostilb apostilb
Apparat apparatus, set, instrument, device, equipment
am Apparat bleiben to hang on, to hold on, to hold the line (621.395)
Apparatebauer set constructor (621.396)
Apparategabel switch cradle, receiver rest (621.395)
Apparateklemme binding post (621.395)
Apparatesatz set (of apparatus), assembly (of apparatus)
Apparateschnur instrument cord
Apparateverteilungsplan apparatus location diagram
Approximationsfehler error of approximation
APT = automatically programmed tool
äquidistant equidistant
Äquipotentialfläche equipotential surface
Äquipotentialkathode equipotential cathode, unipotential cathode
Äquipotentiallinie equipotential line
äquivalent equivalent
äquivalente Eingangsdrift equivalent input voltage (621.382)
äquivalente Eingangsrauschspannung equivalent input noise voltage (621.38)
äquivalenter Erdradius effective radius of the earth, equivalent earth radius
äquivalente isotrope Strahlungsleistung equivalent isotropically radiated power
Äquivalent-Widerstand equivalent resistivity
Araldit araldite
Arbeit labour, work
Arbeit am Band work on the assembly line, line production
Arbeit leisten to accomplish work (53)
Arbeit von der Stange work from the bar, long-bar work (621.9)
Arbeitsablauf processing; sequence of operation
Arbeitsablaufkarte flow sheet
Arbeitsbedingungen working conditions; labour conditions
Arbeitsbereich operation range, operation region
Arbeitsbereich des Ausgangsstromes output current range
Arbeitsbereich der Eingangsspannung input voltage range
Arbeitsbereichgrenze operating limit, overload level
Arbeitsbühne working platform
Arbeitsdruck working pressure, effective pressure
Arbeitsergiebigkeit productivity
Arbeitsfläche working face
Arbeitsfolge operating sequence
Arbeitsfrequenz working frequency, primary frequency, actual frequency
Arbeitsgang working process, shop operation
Arbeitsgang von Hand zu Hand hand operation
Arbeitsgebiet field of activity
Arbeitsgemeinschaft team, working pool
Arbeitsgeschwindigkeit working speed; operating speed
Arbeitshub working stroke; forward traverse (621.7/9)
Arbeitskenngröße performance characteristic
Arbeitskennlinie dynamic characteristic (621.396)
Arbeitskontakt make contact, normally-open contact, NO contact
Arbeitskräfte man power
Arbeitsmaterial work material
Arbeitsmatrix function matrix (621.374)
Arbeitsmodul index of co-operation (621.394.6)
Arbeitsplatz workplace, workstation
Arbeitsplatzbeleuchtung individual workplace lighting
Arbeitsprozeß working process
Arbeitspunkt operating point, working point, quiescent point (621.385)
Arbeitspunkt-... quiescent-operation-..., steady-operation-... (621.385)
Arbeitsregel working rule
Arbeitsregister working register (681.3)
Arbeitsschacht manhole chimney
Arbeitsspannung working voltage, rated operating voltage; on-load voltage
Arbeitsspeicher working storage, temporary storage
Arbeitsspiel working cycle
Arbeitsstellung operating position
Arbeitssteilheit dynamic slope
Arbeitsstrom working current, operating current; load current (621.382)
Arbeitsstromkreis make circuit, circuit-closing connection
Arbeitsstromsystem open circuit system

Arbeitsstunden man hours
Arbeitstabelle einer Digitalschaltung truth table
Arbeitstemperatur working temperatur, operating temperature range
Arbeitstiefe working depth (621.8), reach (621.9)
Arbeitsverfahren working method; (manufacturing) process, production process
Arbeitsvermögen working capacity
Arbeitsvorbereitung operations scheduling, production planning
Arbeitsweise mode of working, working method, function, procedure
Arbeitszyklus working cycle, duty cycle, cycle of operation
arbeiten to work, to labour; to act, to go, to run, to operate, to function
Archimedisches Prinzip principle of Archimedes
Architektur eines Mikroprozessors architecture (of a microprocessor)
arc tan $\tan^{-1}$
Argandbrenner Argand burner
Argon argon
Argonarc-Schweißung argon arc welding
Argonaut-Reaktor Argonaut (research) reactor
Argument argument (51)
Arithmetik arithmetic
arithmetisch arithmetic(al)
arithmetischer Befehl arithmetical instruction
arithmetisch-logische Einheit arithmetic logic unit = ALU
arithmetisches Mittel arithmetic mean(s)
arithmetischer Mittelwert average value, arithmetic mean(s)
arithmetische Reihe arithmetic progression
arithmetischer Spannungsmittelwert average voltage
arithmetischer Strommittelwert average current
Arkatombogen shielded arc (621.7)
Arkatom-Elektrode shielded arc electrode
Arkatomschweißung atomic hydrogen welding
arm an, ...-arm deficient in
Armaturen armatures, fittings, mountings
Armaturenbrett instrument panel; dashboard (629.113)
Armaturteil fitment
armieren to armour; to reinforce
armiertes Kabel armoured cable
armierter Träger reinforced beam
Aronschaltung two-wattmeter circuit
Aron-Gleichstromzähler Aron meter
Aronscher Pendelzähler Aron clockmeter, pendulum meter
arretieren to lock, to stop, to secure
Arretierhebel stop lever, lifter (621.317)
Arretierschraube stop screw
Arretierstift locking pin
Arretierung catch, arrest, locking device
Arsen arsenic
Arsenkupfer copper arsenide
Art mode, modus, way; nature, quality, type
nach Art eines Spannungsteiler in potentiometer fashion
Artgewicht specific weight, specific density, specific gravity
Artwiderstand specific resistance
Arztlampe surgical lamp
Asbest asbestos
Asbestband astbestos ribbon, asbestos tape
Asbestdichtung asbestos packing, asbestos gasket
Asbestfaden asbestos cord, asbestos rope
Asbestfaser asbestos fibre
Asbestliderung asbestos packing
Asbestmantelkabel asbestos-covered cable
Asbestpappe asbestos (mill)board
Asbestschieferplatte asbestos slate (board)
Asbestzementplatte asbestos-cement plate
asbestisoliert asbestos coverd
asbestisolierter Draht asbestos covered wire
A-Schrank "A" switchboard (621.395)
Asphalt asphalt
Asphaltanstrich asphalt coating
Asphaltkabelkanal bituminous conduit
Asphaltkleber bitumastic
Asphaltmasse bituminous compound
Assembler assembler
Assembleranweisung assembler directive
Assoziativspeicher associative storage
astabile Kippschaltung astable trigger circuit
astabiler Multivibrator astable multivibrator, free running multivibrator
astatisch astatic (al)
astatisches Galvanometer astatic galvanometer
astatischer Leistungsmesser astatic wattmeter
astatische Magnetnadel astatic needle
astatisches Nadelgalvanometer Broca galvanometer
astatische Regelung astatic control
astatisches Verhalten floating action (621.316.7)
astatisches Wattmeter astatic wattmeter
astigmatisch astigmatic
Astigmatismus astigmatism
Astro-Navigation astro navigation

Astronomie astronomy
Astrophysik astrophysics (pl.)
asymmetrisch asymmetric (al)
  asymmetrische Leitfähigkeit asymmetrical conductance, unilateral conductance
asymptotisch asymtotic (al)
  asymptotische Darstellung asymptotic representation
  asymptotische Spalterwartung iterated fission expectation
asynchron asynchronous
  asynchroner Anlauf asynchronous starting, induction starting
  asynchron arbeitende Übertragungsschnittstelle asynchronous communications interface (adapter)
  asynchrone Arbeitsweise asynchronous operation
  asynchroner Frequenzwandler induction-frequency converter
  asynchroner Motorgenerator induction motor-generator
  asynchroner Phasenschieber asynchronous condenser
  asynchrone Übertragung asynchronous transmission
Asynchron-... asynchronous
  Asynchrongenerator asynchronous generator, induction generator
  Asynchronmaschine asynchronous machine
  Asynchronmotor asynchronous motor, induction motor
    Asynchronmotor mit Kurzschlußanker squirrel-cage induction motor
    Asynchronmotor mit Schleifringläufer slip-ring induction motor; wound-rotor induction motor (US)
  Asynchron-Synchron-Umformer induction-synchronous motor-generator
  Asynchronverfahren asynchronous operation (621.394.6)
Äthan ethane, ethyl hydride
Äther ether (53), (54)
Äthylalkohol grain alcohol
Äthylen ethylene
Atlantikkabel atlantic cable, transatlantic cable
atmen lassen to vent (621.3)
Atmosphäre atmosphere
  Atmosphärendruck atmospheric pressure
atmosphärisch atmospheric(al)
  atmosphärische Normalbedingungen standard atmospheric conditions
  atmosphärisches Rauschen atmospheric noise
  atmosphärischer Störpegel atmospheric(al) noise, static noise, statics, atmospheric disturbances (621.396.6)
  atmosphärische Störungen spherics, statics, atmospherics, atmospheric disturbances
Atom atom
  Atomabstand interatomic distance
  Atomantrieb atomic propulsion
  Atomart atomic species
  Atomaufbau, Atombau atomic structure, configuration
  Atombatterie atomic battery, atomic pile
  Atombindung atomic bond, atomic linkage
    Atombindungsenergie atomic binding energy
  Atombombe atom bomb, atomic bomb
  Atombrenner atomic pile, atomic reactor, nuclear reactor (539), (621.311)
    Atombrenner mit Graphit als Bremssubstanz graphite-moderated pile
  Atombrennstoff fissile matter
  Atomenergie (sub)atomic energy, nuclear power
  Atomforschung atomic research, nuclear research
  Atomfrequenznormal atomic frequency standard
  Atomgewicht atomic weight, atomic mass
    Atomgewichtsbestimmung atomic-weight determination
  Atomgitter atomic lattice
  Atomhülle atomic shell
  Atomkern atomic nucleus, atomic center
  Atomkraft atomic energy, atomic power, nuclear energy
    Atomkraftwerk atomic-power plant, nuclear-power station
  Atomladung atomic charge
  Atommasse atomic mass
  Atommeiler atomic pile
  Atommodell atomic model
  Atomordnungszahl atomic number
  Atomphysik atomic physics (pl.)
  Atomradius atomic radius
  Atomreaktor atomic reactor, atomic pile
  Atomrumpf atomic torso
  Atomschale atomic shell
  Atomspaltung nuclear fission
  Atomsymbol chemical symbol
  Atomteilchen atomic particle
  Atomtheorie atomic theory
  Atomuhr atomic clock
  Atomumwandlung atomic transformation
  Atomvolumen atomic volume

Atomwärme atomic heat
Atomwertigkeit atomic valence
Atomzahl atomic number
Atomzerfall atomic disintegration, nuclear disintegration
Atomzertrümmerung atom splitting, atom smashing
atomar atomic
atomares Frequenznormal atomic frequency standard
atomarer Zustand atomic state
atomisieren to atomize
atomistisch atomistic
Attrappe dummy, mock-up
Ätzalkalimetalle caustic alkali metals
ätzen to etch
ätzend caustic
Ätzlösung etchant
Ätznatron caustic soda
Audiogramm audiogram
Audiometer audiometer
Audion audion, grid (leak) detector
Audiongleichrichtung grid-leak detection, grid-leak rectification
Audionschaltung leaky grid detector, grid-leak bias
„Auf" "Up", "Raise"; "Higher" (621.313)
Auf- und Ab-Bewegung up-and-down motion
Auf- und Abziehvorrichtung pusher-puller
aufarbeiten to recondition
Aufarbeitung reconditioning; reprocessing (54)
Aufbau structure, composition (54); mounting, installation, assembly, construction, erection, building-up (621)
Aufbau eines Codes code structure
Aufbau eines Feldes setting up to a field, creation of a field
Aufbau der Selbsterregung build-up of self-excitation
Aufbaufaktor build-up factor
Aufbaukurve build-up curve
aufbauen to mount, to construct, to erect, to build-up, to set up
aufbauend up-building (539)
aufgebaute Auslösung series overcurrent tripping (621.316.5)
aufbereiten to process, to treat, to prepare, to purify; to dress (622)
Aufbereitung, Aufbereiten preparation, dressing, pretreatment, processing
Aufbereitungsanlage processing plant, treating plant; reprocessing plant (539); dressing plant (622)
Aufbereiten der Daten data preparation
aufbiegen to bend up, to prize open
aufblitzen (lassen) to flash
aufbocken to jack up
aufbohren to bore, to drill
aufbrennen auf to fire on (66)
Aufbringen deposition, application
Aufbringen einer Einschmelzperle pre-beading
Aufdampfanlage vapourizing equipment
aufdampfen to vapor deposit, to evaporate
aufdrehen to turn on; to unscrew (621.396.62), (628.1)
aufdrücken to impress, to force
aufgedrückte Schwingung forced oscillation
aufgedrückte Spannung impressed voltage
aufeinander abrollen, (sich) to roll together
aufeinander abstimmen to correlate (51), to tune (621.396)
aufeinanderfolgend successive, consecutive
aufeinanderfolgende Impulse successive pulses (621.397)
aufeinanderfolgende Phasen consecutive phases
aufeinander senkrecht mutually perpendicular
auferregen, sich to run up (621.313)
Auferregung build-up of self-excitation
auffädeln to thread
auffallen auf to incide upon (535), to impinge upon (621)
Auffallwinkel angle of incidence
Auffanganode collector plate (621.385. Klystron)
Auffangelektrode collecting electrode, collector plate
Auffang-Flip-Flop D-type flip-flop, latch
Auffangstange elevation rod, terminal
auffangen to intercept, to catch, to collect (621.395); to pick up; to spot (621.396.96)
aufgefangene Nachricht intercepted message
Auffaßbereich pick-up range (621.396.96)
Auffaßbereich eines Funkmeßgeräts coverage of a radar set
auffassen to pick-up
Auffaßreichweite coverage (621.396.96)
auffinden to spot, to locate, to detect
Auffindung der Störstelle location of the trouble, location of fault
auffrischen to vivify, to fortify (54), (77); to regenerate, to rejuvenate, to recondition (621.385); to refresh (681.3)
aufgefrischter Träger reconditioned carrier
Aufgabe task; function; problem (51); feed, feeding (621.396), (621)
Aufgabeband feed conveyor (621.876)
Aufgaberutsche feeder chute

Aufgabesollwert desired value, control point (621.316.7)
Aufgabetrichter feeding funnel
aufgebaucht bulbous
Aufhänge-... suspension, mounting
Aufhängeband ribbon wire, suspension band (621.317)
Aufhängebügel lifting lug
Aufhängedraht suspension (wire); messenger wire (621.33)
Aufhängeeinrichtung hanger
Aufhängefeder suspension spring
Aufhängegestänge suspension rods
Aufhängehöhe mounting height
Aufhängekabel suspension wire
Aufhängelitze lamp cord
Aufhängeöse U-bolt, mounting lug; ball eye, suspension eye (621.315.1)
Aufhängepunkt hinge point, centre of suspension
Aufhängestange suspension rod
Aufhängevorrichtung suspension attachment, suspension mounting
aufhängen to hang up, to suspend
Aufhänger hanger
Aufhängung der Lichtmaschine am Wagen truck generator suspension (629.113)
aufhaspeln to reel in, to wind up, to coil up
aufheben, sich to neutralize, to cancel out each other
sich aufhebende Spannungsabfälle cancelling voltage drops
aufheizen, (sich) to warm up, to heat up
Aufheizung heating
Aufheizzeit heating time, heating-up period
aufhören to cease, to stop
aufkeilen to wedge, to key
aufkleben to stick on, to paste on, to glue on **(auf:** to)
Aufkreis epicycle
aufladen to charge, to load
sich aufladen to charge up
aufladbare Batterie rechargeable battery
Auflage rest, stay, support (621.9); specification, requirement
mit einer Auflage versehen to tip (621.9)
Auflagefläche seating, bearing surface
Auflagekraft vertical force; needle force (681.84)
Auflagepunkt contact point, point of support
Auflagescheibe back plate
Auflagetisch lay-up table (621.9)
Auflager support, fixed location point, reaction point, bearing
Auflagerdruck bearing stress, reactance
Auflagerleiste rib
Auflagerpunkt reaction point, support point
auflaufen to sum up (to) (51); to strike ground (629)
Auflaufpunkt first point of contact
Auflaufschuh side approach, end approach (621.33)
auflegen auf to apply to
Aufleuchten intensity brightening; flashing
aufleuchten to flash up, to light up
auflösen to solve (51); to dissolve (54); to separate (539)
aufgelöste Darstellung exploted view
aufgelöster Sternpunkt open star-point, open neutral
auflösendes Feld separating field (539)
Auflösespalt defining slit (535)
Auflösung resolution, dissolution; definition (621.397), (621.396.96)
Auflösung in Bildelemente scan (621.397)
Auflösungsfehler resolution error
Auflösungsvermögen resolving power (535)
Auflösungszeit resolving time
auflöten to solder on; to braze on
aufmagnetisieren to remagnetize
aufmodulieren auf to modulate upon
Aufnahme absorption (54); record (681.8); transcription, pick-up (621.396); exposure (77)
Aufnahmeentzerrung pre-equalization
Aufnahmegerät recorder
Aufnahmeraum studio (621.396/397)
Aufnahmeröhre camera tube
Aufnahmevermögen susceptibility, receptivity
Aufnahmevorrichtung recording device (681.84)
aufnehmen to absorb (54); to record (681.8); to pick-up, to monitor (621.396); to photograph (77); to survey (526)
aufgenommene Leistung power input, power consumed
Aufnehmer sensor; pick-up, pick-off
aufpolieren to furbish, to shine up, to polish, to refinish
Aufprall impact
aufprallen to impact
aufprallen auf to strike
aufpressen to press on, to force on
aufprüfen to hunt (621.395)
Aufpunkt representative point, selected point (51)
Aufputzinstallation surface mounting
aufrauhen to roughen
sich aufrauhen to gall (621.9)

aufräumen to clear
Aufrecht-Formmaschine straight-draw moulding machine
aufrechterhalten to sustain
aufrechtstehend upright
aufreiben to ream
aufreißen to line out, to protract (621.71)
aufrichten to right, to set up, to erect
aufriegeln to unlock
Aufriß (front) elevation, vertical section
Aufriß im Schnitt sectional elevation
Aufruf call(-in)
Aufruftafel call board
aufrunden to round up (51)
aufsaugen to absorb, to soak up (54)
Aufschalten offering, intrusion, cutting in (621.395)
Aufschaukeln (von Schwingungen) building up (of oscillation), resonant rise
aufschäumen to froth, to effervesce, to foam up, to boil up
aufschichten to arrange in layers, to pile up, to stack
Aufschiebestromwandler slip-on current transformer
Aufschlag percussion, impingement, impact
Aufschlagpunkt point of impact
aufschlagen to impact
aufschlagen auf to knock against, to hit
aufschlitzen to split (open), to slot
aufschrauben auf to screw on to
aufschrumpfen to shrink on
aufschütten to feed (621); to throw up (624)
aufgeschütteter Boden made ground
Aufschütt-Trichter (feeding) hopper
Aufschwemmung suspension (54)
Aufsetzpunktfeuer to uch-down point light
Aufsicht control; supervisor (621.395)
Aufsicht führen über to superintend
Aufsichtsbeamter superintendent
Aufsichtsbehörde board of control, supervisory board
aufsichtführender Obermonteur foreman electrician
aufspalten to split (into) (51); to split open (621)
aufspannen to step up (621.3)
Aufspanner step-up transformer (621.3)
aufspeichern to accumulate; to store (681.3)
aufspringen to fly open (621.3)
aufsprudeln to bubble up, to boil up
aufsprudelnd ebullient
aufspulen to reel up, to wind on reels; to spiralize, to coil up
Aufspulen einer Verbindung tail eating
aufspritzen to spray on
aufspüren to trace (out), to detect, to locate
aufstäuben to sputter
aufsteckbar attachable
aufstecken to attach, to slip on
Aufsteckstromwandler through-type current transformer
aufstellen to erect, to mount, to fit up, to install
Aufstellungsort installation site
Auftasten intensity brightening (621.397)
auftasten to brighten (621.397)
Auftastimpuls gate pulse (621.397)
Auftastschaltung gate circuit
auftauen to thaw; to defrost
Auftrag deposit (621.791); order (33); job (681.3)
Auftragmetall filler metal
auftragschweißen to build up
Auftragschweißung hard facing
Auftragverfahren method of application
Auftragen building up (621.35), (621.79)
auftragen to apply (621.35); to measure off, to plot (621.79)
auftragen über ... als Abszisse to plot against...
Auftreffen impact, impingement; incidence (535)
Auftreffelektrode target (electrode)
Auftreffgeschwindigkeit impact velocity
Auftreffwinkel angle of incidence
auftrennen to open
Auftrieb buoyancy; lift (532/533)
Auftriebswirkung buoyancy effect
auftrommeln to reel up
Aufwachsen growth (621.382)
aufwallen s. **aufschäumen**
Aufwand input (621.3); expenses (pl.) (65)
aufwärmen to warm up, to heat up
Aufwärmung warm-up
aufwärts upward
Aufwärtsbewegung upward movement, upward motion
Aufwärtsblitz upward streamer
Aufwärtshub up-stroke (621.4)
aufwärtskompatibel upward compatible (681.3)
aufwärtsstrebend tending to rise, with a rising tendency
Aufwärtstransformator step-up transformer
Aufwärtsübersetzung step-up (ratio)
aufweiten to splay, to widen out; to flare (621.396.67)
sich aufweiten to enlarge
Aufweitung enlargement; flare (621.396.67); expansion

aufwenden to spend, to use
aufgewandte Energie put-in energy, applied energy
aufgewandte Leistung power input, power consumed
aufwickeln to coil, to reel up
Aufwickelhaspel up-coiler
aufwinden to hoist, to haul up
Aufwindstrom up-current (of air) (55)
aufzählen to enumerate, to count down
aufzeichnen to delineate, to draw (down); to record; to trace out (621.385); to lay off (621.9); to log
Aufzeichnung record, recording (681.84); drawing, graph (51), (62)
Aufzeichnungsdichte recording density (681.84)
Aufzeichnungsspur recording track
aufziehen to wind up; to warp (67)
Aufziehvorrichtung pusher
Auf-Zu open-shut, on-off
Auf-Zu-Regler on-off control, two-position action controller
Aufzug lift (Br.), elevator (US) (621.8); hoist (624); warp, woof (67)
Aufzugfeder main spring
Aufzugkorb cage (622)
Aufzugseil hoisting cable
Aufzugswinde hoist winch
Auge eye
Augenermüdung eyestrain
Augenschutz eye protection
Augenschutzglas eye protection glass
Augenblick moment, instant
Augenblickswert instantaneous value; momentary value
Augenblickswert des Stromes instantaneous current
Aureole flame cap
ausbalancieren to outbalance, to balance out, to equilibrate
ausbauchen to bulge (out)
ausbauen to remove, to detach
ausbaubares Glied detachable link
in ausgebautem Zustand opened up
ausbessern to repair, to mend up, to redress
Ausbesserung repair, mending
Ausbesserungsarbeit repair work
Ausbesserungswerkstatt repair shop
ausbesserungsfähig reparable
ausbeulen to buckle, to bulge
sich ausbeulen to bilge out
ausgebeult baggy
Ausbeute yield; response (54)
Ausbeutekurve yield curve
Ausbildung training
Ausbildungsprogramm training program
Ausbildungsreaktor training reactor
ausblasen to blow out; to blow off, to shut down (669)
Ausbleiben der Spannung loss of voltage
ausblenden to gate
ausblühen to effloresce; to spot out (621.1)
Ausblühung stain spots (621.1)
ausbohren to bore out, to drill out
Ausbrandprobe scaling test (621.791)
ausbrauchen to spend, to use, to deplete; to exhaust (54)
ausgebrauchte Batterie exhausted battery
ausgebrauchte Lösung spent solution
ausbrechen to erupt (55); to break out (614.84)
Ausbrechöffnung knockout (696)
ausbreiten, sich to propagate, to travel (621.396); to spread
Ausbreitung propagation (621.3); spreading
Ausbreitung über den Horizont transhorizon propagation
Ausbreitung über Wasser overwater propagation
Ausbreitungsart mode of propagation
Ausbreitungsbedingungen propagation conditions
Ausbreitungscharakteristik propagation pattern, propagation characteristic
Ausbreitungsdämpfung path loss (621.371)
Ausbreitungsfehler propagation error
Ausbreitungsgeschwindigkeit speed of propagation, velocity of propagation
Ausbreitungsrichtung direction of travel
Ausbreitungsverlust propagation loss
Ausbreitungsweg route, path (of propagation) (621.396)
Ausbreitungsweg in der Ionosphäre ionospheric path
Ausbreitungswiderstand earth resistance
ausbrennen to cauterize (61); to burn out (621.3), (621.791); to cut out (621.791)
ausgebrannte Röhre burnt-out valve
ausbröckeln to chip (off)
ausbügeln to iron out; to equalize, to smooth down
ausdehnen, (sich) to expand, to extend, to stretch
Ausdehnung expansion, extension; magnitude, dimension
Ausdehnungsfähigkeit expansability
Ausdehnungsgefäß expansion tank, expansion vessel (621.314.2)
Ausdehnungsglied expansion joint

Ausdehnungskoeffizient coefficient of expansion
ausdrehen to bore; to turn out, to turn off
Ausdruck expression; term (51); hardcopy, printout (681.3)
Ausdruck zweiter Ordnung second-order term
ausdünsten to exhale
Ausdünstung exhalation, effluvium
auseinander apart, asunder
auseinanderbrechen to break apart, to break up
auseinanderfliegen to fly apart
auseinandergehen to diverge (51)
auseinanderhalten to keep apart
auseinanderklaffen to spring
auseinanderlaufen to diverge (51)
auseinanderlaufend divergent
auseinandernehmen to disassemble, to take to pieces
auseinandergenommen opened up
auseinanderschneiden to separate (621.791)
Ausfädeln vacating (351.814.33)
ausfahrbar removable, collapsible
ausfahrbares Schaltgerät drawout switchgear
ausfahrbarer Teil removable element (621.316.5)
ausgefahrene Antenne extended antenna
Ausfall (manufacturing) loss; deficiency, stoppage, failure, dropout
Ausfallalarm alarm on failure
Ausfalldauer down time
Ausfallhäufigkeit failure frequency
Ausfallkriterien failure criteria
Ausfallquote deficiency rate
Ausfallrate failure rate
Ausfallwinkel angle of reflection (535)
Ausfallzeit down time, fault time, outage time
ausfallen lassen to cancel, to omit
ausfiltern to filter out
Ausfleckung spotting out (621.357)
ausfließen to flow out; to discharge
ausfließende Menge outflow
Ausfluß efflux, outflow; emanation, discharge
Ausflußgeschwindigkeit velocity of efflux, discharge velocity
Ausflußkoeffizient coefficient of discharge
ausfräsen to mill out
ausgefräster Teil milled-out portion
Ausfrierfalle cold trap, low-temperature trap
ausführen to design, to perform, to practise; to export (33)
ausführbar practicable, feasible
Ausführung practice, performance; type, model
Ausführungsbefehl executive instruction (681.3)
Ausführungszeit execution time
Ausführungszyklus executive cycle
Ausfüllmasse packing, stuffing
ausfüttern to line
Ausgabe output (681.3)
Ausgabebefehl actuate message, output instruction
Ausgabeblock output block (681.3)
Ausgabeeinheit output unit, output equipment
Ausgabegerät output equipment (681.31)
Ausgabegeschwindigkeit output rate
Ausgabespeicher output register
Ausgang outlet (62); output, load end, output end, receiving end; sink (621.392)
Ausgangsadmittanz output admittance
Ausgangsatom parent atom
Ausgangsbefehl output signal (621.316.7)
Ausgangsbelastbarkeit output loading capacity
Ausgangsbelastungswiderstand (output) load resistor
Ausgangsbuchse outlet (socket)
Ausgangsdehnung original elongation
Ausgangsdrehmoment torque output
Ausgangsfeld parent field
Ausgangsfunktion trial funktion, original function (51)
Ausgangsgleichstrom/Eingangswechselstrom d. c. to a. c. ratio (621.314.5)
Ausgangsgleichung basic equation, initial equation
Ausgangsgröße output variable, output quantity (621.316.7)
Ausgangskapazität output capacitance
Ausgangskern parent nucleus
Ausgangsklemme output terminal, outlet terminal
Ausgangsklemmenimpendanz terminating impendance
Ausgangskonfiguration output configuration, output pattern
Ausgangskreis output circuit, load circuit
Ausgangsleistung (power) output
Ausgangsleistungsmesser output meter, volume indicator (621.396.62)
Ausgangsnutzleistung useful output power (621.396.61)
Ausgangspegel output level
Ausgangspunkt starting point, basing point
Ausgangsrohr, Ausgangsröhre power tube
Ausgangsruhespannung quiescent output voltage (621.382)
Ausgangsscheinleitwert output admittance
Ausgangsscheinwiderstand output impedance
Ausgangsseite load end, output end; exit side

Ausgangssignal output signal
Ausgangsspannung output voltage
Ausgangsspannung / Eingangswechselspannung degree of voltage rectification (621.314.5)
Ausgangsspannungshub output voltage swing
Ausgangsstellung neutral position, initial position, starting position, home position
Ausgangsstoff primary (54); source material (539)
„Ausgangsstrippe" output cord
Ausgangssubstanz parent substance, primary (54)
Ausgangstransformator output transformer
Ausgangsübertrager output transformer, telephone transformer
Ausgangsübertragungsfunktion output transfer function
Ausgangsverstärker output amplifier, power amplifier
Ausgangswert basic value, initial value; output value (621.316.7)
Ausgangswiderstand load, load resistor; internal resistance, output impedance
Ausgangswirkwiderstand internal resistance, output resistance
Ausgangszustand initial state
Ausgasprüfung outgassing test
Ausgleich balance, compensation, building out
Ausgleichsbrückenschaltung balanced-bridge transition
Ausgleichsdrossel interphase reactor, balance coil (621.314.2); compensating coil (621.314.6)
Ausgleichsdrosselspule balancer-transformer (621.316.1)
Ausgleichsgetriebe balancing gear, diffential gear
Ausgleichsgewicht balance weight, compensating weight
Ausgleichsimpuls equalizing pulse (621.397)
Ausgleichskabel equalizing cable
Ausgleichskammer compensation cell
Ausgleichskondensator building-out capacitor, balanced capacitor
Ausgleichslademaschine battery booster (621.355)
Ausgleichsladung equalizing charge (621.355)
Ausgleichsleiter equipotential connection, equalizer (621.313)
Ausgleichsleitung interconnector, interconnecting main (621.315)
Ausgleichsmaschine d. c. balancer (621.3)
Ausgleichsmembran compensating diaphragm
Ausgleichsregelsatz balancer booster (621.316.1)
Ausgleichsring equalizer ring (621.313)
Ausgleichsschaltung correcting network, balancing circuit (621.3)
Ausgleichsscheibe shim
Ausgleichsschiene equalizing bar
Ausgleichsschwingung transient oscillation
Ausgleichsspannung transient voltage; compensating voltage
Ausgleichsspule compensating coil
Ausgleichsstrom transient current
Ausgleichstransformator a. c. balancer (621.315.1), three-wire transformer, static balancer (621.316); hybrid coil (621.395)
Ausgleichsverbindung equipotential connection, equalizing connection (621.313)
Ausgleichsverbindungen balancing rings (621.313)
Ausgleichsvorgang transient reaction, transient phenomenon
beim Ausgleichsvorgang wiederkehrende Spannung transient recovery voltage (621.316.5)
Ausgleichswicklung balanced winding
Ausgleichswiderstand compensating resistor, balancing resistor
Ausgleichszeit recovery time
ausgleichen to compensate (for), to counter-balance, to equalize; to weight up; to smooth out
ausgeglichene Temperatur equable temperature
ausgeglichener Leerlauf-Telefonstörfaktor no-load balanced telephone interface factor
Ausgleicher equalizer
ausglühen to anneal
Ausgußmasse compound filling, sealing compound
aushaken to unhook, to hook out
aushängen to lift the handset, to go off-hook (621.395)
aushärten to cure; to polymerize
aushärtendes Kunstharz cold-curing synthetic raisin
Aushärtung curing
ausheizen to bake (621.38)
Ausheizung von Wigner-Energie Wigner release (539)
aushöhlen to cup, to cave; to gouge (621.9)
auskehlen to flute, to chamfer, to channel
auskleiden to line
Auskleidung lining
ausklingen to die away, to decay

ausklinken to release, to disengage, to trip, to unlatch
ausklinkbar disengageable
Ausklinkdrehzahl tripping speed
Ausklinkvorrichtung tripping device (621.3); unlatching mechanism, release mechanism
ausknipsen to snap off, to snap out
auskoppeln aus to couple from (621.396)
Auskoppelraum catcher space
Auskoppelungsfeld, Auskoppelungsglied catcher [Klystron]
Auskoppelungsgitter catcher grid
auskragen to corbel (72)
auskreuzen to Vee-groove (621.791); to transpose (621.3)
ausgekreuzte Wicklung transpose winding
Auskreuzschema pole diagram
auskristallisieren to crystallize out
auskristallisierend efflorescent
auskühlen to cool (down)
Auskühlzeit cooling period
auskuppeln to uncouple, to ungear, to declutch, to disengage, to throw out of gear, to swing out of mesh, to disconnect
Ausladung reach; overhung portion, throw (624)
Auslandsvermittlungsstelle international exchange, international terminal exchange
Auslaß exhaust, outlet, port
Auslassungszeichen ignore character (681.3)
Auslauf run, deceleration, running down
Auslaufversuch pouring test (66); decelerating test (run) (621.313)
auslaufen to escape, to run dry; to wear out (621.4); to decelerate (621.3)
Ausläuferleitung dead end feeder, dead ended line
auslegen für to rate
Ausleger suspension bracket (621.33)
Auslegeraufhängung bracket arm suspension (621.33)
Auslegermast bracket pole, bracket type supporting structure
ausleiern, (sich) to wear out
auslenken to deflect
Auslenkung displacement (534); deflection (621.385), (621.397)
Auslesen readout
Auslesen mit Zerstörung des Speicherinhalts destructive readout
auslöschen to extinguish, to quench, to cancel; to delete
sich auslöschen to neutralize (51), (53)
Auslöse-... trip, tripping, trigger; release (621.395)
Auslösen durch den gerufenen Teilnehmer called-party release
Auslösen durch den rufenden Teilnehmer calling party release
Auslösen durch den zuletzt auflegenden Teilnehmer last party release
Auslösedaumen releasing cam
Auslöseeinrichtung tripping device, trigger device (621.397)
Auslösefunke trigger spark, trip spark
Auslösegeschwindigkeit tripping speed
Auslösegestänge trip gear, tripping gear, release gear
Auslösegetriebe tripping gear, tripping mechanism
Auslöseglied tripping piece, disconnecting link
Auslöseimpuls trigger pulse, trigger (621.397)
Auslöseimpulsgenerator trigger oscillator
Auslöseklinke trip latch
Auslösekontakt release contact, tripping contact
Auslösekreis tripping circuit (621.31); trigger circuit (621.397)
Auslöseleitung release wire (621.395.34)
Auslösemagnet tripping magnet, releasing magnet
Auslösemutter quick nut
Auslösequittungszeichen release guard signal (621.39)
Auslösepegel triggering level
Auslösepunkt trip point
Auslöserelais trip relay
Auslöseschalter trip switch
Auslöseschwelle minimum triggering level, threshold
Auslösesignal trigger signal
Auslösespannung firing point, tripping voltage (621.316.5)
Auslösespule trip coil
Auslösestift striker pin
Auslösestrom release current, tripping current
Auslösestromkreis trip circuit, release circuit
Auslösetaste operational key, functional key (621.3)
Auslösewirkung tripping effect, tripping action
Auslösezeichen clear-forward signal, release signal
Auslösezeit tripping time, release time
auslösen to trip, to fly off (621.3); to release, to clear (621.395); to trigger (621.397), (77)
Auslöser trip, tripping piece, tripping device, cut-out (621.3); trigger (621.397), (77)
Auslösung release (621.395); tripping (621.316.5); initiation

Auslösung von Elektronen aus einer Oberfläche liberation of electrons from a surface
Auslösung mit Verzögerung time lag tripping (621.316.5)
Auslösung ohne Verzögerung instantaneous tripping
auslöten to unsolder
auslüften to air, to aerate, to ventilate
ausmachen to locate, to spot
ausmauern to line (with bricks)
ausmessen to measure, to meter, to gauge
Ausmessungen size (sing.), dimensions
ausmultiplizieren to multiply out, to expand
Ausnahmeanschluß foreign exchange line
Ausnahmetarif exceptional tariff, exceptional rate, special rate
ausnutzen to utilize, to turn to account
ausnutzbar usable, useful
ausnutzbarer Querschnitt useful cross-section
Ausnutzung efficiency, utilization
Ausnutzungsfaktor utilization factor, coefficient of utilization, strength factor; (plant) capacity factor
ausprägen to stamp; to mark distinctly
ausgeprägter Pol salient pole (621.313)
Ausprüfen eines Programms program testing (681.3)
Auspuff exhaust
Auspuffableiter expulsion-type lightning arrester
auspumpen to exhaust, to pump out
ausräuchern to fumigate
ausrechnen to calculate, to compute
ausrechnen zu to work out as (51)
ausregeln to control, to stabilize, to correct
Ausregelzeit correction time
ausreiben, sich to wear out
ausrichten auf to realign to, to point towards, (at)
ausgerichtet halten nach to keep pointed at (621.396.67)
Ausrichtung alignment
ausrollen to run out (621.315.2)
ausrücken s. **auskuppeln**
Ausrundung cove, rounding-off radius
ausrüsten to fit out, to equip
ausrüsten mit to fix up with
Ausrüster outfitter (629.12)
Ausrüstungsstück equipment item
ausrutschen to skid
ausschachten to excavate
ausschalten to eliminate; to disconnect, to switch off, to cut off, to open circuit (621.3), to extinguish (621.32); to shut down (62)
Ausschalter cut-out, circuit breaker
Ausschalt... breaking..., opening...
Ausschaltabbrand burning on opening, burning on breaking (621.316.5)
Ausschaltbefehl open order
Ausschaltdauer total break time
Ausschaltkontakt interrupting contact
Ausschaltleistung breaking capacity, interrupting capacity
Ausschaltlichtbogen circuit-breaking arc
Ausschaltspitzenstrom cut-off current (621.316.9)
Ausschaltspule trip coil
Ausschaltstellung disconnect position, off position
Ausschaltstrom breaking current; interrupting current (US)
Ausschaltung elimination; disconnecting, disconnection, opening (of a circuit breaker)
Ausschaltverlustenergie energy dissipation during turn-off time (621.382)
Ausschaltverlustleistung turn-off loss
Ausschaltverlustleistungsspitze peak value of turn-off loss
Ausschaltverzögerung opening delay
Ausschaltzeit break time, opening time, tripping time (621.316.5); operating time (621.316.9)
ausscheiden to eliminate, to rule out; to precipitate (54)
Ausscheidung elimination, precipitation (54); secretion (61)
Ausscheidungshärtung precipitation hardening (66)
ausschlacken to slag out
Ausschlag swing, throw, deflection (621.317)
Ausschlag auf der Skala swing on the dial
Ausschlagsbegrenzung limit of travel, limitation of deflection (621.317)
Ausschlagsverfahren deflectional method, method of direct deflections (621.317)
ausschlagen to deflect, to move
ausschlagen über die Skala hinaus to go off scale
ausschleifen to grind out, to turn out; to rebore (621.43)
sich ausschleifen to wear out
ausschließen, sich gegenseitig to be mutually incompatible
ausschmelzen to smelt, to melt out
ausschneiden to cut (out)

Ausschnitt sector (51); waste piece (621.96); cut-out
ausschrauben to screw out
Ausschuß scrap, wastage, rejection (621)
Ausschußfaktor loss factor
Ausschuß-Gußstück waster casting (621.7)
Ausschußstück waster, wastrel, reject
ausschwenken to swing out
ausschwenkbares Schaltfeld swing-out panel
ausschwingen to swing aside; to die away, to decay (621.37)
Ausschwing... decay...
Ausschwingzeit decay time, dying-down time
ausschwitzen to exude
außen out, without, outside, outward
außenliegende Nebenstelle external extension
nach außen federn to spring out
nach außen führen zum Anschluß an to bring out to
nach außen gerichtet outward
nach außen gerichtete Flächennormale outward normal
nach außen klappen to swing outward
von außen zu betätigen externally operable
von außen her from outside
von außen sichtbar outwardly visible
Außen-... outer, outdoor, exterior, external, outside
Außenabmessungen external dimensions
Außenansicht exterior view
Außenantenne outdoor antenna
Außenarmatur outdoor fitting
Außenbeleuchtung exterior lighting, outdoor lighting
Außenbezirk outskirt
Außendurchführung wall bushing, wall entrance insulator
Außendurchmesser outside diameter, outer diameter
Außenende outer end
Außenerdkopplung sheath unbalance (621.315.2)
Außenfläche surface
Außengewinde external thread, male thread
Außenglimmschutz outer corona protection
Außenhaut shell
Außenläufermotor motor with outer rotor, external rotor motor
Außenleiter live wire, outer cable, outer conductor, outside wire
Außenleiterstrom phase current
Außenleuchte outdoor luminaire
Außenlüfter external fan
Außenlunker surface defect (621.7)
Außenmantel outside coating, outer sheath
Außenmaß outside dimension
Außenpolgenerator external-pole generator, external-field generator, revolving armature generator (621.313)
Außenradius male radius
Außenraumklima outdoor climate, climate in weather protected areas
Außenschicht outer layer
Außenspeicher external memory
Außentaster outside calipers (pl.)
Außenverzahnung external gearing
aussenden to emit, to transmit, to radiate (621.396/397)
außer... out of..., off...
außer Betrieb out of operation
außer Betrieb setzen to put out of operation, to put out of service, to shut down
außer Tritt fallen to fall out of synchronism, to fall out of step
Außertrittfallmoment pull-out torque
äußer (-e, -s) outer, exterior, external, outward
äußerer Anschluß outer end (621.3)
äußere Beflechtung outer braid (69)
äußere Charakteristik external characteristic (621.313)
äußeres Elektron outer electron
äußeres Ende outer end
äußere Flächennormale outward normal
äußeres Fremdfeld extraneous field
äußere Kräfte impressed forces
äußerer Kurzschluß (beim Stromwandler) zero burden (of current transformer) (621.317)
äußerer Leiter outer conductor, external conductor
äußeres magnetisches Feld external magnetic field
äußeres Produkt outer product, vector product, cross product
äußerer Stromkreis external circuit
äußerer Vorwiderstand instrument multiplier (621.317)
äußerer Wärmewiderstand external thermal resistance, thermal resistance case to ambient (621.382)
äußerer Widerstand external resistance (621.3)
äußere Zuführung terminal lead
Äußeres exterior
außerhalb off, out of; outside
außerhalb des Bandes outband
außerhalb des Kerns befindlich extranuclear
außerhalb der Lastspitze off-peak period
außerhalb der Resonanz off resonance
außerhalb der Spitze off peak

äußerlich external
außermittig eccentric
außerordentlich exceptional, extraordinary
  außerordentlicher Strahl extraordinary ray (535)
  außerordentliche Welle extraordinary wave
äußerste Grenze ultimate limit
Aussetzbelastung intermittent loading
Aussetzbetrieb (AB) intermittent duty, periodic duty, intermittent service, periodic service
Aussetzfehler dropout
Aussetzlast intermittent load
aussetzen to stop, to intermit, to interrupt; to break down (621.4)
aussetzend intermittent
  aussetzende Belastung intermittent load
  aussetzender Betrieb intermittent operation
  aussetzende Entladung intermittent discharge
  aussetzender Erdschluß intermittent (earth)
  aussetzender Fehler intermittent (trouble)
  aussetzendes Feuer occulting light (656.7)
  aussetzende Störung intermittent trouble
aussieben to filter out (621.3); to sieve
aussondern to isolate, to separate; to sort, to single out
aussortieren to sort out; to sift out (621.385); to screen (66), (622)
aussparen to spare out, to leave out
aussperren to shut out, to bar; to lock out (33)
ausstechen to gouge (621)
Ausstellung exhibition; display
„Aus"-Stellung "out"-position
Ausstiegsklappe escape hatch
Aussteuerbereich range of control, dynamic range; range of modulation; recording-level range (681.8)
aussteuern to control, to drive
Aussteuerung control, driving
  Aussteuerungsanzeiger percent-modulation indicator (621.396), recording-level indicator (681.8)
  Aussteuerungsbereich range of modulation (621.396), recording-level band (681.8)
  Aussteuerungsgrad depth of modulation, percentage of modulation
  Aussteuerungsgrenze operating limit
  Aussteuerungsmesser peak programme meter, speech-level meter, percent-modulation indicator, programme level meter
Ausstoß output, discharge, ejection
  Ausstoßableiter expulsion-type arrester
  Ausstoßseite discharge side
  Ausstoßstift knock-out bar, ejector (621.7)
ausstoßen to exhaust, to blow off; to eject
ausstrahlen to radiate, to emit
  ausgestrahlte Energie radiated energy
  ausstrahlender Kreis radiative circuit
Ausstrahlung radiation, emission, sending (621.396); radiance (621.32)
  Ausstrahlung unter kleinem Erhebungswinkel low-angle radiation
  Ausstrahlungsvermögen emissivity
  Ausstrahlungswinkel angle of radiation
austasten to blank out, to gate (621.397)
Austast... blanking
  Austastgemisch blanking signal (621.397)
  Austastimpuls blanking pulse
  Austastlücke blanking interval
  Austastpegel pedestel (level), blanking level
  Austastschaltung killer circuit
  Austastschulter porch
  Austastsignal blanking signal (621.396.96)
  Austastung blanking, gating
  Austastwert blanking level
Austausch interchange, exchange
  in Austausch gegen in exchange for
  Austauschbauteile interchangeable components
  Austauschreaktion exchange reaction (54)
  Austauschstoff substitute
  Austauschteile interchangeable components
  Austauschvorgang interaction
austauschen (gegen) to exchange for, to substitute for, to replace; to swap (681.3)
  austauschbare Grenzen interchangeable limits
  Austauschbarkeit pin-for-pin replacement (621.382)
Austauscher interchanger
Austenit austenite
austesten to check out
Austinsche Ausbreitungsformel Austin formula (621.396)
austreiben to expel, to eliminate
austretende Flüssigkeit outpouring
Austritt emission, ejection
  Austrittsarbeit work function (621.385); electron affinity (539)
  Austrittsgeschwindigkeit speed of ejection, emission velocity
  Austrittsöffnung orifice, exhaust port
  Austrittspotential work function (621.385)
  Austrittsspannung work function (621.385)
  Austrittstemperatur outlet temperature
austrocknen to dry out, to desiccate, to exsiccate
auswägen to poise, to weigh out
auswählen to select
  ausgewählter Punkt selected point
auswandern to wander (621.317)

auswechseln gegen to replace with, to interchange with
auswechselbar replaceable, interchangeable
auswechselbarer Kontakt interchangeable contact piece
Auswechslung replacement
Ausweich . . . alternative . . .
Ausweichfrequenz alternative frequency
Ausweichgleis siding, shunting line
Ausweichkanal alternate channel, alternative channel
Ausweichleitweg alternative route
Ausweichleitwegzuteilung alternative routing
Ausweichvermittlung emergency exchange
ausweiten to enlarge; to expend; to widen
Auswerteimpuls strobing pulse, gating pulse
auswerten to reckon, to evaluate; to analyze
auswickeln to unwrap
auswiegen to weigh out; to counterpoise, to counterbalance, to level
auswischen to wipe out
auswittern to weather (-wear); to season
auswuchten to balance, to counterbalance
Auswuchtmaschine balancing machine
ausziehen to extract (51), (54); to exhaust
ausziehbar telescopic; removable, withdrawable, extractable
ausziehbarer Geräteblock drawout equipment unit
ausziehbares Chassis drawout chassis, pull-out chassis
ausziehbares Leitungsstück telescopic section
ausziehbares Rohr telescopic tube
ausziehbares Rohrstück telescopic section
ausziehbarer Schlitten slide-out tray
ausziehbares Teil removable element, pull-out unit
Auszieh . . . pull-out . . ., drawout . . .
Ausziehgriff withdrawal handle
Ausziehschalter drawout circuit breaker, truck-type circuit breaker
Ausziehsicherung pull-out fuse
Ausziehvorrichtung extractor, puller
Auszieher extractor (621.9)
Auszieherkralle extractor claw
Auszug extract (54)
Auto = Automobil car, motor car; automobile (US)
Autoanlasser starter
Autoantenne car aerial
Autobatterie car battery
Autoempfänger car radio, car receiver (set)
Autolampe car sidelamp
Autolichtanlage car lighting
Autoscheinwerfer (car) headlight
Autosteuerung car control
Auto-Suchscheinwerfer (car) searchlight
Autowinker indicator
Autozündleitung (automobile) ignition cable
Auto . . . = Selbst . . . auto . . .
Autoalarmempfänger autoalarm receiver
autogen autogenous
Autogen-. . . autogenous
Autogenschneidemaschine autogenous cutting machine
Autogenschweißdraht gas-welding apparatus
Autogenschweißung autogenous welding
Autoinoisation auto-ionisation
Autokollimationsfernrohr autocollimatory telescope
Automat (automatic) cut-out (621.3); automatic (621.9); automaton (pl.: -ta), robot
Automatenschweißung automatic welding
Automatenstahl free cutting steel (621.9)
Automation automation
automatisch automatic, self-. . .
automatisch gesteuert self-controlled, automatic controlled
automatisches Abfahrtsignal automatic dispatching device (656.25)
automatische Abfahrsteuerung automatic acceleration (621.33)
automatische Abschaltung automatic opening, automatic shut-off; automatic stop (681.84)
automatischer Alarmempfänger automatic alarm receiver, autoalarm receiver
automatisches Amt automatic telephone exchange, dial central office
automatischer Anlasser self-starter
automatische Anlaßsteuerung automatic starting control
automatische Aufschaltung automatic lock-on
automatische Auslösung automatic tripping
automatische Bahnsignalanlage automatic railway-signalling
automatische Bandbreitenregelung A. S. C. = automatic selectivity control
automatischer Batteriespannungsregler automatic reversible battery booster
automatische Betätigung automatic operation
automatisches Blocksystem automatic block system
automatische Bremse self-acting brake
automatische Dynamikregelung automatic volume expansion

automatische Einschaltung automatic connection, automatic closing; automatic starting (621)
automatische Einstellung self-adjustment
automatische Fehlerkorrektur automatic R. Q.
automatische Feldschwächungseinrichtung automatic field suppressor (621.316.9)
automatische Flugüberwachungs- und Steueranlage automatic flight controller
automatische Frequenzregelung automatic frequency control (AFC) (621.396.61)
automatische Gebührenerfassung automatic assessment of message fees
automatische Geräuschsperre squelch (circuit)
automatische Gesprächstrennung calling party release
automatische Gewindeschneidmaschine automatic tapping machine
automatische Gittervorspannung self-bias (621.385)
automatische Hauszentrale private automatic exchange, PAX
automatische Kesselsteuerung automatic combustion control
automatisches Kraftwerk automatic power station
automatischer Kursrechner course-line computer
automatische Kurssteuereinrichtung automatic pilot (629.13)
automatische Kurzschlußsicherung automatic short-circuit protection
automatische Lautstärkeregelung automatic volume control
automatischer Leistungsschalter automatic circuit breaker
automatische Lichtbogenschweißung automatic arc welding
automatische Meldeanlage automatic signalling device, automatic alarm
automatische Nachführung automatic aiming, automatic tracking
automatische Nebenstellenanlage private automatic branch exchange, PABX
automatische Nullstellung automatic reset (681.3)
automatische Pegelregelung automatic level regulation
automatischer Peiler automatic direction finder
automatische Phasenregelung automatic phase control
automatischer Plattenwechsler automatic record changer
automatisches Programmieren automatic programming, automatic coding (681.3)
automatische Prüfeinrichtung automatic routiner (621.395), routine test circuit
automatischer Regelumspanner automatic variable voltage ratio transformer
automatisches Regelungssystem servo system
automatischer Regler automatic regulator, automatic controller (621.3)
automatische Rückführung automatic reset
automatischer Rückruf automatic call booking
automatischer Ruf automatic calling
automatische Schachttürverriegelung automatic gate lock (621.8)
automatischer Schalter automatic circuit-breaker
automatisches Schaltschütz self-acting cut-out
automatische Scharfabstimmung A. F. C. = automatic frequency control
automatische Schwarzwerthaltung automatic black-level control
automatische Schwundregelung automatic gain control
automatische Signalanlage automatic signalling
automatischer Spannungsregler automatic voltage regulator
automatische Steuerung automatic control, self-control
automatische Stockwerksschaltung automatic floor-stop operation (621.876)
automatischer Suchempfänger automatic scanning receiver
automatische Synchronisiereinrichtung automatic synchronizer
automatische Telegraphie-Speichervermittlung automatic message switching system
automatische Telegraphie-Teilstreckenvermittlung telegraphic automatic relay system
automatische Umschalteinrichtung automatic transfer equipment
automatisches Unterwerk unattended substation, automatic substation
automatischer Verschluß automatic shutter
automatische Verstärkungsregelung automatic gain control, AGC
automatischer Vorschub automatic carriage (621.394.6)
automatische Wählwiederholung recorded call
automatische Warnanlage automatic alarm
automatische Wiederzuschaltung automatic reclosing

automatische Zielverfolgung automatic tracking, automatic following
automatische Zugbremsung automatic train deceleration
automatische Zugsteuerung automatic train control
Autopilot autopilot (629.13)
Autotransformator auto-transformer
A-Verstärker class A amplifier
Avogadrosche Zahl Avogadro's number
AW = Amperewindungen a. t. = ampere turns
A-Wellen type "A" waves
A 1-Wellen type "A 1" waves, unmodulated key controlled continuous waves
A 2-Wellen type "A 2" waves
Axial-... axial
Axialbelüftung axial ventilation
Axialdrucklager axial thrust bearing
Axialdrucklaufring ball-thrust washer
Axialfluß axial flow
Axialgebläse axial flow fan
Axialkugellager axial ball-bearing
Axialkühlung axial system of ventilation
Axiallager thrust bearing
Axialpumpe axial flow pump
Axialschnitt axial section
Axialschub axial pressure, axial thrust
Axialspiel end play
Axialteilung axial pitch (621.8)
Axialverdichter axial flow fan
axial axial
axiale Anschlußdrähte axial leads, axial terminations
axiale Lüftungsschlitze axial ducts (621.313)
axialer Schub end pressure, end thrust
axiales Spiel end play
mit axialen Nuten versehen to spline
Azetat acetate
azetatfolienisolierter Draht acetate wire
Azetatseide acetate silk
Azetylen acetylene
Azetylenbrenner acetylene torch
Azetylen-Sauerstoff-Schneidbrenner oxy-acetylene cutting torch
Azetylen- (Sauerstoff-)Schweißung oxy-acetylene welding
Azimut bearing angle, azimuth
Azimutanzeiger bearing indicator
Azimutauflösung bearing resolution
Azimutfehler bearing error
Azimuthöhenbild im Funkmeßgerät C-scan
azimutal azimuthal
azyklisch acyclic

B

babbeln to gabble, to gibber
Babbeln babble (621.392)
Backe jaw (621)
Backenbrecher jaw breaker, jaw crusher
Backenfutter jaw lining, loose jaw
Backen packing, agglomeration (621.395.6)
b-Ader B-wire
Badewannenkurve bathing-tub diagram
Badspannung bath voltage, tank voltage (621.357)
Bagger excavator dredger
Bahn track; railway (Br.), railroad (US); orbit (52)
Bahnbeschleunigung acceleration along the path (53)
Bahnbewegung orbital motion (53)
Bahndrehimpuls orbital angular momentum, path spin
Bahnebene orbital plane (52); trajectory plane (53)
Bahnelektronen orbital electrons (53)
Bahnellipse orbital ellipse (53)
Bahnexzentrizität orbital eccentricity (53)
Bahngenerator traction generator
Bahngleichrichter traction rectifier
Bahnmotor railway motor, traction motor (621.33)
Bahnprofil maximum equipment line, railway loading gauge; railroad clearance (US)
Bahnquantenzahl orbital quantum number
Bahnsteigrufanlage railway call system
Bahnsteuerung path control, straight path control (52); continous path control (621.9)
Bahnstromgenerator traction generator (621.33)
Bahnstromleitung traction supply line
Bahnstromversorgung traction supply
Bahntelegraph railway telegraph
Bahntransformator traction transformer
Bahnunterstation railway substation, traction substation
Bahnwiderstand extrinsic resistance (621); bulk resistance (621.314)
Bajonettfassung bayonet holder (621.32)
Bajonettsockel bayonet cap
Bajonettverbindung bayonet joint
Bake beacon
bakelisieren to bakelize
Bakelit bakelite
Bakelit aushärten to cure bakelite
Bakelitgehäuse molded bakelite jacket

Bakelitharz bakelite resin
Bakelitlack bakelite varnish
Bakelitpapier bakelite paper
Bakelitprodukt bakelite derivative
Bakkerrohr Calrod heating element (621.36)
Balance balance, equipoise
balancieren to balance
balanciertes Drehstromsystem balanced three-phase system
balanciertes Dreileitersystem balanced three-wire system
balanciertes Mehrphasensystem balanced polyphase system
Balkencode bar code
Balkendiagramm bar chart, bar graph
Balkengenerator bar generator
Ballast ballast
Ballastbelastungswiderstand bleeder resistance
Ballastwiderstand ballast resistance, buffer resistance
Ballempfang reception via demodulator repeater
Ballempfänger relay receiver
ballistisch ballistic
ballistischer Ausschlag throw (621.317)
ballistisches Galvanometer ballistic galvanometer
ballistische Methode ballistic method
Ballonreflektor balloon reflector
Ballung bunching (621.385); concentration
Bananenstecker banana plug, banana pin
Bananensteckerbuchse banana jack
Bananensteckerstift pin plug
Band band, tape, ribbon; brace, strap, strip
Bandabstand energy gap (53)
Bandandruck head pressure (681.84)
Bandanfangsmarke begin of tape = BOT
Bandantrieb tape drive, tape-drive servo
Bandaufhängung ribbon suspension
Bandaufnahme tape recording; tape record (534.86)
Bandaufnahme machen to record on tape
Bandblock tape block (621.394)
Bandbreite band (width) (621.39)
Bandbreitenregler bandwith control (621.396.62)
Bandbremse band brake, strap brake (629.113)
Banddehnung band spreading (621.396)
Banddreher speech inverter (621.395.1)
Bandeinziehvorrichtung tape lead-in
Bandeisen strap iron, hoop iron, steel tape
bandeisenbewehrtes Kabel band-armoured cable, steel-tape armoured cable
Bandendmarke end of tape = EOT
Banderder surface earth, iron-strip earth conductor, copper-strip ground conductor
Bandfabrikation assembly-line production
Bandfehler bad spot (681.3)
Bandfilter band (pass) filter (621.396.62)
Bandfilterabstimmung bandpass tuning
Bandfiltereigenschaft bandpass characteristic
Bandfilterkennlinie bandpass characteristic, bandpass response
Bandfilterkopplung zwischen den Stufen interstage tuning unit
Bandfilterwelligkeit bandpass flatness
bandfiltergekoppelter Verstärker bandpass amplifier
Bandförderer belt conveyor
Bandführungssegment vacuum guide (681.84)
Bandführungstrommel tape drum, head drum (681.84)
Bandgenerator van de Graaf generator
Bandgeschwindigkeit tape speed
Bandgrenze band limit
Bandkabel ribbon cable
Bandkassette tape cartridge
Bandkern tape core, tape-wound core
Bandkette band chain
Bandkontakt head-to-tape contact (681.84)
Bandlaufwerk tape drive
Bandleiter strip conductor, stripline, micro strip
Bandleitertechnik strip-line technique
Bandmagnetophon tape recorder (534.86)
Bandmarke tape mark (681.84)
Bandmikrophon ribbon microphone
Bandmitte midband, centre of the band (621.396)
Bandmittenfrequenz midband frequency
Bandpaß bandpass filter, passband
Bandringkern strip-wound toroidal core
Bandrolle tape reel
Bandrollenmotor drum motor
Bandrücklauf rewinding (681.84)
Bandrücksetzung backsparing (621.377), (681.3)
Bandschleifmaschine belt sander
Bandschreiber tape recorder, strip-chart recorder (621.317)
Bandspeicher tape store
Bandsperre band-elimination filter, band stop filter, band rejection filter
Bandspreizung band spread, band spreading
Bandsprosse frame (681.3)
Bandspule ribbon coil, spool
Bandstahl strip steel, steel tape

Bandtrommel tape drum
Bandverständlichkeit band articulation
Bandvorschub tape feed
Bandwalzwerk strip mill
Bandwiderstand ribbon resistance; strip resistor
Bandzuführung tape feed
Bandage strap, (binding) band (621.313)
Bandagendraht binding wire
Bandagenverluste band losses (621.313)
Bändchengalvanometer tape galvanometer, twisted strip galvanometer
Bändchenmikrophon ribbon microphone
Bandenspektrum band spectrum
Barium barium
Barkhausen-Effekt Barkhausen effect
Barkhausen-Kurz-Generator Barkhausen-Kurz oscillator
Barkhausen-Lautstärkemesser alternation phonometer
Barkhausen-Sprung Barkhausen jump (538)
Barlowsches Rad Barlow's wheel
barn barn (539)
Barometer barometer, (weather-) glass
Barometerstand barometric height, barometer reading
Barometerstand auf NN reduzieren to correct barometer readings
Barretter barretter, thermo detector
Barriere barrier
Baryt baryte
Base base (54)
BA-Signal blanked picture signal
Basis base (621.382), basis, fiducial line (51), (52), (72)
Basisadresse base address
Basisadreßregister base register
Basisanschluß base contact; base terminal (621.382)
Basis (-Elektrode) basis, base electrode
Basisbahnwiderstand extrinsic base resistance
Basisband base band
Basisfrequenz base frequency
Basisinformation raw storage (621.316.7)
Basislinie fiducial line
Basisschaltung rückwärts inverse common base
Basisschaltung vorwärts common base
Basissignal basic signal (681.3)
Basisspannung base voltage
Basissperrstrom reverse base-bias current
Basisstrom base current
Basiswert fiducial value (621.317)
Basiswiderstand base resistance
Basiszahl base number
Basiszone base region
Basiszwischenzone interbase region
BAS-Signal composite picture signal
basisch basic
basisches Verhalten basicity
Baß bass (78)
Baßanhebung bas compensation (621.396.62)
Baßlautsprecher subwoofer
Bassin-Reaktor aquarium reactor, swimming-pool reactor
basteln to gadget, to tinker
Bastler amateur (constructor)
Batterie battery (621.3), (623), accumulator; (storage) battery (621.33)
Batterie belasten to load a battery
Batterie entladen to discharge a battery
Batterie laden to charge a battery
Batterieanruf battery call
Batterieantrieb battery traction (625/629.1)
Batteriebetrieb battery operation
Batterieempfänger battery set, battery radio, battery receiver
Batterieentladung battery discharge
Batteriefahrzeug battery-driven vehicle, accumulator car
Batteriegehäuse battery container, battery box
Batteriegerät battery set, battery-operated equipment
batteriegespeist battery-operated, battery powered
Batteriegestell battery rack
Batteriekapazität battery capacity
Batteriekasten battery container, battery box
Batterieklemmen terminal posts (621.355), battery clip
Batterieladegerät battery charger, accumulator charger
Batterieladegleichrichter battery-charging rectifier
Batterieladung battery charging, accumulator charging
batterieloser Fernsprecher sound-powered telephone
Batterieprüfer battery gauge, battery tester (621.317)
Batterieruf battery call, battery ringing
Batteriesäure battery acid
Batterieschaltfeld battery rack
Batterieschalttafel accumulator switchboard
Batteriespeichervermögen battery capacity
Batteriestromverbrauch battery drain
Batteriestromversorgung battery supply
Batteriewahl battery dialling (621.395)

Batteriezähler battery meter
Batteriezündung battery ignition
Batteriezusatzmaschine battery booster
Bau construction, structure; building
in Bau under construction, in process of construction
Bauart make, style, type of construction
Baugewicht construction weight
Baugröße physical size
Baugruppe subassembly
Bauhöhe overall size, head room
Baukastensystem modular system, modular construction
Baulänge overall size
Bauplan component layout (621.3)
Bausatz kit
Bauschaltplan internal wiring and assembly diagram
Baustahl structural steel
Baustein module, building block (621.3); tile (621.316.5)
Baustein-Bauweise brick-building method, modular system
Baustufe stage of construction
Bauteil component
Bauteilchen constituent particle (539.15)
Bauweise form of construction
Bauelement component
Bauelement der Starkstromtechnik heavy-current component
Bauelement unter Überdruck pressurized component
Baugruppe sub assembly, component, module
Baugruppenträger subrack
Bauch anti node, internode (621.37)
Baud Baud
Baudot-Drucktelegraph Baudot multiplex printing telegraph
Baudot-System Baudot telegraph system
bauen to build, to construct, to erect
baumförmig tree-like
Baumharz wood rosin
Baumwolle cotton
Baumwollband cambric tape, cotton tape
Baumwollgewebe cambric, cotton fabric
baumwollisoliert cotton-covered
baumwollisolierter Draht cotton-covered wire
Baumwollumklöppelung cotton-braided covering
Baumwollumspindelung cotton covering, cotton wrapping
baumwollumsponnener Draht cotton-covered wire
Bausatz kit
Bausch wad, wadding
Bauxit bauxite
BB = zweimal baumwollisoliert double cotton-covered
B-Beamter, B-Beamtin "B" operator
B-Betrieb class B operation
BCD-Code BCD = binary coded decimal
beanspruchen to stress, to stain (621), (621.3)
Beanspruchung stress (ing), strain (ing), load, duty, service (621), (621.3)
Beanspruchung, zulässige maximum permissible stress
bearbeiten to process; (spanlos:) to work, (spanabhebend:) to machine
bearbeitbar workable, machinable
Bearbeitbarkeit workability, machinability
bearbeitete Fläche machined face
Bearbeitung process (ing); (spanlos:) working, (spanabhebend:) machining
Bearbeitung von Hand hand process
Bearbeitungsanweisung handling instruction, processing instruction
Bearbeitungsfehler defective workmanship
Bearbeitungsfläche working face
Bearbeitungsfolge tooling sequence
Bearbeitungsgang shop operation, machining operation
Bearbeitungsgenauigkeit tooling accuracy
Bearbeitungsgüte surface quality
Bearbeitungskarte flow sheet
Bearbeitungslehre jig
Bearbeitungsspur (tool) mark (621.9)
Bearbeitungsspuren hinterlassen to mark
Bearbeitungsverfahren working procedure, working method
Bearbeitungszeit handling time, processing time
Bearbeitungszugabe (spanlos:) working allowance, (spanabhebend:) machining allowance
beaufschlagen to admit (steam); to trigger (621.3)
beaufschlagen mit Spannung to apply voltage, to impress voltage
beaufsichtigen to inspect, to control, to superintend, to supervise
Beaufsichtigung inspection, control, superintendence, supervision
beauftragen to charge
beben to vibrate; to quake (55)
Becher can, cup (621.35); bucket (621.86/87)
Becherkondensator encased capacitor
bechern to pot (621.318)
Beck-Lichtbogenlampe Beck arc lamp

bedämpfen to dampen
Bedarf an demand for, need of, want of
Bedarfsautomatik automatic load-response (621.311)
Bedarfshaltestelle conditional stop
Bedarfsspitze maximum demand
bedecken to cover, to clad, to shelter
bedeckt overcast (55)
Bedeckungsbereich coverage area
bedienen to operate, to attend
Bediener operator
Bediengerät terminal, control equipment
Bedienung attendance, control
Bedienungsanweisung instruction book, operating instruction, (owner's) manual, handbook
Bedienungsdeckel inspection cover, servicing cover
Bedienungseinrichtung control equipment
Bedienungsfehler operating error
Bedienungsfeld operator control panel
Bedienungseinfluß the human element
Bedienungsgang operating aisle, operation gangway (621.316)
Bedienungsknopf control
bedienungsloses Unterwerk unattended substation, automatic substation
Bedienungsmann operator
Bedienungsmannschaft operating crew
Bedienungsöffnung access opening
Bedienungsperson operator
Bedienungsperson in Nebenstellenanlagen attendant
Bedienungspult control desk, operators console, control console
Bedienungsschild instruction plate
Bedienungsvorschrift operating instructions
Bedienungswarte control room
bedingter Gegenverkehr semiduplex operation
bedingtes Halt permissive block (625)
bedingter Sprung conditional transfer of control, conditional jump
Bedingung condition, requirement
Bedingung am Leitungsende terminal condition
beeinflussen to affect, to influence, to act on, to control; (störend:) to interfere with
Beeinflussung (gegenseitige) interaction (mutual) interference
Befehl instruction (681.3)
Befehlsadresse instruction address
Befehlsadressenregister program counter, instruction address register
Befehlsaufbau instruction format
Befehlsbibliothek library
Befehlscode instruction code, operating code
Befehlsdecoder instruction decoder
Befehlsfolge control sequence, instruction sequence
Befehlsliste instruction list
Befehlsnockenschalter cam-type control switch
Befehlsregister control register, instruction register
Befehlssatz instruction set
Befehlsschema flow chart
Befehlssprache command language
Befehlsvorrat instruction set
Befehlswort instruction word
Befehlszähler address counter, location counter, program counter
Befehlszählregister instruction counter
Befehlszeit instruction time
Befehlszyklus instruction cycle
befestigen to fasten, to mount, to fix; to fortify
befestigen an to join to, to fasten to, to attach to
befestigen in to secure into
Befestigung fastening, fixing, securing, mounting, attachment; fortification (623)
Befestigungsbolzen fastening bolt
Befestigungsbügel fixing strap
Befestigungsdose support box (621.315.6)
Befestigungselement fastener, mounting
Befestigungsflansch mounting flange
Befestigungskeil wedge cotter, feather (621)
Befestigungsklemme mounting clip, binding clip
Befestigungsloch location hole
Befestigungsöse mounting lug
Befestigungsschelle mounting clamp
Befestigungsschraube attachment screw, fastening screw, fixing screw
Befestigungsstück fastener
Befestigungsvorrichtung mounting device
befeuchten to moisten, to wet, to humidify
Befeuchtung humidification, moistening
Befeuchtungsanlage humidification equipment
Befeuchtungsmittel moistening agent
befeuern to light (351.814)
befeuerte Landebahn flare path (629.13)
Befeuerung navigation lights, airway-lighting installation
beflechten to braid
Beflechtung braid, braiding
befördern to transport, to forward, to convey
Beförderung transportation, forwarding, conveyance

Beförderung untertage underground conveyance
begehbar passable
begehbares flaches Dach platform roof
begehbarer Kabelkanal cable subway
beginnen Strom durchzulassen to go into conduction
Beginnzeichen answer signal, clear forward signal, start signal
Beginnzustand initial condition
beglaubigen to verify, to testify
beglaubigte Normalien Board of Trade standards
Beglaubigungsprüfung referee test
Begleiterscheinung attendant phenomenon
Begleitsubstanz companion (matter), associated matter
begradigen to straighten (out)
begrenzen to outline; to limit, to clip (621.369)
begrenzt abhängige Verzögerung inverse time-lag with definite minimium
begrenzt-abhängige Maximal-Zeit-... inverse time-lag with definite minimum
begrenzt zugänglich restricted
begrenzter Raum confined space, bounded space
begrenzt verzögerte Auslösung definite time-lag tripping (621.316.5)
Begrenzer limiter, clipper (621.396); feed-trip stop (621.9)
Begrenzerschaltung suppressor (circuit), limiting circuit, clipping circuit
Begrenzerstufe limiter stage (621.396.62)
Begrenzerwirkung limiting action
Begrenzung limitation, limiting (action)
Begrenzung bilden to outline
Begrenzung durch Erdkrümmung horizon limitation
Begrenzung durch Gitterstromeinsatz cut-off limiting
Begrenzungsanschlag limiting stop
Begrenzungsdrossel limiting reactor, impeder
Begrenzungseigenschaft cut-off characteristic
Begrenzungsfläche bounding surface (51)
Begrenzungskontakt terminal contact
Behälter container, tank, retaining vessel, receptacle
Behälterinhalt tankage
behämmern to pene
behandeln to process, to manipulate, to handle
sich behandeln lassen nach to go by (51)
Behandlungsweise method of procedure; method of attack (51)
Beharrungsverhalten steady-state behaviour
Beharrungsvermögen inertia
Behelf expedient, auxiliary device
Behelfs-..., Behelfsmittel make-shift
behelfsmäßig improvised; by way of expedient
beidohrig binaural
Beidraht sheath wire (621.315)
Beidrahtschelle bonding clip
beidseitig bilateral, double-sided
beigeordnet co-ordinate
Beilage packing block, packing piece (621); pole shim (621.3)
durch Beilagen ausgleichen to shim
Beilageblech shim (plate)
Beilagekeile wedge cotters; adjustable parallel blocks
Beilagescheibe washer
Beilauf filler, filling (621.315.2)
Beilauffaden tracer
Beilegierung alloying inclusion
Beimengung admixture, additive; impurity (621.382)
beimischen to admix
beiseite aside, apart
beiseite drehen to swing aside
beißend mordant, pungent; caustic
beißender Geschmack pungency
beißender Rauch pungent smoke
Beißzange nippers (pl.)
Beiwert, Beizahl coefficient
Beize mordant, pickle, stain (67), (621)
beizen to pickle, to stain
beklöppeln to braid
Beklöppelung braiding
Bel T.U. = transmission unit, bel
beladen to load, to lade
Belag lining, facting, fur, coating
...belag ... per unit length
Belastbarkeit (load) carrying capacity, rating, permissible load
belasten to weight (621); to load (621.3); s. auch **beanspruchen**
belastet under load, loaded
belastete Antenne loaded antenna
belastetes Kabel loaded cable
belastete Leitung loaded line, line under load
Belastung load, loading, burden, stress; traffic carried (621.39); s. auch **Beanspruchung**
Belastung außerhalb der Spitzenzeit off-peak load
Belastung durch Wirkwiderstand resistive load
bei Belastung on load
auf eine Belastung arbeiten to work into a load

Belastungsfaktor (plant) capacity factor, output factor, load factor
Belastungsfaktor des Ausgangs fan-out (621.382)
Belastungsfaktor des Eingangs fan-in (621.382)
Belastungsgewicht dead weight
Belastungsgrenze load limit
Belastungskennlinie load line (621.396); load characteristic
Belastungskreis load circuit
Belastungskurve load curve
Belastungsprobe load test
Belastungsregler load regulator
Belastungsscheinwiderstand load impedance
Belastungsschwankung load variation
Belastungsschwerpunkt load centre
Belastungsspannung on-load voltage
Belastungsspitze load peak
Belastungsstoß current inrush
Belastungsstrom load current
Belastungsstrom entnehmen to draw a load current
Belastungsteilung load sharing (621.395)
Belastungswiderstand load resistance, load resistor, load impedance, artificial load
Belastungszustand loading conditions (pl.)
belegen to seize (621.395)
belegen mit to line with (621); to plate
belegt busy, occupied
belegte Bandbreite occupied bandwidth
Beleg (stück) proof, document
Belegleser record reader, document reader
Belegsortierer document sorter
Belegung occupation (621.395)
Belegungsdauer holding time, circuit holding time (621.395)
Belegungssignal seizure signal
Belegungsspektrum call mix (621.395)
Belegungsversuch call attempt
Belegungsvorgang seizure (621.395)
Belegungszeichen seizing signal
Belegungszeit busy time, holding time (621.395); cycle time (681.3)
Belegungszustand occupation level
beleuchten to illuminate
beleuchtete Skala illuminated dial
Beleuchtung lighting, illumination
Beleuchtung einschalten to illuminate
Beleuchtung wiedereinschalten to re-illuminate
Beleuchtungsanlage lighting installation
Beleuchtungsarmatur lighting fitting
Beleuchtungsdichte surface brightness, irradiancy (621.32)
Beleuchtungseinrichtung illuminating apparatus
Beleuchtungskabel (elctric) light cable
Beleuchtungskörper lighting fitting
Beleuchtungsmesser luxmeter, luxometer, photometer, illuminometer
Beleuchtungsmessung illumination measurement
Beleuchtungsmuster illumination pattern
Beleuchtungsradar illuminator
Beleuchtungsstärke illumination, lighting intensity, illuminance
Beleuchtungswirkungsgrad coefficient of utilization (621.32)
Beleuchtungszubehör lighting fittings (pl.)
belichten to expose (77)
Belichtung exposure (to light), lighting (77)
Belichtung mit Tageslicht daylight exposure
Belichtungsmesser exposure meter
beliebig arbitrary, optional, random
beliebiger Zugriff random access
Bell-Telefon Bell telephone, Bell receiver
belüften to vent, to ventilate, to aerate
Belüftung ventilation, artificial draught; windage (621.313)
Belüftungsanlage ventilating plant, ventilation system
Belüftungsöffnung ventilating aperture, breathing vent
Belüftungsschacht ventilating shaft
Belüftungssystem ventilating system, system of ventilation
bemannte Station attended station
bemessen to proportion, to rate
bemessen sein für to be rated
Bemessung rating
Bemessungsregeln dimensioning rules
Bendmannschutz Garton lightning arrester
benetzen to moisten, to wet
Benutzer user, subscriber
Benutzerstation user terminal (681.3)
Benutzerzeit serviceable time, up time
Benutzung utilization
Benutzungsdauer useful life; operation period
Benzin petrol, benzine; gasoline (US)
benzin-elektrischer Antrieb petrol-electric drive; gas-electric drive (US)
Benzinmotor petrol engine; gasoline engine (US)
Benzol benzene, benzole
Beobachter observer
Beobachtung observation
Beobachtungsfehler observational error
Beobachtungsplatz observation desk (621.395)

bequem snug, convenient; handy
bequeme Bedienung handiness of operation
bequemer Sitz snug fit
beratender Ingenieur consulting engineer
berechenbar calculable, computable, amenable to calculation
berechnen to reckon, to calculate, to compute, to analyse
berechnen auf, zu to compute at
berechen aus, nach to compute by
Berechnung reckoning, calculation, evaluation
Beregnungsanlage sprinkling plant
Beregnungsversuch wet test, rain test
Bereich sphere, range, scope, region, realm
Bereich des Möglichen realm of possibility
Bereich des negativen differentiellen Leitwerts negative differential conductance region (621.382)
Bereich überdecken (überstreichen) to sweep over a range, to cover a range
Bereich umfassen to enbrace a range, to cover a range
Bereichbreite breadth of range
Bereichsmitte midrange
Bereichsumschalter range switch
Bereitschaft standby, readiness
Bereitschaftsschalter standby switch
Bereitschaftssystem backup system (681.3)
Bereit-Zustand ready status (681.3)
Berg mount, mountain
Bergkristall rock crystal
Bergwerk mine, pit
Bergwerksmaschinen mining machinery (sing.)
Bergmannrohr slip conduit (621.3)
Bernstein amber
Berstdruck bursting limit, explosion limit
bersten to burst
beruhigen to calm
Beruhigung calming
Beruhigungswiderstand smoothing resistance
Beruhigungszeit response time, recovery time (621.317); damping time
berühren to touch, to make contact; to be tangent to
Berührung touch, contact
in Berührung sein mit to engage with
der Berührung zugänglich sein to be exposed (621.3)
Berührungselektrizität contact electricity
berührungsempfindlicher Schalter touch sensitive switch
berührungsempfindliches Terminal touch terminal
Berührungsfläche contact area, contact surface
Berührungsgefahr shock hazard (621.3)
Berührungslinie line of contact, contact line
berührungsloser Schalter proximity switch; magnetically-operated switch
Berührungsnormale common normal
Berührungspunkt point of contact, point of tangency
Berührungsschutz shock (-proof) protection (621.3); protection against contact
berührungsschutzsicher shock-proof (protected) (621.3)
berührungsschutzsicherer Schalter shock-proof switch, all-insulated switch
Berührungsspannung contact voltage
Beryllium beryllium
besäumen to edge, to trim (67)
besäumter Balken square-edged timber
Beschaffenheit quality (621.316)
Beschallung ultrasonic irradiation, exposure to ultrasonic waves
beschalten to wire, to load (621.395)
beschaltete Leitung loaded line (621.395)
Beschaltung wiring, cabling
Beschaltung mit Querdrosseln shunt loading
Bescheiddienst intercept service
Bescheidleitung enquiry circuit (621.395)
beschichten to laminate, to coat
Beschichtung deposition
Beschichtungseinrichtung laminator; coating machine
Beschickung feeding, charging
Beschießung mit Neutronen neutron bombardment
beschlagene Scheibe dimmed pane, hazy pane
beschleunigen to accelerate, to speed up
Beschleuniger promoter (66); accelerator (53), (61)
Beschleuniger mit Gleichspannung constant-potential accelerator
Beschleunigung acceleration
Beschleunigungsanode accelerator (621.385), accelerator anode
Beschleunigungsleistung accelerating power (621.33)
Beschleunigungsmesser accelerometer
Beschleunigungsmoment accelerating torque
Beschleunigungsspannung accelerating potential (621.385)
Beschleunigungsverlust accelerating loss (621.33)
beschmieren mit to smear with
beschneiden to pare, to trim; to clip (621.396)
beschossener Kern target nucleus (539)

Beschreibung description, descriptive literature, manual
Beschriftung lettering, (descriptive) legend
Beschriftungsschild tab
besetzt engaged, busy (621.395)
besetzte Leitung busy line (US), engaged line
besetzter Wähler engaged selector (621.395)
Besetzt ... busy ...
Besetztklappe supervisory signal
Besetztlampe busy lamp, hold lamp
Besetztprüfung busy test
Besetztton busy to ne
Besetztzeichen busy to ne, busy signal, engaged tone, busy-back, busy flash signal, engaged signal
Besetzzustand busy condition
bespinnen to braid
besponnener Draht braided wire
Bespinnung braid, braiding
bespulen to load, to pupinize
bespultes Kabel loaded cable, pupinized cable
bespulte Leitung loaded line (621.395)
Bespulung series loading
Besselsche Funktion Bessel function
Bestand stock, store; duration, consistence; stability
Bestandteil constituent matter, constituent substance, component part, component, inclusion (54)
Bestandteile aggregates
beständig of keeping qualities; settled (55)
...-beständig resistant to ..., ... proof
Beständigkeit fastness, stability, persistency
bestätigen to acknowledge
bestimmen to fix, to govern (51); to determine
bestimmte Fläche definite area
bestimmtes Integral definite integral
bestrahlen to irradiate
Bestrahlung irradiation; raying (61)
Bestrahlungsgerät radiation equipment
Bestrahlungskanal irradiation channel (539)
bestreichen mit to spread with, to coat with
Besuchsschaltung call transfer (621.395)
Bestückung instrumentation; component parts
Beta-Strahl beta ray
Beta-Teilchen beta particle
Betatron betatron
Beta-Zerfall beta decay
betätigen to operate, to actuate, to work
Betätigung operation, actuating, actuation, control
Betätigungsbatterie control battery
Betätigungseinrichtung operating device, handling device
Betätigungsgröße actuating quantity
Betätigungshebel operating lever, handling lever, actuating arm
Betätigungskabel control cable
Betätigungskraft operating force
Betätigungskurbel operating crank
Betätigungsmechanismus operating mechanism
Betätigungsschalter (remote) control switch, trip switch, actuating switch
Betätigungsseil tracker wire
Betätigungssignal actuating signal
Betätigungsspannung control voltage, operating voltage, actuating voltage
Betätigungsstrom actuating current
Betätigungsstromkreis control circuit
Betätigungstaste operational key, functional key
Betätigungsventil valve actuator
Beton concrete
Betonmast pre-cast pile, concrete pole
betonieren to concrete
Betrachtung consideration; viewing
Betrachtungsabstand viewing distance
Betrachtungsapparat viewing lantern (77), (621.386)
Betrachtungsschirm oscilloscope screen
Betrachtungswinkel viewing angle
Betrag modulus (51); amplitude (621.385); magnitude (62)
Betrag eines Vektors magnitude of a vector
Betragbildung rationalization (51)
betreiben to run, to operate; to work (**mit:** by)
betrieben driven
betrieben werden to be in action
Betrieb plant, factory, work (s); operation service
„Betrieb" run
Betrieb einstellen to shut (down); to take out of operation
außer Betrieb out of action, out of service
außer Betrieb kommen to come to rest
außer Betrieb setzen to shut down, to put out of action
im Betrieb gealtert service-aged
in Betrieb in operation, on the go, in action, on service
in Betrieb gehen to go into operation
in Betrieb halten to keep in service
in Betrieb kommen to come into operation
in Betrieb setzen to put into operation
voll in Betrieb sein to work full swing; in full operation
Betrieb eines Netzes network operation

betrieblich operational
betriebliche Anforderung duty
betriebliche Bedingungen operational conditions
betriebliches Verhalten performance
Betriebs-... operating, service
Betriebsabschnitt run, operational phase
Betriebsanweisung operating instructions (pl.)
Betriebsart mode of operation, operation mode, method of operation; type of service, operating duty (621.313)
Betriebsbeanspruchung operating duty, working stress, usage
Betriebsbedingungen operating conditions, service conditions, working conditions
unter Betriebsbedingungen under service conditions
Betriebsbereich service band (621.396)
betriebsbereit operative, ready for service
Betriebsdämpfung composite attenuation; overall loss, overall attenuation (621.392)
Betriebsdauer operating period, working time
Betriebsdrehzahl running speed, working speed, operating speed
Betriebsdruck working pressure
Betriebseigenschaft operating characteristic
Betriebserde system earth; system ground (US); signal ground (UITT)
betriebsfähig in working condition, operable
betriebsfähiger Zustand working condition, operating condition
Betriebsfrequenz operating frequency, normal frequency
Betriebsgüte operational quality
Betriebsingenieur operating engineer, production engineer, shop manager
Betriebskapazität eines Leiters bei Erdung aller übrigen grounded capacitance
Betriebskapazität zwischen zwei Leitern bei Erdnung aller übrigen mutual capacitance
Betriebskapazität zwischen zwei Leitern bei freiem Potential aller übrigen plenary capacitance
Betriebskennlinie operating characteristic
Betriebskraft working power
Betriebs-Leistungsverstärkung transducer (power) gain (621.382)
Betriebsleiter works superintendent, works manager; plant manager
betriebsmäßig operational
betriebsmäßige Beanspruchung working stress
betriebsmäßige Überlastung operating overload
in betriebsmäßigem Zustand in working order
Betriebspause period of rest
Betriebspegel working level, power level (621.395)
Betriebsprogramm executive program (681.3); service program, service routine
Betriebsschaubild diagrammatic display, graphic panel, functional diagram
betriebssicher reliable, fail-safe
Betriebssicherheit service reliability
Betriebssicherung active fuse
Betriebsspannung service voltage, operating voltage (621.316.5)
Betriebsspannung (eines Motors, der mit Anlaßspannung angelassen wird) running voltage
Betriebsspiel duty cycle
„Betriebs"-Stellung running position, operating position
Betriebsstörung service interruption, service derangement; break down
Betriebsstrom working current
Betriebsstromkreis working circuit, service circuit
Betriebsstufe running notch
Betriebsstunde hour of service, operating hour(s)
Betriebsstundenzähler (working) hours counter, elapsed-time indicator, service-time register, operating-hour counter
Betriebssystem operating system (681.3)
Betriebstemperatur operating temperature, working temperature
Betriebsunterbrechung service interruption
Betriebsverfahren operation procedure, mode of operation
Betriebsverstärkung overall amplification (621.396.6)
Betriebsvorschriften working instructions, operating instructions
Betriebswellenlänge operating wavelength
Betriebswerterechner operating data computer
Betriebszeit opertation time, up time
Betriebszustand working conditions, service conditions (pl.)
beugen to diffract (53), (621.396); to bend (62)
beugende Kante diffraction knife edge (621.396)
Beugung diffraction
Beutelelement Leclanché element, sack cell
Beverage-Antenne Beverage aerial, wave aerial
bewegen, (sich) to move

bewegende Kraft motive force
beweglich moving, movable, mobile; flexible
bewegliche Aufhängung flexible suspension
beweglicher Funkdienst mobile radio service
bewegliche Installationsleitung flexible conduit
beweglicher Kontakt moving contact
bewegliche Landfunkstelle land mobile station
beweglicher Plattensatz moving plates, moving vanes (pl.) (621.319.4)
bewegliches System moving system, moving element (621.317)
bewegliche Teile motion work (sing.), moving parts
Beweglichkeit mobility
Bewegung motion, movement, agitation
Bewegung in einem Feld movement across a field
Bewegung nach links left-hand motion
Bewegung (von Ort zu Ort) locomotion
Bewegung übertragen auf to convey movement to
aus Bewegung entstehend motional
in Bewegung setzen to set in motion, to convey movement to
Bewegungsgröße momentum, quantity of movement
bewegungslos motionless
bewehren to reinforce, to armour; to armor (US) (621.315.2)
bewehrtes Kabel armoured cable
Bewehrung armour(ing), metal reinforcement
Bewehrungsschelle armour clamp
bewerten to weight; to rate
bewertete Geräuschleistung psophometric power
bewerteter Strom weighted current
Bewertungsfaktor weighting factor
Bewertungsprogramm benchmark (program) (681.3)
bewickeln (mit) to wind (with); to wrap
bewickelter Rotor wound rotor
Bewicklung (mit Band) taping (621.3)
bewirken to cause, to effect
bezeichnen to indicate, to mark, to denote, to designate
bezeichnen mit to denote by
mit „x" bezeichnen to mark "x"
Bezeichnung designation, term; symbol
Bezeichnungsfahne marking tag
Bezeichnungshülse identification sleeve
Bezeichnungsschild name plate, label
Bezeichnungsweise terminology, notation
beziehen auf, sich to refer to; to be related to
bezogener Blindleitwert "per unit" susceptance
bezogene Leitungseinheit primary constant
bezogener Leitwert "per unit" admittance
Beziehung correlation; relationship; reference
in Beziehung setzen zu to relate to, to relate with, to take in relation to
beziffern to number, to figure
sich beziffern auf to amount to, to total up to
Bezifferung designation, (numeral) notation
Bezirk district, region
Bezirkssender regional broadcasting station
Bezirkswähler code selector (621.395)
Bezug covering; s. auch **Beziehung**
Bezugsachse axis of reference
Bezugsdämpfung reference equivalent; volume equivalent (621.39)
Bezugsebene reference plane
Bezugsfrequenz reference frequency
Bezugsgröße reference value
Bezugskreis reference standard, reference circuit (621.392/6)
Bezugsnormal reference standard
Bezugspegel reference level (621.395/6)
Bezugspegel Null zero reference level
Bezugspotential reference potential
Bezugspunkt reference point
Bezugsspannung voltage reference
Bezugssystem system of reference (51), (53)
Bezugstemperatur reference temperature
Bezugszeit reference time
B-H-Kurve B-curve
Biberschwanzkeule beaver-tail beam (621.396.67)
Bichromat bichromate
Bichromat-Element Fuller cell
bidirektionaler Transistor bi-directional transistor
bidirektionaler Thyristor Triac
Biege-... bending, flexural
Biegebeanspruchung bending strain
Biegefestigkeit bending strength, cross-breaking strength
Biegemaschine folding machine
Biegeschwingung flexural vibration, flexural mode
Biegespannung bending stress
Biegeversuch bend test
Biegezahl bend-test number
Biegezange pliers (pl.)
biegefähig flexile
biegen to bend, to inflect; (Schmiegsamkeitsprüfung:) to flex
biegeprüfen to test by bending
gebogen bent, curved
biegsam flexible, felxile; pliable
biegsamer Leiter flexible conductor

biegsame Welle flexible shaft, flexible drive
Biegsamkeit flexibility
Biegung bend, deflection; curvature
Biegungsspannung bending stress
biegungsstarr, biegungssteif stiff
bifilar bifilar
bifilar gewickelt non-inductively wound
bifilare Aufhängung bifilar suspension
bifilare Wicklung bifilar winding, double spiral coil
bifokal bifocal
bikonvexe Linse double-convex lens
Bild image, picture, frame (621.397), (77)
Bild erzeugen to form an image (535)
Bild festhalten (durch Nachleuchten) to hold the image
Bild im Text inset
Bildabschattung shading
Bildabtastgerät scanner
Bildabtaströhre scanning tube
Bildabtastung scanning (621.397)
Bildauflösung picture resolution
Bildaufnahme (direct) recording
Bildaufnahmeröhre camera tube
Bildaufnahmeröhre mit Photohalbleiterschicht photoconductive camera tube
Bildaufnahmeröhre mit Photokathode photoemissive camera tube
Bildaufzeichnungsgerät video recorder, video tape recorder
Bildaustastsynchronsignal BAS-signal
Bildbreiteregler width control
Bilddauer frame duration
Bilddurchlauf picture slip, frame slip
Bildelement picture element (621.397)
in Bildelemente auflösen to scan
Bildfalle adjacent carrier trap
Bildfernsprecher videophone
Bildfolgefrequenz repetition rate, picture frequency, frame frequency (621.397)
Bildgüte pictorial quality (621.397)
Bildhelligkeit picture brightness, level of illumination, brightness of image (621.397)
Bildimpuls picture pulse, frame pulse
Bildkanal picture channel
Bildkontrollgerät picture and waveform monitor
Bildkontrolloszillograph waveform monitor
Bildlageeinstellung framing (621.397)
Bildmischgerät video mixer
Bildmodulation vision modulation, image modulation
Bildmonitor picture monitor
Bildmuster test pattern
Bildmustergenerator grating and dot generator, pattern generator
Bildpunkt representative point (51)
Bildröhre picture tube
Bildschärfe definition (53), image clearness
Bildschärferegler focusing control
Bildschirm screen, viewing screen, display scope, scope
Bildschirmkonsole display console
Bildschirmtext show cable
Bildsender TV transmitter, television transmitter, picture transmitter
Bildschwarz picture black
Bildsignal picture signal
Bildsignal mit Austastung blanked picture signal
Bildsignalmittelwert average picture signal level
Bildspeicherröhre mosaic tube
Bildspur paint
Bildstandsregelung centering control
Bildsynchronisierung sychronizing of images (621.397)
Bildtelegraphie phototelegraphy, picture telegraphy
Bildträger video carrier
Bildträgerfrequenz picture tone
Bildträgersperre adjacent video carrier trap
bildübertragen to telephotograph
Bildübertragung facsimile transmission, picture transmission
Bildübertragungsdauer picture transmission period
Bildverhältnis aspect ratio (621.397)
Bildvermittlungsplatz phototelegraph position (621.394.6)
Bildverstärkerröhre image-intensifier tube
Bildwandler image converter
Bildwandlermikroskop infrared microscope
Bildwechselfrequenz frame frequency (621.397)
Bildwechselzahl picture frequency
Bildweiß picture white
Bildwiedergaberöhre picture tube
Bildwinkel viewing angle
Bildzeile scanning line, picture strip (621.397)
Bildzerleger image dissector
Bildzerlegung scanning
bildlich darstellen to picture, to depict, to illustrate
bildsam plastic, ductile
Bildung formation, structure, organization; education (3)
Bildungswärme heat of formation

Billiarde quadrillion (US)
Billion billion (Br), trillion (US)
Bimetall bimetal
Bimetallauslöser bimetallic release, bimetallic trip, thermal release
Bimetall-Meßinstrument thermo-bimetallic instrument
Bimetallrelais bimetal relais, thermal relay
Bimetallstreifen bimetal strip, compound strip
Bimsstein pumice (stone)
binär binary
binäre elektrische Größe binary electrical quantity
binäres Speicherelement binary cell
binär verschlüsselt binary coded
binär verschlüsseltes Dezimalsystem binary coded decimal system
binäres Zahlensystem binary notation, binary number system
Binärcode binary code
Binärcode für Dezimalziffern binary coded decimal, BCD-code
Binär-Dezimalumwandlung binary-to-decimal conversion
Binärkomma binary point
Binärkonverter binary converter
Binäroperation binary operation
Binärschaltung binary circuit
binäres Speicherelement binary cell
Binärteiler binary divider, scale of two counter, binary counter
Binärzahl binary number
Binärzähler binary counter, radix-two counter
Binärzeichen binary digit, bit
Binärziffer binary figure
Binde-... binding, bonding
Bindedraht binding wire
Bindegarn binder twine (621.3)
Bindeglied (connecting) link, bond
Bindelack binder
Bindemittel binder, bond, binding agent, bonding agent, binding substance, bonding substance
binden to bind, to bond; to fix, to tie up, to combine (54)
gebundene Energie latent energy
gebundener Kohlenstoff combined carbon
gebundener Sauerstoff fixed oxygen
gebundenes Wasser combined water
Bindung binding, bond(ing); bond (539)
binokular binocular
Binom binomial
Binominalantenne binominal antenna array
binomisch binomial
binomischer Lehrsatz binomial theorem
Biot-Savartsches Gesetz Biot-Savart law, Ampére's law
bipolar bipolar
bipolarer Halbleiter bipolar semiconductor
biquadratisch biquadratic
Birne (lamp-) bulb (621.326)
birnenförmig pear-shaped
Biskuitporzellan biscuit, bisque
bistabil bistable
bistabile Kippschaltung bistable trigger circuit
bistabiler Multivibrator bistable flip-flop, R-S flip-flop, bistable trigger circuit, bistable multivibrator, toggle trigger
Bit bit = binary digit
Bitdichte bit density, recording density
Bitfehlerhäufigkeit bit error rate
Bitfolge bit rate
bitparallel bit parallel
Bitstelle bit location
Bit mit dem niedrigsten Stellenwert least signification bit
Bitumen bitumen
Bitumenüberzug bituminous coating
bituminös bituminous
blank bare (621.3), (621.791); bright (535)
blankes Bleikabel plain leadcovered cable
blanker Draht bare wire, uninsulated wire
blanke Elektrode bare electrode (621.791)
blanker Leiter bare conductor
bis aufs blanke Metall down to the bare metal
blankgemachtes Ende bared end
blankreiben to rub bare
Blas-... blow, blowing
Blaskopf blower
Blasmagnet arc blow-out, arc deflector, magnetic blow-out (621.325)
Blassicherung blow-out fuse
Blasspule blow-out coil (621.316.5)
Blase blow hole (621.791); bleb (66); blister, void (66); bubble
Blase bilden to bubble
Blasen werfen to blister
Blasenbildung blister formation; formation of blow holes; bubbling
blasenfrei non-porous (621.7)
Blasenspeicher bubble memory
blasig auftreiben to blister
Blatt leaf, foil; blade (621.9); scarf (69); sheet (65)
Blatt einer Dickenlehre leaf of a gauge
Blattelektrometer leaf electrometer
Blattfeder spring blade, spring leaf, flat spring, leaf spring

Blattgold leaf gold, gold foil
Blattmetall metal foil
Blattschreiber page printer
blättern to blister
blättrig foliated, leafy
Blaupause blueprint
blaupausen to blueprint
Blauschriftröhre dark trace tube
Blech sheet, sheeting, plate
Bleche walzen to roll sheets
aus Blechen aufbauen to laminate
blechabgedeckter Kabelkanal surface-metal raceway
blechumhüllte Elektrode sheathed electrode
Blechalterung magnetic fatigue (621.313)
Blechbearbeitung sheet metal working, plate working
Blechbeilage shim
Blechbelag sheeting
Blechbiegemaschine plate bending rolls (pl.)
Blechgehäuse sheet-iron casing
Blechisolation lamination insulation
Blechjoch laminated yoke
Blechkern laminated core
Blechrichtmaschine plate-straightening rolls (pl.)
Blechrohr sheet-iron tube
Blechschnitt lamination, iron punching, iron stamping
Blechschnitte stanzen to blank
Blechwand sheet-iron panel, metal panel
Blechzwischenlage separator plate
blechen to laminate
geblechter Pol laminated pole
Blei lead
aus Blei leaden
Bleiakkumulator lead cell, lead accumulator storage battery
Bleiauskleidung lead lining
Bleiazetat sugar of lead
Bleibacken jaw lining (sing.)
Bleibaum lead tree (621.35)
Bleibaumbildung treeing
Bleibelag lead lining
Bleiblende, Bleiblendenanordnung grid technique (621.386)
Bleidichtung lead seal
Bleiform lead mould (621.357)
Bleigitter Potter-Bucky grid (621.386)
Bleiglanz galena
Bleiglanzdetektor galena detector
Bleiglätte litharge
Bleigummi lead rubber
Bleihammer lead hammer, babbitt hammer
Bleihülle lead sheath
Bleikabel lead-covered cable
Bleikabel mit Korrosionsschutz served lead-covered cable
Bleikappe (= Zubleiung) für unter Spannung stehendes Kabel live cable test cap
Bleimantel lead sheath, lead coating, lead covering
mit Bleimantel lead-covered
Bleimantelkabel lead-sheathed cable
Bleimuffe lead jointing
Bleiplatte lead plate
Bleirohr lead pipe
Bleisammler lead cell, lead accumulator
Bleistreifen lead strip
Bleisulfiddetektor galena detector
Bleisuperoxid lead dioxide
Bleiummantelung lead covering, lead sheathing
bleibend permanent, residual
bleibende Dehnung strech, permanent extension (620.1)
bleibende Einstellung permanence of setting
bleibende Regelabweichung (permanent) offset, steady-state error
bleibende Verformung permanent set, permanent deformation
Blende aperture, stop (77), diaphragm (535); blende (54)
Blendenverzerrung aperture distortion
blendfreies Glas anti-glare glass
Blendschutzbrille antiglare spectacles (pl.)
Blendschutzlampe antidazzle lamp
blendungsfrei glareless
blendungsfreie Beleuchtung glareless lighting
Blickfeld field of vision, field of view
Blickfeldanzeige head-up display, field of view display
Blind-... reactive, wattless; mock, dummy (621.396); blind
Blindanflug instrument approach
Blindanteil wattless component
Blindantenne dump aerial, dummy aerial, absorber
Blindfeld blank panel
Blindflug instrument flight
Blindkilowatt, Blind-kVA kilovar
Blindkilowattstunde kilovarhour, kvarh
Blindkomponente wattless component, idle component, reactive component (621.3)
Blindlandeanlage blind landing system, instrument landing system
Blindlandeausrüstung instrument landing equipment

Blindlandung blind landing, instrument landing
Blindlast reactive load
Blindlastzählung reactive metering
Blindleistung wattless power, blind power, reactive power, reactive voltamps (pl.)
Blindleistung der gegenläufigen Komponente negative-sequence reactive power
Blindleistung des mitläufigen Systems positive-sequence reactive power
Blindleistung des Nullsystems zero-sequence reactive power
Blindleistungsausgleich power-factor correction, reactive power correction
Blindleistungsmaschine synchronous reactor, synchronous compensator
Blindleistungsmesser reactice-power meter, idle-current wattmeter
Blindleistung/Scheinleistung reactive factor
Blindleistungszähler reactive volt-ampere (-hour) meter
Blindleitung nondissipative stub, stub
Blindleitwert susceptance
Blindplatte blind
Blindschaltbild mimic diagram, mimic bus diagram
Blindspannung reactive (component of) voltage
Blindspritze matching stub, stub line (621.396)
Blindspritzenanschluß stub attachment (621.396)
Blindstrom idle current, blind current, wattless current, reactive current, reactive amperage
Blindstrom-Amperemeter reactive current meter
Blindstrom-Amperestundenzähler reactive ampere hour meter
Blindstromausgleich power-factor correction
Blindstromkompensation power-factor compensation
Blindstrommesser idle current meter, reactive current meter
Blindstromzähler reactive ampere-hour meter, wattless component meter, varhour meter
Blindverkehr waste traffic
Blindversuch dry running
Blindwatt wattless power
Blindwiderstand, (induktiver) reactance
Blindwiderstand für die gegenläufige Komponente negative-sequence reactance
Blindwiderstand für das Nullsystem zero-sequence reactance
Blindversuch dry running
Blinkanzeige flashing indication
Blinkeinrichtung flasher; direction indicator (629)
Blinkfeuer intermittent light, flashing light (656.7)
Blinklicht flashing light, flicker light
Blinkrelais flicker-light relay
Blinkscheinwerfer signal light
blinken to wink, to glimmer; to flash, to blink
Blitz lightning; flash
Blitzableiter lightning arrester, surge diverter (621.316.9); lightning protector, lightning conductor, lightning rod (69)
Blitzauffangstange lightning rod
Blitzeinschlag lightning stroke
Blitzentladung lightning discharge
Blitzfeuer flashing beacon, flashing light (656.7)
Blitzfrequenz glimpse frequency
Blitzgeschwindigkeit lightning speed
Blitzlicht flash-light
Blitzlichtlampe flash lamp, photoflash, flash bulb
Blitzröhre fulgurite (55); flash tube (621.38)
Blitzschlag lightning stroke, lightning flash, thunderbolt
Blitzschutz lightning protection, lightning protector
Blitzschutzrelais arrester relay
Blitzschutzschalter lightning switch
Blitzstrom lightning current
Blitzüberspannung lightning surge
blitzen to flash, to fulminate (621.394.6)
Block block (section) (621.33); blocking capacitor (621.33); block (681.3)
Blockbeginnzeichen start-of-block signal (681.3)
Blockbetrieb unit-system operation (621.311)
Blockendezeichen end-of-block signal
Blockfehlerquote block error rate
Blockkondensator blocking capacitor, blocking condenser
Blockschaltbild block (schematic) diagram
Blockschwinger blocking oscillator, shock-excited oscillator, Squegger oscillator
Blockschwingerbetrieb intermittent operation (621.396)
Blocksignal interlocking signal, block signal (621.33)
Blockstation isolated generating plant (621.311); blocking signal station (625)
Blockstrecke block (section) (621.33)
Blocksystem block system (621.33)

Blockübertragung block transfer
blockieren to lock, to block; to freeze (621); to inhibit (681.3)
Blockierung congestion, blocking, interlocking, hold
Blockierung des Arbeitsablaufs hangup (681.3)
Blockierungsrelais interlocking relay
Blockierungssignal inhibiting signal (681.3)
blonder Lack laquer (681)
bloßes Auge naked eye
blubbern to bubble
BMZ = begrenzt-abhängig-Maximal-Zeit (s. dort)
Bock (gerüst) gantry, trestle, jack, stand support
Boden soil, earth, land, ground (level); floor
im Boden versickern to sink into the soil
bodenabhängige Navigation ground dependent navigation
bodengesteuertes Anflugsystem G. C. A., ground controlled approach system
bodengesteuerte Jägerleitung ground controlled interception
bodengesteuerte Verkehrskontrolle ground-base traffic control (351)
bodenunabhängige Navigation autonomous navigation, inertial navigation
Boden-Boden Flugkörper surface-surface missile
Boden-Bord-Sprechverkehr ground-to-air radiotelephony
Boden-Bord-Verbindung ground-to-air communication
Bodenanlage ground installation
Bodendämpfung ground absorption
Bodendruck foundation pressure
Bodeneinfluß ground effect (621.396)
Bodenfeuchtigkeit ground damp
Bodenfläche bottom face, floor space
Bodenflansch bottom flange
Bodenkontakt floor contact (621.33), (621.876)
Bodenlast panel loads (pl.)
Bodenleitfähigkeit earth conductivity
Boden-Luft Flugkörper surface-air missile
Bodenmarkierung ground marker (656.7)
Bodenpressung foundation pressure, specific pressure (624)
Bodenradar ground radar (621.396.96)
Bodenreflexion ground reflexion (621.396)
Bodenrückstreuung ground-back scatter
Bodensatz deposit, ground; foots (pl.) (54); residue (66)
Bodenstation ground station (621.396)
Bodentemperatur soil temperature
Bodenwelle ground wave, direct wave, surface wave (621.396)
Bodenwellenrandzone intermittend service area (621.396)
Bogen arc, bend (51), bend (621.6); arc (621.3); arch (72)
90°-Bogen normal bend, sharp bend (621.315.6)
135°-Bogen half bend (621.315.6)
Bogenfußpunkt arc electrode (621.3)
Bogenlampe arc lamp, arc light
Bogenlampenbeleuchtung arc lighting
Bogenlampenkohle arc-lamp carbon
Bogenlicht arc lighting
Bogenminute minute of arc
Bogenschiene curved rail (625)
Bogenspannung arc voltage
Bogenspektrum arc spectrum
Bohnermaschine electric floor polisher
Bohr-... drilling; boring
Bohrfutter drill chuck
Bohrknarre ratchet brace
Bohrmaschine drill(ing) machine
Bohrmehl drillings (pl.)
Bohröl cutting oil, soluble oil (621.9)
Bohrschablone template
Bohrtisch bed (621.9)
Bohrzeug drilling tackle (622)
bohren to drill; to bore
Bohrer drill; borer, boring tool
Bohrung bore
Bohrungsachse bore axis
Boiler electric boiler
Boje buyo
Bold decoy, sham (621.396.96)
Bolometer bolometer
Bolometerkopfverstärker hotwire magnifier, Heurtley magnifier
Boltzmannsche Konstante Boltzmann's constant, constant of entropy equation
Bolzen bolt, pin, gudgeon
Bolzenloch bolt hole
Bolzenverbindung pin-joint
Bombardement, Bombardierung bombardment (621.385)
bombardieren to bomb, to bombard (621.385)
Boolsche Algebra Boolean algebra
Boolsches Schaltglied Boolean gate
Borax borax, tincal
Bord ...; airborne (Flugzeug), shipborne (Schiff)
Bord-Boden ... air-ground
Bordanzeige airborne display
Bordfunk airborne radio
Bordnetz (airborne) power system

Bordradar airborne radar
Bordsprechanlage intercommunication system, internal communication, interphone
Bördelschweißung double-flanged butt-weld
Borium-Zählrohr boron counter tube
Boschhorn electric horn, (electric) hooter
Böschung escarpment, slope; bank
Böschungswinkel slope angle, angle of repose
Boucherot-Motor Boucherot motor
Boucherot-Schaltung Boucherot circuit
Bougie-Kerze decimal candle
Bourdonrohr Bourdon tube
Bowdenzug Bowden control, control cable
Br. = Breite lat. = latitude
Brand blaze, fire
in Brand on fire, alight; ablaze
in Brand geraten to catch fire, to take fire
in Brand setzen to set on fire, to set alight
Brandfleck burn mark
Brandgefahr fire risk, fire hazard
Brauchbarkeitsdauer life utility
Brauneisenstein brown iron ore
Braunkohle brown coal, lignite
Braunkohlenkraftwerk lignite-fired power station
Braunstein manganese ore
Braunsches Rohr CR-tube = cathode-ray tube, Braun tube
Braunsches Rohr mit magnetischer Ablenkung magnetic cathode-ray tube
Brech-... breaking, facturing (620.1); crushing (622); refracting (535)
Brechvorgang breaking (62)
Brechwerk crushing, machine crusher
brechen to break, to fracture (620.1); to crush (622); to refract (535)
Brechung refraction (535), (621.371); sound-wave refraction (534)
Brechungsgesetz law of refraction
Brechungsindex refractive index, index of refraction
Brechungswinkel angle of refraction
Brechungszahl refractive index
Brei paste
breit broad, wide
breites Band broad band, wide band
breite, flache Resonanzkurve broad flat resonance curve
breiter Flansch wide flange
breites Frequenzband wide-frequency range, wide band (621.396)
breite Resonanzkurve broad resonance curve
Breitband wide band, broad band (621.396)
Breitbandantenne wide-band aerial; broad-band aerial
Breitbandauslegung broadbanding (621.396)
Breitbandempfang wide-band reception, broad-band reception
Breitbandempfänger wide-band receiver, broad-band receiver, wide-open receiver
Breitbandfilter wide bandpass filter
breitbandig frequency inselective flat, wide-band (621.396.62/67)
breitbandige Frequenzkurve flat response
Breitband-Richtfunksystem wide-band radio relay system
Breitbandstörung barrage jamming
Breitband-TF-System broad-band carrier system
Breitbandübertrager wide-band transformer
Breitbandverstärker wide-band amplifier
Breitflansch-Schlagwettersicherung flange protection
Breitflanschträger wide-flanged girder
Breitspur broad gauge (625)
Breitstrahler wide-angle lighting fitting; wide-angle luminaire (US)
Bremse brake
Bremse (n) anziehen to put the brake (s) on
Bremse (n) lösen to release the brake (s)
Bremse (n) ziehen to apply the brake (s)
Bremsbelag brake lining
Bremselektrode decelerating electrode
Bremsfeld retarding field
Bremsfeldröhre retardation valve, retarding field tube, inverted tube (621.396)
Bremsfeldsender retarding field generator
Bremsfläche slowing-down area (539)
Bremsgitter anti-secondary grid, suppressor (grid) (621.385)
Bremsgittermodulation suppressor (grid) modulation
Bremskraft retarding force, decelerating force
Bremsleistung b.h.p. = brake horse power
Bremsleitung brake pipe
Bremslüfter brake lifting magnet
Bremsmagnet breaking magnet, magnetic brake (621.317)
Bremsmittel moderator (54), (539)
Bremsmoment retarding torque, brake torque
Bremsmotor brake motor
Bremsprobe brake test
Brems-PS s. **Bremsleistung**
Bremsring brake collar
Bremsschalter braking controller, braking switch, braking notch (621.33)
Bremsscheibe brake disk (621.317)

Bremsschelle brake clamp; brake collar
Bremsschuh brake shoe, (wheel-) drag (621.33)
Bremsstand brake-test bench
Bremsstandprüfung brake test
Bremsstellung retard position
Bremsstrahl brake jet (621.2)
Bremsstrahlung impulse radiation (621.386), radiation due to retarding of particles
Bremsstrecke braking distance
Bremsstrom braking current
Bremssubstanz moderation (54), (539); inhibitor (621.357)
Bremssystem braking system
Bremstrommel brake drum
Bremsverzögerung deceleration
Bremsvorrichtung brake mechanism, braking device
Bremswaage brake dynamometer
Bremswächter brake guard, brake control device
Bremsweg braking distance
Bremswirkung braking effect
bremsen to brake, to retard, to stop; to moderate (54), (539)
Bremsung braking, retardation
Brenn-... combustion, firing, burning
Brenndauer burning time, burning period, duration of combustion
Brennelement fuel element (539)
Brennelementgruft fuel-element cave
Brennelementhülle cladding, clad (539)
Brennelementschaden fuel burn out (539)
Brennelementschleuse fuel air lock
Brennfleck focus (535)
Brennkammer combustion chamber
Brennofen kiln
Brennöl fuel oil
Brennprobe life test (621.326)
Brennpunkt (principal) focus, focal point
in den Brennpunkt stellen to focalize, to focus
vor dem Brennpunkt liegend pre-focus
Brennschneiden cauterization (61); torch-cutting, cut-welding (621.7)
Brennspannung tube voltage drop, arc voltage (621.385)
Brennstelle lighting outlet (696)
Brennstoff fuel, combustible
Brennstoffaufbereitung fuel reprocessing
Brennstoffelement fuel cell (539); carbon-consuming cell, carbon-combustion cell, thermo-chemical generator
Brennstofflagerung fuel storage
Brennstoffstab fuel rod (539)
Brennstoffverbrauch fuel consumption
Brennstunde hour service (621.385)
Brenntemperatur firing temperature (666)
Brennvorgang burning; vitrification (66)
Brennweite principle focus, focal length
mit zwei Brennweiten bifocal
brennbar inflammable, combustible, ignitable
brennbarer Bestandteil combustible constituent
brennen to burn (621); to bake (66); to calcine (622); to distil (66)
brennend on fire, afire, alight; lit
brennende Lampe lit lamp
Brenner burner, torch (621.791)
Brennerdüse burner nipple
Brennermund port, torch orifice
Brett plank, board, panel
Brettschaltung bread-board layout
Brewster-Winkel Brewster angle
Briefaufstellanlage letter-facing equipment (35)
Briefsortieranlage letter-culling equipment (35)
Briggsscher Logarithmus common logarithm
Brikett briquette
Brinellhärte Brinell hardness
Brinell(härte)zahl Brinell-hardness number
Britischer Normenausschuß BESA = British Engineering Standard Association
Britische Wärmeeinheit B. Th. U. = British thermal Unit (1 B. Th. U. = 0, 2522 kcal = 1,056 kJ)
Brodelstörung grinders (pl.) (621.396)
Bronze bronze, gun metal, admiralty metal
Bronzedraht bronze wire, bronze conductor
Bronzeseil stranded bronze conductor
Brotröster electric toaster
Bruch rupture, breakage, fissure; fraction (51); fracture (61); breakdown (621)
bis zum Bruch prüfen to test to destruction
Bruchbelastung breaking load
Bruchdehnung elongation (on break), elongation on fracture
Bruchdicke throat (539)
Bruchfestigkeit breaking strength, rupturing strength, U.T.S. = ultimate (tensile) strength, braking strain, modulus of rupture
Bruchfläche fracture plane, plane of rupture
Bruchgefüge fracture structure
Bruchkupplung breakable coupling
Bruchlast breaking load, ultimate stress (621.1)
Bruchlochwicklung short chord winding (621.313)
Bruchneigung failure tendency
Bruchquerschnitt throat

Bruchspannung breaking stress, ultimate stress
Bruchstelle break (621)
Bruchstrich fracture line, fractional line (51)
Bruchstücke breakup constituents
Bruchteil fraction
bruchzahlig fractional
Bruchziffer fractional number
brüchig brittle
Brüchigkeit brittleness
Brücke bridge (621.1), (621.317), (629.12)
Brückenabgleich bridge balance
Brückenanordnung bridge assembly
Brücken-Duplexschaltung bridge duplex system
Brückeneckpunkt corner of a bridge (621.317)
Brückengleichrichter bridge rectifier (circuit)
Brückenglied bridge section (621.317)
Brückenmessung bridge measurement
Brückenschaltung bridge assembly, bridge circuit; double-way connection (621.314)
Brückenschaltung mit Selbstabgleich self-balancing bridge
Brückenweiche diplexer
Brückenzünder low-tension detonator (623)
Brückenzweig arm of a bridge, leg of a bridge (621.317)
Brückenzweig gegenüber... arm opposite to... (621.317)
Brumm, Brummen hum, humming, ripple
Brumm und Rauschen ripple and noise
Brummabstand signal-to-hum ratio
Brummeinstreuung hum pick-up
Brummentstörungsspule humbucking coil
Brummpegel hum level
Brummspannung hum voltage, ripple voltage
brummen to hum; to buzz, to grumble
Brustmikrophon breast transmitter
Brustsprechsatz head-and-chest set
Brüten breeding
Brüter (breeder) reactor (539)
Brütermaterial fertile
Brutausbeute breeding-processing efficiency
Brutgewinn breeding gain
Brutreaktor breeder reactor
Brutverhältnis breeding ratio
Brutzelle blanket
B-Signal picture signal
Buchholzschutz Buchholz protective device, Buchholz protection
Buchse bush, bushing, socket, jack (621.3); inner lining (621.4); gland, sleeve (621.8)
Buchsenfeld socket panel
Buchsenleiste contact strip, jack strip
Büchse (cylinder) lining (621.4); can (539); can, tin, box
Buchstabe symbol (51); letter, type
in Buchstaben in symbols, alphanumeric
Buchstabenschaltung letter shift (621.394)
Buchstabentelegraph A.B.C.-telegraph
Buckel hump (51); boss (621.9)
Buckelschweißung projection welding
Bügel bow; head band (621.395)
Bügelbelag bow strip (621.33)
Bügelsäge bow saw
Bügelstromabnehmer bow collector, bow trolley (621.33)
Bügeleisen (flat-) iron, electric iron
Bühne platform (625); stage (79)
Bühnenbeleuchtung stage lighting
Bühnenregler stage-light dimmer
Bühnenschalttafel stage switchboard
Bums thump
Bund, Bündel bunch, bundle; group (of trunks) (621.395)
bündeln to bundle, to bunch, to beam, to narrow (621.396); to focus (621.385/396), (535), to collimate
Bündelung bunching, narrowing, focusing; beaming (621.396.6); directivity (534)
Bündelungselektrode focusing electrode, beam-forming plate
Bündelungsröhre beam (power) tube
Bündelungsschärfe sharpness of directivity
bündig flush
Bunsenbatterie Bunsen battery
Bunsenbrenner Bunsen (burner)
Bunsenelement Bunsen cell
Bunsenphotometer Bunsen photometer
Buntmetall yellow metal, non-ferrous metal
Bürde burden, load (621.317)
Bürette burette
Burst burst
Bürste brush (621.3), (64)
Bürsten abheben to lift the brushes
Bürsten auflegen to put down the brushes, to seat the brushes
Bürsten einschleifen to bed brushes
die Bürsten hüpfen the brushes jump
Bürstenabhebeeinrichtung brush lifter
Bürstenandruckfeder brush spring
Bürstenbolzen brush arm, brush spindle, brush stud
Bürstenbrücke brush rocker
Bürstendruck brush pressure (621.313)
Bürstenfeder brush spring
Bürstengestell brush rocker, brush gear, brush yoke

Bürstenhalter brush holder, brush box
Bürstenhalterkasten brush box
Bürstenhalterpatrone cartridge-type brush holder
Bürstenhalterring rocker, brush-rocker ring
Bürstenhalterstift brush-holder stud
Bürstenkontakt brush contact
Bürstenreibungsverlust brush-friction loss
Bürstenring brush ring
Bürstenschalter brush switch
Bürstenstift s. **Bürstenbolzen**
Bürstenstrom brush current
Bürstenverschiebung brush shift, brush lead
Bürstenverschiebung nach vorn angular advance of the brushes
Bürstenverschiebung gegen den Drehsinn backward shift of brushes, backward lead of brushes
Bürstenversteller brush-shifting device
Bürstenvorderkante front of a brush
Bürstenvorverstellung forward lead of the brushes
Bürstenwiderstand brush resistance
Bus bus
Buserweiterung bus extend
Bussystem bus system
Bustreiber bus driver (681.3)
Büschel brush (621.3)
Büschelentladung brush discharge, tuft discharge
B-Vermittlungsschrank "B" switchboard (621.395)
B-Verstärker class B amplifier (621.396.64)
B-Wellen type "B" waves
Byte byte
byteseriell byte serial

C

C = Coulomb C = coulomb
c-Ader "p" wire
Cadmium cadmium
Caesium caesium
Caesium-Photozelle caesium cell
Cambrik cambric
Carey-Foster-Brücke Carey-Foster bridge (621.317)
Carnaubawachs carnauba wax
Cassegrainantenne Cassegrain antenna
Castner-Kellner-Prozeß Castner-Kellner process
Cauchyscher Integralsatz Cauchy's integral theorem
C-Betrieb class C operation
CB-Funksprechgerät citizens-band radio telephon, CB radio
CCD-Bildsensor CCD image sensor
ccm = Kubikzentimeter c.c. = cubic centimetre
Celsius Celsius
Celsiusgrad degree Centigrade
Centigramm centigramme
Centiliter centilitre
Centimeter centimetre
CERDIP-Gehäuse cer-DIP package
Cereisenstein flint
Ceresin ceresin
cgs-Einheiten absolute units, gaussian units
cgs-System C.G.S. system
Charakteristik characteristic (equation) (51); graph, pattern; line characteristic (621.39)
charakteristisch characteristic, representative (**für:** of)
charakteristischer Feldwiderstand characteristic wave impedance (621.396)
charakteristische Strahlung characteristic radiation (621.386), spontaneous radiation (621.385)
charakteristische Verzerrung characteristic distortion
charakteristischer Widerstand characteristic impedance
Charge charge, lot, batch
chargieren to charge
Chassis chassis, car body (629); mounting (621.396.62); chassis (621.396.64)
Chemie chemistry
Chemie der Kohlenwasserstoffe carbohydrate chemistry
Chemiefaser man-made fiber
Chemikalie chemical
chemisch chemical
chemisch inert chemically inert
chemisch reagierender Anteil reactive component
chemisch rein chemically pure
chemisch träge chemically inert
chemisch verwandt chemically allied
chemische Aktivität chemically activity
chemische Analyse chemical analysis
chemischer Aufbau chemical constitution
chemische Einwirkung chemical action
chemische Energie chemical energy
chemische Fabrik chemical plant
chemische Formel molecular formula
chemische Trennung chemical separation

chemische Verbindung chemical combination
chemisches Verhalten mode of reaction
chemische Verwandtschaft affinity
chemisches Zeichen chemical symbol
chemische Zusammensetzung chemical constitution
Chiffre cipher
chiffrieren to cipher, to encipher, to code
Chiffriermaschine cryptographer, cryptosystem, coder
chinesische Tusche Indian ink
Chinhydronelektrode quinhydrone electrode, quinhydrone half cell
Chip chip
Chireix-Antenne Chireix-Mesny array
Chireix-Modulation outphasing-modulation system
Chlor chlorine
Chrom chromium
Chromnickel nichrome
Chromnickeldraht chrome-nickel wire, nichrome wire
Chromsäure chromic acid
Chromsäure-Element bichromate cell
chromatisch chromatic (534), (535)
chromatische Aberration chromatic aberration
Chrominanz chrominance
Chromosphäre chromosphere (52)
Chronograph chronograph
Chronometer chronometer
Citizen-Band citizen band = CB
Clark-Element Clark cell
Clausius-Clapeyron-Beziehung Clausius-Clapeyron relation
Clophen inerteen, clophen
Clophen-Transformator clophen-immersed transformer
cm-Quecksilbersäule cms mercury
cm-Wassersäule = cm WS 0.394 ins. of water
cm-Welle micro wave (621.396)
Cobol common business oriented language, COBOL
Cochenille cochineal
Code code, signal book
Code-Element code element
Code-Erkennung code recognition, code detection
codetransparente Übertragung code-transparent transmission
Code-Umsetzer code converter
Code-Umsetzung code translation
Code-Verwirrung code garbling
Code-Wandler code converter
codieren to encode
Codierer encoder
Codierung coding
Codierungsröhre coding tube
Compilierer compiler
Compton-Effect Compton effect
Compton-Streuung Compton scattering
Concordia-Sicherung [trade name of] carbon tetrachloride fuse
Consol-Drehfunkfeuer Consol beacon
Coolidge-Draht swaged wire
Coolidge-Röhre Coolidge tube, radiator tube (621.386)
cos = Cosinus cos = cosine
cos φ-Ausgleich power-factor correction
cos φ-Kompensation power-factor compensation
cos φ-Messer power-factor indicator
cos φ-Regler power-factor regulator
Cosecans cosec = cosecant
Cosh = Hyperbelcosinus cosh = hyperbolic cosine
Cosinus cosine
Cosinus-Charakteristik cosine directivity (621.396)
Cosinussatz cosine rule
Cotangens cotangent
Coulomb coulomb
Coulomb (sche) Abstoßung coulomb repulsion
Coulomb (sche Anziehungs)-Kraft electric coulomb force
Crookescher Dunkelraum Crooke's dark space
Crownglas Crown glass
ctg = Cotangens cot = cotangent
Cursor cursor
Cursor Anfangsstelle cursor home
Cumberlandverfahren cathodic protection
Cupri-... (= zweiwertige Cu-Verbindung) cupric
Cupro-... (= einwertige Cu-Verbindung) cuprous
Curie Curie
Curiepunkt Curie point, magnetic transition temperature
C-Verstärker class C amplifier
c_w = Widerstandsbeiwert drag coefficient
Cyan cyan, cyanogen
Cyanwasserstoff hydrocyanic acid
Cyanid cyanide

D

Dach roof; flat-top structure (621.396.67)
Dachantenne roof aerial, flat-top aerial
Dachlast panel loads (pl.)
Dachleitung roof conductor (699.83)
Dachluke skylight
Dachneigung pitch of the roof
Dachschräge (eines Impulses) pulse droop, pulse tilt
Dachständerkopf service head (696.6)
Dachstuhl roof truss
Dachterrasse platform roof
Damm dam; dike; barrage
Dämmerung (Morgen-:) dawn, (**Abend-:**) dusk
Dämmerungseffekt night error, polarisation error (621.396.9)
Dämmerungsfrequenz transition frequency (621.371)
Dämmerungsgürtel dawn belt, dusk belt
Dämmerungsschalter photo-electric switch, photo-electric lighting control
Dämmplatte insulation board (69)
Dämmstoff insulating material (69)
Dampf vapour; vapor (US), steam
Dampf abgeben to steam
Dampf ablassen to let off steam, to release steam
Dampf aufmachen to get up steam
Dampf erzeugen to raise steam, to generate steam
Dampf reinigen to purify steam
Dampf überhitzen to superheat steam
Dampfabgabe exhalation; steam output (621.1)
Dampfabsorption vapour absorption
dampfbeheizt steam-heated
dampfdicht steam-proof, steamtight, vapour-proof
Dampfdichte density of steam, vapour density
Dampfdom steam dome
Dampfdruck steam pressure (621.1); vapour tension
Dampfdruckmanometer steam pressure gauge
Dampfeinlaß steam inlet
Dampfentnahme bleeding
Dampfentnahmeventil bleeder valve
Dampfentwicklung formation of steam
Dampferzeuger steam generator, steam raising unit
Dampffeuchtigkeit steam moisture
Dampfhammer steam hammer
Dampfheizung steam heating
Dampfkessel (steam) boiler
Dampfkesselanlage steam boiler plant, steam generator
Dampfkesselarmaturen steam boiler fittings, boiler accessories
Dampfkesselexplosion explosion of a steam boiler
Dampfkraft steam power
Dampfmaschine steam engine
Dampfspannung steam tension, steam pressure
Dampfstrahl steam jet, jet of steam
Dampfstrahlapparat steam ejector
Dampfstrahldiffusionspumpe vacuum diffusion pump
Dampfstrahlgebläse steam-jet blower
Dampfstrahlpumpe (steam jet air) ejector, steam-ejector vacuum pump
Dampfstrahlzerstäuber steam-jet sprayer
Dampftemperatur steam temperature
Dampfturbine steam turbine
dampf-turboelektrische Lokomotive steam-turbine electric locomotive
Dampfventil steam valve
dampfen to steam
dämpfen to damp (64); to damp (down) (534), (621.3); to muffle (534), (681), to attenuate (621.3)
dämpfend dissipative (621.3)
gedämpfte Schwingung damped sinusoidal quantity, damped oscillation, attenuated oscillation
gedämpfte Welle damped wave
gedämpfter harmonischer Schwinger damped harmonic oscillator
gedämpftes Licht subdued light
gedämpftes Tageslicht shaded daylight
Dämpfer damper (534); muffler (681.8)
Dämpferrahmen amortisseur (621.317)
Dämpferwicklung damper winding, amortisseur winding (621.313)
Dämpfung damping, transmission loss (621.394, 621.395, 621.396); attenuation, loss (621.392)
Dämpfung der Bodenwelle ground absorption (621.396)
Dämpfungsbereich attenuation band, rejection band
Dämpfungsglied attenuator, pad
Dämpfungsentzerrer attenuation equalizer, attenuation compensator (621.395)
Dämpfungsentzerrung attenuation equalization
Dämpfungsfähigkeit damping power

Dämpfungsfaktor damping factor, attenuation factor
Dämpfungsflügel vane (621.317)
Dämpfungsglied attenuator, attenuation unit (621.392), attenuation network, pad
Dämpfungskette attenuator set
Dämpfungskonstante (d = 1/Q) damping constant (621.396), attenuation coefficient, attenuation constant (US) (621.392)
Dämpfungskreisel rate gyro
Dämpfungsmaß attenuation (constant)
Dämpfungsmesser "Q"-meter, transmission measuring set (621.395), decibel meter
Dämpfungsmoment damping torque (621.317)
Dämpfungsplan overall-loss plan
Dämpfungsregler attenuator set, variable attenuator
Dämpfungsschaltung im Regelkreis antihunt circuit
Dämpfungsschwert flap attenuator (621.396)
Dämpfungsverhältnis ballistic factor (621.317)
Dämpfungsvermögen damping power
Dämpfungsverzerrung attenuation distortion; frequency distortion (US)
Dämpfungsvierpol attenuation network
Dämpfungszylinder dashpot
Daniell-Element Daniell cell
Darlington-Schaltung Darlington circuit
d'Arsonval-Instrument d'Arsonval instrument
darstellen to feature; to depict, to display
darstellen als to state as (51)
darstellende Geometrie demonstrative geometry
Darstellung (re)presentation, display
Datei file (681.3)
Dateiorganisation file structure
Dateiverzeichnis directory
Dateldienste datel service
Daten data
Datenabruf(-Programm) fetch
Datenanzeige data display
Datenaufbereitung data processing
Datenaufzeichnung data recording
Datenausgabe data output
Datenbank data bank, file, data base
Datenblatt data sheet, specification sheet, leaflet
Datenblock data block
Datenbus data bus
Datendienst data service
Dateneingabe data input
Dateneingabeterminal data-entry terminal
Daten-Ein-und Ausgabegerät data terminal, input-output device
Datenendeinrichtung data processing terminal equipment
Datenerfassung data acquisition, data collection, data recording
Datenfernverarbeitung teleprocessing
Datenfluß data flow
Datenflußplan data-flow chart, dynamic flow diagramm
Datenformat data format
Datenquelle data source
Datenreduzierung data reduction
Datenschutz data protection
Datensenke data sink
Datensicherung data security
Datensichtgerät terminal
Datensichtstation display, data display
Datenspeicherplatz storage location
Datenspeicherung data storage
Datenstation terminal, data station
Datenträger data carrier, data medium, storage medium
Datentransfer data transfer, data movement
Datentransfergeschwindigkeit data-transfer rate
Datentypist punch operator
Datenübertragung data transmission, data communication, data transfer
Datenübertragungsverbindung data link
Datenverarbeitung data processing, data handling
Datenverarbeitungsanlage data-processing system, computer
Datenverdichtung data compression, data compaction
Datenverschiebung data shift
Datenwort data word
Dauer duration, time
Dauer des Vorhandenseins time of persistence
Dauerbefehl persistent command
Dauerbelastung constant load, continuous load, permanent load
Dauerbenutzung continuous use
Dauerbetrieb continuous running, permanent service
Dauerbiegung fatique bending
Dauerbruch fatigue failure, fracture by fatigue
Dauerbruchsicherheit resistance to fatigue
Dauerdurchschlagspannung asymptotic voltage
Dauererdschluß continuous earth; permanent ground (US)
dauererprobt time-tested
Dauerfestigkeit fatigue limit, endurance limit, fatigue strength; fatigue life

Dauerfestigkeitsprüfung endurance test, fatigue test
Dauergeräusch permanent noise
Dauergleichstrom continuous forward current, continuous on-state current
dauerglühen to soak (66)
Dauerglühung soak, soaking
Dauergrenzstrom limiting value of mean on-state current; maximum rated mean forward current
Dauerkennzeichen continuous signal
Dauerkurzschluß substained shortcircuit
Dauerladegerät trickle charger (621.355)
Dauerladung trickle charge
Dauerladungsbatterie a. c. floating storage battery system
Dauerlast continuous load
Dauerlauf continuous running
Dauerleistung constant power, continuous output
Dauermagnet p. m. = permanent magnet
Dauermeldung persistent alarm
Dauerprüfmaschine fatigue testing machine
Dauerschlagversuch repeated impact test
Dauerschmelzstrom minimum fusing current
Dauerschwingung maintained vibration
Dauerspannung constant voltage
Dauerstandfestigkeit (limiting) creep stress, creep limit
Dauerstrich unmodulated signal, unmodulated wave, c. w. = continuous wave (621.396)
Dauerstrich mit Anodenwechselspannung I. C. W. = interrupted continuous waves
Dauerstrom standing current, constant current
Dauerstromhalten continuous current holding
Dauerversuch fatigue test, endurance test
Dauerwechselbeanspruchung repeated stress
Dauerzugfestigkeit endurance tensile strength
Dauerzustand perpetuity
dauernd perpetual, persistent, permanent
dauernde Stromentnahme permanent discharge
Davy-Grubenlampe Davy-lamp
dazwischenkeilen to wedge in between
Debye-Scherrer-Aufnahme powder spectrogram
Decca-Navigations-System Decca Navigator System
dechiffrieren to decode, to decipher
Decibel decibel
Deck-... cover (ing), superficial
Deckanstrich (paint) finish, finishing coat, top coating
Deckfarbe body colour
Decklack coating lacquer
Deckleiste slot wedge (621.313)
Decköl anti-spray film (621.355)
Deckschicht superficial layer, plating
Decke ceiling (72); tire (US), tyre (629.113); blanket (67)
Deckenabstand suspension distance (621.32)
Deckendurchführung floor insulator (621.315.6)
Deckenrosette ceiling rose, ceiling plate
Deckenschalter ceiling switch
Deckenstrahler ceiling-reflecting lamp
Deckel lid, cover (plate), cap
decodieren to decode
Decodiermatrix decoding matrix
Deemphasis de-emphasis
Defekt defect
Defekt-Elektron defect electron, hole, mobile hole
Defektleitung hole conduction, p-type conduction
defekt defective, at fault
definieren to define
ohne definiertes Potential floating (621.3)
deformieren to deform
Deformierung deformation
degenerieren to degenerate
dehnbar expansible, tensile, dilatable
dehnen to extend, to elongate, to strain
sich dehnen to expand, to stretch
gedehnte Skaleneinteilung open scale (621.317)
Dehner expander
Dehnung (Wärme-:) expansion, dilatation; (Längen-:) extension; (Bruch-:) elongation; (elastisch:) stretch; (verformend:) strain
Dehnungsdiagramm strain variation diagram
Dehnungsfuge expansion joint (69)
Dehnungskupplung expansion coupling
Dehnungsmesser dilatometer, extensometer, strain gauge
Dehnungsmeßstreifen strain gauge
Dehnungsriß expansion crack (69)
Dehnungsstoß expansion joint
dehydrieren to dehydrate
Deionisationsschalter deion circuit breaker
Dekade decade
Dekaden-Kurbelkondensator dial dicade condenser
Dekaden-Kurbelwiderstand dial decade box
Dekaden-Widerstandskasten decade (resistance) box (621.317)
dekadischer Frequenzgenerator (decade) frequency synthesizer

dekadische Steuerstufe master oscillator with decadic setting, synthesizer
dekadischer Zähler decade scaler
Dekagramm dag. = decagram (me)
Dekaliter dal. = decalitre
Dekameter dam. = decametre
Deklination declination (52), (538)
Dekoder decoder
dekodieren to decode
Dekontamination decontamination
Dekrement decrement
D-Elektrode Dee (539)
Deltafunktion delta function
Demodulation demodulation, detection (621.396)
Demodulator demodulator
demodulieren to demodulate
Demontage dismantling
denaturieren to denature
Dendrit dendrite
Depolarisation depolarization
Depolarisator depolarizer; depolarizing mix (US)
Depolarisierungsmittel depolarizer
Depolymerisation depolymerization
Dérimotor Déri-motor
Derivat derivative
Desintegrator disintegrator
Desorption desorption
Desoxidierung deoxidation
Destillat distillate
Destillation distillation
Destillier-... distilling
Destillierkolonne still
Destillierofen distilling furnace
destillieren to distil
destilliertes Wasser distilled water
Detail detail
Detailkonstrukteur draftsman
Detailzeichnung detailed drawing
Detektor (wave) detector
Detektorapparat, Detektorenempfänger crystal set
Detektordrähtchen cats whisker
Detektorfassung crystal mounting
Detektorkreis detector circuit
Detektormischkopf crystal mixer (621.396)
Determinante determinant
Deuterium deuterium
Deuteron deuteron
deutliches Bild defined image
Deutsche Industrie-Norm DIN (s. dort)
Deutscher Normenausschuß (DNA) German Engineering Standard Association [counterpart of British Engineering Standard Association (BESA)]
Deviation deviation
Dewarsche Flasche, Dewargefäß Dewar flask
Dextrin starch (gum), dextrin
Dextrose dextrose
dezentral, dezentralisiert distributed, decentralized
dezentralisierte EDV decentralized data processing
dezentrierte Speisung offset feed (621.396.677)
dezentrales System distributed system
Dezibel decibel, dB
Dezimal-... decimal
Dezimal-Binär-Umwandlung decimal-to-binary conversion
Dezimalbruch decimal (fraction)
in einen Dezimalbruch verwandeln to decimalize
Dezimalklassifikation universal decimal classification
Dezimalkomma decimal point; radix point
Dezimalskala decimal scale
Dezimalstelle decimal place
Dezimalstellen decimal points, places of decimals
Dezimalsystem decimal system, decimal notation, decimal scale
Dezimalwaage decimal balance
Dezimalwährung decimal coinage
Dezimalzahl decimal number
Dezimalzählung decimal numeration
Dezimalziffer decimal digit
Dezigramm decigramme
Deziliter decilitre
Dezimeter decimetre
Dezimeterbereich u.h.f. range = ultra-high frequency range
Dezimeterwellen decimetric waves, u. h. f.
D-Flip-Flop D-type flip-flop
Dia slide
Dia-Abtaster dia scanner
Dia-Projektor slide projector
Diagnoseprogramm diagnostic routine
diagonal diagonal
Diagonaldraht incidence wire (629.13)
Diagonale diagonal
Diagonalschnitt angle-cutting
Diagonalverstrebung diagonal bracing
Diagramm diagram, pattern, plot; graph, chart
Diagramm auftragen to plot a diagram
in einem Diagramm ablesen to scale from a diagram

in ein Diagramm eintragen to carry down to a diagram
in ein Diagramm übertragen to bring down to a diagram
zu einem Diagramm verarbeiten to plot (621)
Diagramm-Anhebung uptilt (621.396.677)
Diagrammstreifen chart paper, record
diaktinisch diactinic
Dialogbetrieb interactive mode (681.3)
Dialogsprache interactive language
diamagnetisch diamagnetic
Diamagnetismus diamagnetism
Diamant diamond
Diamantlager jewel bearing
Diamantkohle jet
Diaphragma diaphragm, porous barrier, porous cell (621.35)
Diatomeenerde diatomite
Diathermie diathermy, diathermal heating
Diathermie mit induktiver Kopplung auto-conduction
Diathermie mit kapazitiver Kopplung auto-condensation
Diathermiebehandlung diathermal heating, high-frequency treatment
Dichromatelement dichromate element
dicht dense, thick, intense, compact, solid, massive; tight, sealed
dicht passend tight-fitting
dicht schließender Deckel close lid
dicht unterhalb der Frequenz short of the frequency
dichter Abschluß solid enclosure
dichtes Gefüge dense texture, tight texture
dichtes Papier high-density paper (621.315.2)
dichter Rauch dense smoke
dichter Verkehr congested traffic
dichter Verschluß hermetic seal
Dichte density, thickness, compactness, massiveness; tightness
Dichtemodulation density modulation
dichten to tighten, to seal
Dichtigkeit s. **Dichte**
Dichtschweißung sealing weld
Dichtung seal, gasket, packing
Dichtungsmasse tightening compound, sealing compound
Dichtungsmaterial packing material, gasket board
Dichtungsmittel sealing substance
Dichtungsring ring seal, washer
Dichtungsscheibe washer, gasket
dick thick, big, heavy-gauge
dick werden to congeal (54); to put on weight (61)
dickes Blech heavy(-gauge) sheet
dicker Draht heavy(-gauge) wire, stout wire
dickes Ende butt
dicker Strich heavy line
dickdrähtig thickwire
dickdrähtige Spule thickwire coil
dickwandig thick-walled
Dicke thickness, depth, diameter, gauge
Dickenlehre thickness gauge, feeler gauge
Dickenschwingung thickness vibration, width vibration
Dickentoleranz tolerance on thickness
Dickschichtschaltung thick-film integrated circuit
diebessicher thief-proof
diebstahlsichere Fassung locking lamp holder
Dielektrikum dielectric
Dielektrikum-Meßgerät dielectric parameter measuring set
dielektrisch dielectric
dielektrische Absorption dielectric absorption
dielektrische Antenne dielectric antenna
dielektrisch beschichtetes Hohlkabel dielectric coated waveguide
dielektrischer Draht dielectric wire, dielectric rod
dielektrische Eigenschaft dielectric property
dielektrische Erwärmung dielectric heating
dielektrischer Fluß dielectrical flux
dielektrische Flußdichte dielectric flux density, displacement
dielektrischer Leiter dielectric wire, dielectric guide
dielektrische Mehrfachantenne polyrod aerial
dielektrische Nachwirkung dielectric hysteresis, anomalous displacement
dielektrische Übertragungsleitung dielectric transmission line
dielektrischer Verlust dielectric loss, dielectric absorption
dielektrischer Verlustfaktor dissipation factor
dielektrische Verschiebung dielectric displacement
dielektrischer Wellenleiter dielectric wave-guide
Dielektrizitätskonstante (specific) inductive capacity, dielectric constant, dielectric coefficient, permittivity, permittance
Dielektrizitätszahl relative permittivity
Dienst service, duty
Dienstanweisung service instructions (pl.)

Dienstgespräch service call, service massage
Dienstgüte service quality, grade of service
Dienstkanal order channel
Dienstleistung software, service
Dienstleitung order-wire (circuit), speaker circuit (621.395)
Dienstprogramm utility program (681.3)
Diesel(motor) Diesel (engine), oil engine
diesel-elektrischer Antrieb Diesel-electric drive, oil-electric drive
Dieselaggregat Diesel-driven generating set
Dieselöl Diesel oil
Dieselhorst-Martin-Vierer multiple twin quad (621.395)
Dietze-Anleger clamp ammeter (621.317)
Differential differential, derivative (51); differential (gear) (629.113)
Differentialanalysator differential anlyzer
Differentialbogenlampe differential arc lamp
Differential-Drehkondensator differential capacitor
Differentialdrehmelder differential synchro
Differentialempfänger differential motor, torque-differential synchro receiver (621.316.7); differential repeater (621.39)
Differentialentzerrer derivative equalizer
Differentialfilter lattice network
Differentialflaschenzug differential pulley
Differentialgeber differential transmitter, differential generator
Differentialgegenverkehrssystem differential duplex system (621.394)
Differentialgetriebe differential (gear)
Differentialgleichung differential equation
Differentialkalkül (differential) calculus
Differentialmanometer differential manometer, differential pressure gauge
Differentialmikrophon push-pull microphone
Differentialoperator differential operator
Differentialrechnung (differential) calculus
Differentialrelais balanced relay
Differentialschutz differential protection, core-balance protection (621.316.9)
Differentialschutzwandler biasing tranformer
Differentialtransformator,
Differentialübertrager differential transformer
Differentialverstärker differential amplifier
Differentialzeichen sign of differentiation
Differentiation differentiation
Differentiator differentiator
differentiell differential, incremental
differentieller Hysteresisverlust incremental hysteresis loss
differentielle Induktion incremental induction
differentielle Kapazität small signal capacity
differentielle magnetische Feldstärke incremental magnetizing force
differentielle Permeabilität incremental permeability
differentieller Phasenfehler differential phase
differentieller Verstärkungsfehler differential gain
differentieller Widerstand differential resistance
Differenz difference
Differenz zwischen Kleinst- und Größtwert range
Differenzausgang differential output
Differenz-Ausgangsimpedanz differential-mode output impedance
Differenzeingang differential input
Differenzfrequenz difference frequency
Differenzreihe differences sequence
Differenz-Spannungsverstärkung differential-mode voltage amplification
Differenzstromrelais differential current relay
Differenzton difference tone
Differenzträgerverfahren intercarrier system
Differenzverstärker differential amplifier
differenzieren to differentiate (51), to derive
differenzierendes Netzwerk differentiating circuit
Differenzierentzerrer derivative corrector, derivative equalizer
Differenzierglied advance element
Differenzierkreis differentiating circuit
Differenzierschaltung differentiator, differentiating circuit
Differenzierung differentiation (51)
Differenzierungseinrichtung differentiator
diffundieren nach to diffuse to
diffundierend diffusive
diffundierter Transistor diffused transistor
diffundierter Zonenübergang diffused junction
diffus diffuse
diffuse Beleuchtung diffuse illumination
diffuse Durchlässigkeit diffuse transmission (535)
diffuse Reflexion diffuse reflection
diffuses Reflektionsvermögen diffuse reflecting power
Diffusion diffusion
Diffusion im festen Zustand solid-state diffusion (621.382)
Diffusionskoeffizient coefficient of diffusion, diffusion constant
Diffusionslänge diffusion length

Diffusionsofen diffusion furnace
Diffusionsprodukt diffusate
Diffusionspumpe vacuum diffusion pump, condensation pump
Diffusionsströmung diffusive flow
Diffusionsspannung diffusion potential
Diffusionstechnik impurity diffusion technique
Diffusionstransistor diffused transistor
Diffusionszeit diffusion-rate period
Diffusor diffuser
digerieren to digest (54)
digital digital
digitale Bildschirmdarstellung digital display
digitale Darstellung digital display
digitale elektrische Größe digital electric quantity
digitaler Integrator incremental computer
digitale Prozeßsteuerung digital control
digitales Signal digital signal
digitalisieren to digitize
Digital... digital
Digital-Analogwandler digital-analog converter
Digitalanzeige digital display, digital indication, digital readout
Digitalfrequenzmesser counter, frequency meter
Digitalmeßgerät digital measuring instrument, digital meter
Digitalrechner digital computer
Digitalschaltung digital circuit
Digitaltonband digital audiotape
Digitalumsetzer digital data conversion equipment, digital transducer
Diktaphon dictaphone, phonographic recorder
Diktiermaschine dictograph, dictating machine, recorder
DIL-Steckverbinder DIP connector
Dilatometer dilatometer
Dimension dimension
dimensionslos dimensionless
dimensional dimensional
dimensionieren to proportion
DIN German Engineering Standard [counterpart of British Standard]
DIN-Blatt German Standard Specification [counterpart of B. S. S.]
Diode diode, two-electrode valve; two-element tube (US)
Diodenbegrenzer diode limiter
Diodengleichrichtung diode detection, diode rectification
Dioden-Transistor-Logik diode-transistor logic
Diopter diopter
dioptrisch dioptric
Dioxid dioxide
Diplexbetrieb diplex operation
Diplexer diplexer
Dipol dipole, doublet (53), (621.396)
Dipolanordnung dipole array
Dipolantenne dipole aerial, doublet aerial
dipolartig dipolar
Dipolmoment dipole moment
direkt direct
direkt angetrieben direct driven
direkt anzeigendes Instrument direct-reading instrument
direkt anzeigender Peiler direct-reading direction finder
direkt durchgehend straight-through
direkt erden to earth direct
direkt geheizte Kathode direct-heated cathode, filament(-type) cathode
direkt geheizte Röhre filament(-type) valve
direkt gekoppelter Verstärker direct resistance-coupled amplifier, d.c. coupled amplifier
direkt gesteuertes System direct-control system (621.395)
direkte Adressierung direct addressing
direktes Anlassen direct on-line starting, full-voltage starting
direkter Antrieb direct drive
direkte Aufhängung direkt suspension
direkte Beleuchtung direct lighting
direkte Betätigung direct operation
direkte Betrachtungsweise short-cut method
direkte Datenerfassung direct data capture
direkte Handbetätigung direct manual control
direkte Heizung direct heating
direkte Modulation final modulation
direkte Nullstromauslösung series undercurrent tripping
direkte Proportionalität direct proportion
direkte Speiseleitung direct feeder
direkter Strahl straight-through ray (621.396)
direkte Überstromauslösung series overcurrent tripping (621.316.5)
direkter Zugriff direct access
Direktantrieb direct drive
Direktrufeinrichtung direct key selector
Direktübertragung live broadcast (654.19)
Direktweg direct route, high usage route (621.395)
Direktzugriff direct access
Direktion (general) management, managing board, board of directors

Direktor director (621.396.67)
Discus (Navigationsverfahren) OBOE (621.396)
disharmonisch disharmonious
Disjunktion disjunction
Diskette floppy disk
  Disketten-Laufwerk floppy-disk drive
diskontinuierlich discontinuous
Diskriminator discriminator, frequency discriminator
dispergierbar dispersable
dispergieren to disperse, to intersperse
  dispergierter Metallstaub metal fog, metal mist (621.357)
Dispersion dispersion, dispersal
Dissonanz dissonance
Dissoziation dissociation
  Dissoziationsgrad grade of dissociation
  Dissoziationsprodukt dissociated component
dissoziieren, (sich) to dissociate
Distanz distance
  Distanzadresse displacement (681.3)
  Distanzrelais distance relay, impedance relay
  Distanzring bush, spacer ring
  Distanzschutz distance protection
    Distanzschutz mit steiler Auslösekennlinie continous-curve distance-time protection
    Distanzschutz mit Stufenkennlinie stepped-curve distance-time protection
distributiv distributive (51)
div div (51)
Divergenz divergence (51), (535)
divergieren to diverge
  divergierend divergent
Diversity-Empfang diversity reception
Dividend dividend (51)
dividieren to divide (51)
Division division (51)
  Division durch Zahlen bis 12 short division
  Division durch Zahlen über 12 long division
Divisor divisor
D. K. = Dielektrizitätskonstante s. dort
DME distance measuring equipment
dm-Gebiet, dm-Welle u. h. f. = ultra high frequency
DNA = Deutscher Normenausschuß s. dort
Docht wick
  Dochtöler felt-wick lubricator (621.8)
Doherty-Sender Doherty transmitter
  Dokumentfax documentfax
Donator donor
Donner thunder
Doppel duplicate; double, twin
  doppeladrige Litze twin flexible cord, twin flex
  Doppelader pair, twin wire
Doppelanker double armature
Doppelbindung double bond
Doppelboden double bottom, false bottom
Doppelbogen offset (625)
Doppelbrechung double refraction
Doppelbrücke double bridge, Thomson bridge, Kelvin bridge (621.317)
Doppel-Cosinus-Charakteristik figure-eight pattern
Doppeldiode duodiode, double diode
Doppeldiplextelegraphie four frequency diplex telegraphy
Doppeldrehregler twin induction regulator
Doppeldreieckschaltung double-delta connection
Doppeldreipolröhre duotriode, twin triode, pentatron
Doppelerdschluß polyphase earth
Doppelfahrdraht double-contact wires, twin contact wires (pl.) (621.33)
  Doppelfahrdrahtaufhängung double-catenary suspension, duplex catenary suspension
Doppelfahrleitung double-contact wires (pl.)
Doppelgitterröhre bigrid valve, dual grid tube
Doppelindizes double subscripts
Doppelkäfig-Induktionsmotor double squirrel cage induction motor
Doppelkamm-Magnetron interdigital magnetron
Doppelkeilnut double Vee groove (621.791)
Doppelkontakt twin contact
Doppelkopfhörer telephone headset, earphones
Doppelkurbel Thomas-Varley slide (621.317)
Doppelleiter twin conductor
Doppelleitung two-wire line, two-wire circuit, duplicate lines (pl.), twin conductor, pair (of leads) (621.3); metallic circuit (621.394)
  Doppelleitung mit entgegengesetzt gerichteten Strömen unlike pair
  Doppelleitung mit gleichsinnigen Strömen like pair
  Doppelleitungsschnur twin flexible cable
  Doppelleitungsschutz parallel line protection
Doppellichtbogenofen series-arc furnace
Doppellitze twin flexible cable
Doppelmast "H" pole
Doppelmeldung double-point information (621.398)
doppelpoliger Motor bipolar motor
doppelpolige Speisung insulated return system (621.33)

Doppelrahmen-Peilsystem spaced-loop direction finder
Doppelschicht double layer
Doppelschleifbock two-wheel machine
Doppelschlußgenerator compound generator
doppelseitig gerichtet bidirectional
doppelseitiger Antrieb bilateral drive, double-ended drive (625)
doppelseitig gedruckte Schaltung double-side pc board
Doppelspullautsprecher articulated voice-coil arrangement
Doppelstecker socket-outlet adapter (621.315.6)
Doppelstern double star (52)
Doppelsternvierer quad pair
Doppelsteuerung dual control, dual operation (621.876)
Doppelstrahloszillograph double-beam oszilloscope
Doppelstrombetrieb polar operation, double-current working
Doppelstromgenerator double-current generator
Doppel-T-Anker two-pole armature, "H"-armature, Siemens armature
Doppeltarif double tariff, multiple tariff
Doppeltarifzähler double-tariff meter
Doppeltriode twin triode, pentatron
Doppelversorgung dual service, plural service
Doppelweggleichrichter double-way rectifier, full-wave rectifier
Doppelwendel coiled coil
Doppelwendellampe coiled-coil lamp
Doppelwicklung step winding
Doppelwicklungs-Asynchronmotor Boucherot motor
Doppelzweipolröhre double diode
doppelt double, twofold
doppelt baumwollisoliert double-cotton covered
doppelt geladen doubly charged
doppelt gelascht double covered (621.884)
doppelt isoliert double covered (621.3)
doppelt konzentrisches Kabel twin concentric cable
doppelt seidenisolierter Draht double-silk covered wire
doppelt umsponnen double covered
mit doppeltem Brennpunkt bifocal
Doppler gemittelter Wirkungsquerschnitt Doppler averaged cross section (539)
Dopplerfrequenz Doppler frequency, Doppler shift
Dopplernavigation Dopplernavigation
Dopplersches Prinzip Doppler's principle
Dopplerverschiebung Doppler shift
Dopplung doubling (621.7); reproduction (651)
Dose conduit box (621.315.6), (696); jar, can, capsule
Dosenbarometer aneroid barometer
dosieren to dose
Dosierpumpe proportioning pump
Dosis dose
Dosismesser dosage meter, dosimeter, intensimeter, Roentgen meter, milliampere-second-meter, radiation rate meter (621.396)
Dosisüberschreitung overexposure to radiation (61)
dotieren to dope, doping
Dotiermittel dopant, dope additive
Draht wire, thread, conductor
Draht abisolieren to strip a wire (621.315.3)
Draht aufwickeln (auf eine Rolle) to lace wire (on a reel), to coil up wire
Draht umknicken to bend a wire over on itself
Drähte verlegen to wire
Draht ziehen to draw wire
Drahtarmierung wire armour(ing) (621.315.2)
Drahtbarren wire bar
drahtbewehrtes Kabel wire-armoured cable
Drahtbewehrung wire armour(ing)
Drahtbondung wire bond
Drahtbruch wire break
Drahtbügel wire bow
Drahtbürste (rotary) wire brush
mit der Drahtbürste reinigen to wire-brush
Drahtdrehwiderstand variable wire-wound resistor
Drahtfederrelais wire-spring relay
Drahtfernsehverbindung line television, closed-circuit television, C. C. T. V.
Drahtfestwiderstand fixed wire-wound resistor
Drahtfunk wire broadcasting, wired wireless, line broadcasting
Drahtgaze wire gauze, wire netting, wire screen
Drahtgeflecht wire mesh, wire netting
drahtgewickelt wire-wound
drahtgewickelter Anker wire-wound armature
drahtgewickelter Läufer wire-wound rotor
Drahtgitter wire grid
Drahtglas wired glass
Drahtisolation wire insulation
Drahtkorb wire guard (621.315.1)
Drahtlage layer of wire

Drahtlampe filament lamp, wire lamp
Drahtleitung wire line
Drahtmagnetophon wire recorder
Drahtnachrichtentechnik wire telephony and telegraphy, wire communication
Drahtnadel wire pin
Drahtnetz hard wired netting, wire netting (621.315.1)
Drahtpotentiometer wire-wound potentiometer
Drahtschere wire nippers (pl.), wire cutter
Drahtschleife loop of wire
Drahtschwabbel rotary wire brush
Drahtseil cable, steel rope, stranded rope
Drahtseilbahn cable railway, aerial cable way, aerial ropeway
Drahtsieb gauze filter
Drahtsonde probe (621.396)
Drahtspule wire coil
Drahtstärke wire gauge, size of wire
Drahtstift wire nail, wire tack, sprig
Drahttelefonie line telephony, wire telephony
Drahttelegraphie line telegraphy
Drahtverbindung hard wired connection, hard wired circuit, wire communication circuit
Drahtverspannung wire bracing, bracing wire
Drahtwellenleitung single-wire transmission line, surface-wave transmission line
Drahtwendel helix of wire
Drahtwickelverbindung wire wrap
Drahtwiderstand wire resistance; wire-wound resistor (621.396)
Drähtchen filament
drahten to wire, to telegraph
drahtlos wireless; radio... (US)
drahtlos telegraphieren to wireless, to radiotelegraph
drahtlose Nachrichtentechnik wireless telephony and telegraphy, radio telecommunication
drahtlose Nachrichtenübermittlung wireless communication, radio communication, radio transmission
drahtloser Ruf wireless call signal
drahtloser Sender wireless transmitter
drahtlose Telefonie wireless telephony, radio telephony
drahtlose Telegraphie radio telegraphy
auf drahtlosem Wege by wireless
Drain drain (621.382)
Drain-Anschluß drain terminal
Drain-Elektrode drain electrode
Drain-Reststrom drain cut-off current
Drain-Schaltung common drain
Drain-Source-Spannung drain-source voltage
Drainstrom drain current
Drain-Zone drain region
Drainage drain(age), sewage; drainage (621.33)
Draisine (rail) trolley
Drall twist, spin
D-Regler derivative controller
Dreh-... rotary, rotating, turning
Drehachse axle, spindle (bar), hinge axis of rotation; centre of gyration (52)
Dreharm wiper rotary arm (621.39)
Drehbeanspruchung torsional strain
Drehbewegung rotary movement; spin
Drehbühne revolving stage
Dreheiseninstrument moving-iron instrument, moving-magnet instrument, electromagnetic instrument
Dreheisenspannungsmesser moving-iron voltmeter
Drehfahnenrelais vane-type relay
Drehfeld rotating field, revolving field, rotary field
Drehfeldantenne rotating field aerial
Drehfeldgenerator reaction generator
Drehfeldhysteresis rotary hysteresis
Drehfeldinstrument rotating field instrument, induction instrument, Ferraris instrument
Drehfeldleistungsmesser induction wattmeter
Drehfeldmagnetisierung rotary magnetism
Drehfeldmaschine induction machine
Drehfeldmotor Ferraris motor, induction motor
Drehfeldphasenschieber Helmholtz coil, phase shifter
Drehfeldsystem synchro
Drehfeldrichtungsanzeiger phase-sequence indicator
Drehfeldumformer transformer generator
Drehfeldummagnetisierung rotary hysteresis
Drehfeldwattmeter induction wattmeter
Drehfunkfeuer rotating beacon, omnidirectional range station
Drehgestell bogie, swivel truck
Drehgestellrahmen bogie frame
Drehimpuls moment of momentum, angular momentum
Drehknopf (control) knob
Drehkondensator adjustable condenser, variable condenser, adjustable capacitor, variable capacitor
Drehkondensatoren eines mehrstufigen

Empfängers abgleichen to line up the condensers of a multistage receiver
Drehkontaktwange pivot jaw
Drehkraft torsional force
Drehkreuz turnstile
Drehkreuzantenne turnstile aerial
Drehkupplung rotating joint
Drehmagnet moving magnet
Drehmelder synchro, telesyn
Drehmeldesystem synchro system
Drehmoment turning moment, torque (moment); twist moment
Drehmoment um einen Punkt moment about a point
Drehmoment bei festgebremstem Rotor locked rotor torque, static torque
Drehmomentantrieb torque motor
Drehmomentempfänger synchro-torque receiver
Drehmomentgeber synchro-torque transmitter
Drehmomentenverstärker torque amplifier
Drehofen revolving furnace
Drehpunkt pivotal centre, hinge point, fulcrum, centre of rotation
Drehrahmensender rotating-loop radio-transmitter
Drehregler induction (voltage) regulator, berry transformer, static induction motor (621.313)
Drehrichtstrahl rotary beam
Drehrichtung direction of rotation
Drehrichtung umkehren to reverse
Drehrichtungsumkehr change of rotation, reversal
Drehrichtungsumschalter reversing controller
Drehrohr-Lichtbogenofen revolving-arc furnace
Drehrohrofen rotary kiln
Drehschalter turn switch, rotary switch
Drehscheibe turntable (625)
Drehschritt rotary step (621.395.3)
Drehsinn hand, sense of rotation
Drehspule moving coil
Drehspulaufhängung moving-coil support
Drehspulgalvanometer moving-coil galvanometer
Drehspulgalvanometer mit Fadenaufhängung suspended coil type galvanometer
Drehspulinstrument (permanent-) magnet moving-coil instrument, moving-coil meter
Drehspulrahmen former (621.317)
Drehspulrelais moving-coil relay
Drehspulspannungsmesser moving-coil voltmeter
Drehspultelefon moving-coil telephone
Drehspultragkonstruktion moving support
Drehspulvoltmeter moving-coil voltmeter
Drehstahl tool (621.9)
Drehstrom multiphase (alternating) current, three-phase (alternating) current
Drehstromanlage polyphase system
Drehstromanlage ohne Nulleiter three-phase three-wire system
Drehstromanlasser three-phase starter
Drehstrom-Asynchronmotor three-phase induction motor, polyphase induction motor; asynchronous three-phase motor
Drehstrombahn three-phase railway
Drehstrombahnbetrieb three-phase traction
Drehstrom-Brücken-Schaltung three-phase bridge connection
Drehstrom-Dreileiternetz three-phase three-wire circuit
Drehstromgenerator three-phase generator, multiphase generator, polyphase generator
Drehstrom-Induktionsmotor polyphase induction motor
Drehstrom-Kollektormotor three-phase commutator motor
Drehstrom-Kommutatormotor three-phase commutator motor
Drehstromkreis three-phase circuit, polyphase circuit
Drehstrom-Kurzschlußläufer-Motor (three-phase) squirrel cage motor
Drehstromlokomotive three-phase locomotive
Drehstrommotor three-phase motor, polyphase motor, multiphase motor
Drehstromnetz three-phase supply network
Drehstromnetz mit Nulleiter three-phase four-wire system
Drehstromnetz ohne Nulleiter three-phase three-wire system
Drehstromschalter three-phase switch, polyphase switch
Drehstromschaltung three-phase circuit, polyphase circuit
Drehstrom-Siebenleiternetz seven-wire three-phase system
Drehstrom-Stickstoffofen Kilburn-Scott furnace
Drehstromsystem polyphase system
Drehstromsystem in offener Schaltung three-phase six-wire system

Drehstromtransformator,
Drehstromumspanner three-phase transformer, polyphase transformer
Drehstrom-Vierleiteranlage three-phase four-wire system
Drehstromentwicklung three-phase winding, polyphase winding
Drehstromzähler three-phase meter
Drehstufenschalter rotary stepping switch, rotary multi position switch, rotary switch
Drehstuhl swivel chair
Drehtisch turntable, rotating table, revolving table (621.9)
Drehtransformator induction (voltage) regulator, phase-shifting transformer, berry transformer
Drehvorrichtung turning gear
Drehwähler rotary selector; rotary switch
Drehwählersystem rotary automatic telephone system, rotary dial system
Drehwiderstand variable resistor
Drehwirkung turning effect
Drehzahl speed, number of revolutions
Drehzahl bei Belastung on-load speed
Drehzahl zu hoch treiben to overrev
Drehzahl eines Zählers rate of registering
auf Drehzahl kommen to attain speed, to get up speed
Drehzahlabfall falling-off in speed
Drehzahländerung change of speed
Drehzahlanzeiger speed indicator, revolution indicator
Drehzahlbereich speed range
Drehzahleinstellung speed adjustment
drehzahlgesteuerter Achsdynamo controlled speed-axle generator (621.33)
Drehzahlkennlinie speed torque characteristic
Drehzahlmesser speed indicator, revolution indicator
drehzahlregelbarer Motor variable-speed motor, change-speed motor, adjustable speed motor
Drehzahlregler speed regulator, speed controller, governor
Drehzahlregelung speed regulation, speed control, speed governing
Drehzahlregelung mit Kompoundcharakteristik adjustable varying-voltage control (621.313)
Drehzahlregelung mit Nebenschlußcharakteristik adjustable voltage control
Drehzahlregelwiderstand speed-regulating rheostat
Drehzahlsteuerung speed control
Drehzahlstufe speed
Drehzahlumkehr reversing
drehzahlumschaltbarer Motor single-winding multispeed motor
Drehzahlveränderung durch Polumschaltung pole-changing control
Drehzahlverhalten speed characteristic
drehbar rotary, revolving, turnable, moving
drehbar gelagert sein auf to pivot on
drehbar lagern auf to pivot on
drehbar um pivoted about
drehbarer Plattensatz rotating plates (pl.) (621.319.4)
drehen to rev = to revolve, to rotate; to turn (621.94)
drehen, sich to revolve, to rotate, to turn, to spin
sich drehen um ... to turn round ...
sich drehen um einen Winkel von ... to turn round through ...
drehend rotating, rotational
drehend, sich rotary, rotating
drehend antreiben to spin (621)
drehende Winde veering winds
gedrehte Schraube turned bolt
gedrehte Welle turned shaft
Drehung rotation, turn, turning, gyration; torsion, twisting
Drehung entgegen dem Uhrzeigersinn counterclockwise rotation
Drehung im Uhrzeigersinn clockwise rotation
Drehung um einen Winkel rotation through an angle
durch Drehung bewirkt rotational
dreiadriges Kabel three-core cable
Dreiadresscode three-address code
Drei-Amperemeter-Leistungsmessung three-ammeter method (621.317)
dreiatomig triatomic
Dreibackenfutter three-jaw chuck
dreibegriffiges Signal three-aspect signal
Drei-dB-Koppler 3-dB coupler, hybrid coupler
Drei-Dekaden-Kurbelkondensator three-dial decade condenser
dreidimensional three-dimensional
Dreieck delta (621.3); triangle (51)
dreieckförmig triangular, deltoid
dreieckgeschaltet delta-connected, connected in mesh
Dreieckimpuls triangular (im)pulse
Dreiecksgewinde triangular thread
Dreiecksloch triangular hole

Dreiecksschaltung delta connection, mesh connection, delta circuit
Dreieckspannung delta voltage, phase-to-phase voltage
Dreieckswicklung delta winding
dreieckig triangular, three-cornered
Dreielektrodenröhre three-electrode valve
dreietagig three-storeyed
dreietagiger Schalter three-storeyed switch
dreietagige Wicklung three-range winding, three-tier winding
Drei-Exzeß-Code excess-three code
dreifach thrice, treble, triple
dreifach konzentrisches Kabel triple concentric cable
Dreifach-... triple
Dreifachbindung triple bond
Dreifachexpansion triple expansion
Dreifachkondensator three-gang condenser
Dreifachröhre three-in-on valve
Dreifachuntersetzung triple reduction
Dreifaches treble (51)
Dreifarbenverfahren three-colour process (77)
dreifarbig trichromatic
Dreifußstativ tripod
Dreigangkondensator three-gang condenser
Dreigitterröhre triple-grid valve
dreihalbte Potenz three-halves power
Dreikantauflage Vee-block (621.9)
Dreikantfeile triangular file, three-square file
Dreikantmaßstab triangular scale, scale rule
Dreiklang common chord
Dreileiter-... three wire
Dreileiteranlage three-wire system
Dreileitergenerator three-wire dynamo
Dreileiterkabel triplex cable, three-core cable
Dreileiternetz three-wire system
Dreileiter-Transformator three-wire transformer
Dreileiterverteilung three-wire distribution
Dreileiterverteilungsnetz Edison distribution system
Dreileiterzähler three-wire meter
Dreimantel(blei)kabel "SL" cable = separately lead-covered cable
Drei-Pegel-Modulation three-level modulation
Dreiphasen-... three-phase
Dreiphasenstrom three-phase current
Dreiphasentransformator three-phase transformer
Dreiphasenwechselstrom three-phase alternating current
Dreiphasenwicklung three-phase winding
dreiphasige Läuferwicklung three-phase rotor winding
Dreipol-... three-electrode
Dreipol-Fünfpolröhre triode-pentode
Dreipolröhre three-electrode valve, triode
dreipolig three-pole, three-pin, three-contact, triple-pole, tripolar
dreipolige Abzweigdose three-pole box
dreipoliger Ausschalter three-pole circuit breaker
dreipoliges Kabel triplex cable, three-core cable
dreipolige Klinke three-pole jack
dreipoliger Schalter three-pole switch, three-electrode switch
dreipolige Steckdose three-pole jack
dreipoliger Stecker three-pin plug, triple-pole plug
dreipoliger Umschalter three-pole doublethrow switch
Dreipunktkurs line-of-sight course
Dreipunktschaltung autodyne circuit
Dreipunktschaltung mit induktiver Spannungsteilung Hartley-oscillator circuit
Dreipunktschaltung mit kapazitiver Spannungsteilung Colpitts oscillator circuit
Dreipunktverfahren three-point system
Dreiröhren-... three-valve...
Dreiröhrenempfänger three-valve receiver
Dreisatzrechnung rule of three
Dreischichtglas triplex glass
Dreischleifenwicklung triplex winding
dreiseitig trilateral
Dreistellenrelais three-position relay
dreistellige Zahl number of three ciphers
Dreistiftstecker three-pin plug
Dreistofflegierung three-component alloy
Dreistoffsystem ternary system (669)
Dreistrahlfarbbildröhre three-gun color picture tube, three-gun shadow mask tube
dreisträngige Wicklung three-phase winding
Dreistufenschalter three-stage switch
dreistufig three-stage
dreistufiger Verstärker three-stage amplifier
dreiteilen to trisect (51)
Dreiteilung trisection
dreiviertel Wellenlänge three-quarter wavelength
Dreivoltmeter-Leistungsmessung three-voltmeter method
Dreiwangenbett three-shear bed (621.94)
Dreiwattmeterschaltung three-wattmeter method
Drei-Weg-Box three-way speaker
dreiwertig trivalent

Dreiwickler, Dreiwicklungstransformator three-winding transformer
Drei-Zustands-Ausgang three state output (621.374)
Dreschmaschine thresher, threshing machine
Dreschmotor threshing motor
Drift drift
Driftausfall degradation failure (620.1)
Driftbeweglichkeit drift mobility
Driftgeschwindigkeit drift velocity
Drifttransistor graded-base transistor
Drill-... torsional, twisting
Drillmoment twisting moment, moment of torsion
Drillschwingung torsional vibration
Drillwinkel angle of twist
Drillung twist, twisting, torsion
dritte Harmonische, dritte Oberwelle third harmonic, triple-frequency harmonic
dritte Potenz third power, cube
in die dritte Potenz erheben to cube
dritte Wicklung tertiary winding
dritte Wurzel third root, cube root, cubic root
Drossel choke, choker, kicking coil, impeder (621.3); throttle (621.1)
Drosselerdung inductive leak (621.394)
Drosselflansch choke flange (621.396)
drosselgekoppelt impedance coupled
Drosselklappe throttle valve, butterfly valve
Drosselkopplung choke coupling, impedance coupling, resistance-inductance coupling (621.396.6)
Drosselkupplung reactance tie bars (pl.) (621.311)
Drosselmodulation (saturated) iron-core modulation, choke modulation, choke control
Drosselregelung reactance control method (621.314.5)
Drosselregler throttling-type governor
Drosselschaltung suppressor (circuit)
Drosselspule choke, choker, choking coil, kicking coil, reactance coil, self inductor, reactor (621.316)
Drosselstoß choke joint (621.391); impedance joint (621.33)
Drosselventil throttle (valve), pressure-relief valve
Drosselverbinder impedance bond, reactance bond
Drosselwirkung choking effect
drosseln to choke; to throttle (621.4)
Druck pressure, compression, thrust, push (621); print (655)
Druck aufnehmen to take a pressure
Druck nach außen outthrust
Druck auf der Druckseite fore-pressure (621.6)
Druck nach oben up-thrust
Druck von oben top pressure
Druck von der Saugseite back pressure
Druck auf der Überdruckseite upstream pressure
Druck zusetzen to boost
auf Druck ansprechend pressure-sensitive
auf Druck beanspruchen to compress
unter Druck formen to mould under pressure
Druckabfall pressure drop, fall in pressure
Druckaufnehmer pressure sensor
Druckausgleich pressure balance
Druckbeanspruchung compressive stress, compressive strain
unter Druckbeanspruchung compressed
Druckbelastung compressive load
druckabhängig pressure-responsive
druckbetätigt pressure-actuated
Druckdehnung compressive strain, compression strain
Druckdifferenz pressure difference
Druckdose pressure-sensitive bulb
druckempfindlich pressure-sensitive
Druckfeder compression spring, pressure spring
druckfest pressure-proof, resistant to pressure; flameproof (622); explosion-proof (US)
druckfeste Kapselung flameproof enclosure
Druckfestigkeit compression strength, compressive strength
Druckgaskabel gas pressure cable, gas filled cable
Druckgeber pressure transmitter
Druckgefälle pressure drop, pressure gradient
Druckgeschwindigkeit print speed (681.3)
Druckhebel downward-acting lever
Druckhöhe (pressure) head
Druckhöhe zur Überwindung des Reibungswiderstandes frictional head
Druckhub compression stroke
Druckkabel compression cable, high oil-pressure cable
Druckkammereichung pressure calibration
Druckkammerlautsprecher pressure-chamber loudspeaker, pneumatic loudspeaker
Druckknopf push-button, press-button (621.3)
Druckknopf betätigen to depress a push-button
Druckknopf einschnappen lassen to latch a push-button
Druckknopfabstimmung touch-tuning (621.396.62)

Druckknopfaufzugsteuerung push-button lift control
Druckknopfbirne pear push, pendant push
Druckknopf-Fahrstuhlsteuerung push-button lift control
Druckknopfschaltbirne pressel switch, pear switch, suspension switch
Druckknopfschalter push-button switch, finger-push switch
Druckknopfsteuerung push-button control, press-button control
Druckknopftableau push-button box, push-button panel, push-button station
Druckknopfwähler autodial (621.395.3), push-button selector (621.396.62)
Druckkontakt pressure contact
Drucklager thrust bearing
Druckleitung pressure tubing
Drucklinie line of thrust
Druckluft compressed air; air power
Druckluftanlage compressed-air plant
Druckluftbremse air-pressure brake, compressed-air brake
Drucklufthammer pneumatic hammer
Druckluftkühlung forced air (cooling)
Druckluftlautsprecher pneumatic loudspeaker
Druckluftleitung compressed-air line
Druckluftmotor pneumatic motor
Druckluftschalter gas blast switch
Druckluftschalter mit Selbstblasung auto-pneumatic circuit breaker
Druckluftstellwerk mit elektrischer Betätigung electropneumatic interlocking machine
Druckluftsteuerschalter electropneumatic controller
Druckluftsteuerung pneumatic control
Druckluftversorgung compressed-air supply
Druckmesser manometer, pressure gauge, compression gauge
Druckmeßquarz piezo electric pressure gauge
Druckmeßwertwandler pressure transducer
Druckmikrophon pressure-actuated microphone
Druckminderung depression
Druckminderventil pressure-relief valve
Druckmittelpunkt centre of pressure
Druckplatte end-plate (621.313); platen (655)
Druckprobe compressive test
Druckpumpe pressure pump, compression pump, hydraulic pump, compressor
Druckreduzierung pressure reduction
Druckreduzierventil pressure-reduction valve
Druckregler pressure regulator, pressure governor
Druckrelais pressure relay
Druckrichtung line of pressure
Druckring thrust washer
Druckscheibe thrust washer
Druckschmierung forced (feed) lubrication, pump-fed lubrication
Druckschwankung pressure fluctuation
Druckseite pressure side, compression side (621.5); discharge side, delivery side (621.6)
Druckspannung compressive stress
Drucksteigerung increase in pressure
Druckstelle bruise (61)
Druckstock electrotype
Druckstock galvanisch herstellen to electrotype
Drucktaste push-button, key
Drucktastenauslöser tab
Drucktastenradio push-button radio (set)
Druckverlust loss of pressure, fall of pressure
druckwasserdicht watertight under pressure
druckwassergeschützt pressed waterproof
Druckwasserreaktor pressurized-water reactor
Druckwelle pressure wave, shock wave
Drückband spinning lathe
drucken to print
druckender Empfangslocher printing reperforator
druckender Zähler printometer
gedruckte Leiterplatte printed board
gedruckte Verdrahtung printed wiring
Drucker printer (681.3)
drücken to thrust, to push, to force, to press, to stamp, to emboss, to form, to spin (621.9)
D-Region D-region
Drudenfuß pentagram
Drüse gland (61)
Dtz = Dutzend doz. = dozen
dual dual
Dualsystem binary-digit system, dual code
Dualziffer binary digit
Dübel dowel, cleat
auf Dübeln verlegen to cleat (621.3)
Duddell-Galvanometer thermogalvanometer
Dukt wave duct
duktil ductile
Duktilität ductility
dumpf woolly (534)
dumpfe Wiedergabe woolly reproduction (621.396.62)
Dunkel dark, darkness

Dunkelentladung cold electronic discharge
Dunkelfeld dark field (535)
Dunkelkammer darkroom
Dunkelkammerlampe dark (-room) lamp
Dunkelraum dark space (621.385)
Dunkelschaltung synchronizing dark method (621.311)
Dunkelschirm black scope (621.385), (621.396.96)
Dunkelschriftröhre dark trace tube, skiatron
Dunkelstrahler dark-ray radiator
Dunkelsteuerwert blanking level
Dunkelstrom dark current (621.83)
Dunkeltasten blanking (621.397)
Dunkelwiderstand dark resistance
dunkel dark
dunkle Glut dark heat, dull heat
dunkle Rotglut dark red heat, dull red heat
dunkeltasten to blank out
Dünnfilm-Feldeffekttransistor insulated-gate thin film field effect transistor
Dünnfilmschaltung thin-film circuit
Dünnziehen thin-rod pulling, slim-rod pulling (621.382)
dünn thin, fine, small, delicate
dünner Draht fine wire
dünnes Ende small end
dünne Luft thin air; tenuous air (551.5)
dünne Schicht (thin) film
dünner Strahl narrow beam (535)
dünner Überzug flash plate (621.357)
dünndrähtig thin-wire, fine-wire
dünndrähtige Litze fine strand
dünndrähtige Sicherung fine-wire fuse
dünndrähtige Wicklung thin-wire winding
dünnflüssig mobile
dünnflüssig machen to thin
dünnflüssige Substanz mobile substance
dünnwandig thin-walled
Dunst damp, fume; mist, haze (551)
Duodezimalumsetzung duo-decimal conversion
Duodiode double diode
Duplex-... duplex
Duplexbetrieb duplex operation
Duplexschrank duplex switchboard
Duplextelegraphie duplex telegraphy
duplex schalten to duplex (621.394)
Duplexer duplexer
Düppel chaff, straw, streamer window (621.397)
Düppelband rope
Düppelblüte blossom
Düppelklumpenbildung birdnesting
Düppelstoß chaff burst, window burst
Duralumin(ium) duralumin(i)um
durchbiegen, sich to sag (539.384); to deflect (621)
Durchbiegung sag (539.384); deflection, bend, set (621)
Durchbiegung in der Mitte central deflection
Durchbiegungskoeffizient deflection coefficient
Durchbiegungsmesser deflection indicator
Durchbrennen fusing, fusion, blowing (621.3)
durchbrennen to fuse, to blow, to burn out (621.3)
durchgebrannte Glühbirne burnt-out bulb
durchgebrannte Sicherung blown fuse, gone fuse
durchbrechen to break through; to pierce, to punch, to perforate
Durchbruch rupture; aperture, opening, breakdown
Durchbruchsbereich breakdown region
Durchbruchsfeldstärke dielectric strength, disruptive electric field strength
Durchbruchspannung break down voltage
Durchdrehmotor barring motor
durchdringen to penetrate (through); to pervade, to pierce
durchdringende kosmische Strahlung penetrating cosmic radiation
Durchdringung penetration
Durchdringung von Materie penetration of matter
Durchdringungsfähigkeit penetrating power, radiation hardness (539)
Durchdruck contact-pressure movement (621.316.5)
durchfädeln to thread; to fish (696)
Durchfahrschalter non-stop switch
durchfließen to flow through
Durchflußgeschwindigkeit rate of flow
Durchflußerhitzer flow heater
Durchflußmesser flow meter
Durchflußregelung flow control
Durchflutung flux, ampere turns
Durchflutungsempfindlichkeit magnetomotive force sensitivity
durchführen to carry into effect; to lead through
durchführen nach to bring out to (621.3)
Durchführung (lead-through, feed-through) bushing
Durchführungsdraht lead-in-wire
Durchführungsisolator bushing insulator, insulated bushing
Durchführungsklemme bushing clamp
Durchführungskondensator feed through capacitor

Durchführungsperle beading (621.38)
Durchführungs(strom)wandler bushing current transformer
Durchgang continuity (621.3); pass, passage (621.9)
Durchgang behalten to retain the continuity
auf Durchgang prüfen to test for continuity, to test through
Durchgangsamt tandem exchange (621.395)
Durchgangsdämpfung transmission loss
Durchgangsfernleitung gateway trunk circuit (621.394)
Durchgangsleistung throughput (621.3)
Durchgangsleistungsmesser transmission dynamometer, interconnectable wattmeter
Durchgangsleitung through circuit (621.394/5)
Durchgangsmeßkopf interconnectable probe
Durchgangsmuffe straight joint (621.315.2)
Durchgangsprüfung continuity check, continuity test
Durchgangsregister transit register (621.395)
Durchgangsverbindung through connection
Durchgangsvermittlungsstelle transit exchange
Durchgangszeit transit time (621.38)
Durchgehen overspeed(ing), racing (621.313); blowing (621.316); run-away (539.17)
Durchgehschutz overspeed protection
durchgehen to pass through; to race (621.313); to blow (621.316); to run away (539)
durchgehen durch to permeate through
durchgehender Betrieb permanent service
durchgehende Härte through hardness
durchgehende Leitung through wiring
durchgehend verlegte Leitung continuous run of wiring
durchgegangene Sicherung blown fuse
durchgehende Signalisierung end-to-end signalling
Durchgreifspannung punch-through voltage, penetration voltage, reach-through voltage (621.382)
Durchgriff reciprocal of amplification (621.385)
Durchgriffs-Meßbrücke Miller bridge
Durchhang dip, sag (621.315.1), (67)
Durchhangseinregelung regulation of wires (621.315.1)
durchhängen to sag
Durchimprägnierung mass impregnation
Durchlaß passage, passband
Durchlaßbandbreite filter passband
Durchlaßbereich transparency region (535); passband (621.392); transmission range (621.395); conducting-state range, on state range (621.382)
Durchlaßbreite width of the passband
Durchlaßcharakteristik passband characteristic (621.372.5), conducting state characteristic (621.382)
Durchlaßfrequenzband filter transmission band (621.39)
Durchlaßkennlinie forward characteristic (621.314.5)
Durchlaßrichtung forward direction (621.314.6)
Durchlaßspannung forward voltage (621.314.6), conducting state voltage, on-state voltage, forward bias
Durchlaßstrom forward current, conducting current, on-state current
Durchlaßstrom-Mittelwert mean on-state current
Durchlaßverlustleistung conducting-state power loss, forward power loss
Durchlaßverzögerungszeit forward recovery time
Durchlaßwiderstand forward (d. c.) resistance, on-state resistance
Durchlaßzeit conduction period (621.385)
Durchlaßzustand conducting state, on-state
durchlassen to let through, to pass (621.3); to transmit (535), (621.392), to discriminate in favour of (621.392)
durchlässig für permeable to , pervious to
durchlässig für Röntgenstrahlen radioparent
Durchlässigkeit permeability; transmissivity, transmittance, transmission factor (535)
Durchlauf traverse; run (621.3)
Durchlaufbetrieb continous duty
Durchlaufbetrieb mit Aussetzbelastung continous duty with intermittent loading
Durchlaufbetrieb mit Kurzzeitbelastung continous duty with short-time loading
Durchlaufgalvanisierung semi-automatic plating
Durchlaufkühlung once through cooling
Durchlaufofen bogie furnace, belt furnace, continous belt furnace
durchlaufen to traverse, to run through; to cover; to be continuous
durchlaufende Schweißnaht continuous weld
durchleiten to pass (through)
durchleuchten to X-ray, to screen (621.396)
Durchleuchtung radioscopy, screening, X-raying (621.386)
durchlöchern to perforate, to pierce, to punch
Durchmesser diameter
Durchmesserspannung diametral voltage (621.313)

Durchmesservergrößerung swelling
Durchmesserwicklung full-pitch winding
durchpausen to trace
durchrutschen to slip through
Durchschaltebene junction stage
durchschalten to switch through, to cut through, to connect through
Durchschaltetechnik circuit switching (621.395)
Durchschaltung cut through
Durchschaltzeit gate-controlled rise time (621.382)
durchscheinend translucent, semi-transparent
durchscheuern to fray
durchschicken to pass (through)
durchschieben to push through
sich durchschieben lassen zwischen to pass between
Durchschlag breakdown (671.3), puncture (621.3), (629.113); (pin) punch (621.9); copy (651)
durchschlagfest puncture-proof
durchschlagfester Isolator puncture-proof insulator
Durchschlagentfernung flashover distance
Durchschlagsfestigkeit (specific) dielectric strength, puncture strength, disruptive strength (621.3), breakdown rating
Durchschlagprüfung puncture-voltage test
Durchschlagsspannung breakdown voltage, puncture voltage, potential of puncture; disruptive voltage (621.3)
Durchschlagverzögerung impulse inertia
durchschlagen to punture, to spark through
durchschlagende Wirkung striking effect
Durchschleifeingang feed-through input (621.397)
Durchschnitt average, mean
Durchschnittsausstoß average output
Durchschnittslast average load, mean load
Durchschnittswert standard (value), average
durchsichtig transparent, lucid pellucid
Durchsichtigkeit transparence, coefficient of transparency
durchsickern to trickle through
durchsieben to sieve, to strain, to screen, to riddle
Durchsteckstromwandler bar-type current transformer, stud current transformer
durchstimmbarer Sender variable-frequency oscillator
Durchstimmbereich tuning range
durchstrahlen to X-ray (621.386)
durchstreichen to cross out
durchtropfen (lassen) to percolate
durchverbinden to bond (621.3), to connect through, to build up a connection (621.395)
Durchwahl in-dialling, direct inward dialling, DID
Durchziehband fishing wire, snake (696)
Durchziehloch eyelet
Düse nozzle, jet
Düsenantrieb jet propulsion
Düsenflugzeug jet plane
D-Verhalten derivative action (621.316.7)
DVOR-System DVOR = Doppler Very high frequency Omnidirectional Radio range
Dyn (= c.g.s.-Einheit der Kraft) dyne
Dynamik dynamics (pl.) (53); volume range (621.396.61/62)
Dynamikbegrenzer volume compressor (621.396.61)
Dynamikdehner volume expander (621.396.62)
Dynamikkompression volume compression
Dynamikregler compandor (621.395/6)
Dynamikpresser contractor (621.396.61)
Dynamikverstärker contrast amplifier (621.396.6)
dynamisch dynamic
dynamisch ausgewuchtet dynamically balanced
dynamisch stabil transiently stable
dynamische Auslösung inverse-time overcurrent tripping (621.316.5)
dynamische Auswuchtung dynamic balancing, running balance
dynamische Beanspruchung dynamic stresses (pl.)
dynamische Charakteristik dynamic characteristic
dynamischer Druck dynamic pressure
dynamische Gitterkapazität effective input capacitance
dynamisches Gleichgewicht dynamic equilibrium
dynamische Kennlinie rectification curve (621.385); dynamic characteristic
dynamische Kippleistung transient-power limit, transient-stability limit (621.311)
dynamischer Lautsprecher electrodynamic (loud)speaker, moving-coil speaker
dynamisches Mikrophon moving-coil microphone, electrodynamic microphone
dynamische Stabilität transient stability
dynamische Schleifenverstärkung closed-loop gain
dynamischer Speicher dynamic storage
dynamische Streuung dynamic scattering mode
dynamisches Telefon moving-coil telephone, moving-coil headphone

dynamischer Tonabnehmer moving-coil pick-up
Dynamo(maschine) dynamo (generator)
Dynamoblech dynamo sheet, lamination
dynamoelektrisch dynamoelectric
Dynamometer dynamometer
dynamometrisch dynamometric(al)
dynamometrisches (Maß-)Instrument dynamometer-type instrument, moving-conductor instrument
dynamometrisches Wattmeter dynamometer-type wattmeter, moving-coil wattmeter
Dynamotor dynamotor
Dynatron dynatron, transitron (621.396.61)
Dynatronsender negative transconductance oscillator, dynatron oscillator
Dynode dynode
Dyne s. **Dyn**

E

e [Ladung des Elektrons] e [charge of the electron]
eben even, flat, plane, planar; level, flush (**mit:** with)
eben polarisiert plane polarized
eben polarisierte Welle plane-polarized wave
ebener Boden flat ground, flat bottom
ebene Erdoberfläche flat earth (621.396)
ebene Fläche plane (51); flat surface
ebene Polarisation plane polarization
ebene Welle plane wave, plane polarizedwave
Ebene plane (51), plain (91)
in einer Ebene liegend co-planar, flush with a face
Ebonit ebonite
Eccles-Jordan-Trigger-Schaltung Eccles-Jordan trigger circuit
Echelongitter echelon grating
Echo echo, repercussion
Echobild multiple image, double image, ghost
Echodämpfung active balance return loss (US); echo attenuation
Echo geben to echo
Echolaufzeit echo-transmission time
Echolot fathometer, echo sounder, sonic depth finder (629.12); terrain-clearance indicator (629.13)
Echolotung acoustic sounding, reflection sounding, sonic sounding
Echoortung echo ranging
Echosperre echo eliminator, echo suppressor (621.395)
Echosperrung echo elimination
Echostärke echo strength (621.397)
Echozeichen echo signal (621.397), blip, pip (621.396.96)
echt genuine, true, real, proper, actual
echte Adresse actual address
echtes Bild real image
echter Bruch proper fraction (51)
echter Ladestrom normal charging current
echte Reflexion regular reflection (535)
Echtwiderstand actual resistance, effective resistance (621.3)
Echtzeit real time
Echtzeitbetrieb real-time operation, real-time processing
Ecke corner; plane angle, solid angle (513); (des Dreiecks:) apex (pl.: apices)
Ecke scharf ausdrehen to square out a corner
Eckenabrundung corner-rounding
Eckfrequenz corner frequency (621.372.5)
eckig angular
eckige Klammern square brackets
Edelgase rare gases, noble gases
Edelmetall noble metal, precious metal
Edelpassung close fit
Edelstein precious stone, jewel, gem
Edison-Akkumulator Edison accumulator, alkaline cell, alkaline-type battery
Edison-Batterie Edison battery, steel alkaline battery
Edison-Effekt Edison effect
Edison-Element Edison cell
Edison-Fassung screwed-type socket, screwed holder
Edison-Goliathfassung Goliath Edison screw holder
Edison-Lampe Edison lamp
Edison-Sammler Edison cell, Edison accumulator, alkaline cell
Edison-Sockel Edison screw cap
Editor editor
EDV = elektronische Datenverarbeitung electronic data processing = EDP
Effekt effect
Effektbogenlampe flame-arc lamp, luminous-arc lamp
Effektdochtkohle flame-cored carbon
Effektkohle flame carbon, flaming carbon, mineralized carbon
effektiv effective
effektive Adresse effective address

effektive Antennenhöhe effective aerial height
effektive Höhe effective height (621.396.67)
effektive Höhe der Ionosphärenschichten effective height of ionospheric layers
effektive Kadmium-Schwellenenergie effective cadmium cut-off
effektiver Laufwinkel effective bunching angle (621.385)
effektive Leistung effective power
effektive Strahlungsleistung effective radiated power
Effektiv-... effective
Effektivamplitude des Videosignals picture-signal amplitude (621.397)
Effektivspannung effective voltage, virtual voltage, root-mean-square voltage
Effektivstrom root-mean-square current, r.m.s. current
Effektivstrommesser current square meter
Effektivwert effective value, virtual value, r.m.s. = root-mean-square
Effektivwert der Spannung root-mean-square voltage
Effektlautsprecher effect loudspeaker, auditorium loudspeaker
e-Funktion exponential function
egalisieren to equalize
Eich-... calibration, standardizing
Eichamt standardizing laboratory, standards institution, Bureau of Standards
Eichdiagramm calibration chart
Eichfehler calibration error (621.317)
Eichgenauigkeit calibration accuracy
Eichgerät calibrator
Eichkurve calibration curve
Eichkurvenblatt calibration chart
Eichleitung standard transmission line, reference circuit
Eichmarke calibration mark
Eichpunkt calibration point
Eichraum standardizing room
Eichsender standard-frequency generator
Eichskala calibration scale (621.318.5)
Eichspannung calibration voltage (621.317)
Eichtabelle table of correction
Eichteiler calibrated attenuator
Eichelröhre acorn (type) tube
eichen to gauge, to calibrate, to standardize
Eichung calibration, standardizing
Eichung im Feld fortschreitender Wellen field calibration (534.86)
Eidophorröhre eidophor tube
Eierisolator globe-strain insulator, egg insulator
„eiern" to wobble, to run out of truth

Eigen-..., eigen-... self, natural
Eigenantrieb self-propulsion
Eigenbedarfsanlage auxiliary generating station; station-service plant (621.311)
Eigenbedarfsleistung station auxiliary power (621.311)
Eigenbedarfsmaschine station-service machine
eigenbelüfteter Motor fan-cooled motor
Eigenbewegung intrinsic mobility
eigenerregter Schwinger self oscillator
eigenerregter Sender self-oscillator transmitter, self-excited transmitter
Eigenerregung self oscillation (621.37); direct-connected excitation (621.313)
Eigenfehler residual error
Eigenfeld self field (621.382)
Eigenfrequenz natural frequency, resonant frequency, characteristic frequency
eigengespeiste Verstärkerstelle self-contained repeater station
Eigengewicht self-weight, dead weight
Eigenhalbleiter intrinsic semiconductor
Eigeninduktivität self-inductance
Eigenkapazität self-capacitance
Eigenkorrektur autocorrection (621.317)
Eigenkühlung natural cooling
Eigenleitung intrinsic conduction
Eigenmodulation internal modulation
Eigenoszillator local oscillator
Eigenrauschen inherent noise (621.396.62), inherent random noise
Eigenschaltanlage station-service switchgear
Eigenresonanz natural resonance, self-resonant frequency
Eigenschwingung natural resonance, natural oscillation, self-resonance
Eigenschwingungszahl natural frequency
Eigenschwingungszeit natural period of vibration
eigensicher intrinsically safe (622)
Eigenverkettung self-linkage
Eigenverluste inherent losses; natural leak (621.396.69)
Eigenverständigung intercommunication (629.13)
Eigenwellenlänge natural wavelength, fundamental wavelength
Eigenwellenlänge der Antenne ohne Abstimmelemente unloaded wavelength
Eigenwert inherent value
Eigenwiderstand inherent resistance, resistivity
Eigenzeit inherent response time; operating time (621.316.5)
Eigenzeitkonstante residual time constant

Eimerkettenschaltung bucket-brigade devices
„Ein" "ON"
einachsig uniaxial
Einadressenbefehl one-address instruction, single address instruction
einadrig single-core
einadriges Kabel single-core cable, single-wire cable
Einankerumformer single-armature converter
Einankerumformer-Zusatzmaschine synchronous booster converter
Einanodengleichrichter single-anode rectifier
einatomig monatomic
Ein-Ausgabegerät terminal, input-output device
Ein-Aus-Regler on-off controller, two-position action controller
ein- und ausschalten to turn on and off (621.3)
Ein- und Ausschalter make-and-break key, single-throw switch
Ein-Aus-Tastung on-off keying (621.394)
Einbau installation, insertion, application, building-in, mounting
einbaufertig ready to install
Einbauhöhe mounting height
Einbauinstrument flush-mounting meter, flush-type instrument (621.317)
Einbaulage mounting position
Einbaupuffer cushion mounting
Einbausatz installation kit
Einbauschalter recessed switch, flush-mounting switch
Einbauspannung mounting stress
Einbautype flush pattern
Einbauzeichnung installation plan; locating plan
Einbauzeit installation time
einbauen to build-in, to locate, to mount; to house; to sink (into the wall)
eingebaut built-in, self-contained
eingebaute Antenne built-in aerial
eingebautes Kontrollthermometer embedded temperature detector (621.313)
eingebauter Nebenwiderstand load-indicating resistor (621.317)
eingebaute Prüfeinrichtung built-in check, self check
einbeschreiben to describe in, to inscribe (51)
einbeschriebener Kreis in-circle
einbetten to embed, to imbed
Ein-Bit-Verzögerungsglied digit delay, digit delay element
einbördeln to bead in
Einbrand penetration (621.791)
Einbrandstelle flat (621.313)
einbrennen to burn in, to stove, to fire
Einbrennlack baking varnish, hard-store enamel
Einbrenntest burn-in test
Einbruchsicherung burglar alarm
Einbuchtung recess, inflection
Einbuße loss, damage
Einbuße an Zeichenschärfe loss of definition
Einchip-Element one-chip device
eindeutige Funktion single-valued function
Eindeutigkeit einer Lösung uniqueness of a solution (51)
eindimensional unidimensional
eindimensionaler Hornstrahler sectoral horn (621.396.67)
Eindraht-... single wire, single conductor, single trolley (621.33)
Eindrahtantenne single-wire aerial
Eindrahtinstallation single-conductor wiring
Eindraht-Oberleitungssystem single-trolley system
Eindringtiefe throwing power (621.357); depth of penetration, skin depth (54); (621.3)
Eindringvermögen penetrating power
eindrücken to force in, to insert; to crush
sich eindrücken in to get imprinted upon
einebnen to level, to smooth out
einetagiger Wickelkopf one-range winding
Einer-Zählröhre units-counting tube
einfach simple, plain, elementary
einfach aufgelagert simply supported
einfach wirkend single-acting
einfach wirkende Maschine single-acting engine
einfacher Ausschalter plain cut-out
einfache Abschrägung single bevel
einfacher Einschalter single-way switch
einfacher Guß gravity casting
einfache Maschine simple machine
einfache Muffe straight joint (621.315.2)
einfacher Schalter one-way switch
einfache Viererverdrillung twist system
Einfach-... single
Einfachkabelstutzen single-cable sealing end
Einfachreflexionübertragung single-hop transmission
Einfachspeiseleitung unit feeder
Einfachstrom single current (621.394)
Einfachstrom-Arbeitsstrombetrieb inverse neutral telegraph transmission
Einfachstrom-Betrieb neutral operation, single-current working
Einfachthermoelement single-junction thermocouple
Einfachzählung single metering (621.395.1)

mit Einfachunterbrechung single-break
Einfachwendel single-coil filament
Einfachwendellampe single-coil lamp
einfädeln to thread (in); to channel in, queuing in, to line up, to intercept and proceed the track (351.814.33)
eingefädelte Wicklung threaded-in winding
Einfadenaufhängung unifilar elastic control
Einfaden-Elektrometer unifilar electrometer
einfädig unifilar
Einfahrschalter slowing switch (621.876)
Einfall von Strahlung impact to radiation
Einfallpunkt point of incidence
Einfallswinkel angle of incidence, angle of arrival, wave angle
einfallen to incide (535)
einfallend incident (**in:** in)
einfallende Strahlungsdichte incident intensity
Einfang capture
eingefangenes Elektron trapped electron, captured electron
Einfang eines K-Schalen-Elektrons K-electron capture
Einfangquerschnitt capture cross section (539)
einfarbig self-coloured, monochromatic
einfetten to grease
Einfluß influence, effect, bearing; influx
Einfluß der Enden end effect, effects of ends
Einflußfunktion importance function (539)
Einflußgröße parameter, factor (53)
einfrieren (lassen) to freeze in
eingefrorene Spannung locked-in stress
Einfügungsdämpfung insertion loss
Einfügungsgewinn insertion gain (621.392)
Einfügungs-Leistungsverstärkung insertion power gain
Einfügungsverlust insertion loss
einführen to lead in, to inject (621.3)
Einführung lead-in, lead wire; wall-entrance insulator (621.315.6); injection
Einführungsisolator wall-entrance insulator, wall-entrance bushing (621.315.6)
Einführungskabel leading-in cable
Einführungsleitung leading-in wire (621.315), drope wire (621.395)
Einführungspfeife entrance ell (696.6)
Einfülltrichter filling funnel, charging funnel
Eingabe input
Eingabegabeblock input block (681.3)
Eingabegerät input equipment, input device
Eingabeinformation order, input information
Eingabeprogramm input routine
eingabeln to stradle
Eingang input end, sending end, source (621.3); generator end (621.394/396)
Eingangsadmittanz input admittance
Eingangsanschluß input terminal
Eingangsfehlspannung input offset voltage (621.382)
Eingangsgröße input variable
Eingangsimpedanz input impedance
Eingangskapazität input capacitance
Eingangsklemme input terminal
Eingangskonfiguration input configuration
Eingangskoppelblock input coupling capacitor
Eingangslastfaktor input loading factor (621.374/92); fan-in (681.3)
Eingangsleistung (power) input
Eingangsleitwert input admittance
Eingangsoffsetspannung input offset voltage
Eingangsruhespannung quiescent input voltage
Eingangsruhestrom quiescent input current
Eingangsspannung input voltage; sending-end voltage, signal voltage (621.39)
Eingangsstrom input current, entering current
Eingangsstufe input stage, preceding stage (621.396.62/64)
Eingangstransformator input transformer
Eingangsverstärker low-level amplifier
Eingangswiderstand input impedance, input resistor, sending-end impedance
Eingangswiderstand bei Leerlauf an allen anderen Klemmen self-impedance
eingefahrene Antenne retracted antenna
eingehende Untersuchung thorough examination; complete analysis (51)
eingekuppelt in gear, coupled
eingeregelter Zustand balanced condition
eingeschwungener Zustand steady state
eingleisige Strecke single-track line
eingreifen in to intervene in; to interfere with, to infringe upon (34), to gear into, to mate with, to mesh with, to register with (621.8)
eingreifendes Rad mating gear
Eingriff engagement, gear, gearing, mesh; register (621.8)
außer Eingriff bringen to push out of gear, to swing out of mesh
in Eingriff bringen to bring into engagement, to engage
Eingriffsfläche mating surface
Eingriffslinie line of force, line of contact, pressure line
Eingriffszeit period of contact
eingruppieren to group (together)
Eingruppierung grouping
Einguß feeder, runner (621.7)

einhaken to hook
einhalten to stop, to check, to arrest; to keep (up), to observe
Einhaltung der vorgeschriebenen Toleranzen überprüfen to check for specified limits
einhängen to hook to hang in; to hang up, to replace the handset, to go on-hook (621.395)
Einhängedraht slinging wire (621.357)
Einheit unit (53) (621); set (621.3)
Einheit des elektrischen Flusses unit electric flux
Einheit der Kraft unit (of) force
Einheit der Ladung unit (of) charge
Einheit der Leuchtdichte unit of brightness
Einheit des Rauminhalts unit of volume
Einheit der Wärmemenge thermal unit
Einheit der Zeit unit (of) time
die Einheiten betreffend unitary
auf die Einheit bezogen per unit
Einheitsbauweise standard design, standard equipment practice
Einheitsbohrung-System hole-basis system
Einheitsfluß unit electric flux
Einheitskreis unit circle
Einheitsleistung specific output
Einheitspol unit pole
Einheitsröhre unit tube of force (537)
Einheitssprung unit step function, unit discontinuous function
Einheitsstoß unit impulse
Einheitsstrom unit current
Einheitsvektor unit vector
einkeilen to wedge
einkitten to putty
Einklang consonance, harmony
einklinken to latch
sich einklinken to fall into a notch
Einknopfabstimmung ganged tuning
Einknopfbedienung single-dial control (621.396.62)
Einkreis-Triftröhre single-cavity velocity-modulated tube
Einkristall single crystal
Einkristallschaltung monolithic circuit
einkuppeln to gear, to throw in, to bring into engagement
eingekuppelt in gear
Einlagerungstelegraphie intraband telegraphy
einlagig single-layer
einlagige Spule single-layer coil
einlagige Wicklung, einlagige Zylinderspule single-layer winding (621.396)
Einlaß admission, inlet (621.1/4)
Einlaßventil inlet valve, intake valve, admission valve
einlassen to admit
Einlaufen, Einlaufvorgang break-in (621)
Einlaufverschleiß break-in wear
Einlaufzeit warm-up time
Einlegen insertion (621.9)
Einlegebacke loose jaw
einlegen to insert
Einleiterkabel single (-wire) cable
Einleiterwandler single-turn transformer, bar-type current transformer (621.317)
einlesen to read in
Einlochbefestigung one-hole fixing
Einmannbedienung one-man operation
einmessen to calibrate
einmontieren to fit (in)
einnebeln to fog
einpassen to fit in, to nest, to register
einpassen in to match into
Einpegelung alignment, level adjustment
einpendeln to hunt to a steady state
einphasen to phase (621.394.6)
Einphasen-... s.p. = single-phase, single-pole; uniphase, monophase
Einphasenbahn single-phase railway
Einphasenbahnbetrieb single-phase traction
Einphasenbahnmotor single-phase traction motor
Einphasen-Dreiphasenverteilung monocyclic system
Einphasengenerator single-phase generator, single-phase alternator
Einphasen-Induktionsmotor single-phase induction motor
Einphasen-Kommutatormotor mit Kompensationswicklung neutralized series motor, series-compensated motor
Einphasenlokomotive single-phase locomotive
Einphasenmotor single-phase motor
Einphasenreihenschlußmotor mit kurzgeschlossener Kompensationswicklung single-phase series commutator motor with short-circuited compensation winding
Einphasenstrom single-phase current
Einphasen-Synchrongenerator single-phase synchronous generator
Einphasensystem single-phase system
Einphasentransformator single-phase transformer
Einphasenwechselstrom single-phase a. c. = single-phase alternating current
Einphasenwicklung single-phase winding
Einphasenzähler single-phase meter

einphasig one-phase, single-phase, monophase, uniphase
einpolig single-pole, single-contact
single-conductor (s. auch **einphasig**)
einpolige Installation single-conductor wiring
einpoliger Kurzschluß one-line-to-earth short circuit
einpolige Leitung single-wire line, single-wire conductor
einpoliges Relais single-contact relay
einpoliger Schalter single-contact switch, single-pole circuit breaker
einpoliges Schütz single-pole cut-out
einpolige Sicherung single-pole fuse
einpolige Signalleitung signalling wire
einpoliger Umschalter single-pole double-throw switch
einprägen to imprint, to impress
eingeprägte E.M.K. impressed e.m.f.
eingeprägte Kraft impressed force
eingeprägte Spannung impressed voltage
einrasten to snap in, to lock in place
Einraststrom latching current (621.382)
einregulieren to regulate, to adjust (621), (681)
einreihiges Kugellager single-race ball bearing
einrichten to adjust, to set right; to arrange, to establish, to install
Einrichtung facility, installation
Einrillenscheibe single-groove sheave
einrollen to roll up, to bead (621.7)
eingerollter Rand bead (621)
mit eingerolltem Rand versehen to wire (621.396)
Eins one
Einszustand one-condition
Einsfrequenz frequency of unity, current transfer ratio
Einsäulentrenner (single-post) pantograph-type isolating switch
Einsattelung fold (55); dip (621.396)
Einsatzhärtung case hardening
Einsatzpfeifen threshold howl (621.392), fringe howl (621.396)
Einsatzpunkt cut-off (621.385); zero of operating range (621.316.7)
Einsatzsicherung switch fuse
Einsatzwert threshold value
einsaugen to imbibe; to draw (621.74)
Einschalt-... make, making, operating
Einschaltfolge starting sequence
Einschaltfunke make spark
Einschaltleistung making capacity
Einschaltstellung on-position, closed position
Einschaltstoß make impulse
Einschaltstoß einer Lampe overshooting of a lamp
Einschaltstrom inrush current
Einschaltstromstoß starting current inrush (621.313); magnetizing-current inrush (621.314.2); making-current inrush
Einschaltverlustenergie energy dissipation during turn-on time (621.382.3)
Einschaltverlustleistung turn-on loss (621.382.3)
Einschaltverlustleistungsspitze peak value of turn-on loss
Einschaltvorgang circuit closing; transient; excursion (539)
Einschaltvorrichtung closing device
einschaltbereite Reserve hot reserve
einschalten to switch in, to switch on, to turn on, to snap on; to power-up
(plötzlich) einschalten auf to throw on to
eingeschaltet lassen to leave switched on
eingeschalteter Zustand closed-circuit condition, switched-on condition
Einschaltung making, circuit-closing (621.316); starting (621.313); cut-over (621.395)
einschichtige Wicklung single-layer winding
Einschienenbahn monorail
Einschlagwecker single-stroke bell
Einschleifen looping-in (621.316)
einschleifen to loop in
Einschleifenwicklung singly re-entrant winding, simplex winding
Einschleifung loop service (621.316)
einschließen to enclose, to shut in; to include (51)
eingeschlossener Winkel included angle
Einschlüsse crystal spots (621.357)
einschmelzen to melt down; to seal in
Einschmelzglas beading glass (621.385)
Einschmelzperle beading
Einschmelzung (glass-to-metal) seal, sealing-in joint (621.385)
einschnappen to snap in
einschneiden to incise, to indent, to slot, to cut
Einschnitt incision, slot, notch, cut
mit Einschnitten versehen to notch
Einschnüreffekt pinch effect (621.382)
einschrauben to screw (in)
einschreiben to write-in (681.14)
Einschreibesignal writing signal
Einschub slide-in chassis, plug-in unit
Einschubsystem equipment practice; plug-in-unit type
einschweißen to fuse in (621.791)
Einschwingfrequenz natural frequency

Einschwingverhalten transient response
Einschwingverzögerung delay
Einschwingvorgang transient (phenomenon)
Einschwingverzerrung characteristic distortion (621.394)
Einschwingzeit rise time, build-up time, response time, settling time
Einseitenband... single sideband
  Einseitenhandsender single sideband transmitter
  Einseitenbandübertragung single sideband transmission
einseitig unilateral
  einseitiger Antrieb unilateral drive (625)
  einseitig beschreibbare Diskette single-sided diskette
  einseitig eingespannter Träger cantilever
  einseitig gerichtet unidirectional
  einseitig unbegrenzt semi-infinite
  einseitiger magnetischer Zug unbalanced magnetic pull
  einseitige Richtcharakteristik unidirectional characteristic
  einseitige Verzerrung bias distortion
einsenken to recess
einsetzen to insert, to place; to set in; to substitute (51)
  Einsetzklima set-in climate
Einsfrequenz frequency of unity current transfer ratio
einsickern (lassen) to infiltrate
einsinnig unidirectional
  einsinnig verknüpft unidirectional
  einsinniger Strom unidirectional current
einsortieren to file (681.3)
einspeisen to feed (in)
Einspeisepunkt supply point, feed-in point
Einspeisung feed-in, incoming supply, incoming feeder; line entry (621.315.1)
Einsprache, Einsprechöffnung mouthpiece, transmitter (621.395)
einspringend re-entrant
  einspringende Vertiefung recess
Einspritz-... injection
  Einspritzdüse injection nozzle
  Einspritzmotor injection motor
einspritzen to inject
Einspritzung injection
Einsprungstelle entry point (681.3)
Einspurtechnik single-track recording (681.84)
einstecken in to slip into, to engage into (621.3)
Einsteckführung entering guide
Einstein-de Haas-Effekt gyromagnetic effect
Einstell-... setting, adjusting, positioning
  Einstellbereich timing, adjustment area; adjustment range, setting range
  Einstelleinrichtung positioning device, adjusting device
  Einstellgenauigkeit accuracy of adjustment, setting accuracy
  Einstellknopf adjusting knob, setting knob
  Einstell-Lehre setting gauge
  Einstellmechanismus setting mechanism
  Einstellmutter adjusting nut
  Einstellschraube set screw, adjusting screw
  Einstellskala calibration scale (621.318.5); setting scale
  Einstellvorrichtung adjusting device
  Einstellwert setting, set point
  Einstellzeit setting time
einstellbar adjustable
  einstellbarer Kondensator variable capacitor
  einstellbare Drehzahl adjustable speed
  einstellbarer Spannungsregler adjustable voltage regulator
einstellen to focus (535); to set, to adjust (621); to calibrate (621.318); to modulate (621.396)
  einstellen auf to set at (621); to tune in for (621.396)
Einstellen-Schweißumformer single-operator motor-generator set
Einstellung setting, adjustment; calibration
  Einstellung auf Null zero setting
  Einstellung der Nullinienhöhe position control (621.385)
„Ein"-Stellung "on" position
einstrahlen to irradiate
Einstrahlpeilung single-beam direction finding
Einstrahlröhre mit Farbsteuergitter chromatron, lawrence tube (621.397)
Einstrahlung irradiation
einstreuen to intersperse, to strew in
Einstreuung stray pick-up (621.396)
einströmen bei to flow in at
Einstromsystem single-current (telegraph) system
Einstromtaste single-current key (621.394)
Einströmung admission (621.4)
einstufig single-stage
  einstufiger Verstärker single-stage amplifier
Eintaktausgang single-ended output (621.375)
Eintakteingang single-ended input (621.375)
einteilbar divisible
einteilen to divide, to portion; to graduate (531)
  einteilen in to mark off in
  einteilen nach to graduate in terms of
  sich einteilen lassen in to separate into
einteilig one-piece
  einteiliger Aufbau one-piece construction

einteiliger Läufer solid rotor
einträufeln to instil
eintreten bei to flow in at (621.3)
eintreten in to enter (621.3); to cut in (621.395)
eintreten lassen in eine Fläche to insert into a face
Eintrittsöffnung intake
Eintrittsstelle point of entry
Einweggleichrichter one-way rectifier, half-wave rectifier
Einweggleichrichtung one-way rectification, half-wave rectification
einweichen to soak
Einwellenbetrieb single-wavelength working, single-wavelength operation
Einwellenempfänger single-signal receiver, selective receiver
einwellig sine wave
einwellige EMK simple sinusoidal e.m.f.
einwellige Spannung sine-wave voltage
Einwelligkeit selective resonance
einwerfen to throw in (621.316.5); to enter, to insert
einwertig univalent (54); single-valued (51)
einwertige Funktion single-valued function
Einwicklungstransformator autotransformer
Einwindungsrahmen one-turn loop, single-turn loop (621.396)
Einwirkung effect, effectiveness
Einwirkung von Streufeldern stray pick-up
der Einwirkung eines Feldes unterwerfen to set in a field
einzeichnen to draw in
einzeilig one-lined
Einzel-... single, separate, individual
Einzeladresse single address
Einzelanschluß individual line, single line (621.395)
Einzelantrieb separate motor drive, individual drive (621.9)
Einzelbild single picture, frame (621.397), (77)
Einzeleinbau separate installation
Einzelgerät separate unit
Einzelinstallation separate installation
Einzelkippschaltung one-shot multivibrator
Einzellast single-point load (624); individual load (621.3)
Einzelleiter isolated conductor
Einzelleitung single conductor
Einzelpol pole unit (621.316.5)
Einzelschrittbetrieb one-shot operation; step-by-step operation (681.3); single step operation
Einzelspeiseleitung single feeder
Einzelteil separate part, individual part
in Einzelteile zerlegen to break down
Einzelzeichendrucker character-at-a-time printer
Einzellen-Lademaschine milking generator (621.355)
einzeln isolated (621.3); individual, separate, single (s. auch **Einzel-...**)
einzeln angetriebene Achse individual gear-driven axle
einzelne Lage single layer
einzelne Leitung single conductor
einzelne Windung single turn
Einziehdraht fish wire, snake (696)
Einziehschacht draw-in box (621.315.2)
Einzylindermotor single-cylinder engine
Eis ice
Eisbelastung, Eislast ice loading (621.315.1), (621.396.67)
Eislastbildung icing
Eisschutzhaube sleet hood
Eisüberzug sleet
Eisen iron
Eisenbahn Ry. = railway; railroad (US)
Eisenbahnsignal railway signal
Eisenbahnsignalwesen railway signalling
Eisenbahnwagen railway carriage, railway coach; wagon (US)
Eisenbeton ferro-concrete, reinforced concrete
Eisenbetonmast reinforced-concrete pole
Eisenblättchen iron vane (621.317)
Eisenblech sheet iron
Eisenblechgehäuse sheet-iron casing
Eisendraht iron wire
Eisendrossel saturable reactor, iron-core reactor
Eisenerz iron ore
Eisenfahne iron vane (621.317)
Eisenfeilicht, Eisenfeilspäne iron filing(s)
eisengekapselt iron-clad
Eisengerüst iron framework
eisengeschirmtes Wattmeter shielded wattmeter
eisengeschütztes Wattmeter iron-clad wattmeter
Eisengitter iron grating
Eisenglanz iron glance
Eisengleichrichter steel-cylinder mercury-arc rectifier, steel-tank rectifier
Eisenguß iron castings (pl.)
Eisenkarbid iron carbide
Eisenkern iron core
mit Eisenkern iron-cored

Eisenkernspule iron-cored coil, iron-cored inductor; dust-cored coil (621.396)
Eisenkerntransformator iron-cored transformer, saturable transformer
Eisenlegierung iron alloy
Eisenleiter steel conductor
eisenlose Drosselspule air-core chocke
Eisenmast (steel) tower (621.315)
Eisenmastleitung tower line
Eisenmetalle iron metals, ferrous metals
Eisen-III-Oxid ferric oxide
Eisenoxidul ferrous oxide
Eisenprobe iron specimen
Eisenprüfgerät ferrometer
Eisenspulver iron dust, powdered iron, pulverized iron, dust iron (621.396)
Eisenpulverkern iron-dust core, iron-dust slug, powdered iron core (621.396)
Eisenpulverkernspule dust-cored coil
Eisenpulvertopfkern pot-type iron-dust core
Eisenpyrit iron pyrite
Eisenrost iron grating (621.1), (624); iron rust (54)
Eisensäge iron-cutting saw
Eisenschluß keeper (538)
Eisenstaub pulverized iron (621.396); dust iron
Eisenstein ironstone
Eisentransformator iron-cored transformer
Eisentransformator mit Luftspalt open-core transformer
Eisenverlust iron loss (621.3)
Eisenverlustmessung watt-meter method (621.318)
Eisenvitriol green copperas
Eisenwasserstoffwiderstand ballast tube, iron-hydrogen resistance, barretter
eisern iron(-bound)
eisernes Wellblech corrugated sheet iron
Eivau = Eigenverständigung intercom. = intercommunication (629.13)
Ejektor ejector
Ekonomiser economizer, feed heater (621.1)
elastisch elastic, flexible
elastische Aufhängung resilient suspension, resilient mounting, flexible mounting
elastische Beanspruchung elastic strain
elastischer Bereich dimensional elastic range (620.1)
elastische Dehnung elastic strain, elastic stretch
elastische Kontraktion elastic contraction
elastische Nachwirkung creep (620.1)
elastischer Stoß elastic collision
Elastizität elasticity
Elastizitätsgrenze elastic limit, limit of elasticity
Elastizitätsmodul modulus of elasticity, elastic modulus, Young's modulus
Elektret electret
elektrifizieren to electrify
elektrifizierte Strecke electrified line
Elektrifizierung electrification
Elektriker electrician
elektrisch electric, electrical (siehe auch **Elektro . . .)**
elektrisch aufladen to electrify
elektrisch betätigte Bremse electrically operated brake
elektrisch betätigtes Druckluftventil electro-pneumatic valve
elektrisch leitend electrically conductend, electroconductive
elektrisch leitend verbinden to bond together
elektrischer Abgriff sonde (621.396); tap (621.31)
elektrische Abtastung electrical scan (621.397)
elektrische Achse electrical axis (537.228)
elektrische Anlage electric installation
elektrischer Anschluß electric connection; electric supply
elektrischer Antrieb electric drive; electric propulsion (629)
elektrische Auslösung electromagnetic release
elektrische Ausrüstung electrical equipment
elektrischer Bahnbetrieb elctric traction
elektrische Belastung electrical load
elektrische Beleuchtung electric lighting
auf elektrischen Betrieb umstellen to electrify
elektrische Bremsung electric braking, dynamic braking
elektrisches Bügeleisen electric iron
elektrischer Dipol electric dipol, electric douplet
elektrische Doppelschicht electric double layer
elektrischer Drehzahlmesser magneto speed indicator
elektrisches Echolot electric depth finder, fathometer
elektrische Eigenschaft electrical property
elektrische Einheit absolute electrical unit
elektrische Energie erzeugen to generate electrical energy
elektrische Energie (nutzlos) in Wärme umsetzen to waste watts
elektrische Entladung electric(al) discharge
elektrische Fahrstraßenverriegelung electric route locking (625)
elektrisches Feld electric field

elektrisches Fernmeßgerät electric telemeter
elektrische Festigkeit electric strength
elektrischer Fluß electric(al) flux
elektrische Grenzlehre electro-limit gauge
elektrische Größe electrical quantity
elektrische Grubenlampe electric safety lamp, miner's electric lamp
elektrische Handbohrmaschine portable electric drill
elektrische Heizung electric heating
elektrischer Höhenmesser radioaltimeter, terrain clearance indicator, height indicator
elektrische Hörhilfe hearing aids
elektrische Induktion electric induction, dielectric-flux density
elektrische Installation electrical installation, electrical wiring
  elektrisches Installationsmaterial electric-installation material, wiring material
elektrische Kaffeemühle electric coffee grinder
elektrischer Kontakt electrical contact
elektrische Kraftübertragung electric transmission
elektrische Länge electrical length (621.396.67)
elektrischer Laufkran overhead electric traveller
elektrische Leistung electrical power
elektrischer Leiter electric conductor
  elektrische Leitfähigkeit electric(al) conductivity
elektrische Leitung electric line
  elektrische Leitungen electric wiring (sing.)
  elektrische Leitungsanlage electrical installation
  elektrischer Leitungsmast electricity standard, pole, tower, pylon
elektrischer Lichtbogen electric arc
  elektrischer Lichtbogenofen electric-arc furnace
  elektrische Lichtbogenschweißung electric-arc welding
elektrisch löschbares ROM electrically erasable ROM = EEROM
elektrische Luftdruckbremse electropneumatic brake
elektrisches Meßinstrument electrical (measuring) instrument
elektrische Meßtechnik electrical measurements (pl.)
elektrische Messung electrical measurement
elektrisches Moment electric moment
elektrische Nachrichtenübertragung electrical communication
elektrische Nebenuhr electric-impulse clock, slave clock, (electric) secondary clock
elektrisches Netz electric mains (pl.)
elektrisches Photometer photoelectric photometer
elektrischer Plattenspieler electric gramophone, record player
elektrisches Potential electric potential
  elektrische Potentialdifferenz electric potential difference
elektrische Prüfung electric test(-ing)
elektrische Rangierlokomotive electric switching locomotive
elektrische Registrierung electrical recording
elektrische Rückstellung electrical reset
elektrische Schaltung electric network, electric circuit
elektrischer Schlag electric shock
elektrisches Signal electric signal
  elektrisches Signalwesen electric signalling
elektrische Spannung electric voltage, electric tension, electric potential difference
elektrischer Speisenaufzug electric dumbwaiter
elektrische Staubabscheidung electrical precipitation, electrostatic precipitation
elektrischer Staubsauger electric vakuum cleaner, electric sweeper
elektrisches Stellwerk electric signal box, electric interlocking machine, electromechanical interlocking machine
elektrischer Strahlofen electric fire; electric radiator (US)
elektrischer Streustrom stray electric current
elektrischer Strom electric current
  elektrischer Stromkreis electric(-al) circuit
elektrische Suszeptibilität dielectric susceptibility
elektrische Taschenlampe electric torch
elektrischer Torpedo Simms-Edison torpedo
elektrischer Triebwagen battery-driven vehicle
elektrische Trocknung electro-desiccation
elektrischer Turmwagen electric tower car (621.33)
elektrische Übertragungsschaltung electric transducer
elektrisches Unterwerk electric (power) substation
elektrisch veränderbarer Festwertspeicher electrically alterable ROM = EAROM
elektrische Verblockung electric section locking, electromagnetic locking
elektrische Versenkung electric stage lift
elektrische Versorgungsleitungen electric supply lines

elektrische Weichenstellverriegelung electric switch-lever lock
elektrische Weichenverriegelung electric switch lock
elektrischer Weidezaun electric fence
elektrische Welle electric wave (537), (621.396); synchronous tie (621.87)
elektrischer Widerstand electrical resistance, electrical resistor
elektrische Wiedergabe electrical reproduction
elektrischer Wind ionic wind, static breeze (537)
elektrische Zeitsperre electric time locking (65)
elektrische Zündanlage electrical ignition system
elektrischer Zünder electric detonator, electric igniter (622); electrical fuse (623)
elektrische Zündung electric ignition (621.4)
elektrisieren to electrize, to electrify (61); to faradize (621.384)
Elektrisierung electrization, electrification; faradization
Elektrizität electricity
Elektrizitätsdurchgang durch Gase passage of electricity through gases
Elektrizitätserzeugung generation of electricity
Elektrizitätslehre science of electricity
Elektrizitätsmenge quantity of electricity
Elektrizitätsträger carrier of current
Elektrizitätsversorgung electric supply, electricity supply
Elektrizitätsversorgungsunternehmen electric light and power company
Elektrizitätswerk electricity works (pl.), electric plant
Elektrizitätszähler electricity meter, electric (supply) meter
elektro-... electro, electric(al)
elektro-akustisch electro-acoustic, electro acoustical
elektro-akustische Geräte sound system equipment
elektro-akustischer Übertragungsfaktor force factor, transmission factor
elektro-akustisches Übertragungsmaß pressure rating, response level
elektro-akustischer Wandler electro-acoustic transducer, audio receiving device
Elektroartikel electrical goods
Elektrobrand electrical fire
elektrochemisch electrochemical
elektrochemisches Äquivalent electrochemical equivalent
elektrochemische Passivierung electrochemical passivation
Elektrode electrode (621.3); tip (621.791)
Elektrode mit gasumhülltem Lichtbogen shielded-arc electrode
Elektrodenkapazität electrode capacity, interelectrodic capacitance (621.385)
Elektrodenkessel electrode boiler
Elektrodenkragen electrode economizer (621.357)
elektrodenlose Glimmlampe neon induction lamp
Elektrodenmetall electrochemical valve metal (621.385)
Elektrodenplatte electrode plate
Elektrodenpotential electrode potential, electrolytic potential; (gegen Kathode:) electrode voltage
Elektrodenpotential bei Stromfluß dynamic electrode potential
Elektrodenschweißung metal-arc welding
Elektrodenspitze electrode tip
Elektrodensystem electrode system
Elektrodenverlustleistung electrode dissipation
Elektrodenwiderstand electrode resistance
Elektrodynamik electrodynamics (pl.)
elektrodynamisch electrodynamic
elektrodynamische Blaseinrichtung solenoid blow-out (621.316.5)
elektrodynamisches Instrument electrodynamic instrument
elektrodynamische Kraft electrodynamic force
elektrodynamischer Lautsprecher electrodynamic (loud)speaker, moving-coil loudspeaker
Elektroerosion electroerosion, spark erosion
Elektrofilter electrofilter, electrostatic precipitator
elektrogalvanisch electro-galvanic
elektrogalvanische Lösung electro-dissolution
Elektrogeräte electrical appliances; electric equipment (sing.)
Elektro-Hartlötung electric brazing
Elektroheißwasserspeicher electric water heater, thermal-storage water heater
Elektroheizung electric heating
Elektroherd electric range, electric cooker
elektrohydraulisch electrohydraulic
Elektroindustrie electrical industry
Elektroingenieur electrical engineer
Elektro-Installateur electrical contractor; electrician
Elektro-Installation electric wiring, electrical installation

Elektrokardiogramm electrocardiogram
Elektrokarren electric truck
Elektro-Kauterisation hot wire cautery (61)
Elektrokessel electric boiler
Elektrokran electric crane
Elektrolokomotive electric locomotive
Elektro-Lötkolben electric soldering iron
Elektrolyse electrolysis
  Elektrolysebad electrolytic bath
Elektrolyseur Allan cell (621.357)
elektrolysieren to electrolyse
Elektrolyt electrolyte
  Elektrolytableiter electrolytic arrester
  Elektrolyteisen electrolytic iron
  Elektrolytgleichrichter electrochemical valve, aluminium rectifier, electrolytic rectifier
  Elektrolytkondensator electrolytic condenser, electrolytic capacitor
  Elektrolytkupfer electrolytic copper, standard copper
  Elektrolytventil aluminium rectifier
  Elektrolytzähler mercury meter, electrolytic meter
elektrolytisch electrolytic(al)
  elektrolytisch verstärken to increase by electrodeposition
  elektrolytische Abscheidung electrodeposition
  elektrolytische Analyse electrometric analysis
  elektrolytisches Bad electrolytic bath
  elektrolytische Dissoziation electrolytic dissociation
  elektrolytisches Element electrolytic cell
  elektrolytische Korrosion autogalvanic corrosion, local corrosion
  elektrolytischer Mittelwert rectified value
  elektrolytisches Photoelement photo-chemical cell, light-voltaic cell
  elektrolytische Reduktion electrolytic reduction
  elektrolytischer Widerstand electrolytic resistance
  elektrolytische Zelle electrolytic cell
  elektrolytischer Zerfall electrolytic dissociation
Elektromagnet electromagnet
elektromagnetisch electromagnetic
  elektromagnetische Ausstrahlung electromagnetic emission
  elektromagnetische Auslösung electromagnetic release, electromagnetic tripping
  elektromagnetische Bremse electromagnetic brake
  elektromagnetische Einheit E.M.U. = Electromagnetic Unit
  elektromagnetische cgs-Einheit der Induktivität abhenry
  elektromagnetische cgs-Einheit der Kapazität abfarad
  elektromagnetische cgs-Einheit der Ladung abcoulomb
  elektromagnetische cgs-Einheit des Leitwertes abmho
  elektromagnetische cgs-Einheit der Spannung abvolt
  elektromagnetische cgs-Einheit des Stromes abampere
  elektromagnetische cgs-Einheit des Widerstandes abohm
  elektromagnetisches Feld electromagnetic field
  elektromagnetische Fokussierung electromagnetic focusing
  elektromagnetische Induktion electromagnetic induction
  elektromagnetischer Lautsprecher electromagnetic (loud)speaker
  elektromagnetische Linse electromagnetic lense
  elektromagnetisches Relais electromagnetic(-type) relay
  elektromagnetische Strahlung electromagnetic radiation
  elektromagnetische Verträglichkeit electromagnetic compatibility, EMC
  elektromagnetische Verzögerungsleitung electromagnetic delay line
  elektromagnetische Welle electromagnetic wave, Hertzian wave
Elektromaschinen electrical machinery (sing.)
elektromechanisch electromechanical
  elektromechanischer Kopplungsfaktor electromechanical coupling factor
elektromedizinisch electromedical
  elektromedizinische Geräte electromedical appliances
Elektrometallurgie electrometallurgy
Elektrometer electrometer
  Elektrometernadel versorium, electrometer needle
  Elektrometerröhre electrometer valve
Elektromonteur electrician; line man (621.315)
Elektromotor electric motor
elektromotorisch electromotive
  elektromotorische Kraft electromotive force
Elektron electron (particle)
  Elektron der äußeren Schale peripheral electron
  Elektronen abgeben to lose electrons
  Elektronen aufnehmen to gain electrons

Elektronen auslösen to set electrons free
Elektronen austauschen to interchange electrons
Elektronen befreien to dislodge electrons, to set electrons free
Elektronen einfangen to capture electrons, to collect electrons, to trap electrons
Elektronen emittieren to emit electrons, to eject electrons
Elektronen paketieren to bunch electrons
Elektronen treiben to urge electrons
Elektronen übernehmen to collect electrons
Elektronen auf eine gekrümmte Bahn zwingen to curve electrons around along a path
elektronenarm deficient in electrons
Elektronenausbeute emission efficiency
Elektronenbeschleuniger electron accelerator
Elektronenbahn electronic path, electron trajectory
Elektronenbeugung electron diffraction
Elektronenbewegung electron motion
Elektronenblitz electronic flash, photoflash
Elektronenbombardement electron bombardment
Elektronenbremsung electron deceleration
Elektronenbündelröhre electron beam valve
Elektronendichte electron concentration
Elektronendrall electron spin
Elektroneneinfang electron capture
Elektronenemission emission of electrons, electron emission
elektronenemittierend electron-emissive
Elektronenfehlstelle electron trap
Elektronenfluß electron flow
Elektronengas electron gas
Elektronengehirn electronic brain
elektronengekoppelt electron-coupled (621.396)
  elektronengekoppelter Frequenzumsetzer electron-coupled frequency changer
  elektronengekoppelter Oszillator electron-coupled oscillator, ECO
Elektronenhülle electron shell
Elektronenkanone electron gun
Elektronenkonzentration electron concentration
Elektronenlaufzeit electron transit time, time of flight (621.385)
Elektronenlawine electron avalanche
Elektronenleitung electron conduction, N-type conduction
Elektronenlinearbeschleuniger linear electron accelerator
Elektronenlinse electronic lense
Elektronenmangelleitung hole conduction
Elektronenmangelstelle hole
Elektronenmikroskop electronic microscope
Elektronenpaket bunch of electrons
Elektronenplasma electron gas, electron plasma
Elektronenquelle electron emitter, electron gun (621.385)
Elektronenrechner electronic computer
Elektronenröhre electronic valve, thermionic valve, electron tube
Elektronenschalter electronic switch
Elektronenschleuder betatron
Elektronenschale electron shell
Elektronenspiegel electron mirror
Elektronenspin spin of the electrons, electron spin
Elektronenstrahl electron ray, electron beam
  Elektronenstrahloszilloskop cathode-ray oscilloscope
  Elektronenstrahlröhre electron ray tube, oscilloscope tube
Elektronenstreuung electron scattering
Elektronenstrom electron flow, electron stream, electron current
Elektronenübergang electron transition
Elektronenüberschußzone electron-excess region
Elektronenvervielfacher electron multiplier
Elektronenvolt electron volt
Elektronenwolke electron sheath
elektro-negativ electro-negative
Elektronik electronics (pl.)
elektronisch electronic
  elektronische Abstimmung electronic tuning
  elektronische Abtastung electronic scanning
  elektronisches Bild electron image
  elektronische Bindung electrovalent link
  elektronische Datenverarbeitung electronic data processing = EDP
  elektronischer Herzschrittmacher electronic pace maker
  elektronische Kampfführung electronic warfare
  elektronische Rechenmaschine electronic computer
  elektronischer Stift electronic pen
  elektronisches Testbild electronic test pattern
  elektronischer Umrechner electronic director
Elektroofen electric furnace
  Elektroofen mit direkter Beheizung des Heizguts charge-resistance furnace
elektro-optischer Effekt electro-optical effect
Elektroosmose electro-osmosis

Elektrophor electrophorus
Elektrophorese electrophoresis
elektro-pneumatisch electropneumatic
elektropositiv electropositive
Elektrorasierer electric shaver
Elektroschleifer electric grinder
Elektroskop electroscope
Elektro-Schlacke-Umschmelzofen electro-slag remelting furnace
Elektroschweißer electric welder
Elektroschweißung electric welding
Elektrostatik (theory of) electrostatics (pl.)
elektrostatisch electrostatic
elektrostatische Ablenkung electrostatic deflection
elektrostatische Adhäsion electrostatic adhesion
elektrostatische Anziehung electrostatic attraction
elektrostatischer Ausscheider electrostatic precipitator
elektrostatische Beanspruchung electrostatic stress
elektrostatische Bündelung electrostatic focusing
elektrostatischer Drucker electrostatic printer
elektrostatische Einheit electrostatic unit
elektrostatische cgs-Einheit der Kapazität abstatfarad
elektrostatische cgs-Einheit der Ladung abstatcoulomb
elektrostatische cgs-Einheit der Spannung abstatvolt
elektrostatische cgs-Einheit des Stromes abstatampere
elektrostatisches Feld electrostatic field
elektrostatische Feldstärke electrostatic fieldstrength, electrostatic stress
elektrostatisches Filter electrostatic filter
elektrostatisches Instrument electrostatic instrument
elektrostatische Ladung electrostatic charge
elektrostatischer Spannungsmesser electrostatic voltmeter
elektrostatische Speicherung electrostatic storage
elektrostatisches Streufeld stray electrostatic field
elektrostatisches Wattmeter electrostatic wattmeter
Elektrotechnik electrical engineering, electrotechnology
elektrotechnisch electrotechnic(al), electrical
Elektrotherapie physiotherapy
elektrothermisch electrothermic, electrothermal
elektrothermische Aufzeichnung electrothermal recording
elektrothermischer Auslöser electrothermal release
elektrothermisches Relais electrothermal relay
elektrothermischer Wirkungsgrad electrothermal efficiency
Elektrotypie galvanograph (655)
Elektro-Unternehmer electrical contractor
Elektroventil solenoid valve
Elektrowärme electric heat
Elektro-Weidezaun electrical fence
Elektrowerkzeug electric tools (pl.)
Elektrozug electric pulley block, (overhead) electric hoist (621.387)
Element element (54); single body, member; cell (621.352)
Element des Abstimmkreises tuned circuit element
Element bricht zusammen cell runs down (621.352)
auf ein Element bezüglich elemental
Elementar-..., elementar elementary, fundamental
elementare Verknüpfungsschaltung elementary logical circuit
Elementarbereich domain (538)
Elementardipol elementary dipole (621.396.67)
Elementarladung electronic charge, elementary charge
Elementarteilchen elementary particle, fundamental particle, ultimate particle (439)
Elementarwelle elementary wave
Elevation (angle of) elevation
Elevator elevator, hoist
Ellipse ellipse
Ellipsoid ellipsoid, spheroid
ellipsoidähnlich spheroidal
elliptisch elliptic(al)
elliptisch polarisiert elliptically polarized
elliptische Funktion elliptic function
Ellira-Verfahren Unionmelt process (621.791)
Elmsfeuer Saint Elmo's fire, corposant
„Elok" electric locomotive
elongieren to elongate
eloxieren to anodise; to anodize (US)
Eloxierung anodization, electrolytic oxidation (of aluminium)
Em = Emanation Nt = Niton, Rn = Radon
Email, Emaille (vitreous) enamel; porcelain (US)
Emaildraht enamel-insulated wire
Emailfritte enamel frit

emaillieren to enamel
Emaillierung (vitreous) enamelling
Emanation emanation
emanieren to emanate
EMD-Schalter = Edelmetall-Motor-Drehwähler EMD-switch, noble-metal uniselector switch
Emission emission
Emissionsfläche emitting surface
Emissionskennlinie emission characteristic; cathode characteristic (621.385)
Emissionsrauschen emission noise
Emissionsspektrum emission spectrum
Emissionsstrom cathode stream, cathode current
Emissionssubstanz active material
Emissionsvermögen emissive power, emissivity
Emitter emitter
Emitteranschluß emitter terminal
Emitterbahnwiderstand emitter series resistance
Emitter-Basis-Spannung emitter base voltage
Emitter-Basis-Zonenübergang emitter junction
Emitter-Elektrode emitter electrode
Emitterfolger emitter follower
emittergekoppelt emitter coupled
Emitter-Reststrom emitter-base cut-off current
Emitterschaltung rückwärts inverse common emitter
Emitterschaltung vorwärts common emitter
Emitter-Sperrschicht emitter depletion layer, emitter junction
Emittersperrschichtkapazität emitter depletion layer capacitance
Emitterstrom emitter current
Emitterübergang emitter junction
Emittervorspannung emitter bias
Emitterwirkungsgrad emitter-injection efficiency
Emitterzone emitter region
emittieren to emit
emittierendes Material emissive material
EMK = elektromotorische Kraft e.m.f. = electromotive force
EMK der Drehung rotational electromotive force (621.313)
Empfang reception (621.394/396)
Empfang mit Trägeranhebung exalted carrier reception
Empfang mit Trägerzufügung local carrier reception
Empfang zudecken to blot out reception
Empfangsanlage receiving installation (621.396)
Empfangsantenne receiving aerial
Empfangsapparat receiving apparatus, receiver (621.394)
Empfangsband(breite) received band
Empfangsbereich coverage
Empfangsfeldstärke signal strength
Empfangskreis receiving circuit
Empfangslocher reperforator, receiving perforator
Empfangslücke dead spot (621.396)
Empfangsnullstelle zero response (621.396.677)
Empfangsseite receiving end, output end, load end
Empfangsspannung signal voltage
Empfangsstation receiving station
Empfangsstelle head end (621.396.67)
Empfangsstellung receive position (621.396)
Empfangszeichen signal received
empfangen to receive
Empfänger receiver
Empfängerchassis receiver chassis
Empfänger-Erdanschluß receiver earth
Empfänger-Erdklemme receiver-earth terminal
Empfängergehäuse receiver cabinet
Empfängerlautstärke receiver-output volume
Empfänger-Oszillator receiver local oscillator
Empfängerröhre receiver valve, receiving valve
Empfängerschaltung receiver circuitry
Empfängerschwelle minimum receiver input
Empfängersperröhre T. R. tube
Empfängerspule receiving coil
empfindlich für sensitive to , responsive to
empfindliche Waage delicate balance
Empfindlichkeit response, figure of merit (621.317); sensitivity (621.396.62); susceptibility
Empfindlichkeit einer Bildaufnahmeröhre radiant sensitivity of a camera tube
Empfindlichkeit bei Glühlampenlicht tungsten sensitivity (621.385)
Empfindlichkeit gegen Versorgungsspannungsänderungen supply-voltage sensitivity
Empfindlichkeitsgrenze limit of sensitivity (621.317)
emulgieren to emulsify
Emulsion emulsion
Ende end; close; sink (621.392)
Ende einer Kolonne rear of a column (355)
Ende Rohr length of piping
am Ende kurzgeschlossene Leitung closed-end line
am Ende offen open-ended

am Ende offene Leitung open-end line (621.3)
am Ende von at the foot of
mit dem Ende vorneweg end on
zu Ende führen to carry through
Endanflug final approach
Endanschlag (end)stop, (stop)peg; feed-trip stop
Endausschalter terminal switch, final limit switch (621.9)
Endbohrer plug tap (621.9)
Endbund-Klemme terminal lug
Enddruck terminal pressure (624)
Endeffekt net effect
Endeinrichtung terminal equipment
Endeinspannung end restraint
Endfläche end plane
Endgerät terminal (device)
Endkappe end bell
Endkern resultant nucleus (539)
Endkontakt end contact, terminal contact
Endladestrom finishing rate
Endlagenschalter limit switch
Endlosformular continous form, fan-fold paper
Endleistung ultimate capacity
Endmaß parallel end gauge, slip gauge, block gauge
Endmaße ansprengen to wring gauge blocks together
Endmast dead-end tower, terminal pole (621.315.1)
Endplatte end plate (621.313)
Endpunkt final point
Endröhre high-power tube, output valve
Endschalter limit switch, terminal limit switch (621.876)
Endschaltung terminal equipment (621.39)
Endspannung final voltage, terminal voltage
Endstation terminal station, final position
Endstelle terminal, terminal equipment; station (621.395)
Endstelleneinrichtung subscriber's apparatus
Endstellung end position, final position
Endstrom final current
Endstück terminal
Endstufe final stage, output stage, power stage
Endstufe in B-Betrieb class B output stage
Endstufenmodulation high-power modulation, high-level modulation
Endteil terminal part
Endtemperatur final temperature, ultimate temperature
Endtetrode mit Elektronenbündelung beam-power tetrode
Endumschalter travel-reversing switch
Endverbleiung test cap (621.315.2)
Endverschluß terminal box, sealing chamber, sealing end, pothead (621.315.2)
Endvermittlung terminal exchange
Endverstärker terminal repeater (621.395); power stage (621.396)
Endverstärkerröhre power tube, output tube
Endziffern terminal figures
Endziffernwähler final selector
enden bei to end up with
endigen mit to terminate in
endotherm endothermic
energetischer Wirkungsgrad energy efficiency (621.355)
Energie energy
energielos de-energized
energiereich high-energy
Energie abgeben an to give energy to
Energie aufnehmen to gain energy, to consume energy
Energie austauschen to interchange energy
Energie einsparen to save energy
Energie entziehen to absorb energy, to abstract energy, to remove energy, to rob energy (**von, aus:** from)
Energie hineinstecken in to put energy into
Energie liefern to supply energy, to contribute energy
Energie liefern an to energize
Energie rückliefern to give back energy
Energie speichern to store energy
Energie übertragen auf to give energy to
Energie verbrauchen to expend energy, to dissipate energy, to consume energy
Energieabgabe delivery of energy
energieabhängiger Speicher volatile storage
Energieaufnahme uptake of energy, energy intake, energy consumption, energy input
Energieaufwand expenditure of energy
Energieausbeute energy yield
Energieaustausch interchange of energy
Energieband energy band, band-edge energy (621.382)
Energiefluß energy flow
Energieflußdichte energy-flux density (539)
Energieinhalt energy
Energieleitung power-transmission line (621.396), feedline (621.396.67); waveguide (621.372)
Energielieferung delivery of energy, energy supply

Energieniveau energy level
Energiepaket energy packet
Energiequelle source of energy, source of power, source of supply
Energiesatz law of conservation of energy, work law
energiereich energy-rich, rich in energy
Energiespeicher storehouse of energy, warehouse of energy
Energiespektrum energy spectrum
Energiestrom energy flow, stream of energy
Energietransport energy transfer
Energieterm term, energy term
Energieübertragung transmission of energy (621.315/396)
Energieumwandlung conversion of energy
energieunabhängiger Speicher nonvolatile memory
Energieverbrauch consumption of energy, expenditure of energy
Energieverlust energy dissipation, loss of energy
Energieversorgung power supply
Energieversorgungsunternehmen public utility
Energieverteilung distribution of power
Energiewandler energy converter, (active) transducer
Energiewirtschaft power economy
Energiezufuhr supply of energy, power supply
eng narrow, close; intimate
eng anliegend close-fitting
eng verbunden intimately connected
enge Beziehung intimate relation
enge Bohrung small bore
enges Bündel narrow beam (621.396.677)
enge Grenzen narrow bounds
enge Kopplung close coupling
enge Toleranz close tolerance
engere Toleranzen closer limits
„Engländer“ monkey, adjustable wrench, adjustable spanner
entblocken to unblock
entblockter Streckenabschnitt open block (625)
entbrummen to dehum
Entbrummerspule hum-bucking coil
Entdämpfung deattenuation
Entdämpfungsgrenzfrequenz resistive cut-off frequency
enteisen to de-ice
Enteisungseinrichtung de-icer
Entfernung (linear) distance
Entfernung messen to range
Entfernung überbrücken to bridge a distance
Entfernungsauflösung range resolution, range discrimination
Entfernungsbereich search capability (621.397)
Entfernungsfeinmeßstufe precision range unit
Entfernungsfeuer distance-marking light
Entfernungs-Höhensichtgerät height-position indicator, H. P. I.
Entfernungsmeßmarkengeber range marker
Entfernungsmesser range finder, distance measuring equipment
Entfernungstorschaltung range gate (621.396.96)
entfetten to degrease
Entfettungsmittel detergent
entfeuchten to dehydrate
Entfeuchtung dehydration
entflammbar ignitable, inflammable
Entflammbarkeit , ignitability, inflammability
entflammen, sich to burst into flames
Entflammungstemperatur flash point
entfokussieren defocusing
entgasen to de-gas
Entgasung de-gasification
entgegengesetzt opposite
entgegengesetzt gerichtet opposite in direction, unlike
entgegengesetzt gerichtete Ströme unlike currents, opposing currents
entgegengesetztes Feld opposing field
entgegengesetzte Richtung opposite direction
entgegengesetztes Vorzeichen haben to be opposite in sign
entgegenwirken to counterbalance, to counteract, to oppose
entgraten to debur
enthalten to contain, to hold, to house, to include; to involve
enthalten in implicit
enthärten to soften
entionisieren to de-ionize
Entionisierung de-ionization, de-ionizing
Entionisierungsgeschwindigkeit de-ionization rate
Entionisierungsspannung de-ionization potential (621.385)
Entionisierungszeit de-ionization time, recovery time
entkoppeln to decouple (621.396)
entkoppelte Zweige conjugate branches
Entkopplung decoupling
Entkopplungskondensator balancing capacitor
Entkopplungsschaltung decoupling network
Entkopplungswiderstand plate decoupling resistor (621.396.64), isolating resistor

entkuppeln to uncouple, to declutch, to disconnect (625)
entladen to unload
entladen, sich to discharge (621.3)
Entladekreis discharge circuit
Entladeschalter discharge key
Entladespannung discharge voltage
Entladestrom discharging current, discharge current
Entladeverzug carrier storage time (621.382)
Entladewiderstand bleeder (resistor)
Entladung unloading; discharge (621.3)
Entladung abdrosseln to quench a discharge
Entladung in Gang setzen to set off a discharge
Entladung löschen to quench a discharge
Entladung setzt ein discharge sets in
Entladung zünden to strike a discharge, to set off a discharge
Entladungsbahn discharge path
Entladungsdauer service life (621.355); time of discharge
Entladungsfunke discharge spark
Entladungskanal discharge channel
Entladungslampe discharge lamp, flash tube
Entladungslampe zünden to light a discharge lamp
Entladungsraum interelectrode space
Entladungsstoß discharge surge
Entladungsstörung precipitation static, interference, precipitation noise
Entladungsweg discharge path
entlang along
entlang an ... alongside ...
entlanglaufen an ... to travel down ...
entlangschwingen an ... to swing along ...
entlasten to discharge, to unload; to relieve, to remove the stress (624)
entlüften to de-air (621.1), to vent (621.7)
Entlüfter de-aerator (621.1); ventilator
Entlüftung de-aeration (621.1); ventilation (69)
Entlüftungskammer de-aeration reservoir (621.1)
Entlüftungsleitung vent line
Entlüftungspfropfen vent plug (621.355)
Entlüftungsschraube breather plug (621.355)
entmagnetisieren to demagnetize, to de-gauss (623)
Entmagnetisierung demagnetization; de-gaussing (612.12)
Entmagnetisierungseinrichtung de-gaussing equipment
Entmagnetisierungskabel de-gaussing belt, de-gaussing cable, demagnetizing cable
Entmagnetisierungskurve demagnetization curve
Entmagnetisierungsschleife de-gaussing belt
entmischen, sich to segregate (66)
entnehmen to draw off, to tap off; to drain
entnebeln to defog, to demist
entprellen to debounce
entriegeln to unlock
Entropie entropy
entrosten to descale, to remove rust
entsäuern to deacidify
Entscheidungsschwellwert decision level
Entscheidungssymbol decision box (681.3)
entschlüsseln to decode, to decipher
Entschlüßler decoder
entschwefeln to desulphurize
entsilbern to desilver
entspannen to remove the stress, to relieve
entsperren to unblock
entstauben to dedust
Entstaubungsanlage dust-collection equipment; precipitation equipment
entstehen to rise
entstehender Strom current response
entstören to radio-shield, to screen; to suppress noise, to eliminate interference
Entstörung interference suppression
Entstörungsdrossel anti-interference choke
Entstörkondensator anti-interference capacitor
Entstörungswiderstand suppressor resistance (621.396), (629.113)
Enttrübung minimum clearing, compensation, anticlutter rain (621.396.96)
entwässern to dewater, to drain
Entwässerung dewatering, drainage, draining; dehydration
Entweichen von Dampf escape of steam
entwerfen to plan, to scheme, to line out, to design, to lay out
entwickeln to evolve; to liberate (54); to develop (77) (62)
entwickeln, sich to start up, to develop
entwickeln und fixieren to process (77)
Entwickler developer (77)
Entwicklung development, evolution; research and development (department)
Entwicklungssystem development system
Entwurf lay-out, design, draft, outline, scheme; artwork
Entwurf von Platinen circuit board artwork
Entwurfsexemplar draft copy
Entwurfpapier design drawing paper
Entwurfsrichtlinie design criterion

entzerren to correct (621.39); to equalize (621.372.4)
Entzerrer equalizer; compensator (621.375)
Entzerrerschaltung corrector circuit, correcting network, equalizing network
entzundern to descale
entzünden, (sich) to inflame, to ignite
entzündlich inflammable, ignitable
Entzündung ignition
Entzündungstemperatur temperature of ignition
Epidiaskop epidiascope
Epikadmiumneutronen epicadmium neutrons
Episkop reflectoscope, aphengescope (535)
Epitaxial-Transistor epitaxial transistor
Epitaxie epitaxy
epithermische Neutronen epithermal neutrons
epithermischer Reaktor epithermal reactor
Epizykel epicycle
Epicykloide epicycloide
Epoxyd-Gießharz epoxy resin
Epoxyd-Glashartgewebe epoxy-glass
EPROM EPROM
Eprommer EPROM Programmer
Epsteinapparat Epstein hysteresis tester, magnetic square
Erd-... earth, ground (621.3)
Erdalkali alkaline earth
Erdanschluß earth(ing) connection; ground lug (US)
Erdantenne buried aerial, ground aerial
Erdanziehung earth's pull
Erdausbreitungswiderstand earthing resistance, earth's resistance
Erdball terrestrial globe
Erdbeanspruchung earth pressure (624)
Erdbeschleunigung acceleration due to gravity, gravitational constant
Erddruck earth pressure
Erdefunkstelle earth station
Erdelektrode earth electrode
erdfrei free of ground, floating
erdfreier Eingang floating input
Erdgas natural gas, rock gas
Erdinduktor earth inductor (538)
Erdkabel buried cable, underground cable
Erdkapazität capacity to ground
Erdklemme earthing terminal; ground terminal (621.315) (US)
Erdkrümmung earth curvature
Erdleiter earthing conductor
Erdleitung earth wire, earth lead
Erdmagnetismus terrestrial magnetism
Erdmikrophon seismicrophone
Erdoberfläche earth's surface, ground (surface)
Erdöl petroleum, naphtha
Erdpech earth pitch
Erdpotential earth potential
auf Erdpotential earthy, at zero potential (621.3)
Erdradius earth's radius
Erdrinde earth's crust
Erdrückleitung ground return, earth return (circuit)
Erdschicht layer of earth
Erdschluß earth-(leakage) fault; ground fault, line-to-ground fault (US), accidental ground, (satter:) dead earth, (schleichender:) earth leakage
Erdschlußanzeige ground indication, earth indication
Erdschlußanzeiger ground-leakage indicator, earth-leakage indicator
Erdschlußautomat ground-leakage trip, earth-leakage trip
Erdschlußlöscheinrichtung arc suppressor (621.316.9)
Erdschlußmesser earth leakage current meter
Erdschlußprüfer earth-leakage indicator
Erdschlußrelais earth-leakage relay, residual relay, zero-phase sequence relay
Erdschlußschutz ground protection
Erdschlußspule Petersen coil
Erdschlußwächter earth-leakage monitor; ground-fault monitor (US)
Erdschlußstrom earth current, earth-leakage current, loss current to earth; ground-fault current, fault-to-ground current (US)
Erdschlußwischer passing earth, transien tearth, transitory line-to-ground fault
Erdschlußzeiger earth detector
Erdschutzleiter earth-continuity conductor
Erdseil earth wire, earth conductor, shield wire
Erdstrom earth current, stray current, vagabond current
Erdstromausgleicher earth-current equalizer
Erdstromrelais residual relay
erdsymmetrisch balanced to ground
Erdtrennschalter earthing switch; grounding switch (US)
Erdumlaufecho round-the-world echo (621.396)
Erdumlaufzeichen around-the-world signal (621.396)

erdunsymmetrische Ablenkung unbalanced deflection (621.385)
Erdverbindung earth(ing) connection
Erdverlegung underground laying (621.315.2)
Erdwärme terrestrial heat
Erdwärmekraftwerk geothermic power station
Erde earth, ground (US) (621.3); soil (624)
an Erde legen, gegen Erde kurzschließen to (connect to) ground, to short to earth, to (connect to) earth
gegen Erde isoliert insulated from earth
in die Erde verlegen to lay underground
über Erde by way of the ground (621.3); above ground level
Erde als Rückleitung earth return
erden to (short to) earth; to (connect to) ground (US)
geerdete Abschirmung earth-protective sheath
geerdeter Draht earthed wire
geerdeter Leiter earthed conductor
geerdeter Mittelleiter earthed neutral conductor
geerdetes Netz earthed neutral system
geerdeter Sternpunkt earthed neutral
geerdeter Stromkreis earthed circuit
Erder earth(ing) electrode
Erdung earthing, grounding, earth connection, ground connection
Erdungsanlage earth system, ground system (621.396.67)
Erdungsanschluß earth connection point, grounding lug
Erdungsband earthing strip
Erdungsblech earth plate
Erdungsdraht earth wire
Erdungsdrossel earthing (auto-)transformer, neutral (auto-)transformer (621.316.9)
Erdungsklemme earthing clip, earthing terminal, ground terminal
Erdungskreis leak circuit
Erdungslasche ground strap
Erdungsmantel earth sheath, test shield (621.315.2)
Erdungsmesser earth-resistance meter
Erdungsplatte earth plate
Erdungspunkt earthing point, earth connection point
Erdungsring bonding ring
Erdungsrohr earth rod
Erdungssammelleitung ground bus
Erdungsschalter earthing switch, grounding switch
Erdungsschelle bonding ring, bonding clip
Erdungsschiene fault bus
Erdungsschraube grounding screw
Erdungsstange ground rod
Erdungsstöpsel earthing plug
Erdungstransformator s. **Erdungsdrossel**
Erdungstrennschalter earthing isolator
Erdungswandler s. **Erdungsdrossel**
Erdungswiderstand earth(ing) resistance, ground resistance; resistance to earth, resistance of earthed conductor; discharging resistor (621.315)
Ereignis-Zähler event counter (681.3)
erfassen to detect (623); to register, to record; to cover; to monitor (a fault)
Erfassungsradar acquisition radar
Erfassungs- und Verfolgungsradar acquisition and tracking radar
Erfassungsreichweite detection range, coverage
erfinden to invent
Erfinder inventor
Erfindung invention
erforschen to explore; to research
Erforschung exploration; research
Erg erg
ergänzen to complete; to supplement
ergänzende Unterlagen supplementary data
Ergänzung completion; supplement
Ergänzungsnetzwerk building-out network
Ergänzungsspeicher backing store (681.3)
Ergonomie human engineering
Erhaltung preservation, conservation; maintenance, upkeep
Erhaltung der Energie conservation of energy
Erhaltungsladung trickle charging, compensation charge (621.355)
Erhaltungszustand (state of) preservation (621)
Erhebung (angle of) elevation
Erhebungswinkel elevation angle
erhellen to illuminate
erhellen aus be(come) evident from
erhitzen to heat, to warm
Erhitzung heating, warming
bei Erhitzung on heating, on warming
erhöhen to raise, to elevate; to increase (**um:** for, by); to increment (681.3)
erhöhte Sicherheit improved safety
erhöhte Temperatur raised temperature, elevated temperature
Erholzeit recovery time (621.382)
Erkennung recognition (623)
Erkennungsgerät identification equipment (623)
Erkennungssignal identification sign(al)
erläutern to illustrate, to interpret, to explain

erläuternde Aufschrift descriptive legend
Erlenmeyer-Kolben Erlenmeyer flask
erleuchten to lighten, to illuminate
ermitteln to find (out), to detect, to locate
Ermittlung der Schadenstelle location of the trouble
Ermüdung fatigue (620.1), (621.385)
Ermüdungsbruch fracture by fatigue, fatigue failure, fatigue fracture
Ermüdungsfestigkeit fatigue strength
Ermüdungsgrenze fatigue limit, endurance limit
Ermüdungszuschlag fatigue allowance
erneut abgleichen to readjust
erneut prüfen to re-check, to re-examine
erodieren to erode
Erosion erosion
Erdproben to try, to test (**auf, hinsichtlich:** for)
Erprobung trial, proving, test, trial run
Erprobungsstelle proving ground
erratisch erratic
erregbar excitable
erregen to excite; to energize (621.318)
Erreger exciter, exciting cause; aerial feed (621.396.67)
Erregeranode exciter anode, exciting anode, excitation anode (621.314.5)
Erregerdipol excitative douplet, driven dipol, active dipol
Erregerfeld field system, excitation field, exciting field
Erregerfluß excitation flux
Erregerkreis exciter circuit, exciting circuit
Erregermaschine exciter, exciting generator
Erregerreflektor vertex plate (621.396.67)
Erregerspannung exciting voltage
Erregerstrom exciting current, inductor current, field current
Erregerstromregler field-current regulator
Erregerwicklung exciting winding, field winding
Erregerwiderstand exciter resistance
Erregung excitation
Erregungsstadium stage of excitation
Erregungsverluste excitation losses, field losses
Erreichbarkeit availability, accessibility
errichten to erect, to build, to construct
Ersatz substitute, equivalent; reserve, spare, standby
Ersatz leisten für to compensate for
Ersatz von x durch y substitution of y for x
Ersatzader spare conductor
Ersatzdämpfung equivalent articulation loss, articulation reference
Ersatzkettenleiter equivalent periodic line, equivalent ladder network
Ersatzmaschinen spare machinery (sing.)
Ersatzmittel substitute
Ersatzröhre spare tube
Ersatzschaltbild equivalent circuit
Ersatzschaltung equivalent circuit, equivalent network, lumped circuit
Ersatzsicherung spare fuse
Ersatzsperrschichttemperatur virtual junction temperature, equivalent junction temperature (621.382)
Ersatzspule replacement bobbin, rewind
Ersatzstoff substitute, surrogate
Ersatzstromkreis equivalent circuit
Ersatzteil replacement (part), spare (part)
Ersatzteilausstattung service kit
Ersatzteilliste spare-parts list
Ersatzteilnummer spare-part number
Ersatzwiderstand substitutional resistance, equivalent resistance; slope resistance (621.382.2)
Ersatzwiderstand für die mechanische Rückwirkung motional impedance
Ersatzwiderstand für mechanische Verluste motional resistance
erschöpfen to exhaust, to deplete
erschöpfte Batterie exhausted battery, run-down battery
erschüttern to shock, to jar
Erschütterung percussion, concussion, vibration, jar
erschütterungsfreie Befestigung anti-vibration mounting, vibration-proof mounting
erschütterungssichere Fassung anti-vibration lamp-holder
ersetzbar replaceable, renewable
ersetzen durch to replace by, to exchange by, to substitute by
erst first; prime, superior
erster Abstich first run
erster Hauptsatz der Thermodynamik first law of thermodynamics
erste Hilfe first aid
erste Wahl first choice
erster Wähler first selector (621.395)
Erstarren, Erstarrung freezing, congealment, solidification
Erstarren der Schweißraupe freezing of the fillet
Erstarrungspunkt freezing point, point of congealment, solidification point
erstarren to freeze, to congeal, to solidify
ersticken to quench

ertönen (lassen) to sound
erwärmen to heat
sich erwärmen to warm (up)
Erwärmung heating; temperature rise (621.313); overtemperature
Erwärmungsversuch heating test, heat run
Erwärmungszyklus heat(ing) cycle
sich bei Erwärmung ausdehnen to expand with heat
erweichen to soften
Erweichungspunkt softening point; fusion point, sintering point
erweitern to widen, to enlarge, to extend; to multiply throughout (51)
sich erweitern to grow larger, to get wider
Erweiterung enlargement, widening, extension
Erythem erythem
erythembildend erythemal
Erythemdosis unit of erythemal flux, finsen
Erythemgefährdung erythemal exposure
Erythemwert der Strahlung erythemal flux
Erz ore
erzeugen to generate, to induce (621.3); to produce, to manufacture, to give rise to
Erzeuger producer, manufacturer
Erzeugerpreis producer's price
Erzeugung generating, generation; manufacture, manufacturing
Erzeugungsanlage power-generating plant
Erzeugungswälzkurve generating rolling curve (621.8)
erzwingen to force; to impose
erzwungener Luftstrom forced draft
erzwungene Randbedingungen imposed boundary conditions
erzwungene Schwingung forced vibration, forced oscillation
erzwungener Strom forced current
E-Schicht E-layer, Heaviside layer (621.396)
E.S.E. = elektrostatische (cgs-) Einheit E.S.U. = Electrostatic Unit, abstat-...
Esse funnel, furnace offtake, flue, chimney (stack)
Etikett label, tag, ticket
Etikett mit Gebrauchsanweisung labelled instructions (pl.)
Eutektikum eutectic (mixture)
eutektisch eutectic
eutektisches Gemisch eutectic mixture
eutektische Legierung eutectic alloy
evakuieren to evacuate, to exhaust
Evakuierung evacuation (621.315.2)
eV (= Elektronenvolt) electron volt
Evolvente evolute, involute
Evolventenverzahnung involute gearing
EVU = Elektrizitäts-Versorgungs-Unternehmen power-supply company
E-Welle E-wave, transverse magnetic wave
exakt exact
exakte Wissenschaften exact sciences
Excelsiorband varnished cambric
Excitron excitron
Ex-geschützt explosion proof
Exhaustor exhauster
exklusives Oder exclusive OR, nonequivalence (621.374)
exotherm exothermal, exothermic(al)
Expander expander
expandieren to expand
Expansion expansion
Expansionskraft expansive force
Expansionsstufe expansion stage (621.395.3)
exotherm exothermic
exotherme Kernreaktion exothermic nuclear reaction
expandieren to expand
Expansion expansion
Expansionsschalter expansion(-type) circuit breaker
Experiment experiment
Experimentalphysik experimental physics
Experimentator experimenter
experimentell experimental
experimentell nachweisen to verify
experimenteller Befund experimental findings (pl.)
experimentieren to experiment
explizit explicit
explodieren to explode, to detonate, to burst, to fulminate
explosible explosive
Explosion explosion, detonation
explosionsartig explosive
explosionsfähiges Gas explosive gas
Explosionsansicht exploded view
Explosionsfähigkeit explosiveness
Explosionsgefahr explosion hazard
explosionsgefährdete Räume hazardous locations
explosionsgeschützt explosion proof
explosionsgeschützte Betriebsmittel apparatus for use in explosive gas atmospheres
explosionsgeschützte Motoren motors for hazardous locations, motors for operation in explosive gas atmospheres
Explosionskraft explosive force

Explosionsschutz explosion protection, flameproof enclosure; explosion vent (621.314.2)
Explosiv-..., Explosivstoff, explosiv explosive
Exponent exponent, index (51)
Exponential-... exponential
Exponentialbatterie exponential battery (539)
Exponentialcharakter exponential character
Exponentialfunktion exponential (function)
Exponentialgesetz exponential law
Exponentialhornstrahler exponential horn
Exponentialkennlinie variable mu characteristic (621.385)
Exponentialröhre variable mu valve, supercontrol tube, remote cut-off tube (621.385)
Exponentialschreibweise exponential notation
Exponentialrichter exponential horn (534), (621.396.67)
Exponentialwirkung variable mu action (621.385)
exponentiell exponential
exponentiell gedämpfte sinusförmige Größe exponentially damped sinusoidal quantity
exponentiell verlaufende Größe exponential quantity
exponentieller Verlauf exponential character
exponentielle Zeitablenkung exponential sweep (621.397)
Exsikkator exsiccator, desiccator
Extinktion extinction
extrahieren to extract
Extraktion (solvent) extraction, stripping (54)
Extrapolation, Extrapolierung extrapolation
extrapolieren nach to extrapolate towards
extrapolierte Reaktorbegrenzung extrapolated boundary
extrem extreme
extreme Verkleinerung miniaturization
Extremum, Extremwert extreme (pl.: -ma)
Exzenter eccentric
Exzenterstange eccentric rod
Exzenterstift eccentric stud
Exzenterwelle eccentric shaft
exzentrisch eccentric, off-centre, out of centre
exzentrische Rundsichtdarstellung off-centre plan display (321.396.96)
Exzentrizität eccentricity
exzerpt excerpt

F

F = Farad f. = farad
Fabrik factory, mill, plant; works (pl.)
Fabrik betreiben to run a factory
Fabrikant manufacturer, maker, producer
Fabrikat make, brand
Fabrikation manufacture, production
Fabrikationsart method of production
Fabrikationsprogramm manufacturing programme, product range
Fabrikationsverfahren manufacturing process
fabrizieren to manufacture, to produce, to make
Fach branch, profession, line: pigeon hole (656.8); bin (651)
Facharbeiter specialized worker, skilled worker, trained worker
Fachausdruck (technical) term
Fachgebiet branch, profession, line
Fachingenieur expert engineer
Fachschule vocational school
Fachwissen knowhow
n-fach-Umschalter n-way switch
Fächer fan
Fächerantenne fan aerial, fanned-beam antenna
Fächerfunkfeuer fan marker beacon
Fächerkeule fan beam
Fackel torch(-light); solar flare (52)
fädeln to thread
gefädelter Anker tunnel-wound armature (621.313)
gefädelte Wicklung threaded-in winding
Fädelwicklung tunnel winding, pin winding
Faden hair line (535.8); filament (621.32); string, thread (67)
Faden durch ein Loch fädeln to run a thread through a hole
Fadenaufhängung unifilar (elastic) suspension
Fadenformierung flashing of filament (621.396)
Fadengalvanometer thread galvanometer
Fadenkreuz cross hairs, spider lines, hair lines (pl.)
Fadenkreuzblende cross-hair diaphragm
Fadenwiderstand filament resistance (621.32)
fadenziehend ropy
Fading fading (621.396)
Fadingregulierung automatic volume control (= a. v. c.) automatic gain control
Fahne lug (621.3), riser (621.315); vane (621.317)
Fahneneffekt streaking (621.397)
fahrbar mobile, portable, travelling

fahrbares Gerät mobile set, mobile equipment
fahrbarer Sender mobile transmitter
fahrbarer Untersatz dolly
fahrbares Unterwerk travelling substation
Fahrdraht trolley wire, overhead line, contact wire (621.33)
Fahrdrahtabspannung pull-off
Fahrdrahtabstützung push-off
Fahrdrahtanlage overhead-wire system
Fahrdrahtaufhängedraht catenary hanger
Fahrdrahtlokomotive trolley-electric lokomotive
Fahrdrahtspannung trolley voltage
Fahrdrahtspeiseleitung trolley-wire feeder
Fahrdrahtverbinder splicing ear
Fahrdrahtweiche trolley frog
Fahrgestell bogie (625), chassis (629.113), undercarriage, landing gear (629.13)
Fahrkette track
Fahrkorb car (621.3)
Fahrkorbschaltung car-switch (lift) control
Fahrlässigkeit negligence, carelessness
Fahrleitung overhead line, overhead conductor, trolley line (621.33)
Fahrleitungs-Abspannisolator buckle insulator
Fahrleitungsausleger bracket arm
Fahrleitungsmast tram standard
Fahrleitungsschalter line breaker
Fahrleitungsstütze steady brace
Fahrleitungssystem overhead-contact system
Fahrmotor traction motor, railway motor
Fahrplangeschwindigkeit scheduled train speed
Fahrprofil railway profile, railroad profile
Fahrraddynamo magneto alternator, bicycle magneto
Fahrrinnenkabel leader cable, guide cable (527)
Fahrschalter traction switch, master controller, power switchgroup
Fahrschaltung vom Anhänger aus trailer control (621.33)
Fahrstuhl lift (Br.), elevator (US)
Fahrstuhl mit automatischer Druckknopfsteuerung electric-automatic push-button lift
Fahrstuhl mit Selbststeuerung electric-automatic self-service lift
Fahrstuhlantriebsmaschine lift machine
Fahrstuhl-Nullspannungsschalter lift potential switch
Fahrstuhlschacht lift shaft, hoistway
Fahrstuhlschachttür lift-shaft gate
Fahrstuhl-Schlaffseilschalter lift slack-cable switch
Fahrstuhlsteuerung lift control
Fahrstuhl-Stockwerksdruckknopf landing-call push
Fahrstuhltür lift door
Fahrstuhltürverriegelung lift-door interlock, hoistway-door interlock
Fahrt speed
Fahrtbeschränkung speed restriction
Fahrtmesser speed indicator (629.13)
Fahrtregler winding controller (622)
Fahrtrichtungsschalter reverser (621.33)
Fahrzeug vehicle
Fahrzeugakkumulator vehicle cell
Fahrzeugbatterie vehicle battery, traction battery
Fahrzeugfunk mobile radio
Fahrzeugmotor traction motor
Faksimile facsimile
Faksimileschreiber facsimile recorder
Faksimiletelegraphie facsimile telegraphy
Faktor factor (51)
in Faktoren zerlegen to factorize
Fakturiermaschine invoicing machine
Fakultät faculty
n Fakultät = n! factorial n
Fall fall, drop; case, act
Fallbügel dropping loop
Fallbügelinstrument relay-type recording instrument, instrument with locking device; chopper-bar recorder
Fallbügelregler chopper-bar controller
Fallbügelrelais hoop-drop relay
Fallbügelschreiber intermittent contact recorder, thread recorder, chopper-bar recorder
Fallgewicht drop weight, falling weight
Fallgewichtsprüfung falling-weight test (622)
Fallhöhe height of fall
Fallklappe drop shutter (621.318); drop indicator, drop signal (621.395)
Fallklappenrelais trigger relay
Fallklappentafel announciator (621.3)
Fallzeit fall time
Falle trap
Fallen des Barometers drop of barometer, sinking of the barometer
Fallen des Stromes fall of current
fallen to fall, to drop; to decline, to diminish
fallen lassen to drop, to release; to eliminate
fallende Charakteristik falling characteristic, negative resistance characteristic
fallender Luftdruck drop of barometer
fallende Potenzen inverse powers
fallende Potenzreihe inverse power series
fallende Tendenz downward trend

fällen to precipitate (54)
falsch false; improper, spurious
falsch berechnen to miscalculate
falsch deuten to misinterpret
falsch eingebaut misplaced
falsch hören to mishear
falsch sitzend out of truth
falsch zählen to miscount
falsche Berechnung miscalculation
falsche Bezeichnung misnomer
falsche Darstellung misrepresentation
falsche Form oddside (621.74)
falsche Nummer wrong number (621.395)
falsche Zählung miscount
falschphasige Komponente out-of-phase component
an falscher Stelle out of place
in falscher Stellung out of position
Falschabstimmung mistuning
Falschverbindung wrong connection
Falschwahl faulty selection (621.395)
Faltdipol folded doublet, folded dipole antenna
Falte fold, crease, wrinkle
Falten werfen to wrinkle
falten to fold
gefaltetes Koppelnetz folded network
faltig creasy
Faltung convolution (51)
Falz joggle, rabbet
Falzrohr slip conduit, slip joint tube (621.3)
falzen to joggle, to rabbet
Fang capture, catch(ing); pulling
Fangbereich pulling range
Fangdamm coffer dam
Fangeinrichtung interception circuit (621.395)
Fangentladung (upward) streamer
Fangnetz catch net (621.315.1)
Fangschaltung interception circuit; call-tracing circuit (621.395)
Fangstelle trap (621.382)
Fangvorrichtung cage arrester (622)
fangen to catch, to trap
Farad farad
Faraday-Käfig Faraday cage
Faraday-Kammer Faraday cell
Faradayscher Ring Faraday's ring
Faradisiergerät faradizer
Farbe colour, paint, ink; dye (67)
Farbabschalter colour killer
Farbanpassung colour match
Farbanstrich paint, painting, paint coat
Farbart chromaticity
Farbartsignal chrominance signal
Farbbalkengeber colour-bar generator
Farbband ink ribbon
Farbbereich colour range, colour gamut
Farbbild colour picture
Farbbildröhre colour-picture tube
Farbbildsignal colour-picture signal
Farbbildsignalgemisch colour video signal; composite color (US)
Farbbildübertragung colour-picture transmission
Farbcode colour code
Farbcoder colour encoder, colourplexer
Farbdecoder colour decoder
Farbdemodulator chrominance demodulator
Farbdia colour slight
Farbdreieck colour triangle
Farbempfindung colour sensation
Farbfernsehen colour television
Farbfernsehkamera colour-television camera
Farbfernsehröhre TV-colour tube
Farbfestigkeitsprüfer fadometer
Farbfleck stain (spot)
Farbfolge colour sequence (621.397)
Farbinformation chroma information
Farbkennzeichnung colour code
mit Farbkennzeichnung colour-coded
farblos colourless
farbloser Lack lacquer (681)
Farbmaske colour mask
Farbmatrixschaltung colour-matrix unit
Farbmetrik colourimetry
Farbmischung colour mixture
Farbmodulator chrominance modulator
Farbröhre = Farbfernsehröhre
coloursaturation
Farbschirm colour-picture screen
Farbspritzanlage spraying plant
Farbspritzpistole paint-spray gun
Farbstoff colour substance, pigment; dye (stuff) (677)
Farbstreifen stripe
Farbsynchronimpuls burst, colour burst
Farbtafel chromaticity diagram
Farbtemperatur colour temperature
Farbton tinge, tint, hue
Farbträger chrominance subcarrier (621.397)
Farbtripel colour triad, phosphor triple
Farbvalenz colour-stimulus specification
Farbverfälschung colour-registration error, colour-purity error
Farbwert tristimulus value
Farbwertsignal primary signal
Farbwiedergabe colour rendering
Farnsworth-Röhre image dissector
Fase chamfer (674.2)

fasen to chamfer
Faser fibre, fiber
Faserausrichtung fibre alignment
Faserform fibre shape
Faseroptik fibre optic
Faserplatte mill boarding
Faser(richtung) grain (66), (67)
gegen die Faserrichtung against the grain
Faserstoff fibrous material
Faserstruktur fibrous structure
faseriger Rand fringing
Faß barrel, drum; kilderkin
Faßlampe inspection lamp, inspection torch
Faßwicklung barrel winding (621.313)
fassen to hold, to contain, to include; to set, to mount (671), (621.396)
Fassung socket, (lamp) holder, bulb holder; mount, mounting, setting
Fassung für Leuchtröhren tubular-lamp holder
Fassungsader fixture wiring; flexible wire
Fassungslitze lamp cord
Fassungsnippel cord grip
Fassungsschalter socket switch, key switch
Fassungsstecker lamp-holder plug
Fassungsvermögen cubic capacity, storage capacity; pondage (627.8); (gewichtsmäßiges:) poundage
fast kreisförmig subcircular
Faure-Platte Faure plate
Faustregel rule of thumb, rough and ready rule
Feder feather (59), (621); spring (621.888); sunk key (621.886); tongue (694); (ink) pen (621.317)
Feder wickeln to wind a spring
federangetrieben spring-operated
Federantrieb spring motor
federbelastet spring-loaded, spring-controlled
federbetätigt spring-operated
Federblatt spring leaf
Federbolzen spring bolt
Federbuchse spring bush
Federdruck spring pressure
Federgehäuse spring case
federgelagert spring-mounted
Federhaken spring hook
Federklammer spring clip
Federkonstante force per unit elongation; torque per unit twist
Federkolben spring plunger
Federkontakt spring contact
Federkraft spring force, spring elasticity
Federleiste edge connector
Federmanometer spring pressure gauge
Federmotor spring motor
Federring spring ring
Federrückstellung spring control
mit Federrückstellung spring-controlled
Federscheibe spring washer
Federsockel spring suspension
Federstahl spring steel
Federsteife elongation per unit force
Federteller spring plate, spring seat
Federventil spring-operated valve
Federwaage spring balance
Federwickelmaschine spring-winding machine, spring winder
federn to spring, to yield (539.3)
federnd springy, snappy, resilient
federnd lagern to spring
federnd nachgeben to yield
federnde Aufhängung spring suspension
federnder Schlag snappy blow
federnde Unterlegscheibe spring washer
gefederte Achse spring axle (625.2)
Federung spring, springing suspension; elasticity
Fehl-... mis-..., spurious
Fehlalarm false alarm
Fehlabstimmung mistuning
Fehlanflug missed-approach procedure
fehlanpassen to mismatch (621.396)
Fehlanpassung standing-wave ratio (621.317/396); mismatch (621.396)
Fehlanpassungsmesser reflectometer
Fehlansprechen misoperation
Fehlauslösung random triggering
Fehleinstellung misadjustment
Fehlkonstruktion misconstruction
Fehlordnung imperfection (621.382)
Fehlpeilung spurious bearing
Fehlwinkel impurity phase angle (621.313)
fehlzünden to misfire (621.43)
Fehlzündung misfiring
Fehler fault, error (51), (621); defect, flaw (621.7); fault, failure (621.3); mistake, blunder
Fehler beseitigen to debug (681.3); to remove the fault
Fehler festbrennen to break down a fault (621.315.2)
Fehler durch nachfolgenden Betriebsstrom secondary fault
fehleranfällig fallible
Fehlerausgleich correction of error
fehlerbehaftet faulty, faulted, defective
fehlerbehaftete Leitung faulty line
Fehlerbereich range of error
Fehlerbeseitigung debugging (681.3)
Fehlerbit erroneous bit

Fehlerdämpfung return loss between line and network (US), balance return loss
Fehlerdämpfungsmeßgerät impedance unbalance measuring set, reflection measuring set
Fehlerdiagnose fault diagnosis
Fehlereingrenzung broad fault location, fault localization
fehlerempfindlich fail soft
Fehlererkennbarkeit flaw discrimination (621.386)
Fehlererkennung error detection
fehlererkennender Code self-checking code, error-detecting code
Fehlererkennungssystem mit Signalwiederholung request repeat system
fehlerfrei faultless, error free
Fehlergrenze margin of error, limit of error, accuracy
Fehlergrenzen-Klassifizierung accuracy rating
fehlerhaft defective, faulty; erroneous
fehlerhafte Funktion malfunction
fehlerhafte Peilung spurious bearing
Fehlerhäufigkeit error rate
Fehlerkorrektur error correction
fehlerkorrigierender Code self-correcting code
Fehlermöglichkeit likelihood of errors
Fehlerort point of fault
Fehlerortsbestimmung location of faults, fault localization, troubleshooting
Fehlerortsbestimmung für Kabel Varley's test
Fehlerortsbestimmung durch Schleifenmessung Murray loop test
Fehlerortsbestimmung durch Spannungsabfallmessung overlap test
Fehlerprüfschalter feeler switch (621.315.1)
Fehlerquelle source of defects, source of errors (51), (621.317)
Fehlerrate error rate
fehlersicher errorproof, fail safe
Fehlerspannung potential to ground
Fehlerspannungsschutzschalter voltage-operated leakage trip, fault-voltage circuit breaker
Fehlerstrom leakage current, fault current
Fehlerstrom-Differentialschutz Merz-Hunter protective system
Fehlerstromschutz leakage-protective system; core-balance protective system
Fehlerstromschutzeinrichtung residual current-operating system
Fehlerstromschutzschaltung mit Überstromauslösung current-operated earth-leakage circuit breaker with over current protection
Fehlersuchcode error-detecting code
Fehlersuche fault finding, troubleshooting, fault locating
Fehlersuche in einem Programm programm debugging (681.3)
Fehlersucher trouble shooter (621.3), (621.396)
Fehlersuchprogramm diagnostic routine, debug program (681.3)
Fehlerwiderstand fault resistance
Fehlstelle imperfection, defect (621.382)
Fehlstellenerzeugung disordering (621.382)
Fehlstellenhalbleiter extrinsic semiconductor
Feilspäne, Feilicht filings (pl.)
Feilspanbild magnetic figures (pl.)
Fein-... fine, smooth; micro-...
Feinabgleich fine adjustment
Feinabstimmung fine tuning
Feinbearbeitung finish machining (621.9)
Feinblech thin sheet
Feindraht fine wire
Feineinstellung fine setting, fine adjustment
Feineinstellskala vernier scale
Feinendtaster precision momentary-contact limit switch
Feinfeile smooth file
Feingehalt degree of purity (669)
Feingewinde fine thread
Feinkohle coal fines (pl.)
Feinkontrast detail contrast (621.397)
Feinlunker pin hole (621.7)
Feinmechanik fine mechanics, light engineering
Feinmechaniker precision mechanic
Feinmeßdiagramm stress-strain diagram
Feinmeßfunkenstrecke micrometer gap
Feinmeßinstrument fine measuring instrument, precision-type instrument
Feinmeßuhr micrometer dial
Feinporigkeit microporosity
Feinschnitt microtome
Feinsicherung fine-wire fuse, minifuse
Feinsitz snug fit, close fit
Feinstellknopf fine-regulating knob
Feinstellschraube micrometer screw
Feinsteuerung fine control
Feinstruktur microstructure, fine structure
Feintrieb fine setting, fine adjustment, precision adjustment
Feintriebskala slow speed dial (621.396.62)
Feinwanderung fine migration

Feinwanderungskonstante transport ratio (621.316.5)
Feinzerstäubung atomization
fein fine, delicate; first-class, first-rate; sharp (61)
fein bearbeiten to finish
fein verteilen to sprinkle, to disperse, to intersperse
fein verteilt finely divided
feines Gewinde fine thread
feine Linie delicate line
feindrähtig multi-strand
feindrähtige Litze multi-strand wire
feinfaserig close-grained
feingeschliffen precision-ground
feinkörnig fine-grained, close-grained
feinkörniger Bruch fine-grained fracture
feinmaschig fine-mesh
feinregelbarer Motor infinitely variable speed motor
feinrissig werden to gall
feinschleifen to hone
feinen to fine, to refine
Feinheit fineness
feinst verteilen to atomize
Feld field (53), (63), (621.3) (623)
ein Feld aufbauen to build up a field, to make a field
ein Feld aussetzen to set in a field
ein Feld begrenzen to bound a field
ein Feld breitet sich aus a field expands
ein Feld schwächen to weaken a field
ein Feld verstärken to aid a field
Feldänderung field change
Feldbild field pattern, field configuration
Felddichte field density
Feldeffekt-Transistor field-effect transistor, unipolar transistor, FET
Feldeffekttransistor mit isolierter Steuerelektrode insulated-gate field-effect transistor
Feldeffekttransistor mit Metall-Nitrid-Halbleiter-Aufbau metal-nitride semiconductor field-effecttransistor
Feldeffekttransistor mit Metalloxid-Halbleiter-Aufbau metal-oxide semiconductor field-effect transistor
Feldeffekttransistor mit PN-Übergang junction-gate field-effect transistor
Feldeffekttransistortriode triode field-effect transistor
Feldemission field-enhanced emission
Felderregung field excitation (621.313)
Feldkabel field wire
Feldlinie line of a vector, characteristic line (of a vector field), line of force
Feldmagnet field magnet
Feldmühle rotary voltmeter, generating voltmeter
Feldplatte magneto-resistive element
Feldpol field pole
Feldregelstab fine-regulating rod (539)
Feldregelung field control (621.34)
Feldregler field regulator, field rheostat, shunt regulator, exciter resistance
Feldrichtung field direction
Feldröhre field tube
Feldschwächung weakening of the field, flux shunting (621.313)
Feldschwächungseinrichtung field suppressor
Feldspannung inductor voltage
Feldspannungsteiler potentiometer-type field rheostat
Feldspat feldspar, felspar, orthoclase
feldspatartig, feldspathaltig felspathic
Feldspule stationary coil, field (winding) coil (621.313)
Feldspulenkasten field spool
Feldstärke field strength, stress, field intensity; signal strength (621.396)
Feldstärkemeßgerät field-strength measuring set, field-intensity meter (621.396)
Feldstörung field obstruction (621.396)
Feldstrom field current, inductor current
Feldstromkreis shunt circuit (621.313)
Feldtelefon field telephone
Feldtrenner field break-up switch
Feldtyp mode (621.372)
Feldüberbrückungswiderstand field-diverter rheostat (621.313)
Feld-Übertragungsfaktor field sensitivity (539)
Feldverteilung field distribution
Feldwellenwiderstand characteristic wave impedance (621.371)
Feldwicklung field (magnet) winding
Felge rim
Fenster window; aperture (53)
Fensterscheibe (window) pane
Fensterventilator window fan
Ferment ferment
Fermi-Funktion Fermi function
FERMI-Niveau FERMI level
Fern-... remote, tele-..., long-distance...
Fernablesung distant reading, remote reading
Fernabstimmung remote tuning

Fernamt trunk exchange (Br.), long-distance exchange (US)
Fernamtsbeamtin trunk operator
Fernanruf trunk call (Br.), long-distance call (US)
Fernanzeige teleindication, remote indication
Fernanzeigevorrichtung remote-indicating device
Fernbeamter, Fernbeamtin trunk operator
Fernbediengerät remote control device
Fernbetätigung remote control
Fernbildübertragung telephotography, picture transmission
Ferndrucker teleprinter
Ferneinstellung remote setting, teleadjusting
Fernempfang long-distance reception, long-range reception (621.396)
Fernempfangszone secondary coverage
Fernfeld distant field
Ferngeber teletransmitter
Ferngespräch s. **Fernanruf**
Fernkabel trunk cable, toll cable
Fernleistungsmesser remote power indicator
Fernleitung intertoll trunk, long line, long-distance (transmission) line (621.315); trunk line (Br.), toll line (US) (621.395)
Fernleitungsnetz transmission system, grid system
Fernleitungsübertrager repeater coil, phantom coil
Fernleitverfahren Loran = Long Range Navigation
fernlenken to guide
Fernlenksystem guidance system
Fernmeldetechnik telecommunication
Fernmeldetechniker communication engineer
Fernmeldenetz telecommunications network
Fernmeldesatellit communication satellite
Fernmeldeturm communications tower
Fernmeldeverbindung telecommunication line
Fernmeldewesen telecommunication, telephony and telegraphy
Fernmessen telemetering, remote measuring, remote measurement
Fernmeßgerät mit Spannungsübertragung voltage-type telemeter
Fernmeßinstrument telemeter, tallymeter
Fernmessung remote metering
Fernmessung mit Frequenzumsetzung frequency-type telemeter
Fernnebensprechen far-end crosstalk
Fernnetz long-distance network
Fernplatz trunk position
Fernschalten teleswitching
Fernschrank trunk switchboard
Fernschreibamt telex exchange
Fernschreibbild telephotograph
Fernschreiber teleprinter, teletypewriter
Fernschreibentzerrer regenerative repeater
Fernschreiblinienstrom telegraph line current
Fernschreibnetz teleprinter network
Fernschreibsystem printing telegraph system, teleprinter system
Fernschreibteilstreckensystem telegraph retransmission system
Fernschreibvermittlung teleprinter exchange, telex exchange
Fernseh-... television, TV
Fernseh übertragen, durch to televise, to telecast
Fernsehanlage television system
Fernsehaufzeichnung telerecording, video record
Fernsehaufzeichnungsgerät VTR, video-tape recorder
Fernsehbild television picture, image
Fernsehbildröhre siehe **Fernsehröhre**
Fernsehempfang television reception
Fernsehempfänger, Fernsehgerät television receiver, television set
Fernsehfernleitung intercity television link
Fernsehfilm telefilm
Fernsehfrequenz television frequency, television channel
Fernsehfüllsender gap filler
Fernsehfunk television = TV
Fernseh-Gemeinschaftsempfangsanlage multiple-outlet television system, television system
Fernsehgerät television set, TV set
Fernsehkamera television camera
Fernsehkanal television channel
Fernsehkanalabstand television-channel spacing
Fernsehleitung television line
Fernsehleitungskette chain of television lines
Fernsehnorm television standard
Fernsehnormwandler TV system converter, TV standards converter
Fernsehortsleitung local television link
Fernsehprojektionsröhre kinescope
Fernsehprüfsignale television test signals
Fernsehröhre (cathode ray) television tube, picture tube
Fernsehrundfunk television broadcasting
Fernsehrundfunkempfänger TV-set, TV home receiver

Fernsehsatellit TV satellite
Fernsehschaltstelle television program center
Fernsehsender television (broadcasting) transmitter, TV station, television broadcast station
Fernsehsendung TV transmission
Fernsehsignal signal
Fernsehsignalgemisch composite picture signal
Fernsehstudio television studio
Fernsehtechnik video technique
Fernsehtelefon picturephone, viewphone
Fernsehumsetzer television translator, rebroadcasting transmitter, television transposer
Fernsehverstärker video (frequency) amplifier
Fernsehen (wireless) television = TV, radiovision; video (US); (closed circuit) television = CCTV
Fernsehen mit geringer Auflösung (Zeilenzahl unter 200) low-definition television
Fernsehen mit hoher Zeilenzahl high-definition television
Fernsprech-... telephone... (s. auch **Telefon**)
Fernsprechamt telephone exchange
Fernsprechanschluß telephone connection
Fernsprechbeamter telephone operator
Fernsprechbuch (telephone) directory
Fernsprech-Endvermittlung telephone terminal exchange
Fernsprechfernvermittlung telephone toll exchange, telephone trunk exchange
Fernsprechgerät telephone equipment
Fernsprechhäuschen public call box, telephone booth
Fernsprechkabel telephone cable
Fernsprechleitung telephone wire, telephone line, telephone circuit
Fernsprechnetz telephone network
Fernsprechsaal operating room
Fernsprechschrank switch box
Fernsprechtechnik telephone engineering
Fernsprechverkehr telephone transmission
Fernsprechvermittlung central office, telephone exchange
Fernsprechverstärker telephone repeater
Fernsprechwesen telephony
Fernsprechzelle telephone box, telephone booth
Fernsprecher telephone
Fernsteuerung remote control, telecontrol
Fernsteueranlage remote-control system
Fernsteuergerät remote-control apparatus, remote control equipment
Fernthermometer telethermometer
Fernübertragung teletransmission
Fernverbindung trunk connection (621.395)
Fernverkehr long-distance traffic
Fernvermittlung long-distance switching, toll exchange
Fernvermittlungsstelle to ll office, long-distance center
Fernvermittlungsamt trunk exchange
Fernvoltmeter pilot voltmeter
Fernwahl long-distance dialing, intertoll dialing
Fernwahlmünzfernsprecher coin box, telephone for long-distance calls
Fernwahlvermittlungssystem central-office toll dialing
Fernwirkstation remote control and telemetering control room
Fernwirkunterstation out station
Fernwirkzentralstation master station
Fernwirksystem supervisory remote control and telemetering system, telecontrol system
Fernzählung telemetering, telerecording
fern remote
fernbetätigt remote controlled
fernmündlich telephonic
fernschreiben to teleprint
fernsehen to televise
fernsteuern to guide (623), to telecontrol, to telecommand (621.3)
ferngesteuertes Geschoß guided missile
fernwirken to telecontrol
Ferranti-Effekt Ferranti effect
Ferrarisinstrument Ferraris instrument, shaded-pole instrument, shielded pole instrument, induction-type instrument
Ferrarismotor shaded-pole motor
Ferrarisrelais induction relay
Ferraris-Voltmeter induction voltmeter
Ferrariszähler induction (motor) meter
Ferri-... (= dreiwertige Eisenverbindung) ferric
Ferrit ferrite
Ferrit-Antenne antenna core, ferrite-rod aerial, loopstick (antenna)
Ferrit-Ringkern ferrite ring core
Ferritperle ferrite bead
Ferro-... (= zweiwertige Verbindung) ferrous
Ferrocart ferrocart

ferrodynamisch ferrodynamic
ferroelektrisch ferroelectric
ferromagnetisch ferromagnetic
Ferromagnetismus ferromagnetism
Ferrometer ferrometer
Fertigware finished article, finished product
Fertigung production (rate)
Fertigungsbreite production range
Fertigungskontrolle production control
Fertigungsstraße production line
Fertigungsstufe stage of completion
Fertigungsunterlagenerstellung artwork generation
Fest-... solid, fast, fixed
Festantennenpeiler fixed aerial direction-finder
Festbremsleistung locked kilovolt-amperes
Festbremsmoment stalled to rque
Festfrequenzsender fixed frequency transmitter
Festfressen seizure
Festkomma fixed point
Festkondensator fixed capacitor
Festkörper solid (substance)
Festkörperphysik solid-state physics
Festkörperschaltung solid-state circuit, monolithic circuit
Festkörperspeicher solid state memory
Festplatte hard disk
Festpunkt fixed point
Festsitz snug fit
Festspannungsregler fixed voltage regulator
Festspeicher read-only memory, ROM, fixed store, nonerasable storage
Feststation base station, fixed station
Feststellhebel locking lever
Feststellschraube locking bolt, locking screw
Festverzögerung absolute delay (621.397)
Festwertspeicher read only memory, ROM
Festwertspeicher mit Schmelzbrücken fusible ROM = FROM
Festwiderstand fixed resistance, fixed resistor
Festzeichen bench mark (621.397); clutter, permanent echo, ground return (621.396.96)
Festzeichenaustastung clutter gating
Festzeichenunterdrückung permanent echo cancellation, fixed echo elimination
Anzeige mit Festzeichenunterdrückung moving-target indication (621.396.96)
Festzeitgespräch fixed time call (621.395)
fest fast, fixed; firm, strong, solid, compact
...fest ...proof, resistant to ...
fest abgestimmt permanently tuned, fixed-tuned
fest verlegte Leitungen fixed installation (sing.)
fest verschraubt securely screwed
feste Anlage stationary structure (621.316.5)
feste Betriebsbedingungen fixed service conditions
festes Dielektrikum solid dielectric
festes Fundament strong foundation
fester Funkdienst fixed radio service
feste Funkstelle fixed station, base station
feste Funkverbindung point-to-point radio communication
feste Gittervorspannung fixed (grid) bias (621.396)
festes Isoliermaterial firm insulating material, solid insulating substance (621.3)
feste Kopplung tight coupling, close coupling
feste Kosten fixed cost (sing.)
feste Landfunkstelle land-base station
feste Lösung solid solution
festes Material solid substance, strong material
feste Phase solid phase (54)
fester Plattensatz fixed plates (pl.) (621.396)
fester Schiebesitz tight push fit
fester Sitz tight fit
feste Trägerfrequenz resting-carrier frequency
festes Widerlager fixed location point
fester Widerstand fixed resistance, fixed resistor
fester Zustand solid state
festbinden an to lash to
festbremsen to hold fast (621.313)
in festgebremstem Zustand at stall
festfressen to seize (621.82)
sich festfressen auf to seize fast against
festhalten to retain, to hold fast, to hold down
Anzeige festhalten to freeze readings
festkeilen to key, to wedge tightly, to fasten
festkeilen in to wedge into
festklemmen to secure under terminals (621.3); to clamp (621)
festlegen to fix, to locate, to key; to govern (51)
festmachen to fasten, to fix (**an:** to)
festmachen in to secure into
festrosten to rust fast
festschnallen to strap
festschrauben to screw up, to bolt
festsetzen, sich to stick, to foul
festsitzend secure
festspannen auf to clamp to; to secure on (621.9)
feststehend axiomatic (5); stationary (6)
feststehendes Bild still picture (77)

feststehende Platte stator plate, stator vane (621.319.4)
feststehende Spule stationary coil
feststehender Teil stationary element
festverdrahtete Logik hard-wired logic
Festigkeit stability, strength (66)
Festigkeit verringern to weaken the strength
Festigkeitsprüfung strength testing
FET field effect transistor, FET
Fett (lubricating) grease
Fettbüchse grease cup
Fetton fatty clay
fett, fettig fat, fatty, oily, greasy
fettfrei grease-free
fettgeschmiert grease-lubricated
fettlösend detergent
fetten to grease, to lubricate
feucht moist, damp, wet, humid
feucht halten to keep moistened
feucht machen to moisten
feuchte Luft damp atmosphere, moist air, humid atmosphere
feuchtes Wetter damp weather
Feuchtemesser hygrometer
Feuchtgewicht wet weight (67)
Feuchtigkeit moisture, humidity, dampness
Feuchtigkeit abtreiben to eliminate moisture
Feuchtigkeit anziehen to attract moisture
Feuchtigkeit aufnehmen to absorb moisture, to take up moisture
Feuchtigkeit Null zero humidity
gegen Feuchtigkeit abdichten to seal against moisture
vor Feuchtigkeit geschützt moisture-proof
feuchtigkeitsabstoßend moisture-repellent
Feuchtigkeitsaufnahme absorption of moisture
feuchtigkeitsbeständig moisture-resistant, moisture-proof
feuchtigkeitsdicht moisture-tight
Feuchtigkeitseinfluß influence of moisture
Feuchtigkeitsentzug moisture removal
Feuchtigkeitsgehalt moisture content
feuchtigkeitsgeschützte Maschine moisture-proof machine
Feuchtigkeitsgrad degree of humidity
Feuchtigkeitsgradient moisture gradient (551.5), (621.396)
Feuchtigkeitsschutz moisture proofing
Feuchtraum damp room
Feuchtrauminstallation weather-proof wiring, damp-proof wiring
Feuchtraumleuchte vapour-proof fitting (621.32)
Feuer fire
feuerbeständig refractory, fire-proof, fire-resisting
Feuerbeständigkeit refractoriness
Feueremaillierung vitreous enamelling
feuerfest fire-proof, refractory
feuerfester Tiegel heat-resisting crucible
Feuerfestigkeit refractoriness, fireproofness
feuergefährliches Material inflammable matter
feuerhemmend fire-resisting, fire-retarding, flame-retarding
feuerhemmende Isolation slow-burning insulation
Feuerleiteinrichtung gun director
Feuerleitgerät fire-direction device
Feuerleitnetz fire-direction net
Feuerleitradar fire-control radar, gun-laying radar
Feuerleitung fire control, fire direction, gun-laying
Feuerlöscher fire extinguisher
Feuermelder fire alarm (box), fire detector
Feuermeldung fire alarm
Feuerschiff light ship, floating light
feuersicher fire-proof
feuerverzinken to galvanize
Feuerverzinkung (hot) galvanizing, galvanization
feuerverzinnen to fire-tin, to hot-dip
Feuerverzinnung fire-tinning, hot-dipping
feuern to flash (621.3), (623.4)
Feuerung firing, furnace
Fiber fibre
Fiberoptik fibre optic
Fiberunterlegscheibe fibre washer
Fibrille fibril
fiederförmig feathery
figürlich darstellen to figure
Film film (77); (movie) film, (moving) pictures (pl.) (77)
Film vorführen to feature a film
Filmabtaster telecine equipment, (flying spot) film scanner
Filmdosimeter film badge (539)
Filmempfindlichkeit speed of a film
Filmplakette film badge (539)
Filmprojektor film projector
Filter filter (53), strainer, screen (54); frequency filter (621.392)
Filter aus abgestimmten Leitungen line filter
Filtergaze gauze filter
Filterpresse filter press
Filterquarz filter crystal, resonator crystal
Filterrechen strainer rack (621.2)

Filtersperrbereich filter stop band, filter attenuation band (621.395)
Filterspule filter coil, filter reactor
Filterweiche notch diplexer
Filterwirkung filtering action
Filterzelle absorption cartridge (621.314.2)
filtern to filter
Filterung, Filtrierung filtering, filtration
Filtrierpapier filter paper
Filz felt
Filzdichtung felt packing
Filzscheibe, Filzschwabbel rotary felt
Fingeranschlag finger stop
Fingerspitzengefühl delicacy of touch, feel
Firma firm, company
Firmenschild (maker's) name plate
Firmenzeichen identification badge, company initials, trade name
Fischgrätenantenne fish-bone aerial
Fitsche, Fitschenband pin hinge
Fixativ fixative
Fixierbad fixing bath
fixieren to fix (77); to locate (621.9); to key
Fixierung location (621.9)
Fixierungsloch fixation hole
Fixpunkt fixed point
Flach-... flat
Flachbahnanlasser face-plate starter (621.3)
Flachbahnstufenschalter face-plate step-switch
Flachbandkabel ribbon cable
Flachbauweise low-level design, ground type (621.316.5)
Flachbogen segmental arch
Flacheisen flat bar
Flachgewinde parallel thread, square thread
Flachglas plate glass, sheet glass
Flachkabel flat cable (621.395)
Flachkopf pan head (621.884)
Flachkopfniet flat-headed rivet
Flachkopfschraube flat-head screw, cheese-headed screw
Flachkupfer flat-bar copper
Flachkupferwicklung strip-wound armature (621.313)
Flachleitung flat twin cable
Flachlitzenschnur tinsel cord
Flachprofilgerät flat edgewise pattern instrument
Flachrelais flat-type relay
Flachschienenwählersystem panel dial system
Flachspule pancake coil (621.3)
Flachspul-Amperemeter attraction-type ammeter
Flachstahl steel flat, flat-bar iron
Flachwähler rack selector
Flachzange flat-nose pliers
flach flat, flat-topped, flush
flach abbiegen to flatten off
flach auslaufen to level off
flach geschichtet face to face
flach machen to flatten
flach werden to level off
flacher Kopf (einer Kurve) flat top (of a curve)
flache Kurve flat curve
flaches Maximum flat maximum
flacher Schnitt shallow cut
Fläche face, area, surface (area), plane
Fläche abtasten to explore a surface
Fläche gleicher Phase equiphase surface
mit der Fläche nach oben face up
zwischen den Flächen interplane
Flächenausdehnung areal extension
Flächenblitz sheet lightning
Flächendiagonale face diagonal
Flächendiode junction diode
Flächeneinheit square unit, unit area
Flächenhelligkeit surface brightness
Flächeninhalt der Oberfläche surface area
Flächenmaß superficial measure, square measure
flächenmäßig areal
Flächenpressung contact stress
Flächenscherungsschwingung face-shear mode
Flächenschleifmaschine face grinder
Flächenstrahler large-area radiator
Flächentransistor junction transistor
flächentreues Abbildungsnetz equal-area web (51)
flächenzentriert face centred (548)
Flackern flickering, jitter
Flackerfeuer, Flackerlicht blinker light, undulating light (656.7); flicker light, flashing light
Flackerzeichen winking signal, flickering (621.395)
Flackerzeichen geben to call the operator
flackern to flare, to flicker
flackernder Brand flare
Flamme flame
in Flammen afire, aflame
in Flammen aufgehen to burst into flame
Flammbogen flaming arc, arc flame
Flammenfärbung flame coloration
Flammenrückschlag flash (621.791)
Flammofen reverberatory furnace
Flammpunkt flash point, flashing point, temperature of ignition, flaming point
flammen to flame

Flanke flank, side, slope
Flanken-Abfallzeit fall time
Flanken-Anstiegszeit rise time
Flankenspiel backlash (621.831)
Flankensteilheit edge steepness
Flankensteilheit der Ausgangsspannung rate of change of the output voltage
Flankenzeit slope time
Flansch flange
mit Flansch versehen to flange
mit Flansch (versehen) flanged
Flanschbreite flange breadth, flange width
Flanschdichtung rim gasket
Flanschmotor flange motor
Flasche flask (54); cylinder (621.79); bottle
Flaschendruckmanometer contents gauge
Flaschenzug pulley (block), block and tackle, lifting tackle
Flatpack flat pack
Flattern flutter, jitter, chattering
flechten to braid, to lace, to plait
geflochtene Litze braided wire
Fleck spot, stain
Fleckform spot shape (621.385)
Fleckgeschwindigkeit scan speed (621.397)
Fleckgröße spot size (621.385)
Fleckhinlauf forward sweep (621.385)
Fleckrücklauf flyback, return trace
Fleckrücklaufzeit flyback time, return time
Fleckschärfe sharpness of the spot (621.385)
Fleckunschärfe deflection defocusing
fleckenlos spotless, stainless
fleckig stained; patchy (621.32)
fleckig machen to stain
flexibel flexible
flicken to mend, to patch
Flickereffekt flicker effect
fliegen to fly; to plane
fliegender Anschluß jumper
fliegende Installation temporary wiring
fliegende Leitung flying lead, temporary wire
Fliehkraft centrifugal force
Fliehkraftanlasser centrifugal starter, inertia starter (621.4)
Fliehkraftkupplung centrifugal clutch
Fliehkraftregler centrifugal governor
Fliehkraftschalter centrifugal switch
Fließ . . . flow
Fließband assembly line
Fließeigenschaft flow property
Fließfertigung streamlined production
Fließfiguren waving (sing.) (539), rippling (sing.) (66)
Fließgeschwindigkeit rate of flow
Fließgrenze yield point
Fließkomma floating point
Fließpapier absorbent paper
fließen to flow, to well
Flimmern shimmer
flimmern to flicker, to glint, to glisten
Flimmerphotometer flicker photometer, Simmance-Abady photometer
flinke Sicherung quick-break fuse, quick(-response) fuse
Flintglas flint-glass
Flip-Chip-Transistor flip-chip transistor
Flip-Flop flip-flop circuit, to ggle circuit
Flip-Flop-Auslöser Eccles-Jordan trigger
Flip-Flop-Speicher flip-flop storage
Flocke flake, flock
flocken to flake
flockige Masse flocks (pl.)
Floppy-Disk floppy disk
fluchten to align
flüchtig volatile (54), transient
flüchtiger Bestandteil volatile (constituent)
die flüchtigen Bestandteile abtreiben to expel the volatiles (54)
flüchtige Größe transient quantity
flüchtige Komponente transient component
flüchtiger Speicher volatile memory
flüchtiger Stoff volatile matter
flüchtiger Strom transient current
Flüchtigkeit volatility (54)
Flug fly, flight
Flugasche flue dust, smoke dust, suspended ash; light ashes (pl.)
Flugbahn trajectory
Flugbetriebslenkung operational control
Flugdatenrechner air-data computer
Flugdatenschreiber flight-data recorder, crash recorder
Flugdienstleitung control tower
Flugelektronik avionics
Flugfeldüberwachungsradar airport surface-movement indicator
Fluggeschwindigkeit air speed
Fluggewicht all-up weight
Flughafenbefeuerung airport lighting
Flughafenleuchtfeuer airport beacon
Flughafenüberwachungsradar ASR = airport surveillance radar
Flughöhe aircraft altitude
Flugfunkbake radio range
Flugfunkvollzugsordnung aircraft radio regulations
Flugkörper (guided) missile, spacecraft
Flugkörperlenkung missile guidance

Flugkörperverfolgung missile tracking
Flugleitung flight control
Flugmotor aero engine
Flugnavigation air navigation
Flugnavigationshilfe aid-to-air navigation
Flugplatz airport, landing ground
Flugplatzkennfeuer airport beacon
Flugregler flight-control system
Flugschreiber crash recorder
Flugsicherung ATC, airtraffic control, airway traffic control
Flugsicherungsleuchte airway beacon
Flugsimulator flight simulator
Flugstrecke route of flight
Flugstreckenfeuer route beacon
flugtüchtig flyable
Flugverkehrskontrolle ATC, air traffic control
Flugwarnfeuer obstruction light, hazard light
Flugweg track
Flugwegdarstellung track display
Flugwegrechner flight-path computer
Flugzeug aircraft, airplane, aeroplane, plane
Flugzeug mit Druckpropeller pusher
Flugzeuge orten to locate planes
Flugzeugeffekt aeroplane effect
Flugzeugführer pilot
Flugzeugsteuerung aircraft control system
Flügel wing (59), (629); grand (piano) (78); vane (533)
Flügelmutter hand nut, wing nut, thumbnut
Flügelpumpe wing pump
Flügelrad-Anemometer windmill-type anemometer
Fluidik fluidic
Fluor fluorine
Fluoreszenz fluorescence
Fluoreszenzlicht fluorescent light
Fluoreszenzschirm fluorescence screen, fluorescopic screen, fluoroscope, roentgenoscope
fluoreszieren to fluoresce
fluoreszierend fluorescent
Fluß flow (532/533), (536); flux (535...539); river (551)
Fluß tritt aus einer Fläche aus a flux is taken out of a face
Fluß tritt in eine Fläche ein a flux is inserted into a face
im Fluß in a state of flux
in Fluß halten to keep flowing
mit einem Fluß verkettet sein to link a flux
Flußbündel, Flußbüschel flux tuft (538)
Flußdiagramm flowchart, flow diagram
Flußdichte flux density
Flußempfindlichkeit flux sensitivity (621.382)
Flußeisen ingot steel, mild steel, low-carbon steel
Flußlinie flux line
Flußmittel (welding) flux, soldering flux
mit Flußmittel betreichen to flux
Flußröhre tube of flux
Flußsäure hydrofluoric acid
Flußsäurebehandlung hydrofluorination
Flußspat fluor = fluorspar, Derbyshire spar
Flußstahl mild steel, ingot steel, common hardening steel
Flußstahldraht mild steel wire
Flußverdrängung magnetic skin effect
Flußverkettung flux linkage
Flußwechseldichte density of flux changes (537)
flüssig liquid, fluid
flüssig werden to deliquesce, to liquify
flüssiger Brennstoff liquid fuel
flüssige Kathode pool (621.314.6)
flüssige Luft liquid air
flüssiger Sauerstoff liquid oxygen
flüssige Schweiße molten pool (621.791)
flüssiger Treibstoff liquid fuel
Flüssigkeit liquid (matter), liquid substance, fluid
Flüssigkeitsanlasser liquid starter
Flüssigkeitsaufnahme imbibition
Flüssigkeitsdämpfung viscous damping
flüssigkeitsdicht fluid-tight
Flüssigkeitsdruck hydraulic pressure, liquid pressure
Flüssigkeitsmelder fluid detector
Flüssigkeitssicherung liquid fuse unit
Flüssigkeitsstand liquid level, fluid level
Flüssigkeitsstandsmelder level control detector
Flüssigkeitsstandregelung fluid level control, liquid level control
Flüssigkeitswiderstand liquid resistance
Flüssigkristallanzeige liquid-crystal display, LCD
Flüssigspaltstoff-Reaktor fluidized(-fuel) reactor
Flut (high) tide, flood (tide)
Flutkraftwerk tidal-power plant
Flutlicht(leuchte) floodlight
Flutlichtbeleuchtung flood lighting
Flutlicht-Platzbeleuchtung general field floodlight system (656.7)
Fluxmeter fluxmeter
FM = Frequenzmodulation f.m. = frequency modulation
fokal focal

fokussieren to focalize, to (bring into) focus
Fokussierung focusing
Fokussierungseinrichtung focusing device
Fokussierungsspule focus(ing) coil (621.385)
Folge succession, sequence; sequel, consequence
Folge der Operationen train of operations
Folgeausfall secondary failure
Folgekontakt sequence contact
Folgemotor slave motor, secondary motor
Folgeprodukt derivative (54)
Folgeschaltung sequential circuit
Folgesteuerung sequential control
Folie foil
forcieren to force
forcierte Luftkühlung forced draft (621.396)
Förder-... conveying, winding, hoisting, hauling
Förderband conveyor belt, belt conveyor
Förderbandofen conveyor furnace
Förderbandtragrolle troughing roller
Fördereinrichtung haulage device, conveying device
Fördergerüst derrick frame, pit-head gear
Fördergeschwindigkeit conveying speed, winding speed
Fördergetriebe winding gear
Förderkette chain conveyor
Förderlokomotive haulage mining locomotive
Förderscheibe drive wheel, hoisting pulley
Förderschnecke conveyor screw, worm conveyor, spiral conveyor
Förderseil winding rope, hauling rope
Förderspiel winding cycle
Förderturm winding tower, derrick frame, shaft derrick
fördern to mine, to haul, to hoist, to wind
Form form, shape; mould (621.7)
Form geben to (form to) shape
Form verlieren to lose contour
von der Form... the shape of...
Formänderung deformation; strain
Formdraht fashioned wire (e. g. square-sectionwire, ribbon wire)
Formeisen fashioned iron
Formfaktor form factor (621.3)
Formfaktormeßbrücke form-factor meter, form-factor indicator
Formguß gravity casting
Formmasse moulding compound
Formraum mould cavity
Formstein special brick
Formstück sectional form
Formaldehyd formaldehyde
Formalin formalin
Formatierer formatter (681.3)
Formation formation (55), (355.3), Age (55)
Formel formula (pl. -ae), (51), (54)
die Formel gilt the formula holds good
Formelausdruck expression, term
Formelzeichen symbol
formen to form, to shape, to frame, to mould (621.7)
Formerei moulding shop
Formgebung designing, styling
formieren to form (621.385), to bake out (621.314.6)
formierte Platte formed plate (621.385)
formierter Sammler Planté accumulator
Formierung forming (621.355), (621.385), electrical formage
Formierungsspannung formation voltage (621.35)
Formierungsvorgang forming process
Formierungsvorschrift forming schedule
... förmig ...-shaped
Formspule formed coil
Formverzerrung envelope distortion
Forscher scientist
Forschung research (work)
Forschungsabteilung research department
Forschungsanstalt research establishment, research institute
Forschungsarbeit exploratory work, research work
Forschungsingenieur scientist, research engineer
Forschungsreaktor research reactor (539)
fortlaufend running, consecutive, continuous, serial, progressive
fortlaufend wiederholt repetitive
fortlaufende Nahtschweißung progressive induction seam welding
fortpflanzen, (sich) to propagate
Fortpflanzung propagation
Fortpflanzungsgeschwindigkeit speed of propagation
Fortpflanzungskonstante propagation constant
Fortpflanzungsrichtung direction of propagation
Fortran **for**mula **tran**slation (Programmiersprache)
Fortschaltung step-by-step switching, sequence control
Fortschreiten progression
Fortschreitgeschwindigkeit velocity of progression
fortschreiten to proceed, to advance, to progress
fortschreiten zu to step on to
fortschreitende Bewegung progressive motion

fortschreitende Welle progressing wave, travelling wave
Foto-... s. **Photo-...**
Foucaultsches Pendel Foucault's pendulum
Fourier-Integral Fourier integral
  Fouriersche Reihe Fourier series
  Fouriertransformierte Fourier transform
Frahmscher Zungenfrequenzmesser Frahm frequency meter
Fraktion fraction (54)
fraktionieren to fractionate
  fraktionierte Destillation fractional destillation
Franse fringe
„Franzose" adjustable wrench, monkey wrench
fräsen to mill
Fräser milling wheel, milling to ol
Fräsmaschine miller, milling machine
Fräsvorgang milling
Frauenhofersche Linien Frauenhofer's lines
„Fräulein vom Amt" operator (621.395)
frei free; uncombined (54), (66); disconnected, disengaged (621.395)
  frei aufgelagert simply supported, pin-jointed
  frei beweglich free-moving
  frei drehen auf, sich to run freely on
  frei fallen to fall freely
  frei laufen to run freely
  frei liegen to be exposed
  frei verlegte Installation open wiring
  freies Elektron free electron, unattached electron
  freie Fettsäure free fatty acid
  freies Gitter disconnected grid, floating grid
  freie Größe transient quantity
  freie Höhe clearance
  freier Kohlenstoff uncombined carbon
  freie Leitung disengaged line, vacant line (621.395)
  freie Leitwegzuteilung free routing (621.395)
  freier Luftzutritt free airway
  freie Oberfläche free surface
    freies Oberflächenwasser free surface water
  frei programmierbar free programmable
  freier Raum free space, clear space
  freie Schwingung free oscillation, free vibration
    frei schwingender Multivibrator free running multivibrator
  freie Spannung transient voltage
  freier Strom transient current
  freier Vorgang transient reaction
  freier Wähler disengaged selector (621.395)
  freie Weglänge free path
  freie Wellen free waves
  freigeben to release; to enable (681.3)
  freilaufend free-running
    freilaufender Oszillator free-running oscillator
  freiliegend exposed
  freimachen to (set) free, to liberate, to disengage (54); to stamp (383)
  freischalten to isolate (621.315)
  freistehend isolated
    freistehender Kran isolated crane
  freitragend self-supporting
    freitragender Stahlmast self-supporting steel tower
  freizügige Rufnummernzuteilung flexible numbering
  auf freiem Potential liegen to float
Freiauslösung trip-free release
  Freiauslösungsrelais trip-free relay
  ... mit Freiauslösung trip-free ..., non re-entrant
Freigabe... enable... (681.3); release (621)
  Freigabepuls enable pulse
Freiheitsgrad degree of freedom
Freilauf(grad) free wheel (629.118)
  Freilaufkupplung free wheeling clutch
Freileitung overhead (transmission) line, air line, aerial conductor, open-wire line, open-wire circuit, pole line
  Freileitung auf Holzmasten open-wire pole line
  Freileitungsanlage overhead-line system
  Freileitungsdraht overhead wire
  Freileitungsisolator line insulator
  Freileitungsmonteur lineman
  Freileitungsnetz overhead-line system
Freiluft-... open-air, outdoor
  Freiluftanlage open-air plant, outdoor installation
  Freiluftklima open-air climates
  Freiluftleuchten weather-proof fittings
  Freiluftschalter outdoor circuit breaker
    Freiluftschaltanlage switchyard
    Freiluftschaltgerät outdoor-type switchgear (sing.)
  Freiluftstation outdoor station
  Freilufttype outdoor type
  Freiluftumspanner outdoor transformer
Freiraumausbreitung free-space propagation
Freiraumdämpfung free-space attenuation
Freiraumwelle space wave
Freischalten clearing, forced release (621.395)
Freisprechtelephon hands-free telephone
Freistrahlwindkanal open jet tunnel

Freiton ringing tone
Freiträger cantilever
Freiwahl hunting (621.395)
Freiwahl unterbinden to stop hunt
Freiwerdezeit circuit commuted recovery time (621.382.3)
Freizeichen ringing to ne, free-line signal (621.395)
Freizustand idle condition (621.395)
Fremd-... foreign; separate, external
Fremdatom impurity, foreign atom
Fremderregung separate excitation, external excitation
Fremdmaterial foreign material
Fremdmodulation external modulation
Fremdspannung injected voltage; unweighted noise; external voltage (621.392); extraneous voltage
Fremdspannungsabstand unweighted signal-to-noise ratio
Fremdstoff, Fremdsubstanz foreign matter, foreign substance (54)
Fremdstrom parasitic current
fremd foreign
fremde Marke foreign brand
fremdbelüftet pipe ventilated, separately ventilated, forced ventilated
fremdbelüftete Maschine separately ventilated machine
fremdbelüfteter Motor forced-ventilated motor, forced-air-cooled motor
fremdbelüfteter Transformator air-blast transformer
fremderregter Generator separately excited generator
fremdgezündete Funkenstrecke timed spark gap
Frequenz frequency, periodicity (621.3)
Frequenz in der Bandmitte midband frequency
Frequenz fahren to hold frequency, to take the swings (621.311)
Frequenz sperren to reject a frequency
Frequenz der Welligkeit ripple frequency
Frequenz zuteilen to pre-assign frequencies
frequenzabhängig frequency selective, frequency sensitive, frequency discriminating
Frequenzabhängigkeit frequency response, frequency dependence
Frequenzablage frequency departure, frequency drift
Frequenzabweichung frequency devitation, frequency departure
Frequenzanalysator frequency analyzer, wave analyzer
Frequenzanalyse frequency analysis
Frequenzanzeiger frequency indicator
Frequenzauswanderung frequency drift
Frequenzband frequency band, frequency range
Frequenzbereich frequency coverage, frequency range, frequency band
Frequenzbereich verarbeiten to handle a frequency range
Frequenzbewertungsfaktor eines Stromes current-weighting factor
Frequenzdiversität frequency diversity
Frequenzeinfluß frequency error (621.317)
Frequenzeinstellung frequency setting
frequenzempfindlich s. **frequenzabhängig**
Frequenzfahren holding frequency (621.311)
Frequenzgang frequency response, response curve, response characteristic
Frequenzgangausgleich frequency compensation, balance control (621.395.6)
Frequenzgemisch composite signal
Frequenzgenauigkeit frequency accuracy
frequenzgerader Kondensator straight-line frequency capacitor
Frequenzhub frequency deviation, frequency swing
frequenzkonstant stable in frequency
Frequenzkonstanz frequency stability, circuit stability
Frequenzmeßbrücke frequency-measuring bridge
Frequenzmesser frequency meter, frequency indicator
Frequenzmodulation frequency modulation
frequenzmoduliert frequency-modulated
Frequenzmultiplex frequency-division multiplex
Frequenznormal frequency standard
Frequenzplan frequency allocation
Frequenzregler frequency regulator
Frequenzrelais frequency relay
Frequenzschwankungen frequency fluctuation
Frequenzspektrum frequency spectrum
Frequenzspringen moding
frequenzstabil stable in frequency
Frequenzstabilität frequency stability
Frenquenzsteuerung frequency control
Frequenzteiler frequency divider
Frequenzteilerschaltung frequency-reducing system, counting-down device
Frequenzteilung frequency division
Frequenztoleranz frequency tolerance (621.396)

Frequenzuhr master frequency meter, crystal clock
Frequenzumkehrung frequency inversion
Frequenzumsetzer frequency changer, converter
Frequenzumsetzung frequency shift, frequency translation, frequency conversion
Frequenzumtastung frequency-shift keying
Frequenzverdoppler frequency doubler
Frequenzverdopplung frequency doubling
Frequenzverdreifacher frequency tripler
Frequenzverdreifachung frequency tripling
Frequenzverhalten dynamic response, frequency response
Frequenzversatz field offset (621.397)
Frequenzverschachtelung frequency interlace (621.397)
Frequenzverteilung frequency allocation
Frequenzverteilungsplan frequency allotment plan
Frequenzvervielfacher frequency multiplier, harmonic generator
Frequenzvervielfachung frequency multiplication
Frequenzverwerfung shift in frequency
Frequenzvorhersage frequency prediction
Frequenzwanderung (frequency) drift
Frequenzwandler frequency changer, frequency converter
Frequenzweiche branching filter, frequency selective filter; crossover network
Frequenzwobbler wobbulator, sweep generator
Frequenzwobblung wobbulation
Frequenzzählerschaltung frequency counter (circuit)
Frequenzziehen frequency pulling
Fresnel-Zone Fresnel zone, Fresnel region
fressen to bind, to seize (621.8)
Freund-Feind-Kennung IFF, identification friend-foe
Frigen (CCl_2F_2) freon
Friktionsscheibe friction disk
Frischdampf live steam
Frischluft fresh air
Frischluftgebläse forced-draught fan; forced-draft fan (US)
Frischmetall virgin metal
fritten to frit, to sinter
Fritter coherer
Frontlader front loading system (681.84)
Frontplatte front panel
Frontspeisung front feed (621.396.67)
Froschbeinverbindung butterfly connection (621.313)
Froschbeinwicklung frog-leg armature winding
Frostriß frost crack
Frostschutztransformator anti-freeze transformer
frostsicher, frostgeschützt frost-proof
Frühausfälle early failures
Frühwarnung early warning (623)
frühzünden to pre-ignite
Frühzündung pre-ignition, advanced ignition
F-Schicht F-layer, Appleton layer
Fuchsin fuchsin
Fuge kerf (621.791), (621.9); junction (621.3); joint (69)
fugen to joint (69)
fügen to joint (67)
fühlbar sensible, perceptible, palpable
fühlbare Wärme sensible heat
Fühler feeler, sensor, detector; tracer
Fühlersteuerung tracer control
führen to lead, to guide; to locate (621.9)
mit sich führen to carry
geführter Generator driven oscillator (621.396)
Führer leader; driver
Führerraum cockpit (629.13); driving cab
Führerstand (driver's) cab, box (625.282), driving position
Führung guide, guidance, lead (621); direction (65)
Führungsantenne director aerial (621.396)
Führungsbett bed slide (621.9)
Führungsbuchse pilot bush, guide bushing
Führungsdirektleitung command hot line (623)
Führungsgröße control input, command variable, reference input (621.316.7)
Führungsrolle guide roller
Führungssäule guide pillar (621.9)
Führungsschaft guide stem
Führungsschiene guide bar, guidance bar, guide rail; bed slide
Führungsschlitz guide slot, cam groove
Führungsstange guide bar
Führungsstift guide pin, pilot pin, entering guide, follower, alignment pin
Führungsstreifen guide strip
Führungsstück guide block, guide piece
Führungswagen command car
Führungszylinder guide cylinder
Fulgurit fulgurite
Füll-... fill, filler, filling, feeding, charging, stuffing
Füllansatz fill piece
Füllfaktor space factor (621.313), (621.314.2)
Füllgas filler (621.32)
Füllmittel filling medium
Füllrutsche chute feed

Füllsignal dummy (681.3)
Füllstandsanzeiger filling-level meter
Füllstoff filler, filling (67)
Füllstück fill piece
Fülltrichter feeding magazine (621.9); coal hopper (622)
Füllventil filler valve
Füllzeichen stuffing (621.39); fill character (681.3)
füllen to fill, to feed
gefüllter Preßstoff loaded plastic
Fullererde fuller's earth, bentonite
Füllung filling, feeding, charge
als Füllung verwenden to tie up (54)
Fundament foundation, groundwork, base
Fundament setzt sich foundation sinks
Fundamentsgraben foundation trench
Fundamentmauer foundation wall
Fundamentplatte foundation slap
Fundamentrost compound grillage
Fundamentschraube foundation bolt
Fünfeck pentagon
fünfeckig pentagonal
Fünfergruppe word (621.396)
Fünfgitterröhre pentagrid valve
Fünfkanal-Lochstreifen five-track punched tape
Fünfleitersystem five-wire system
Fünfleiter-Zweiphasensystem five-wire two-phase system
fünfpolige Klinke five-point jack
Fünfpolmischröhre pentode mixer tube
Fünfpolröhre pentode, triple-grid tube, five-electrode valve, quintode
fünfstelliger Logarithmus five-place logarithm
fünfstellige Zahl five digits figure
Funk, Funk-... wireless (Br), radio (US)
Funkamateur radio amateur
Funkanlage radio installation, radio plant
Funkaufklärung radio reconnaissance
Funkausbreitung radio propagation
Funkbake wireless beacon, radio beacon
Funkbereitschaft radio guard, radio alert
Funkbeschickung site error correction, bearing correction, error curve
Funkbild radiophotogram, photoradiogram
Funkbildnavigation Teleran = television radar air navigation
Funkbildübertragung radiophotography
Funkdienst radio communication service
Funkdienstband service band
Funkdisziplin radio discipline
Funkempfang radio reception
Funkempfänger wireless receiver, R. F. receiver, radio receiver
Funkempfangseinrichtung wireles receiving equipment, radio receiving equipment
Funkempfangsstelle radio receiving station
Funkentfernungsmessen radio range finding
Funkentstörung suppression of statics, radio noise suppression, radio interference suppression
Funkerfassung radio detection, radio intercept
Funkfeinhöhenanzeiger terrain clearance indicator
Funkfeld hop
Funkfelddämpfung path loss
Funkfeldlänge path length
Funkfernmessen radio telemetry
Funkfernschreiber radio teleprinter
Funkfernsehen (wireless) television
Funkfernsteuerung radio telecontrol
Funkfeuer radio beacon
Funkgerät radio equipment
Funkgitter-Navigationssystem radio-mesh navigation
Funkhöhenmesser radio altimeter
Funkhorizont radio horizon
Funklandehilfe radio approach aid, radio landing aid
Funkleitstrahl radio beam
Funkmeß radar = Radio Detecting and Ranging (siehe Radar)
Funknachrichtenkanal radio channel
Funknachrichtenwesen wireless telephony and telegraphy, radiocommunication
Funknavigation radio navigation
Funknavigationsgerät radio-aid to navigation
Funknavigationshilfe radio-navigation aid
Funknavigationssatellit radio navigation satellite
Funknetz radio net(-work)
Funkortung radio location, wireless fix
Funkpeiler radio compass, wireless direction finder
Funkpeilung radio direction finding, radio bearing
Funkrichtstrahl wireless beam
Funksender wireless transmitter, radio transmitter
Funksicherungsbake radio-range beacon
Funksprechen wireless telephony, radiotelephony
Funksprechgerät radiotelephone
Funksprechverkehr radiophony, radiotelephony
Funkspruch radio message
Funkstandlinie position line

Funkstation wireless station, radio station
Funkstelle radio station
Funksteuerung radio control
Funkstille radio silence, silent period
Funkstörungen atmospheric, statics, radio interference (sing.); jamming (623)
Funkstörmesser radio-interference meter
Funkstörpegel radio noise level
Funksuchempfang radio search
Funktäuschung radio deception
Funktechnik wireless engineering, radio engineering
Funktelefon radiophone, transceiver
Funktelefonie radiophony, radiotelephony, space telephony
Funktelegramm radiogram, marconigram, wireless message
Funktelegraphie Hertzian telegraphy, wireless telegraphy, radiotelegraphy
Funkübertragung wireless transmission, radio transmission
Funkübertragungsweg radio link
Funkverbindung, Funkverkehr radio communication, wireless communication, radio link, radio circuit
Funkwarnung radio warning
Funkwellen radio waves
Funkwettervorhersage ionospheric forecast, radio-propagation prediction
Fünkchen sparklet
Funke, Funken spark
Funken bilden to spark
Funken springen über sparks jump
einen Funken ziehen to strike spark
Funkenableiter gap arrester
Funkenabtragung spark erosion, spark machining
Funkenauswurf sparking (625)
Funkenbildung sparking
Funkenentladung spark discharge
Funkenfang spark catcher, spark arrester (625)
Funkenflug firespread
funkenfrei sparkless
funkenfreie Kommutierung sparkless commutation, sparkless collection
Funkenfrequenz spark frequency
Funkeninduktor induction coil
Funkenlöschung spark quenching (621.318.5), arc suppression; spark absorbing (621.318.5)
Funkenlöscheinrichtung spark quenching device
Funkenmikrometer spintharometer, micrometer gap
Funkenphotographie spark photography
Funkenschlagweite spark length
Funkenschreiber spark recorder
Funkensender spark transmitter
Funkenspannung sparking voltage, sparking potential
Funkenspektrum spark spectrum
Funkenspektroskopie spark spectroscopy
Funkenstrecke spark gap
Funkenstrecken-Überspannungsschutz spark-gap protector
Funkentelegraphie radiotelegraphy, spark system of wireless telegraphy
Funkenwiderstand spark resistance
funkeln to flicker, to twinkle, to sparkle, to scintillate, to glisten
funken to radio, to wireless (621.396); to spark (621.31)
Funktion function
Funktion ausüben to perform a function
Funktion der Zeit time function
funktionsbestimmendes Element resolver (681)
Funktionsdifferentialgeber resolver differential transmitter
Funktionsdrehmelder resolver, resolver synchro
Funktionsempfänger resolver receiver, resolver transformer
Funktionsgenerator function generator, waveform generator (621.397)
Funktionstabelle truth table (621.374)
Funktionstaste function key
Funktionsteilung function sharing
Funktionsverstärker operational amplifier
funktional, funktionell functional (54)
funktioneller Baustein tile with electrical function (621.316.5)
funktionieren to function, to go round
Fusion amalgamation (54)
fusionieren to amalgamate (54); to merge (33)
Fuß foot (pl. feet) (61); pedestal, base, socket (621)
Fußantrieb treadle operation
Fußbetätigung actuation by foot
mit Fußbetätigung, fußbetätigt foot-operated
Fußboden floor
Fußbodenschalter floor contact
Fußbremse foot brake, pedal brake
Fußgänger pedestrian
Fußgänger-Verkehrszeichen Belisha beacon
Fußhebel foot lever, foot treadle
Fußhülse socket
Fußlager socket, step bearing
Fußpuffer rubber foot

Fußpunkt, am at base
Fußpunkt des Lotes foot of the perpendicular
Fußraste foot rest
Fußschalter foot switch, switch pedal
Fußschaltung foot control
Fusseln fluff (sing.)
Futter lining (66), chuck (621.9)
Futter mit Einzelbacken independent jaw chuck
füttern to line

G

g = Gramm gr. = gram, gramme
g = Güte, Spulengüte Q = quality (621.396)
GaAs-Laser = Gallium-Arsenid-Laser Ga-As laser, gallium-arsenide laser
Gabel fork; cradle, receiver rest (621.395), hybrid (621.392)
Gabelabgleich hybrid balance
Gabeldämpfung effective attenuation of a termination circuit
Gabeldrossel hybrid coil (621.395)
Gabelkontakt bifurcated contact
Gabelleitung tied feeder
Gabelmuffe bifurcating box
Gabelschalter switch cradle, cradle switch
Gabelschaltung hybrid termination (circuit), hybrid four wire terminating set
Gabelschlag hook flash
Gabelspule hybrid coil
Gabelstapler fork truck
Gabelübergangsdämpfung transhybrid loss, insertion loss
Gabelübertrager hybrid transformer
Gabelumschalter switch hook
Gabelverstärker hybrid repeater, termination amplifier
gabeln, sich to furcate, to bifurcate, to branch
gegabelt furcate
Gabelung furcation, bifurcation, fork
Gallert jelly
Gallium gallium
Gallone gallon (s. Anh.)
Gallsche Kette Gall chain, roller chain
Galvanik-Gleichrichter electro-plating rectifier
Galvanisation galvanization
galvanisch galvanic, voltaic
galvanisch behandelt plated
galvanisch gekoppelt d.c. coupled, direct coupled
galvanisch vernickelt nickel-plated
galvanisch verstärken to increase by electrodeposition
galvanisch verzinkt electrogalvanized
galvanische Abscheidung electrodeposition
galvanisches Bad electroplating bath
galvanische Batterie galvanic battery
galvanische Beizung electrolytic pickling
galvanischer Bildtelegraph fultograph
galvanisches Element galvanic element, voltaic cell
galvanische Kopplung direct coupling
galvanischer Korrosionsschutz cathodic protection
galvanische Metallabscheidung electroplating
galvanische Spannungsreihe electromotive series
Galvaniseur galvanist; galvanizer, galvanizing device (621.35)
Galvanisieranstalt electroplating plant
Galvanisierbottich plating vat
galvanisieren to galvanize (61); to plate, to electroplate (621.3)
Galvanisiergehänge plating rack
Galvanisierung galvanization (61); (electro)plating (621.3)
Galvanisierung mit bewegten Elektroden mechanical plating
Galvanismus galvanism
Galvanometer galvo = galvanometer
Galvanometer mit Nullpunkt in der Mitte zero-centre galvanometer
Galvanometer mit photoelektrischer Verstärkung microgalvanometer
Galvanometer-Nebenwiderstand galvanometer shunt, galvanometer resistance
Galvanoplastik galvanoplastiy, electrotyping
galvanoplastisch galvanoplastic
Galvanoskop galvanoscope
Galvano-Therapie galvano-therapy
Gammaquant gamma quantum
Gammastrahl gamma ray
Gammastrahlung gamma rays (pl.), gamma radiation
Gammazerfall gamma decay
Gang gear (621.8); speed (629.113); process, procedure, run, operation (621.7); travel, pass (621.7); thread (621.8); rate (51), (621.3); watch rate (681.1); corridor, passage (72); gangway (629.13)
Gang des Prüfverfahrens test procedure

in Gang setzen to set into action, to start, to actuate
Ganggenauigkeit time-keeping quality (681)
Ganghöhe rise, pitch (621.8)
Ganghöhe einer Schraube pitch of a screw
Gangregulierung timing (of a clock)
gängige Größe standard size
ganz whole, entire; integer, integral (51); intact, unhurt
ganz nach ... drehen to turn all the way to ...
ganz durchmultiplizieren to multiply throughout
ganz eingeschaltet all in
ganz füllen to fill to capacity
ganz gekapselt fully enclosed, totally enclosed
ganz geschweißt all-welded
ganz aus einem Stück all of a piece
ganze Zahl integer, whole number, cardinal number
ganzzahlig integer, integral
ganzzahliger Bruchteil (integral) submultiple, subharmonic
ganzzahliges Vielfaches multiple
ganzzahliger Wert integral value
Ganzlochwicklung integral slot winding
Ganzmetall ... all-metal
Ganzwellendipol full-wave dipole
garantieren to guarantee, to warrant
Garn yarn, thread, twine
gären (lassen) to ferment
Gas gas
Gase abgeben to release gases
Gas absorbieren to absorb gases, to clean up gases
Gase austreten lassen to release gases
Gase entwickeln to evolve gases
Gasabflußrinnen gas grooves (621.357)
Gasabschluß gas seal (621.314.2)
Gasansammlung gas accumulation
Gasanzünder gas lighter
Gasaußendruckkabel external gas-pressure cable
Gasaustritt emission of gas, gas leakage
Gasbeladungselement gas cell
Gasblase (in festen Körpern:) gas blebb, gas pocket; (in Flüssigkeiten:) gas bubble
gasdicht gas-tight, gas-proof
Gasdichte gas density
Gasdiffusion gaseous diffusion
Gasdruck gas pressure
Gasdruckkabel gas-pressure cable
Gasdruckmesser gas-pressure meter
Gaselektrode half cell glass
Gaselement gas cell
Gasentladung gas discharge, gaseous discharge (621.385)
Gasentladungsanzeige gas-discharge display
Gasentladungsgleichrichter gas-filled tube rectifier
Gasentladungslampe vacuum-tube lamp, luminous-gas lamp, gas-discharge lamp
Gasentladungsrohr luminous discharge tube
Gasentladungsrohr zünden to trip a gas tube
Gasentladungsröhre ionic valve, gas (discharge) tube
Gasentladungsventil gas-filled rectifier, thermionic rectifier
Gasentwicklung gas formation, gas development, evolution of gas
gaserfüllter Hohlraum gaseous void (621.315.2); gas-filled cavity
Gasexplosion gas explosion
Gasflasche gas cylinder, storage bin for compressed gases
Gasfokussierung gas focusing (621.385)
gasförmig gaseous, gasiform
gasförmige Materie gaseous matter
gasförmige Verbindung gaseous compound
gasförmiger Zustand gaseous state
in den gasförmigen Zustand überführen to gasify
gasfrei gas-free, clear of gas
gasführend gassy
gasgefüllt gas-filled, gaseous
gasgefüllte Dreipolröhre gas triode
gasgefüllter Gleichrichter soft-tube rectifier
gasgefüllte Lampe gas-filled lamp
gasgefüllte Photozelle soft photoelectric cell
gasgefüllte Röhre gas tube, soft tube, ionic valve
gasgefüllte Röntgenröhre gas X-ray tube
Gasgehalt gas contents (pl.)
gasgekühlter Reaktor gas-cooled reactor
Gashülle gas envelope
Gasinnendruckkabel internal gas-pressure cable
Gasinnendruck-Rohrkabel pipe-type internal gas-pressure cable
Gaskonstante gas constant
Gaskonzentrierung gas focusing (621.385)
Gasmantel gas envelope, gas shield (621.791)
Gasmelder gas alarm
Gasstrahlungsbrenner radiant gas burner
Gastriode gas triode, gas-filled three-electrode valve
Gasturbine gas turbine

Gasverlust loss of gas, gas leakage
Gaszelle gas bag (629.13)
gasen to gas, to give off gas
Gate gate
Gate-Drain-Spannung gate-drain voltage, gate-collector voltage
Gate-Isolierschicht insulating layer of an insulated-gate field-effect transistor
Gate-Leckstrom gate-leakage current
Gate-Reststrom gate cut-off current
Gate-Schaltung common gate
Gatter gate
Gaußmeter gaussmeter
Gaußscher Satz Gauss's theorem
Gaußsche Verteilung Gaussian distribution
Gaze gauze, tissue
Gazebürsten gauze brushes (621.313)
Gazefilter gauze filter
durch Gazeabdeckung schlagwettersichere Maschine gauze-protected machine
geätzte Schaltung etched circuit, etched wiring
Gebäude building, edifice, structure
Gebäudeblitzschutz lightning protection of a building
Geber transmitter (621.3); pick-up (534)
Gebläse fan, (air) blower
Gebläseflügel fan blade
geblechter Kern laminated core
Gebrauch use, usage
Gebrauchsanweisung instructions (pl.) for use, instruction book, user's handbook, owner's manual
gebrauchsfertig ready for use, fit for use, ready-made
Gebühr rate, due, fee, charge
Gebührenanzeige tax indication (621.395)
Gebührenanzeiger charge indicator
Gebühreneinheit unit fee, tariff unit, unit charge
Gebührenerfassung charging
Gebühreninformation tax rate (621.395)
Gebührenminute chargeable minute (621.395)
Gebührenzählung tax metering
Gebührenzone metering zone, tariff zone
gedämpft damped, subdued
gedämpftes Licht subdued light, dimmed light
gedämpfte Schwingung damped oscillation
gediegenes Metall native metal
gedruckt printed
gedruckte Leiterplatte printed board
gedruckte Schaltung printed circuit
Gefahr danger, hazard, risk
Gefahrenquelle hazard
Gefahrschwelle danger threshold, danger coefficient
Gefahrenzone area of risk
gefährden to endanger, to imperil
gefahrlos without danger, safe
Gefälle slope, (down) grade; gradient (51); fall (621.2)
gefalteter Trichter re-entrant horn (621.396.67)
Gefäß vessel, tank, container
Geflecht braid, mesh
Gefrier-... freezing, congealing
Gefrierpunkt freezing point, congealing point
Gefrierpunkterniedrigung depression of the freezing point
Gefrierpunktlötstelle ice-bath cold junction
gefrieren (lassen) to freeze, to congeal
Gefüge structure, texture (53), (66)
Gegenadmittanz counteradmittance (621.372.5)
Gegenamperewindungen back ampere-turns, field ampere-turns
Gegenamt distant exchange (621.395)
Gegenbremse counter brake
Gegendruck counter pressure, backpressure
Gegenecho talker echo (621.395)
gegeneinander abgeschirmt mutually screened
Gegenelektrode backplate electrode
Gegen-EMK back e.m.f. = back electromotive force, counter e.m.f.
Gegengewicht counterpoise, compensating weight, balance weight, counterbalance, earth mat, earth system; capacitance earth, balancing capacitance (621.396)
durch Gegengewichte ausbalancieren to counterpoise
Gegenimpedanz counter impedance (621.372.5)
Gegeninduktivität mutual inductance, mutual inductivity
Gegenkompoundmotor decompounded motor
Gegenkompoundwicklung reverse compound winding, countercompound winding
gegenkoppeln to degenerate
gegengekoppelter Sender antidyne circuit
gegengekoppelter Verstärker stabilized feedback amplifier
Gegenkoppelspannung degenerative voltage
Gegenkoppelung negative feedback, degeneration, degenerative feedback, inverse feedback, reverse feedback
durch Gegenkopplung stabilisierter Verstärker stabilized feedback amplifier
gegenläufig negative-sequence, negative-direction (621.392); returning, backward travelling (621.394); moving in opposite direction(s) (621)

gegenläufiger Antrieb counterdirection drive (621.316.5)
gegenläufige Bürste opposed brush
gegenläufiges Drehfeld contra-rotating field
gegenläufiger Drehfeldsinn reversed phase sequence
gegenläufiges Echo talker echo (621.395)
gegenläufiges Feld negative-sequence field
Gegenmutter checknut
Gegennebenschlußwicklung differential-shunt winding
Gegennebensprechen far-end cross talk
Gegenphase phase opposition
in Gegenphase in phase opposition
gegenpolige Schaltung subtractive polarity
Gegenprobe crosschecking
Gegenreihenschlußwicklung differential-series winding
Gegenrichtung reverse direction
gegenschalten to counterconnect
gegengeschaltete Regelzelle counter electromotive force cell (621.395)
Gegenschaltstück countercontact, mating contact
Gegenschaltung, in in opposition
Gegenschreiben duplex teletypewriter communication
Gegenschreiben ohne Mitlesemöglichkeit blind duplex teletypewriter
Gegenseite opposite (51)
gegenseitig reciprocal, mutual, one another, each other
gegenseitig austauschbar pin compatible (621.374)
gegenseitig auslöschen, sich to cancel out each other
gegenseitig beeinflussen, sich to interact
gegenseitig nicht berühren, sich to clear each other
gegenseitig verriegeln to interlock
gegenseitige Abhängigkeit interdependence
gegenseitiger Abstand distance apart
gegenseitige Abstoßung mutual repulsion
gegenseitige Anziehung mutual attraction, reciprocal attraction
gegenseitige Austauschbarkeit reciprocal replacement
gegenseitige Beeinflussung interaction
gegenseitige Induktion mutual induction
gegenseitige Kapazität mutual capacitance
Gegenseitigkeit reciprocity
gegensinnige Hintereinanderschaltung series opposing
Gegenspannung countervoltage, back potential, back bias, reverse bias
Gegenspannungsregelzelle counter-e.m.f. cell
Gegensperröhre ATR-tube, antitransmit-receive tube
Gegensprechen duplex voice communication, duplex operation
Gegensprechanlage intercommunication equipment (two way), two-way telephone system
Gegenstelle distant station, far station
Gegenstrom counterflow, countercurrent (flow)
im Gegenstrom countercurrently
Gegenstrombremsung plugging
Gegenstromprinzip countercurrent principle
Gegenstromschütz plugging contactor
Gegenstromverfahren countercurrent process
Gegentakt push-pull
Gegentaktausgang push-pull output
Gegentakt-C-Betrieb Q.P.P. = quiescent push-pull (621.396.64)
Gegentaktkraftverstärker push-pull power amplifier
Gegentaktmodulator balanced modulator
Gegentaktneutralisation cross neutralization
Gegentaktröhre push-pull valve
Gegentaktschaltung push-pull circuit, push-pull connection
Gegentakttransformator push-pull transformer
Gegentaktverstärker push-pull amplifier, balanced amplifier
gegenüber opposite (to)
gegenüberliegender Winkel opposite angle
Gegenverbunderregung differential excitation
Gegenverbundmotor decompounded motor
Gegenverbundwicklung differential compound winding
Gegenverkehrsverbindung duplex circuit
Gegenwirkleitwert transconductance
Gegenwirkung counteraction
gehämmert hammer-finished
Gehänge shackle, suspension
Gehäuse cabinet, housing, case, casing, jacket, box, can, package, cage, chassis
Gehäuseerdung equipment earth, frame-grounding circuit
Geheim-. . ., geheim secret
Geheimhaltung secrecy, privacy
Gehörprüfung intelligibility test
gehörrichtig tone-compensated (621.396.62)
gehörrichtiger Lautstärkenregler tone-compensated volume control
Gehörschutz click absorber (621.395)
Geh-Steh start-stop
Geh-Steh-Schreiber start-stop printer
Geh-Steh-Telegraphie start-stop telegraphy

Geiger-Müller-Zählrohr Geiger-Müller tube amplifier
Geigerzähler Geiger counter
Geißlersche Röhre Geissler tube
Geist ghost, ghost image, multiple image (621.397)
Geisterecho ghost echo
gekapselt encapsulated; totaly enclosed (621.313)
gekoppelte Kreise coupled circuits
gekreuzte Wellenwicklung retrogressive wave winding
Gel gel; gelly
Gelände terrain
Geländeschnitt path profile (621.396)
Gelatine gelatine
Gelbmetall yellow metal
Geleise track (s. auch **Gleis**)
Gelenk joint (61), (621)
mit Gelenk verbunden jointed
Gelenkbalken hinged girder
Gelenkkette band chain
Gelenkkupplung jointed coupling
Gelenkverbindung pin joint, knuckle joint
gelieren to congeal
gemeiner Bruch vulgar fraction
Gemeinkosten overheads, indirect costs
gemeinsam joint, common
gemeinsam eine Last speisen to share a load (621.3)
gemeinsame Abnehmerleitung common trunk
gemeinsamer Anschlußpunkt common point (621.3)
gemeinsamer Antrieb group operation (621.316.5)
gemeinsamer Pol common leg (621.372.5)
gemeinsamer Schenkel common limb (621.3)
gemeinsamer Schnittpunkt (common) point of concurrency
mit gemeinsamem Schittpunkt concurrent
gemeinsamer Teiler common factor, common measure, common multiplier
mit gemeinsamem Teiler equimultiple
gemeinsamer Zweig common branch (621.3)
Gemeinschaft community
Gemeinschaftsanschluß apartment satellite exchange, party line, shared line
Gemeinschaftsanschlußleitung party line
Gemeinschaftsantenne communal aerial, apartment-house aerial, community-antenna system, common antenna
Gemeinschaftsarbeit team work
Gemeinschaftseinrichtung shared-line equipment (621.395)
Gemeinschaftsleitung party line, shared line (621.395)
Gemeinschaftswellen-Rundfunk S.B. = simultaneous broadcasting
Gemenge, Gemisch mixture (53), (621.43), (621.396), (666)
gemischtadriges Kabel composite cable
genau exact, precise, accurate, sharp, close
genau festliegend definite
genau gehen to keep time (681.11)
genau nach Muster true to type
genau passend close-fitting
genauer Abgleich exact balance
genaue Abstimmung sharp tuning
genaue Analyse close analysis
genaue Durchsicht perusal
genaue Lage truth
genaue Lösung exact solution (51)
genaue Messung close measurement
genaue Untersuchung scrutiny
Genauigkeit exactitude, precision, accuracy, truth
auf Genauigkeit prüfen to check for truth
mit einer Genauigkeit von within a margin of
Genauigkeitsprüfung test of accuracy
genehmigte Frequenz authorized frequency
Generaldirektor general manager, managing director, president
Generalnenner, (kleinster) L.C.M. = lowest common multiple, least common multiple
Generator generator (621.3); producer, generator (66)
Generator mit eingeprägter Spannung constant-voltage generator
Generator mit eingeprägtem Strom constant-current generator
Generator mit feststehendem Feld revolving armature alternator
Generator-AS = Generator Antriebsseite back of a generator
Generator-ES = Generator-Erregerseite front of a generator
Generatorbremsung regenerative braking
Generatorgas producer gas
Generatorklemme generator terminal
Generatorleistung generating capacity, generator output
Generator-Sammelschiene generator bus-bar
Generatorensatz generating set (621.31)
Generatorseite sending end (621.392)
Generatorschutz generator protection
Generatorstoßstreureaktanz subtransient reactance (of generator)

Generatorstromregelung total current regulation (621.32; 621.33)
Generatorwicklung generator winding
genulltes Netz multiple-earthed system; multiple-grounded system (US)
geo-elektrisch geo-electric
geomagnetisch geomagnetic
geomagnetische Erzsuche geomagnetic surveying
Geometer geometrician
Geometriefehler geometrical distortion (621.397)
geometrisch geometric(al), physical
geometrische Abmessung physical dimension
geometrisches Mittel geometric mean
geometrischer Ort (geometric) locus
geometrischer Punkt geometrical point
geometrische Reihe geometrical progression
Geophysik geophysics
geordneter Absuchvorgang sequential hunting (621.395)
geordnete Belastungskurve load-duration curve
geothermisches Kraftwerk geothermal power station
geozentrisch geocentric
gepolter Kondensator polar capacitor
gepoltes Relais polarized relay
gepolter Stecker polarized plug
Gerade straight line (51)
gerade even, rectilineal, rectilinear, straight (51)
gerade Kante straight edge
gerades Leitungsstück straight run
gerade Linie straight line
gerade Parität even parity (681.3)
gerade Zahl even number
geradebiegen to unbend
geradflankig straight-sided
geradlinig straightline, rectilinear
geradlinige Bewegungen straightline motion, rectilinear motion
geradliniger Teil straightline portion
geradliniger Teil der Leerlaufkennlinie air-gap line (621.313)
geraderichten to straighten
geradzahlig even(-numbered)
geradzahlige Oberwellen even-order harmonics
von geradzahliger Ordnung even order
Geradeaus-Empfänger straight receiver, direct receiver, tuned R.F. receiver
geraffte Lebensdauerprüfung accelarated life test
Gerät device, set, equipment; implement; appliance (643); instrument (621.317)
Gerät in Reserve standby equipment (sing.)
Gerät mit guter Tonwiedergabe high-fidelity set = hi-fi set
Geräteanschlußschnur flexible lead
Gerätebaugruppe subrack, package
Gerätefehler instrumental error
Gerätegestell rack
Geräteinnenraumklima equipment internal climate
Gerätekasten equipment box (621.315.6)
Geräteschrank cabinet
Gerätesteckdose coupler socket, (appliance) connector
Gerätestecker coupler plug, connector, appliance inlet
Gerätesteckvorrichtung coupler
Gerätestromkreis appliance branch circuit
Geräteteil equipment item
Geräteuntersuchung set analysis (621.396)
Geräusch noise
Geräusch durch Einschwingvorgänge transient noise
Geräuschabsorber noise absorber
Geräuschabstand signal-to-noise ratio
Geräuschanalyse noise analysis
Geräuschbewertungsfaktor psophometric weight
Geräuschbewertungskurve psophometric weighting curve
Geräuschdämpfung noise absorption
geräuschlos noiseless, quiet, silent
geräuschloser Betrieb quiet operation
geräuschloser Lauf silence in operation
Geräuschlosigkeit silence
Geräuschmeßgerät noise meter
Geräuschpegel noise level, audible sound level
Geräuschspannung noise voltage; weighted noise, psophometric noise (621.392)
Geräuschspannungsabstand signal-to-noise ratio
Geräuschstärke audible sound level
Geräuschunterdrückung noise suppression
geräuschvoll noisy
geregelt regulated, controlled, stabilized
geregelte Größe controlled variable
geregeltes Netzgerät stabilized power supply
geregelte Spannung regulated voltage
gerichtet directional; pointing (at)
gerichtet sein nach to point at
gerichteter interrupt vectored interrupt
gerichtete Strecke vector (line)
gerichtet betriebene Leitung oneway trunk (621.395)
gering small, little; feeble; close; low
geringer Abstand close spacing

geringe Bandbreite narrow band
geringer Elektrodenabstand close electrode spacing
geringe Entfernung short distance
geringe Geschwindigkeit low speed
geringe Qualität low grade, poor quality
geringe Spannung feeble voltage
geringer Widerstand small resistance
geringster Widerstand least resistance
geringfügig faint, slight, minor
geringfügige Verschiebung slight displacement
gerippt ribbed, finned
Germanium germanium
geröbelte Wicklung transposed winding
Geruch smell, odour (**nach**: of)
geruchlos smell-less, odourless
Gerumpel rumbles (pl.), rumble, rumbling (621.396)
Gerüst scaffold, structure, skeleton (693)
Gesamt-... total, grand, overall; overhead, general
Gesamtansicht general view
Gesamtausfall total failure
Gesamtbelastung total load
Gesamtbestand total stock
Gesamtbezugsdämpfung total-loss reference-equivalent
Gesamtdämpfung total loss, attenuation lenght (621.395)
Gesamtdicke overall thickness
Gesamtfläche aggregate area
Gesamtfluß total flux
Gesamtgehalt an Oberwellen total harmonic content
Gesamtheizfläche total heating surface
Gesamthöhe overall depth
Gesamtlänge overall lenght, pitch
Gesamtlaufzeit eines Transponders interrogation suppressed time delay
Gesamtleistung total output
Gesamtleistungsfeld total output panel (621.311)
Gesamtlichtbogenspannung true arc voltage
Gesamtmaß overall dimension
Gesamtreaktionszeit total response time
Gesamtspannung total voltage, overall voltage
Gesamtspannungsabfall impedance voltage
Gesamtstärke overall thickness
Gesamtstrahlung total emissivity
Gesamtstrom line current
Gesamtsumme (sum) total
Gesamt- und Überverbrauchszähler excess and total meter
Gesamtverlust total loss
Gesamtverlustleistung total power loss
Gesamtverstärkung overall amplification, overall gain
Gesamtverzerrung total distortion
Gesamtvolumen bulk
Gesamtwert integral value
Gesamtwiderstand total resistance (621.3)
Gesamtwirkungsgrad total efficiency, overall efficiency
Gesamtzahl (grand) total
geschaltete Fernsprechleitung leased telephone circuit
geschirmt screened
geschirmtes Kabel screened cable, shielded cable
geschirmter Meßraum test cage, screened cabin
Geschirr tackle, gear
geschlitzte Hohlrohrleitung slotted line
geschlossen closed, enclosed
geschlossene Dämpferwicklung interconnected damper winding
geschlossene (Regel-)Schleife closed loop, feedback control loop
geschlossene Sicherung fuse with enclosed fuse element
geschlossene Teilnehmergruppe closed subscriber group
geschlossene Wicklung closed-coil winding, ring winding
geschmeidig ductile
Geschoßverfolgung shell tracking
geschützt protected
geschützt gegen Tropfwasser und Berührung drip-proof screen-protected
geschützt gegen zufällige Berührung protected from accidental contact
geschützte Datenverbindung protected data connection
geschützte Frequenz protected frequency
Geschwindigkeit rapidity, speed, velocity, rate (of motion)
Geschwindigkeit Eins unit velocity
Geschwindigkeit der Erregerspannungsänderung exciter response
Geschwindigkeit über Grund ground speed (629.13)
Geschwindigkeit der Sedimentierung sedimentation range
Geschwindigkeit des Temperaturanstiegs rate of temperature rise
Geschwindigkeit des Zerfalls velocity of decay
mit einer Geschwindigkeit von at a rate of

Geschwindigkeitsbeschränkung speed limit
Geschwindigkeitskomponente component velocity
Geschwindigkeitsmesser speed indicator, speedometer, tachometer, speed meter
Geschwindigkeitsmodulation velocity modulation
Geschwindigkeits-Modulationsgitter bunched electrode
geschwindigkeitsmoduliert velocity modulated
Geschwindigkeitsregelbereich range of speed control
Geschwindigkeitsregler speed controller
Geschwindigkeitsschwankung speed variation
Geschwindigkeitssteuergröße rate signal
Geschwindigkeitsverteilung velocity distribution (53); energy spectrum (539)
Geschwindigkeit-Weg-Diagramm speed-distance curve
gesehnte Wicklung chord winding (621.313)
Gesellschaftsanschluß party line
Gesenk die, swage block
gesenkschmieden to swage, to drop-forge
Gesetz law, principle, theorem
Gesetz von der Massenkonstanz law of mass conservation
Gesetz der Wahrscheinlichkeit probability laws
gesetzlich legal
gesetzlich festlegen to legalize
gesetzliche Einheit legalized unit
gesetzliches Ohm legal ohm, Board of Trade ohm
gesicherte Datenübertragung protected data transmission
Gesicht face
Gesichtsfeld field of vision, visual field
Gesichtskreis horizon, range (of sight)
aus dem Gesichtskreis entschwinden to drop out of sight
Gesichtssinn sight, vision
gespeichert stored
gespeistes Antennenelement driven element
Gespräch call (621.395); talk
Gespräch zur Leitungsprüfung test call
Gespräch mit Voranmeldung preavis call, prearranged call
Gespräch ohne Vorrang nonprecedence call
Gesprächsanmeldung call booking
Gesprächsbeginn beginning of conversation
Gesprächsdauer conversation time
Gesprächsdauerzähler calculagraph
Gesprächseinheit unit call
Gesprächsende end of conversation
Gesprächsgebühr call fee
Gesprächsteilnehmer partner
Gesprächszähler message register, call meter, telephone meter, conversation counter, call counter
Gesprächszähltaste register key, meter key
Gesprächszeitmesser call meter, timing register
Gesprächszettel (call) ticket
Gestalt shape, form
gestalten to shape, to form, to frame
Gestell pedestal, stand; frame; bay, rack (621.3)
Gestellbock trestle; derrick
Gestelldrossel earth-fault reactor (621.313)
Gestellisolator accumulator insulator (621.35)
Gestell-Lampe bay lamp
Gestellschlußschutz Howard protective system
gesteuert guided; controlled (621.3)
gesteuerter Flugkörper guided missile
gesteuerter Siliziumgleichrichter silicon controlled rectifier, SCR
gestreckter Dipol plain dipole
gestrecktes Koppelnetz straight-forward network
gestreute Beleuchtung diffused lighting
gestreutes Licht scattered light, diffused light
getastete Rückstrahlantenne coded passive reflector antenna
geteilter Bildschirm split screen
geteiltes Gehäuse split case
geteilter Kontakt bifurcated contact
getränkte Isolation impregnated insulation
geträufelte Wicklung drop-in winding
getreu faithful, true
getreu aufzeichnen to record faithfully
getreue Wiedergabe fidelity (534.86), (621.396.62)
Getriebe gear, gearing
Getriebe mit Gangvorwahl preselective gearing
Getriebe schalten to shift the gear
Getriebemotor geared motor
Getriebeöl gear-box oil
Getriebewelle transmission shaft
Getter getter (621.385)
Getter abschließen to flash a getter
Getucker chug (621.4); put-put, motor-boating (621.396)
Gewährleistung warranty
gewaltsam by force
gewaltsam trennen to dislocate
gewaltsame Trennung dislocation
Gewebe tissue (61), (67); fabric, cloth (67)
gewellt waved, corrugated
Gewicht weight, poundage (531.5); bob (621)
Gewichte auflegen to weight
Gewicht und Raumbedarf weight and space
Gewichtsabnahme decrease in weight

Gewichtsausgleich equilibration
gewichtsbehaftet poised
gewichtslos weightless, impalpable
Gewichtsverlust weight loss, loss of weight
Gewichtszunahme increase in weight
Gewinde thread (621.8)
Gewinde hinterdrehen to neck down the thread
mit Gewinde versehen to thread
Gewindeanschnitt start of a tap
Gewindebohrer tap
Gewindeflanke thread flank
Gewindefräser thread milling cutter
Gewindeloch tapped hole
Gewindemuffe threaded collar, threaded joint
Gewinderollmaschine screwing machine
Gewindeschneider thread tap
Gewindeschneidkluppe screw plate stock
Gewindesockel screwed holder
Gewindespindel screw spindle
Gewindestift threaded pin, grub screw
Gewindeteil threaded part
Gewinn (power) gain (621.396.67), profit (33)
gewinnen to extract, to mine (622), to recover (66)
Gewinnung extraction, mining (622), recovery (66)
Gewitter storm, thunderstorm
Gewitterortung thunderstorm location
Gewitterwolke thunder cloud
Gewohnheit routine, practice, custom
gewöhnlich ordinary, customary, plain
gewöhnliches Gleitlager plain bearing
gewöhnliches Gußeisen plain cast iron
gewöhnliche Sicherung plain fuse
gewöhnliche Temperatur ordinary temperature
gewogenes Mittel weighted average, effective average
gezackt, gezahnt dented, serrate, serrated
Gezeiten tide (sing.)
Gezeitenenergie tidal power
Gezeitenkraftwerk tidal-power plant, tidal-power station
gezogener p-n Übergang grown junction (621.382)
Gichtgas waste gas, blast-furnace gas, furnace mouth gas
Gießbeton poured concrete
Gießharz epoxy resin, epoxyde (resin), cast resin
Gießmasse moulding compound
Gießtemperatur pouring temperature
gießbar castable
gießen to cast, to found, to pour (621.74)
Gießerei foundry, moulding shop
giftig poisonous, toxic
Gill gill (= 1/4 pint = 142 cm^3)
Gi-Metall rubber-bonded to metal
Giorgisches Maßsystem Giorgi system of units
Gipfel crest, summit
Gipfelhöhe ceiling
Gipfelpunkt culmination, climax, turning point; peak point (621.382)
Gipfel-Tal-Stromverhältnis peak-to-valley point current ratio
gipfeln to culminate
Gips gypsum, plaster (693); plaster of Paris (61)
Gitter grid (621.385); lattice (72), (54), (624); grate (68); grating (535), (72); fence (621.9)
Gitter beaufschlagen to drive a grid
Gitter negativ machen to place a grid negative
Gitter steuern to drive a grid
Gitter ins stromlose Gebiet aussteuern to drive a grid below cut-off
dem Gitter zuführen to inject on the grid (621.385)
Gitterableitung grid leak, grid return
Gitterableitwiderstand grid-leak resistance
Gitterablenkröhre Lawrence tube, chromatron (621.397)
Gitterabstand lattice pitch
Gitterabstimmkreis tuned grid
gitterähnlich grid-like
Gitteramplitude grid-voltage swing, grid sweep
Gitteranodenkapazität (innere) grid-plate tube capacity
Gitteranschluß grid terminal
Gitterbasisschaltung grounded-grid operation
Gitterbatterie grid (bias) battery, C-battery (621.396.64)
Gitterblock grid capacitor, grid-blocking condenser
Gitterblockierung wipe-out (621.396.62)
Gitterdämpfungswiderstand (series) grid resistor, grid suppressor
Gitterdraht grid wire
Gitterdrossel grid choke
Gitteremission grid emission
Gittererder ground mat
Gitterfehler lattice defect, lattice imperfections (621.382)
Gittergleichrichter (leak) grid detector, grid rectifier
Gitterkappe grid cap, to p cap (621.385)
Gitterkathodenkapazität grid-cathode capacity
Gitterklemme grid terminal
Gitterkondensator grid capacitor

Gitterkonstante grating constant, grating space (535); lattice parameter (548)
Gitterkreis grid branch, grid circuit
Gitterlücke vacancy (621.382)
Gittermast lattice mast (621.315)
Gittermodulation grid modulation
gittermoduliert grid modulated
Gittermuster cross hatch (621.397)
Gitterplatte grid plate, accumulator, grid (621.35)
Gitterpotential grid potential
Gitterschwingkreis tuned-grid circuit, grid tank
Gitterspannung grid bias, grid voltage (621.385); exciting voltage (621.314.6), signal (621.385)
  durch Gitterspannung sperren to drive the grid below cut-off
  Gitterspannungserzeugung grid-bias supply
Gittersperrspannung cut-off bias, cut-off grid potential (621.385)
Gitterstabwicklung transported winding
Gittersteuerleistung grid-driving power
gittergesteuert grid controlled
  gittergesteuerter Gleichrichter grid-controlled rectifier
Gitterstörstelle lattice deformation, lattice imperfection (621.382)
Gitterstrich ruling (535)
Gitterstrom grid current
  Gitterstromkennlinie grid-current characteristic, grid-current curve
  Gitterstrom-Modulation grid-current modulation
Gittertastung grid keying (621.385/396)
Gitterverformung deformation of the lattice (54)
Gitterverlustleistung grid dissipation
Gittervorspannung grid biasing voltage, grid potential
Gittervorwiderstand (series) grid resistor
Gitterwechselspannung signal voltage, grid sweep
  Gitterwechselspannung anlegen to apply a signal to the grid
Gitterwechselstrom grid a.c.
  Gitterwechselstromleistung grid-driving power
  Gitterwechselstromwiderstand grid a. c. resistance
Gitterwiderstand grid resistance; grid suppressor, grid resistor
Gitterzündspannung critical grid voltage
Gitterzweig grid branch; grid circuit
Gl. = Glchg. = Gleichung eq. = equation
Glanz gloss, lustre, glare; brilliance, brilliancy (77), (535)
  Glanzgalvanisieren bright electroplating
  Glanzgarn rayon (yarn), mercerized cotton
  Glanzmeßgerät glossimeter
  Glanzplattieren bright electroplating (621.35)
  Glanzreflex reflected glare (535)
  Glanzstellen high lights (621.357)
  Glanzwinkelreflexion specular reflection
Glänzbad bright dip (621.35)
  Glänzblech glazing sheet (77)
glänzen to flash, to brighten, to glance; to gloss, to be bright
  glänzende Oberfläche high finish, bright finish (621.7)
  glänzende Politur high polish
Glas glass
  Glas blasen to blow glass
  Glasbehälter glass container, glass jar
  Glasdurchschlag puncture (621.385)
  Glaselektrizität vitreous electricity
  Glaselektrode glass electrode, glass half cell
  Glasfaser optical fibre, glass fibre
    Glasfaserverbindung fibre link
  Glasflasche glass flask; glass jar
  Glasgewebe woven glass
  Glaskolben glass bulb
  Glaskörper vitreous humour (61)
  Glaspapier glass paper
  Glasperle glass bead
  Glasrohrsicherung tube fuse
  Glasscheibe glass pane, glass plate
  Glasseide spun glass
  Glaswalze, Glaszylinder cannon (66)
  in eine Glaswand einschmelzen to seal through glass (621.385)
  Glaswolle glasswool, spun glass
glasieren to glaze
Glasierung glassivation
glasig glassy, vitreous
  glasig werden to glaze
Glasur glaze
glatt even, smooth
  glatter Anker smooth core armature (621.313)
  glatte Oberfläche flat surface, smooth surface
  glatte Verbindung straight-through joint
Glattschirmisolator plain-shed insulator
Glätte smoothness; polish, finish
glätten to smooth (621.3), (621.9), to finish; to iron out, to equalize
  glättende Wirkung ironing effect
Glättung smoothing (621.3), (621.9)

Glättungsdrossel smoothing choke, smoothing coil
Glättungsfaktor ripple rejection (621.314)
Glättungskondensator smoothing capacitor, filter capacitor
Glättungsschaltung filter circuit, smoothing circuit, stabilizer
Glättungswiderstand smoothing resistance
Glättungswirkung smoothing effect
Glaubersalz Glauber's salt, sodium sulphate
gleich equal, like; same; is, makes (51)
gleich hoher Ton same-pitched note
gleich schwer bleiben to keep the same weight
gleich weit entfernt equidistant
gleichen Abstand haben to be equidistant
gleiche Beleuchtungsstärke isolux
gleiche Pole stoßen sich ab like poles repel
gleiches Vorzeichen haben to be the same polarity
in gleichem Abstand equidistant, equally spaced
auf gleicher Ebene in the same plane, co-planar
auf der gleichen Geraden co-linear
in gleichem Maße wie in sympathy with
zu gleichen Teilen volume for volume, share and share alike
gleichabständig equally spaced
gleichachsige Anordnung in-line construction
gleichen to resemble, to be equal (to)
Gleichenergieweiß equal-energy white (621.397)
gleichfarbig isochromatic
gleichförmig uniform
gleichförmig beschleunigt uniformly accelerated
gleichförmig verzögert uniformly retarded
gleichförmige Bewegung uniform motion
Gleichförmigkeit uniformity
Gleichförmigkeitsgrad degree of uniformity
gleichgerichtet unidirectional (621.3), rectified (621.314)
gleichgerichteter Strom unidirectional current, rectified current (621.314)
gleichgerichtete Ströme like currents
gleichgestaltet isomorphous
Gleichgewicht equilibrium, balance
Gleichgewicht gegen Verdrehung rotational equilibrium
Gleichgewicht gegen Verschiebung positional equilibrium
aus dem Gleichgewicht out of balance
aus dem Gleichgewicht bringen to unbalance
im Gleichgewicht in a state of equilibrium
ins Gleichgewicht bringen to equilibrate, to balance
Gleichgewichtsbedingung condition of balance
Gleichgewichtsgeschwindigkeit balancing speed, free running speed (625)
Gleichgewichtskonzentration equilibrium concentration
Gleichgewichtslage position of equilibrium poise
Gleichgewichtszustand steady state
Gleichheit equality
Gleichheitszeichen equality sign
Gleichkanal... common-channel
Gleichkanalbetrieb same channel operation, co-channel operation
Gleichkanalsender co-channel transmitter
Gleichkanalstörung common-channel interference
Gleichklang unison
Gleichlauf synchronism, isochronism (621.31), alignment (621.396)
auf Gleichlauf abstimmen to align, to gang
Gleichlaufabstimmung alignment of tuned circuits, tracking (621.396.62)
Gleichlaufabstimmelement two-circuit tuner
Gleichlaufkondensator padding capacitor (621.396)
Gleichlaufkontrolleinrichtung scrutineer (621.394)
Gleichlaufpilot synchronizing pilot (621.395.4)
Gleichlaufschwankung flutter and wow (681.84)
Gleichlaufselbststeuerung autoscrutineer
Gleichlauftrimmer tracking capacitor, padding capacitor
Gleichlaufzusatz tracking capacitor, padding capacitor (621.396)
gleichlaufen to track
gleichlaufend concordant (51), synchronous (6)
gleichmachen to equalize
gleichmäßig steady, even, uniform
gleichmäßig aufteilen to proportion uniformly
gleichmäßig geteilte Skala scale of equal parts
gleichmäßig im Raum verteilt evenly spread throughout the space
gleichmäßig wiederkehrender Arbeitsvorgang repetition job
gleichmäßige Abnutzung even wear
gleichmäßiges Andauern permanence
gleichmäßige Last steady load
gleichmäßiger Lauf smooth running
gleichmäßiges Tempo steady pace
in gleichmäßigen Abständen evenly spaced

Gleichmäßigkeit uniformity, permanence
gleichnamig like
gleichnamige Pole like poles
gleichnamige Pole stoßen sich ab like poles repel
gleichphasig in-phase, co-phase, equiphase
gleichphasige Komponente co-phase component, in-phase component
gleichphasige Spannung in-phase voltage
gleichphasiger Strom in-phase current
gleichpolig homopolar
Gleichpolmaschine homopolar generator, homopolar alternator
gleichrichten to rectify (621.3)
Gleichrichter rectifier (621.3), detector (621.396), straightener (621.63)
Gleichrichter mit Außengittersteuerung cathetron
Gleichrichter in Brückenschaltung bridge connected rectifier
Gleichrichter mit hoher Leistung power rectifier
Gleichrichter mit reversibler Durchbruchcharakteristik avalanche rectifier
Gleichrichter mit Zündstift ignitron
Gleichrichteranode rectifier anode
Gleichrichterbrücke bridge rectifier
Gleichrichterdiode rectifier diode, diode, detector
Gleichrichtergerät rectifier, rectifying device, rectifier cubicle
Gleichrichterkathode rectifier cathode
Gleichrichterkennlinie rectification curve, rectifier characteristic (621.385)
Gleichrichterkreis rectifier circuit
Gleichrichtermeßinstrument rectifier(-type) instrument
Gleichrichterphotozelle rectifier photo-electric cell
Gleichrichterröhre rectifier valve, vacuum-tube detector, anode converter
Gleichrichtersatz rectifier unit, rectifier set
Gleichrichtersäule rectifying stack, rectifier stack
Gleichrichterschaltung rectifier circuit, rectifier connection
Gleichrichterscheibe rectifier disc
Gleichrichterstation rectifier substation
Gleichrichterstufe rectifier stage
Gleichrichtertransformator rectifier transformer
Gleichrichterventil valve (621.386)
Gleichrichterwirkung rectifying effect
Gleichrichterzelle rectifier cell
Gleichrichtung rectification (621.3), detection (621.396)
gleichschenklig isosceles
gleichschenkeliges Dreieck isosceles triangle
gleichschenkliges L-Eisen equal angle
gleichseitig equilateral
gleichseitiges Dreieck equilateral triangle
gleichsetzen to equate, to equal (51)
gleichsinnig concordant, unidirectional (621.3)
gleichsinnige Hintereinanderschaltung series aiding connection
Gleichspannung d.c. (= direct current) voltage, d.c. potential, c.c. (= continuous current) voltage
Gleichspannungsverstärker d.c. amplifier
Gleichsperrspannung continous reverse voltage, direct off-state voltage
Gleichstrom d.c. = direct current, c.c. = continuous current
bei Gleichstrom under d. c. conditions
Gleichstromanlage d. c. plant, continuous current plant
Gleichstromanteil steady component, d. c. component
Gleichstrombetrieb d. c. operation
Gleichstromdoppelschlußgenerator d.c. compound-wound generator
Gleichstrom-Dreileiternetz d. c. neutral grid, d.c. three-wire system
Gleichstrom-EMK direct electromotive force
Gleichstromempfänger d. c. receiver
Gleichstromfernwahl d. c. dialling
gleichstromfreie Übertragung a. c. transmission (621.397)
Gleichstromgenerator d. c. generator, d. c. dynamo
Gleichstrom-Gleichstrom-Umformer rotary transformer, motor-transformer
Gleichstrom-Hauptschlußmotor d. c. series-wound motor
Gleichstromklingel trembler bell (621.395), battery bell
Gleichstromkomponente d. c. component, steady component
Gleichstromleistung d. c. power
Gleichstromlichtbogen d. c. arc
Gleichstrom-Lichtbogenschweißung d. c. arc welding
Gleichstrommaschine d. c. machine
Gleichstrommotor d. c. motor
Gleichstromnebenschlußgenerator d. c. shunt-wound generator
Gleichstromnetz d. c. mains

Gleichstrom-Potentialdifferenz direct potential difference
Gleichstromreihenschlußgenerator d. c. series-wound generator
Gleichstrom-Schweißgenerator d. c. arc-welding generator
Gleichstromschweißung d. c. arc welding
Gleichstromtelegraphie d. c. telegraphy, continous current telegraphy
Gleichstromübertragung d. c. transmission
Gleichstromverhältnis static value of the forward current transfer ratio (621.382)
Gleichstromverstärker d. c. amplifier, direct resistance-coupled amplifier
Gleichstromverteilung d. c. distribution
Gleichstrom-Vormagnetisierung polarizing magnetization
Gleichstromwandler d. c. static converter
Gleichstromwecker trembler bell (621.395), battery bell
Gleichstromwicklung d. c. winding, closed-coil armature
Gleichstromwiderstand d. c. (plate) resistance
Gleichstromzähler d. c. meter
Gleichstromzuführung steady supply
Gleichtakt unison, in-phase
Gleichtakteingang common mode input
Gleichtakt-Signal common-mode signal, in-phase-signal
Gleichtaktunterdrückung common mode rejection ratio
Gleichtaktverstärker in-phase amplifier
Gleichung equation
Gleichung ansetzen to write down an equation
Gleichung höheren Grades equation of higher degree
Gleichungen lösen to solve equations
Gleichung in Polarkoordinaten polar equation
Gleichungen mit 6 Unbekannten simultaneous equations in 6 unknows
Gleichungssystem set of equations
Gleichwellenfunk S.B. = simultaneous broadcasting
Gleichwert mean value (of a periodic quantity)
gleichwertig equivalent, alternate
gleichwinklig equiangular
gleichzeitig simultaneous, concurrent
gleichzeitig ereignen, sich to concur
Gleichzeitigkeit simultaneity, concurrence
Gleichzeitigkeitsfaktor diversity factor (621.396), load diversity (621.31)
Gleis track, rail tracks (pl.)
Gleisbildstelltisch (für Eisenbahngleisanlage) track-diagram control panel (for railway interlockings)
Gleisbildstellwerk track-diagram switchbox
Gleisbildtafel track-indicator panel
Gleisbremsanlage car-retarder mechanism
Gleiskontakt rail contact
Gleismagnet track magnet
Gleisrelais track relay
Gleisrückleitung track return circuit
Gleisverbinder intertrack bond
Gleit-... sliding
Gleitbacke slide jaw
Gleitbügel sliding bar (621.3)
Gleitebene sliding plane (54)
Gleitentladung corona, creeping discharge
Gleitfunkendurchschlag creep flashover
Gleitfläche slip plane (66)
Gleitkomma floating point
Gleitkontakt sliding contact
Gleitkufe skid bar
Gleitlager plain bearing, journal bearing, sliding-surface bearing
Gleitmittel grease, lubricant
Gleitmodul shear modulus, shearing modulus
Gleitpassung sliding fit
Gleitreibung sliding friction, slipping friction
Gleitschiene slide bar, skid bar (621.9)
Gleitweg glide slope (629.13)
Gleitwegsender glide-slope transmitter
Gleitwiderstand an den Korngrenzen slip resistance of grains
gleiten to slip, to slide
gleitendes Mittel moving mean
gleitende Reibung slipping friction
gleitende Skala sliding scale
gleitende Spannung sliding voltage
Glied limb (61); member, organ (621); section (621.392); link (67)
Glied einer Reihe term of a sequence (51)
Glimm-... glow, corona
Glimmentladung glow discharge, corona discharge, luminous glow
Glimmgleichrichter neon-tube rectifier, helium valve
Glimmgleichrichterröhre anotron
Glimmlampe glow lamp, neon (-glow) lamp, neon tube
Glimmlicht glow, corona
Glimmlichtanzeigeröhre neon indicator
Glimmlichtdetektor ionized-gas detector
Glimmlichtgleichrichter neon-tube rectifier
Glimmlichtmikrophon glow-discharge microphone

Glimmlichtoszillograph Gehrcke oscilloscope, ondoscope
Glimmlicht-Scheitelspannungsvoltmeter neon-tube peak voltmeter
Glimmlichtstroboskop stroboglow, neon-electric stroboscope
Glimmrelais grid glow relay, grid glow tube
Glimmröhre neon bulb, glow tube, glow (-discharge) lamp
Glimmröhrenanzeiger neon glow tube, h.f. voltage indicator
Glimmröhren-Scheitelspannungsmesser neon-tube crest voltmeter
Glimmröhrenstabilisator glow-tube regulator
Glimmschaltdiode glow switching diode
Glimmstabilisator stabilovolt
Glimmstrecke telephone-type arrestor
Glimmstrom-Anemometer glow-discharge anemometer
Glimmzünder glow starter switch
glimmen to glow (621.3); to smoulder (662)
Glimmer mica
Glimmerblättchen mica sheet, mica foil
Glimmerblock foil-mica capacitor
Glimmer-Dielektrikum mica dielectric
Glimmererzeugnisse reconstructed mica (sing.)
Glimmerkondensator mica (dielectric) capacitor
Glimmerplatte mica slab
Glimmerscheibe mica lamination
Glimmerschicht mica layer
glitzern to glitter, to glisten, to glint, to twinkle
Glocke bell (534); shell (621.315.6)
glockenförmig bell-shaped
glockenförmiger Impuls bell-shaped pulse (621.397)
Glüh-... incandescent, hot; annealing (621.7)
Glühbehandlung annealing (process)
Glühbirne (electric) bulb, incandescent lamp
Glühdrahtkochplatte open-type hot plate, open-type boiling-plate
Glühfaden filament
Glühfadenpyrometer monochromatic pyrometer
Glühhärtung temper hardening
Glühkathode thermionic cathode, hot cathode, cathode-filament emitter
Glühkathode mit Emissionsfleck lime-spot cathode
Glühkathodenanschluß filament terminal
Glühkathodengleichrichter thermionic rectifier, hot-cathode rectifier
Glühkathoden-Quecksilberdampfgleichrichter hot-cathode mercury-vapour rectifier, phanotron
Glühkathodenröhre hot-cathode tube
Glühkathoden-Röntgenröhre hot-cathode X-ray tube
Glühkaustik hot-wire cautery (61)
Glühkerze heater plug (621.4)
Glühkopfmotor semi-diesel engine
Glühlampe bulb, filament lamp, incandescent lamp
Glühlicht incandescent light
Glühofen annealing furnace, reverberatory furnace
Glühstab glower
Glühstrumpf incandescent gas mantle
Glühtemperatur normalizing temperature
glühen to glow (nicht 621.326); to incandesce (621.32); to anneal (621.7)
glühend incandescent
Gold gold
Goldblättchen gold leaf
Goldblatt-Elektroskop gold-leaf electroscope
Golddrahtdiode gold-bounded diode
Goldfolie gold foil, gold leaf
Goldschlägerhaut gold beater's skin
Goldstaub gold dust
Goldschmidt-Maschine Goldschmidt alternator
Goliath-Edison-Gewinde G.E.S. cap = Goliath Edison screw cap
Gon grade (514)
Goniometer goniometer, radiogoniometer, Helmholtz coil arrangement
Goniometrie goniometry
grad = Gradient grad = gradient (51)
Grad deg. = degree (536), (51); grade (514)
30 Grad Celsius 30 ° Centigrade = 30 ° C
5 Grad wärmer 5 points up
um 90 Grad verschoben 90 ° out of phase, in quadrature
Gradeinteilung graduation
Gradring graduated ring
Gradstrich degree mark
gradweise by degrees
Gradation gradation (77)
Gradationsentzerrung gamma corrector
Gradationsfehler gradation distortion
Gradationsverzerrung half-tone distortion
Gradient gradient (51), (53), potential gradient (621.3), (55)
Gradient des Brechungsexponenten refractive gradient
Gradientenfaser gradient fiber
graduieren to graduate

Graetzschaltung full-wave bridge circuit, (full-wave) bridge-connected rectifier
Grafikkarte graphic board (681.3)
Gramm gm = gram, gramme
  Grammäquivalent gram equivalent
  Grammolekül mole (54)
Grammescher Ring Gramme ring, closed-coil winding
Grammophon gramophone
  Grammophonnadel (gramophone) needle, stylus
Granalien grained metal, granulated metal (66)
granulieren to granulate
graphisch graphic (al)
  graphische Anzeige graphic display
  graphisch darstellen to depict graphically, to diagram
  graphische Darstellung graphic(al) representation, diagram, chart, graph
  graphische Differentiation graphic(al) differentiation
  graphisches Verfahren graphical method
  auf graphischem Wege by graphics
Graphit graphite, black lead, plumbago
  Graphitbelag graphite coating
  Graphitbremse carbon moderator (539)
  Graphitbürste graphite brush
  Graphitelektrode graphite electrode
  Graphitflocken flakes of graphite (669)
  graphitgebremster Kernmeiler, (-reaktor) graphite-moderated pile, graphite-moderated reactor (539)
  graphitgebremster Uranbrenner graphite-moderated uranium pile
  Graphittiegel graphite crucible
graphitieren to graphite, to graphitize
  graphitierter Glühfaden graphitized filament
graphitisch graphitic
Grat burr (621)
grau grey
  grau machen to grey
  grauer Körper grey body
Grauguß grey iron
  Graugußteile grey castings
Graukeil gray scale, neutral wedge
  Graukeilphotometer wedge photometer
Graustrahler grey body
Graustrahlung grey-body radiation
gravieren to grave, to engrave (681.9); to aggravate
Graviermaschine engraving machine
Graviernadel graving chisel (76), engraving needle (621.9)
Gravierung engraving
Gravitation gravitation
  Gravitationskonstante gravitational constant
gravitieren to gravitate
Greenwich-Zeit Greenwich time
greifen to seize, to grip, to grasp, to clamp
Greifer grip, gripper, grab
Grenz-... limiting, bounding, boundary, critical
  Grenzbeanspruchung maximum limited stress, to lerated stress
  Grenzbedingung boundary condition
  Grenzdämpfung ultimate damping (539); aperiodic damping, critical damping (621.317)
  Grenzdrehzahl limit speed, ceiling speed
  Grenzempfindlichkeit ultimate sensitivity, noise figure, limiting sensitivity (621.396.62)
  Grenzfall border-line case, limiting case
  Grenzfrequenz critical frequency (621.37), cut-off frequency (621.372), barrier frequency, limiting frequency
  Grenzgebiet interface region (53)
  Grenzlast limiting load
    Grenzlast für funkenfreie Kommutierung sparking limit
  Grenzlinie border line, boundary line, line of demarcation
  Grenzmaß limit dimension
  Grenzpegel overload level (621.395)
  Grenzreibung limiting friction; boundary friction
  Grenzschicht boundary layer, junction, barrier layer (621.382)
  Grenzstockwerkschalter terminal limit switch (621.87)
  Grenzstrom limiting current, carrying current (621.316.5)
    Grenzstromkennlinie limiting overload characteristic; rated overload characteristic (621.314.5)
    Grenzstromrelais marginal relay
  Grenzvakuum ultimate vacuum
  Grenzwellen intermediate waves
    Grenzwellenlänge quantum limit (621.386)
  Grenzwert limit, limiting value, absolute maximum rating
    über den Grenzwert hinaus beyond the limit
    Grenzwertmelder limit sensor
    Grenzwertprüfung marginal check (681.3)
Grenze border, boundary; limit (51); frontier (91)
  Grenze der dynamischen Stabilität transient stability limit
  Grenzen einsetzen to substitute limits (51)
  Grenze des Empfangsbereichs fringe of the operating range (621.396)

Grenze der Möglichkeiten bounds of possibility
Grenze des Zündbereichs limits of inflammability
bis an die Grenze gehen to go near the margin
in Grenzen halten to limit
grenzüberschreitender Leitungsabschnitt international telephone line (621.39)
grenzen an to border on, to verge on
griechisches Feuer Greek fire
Grieß snow (621.396.96), (621.397)
Griff handle
Griffsicherung handle-type fuse (621.316)
Griffverlängerung extension handle
Grill gridiron; electric grill
Grob-... rough, coarse
Grobabgleich rough adjustment, coarse setting
Grobabstimmung rough tuning, flat tuning
Grobblech plate
Grobblitzschutz lightning guard (621.394)
Grobeinstellung s. **Grobabgleich**
Grobgewinde coarse thread
Grobkontrast large-area contrast
Grobkorn large grain, coarse grain
Groblunker cavity
Grobschliff coarse grinding
Grobsteuerung coarse control
Grobstruktur bulk structure
Grobzerkleinerung coarse crushing, pre-grinding
grob rough, coarse
grobe Fahrlässigkeit serious carelessness, perversity
grobes Gefüge coarse texture
grobe Schätzung rough estimate
grobe Zeichnung rude drawing
grobgängig coarse-pitch
Groß-... large (-scale); high-power, high-capacity
Großbasispeiler wide-base direction finder; wide-aperture direction-finder
Großelektromotor high-capacity electrical motor
Großflächengleichrichter broad-area rectifier
Großintegration large-scale integration = LSI
Großkraftwerk high-power station
Großkreis great circle
Großkreisausbreitung straight-line path (621.396)
Großkreispeilung long-path bearing
Großlampe high-wattage lamp
Großlautsprecher high-power loudspeaker
Großmotor high-capacity motor
Großoberflächenplatte large-area plate (621.355)
Großsender high-power transmitter
Großserienfertigung large-scale production
Großsignal large signal
Großzahlforschung statistical research
groß great, big, large, wide
große Abmessungen large size (sing.), large dimension
große Amplitude large amplitude
große Auflösung high definition (621.397)
große Bandbreite wide band
großer Bereich wide range, long range
große Halbachse major semi-axis
große Kraft big force
große Lautstärke large volume (621.396)
großer Maßstab large scale, open scale
große Steilheit high slope (621.385)
große Toleranz wide to lerance
großes Volumen large volume
in großer Höhe at high altitude
in großem Maßstabe large-scale
mit großer Leistung high-powered
größer als greater than, superior to (51)
größte Abmessung overall dimension
größter gemeinsamer Teiler G.C.M. = Greatest Common Measure, H.C.F. = Highest Common Factor, greatest common divisor
größte Keule major lobe (621.396.677)
größter Kugelkreis great circle
größte Leistungsübertragung maximum power transfer
größte prozentuale Abweichung von der Sinusform deviation factor of a wave
größter Scheitelwert crest amplitude
Größe size; magnitude, quantity (51)
Größe zweiter Ordnung second-order term, second-order quantity (51)
auf Größe schneiden to trim to size
der Größe nach konstant constant in magnitude
Größtwert maximum value, amplitude
Grübchen pits (pl.), pitting (621.357)
Grube pit, mine (622)
Grubenbetriebsanforderungen mining duties
Grubenbahn mine railway
Grubenbewetterung mine ventilation
Grubenlampe miner's lamp, safety lamp
Grubenlokomotive mining locomotive
Grubenventilator mine ventilator
Grummeln rumbles (pl.) (621.396)
Grund ground, base, basis; bottom; reason, cause

Grundanstrich prime coat, primary coating, undercoat (69)
Grundausbildung basic training
Grundausrüstung basic equipment
Grundbedingung main condition
Grundbegriff basic term
Grundbelastung base load, permanent load
Grunddrehzahl base speed
Grundelement element, fundamental unit, independent unit (53)
Grundfarbe primary colour (535); rimer colour (655); primary (621.397)
Grundfarbeneinheit unit quantity of primary colour
Grundfläche bottom face (621), floor space (69); base (51)
Grundfrequenz fundamental frequency, base frequency, basic frequency
Grundgebühr fixed charge
Grundgeräusch basic noise, background noise, background hiss
Grundgeräuschabstand noise-power ratio
Grundgleichung basic equation
Grundhelligkeit background brightness
Grundlagen der Elektrizität basic electricity
Grundlagenforschung fundamental research
Grundlagenkenntnisse basic knowledge (sing.)
Grundlast base load
Grundlastfahren holding load (621.311)
Grundlinie lower side (51), base (51), (52)
Grundmasse matrix (66)
Grundmetall parent metal, original metal (621.791), base (metal) (621.35)
Grundpegel background count (539), background noise level
Grundplatte base (plate), bed plate, shoe
Grund-Primärgruppe basic group (621.395.4)
Grund-Quartärgruppe basic super-master group (621.395.4)
Grundriß ground plan, plan (view)
Grundschaltbild essential connections (pl.), basic circuit diagram (621.3)
Grundschwingung fundamental oscillation, basic frequency
Grundschwingungsquarz fundamental crystal
Grund-Sekundärgruppe basic soupergroup (621.395.4)
Grundstoff base (54)
Grund-Tertiärgruppe basic master group (621.395.4)
Grundton fundamental note
Grundtypus principal mode (534), 621.396)
Grundüberholung rebuild (623)
Grundübertragungsdämpfung basic transmission loss (621.371)
Grundwasser ground water, level water, subterranean water
Grundwasserspiegel ground-water level
Grundwelle fundamental wave, 0-wave; dominant wave
Grundwellenanteil fundamental frequency component
Grundwellenblindleistung der gegenläufigen Komponente fundamental negativ-sequence reactive power
Grundwellenblindleistung des k-Systems fundamental k-sequence reactive power
Grundwellenblindleistung der mitläufigen Komponente fundamental positive-sequence reactive power
Grundwellenblindleistung der Nullkomponente fundamental zerosequence reactive power
Grundwellenbrumm fundamental hum
Grundwellenlänge fundamental wavelength
Grundwellenwirkleistung fundamental active power
Grundwert des Nebensprechens signal to crosstalk ratio
Grundzahl base cardinal number (51)
Grundzyklus basic machine cycle (681.3), major cycle
grundieren to prime (69)
Grundierung priming, undercoat, body colour
Gründung foundation (624); establishment (65)
Gründungen footings (621.315.1)
Grünspan verdigris
Gruppe group (51), (621.396), array (621.396.67)
Gruppen pro Minute w.p.m. = words per minute (621.394)
in Gruppen zerfallen to fall into groups
Gruppenantrieb group drive, group driving (621.9)
Gruppenfaktor array factor (621.396.67)
Gruppengeschwindigkeit group velocity (621.392)
Gruppenlaufzeit envelope delay, group delay (time)
Gruppenlaufzeitverzerrung group-delay distortion, envelope-delay distortion
Gruppenlaufzeitverzögerung group retardation
Gruppenpilot group-reference pilot
Gruppenrichtcharakteristik group directional characteristic
Gruppenschalter gang switch
Gruppenschaltung array network

Gruppenträgerfrequenz sub-carrier frequency (621.395)
Gruppenübertragungsquerschnitt group transfer scattering cross section (539)
Gruppenumsetzer group translator
Gruppenverlustquerschnitt group removal cross section (539)
Gruppenwähler group selector
gruppieren to group, to array
Guckloch peep hole, eye hole
Guldinsche Regel Guildinus theorem
Gummi rubber, gum
Gummi auf Metall fest aufbringen to bond rubber to metal
Gummiaderleitung rubber-insulated wire, rubber cable
gummiartig rubbery
Gummiband elastic; rubber belt (629.1)
Gummibleikabel rubber-insulated lead covered cable
Gummidichtung rubber washer, rubber gasket, rubber packing
Gummifeder rubber spring
Gummifüßchen rubber foot
Gummihandschuh rubber glove
gummi-isoliert rubber-insulated
Gummi-Isolierung rubber insulation
Gummikabel rubber cable
Gummipuffer rubber bumper, rubber cushion
Gummischeibe rubber washer
Gummischlauch rubber hose, rubber tubing
Gummischlauchleitung t. r. s. cable = tough rubber sheathed cable, non-kinkable flex, "cab-tyre" cable (621.3)
Gummischlauchmantel "cab-tyre" sheathing
gummieren to gum; to rubber-coat
Gunn-Effekt gunn effect
günstig beeinflussen to promote
günstige Last optimum load
Gürtel belt
Gürtelkabel belted cable
gürtelloses Kabel separately leadcovered cable, H-cable
Guß founding, casting (process)
Gußaluminium cast aluminium
Gußbarren foundry ingot
Gußbeton poured concrete
Gußbetonpfahl pre-cast pile
Gußblase blow hole
Gußeisen C. I. = cast iron
Gußgehäuse cast-iron casing
Gußhaut skin (621.7)
Gußkapselung iron-clad case
gußgekapselte Verteilung iron-clad distribution board
Gußlegierung casting alloy
Gußmodell pattern
Gußstahl cast steel
Gußstück casting
gut good; well
gut ausgeprägt clean-cut (621.9)
gut definiert well-established
gut eingebrannt properly fused
gut erhalten in good repair
gut gebaut well-constructed
gut gekühlt well-cooled
gut gelüftet well-ventilated
gut isoliert well-insulated
gut leiten to conduct heavily
gut passen to be a good fit, to make a good fit
gut proportioniert well-proportioned
gute Annäherung close approach
guten Empfang haben to get a strong signal
guter Gleichlauf perfect alignment
gute Klangtreue high (-grade) fidelity
gute Näherung close approximation
gute Passung close fit, exact register
guter Schwingeinsatz rapid starting (621.396.61)
guter Sitz good fit
gute Tonqualität high (-grade) fidelity
gute Zeichnung sharp definition (621.397)
Gutachten expert opinion
Güte quality (621.396); degree of perfection, merit (621.317)
der Güte nach (geordnet) in order of merit
Gütebestätigung quality assessment
Gütefaktor Q-factor, quality factor
Gütefaktormeßgerät Q-factor meter, quality-factor meter
Gütegrad (quality) standard
Güteklasse grade, standard of accuracy, class of accuracy
Güteprüfung quality inspection
Gütesicherung quality assurance
Güteverlust degradation
Gütezahl quality, Q-value, (magnification) factor (621.396); figure of merit (621.317)
Gütezahl-Meßgerät "Q" meter = quality meter
Guttapercha guttapercha
Gyrofrequenz gyro frequency
gyromagnetischer Effekt gyromagnetic effect

H

H = Henry H = henry
Haarnadel hairpin
Haarnadel-Glühfaden horseshoe filament
Haarriß fine crack
Haarsieb hairsieve, fine sieve
Haarstrich hair line
Haften adherence (**an:** to)
haften an to adhere to, to cling to
Haftfestigkeit adhesive strength
Haftreibung sticking friction
Haftrelais retentive-type relay, remanent relay
Haftsitz wringing fit, interference fit
Haftstelle deathnium center, trap (621.382)
Haftvermögen adhesion, adhesiveness (621.357)
Hahn cock (59), (623), (696); faucet (US), tap (696)
Hahnfassung key holder, switch lampholder (621.32)
Haken hook
Hakenschalter gravity switch, cradle switch (621.395.6)
Hakenschlüssel hook spanner
Hakenumschalter hook (-type) switch (621.395)
haken to hook
Halb-... semi-..., half
Halbachse semi-axis
Halbaddierer half adder
Halbautomat semi-automatic (621.9)
Halbbild field (US); frame (621.397)
Halbbilddauer field duration
Halbbildfolge field sequentiel (621.397)
Halbbyte nibble
Halbduplexbetrieb half duplex
Halbfabrikat semi-finished product
Halbierschaltung scale-of-two circuit, divide-by-two circuit
Halbkreis semi circle
Halbkugel hemisphere
halbleitende Verbindung intermetallic compound (621.382)
Halbleiter semiconductor
Halbleiter-Bildsensor solid-state image sensor
Halbleiterbauelement semiconductor component
Halbleiterdiode crystal diode, semiconductor diode
Halbleitereffekt barrier-layer effect (621.314.6)
Halbleiterkühlelement semiconductor cooling element
Halbleiterscheibe semiconductor wafer
Halbleiterübergang semiconductor junction
Halbleiterzone semiconduction region, semiconductor region
Halbperiode $1/2$ period, half period (539)
Halbquerschnitt half section
Halbton half tone, semitone
Halbtonbild half-tone picture
Halbwattlampe half-watt lamp
Halbwelle half wave, half cycle, half period, alternation (621.3)
Halbwellenantenne half-wavelength aerial
Halbwellendipol half-wave dipole, half-wave radiator
Halbwellendifferentialschutz half-wave differential protection
Halbwellengleichrichter half-wave rectifier, one-way rectifier
Halbwellengleichrichtung half-wave rectification
Halbwellenmittelwert half-period average value
Halbwellenschleife half-wave frame (621.396)
Halbwellenübertragung half-wave transmission (621.315)
Halbwert $1/2$ max. value = half maximum value
Halbwertsbreite beam width at half power, beam width at 3 dB, half-power beam width (621.396.67); 3-dB band-width (621.39)
Halbwertsdicke half-value thickness, half-value layer (621.386)
Halbwertspunkt half-power point (621.395)
Halbwertsverfahren half-deflection method
Halbwertszeit $1/2$ period = half period, (period of) half-life (539)
Halbwertwinkel half-power angle (621.396)
Halbzeug half-finished products (pl.)
halb half
halbe Größe, in halber Größe half size
halbe Höhe mid-height
auf halber Höhe at mid-height
halber Querschnitt half section
halb ausgesteuert controlled to half value
halbamtsberechtigte Nebenstelle partially restricted extension
halbautomatisch semi-automatic
halbautomatisches Amt auto-manual (telephone) exchange
halbautomatische Fahrstuhlsteuerung semi-automatic lift control

halbautomatische Handtaste bug
halbautomatische Station partial automatic station (621.316)
halbautomatische Telefonie semi-automatic telephone system, auto-manual telephone system
halbautomatische Verfolgung aided tracking (623), (621.316.96)
halbautomatisches Vermittlungssystem auto-manual telephone system
halbdurchlässig semi-permeable, semi-porous
halbdurchsichtig semi-transparent
halbempirisch semi-empirical
halbfertig half-finished, semi-finished
halbgeschlossen semi-closed, semi-enclosed
halbgeschlossene Maschine semi-enclosed machine, semi-protected machine
halbgeschlossener Motor screen-protected motor
halbgeschlossene Nuten half-closed slots
halbgeschlossene Sicherung semi-enclosed fuse
halbindirekt semi-indirect
halbindirekte Beleuchtung semi-indirect lighting
halbindirekte Leuchten semi-indirect fittings
halbkreisförmig semi-circular
halbkugelförmig semi-spherical, hemispherical
halblogarithmisch semi-logarithmic(al)
halblogarithmisches Koordinatensystem semi-logarithmical scale
halbsinusförmig half-sinusoidal
halbstarr semi-rigid
halbstarre Aufhängung semi-rigid suspension (621.33)
halbversenkt semi-sunk (621.8/884)
halbversenkter Schalter semi-sunk switch, semi-recessed switch
Hälfte half (pl.: -ves)
Hall tang (534)
Hallraum strong-echo room, reverberatory chamber, echo room
Hall-Beweglichkeit Hall mobility
Hall-Effekt Hall effect
Hallkonstante Hall constant, Hall coefficient
Hallsonde Hall probe
hallen to tang, to echo, to reverberate
hallend reverberatory
Halogen halogen, member of the halogen group
Halogensalz halide
Hals neck (61), (621)
Halsansatz throat (621.396.677)
Halsquerschnitt throat area (621.396.677), (534)
Halsriemen neck strap (621.395.61)
Halsschraube body screw, recessed-collar head screw
Halte-... holding, securing
Halteanode excitation anode, keep-alive anode
Haltebereich hold range
Haltebolzen securing bolt
Haltedraht supporting wire
Halteerregung hold excitation (621.318.5)
Haltekontakt holding contact, hold-on contact
Haltekreis holding circuit (621.395/621.318)
Halteleitung holding line
Haltemagnet holding magnet
Haltepunkt breakpoint (681.3)
Halterelais holding relay
Haltespannung holding voltage
Haltespule holding coil
Haltestrom holding current (621.318.5)
Halteverhältnis holding ratio
Haltevorrichtung holding appliance, holding device
Haltezeit duration of hold
halten to hold; to stop
Halter holder, bracket, support
Halterung holder, securing, mounting, support; jig (621.317)
...-haltig ... containing
Hämatit haematite, hematite
Hammer hammer
hammerförmig hammer-shaped
Hammerkopf head of a hammer
Hammerkopfschraube T-head bolt
Hammerschlaglack hammer-effect varnish
Hammerunterbrecher hammer break
hämmern to hammer, to swage, to pene
gehämmerter Draht swaged wire
Hanauer Quarzlampe Heraeus lamp
Hand hand
von Hand manual
von Hand transportieren to man-handle
Handabstimmung manual tuning
Handamt manual exchange, hand-operated exchange, manual central office
Handantrieb manual drive, hand motion
mit Handantrieb hand-driven, hand-geared
Handapparat hand (telephone) set, (hand) microtelephone
Handaufzug handwinding (681.11)
handbetätigt, handbetrieben hand-driven, hand-actuated, hand-operated
Handbetätigung, Handbetrieb manual

operation, hand actuating, hand motion, human operation
Handbohrmaschine hand drill, portable (electro) drill
Handbuch (technical) manual
Handbuch-Änderungsdienst manual update service
Handeinstellung hand setting, hand adjustment, manual setting
Handfunksprecher walkie-talkie
Handgriff handle; manipulation
Handhebel hand lever
Handkurbel crank-handle, cranked handle
Handlampe hand lamp, inspection lamp, portable lamp
Handlocher keyboard perforator, keypunch, hand punch
Handmikrophon hand-type microphone
Handrad handwheel
Handregel Fleming's rule (621.3)
Handregler hand-operated regulator
Handregelung hand regulation, manual control
Handruf manual ringing
Handschalter handswitch
Handschrank manual switchboard (621.395)
Handschweißung manual weld
handgeschweißt hand-welded
Handsignalscheinwerfer pistol light
Handsteuerung manual control
Handvermittlung manual exchange, manual central office, manual telephone system, hand-operated exchange (621.395)
Handwerkszeug small to ols, working tools, hand tools (pl.)
Handwicklung hand winding (621.313)
handlich handy, manageable
Hanf hemp
Hanfpackung hemp packing
Hanfseele hempen core, fibre core
Hänge-... suspension, pendant
Hängeanordnung underhung mounting (621.316.5)
Hängedose pendant socket outlet
Hängedruckknopf pear push, pendant push
Hängeisolator, Hängekette suspension (-type) insulator, chain insulator
Hängeklemme suspended clamp (621.315.1)
Hängelampe pendant (lamp), suspended lamp, hanging lamp
Hängeschalter suspension switch, pendant switch, pear switch
hängen to hang
hängen an to attach to, to fix to, to suspend on
hängenbleiben to stick
hängengebliebenes Relais stuck relay
hängend pendant
hängendes Ventil overhead valve
Hankelsche Funktion Hankel function, modified Bessel function
Hapugmodulation floating carrier system, carrier control
„Harke" polyrod aerial (621.396)
harmonisch harmonic, harmonious
harmonisch gliedern to proportion
harmonischer Analysator harmonic analyser, wave analyser
harmonische Bewegung harmonic motion
harmonische Oberwelle harmonic (component)
harmonische Reihe harmonic progression
harmonische Strahlung harmonic emission
harmonische Unterwelle subharmonic
Harmonische harmonic
harmonisieren to harmonize
Harnstoff urea
Harnstoffharz urea resin
Harnstoff-Kunstharzleim urea-resin glue
Harnstoffleim urea-base glue
Hart-... hard
Hartblei hard lead
Hartfaserplatte hard-wood fibre slab, fibre wall board
Hartgasschalter hard-gas circuit breaker
Hartglasgewebe hard-glass fibre
Hartgummi hard rubber, vulcanite, ebonite
Hartholz hardwood
Hartlot hard solder, (brazing) spelter
Hartlötung hard soldering, brazing
Hartmetall cutting metal, tungsten carbide; hard alloy
Hartmetallauflage tool tip, carbide tip (621.9)
Hartmetallschneidplättchen tungsten-carbide bit
Hartmetallwerkzeug tipped tool, carbide tool, tungsten-carbide tool
Hartpapier bakelized paper, hard paper, pressed paper, resin bonded paper, phenolic paper
Hartpapierplatte synthetic resin paper board
Hartporzellan hard porcelain
Harttastung hard keying
hart hard; contrasty (77)
hart werden to harden
harte Beanspruchung severe strain
harte Legierung hard alloy
hartmagnetisch hard magnetic

harter Nickelstahl Ni-hard
harte Röntgenstrahlung hard X-rays (pl.)
harter Supraleiter hard superconductor
hartes Wasser hard water (54), (621.18)
hartlöten to hard solder, to braze
hartsektoriert hard-sectored
härtbar hardenable
Härte hardeness (54), (535), (621.386), (628), (66); temper (621.386), (66)
Härtegrad grade of hardness, degree of hardness, temper
Härtemeßgerät penetrometer, qualimeter; radiosclerometer (621.386)
Härteprüfer hardness tester
Härteskala hardness scale
härten to harden
Härter hardener
Härtung hardening
Harz resin, rosin
harzähnlich resinoid
Harzelektrizität resinous electricity
harzig resinous
Haube hood; shell, dome, cover
Hauch (auf einer Oberfläche) cloud
Häufigkeit frequency, abundance
Häufigkeitsbestimmung abundance determination (54)
Häufigkeitsverteilung frequency distribution (51)
Häufungswert mean
Haupt-... principal, chief, grand, main, cardinal, master
Hauptamt main exchange, master office
Hauptamtvermittlung single-office exchange
Hauptanschluß main (station), subscriber's main station (621.395), main terminal (621.382)
Hauptanschlußrelais network master relay
Hauptbedingung main condition
Haupteinflugzeichen middle marker
Haupt-Endausschalter main limit switch
Haupterregermaschine main exciter
Hauptfeld main field
Hauptfeldstrom series field
Hauptfeuermelder master fire-alarm box
Hauptfluß working flux, core flux (621.3)
Hauptfrequenz primary frequency (621.396)
Hauptinduktivität primary inductance
Hauptinstandsetzung major repair (623)
Hauptkabel main cable
Hauptkeule main lobe, main beam (621.396.677)
Hauptkontakt main contact, current-carrying contact (621.316.5)
Hauptleitung main line (621.3)
Hauptlinie main line (625)
Hauptmast main mast (621.315)
Hauptnetzsicherung main-power fuse
Hauptnormal primary standard
Hauptpol main pole (621.315)
Hauptprogramm master routine (681.3)
Hauptprozessor master processor
Hauptraster main pattern (621.396.2)
Hauptschalter master switch, main switch, power-supply switch
Hauptschaltanlage main switchgear connections (pl.)
Hauptschalttafel main switchboard, power switchboard
Hauptschluß... series...
Hauptschlußbogenlampe series arc lamp
Hauptschlußerregung series excitation
Hauptschlußfeld series field
Hauptschlußfeldregler shunted-field control
Hauptschlußfeldspule series coil
Hauptschlußgenerator series-wound generator
Hauptschlußmotor series (-wound) motor
Hauptschlußwicklung series winding
Hauptschwingkreis tank-circuit oscillator
Hauptsicherung main fuse
Hauptspannung principal stress (539.4); principal voltage
Hauptspeicher main storage (681.3)
Hauptspeiseleitung main-supply line, trunk feeder
Hauptstelle subscriber's main station (621.39)
Hauptsteuerschalter master controller
Hauptstrahl main beam (621.396.67)
Hauptstrahlrichtung direction of maximum radiation
Hauptstrahlrichtung in der Elevationsebene uptilt (621.396.67)
Hauptstrahlunterdrückung main-beam killing
Hauptstrahlwinkel (in der Vertikalebene) directivity angle (621.396.67)
Hauptstrahlzipfel main lobe of radiation (621.396.67)
Hauptstrecke main line, trunk line (625)
Hauptstrom main current
Hauptstromkreis main circuit
Hauptstrommotor series motor
Hauptstromregelung armature control (621.34)
Hauptstromregler series-field regulator

Hauptstromrelais main-current relay, primary relay
Haupttelegraphenamt central telegraph office
Haupttyp dominant mode (621.396)
Hauptuhr master clock
Hauptverkehrsstunden rush hours (621.395), (625), crowded hours (625), busy hours (621.395), main-traffic hour (621.395)
Hauptversorgungsgebiet primary service area (621.396)
Hauptverteiler main distributor, main distribution frame (621.395)
Hauptverteilertafel branch-circuit distribution centre
Hauptverteilungspunkt sub centre
Hauptweiche king point (625)
Hauptwelle dominant wave (621.37)
Haus house
Hausamt private exchange
Hausanschluß service entrace, service line
Hausanschlußkabel leading-in cable
Hausanschlußkasten transfer box, service box (621.315.6)
Hausanschlußleitung connecting lead, house-connection line, supply mains to a house, lighting feeder (621.315.1), service equipment (696)
Hausanschlußschalter service switch (696)
Hausanschlußzähler house-service meter
Hausbedarf domestic purposes (pl.); domestic demand
Hauseinführungspfeife entrance ell (696.6)
Haushalt household; domestic...
Haushaltsbedarf household demand
Haushaltsgebrauch domestic application
Haushaltsgeräte household facilities, household appliances
Haushaltsinstallation domestic installation, domestic wiring
Haushaltskühlschrank domestic refrigerator
Haushaltsverbraucher domestic consumer
Haushaltszähler house-service meter
Hausinstallation consumer's installation, house wiring, domestic installation
Hauspostanlage inter-office mail
Hausrohrpost pneumatic house tube
Haustelefon interphone, domestic telephone
Haustelefonanlage interphone system, intercommunication telephone
Hausvermittlung private exchange
Haut skin (61), (621.7), (629.12/13); film (621.7); hide (63)
Hautbildung film formation (621.6)
Hauteffekt s. **Hautwirkung**
Hautwirkung skin effect (621.3)
Häutchen pellette, pellicle
HD = Hochdruck HP = High Pressure
Heaviside-Einheitsspannungsstoß Heaviside unit function
Heaviside-Kennelly-Schicht Kennelly-Heaviside layer
Heaviside-Operatorenrechnung Heaviside operational calculus
Heavisidesche Regel Heaviside expansion theorem
Hebdrehwähler two-motion selector, Strowger selector, step-by-step selector; bank-and-wiper switch
Hebel lever
Hebel betätigen to press a lever
Hebelanlasser lever-type starter (621.313)
Hebelantrieb lever operation, lever drive
Hebelarm lever arm, to rque arm
Hebelarm des Moments arm of the moment
Hebelbetätigung lever operation
Hebeldrehpunkt lever fulcrum
Hebelgesetz principle of the lever
Hebelschalter knife switch
Hebelschere shearing machine (621.9)
Hebelsteuerung lever control
hebeln to lever
heben to lift
gehobene Kathode biased cathode
Hebschritt vertical step (621.395.34)
Hefnerkerze Hefner candle, Hefner unit
Heft haft, helve, tack
Heftniet tack, rivet
Heftschweißung tack weld
heften to stitch, to sew, to staple (65)
Heimkino home movie
Heinisch-Riedl-Schutzschalter earth-leakage trip
Heisingmodulation Heising modulation, plate-current modulation, choke modulation
Heising-Modulationsröhre Heising modulator
Heiß-... hot, superheated
Heißisolation class H insulation, high-temperature insulation
Heißleiter thermistor, high-temperature conductor, pyroconductive material
Heißleitfähigkeit pyroconductivity
Heißwasserspeicher electric water heater
heiß hot
„heiß" live (621.3); hot, active (539.16)
heiß werden to grow hot
heiße Zone torrid zone
heißgewalzt hot-rolled
heißlaufen to run hot
heißverleimen to heat-bond

heiter serene (55)
Heiz-... heating, calorific
Heizbatterie filament battery, A-battery, filament supply
Heizdecke electric blanket
Heizdraht heating wire (621.365)
Heizelement heater element, heating element, resistance element; fuel element (539.17)
Heizfaden heater, filament (621.385)
Heizfläche heating surface
Heizkammer oven
Heizkissen hot pad, heater pad
Heizkörper heater element, heating element, radiator
Heizleistung heating effect, heating power
Heizleistung der Kathode cathode-heating power
Heizleiter heating conductor (621.3)
Heizplatte electric hotplate
Heizraum heating compartment, combustion chamber
Heizrohr heating pipe
Heizsonne electric fire
Heizsonnenreflektor electric fire bowl
Heizspannung heating voltage, filament voltage (621.365)
Heizspannungsquelle filament power supply
Heizspirale heating coil
Heizstrom heater current, heating current
Heizstromversorgung filament supply
Heizteppich electric carpet
Heiztransformator filament-heating transformer, filament transformer
Heizwert calorific value, heat of combustion
Heizwertbombe bomb calorimeter, calorimeter bomb
Heizwicklung heating coil; heater winding (621.396.62)
Heizwiderstand heating resistor, filament rheostat (621.385)
Heizwirkung calorific effect, heating value
Heizzyklus heating cycle
heizen to heat, to fire; to stoke
Heizung heating
Heizungsanlage heating installation
Hektoliter hectol. = hectolitre
Hektometer hectom. = hectometer
Heliograph heliograph
Helium helium
heliumgekühlter Reaktor helium-cooled reactor
Helium-Neon-Gas-Maser helium-neon gas maser
hell bright, brilliant
helle Beleuchtung intense illumination, intense lighting
helle Glut bright heat
helles Licht intense light, bright light
helle Rotglut bright red heat
hellrot bright red
helltasten to intensify (621.397)
Hellfeldbeleuchtung light-ground illumination (535)
Hellsteuerimpuls unblanking pulse
Helligkeit brightness, luminosity (535); intensity (621.397)
Helligkeitsmodulation brilliance modulation, intensity modulation, beam trigger
Helligkeitsignal luminance signal
Helligkeitssteuerung brightness control, dimmer control
Helligkeitsverhältnis brightness ratio
Helltastung intensity modulation
hemmen to drag, to skid, to check
Hemmschuh scotch block, slipper, car retarder
Henry = H henry, secohm
Heptode heptode, pentagrid
herabdrücken to depress
herabsinken to sink
heranbringen to bring
herankommen an to get at
heraus...... out, ... outwards
herausarbeiten to work out, to evolve
herausdrücken to force out, to expel
herausdrücken aus der Achse to throw out of alignment
herausführen durch to take out through (621.385)
herausholen to withdraw, to evacuate; to educe (54)
herauskommen to result (51)
herauslösen to extract, to set free
herausschleudern to eject
herausschrauben to screw out
herausspulen to reel out
heraustreiben to force out
herausziehen to extract, to withdraw, to slip out
herbeiführen to effect, to incur
Herd range (643.3); hearth (669)
hermetisch hermetic(al)
hermetisch abgeschlossen hermetically sealed
Héroult-Ofen Hérould furnace, direct arc furnace
herrühren von to proceed from, to originate from
herstellen to make, to manufacture, to produce; to prepare (54)
Hersteller maker, manufacturer, producer

Herstellung manufacture, manufacturing, production; preparation, formation, extraction (54)
Herstellungsverfahren method of production, manufacturing process
Hertz cycles per second = c. p. s., c/s; cps (US)
Hertzsche Antenne Hertz aerial
Hertzscher Dipol Hertz oscillator, half-wave dipole, doublet
Hertzsche Wellen Hertzian waves
herumdrehen (sich) to spin, to swivel (round)
herumgleiten lassen um to nurse past
herumwickeln to wrap around
herumwirbeln to spin, to whirl
herunterdrehen to turn down
herunterdrücken to push down, to depress; to reduce
herunterkühlen auf to cool down to
herunterschmelzen to melt down
heruntertransformieren to step down
herunterübersetzen to step down (621.3), (621.83), to gear down (621.83)
hervorstehen to project, to protrude, to be salient
hervorstehend protruding, salient
Herz heart; core (539.16)
herzförmig heart-shaped
Herzkurve cardiod, cardioid
Herzschrittmacher (electonic) pacemaker
Herzspannungskurve electrocardiogram
heterogen heterogeneous
Heulen squeal, howl, whine
heulen to squeal, to howl, to whine
Heuslersche Legierung Heusler alloy
Hewlett-Isolator link-type insulator, Hewlett insulator
hexadezimal hexadecimal
Hexan hexane
Hexode hexode
HF = Hochfrequenz h. f. = high frequency, r. f. = radio frequency (s. auch unter Hochfrequenz)
HF-Abschirmung h. f. screen, radio shielding
HF-Bake radio beacon
HF-Bauteile radio-frequency components
HF-Behandlung h. f. treatment
HF-Drossel radio-frequency choke
HF-Eisen electric metal
HF-Erwärmung dielectric heating
HF-Gleichrichter first detector, radio detector
HF-Goniometer radio goniometer
HF-Heizung dielectric heating
HF-Höhenmesser radio altimeter
HF-Induktionsofen h. f. furnace, Northrup furnace
HF-Litze litz, litzendraht (wire), stranded wire
mit HF-Litze gewickelt litz-wound
HF-Messung radio-frequency measurement
HF-Nachrichtenverkehr radio communication
HF-Pentode radio-frequency pentode
HF-Schirm h. f. screen
HF-Spannung r. f. voltage, h. f. voltage, carrier voltage
HF-Störpegel radio-interference level
HF-Stoß oscillatory surge (621.316.9)
HF-Strom h. f. (alternating) current, r. f. current
HF-Strommesser radio-frequency ammeter
HF-Trägertelegraphie h. f. carrier telegraphy
HF-Übertrager radio-frequency transformer
HF-Vormagnetisierung a. c. magnetic biasing
HF-Vorstufe (first) h. f. stage (621.396)
HF-Widerstand radio-frequency resistance
hierarchisches Netz hierarchical network
Hilfs-... subsidiary, auxiliary; ancillary, supplementary; accessory
Hilfsader pilot core; pilot wire (US)
Hilfsanode auxiliary anode
Hilfsbetrieb auxiliary establishment
Hilfsbürste pilot brush (621.313)
Hilfsdraht pilot wire
Hilfsdrahtrundsteuerung pilot-wire control
Hilfsdrahtschutzsystem pilot-wire protection system
Hilfseinrichtung gadget, facility, ancillary equipment
Hilfselektrode auxiliary electrode
Hilfserregermaschine pilot exciter
Hilfsfunke trigger spark
Hilfsfunkenstrecke auxiliary spark gap
Hilfsgittermodulator injection modulator
Hilfsklinke ancillary jack (621.395)
Hilfskontakt auxiliary contact
Hilfsleitungstrennstelle secondary disconnecting devices
Hilfsmaschine donkey engine; auxiliary machine
Hilfsmittel auxiliary means, implement, facility, expedient
Hilfsphase auxiliary winding, split phase
Hilfsphasengenerator monocyclic alternator
Hilfspol auxiliary pole
Hilfspolmotor split-pole motor, shaded-pole motor
Hilfsprogramm auxiliary program (681.3)
Hilfsreihenschlußwicklung auxiliary stabilizing series winding
Hilfsrelais auxiliary relay
Hilfsröhre accessory valve

Hilfssammelschiene hospital busbar
Hilfsschalter auxiliary switch
Hilfsspannung auxiliary voltage
Hilfsspeicher backing storage, auxiliary storage
Hilfsspule auxiliary coil
Hilfsstromauslösung independent trip
Hilfstaste auxiliary button
Hilfsträger subcarrier
Hilfsträgermodulation subcarrier-frequency modulation
Hilfstransformator auxiliary transformer
hin und her to and fro, in and out
hin- und herbiegen to bend to and fro
hin und her auf der Leitung up and down the line (621.3)
hin- und herbewegen to seesaw
sich hin- und herbewegen to sway, to approach and recede, to move back and forth
hin- und hergehende Teile reciprocating parts
hin- und herschwanken to oscillate, to vacillate
hin- und herschwenken to switch back and forth (621.396.67)
hinauftransformieren to step up (621.3)
hinaufübersetzen to gear up (621.8), to step (621.3)
Hindernis obstacle; restriction
Hindernisbefeuerung obstruction lighting, hazard lighting
Hindernisfeuer obstruction light (656.7)
Hindernis-Warnanzeige terrain-clearance warning indicator (656.7)
Hinderniswarn-Radar terrain-avoidance radar
hindurchgehen durch to pass through; to permeate through
hindurchtreiben durch to force through
hindurchziehen to pull through
hineindrücken to force into
hineinfügen to fit in
hineinheben to lift into
hineinleuchten in to throw light upon, to throw light into
hineinpressen to force into
hineinstecken in to put into
hineingesteckte Energie put-in energy, input energy
hineingesteckte Leistung power input
Hineinwahl dialling in (621.395)
Hingang foreward motion
Hinlauf foreward sweep (621.385)
hinlaufende Welle forward wave, progressing wave
hinlegen to lay, to place
hinleiten nach to channel to (621.3)
hinneigen nach to trend to
Hin- und Rückleitung flow and return circuit, lead and return circuit
hinten in the back, in the rear, behind
hinten sichtbar werden to show through (655)
hintere Antennenwand rear curtain (621.396.677)
hintere Austastschulter back porch (621.397)
hintereinanderschalten to connect in series
hintereinandergeschaltet in series, in tandem
Hinter-... rear, back
Hinterachse rear axle
Hinterbandkontrolle read after write (681.3)
Hintereinanderschaltung series connection, series combination
Hintergrundrauschen background noise, snow (621.397)
Hintergrundstrahlung background radiation
Hintergrundverarbeitung background processing
Hinterkante trailing edge (621.3), (629.113)
Hinterlage backing material
Hintermotor tandem motor
Hinterradantrieb rear wheel drive
Hinterschneidungsfläche land (621.9)
Hinterschneidungswinkel clearance angle (621.9)
Hinweis pointer, indication, tip, guide
Hinweisdienst intercept service (621.395)
Hinweiston special information tone (621.395)
hinweisen auf to point to, to point at, to intimate
hinweisend auf indicative of, suggestive of
hinzielen auf to trend to, to aim
hinziehen, sich to draw on
hin- und zurücklaufen to travel forwards and backwards (621.392)
Hittorfsche Röhre Hittorf tube
Hitze heat
Hitzbandinstrument hot-strip instrument
Hitzdraht hot wire (621.317)
Hitzdrahtamperemeter hot-wire (expansion) ammeter
Hitzdrahtanemometer hot-wire anemometer, catharometer
Hitzdrahtausbrennung hot-wire cautery (61)
Hitzdrahthöhenmesser hot-wire altitude meter, hot-wire altimeter (629.13)
Hitzdrahtinstrument thermo-expansion instrument, hot-wire instrument
Hitzdrahtleistungsmesser hot-wire expansion wattmeter (621.396)
Hitzdrahtoszillograph hot-wire oscillograph

Hitzdrahtspannungsmesser hot-wire voltmeter, Cardew voltmeter
Hitzdrahtspulensicherung heat coil (621.395)
Hitzdrahttelefon thermal telephone
hitzebeständig heat resistant, heat resisting
Hitzebeständigkeit resistance to heat, immunity to heat
H-Kabel H-cable, Hoechstaedter cable
hl = Hektoliter hl = hectolitre
hm = Hektometer hm = hectometre
hobeln to plane
Hoch-... high; overhead
Hochantenne high aerial, overhead aerial, outdoor aerial
Hochantenneneffekt aerial effect, capacity effect
Hochbahn overhead railway
Hochbehälter high-level tank
Hochdruck high pressure
Hochdruck-Quecksilberdampflampe high-pressure mercury vapour lamp
Hochflußreaktor high-flux reactor
hochfrequent of high frequency (s. auch unter HF)
hochfrequenter Drahtfunk carrier frequency wire broadcasting, wired wireless
hochfrequenter Wechselstrom high-frequency alternating current
Hochfrequenz high frequency, radio frequency (s. auch unter HF)
Hochfrequenzchirurgie surgical diathermy
Hochfrequenzdrossel radio-frequency choke
Hochfrequenzfeldstärke radio field intensity, radio field strength
Hochfrequenzgenerator radio-frequency alternator; high-frequency generator, radio-frequency oscillator
Hochfrequenzglimmlampe induction lamp
Hochfrequenz-Induktionsofen high frequency induction furnace
Hochfrequenzlinse wave lens
Hochfrequenzmeßbrücke radio frequency bridge
Hochfrequenzsender radio transmitter
Hochfrequenzsperre r. f. filter
Hochfrequenzspule radio coil
Hochfrequenzstufe radio-frequency stage
Hochfrequenztechnik radio (-frequency) engineering, high-frequency engineering
Hochfrequenzverhalten high-frequency response
Hochfrequenzverstärker high-frequency amplifier, radio-frequency amplifier
Hochglanz mirror finish
Hochlastwiderstand power resistor
Hochleistungs-... high-capacity, high-power, high-efficiency, high-level
Hochleistungsbatterie high-level pile
Hochleistungslampe high-wattage lamp
Hochleistungslautsprecher high-power (loud)speaker
Hochleistungsmaschine high-capacity machine, high-efficiency machine
Hochleistungsmotor high-power motor (621.3); high-power engine (621.4)
Hochleistungsreaktor high-power reactor
Hochleistungssicherung high-rupturing-capacity fuse, HRC fuse, high-breaking-capacity fuse, HBC fuse
Hochleistungsstahl high-duty steel
Hochleistungswerkzeug high-duty to ol
hoch liegender Wellenleiter elevated duct
Hochpaß high-pass filter
Hochpegelwahl high-level selection (621.395)
Hochspannung high tension = h. t., high voltage = h. v., high pressure
Hochspannungsanlage high-tension power plant (= h. t. power plant)
Hochspannungsbatterie high-tension battery
Hochspannungsdurchführung h. t. bushing
Hochspannungselektrometer h. t. voltmeter; aislometer
Hochspannungsfreileitung high-tension overhead line
Hochspannungs-Gleichstrombahn high-tension d. c. railway
Hochspannungsisolator high-tension insulator
Hochspannungskabel h. t. cable
Hochspannungs-Kommutatorgleichrichter h. t. transverter
Hochspannungsleitung h. t. (transmission) line, h. t. (power) line
Hochspannungsmast transmission tower
Hochspannungsmessung high-tension measurement
Hochspannungsnetz transmission system, primary network
Hochspannungsprüfung high-voltage test
Hochspannungssammelschienen high-tension busbars
Hochspannungsschalteinrichtung high-tension switchgear
Hochspannungssicherung high-tension fuse
Hochspannungsspeiseleitung power feeder
Hochspannungsstababgleichrichter EHT-stack

Hochspannungstransformator high-tension transformer
Hochspannungs-Übertragungsleitung primary transmission feeder
Hochspannungsversorgungsnetz primary distribution network
Hochspannungsverteilungsschiene primary distribution main; backbone (US)
Hochspannungswicklung high-tension winding
Hochspannungszange fuse to ngs (pl.)
Hochspannungszündmaschine high-tension magneto (621.3)
Hochstromschalter heavy-current switch
Hochtemperaturisolation class H insulation = high-temperature insulation
Hochtonblende treble control (621.396.62)
Hochtonlautsprecher tweeter, treble loudspeaker
Hochtonwiedergabe high-note response
Hochvacuum high vacuum
Hochvacuumgleichrichter high-vacuum thermionic rectifier, hard-tube rectifier
Hochvakuumgleichrichterröhre vacuum-tube detector
Hochvakuumkathodenstrahlröhre hard-type cathode-ray tube
Hochvakuumpumpe high-vacuum pump
Hochvakuumröhre high-vacuum valve, high vacuum tube; hard valve
Hochvolt high-voltage = h. v., high tension = h. t.
Hochvoltdurchführung h. v. bushing
Hochvoltklemme h. v. terminal
Hochvoltsammelschiene h. v. busbar
Hochvoltschaltanlage h. v. switchgear
Hochvoltstation h. v. substation
Hochvolttransformator h. v. transformer
Hochvoltwicklung h. v. winding
hoch, hohe, hoher high (-grade), tall
zu hoch absichern to overfuse
hochaktiver Abfall high activity waste
hochangeregtes Atom hot atom
hochauflösendes Mikroskop high-resolution microscope
hochbeansprucht heavily loaded; heavily stressed
hochbelastet heavily loaded (621.3); heavily stressed
hochbelasteter Stromkreis heavy circuit
hochbocken to jack up
hochempfindlicher Film high-speed film
hochenergetischer Prozeß high-energy reaction
hochführen to carry up
hochglanzpoliert high-polished
hochgradiger Spaltstoff high-grade nuclear fuel
hochheizen to warm up, to heat (up)
hoch ionisiert highly ionized, strongly ionized
hochkant edgewise, on end
hochkant gewickelt edgewise wound
hochkant stellen to set on edge
hochlegiert high-grade alloy
hochmolekularer Stoff high molecular weight substance
hochohmig high-resistive, high-impedance
hochorganisiert highly organized
hochpermeables Material high-permeability material
hochprozentig high percentage
hochverstärkender Verstärker high-gain amplifier
hochwertiges HF-Eisen electric metal
hochwertiges Material superior material, good-quality material
hochwertige Tonwiedergabe high fidelity = hi. fi.
hohe Anforderungen exacting requirements
hohe Auflösung high resolution (535), (621.397)
hoher Außenwiderstand high (-load) resistance (621.385)
auf hohen Außenwiderstand arbeiten to work through a high resistance
hohe Beanspruchung heavy loading conditions (pl.)
hohe Drehzahl high speed
hoher Druck high pressure (auch 551.5)
auf hohen Druck bringen to pump to high pressure
hoher Feuchtigkeitsgehalt high moisture content
hoher Heulton high-speed whine
hohe Leistung high wattage, high power
hoher Luftdruck high barometer, high pressure
hoher Mast tall pylon, tall tower
hohes Niveau high level
hohe Nummer high number
hoher Scheinwiderstand high impedance
hohe Spannung high voltage, intense voltage, high tension, high pressure
hoher Ton high-pitch note
hohe Verstärkung high gain
hoher Widerstand high resistance
über hohen Widerstand arbeiten to work through a high resistance (621.385)
hohe Wiederabgabegüte high-grade fidelity
hoher Wirkungsgrad high efficiency

höhere Harmonische, höhere Oberwellen higher harmonics
Höchst-... maximum, top, super
Höchstbeanspruchung maximum stress
Höchstbedarf maximum demand
Höchstdrehzahl top speed, maximum speed, full speed
Höchstdruck super pressure
Höchstdruckbeanspruchung maximum compression stress
Höchstfrequenz hyperfrequency
Höchstgewicht maximum weight
Höchstinduktion normal induction (621.314); maximum induction
Höchstintegration very large scale integration, VSLI
Höchstlast maximum load, maximum demand
Höchstlastanzeiger maximum-demand indicator
Höchstleistung maximum capacity, maximum power
Höchstspannung maximum stress (620.1); ceiling voltage (621.3), very high tension, supertension (620.1), (621.3)
Höchstspannungskabel super-tension cable
Höchstspannungsleitung very high tension line, extra-high-voltage line
Höchststrom maximum current
Höchsttemperatur maximum temperature
Höchstwert peak (value), maximum (value), crest, extreme
höchst, höchst-... highest, maximum, ceiling
höchste Ablesung highest reading
höchste Ausgangsleistung maximum output (621.396)
höchste Betriebsspannung ceiling voltage
höchste Dringlichkeitsstufe first priority
höchste Einstellung highest reading
höchste nichtzündende Steuerspannung gate non-trigger voltage (621.314.63)
höchstes Potential maximum potential
höchste Spitze crest amplitude
höchste Übertragungsleistung maximum power transfer
höchster Wirkungsgrad maximum efficiency
höchste Zugbeanspruchung maximum tensile stress
höchstmögliche Betriebsfrequenz maximum usable frequency (621.396)
höchstmögliche Leistung maximum capacity
höchstverdichtetes Papier superdense paper
höchstwertige Ausführung top quality
in höchstwertiger Ausführung top quality
höchstwertiges Bit most significant bit, MSB
höchstzulässig rated maximum
höchstzulässige Beanspruchung maximum safety stress (620)
höchstzulässige Drehzahl maximum speed (potential)
höchstzulässiger Modulationsgrad modulation capability
höchstzulässige Spannung maximum potential, rated (maximum) voltage
höchstzulässige Zuglast maximum safe pull
Höchstädterkabel H-cable, Hoechstaedter cable
Höhe height, altitude; perpendicular, vertical height (51), pitch (534), (621.8)
Höhe des Vakuums hardness (621.385)
Höhenbedeckung vertical coverage (621.396.96)
Höhenentzerrung treble equalization (621.396.62)
Höhenfestigkeit altitude resistance
Höhenmesser altimeter, altitude meter, height indicator
Höhensonne Heraeus lamp
Höhenspannungsgenerator teaser (621.314.2)
Höhenspannungstransformator teaser transformer
Höhenstrahlung cosmic radiation
Höhenwicklung teaser winding
Höhenwinkel angle of elevation
Höhenwinkel-Entfernungssichtgerät elevation-position indicator, E.P.I.
Höhepunkt peak, zenith, climax, culmination
höhere Programmiersprache high-level language
Höckerpunkt peak point
Höckerstrom peak-point current
Hohl-..., hohl hollow, concave
Hohlkabel waveguide
Hohlkabeldispersion wave-guide dispersion
Hohlkathode concave cathode
Hohlkegel hollow cone
Hohlkugel spherical cavity, holow sphere
Hohlkugelresonator spherical resonator
Hohlleiter wave guide
Hohlleiterabstimmkolben wave-guide plunger
Hohlleiterantenne radiating guide
Hohlleiterblende baffle plate
Hohlleiterdämpfungsglied wave-guide attenuator
Hohlleiterdifferentialübertrager magic tee
Hochleitergrenzfrequenz wave-guide cutoff frequency
Hohlleiterknie wave-guide elbow
Hohlleiterlinse wave-guide lense

Hohlleiterschlitzantenne leaky wave guide antenna
Hohlleiterverdrehung wave-guide twist
Hohlmaß measure of capacity, liquid measure
Hohlraum void, vacuous space, cave, cavity, interstice
Hohlraumbildung caviation
Hohlraumleitung hollow-wave guide (621.396)
Hohlraumresonator resonant cavity, cavity resonator, resonance chamber, hollow space resonator
Hohlraum-Schlitz-Speisung cutler feed
Hohlrohr (leitung) hollow pipe (621.643), hollow conductor, wave-guide (621.396)
Hohlrohrmeßleitung wave-guide test bench
Hohlrohrwelle hollow-pipe wave, confined-space wave (621.396)
Hohlseil hollow conductor (621.315)
Hohlspiegel concave mirror
Hohlwand hollow (partition) wall
Hohlwellenleiter hollow-wave guide
Hohlwellenantrieb hollow-shaft motor drive
Holz wood; timber
Holzfaserplatte wood-fibre slab
Holzmast (wooden) pole
Holzmastleitung pole line
Holzschraube wood screw, common screw
hölzern wooden
Holzyklus fetch cycle (681.3)
homogen homogeneous; uniform
homogene Gleichung homogeneous equation
homogene Leitung uniform line
homogene Mischung homogeneous grading (621.395.3)
homogener Reaktor homogeneous reactor
Homogenität homogeneity
homolog homologous
homöopolare Bindung non-polar link (54)
honen to hone
homopolarer Generator homopolar generator
Hookesches Gesetz Hooke's law
Hör-... audio, audible
Hörabgleich aural setting
Hörapparat deaf aid, hearing aid
Hörbarkeit audibility
Hörbarkeitsdiagramm audiogramm
Hörempfindung audible sensation
Hörempfang audible response (621.396); sound reception
Hörempfangspeiler aural direction finder
Hörfrequenz audible frequency, audio frequency
Hörgerät hearing aid
Hörfunk radio transmission
Hörhilfe hearing aid
Hörkapsel receiver capsule
Hörorgan organ of hearing
Hörsamkeit acoustical quality
Hörschwelle threshold of audibility, threshold of hearing
Hörschwelle im freien Schallfeld minimum audible field
Hörschwelle des Schalldrucks im Ohr minimum audible pressure
Hörschwellenstrom audibility current
hörbar audible
hörbar machen to render audible
hörbarer Oberton audible harmonic
hörbarer Schwebungston audio-frequency beat note
hörbare Töne tonalities
horchen to listen
Hörer receiver (621.395); listener (654.19); headphone, headset (621.396.62)
Hörer abnehmen to lift the receiver
Hörer auflegen to replace the receiver
Horizont horizon
Horizont absuchen to sweep the horizon
Horizontalablenkung line sweep, horizontal deflection
Horizontalablenktransformator line-scan transformer
Horizontal-Austastlücke line-blanking interval
Horizontaldiagramm horizontal pattern (621.396)
Horizontaldruck horizontal thrust
Horizontalebene horizontal plane
Horizontalfeldstärke horizontal field strength
Horizontalfrequenz line frequency (621.397)
Horizontalfrequenzteiler line divider
Horizontalintensität horizontal intensity (538)
Horizontalschnitt horizontal section
Horizontal-Synchronisation horizontal synchronizing, line synchronizing
Horizontal-Synchronsignal line synchronizing signal, line sync pulse
horizontal horizontal
horizontal polarisiert horizontally polarized
horizontale Bewegungsfreiheit freedom of horizontal movement
horizontale Bündelung horizontal directivity
horizontale Langdrahtantenne flat-top antenna
horizontale Polarisation horizontal polarization
horizontale Reichweite ground range (621.396)
horizontale Richtwirkung horizontal directivity
horizontieren to level up

Horizontierungsfläche levelling pad
Horizontierungskreisel stable vertical gyro
Horn horn (629.113), (67), (681.8)
Hornparabolantenne horn-parabolic antenna
Hornstrahler horn (antenna), electromagnetic horn (621.396.67)
Hornstrahler mit zwei Seitenwänden bisectional horn antenna
Horntrichter horn (621.396.67), horn gate (621.7)
Hörnerableiter gap arrester (621.3), horn arrester (69)
Hörnerfunkenstrecke horn gap
Hörnerschalter horn break, horn-gap switch
Hörnersicherung horn break fuse
Hoyt-Nachbildung special balancing network
Hub stroke (621.4), (621.9); swing, deviation (621.396); shift (621.394)
Hubeinstellung stroke adjustment (621.9)
Hubkraft lifting capacity (621.318)
Hubmagnet lifting magnet
Hubmeßgerät deviation meter (621.396)
Hubraum piston displacement (621.4)
Hubweg length of (the) stroke
Hufeisen horseshoe
hufeisenförmig horseshoe-shaped
Hufeisenmagnet horseshoe magnet
Hughes-Schreiber Hughes print
Hülle sheath, covering; envelope, wrapper, jacket; shell (539), skin (629.12/.13)
Hüllenaufbau extranuclear structure
Hüllenelektron shell electron, extranuclear electron, orbital electron
Hüllkurve envelope (51)
Hülse shell, tubular, cover, bush, sleeve (621.8)
Hülsendipol sleeve dipole
Hülsenklemme, Hülsenverbindung sleeve joint
Hunderterteilung centigradation
Hunderterwähler hundreds selector
Hundertsatz percentage
Hupe horn, klaxon
hupen to hoot, to toot, to blow the horn
hüpfen to skip
Hurwitz-Bedingung Hurwitz relation (51)
Hütchenkondensator cup-shaped condenser
HV = Hochvolt, Hochspannung s. dort
H-Wächter H-type supervisory units (for high-voltage feeders in mines)
H-Welle H-wave, transverse electric wave, TE-wave
Hybrid-... hybrid
Hybridrechner hybrid computer
Hybridschaltung hybrid circuit
Hydraulik hydraulics (pl.)
hydraulisch hydraulic
hydraulisch betrieben hydraulically operated, oil-pressure operated
hydrieren to hydrogenate, to hydrogenize
Hydro-... hydro
Hydrochinon hydroquinone, quinol (77)
Hydrochinonelektrode quinhydrone half cell
Hydrodynamik hydrodynamics (pl.)
hydrodynamisch hydrodynamic
hydroelektrisch hydro-electric
Hydrophon hydrophone
Hydrostatik hydrostatics (pl.)
hydrostatisch hydrostatic
hydrostatischer Druck hydrostatic pressure
Hygrometer hygrometer
Hygroskop hygroscope
hygroskopisch hygroscopic
Hyperbel hyperbola
Hyperbelcosecans cosech = hyperbolic cosecant
Hyperbelcosinus cosh = hyperbolic cosine
Hyperbelcotangens tanh = hyperbolic cotangent
Hyperbelfeld hyperbolic pattern (621.396.93)
Hyperbelfunktion hyperbolic function
Hyperbelhornstrahler hyperbolic horn
Hyperbelnavigation GEE (navigation), hyperbolic navigation
Hyperbelsinus sinh = hyperbolic sine
Hyperbeltangens hyperbolic tangent
hyperbolisch hyperbolic
Hyperboloid hyperboloid
Hypernik hypernik, hipernik
Hypotenuse hypotenuse, long side, base
hypothetisch hypothetical
hypothetischer Bezugskreis nominal maximum circuit, hypothetical reference circuit
Hysteresis, Hysterese hysteresis
Hysteresisexponent hysteretic index
Hysteresiskonstante hysteretic constant
Hysteresismoment hysteretic drag
Hysteresismotor hysteresis motor
Hysteresisschleife hysteresis loop, B-H loop
Hysteresisschleife durchlaufen to go through a hysteresis loop
Hysteresisverlust hysteresis loss, hysteretic loss
Hysteresisverzerrung hysteresis distortion
Hysteresiswärme hysteretic heat
Hz = Hertz c. p. s. = cycles per second
50-Hz-Brumm fundamental hum

I

IC IC, Integrierte Schaltung
IC-Fassung IC socket
IC-Fotomaskierung IC-photomasking
ideal ideal, perfect
ideales Dielektrikum ideal dielectric, perfect dielectric
idealer Gleichrichter perfect rectifier
idealer Kondensator perfect condenser
ideales Metall perfect metal
idealer Wirkungsgrad absolute efficiency
Idealkristall perfect crystal
ideelle Erde reference earth
ideeler Kurzschluß virtual short circuit
Identifizierung des Rufers number identification
Identifizierungszeichen identification mark
idiostatisch idostatic
idiostatische Schaltung idiostatic method (621.317)
Idometer idometer
ID-Regler integral and derivative action controller
Ignitron ignitron
Ignitronschweißtakter ignitron-welding timer
Ikonoskop iconoscope, electron camera
Ilgner-Maschine buffer machine
Ilgner-Rad Ilgner wheel
Ilgner-System Ilgner system
Illumination illumination
illuminieren to illuminate
imaginär imaginary
imaginäre Achse axis of imaginaries
imaginärer Anteil imaginary (term)
imaginäre Einheit unit of imaginaries
imaginärer Teil imaginary part
Imaginärteil imaginary component
Immersion immersion
Impedanz impedance
Impedanz des gegenläufigen Feldes negative-sequence field impedance
Impedanzanpassung impedance matching
Impedanzfehlanpassung impedance mismatch
Impedanzkopplung impedance coupling
Impedanzmeßbrücke impedance bridge
Impedanzrelais impedance relay
Impedanzwandler impedance transformer
impfen to inject (539); to dop
geimpfter Reaktor injected pile (539)
Impfkristall seed crystal
implizite impliedly, implicitly
implizite Adressierung implied addressing
Implosion implosion (621.385)
imprägnieren to impregnate, to waterproof; to compound (621.3); to preserve (67)
Imprägnierbottich impregnating tank
Imprägniertank cable tank
Imprägnierung impregnation, (moisture) proofing (66); preparation (621.3); preservation (67)
Imprägnierungsmasse impregnation compound
Imprägnierungsmaterial impregnating material
Imprägnierungsmittel impregnant
Impuls pulse (621.3), (621.397); momentum, impetus (531), (539); impulse
Impulse formen to wave-shape (621.397)
Impulsabfallzeit pulse decay time
Impulsabstand pulse-digit spacing, pulse period
Impulsabtrennstufe clipper stage, pulse clipper (621.397)
Impulsabstand pulse period, pulse spacing
Impulsamplitude pulse amplitude
Impulsamplituden-Analysator pulse-amplitude analyzer
Impulsanstiegszeit pulse rise time
Impulsausleseschaltung pulse separator, impulse separator
Impulsbandbreite pulse bandwidth
Impulsbefehl pulse command
Impulsbreite pulse width
Impulsdauer duration of a pulse, pulse duration (621.397), pulse length
Impulsdiagramm timing diagram
Impulsenergie pulse energy
Impulserneuerung sync regeneration, pulse regeneration
Impulsfernmeßgerät impulse-frequency telemeter
Impulsfolge pulse repetition, pulse train (621.394)
Impuls(folge)frequenz P. R. F. = pulse-repetition frequency, repetition rate (621.397), impulse frequency
Impulsform pulse envelope
Impulsformerschaltung shaping circuit, pulse-forming circuit, pulse-forming network, pulse shaper
Impulsgeber pulse transmitter, pulse generator; (selective) digit emitter (681.3); synchronization-signal generator, synchronization-clock system (621.397)
Impulsgeberschaltung trigger circuit (621.397)
Impulsgenerator pulse generator
Impulsgruppe pulse group, pulse train

Impulshöhe pulse amplitude
Impulskette pulse-forming network
Impulslänge pulse length
Impulsmagnetisierung flash magnetization
Impulsperiodendauer pulse interval
Impulsrauschen impulse noise
Impulsregenerierung pulse regeneration
Impulsserie pulse train
Impulsspannungsmesser pulse voltmeter
Impulssperrung gating (621.397)
Impulsspitze pulse spike
Impulsspitzenleistung peak power
Impulstastverhältnis mark-to-space ratio
Impulstransformator pulse transformer (621.397)
Impulstrennstufe pulse separator
Impulsübertrager impulse repeater (621.395.34)
Impulsverhältnis impulse ratio (621.395)
Impulsverstärker pulse amplifier
Impulsverteilerschaltung synchronizing pulse distribution amplifier
Impulsverzerrung impulse distortion
Impulsverzögerungsgenerator delayed pulse generator
Impulsverzögerungszeit pulse-delay time
Impulswahl impulse action (621.395)
Impulswärmewiderstand thermal impedance for one single pulse
Impulswiederholer pulse repeater; responder beacon (621.396.96)
Impulszähler pulse counter, pulse meter, impulse summator (621.317)
Impulszählschaltung counting circuit, counting train
Impulszentrale pulser, timer, keyer
Impulszug pulse group, pulse train
inaktiv inactive, idle, inert
inaktiver Draht idle wire
Inbetriebsetzung einer Maschine starting of a machine
Inbus-Schraube Allen screw, hollow screw, socket-head cap screw
Index index, affix
Indexloch index hole (681.3)
Indexregister base register, index register (681.3)
Indexstreifen index stripe (621.397)
Indexträger pilot carrier (621.397)
mit einem Index versehen to index
Indikator indicator (54); (621.1/.4)
indirekt indirect
indirekt geheizte Kathode indirectly heated cathode, heater-type cathode
indirekt gesteuertes System indirect control system (621.395.3)
indirekte Adressierung indirect addressing, indexed addressing
indirekte Auslösung indirect tripping, transformer overcurrent tripping
indirekte Beleuchtung indirect lighting, indirect illumination
indirekte Betätigung indirect operation
indirekte Heizung indirect heating (621.385)
indirekte Leuchte ceiling reflecting lamp
indirekte Nullstromauslösung transformer undercurrent tripping
indirekter Regler pilot-operated controller
indirektes Verfahren indirect mode
Indium indium
indizieren to indicate (621.1)
indizierte Adressierung indexed addressing
indizierte Leistung i. h. p. = indicated horsepower
individuelle Abnehmerleitung individual trunk (621.395)
Individueller Verzerrungsgrad individual distortion (621.394)
Induktion induction, flux density
induktionsarm low-inductance
induktionsfrei inductionless, noninductive, non-reactive
induktionsfrei gewickelt noninductively wound
induktionsfreie Belastung nonreactive load, ohmic load
induktionsfreies Kabel noninductive cable
induktionsfreier Kreis noninductive circuit, nonreactive circuit
induktionsfreie Wicklung Ayrton-Perry winding, inductionless winding
Induktionsfühler inductive pick-off, induction pick-up (coil)
Induktionshärtung induction hardening
Induktionsheizung induction heating
Induktionskonstante space permeability
Induktionslötung induction brazing
Induktionsmaschine induction machine
Induktionsmotor induction motor
Induktionsofen induction furnace, Kjellin furnace
Induktionsofen mit Röhrensender valve-operated induction furnace
Induktionsschutz anti-inductive
Induktionsspannung induced voltage
Induktionsspule induction coil, inductor
Induktionswirkung inductive action
durch Induktionswirkung by induction

Induktionszähler induction (motor) meter
induktiv inductive, reactive
induktiv-galvanische Kopplung autoinductive coupling
induktiv gekoppelt inductively coupled
induktive Abstimmung permeability tuning, inductance tuning
induktive Belastung inductive load(ing)
induktive Blindspannung reactive voltage
induktiver Blindwiderstand inductive reactance, magnetic reactance
induktive Erwärmung induction heating
induktive Hohlleiterkopplung choke joint (621.372)
induktive Komponente reactive component
induktive Kopplung inductance coupling, mutual inductance coupling, transformer coupling
induktive Last inductive load, reactive load, lagging load
induktiver Leistungsfaktor lagging powerfactor
induktives Potentiometer ipot
induktive Restfläche des Hallstromkreises effective induction area of the output loop
induktive Restfläche des Steuerstromkreises (eines Halleffekt-Bauelementes) effective induction area of the control current loop (621.382)
induktiver Scheinwiderstand inductive impedance, reactive voltamps (pl.)
induktiver Spannungsabfall inductance (voltage) drop, reactance voltage
induktiver Spannungsanstieg inductive rise
induktive Störbeeinflussung inductive interference
induktive Stoßschweißung electromagnetic percussive welding
induktiver Strom inductive current, lagging current, reactive current
induktiver Stromkreis inductive circuit
induktiver Widerstand inductive resistance, inductance, reactance
induktive Widerstandsschweißung induction resistance welding
induktiver Zweig inductive branch
Induktivität inductivity, inductance, coefficient of self-induction
Induktivität für überlagerten (kleinen) Wechselstrom incremental inductance
Induktivität der Zuführungen lead inductance
Induktivitätsbelag inductance per unit length
Induktivitätsmeßbrücke inductance bridge
Induktivitätsmesser henrymeter, secohmmeter, inductance meter
Induktivitätsnormal inductance standard
Induktor inductor, magneto ringer, magneto generator
Induktoranruf magneto ringing, manual ringing, magneto call (621.395)
Induktorballen inductor core
Induktorkreis rotor circuit
Induktorpol field pole (621.313)
Induktorruf s. **Induktoranruf**
Induktorspannung inductor voltage
Induktorspule inductor coil
Induktorstrom inductor current, rotor current
Induktorstromkreis inductor circuit
Induktorwecker magneto bell
Induktorwicklung inductor winding, rotor winding
Industrie industry
industriell industrial
industrielles Fernsehen closed-circuit television
industrielle Funkstörung man-made noise
induzieren to induce
induzierte elektrische Ladung induced electricity
induzierter Kernzerfall induced nuclear disintegration
induzierte Radioaktivität induced radioactivity
induzierte Ständerspannung stator e. m. f.
induzierter Strom induced current
ineinander into one another
ineinander eingreifen to mate
ineinander übergehen to slur (51); to coalesce
ineinanderfügen to interlace, to spigot into one another
ineinanderlaufen to run together
ineinanderwickeln to interweave (621.314.2)
ineinanderziehen to slur (51)
inert inert
inertes Gas inert gas
infiltrieren to infiltrate
Infinitesimalrechnung infinitesimal calculus
Influenz influence
Influenzmaschine influence machine, electrostatic generator
Informatik computer science
Information information
Informationsbit information bit
Informationsdichte information density
Informationsdurchsatz information througput
Informationskanal information channel
Informationskapazität information volume
informationsloses bit overhead bit

Informationstheorie information theory (621.39)
Informationsträger (information) storage medium
Informationsverarbeitung information processing
Infrarot infrared
Infrarot-Bildwandler infrared converter, noctovisor
Infrarotmikroskop infrared microscope
Infrarot-Sehen noctovision
Infrarotspektroskopie infrared spectroscopy
Infrarotstrahlung infrared rays (pl.)
Infrarot-Therapie infrared therapy, radiant heat therapy
Infraschall infrasound, subaudio oscillation
Infraschallfrequenz infra-acoustic frequency, subaudio frequency, infra-audible frequency, subaudible frequency, infrasonic frequency, subsonic frequency, infrasound frequency, subsound frequency
Ingenieur engineer
Ingenieur der Praxis practicing engineer
Ingenieurschule polytechnic school, engineering college, technical college
Ingenieurwesen engineering
Inhalt contents (pl.), content; pondage (627.6)
Inhaltsverzeichnis index, table of contents
inhärent inherent
inhärentes Gleichstromverhältnis inherent forward current transfer ratio
Inhibitor inhibitor (621.357)
inhomogen inhomogeneous, non-uniform
inhomogenes Feld non-uniform field
Inhomogenität inhomogeneity
Initialzündung initiation, primer detonation
initiieren to prime (66)
Injektion injection
Injektionsfolgespannung post injection voltage (621.382)
Injektionslaser injection laser
injizieren to inject
Inklination inclination, dip (538)
Inklinationsnadel inclinatorium
inklusives Oder inclusive OR
inkohärent incoherent
Inkohärenz incoherence
Inkohlung (in-)carbonization
inkommensurabel incommensurable, disproportionate, inadequate
inkommensurable Eigenschaft incommensurability
inkompressibel incompressible
Inkompressibilität incompresibility
inkonstant inconstant
Inkonstanz inconstancy
Inkreis in-circle
Inkrement increment
Inland inland, home country
Inlandsferngespräch inland trunk call
Inlandsverbindung internal circuit, domestic circuit
Inlay-Verfahren inlay
Innen-... inner, interior, internal, inside, indoor
Innenabmessungen inside measurements, inside dimensions
Innenansicht interior view
Innenantrieb to tally enclosed drive (621.9)
Innenbeleuchtung interior lighting
Innendurchmesser ID = internal diameter, inside diameter
Innenfläche inside face
Innengewinde internal thread, female thread
Innenlehre feeler gauge
Innenleiter inner conductor, inside conductor, centre conductor (621.315.2), (621.396)
Innenlunker blow hole
Innenmaß inside dimension
Innenmattierung internal frosting (621.326)
Innenpolgenerator rotating-field-type alternator
Innenpolmaschine external armature alternator (621.313)
Innenradius inside radius
Innenraum interior
Innenraumantenne indoor aerial
Innenraumisolator indoor insulator
Innenraumklima room climates, indoor climates
Innenraumumspanner indoor transformer
Innenreflektor internal reflector
Innenschicht inner layer, inner core (674)
Innensechskantschraube Allen screw, socket-head screw
Innenwiderstand internal resistance; plate resistance (621.385)
innen in, inside, within
nach innen gerichtet inward, inwards
von innen her from within
innenmattiert internally frosted
innenmattierte Lampe internally-frosted bulb
innenopalisierte Lampe lamp with internal diffusing coating
inner s. **Innen-...**
inneres Driftfeld internal drift field (621.382)
inneres Elektron inner-shell electron

innere Gitteranodenkapazität grid-plate tube capacity
innere Induktivität inner self-inductance
innere Kräfte erection stresses
innerer Leiter s. **Innenleiter**
inneres Produkt inner product, scalar product, dot product
innere Reibung molecular friction
innere Rückkopplung internal feedback
innere Selbstinduktion internal self-inductance
innere Spannung internal voltage (621.3); locked-in stress (620.1)
innere Spannungen internal stresses, erection stresses (624)
innerer Verstärkungsfaktor gas amplification factor (621.383)
innere Wärme heat equivalent of internal work
innerer Wechselstromwiderstand internal impedance
innerer Widerstand internal resistance, inherent resistance (621.3); anode a. c. resistance (621.385)
innewohnend intrinsic, inherent
innig intimate
innige Berührung intimate contact
innige Mischung intimate mixture
Inschrift inscription, lettering
Inselbildung island effect
Inspektion inspection, examination
inspizieren to inspect, to examine; to superintend
instabil instable
Instabilität instability
Installateur electrical contractor, electrician (621.3); plumber (696.2), (628)
Installation installation, wiring (arrangement); plumbery, plumbing (69)
Installation im Freien outside run, outdoor wiring
Installation auf Porzellanisolatoren knob wiring
Installationsarbeit installation work
Installationserdung interior wiring system ground
Installationsgerät installation equipment
Installationsrohr conduit
Installationsvorschrift wiring regulation, wiring code
installieren to install; to pipe (696); to wire (621.3)
installierte Erzeugerleistung electrical generating capacity
installierte Kraftwerksleistung generatingstation capacity
installierte Leistung plant capacity, prime power, installed capacity
instandhalten to maintain
Instandhaltung maintenance
instandsetzen to repair, to recondition, to overhaul, to service
instandsetzende Unterhaltung corrective maintenance
Instandsetzung repair, reconditioning, overhauling, servicing
Instandsetzungsarbeit repair work, service
Instandsetzungszeit (active) repair time
Institut institute, establishment
Instruktion instruction
Instrument instrument; meter (621.317); implement, device
Instrument mit Federrichtkraft spring-control instrument
Instrument mit Nullpunkt in der Mitte zero-centre instrument
Instrument mit unterdrücktem Nullpunkt suppressed-zero instrument
an einem Instrument ablesen to read off an instrument
Instrumenten-Anflugverfahren instrument approach
Instrumentenanzeige meter indication, meter deflection, meter display
Instrumentenbestand instrumentation
Instrumentenbrett dashboard (629)
Instrumentenflugregeln instrument flight rules, IFR
Instrumentenfuß instrument base
Instrumentengehäuse instrument housing
Instrumentenkorrektur instrument correction
Instrumenten-Landesystem ILS, instrument landing system
Instrumenten-Navigation instrument navigation
Instrumentenraum instrument shop
Instrumentenschalter instrument switch
Instrumententafel instrument panel
instrumentelle Ausrüstung instrumentation
intakt intact, unhurt
Integral integral
integral wirkender Regler integral-action controller, reset-action controller
Integralgleichung integral equation
Integralzeichen integral sign, sign of integration
Integrand integrand
Integration integration
Integration ausführen to perform an integration

Integration im Komplexen complex integration
Integrationsfläche surface of integration
Integrationsgrenzen limits of integration
Integrationskonstante constant of integration
Integrationsschaltung integrator circuit
Integrationsvariable variable of integration
Integrationsverfahren integration method
Integrationsweg path of integration
Integrator integrator, integrating network, integrating circuit
integrieren to integrate
integrierender Bestandteil integral part, integrant
integrierender Fehler cumulative error
integrierte Halbleiter-Großschaltung LSI, large scale integration
integrierter Halbleiterkreis semiconductor integrated circuit, IC
integrierendes Instrument integrating instrument
integrierter Mikrowellenschaltkreis microwave integrated circuit
integriertes Netz integrated network
integrierendes Netzwerk integrating circuit
integrierende Regelung floating action
integrierte Schaltung integrated circuit, IC
Intensimeter dose rate meter (539)
Intensität intensity
intensiv intense, intensive
intensive Hitze intense heat
interaktiv interactive
Intercarrierbrumm intercarrier noise
Intercarrierverfahren intercarrier system
Interface interface
Interferenz interference (535)
Interferenzbild interference pattern, interference figure
Interferenzgebiet fringe area, mush area (621.371)
Interferenzphotometer interference photometer
Interferenzschwund phase fading, interference fading
Interferenzstreifen interference fringe
interferieren to interfere (535)
Interferometer interferometer
Interferometrie interferometry
interkristallin intercrystalline
interkristalline Rißbildung intercrystalline cracking
intermittierend intermittent
intermittierender Erdschluß intermittent (earth) (621.3)
intermittierender Fehler intermittent (trouble)
intermittierender Strom intermittent current
intermittierende Zählung intermittent integration
Intermodulation intermodulation
Intermodulationsverzerrung intermodulation distortion
intermolekular intramolecular
international international
internationale Einheiten international units
internationale Norm international standard
Interpolation, Interpolierung interpolation, intercalation
interpolieren to interpolate, to intercalate
Interpretierer interpreter (681.3)
interstellar interstellar
interstellares Rauschen interstellar noise
Intervall interval
Intrinsicdichte inversion density
Invar invar
invariant invariant
Invarianz invariancy
Inventar (verzeichnis) inventory (sheet)
Inventur machen to take stock
Inversion inversion (621.3)
inverser Betrieb inverse direction of operation (621.382.3)
Inversionsdichte inversion density
Inverter inverter
invertierender Verstärker inverting amplifier
invertierte Sprache inverted speech
in vitro in vitro
Ion ion
Ionen-... ion, ionic
Ionenausbeute ion yield
Ionenaustausch ion exchange (54)
Ionenaustauscher ion exchanger
Ionenbeweglichkeit ionic mobility
Ionenbildner ionogen
Ionenbombardement ion bombardment, ionic bombardment
Ionenemission emission of ions
Ionenfalle ion trap
Ionenfleck ion spot, ion burn (621.385)
Ionengeschwindigkeit ionic speed, ionic velocity, velocity of ions
Ionenhalbleiter ionic semiconductor
Ionenimplatation ion implantation
Ionenleitung ionic conduction
Ionenlehre ionics (pl.)
Ionenrohr ionic valve
Ionenschleuder ionic centrifuge (539)
Ionenstrahl ion beam, ionic jet, ray of ions
Ionenstrahlen ion rays, anode rays
Ionenstrom anode-ray current, ionic current
Ionenwanderung ion migration

Ionenwind ionic wind, ionic breeze
Ionenwindvoltmeter ionic-wind voltmeter
Ionisation ionization
Ionisationsansprechspannung discharge inception voltage
Ionisationsgeschwindigkeit ionization rate
Ionisationskammer ionization chamber
Ionistionsmanometer ionization gauge
Ionisationsmikrophon cathodophone
Ionisationsstrom ionization current
Ionisationsstufe ledge (621.372)
ionisieren to ionize
ionisiertes Atom ionized atom
ionisiertes Gas ionized gas
ionisierte Schicht ionized layer
Ionisierung ionization
Ionisierungsknick ionization threshold
Ionisierungsspannung ionization potential
Ionosphäre ionosphere, Kenelly-Heaviside layer
Ionosphärenhülle ionospheric shell
Ionosphärenlotung ionospheric sounding
Ionosphärenschicht ionosphere layer, ionospheric layer
ionosphärisch ionospheric
ionosphärische Fokussierung ionospheric focusing
ionosphärische Kreuzmodulation ionospheric cross modulation, Luxemburg effect
ionosphärische Streuausbreitung ionospheric scatter
IPD-Regler IPD-action controller
IPD-Verhalten integral proportional derivative action
IP-Verhalten integral proportional action
I-Regelung integral control
I-Profil I-beam, I-section
irden earthenware
Iridium iridium
Iris iris
Irisblende iris diaphragm
irisierend iridescent
I-Rohr surface metal raceway, thinwalled tubing (696)
irrational irrational, surd (51)
irrationale Zahl irrational number, improper number, surd (number)
irreversibel irreversible
irreversible Umwandlung irreversible transformation
ISB-System independent sideband system
Isobare isobar, isobaric line (55)
Isobarenausbeute chain fission yield (539)
Isobutan isobutane
Isochron-Verzerrungsgrad isochronous distortion (621.394)
Isodromregler proportional reset controller
isoklin isoclinal
Isoklinal-... isoclinal (54)
Isokline isocline, isoclinic line
Isolation insulation (621.3), (69)
Isolationsdurchbruch primary fault, insulation failure, breakdown of insulation
Isolationsfehler insulation failure, leak, leakage
Isolationsfestigkeit dielectric strength
Isolationsnennspannung rated insulation voltage
Isolationsprüfer insulation tester, megger, insulation indicator
Isolationsprüfung insulation test
Isolationspunktierung puncture of insulation
Isolationsschaden insulation defect
Isolationsstrom leakage current
Isolationsverluste insulator losses (621.396.67)
Isolationswiderstand insulation resistance, leak resistance, insulance
Isolator insulator
Isolator mit Schwanenhalsstütze swan-necked insulator
Isolatorenaufhanfkluppe hemp-fixing clamp (for insulators)
Isolatorenaufhängung insulator suspension
Isolatorenkette (insulator) string (621.315.6)
Isolatorschirm petticoat, shed
Isolatorenstütze insulator pin
Isolatorenteller porcelain cup
Isolier-... insulated, insulating
Isolieranstrich insulating coating
Isolierband insulating tape, rubber tape
mit Isolierband umwickeln to tape (69.66)
Isolierbuchse insulating bush, insulated bushing
Isolierfähigkeit insulating power, insulativity
Isolierfilz insulation felt
Isolierflüssigkeit liquid insulation, insulating liquid
Isolierhülse slot liner (621.313); insulating sleeve
Isolierklemme insulated bolt, insulated terminal
Isolierkunststoff synthetic insulating material, insulating plastic
Isolierlack insulating lacquer, insulating varnish, coil varnish
Isolierlage insulating course (693)
Isoliermantel insulating jacket

Isoliermantelkabel non-metallic sheathed cable
Isoliermasse insulating compound
Isoliermaterial insulating material, insulator
Isoliermuffe insulating joint (621.315.2)
Isolieröl insulating oil
Isolierpapier vable paper
Isolierpappe insulating cardboard
Isolierperle insulating bead
Isolierpreßstoffe (insulating) plastics, moulded plastics (621.3)
Isolierrohr insulating tube, insulating guide
Isolierschicht insulating layer, insulating course
Isolierschicht-Feldeffekttransistor insulated-gate field-effect transistor
Isolierschlauch insulating hose
Isolierschnalle buckle insulator
Isolierstoß insulated rail joint (625)
Isolierstütze insulating support
Isoliertransformator isolating transformer, one-to-one transformer
Isolierumhüllung insulation sleeve
Isolierunterlage insulating stand
Isoliervermögen insulating power
Isolierwand insulating screen
isolieren to insulate (621.3); to isolate (= trennen)
isolierend insulating, insulant
isolierende Eigenschaft insulating property, dielectric property
isolierende Führung insulating guide
isolierender Untersatz insulating stand
isoliert insulated, covered, disconnected (621.3)
isolierter Bestandteil insulated component
isolierter Draht insulated wire, covered wire
isoliertes Kabel insulated cable
isolierter Leiter insulated conductor
isolierte Leitung insulated cable
isolierte Leitungsklemme insulated clip (696)
isolierter Leitungsträger insulated hook (696)
isolierter Mittelleiter insulated middle wire, insulated neutral (conductor)
isoliertes Netz isolated-neutral system
isolierter Sternpunkt isolated neutral
isolierter Teil insulated part, insulated component
isolierte Wicklung insulated winding
Isolierung insulation; isolation (= Trennung)
isomorph isomorphous
Isoperm isoperm, perminvar
isotherm isothermal
Isotop isotope
Isotopenbatterie isotope battery
Isotopenhäufigkeit isotopic abundance
Isotopenhäufigkeitsverhältnis abundance ratio
Isotopentrenner mass separator (539)
Isotopentrennung isotope separation
isotop isotope
isotrop isotropic
isotroper Isolierstoff isotropic dielectric
isotroper Körper isotropic body
isotroper Strahler isotropic radiator
Isthmusmethode, Isthmusverfahren isthmus method
Istfrequenz actual frequency
Istwert actual value, real value; control print (621.316.7)
Iteration iteration
Iterationsverfahren cut-and-try process, process of trial and error
iterieren to iterate
I-Träger I-beam, joist
I-Verhalten integral action

J

J = Joule J = joule
Jablochkoff-Kerze Jablochkoff candle
Jägerleitradar air-interception radar
Jahr year
Jahresbelastungskurve annual load curve
Jahresgang annual variation
Jahresmittel annual mean
Jahresspitze absolute peak (621.315)
Jahreswirkungsgrad annual efficiency
Jahreszeit season
jahreszeitlich seasonal
jahreszeitliche Belastung seasonal load
jährlich yearly, annual
jährlich wiederkehrend perennial
jährlicher Gang annual variation
J-K Flip-Flop J-K flip-flop
Jedermann-Funkgerät citizen-band radio-telephone
Joch yoke
Joch umschlingen to link a yoke
Jochblech yoke lamination
Jochbolzen yoke-pressure bolt
Jochmagnetisierung yoke induction
Jochverfahren bar and yoke method (538.22)
Jochverlust yoke losses (pl.)

Jod iodine
Joubert-Oszillograph ondograph
Joubertscheibe revolving contact maker
Joulesches Gesetz Joule's law
Joulesche Wärme Joulean dissipation, resistance loss, rheostatic loss
Joulesche Wärmeentwicklung Joule heating effect
justierbar adjustable
justieren to adjust
Justiermast collimation mast
Justierschraube adjusting screw
Justierzange spring-adjusting plier
Jute jute
Juteumspinnung bedding of jute

K

KA = Kabelaufführungspunkt cable-distribution point
kA = Kiloampere kiloamp
Kabel cable
Kabel abbinden to tie off a cable
Kabel mit abgestufter Isolation graded cable
Kabel mit Aluminiummantel aluminium-sheathed cable
Kabel mit Bleimantel lead-covered cable, lead-sheathed cable
Kabel mit Gürtelisolation belted cable
Kabel mit Juteisolation jute-insulated cable
Kabel mit Lackleinenisolation varnished-cambric cable
Kabel mit längsveränderlicher Bespulung tape-loaded cable
Kabel legen to run a cable
Kabel mit ovalem Mantel oval cable
Kabel pupinisieren to load a cable (621.395)
Kabel verlegen to lay a cable
Kabelabschluß cable termination
Kabelabschnitt cable section
Kabelader cable conductor, cable core
Kabelanschluß cable terminal, cable connection
Kabelaufbau cable construction
Kabelaufführung cable post
Kabelaufführungspunkt cable distribution point
Kabelaufhänger cable suspender, cable sling
Kabelaufteilung cable distribution
Kabelbaum cable tree, cable harness, wiring harness
Kabelbeilauf cable filler
Kabelbewehrung cable armour(ing)
Kabelbündel bunched cables (pl.), bunch of cables
Kabeleinführung cable entry, cable inlet
Kabelendverschluß cable sealing end, pot head (621.315.2)
Kabelerdungsschelle ground cable bond
Kabelerdstrom-Differentialschutz Ferranti-Hawkins protective system
Kabelfehler cable fault
Kabelfehlersuchgerät electrode cable detector (621.315.2)
Kabelfernsehen cable television, closed-circuit television
Kabelführung cable running
Kabelführungsplan cabling plan
Kabelgeschirr cable gear
Kabelgestell cable rack
Kabelgraben cable trench
Kabelgraben ausheben to trench
Kabelhalterung cable support bracket
Kabelinstallationsplan cable layout
Kabelkanal cable tunnel, cable throughing, cable duct, raceway duct bank, cable conduit
Kabelklemme cable terminal
Kabelkupplung cable coupling
Kabellöter jointer (621.395)
Kabelmantel cable sheath (ing)
Kabelmasse sealing compound, insulating compound
Kabelmast cable post
Kabelmerkstein cable marker
Kabelmesser cable stripper
Kabelmuffe cable box, cable joint, cable fitting, sealing box
Kabelnetz grid (621.315)
Kabelplan wiring diagram, circuit lay-out card
Kabelöse cable eye
Kabelrost raceway
Kabelrückleitung return cable (621.33)
Kabelschacht manhole chimney; cable chamber
Kabelschelle cable clamp, cable clip
Kabelschrank cable terminal box
Kabelschuh cable socket, forked terminal, cable lug
Kabelschutzbinde teredo tape, cable bandage (621.395.3)
Kabelschutzrohr cable conduit (621.395)
Kabelseele cable core
Kabelstromwandler cable current transformer
Kabeltrasse cable running

Kabeltrenner, Kabeltrennschalter cable isolator
Kabeltrommel cable drum
Kabeltunnel cable tunnel, cable subway
Kabelverbinder cable fitting, cable coupler
Kabelverbindung cable connection, cable jointing, cable coupling
Kabelvergußmasse filling material
Kabelverlegung cable laying
Kabelverlustwinkel cable PF = cable power factor
Kabelverzweiger cable distribution box
Kabelzubehör cable equipment
Kabine cabin; cage, car (621.876), telephone box (621.395)
Kabinenansage-Gerät cabin address equipment
Kabinenkabel lift electric travelling cable
Kabinenschaltung car-switch operation (621.876)
Kabinensteuerung car-switch (lift) control
kadmieren to cadmium-plate
Kadmierung cadmium plating
Kadmium cadmium
Kadmiumbronze cadmium copper
Kadmiumdampflampe cadmium-vapour lamp
Kadmium-Normalelektrode cadmium electrode
kadmiumplattiert cadmium-plated
Kadmiumregelstab cadmium control rod
Kadmiumschirm cadmium screen
Kadmium-Selen-Zelle cadmium-selenide cell
Kadmiumsteuerstab cadmium control rod
Käfig cage
Käfiganker squirrel cage (621.313)
Käfigläufer squirrel-cage rotor
Käfigwicklung squirrel-cage winding
Kali potassium, potash
Kalilauge potassium hydroxide, potash lye
Kaliber calibre (623.4)
kalibrieren to calibrate
Kalium potassium
Kalk lime (54)
Kalomel calomel
Kalomelelektrode calomel electrode, calomel half-cell
Kalorie calorie, calory
Kalorimeter calorimeter
Kalorimeterbombe bomb calorimeter, calorimetric bomb
kalorimetrisch calorimetric
Kalt-... cold, low-temperature
Kaltbrüchigkeit low-temperature brittleness
Kaltdruckfestigkeit cold-crushing strength (66)
Kaltgerätestecker appliance plug
Kaltkathode cold cathode
Kaltkathodenröhre cold-cathode tube
Kaltleim cold-setting glue
Kaltleiter PTC thermistor
kalt cold, cool, frigid
kalt anfahren to start up cold
kalt einschalten to start up cold (621.314.6)
kaltgereckter Draht cold-strained wire
kaltgewalzt cold rolled
kaltgezogen cold-drawn
zu kalt deficient in temperature
kalte Kathode cold cathode
kalte Lötstelle cold junction, dry joint, rosin connection
kalte Zone frigid zone
Kälte cold
kältebeständig resistant to cold
Kältemaschine refrigerating machine
Kältemischung freezing mixture
Kalutron calutron
kalzinieren to calcine
Kalzium calcium
Kalziumkarbid calcium carbide
Kamera camera
Kameraröhre camera tube
Kamin funnel, chimney
Kamm comb
Kammpol comb-shaped pole, comb-type pole
Kammrelais cradle relay
kämmen mit to mate with, to gear with
kämmendes Rad mating gear
Kammer chamber, cabinet
Kammerofen chamber kiln
Kammerton standard musical pitch, a above middle c
Kammertrockner cabinet dryer
Kanal canal (626.1), (621.385), duct; channel (621.39); trunk (621.395)
Kanal bilden to channelize (621.395)
Kanalabstand channel spacing, channel separation (621.395.6)
Kanalanfang duct entrance
Kanalbreite channel division, communication band (621.396), bandwidth (621.395)
Kanalcodierung channel coding
Kanalgerät line-control unit (621.398)
Kanalkapazität channel capacity
Kanalraster channel spacing, channel pattern
Kanalstrahl canal ray, anode ray, ion beam
Kanalstrom anode-ray current
Kanalumschalter telegraph distributor, channel selector (621.394)

Kanal-Umsetzer channel modulator, channel translator, channel converter (621.395)
Kanal-Umsetzung channel translation (621.395)
Kanalverguß duct sealing (621.315.2)
Kanalverlegung (von Kabeln) solid system of cable laying
Kanalwähler channel selector, channel switch, tuner (621.396)
Kanalwiderstand channel resistance (621.382)
Kandelaber candelabrum
Kante edge, verge
Kanten behobeln to trim the edges
Kanten brechen to chamfer
auf der Kante stehend on edge
Kantenabtaster edge scanner (621.316.7)
Kantenbeugung knife-edge diffraction
Kantenbruch chamfer
Kantenverstärkung webbing (67)
kanten to tilt (over)
kantig angular
Kaolin kaolin, China clay
Kapazität capacity, storage capacity, output capacity, capacitance
Kapazität gegen Erde capacity to ground, capacity to earth
Kapazität pro Flächeneinheit unit-area capacity
Kapazität der Leitunge wiring capacity
Kapazität Wicklung gegen Wicklung interwinding capacity (621.314.2)
kapazitätsarm low-capacity, anticapacitance, bank-wound
kapazitätsarme Verdrahtung low capacity wiring
kapazitätsgerade Charakteristik linear capacity law
kapazitätsgerader Kondensator straightline capacity capacitor
Kapazitätsbelag capacitance per unit length
Kapazitätsdiode capacitance diode, varactor diode, varicap
Kapazitätsmeßbrücke capacitance bridge
Kapazitätsmesser capacitance metering device, capacitance meter, faradmeter
Kapazitätsnormal capacitance standard
Kapazitätsprüfung service test (621.35)
Kapazitätsprüfung nach Lagerung delayed test (621.35)
Kapazitätsprüfung im neuen Zustand initial test (621.35)
Kapazitätsprüfung bei Programmentladung intermittent test
Kapazitätsschwankung variation in capacity
Kapazitätsverhältnis (C_{min}/C_{max}) capacity ratio
Kapazitätswert coefficient of capacity
kapazitiv capacitive
kapazitiv-galvanische Kopplung autocapacitive coupling
kapazitiv zusammenschalten to join capacitively
kapazitiver Blindwiderstand capacitive reactance, capacitance
kapazitives Dach capacity to p (621.396.67)
kapazitive Dreipunktschaltung Colpitt's circuit (621.396.61)
kapazitive Erdung capacitance earth
kapazitive Erwärmung dielectric heating
kapazitiver Höhenmesser electric capacity altimeter
kapazitive Kabelfehlerortsmessung Gott's method
kapazitive Kopflast capacity to p (621.396.67)
kapazitive Kopplung capacitive coupling, capacitance coupling
kapazitive Last capacitive load
kapazitiver Leistungsfaktor leading power factor
kapazitiver Spannungswandler capacitor transformer
kapazitiver Strom capacitance current
kapazitiver Stromkreis capacitive circuit
kapazitiver Teiler capacitance divider
kapazitiver Widerstand capacitance
mit kapazitiven Eigenschaften condensive
kapillar capillary
Kapillare capillary (tube)
Kapillardetektor capillary detector, Armstrong-Orling detector
Kapillarelektrometer capillary electrometer
Kapillarkraft capillary force, capillary attraction
Kaplanturbine Kaplan turbine
Kapok kapok
Kappe cap (621.315.6); cover, hood
Kappenisolator disc insulator; suspension insulator (US)
kappen to lop
Kapsel cap, capsule, encapsulation
kapseln to enclose, to enclose totally, to enclose fully, to confine, to encapsulate
gekapselter Blitzschutz confined discharge arrester
gekapselte Maschine to tally enclosed machine
gekapselter Motor to tally enclosed motor
gekapselter Motor mit Eigenlüftung fan-ventilated motor

gekapselte Schaltanlage metalclad switchgear
gekapselte Schaltanlage mit Massefüllung solid system of switchgear
gekapselte Sicherung enclosed fuse cut-out
Karabinerhaken clipper, snap hook
Karat carat
Karbid carbide
Karborundum carborundum
Karbowid(widerstand) carbon-rod resistor, carbon-stick resistor, rodtype carbon resistor
Kardan gimbal, cardan
Kardanachse gimbal axis
Kardangelenk universal joint, spider
Kardanlager gimbal bearing
Kardanrahmen gimbal ring
Kardanreibung gimbal friction
Kardanwelle gimbal shaft
kardanisch gimbal
kardanisch aufgehängt, kardanisch gelagert gimbally mounted
kardanische Aufhängung gimbals (pl.)
Kardiogramm cardioid diagram
Kardioide cardiod, cardioid
Kardioidendiagramm cardioid pattern, heart-shaped diagram (621.396)
karieren to square
kariertes Papier squared paper
Karolus-Bildtelegraphie Karolus (system of) phototelegraphy
Karoluszelle Karolus cell
Karosserie car body, automobile body
Karte card (658); map (912); chart (51); printed board (621.3)
kartengesteuerter Streifenlocher card-to-tape converter
Karte stechen to clock one's card
Kartenablage card stacker
Kartenanschlag card aligner
Kartendarstellungsgerät map display (623)
Kartendoppler (card)reproducer
Karteneinblendung chart-comparison unit, video mapping
Kartenhinterkante card-trailing edge
Kartenleser card reader
Kartenlocher card punching equipment, card punch, card puncher
Kartenmagazin card bed, hopper
Kartenstapel deck
Kartenvorderkante card-leading edge
Kartenzuführung card feed
kartesisch Cartesian
kartesische Koordinaten Cartesian co-ordinates
kartesisches Koordinatensystem Cartesian grid
Karussell revolving table, rotating table (621.9)
Karusselldrehbank turning and boring mill, vertical lathe
Kaskade cascade
in Kaskade schalten to connect in cascade, to connect in tandem
Kaskadenanlasser cascade starter
Kaskadenmotor cascade motor
Kaskadenregelung cascade control (621.313)
Kaskadenschaltung cascade connection, tandem connection, concatenation connection (621.313)
Kaskadenumformer cascade converter, Lacour converter, Bragstad converter
Kaskadenverstärker cascade amplifier, cascode amplifier
Kaskadenwandler cascade transformer
Kassenterminal point of sales terminal
Kassette cartridge (681.84); cassette
Kassettenabspielgerät cartridge playback
Kassettenlaufwerk cartridge drive
Kassettenrekorder cassette recorder
Kassettentonbandgerät cartridge tape recorder, cassette recorder
Kasten box, kit, case; cage
Kastenbürstenhalter box-type brush-holder
Kastenplatte cape-type negative plate (621.355)
Kastenrahmen box-type frame (621.313)
Kastenwicklung spool winding
Katalog catalogue (Br.), catalog (US)
Katalysator catalyst, promoter
Katalysatormasse catalyst (material)
Katalyse catalysis
katalytisch catalytic
Kataphorese cataphoresis
Kathode cathode, filament, emitter, electron gun
kathodengekoppelte Gegentaktstufe longtailed pair
kathodenseitig steuerbarer Thyristor P-gate thyristor
Kathodenabbau cathode disintegration
Kathodenanheizzeit cathode-heating time
Kathodenanschluß cathode terminal
Kathodenbasisschaltung grounded-cathode circuit, cathode-basis connection
Kathodenblock cathode-bypass capacitor
Kathodenbombardement filament bombardment
Kathodendunkelraum cathode dark space
Kathodenfall cathode fall, cathode drop

Kathodenfallableiter cathode arrester, valve-type lightning arrester, multigap arrester (621.316.9)
Kathodenfläche cathode area
Kathodenfleck cathode spot
Kathodenfleck in Mittelstellung bringen to centre the cathode spot
Kathodenfolgeschaltung (cathode) follower
Kathodenform matrix (621.357)
kathodengekoppelt cathode-coupled
Kathodenheizung cathode heating
Kathodenheizleistung cathode-heating power
Kathodenklammer filament clamp
Kathodenklemme cathode terminal
Kathodenoberfläche cathode area
Kathodenröhrchen cathode sleeve
Kathodensättigung emission saturation (621.385)
Kathodensaum cathode border
Kathodenschicht cathode layer
Kathodenstrahl cathode ray
Kathodenstrahlbündel cathode-ray pencil
Kathodenstrahlfleck cathode-ray spot
Kathodenstrahlkardiograph tyriograph
Kathodenstrahloszillograph (ohne Registrierung) cathode-ray oscilloscope
Kathodenstrahloszillograph (mit Registrierung) cathode-ray oscillograph = c. r. o
Kathodenstrahlpeilung cathode-ray direction finding
Kathodenstrahlspeicherröhre cathode-ray storage tube
Kathodenstrahlröhre cathode-ray (oscillograph) tube
Kathodenstrahl-Schirmbilddarstellung cathode-ray tube display
Kathodenstrahlstrom cathode-ray current
Kathodenstrahlteilchen cathode-ray particle
Kathodenstrom cathode current, cathode stream
Kathodensumpf mercury pool
Kathodenverstärker cathode follower
Kathodenvorspannung cathode bias
Kathoden(vor)widerstand cathode-bias resistor
durch Kathodenwiderstand gekoppelt cathode-coupled
Kathodenzerstäubung cathode sputtering, cathode disintegration
kathodisch cathodic
kathodische Beizung cathode pickling
kathodischer Dunkelraum cathode dark space
kathodischer Elektrolyt catholyte
kathodisches Glimmlicht cathode glow
kathodische Polarisation cathodic polarization
kathodische Reinigung cathode cleaning (621.35)
kathodischer Schutz cathodic protection, Cumberland protection
Kation cation
katoptrisch catoptric
kaustisch caustic
Kautschuk caoutchouc, India rubber
Kavernenkraftwerk underground power station, cavern-type power station
Kavitation cavitation
KB = Kurzzeitbetrieb short-time duty
kcal = Kilokalorie kcal, kg.cal. = kilogramm calorie
Kegel taper (621), cone (51)
Kegelankermagnet long-pull magnet
Kegelgetriebe bevel gear
Kegelkopf cone head
Kegelkopfniet cone-headed rivet
Kegelreibungskupplung cone-friction clutch
Kegelrolle tapered roller
Kegelrollenlager taper-roller bearing
Kegelschale taper bush
Kegelschnitt conic section
Kegelstift tapered pin
Kegelstumpf truncated cone, frustrum of a cone
Kehlkopfmikrophon laryngaphone, throat (-type) microphone
Kehlschweißung fillet weld
Kehrlage reverse-frequency position (621.395.4)
Kehrwert reciprocal (value)
Kehrwert der Kapazität elastance
Kehrwert der Permeabilität reluctivity
Kehrwert des Isolationswiderstandes leakance
Keil wedge, cotter, key (621)
Keil vorlegen to scotch
mit Keilen öffnen to wedge open
Keilform wedge shape
keilförmig wedge-shaped, tapered
Keilkupplung wedge coupling
Keilnut Vee-groove (621.79), cotter way, keyway (621.8)
Keilriemen V-rope, V-belt
Keilriemenantrieb V-belt drive, V-belting
Keilriemenscheibe V-grooved wheel
Keilrille V-groove
mit Keilrille V-grooved
Keilschlitz cotter way
Keilschloß wedge lock
Keilverbindung wedging
keilen to wedge
Keim (für Kristallbildung) seed crystal

Keinathschreiber recording potentiometer, potentiograph
K-Einfang K-electron capture
Kellergeschoß basement
Kellerstationstransformator vault type transformer
Kellerspeicher stack, push-down storage, LIFO = Last In - First Out
Kellfaktor Kell factor (621.397)
Kellogschalter Kellog switch, to ggle switch, tumbler switch
Kelvin-Quadrantelektrometer Kelvin quadrant electrometer
Kelvinsche Stromwaage Kelvin balance, Kelvin scale (of temperature)
Kenn... significant, identification
Kennabschnitt significant interval (621.394.6)
Kennfaden tracer
Kennfarbe identification colour
Kennfrequenz assigned frequency
Kennlampe pilot lamp
Kennlinie characteristic (curve), response curve
Kennlinie einer Sicherung time-current characteristic of a fuse
Kennlinie einer Steuerspannung lumped characteristic (621.385)
Kennlinie eines Thermoelements thermo-electric characteristic
Kennlinienfeld family of characteristics
Kennlinienfläche characteristic surface (621.385)
Kennlinienknick turn-over characteristic
Kennlinienschar field of characteristics
Kennmarke distinctive mark
Kennung identity
Kennungsbake, Kennungsfeuer code beacon (656.7)
Kennungsabfrage IFF, identification of friend or foe
Kennungsgeber identity keyer
Kennungszeichen identification (sign); code (621.395.3)
Kennwiderstand image impedance
Kennwort pass word (681.3)
Kennzahl coefficient; prefix (654); code (621.395.3)
Kennzeichen characteristic, feature, mark, marking; flag (681.3)
Kennzeichenbit flag bit
kennzeichnen to mark, to designate; to flag (621.395), (681.3)
gekennzeichnete Verbindung flagged call (621.395)
Kennzeichnung designation, identification
Kennzeitpunkt significant instant (621.394), (681.3)
Keramik ceramics (pl.); ceramic material
Keramikröhre ceramic tube
keramisch ceramic
keramisches Dielektrikum ceramic dielectric
keramisches Gehäuse ceramic package
keramischer Isolator ceramic insulator
keramisches Isoliermaterial ceramic insulating material
keramischer Kondensator ceramic condenser
keramischer Sperrschichtkondensator barrier-layer capacitor
Kerbe notch, indent
Kerbfestigkeit notch toughness
Kerbschlagfestigkeit notched bar impact strength
Kerbschlagversuch notch bar impact test
Kerbwirkung stress concentration, notch effect
Kerbzähigkeit notch impact strength
kerben to notch, to indent
Kern core (52), (53), (539), (621.74), (621.314.3); nucleus (pl.: -lei) (539)
Kerne einlegen to core, to set cores (621.74)
mit einem Kern versehen to core (621.3)
Kernader core wire
Kernansatz core print (621.74)
Kernart atomic species, nuclid
Kernbau nuclear structure
Kernbauteil subatomic particle, nucleon
Kernbereich core of network (621.395)
Kernbindungsenergie nuclear binding energy
Kernblech core plate, core lamination (621.314.2)
Kernbrennstoff nuclear fuel
Kernchemie nuclear chemistry
Kerndurchmesser core diameter (621.3); (621.74)
Kernenergie nuclear energy
Kernenergieniveau nuclear energy level
Kernfluß core flux (621.3)
Kernforschung nuclear research
Kernfusion nuclear fusion
Kerngeschoß atomic projectile
Kernkettenreaktion nuclear chain reaction
Kernkonstante nuclear constant
Kernkraft nuclear force
Kernkraftwerk nuclear power plant, nucleonic power plant
Kernladung nuclear charge
Kernladungszahl atomic number, nuclear charge number

Kernleitwert short-circuit transmittance (621.39)
Kernmasse nuclear mass
Kernmatrix core array, core matrix
Kernphysik nuclear physics (pl.)
Kernquerschnitt core cross-section (621.3)
Kernreaktion nuclear reaction
Kernreaktor nuclear reactor, nuclear pile
Kernspaltstoff fissionable material
Kernspaltung atomic fission, nuclear fission
für Kernspaltung brauchbares Material fissible matter
Kernspeicher magnetic-core storage, core memory
Kernspin nuclear spin
Kernstopfen centre core (621.396)
Kernstruktur nuclear structure
Kerntechnik nuclear engineering
Kerntransformator core-type transformer
Kerntrocknung core baking (621.7)
Kernumwandlung nuclear transmutation, nuclear transformation
Kernverschmelzung nuclear fusion, atomic fusion
Kernwiderstand open-circuit transimpedance (621.372.5)
Kernzerfall nuclear disintegration, nuclear decay
Kernzertrümmerung fragmentation of nucleus
Kerrzelle Kerr cell
Kerze candle
Kerzenlampe candle lamp, flame lamp (621.32)
Kerzenstärke candle power
Kessel boiler (621.1); vessel, tank
Kesselspeisepumpe boiler-feeding pump
Kesselstein incrustation, fur, scale (621.1)
Kesselstein ansetzen, Kesselstein entfernen to scale, to descale
Kette string (621.315.6); chain (621.855), (67)
Kettendämpfung attenuation constant, iterative attenuation constant (US)
Kettendrucker chain printer
Kettenflaschenzug chain block
Kettenförderer chain conveyor
Kettengespräch serial call, chain call
Kettenglied section (621.392); chain link (621)
Kettenisolator chain insulator
Kettenleiter ladder network, lattice network, lattice circuit, wave filter, lumped-constant transmission line, recurrent networks
Kettenlinie catenary curve
Kettenreaktion chain reaction, cumulative process, self-propagating reaction
als Kettenreaktion verlaufend chain-reacting
Kettenrad chain sprocket
Kettenrolle plain pulley
Kettenschaltung concatenation connection, tandem connection
in Kettenschaltung in sequence
Kettenübertragungsmaß iderative propagation constant (US), propagation coefficient
Kettenverstärker distributed amplifier
Kettenwiderstand iterative impedance
Kettenwinkelmaß phase constant, iterative phase constant (US) (621.372)
Kettenwirkungsgrad string efficiency (621.315.6)
ketten an to lock to, to link to
Keule beam, lobe (621.396.67)
Keule schwenken to shift the lobe
Keulenachse beam axis
Keulenbreite beam width, beam angle, spread
Keulenschwenkung, Keulenumtastung lobe switching
Keulenumtastung beam switching, beam-lobe switching
Keulenumtastfrequenz lobe frequency
Keulenwinkel beam angle
kg = Kilogramm kg. = kilogram(me)
kHz kc. (p. s.) = kilocycles (per second)
9-kHz-Sperre 10 kc. trap, heterodyne filter
KI = künstliche Intelligenz AI, artificial intelligence
Kiesel(stein) pebble
Kieselgur kieselguhr
Kieselsäure silica; silicid acid
Kiloampere kiloamp
Kilohertz kilocycles (per second); (s. auch kHz)
Kiloliter kilolitre
Kilometer kilometre
Kilometerzähler milometer, cyclometer
Kiloohm kiloohm
Kilovolt kV = kilovolt
Kilovoltampere kVA = kilovoltampere
Kilowatt kW = kilowatt
Kilowattstunde kWh = kilowatthour, B. T. U. = Board-of-Trade Unit
Kilowattstundenzähler kilowatthour meter
kinematisch kinematic
Kinematograph cinematograph, cinema
kinetisch kinetic
kinetische Energie kinetic energy, velocity energy
kinetische Gastheorie kinetic theory of gases
Kino movies (pl.), (moving) pictures (pl.); cinema; motion pictures (US)
Kinoaufnahme, Kinobild motion picture

Kinokamera cine camera
Kinoprojektor cinema projector, film projector
Kinotechnik motion picture engineering
Kipp-... tilting, tipping
kippbar tiltable, inclinable
Kippbewegung tilting movement
Kippbewegung des Kolbens tilting movement of the bulb (621.314.6)
Kippbühne tipping stage
Kippfrequenz sweep-repetition rate, relaxation frequency
Kippfrequenzregelung hold control (621.317)
Kippgenerator relaxation oscillator (621.396), scanning generator, scan current generator (621.397)
Kipphebel rocker (621.4), to ggle (621.396.69)
Kippkreis trigger circuit, time-base circuit
Kippleistung stability limit (621.311)
Kippmoment tilting moment, overturning moment (531); breakdown to rque, pull-out torque (621.313)
Kippofen tilting furnace
Kipprelais trigger-action relay; throw-over relay
Kippschalter to ggle switch, tappet switch, tumbler switch
Kippschaltung trigger circuit; relaxation circuit, relaxation oscillator (621.397)
Kippschaltung für einmalige Ablenkung Eccles-Jordan circuit, one-shot flip-flop
Kippschwinger multivibrator, relaxation-type oscillator
Kippschwingungen relexation-type oscillations, saw-tooth oscillation
Kippschwingungsgenerator s. **Kippschwinger**
Kippspannung breakover voltage (621.394.63); triggering voltage
Kippunkt breakover point
Kippvorgang breakdown (621.313); sweep (621.373)
Kippvorrichtung tumbler device
Kippwelle tumbler shaft
Kippzündung tilting ignition (621.314.6)
kippen to tilt, to tip (531); to pull out (621.313)
Kippscher Apparat Kipp's apparatus
Kirchhoffsche Gesetze Kirchhoff's laws
Kissenentzerrung pincushion correction (621.397)
Kissenverzeichnung pincushion distortion (621.397)
Kiste box, chest, case; crate
in Kisten verpacken to box, to case; to crate
Klammer clamp, cramp; clip; paranthesis, bracket
Klammer auf bracket up
Klammer zu bracket closed
in Klammern in parentheses, in brackets
Klammerausdruck term in parantheses, term in brackets (51)
klammern to cramp, to clamp
Klang sound (534); note; peal, tang
Klanganalysator sound analyzer
Klangcharakter distinctive note
Klangfarbe timbre, quality (of a sound)
Klangfarbenregler to ne-corrector, tone control; bass control
Klappanker hinged armature
Klappdeckel hinged cover, hinged lid
Klappe flap, tab; stop (681.8)
Klappenschrank drop switchboard; magneto-board (621.395)
Klapper rattle
Klapperschaltung jogging (621.34)
Klapperventil self-acting valve
klappern to rattle, to clatter, to jingle
klar clear, defined, lucid; evident
klares Bild defined image
klarer Ton clear ring (78)
Klarglas clear glass
Klarschriftleser character reader
Klartext clear text, plain text
klären to clarify, to purify; to clear
Klasse standard of accuracy, accuracy rating (621.317); class; form (35)
Klasse 0,5 substandard grade (621.317)
Klasse 1 first grade (621.317)
Klasse 2 second grade (621.317)
klassieren to grade, to size, to classify
Klassierung grading, sizing, classification
klassifizieren to classify
Klassifizierung classification, rating, grading
Klassifizierungsvorschrift grading rule
klassische Mechanik Newtonian mechanics
Klaue claw
Klauenkupplung claw clutch
Klauenpolgenerator claw-pole stationary-field generator
kleben to glue, to paste, to gum; to stick
Klebfähigkeit adhesiveness
Klebstoff bonding medium, adhesive substance, gum, cement
Klebestreifen adhesive tape
Klebwachs adhesive wax, universal wax
klebrig gluey, gummy, sticky
Kleeblattantenne cloverleef antenna
Klein-... small, little; miniature; light
Kleinabnehmer domestic consumer
Kleinbahn light railway, narrow gauge railway

Kleinflächenkontakt small-area contact
Kleinkondensator miniature capacitor
Kleinmotor fractional horsepower motor
Kleinrechner minicomputer
Kleinrelais midget relay
Kleinsicherung littlefuse
Kleinsignal small signal
Kleinsignal-Kurzschlußstromverstärkung small-signal short-circuit forward-current transfer ratio
Kleinsockellampe s. b. c. lamp = small bayonet cap lamp
Kleinspannung first voltage range
Kleinspannungsmotor low-voltage motor
Kleinspannungstransformator single-lamp transformer
Klein-Swansockel s. b. c. = small bayonet cap
Kleinteile bits and pieces; mounting material, wiring material
Kleinverbrauch individual consumption, domestic consumption
klein small, little, minor
klein gegen... small in relation to
klein zweiter Ordnung second-order small
kleinstellen to turn down
zu klein undersized
kleiner Abstand short distance, short range
kleine Bohrung small bore
kleine Halbachse minor semi-axis
kleiner konzentrischer Sockel s. c. c. cap = small central contact cap
kleiner Prüfstrom minimum fusing current
kleine Reichweite short range
mit kleiner Reichweite short-range
kleine Schwankung ripple
kleine Terz minor third
kleiner Vorsprung pip
Kleinst... micro..., minimum..., miniature, extra low
Kleinstdurchmesser minimum diameter
kleinste Last power minimum
kleinstes wahrnehmbares Signal minimum discernible signal
Kleinstmotor miniature motor
Kleinstschalter miniature switch, microswitch
Kleinstspannungsbeleuchtung extra low voltage lighting
Kleinstwert minimum value
Kleistsche Flasche electric jar, Leyden jar
K-Leiter fault bus (621.316.9)
Klemmbrett terminal block, terminal board (621.313)
Klemme terminal, binding post (621.3), terminal insulator (621.314.2); lock, clamp (621.8)
Klemme für Nennübersetzung rated kva tap
Klemmenkasten terminal box
Klemmenkopf terminal head
Klemmenleiste terminal strip, connecting block, terminal board
Klemmenpaar set of terminals (621.3)
Klemmenspannung working voltage (621.352); terminal voltage, pole voltage
Klemmenscheinwiderstand terminal impedance
Klemmenschraube terminal screw, binding screw (621.3); clamping screw
Klemmenzuleitung terminal lead (621.385)
Klemmhülse clamp bushing
Klemmleiste terminal strip, terminal board
Klemmimpuls clamping pulse
Klemmschaltung clamp circuit, clamping circuit
Klemmspannung clamp voltage
Klemmverbindung clamped connection
Klemmverstärker clamped amplifier
Klemmverzerrung clamping distortion
Klempner plumber, tinsmith, whitesmith
Klettereisen pole step; climbing irons
Klima climate
Klimaanlage air-conditioning equipment
klimafest all-climate proof
klimatisieren to condition (628)
klimatisierter Betrieb air-conditioned workshop
klimatisierte Luft conditioned air
klimpern to jingle, to chink
klingarm non-microphonic (621.385)
Klinge blade
Klingel bell
Klingel mit Unterbrecher trembler bell
Klingeldraht ringing wire, bell wire
Klingelknopf bell button, push button
Klingeltransformator bell transformer
Klingelzug bell pull
klingeln to ring (the bell)
klingen to sound, to ring, to chink
klingende Röhre microphonic valve
Klingerit gasket board
klingsicher anti-microphonic, anti-vibration (621.396)
Klinke latch; pawl, release (621); jack (621.395)
Klinkenreihe, Klinkenstreifen jack strip
Klinkenschaltwerk ratched mechanism
Klipper clipper
klirren to clank, to clink, to clatter
Klirrdämpfung harmonic distortion
Klirrfaktor (non-linear) distortion factor, harmonic distortion coefficient, k-rating
Klirrfaktormeßbrücke distortion bridge
Klirrfaktormeßgerät distortion-factor meter

klopfen to tap; to knock, to pink (621.43)
Klopfer sounder (621.395)
klopffest non-pinking (621.4)
Klöppel hammer, clapper
Klotzkontakt block contact
kluckern to purl
Klydonograph klydonograph, surge recorder
Klystron klystron
Klystronsender klystron oscillator
Knackgeräusche clicks
Knackprobe click control (621.395)
Knackschutz acoustic shock reducer, click suppressor
Knall report, bang
Knallfunkensender Mie oscillator
Knallgas detonating gas
Knallgasflamme oxy-hydrogen flame
Knallquecksilber mercury fulminate, fuminating mercury
knallen to bang
Knarre ratchet
Knebelschalter toggle switch
Kneifzange nippers, pincers (pl.)
kneten to knead
Kneter, Knetmaschine kneading machine, kneading mill
Knetlegierung forging alloy (66)
Knick kink, crease, fold, break
Knick einer Kurve break
Knicke bekommen to knick
Knickbelastung buckling load
Knickfestigkeit buckling strength, breaking strength
Knickpunkt inflection (51); break point
Knickspannung buckling stress
Knickversuch buckling test
knicken to fold, to buckle
Knickung buckling, bending
Knie knee (61), (621.6); elbow, bend (621), (66)
Kniehebel toggle (lever)
Kniehebelwirkung to ggle action, knuckle action
Kniestück knee (621.6)
knipsen to snap (621.3); to snapshot (77); to punch
„Knipser“ snap switch
Knipszange punching tongs (pl.)
knistern to crackle
knittern to crease
Knochentelefon osophone, ossiphone
Knolle bulb
Knopf knob, button
auf den Knopf drücken to push the button, to press the button
Knopf-Kondensator button-type capacitor
Knopfloch buttonhole
Knopflochmikrophon lapel microphone
Knopfröhre door-knob tube, acorn tube
Knoten node (532), (537); knot (67)
Knotenamt main centre office, tandem central office
Knotenblech junction plate, connection plate, gusset (plate)
Knotenfläche nodal plane
Knotenlinie nodal line
Knotenpunkt nodal point (532), junction (621.3)
Knotenpunktslast joint load
Knotenvermittlung tandem exchange, toll centre, nodal exchange
K. O. = Kathodenstrahl-Oszillograph C. R. O. = Cathode Ray Oscillograph
koaxial coaxial
Koaxialantenne mit Gegengewicht sleeve-stub antenna
Koaxialantenne mit Sperrkopf sleeve-dipole antenna
Koaxial-Hohlleiter-Übergang coaxial wave-guide junction
Koaxialkabel coaxial cable
Koaxialkabel mit Abstandshaltern beaded transmission line
Koaxialrelais coaxial relay
Kobalt cobalt
Koch-... cooking, boiling
Kochgerät cooker, cooking appliance
Kochherd cooking range
Kochplatte hot plate, boiling plate
Kochplattenumschalter three-heat switch, five-heat switch
Kochpunkt boiling point
Kochsalz common salt, rock salt, sodium chloride
kochen to boil
kochend ebullient
kochender Akkumulator gassing accumulator
kochendheiß steaming-hot
Kochwasser-Reaktor water-boiler reactor
Kode (siehe Code)
kodieren (siehe codieren)
Koeffizient coefficient, factor
Koeffizient der diffusen Durchlässigkeit diffuse transmission factor
Koeffizient der diffusen Reflexion diffuse reflection factor
Koeffizient der Gesamtreflexion total reflection factor

Koeffizient der regulären Durchlässigkeit regular transmission factor
Koeffizient der regulären Reflexion regular reflection factor
koerzitiv coercive
koerzitive Eigenschaft coerciveness, coercitivity
Koerzitivfeldstärke coercive field
Koerzitivkraft coercive force, coercive intensity, coercitivity
Kofferempfänger portable receiver, self-contained radio set
Koffersender self-contained transmitter, portable transmitter
Kohärer coherer
Kohärerwirkung coherer effect
kohärent coherent
Kohäsion cohesion
Kohäsionskraft cohesive force
Kohle carbon (54), coal (622); brush (621.3)
Kohleblockmikrophon solid back microphone
Kohlebogenlampe carbon-arc lamp
Kohlebürste carbon brush, graphite brush (621.3)
Kohlendioxid carbon dioxide
Kohledruckregler carbon rheostat, carbon regulator, carbon-pile regulator
Kohledrucksäule carbon pile (621.3)
Kohlenelektrode carbon electrode
Kohlefadenlampe carbon (-filament) lamp
Kohlegrieß carbon granules (pl.)
Kohlehydrierung hydrogenation of coal
Kohlekontakt carbon contact (621.316)
Kohlekörner carbon granules
Kohlelichtbogen carbon arc
Kohlelichtbogenschweißung carbon-arc welding
Kohlemikrophon carbon (-type) microphone
Kohlemikrophon-Tonabnehmer carbon-contact pick-up
Kohlenoxid carbon oxide, carbon monoxide
Kohlensäure carbonic acid, carbon dioxide
Kohlestab, Kohlestift carbon rod
Kohlenstoff carbon
Kohlenwasserstoff hydrocarbon, carbohydrate
Kohleschichtfestwiderstand fixed-carbon film resistor
Kohleschichtwiderstand carbon-film resistor
Kohlewiderstand carbon resistor, graphite resistance
Kohlrauschbrücke Kohlrausch bridge
koinzident coincident
Koinzidenz coincidence
Koinzidenz-Entfernungsmesser coincidence range finder
Koinzidenzschaltung coincidence circuit
Koinzidenzspeicher coincident-current memory
Koinzidenzsteuerung coincident-current selection
Koinzidenz-Tor coincidence gate
Koinzidenzzähler coincidence counter
Koks coke
Kolben piston (621.43); bulb, envelope, (621.385); plunger (621.65), (621.396)
Kolbenhub length of the stroke, travelling of the piston
Kolbenmembran piston diaphragm (534.86)
Kolbenring piston ring
Kolbenweg (piston) stroke
kolinear co-linear
Kollektor collector (539), (621.313), (621.382); commutator (621.313)
Kollektor abdrehen to turn a commutator
Kollektoranschluß collector terminal
Kollektorbahnwiderstand collector series resistance (621.382)
Kollektor-Basis-Reststrom collector-base cut-off current
Kollektor-Basis-Spannung collector-base voltage
Kollektor-Basis-Zonenübergang collector junction
Kollektor-Brandstelle flat
Kollektorbürste commutator brush
Kollektor-Emitter-Reststrom collector-emitter cut-off current
Kollektor-Emitter-Spannung collector-emitter voltage
Kollektorfahne commutator connector; riser (621.315)
Kollektorglimmer commutator mica
Kollektorlamelle commutator section, commutator segment
Kollektormotor commutator motor
Kollektorreststrom collector-base cut-off-current
Kollektorschaltung vorwärts common collector
Kollektorschaltung rückwärts inverse common collector
kollektorseitiger Wicklungsschritt front pitch
Kollektorspannring „V" ring
Kollektorsperrschicht collector depletion layer, collector junction
Kollektorsperrschichtkapazität collector depletion-layer capacitance

Kollektorstörungen (pl.) interference caused by the commutator
Kollektorstrom collector current
Kollektortöne (pl.) commutator ripple
Kollektorübergang collector junction
Kollektorzone collector region
kollidieren to collide
Kollimator collimator
Kollision collision
Kollisionskursmarkierung collision-curse indication
Kollisionsschutzradar anticollision radar
Kollisionswarnung collision warning
Kollodium collodion
Kollodiumhäutchen collodion film
Kolloid colloid
kolloidal colloid(al)
Kolophonium colophony (rosin)
Kolophoniumlötdraht resincored solder
Kolorimeter colorimeter
Kombination combination
Kombinationsfrequenz combination frequency
Kombinationsrechnung combinatorial analysis
Kombinationsschwingungen intermodulation products generated in the transmitter (621.396)
Kombinationszange combination pliers (pl.)
kombinatorische Schaltung combinatorial circuit (621.374)
kombinieren to combine
kombinierte Wellen-Schleifen-Wicklung mixed winding (621.313)
Komma (im Dezimalbruch) decimal point
Kommando order; (control) signal, command (621.398)
Kommando Empfänger control receiver
Kommandogabe control initiation
Kommando-Pult command console, control desk
Kommandospeicher control-impulse store
Kommunikationslehre communication theory
kommunizieren to communicate
kommutativ commutative
Kommutator commutator (siehe auch unter Kollektor)
Kommutator nachdrehen to true up a commutator
Kommutatoranker commutator armature
Kommutatorbürste commutator brush
Kommutatorfahne commutator lug; commutator riser (US)
Kommutatorgleichrichter permutator
Kommutatorlamelle commutator segment
Kommutatormotor commutator motor, brush motor
Kommutatorring metal „V" ring
Kommutator-Ringisolierung mica „V" ring
Kommutator-Rundfeuer commutator flashing
Kommutatorsegment commutator segment
kommutieren to commutate
Kommutierung commutation
Kommutierungsachse axis of commutation (621.313)
Kommutierungsanode transition anode, discharge path
Kommutierungsfeld reversing field
Kommutierungswinkel commutating angle (621.314.5)
Kommutierungszone reversing zone
kompakt compact, massive
Kompander compandor (621.395/396)
Kompandierung companding
Komparator comparator
Kompaß compass
Kompaß kompensieren to box the compass
Kompaßnadel compass needle
Kompaßpeilung compass bearing
Kompaßpunkt, Kompaßstrich point of the compass
Kompaßscheibe dial, compass card
Kompatibilität compatibility
Kompensation compensation
Kompensationsantenne balancing aerial
Kompensationsbogen expansion joint (621.6)
Kompensationshalbleiter compensated semiconductor
Kompensationsinstrument self-balancing instrument
Kompensationsmethode potentiometer method
Kompensationsspannungsmesser compensation voltmeter
Kompensationsspule coumpound coil (621.317)
Kompensationstransformator neutralizing transformer
Kompensationswicklung compensating winding
kompensieren to compensate (for); to balance
kompensieren, sich to neutralize (51), (53)
kompensierter Motor compensated motor
kompensierter Motor mit Kondensator capacitor motor
Kompilierer compiler
Komplement complement
komplementär complementary
komplementäre Transistoren complementary transistors
Komplementärfarbe complementary colour

komplett complete, entire
komplettes Lagerschild end-shield assembly
komplette Lösung complete solution (51)
komplettieren to complete
komplex complex
komplexe Darstellung complex quantity representation, complex notation
komplexe Ebene complex plane
komplexe Größe complex quantity
komplexe Rechnung complex analysis
komplexe Schreibweise complex quantity notation
komplexe Veränderliche complex variable
komplexer Verstärkungsfaktor vector amplification (factor)
komplexer Widerstand complex impedance, vector impedance
komplexe Zahl complex, complex number, complex quantity
Komponente component
Komponentenzerlegung resolution
Kompound-... compound (-wound)
Kompounderregung compound excitation
Kompoundgenerator compound (-wound) generator
Kompoundmotor compound (-wound) motor
Kompoundwicklung compound winding (621.313)
mit Kompoundwicklung compound wound
kompoundieren to compound (621.3)
kompoundierter Nebenschlußmotor stabilized shunt-wound motor
kompoundierter Wechselstromgenerator compound alternator, compensated alternator
Kompoundierung compounding (621.313)
kompressibel compressible
Kompression compression (strain), compression ratio (621.4)
Kompressionskühlschrank compression-type refrigerator
Kompressionspumpe compression pump
Kompressionsverhältnis ratio of compression
Kompressor compressor (621.2), (621.395); blowing engine, blower (621.2)
komprimieren to compress
Kondensation condensation
Kondensationsdampfmaschine condensing steam engine
Kondensationsdampfturbine condensing steam turbine
Kondensationskern nucleus
Kondensator condenser (621.1), capacitor (621.3)
Kondensator mit hoher D. K. hicap
Kondensator entladen to discharge a capacitor
Kondensator für Blindleistungskompensation power-factor correction capacitor
Kondensatorbatterie bank of capacitors
Kondensatorelektroskop capacitor electroscope
Kondensatorenabgleich ganging, aligning
Kondensatorenbank capacitor bank
Kondensatorenbatterie battery of capacitors
Kondensatorbecher capacitor case
Kondensatordurchführung capacitor (-type) bushing
Kondensatorengruppe capacitor bank, bank of capacitors
Kondensatorgüte capacitor constant (621.3)
Kondensatorkasten subdivided capacitor
Kondensatorlautsprecher electrostatic loudspeaker
Kondensatormikrophon electrostatic microphone, capacitor microphone
Kondensatormotor capacitor motor
Kondensatorspeicher capacitor store
Kondensatorstoßschweißung electrostatic percussive welding, capacitive storage welding
Kondensatorwanne capacitor case (621.396)
Kondensatorwickel capacitor coil
kondensierbar condensable
kondensieren to condense
Kondensor condenser (535)
Kondensorlinse condensing lens, condenser
konditionieren to condition (67)
Konduktanz conductance
Konferenzgespräch conference call
Konferenzschaltung conference circuit, conference service
Konfiguration configuration
konfokal confocal
konform conform(al)
konforme Abbildung conform(al) representation
Konglomerat conglomerate, conglomeration
kongruent congruent
Kongruenz congruence
Königswasser aqua regia
konisch conic(al) (51), taper, tapered (621)
konisch erweitern, (sich) to flare, to taper
konische Abtastung conical scan (621.396.96)
konische Bohrung taper hole
konische Gewinde tapered thread
konischer Gewindebohrer taper tap
konisches Loch taper hole
konische Reibahle taper reamer
konische Trommel conic drum
Konizität taper, conicity
konjugiert komplex conjugate complex

konjugiert komplexe Dämpfung conjugate attenuation constant, conjugate attenuation coefficient
konjugiert komplexes Übertragungsmaß conjugate transfer coefficient
konjugiert komplexe Zahl conjugate (complex number)
konjugierter Widerstand conjugate impedance (621.3)
Konjunktion conjunction (52)
konkav concave
konkave Tastatur sculptured keyboard
Konkav-Konvexlinse meniscus (pl.: -ci)
Konkavlinse concave lens
Konsekutivmodulation differential modulation
konservieren to preserve, to cure
konsistent consistent
Konsole bracket; console
Konsonanz consonance
konstant constant, steady, permanent
konstanter Druck steady pressure
bei konstantem Druck at constant pressure
konstante Einstellung permanent setting, permanence of setting
konstante Last constant load
konstante Lichtquelle steady light
konstanter Strom constant current
bei konstantem Volumen at constant volume
Konstanthaltung maintaining constant, stabilization
Konstantspannungsgeber constant-voltage generator
Konstantspannungsgenerator constant-voltage generator
Konstantspannungsgerät constant-voltage (regulating) unit
Konstantan constantan
Konstante constant (51)
Konstante eines Galvanometers bestimmen to take the constant of a galvanometer
Konstante des Stromkreises circuit constant
Konstanz constancy, permanence, steadiness
Konstanz über kurze Zeit short-term stability
konstruieren to design; to construct (nur 51)
Konstrukteur designer, design engineer
Konstruktion design, structure, building; construction (nur 51)
Konstruktionsbüro drawing office, design office
Konstruktionsdaten design data
Konstruktionselement structural part
Konstruktionsgrundlagen design principles
Konstruktionsgüte design merit
Konstruktionsmerkmal constructional feature
Konstruktionsprinzip principle of design
Konstruktionsregel design formula
Konstruktionsspannung design voltage (621.3); internal stress (624)
konstruktiv constructive; design
Konsument (current) consumer
Konsumgut consumer appliance, consumer goods
Kontakt contact
Kontakt berühren to engange with a contact, to strike a contact
Kontakt herstellen, Kontakt machen to make contact
Kontaktabbrand contact pitting, contact burn
Kontaktabstand contact separation, break distance (621.318.5)
Kontaktarm contact arm, wiper, brush (621.395)
Kontaktbacke contact jaw
Kontaktbelastung contact duty
Kontaktbemessung contact rating
Kontaktbürste contact brush
Kontaktdauer period of contact, contact time
Kontaktdruck contact pressure, contact force
Kontaktelektrizität voltaic electricity
Kontaktfeder contact spring
Kontaktfeld contact bank (assembly)
Kontaktfinger contact finger, contact stud, contact pin
Kontaktfläche contact surface, contact area
Kontaktgeräusch contact noise, scratches
Kontaktgleichrichter mechanical rectifier
Kontakthammer trembler
Kontakthebel contact lever
Kontaktmaterial contact material
Kontaktmesser contact blade
Kontaktmikrophon contact microphone
Kontaktmitnahme contact follow
Kontaktpotential contact potential; liquid-junction potential (621.35)
Kontaktprellung bounce, contact chatter
Kontaktpunkt point of contact
Kontaktrauschen contact noise
Kontaktrolle contact roller
Kontaktsatz bank of contacts
Kontaktscheibe contact disk
Kontaktschieber sliding contact
Kontaktschiene contact rail
Kontaktschraube contact screw
Kontaktsegment contact segment
Kontaktspannung contact voltage, Volta effect; contact-operating voltage
Kontaktspiel contact float
Kontaktstelle point of contact, junction

Kontaktstift s. **Kontaktfinger**
Kontaktstück contact piece, contact plate
Kontakttrennung contact separation
Kontaktweg contact travel
Kontaktwiderstand contact resistance
kontern to lock
Kontermutter lock-nut, jam nut
kontinuierlich continuous
kontinuierliche Regelung infinite regulation, stepless regulation
kontinuierliche Reglerwirkung continuous controller action
kontinuierliches Spektrum continuous spectrum
Kontinuität continuity
Kontinuitätsprüfung continuity check
Kontinuum continuum
Kontraktion contraction
Kontrast contrast; brilliance, brilliancy (621.397)
Kontrastphotometer contrast photometer
kontrastreich contrasty
kontrastrieren to contrast
Kontrolle inspection, check; control; proof test (621.317)
Kontrollablesung check reading
Kontrollabschnitt counterfoil, check slip; control section (625)
Kontrollanalyse check analysis
Kontrollantenne monitor aerial
Kontrolleinrichtung monitoring device
Kontrollempfänger check receiver, monitoring (radio) receiver, monitor
Kontrollgerät (check) gauge, monitoring device, monitor
Kontrollhörer monitor headphone
Kontrollampe monitor lamp, telltale lamp
Kontrollautstärke sidetone level (621.395)
Kontrollicht telltale lamp, pilot lamp
Kontrollichtöffnung inspection window; back light (629.13)
Kontrollmessung check measurement
Kontrollraum inspection room
Kontrollschacht inspection shaft
Kontrollsignal check signal
Kontrollstelle control station
Kontrollthermometer temperature detector
Kontrollturm control tower
Kontrolluhr control clock, time clock
Kontrolluhr stechen to clock
Kontrollversuch check test, verificatory test
Kontrollwiedergabe playback (654.19)
Kontrollzeichen check mark
Kontrollzettel control sheet, inspection ticket, counterfoil
Kontroller controller; camshaft controller (621.316.5)
Kontrolleur supervisor
kontrollieren to control, to check, to monitor, to inspect
Kontur contour, outline
Konturenbeleuchtung outline lighting
Konus cone (51), taper (621)
Konusantenne cone antenna
Konuslautsprecher cone speaker
Konusmembran speaker cone
Konuswinkel angle of flare
Konvektion convection
Konvektionskühlung convection cooling
Konvektionsschalter convection circuit-breaker, blast contact circuit-breaker
Konvektionsstrom convection current
Konvektionstrocknung convection air drying
Konvektionsverlust convection loss (536)
Konvektionswärme convection heat
Konvention convention
Konventionalstrafe (bei Lieferverzug) penalty (for nonfulfillment of contract)
konvergent convergent
Konvergenz convergence, convergency
Konvergenzbereich region of convergence
Konvergenzelektrode convergence electrode
konvergieren to converge (**nach:** upon)
Konversionsreaktor converter reactor
Konverter converter (669)
Konverterbrennelement converter fuel element
konvex convex
Konvexlinse convex lens
Konzentrat concentrate
Konzentration concentration; busy hour today ratio (621.395)
Konzentrationselement concentration cell (621.35)
Konzentrationsstufe concentration stage (621.395)
Konzentrator concentrator (621.395)
konzentrieren to concentrate; to focus (535), (621.385)
konzentriert concentrated (54), (621.3), focussed (535), lumped (621.392); intense
konzentriert anordnen to concentrate, to lump
konzentrierte Induktion concentrated inductance
konzentrierte Induktivität lump inductance
konzentrierte Kapazität lump capacitance
konzentrierte Last concentrated load
konzentriertes Schaltelement lumped element, lumped constant

konzentrierte Wicklung concentrated winding
Konzentrierung concentration; focussing
konzentrisch concentric(al), coaxial
konzentrisches Kabel concentric cable, concentric line, coaxial cable, single-wire cable
konzentrische Kreise concentric circles
konzentrisches n-Leiterkabel n-conductor concentric cable
konzentrische Muffe coaxial cable fit
konzentrischer Stecker coaxial connector
Koordinate co-ordinate
Koordinatenachse axis of reference
Koordinatennetz grid
Koordinatenschalter crossbar selector, crossbar switch (621.395)
Koordinatenschreiber xy-recorder, xy-plotter
Koordinatensteuerung point-to-point control
Koordinatensystem system of co-ordinates, set of co-ordinates
in ein Koordinatensystem eintragen to plot upon a co-ordinate system
Koordinatenwähler two-motion selector (621.395)
Koordinatenwandler synchro resolver (621.316.7)
Koordinierung co-ordination
Koordinierungsentfernung co-ordination distance (621.396)
Koordinierungsgebiet co-ordination area (621.396)
Koordinierungskontur co-ordination contour (621.396)
Kopalharz copal resin
Kopf head
Kopfanker end-on armature
Kopfbogen return bend
Kopfbügel head band (621.395)
Kopfhörer headphone, headset; ear-piece, earphone
Kopfhöreranschluß headphone connection
Kopfhörerbetrieb headset operation
Kopfhörerkreis head circuit
Kopfhörerlautstärke headset level
Kopfrad head wheel (681.84)
Kopfschuh vacuum guide (681.84)
Kopfspiel tip clearance (621.8)
Kopfüberstand head-tip projection (681.84)
Kopfumschalter head-switching circuit
Kopfwelle explosion front
Kopie copy, hardcopy
Kopierautomat automatic printing machine
Kopiergerät copying apparatus
kopieren to copy, to replicate; to print (77)
koplanar co-planar
koppeln to couple, to link
Koppel... switching... (621.395.3)
Koppelanordnung switching network, connecting network
Koppelelement crosspoint, switching element (621.395)
Koppelfeld crosspoint matrix
Koppelkondensator coupling capacitor
Koppelnavigation lead reckoning navigation
Koppelnetz switching network, connecting network (621.395)
Koppelpunkt crosspoint (621.395)
Koppelreihe switching row, connecting row
Koppelschleife coupling loop (621.37)
Koppelspule link coil, coupling coil
Koppelstufe switching stage, connecting stage (621.395)
Koppelvielfach connecting matrix, switching matrix (621.395)
Koppelwirkung coupling effect
Kopplung coupling (621.396)
Kopplung von Wärme- und Kraftversorgung combined power and heating system
Kopplungsfaktor coefficient of coupling (621.396); leakage factor (621.313)
Kopplungsglied coupling element, linking element
Kopplungsgrad degree of coupling
Kopplungskondensator coupling capacitor
Kopplungsvariometer vario coupler
Kopplungs-Wechselstromwiderstand mutal impedance, coupling impedance, interaction impedance (621.372.5)
Köpsel-Apparat Koepsel permeameter
Korb basket
Korbspule basket (wave) coil (621.396)
Korbwicklung chain winding (621.313), basket winding (621.396)
kordeln to knurl
Kordelmutter knurled nut
Kordelring knurled ring
Kork cork
Korkdichtung cork gasket
Korkmehl, Korkstaub cork dust, cork meal
Korkenzier corkscrew
Korkenzieherregel corkscrew rule
Korn grain (53), (63), (77)
Kornfeuchtigkeit moisture in grain (63)
kornfrei grainfree
Korngrenzkorrosion intercrystalline corrosion, intergranular corrosion
Korngröße grain size

Korngrößenbestimmung grading test
Kornmaterial aggregate
kornorientiertes Transformatorblech grain-oriented transformer sheet
Kornverfeinerung grain refinement
Kornvergröberung grain coarsening
Körnchen granule
körnen to granulate, to grain, to corn
gekörntes Metall granulated metal
Körner punch, (centre) pop
Körnermarke, Körnerschlag punch mark
körnig granular, grainy
körnig machen to corn
Körnung granulation, graining
Korona corana (52), (621.3)
Koronadämpfung corona attenuation
Koronaentladung corona discharge
Koronograph coronograph
Körper body, solid (substance); casing (621.3)
einem Körper Energie zuführen to do work on a body
Körpererdung equipment earth
körperlos bodiless, disembodied
Körperschallmikrophon contact microphone
Körperschluß insulation fault
Körperstrom fault current, leakage current
Körperwärme warmth of body
Körperchen corpuscle, particle
körperlich physical (51), (61)
körperliche Gewalt physical force
Korpuskel corpuscle, particle
korpuskular corpuscular
Korrektionsglied correction term
Korrektur correction, amendment; emendation (655)
Korrekturen anbringen für to correct for
Korrektur für endliche Länge end correction
Korrektur lesen to read the proofs
Korrekturbefehl partel, patch (681.3)
Korrekturbogen, Korrekturfahne proof (-sheet)
Korrekturfaktor correction factor
Korrekturfaktor für das Übersetzungsverhältnis ratio correction factor (621.317)
Korrekturkurve correction curve
Korrekturroutine patch (681.3)
Korrelat correlate
Korrelation correlation
Korrelationskoeffizient correlation coefficient (519)
Korrelationsrechnung correlation analysis
korrelieren to correlate
korrespondieren to correspond
korrigieren to correct, to amend
korrodieren to corrode
korrodierende Substanz corrosive
Korrosion corrosion
Korrosionsbeständigkeit resistance to corrosion
korrosionsbeständiger Stahl stainless steel, non-corroding steel
Korrosionsermüdung corrotion fatigue
korrosionsfest resistant to corrosion, corrosion-proof, anti-corrosive
Korrosionsmittel corrosive (agent), corrosion medium
Korrosionsschutz corrosion protection, protective coating
Korund, Korundum corundum
Korundbohrkrone corundum bit
kosmisch cosmic
kosmisches Rauschen interstellar noise, extra-terrestrial noise (621.396)
kosmische Strahlung cosmic rays, ultra-X-rays (pl.)
Kosmotron cosmotron, proton-synchrotron
Kosten cost; costs (pl.), expenses
Kostenanschlag (cost) estimate, tender
Kostenberechnung costing
Krach bang, noise
krachen to bang (534); to crack, to break down
krachtöten to squelch, to mute, to quiet (621.396.62)
Krachtöter noise suppressor, interference suppressor; silencing device, noise killer, static eliminator
Krachtöterschaltung Q-circuit, quieting circuit, squelch circuit
Krachtötung noise suppression
Kraft force (53); power (621), (621.3)
Kraft Eins unit force
Kraft pro Flächeneinheit force on unit area
Kraft und Gegenkraft action and reaction
Kraft am Zughaken draw bar pull
Kraftanlage power plant
Kraftbedarf power demand
Kraftbelastung power load
Kraftbetätigung power operation
Krafteck polygon of forces, force polygon, force diagram
Krafteinheit unit (of) force
Krafterzeugung power generation
Kraftfahrzeug motor vehicle
Kraftfeld field of force
Kraftkomponente component force
Kraftlinie line of force, flux line
Kraftlinien schneiden to cut (across) lines of force

Kraftlinien treten aus dem Pol aus lines of force proceed from the pole
Kraftlinienzahl number of lines of force
Kraftmaschine engine, prime mover
Kraftmeßdose pressure gauge
Kraftnetz power mains
Kraftquelle power source
Kraftröhre field tube, unit tube of force (537)
Kraftsteckdose power receptacle
Kraftstoff fuel, petrol (Br.); gas, gasoline (US)
Kraftstrom electric power
Kraftstromkreis power circuit
Kraftstromschalttafel power switchboard
Kraftstromstecker power plug
Kraftübertragung power transmission
Kraftverbrauch power consumption, power absorption
Kraftversorgung power supply
Kraftverstärker power amplifier (621.375)
Kraftverstärkerröhre power valve
Kraftverstärkerstufe power stage
Kraftverteilung power distribution
Kraft-Weg-Diagramm force-displacement characteristic
Kraft-Wärme-Kopplung combined power and heating system
Kraftwerk power (generating) plant, power house, power station
Kraftwerkseinrichtung power-station equipment
Kraftwerkshilfsmaschinen generating-station auxiliaries
Kraftwerksleistung generating capacity
Kraftwerksmaschinen power-station machinery (sing.)
Kraftwerkstransformator power-station transformer
Kraftwerkswirkungsgrad generating station efficiency
Kraftzentrale power station
Kräfte-... force
Kräftedreieck triangle of forces
Kräftegruppe set of forces
Kräftemaßstab force scale
Kräftepaar (turning) couple
Kräfteplan stress diagram
Kräftepolygon polygon of forces
Kräftesystem, Kräfteverteilung set of forces
Kräftezug load line
kräftig strong, powerful, intense, intensive
kräftig leiten to conduct heavily
kräftiger Empfang strong signal (621.396)
kräftige Farbe intense colour
kräftige Feder powerful spring
kräftiges Feld intense field
kräftige Hitze good heat
kräftiger Strom strong current
Kran crane
Kranfernsteuerung crane telecontrol
Kranmotor crane motor
Kranschalter crane controller
Kranschiene crane rail
Krarup-Kabel Krarup cable
Krater crater
kratzen to scratch, to scrape
Kratzer, Kratzspur scratch, scratching
kratzfest scratch-proof, resistant to abrasion
Kratzfestigkeit resistance to scratching, abrasion resistance
Kratzgeräusche line scratches, contact noise (US) (621.395)
Kräusel-... ripple
Kräuselanteil ripple ratio
Kräusel-EMK ripple electromotive force
Kräuselfalte crinkle
Kräuselgröße ripple quantity
Kräuselspannung ripple (voltage), ripple potential difference
kräuseln to ripple; to crinkle, to goffer
Kräuselung ripple; goffering
Kreis circle (51); orbit (52), (539); circuit (621)
Kreis schlagen um to draw a circle around
Kreisabschnitt segment of a circle
Kreisantenne circular antenna
Kreisausschnitt sector (of a circle)
Kreisbahn circular trace; orbit
sich auf Kreisbahnen bewegen to circle
Kreisbewegung circular motion
Kreisbogen arc of a circle, circular arc
sich auf einem Kreisbogen drehen to revolve through the arc of a circle
Kreisdiagramm circle diagram (621.3)
kreisförmig circular
kreisförmige Zeitablenkung circular sweep (621.396.96)
Kreisfrequenz angular frequency, radian frequency
Kreisfunktion periodic function, circular function
Kreisgröße circuit constant, circuit element (621.3)
Kreisgüte Q-factor
Kreiskegel circular cone
Kreislauf circuit, circulation, recirculation
im Kreislauf führen to recirculate, to recycle (54), (66)
Kreislaufkühlung closed-circuit cooling
Kreislaufleitung recirculation line

Kreislaufreaktor circulating reactor
Kreisprozeß complete cycle (536)
Kreisprozeß durchlaufen to perform a cycle
Kreisquerschnitt circular section
Kreisringspule single ring coil
Kreisspur circular trace
Kreisstrom circulating current
Kreisel gyro, gyroscope
Kreiselanlasser centrifugal starter
Kreiselbewegung gyratory motion, gyration
Kreiselgebläse turbo blower, centrifugal fan
Kreiselkompaß gyro compass, gyrostatic compass
Kreiselkompressor turbo compressor, centrifugal compressor
Kreiselpumpe rotary pump, centrifugal pump, turbo-pump
Kreiselrad gyro wheel, gyroscopic wheel; turbo-pump wheel
Kreosot creosote
Kreuz cross
Kreuzdipolantenne turnstile
Kreuzgelenk universal joint
Kreuzgriff star handle, cross handle
Kreuzkoppler cross guide
Kreuzmodulation intermodulation, crossmodulation (621.395/6)
Kreuzpeilung cross bearing
Kreuzrahmenpeiler Bellini-Tosi direction finder
Kreuzschaltung cross connection; back-to-back connection (621.314.6)
Kreuzschienenverteiler crossbar distributor
Kreuzschienenwähler crossbar switch
Kreuzschraffur cross hatching
Kreuzspulinstrument cross-coil instrument
kreuzweise cross-wise
Kreuzzeiger cross pointer
kreuzen (sich) to cross
gekreuzte Nicols crossed Nicols
gekreuzter Riemenantrieb crossed belt drive
Kreuzung cross, crossing; transposition (621.315), (621.395); junction (625)
Kreuzungsebene crossing plane
Kreuzungsgleis crossing loop (625.1)
Kreuzungsherz frog (625)
Kreuzungslinie crossing line
Kreuzungspunkt crossing point, junction, cross point
Kreuzungsweiche crossing point
Kriech-... creep (620); leakage (621.3)
Kriechanlasser inching starter
Kriechdehnung creep strain, rate of stretch
Kriechgang inching speed, crawling speed
Kriechentladung dissipation of charge
Kriecherholung creep recovery
Kriechgalvanometer quantometer
Kriechgeschwindigkeit rate of creep, creep rate (66)
Kriechspurbildung tracking (621.3)
Kriechstrecke (current) leakage path, tracking distance (621.3); creeping distance, creepage distance
Kriechstrom leakage current, surface leakage, creepage current
kriechstromfest non-tracking, creepage proof
Kriechvorgang creeping
Kriechweg s. **Kriechstrecke**
kriechen to creep
kriechend sluggish (621.317)
Kristall crystal, piezoelectric...
Kristalle ansetzen to effloresce
Kristalle züchten to breed crystals
Kristallanalyse crystal analysis
Kristallanalysenkamera crystal-analysis camera
Kristallbau crystal structure
Kristalldetektor crystal detector, crystal rectifier
Kristalldiode crystal diode
Kristalldiodenmischstufe crystal mixer
Kristalldrehungsschwingung torsional vibration of a crystal
Kristalldruckabnehmer piezoelectric pressure gauge
Kristallfilter crystal filter
Kristallform crystalline form
Kristallgitter crystal lattice
Kristallautsprecher crystal speaker, piezoelectric loudspeaker
Kristallmikrophon crystal microphone, piezoelectric microphone
kristallorientiertes Blech oriented sheet (621.314.2)
Kristallscheibe crystal wafer
Kristalltemperatur crystal temperature (621.382)
Kristalltonabnehmer crystal pickup
Kristallwachstum crystalline growth
Kristallziehen crystal pulling
Kristallziehvorrichtung crystal puller
Kristallzwillinge twinned crystals
kristallin crystalline
kristalline Form crystalline form
Kristallisation, Kristallisierung crystallization
Kristallisationskern nucleus (548)
Kristallisationspunkt crystallizing temperature

kristallisieren to crystallize
kristallischer Bestandteil constituent crystal
Kristallit crystallite
Kristallode transistor (621.382)
Kristallographie crystallography
kristallographisch crystallographic
Kriterium criterion (pl.: -ria)
kritisch critical; divergent
kritisch gedämpft critically damped
kritische Drehzahl critical speed
kritischer Druck critical pressure
kritische Frequenz critical frequency, optimum coupling
kritische Gitterspannung critical grid voltage
kritische Größe critical magnitude
kritische Kopplung critical coupling
kritische Masse critical mass
kritischer Reaktor critical reactor
kritische Spannungssteilheit critical rate of a rise of off-state voltage (621.382)
kritische Stromsteilheit critical rate of a rise of on-state current (621.382)
kritische Temperatur critical temperature
kritischer Zustand critical state
Krokodilklemme crocodile clip, alligator clip (621.396)
Krone crown
Kronleuchter chandelier
kröpfen to crank
krumm curved, warped, out of straight
krumme Linie curve, curved line
krummlinig curvilinear
krummliniges Koordinatensystem curvilinear system of coordinates
krummlinige orthogonale Koordinaten orthogonal curvilinear coordinates
krümmen to curve, to crook, to bend
gekrümmte Erdoberfläche round earth
Krümmer elbow, bend, knee (621.64)
Krümmerradius radius of a bend
Krümmung curvature (51); bend, turning, flexure
Krümmung der Anodenkennlinie anode bend
Krümmungshalbmesser radius of curvature, radius of bend
Krümmungsmittelpunkt centre of curvature
Krümmungsradius radius of curvature
Kryotron cryotron
Krypton krypton
Kubik-... cubic
Kubikfuß cu. ft. = cubic foot = 28, 32 1
Kubikmeter cubic metre, metre cube = 1 cu. yd. 8 cu. ft.
Kubikwurzel cube root, third root
Kubikzentimeter c. c. = cubic centimetre, centimetre cube
kubisch cubic(al)
kubische Dehnung cubical expansion
kubische Gleichung cubic equation, third degree equation
Kubus cube
Küchenherd cooker (B); (cooking) range, kitchen range (US)
Kufe skid
Kugel globe (55); sphere (51); ball (621)
Kugelblitz globe lightning, globular lightning, fire ball, ball lightning
Kugeldreieck spherical triangle
Kugeldruckhärte ball hardness
Kugelelektrode ball electrode, spark ball (621.38)
kugelförmig globular, spherical
kugelförmig gelagert spherically seated
kugelförmiger Hohlraum spherical cavity
Kugelfunkenstrecke sphere gap
Kugelfunktion spherical function
Kugelgelenk ball and socket, ball joint, socket joint
Kugelgriff ball handle
Kugelkoordinaten spherical polars
Kugellager ball bearing
Kugelneigungsmesser ball-bank indicator
Kugelphotometer globe photometer, sphere photometer
Kugelstrahler isotropic radiator (621.396.67)
Kugelviskosimeter ball and bucket viscosimeter
Kugelwelle spheric wave
Kügelchen globule
Kühl-... cooling, ventilating, refrigerating
Kühlanlage cooling plant, refrigerating plant, freezer
Kühlbehälter cooling tank, cooling vessel
Kühlelement cooler element
Kühlfahne cooling vane
Kühlfalle cold trap, low-temperature trap
Kühlfalle mit flüssiger Luft liquid-air trap
Kühlflügel electrode radiator (621.396.69)
Kühlgebläse ventilating fan, cooling blower (621.314)
Kühlgeschwindigkeit cooling velocity
Kühlkörper heat sink
Kühlmantel cooling jacket
Kühlmaschine refrigerating machine
Kühlmittel coolant, refrigerant
Kühlmitteltemperatur temperature of cooling medium
Kühlofen cooling oven, cooling tunnel

Kühlöl cooling oil
Kühlraum cooling chamber
Kühlrippe cooling fin
Kühlschlange serpentine cooler, condenser coil
Kühlschlitz air duct, ventilating duct (621.313)
Kühlschlitzbreite ventilating space (621.313)
Kühlschrank refrigerator, refrigeration cabinet, freezer
Kühlsystem system of ventilation (621.313)
Kühltank cooling tank, cooling vessel
Kühlturm cooling tower
Kühlvorlage condenser
Kühlwasser cooling water
kühl cool, frigid, gelid
kühle Flamme cool flame, soft flame
kühlen to cool, to refrigerate
Kühlung cooling, refrigerating, ventilation
kulminieren to culminate
Kumaronharz coumaric resin
kumulativ cumulative
Kunde purchaser, user, customer
Kundendienst service
Kundendienst betreiben to service
Kunst art; artificial..., synthetic...
Kunstantenne dummy aerial, dummy load, artificial aerial
Kunstbeleuchtung artificial lighting
Kunstfaser artificial fibre
Kunstgummi synthetic rubber
Kunstharz synthetic resin; thermoplastic material
Kunstharz getaucht resin dipped
Kunstharz umpreßt resin molded
Kunstharz vergossen resin sealed
Kunstharzkitt bakelite paste
Kunstharzlack synthetic resin varnish
Kunstharzleim synthetic resin glue
Kunstlast dummy load (621.396.61)
Kunstleitung artifical line
Kunstlicht artifical light
Kunstpreßstoffe synthetic plastics
Kunstpreßstoffteile plastic mouldings
Kunstseide artificial silk
Kunststoff plastic (compound), synthetic material
Kunststoffilm plastic film
Kunststoffisolation plastic insulation
Kunststoffolien-Kondensator KP-capacitor, polypropylene capacitor; plastic foil capacitor
Kunststoffkapselung plastic encapsulation
Kunststoffmantelkabel non-metallic sheathed cable
Kunststoffschienenverbinder plastic (rail) bond
künstlich artificial, synthetic, made
künstlich altern to age
künstliche Alterung artificial ageing
künstliche Antenne dummy aerial, artificial aerial, mute aerial
künstliche Belüftung induced ventilation
künstliche Kernumwandlung, künstlicher Kernzerfall artifical nuclear disintegration
künstliche Leitung artificial circuit
künstlicher Nullpunkt artificial neutral point
künstliches Ohr artificial ear
künstliche Radioaktivität induced radioactivity
künstliches Schwarz (Weiß) artificial black (white) signal, nominal black (white) signal
künstliches Sonnenlicht artificial sunlight
künstlicher Sternpunkt artificial neutral point
Kupfer copper
Kupferband copper tape
Kupferbergwerk copper mine
Kupferblech sheet copper
Kupferbürsten copper brushes, gauze brushes
Kupferdochtkohle copper-cored carbon
Kupferdraht copper wire
Kupferglanz copper glance
kupferkaschierte Leiterplatte printed circuit board, copper-clad laminate
Kupferkontakt copper contact
Kupferleiter copper conductor
Kupfermanteldraht copper-clad wire (621.385)
Kupfer-Nickel-Legierung cupronickel alloy
Kupferoxidul cuprous oxide, copper oxide
Kupferoxidulgleichrichter copper-oxide rectifier
Kupferplatte copper plate
kupferplattiert copper-clad, copper plated
Kupferrohr copper tube
Kupferseil copper (wire) cable
Kupferschweißung copper welding
Kupfersulfat copper sulphate
kupferummantelt copper-clad
Kupferverluste (pl.) copper loss (621.3)
Kupfervitriol copperas, blue vitriol
kupfern copper
kupferner Abschirmbecher copper can, copper screen
Kupol cupola
Kupolofen cupola (furnace)
Kuppel dome; cupula
Kuppel-... coupling
Kuppelgestänge linkage gear

Kuppelleitung trunk feeder, tie feeder, interconnection tie
Kuppelschalter coupling switch
Kuppelstange coupling rod, link (625.282)
Kuppeltransformator coupling transformer
Kuppelwelle coupling shaft
kuppeln to link, to interlink, to connect (621.3); to couple, to connect, to clutch, to engage (621.8)
gekuppelte Schalter linked switches
Kupplung linkage (537); plug-and-socket connector (621.315.6); coupler, coupling (625); clutch (621.8)
Kupplungsbelag clutch lining, friction facing
Kupplungsdose coupler socket, jumper receptacle
Kupplungsgabel clutch fork
Kupplungsglied engaging link
Kupplungsleitung interconnecting feeder, interconnection
Kupplungsscheibe clutch disk
Kupplungsschlitz coupling way
Kupplungsseite coupling end, pulley end (621.313)
Kupplungsseite eines Generators back of a generator
Kupplungsstecker coupler plug
Kupplungswelle coupling shaft
Kurbel crank
Kurbelantrieb crank drive
mit Kurbelantrieb crank-driven
Kurbelbewegung crank motion
Kurbeldekade decade dial (621.317)
Kurbelgehäuse crank case
Kurbelinduktor crank inductor, magneto generator, hand generator
Kurbelinduktor für Widerstandsmessung magneto-ohmmeter
Kurbellager crank bearing
Kurbelscheibe crank plate
Kurbeltrieb crank mechanism, crank drive, crank gear
Kurbelwanne crank case
Kurbelwelle crank shaft (621.43), crank axle (625.2)
Kurbelwiderstand rotary rheostat, decade resistance box (621.317)
Kurbelzapfen crank pin
Kurs course, bearing
Kurs über Grund track
Kursabweichung heading displacement, course deviation
Kursanzeiger course-deviation indicator
Kursfeuer route beacon
Kurskoppler dead reckoner
Kurspeilung course bearing
Kursrechnen dead reckoning (629.12.13), (527)
Kursrechner dead-reckoning device, course calculator, course-line computer
Kurssender localizer beacon
Kurswähler omnibearing selector
Kurve curve, curved line (51); bend (625)
Kurve auftragen to plot a curve
Kurve ausgleichen to smooth a curve
Kurve beschreiben to trace a curve
Kurve biegt flach ab a curve flattens off
Kurve ermitteln to take a curve
Kurve weist Buckel auf a curve displays humps
Kurvenanstieg built-up curve
Kurvenblatt graph, diagram, chart
Kurvenform wave (-form), wave shape
Kurvenform des Empfangszeichens arrival curve
Kurvenform des Stromes current undulation
Kurvenformmeßgerät waveform instrument
Kurvenformuntersuchung wave analysis
Kurvenknick inflexion
Kurvenschar family of curves
Kurvenschlitz cam groove
Kurvenschreiber graphic recorder, x-y recorder, graphic plotter
Kurvenvergleich signal comparison
Kurvenverlauf curve trace
kurven to curve
Kurz-... short
Kurzdarstellung outline, summary
Kurzhub-... short-stroke
Kurzimpulsmeldung fleeting alarm
Kurzrufnummer abbreviated number
Kurzschließen short circuiting
Kurzschließer short-circuiting device
Kurzschluß short circuit
Kurzschluß hinter... short circuit beyond...
Kurzschluß von... nach... short circuit across...
auf Kurzschluß arbeiten to be on short circuit
Kurzschlußabschaltung short-circuit interruption
Kurzschlußanker short-circuit armature, squirrel-cage rotor
Kurzschluß-Ausgangsadmittanz short-circuit output admittance (621.382)
Kurzschlußauslöser short-circuit trip
Kurzschlußberechnung short-circuit calculation, short-circuit analysis
Kurzschlußbrücke (shorting) bridge

Kurzschlußbügel jumper wire, shorting bar (621.396)
Kurzschlußcharakteristik short-circuit characteristic
Kurzschlußdauer short-circuit period, duration of short(-circuit)
Kurzschlußdrossel (short-circuit) limiting reactor
Kurzschluß-Eingangsadmittanz (small signal) short-circuit input admittance
Kurzschluß-Eingangs-Kapazität input-capacitance short-circuit output (621.382)
Kurzschlußersatzschaltbild constant current equivalent circuit (621.385)
kurzschlußfest, kurzschlußsicher short-circuit proof
Kurzschlußfortschaltung automatic reclosing (621.316)
Kurzschlußkäfig squirrel cage (winding)
Kurzschlußkennlinie s. **Kurzschlußcharakteristik**
Kurzschlußkolben shorting plunger (621.396)
Kurzschlußlasche shorting strip, shorting strap
Kurzschlußläufer (squirrel-) cage rotor
Kurzschlußleistung short-circuit power
Kurzschlußleitung shorting link
Kurzschlußmotor squirrel-cage motor
Kurzschlußring shading ring (621.317)
Kurzschlußschieber shorting plunger
Kurzschlußschleife short-circuited loop
kurzschlußsicher blow-out proof (621.382)
Kurzschlußspannung short-circuit voltage
Kurzschlußspannung eines Transformators impedance voltage of a transformer
Kurzschlußstecker shorting plug
Kurzschlußstrom short-circuit current
Kurzschlußstromverstärkung short-circuit forward current transfer ratio, current amplification
Kurzschluß-Übertragungsadmittanz short-circuit transfer admittance
„Kurzschluß"-Verfahren short-cut method (nicht 621.3)
Kurzschlußverhältnis short-circuit ratio
Kurzschlußvorgang short circuiting
Kurzschlußvorrichtung short-circuiting device
Kurzschlußwicklung short-circuited winding, squirrel cage
Kurzschlußwiderstand short-circuit impedance
Kurzschlußwindung short-circuited turn, shorted turn; slug (621.318.5)
Kurzschlußzeit s. **Kurzschlußdauer**
Kurzstart STOL, short take-off and landing (629.13)
Kurzunterbrechung automatic reclosing (621.396)
Kurzwahl abbreviated dialling
Kurzwellen short waves
Kurzwellenbereich short-wave range
Kurzwellengerät short-wave set
Kurzwellentherapie short-wave therapy
Kurzwellenverbindung short-wave connection, short-wave link
Kurzwellenvorsatz short-wave adapter (621.396)
Kurzzeit short time; short term
Kurzzeitbetrieb short-time service, short-time duty
Kurzzeitleistung short-time rating
Kurzzeitmessung short-time metering
Kurzzeitnennstrom rated short-time current (621.316.5)
Kurzzeitschalter fast-action switch, short-time switch
Kurzzeitstrom rated short-time current
kurz short
kurz darstellen to outline
kurzer Hub short stroke
kurze Lebensdauer short life
kurze Reichweite short range
kurze Wellen short waves
für kurze Belastung ausgelegt short (-time) rated
kurzlebig short-lived, short-living
kurzschließen to short circuit, to short
kurzgeschlossen sein to be on short-circuit, to be shorted
kurzgeschlossene Leitung short-circuited line
kurzgeschlossene Wicklung short-circuited winding
kurzwellig short-wave
kurzwellige Grenze minimum wavelength, quantum limit (621.386)
kurzwelliges Infrarot near infrared
kurzzeitig short-time, temporary
kurzzeitige Betriebsunterbrechung temporary stoppage
kurzzeitiger Betrieb temporary service
kurzzeitige Frequenzkonstanz short-term stability (621.396.61)
Küste coast
Küstenbrechung coastal refraction (621.396)

Küsteneffekt shore-line effect (621.396)
Küstenfunkstelle coast station
Küstenkabel shore-end cable
Küstenstation shore-based equipment
kV = Kilovolt kV = kilovolt
kVA = Kilovoltampere kva = kilovoltampere
kVA-Zähler kilovolt ampere-hour meter
kVar = Kilovar kvar = kilovoltamps reactive (537)
kW = Kilowatt kw = kilowatt
kWh = Kilowattstunde kwh = kilowatthour, B. T. U. = Board of Trade Unit
kWh-Zähler kwh meter
Kybernetik kybernetics (pl.), communication theory

L

L [Induktivität] [inductivity]
Lab. = Labor = Laboratorium lab. = laboratory
labil unstable, instable
labiles Gleichgewicht unstable equilibrium
Labilität instability
Laboratorium laboratory
Laboratoriumsarbeit laboratory work
Laboratoriumseinrichtung laboratory equipment
Labortisch laboratory desk, bench
Labyrinth labyrinth
Labyrinthdichtung labyrinth seal
Labyrinth-Explosionsschutz labyrinth protection
Labyrinthpackung labyrinth packing
Lack lacquer, varnish, enamel
Lackanstrich (application of) varnish; varnish coating, lacquer coating
Lackband varnished cambric
Lackdraht lacquered wire, enamel-insulated wire
Lackfilm lacquer film
Lackimprägnierung varnish impregnation
Lackleinen varnished cambric
Lackpapier varnished paper
Lackpreßpappe varnished pressboard
Lackschlauch varnished flex tubing, flexible varnished tubing
Lackschicht resin finish
Lackseide varnished silk
Lackstrich enamel mark
lackieren to varnish, to lacquer, to finish
Lackierung varnish, finish(ing), lacquer coat
lackisoliert enamelled
lackisolierter Draht enamel-insulated wire
Lackmus litmus
Lade-... loading, charging
Ladeaggregat charging set, charging generator
Ladeband loading belt (621.876)
Ladebefehl load instruction (681.3)
Ladedruck boost pressure
Ladefähigkeit loading capacity
Ladegerät battery charger
Ladegleichrichter battery charger, charging rectifier
Ladekondensator charging capacitor (621.3); (s. auch **Glättungskondensator**)
Ladekreis charging circuit
Lademaschine charging generator (621.355)
Ladeprogramm bootstrap, initial instruction, initial program load (681.3)
Laderegler lamp regulator (621.32/33); charging regulator (621.355)
Ladeschaltung charging circuit
Ladespannung charging voltage
Ladestrom charging current, capacitance current
Ladestromstärke charging rate
Ladevorgang charging
Ladewiderstand charging resistor
laden to load, to charge; to super-charge (621.4)
Lader loader (621.3)
Lader für verschiebbare Programme relocating loader
Ladung charge (621.3), (537); charging; load, cargo (65)
Ladung abführen to drain away the charge, to bleed off the charge
Ladung beibehalten to keep the charge
Ladung auf einen Kondensator aufbringen to put a charge on a capacitor
Ladung mit konstanter Spannung constant-voltage charge (621.355)
Ladung mit konstantem Strom constant-current charge
Ladungsdichte density of charge
Ladungseinheit unit charge
ladungsgekoppelt charge coupled
Ladungsträger charge carrier
Ladungsträgerbeweglichkeit charge-carrier mobility (621.382)
Ladungsträger-Diffusion charge-carrier diffusion
Ladungsträgerinjektion charge-carrier injection

Ladungsträgerspeicherung carrier storage (621.382.2)
Ladungstransport charge transport
Ladungsverlust loss of charge
Lage situation; location; site, position, layer (621); ply (674); thickness
Lage wechseln to interchange position
Lage zu ... location with reagard to ...
Lagekreisel directional gyro
Lageplan site plan, plan view; space diagram
Lager bearing (621); magazine, stock, store, depot
Lager mit Lagerbock bearing bracket assembly
Lager räumen to clear out
ab Lager ex store, from stock
ab Lager lieferbar delivery from stock
auf Lager nehmen to stock
nicht (mehr) am Lager out of stock
Lagerbock bearing bracket, bearing support
Lagerbuchse lining of bearing, bearing bush
Lagerdruck journal load, bearing pressure
Lagerfähigkeit shelf life
Lagerfuß bearing pedestal
Lagerfußanordnung bearing pedestal assembly
Lagergefäß storage bin, storage tank
Lagerkörper body of the bearing
Lagerliste inventory
Lagermetall bearing metal, white metal, anti-friction metal, babbitt
Lagerplatz storing place
Lagerprüfung shelf test (621.353)
Lagerreibung journal friction, bearing friction
Lagerschale bearing shell, bearing cup, bush, brass
Lagerscheibe bearing washer
Lagerschild end frame, end shield (621.313)
Lagerschmierung bearing lubrication
Lagerstätte deposit, bed (55)
Lagerstein jewel (621.317), (681)
Lagersteine einsetzen to jewel
Lagerstift pivot
Lagerzapfen stud, journal
lagern to store, to stock; to bed (621)
Lagerung storage, storing; support, bedding (621)
Lagerung der Drehspule moving-coil support
Lagerung unter freiem Himmel open-air storage
Lagerungsfähigkeit storage life
Lagerungstemperatur storage temperature
Lahnlitzenschnur tinsel cord
Lambda = [Wellenlänge] lambda
Lambda-Viertel quarter-wave
Lambda-Viertel-Abschluß quarter-wave termination
Lambda-Viertel-Antenne Marconi antenna
Lambda-Viertel-Dipol quarter-wave aerial
Lambda-Viertel-Glied quarter-wave sleeve
Lambda-Viertel-Leitung quarter-wave line, quarter-wave transmission line
Lambda-Viertel-Sperrtopf quarter-wave shield
Lambda-Viertel-Stütze metallic insulator, quarter-wave stub, support (621.396)
Lambda-Viertel-Transformator quarter-wave transformer
Lambertsches Gesetz Lambert's law (535)
Lamelle segment, commutator section (621.31); lamination
Lamellenstruktur laminated form of structure
lamellieren to laminate
lamelliertes Blech lamination
lamellierte Bürste laminated brush (621.3)
lamelliertes Joch laminated yoke
lamellierter Magnet laminated magnet
lamellierter Pol laminated pole, stamped pole
lamellierter Polschuh laminated poleshoe
laminar laminar
laminare Strömung laminar flow, viscous flow
Lampe lamp
Lampe einschalten to light a lamp
Lampe mit Edisonsockel screw-type lamp
Lampe mit Swanfassung b. c. lamp = bayonet-cap lamp
Lampe mit kleiner Bajonettfassung s. b. c. lamp = small bayonet-cap lamp
Lampenabdeckung lamp cap
Lampenfassung lamp holder, bulb holder, lamp socket
Lampenfuß stem (621.326), lamp stand (64)
Lampenkolben (lamp) bulb
Lampenkugel globe
Lampenleisten battens
Lampenlicht lamp light
Lampennormal standard lamp, secondary luminous standard
Lampenschirm lamp cap, shade
Lampensockel lamp cap, lamp base
Lampensockelstecker plug adapter
Lampenträger hickey
Lampenwiderstand lamp resistance
Landebahn runway (629.13)
Landebahnanflug runway approach
Landebahnbeleuchtung runway floodlight
Landebahnfeuer runway lights
Landebahnleuchte landing-area floodlight
Landefunkbake mit Punkt-Strich-Tastung equisignal localizer

Landefunkfeuer runway-localizing beacon
Landekurssender localizer (beacon)
Landepfadbefeuerung landing flares
Landeplatz landing ground
Landerichtungsanzeiger landing-direction indicator
Landeskennzahl country code (621.395)
Landmarke landmark (527)
Landungsscheinwerfer landing floodlight
Landwirtschaft agriculture
Landwirtschaftstransformator agricultural service transformer
Landzentrale rural exchange
Lang-..., lang long
lange Haltbarkeit long-wearing quality
lange Kathete long side
lange Lebensdauer long life
lange Leitung long line
lange Nachleuchtzeit long persistence (621.396.96)
lange Seite (eines Rechtecks) broad face
lange Wellen long waves
Langdrahtantenne wave aerial, Beverage aerial, long-wire aerial
Langdraht-Leuchtdraht straight-up-and-down filament (621.32)
Langhubsicherheitsventil highlift safety valve
Langlebensdauer longevity
langlebig long-lived
langlebiges Spaltprodukt long-lived fission product
Langstabisolator rod(-type) insulator
Langwelle long wave, kilometric wave
Langwellenbereich long-wave range
Langwellenempfänger long-wave receiver
Langwellenverbindung long-wave channel, long-wave connection
Langwellenverkehr long-wave communication
Langzeitprüfung long-term test
Länge length; longitude (527)
Länge über alles total length, overall length
Länge Eins unit length
der Länge nach in order of length
auf Länge schneiden to cut to length
Längeneinheit unit length
Längenelement element of length, elementary length
Längengrad degree of longitude, meridian
Längenlehre length gauge
Langevin-Ion Langevin ion
Längs-... longitudinal
Längsbeanspruchung longitudinal stress
Längsbewegung length movement, travelling movement, longitudinal movement
Längsdifferentialschutz biased differential protective system
Längsdruck axial pressure
Längsdurchflutung direct-axis component of mmf (621.313)
Längs-EMK longitudinal electromotive force
Längsfeldkomponente direct-axis component (621.313)
Längsmagnetisierung solenoidal magnetization
Längsnaht longitudinal seam
Längsneigungsregler pitch control (629.13)
Längsreaktanz direct-axis subtransient reactance, positive-sequence reactance
Längsregelung horizontal control (621.316.7); in-phase regulation (621.314.2)
Längsresonanz longitudional resonance
Längsrippe longitudinal rib
Längsschnitt longitudinal section, axial section
Längsschnittwelle longitudinal-section wave (621.396)
Längsschweißung longitudinal weld(ing)
Längsschwingungen extensional mode
Längsspannung longitudinal voltage
Längsspurverfahren longitudional recording
Längsstrahler end-on directional aerial, end-fire array
Längsstrahlerreihe end-fire array (621.396.677)
Längsstrahlrichtantenne end-on directional aerial
Längsstrich down rule
Längsträger longitudinal girder (624)
Längsverbindung straight-through joint
längs along, alongside of
längs der Leitung along the line, down the line
längsdrehen to slide (621.94)
längsgetrennte Sammelschiene sectionalized bus bar
Langsam-... slow, slowing
Langsamanlasser slow-motion starter
Langsamschalter slowing switch (621.876)
Langsamsteuerung inching control (621.9)
langsam slow
langsam drehen, sich to run at slow speed
langsam fahren to slow
langsam laufende Maschine slow-speed machine
langsame Änderung slow change, flat variation
langsamer Brutreaktor thermal breeder
langsame Entwicklung slow development
langsamer Gang low gear

langsames Neutron thermal neutron, slow neutron
langsamer Reaktor slow reactor, thermal reactor
langsamer werden to slow down, to slow off
Längstwellen myriametric waves, very low frequency
Laplacesche Gleichung Laplace's equation
Laplace-Transformation Laplace transformation
Laplace-Transformierte Laplace transform
läppen to lap
Lappnaht lap joint
Larmorpräzession Larmor precession, Larmor theorem
Lasche cover plate, (cover) strap (621.8); tongue (68); clip, lug
Laschenschraube fish bolt
Laschenverbindung strap connection (621.3), lap joint (621.791)
Laser light amplification by stimulated emission of radiation = laser
Laserdrucker laser printer
Last load, power-load (621.3); burden (621), (621.317); weight (531)
Last im Mittelpunkt central load
Last an der Welle shaft load
unter Last on load
unter Last anlassen to start under load
Lastabschaltung load break
Lastabschaltversuch load-rejection test, load-shedding test
Lastabwurf load shedding, load rejection
Lastanpassung load matching
Lastarm weight rod
Lastaufzug goods elevator
Lastbandtragrolle troughing idler
Lastbegrenzungsrelais load-limiting relay, load-levelling relay, load-limiting relay
Lastkreis load circuit
Lastmagnet lifting magnet
Lastminderung derating
Lastminimum power minimum, base load (621.31)
Lastrichtungswechsel reversal of stress
Lastschalter power circuit breaker, power cut-out; on-load tap changer (621.314.2)
Lasttrenner load break switch
Lastschwankung variation in load, load variation
Lastschwerpunkt load center, center of distribution (621.316)
Lastspitze peak load, load peak, maximum demand
Laststoß shock load, load shock
Lasttal minimum power, minimum load
Lastteilung load sharing
Lastumschaltrelais power-transfer relay
Lastverteiler load dispatcher, control engineer (621.316), load divider (621.316.5)
Lastverteilungsstelle load-dispatching office
Lastvielfaches load factor (629.13)
Lastwagen truck
Lastwechsel reversal of stress
Lastwiderstand ballast resistance
Lastwinde windlass
Lastzuschaltung throwing on the load
latent latent
latente Wärme latent heat
Laterna magica magic lantern
Laterne lantern
Laternenpfahl lamp post
Lattenverschlag, Lattenkiste crate
Laue-Film non-screen film (621.386)
Lauf course; run, running, operation
Laufachse straight axle (625)
Laufbüchse bushing
Laufgestell running gear, trailing bogie (625)
Laufgewicht travelling weight
Laufkante running edge
Laufkatze travelling crab, trolley, monkey (621.87)
Laufkraftwerk run-of-river plant, river-run plant
Laufkran travelling crane, overhead crane, movable crane
Lauflager rotor bearing
Laufprobe running test
Laufrichtung direction of propagation
Laufrille groove
Laufring ball race (621.8)
Laufsitz running fit, clearance fit
Laufwerk drive (mechanism) (681.84)
Laufzeit time delay (621.392), (621.396.62), time of flight, (electron) transit time, transition interval (621.385), progagation time (621.377), (621.392); run time (681.3)
Laufzeitempfängerröhre velocity-modulated receiver tube
Laufzeitentzerrer delay equilizer, phase equalizer
Laufzeitfehler phase-delay distortion
Laufzeitgang delay vs. frequency
Laufzeit-Isotopentrenner isotron
Laufzeitkette delay line, retardation network
Laufzeitkettentastung line pulsing
Laufzeitleitung delay line

laufzeitmoduliert velocity-modulated (621.396)
Laufzeitoszillator velocity-modulated oscillator
Laufzeitröhre velocity-modulated tube, klystron, transit-time tube
Laufzeitspeicher delay-line memory, delay-line store
Laufzeitunterschied delay difference
Laufzeitverstärker velocity-modulated amplifier
Laufzeitverzerrung phase distortion, delay distortion, transient effect (621.392)
Laufzeitverzögerung propagation delay
Laufzeitwinkel transit (phase) angle
laufen to run; to be on the go
laufen lassen to rev = to revolve
laufender Bedarf current demand
laufende Länge foot run, footage
laufende Prüfung routine inspection, routine test
laufende Schmierung stream lubrication, permanent lubrication
laufende Unkosten running costs, running expenses
laufende Welle travelling wave (621.3), (534); running shaft (621.8)
Läufer rotor, inductor (621.313); cursor (51)
läufererregte Kommutatormaschine machine with inherent self-excitation
läufergespeist rotor-fed
läufergespeister kompensierter Induktionsmotor (rotor fed) self-compensated motor
Läuferanlasser rotor starter
Läuferblech rotor lamination
Läuferblechpaket laminated rotor core
Läufer-EMK rotor e. m. f.
Läuferkranz rotor rim
Läuferkreis rotor circuit
Läufernut inductor slot
Läuferspannung rotor voltage, inductor voltage
Läuferstrom rotor current, inductor current
Läuferwicklung rotor winding, inductor winding
Lauge lye, alkali
. . . -lauge . . . -hydroxide
Laut sound; tone; note
Lautheit loudness
Lautsprecher loudspeaker, speaker
Lautsprecheranlage public address system
Lautsprechererregung speaker field
Lautsprechermembran speaker diaphragm, speaker cone
Lautsprecherspule voice coil, moving coil
Lautstärke intensity of sound, loudness level, sound level; (output) volume (621.396.62)
Lautstärke aufdrehen to advance the volume control
Lautstärke vermindern to reduce the volume control
Lautstärkemesser sound-level meter, phonometer
Lautstärkenregler, Lautstärkenreglung volume control
Lautstärkepegel volume level, loudness level
Lautstärkeschwankung im Empfang variation in signal strength, fading (effect)
Lautverständlichkeit sound articulation
laut loud, noisy
lauter Empfang strong signal
lauter Raum noisy room
läuten to ring (the bell)
läutern to clarify, to refine, to purify
lawinenartige Ionisation cumulative ionization
Lawinendurchbruch avalanche breakdown
Lawinendurchbruchspannung avalanche voltage
Lawineneffektgleichrichter avalanche rectifier
Lawinen-Gleichrichterdiode mit eingegrenztem Durchbruchsbereich controlled avalanche rectifier diode
LC-Schaltung coil-and-capacitor circuit
LC-Kopplung impedance coupling, choke-capacitance coupling
Leben life
Lebensdauer life (period), longevity, useful life, service life, service limit, course of life, operational life, lifetime (539), (621.382)
Lebensdauererwartung life expectancy
Lebensdauer in unbenutztem Zustand shelf life
Lebensdauerprüfung life test, test of life, reliability test
Lebensdauerprüfung mit steigender Belastung progressive stress testing
Lebensgefahr danger of life
LED LED
LED-Bandanzeige bargraph
lebendige Kraft velocity energy
Lecherdrähte Lecher wires
Lecherleitung Lecher circuit, parallel-wire line
Lechersystem s. **Lecherleitung**
Leckstrom leak(age) current
Lecksucher leak detector

Leclanché-Element Leclanche cell, sack cell
Leer-... idle, neutral, zero
Leeranlauf no-load starting
Leerausschaltung off-load interruption
Leerbandtragrolle return idler
Leerbefehl blank instruction, do-nothing instruction, skip, waste instruction (681.3)
Leerfeld blank panel, empty panel
Leergang neutral gear
Leerhub idle stroke (621.43)
Leerlauf no-load, open-circuit (621.3); idle motion, lost motion, idling (speed) (621)
Leerlauf haben to creep (621.317)
Leerlauf eines Spannungswandlers zero burden (of a potential transformer)
bei Leerlauf on open-circuit, on no-load, under no-load conditions
auf Leerlauf schalten to open circuit
Leerlaufarbeit wasted energy
Leerlauf-Ausgangsadmittanz open-circuit output admittance
Leerlauf-Ausgangsimpedanz open-circuit output impedance
Leerlaufcharakteristik no-load characteristic, open-circuit characteristic
Leerlaufdrehzahl idle-running speed, idling speed, no-load speed
Leerlaufersatzschaltbild constant-voltage equivalent circuit (621.385)
Leerlaufgleichspannung floating voltage (621.314.6)
Leerlaufhemmung anti-creep device (621.371)
Leerlaufkurzschlußverhältnis short-circuit ratio
Leerlaufmagnetisierungsstrom no-load magnetizing current
Leerlaufmessung no-load measurement, open-circuit measurement
Leerlaufrolle idler
Leerlaufschaltung no-load switching
Leerlaufspannung no-load voltage, open-circuit voltage
Leerlauf-Spannungsrückwirkung value of the open-circuit reverse voltage transfer ratio
Leerlaufspannungsverstärkung open-loop voltage gain, voltage amplification, output open-circuit to a. c.
Leerlaufstrom no-load current
Leerlauftrommel idle drum
Leerlaufübersetzungsverhältnis voltage ratio, no-load ratio
Leerlaufverluste no-load losses, open-circuit losses
Leerlaufvortrieb shunt running (621.317)
Leerlaufwiderstand open-circuit impedance
Leerlaufzeit idle time, waste time
Leerlaufzustand no-load state, no-load condition, open-circuit condition
Leerschalter no-load isolator
Leerscheibe idle pulley
Leerstelle blank (681.3); vacancy, hole (539)
Leertakt idle stroke (621.43)
Leertaste spacing key
Leerzeichen blank
Leerzeile blank line
leer empty, void
leer mitlaufende Reserve spinning reserve
leer mitlaufende Tragrolle idler
leer werden to run dry, to run low
leerer Raum (empty) space
leerlaufen to be on open-circuit; to idle, to run loosely
leerlaufende Führungsrolle idle wheel
leerlaufende Leitung open-ended line
leerpumpen to evacuate
Leere void
leeren to void, to empty, to clear
legen to lay, to put
legieren to alloy
legiertes Blech silicon steel (lamination)
legierter Stahl alloy steel
legierter Transistor alloyed transistor
legierter Übergang alloyed junction (621.382)
Legierung alloy
Legierungsbestandteile alloying elements
Legierungstransistor alloyed transistor
Lehre gauge (Br.), gage (US); science, theory; apprenticeship
Lehre vom Magnetismus magnetics (pl.)
Lehrdorn plug gauge
Lehrplan curriculum
Lehrsatz theorem, rule
Lehrling apprentice
Leicht-... light, light-weight
Leichtbau light-weight construction
Leichtmetall light metal
Leichtmetalle light alloys
Leichtmetallegierung light alloy
Leichtöl light oil, light-grade naphtha
leicht light; slight, easy
leicht ansäuern to acidulate
leicht anstoßen to give a tap
leicht auswechselbar easy to exchange, easy to replace
leicht bedienbar easy to operate

leicht berühren to graze
leicht brennbar readily combustible
leicht entzündlich highly inflammable
leicht konisch slightly taper
leicht sauer acidulous
leicht schmelzbar easy-to-melt
leicht ummantelte Elektrode lightly coated electrode
leicht verdampfbar volatile
leicht wellen to ripple
leicht zugänglich easy to get at, get-at-able
leichter Anschlag light stroke
leichter Kern light nucleus
leichter Lauf smooth running
leichte Löslichkeit ready solubility
leichter Schlag tap
leichtschmelzendes Lot soluble solder
sich leicht bearbeiten lassen to machine easily
zu leicht underweight
Leim glue, size
Leim zum Abbinden bringen to set glue
Leimabbindung glue setting
Leimfläche, Leimfuge glue line
leimen to glue, to size
Leinöl linseed oil
Leinwand canvas; screen (77)
leise low, faint, weak
leises Echo weak echo
leises Geräusch faint noise
leise Stelle weak passage (78)
leiser Ton low sound, weak note
leise zischen to fizzle, to hiss faintly
Leiste lath, batten, batton
leisten to do, to work, to effect, to accomplish
Leistung performance, effect; (output) power, wattage (53), (621); capacity; traffic capacity (621.395)
Leistung abgeben to give out power
Leistung abnehmen to consume power
Leistung aufnehmen to take watts, to absorb power, to consume power
Leistung auskoppeln to couple out energy
Leistung bei Aussetzbetrieb intermittent rating
Leistung benötigen to require power
Leistung beziehen von to take power from
Leistung im Endausbau ultimate capacity
Leistung entnehmen, Leistung entziehen to draw power
Leistung erzeugen to generate power
Leistung verarbeiten to handle power
Leistung verbrauchen to spend power, to consume power, to absorb power
an Leistung übertreffen to outperform
mit Leistung versehen to power(-up)
Leistungsabfall fall in output, drop in power
Leistungsabgabe power output
Leistungsanpassung load matching
Leistungsanstieg power increase (621.396.667)
Leistungsaufnahme power consumption, power drain, power absorption, energy intake
Leistungsaustausch power interchange
Leistungsbandbreite power bandwith
Leistungsbedarf power demand, required power, power consumption
Leistungsbedarfsspitze maximum power demand
Leistungsbegrenzung load limitation
Leistungsbegrenzungsschutz over-power protection
Leistungsbemessung power rating
Leistungsbremse dynamometric brake, (absorption) dynamometer
Leistungsfähigkeit (power handling) capacity, capability
Leistungsfaktor (vector) power factor = p. f., phase factor
Leistungsfaktor Eins unity power factor
Leistungsfaktormeßinstrument phase meter
Leistungsfaktorverbesserung power-factor compensation
Leistungsfluß power flow
Leistungsgebühr demand rate
Leistungsgewinn power gain
Leistungsgleichrichter power rectifier
Leistungsgrad efficiency
Leistungsgröße output size
Leistungskonstante specific output
leistungsloser Speicher non-volatile memory, permanent memory
Leistungsmangel power shortage (621.311)
Leistungsmerkmal performance characteristics
Leistungsmesser (horse-)power meter, power indicator, dynamometer, wattmeter
Leistungsmeßsender high-power signal generator
Leistungsoszillographenschleife puissancegraphe
Leistungspegel power level (621.375)
Leistungspentode output pentode, power pentode
Leistungsquelle source of power, power source
Leistungsreaktor power reactor
Leistungsreflexionsfaktor pswr.; power standing wave ratio
Leistungsregelung power regulation, power control
Leistungsrelais power relay

Leistungsrichtungsrelais power directional relay
Leistungsröhre power valve, power tube
Leistungsrückgang fall in output, decrease in output
Leistungsschalter power switch, (power) circuit breaker
Leistungsschild nameplate (621)
Leistungsschreiber power-level recorder (621.39)
Leistungsschwankung fluctuation of power
Leistungsspitze power peak
Leistungssteigerung increase of output
Leistungsstufe power stage (621.396.64)
Leistungsteiler power divider
Leistungstransformator power transformer
Leistungstransistor power transistor
Leistungsübertragung power transfer, power transmission
Leistungsumkehr power reversal
Leistungsverbrauch power consumption, power absorption
Leistungsverlust power loss
Leistungsverminderung power derating (621.396.69)
Leistungsvermögen (power handling) capacity, capability
Leistungsverstärker power amplifier (621.375)
Leistungsverstärkerröhre power-amplifier tube
Leistungsverstärkung power amplification, power gain (621.375)
Leistungswaage horse-power meter
Leistungswandler power converter (621)
Leistungswelligkeitsfaktor power standing wave ratio
Leistungszunahme, Leistungszuwachs power increase
Leit-... guide, lead (531); conduction (537)
Leitantenne director (621.396)
Leitapparat guiding vanes (pl.), gate apparatus (621.2)
Leitblech vane
Leitfähigkeit conductivity
Leitfähigkeitsmesser conductivity meter
Leitfähigkeit des Erdbodens soil conductivity
Leitfernrohr finder
Leitkabel guide cable (527)
Leitlinie directrix
Leitliniensystem track-guidance system
Leitmotor master motor, prime motor, pilot motor (621.316.7)
Leitradar guidance radar
Leitregister originating register (621.395)
Leitrolle guide pulley
Leitschicht wave duct (621.371)
Leitschiene check rail
Leitstrahl (guide) beam; equisignal
Leitstrahl eines Kursfeuers radio-range leg
Leistrahlanflug-Funkfeuersystem beam-approach beacon, BABS
Leitstrahlführung control by guide beam
Leitstrahlfunkfeuer radio range beacon
Leitstrahlreiter beam rider
Leitstrahllenkung beam-rider guidance
Leitstrahlverfahren interlocking equisignal system
Leitvermittlung control exchange (621.395)
Leitvermögen conductivity
Leitweg route, routing (621.395)
Leitwegaufteilung routing-line segregation
Leitweginformation route data
Leitweglenkung call routing, alternate routing
Leitwerk computer control (681.3)
Leitwert conductance, admittance
Leitwertbelag „per unit“ admittance
Leitwertdiagramm admittance chart
Leitwertoperator (vector) admittance
leiten to conduct (621.3); to manage (33)
leitend conducting, conductive
leitend werden to go into conduction
leitender Mantel external conductor
leitende Verbindung conductive connection, conductive path
leitfähig conducting
leitfähige Tinte electrographic ink
Leitfähigkeit conductivity, conductance
Leitfähigkeitsmodulation conductivity modulation
Leiter conductor (621.3); manager, managing director (33); ladder
Leiter erster Klasse prime conductor, first-class conductor
Leiter zweiter Klasse second-class conductor
Leiterbahn trace
Leiterband strip conductor
Leiterbündel bunched conductors (pl.)
Leiterdicke conductor size
Leiterplatte printed circuit board
Leiterseele core conductor
Leiterseil core, cable (621.3)
Leiterspannung line voltage, delta voltage, voltage between lines, voltage between phases, voltage of the system, system voltage
Leitung line, conductor, cable (621.3); duct (624); management

Leitung anschließen an to bring a line to
Leitung mit hohem Verlust lossy line
Leitung im Landbezirk rural line (621.39)
Leitung auf Metallisolatoren stub-supported line (621.396.67)
Leitung mit offenem Ende open-end line
Leitung prüfen to examine a line (621.395)
Leitung in Stahlpanzerrohr metal-cased conductor
Leitung verlegen to run a cable
Leitungen neu verlegen to rewire
Leitung mit Verlust line with dissipation
Leitung ziehen to wire
an der Leitung bleiben to hang on, to hold the line (621.395)
Leitungsabschaltung line disconnection
Leitungsabzweig stub line, branch line
Leitungsanfang sending end (621.392)
Leitungsanlage electric installation
Leitungsanordnung wiring arrangement (696); line configuration (621.315)
Leitungsarbeiter lineman (621.315.1)
Leitungsausgleich line equalization
Leitungsband conduction band (621.382)
Leitungsbogen elbow
Leitungsbruch phase failure, circuit break
Leitungsbündel bunch of pairs (621.394); group of trunks (621.395); conductor bundle
Leitungsdämpfung transmission loss (621.395)
Leitungsdraht electric wire, line wire
Leitungsdrossel feeder reactor, choke coil
Leitungsdurchschalter line concentrator (621.395)
Leitungseinführung (cable) inlet
Leitungselektron conduction electron
Leitungsende receiving end, receiver (621.3)
Leitungsentzerrer line balancer (621.395)
Leitungserde transmission-line earth
Leitungsfeld feeder panel (621.316.34); line section (621.315)
Leitungsführung cable running; routing
Leitungsgeräusch line noise
leitungsgerichtete Telefonie carrier telephony (along lines)
leitungsgerichtete Welle guided wave (621.396)
Leitungshalterung wire guide
Leitungsinstallation wiring installation
Leitungskanal busway
Leitungskonstanten constants of the line
Leitungskonzentrator line concentrator (621.39)
Leitungskreuzung cross (621.315)
Leitungskupplung bus-line coupler (625)
Leitungsmast mast, structure; pole (wood or steel); steel tower
Leitungsnachbildung (duplex) artificial line, balancing network, line simulator (621.395)
Leitungsnetz line system, grid (621.315)
Leitungspaar pair of leads (621.3)
Leitungsplan wiring diagram, connection diagram, wiring scheme
Leitungsprobe test call, talking test, line check (621.395)
Leitungsprüfer continuity checker
Leitungsrelais line relay
Leitungsrohr duct, conduit, raceway
Leitungssaal terminal room (621.395)
Leitungssammelschiene feeder bus bar
Leitungsschacht duct
Leitungsschalter line switch, line circuit breaker
Leitungsschelle cord grip; clip, clamp
Leitungsschnur twin flexible cable, cord, flex
Leitungsschwingkreis resonant line, resonator
Leitungssignal line signal
Leitungsstrom conduction current, convection current
Leitungssuchen trunk hunting (621.395)
Leitungssucher line selector
Leitungssymmetrie line balance (621.395)
Leitungstheorie transmission-line theory
Leitungstransformator matching line, matching section (621.396)
Leitungsübertragung line transmission
Leitungsunterbrechung dislead (621.3)
Leitungsverlegung wiring
Leitungsverlegung auf Putz surface wiring
Leitungsverlegung unter Putz buried wiring
Leitungsverlust line loss, transmission loss (621.315)
Leitungsverstärker (telephone) repeater, intermediate repeater, transmission-line amplifier, line amplifier
Leitungsvoltmeter feeder voltmeter
Leitungswähler line selector, final selector (621.395)
Leitungswächter earth-leakage monitor
Leitungsweg transmission route (621.315.1)
Leitungswiderstand line resistance
Leitungszeichen line signal (621.395.3)
Lenard-Fenster Lenard window
Lenard-Röhre Lenard tube
Lenard-Strahlen Lenard rays
lenken to guide, to steer, to drive (621.113)
Lenkrakete guided rocket, guided weapon, guided missile
Lenksystem guidance system

Lenkung steering
Lenzsche Regel Lenz's law
Leonard-Schaltung Ward-Leonard control
Lesekopf read head, reading head
  Leselocher read punch
  Leseschaltung read out circuit
  Lesesignal sense signal
  Lesestift scanner (681.3)
  Leseverstärker sense amplifier
  Lesevorimpuls preread disturb pulse
lesen to sense, to read out (681.3)
lethal lethal
  lethale Dosis lethal dose
letzt last, ultimate, final
  letzte Bestandteile ultimate constituents
  letzte Stellen weglassen to strike off terminal figures (51)
Letztweg last-choice route (621.395.3)
Leucht-... luminous, light(ing)
  Leuchtbetriebsstoff illuminant
  Leuchtbildschalter luminous-annunciator switch
  Leuchtboje floating light
  Leuchtdecke luminous ceiling
  Leuchtdichte lighting density. lighting intensity; (surface) brightness, brilliancy; luminance (621.397)
    Leuchtdichteschwankung variation range (621.32)
    Leuchtdichtesignal luminance signal
  Leuchtdiode light emitting diode, LED
  Leuchtfaden streamer (55); filament (621.32)
  Leuchtfarbe, Leuchtfarbenanstrich luminous paint, radium paint
  Leuchtfeuer pilot light; airway beacon (351); lighthous beacon (626)
  Leuchtfleck spot, fluorescence spot, beam spot, spot of light (621.385)
    Leuchtflecküberhellung, Leuchtfleckaufweitung blooming
  Leuchtgas illuminating gas, coal gas
  Leuchtkraft luminosity, lighting power, brilliance, brilliancy
  Leuchtmarke cursor (621.396.96)
  Leuchtmasse luminous substance
  Leuchtöl kerosene, kerosine; illuminant
  Leuchtplasma incandescent column
  Leuchtquarz luminator crystal, luminous crystal
  Leuchtreklame electric sign, sign lights, neon lights
  Leuchtröhre neon tube
  Leuchtschaltbild luminous (circuit) diagram
  Leuchtschirm light screen; fluorescent screen, luminescent screen
  Leuchtschrifttafel luminous (annunciator) board
  Leuchtskala illuminated dial
  Leuchtstofflampe, Leuchtstoffröhre fluorescent lamp
  Leuchtstoffpunkte phosphor dots
  Leuchtturm lighthouse
  Leuchtzone zone of luminosity
Leuchte lamp, lighting fitting, luminaire
  Leuchtenhalter fixture stud (696)
Leuchten luminescence
leuchten to light, to shine; to phosphoresce
  leuchtend luminous
    leuchtende Gasentladung luminous gas discharge
    leuchtender Punkt high light
Leydener Flasche Leyden jar
lfd. Meter = laufende Meter metre run
  je lfd. Meter per metre run
Lg. = Länge long. = longitude (91)
L-Glied L-section (621.392)
Libelle cross-level
Licht light
  Licht abgeben, Licht ausstrahlen to give off light
  Licht abstrahlen to emit light
  Licht ist an light is on
  Licht anknipsen to snap the light on
  Licht ist aus light is off
  Licht ausknipsen to snap the light out, to snap the light off
  Licht und Kraft light and power
  Licht werfen auf to throw light on
  Lichtausbeute luminous efficiency
  Lichtbandbreite width of band (621.396.96)
    Lichtbandinstrument luminous-band instrument
  Lichtbatterie light battery
  Lichtbild photo = photograph; (lantern) slide
  Lichtblitz flash (of light)
    Lichtblitzfeuer flashing light (656.7)
  Lichtbogen (electric) arc, flame
    Lichtbogen bilden to arc
    Lichtbogen löschen to quench an arc
    Lichtbogen reißt ab an arc breaks
    Lichtbogen zünden to strike an arc
    Lichtbogenbildung arcing
    Lichtbogenbrenndauer arcing time (621.316.5)
    Lichtbogenentladung arc discharge
    Lichtbogenfußpunkt foot(ing point) of an arc, arc electrode

Lichtbogengleichrichter arc rectifier
Lichtbogengleichrichter mit Funkensteuerung pilot spark arc rectifier
Lichtbogenkammer flash guard, arc chute (621.316.5)
Lichtbogenkohle arc-lamp carbon
Lichtbogenlampe arc lamp
Lichtbogenlöscheinrichtung quencher
Lichtbogenlötung arc brazing
Lichtbogenofen arc furnace
Lichtbogenperle blob
Lichtbogenplasma arc stream (537.5)
Lichtbogensäule arc column
Lichtbogenschneiden arc cutting (621.791)
Lichtbogenschutzhorn arcing horn (621.315.6)
Lichtbogenschutzkammer arc shield, arc deflector (621.316.5)
Lichtbogenschutzring arcing ring, insulator-grading shield
Lichtbogenschweißung arc welding
Lichtbogensender arc generator, arc transmitter, arc converter (621.396)
Lichtbogenspannung arc voltage, arcing voltage, arc-drop voltage
Lichtbogenspannungs(ab)fall arc drop
Lichtbogenüberschlag arcing
Lichtbogenwiderstandsofen smothered arc furnace
Lichtbündel light pencil (535)
Lichtbüschel beam of light
lichtdicht light-proof, light-tight
lichtdicht abgeschlossen light-trapped
lichtecht light-proof
lichtelektrisch photo-electric
lichtelektrisches Relais light relay
lichtelektrische Zelle photoelectric cell
lichtempfindlich photosensitive, lightsensitive
lichtempfindliche Zelle lightsensitive cell
Lichtempfindlichkeit luminous sensitivity (621.385)
Lichtempfindlichkeit eines Filmes speed of a film
Lichtempfindung sensation of light
lichtemittierende Diode light-emitting diode, LED
Lichtfleck lightspot, luminous spot
Lichtfleckinstrument projected scale instrument
Lichtfluß luminous flux
Lichtgeschwindigkeit velocity of light
Lichtgriffel light pen (681.3)
Lichtjahr light year
Lichtkabel light cable
Lichtkorpuskel photon, light corpuscle
Lichtleitfaser optical fibre
Lichtleitfaseroptik fibre optic
Lichtleitung fibre link
Lichtmarke spot light, luminous pointer (621.317)
Lichtmaschine generator (629.113); train-lighting generator (625)
Lichtmenge quantity of light
Lichtmesser light meter
Lichtnetz lighting circuit, electric wiring, electric (light) installation
Lichtpause tracing, blueprint
lichtproportionale Lichtmodulation positive light modulation
Lichtpunktabstand light centre length (621.326)
Lichtpunktabtastung spot-light scanning, flying-spot scanning
Lichtquant photon, quantum of light
Lichtquelle (source of) light, light source, illuminant (53)
Lichtreklame illuminated advertisement, sign light, electric sign (advertising)
Lichtrelais light relay, photo-electric relay
Lichtrufanlage luminous call system, light signalling installation
Lichtschalter lighting switch
Lichtschranke light barrier, light relay
Lichtschreiber light pen
lichtschwach dim
Lichtsignal flash signal, position light signal
Lichtspitze lighting peak
Lichtstärke luminous intensity
Lichtsteckdose lamp receptacle, wall socket
Lichtstift light pen, electronic pen
Lichtstrahl light beam
Lichtstrahlen, Lichtstrahlung light rays (pl.)
lichtstreuende Fläche diffusing surface
Lichtstreuung light diffusion, diffusing effect
Lichtstreustoff scattering medium
Lichtstrom luminous flux
Lichtstromdiagramm isocandle diagram
Lichtstromkreis lighting (branch) circuit
Lichtstromzähler lighting meter
Lichttelefonie optical telephony, phototelephony
Lichttelefoniegerät phototelephone
Lichttelegraph heliograph
Lichttransformator lighting transformer
lichtundurchlässig opaque
Lichtverlust light loss
Lichtverstärkerröhre intensifier tube

Lichtverteilungsdiagramm isolux diagram
Lichtwelle light wave
Lichtwellenleiter optical waveguide
Lichtzähler lighting meter
Lichtzeiger optical lever (621.317)
Lichtzeiger-Instrument light-beam instrument
lichte Breite clear width
lichte Höhe free headroom, clearance
lichte Weite ID = internal diameter, inside diameter, clear width
in der lichten Weite in the clear
Lichtenbergfiguren Lichtenberg figures
Liderungsring packing ring
Lieben-Röhre Lieben-Reiss relay
liegende Maschine horizontal machine
Lift lift; elevator (US)
Lignit lignite
Liliput-... liliput, miniature, midget
limitieren to limit
linear linear
linear polarisierte Welle linearly polarized wave
linear verknüpftes System linear connected system
linearer Ausdehnungskoeffizient coefficient of linear expansion
linearer Ausdruck linear combination (51)
lineare Dipolgruppe collinear antenna array, linear antenna array
lineare Funktion linear function
linearer Gleichrichter linear rectifier, linear detector
lineare Gleichung simple equation
lineares Gleichungssystem first degree algebraic equations (pl.)
linearer Koeffizient der Wärmedehnung coefficient of linear expansion
lineare Modulation linear modulation
lineares Nebensprechen linear crosstalk
lineare Superposition linear superposition, linear combination (51)
lineare Überlagerung linear superposition, linear combination (621.3)
lineare Verstärkung linear amplification (621.396.64)
lineare Verzerrung linear distortion
lineare Wärmedehnung linear expansion
lineare Zeitablenkung linear sweep (621.385)
linearisieren to linearize
Linearbeschleuniger linear accelerator
Linearität linearity
Linearitätsfehler linearity error
Linearverstärker linear amplifier
Linearitätsmaß nonlinearity
Linie line (51), (53), (625)
Linie anreißen to set off a line
Linie gleicher Feldstärke equisignal line
Linien gleicher Horizontalfeldstärke isodynamic lines (538.7)
Linie gleicher Neigung isocline (51)
Linie verlängern to produce a line (51)
Linie ziehen to rule a line
in eine Linie bringen mit to move into line with (51); to align with (621)
Linienanordnung line grouping, omnibus configuration
Linienbüschel set of lines
Liniengitter rod lattice
Linienintegral line integral
Linienschreiber line recorder, strip-chart recorder, graphic recorder
Linienspektrum line spectrum, selective radiation
Linienverstärker line amplifier
Linienzahl scanning density (621.39)
Linke-Hand-Regel left-hand rule
Links-... left-hand, left-handed, counter-clockwise
Linksantrieb left-hand drive
Linksdrehung left-handed rotation
Linksgewinde left-handed thread
Linksschraube left-handed screw
Linksschweißung leftward welding
Linkssteuerung left-hand drive
links left
nach links gerichtet leftward
von links nach rechts gerichtet left-to-right
linksdrehend, linksgängig left-handed, anticlockwise, counter-clockwise
linksgängiges Gewinde left-handed thread
linksgängige Schraube left-handed screw
linksläufig negative-sequence (621.3)
linksläufige Maschine left-hand machine
Linse lens (535)
Linsenantenne lens antenna
Linsenkopfschraube domed head screw
Linsensystem lens system
Lissajou-Figur Lissajou's figure
Liter litre; liter (US)
Lithium lithium
Litze, Litzendraht litz wire, strand, stranded wire, flexible
Litzenkabel flexible cable
Litzenleiter mit Hanfseele cased cord
Litzenleitung flexible conductor
Litzenseil strand, stranded rope
Lizenz licence, permit

Lizenzgebühr royalty, licence fee
Lizenzträger licensee
lizenzieren to license
Ljungströmturbine radial turbine, double-rotation steam turbine
Loch hole (621.382); leakage (621.6); pit (621.357)
Loch mit flachem Boden flat-bottom hole
Loch stechen to prick a hole
Loch zustopfen to plug a hole
Lochblech punched plate, perforated plate
Locheisen punch, puncher (621.9)
Lochfraß perforation (621.315.2), (696); pitting (621.8)
Lochkarte punched card
lochkartengesteuert card-controlled
Lochkartenleser punched-card reader, punched-card interpreter
Lochkartenprüfgerät key-verifying unit, punched-card verifier
Lochkartenstanzer card puncher
Lochkreis pitch circle
Lochkreisdurchmesser P. C. D. = Pitch Circle Diameter, pitch diameter
Lochmaske shadow mask, aperture mask
Lochmaskenröhre shadow-mask tube
Lochscheibe washer; hole disc; Nipcow disc (621.397)
Lochschweißung plug weld
Lochstelle punching position (681.3)
Lochstreifen punched tape, tape
Lochstreifenempfänger reperforator
Lochstreifengerät punched-tape equipment
lochstreifengesteuert tape controlled
Lochstreifenleser (punched-)tape reader
Lochstreifensender (paper) tape transmitter, transmitter distributor
Lochstreifenspeicher paper-tape store
Lochstreifenstanzer paper-tape punch
Lochstreifenvervielfältigungsanlage tape-copying system
lochen to punch, to pierce, to perforate
gelochtes Blech punched plate; perforated plate
gelochte Metallfolienabschirmung perforated metal screen (621.315.2)
Löcher pits (621.357); holes (621.38)
Löcherbeweglichkeit hole mobility (621.382)
Löcherleitung hole conduction, P-type conduction (621.382)
Löcherstrom hole current
locker slack, loose
locker werden to slacken, to work loose
lockern to loosen; to slacken (a screw)
Löffelkontakt spoon-type contact
log = Logarithmus log = logarithm
Logarithmenpapier logarithmic paper
Logarithmentafel table of logarithms
logarithmisch logarithmic
logarithmische Auftragung logarithmic plot, logarithmic chart
logarithmisches Dekrement logarithmic decrement
logarithmisches Diagramm log plotting
logarithmische Spirale logarithmic spiral
logarithmischer Trichterlautsprecher logarithmic horn (speaker)
logarithmischer Verstärker logarithmic amplifier
in logarithmischer Stufung logarithmically spaced
logisch logic, logical
logische Addition logical add
logischer Befehl logical instruction
logische Entscheidung logical decision
logische Mehrfachschaltung logic array
logische Schaltung logical circuit, logic network, gate
logische Verknüpfung logical operation
logische Verschiebung logic shift
logischer Zustand logic level, (logic) state
Logistik logistics
Lokomotive locomotive
Lokomotive mit Rollenstromabnehmer trolley locomotive
Longitudinalwelle longitudinal wave
los-... un-..., dis-...
losarbeiten, sich to work loose (621.3)
loshaken to unhook
loskuppeln to uncouple
loslöten to unsolder, to unsweat
losnieten to unrivet
losschrauben to unscrew, to unbolt
lösbar solvable (51); detachable (621)
lösbarer Teil detachable link, detachable component
Lösch-... (fire) extinguishing; quenching (621.3); clearing (681.3)
Löschen des Bildschirms clearing of screen
Löschbefehl clear instruction (681.3)
Löschfunken quenched spark
Löschfunkensender quenched spark-gap transmitter
Löschfunkenstrecke quenched spark gap
Löschgeräte fire-fighting equipment (sing.)
Löschkammer quenching pot, arc chamber
Löschkammerkontakt boxed contact, blast contact (621.316.5)

Löschkammerschalter oil-blast breaker, pot-type breaker
Löschkondensator quenching capacitor
Löschkopf erasing head, wiping head (681.84)
Löschkreis quench circuit
Löschmittel extinguishing medium, quenching medium
Löschrohrableiter expulsion-type arrester
Löschrohrsicherung expulsion-fuse unit
Löschschaum suds (pl.), fire foam
Löschspannung extinction potential, de-ionizing potential; reseal voltage (621.316.9)
Löschspule blow-out coil (621.316.5)
Löschtaste erase button (681.84); cancel key, release key; clearing key (681.3)
Löschung magnetic biasing (621.377); erasing (681.3), (681.84); clearing (681.3)
Löschwinkel communication margin (621.314.5)
Löschzeit gate controlled turn-off time (621.314.63); arcing time (621.316.9)
löschen to extinguish, to quench, to erase; to clear (681.3); to delete (681.3)
löschendes Lesen destructive reading, destructive readout
löschbarer Speicher erasable storage
lösch- und programmierbarer Speicher erasable programmable ROM = EPROM
lose loose, slack
loses Ende slack end
loses Gefüge loose texture
lose gekoppelt lightly coupled, loosely coupled
lose hängen to dangle
loser Kern loose piece, loose core (621.7)
lose Kopplung loose coupling (621.396)
loser Lagerdeckel loose cap
loser Reibsitz loose friction fit
loser Sockel loose socket
lose Strippe loose cord
loser Teil detachment
lose Verbindung loose connection, slack joint
lose Zuführungsleitung trailing cable
lösen to solve, to resolve (51), (53); to dissolve (54); to loose, to slacken, to disconnect, to detach; to unfasten (621)
sich lösen lassen nach... to go by... (51)
sich nicht lösen lassen to defy solution
lösend solvent
gelöster Stoff dissolved matter
gelöste Substanz solute
löslich soluble, dissoluble (54)
Löslichkeit solubility, dissolubility (54)
Lösung answer (51); solution (51), (54); dissolution (54)
in Lösung gehen to pass into solution
Lösungsdruck solution tension
Lösungsflüssigkeit solvent (liquid)
Lösungsmittel solvent, dissolvent
Lösungspotential gegen... solution potential against...
Lösungsspannung (electrolytic) solution voltage
Lösungsweg (method of) attack, (method of) approach (51)
Lot normal, vertical, perpendicular (line) (51); sound (629.12); plumb (line), bob (69); solder (metal) (621.791)
Lot fällen to erect a perpendicular line
aus dem Lot out of plumb
ins Lot bringen to set plumb
Lotabweichung plumb-bob deflection
Loteinrichtung sounding device (629.12)
lotrecht perpendicular, vertical plumb
lotrecht einrichten to set plumb
Löt-... soldering
Lötanschluß soldering tag
Lötbad floating solder bath
Lötdraht solder wire
Lötdraht mit Flußmittel cored solder
Lötdrähtchen pigtail lead
Lötfahne soldering tag, lug (621.3)
Lötfett soldering paste
Lötflußmittel soldering flux
lötfreie Drahtverbindung wire wrap, solderless connection
Lötgeräte soldering appliances
Löthülse soldering sleeve
Lötkolben solder iron, soldering iron
Lötkragen, Lötmuffe wiping gland, wiping sleeve
Lötlampe blow lamp, blow torch, soldering blow pipe
Lötmittel solder (metal)
Lötöse lug (621.3)
Lötösenstreifen soldering-tag, terminal strip, tag block
Lötpaste solder cream
Lötpastendosierer solder-cream dispenser
Lötperle blob, bead of solder
Lötperlenprobe borax-bead testing
Lötrohr (mouth) blow pipe, soldering blow pipe
Lötsäure soldering solution
Lötschuh sweating thimble, soldering lug
Lötschwanz pigtail lead
Lötsicherung heat coil (fuse)

Lötstab stick of solder, solder stick
Lötstelle junction, soldered joint
Lötstelle des Thermoelements thermocouple junction
Lötstützpunkt binding post
Lötverbinder splicing sleeve
Lötvorgang soldering (process)
Lötwasser soldering solution
Lötzinn solder metal, solder tin
loten to fathom, to sound (656); to plumb (621)
löten to solder, to sweat
Lotung sounding (656)
L-Schaltung L-network
Lücke valancy (621.382)
Luft air; clearance (621.8)
Luftabschluß exclusion of air
Luftansaugung air intake
Luftbefeuchtung air moistening
Luft-Boden Flugkörper air-to-surface missile
luftdicht air-tight
Luftdichte air density
Luftdrähte air wires (621.396)
Luftdrehkondensator air-dielectric variable capacitor, variable air capacitor
Luftdrosselspule air-core choke, air-core reactor
Luftdruck air pressure, barometric pressure, atmospheric pressure
luftelektrisches Feld earth's electric field
luftelektrische Störungen atmospherics
Luftelektrizität atmospheric(al) electricity
Luftentfeuchter air dehumidifying device
Lufterhitzer air heater, airheater
Luftfeuchtigkeit atmospheric humidity, air humidity
Luftfilter air cleaner
Luftisolation air insulation
luftisoliertes Kabel air-space cable
Luftkabel aerial cable
Luftkabelleitung aerial-cable line, cable-pole line
Luftkanal air duct
Luftkondensator air capacitor (621.3)
Luftkühlung air cooling
luftgekühlt air-cooled, ventilated (621.313)
luftgekühlter Transformator air-cooled transformer, ventilated transformer
Luft-Luft-Flugkörper air-to-air missile
Luftnavigation aerial navigation
Luftraum atmosphere, airspace
Luftraumüberwachung airway traffic control, airtraffic control, air surveillance
Luftreibung air friction
Luftreibungsverluste windage losses
Luftreibungswiderstand drag (629.1)
Luftringlaufkühlung closed-circuit air cooling
Luftsauerstoff atmospheric(al) oxygen
Luftsauerstoffbatterie air-cell battery
Luftsauerstoffzelle air cell (621.35)
Luftschall airborne sound
Luftschalter air-break switch, air-blast breaker
Luftschleuse air lock
Luftschütz air-break contactor
Luftspalt air gap, air space, gap clearance; head gap (681.84)
Luftspaltbreite width of the gap, gap clearance
Luftspaltfluß air-gap flux
Luftspaltinduktion air-gap flux density
Luftspaltstreuung main leakage, peripheral leakage (621.313)
Luftspaltwiderstand gap reluctance (621.313)
Luftspule air-cored coil, air-cored choke, air-cored reactor
Luftstreckentrenner air-section break (621.33)
Luftstrom draught, air current, air blast
im Luftstrom leer mitlaufen to windmill (629.13)
Luftströmung air flow
Lufttemperatur air temperature
Lufttransformator air-cooled transformer
Lufttrennstrecke air break, isolating air gap
Lufttrimmer air-type trimmer
Luftwellenlänge free-space wavelength (621.396)
Luftwerterechner air-data computer (629.13)
Luftwiderstand drag, air resistance
Luftzutritt air admission
lüften to vent, to ventilate, to air, to aerate
Lüfter ventilator; blower (621.313)
Lüfterrad ventilating fan, ventilating wheel
Lüftung ventilation
Lüftungsgitter ventilating grating
Lüftungskanal ventilating (air) channel, ventilating duct
Lüftungsloch vent (hole)
Lumen lumen (621.32); i. d. = internal diameter (621.6)
Lumenstunde lumen hour
Luminanzsignal luminance signal
Lumineszenz luminiscence
Lumineszenzdiode light-emitting diode, LED
Luminophor luminescent material, phosphor
Lummer-Brodhun Photometer Lummer-Brodhun photometer
Lunker cavity, shrinkhole, void
Lunkerbildung shrinkage

Lupe magnifier, magnifying glass
Lüster gloss
Lüsterklemme lamp-wire connector
Lux = Lumen/qm lux, meter-candle
Luxmeter (illumination) photometer
Luxemburg-Effekt ionospheric cross-modulation, Luxemburg effect

M

M = Meg(a)-... M = meg(a)-...
M = Mille M = mille
m = Meter m = meter
Madenschraube grub screw, headless screw
Magazin laminated magnet (621.317); store (621.9); cartridge (681.84)
Magenlampe photophore
magisches Auge magic eye, electron ray tube, cathode ray tuning indicator, shadow tuning indicator
magischer Fächer electron-ray indicator tube
Magnesium magnesium
Magnet magnet
Magnetanker armature of a magnet
Magnetanlasser solenoid starter, magneto (621.386)
Magnetaufzeichnung magnetic recording
Magnetband magnetic tape, magnetic recording tape
Magnetbandaufnahme magnetic recording, tape recording
Magnetband-Fernsehaufzeichnung video-tape recording, VTR
Magnetbandgerät magnetic recorder, tape recorder, magnetic tape device, magnetic tape recording and reproducing device
Magnetbandgerät in Kassettentechnik cartridge tape recorder
Magnetbandgerät in Spulentechnik reel-to-reel tape recorder
Magnetbandgerät mit endlosem Band continuous loop recorder
Magnetbandkassette tape cartridge
Magnetbandlaufwerk tape drive
Magnetbandleser magnetic tape reader
Magnetband-Normwandlung standards conversion of taped programs
Magnetbandspeicher magnetic tape store
Magnetbetätigung solenoid operation (621.316)
Magnetblasenspeicher magnetic-bubble memory
Magnetbremse magnetic brake, solenoid brake (621.33)
Magnetdrahtspeicher magnetic-wire store
Magneteisenerz magnetite
Magneteisenstein lodestone, loadstone
magnetelektrischer Generator magneto-electric generator
Magnetfeld magnetic field
Magnetfeldröhre magnetron
Magnetfeldröhrenkennlinie critical-voltage parabola
Magnetfluß magnetic flux
Magnetfutter magnetic chuck (621.9)
Magnetgestell magnetic frame
Magnetglocke magnet shell
Magnetinduktor magneto inductor
Magnetjoch magnet yoke
Magnetkarte magnetic card
Magnetkern magnet core, magnetic core
Magnetkernspeicher magnetic core storage
Magnetkompaß magnetic compass, compass magnet
Magnetkompaßpeilung magnetic bearing
Magnetkopf magnetic head; pick-up magnet; recording magnet
Magnetkupplung magnetic clutch
Magnetlautsprecher balanced-armature loud-speaker, magnetic-armature speaker
Magnetlegierung magnetic alloy
Magnetmikrophon magnetic-type microphone; variable-reluctance microphone
Magnetmine magnetic mine (623)
Magnetnadel magnetic needle
Magnetplatte hard disk, Winchester
Magnetplattenspeicher (magnetic) disc memory
Magnetpol (magnetic) pole (538)
Magnetprüfgerät permeameter
Magnetscheider magnetic (ore) separator, drum apparatus
Magnetscheidung magnetic separation
Magnetspeicher magnetic memory, magnetic store
Magnetspur magnetic track
Magnetstahl magnet steel
Magnetstreifen magnetic strip
Magnettaste magnetic key
Magnetton-... magnetic recording
Magnettonabnehmer moving magnet (pick-up)
Magnettrommel magnetic pulley, magnetic drum

Magnettrommelspeicher magnetic drum (storage)
Magnetventil solenoid valve
Magnetverstärker magnetic amplifier, transductor
Magnetwaage magnetic balance
Magnetwicklung magnet winding
Magnetzeichenleser magnetic character reader
Magnetzündung magneto ignition (621.43)
magnetisch magnetic
magnetisch betätigt solenoid-operated
magnetisch betätigter Schalter solenoid switch
magnetisch gesteuertes Ventil solenoid valve
magnetisch Nord magnetic north
magnetische Ablenkung magnetic deflection (621.385)
magnetische Abschirmung magnetic shield, magnetic screen
magnetische Abstoßung magnetic repulsion
magnetische Achse magnetic axis (538)
magnetische Anziehung magnetic attraction
magnetische Anziehungskraft magnetic pull
magnetischer Äquator aclinic line
magnetische Aufbereitung magnetic separation
magnetische Aufspaltung magneto-ionic splitting
magnetische Aufzeichnung magnetic recording
magnetische Auslösung magnetic release, magnetic trip
magnetischer Ausscheider magnetic separator
magnetische Beschichtung magnetic coating
magnetische Blaseinrichtung magnetic blow-out (621.316.5)
magnetisches Blatt magnetic shell
magnetische Breite magnetic latitude
magnetische Deklination magnetic declination, magnetic variation
magnetische Doppelbrechung magneto-ionic double refraction
magnetisches Drehfeld rotating magnetic field
magnetische Drehung magnetic rotation (535)
magnetische Eigenschaft magnetic property
magnetischer Eigenschutz degaussing equipment (629.12)
magnetischer Einheitspol unit magnetic pole
magnetischer Erzscheider magnetic ore separator
magnetisches Feld magnetic field
magnetische Feldlinie line of magnetism
magnetische Feldstärke magnetic fieldstrength, magnetic force, magnetic intensity
magnetischer Fluß magnetic flux
magnetischer Fluß drängt sich zusammen magnetic flux crowds
magnetische Flußdichte magnetic flux density, magnetic induction
magnetische Flußlinien lines of magnetic flux
magnetische Flüssigkeitskupplung electromagnetic fluid clutch
magnetisches Fremdfeld external magnetic field
magnetischer Fremdfeldeinfluß effect of external magnetic field
magnetisches Futter magnetic chuck
magnetische Größe magnetic quantity
magnetische Hysteresis magnetic hysteresis
magnetische Induktion magnetic induction, magnetic-flux density
magnetische Inklination magnetic inclination, dip
magnetische Karte magnetic map
magnetische Klebkraft magnetic adhesion
magnetischer Konstanthalter magnetic regulator
magnetische Kraftlinie line of magnetic force, line of magnetism
magnetische Kraftröhren tubes of magnetic force
magnetischer Kreis magnetic circuit
magnetische Ladung charge of magnetism
magnetische Längsaufzeichnung longitudional magnetization
magnetischer Lautsprecher magnetic speaker, magnetic-armature speaker
magnetische Leitfähigkeit permeability
magnetischer Leitwert permeance
magnetische Linse magnetic lens
magnetischer Luftspaltwiderstand gap reluctance
magnetischer Meridian magnetic meridian
magnetisches Mikrophon magnetic (-type) microphone
magnetische Mißweisung magnetic variation
magnetisches Moment magnetic moment
magnetisches Moment des Kerns nuclear magnetic moment (539)
magnetische Nachwirkung magnetic lag, magnetic viscosity
magnetischer Nebenschluß magnetic shunt
magnetische Permeabilität magnetic permeability
magnetischer Pol magnetic pole
magnetische Pole magnetic foci
magnetische Polstärke magnetic-pole strength

magnetisches Potential magnetic potential
magnetische Potentialdifferenz magnetic-potential difference
magnetische Prüfung magnetic test(ing)
magnetische Quantenzahl magnetic quantum number
magnetische Queraufzeichnung perpendicular magnetization
magnetische Remanenz magnetic remanence
magnetische und elektrische Rückkopplung mixed retroaction
magnetische Sättigung magnetic saturation
magnetische Spannung magnetic-potential difference
magnetischer Spannungskonstanthalter magnetic voltage stabilizer
magnetische Stoßschweißung electromagnetic percussive welding
magnetische Strahlkonzentrierung magnetic focusing
magnetisches Streufeld stray magnetic field
magnetischer Streufluß stray magnetic flux
magnetische Streuung magnetic leakage
magnetischer Streuweg magnetic-leakage path
magnetischer Sturm magnetic storm (538.7)
magnetische Suszeptibilität magnetic susceptibility
magnetische Teile magnetics
magnetische Tinte magnetic ink
magnetischer Tonabnehmer magnetic pick-up
magnetische Umwandlungstemperatur magnetic transition temperature, Curie point
magnetische Verriegelung magnetic interlocking
magnetischer Verstärker magnetic amplifier
magnetische Videosignalaufzeichnung video-tape record(ing)
magnetische Vorgeschichte previous magnetic history
magnetischer Weg magnetic path
magnetischer Widerstand magnetic resistance, magnetic reluctance
magnetischer Widerstand des Luftspalts gap reluctance
magnetisierbar magnetizable
Magnetisierbarkeit magnetizability
magnetisieren to magnetize
Magnetisierung (intensity of) magnetization, ferric induction
Magnetisierungsbereich magnetization range
Magnetisierungskurve curve of cyclic magnetization, BH curve
Magnetisierungskurve des Eisens iron curve
Magnetisierungsstrom magnetizing current, exciting current
Magnetisierungswicklung magnetizing winding
Magnetismus magnetism; magnetics
Magnetismusmenge quantity of magnetism
Magnetit magnetite
Magnetohydrodynamik magnetohydrodynamics, MHD
magnetohydrodynamischer Generator magnetohydrodynamic generator
Magnetometer magnetometer, magnetic tester, gaussmeter
Magnetometerverfahren magnetometer method
Magnetometrie magnetometry
magnetomotorische Kraft magneto-motive force = m. m. f., magnetizing force
Magneton magneton
Magnetooptik magneto-optics
magnetooptisch magneto-optic(al)
magnetooptische Drehung magneto-optical rotations (pl.)
Magnetophon magnetophone, magnetic tape recorder
Magnetostriktion magnetostriction
magnetostriktiv magneto-strictive
magnetostriktiver Effekt magnetostrictive effect
magnetostriktives Filter mechanical filter
magnetostriktiver Lautsprecher magnetostriction loudspeaker
magnetostriktiver Schwinger magnetostrictive oscillator
Magnetron magnetron
Magnetronfrequenzziehen magnetron pulling
Magnetronschwingungen magnetron oscillations
Majoritätsträger majority carrier (621.382)
Makelverbindung broker's call
Makro . . . macro . . .
Makrobefehl macro instruction
Malachit malachite
Malteserkreuz Maltese cross
Mangan manganese
Mangel defect; deficiency
Mangel an lack of, shortage of; deficiency in
Mangel an Biegsamkeit inflexibility
Mangel an Elastizität inelasticity
Mangel an Reaktionsfähigkeit inertness
Mangel an Widerstandsfähigkeit frailness
Mangelleitung p-type conduction, hole conduction (621.382)
mangelhaft defective; deficient
mangelnder Gleichlauf misalignment

manipulieren to manipulate
Mannloch manhole, access opening
Manometer manometer, pressure gauge, compression gauge
Manschette cuff (621)
Mantel coat, jacket (621); sheath (621.315.2); shell (623.4)
mit einem Mantel versehen to jacket; to can
Manteldraht bimetallic wire
Mantelelektrode sheated electrode, coated electrode
Mantelerdung earthing cable bond
Mantelfläche lateral area, lateral surface (513)
mantelgekühlt frame-cooled (621.313)
Mantelleiter external conductor
Mantelleitung non-metallic sheathed cable, plastic-sheathed cable
Mantelspannung induced voltage (621.315.2)
Mantelstrom sheath current
Manteltransformator shell-type transformer
Mantelverbinder cable bond (621.315.2)
Mantelverbindung cable-sheath bond
Mantelverluste sheath loss (sing.)
Mantisse mantissa
manuell manual
manueller Betrieb manual operation (681.3)
manuell eingegebene Daten manual data
maritim maritime
Marke mark, strobe (621.317); make, brand; stamp; label (681.3)
Markenartikel standard product
markieren to mark; to flag (681.3)
Markierer marker (621.395.3)
Markierung mark, marking; marker; identification (621.396.93); flag (681.3)
Markierungsfarbe marking substance
Markierungsfunkfeuer marker (beacon)
Markierungsimpuls marker pip, strobe pulse (621.397)
Markierungsleser document mark reader
Markierungslinie marking, line of demarcation
Markierungsstrich (graduation) mark
Marmor marble
Marmorplatte marble slab
Martensit martensite
Marx-Gleichrichter Marx rectifier
Masche mesh
in einer Masche round a mesh
Maschendraht screen wire, wire mesh
Maschennetz mesh, mesh network (621.3)
Maschenstrom mesh current
Maschenweite mesh, width of meshes, sieve spacing
Maschine machine (621.3), (621.9); engine (621.1), (621.4), (625); motor (621.313), (621.4); generator (621.313.12)
Maschine mit ausgeprägten Polen salient-pole machine
Maschine mit Berührungsschutz protected machine
Maschine leer laufen lassen to run a machine idle
Maschinen warten to maintain machinery (sing.)
Maschinenadresse machine address
Maschinenbau mechanical engineering
Maschinenbauingenieur, Maschinenbauer mechanical engineer
Maschinenbefehl computer instruction, machine code instruction (681.3)
Maschinencode machine code, object code
Maschineneinrichter tool setter
Maschinengestell machine frame, machine base
Maschinenhalle engine house
maschinenlesbar machine readable
Maschinenmoment machine torque
Maschinenprogramm machine program, object program (681.3)
Maschinenraum machine shop; engine room; power house
Maschinensatz generating set (621.31)
Maschinenschraube machine screw
Maschinensender alternator transmitter (621.396)
Maschinensockel machine base
Maschinenspannungsregler generator-voltage regulator
Maschinensprache computer language, machine language, object language (681.3)
Maschinentelegraph engine room telegraph
Maschinentelegraphie machine telegraph system, automatic transmission (621.391)
Maschinenwärter attendant
Maschinenwicklung machine winding
Maschinenzyklus machine cycle
maschinell mechanical
maschinell hergestellt machine-made
maschinenorientierte Programmiersprache computer-oriented language
Maser (microwave emission by stimulated emission of radiation), maser
Maske mask
Maskenmikrophon mask microphone
Maskenröhre shadow-mask tube
Maskentechnik mask engineering (621.382)

maskenprogrammierbarer Festwertspeicher mask-programmable ROM
Maskieren masking (681.3)
Maß measure, dimension, rate; gauge
Maß und Gewicht weights and measures (pl.)
Maß- und Gewichtswesen metrology
Maß nehmen to take measure
Maß nehmen an to take the gauge of
auf Maß formen to form to shape
auf Maß schneiden to trim to size, to true
nach Maß to measure, true (621)
nach Maß gefertigt made to measure
wieder auf Maß bringen to true up
Maßanalyse quantitative analysis
Maßband tape measure
Maßeinheiten units of measure; absolute units
maßgerecht true to size
maßhaltig true, correct, to measure
Maßhaltigkeit correctness, accuracy
auf Maßhaltigkeit prüfen to check for specified limits (pl.)
Maßhaltigkeitsprüfung test of accuracy
Maßstab measure, scale; scale rule, yardstick
Maßstab 1 : 1 full size, one-to-one correspondence
Maßstab 1 : 2 half size
Maßstab 1 : 4 quarter size
maßstäblich to scale
maßstäblich abändern to scale
maßstäblich abgreifen to scale off
maßstäblich vergrößern to scale up
maßstäblich verkleinern to scale down
maßstäbliche Zeichnung scale drawing
Maßsystem system of units
Maßtafel table of measures
Masse mass, crowd, bulk; compound (621.315.2); frame chassis ground (621.31), (621.396), (621.4); body (66)
an Masse legen to ground, to earth, to bond to frame (621.3)
mit Masse füllen to compound (621.3)
Masseanschluß earthing, grounding, frame (bonding) connection
massebehaftet massive, poised, heavy
Massefestwiderstand composition resistor
Masseisolierung insulating compound, solid system
Massekabel solid-type cable, bitumen cable (621.315); ground wire (621.396.6)
Massekern compressed iron core, powder core
masselos massless, weightless, bodiless, disembodied
Masselosigkeit lightness
Massenabsorption mass absorption
Massenabsorptionskoeffizient mass-absorption coefficient (621.386)
Massenanziehung mass attraction, gravitation
Massendefekt mass difference, mass defect (539)
Masseneinheit mass unit; unit mass
Massenerhaltungssatz law of mass conservation
Massenfertigung mass production
Massenkonstanz mass conservation
Massenprüfung mass-production testing
Massenspeicher mass storage
Massenspektrograph mass spectrograph
Massenträgheit inertia
Massenzahl mass number (539)
Massewiderstand composition-type resistor
Masseleisen pig iron
mäßig moderate, temperate
mäßig groß moderate-size
mäßig schwer moderate-weight
mäßig stark moderate-strength
mäßige Belastung moderate load
mäßige Entfernung moderate distance
für mäßigen Abstand moderate-distance
mäßigen to moderate
Massiv-. . . massive, solid
Massivdraht solid wire
Massivinduktor one-piece inductor
Massivkohle (pure) solid carbon
Massivpol solid pole
massiv massive, solid
massiv rund solid circular
massiver Leiter solid conductor
massives Rohr continuous tube
massives Silber solid silver
Mast pole, post, tower, pylon (621.315.1); mast (621.396.67)
Mast setzen to sink a pole
Mastabspannung pole guy
Mastabstand span (621.315.1)
Mastanker stay
Mastaufführung post head (621.33)
Mastaufhängung pole mounting
Mastbelastung tower loading
Masterdungswiderstand tower-earthing resistance
Mastfuß mast base, tower base, tower footing; pole butt
Mastkopf head of the mast
Mastschalter mast switch, pole switch
Mastspitze point of a pole
Maststrahler to wer radiator
Masttransformator pole (-type) transformer

Masttrennschalter mast isolator, pole-type isolator
Mastic mastic
Material material, materials
Materialausnutzung utilization of material
Materialfehler defect in material
Materialkonstante characteristic constant of material
Materialprüfreaktor materials testing reactor (539)
Materialprüfung materials testing, testing of materials
Materialsammlung collection of data
Materialwahl selection of materials
Materialwanderung transportation of contact material (621.316.5)
materialisieren to materialize
Materie matter
in Materie verwandeln, (sich) to materialize
Materieteilchen material particle, constituent particle
Mathematik mathematics (pl.)
mathematisch mathematical
mathematisch behandeln to analyse, to apply mathematics to, to deal mathematically with
mathematische Ableitung mathematical deduction
mathematische Aufgabe mathematical problem
mathematischer Beweis mathematical proof
mathematische Darstellung mathematical representation
mathematische Fiktion mathematical fiction
mathematische Formulierung mathematical formulation
mathematisches Problem mathematical problem
mathematische Schreibweise mathematical representation
mathematisches Zeichen mathematical symbol, code symbol
Mathieusche Funktion Mathieu function
Matrix (pl.: -izen) matrix (pl.: -ices) (51)
Matrixdrucker matrix printer
Matrixzeichen matrix character
Matrize female die (621.9); matrix (51)
Matrizenrechnung calculus of matrices
Matrizenstahl die steel
matt mat, dull (535); frosted (69)
matte Farbe dull colour
matter Glanz dull lustre
mattes Glas frosted glass
matte Oberfläche dead surface
matte Säure depleted acid
Matte mat
mattieren to frost; to grey (77)
mattiertes Glas mat-surface glass
mattierte Lampe frosted lamp, pearl lamp
Mattierung frosting
Mattierungsbad matte dip
Mattscheibe diffusing screen, ground glass
Maulklemme jaw-type terminal
Maus mouse (681.3)
Maxigraph maxigraph
Maximal-... maximum
Maximalauslöser overload release
Maximalauslösung overcurrent circuit breaking
Maximalkontrast contrast range (621.317)
Maximalrelais maximum (current) relay, overcurrent relay, overload relay
Maximal-Minimalrelais over-and-undercurrent relay
Maximalschalter overload circuit breaker, overcurrent switch
Maximalschalter mit Nullspannungsauslösung overcurrent undervoltage circuit breaker
Maximalspannung maximum voltage
Maximalthermometer maximum thermometer
Maximalverstärkung maximum amplification, ceiling of amplification
Maximalwert maximum value
maximal maximum
maximal abgebbare Leistung available power
maximale Datenübertragungsgeschwindigkeit peak data transfer rate
maximal erreichen to top
maximaler Spanungswert maximum voltage, ceiling voltage
maximaler Strom maximum current
Maximum maximum (pl.: -ma)
Maximumanzeiger maximum-demand indicator
Maximumtarif (fixed) charge for maximum demand
Maxwell maxwell (538)
Maxwellsche Brücke Maxwell bridge (621.317)
Maxwellsche Gleichungen Maxwell field equations
Maxwell-Statistik Maxwell statistics (pl.)
MAZ = magnetische Videosignal-Aufzeichnung video tape recording
Mechanik mechanics (531); mechanism (621)
Mechaniker mechanic, mechanician (681)
mechanisch mechanical
mechanisch beanspruchen to put mechanical stresses on ...

mechanisch beanspruchte Verbindung structural joint
mechanisch betätigt mechanically operated
mechanisch kuppeln to gear together, to gang together
mechanische Achse mechanical axis (537.228), (548)
mechanischer Anschlag mechanical stop
mechanischer Antrieb mechanical drive
mechanisches Antriebssystem mechanical-drive system
mechanische Auslösung mechanical trip
mechanische Belastung shaft load (621.313.13)
mechanische Betätigung mechanical operation
mechanischer Blindwiderstand mechanical reactance
mechanische Eigenschaft mechanical property
mechanische Energie mechanical energy
mechanische Festigkeit mechanical strength
mechanischer Filter mechanical filter
mechanischer Gleichrichter mechanical rectifier
mechanische Größe mechanical quantity
mechanische Impedanz mechanical impedance
mechanische Leistung mechanical power
mechanisches Lichtäquivalent mechanical equivalent of light
mechanische Mischung mechanical mixture
mechanisches Ohm mechanical ohm
mechanischer Schutz mechanical protection
mechanisches Stellwerk lock machine (625)
mechanische Teile mechanical parts
mechanische Verriegelung mechanical blocking; mechanical interlocking
mechanisches Wärmeäquivalent mechanical equivalent of heat, thermal equivalent of work
mechanische Werkstatt machine shop
mechanischer Wirkwiderstand mechanical resistance
mechanisieren to mechanize
Mechanisierung mechanization
Mechanismus mechanism; gadget
Median(wert) median
Medium medium
Meer ocean, sea
Meereshöhe sea level
Meeresspiegel sea level, normal zero, ordnance datum
Mega-... mega
Mega-Elektronenvolt mega electron volt
Megahertz megacycles (per second) = mc. p. s.
Megaphon megaphone, speaking trumpet
Megavoltampere = MVA megavoltamperes (pl.) = M. V. A.
Megohm megohm
mehradriges Kabel multi-conductor cable, multi-core cable
mehradrige Leitung multi-conductor line, multi-core line
Mehrbelastung surplus load
Mehrbenutzersystem multiuser system
Mehrdeutigkeit ambiguity
mehretagiger Schalter multi-storey switch
Mehrfach-..., mehrfach multiple, multiplex
mehrfach konzentriertes Kabel concentric-lay cable, multiple concentric cable
Mehrfachanschluß multiple connection; party line (621.395)
Mehrfachantenne multiple tuned aerial, aerial array
Mehrfachausnutzung, Mehrfachbetrieb multiplex system, channel working (621.394)
Mehrfachausnutzung von Leitungen multiplexing of lines
Mehrfachecho multiple echo, mirror-reflection echo, round-trip echo
Mehrfachempfang diversity reception; multiple reception
Mehrfachkondensator gang capacitor
Mehrfachleiterkabel split-conductor cable
Mehrfachleitung multiple line
Mehrfachreflexionsweg multi-hop path (621.396)
Mehrfachröhre multi-unit tube, multisection tube
Mehrfachschaltung intermediate wiring; multiple connection
Mehrfachschleifendipolantenne multiple-folded dipole antenne
Mehrfachschreiber multiple recorder
Mehrfachspeiseleitung multiple feeder, parallel feeder
Mehrfachstecker multi-contact plug, multiple plug, socket-outlet adapter
Mehrfachstoßpunktschweißung interrupted spot welding
Mehrfachtarif multiple tariff
Mehrfachtarifzähler multirate meter
Mehrfachtelegraph multiplex telegraph
Mehrfachtelegraphiesystem multiple telegraph system
Mehrfachumschalter multi-way switch, multiple change-over switch
Mehrfachwegausbreitung multipath propagation
Mehrfachwurzel repeated root (51)

Mehrfachzählung multimetering (621.395)
Mehrfachzeichen multiple signal (621.396)
Mehrfachzugriff multi access, multiple access
mehrfeldrige Tafel sectional panel
Mehrfrequenzcode-Wählverfahren MFC-signalling (multi-frequency code)
Mehrganggetriebe multi-change gear
Mehrgangkondensator gang capacitor
mehrgängiges Gewinde multiple thread, multistart thread
mehrgängige Schleifenwicklung multiple winding
mehrgängige Wellenwicklung multiplex winding
Mehrgitterröhre multi-grid valve
Mehrkammerklystron multicavity klystron
Mehrkanal-... multi-channel, multi-way
Mehrkanalanlage multi-way system
Mehrkanalausnützung multiplexing
Mehrkanalfernsehen multi-channel television
Mehrkanalrundfunkgerät multi-channel radio set
Mehrkanaltelegraphie multi-channel telegraphy
Mehrkanal-Tonfrequenztelegraphie multi-channel voice-frequency telegraph system
in Mehrkanalordnung schalten to multiplex
Mehrkosten additional costs
Mehrkreis-Triftröhre multicavity v. m. tube
Mehrlagenschaltung multilayer circuit
mehrlagiges Gewebe plied fabric
mehrlagige gedruckte Schaltung multilayer printed board
mehrlagiges Seil concentric-lay cable
mehrlagige Spule multi-layer coil
mehrlagige Wicklung multiple-layer winding
Mehrleiterkabel multi-conductor cable
Mehrleitersystem multi-conductor system, multiple-conductor system
mehrpaariges Kabel multiple-twin cable, paired cable
mehrpaariges Kabel ohne Vierer non-quadded cable
Mehrphasen-... multiphase, polyphase
Mehrphasengenerator multiphase generator, polyphase alternator
Mehrphasenmotor multiphase motor
Mehrphasennebenschlußmotor polyphase shunt commutator motor
Mehrphasenschaltung multiphase circuit, polyphase circuit
Mehrphasenspannungssystem polyphase set of potential differences
Mehrphasenstrom polyphase (alternating) current
Mehrphasenstromsystem polyphase set of alternating currents
Mehrphasensynchrongenerator polyphase synchronous generator
Mehrphasentransformator multiphase transformer
Mehrphasenwechselstrom multiphase alternating current
Mehrphasenwicklung multiphase winding
Mehrphasenzähler polyphase meter
mehrphasig multiphase, polyphase
mehrphasiges Aggregat polyphase aggregate
mehrphasiger Erdschluß polyphase earth
mehrphasiger Schalter polyphase circuit-breaker
mehrphasiger Wechselstrom polyphase alternating current
mehrphasige Wicklung polyphase winding
Mehrplatzsystem multiuser system
Mehrplatzvermittlung multiposition switchboard
mehrpolig multipole, multipolar
mehrpoliger Generator multipolar generator
mehrpoliger Messerschalter tandem knife switch
mehrpoliger Schalter multipole switch, multiple-pole switch
mehrpolige Steckdose multicontact jack
mehrpoliger Stecker multiple-pin plug
mehrpolige Wicklung multipolar winding
Mehrpolröhre polyode
Mehrprogrammbetrieb multiprogramming mode (681.3)
Mehrprozessorsystem multiprocessor system
Mehrrechnersystem multicomputer system
Mehrrillenscheibe multiple-groove sheave
Mehrröhrengerät multivalve set
Mehrröhrenverstärker multistage amplifier
Mehrschicht... multilayer
Mehrschichtenbildschirm double screen
Mehrschicht-Leiterplatte multilayer printed board
Mehrspeicherprozeß multiple capacity process
Mehrspuraufnahmeverfahren multitrack recording system
Mehrspurkopf head stack
Mehrstellenreihenschweißung series welding
Mehrstellenschweißumformer multiple operator motor-generator set
Mehrstrahlkathodenstrahlröhre multigun tube, multibeam CRT
Mehrstrahloszillograph multibeam oscillograph

Mehrstrahlpeilung multiple beam direction finding
Mehrstromgenerator multiple-current generator
Mehrstufen-Stichprobenverfahren multistage sampling
mehrstufige Turbine multistage turbine
mehrstufiger Verstärker multistage amplifier, cascade amplifier
mehrteiliger Isolator multipart insulator
Mehrverbrauch excess consumption
Mehrweg... multipath
Mehrwegausbreitung multipath propagation, multipath transmission
Mehrwegreflexion multipath reflection
mehrwertige Funktion multiple-valued function
Mehrzellenanlage multiple system (621.357)
Mehrzweckregister general register
Meilen pro Stunde m. p. h. = miles per hour
Meiler pile (539.17)
Meißner-Schaltung Meissner circuit (621.396.61)
Meister foreman, master
Melde... answering..., warning..., signalling..., indicating
Meldeklingel alarm bell
Meldelampe warning lamp, indicating lamp; pilot signal (621.395)
Meldeleitung telling circuit
Meldeplatz record position (621.395)
Melderelais pilot relay
Meldesignal answering signal
Meldetafel indicator board; diagram panel (625)
Meldeverzug answering delay
melden to answer (621.398)
Melkmaschine milker, milking machine
Membran membrane, diaphragm
Membranventil diaphragm valve
membranloses Mikrophon diaphragmless microphone
Menge quantity, quantum (pl.: -ta); poundage
Mennige red lead
menschliche Arbeitsleistung manpower; human work
menschliche Einfluß, der the human element
Mensur glas measure, gauge glass, graduate (54)
Meridian meridian
in den Meridian einstellen to set in the meridian
Meridiandurchgang culmination
Merkatorkarte Mercator (chart)
Merkatorprojektion Mercator projection
Merklampe telltale lamp; pilot signal (621.395)
merklich substantial
merklich linear substantially linear
Merkmal feature, sign, characteristic
Merkuri-... mercuric
Merkuro-... mercurous
MES = magnetischer Eigenschutz degaussing belt, degaussing equipment (629.12)
Mesatechnik mesa technique (621.382)
Mesny-Schaltung Mesny circuit
Meson meson
Mesotron mesotron
Meß-... measuring, metering, gauging
Meßabschnitt measuring section
Meßanordnung (measuring) set-up, measuring circuit, measuring device
Meßbasis basis for measurement, fiducial line
Meßbereich (measuring) range, effective range (621.317)
den Meßbereich erweitern to extend the range (621.317)
mit mehreren Meßbereichen multirange
auf einen Meßbereich umschalten to adjust for a scale
Meßbereichserweiterung measuring-range extension, multiplying power
Meßbereichumschalter meter switch
Meßbereichwähler range selector (switch), meter switch
Meßbuchse measuring bush
Meßbrücke measuring bridge
Meßdorn gauge plug
Meßeinrichtung measuring device, measuring equipment
Meßelektrode half cell (621.35); sounding electrode, sonde
Meßfehler measuring error
Meßfläche gauging face
Meßflasche graduated flask
Meßfrequenz test frequency, frequency used in the test
Meßfühler (measuring) pick-up, sensor, detecting element, probe
Meßgefäß measuring container
Meßgenauigkeit accuracy (621.317), measuring precision
Meßgerät meter, measuring device, metering device, measuring instrument
Meßgrenze limit of measurement
Meßgröße quantity to be measured, measured quantity
Meßinstrument s. **Meßgerät**
Meßinstrument mit Hängelager unipivot instrument, single pivot instrument
Meßinstrument mit Nullpunkt in der Skalenmitte zero centre instrument
Meßkette measuring chain (526)

Meßklima measuring climate
Meßkopf gauging head; measuring head
Meßlänge gauge length
Meßlatte measuring pole
Meßleine measuring line (526)
Meßleitung C-wire, measuring line (621.396), slotted line (631.372)
Meßmarke strobe
Meßmarkengeber strobe marker (621.396.96)
Meßmethode method of measurement
Meßobjekt measured object
Meßpegel through level, expected level (621.39)
Meßplatz test board, test desk
Meßprotokoll record of measured values
Meßraum test room, rest shop
Meßreihe series of measurements
Meßsender signal generator (621.396)
Meßsicherheit accuracy (621.317)
Meßsonde experimental probe
Meßspannung measuring voltage
Meßspule search coil, search winding (538)
Meßstange buzz stick (621.315.1/.6)
Meßstrecke measuring section
Meßtechnik measurements (pl.), measuring technique
Meßtemperatur test temperature
Meßtisch measuring table, surveyor's table
Meßtrupp test team
Meßuhr clock gauge, dial indicator
Meßverfahren technique of measurement, method of measurement
Meßwagen dynamometer car (625.1)
Meßwandler instrument transformer, transducer
Meßwerk movement, measuring element, measuring means
Meßwert measured value
Meßwertaufnehmer transducer
Meßwertschreiber event recorder
Meßwertumformer measuring transducer, transducer
Meßwertumformer mit digitalem Ausgang digital transducer
Meßwesen metering, metrology
Meßwinkel square
Meßzapfen stud
meßbar measurable
messen to measure, to meter, to gauge
Messer knife, blade (621.316.5); meter (621.317)
Messerführung blade guide (621.316.5)
Messerkontakt V-type contact, knife-blade contact
Messerleiste blade-contact connector
Messerschalter knife switch
Messersteckverbinder knife-blade connector
Messerverriegelung blad latch
Messing brass
Messingblech sheet brass
Messingdraht brass wire
Messingguß cast brass
Messingklotz brass block
Messingniet brass rivet
Messung measurement, measuring
Metadyne metadyne (621.313)
Metall metal
Metall bearbeiten to work metal
Metall sägen to cut metal
Metall verarbeiten to work metal
Metallbeflechtung metal braid
Metallblech sheet metal, plate metal
Metalldach metal roof
Metalldampf metal vapour
Metalldampfgleichrichter metal-vapour rectifier
Metalldampflampe metal-vapour lamp
Metalldochtkohle metal-cored carbon
Metalldraht metal filament (621.32); metal wire
Metalldrahtlampe metallic-filament lamp, tungsten lamp
Metalleinlage metal insert
Metallelektrode metal electrode
Metallfadenlampe metallic-filament lamp, tungsten lamp
Metallfolie metal foil
Metallfolien-Papierkondensator foil and paper capacitor
Metallgeflecht metal braid
Metallgehäuse metal case
metallgekapselt metal-clad
metallgekapselte Schaltanlage metal-clad switchgear
Metallgestell metal stand; metal frame (621.31)
Metallglanz metallic lustre
Metallglasurfestwiderstand cermet resistor
Metallhülse metal shell, metal envelope
Metallisolator metallic insulator (621.396)
Metallkeramik metal ceramic
Metallkohlebürste compound brush
Metalliderung metallic packing
Metall-Nitrid-Feldeffekttransistor metal-nitride semiconductor field-effect transistor
Metalloxid metallic oxide

Metall-Oxid-Feldeffekttransistor metal-oxide semiconductor field-effect transistor
Metalloxidschicht-Festwiderstand fixed metal oxide film resistor
Metall-Papier-Kondensator metallised paper capacitor
Metallpulver metal powder
Metallrohr metal pipe
Metallröhre metal tube, metal valve
Metallsäge hack saw
Metallschicht-Festwiderstand fixed metal film resistor
Metallschirm metal screen
Metallschlauch flexible metal tube
Metallspritzverfahren metal spraying
Metallschichtwiderstand metal-film resistor
Metallstativ metal stand
Metallstreifen metal strip
Metallsuchgerät metal detector, metal locator
Metalltafel metal panel (621.315)
metallisch metallic; metalloid
metallischer Belag metal coating
metallischer Glanz metallic lustre
metallischer Leiter metal conductor, first-class conductor
metallischer Zustand metallic state
metallisieren to metallize
metallisierter Leuchtdraht metallized filament
metallisierte Röhre metallized valve
Metallisierung metallization, metallizing
Metallographie metallography
metastabil metastable
metastabile Ausgangskonfiguration metastable output configuration
metastabiles Gleichgewicht metastable equilibrium (54)
Meter meter
Meterwellen very high frequency, metric waves
Methan methane
Methode method
Methode der kleinsten Quadrate method of least squares
Methylalkohol methylalcohol, wood-alcohol
metrisch metric
metrisches Gewinde metric thread
in metrischem Maß in metric (measurement)
in metrischem System in metric (system)
Metronom metronome
MeV = Mega-e-Volt MeV = mega electron volt
Mg Mg (54); G (US) (66)
mg = Milligramm mg = milligram (-me)
mH mh. = millihenry
Michelson-Versuch Michelson-Morley experiment (53)
Mietleitung leased circuit
Mignonbatterie penlight battery
Mignonfassung miniature base, midget base
Mignon-Gewindesockel s. E. s. cap = small Edison screw cap
Mikanit micanite, sheet mica, reconstructed mica
Mikartit micarta
Mikro-..., mikro-... micro
Mikroamperemeter microammeter
Mikrobefehl microinstruction
Mikrocomputer microcomputer
Mikroelektronik microelectronics
Mikrofarad microfarad
Mikrofilm microfilm
Mikrofon s. **Mikrophon**
Mikrogramm microgram
Mikrohenry microhenry
Mikroklima microclimate
Mikrokosmos microcosm
Mikromanipulator micromanipulator
Mikrometer micrometer (caliper)
Mikrometereinteilung micrometer-graduated
Mikrometerfunkenstrecke micrometer gap
Mikrometerschraube micrometer screw, screw caliper
Mikro-Ohm microhm
Mikrophon microphone (transmitter), telephone transmitter, "mike"
Mikrophon mit kleiner Ausgangsleistung low-power microphone
Mikrophon besprechen to speak into a microphone
Mikrophoneffekt microphonic action, microphonism
Mikrophongeräusch frying, microphonics (pl.), transmitter noise (US) (621.395.61)
Mikrophonkabel microphone cable
Mikrophonkapsel microphone capsule, microphone button
Mikrophonleitung microphone cable
Mikrophonstativ microphone stand
Mikrophonverstärker microphone amplifier
Mikrophonie microphonic effect, microphonism, microphony
Mikrophotographie microphotography
Mikroporosität microporosity
Mikroprogramm microprogram
Mikroprozessor microprocessor
Mikrorille microgroove
Mikroschalter microswitch

Mikroschaltung microcircuit
Mikrosekunde microsecond
Mikroskop microscope
an einem Mikroskop ablesen to read in a microscope
Mikroskoplinse microscopic lens
Mikroskopschnitt microscopic section, microscope pillar
Mikrosynchronisation microsynchronisation (621.395.3)
Mikrotom microtome
Mikrowaage microbalance
Mikrowelle microwave (621.396)
Mikron micron
Milch milk
Milchglas opal gals
Milchglasscheibe opal pane
milchig milky
Milchigkeit milkiness (54); opalescence (535)
Milli-... milli-...
Milliampere milliampere
Milliamperemeter milliammeter, milliamperemeter
Milligramm milligram(me)
Milli-Henry mh. = millihenry
Millilambert millilambert
Milliliter ml. = millilitre
Millimeter millimetre
Millimeterpapier graph paper
Millimeter Quecksilbersäule millimeter of mercury
Millimeterwellen EHF, extremely high frequency
milli-micro... (10^{-9}) milli-micro...
Millivolt millivolt
Millivoltmeter millivoltmeter
Milliarde milliard (Br.), billion (US)
Million million
millionstel ppm., part per million
minderwertig poor, low-grade, sub-standard, inferior
minderwertiger Brennstoff inferior fuel, low-grade fuel
minderwertige Konstruktion poor design
minderwertige Qualität substandard quality
Minderwertigkeit inferiority, inferior quality
Mindest-... minimum
Mindestabstand minimum distance, minimum clearance
Mindestabschaltspannung gate turn-off voltage (621.314.6)
Mindestbedarf minimum demand
Mindestbetätigungsspannung working value (621.318)
Mindestbetrag minimum amount, minimum value
Mindestgehalt minimum percentage (66)
Mindestgrenze minimum (limit)
Mindestlast minimum power (621.3)
Mindestleistung minimum output (621.3)
Mindestmaß minimum (size); minimum measure
aufs Mindestmaß herunterdrücken to minimize
Mindestpotential minimum potential
Mindestspannung minimum voltage
Mindeststrom minimum current
Mindestwert minimum (value), minimum amount
Mindestzersetzungsspannung equilibrium reaction potential
Mindestzündstrom striking current
Mineral mineral
Mineralöl petroleum
Mineralsäure mineral acid
Mineralwachs mineral wax
mineralisch mineral
mineralischer Bestandteil mineral constituent
Mineralogie mineralogy
Mini... mini... (siehe auch **Miniatur-...**)
Minikassette minicartridge
Minicomputer minicomputer
Miniatur-... miniature, midget miniaturized
Miniaturausführung miniature design
Miniaturröhre miniature valve, midget valve
miniaturisierte Schaltung microcircuit
Minimal-... minimum, underload, undercurrent
Minimalanreicherung minimum enrichment (539)
Minimalauslöser unterload trip
Minimalauslösung underload circuit-breaking, undercurrent circuit-breaking
Minimalbedarf minimum demand
Minimallast minimum power
Minimalleistung minimum load, trough demand; minimum output (621.39)
Minimalrelais minimum relay, underload relay, undercurrent relay
Minimalschalter minimum-circuit breaker, underload circuit breaker, undercurrent circuit breaker
Minimalspannung minimum voltage, minimum potential; undervoltage, trough voltage
Minimalspannungsmesser trough meter (621.317)
Minimalthermometer minimum thermometer

Minimalverbrauch minimum demand
minimal minimum
Minimum minimum (pl.: -ma)
Minimum durchlaufen to pass through a minimum
Minimumpeilung minimum bearing, minimum direction finding
Minoritätsträger minority carrier
minus minus
Minusklemme minus terminal, negative terminal
Minusleiter negative lead, negative conductor
Minusleiter des Netzes negative main
Minuspol minus pole, negative terminal
Minuszeichen minus sign, negative sign
Minute minute
Minutenkontakt minute contact
Minutenzeiger minute hand (681.11)
Misch-... mixing
Mischbatterie mixer (621.6)
Mischbühne mixing stage (69)
Mischdämpfung conversion loss (621.396)
Mischdetektor converter detector, crystal mixer
Mischdiode mixer diode
Mischgerät scrambler (621.39)
Mischfarbe secondary colour (535)
Mischgitter injection grid
Mischgittermodulator injection modulator
Mischkammer mixing chamber (621.1); mixer (621.6)
Mischkopf mixer (621.396)
Mischkristall solid solution
Mischlichtlampe incandescent arc lamp
Mischmeßkopf mixing probe
Mischpult mixing unit (621.396.7); mixing console (654.19); mixing board (621.39)
Mischröhre converter valve, modulator
Mischsteilheit conversion transconductance, conversion mutual conductance
Mischstufe mixerstage, modulator (stage), converter, first detector (621.396.62)
Mischventil mixing valve (621.6)
Mischverstärker mixer amplifier
Mischverstärkung mixing gain, conversion gain
mischen to mix; to mingle, to blend, to interblend
mischen mit, (sich) to mix (up) with
Mischung mix, mixture; conversion (621.396); grading (621.395)
Mischungsverhältnis ratio of components (54); mean interconnecting number (621.395)
mißweisender Kurs (Peilung) magnetic bearing
mitangeschlossene Stromkreise associated electric circuits
Mitecho listener echo
miteinander mischen to interblend (54)
miteinander multiplizieren to multiply together
miteinander verdrillen to intertwist
miteinander verknüpfen to interrelate
mitgeschleppter Fehler inherent error; inherited error (681.3)
Mithör-... monitoring, sidetone, branching
Mithörklinke branching jack (621.395)
Mithörlautstärke sidetone level
Mithörschaltung monitoring circuit
Mithörtaste monitoring key
Mithörtelefon sidetone telephone set
Mithörton sidetone
mithören to intercept, to listen in, to overhear, to monitor (621.395)
mitgehörte Nachricht intercepted message
Mitkopplung positive feedback, regenerative feedback (621.396); feedforward
mitlaufende Reserve spinning reserve (621.311)
mitläufig positive-sequence, positive-direction (621.392)
mitläufige Bürste trailing brush
mitläufiges Echo listener echo (621.395)
Mitnahmebereich pulling range
Mitnahmeeffekt pulling effect
Mitnahme-Oszillator locked oscillator
Mitnahmeschaltung intertripping (621.318.5)
mitnehmen to pull, to lock-in (621.396.61); to take along (621)
Mitnehmer driver, carrier, catch, dog
Mitnehmerstift driving pin
mitschwingende Masse virtual mass
Mitsprech-... sidetone
Mitsprechdämpfung sidetone attenuation
mitsprechfreie Schaltung sidetone reduction wiring (621.395)
Mitte centre, middle
Mitte - Mitte centre to centre
in der Mitte abgreifen to centre-tap
in der Mitte liegend central
mitteilen to communicate, to inform, to notify
Mitteilung communication, statement
Mittel means, medium agent; average (51)
im Mittel as an average
Mittel-..., mittel-... mean, medium, middle, centre; average
Mittelabgriff centre tap, mid-point (621.3)
Mittelabschnitt middle section
Mittelachse central axis
Mittelantrieb central drive
Mittelbereich middle range

Mittelblech medium plate
Mitteldraht middle wire
Mitteldruck medium pressure, intermediate pressure (621.18)
Mitteldurchbiegung central deflection
mittelfeines Korn medium-grained texture (669)
Mittelfrequenz medium frequency
Mittelhalbierende median
Mittelkern centre core (621.396)
Mittellappen central lobe (621.396.677)
Mittellast central load
Mittelleiter middle conductor, third wire, middle wire, neutral conductor; centre conductor, core conductor
Mittelleiterstrom neutral-conductor current
Mittelleitersystem d. c. neutral grid
Mittellinie C. L. = centreline, median line
Mittelmastaufhängung centre-pole suspension
auf Mittelpotential mid-way in potential
Mittelpunkt centre, center; middle point, neutral point; mid-point (621.3)
Mittelpunktsabstand centres (pl.), centre-to-centre distance
Mittelpunktspeisung apex driven (621.396.67)
mittelpunktstrebig centripetal
Mittelschenkel middle limb, common limb (621.3)
Mittelschicht inner layer (67)
Mittelspannung medium voltage, intermediate high voltage
mit Mittelspeisung centrefed (621.396.67)
Mittelstellung mid position, central position, neutral position
Mittelstift centre pin
Mittelstromschiene centre contact rail
Mittelteil middle (section), central portion
Mitteltonlautsprecher midrange speaker
Mittelwasser average flow (626)
Mittelwelle medium wave
Mittelwellenfunkfeuer (radio frequency) nondirectional beacon
Mittelwert mean (value), average (value)
Mittelwert der Ablesungen mean of the readings
Mittelwert der Leistung mean power
Mittelwertmesser mean-value meter
Mittelwertphotometer integrating photometer
mitten zwischen mid-way between
Mittenanzeige band-centre indication (621.396)
mittenbelastete Antenne center-loaded antenna
mittengespeist centerfed (621.396.67)
Mittenfrequenz centre frequency
mittlere middle, central (513); mean, average (512)
mittlere Anschlußleistung demand of an installation
mittlere ausfallfreie Zeit mean-time-between-failures, MTBF
mittlere Ausgangsleistung mean output, average output (621.396)
mittlere Dauerleistung mean continuous output (621.313)
mittleres Drittel middle third
mittlere Durchlaßverlustleistung mean conducting state power loss (621.382)
mittlere freie Weglänge mean free path
mittlere Größe middle size, middle height
mittlere hemisphärische Lichtstärke mean hemispherical luminous intensity
mittlere horizontale Lichtstärke mean horizontal luminous intensity
mittlerer Hub mean deviation
mittlerer Integrationsgrad medium scale integration
mittleres Korn middle grain
mittlere Korngröße medium grain
mittlere Last mean load; central load, average load
mittlere Lebensdauer mean life, average life
mittlere Leistung mean power, average power; meter watts (621.317)
mittlere Lichtstärke in der Nutzzone mean zonal luminous intensity
mittlere Lichtstärke im Nutzkegel mean conical luminous intensity
mittlere Netzleistung mean system capacity
mittlere obere hemisphärische Lichtstärke mean upper hemispherical luminous intensity
mittlere Proportionale mean proportional
mittlere Qualität medium quality
mittlere Seehöhe mean sea level
mittleres spezifisches Gewicht average density
mittlere sphärische Lichtstärke mean spherical luminous intensity
mittlere Temperatur mean temperature; temperature average
mittlere Übertragungsgeschwindigkeit average transfer rate
mittlere untere hemisphärische Lichtstärke mean lower hemispherical luminousintensity
mittlere Verkehrsdichte average traffic flow (65)
mittlere Wartezeit average delay (621.395)

mittlere Zeit zwischen Ausfällen mean-time-between-failures = MTBF
mittlere Zugriffszahl readaround number
mitziehen to drive (621.396)
Mitziehen von Sendern synchronization of oscillators (621.396)
M. K. S.-Einheiten m. k. s. units
M. K. S.-System Giorgi system
ml ml = millilitre
mm mm = millimetre
MMK = magnetomotorische Kraft magnetizing force, m. m. f. = magnetomotive force
Mn Mn (54); M (US) (669.14)
mnemonischer Code mnemonic code (681.3)
MNOS-Speicher = Metall-Nitrid-Oxid-Silizium-Speicher MNOS memory
Modell model, pattern, mock-up
Modellanlage model plant
Modellbau pattern making
Modellstück master part
modellieren to model
modeln to modulate
Modelung modulation
Modem modem = modulator demodulator
Moderator moderator (54), (539)
moderierter Reaktor moderated reactor
Modifikation modification, variant
modifizieren to modify
modifizierte Brechzahl modified refractive index
modifizierter Brechwert refractive modulus
Modul module, modulus, reciprocal of diametral pitch (621.833); mode
modular modular
Modulation modulation (621.396); speech signal, audio current (621.395)
Modulation aufs Gitter geben to apply a signal to the grid
modulationsabhängig gesteuerter Träger controlled carrier
Modulation mit Trägerunterdrückung in den Pausen quiescent-carrier modulation
Modulationsart type of modulation
Modulationsfeld bunching field, buncher (621.385)
Modulationsfrequenz modulating wave, modulation frequency
Modulationsgitter modulated grid
Modulationsgrad degree of modulation, modulation depth, modulation factor
Modulationshüllkurve modulation envelope
Modulationsindex modulation index
Modulationskennlinie modulation characteristic (621.396.61)
Modulationsklirrfaktor modulation distortion
Modulationskontrollgerät modulation monitor
Modulationsleistung a. f. driving power
Modulationsmeßgerät modulation meter
Modulationspause quiet interval (621.396)
Modulationsrauschen modulation noise
Modulationsröhre modulator tube, mixer tube
Modulationsspannung modulator voltage, modulating voltage, speech signal
Modulationsspule modulation coil
Modulationsstufe modulator stage, mixer stage (621.396.21)
Modulationssystem method of modulation (621.385)
Modulationstiefe modulation depth
Modulationsträger carrier
Modulationsunterdrückung modulation suppression
Modulationsverstärker modulation (voltage) amplifier, a. f. amplifier
Modulationsverstärker eines Rundfunksenders program amplifier
Modulator modulator
Modulator mit unterdrücktem Träger balanced modulator (621.396.61)
Modulatorröhre modulator valve, mixer valve
Modulatorstufe modulator (stage)
modulieren mit to modulate with
modulierende Schwingung modulating wave
modulierter Dauerstrich M. C. W. = modulated continuous wave
modulierte Größe modulated quantity
modulierte Sinusgröße generalized sinussoidal quantity
modulierter Träger modulated carrier (621.396)
Modulus modulus (51), (621.833)
Mögel-Dellinger-Effekt Dellinger fade out, sudden short-wave fade out, radio fade out
Mohs'sche Härteskala Mohs scale of hardness
Moiré moiré (621.397)
Mol gram molecule, mole (54)
molearer spezifischer Widerstand molar resistivity
Molekel, Molekül molecule
Molekülbindung molecular bond
Molekülgröße molecular size
1 Moleküllage dick one molecule thick, one molecule deep
Molekülverkleinerung durch mechanische Beanspruchung shear breakdown (665)
molekular molecular

molekulare Reibung molecular friction
Molekulargewicht molecular weight
Molekularkristall molecular crystal
Molekularpumpe molecular pump
Molybdän molybdenium
Moment moment, torque (531); moment
Moment beim Anlauf torque at starting
Momente ansetzen, Momente berechnen to take moments
Moment einer Kraft moment of a force
Moment einer Welle torque on a shaft
Momentauslösung instantaneous tripping
Momentausschaltung rapid make-and-break mechanism (621)
Momentdruckknopfschalter strap key
Momentempfänger synchro-torque receiver, torque receiver
Momentendiagramm bending-moment diagram
Momentengleichung moments equation, equation of moments
Momentenmotor torque motor
Momentenverstärker torque amplifier
Momentenwaage torque meter, torsiometer (621.313)
Momentgeber synchro-torque transmitter, torque transmitter
Momentrelais instantaneous relay
Momentschalter quick make-and-break switch, snap-type switch (696.6)
Momentschaltung instantaneous connection
Momentschlüssel torque wrench
Momentzünder instantaneous fuse (662)
monaural monaural
Monazitsand monazite sand
Monelmetall Monel metal
Monitor monitor
Monitorprogramm monitor program
Monochord monochord
monochrom monochrome
monochromatisch monochromatic
monoklin monoclinic
Monokristall monocrystal
monokular monocular
Monolith monolith
monolithisch monolithic
monolithische integrierte Schaltung semiconductor monolithic integrated circuit
Monomode-Faser monomode fiber
monomolekular monomolecular, one molecule deep, one molecule thick
Monoskop monoscope
monostabil monostable
monostabiler Multivibrator monostable multivibrator, monostable trigger circuit, single-shot multivibrator
monostabile Schaltung monostable circuit
monoton monotonous; monotonic
monotrop monotropic
Montage mounting, erection, construction; assembly, fitting
Montageanleitung installation instructions, assembly instructions
Montagearbeit fitting work; construction work; assembly work, installation work
Montageband assembly line
Montagebock fitting stand
Montagedurchhang erection sag (621.315.1)
Montagegerüst assembly jig
Montagegrube inspection pit
Montagehalle assembly shop
Montagekran erection crane, assembly crane
Montagemeister construction foreman; site foreman
Montageschlosser fitter
Montagewerkstatt fitting shop
Montagezeichnung assembly drawing
Monte-Carlo-Methode random-walk method
Monteur fitter, mechanician; electrician
montieren to mount, to erect, to construct; to fit, to assemble (parts), to install
Moorelichtbeleuchtung Moore lighting
Morse Morse
Morsealphabet Morse alphabet, Morse code
Morse-Doppelstromsystem polar d. c. telegraph system
Morse-Einstromsystem neutral d. c. telegraph system
Morseempfang code-signal reception
Morsekegel Morse cone (621.9)
Morsepunkt dot
Morseschreiber Morse printer
Morseschrift Morse code
Morsestreifen Morse tape, Morse slip
Morsestrich dash
Morsetaste(r) Morse key, telegraph key
Morsetaste mit selbsttätiger Punktgebung semi-automatique key; bug
Morsetelegraph Morse telegraph
Morseticker Morse sounder
Morseticker mit Relaissatz relaying sounder
Morsezeichen Morse code, code signal
MOS Metal Oxide Semiconductor
MOS-Transistor metal-oxide semiconductor field effect transistor
MOS-Transistor vom Anreicherungstyp MOS-transistor enhancement-mode

MOS-Schaltung mit Silizium-Gate silicon-gate MOS
Mosaik mosaics (pl.)
mosaikartig unterteilt tesselated
mosaikartige Unterteilung tesselation
Mosaikbaustein tile (621.316.5)
Mosaikbetriebsschaubild mosaic-diagram display
Mosaikdrucker mosaic printer, matrix printer
Mosaiktelegraphie mosaic telegraphy
Motor engine (621.4); motor (621.313); traction motor (621.33)
Motor anlassen to start a motor
Motor einbauen to mount an engine
Motor umsteuern to reverse a motor
Motor warmlaufen lassen to warm an engine
Motor mit Achsaufhängung axle-hung motor (625)
Motor mit Doppelschlußverhalten motor with compound characteristic
Motor mit Drehzahlregelung variable-speed motor
Motor mit Eigenbelüftung fan-cooled motor
Motor mit Eigen- und Fremdbelüftung motor with combined ventilation
Motor mit Fremdbelüftung separately air-cooled motor, forced ventilated motor
Motor mit Fremderregung separately excited motor
Motor mit Hauptstromcharakteristik inverse-speed motor
Motor mit Hilfsphase split-phase motor
Motor mit kapazitiver Hilfsphase capacitor-start motor
Motor mit Kompensationswicklung compensated motor
Motor mit Kompoundwicklung compound-wound motor
Motor mit Kondensatorenanlauf capacitor-start motor
Motor mit konstantem Drehmoment torque motor
Motor mit konstanter Drehzahl constantspeed motor
Motor mit Kurzschlußläufer motor with squirrel-cage armature
Motor mit Mantelkühlung ventilated totally-enclosed motor
Motor mit mehreren Drehzahlstufen multi-speed motor, change-speed motor
Motor mit Nebenschlußwicklung shunt-wound motor
Motor mit Reihenschlußverhalten motor with series characteristic
Motor mit senkrechter Welle vertical-spindle motor, vertical-shaft motor
Motor mit Vorverdichtung supercharged engine (621.4)
Motor mit Widerstandsanlasser resistance-start motor
Motorachse motor axle (625.2)
Motoranker motor armature
Motoranlasser motor starter
Motoranschlußklemme motor terminal
Motorantrieb motor drive
mit Motorantrieb motor-driven, motor-operated, power-driven
Motordrehmoment motor torque
Motordrehzahl motor speed
Motorfuß motor base
Motorgehäuse motor casing, motor frame
Motorgenerator motor generator, motor converter
Motorgeneratorsatz motor-generator set
Motorgestell motor frame
Motorgetriebe motor gear
Motorkennlinie motor characteristic
Motorklemme motor terminal
Motorklemmbrett motor-terminal board
Motorkupplung motor clutch
Motorleistung motor output
Motorpumpe motor pump, motor-driven pump
Motorschalter motor switch
Motorschild end bell
Motorschleifstein electric grinder
Motorschutz motor protection
Motorschutzschalter termal circuit breaker, open-phase circuit breaker, motor-protection circuit breaker
Motorspannung motor voltage
Motorspule motor coil
Motorstrom motor current
Motorstromkreis motor (branch) circuit
Motortrennschalter motor isolator; isolating switchgroup (625)
Motorwagen motor coach (625); tractive power unit (621.33)
Motorwähler motor uniselector (621.395)
Motorwelle motor shaft
Motorwicklung motor winding
Motorwiege cradle
Motorwirkungsgrad motor efficiency
Motorzähler motor meter
motorisch motory
motorisch angetrieben motor-driven, power-driven

motorischer Betrieb motoric action, motive action
MP-Kondensator metallized-paper capacitor
MPF-Kondensator metallized-paper capacitor with plastic-film dielectric
Muffe box (621.3); straight joint (621.315.2); sleeve, collar; joint (621.6)
Muffenkammer handhole (621.315.2)
Muffenkupplung sleeve coupling
Muffenschacht splicing chamber (621.3)
Muffenüberbrückung continuity sheath, continuity bond
Muffenverbinder continuity cable bond (621.315.2)
Muffenverbindung sleeve joint
Muffenwiderstand joint resistance
Muffel muffle (669)
Muffelofen muffle furnace
Mühle mill; grinder, grinding mill
Mulde trough
muldenförmig trough-shaped
Müllofen destructor, refuse destructor furnace, incinerator
Multiburst multiburst signal
Multichip multi-chip integrated circuit
Multi-Instrument multimeter
Multimode-Faser multimode fiber
Multiplexbetrieb multiplex operation, multiplex system (621.394)
Multiplexer multiplexer
Multiplexübertragung multiplex transmission
Multiplikand multiplicand
Multiplikation multiplication (51), (54), (539)
Multiplikationsfaktor multiplication factor (54), (539)
Multiplikator multiplier (51)
multiplizieren to multiply (**mit:** by)
Multivibrator multivibrator
Multizellularvoltmeter multi-cellular voltmeter
Mumetall Mumetal
Mundloch, Mundstück mouth (piece), orifice
Mündungsquerschnitt mouth area (621.396)
Münzfernsprecher coin (operated) telephone, taxi-phone, coin collector, coin-box telephone
Münzkassierer coin box (621.395)
Münzzähler prepayment meter, slot meter (621.317)
Musa-Antenne musa aerial
Muschelantenne shell antenna
Muskowit muscovite
Muster pattern, model; sample, sampling, specimen
Muster entnehmen to draw samples
Musteranlage model plant
Musterbeispiel representative
Mustermesse fair, exhibition; exposition (US)
Musterstück master copy, standard
Mutation mutation
Mutter nut (621)
Mutter-... parent; female; master
Mutteratom parent atom
Mutterband mother tape
Muttergewinde female thread
Mutterkern parent nucleus
Mutterkreisel transmitter gyro
Mutternuklid precursor (of a nuclide) (539)
Mutterplatine motherboard
Mutterschraube nut, female screw
Muttersubstanz parent substance, parent compound, parent matter
Muttervermittlungsstelle master exchange
MVA = Megavoltampere M. v. a. = megavoltamperes (pl.)
Mycalex mycalex

N

0, 1 n = zehntelnormal 0, 1 = point one normal (54)
Nabe hub, nave (621)
Nabla Operator del, nabla (51)
Nachbar neighbour
Nachbaramtstaste junction key (621.395)
Nachbarbildfalle adjacent carrier trap
Nachbarkanal adjacent channel (621.395.4)
Nachbarkanalselektion near-channel selection
Nachbarwindungen adjacent turns
Nachbehandlung subsequent treatment
nachbeschleunigen to accelerate subsequently
Nachbeschleunigung post acceleration, after acceleration
Nachbeschleunigungselektrode post-accelerating electrode, post-deflection accelerating electrode (621.385)
nachbilden to reproduce, to replicate, to copy; to simulate (621.3)
nachbilden (im Ersatzschaltbild) to simulate
Nachbildung replication, copy; balancing network, simulating network, equivalent network, circuit simulation (621.3); mock-up
nachbleibend residual
nachbohren to rebore
Nachdrehen eines Kollektors trueing up of a commulator

nacheichen to recalibrate
nacheilen gegen to be behind, to lag (behind) (621.3)
nacheilen um 30° to lag by 30°
nacheilend um 90° reactive (621.3)
nacheilende Last lagging load, reactive load
nacheilender Leistungsfaktor lagging power factor, reactive power factor
nacheilende Phasenverschiebung phase lag
Nacheilung lag (in phase), time lag (621.3); retardation
Nacheilung durch Trägheit capacity lag
Nacheilungswinkel angle of lag
Nacheinanderverfahren one-at-a-time principle
Nachentladung soaking out
nacherhitzen to re-heat
Nacherhitzung re-heating
nachfolgen to follow, to succeed
nachfolgender Betriebsstrom follow current
nachfolgende Röhre succeeding valve (621.396.64)
nachfüllen to refill, to fill up, to top up, to replenish
Nachfüllmasse replenisher
Nachfüllung replenishment; refill
nachgesetzte Kennung suffix (621.395)
Nachgleich resetting, readjustment
Nachhall reverberation
Nachhallzeit reverberation time
nachhallen to reverberate
nachhallarmes Studio dead studio
nachindizierte Adressierung post-indexed addressing
Nachklingeffekt memory effect
nachladen to recharge, to replenish (621.3); to refill
Nachladung replenishment, boost charge (621.355)
nachlassen to decline, to drop off, to weaken, to relax
Nachlassen eines Permanentmagneten weakening of a permanent magnet
nachlaufen to lag; to follow; to hunt
Nachlaufinstrument servo-operated instrument
Nachlaufregelung servo governor; servo control (US)
Nachlaufwerk follower
Nachlaufwiderstand self-balancing potentiometer
Nachlaufzeit run-out time
Nachleuchten afterglow, persistence (621.385), (621.397)
Nachleuchtbild residual image
Nachleuchtcharakteristik screen luminous persistence characteristic, decay characteristic
Nachleuchtdauer time of persistence
Nachleuchteffekt screen persistence
Nachleuchtfreiheit freedom from afterglow
Nachleuchtschirm persistence screen
nachleuchten to afterglow, to linger
nachleuchtend persistent (621.385)
Nachlieferung replenishment (33); dispensation (621.385)
nachlöten to resolder
nachmachen to counterfeit, to copy
nachmessen to re-measure, to check
Nachmessung check measurement
nachpolieren to repolish, to shine up
nachprüfen to check
Nachprüfung check; review
nachrechnen to recalculate, to (re-)check
Nachrechnung recalculation, re-check
Nachregulierung readjustment
Nachricht message, communication; intelligence; information (621.39)
Nachricht durchtelegraphieren to flash a message
Nachrichten übermitteln to communicate
Nachrichtendienst intelligence (327.8)
Nachrichteneinheit bit (621.39)
Nachrichteningenieur communication engineer
Nachrichtenkabel communication cable
Nachrichtenkanal (communication) channel
Nachrichtenmittel communication means
Nachrichtenpfad information path, transmission path
Nachrichtenquelle information source
Nachrichtenreichweite communication range
Nachrichtensinke information drain, information sink
Nachrichtenspeicher memory, buffer
Nachrichtenspeicherung information storage
Nachrichtentechnik telecommunication
Nachrichtentruppe signals (pl.), signal corps
Nachrichtenübertragung transmission of communications
Nachrichtenübertragung mit unterdrücktem Träger suppressed carrier transmission
Nachrichtenwesen communication (service)
Nachrufen rering (621.395)
Nachrüstprogramm retrofit program
nachschärfen to touch up (621.9)
nachschieben to boost
nachschleifen to regrind, to touch up
nachschleppen to drag, to trail
Nachstellzeit integral action time (621.316.7)
nachstimmen to retune

Nachstrom post-arc current; follow current
Nacht night
Nachtarbeit night work
Nachtbeleuchtung night light (ing)
Nachtdienst night duty, night shift, after office hour service
Nachteffekt night effect (621.396.677)
Nachteffekt-Fehler night error
nachteffektfreies Funkfeuer radiophare
Nachtfernsehen low light-level television, night vision
Nachtfrequenz night frequency (621.396)
Nachtplatz concentration position (621.395)
Nachtrufnummer night-service number
Nachtschaltung night service
Nachtschicht night shift
Nachtsichtgerät night-vision equipment
Nachttarif night tariff, night rate
nachtrimmen to retrim, to readjust
nachtrocknen to redry (67)
nachverzinnen to retin
nachweisen to trace (54)
Nachwirkung after effect
Nachwirkungsbild retained image
Nachwirkungsladung soaking-in
Nachwirkungsstrom absorption current, reversible current (537)
Nachwirkungsverlust dielectric absorption
Nachwirkzeit hangover time
nachziehen to retighten (621); to re-draw; to add
Nachzieheffekt smearing (621.397)
Nadel needle (534), (538); pin
Nadelabnutzung needle wear, stylus wear
Nadeldrucker wire printer, pin writer
nadelförmig needle-shaped, acicular (54)
Nadelfunkenstrecke needle-point gap
Nadelgalvanometer needle galvanometer, moving-magnet galvanometer
Nadelgeräusch needle scratch
Nadelgeräuschfilter scratch filter
Nadelgleichrichter snook rectifier
Nadellager needle bearing
Nadelnachgiebigkeit compliance (681.84)
Nadelschaltung heterostatic method (621.317)
Nadelspitze needle point
Nadelstrahl sting ray (621.396.67)
Nadelventil needle valve
Nagel nail (61), (621), (69)
Nah-... short-range
Nahbereich-Feuerleitradar close-range fire control radar
Nahbereichsradar short-range radar
Nahechodämpfungsschaltung anti-clutter gain control (621.396.96)
Nahfeld proximity zone
Nahkraft short-range force (539)
Nahnebensprechen near-end crosstalk (621.395)
Nahschwund short-range fading
Nahschwundantenne low-angle fading antenna
Nahverkehr short-haul traffic (656); toll service (621.395)
Nahverkehrsamt toll exchange
Nahverkehrsbereichsradar TAR, terminal area surveillance radar
Nähe nearness, proximity
nähen to sew
Nähmaschine sewing machine
Näherung approach, approximation (51); proximity (621.39)
Näherungsausdruck approximate form
Näherungsformel approximate formula
Näherungslösung trial-and-error solution
Näherungswert approximate value
Näherungszünder proximity fuse
Naht seam, junction
Naht löten to solder a seam
Nahtschweißung seam welding
nahtlos seamless
nahtlos gezogen solid drawn
nahtloses Rohr seamless pipe
Name name
Namengeber answer-back unit
Namengeberzeichen answer-back code
NAND-Schaltung NAND circuit; NAND element, NAND gate
NAND-Verknüpfung NAND operation
nano-... 10^{-9} milli-micro-..., billi...
nano-Ohm = 10^{-9} Ohm billohm
Naphta naphta
Naß-... wet
Naßdampf wet steam
Naßüberschlagsspannung wet flashover voltage
Naßverfahren wet process
naß wet
nasse Batterie wet battery (621.35)
nasser Elektrolytkondensator wet capacitor
nasses Element wet cell (621.35)
naszierend nascent (54)
Natrium sodium
Natriumdampf sodium vapour
Natriumdampfentladung sodium (vapour) discharge
Natriumdampflampe sodium (vapour) discharge lamp

natriumgekühlter Reaktor sodium-cooled reactor
Natriumgraphitreaktor sodium (-graphit) reactor
Natron (baking-) soda
Natronlauge soda lye
Natur nature
Naturglimmer natural mica, block mica
Natururanreaktor natural uranium reactor
Naturseide real silk
Naturwissenschaft natural science
natürlich natural
natürliche Belüftung natural ventilation
natürliche Eigenschaft inherent property
natürliche Größe full size
in natürlicher Größe full-sized, in large
natürlicher Logarithmus ln = naturallogarithm, Naperian logarithm, hyperboliclogarithm
natürliche Magnet natural magnet
natürlicher Meßstab (1:1) full size, one-to-one correspondence
natürlicher Zustand natural state, original state
nautisch nautic(al)
nautisches Instrument nautical instrument
Navigation navigation
Navigation nach Polarkoordinaten omni bearing-distance navigation
Navigationsfunkdienst radio-navigation service
Navigationsfunkfeuer radio beacon
Navigationsfunkhilfsmittel radio aids to navigation
Navigationsgerät navigational instrumentation
Navigationshilfe navigational aid
Navigationsradarbake radar beacon
Navigationssatellit navigation satellite
Nb Cb (54)
NC-Akkumulator alkaline accumulator, nickel-cadmium accumulator, Edison cell (621.356)
NC-Batterie nickel-iron battery, Edison battery
NC-Sammler steel-alkaline battery, Edison cell
ND = Niederdruck LP = Low Pressure
NDB non-directional beacon
n-dotiert n-doped (621.382)
NE = Nichteisen... non-ferrous
Nebel nebula (pl.: -ae) (52); mist, fog
Nebelbildung formation of fog (55)
Nebelhorn fog horn
Nebelkammer cloud chamber, cloud apparatus, condensation apparatus, Wilson chamber
Nebelkammerbahn cloud track
Nebelscheinwerfer fog light
Neben-... auxiliary, secondary, ancillary; by-pass
Nebenanschluß extension line (621.395)
Nebenentladungsweg by-pass discharge path
Nebengebühr accessory charge (621.317); extras
Nebengeräusch background noise
Nebengleis side track, siding, shunt, branch line
Nebenhörer bridging telephone
Nebenkapazität stray capacity
Nebenkosten extras, accessory charge (sing.)
Nebenlautsprecher extension loudspeaker
Nebenlicht ambient light
Nebenprodukt by-product
Nebenresonanz subsidiary vibration; (unerwünschte:) spurious resonance
Nebenschluß shunt (resistance), shunt circuit
im Nebenschluß connected in parallel, shunt connected (621.395)
ohne Nebenschluß unbypassed (621.3)
Nebenschlußbogenlampe shunt arc lamp
Nebenschlußcharakteristik shunt characteristic
Nebenschlußdynamo shunt-wound dynamo
Nebenschlußerregung shunt excitation
Nebenschlußfeld shunt field (621.313)
Nebenschlußfeldregler shunt field rheostat
Nebenschlußgenerator shunt dynamo
Nebenschlußmotor shunt(-wound) motor
Nebenschlußregler shunt regulator, shunt rheostat
Nebenschlußregelwiderstand shunt regulating resistance
Nebenschlußverhalten shunt characteristic
Nebenschlußwicklung shunt winding, shunt coil (621.313)
mit Nebenschlußwicklung shunt-wound
Nebenschlußwirkung shunting effect
Nebensender relay station (621.396)
Nebensprechen cross talk
Nebensprechen aus dem Nachbarkanal adjacent-channel interference
Nebensprechdämpfung crosstalk attenuation
Nebensprechkopplung crosstalk coupling
Nebenstelle extension, extension station
Nebenstellenanlage private branch exchange (621.395), PBX

Nebenstellenanlage mit Direktzugriff intercom system
Nebenstromkreis branch circuit, subcircuit
Nebenuhr secondary clock, slave clock
Nebenwellenausstrahlung spurious emission, spurious radiation
Nebenwellendämpfung spurious-emission suppression
Nebenwiderstand shunt (resistance), parallel resistance (621.3), instrument shunt (621.317)
ohne Nebenwiderstand unshunted
Nebenwiderstandskasten shunt box (621.317)
Nebenwiderstandsklemme shunt terminal
Nebenzipfel secondary lobe, side lobe, minor lobe, parasitic lobe (621.396.677)
Nebenzipfeldämpfung side-lobe attenuation
Nebenzipfelecho side echo
Nebenzipfelfreiheit freedom from secondary lobes
Nebenzipfelunterdrückung side-lobe suppression
Nebenzuleitung shunt lead
neben beside, next to ; in addition to, besides
nebeneinandergeschaltet in shunt, shunt connected, in parallel
nebeneinandersetzen to juxtapose
nebenschlußerregt shunt excited
Negation negation
Negationsglied negator
Negativ negative (77)
Negativ-Amplitudenmodulation negative-amplitude modulation
Negativform negative matrix (621.357)
negativ negative
negativ geladen negatively charged
negativ gerichtet negative-going
negativ machen to place negative
negativ vorspannen to place (a grid) negative, to apply a negative bias
negativ vorspannen bis zur Stromlosigkeit to bias off
negativ werden to go negative
negative Bürste negative brush
negative Charakteristik negative-resistance characteristic
negativer differentieller Leitwert negativ differential conductance (621.382)
negative Elektrizität negative electricity
negative Elektrode negative electrode, cathode
negatives Glimmlicht negative glow
negative Größe negative quantity
negative Halbwelle negative swing
negative Ladung negative charge
negative Leitung negative wire
negativer Pol negative pole, negative element
negative Rückführung negative feedback (621.316.7)
negative Rückkopplung negative feedback, inverse feedback, negative retroaction
negative Steilheit negative transconductance
negative Temperatur sub-zero temperature
negatives Vorzeichen negative sign
negativer Wert negative value
negativer Widerstand negative resistance
Negatron negatron
neigen, (sich) to incline, to tilt, to tip, to bank
neigen zu to tend to
Neigung incline, inclination; declivity, slope, gradient; tilt, tilting
Neigung der Wellenfront wave tilt
Neigung zum Kippen tilting tendency
Neigungsmesser clinometer, inclinometer
Neigungswinkel tilt angle; angle of slope, angle of depression
Nein-Schaltung NOT circuit
Nenn-... nominal, rated
Nennabgabe rated output
Nennabschaltleistung rated interrupting capacity
Nennabschaltstrom rated interrupting current
Nennausgangsleistung rated output power
Nennbeanspruchung designed rating
Nennbelastung rated load
Nennbereich nominal range of use
Nennbetrieb rated service, normal service
Nennbürde rated burden (621.317)
Nenndrehzahl rated speed
Nennerregung rated excitation
Nennfrequenz rated frequency
Nenngeschwindigkeit rated speed
Nennkurzschlußspannung impedance voltage (621.314.2)
Nennkurzschlußverhältnis full-load short-circuit ratio
Nennlast rated load, designed rating
Nennleistung (power) rating, rated output (power), nominal output, normal output
Nennleistung für Aussetzbetrieb periodic rating, load factor rating
Nennleistung eines Umspanners rated kva of a transformer
Nennleistungsfaktor rated power factor
Nennleistungsklemme rated kva tap (621.314.2)
Nennmeßbereich des Instruments rating of the instrument
Nennmoment rated torque

Nennquerschnitt rated cross section, nominal cross section
Nennsekundärspannung rated secondary voltage
Nennspannung (nominal) voltage rating, rated voltage, design voltage
Nennspannungsabfall rated voltage drop
Nennspanungsbereich range of nominal voltage
Nennsperrspannung recommended crest working reverse voltage (621.382.2)
Nennstrom rated current, nominal current; nominal value of mean forward current (621.382.2)
Nennübersetzungsverhältnis marked ratio (621.317)
Nennwert nominal value, rated value
Nennwertbelastung rated working condition
Nennwerte festlegen to rate
über den Nennwert hinaus above the rated value
Nenner denominator
Nenner eines komplexen Bruches reell machen to clear a complex
Neon neon
Neonlampe neon lamp
Neonreklame neon sign
Neonröhre neon bulb
Neopren neoprene
Neper neper (621.39)
Nernstlampe Nernst lamp
Nernststäbchen glower
Nester rosettes (621.7); clusters (55)
Netz net, netting; network (621.3), (537); mains, supply (mains), power mains, grid (system) (621.315); (line) system (625); pipe system (628)
Netz mit drahtferngesteuerten Schalteinrichtungen pilot-wire controlled network
Netz mit Erde als Rückleitung earth-return system
Netz mit geerdetem Mittelpunkt earthed-neutral system
Netz mit isoliertem Sternpunkt network with insulated neutral
Netz mit unmittelbar starr geerdetem Stern network with earth-connected neutral
unabhängig vom Netz self powered
Netzanschluß main circuit connection, power supply
Netzanschlußempfänger mains receiver
Netzanschlußgerät mains set, mains power unit, power supply set, power pack (621.396)
Netzanschlußleitung supply mains
Netzanschlußtransformer mains transformer
Netzantenne mains aerial
Netzausfall power failure, line failure (US); a. c. dump (681.3)
Netzausfallschutz durch Batterien battery backup
Netzbild mimic-system diagram (621.316)
Netzbrumm (power line) hum (621.396.62/64)
Netzdrossel power filter
Netzebene network level (621.395)
Netzerder ground mat (US); earthing mat
Netzfilter low-frequency filter, power filter
Netzflimmern power-line ripple (621.397)
Netzfrequenz mains frequency, power frequency, industrial frequency
Netzgerät power pack, power supply
Netzgleichrichter power rectifier
Netzgruppenamt group centre
Netzkabel main cable, power cable
Netzkennlinienregelung load-frequency control of interconnected systems
Netzklemme main terminal
Netzknotenpunkt network junction point; nodal point of plant (621.315)
Netzkupplung interconnection, system linking
Netzleistung system capacity
Netzleitstelle system dispatching center
Netzleitung main
Netzmasche mesh (631.3)
Netzmantel-Elektrode spiral-wound electrode
Netzparallelbetrieb system interconnection
Netzplantechnik network planning, critical path scheduling
Netzreserve system reserve (621.311)
Netzschalter power switch, line switch
Netzschaltfeld mains-distribution rack (or cubicle)
Netzschutz system protective equipment
Netzschutzschalter network protector
Netzschwankung mains variation
Netzspannung mains voltage, line voltage, supply voltage
Netzspannungskonstanthalter voltage stabilizer, voltage regulator
Netzspannungsschwankung mains fluctuation; line-voltage variation
Netzspeisegerät battery eliminator, (mains) power supply, power pack
Netzspeiseleitung network feeder
Netzsperre mains filter

Netzsteckdose power receptacle, power outlet, wall socket
Netzstecker power plug, mains plug
Netzstörung system fault
Netzstrom main current, line current, supply current
Netzstruktur network structure
Netztransformator mains transformer, distribution transformer, power transformer
Netzwerk network
Netzwerkebene network level; network layer (681.3)
Netzwerkknoten network node
Netzwirkungsgrad system efficiency
Netzzähler supply meter
neu new, fresh, re...
neu auftragen to replot
neu formieren to re-form (621.356), (621.385)
neu installieren to rewire
neu konstruieren to re-design
neu laden to recharge
neu ordnen to rearrange
neu speichern to re-store
neu verteilen to redistribute
neu wickeln to rewind, to re-coil
neu zünden to re-strike, to re-ignite (621.385)
Neue Kerze platinum standard
Neuneck nonagon
Neusilber German silver, nickel-silver
neutral neutral (54), (621.3)
neutrale Achse neutral axis, neutral line (621.313.1)
neutrales Atom neutral atom
neutrale Faser neutral layer (539.384)
neutrale Zone neutral zone (621.313)
Neutralisation neutralization (54), (621.396)
neutralisieren to neutralize (54), (621.396); to decouple (621.396)
neutralisierter Empfänger neutrodyne receiver
Neutralisierungskondensator neutralizing capacitor, neutrodon
Neutralleiter neutral conductor
Neutralpunkt neutral temperature (621.362)
Neutrino neutrino
Neutron neutron
Neutronen emittieren to eject neutrons
Neutronen liefern to yield neutrons
Neutronenabsorbermaterial neutron absorper
Neutronenabsorption neutron absorption
Neutronenausbeute yield of neutrons, neutron yield
Neutronenausfluß leakage (539)
Neutronendiffusion neutron diffusion
Neutroneneinfang neutron absorption, neutron capture
Neutronenfluß neutron flux
Neutronenquelle source of neutrons, neutron source
Neutronenstrahlung neutron radiation
Neutronenstrom neutron flux
Neutronenüberschuß neutron excess
Neutronenvervielfachung neutron multiplication
Neutronenzähler, Neutronenzählrohr neutron counter
Newtonsche Mechanik Newtonian mechanics (pl.)
Newtonsche Ringe Newton's rings
NF = Niederfrequenz l. f. = low frequency; a. f. = audio frequency (siehe auch **Niederfrequenz**)
N-Halbleiter N-type semiconductor
Ni Ni (54); N (US) (669.14)
nicht abgeschlossen unconfined
nicht abgestimmt untuned (621.3); atonic (531)
nicht abgestimmter Unterbrecher atonic interrupter
nicht abschmelzende Elektrode nonfusing arcing electrode
nicht abtrennbar inseparable
nicht amtsberechtigte Nebenstelle restricted extension
nicht auf der Achse liegend off the axis
nicht aktiv inactive
nicht angepaßt unmatched
nicht angetrieben free-running
Nichtantriebsseite front
nicht angezogen slack (621.88)
nicht ansprechend auf irresponsive to
nicht ausbalanciert out of balance
Nichtausbalancierung want of balance
nicht besetzt disengaged (621.395)
nicht betriebsfähig inserviceable, inoperative, out of order
nicht brennbar noninflammable
nichtdruckendes Steuerzeichen nonprinting control character (681.3)
nicht einlaufend unshrinkable, unshrinking
Nichteisenmetalle nonferrous metals
nichtelektrisch nonelectric
nicht fertig bearbeitet unfinished
nichtflüchtiger Speicher nonvolatile storage
nicht geerdet nonearthed, unearthed, ungrounded
nicht geerdeter Mittelleiter insulated middle wire, nonearthed neutral
nicht geerdeter Nulleiter floating neutral

nicht geschlossener magnetischer Kreis imperfect magnetic circuit
nicht gesteuert uncontrolled
nicht harmonisch inharmonic (51), (621.3)
nicht isoliert bare
nicht kondensierbar incondensable
nicht korrodierend incorrosible
nichtleitend nonconducting
nichtleitende Abdeckung resist (621.357)
Nichtleiter nonconductor
nichtlinear nonlinear
nichtlineares Koordinationssystem curvilinear system of co-ordinates
nichtlineares Nebensprechen nonlinear crosstalk
nichtlineares Schaltelement nonlinear circuit element
nichtlinearer Stromkreis nonlinear circuit
nichtlineare Verzerrung nonlinear distortion, amplitude distortion
nichtlinearer Widerstand nonlinear resistor
Nichtlinearität nonlinearity
nichtlöschendes Lesen nondestructive reading, nondestructive readout
Nichtmetall nonmetal
nichtmetallisch nonmetallic
nicht mischbar immiscible
nicht normgerecht not standard(ized), not to standards (pl.)
nicht oxidierende Atmosphäre nonoxidizing atmosphere, noncorrosive atmosphere
nicht polarisiertes Relais nonpolarized relay, unpolarized relay
nicht potentialbezogener Ausgang output not referred to a potential, floating output
nicht reaktionsfähig unreactive (54)
nichtrostend rust resisting, rust-free, rust-proof
nichtrostender Stahl stainless steel
NICHT-Schaltung NOT-circuit, NOT-element, NOT-gate
nicht schienengebunden trackless
nicht schmelzbar infusible
nicht selektiv unselective
nicht sinusförmig nonsinusoidal, nonsine
nicht sinusförmige Wellen nonsine waves
nichtspannungsführender Leiter dead conductor
nichtspannungsführender Teil noncurrent carrying part (696.6)
nichtstromführend non-current carrying
nichtzugelassene Leitung nonapproved circuit
nichtüberbrückende Kontakte nonbridging contacts; break before make (621.316.5)
nicht umkehrbar irreversible
nicht wesentlich zugehörig extrinsic
Nickbewegung dipping movement (629.11); pitch (629.13)
Nickel nickel
Nickelbronze nickel bronze
Nickeleisen nickel iron
Nickel-Eisen-Akkumulator iron-nickel storage battery, nickel-iron alcaline accumulator
Nickel-Kadmium-Akkumulator cadmium-nickel storage battery
Nickellegierung nickel alloy
nickelplattiert nickel-clad
Nickelstahl nickel steel
Nickelstaubkern nicore
Nickel-Zink-Akkumulator Drumm accumulator
Nickelin nickeline
Nicolsches Prisma Nicol (prism)
Niederdruck low pressure
Niederflußreaktor low-flux reactor
niederfrequenter Drahtfunk audio-frequency wire-broadcasting
niederfrequente Drahtfunkverteilung audio-frequency rediffusion
Niederfrequenz = NF low frequency = l. f., audio frequency = a. f.
Niederfrequenzausgang low-frequency terminals (pl.)
Niederfrequenzdrossel audio-frequency choke
Niederfrequenz-Fernkabel audio trunk cable
Niederfrequenzfilter low-frequency filter
Niederfrequenzgenerator audio generator, audio-frequency generator, audio-frequency oscillator
Niederfrequenzleistung low-frequency output, audio-frequency output
Niederfrequenzsieb low-frequency filter
Niederfrequenzspannung low-frequency voltage
Niederfrequenzstufe low-frequency stage
Niederfrequenz-Tonleitung audio-frequency circuit
Niederfrequenztransformator low-frequency transformer
Niederfrequenzverstärker low-frequency amplifier, audio amplifier
Niederfrequenzverstärkung low-frequency amplification
Niederfrequenzwiderstand low-frequency resistance
Niederführungskabel downlead cable
Niederleistungsfunkfeuer low-power beacon
Niederleistungsreaktor low-power reactor
niederohmig low-impedance, low-resistive

niederohmiger Weg low-impedance path
Niederschlag deposit (54); precipitate, precipitation (54), (551.5)
Niederschlagselektrode starting sheet (621.357)
niederschlagen, sich to precipitate, to deposit
niederschmelzen to flux, to melt down
Niederspannung low tension, low voltage
Niederspannungsabgabe low-voltage demand
Niederspannungsabnehmer low-voltage consumer, secondary consumer
Niederspannungsanlage low-voltage system, low-voltage installation
Niederspannungs-Anschlußleitung secondary distribution feeder
Niederspannungsdurchführung low-voltage bushing
Niederspannungsinstallation low-voltage wiring
Niederspannungsisolator low-tension insulator, low-voltage insulator
Niederspannungskabel low-voltage cable
Niederspannungsklemme low-voltage terminal
Niederspannungskraftanlage low-voltage power plant
Niederspannungsleitung low-voltage line, low-voltage bus(bar)
Niederspannungsnetz low-voltage grid
Niederspannungsschalter low-voltage switch
Niederspannungsschaltgeräte low-voltage switchgear (sing.)
Niederspannungsversorgungsnetz secondary distribution network, low-voltage supply system
Niederspannungsverteilung low-voltage distribution
Niederspannungsverteilungsanlage secondary distribution system
Niederspannungsverteilungsschiene low-voltage bus(bar)
Niederspannungswicklung low-voltage winding, secondary winding (621.314.2)
Niedervolt low voltage (s. auch **Niederspannungs-...**)
niedrig low
niedrige Drehzahl low speed
niedrige Einstellung low setting
niedrige Kapazität low capacity
niedriges Niveau, niedriger Pegel low level
niedrige Permeabilität low permeability
niedrige Programmiersprache low level language
niedrige Spannung low tension, low voltage, low potential
niedriger Verlust low loss
niedriger Widerstand low impedance, low resistance
niedrigste Ablesung lowest reading
niedrigste Einstellung lowest setting
niedrigstes Moment während des Anlaufs pull-up torque
niedrigwertigste Stelle least significant digit
recht niedrig halten to keep to the lowest figure
zu niedrig absichern to underfuse
Nierencharakteristik cardioid diagram, appleshaped diagram (621.396.67)
Niet rivet
Niet setzen to buck a rivet
Nietarbeit riveted work
Nietkonstruktion riveted design
Nietkopf rivet head
Nietkopfsetzer rivet set
Nietkörner rivet punch
Nietmaschine riveting machine, riveter
Nietschienenverbinder Chicago rail bond (621.33)
Nietverbindung rivet joint
nieten to rivet
genietetes Gehäuse girder-type frame (621.313)
genietete Stahlkonstruktion riveted work
Nife-Akkumulator = Nickel-Eisen-Akkumulator iron-nickel storage battery
Niob, Niobium columbium, niobium
Nippel nipple, grip
Nippelelektrode noncontinuous electrode (621.357)
Nirosta rustless steel, stainless steel
Niton niton, radium emanation
nitrierhärten to nitrard
Nitrierhärtung nitrogen hardening
Nitrolack nitrocellulose laquer
Nitroverbindungen nitro compounds
nitros nitrous
nitroses Gas nitrous gas, nitrous fume
Niveau level
Niveaufläche equipotential surface
Niveaulinie equipotential line
nivellieren to level
Nixie-Röhre nixie tube
N. N. = Normal Null mean sea level, ordnance datum
über N. N. above sea level
N-Kanal-Feldeffekttransistor N-channel field-effect transistor
N-Leitung n-type conduction (621.382)
n-leitend n-conducting
NLT-Verstärker negative repeater
Nocken cam
Nockenendschalter cam-limit switch

Nockenfahrschalter cam controller, cam-operated master controller
Nockenkontakt cam contact
Nockenring cam roller
Nockenschalter cam (operated) switch, camshaft switch
Nockenscheibe cam drum
Nockenscheibenschalter cam-shaft controller
Nockensteuerung cam control
Nockentrieb cam gear
Nockenwelle cam shaft, tumbler shaft
Nöggerathmaschine acyclic machine
Nomenklatur nomenclature
Nomogramm nomograph, chart diagram, calculator, abac
Nonius vernier
Noniusskala vernier scale
Nord, Norden north
Nordpol north pole
Nordpol der Magnetnadel north seeking pole, marked pole
nordweisend north seeking
NOR-Glied NOR circuit, NOR gate, NOR element
NOR-Verknüpfung NOR operation
Norm norm, standard
als Norm festlegen to standardize
Normenausschuß standards committee
Normbau standard building
Normbauteil standard component
Normdraht code wire
Normgewinde standard thread
Normkabel standard cable, Association cable
Normspannung standard voltage
Normteil standard part
Normstimmton standard tuning frequency
Normtype standard type
Normvorschriften standard book (sing.), code regulations
Normal-... normal, standard; full-sized
Normalabgabe normal output
Normalausbreitung normal propagation (621.396)
Normalausführung standard design, standard pattern
Normalbauart standard type
Normalbeanspruchung standard stress (539); design stress
Normalbetrachter standard observer
Normaldrehzahl base speed; standard speed
Normaldruck normal pressure
Normalelektrode half cell (621.35)
Normalelement standard cell (621.3)
Normalformat standard size
Normalfrequenz normal frequency, standard frequency
Normalfrequenzaussendung standard frequency transmission
Normalfrequenzgenerator standard frequency generator
Normalgenerator 1-milliwatt-generator, signal source (621.395)
Normalgeschwindigkeit normal speed
Normalgewicht standard weight
Normal-Glimmerkondensator standard mica condenser
Normalinduktivität standard inductance
Normalkern standard frame (621.313)
Normalkerze primary luminous standard
Normalklima standard atmosphere
Normalkondensator standard capacitor
Normalkonstruktion standard design
Normallampe standard lamp
Normalleistung normal output
Normallichtquelle s. **Normalkerze**
Normallösung normal solution (54)
Normalmaß standard size
Normalmeter standard meter
Normalmuster standard pattern
Normal Null = N.N. mean level, normal zero
Normal-Ohm standard ohm
Normalpegel standard level (621.392)
Normalpotential standard electrode potential
Normalprüfgefäß standard test vessel
Normalsockel standard socket (621.32), (621.385)
Normalspannung standard voltage (621.3); standard stress (539)
Normalspur standard gauge (625)
Normalspurbahn standard gauge railway
Normalstrom normal current
Normaltemperatur normal temperature, standard of temperature
Normalton reference tone, reftone; standard musical pitch
Normalverteilung characteristic spread, Gaussian distributian
Normalwert normal value
Normalwiderstand normal resistance; resistance standard, standard resistor
Normalzustand normal (state)
Normalzustand (760 mm Hg und 0 °C) N.T.P. = Normal Temperature and Pressure
normal normal, regular; routine
normal polarisierte Wellen normally polarized waves

normale Bauart standard design
normale Nennlast normal rating
normalspuriges Gleis standard gauge track
Normalien, Normen standards
normalisieren to normalize, to standardize
Normalisierung normalization, standardization
normen to standardize
normieren to scale
Normierungsfaktor scale factor, scaling factor
Normung standardization, standardizing
Normfarbtafel chromaticity diagram
Normlichtart C illuminant C (621.397)
Not distress; emergency
Notabschaltung emergency shutdown
Notausschalter off emergency (switch)
Notbehelf shift, makeshift
Notbeleuchtung emergency lighting, pilot lighting
Notbetrieb emergency service
Notbremse emergency brake
Notdienst emergency service
Notfrequenz emergency frequency, emergency radio channel
Notlampe pilot lamp
Notlicht emergency light
Notruf distress signal, emergency call
Notschalter emergency switch
Notsignal distress signal
Notstromaggregat stand-by set
Notstromversorgung emergency power supply, standby generating set, mains-failure generating set
Notverkehr distress traffic
Notzeichen distress signal, SOS call
Notizbuchspeicher scratchpad memory
Np = Neper neper
NPN-Transistor NPN transistor
NP-Übergang NP junction
N.Sp. = Niederspannung L.T. = Low Tension (s. **Niederspannung)**
NTC-Widerstand NTC resistor
N-Typ-Halbleiter N-type semiconductor
Nuklearstromversorgung nuclear-power supply
Nuklid nuclide
Null null, naught, zero; (im Dezimalbruch:) cipher, cypher
auf Null abgleichen to restore to zero, to set to zero
auf Null absenken to reduce to zero, to bring to zero, to depress to zero
auf Null absinken to sink to zero, to decay to zero
auf Null stellen to set at zero
auf Null zeigen to register zero
gegen Null gehen to vanish (51)
unter Null sub zero
Null Komma Eins = 0, 1 point one = .1
Nullabgleich nullification
Nullbezugszeit zero-time reference
Nulldetektor zero-crossing detector, nulldetector
Nulldurchgang crossover (51); passing through zero, zero cross
Nulldurchgang der Spannung voltage zero
Nulleffekt background counting rate
Nulleinstellung zero setting, zero adjustment
Nullenunterdrückung zero suppression (681.3)
Nullfrequenz zero frequency (621.395.4)
Nullfrequenzton-Methode zero-beat methode (621.317)
Nullinstrument (balance) indicator, null indicator
Nulleiter third wire, return wire (d.c.); neutral main, neutral conductor, identified conductor
Nulleiter markieren to polarize
Nulleiterdraht neutral wire
Nulleiternetz secondary neutral grid
Nulleiterstrom neutral conductor current
Nulleitersystem secondary neutral grid
Nulleiterzuführung neutral distributor
Nullinie base line, time base
Nullinien-Festhalteschaltung clamping circuit (621.397)
Nullinien-Stabilisiereinrichtung base-line stabilizer
Nullkegel cone of nulls, cone of silence (621.39)
Nullkennzeichen zero flag
Nullmethode null method, zero method (621.317)
Nullmomentenstellung no-torque position
Nullniveau zero level
Nullode ATR-tube, antitransmit-receive tube
Nullpegel zero level
Nullpegeldiode d. c. restorer diode, d. c. clamp diode
Nullpotential zero potential
Nullpunkt neutral point, earthed (neutral) terminal; origin, reference point (51)
Nullpunktdrossel absorption coil
Nullpunkteinstell-... zero-adjusting
Nullpunkteinstellung setting of zero reference, zero adjustment
Nullpunktserdung neutral earth (Br.), neutral ground (US)
Nullpunktfehler zero error
Nullpunktstabilisierung drift stabilisation
Nullpunktunterdrückung suppressed zero

Nullpunktverschiebung zero offset, zero shift
Nullpunktswanderung zero shift, drift (621.317)
Nullspannung zero voltage, novoltage
Nullspannungsauslöser novoltage release
Nullspannungsauslösung under-voltage circuit breaking, under-voltage tripping
Nullspannungsrelais under-voltage relay, novoltage relay
Nullspannungsschalter novoltage circuit breaker
Nullstelle (zero)root (51); gap (621.396.677)
Nullstellung zero adjustment; null position, point of rest, neutral position
Nullstrich zero mark
Nullunterdrückung zero suppression, zero compression
Nullwert zero value, zero root (51)
Nullzustand zero condition, zero state; nought state (681.3)
Nullzweig detector branch (621.317)
nullen to neutralize (621.316)
nullstellen to reset
Nullung m. p. e. = multiple protective earthing, neutralizing, neutralization
Nullungsleitung neutral feeder
numerieren to number
Numerierungsbereich numbering area (621.395)
Numerierwerk numbering box
numerisch numerical
numerischer Abstand numerical distance
numerische Anzeige numerical display
numerische Apertur numerical aperture (535)
numerische Steuerung numerical control, NC
numerisches Wort numeric word (681.3)
Numerus number, antilogarithm, characteristic
Nummer number (51); size (621)
Nummernschalter dial switch (621.395)
Nummernschalter-Impulskontakt dial-pulse spring, pulse spring
Nummernschalterwahl dial pulsing
Nummernscheibe (calling) dial (621.395)
Nummernschild number plate
Nummernumsetzer director, register translator (621.395)
Nummernwähler numerical selector
Nußisolator egg insulator
Nut, Nute groove, flute; slot
Nutauskleidung slot lining
Nutbreite slot width (621.313)
Nutbrücke slot bridge (621.313)
Nutfrequenz tooth-pulsation frequency (621.313)
Nutfüllfaktor slot-space factor
Nuthülse slot liner
Nutisolation slot insulator, slot liner
Nutenkeil keypiece, slot wedge
Nutleitwert slot permeance
Nutenobertöne slot ripples, to oth ripples
Nutenstreuung slot leakage
Nutteilung slot pitch
Nutenverschlußkeil slot key (621.313)
Nutation nutation
nuten to slot; to spline, to flute, to groove
genuteter Anker slotted-core armature (621.313)
genutete Welle splined shaft
Nutz-... useful, usable
Nutzausgangsleistung power output
Nutzbildfläche useful screen area (621.397)
Nutzbremsung regenerative breaking (621.33)
Nutzfeldstärke useful field intensity
Nutzfluß useful flux, working flux (621.313)
Nutzinformation useful information
Nutzkapazität service capacity (621.35)
Nutzkilometer effective kilometre (625)
Nutzlast useful load, carrying capacity
Nutzleistung useful effect, effective capacity
Nutzpegel useful level
Nutzquerschnitt useful cross section (621.313)
Nutzstrom useful current
Nutzsignal useful signal, wanted signal
Nutzträger wanted carrier
nutzbare Maschinenzeit available machine time
Nutzung utilization, use, exploitation
Nutzungsdauer useful life, service life
Nuvistor nuvistor
NV = Niedervolt l. t. = low tension, low voltage (siehe **Niederspannung**)
Nylon nylon
Nyquist-Charakteristik Nyquist slope
Nyquistdemodulator vestigal-sideband demodulator

O

OB = Ortsbatterie local battery
OB-Amt magneto central office, local-battery exchange
OB-Apparat local-battery telephone set
OB-Telefonie mit Induktorruf magneto-telephone system
OB-Vermittlung local-battery exchange, magneto-switchboard exchange

oben above
oben auf on the top of
oben flach flat-topped
nach oben gerichtet upward
nach oben gerichteter Lichtstrom upward flux
obengespeiste Antenne top-fed antenna
ober upper, superior; high
oberer Flansch upper flange
oberer Heizwert high heat value, gross calorific value
oberer hemisphärischer Lichtstrom upper-hemispherical flux
oberes Seitenband upper sideband
oberer Totpunkt top of the stroke
oberer Umwandlungspunkt upper change-point
oberhalb der Resonanz above resonance
oberirdisch above ground; over ground
oberirdische Anlage surface plant
oberirdische Kabelanlage aerial-cable system
oberste Platte top vane (621.319.4)
Oberband high band, upper band
Oberbau permanent way (625.1)
Oberfläche surface (area)
Oberflächenbehandlung surface treatment
Oberflächenbelüftung surface cooling
Oberflächenbeschaffenheit surface finish
Oberflächeneffekt surface effect
Oberflächenerder surface earth (electrode)
Oberflächenerdung surface earthing, surface grounding
Oberflächenfehler surface defect
Oberflächengüte surface quality, surface finish, surface polish
Oberflächenhärtung surface hardening
Oberflächenhelligkeit surface brightness
Oberflächenkriechstrom surface leakage
Oberflächenkühlung surface cooling
Oberflächenladung surface charge
Oberflächenladungsdichte electric surface density
Oberflächenleitfähigkeit surface conductivity
Oberflächenleitschicht surface duct (621.371)
Oberflächenmontierte Bauelemente surface mounted devices
Oberflächenpassivierung (IC) surface passivation
Oberflächenreflexion surface reflectance
Oberflächenreibung surface friction
Oberflächenrekombination surface recombination (621.382)
Oberflächenschicht superficial layer, surface skin
Oberflächenspannung surface tension
Oberflächenspannungen skin stresses
Oberflächensperrschichttransistor surface-barrier transistor
Oberflächenwelle surface wave (621.371)
Oberflächenwellenleiter ground-based duct, surface wave guide
Oberflächenwiderstand surface resistance, surface resistivity
Oberflächenwirkung surface effect
oberflächlich skin-deep, superficial
oberflächlich berühren to glance
Oberingenieur chief engineer, senior engineer
Oberkante top edge
Oberleitung overhead line, overhead conductor (621.3)
Oberleitungsbus s. **Obus**
Oberleitungsdraht overhead wire
Oberleitungsnetz local-line network (621.395)
Oberleitungsomnibus s. **Obus**
Oberlicht skylight, roof light
Oberschmierung upper lubrication
Oberschwingung overtone, harmonic (534)
Oberschwingungsquartz overtone crystal, harmonic-mode crystal
Oberspannung high tension (s. auch **HV**); high-side voltage (621.314.2)
Oberspannungsklemme high-tension terminal
Oberspannungsstation h. t. sub-station
Oberspannungswicklung high-tension winding = h. t. w.
Oberton harmonic (tone), overtone (534), (621.3)
Oberwelle harmonic (wave)
Oberwellen herausheben to accentuate harmonics (621.385)
Oberwellenanalysator harmonic analyzer
Oberwellenantenne harmonic antenna
Oberwelleneisenverluste incremental hysteresis loss
Oberwellenerzeuger harmonics generator
Oberwellenfilter ripple filter, harmonic filter, static harmonic absorber
Oberwellenfreiheit freedom from harmonics
Oberwellengehalt der Spannungskurve ripple content of the voltage wave
Oberwellenleistung harmonic power
oberwellenreich rich in harmonics
Oberwellensieb s. **Oberwellenfilter**
Oberwellensiebschaltung harmonic stopper
Oberwellenstörung harmonic interference
Oberwellenverminderung suppression of harmonics
Object object, objective
Objektcode object code
Objekttisch stage (535.8)

Objektträger object carrier, (glass) slide (535)
Objektiv objective, object glass, object lens
objektiv objective
objektive Anzeigeneinrichtung visual detector (621.317)
Obus = Oberleitungsomnibus trolley (bus)
Obuslinie trolley-bus line
ODER-Schaltung OR circuit, OR element, OR gate
ODER-Verknüpfung inclusive OR, OR operation
Oersted oersted
Ofen oven (647); stove (697); kiln (666); furnace (621.3), (669)
Ofentrocknung kiln drying, oven drying
offen open; exposed work (621.3)
offene Amtsabfrage unassigned answer
offene Bogenlampe open arc lamp
offenes Dreiphasensystem six-wire three-phase system
offene Funkenbildung open sparking
offenes Gitter floating grid, disconnected grid
offener Kabelschuh space end (621.3)
offene Leitung open-ended line
offene Leitungsverlegung exposed wiring
offenes Licht naked light
offene Maschine open machine (621.313)
offene Numerierung open numbering (621.395)
offene Nut open slot
offener Riemenantrieb open belt drive
offene Schaltanlage open-type switchgear
offene Sicherung open fuse, plain fuse, renewable fuse unit
offene Wicklung open-circuit winding, open-coil armature winding (621.313)
offenliegend exposed
öffentlich public; in public, commercial
öffentliche Beleuchtung street lighting
öffentlich beweglicher Landfunk public mobile land radio service
öffentliche Energieversorgung public power supply
öffentliches Fernmeldenetz public communication network, public network
öffentliches Fernsprechnetz switched public telephone network (commercial network)
öffentlicher Funkdienst public radio service
öffentliches Rundfunknetz (commercial) radio network
öffentliche Sprechstelle public call box, public telephone station
öffentlicher Versorgungsbetrieb public utility (works)
öffentliches Versorgungsnetz public supply mains
öffnen to open, to break; to dismantle
Öffnen eines Schalters opening of a circuit-breaker
öffnen, sich to open; to fly open (621.3); to separate (621.316.5)
Öffnen vor Schließen break before make
Öffner (Kontakt) break contact, NC, normally closed contact
Öffnung opening, aperture; mouth, orifice (621)
Öffnungsbreite aperture angle
Öffnungsgeschwindigkeit opening speed (621.316.5)
Öffnungsimpuls break impuls
Öffnungskontakt normally-closed contact, NC-contact, break contact
Öffnungsstromstoß break pulse
Öffnungswinkel divergence (535); flare angle, angle of beam (621.396.67)
Öffnungszeit gating time, break periode
Offsetbetrieb carrier offset (621.396)
Offset-Spannung offset voltage
OF-Transformator forced air-cooled oil-immersed transformer
OH-Gruppe hydroxyl group
Ohm ohm
Ohmmeter ohmmeter
Ohmsekunde secohm
ohmsch, ohmisch ohmic, resistive
ohmsche Belastung ohmic load, resistive load, non-inductive load
Ohmsches Gesetz Ohm's law
ohmscher Spannungs(ab)fall ohmic drop, i. r. drop, resistance voltage
ohmscher Spannungsteiler resistance voltage divider, potentiometer
ohmscher Verlust ohmic loss, resistance loss, rheostatic loss
ohmscher Widerstand ohmic resistance, true resistance
ohmscher Zweig resistive branch
Ohnelastschalter off-load switch
Ohr ear
das Ohr betreffend aural
Ohrhörer earphone
Ohrkurve ear-response characteristic
Oktalschreibung octal notation (681.3)
Oktan octane
Oktananzahl octane number
Oktave octave (534)
Oktavsieb sound analyser, octave filter
Oktode octode
Okular eye lens, eye piece

Okularmikrometer ocular micrometer
Öl oil
in Öl abgeschreckt oil-quenched
unter Öl oil-immersed, immersed in oil
unter Öl laufende Pumpe oil-immersed pump
Ölanlasser oil starter, oil-cooled starter, oil-immersed starter
ölarmer Schalter low-oil content circuit breaker, minimum-oil circuit breaker
Ölausgleichsgefäß oil feeding reservoir (621.315.2); oil conserver (tank) (621.313); oil-expansion tank (621.314)
Ölaustritt drainage of oil, oil leakage (621.315.2)
Ölbehälter oil (storage) tank, oil container
Ölbelüfter oil breather
ölbenetztes Filter viscous impingement-type filter
Ölbrenner oil burner
öldicht oil-tight
Öldrehkondensator variable oil capacitor
Öldruck oil pressure, hydraulic pressure
Öldruckanzeiger oil-pressure gauge
Öldruckkabel oilostatic cable
Ölfalle, Ölfangraum separating chamber (621.316.5)
Ölfarbe oil colour, oil paint
Ölfarbenanstrich oil painting
ölfest oil-proof
ölfrei, öllos oil-less
ölfreier Schalter oil-less breaker
ölgetränkt oil-impregnated
Ölharz oleo-resin, slum
Ölhärtung oil quenching
Ölisolation oil insulation
Ölkabel oil-impregnated cable, oil-filled cable
Ölkammer oil chamber (621.8)
Ölkondensator oil capacitor, oiled-paper capacitor
Ölkühlung oil cooling
ölgekühlt oil-cooled
Ölumlaufkühlung forced-oil cooling
Öllack oil varnish
öllöslich oil-soluble
Ölpapier oil paper, oiled paper
Ölpapierkondensator s. **Ölkondensator**
Ölringschmierung oil-ring lubrication
Ölrinnenisolator oil insulator
Ölsäure oleic acid
Ölschalter oil(-break) switch, oil circuit breaker
Ölschauglas oil gauge(glass)
Ölschicht oil film
Ölschlamm slum, oil sludge
Ölschütz oil-immersed contactor, oil-break contactor
Ölsicherung oil-break fuse, liquid-fuse unit
Ölsieb oil strainer
Ölspiegel oil level, divide line
Ölstand oil level
Ölstandsrohr oil gauge
Ölsumpf sump (621.3)
Öltransformator oil(-cooled) transformer, oil-immersed transformer
Öltransformator mit Anblasung forced air-cooled oil-immersed transformer
Öltransformator mit Fremdlüftung (OF) oil-immersed forced air-cooled transformer
Ölumlaufschmierung circulating-oil lubrication
Ölverlust drainage of oil, loss of oil
Ölwanne oil sump (621.9), (629.113)
Ölzufuhr oil supply, oil feed
Oleinsäure oleic acid
ölen to oil, to lubricate
Öler lubricator
ölig oily
Opal opal
Opalglas opal glass, opalescent glass, frosted glass
Opallampe opal lamp, pearl lamp
Opalschirm opal shade
Operand operand
Operationspfad operation path
Operationsverstärker operational amplifier, op-amplifier
Operator operator (51)
Operatorenrechnung operational calculus
operieren to operate (61), (355.4)
Ophthalmoskop ophthalmoscope
Oppanol polithion
Optik (science of) optics
optimale Betriebsfrequenz optimum working frequency, F. O. T.
Optimalwert optimum value
Optimierung optimization, optimalizing
Optimum optimum (pl.: -ma)
optisch optical
optisch eben optically flat
optisch linksdrehend laevo-rotary (54), (535)
optische Abstimmung visual tuning
optische Achse optical axis (548)
optische Anzeige visual indication, visual display
optische Bank optical bench
optisches Bild optical image
optischer Bildausgleich optical compensation

optische Drehung optical rotations (pl.)
optisches Gitter optical grating
optischer Markierungsleser optical-mark reader, OMR
optische Meldung visual signal
optische Personensuchanlage visual-paging system
optisches Pyrometer optical pyrometer
optisches Rufzeichen visual-call sign
optische Sicht(weite) optical range, range of optical vision, line of sight
in optischer Sicht in sight of each other (621.396.67)
optischer Spannungsprüfer strain viewer (666.1)
optische Strahlung visible radiation
optische Täuschung optical illusion
optische Untersuchung optical inspection, visual check
optische Zeichenerkennung optical character recognition, OCR
Optokoppler optical coupler
Optronik opto-electronics (pl.)
ordentliche Welle ordinary wave
Ordinate ordinate
Ordinate auftragen to erect an ordinate
Ordinatenachse axis of ordinate
ordnen to classify, to arrange; to gather together (51)
geordnete Belastungskurve load duration curve
Ordnungszahl ordinal (number) (51); electron number (54); atomic number, nuclei number (539)
Organisation organization
organisch organic
organische Chemie organic chemistry
organischer Farbstoff organic dye
organische Verbindung organic compound
Original original
Originalgröße life size
Originalverpackung original packing
Ornamentlampe decorative lamp
Ort site, spot, place; locus (51); location (621.315)
an Ort und Stelle prüfen to spot check, to check on sight
Ortsamt local office, local exchange
Ortsanruf local call
Ortsbatterie (= OB) local battery
Ortsbeamter, Ortsbeamtin local operator (621.395)
Ortsbedienung local control
Ortsbereich local range
Ortsbestimmung location
ortsbeweglich portable
ortsbeweglicher Sender portable transmitter
ortsfest stationary, fixed, land-based
ortsfeste Batterie stationary battery
ortsfeste Funkstelle land station, fixed radio station
ortsfeste Lampe stationary lamp
ortsfester Transformator stationary transformer
Ortsgespräch local call (621.395)
Ortskabel local cable
Ortskennzahl trunk code
Ortskoordinaten coordinates
Ortskurve locus (of points) (621.3); geometric locus (538.55)
Ortskurve des Frequenzgangs frequency-response locus
Ortsleitung local circuit
Ortsmünzfernsprecher coinbox telephone for local calls
Ortsnetz local network
Ortsnetzbereich local telephone area
Ortssender regional broadcasting station
Ortstarif local tariff
Ortsveränderung locomotion
Ortsverbindungsleitung interoffice trunk, local junction circuit
Ortsvermittlungsstelle local exchange, local office
orten to locate, to spot
Orthicon image orthicon
orthochromatisch orthochromatic
orthogonal orthogonal
örtlich local
örtlicher Bauleiter resident engineer
örtliche Gegebenheiten site characteristics
örtliche Überhitzung localized heating
örtliche Verhältnisse local conditions
örtliche Vorschrift by-law, bye-law
Ortung location, position finding, location finding
Ortungsfunkdienst radio navigation service
Ossanakreis Ossana's circle diagram
OS-Transformator self-cooled oil-immersed transformator
Öse eye, eyelet, ear; lug (621), (621.3)
mit Ösen versehen to eyelet
Ösenschraube eye bolt, ring bolt, screw eye
Ösenverbindung eyeletting
Osmium osmium
Osmiumlampe osmium lamp
Osmose osmose, osmosis
osmotisch osmotic
osmotischer Druck osmotic pressure

Oszillator oscillator
Oszillator mit kapazitiver Rückkopplung Collpitts oscillator
Oszillator mit negativem Widerstand negative resistance oscillator
Oszillatorquarz oscillator crystal
Oszillator-Rauschen oscillator noise
Oszillatorschwingkreis tank circuit
Oszillogramm oscillogram
Oszillograph (mit Registrierung) oscillograph
Oszillographenröhre cathode ray tube, oscilloscope tube
Oszillographenschirm oscilloscope screen
Oszilloskop oscilloscope, cathode ray oscilloscope
Otometer audiometer
Outputmeter output meter
Oval, oval oval
Overlay-Verfahren overlay
Oxalsäure oxalic acid
Oxid oxide, calx
...-oxid flower of... (54)
mit Oxid bedeckt oxide coated
oxidbildend oxidative
Oxidfilm, Oxidhaut oxide film
Oxidhaut bilden to scale
Oxidkathode oxide-coated cathode, oxide-coated emitter
Oxidschicht film of oxide
Oxidschichtglühlampe ultra-incandescent lamp
Oxidation oxidation, oxidization
Oxidationsbeständigkeit oxidation resistance
Oxidationsschutzmittel oxidation inhibiter
Oxidationsstufe oxidation state
oxidieren to oxidize
oxidierend oxidative
Ozokerit ozocerite, earth wax
Ozon ozone
Ozonisator ozonizer

P

Paar pair, couple; couplet (621.396)
Paarbildung pair production, pair generation
Paarigkeit (der Zeilen) pairing, twinning (621.397)
Paarverzahnung pair gearing
paaren, paarig abteilen to pair
Pacco-Schalter packet-type switch
packen to pack
Packmaterial packing (material)
Packung packing
Packungsdichte packing density, package density
Padding padding (621.396)
Paddingskondensator padding condenser, padder, tracking condenser
Paketschalter gang switch (621.3)
panchromatisch panchromatic
Panikbeleuchtung emergency lighting, panic lighting
Panoramaempfänger panoramic receiver, panoramic monitor
Panoramagerät plan position indicator, PPI; radar landscape scanner
Pantograph pantograph
Panzerader rubber insulated steel-wire-braided cable
Panzerkabel armoured cable
Panzerrohr steel pipe, steel conduit
Panzerschlauch armoured hose
Papier paper
Papieraufspulvorrichtung paper winder
Papierauskleidung paper lining
Papierbleikabel paper-insulated lead-covered cable = PILC cable
Papierisolation paper insulation, paper dielectric
papierisoliert paper-insulated
papierisoliertes Dynamoblech core-plated punches
Papierkabel paper-insulated cable, paper-core cable, dry-core cable
Papierkondensator paper capacitor
Papierpuppe paper-lined construction (621.35)
Papierspinner paper spinner (621.315.2)
Papiervorschub paper throw, paper feed
Pappe paste board, card, cardboard
Pappkarton cardboard box
Pappröhre cardboard tube
PAR precision approach radar
Parabel parabola
Parabelscheitel vertex of the parabola
Parabelschnitt section of a parabola
parabolisch parabolic
Paraboloid paraboloid
Parabolantenne parabolic aerial, dish antenna
Parabolspiegel parabolic mirror, searchlight-type reflector, parabolical reflector (621.396.677)
Paraffin paraffin
paraffiniert paraffined
paraffiniertes Papier paraffin (-impregnated) paper

Parallaxe parallax
Parallel-... parallel, shunting, bypass (621.3)
Paralleladdition parallel addition
Parallelbetrieb parallel running (621.311); parallel operation
Parallelbetrieb von Netzen interconnection of power systems
Parallelcode parallel code
Paralleldrahtleitung parallel wire line, parallel conductor line, twin lead
Paralleldrossel shunt inductor
Paralleldrucker parallel printer, line-a-time printer, line printer
Parallelkapazität shunt capacity
Parallelklinkenstreifen bunching strip
Parallelkondensator parallel capacitor, shunt capacitor, bypass capacitor
Parallelnuten parallel slots
Parallelperspektive perpendicular projection
Parallelreißer surface gauge
Parallelresonanz parallel (phase) resonance
Parallelresonanzkreis parallel circuit
Parallelschaltstufen parallel notches (621.33)
Parallelschaltung parallel connection, shunt connection, parallel combination; synchronization (621.311)
in Parallelschaltung in shunt (connection), in parallel
Parallelschieber parallel-slide valve
Parallelschraubstock parallel-jaw vice
Parallelschwingkreis parallel resonant circuit, tank circuit (621.396)
Parallel-Serien-Umsetzer parallel-serial converter
Parallelspeisung shunt feed (621.396.61)
Parallelstromkreis parallel circuit, shunt circuit
Parallelübertragung parallel transmission
Parallelwiderstand parallel resistance, shunt resistance, shunt resistor
Parallelzahnnut taper slot
Parallelzugriff parallel access
Parallelzweig parallel branch, shunt branch
Parallelzweigschutz split-conductor protective system
parallel parallel
parallel anschließen to join in parallel
parallel betreiben to run in parallel
parallel zur Faser parallel to fibre
parallel laufen mit to run in parallel with, to run in step with
parallel laufend in parallel; concurrent
parallel liegen zu to parallel with
parallelgehend concurrent
parallelschalten to connect in parallel, to combine in parallel; to parallel, to synchronize (621.311)
parallelgeschaltet zu in shunt across, shunt connected to, as a bypass to
parallelgeschaltete Induktivität shunt inductor
parallelgeschalteter Widerstand shunt resistor
parallelgeschalteter Zweig shunt circuit
parallel-gespeist shunt fed
Parallele parallel
Parallele ziehen to draw a parallel
in Parallele setzen zu to place in parallel to
Parallellineal parallel ruler
Parallelepiped parallelepiped
Parallelismus parallelism
Parallelogramm parallelogram, rhomboid
Parallelogramm der Kräfte parallelogram of forces
paramagnetisch paramagnetic
paramagnetisches Material paramagnetic material
Paramagnetismus paramagnetism
Parameter parameter
Parameterdarstellung parametic form
parametrischer Verstärker parametric amplifier, variable-reactance amplifier
parasitär parasitic, spurious
parasitäre Strahlung parasitic emission, spurious emission
Pardune guy wire
Pardunenisolator guy (shackle) insulator
Parität parity
Paritätsbit parity bit
Paritätskontrolle parity check
Partialbruch partial fraction
Partialbruchzerlegung partial fraction expansion
Partialdruck partial pressure
partiell partial
partiell integrieren to integrate by parts
partielle Ableitung partial derivate (51)
partielles Differential partial differential
partielle Differentialgleichung partial differential equation
partielle Integration integration by parts
Partikel particle
partikuläres Integral particular integral
partikuläre Lösung particular solution (51)
Paschensches Gesetz Paschen's law
Paß filter (621.39)
Paß-... matched, locating, mating
Paßfläche locating face, mating surface

Paßlehre setting gauge
Paßloch location hole
Paßmarke locating mark, register mark
Paßpaar matched pair
Paßpunkt locating point
Paßstift alignment pin
Paßstück match, fitting, gadget, adapter
Paßteile fitments
passen in to fit into
passen zu to check with, to be a match for
passen zwischen to pass between
passendes Stück match
passiv passive (54), (621.3), (669); unreactive (54)
passive Electrode collecting electrode (621.35)
passive Relaisstelle passive repeater
passive Reserve standby equipment
passive Schaltung passive electric network
passiver Strahler passive reflector, passive aerial
passiver Vierpol passive quadripole, passive transducer
passiver Zustand passive state (54)
passiver Zweipol two-terminal impedance (621.3)
Passivierungsschicht passivation layer
Passung fit, seat(ing)
Passungsklasse class of fit
Passungssystem (classes of) fits (pl.)
Passungssystem-Einheitswelle shaft-basis system (of fits)
Paste paste
Pastekathode pasted filament (621.385)
Pasteplatte pasted plate (621.356)
Pasteplatte-Akkumulator Faure accumulator
pastieren to paste (621.356), (621.385)
pastierte Kathode coated cathode
pastierte Platte pasted plate (621.356)
Patent patent
Patent erteilen to issue a patent
Patent nehmen to take out a patent
Patent umgehen to circumvent a patent
Patent verletzen to infringe a patent
Patentamt patent office
Patentanmeldung patent application
Patentanspruch claim, patent-right claim
Patentbeschreibung patent specification
Patentbrief letters patent
Patentinhaber patentee
Patentschlüssel monkey, adjustable wrench
Patentschrift letters patent
Patentumgehung patent evasion
Patentverletzung patent infringement
Paternoster rotary lift (621.3), (621.876)
Patrone cartridge
Patronensicherung cartridge (-type) fuse
Pauling-Stickstoffofen Pauling furnace
pauschal zusammenfassen to lump
Pauschale flat rate
Pauschalbetrag lump sum
Pauschtarif flat-rate tariff
Pause tracing; print, blueprint (62); space (621.394); interval, break
Pausen-Impulsverhältnis break-make ratio, mark-space ratio
Pausencodierung gap coding
Pausenzeichen identification signal
Pausleinwand tracing cloth
Pauspapier tracing paper
pausen to trace; to print, to blueprint
PD-Regler proportional plus derivate action controller
Pech pitch (665.5)
Pechblende pitch blende
Pedal pedal, (foot) treadle
Pedal betätigen, Pedal treten to press a pedal
Pedalschalter foot switch
Pegel level (621.395), (621.396), (628); power level (621.395); tide gauge (626/627); volume
Pegelbildgerät level tracer
Pegeldiagramm level-recorder chart, hypsogram
Pegelhaltung clamping (621.397)
Pegelmeßgerät level meter, level indicator, transmission measuring set
Pegelregler level controller (621.395), (621.6)
Pegelregler mit Hilfsader pilot-wire regulator
Pegelschreiber recording transmission measuring set, level recorder
Pegelsender signal generator, level generator
Pegelsteuerung level control
Pegelumsetzung level conversion
Pegelzeiger volume indicator, level indicator (621.395)
Peil-... direction finding (D. F.)
Peilantenne direction finding aerial
Peilausrüstung, Peilgerät D. F. equipment
Peilauswertung bearing evaluation
Peilempfänger directional receiver, D. F. receiver
Peilergebnis bearing
Peilerkorrekturkurve bearing correction
Peilfehler durch Aufstellungsort bedingt site error
Peilfehler durch Großkreisausbrechung lateral deviation error

Peilfehler durch Mehrwegeausbreitung wave-interference error
Peilhilfe direction-finding aid
Peilstandlinie position line
Peilstation D. F. station
Peilstrahl (radio) bearing beam, direction-finding beam
Peilwinkel bearing angle, radio bearing
Peilwinkelfehler error in bearing
peilen to take bearings; to sound
Peiler direction finder, DF
Peilung D. F. = direction finding; bearing
Peitschenantenne whip aerial
Peitschenmast upsweep columm, davit-arm pole
Peltier-Effekt Peltier effect
Peltier-Kühlblock Peltier cooling block
Peltonrad Pelton runner, Pelton wheel, impulse-water turbine
Pelzbildung furring
Pendel pendulum (531); pendant (621.32)
Pendelausschlag swing of the pendulum
Pendelfrequenz quenching frequency (621.396.61)
Pendelgenerator cradle dynamometer (621.317); governor generator (621.1)
Pendelkugellager self-aligning ball-bearing
Pendelleuchte pendant (lighting fitting)
Pendellinse bob (681)
Pendelmühle pendulum mill, centrifugal grinder
Pendeloszillator self-quenching oscillator, squegging oscillation, squegger
Pendelrückkopplung supperregeneration
Pendelrückkopplungsempfang super-retroactive reception
Pendelrückkopplungsschaltung superregenerative circuit, flewelling circuit
Pendelschnur pendant cord (631.32)
Pendelschwingung swing of the pendulum; quenching oscillation (621.396.617)
Pendelverkehr shuttle traffic
Pendelzähler oscillating meter (621.317)
Pendelzug shuttle train (625.1)
pendeln to swing, to oscillate, to hunt (621.311), (621.316) to pump (621.313.3)
Pendelung phase swinging (621.311); hunting, cycling (621.316.7)
Pentagon pentagon
Pentagramm pentagram
Pentan pentane
Pentode pentode, quintode, triple-grid tube
perforieren to perforate
Perforierung, Perforation (pin-hole) perforation
Pergament parchment
Per. = Periode c. = cycle
Periode period; cycle (51), (53), (621.3)
Periode durchlaufen to traverse a cycle, to perform a cycle, to go through a cycle
Periode der Eigenschwingung natural period of vibration (531); natural period of self-oscillation (621.37)
Periode des Geschehens cycle of events
Periode der Grundwelle primitive period
Perioden pro Sekunde cycles per second = c. p. s.
Periodendauer (T) periodic time (T), period
Periodendauer der Punktfolge dot-cycle (621.394)
Periodendauer der Temperaturwelligkeit period of temperature cycling
Periodensystem periodic table (54)
Periodenzahl number of cycles, periodicity
Periodenzähler cycle-rate counter
periodisch periodic(al), cyclic(al), repetitive
periodisch verlaufen to cycle
periodisch wiederkehrend mit ... cyclic with ...
periodische Dämpfung periodic damping
periodische EMK periodic e. m. f.
periodische Funktion periodic function
periodische Größe periodic quantity
periodische Potentialdifferenz, periodische Spannung periodic potential difference
periodische Spitzensperrspannung repetitive peak reverse voltage, repetitive peak off-state voltage (621.382)
periodischer Spitzenstrom repetitive peak forward current
periodische Spitzenverlustleistung repetitive peak reverse power dissipation
periodischer Strom periodic current
periodische Überprüfung periodic test (621.317)
periodische Veränderung cyclic variation
periodische Welle periodic wave
Periodizität periodicity
peripher peripheric, peripheral
periphere Einheit peripheral unit
Peripherie periphery
Peripheriegerät peripheral device
Periskop periscope
Perle pearl; bead
Perlfeuer brush sparking, slight sparking
Perlmutter mother-of-pearl
Perlenprüfung bead test (54)
Perlensäureprüfer Hick's hydrometer (621.355)
Perlschnurblitz bead lightning
perlen to effervesce

Perlit pearlite
Perlon perlon
Permalloy permalloy
permanent permanent
permanent-dynamischer Lautsprecher p. m. speaker = permanent-magnet speaker
permanenter Magnet, Permanentmagnet p. m. = permanent magnet
permanenter Magnetismus permanent magnetism
permanentes magnetisches Feld permanent magnetic field
permanenter Speicher permanent memory
Permanenz permanency
permeabel permeable
Permeabilität permeability (538)
Permeabilität des leeren Raumes space permeability
Permeabilitätsabnahme durch Alterung time-decrease of permeability
Permeabilitätsabstimmung permeability tuning
Perminvar perminvar
permissiver Block permissive block
Permutation permutation
Permutit permutite
Perpetuum mobile perpetual motion machine
Personal Computer Personal Computer
Personenaufzug passenger lift
Personensuchanlage selective paging system, staff-locator system
Personenzug passenger train
Perspektive perspective
perspektivisch verkürzt foreshortened
perspektivische Verkürzung foreshortening
Pertinax pertinax
Petersen-Spule Petersen coil
Petrographie petrography
Petroleum petroleum, kerosene
Pfahl pole, post; pile
Pfannkuch-Kabelschutz fault-signalling network
Pfannkuch-Schutz Bowden-Thomson protective system
Pfeif-... whistle, whistling
Pfeifeinsatzpegel singing point
Pfeiffrequenz singing-point frequency
Pfeifton whistle, whistling note; beep tone (621.39)
Pfeifpunkt singing point
Pfeifpunktabstand singing margin
Pfeifsicherheit stability
Pfeife whistle, pipe (681)
pfeifen to whistle, to squeal, to howl (621.39)
Pfeil arrow, dart
Pfeilrad, pfeilverzahntes Rad double-helical gear
Pfeilzähne herringbone teeth (621.8)
Pferdestärke horse power (s. Anh.)
Pflege maintenance, tendance (621)
pflegen to maintain, to tend (621)
Pflichtenheft standard specifications, standard requirements
Pflichttempo obligatory keying speed (621.394)
Pflock peg, bung
Pfropfen wad, slug
Pfund lb = libra; pound (s. Anh.)
pH-Wert pH-value, hydrogen ion concentration
P-Halbleiter P-type semiconductor
Phantastron phantastron
Phantom phantom (621.396.67)
Phantomantenne phantom aerial, dummy aerial
Phantombetrieb ghost working (621.394)
Phantomkreis phantom circuit
Phantomkreisübertrager phantom repeating coil
Phantomleitung phantom circuit
Phantom-Oder wired or (621.374)
Phase phase
Phase drehen, Phase schieben to shift the phase
aus der Phase bringen to outphase
in Phase in phase, in step
in Phase bringen to bring into phase
in der Phase einstellen to phase
in der Phase verschoben sein to be off(set in) phase
Phasenabstand phase spacing
Phasenabweichung phase difference, phase displacement
Phasenanschnittsteuerung phase control, phase-shift control
Phasenausgleich power-factor correction
Phasenausgleichkette delay equalizer
Phasenbelag phase constant (621.392)
Phasenbeziehung phase relation, phase relationship
in Phasenbeziehung setzen to phase
Phasenbruch phase failure
Phasenbruchrelais phase-failure protection
Phasenbruchschutz open-phase protection
Phasendifferenz phase difference
Phasendifferenzmodulation differential phase-shift keying
Phasendreher phase shifter, phase-shifting transformer
Phaseneinstellung phasing, phase timing
Phasenentzerrer phase equalizer

Phasenfehler phase angle; phase error (621.317)
Phasenfläche equiphase surface, surface of constant phase (535), (621.396)
Phasenfokussierung bunching (621.385)
Phasenfolge phase sequence, succession of phases
  Phasenfolgerelais phase-rotation relay, phase-sequence relay
  Phasenfolgewinkel (der symmetrischen Komponente) characteristic angular phase difference
  Phasenfolgezeiger phase sequence indicator
Phasengang phase response
Phasengeschwindigkeit phase velocity, wave velocity, speed of phase (621.392)
phasengesteuerter Gleichrichter phase-controlled rectifier
phasengleich coincident in phase
Phasengleichheit phase balance
Phasengleichrichter phase selective rectifier; phase discriminator
Phasenhub modulation index, phase deviation, swing
phasenkompensierter Asynchronmotor allwatt motor
Phasenkonstante phase coefficient
Phasenkontrast phase contrast (535)
Phasenlage phase position, phasing, phase relationship
Phasenlampe phase lamp
Phasenlaufzeit phase-delay time, phase delay
Phasenleiter outer conductor
Phasenmaß phase constant
  Phasenmaß bei stoßfreiem Abschluß image phase constant
Phasenmesser phase meter, power-factor indicator
Phasennacheilung phase lag, lagging phase angle
Phasenmodulation phase modulation
  phasenmoduliert phase modulated
Phasenopposition phase opposition
Phasenregelschleife phase-locked loop
phasenreiner Widerstand unity power-factor impedance, ohmic resistance
Phasenrelais phase balance relay
Phasenresonanz parallel (phase) resonance
Phasenschieben power-factor correction; phase shifting
Phasenschieber phase shifter, phase advancer, phase turner, phase modifier
  Phasenschieber-Hintermaschine slip-ring phase advancer
  Phasenschieberkondensator power-factor capacitor
  Phasenschiebermaschine rotary phase converter
  Phasenschieberschaltung phasing arrangement
  Phasenschiebertransformator phase-shifting transformer
Phasenschwankung angular variation
Phasenspalter phase splitter, phase splitting device
Phasenspannung phase voltage, star voltage, "X" voltage, voltage to neutral
Phasensicherheit phase margin (621.375)
phasenstarr phase locked
  phasenstarrer Verstärker lock-in amplifier
Phasenstellglied phase shifter
Phasenstoß phase (section) break
Phasenstrom phase current
Phasentransformator phase transformer
Phasentransformierung phase transformation
Phasentrennung phase break
Phasenumkehr phase inversion, phase reversal
  Phasenumkehrröhre phase inverter (tube)
  Phasenumkehrschaltung phase inverter (circuit)
  Phasenumkehrverstärker paraphase amplifier
Phasenumtastung phase-shift keying
Phasenumwandlung phase transformation
Phasenunsymmetrie phase unbalance
Phasenunterbrechungsrelais open-phase relay
Phasenunterschied phase difference, difference in phase
Phasenvergleich phase comparison
  Phasenvergleichschaltung phase-comparison circuit
Phasenverschiebung phase shift, phase displacement
  Phasenverschiebung zwischen korrespondierenden Klemmen eines Transformators angular displacement of a polyphase transformer
  Phasenverschiebungswinkel phase-displacement angle
  phasenverschoben sein to be out of phase, to be offset in phase
Phasenverzerrung phase (frequency) distortion, phase delay distortion
Phasenvoltmeter phase voltmeter
Phasenvoreilung leading phase angle, phase lead
Phasenwicklung phase (winding)
Phasenwinkel phase angle

Phasenwinkel Null unity power factor
Phasenzahl number of phases, phase number
Phasenzeiger phase indicator
Phasenzeit phase timing
Phenol phenol, carbolic acid
mit Phenol behandeln to carbolize
Phenol-Formaldehyd-Kunstharz phenol formaldehyde
Phenolphthalein phenolphthalein
phi [Leistungsfaktor] phi
Phon phon
Phonograph phonograph
Phosphor phosphor (53); phosphorus (54)
Phosphorbronze phosphor bronze
Phosphorsäure phosphoric acid
Phosphoreszenz phosphorescence, afterglow
phosphoreszieren to phosphoresce, to phosphorate
phosphoreszierend phosphorescent
Photo-. . ., photo-. . . photo
Photoapparat camera
Photoblitzgerät foto flashlight
Photodiode photodiode
Photoeffekt photoelectric effect
photoelektrisch photoelectric
photoelektrische Ausbeute photoelectric yield
photoelektrischer Effekt photoelectric effect
photoelektrische Emission photoelectric emission
photoelektrischer Kompensator photoelectric compensator, photoelectric balance selenium magnifier
photoelektrischer Leser photoelectric reader
photoelektrisches Relais photoelectric relay
Photoelektrizität photoelectricity
Photoelektronen photo electrons
Photoelektronenvervielfacher photomultiplier
Photoelement photovoltaic cell, photronic cell
Photoemission photo-emissivity
photogeätzt photo etched
Photokathode photosensitive cathode, photo(-electric) cathode
Photokopie photographic copy, photoprint, photostat
Photokopiereinrichtung photoprinting machinery
photokopieren to photostat
Photolampe photolamp
Photoleitfähigkeit photoconductivity
Photolitografie photolitography
Photoluminiszenz photo-luminescence
Photometer photometer, light meter
Photometerschirm photometer screen
Photometrie photometry
photometrisch photometric
Photospannung photoelectric voltage
Photosynthese photosynthesis
Phototransistor phototransistor
Photovervielfacher photomultiplier, photomultiplier tube
Photowiderstand photoresistor, photoconductive resistor, photoconductive cell
Photozelle photocell, photoemissive cell, lightsensitive cell
Photon photon, corpuscle of radiation
Physik physics
physikalisch physical
physikalische Chemie physical chemistry
physikalische Eigenschaft physical property, physical characteristic
physikalische Größe physical magnitude, physical quantity
physikalische Konstante physical constant
physikalische Mischung mechanical mixture
physikalischer Zustand physical state
Physiker physicist
physiologisch physiological
physiologische UV-Strahlung vita rays (pl.)
physiologische Wirkung physiological effect
Pickup-Spule pick-up coil
PI-Regler proportional plus integral controller
PID-Regler proportional plus integral plus derivative controller
piepsen to squeak
piezoelektrisch piezoelectric
piezoelektrischer Effekt piezoelectric effect
piezoelektrischer Indikator piezoelectric indicator
piezoelektrischer Kristall piezoelectric crystal
piezoelektrischer Lautsprecher piezoelectric (loud)speaker
piezoelektrisches Mikrophon crystal microphone
piezoelektrisches Relais piezoelectric relay, crystal relay
piezoelektrischer Tonabnehmer piezo-electric pick-up, crystal pick-up
piezoelektrischer Schwinger piezoelectric oscillator, crystal oscillator
Piezoelektrizität piezoelectricity
Piezo-Keramik piezoelectric ceramics
Piezoquarz piezoelectric crystal
Pilgerschrittschweißung step-back welding
Pilot pilot
Pilotänderung carrier compression

Pilotfrequenz pilot carrier, pilot frequency
Pilottechnik pilot-frequency technique
Pilotton pilot, reference pilot
Pilotüberwachung pilot control
Pilz fungus (pl.: -gi); mushroom
pilzförmig mushroom-shaped
pilzverhindernd, pilztötend fungicide
Pimpelluft armature clearance (621.318.5)
pinkompatibel pin compatible
Pinsel brush
Pinzette (pair of) tweezers, pincette
Pipette pipette
Pips blip (621), (621.397)
PI-Regler proportional plus integral controller
Pitotrohr Pitot tube, Pitot static
P-Kanal-Feldeffekttransistor P-channel field-effect transistor
P-Kanal MOSFET vom Anreicherungstyp P-channel enhancement-mode MOSFET
Plan plan, schedule, scheme; drawing, design
Plandrehen facing (621.94)
Planfräsen (flat-) surface milling
planparallel plane-parallel, parallel-faced
planparallele Glasplatte parallel-faced glass slab, optical flat
Planscheibe face plate (621.94)
Planzeit schedule time
Planartechnik planar technique
Planartransistor planar transistor
Plancksches Strahlungsgesetz Planck radiation law
planen to plan, to scheme, to schedule
Planet planet
Planetengetriebe planetary gearing, epicyclic gearing
planieren to level
Planimeter planimeter
Planté-Akkumulator Planté accumulator
Planung planning, layout
Plasma plasm, plasma; arc column, arc stream (537)
Plasmaspannung arc-stream voltage (537.5)
Plasmastrahl plasma beam, plasma jet
plastifizieren, plastizieren to plasticize; to masticate
Plastikeffekt plastic effect (621.397)
plastisch plastic
plastisches Hören auditory perspective, 3 D-hearing
plastisches Material plastic material, plastic matter
Plastizität plasticity
Platin platinum
Platincyanür platinocyanide
Platindraht platinum wire
Platin-Iridium platinum iridium
Platinmohr platinum black, platinum sponge
Platinrhodium platinum rhodium
Platinschwamm platinum sponge
Platinschwarz platinum black
Platinsilber platinum silver
Platine blank (621.9); (circuit) board (621.39), (681.3)
Platinenentwurf circuit board artwork
Platte slab (624): sheet; plate (77), (621.355/6); vane (621.319.4); record, disc (681.84); disc, disk (681.3)
Platte formieren to form a plate (621.355)
zwischen zwei Platten legen to sandwich
Plattenbelag slab covering
Plattenbetriebssystem disk operating system
Plattendatei disk file
Plattenelektrometer attracted disc electrometer
Plattenerder earth plate
Plattenfläche plate area (621.356)
Plattengeräusch record noise
Plattenkondensator plate capacitor, disc capacitor (621.319.4)
Plattenlaufwerk disk drive
Plattensatz set of plates, system of plates (621.319.4), (621.356)
Plattenschutz flange protection (612.3)
Plattenspeicher disk memory, disk storage
Plattenspieler record player, phonograph, disk reproducing equipment, gramophone (681.84)
Plattenspielerlaufwerk disc drive
Plattenspielermotor gramophone motor
Plattenteller (record) turntable
plätten to iron
Plattform platform
plattieren to plate, to clad
plattierte Kohle plated carbon
plattiertes Metall clad metal
plattierter Stahl clad steel
Plattierung plating
Platzeinflugzeichen inner (boundary) marker
Platzgrenze-Einflugzeichen inner marker beacon
Platzverbindungsleitung interposition trunk
P-Leitung P-type conduction
PLL-Schaltkreis phase-locked loop
Plombe (leaden) seal
Plombendraht safety wire (621.317)
plombieren to seal
Plotter plotter
plötzlich sudden, abrupt
plötzlich abkühlen to quench
plötzliche Änderung abrupt change

plötzlicher Ausschlagsrückgang dip (621.317)
plötzlicher Durchschlag disruptive discharge
plötzlicher Stoß flash
plötzliche Temperaturänderung thermal shock
Plumbikon plumbicon
Plus... Plus..., positive
Plus-Drei-Code excess-three code
Plusklemme, Pluspol plus pole, positive element (621.3)
Plusleiter positive main
Plusleitung plus wire
Plusseite plus side
Pluszeichen plus sign, positive sign, sign of addition
Plutoniumbrenner plutonium pile, plutonium reactor
PN-Grenze PN boundary
PN-Schichtgleichrichter PN junction diode
PN-Übergang PN junction
pneumatisch pneumatic (s. auch **Druckluft...**)
pneumatische Lehre pneumatic micrometer (621.9)
pneumatische Presse pneumatic press
pneumatischer Regler pneumatic controller
pneumatisches Stellglied pneumatic operator
PNP-Flächentransistor PNP junction transistor
Poissonsche Gleichung Poisson's equation
Poissonsche Magnetisierungstheorie Poisson's theory of magnetization
Poissonsche Zahl Poisson ratio
Pol pole (51), (91), (538), (621.3); field pole, pole piece (621.313)
Pole field system (sing.) (621.313)
Pol eines Strahlenbüschels point of concurrency
Polabgleichblech pole shim
Polbogen pole arc, polar arc (621.313)
Polende pole tip
Polfläche pole face (538)
Polgestell carcase, (pole) frame (621.313)
Polkern pole shank
Polklemme pole (terminal) (621.3)
Polkreuz armature spider
polnahe circumpolar
Polpaar pair of poles
Polpaarzahl number of pairs of poles
Polprüfer pole tester, polarity indicator
Polprüfpapier polarity test paper, pole-finding paper
Polrad (pole) spider, magnet wheel (621.313)
Polschuh pole shoe, pole piece, pole tip
Polspitze pole tip, pole horn
Polstärke pole strength, quantity of magnetism
Polstern pole spider, field spider, armature spider, rotating field magnet
Polteilung pole pitch (621.313)
polumschaltbar pole-changeable
polumschaltbarer Generator multifrequency generator
polumschaltbarer Motor pole-change motor, change-pole motor
polumschaltbare Wicklung consequent pole winding
Polumschalter, Polwechsler pole-changing switch
Polwender reversing switch, reversing key
Polzahl number of poles
Polar-... polar
Polaranzeige polar indication
Polarbild I-scan (621.397)
Polardiagramm polar diagram
Polargleichung polar equation
Polarkoordinaten polar co-ordinates
Polar(koordinaten)papier polar paper
Polarkoordinatenwandler angle component solver
polar polar
polare Abtastung I-scan (621.397)
polare Bindung polar link (54)
polares Trägheitsmoment polar moment of inertia
Polarisation polarization
Polarisationsapparat saccharimeter
Polarisationsebene plane of polarization
Polarisationsfehler polarization error (621.396)
Polarisationsfilter polaroid screen
Polarisationseinrichtung direction of polarization
Polarisationsentkopplung depolarization loss
Polarisator polarizer
polarisieren to polarize
polarisiertes Amperemeter polarized ammeter
polarisierter Anker polarized armature (621.318)
polarisiertes Dreheiseninstrument polarized moving-iron instrument
polarisiertes Licht polarized light
polarisiertes Relais polarized relay
polarisiertes Richtungsrelais polarity directional relay
polarisierte Streustromableitung polarized electric drainage
polarisierter Ticker polarized sounder
polarisierter Wecker biased bell
polarisierte Wellen polarized waves
Polarität polarity

Polarität der Klemmen lead polarity
Polaritätsprüfung polarity test
Polaritätsumschlag der Thermospannung thermo-electric inversion
Polarlicht aurora
polieren to polish, to shine, to furbish
Polierfilz rotary felt
Polierscheibe polishing wheel
polig p = pole
n-poliger Schalter n-p switch
Polizei police
Polizeifunk police radio
Polonium polonium
polstern to cushion; to upholster
gepolsterter Einbau cushion mounting
Polyäthylen polyethylene
polychromatisch polychromatic
Polyestergießharz polyester cast resin
Polygon polygon
Polygonschaltung polygonal connection, mesh connection (621.3)
Polymer, Polymeres polymer
Polymerisation polymerization
polymerisieren to polymerize
Polynom polynom
Polynomrechner equation solver
polynom(ial) polynomial
Polystrol polystrol, polystyrene
Polytechnikum polytechnic (school)
Polyvinylchlorid polyvinyle chloride
Pond gram (me) [as a unit of force]
PPI-Anzeigegerät plan-position indicator
Pore pore
porös porous
Portal portal (72)
Portalgerüst gantry
Portalbauweise portal design
Portion portion
portionsweise abteilen to portion
Porzellan porcelain (621.3); china (64)
Porzellananschlußklemme rosette (696)
Porzellaneinsatz porcelain interior (621.315.6)
Porzellanisolator porcelain insulator
porzellanisoliert porcelain insulated
Porzellankörper porcelain former
Porzellanteller porcelain cup (621.315.6)
Porzellanwandler porcelain-insulated transformer
Posaune variable length line, variable length section, telescoping section, line stretcher (621.396); trombone (681.8)
Position position
Positionierungssteuerung positioning control
Positionsfehler positional error
Positionsfeststellung position fix
Positionslampe, Positionslicht position light, wing-tip light (629.13)
Positionsmessung position fix
positiv positive (going), +ive, plus (621.3)
positiv aufgeladen positively charged
positiv gegen(über) positive relative to
positiv geladen deficient in electrons
positiv gerichtet positive going
positiv reagieren to respond positively
positiv werden to go positive
Positiv-Amplitudenmodulation positive amplitude modulation
positive Bürste positive brush (621.313)
positive Elektrizität plus electricity, vitreous electricity
positive Größe positive quantity (51)
positive Klemme positive terminal
positive Kohle positive carbon
positive Ladung positive charge (621.3)
positive Leitung positive wire, positive lead
positive Platte positive (621.356)
positiver Pol positive pole, positive element (621.35); anode
positive Richtung forward direction (537)
positive Rückführung positive feedback, regenerative feedback (621.316.7)
positive Rückmeldung acknowledge (681.3)
positive Säule positive column (621.385), arc stream (537.5)
positive Speiseleitung positive feeder
positive Sperrspannung positive off-state voltage (621.314.63)
positiver Strom forward current
positives Vorzeichen positive sign
positive Zahl positive number
Positive positive (plate) (621.355)
Positivempfang positive reception
Positivform positive matrix (621.357)
Positron positron
Potential potential
auf ein Potential bringen to carry to a potential
auf einem Potential halten to maintain at a potential
im Potential nicht definiert sein to float
potentialbezogen referred to a potential
Potentialausgleich potential equalization
Potentialdifferenz p. d. = potential difference
Potentialfeld potential field
Potentialfunktion potential function
Potentialgefälle, Potentialgradient potential gradient (551.5), (621.3)
Potentialklemme calibration terminal

Potentialschwelle minimum potential (621.385); potential barrier
Potentialsprung zwischen zwei Elektrolyten liquid-junction potential
potentiell potential
potentielle Energie potential energy
Potentiometer potentiometer, voltage dividing resistor
Potentiometerabgriff potentiometer pick-up
Potentiometerarm moving arm of the potentiometer
potentiometrisch potentiometric
potentiometrische Messung potentiometric measurement
potentiometrischer Meßwertwandler potentiometric transducer
Potenz power (51)
Potenzgesetz power law
Potenzreihe power series
Potenzschreibweise index notation
potenzieren to power, to raise to a power
Potenzierung involution, raising to a power
Potierdreieck Potier diagram
Pottasche potash
Poulsen(licht)bogen Poulsen arc
Poulsensender arc transmitter
Poyntingscher Vektor Poynting's vector
prägen to emboss, to impress (621)
Praktiker practitioner
praktisch practical
praktisch anwenden to practise
praktisch ausführen to put into practice
praktisch linear substantially linear
praktisch möglich practicable
praktische Anwendung practical application
praktische Ausführbarkeit practicability
praktische elektrische Einheiten practical electrical units
praktische Erfahrung practical experience
Prallanode dynode
prasseln to sizzle
prasselndes Geräusch sizzling noise
Praxis practice
für die Praxis for practical purposes (pl.)
in der Praxis in the field
in die Praxis umsetzen to realize, to put into practice
Präzision precision
Präzisionsanflugradar PAR, precision approach radar
Präzisionsguß precision casting
Präzisionsinstrument precision(-type) instrument
Präzisionswellenmesser precision wavemeter
Präzisionswerkzeug precision tool
Preemphasis pre-emphasis, accentuation
P-Regelung proportional control
prellen to pounce, to chatter
prellfreie Kontakte non-bounce contacts
Prellung (Kontakt) pounce
Preß-... compressed; moulding
Preßarbeit press work, press job (621.97)
Preßbär ram
Preßdruck moulding pressure
Preßform mould (621)
Preßgas compressed gas
Preßgaskondensator compressed-gas condenser
Preßgasschalter gas blast switch
Preßglas press-glass
Preßglasteller press-glass base (621.385)
Preßluft compressed air (s. auch **Druckluft**)
Preßluftanlasser air starter (621.436)
Preßluftarmatur compressed air fitting
Preßluftkondensator compressed-air condenser
Preßluftmotor pneumatic motor
Preßluftschalter air-blast circuit breaker
Preßluftschlauch air hose
Preßluftstrom (gas) blast
mit Preßluftantrieb air-driven
Preßmasse moulding compound
Preßpapier pressed paper
Preßpappe pressboard
Preßschweißung pressure welding
Preßsitz press fit, force fit (621.9)
Preßspan pressboard, fuller board
Preßstempel press die, squeeze head, punch, platen (621.9)
Preßstoff plastic, plastic compound, plastic material, laminate
Preßstoffgehäuse moulded bakelite jacket
Preßstoffisolator moulded insulator
Preßstoffteile plastic mouldings
Preßstück press work
Preßtechnik moulding technique (67)
Preßverbinder compression fitting
Preßwasser-Reaktor pressurized-water reactor
Preßzeit moulding time
Presse press (621.97), (655)
Pressefax pressfax, pagefax
pressen to press, to mould, to stamp (621.9)
Presser compressor (621.395)
Presserei press shop
Primär-... primary
Primäranschluß primary terminal
Primärausfall primary failure

Primärbatterie primary battery, galvanic battery
Primärelektron primary electron (621.385)
Primärelement primary cell (621.35)
Primärfarbe non physical primary (621.397)
Primär-Farbauszug primary colour picture
Primärgruppe group (621.395.4)
Primärgruppendurchschaltefilter through-group filter
Primärgruppenumsetzer group modulator
Primärgruppenverbindung group link
Primärklemme primary terminal
Primärleistung input (power)
Primärprogramm source program (681.3)
Primärrelais main-current relay
Primärspannung primary voltage
Primärspule primary coil
Primärstrahler primary radiator, driven aerial, active aerial, exciter
Primärspannung primary voltage, input voltage
Primärstrom primary current
Primärvalenzen reference stimuli
Primärwicklung primary (winding) (621.314.2)
primär primary
primäre Nennspannung rated primary voltage
primärer Nennstrom rated primary current
primäre Transformatorwicklung transformer primary (winding)
Primzahl prime number
Printometer printometer, demand printer (621.317)
Prisma prism (51), (53)
prismatisch prismatic
Prisonstift locking pin, dowel pin
Privattelefonanlage private exchange
privater Nebenanschluß private extension (621.395)
Probe proof, test, trial; sample, specimen; check (51)
Proben entnehmen to sample, to take samples, to draw samples
Probebelastung proof load, test loading
Probebetrieb experimental run
Probelauf trial run
Probestück sample, sampling, test piece, test specimen (620)
probieren to try, to test
Probiermethode cut-and-try process
Produkt product, produce
Produktion manufacturing, production; output
Produktionskapazität manufacturing capacity
Produktionskosten production costs
Produktionsreaktor production reactor
produktiv productive
produktive Zeit productive time
Produktivität productivity
Profil profile; outline, (structural) section (624)
Profilbeleuchtung outline lighting
Profildraht profiled wire
Profileisen section iron, structural section, sections (pl.)
Profilfahrdraht grooved (trolley) wire, figure eight wire (621.33)
Profilfräser form cutter
Profilinstrument edgewise instrument
Profilrahmen maximum structure gauge (625.2)
Profilstab sectional bar
Profilstahl section (al) steel, steel section
Profiltiefe chord (533)
Programm program, routine
Programmablauf computer operation, routine, program flow (681.3)
Programmablaufplan program flowchart
Programmabschnitt segment (681.3)
Programmabsturz crash
Programmbaustein program unit
Programmbibliothek program library
Programmfortschaltung program advance
Programmkarte program card
Programmlauf program run
Programmpflege maintenance of programs
Programmprüfung program checking
Programmregelung program control, process-cycle control, time-pattern control
Programmschalter program switch, sequence switch; channel switch (691.397)
Programmschleife loop
Programmschritt program step
Programmschrittzähler program counter
Programmstecker plug-in program, coded plug
Programmsteuerung program control, automatic sequencing
Programmstopp breakpoint (681.3)
Programmumsetzer assembler
Programmverzweigung program jump
Programmzähler program counter
Programmzustandswort program status word
programmieren to program
programmierbarer Festspeicher programmable read only memory, PROM
programmierter Unterricht programmed instruction, programmed learning
Programmierer programmer
Programmierfehler program fault
Programmiersprache programming language

Programmierung programming
Programmierung in Maschinensprache computer-oriented language programming, machine language programming
Projekt project, plan, scheme
projektieren to project, to plan, to scheme
Projektion projection (51), (535)
Projektionsapparat projection lantern, projector
Projektionsfläche projected surface
Projektionsinstrument projected-scale instrument (621.317)
Projektionslampe projector lamp, projection lantern
Projektionsschirm, Projektionswand screen (778)
projizieren to project (51), to screen (535), (77)
Pronyscher Zaum Prony brake, brake dynamometer
Propan propane
Propeller propeller, airscrew
Propellerblatt vane
problemorientierte Programmiersprache problem-oriented language
Proportion proportion
proportional proportional
proportionale Skala scale of equal parts
Proportional-Differential-Regelung = PD-Regelung proportional plus derivative action control, rate action
proportional-differenzierende Regelung rate action
Proportional-Fernmeßgerät direct-relation telemeter
Proportional-Integral-Differential-Regelung = PID-Regelung proportional plus integral plus derivative action control
Proportional-Integral Regelung = PI-Regelung proportional plus integral control
Proportionalzähler proportional counter (539)
Proportionalität proportionality
Proportionalitätsfaktor factor of proportionality
Proportionalitätsgrenze limit of proportionality
proportionieren to proportion
Prospekt descriptive booklet, brochure, leaflet
Prospektmaterial advertising literature
Protodurisolation Protodur insulation
Protoflexrohr Protoflex tube
Protokoll minutes (pl.), record
Protokoll führen to keep the minutes, to minute; to log
zu Protokoll nehmen to record
Proton proton
Protonenstrahl ray of protons, beam of protons
Prototyp prototype
provisorisch provisional, temporary
provisorischer Aufbau temporary set-up
provisorische Installation provisional wiring, temporary wiring
provisorisch verlegte Leitung temporary wire
Prozent per cent
Prozentdifferentialrelais percentage differential relay
Prozentsatz percentage
100-prozentige Modulation complete modulation, total modulation
prozentische Abweichung fractional deviation
prozentual (expressed as) percentage
prozentuale Benutzungsdauer operation factor
prozentualer Drehzahlabfall percentage speed reduction
prozentualer Fehler percentage error, percentage of accuracy (621.317.8)
prozentuale Kräuselspannung percent ripple
prozentuale Zusammensetzung percentage composition
Prozeß... process
Prozeßautomation process automation
Prozeßleittechnik computer-process control
Prozeßrechner process(-control) computer
Prozeßsteuerung process control
Prozeßsteuersprache process-control language
Prozessor processor
Prozessorunterbrechungsmöglichkeit processor interrupt facility
Prüf-... monitoring, testing, proving, checking
Prüfabgriff probe
Prüfader C-wire
Prüfamt standardizing laboratory, testing laboratory
Prüfanschluß probe
Prüfanstalt testing station
Prüfaufbau bench test set up
Prüfbelastung test load, test loading
Prüfbericht test report
Prüfbit check bit
Prüfbuchse test point, test socket
Prüfbürste exploring brush, pilot brush
Prüfdauer time rating, test time
Prüfdraht test cord, testing wire, C-wire, holding wire
Prüfdruck testing pressure
Prüfeinrichtung test equipment, test stand

Prüffeld test department, testing centre; test floor, proving ground
Prüffrequenz test frequency
Prüfgefäß test vessel
Prüfgenauigkeit testing accuracy
Prüfgerät test set, test unit, tester
Prüfgestell test rack (621.3)
Prüfhahn test cock, gauge, cock, control cock
Prüfklemme test terminal, calibrating terminal
Prüfklima test atmosphere
Prüfklinke test jack (621.395)
Prüflampe test lamp (621.32)
Prüflast test load
Prüfmuster (check) sample
Prüfobjekt test piece, specimen
Prüfplatz test desk (621.395); test board
Prüfprogramm check routine (681.3)
Prüfprotokoll test log
Prüfraum testing room, control room
Prüfreaktor (materials) testing reactor
Prüfring testing bush (621.395)
Prüfsatz testing set, testing unit
Prüfschalter feeler switch (621.315.1); test switch
Prüfschein test certificate
Prüfsender test oscillator, signal generator (621.396)
Prüfsignalgeber test-signal generator
Prüfsonde (experimental) probe
Prüfspannung testing potential, proof test voltage (621.3); testing pressure (620)
Prüfspitze test-lead point
Prüfspule search coil, (magnetic) proof-plane (621.3); search winding
Prüfstand test bench, test stand
Prüfstecker test connector, test plug
Prüfstöpsel test plug (621.395)
„Prüfstrippe" test cord
Prüfstrom testing current, C-wire current
Prüftafel test box (621.316); Board of Trade panel (656.25)
Prüftisch test desk
Prüfton test to ne
Prüftransformator testing transformer
Prüfverfahren test method
Prüfzeichen check digit (681.3)
Prüfzeile test line (621.397)
Prüfzeilenmeßsignal vertical-interval test-signal
Prüfzeilenreferenzsignal vertical-interval reference-signal
prüfen to test, to probe, to inspect
prüfen auf to test for, to check for
Prüfung test, testing, check, inspection
Prüfung aushalten to stand a test
Prüfung auf Durchgang continuity test
Prüfung auf Isolation insulation check
Prüfung auf Lagerfähigkeit shelf test (621.353)
Prüfung auf Schluß closure check, leakage test
Prüfung auf Windungsschluß short-circuited turns test
einer Prüfung unterziehen to put to test
Prüfungsergebnis test result
pruzzeln to fizzle, to fry
Pseudosättigung (der Pentode) plate-current saturation
Pseudo-Winkel Brewster angle
pseudostabil pseudo-stable
Psophometerkurve psophometric curve
P-Typ-Halbleiter P-type semiconductor
P-Übergangsfunktion proportional response (621.316.7)
Puddeleisen puddled iron, wrought iron
Puder powder
Puffer buffer; cushion, knuckle
Pufferbatterie buffer battery, floating battery, booster battery
Pufferfeder buffer spring
Pufferglied buffer element
Pufferlösung buffer solution
Puffermittel buffers (pl.) (621.35)
Puffersalze buffer salts (621.35)
Pufferspeicher buffer (store), buffer storage
Pufferstempel plunger (625.2)
Pufferung float loading
puffern to buffer; to float-charge (621.355)
Puls pulse (train)
Pulsamplitude pulse amplitude
Pulsamplitudengang step-function response, indicial response
Pulsamplitudenmodulation pulse-amplitude modulation
Pulscodemodulation pulse-code modulation
Pulsdauermodulation pulse-width modulation, puls-duration modulation, pulse-length modulation
Pulsfolgefrequenz bit rate (681.14); pulse rate, pulse frequency (621.39)
Pulsform pulse shape
Pulsformer pulse shaper
Pulsformung pulse shaping
Pulsfrequenz pulse recurrence frequency, pulse-repetition frequency
Pulsfrequenzmodulation pulse-frequency modulation
Pulsfreies Rückstrahlverfahren continuous wave radar
Pulshöhenanalysator pulse-height analyzer

Pulskennlinie pulse characteristic
Pulskohärenzradar coherent-pulse radar
Pulslängenmodulation pulse-duration modulation
Pulsmikrophon sphymophone
Pulsmodulation pulse modulation
Pulsmodulationsradar pulse-modulated radar
Puls-Pausenverhältnis pulse-duty factor (621.39); high-low ratio (621.374)
Pulsperiodendauermodulation pulse-spacing modulation
Pulsphasenmodulation pulse-phase modulation, pulse-position modulation
Pulsregenerierung pulse regeneration
Pulsschreiber sphygmograph
Pulswärmewiderstand thermal inpedance under pulse conditions (621.382)
Pulszeitmodulation pulse-time modulation
Pulsation pulsation
pulsieren to pulsate
pulsierende EMK pulsation e. m. f.
pulsierendes Feld pulsating field
pulsierender Gleichstrom pulsating direct current
Pultdach shed roof
Pulver powder; dust
Pulveraufnahme ring-powder pattern (621.386)
Pulverkern dust core
Pulverkernstopfen iron-dust slug (621.396)
pulverisieren to pulverize, to powder
Pump-. . . pumping
Pumpspeicherwerk pumped-storage hydroelectric station
Pumpe pump
an der Pumpe liegender Gleichrichter maintained-vacuum rectifier
pumpenloser Gleichrichter pumpless rectifier, sealed rectifier
Pumpenmotor pump motor
pumpen to pump
gepumpt maintained vacuum
Pungsdrossel magnetic modulator
Punkt point (51); spot (51), (621.3), (621.79); dot (621.394)
Punkt (als Zeichen der Differentiation nach der Zeit) dot
Punkt für Punkt point by point
Punkt-zu-Punkt-Verbindung point-to-point-connection
durch einen Punkt gehen to pass through a point
Punktanstrahlung spot lighting
Punktbelastung point load (621.3), (624)
Punktberührung point contact
Punktfolge dot sequential (621.397)
Punktfrequenz dot frequency
punktförmig punctiform
punktförmig angefressen pitted
punktförmige Last single-point load (621.3), (624)
punktförmige Lichtquelle point source (of light)
Punktgitter point lattice
Punktkontakt point contact (621.382)
Punktlichtlampe spot lamp, pointolite lamp
punktlöten to spot solder
Punktmatrixdrucker dot matrix printer
Punktnahtschweißung progressive spot welding
Punktquelle point source (621.3)
Punktrasterbild half-tone picture
Punktschärferegler astigmatism control
Punktschreiber intermittent contact recorder, chopper bar recorder
punktschweißen to spot weld
Punktschweißmaschine spot welder, spot-welding machine
Punktschweißung spot welding
Punktsprungsverfahren dot interlace (621.397)
Punktsteuerung point-to-point control
punkten to spot weld
punktieren to dot; to puncture (621.3)
punktierte Linie dotted line
Punktierung needlepoint puncture, primary fault (621.3)
Pupille pupil
pupinisieren to load, to pupinize
pupinisiertes Kabel loaded cable, pupinized cable
pupinisierte Leitung loaded line
Pupinisierung series loading, coil loading
Pupinspule loading coil
Pupinspulenkasten loading-coil pot, loading-coil box
purpurfarben anlaufen to purple
Purpurlinie purple boundary (621.397)
Putzwolle cotton waste
PVC polyvinyl chloride
P-Verhalten proportional action
P-Verunreinigung P-type impurity (621.382)
Pyknometer pycnometer
Pyramide pyramid
Pyramiden-Horn pyramidal horn
Pyrheliometer pyrheliometer
Pyrit pyrite
pyroelektrischer Kristall pyroelectric crystal
Pyrometer pyrometer

pyrometrisch pyrometric
pyrometrische Messung pyrometric measurement
PZ-Regler proportional controller with disturbance variable feedforward system

Q

Q [Elektrizitätsmenge] Q
qm = Quadratmeter sq. m. = square metre
Q-Meter Q-meter
Quader(stein) rectangular block, square stone
Quadrant quadrant (51), (692.12)
Quadrantenelektrometer quadrant electrometer
Quadrantenschaltung symmetrical heterostatic circuit
Quadrantruderanlage quadrant-steering gear
Quadrat square
Quadrat über einer Linie square on a line
ins Quadrat erhoben, zum Quadrat squared
Quadratkilometer square kilometre
Quadratmeter square metre, metre square
Quadratmetergewicht grammes (pl.) per square metre
Quadratnetzverfahren co-ordinate grid reference system (527)
Quadratwiderstand sheet resistivity
Quadratwurzel square root
Quadratzahl square
Quadratzentimeter centimetre square, square centimetre
quadratisch square, quadratic
quadratischer Ausdruck quadratic term
quadratischer Detektor square-law detector (621.396)
quadratisches Gesetz square law (51)
quadratische Gleichrichtung square-law detection
quadratische Gleichung quadratic equation
quadratische Grundfläche square base
quadratischer Mittelwert r. m. s. = root mean square, effective value
quadratischer Querschnitt square section
quadratische Skala quadratic scale
Quadratur (phase) quadrature
Quadratur des Kreises quadrature of the circle, squaring of the circle
Quadraturmodulation quadrature modulation
Quadraturverzerrung quadrature distortion
quadrieren to square (51)
Quadrophonie quadrophony, quadrophonics
Quadruplex-System quadruplex (telegraph) system
Quadrupol quadripole
Qualität quality, grade
Qualität verbessern to upgrade, to improve the quality
Qualitätserzeugnis high-quality product, high-grade product
Qualitätsprüfung quality control
Qualitätssicherung quality assurance
qualitativ qualitative
qualitative Analyse qualitative analysis
qualitative Untersuchung qualitative investigation
Qualm smother, thick smoke
Quant quantum (pl.: -ta)
Quantenmechanik quantum mechanics
quantenmechanisch quantum-mechanical
Quantensprung quantum jump
Quantentheorie quantum theory
Quantisierung quantization
Quantisierungseinrichtung quantizer
Quantisierungsgeräusch quantization noise
Quantität quantity
quantitativ quantitative
quantitative Analyse quantitative analysis
quantitatives Ergebnis quantitative result
Quantum quantum (pl.: -ta)
Quartal quarter (of the year)
Quartärgruppe supermaster group
Quartärgruppendurchschaltefilter through supermaster groupfilter
Quarz quartz (54), (621); (piezoelectric) crystal (621); silica (54)
Quarzfaden quartz fibre
Quarzfassung crystal mounting
Quarzfilter crystal filter (621.395)
Quarzfrequenz crystal frequency
Quarzglas quartz glass
Quarzhalterung crystal holder, crystal mounting
Quarzlampe quartz lamp, silica lamp
Quarzkristall quartz crystal
Quarzoszillator quartz oscillator, crystal oscillator
Quarzröhre quartz valve, silica valve
Quarzrohrheizelement quartzolite heating elements
Quarzschwinger piezoelectric resonator
Quarzsender quartz oscillator, crystal(-controlled) oscillator
Quarzschnitt crystal cut
Quarzsteuerung crystal control (621.396)

quarzgesteuert crystal controlled
Quarzthermostat crystal oven, crystal thermostat (621.396)
Quarzuhr quartz clock, crystal clock
quasi-optisch quasi-optical
quasi-optische Ausbreitung quasi-optical propagation
quasi-stationär quasi-stationary
quasi-stationärer Stromkreis quasi-stationary circuit, lumped network, localized circuit
quasiternäre Tastung quasi-ternary keying
quaternär quaternary (669)
Quaternion quaternion
Quaternionenrechnung quaternionics, quaternion calculus
Quecksilber mercury, quicksilver
mit Quecksilber genetzter Kontakt mercury-wetted contact
Quecksilberdampf mercury vapour
Quecksilberdampfgleichrichter mercury (arc) rectifier, mercury (vapour) rectifier
Quecksilberdampfgleichrichter mit flüssiger Kathode mercury-pool type rectifier
Quecksilberdampfgleichrichter mit Glasgefäß mercury rectifier with glass bulb
Quecksilberdampfgleichrichter mit Eisengefäß mercury rectifier with iron tank
Quecksilberdampfhöchstdrucklampe extra-high-pressure mercury-vapour lamp
Quecksilberdampflampe mercury-arc lamp
Quecksilberdampflichtbogen mercury arc
Quecksilberdampfniederdrucklampe low-pressure mercury-vapour lamp
Quecksilberdampfröhre mercury-vapour tube
Quecksilberdampfumrichter mercury-vapour frequency changer, mercury-arc converter
Quecksilberdampfwechselrichter mercury-arc inverter
Quecksilberelektrode mercury electrode
Quecksilberentladung mercury-vapour discharge
Quecksilberfaden mercury thread
Quecksilberhochdrucklampe high-pressure mercury-vapour lamp, capillary lamp
Quecksilberkathode mercury pool
Quecksilbermanometer mercury gauge
Quecksilberrelais mercury relay, mercury-wetted contact relay
Quecksilbersäule mercury column, mercury thread
Quecksilberschalter mercury switch
Quecksilberschaltröhre mercury-in-glass switch
Quecksilberstrahl mercury jet
Quecksilberstrahlgleichrichter mercury-jet rectifier
Quecksilbersublimat corrosive sublimate
Quecksilberthermometer mercury thermometer
Quecksilberunterbrecher mercury interrupter, mercury break
Quecksilberzähler mercury meter
Quelle source (53); well, spring
Quellenbereich source range (539)
quellenfreies Feld solenoidal field, tubular vector field
Quellenkodierung source coding
Quellenwiderstand source impedance
Quellprogramm source program
Quellsprache source language
Quer-... cross, transverse; shunt (621.3)
Querabmessung transverse dimension
Querabspannung side anchor (621.315.1)
Querachse quadrature axis (621.313)
Querbalken cross bar, cross beam
Querbewegung cross-wise movement, transverse motion
Querdifferentialschutz transverse differential protection, split-conductor protective system
Querdruck transverse pressure
Querdurchflutung quadrature-axis component of the m. m. f.
Querfeld transverse field, quadrature-axis field, cross-field (621.313)
Querfeldamperewindungen cross ampere turns
Querfelddynamo, Querfeldgenerator cross field dynamo
Querfeldkomponente quadrature-axis component
Querfeldmaschine armature-reaction excited machine
Querfeldreaktanz quadrature-axis transient reactance
Querfeldspannung quadrature-axis voltage
Querfeldverstärkermaschine amplidyne
Querglied shunt (621.3)
Querinduktivität primary inductance
Querkapazität series capacitance (621.314.2)
Querkondensator shunt capacitor (621.392)
Querkopplung cross feed
Querlageanzeiger cross level (indicator) (629.13)
Querleitweg transverse conductance

Querlochstromwandler window-type current transformer
Quermagnetisierung lamellar magnetization
Querneigung bank
Querprüfung vertical redundancy check, VCR (681.3)
Querrichtung cross direction
in Querrichtung crosswise
Querschiene cross rail
Querschnitt (cross) section
Querschnittsänderung cross-sectional deformation; change of section
Querschnittsebene sectional plane
Querschnittsform sectional form
Querschnittsschrumpfung sectional shrinkage
Querspannung transverse voltage; quadrature-axis component of the voltage
Querspurverfahren transversal recording, quadruplex recording (681.84)
Querstab, Querstange crossbar
Querstrahler broadside aerial, broadside array, billboard antenna
in Querstrahlanordnung broadside (621.396.67)
Querstrahlantennenwand broadside curtain of aerials
Querstrahlgruppe broadside array
Querstrahl(richt)antenne broadside (directional) aerial
Querstrahlschalter orthojector circuit-breaker
Querstrom transverse current; shunt current
Querstromlöschkammer turbulator, cross-jet pot
Querstrommikrophon transverse-current microphone
Querstromlüfter lateral blower
Querträger crossarm, cross-arm (621.315.1); pole arm (621.395)
Querverbindung tie trunk, tie line (621.395)
Querverspannung cross span
Querweg high usage route (621.395)
quer cross, transverse, diagonal; crosswise, across
quer zur Achse across axis
quer zur Faser across fibre
quergerichtet transverse
quergerichtetes Feld cross field
querschraffiert cross-shaded
Quetsche squeeze; crushing mill, grinder
Quetschfuß pinch (621.326), (621.385); press squeeze, pressed stem (621.385)
Quetschfuß drücken to pinch
Quetschfußautomat pinch-making machine
Quetschfußdurchführung pin-type lead-in
Quetschhahn snap valve
Quetschhohlleiter split waveguide, squeezable waveguide
Quetschkondensator book capacitor
Quetschkopf squeeze head
Quetschvorrichtung squeezing device
Quetschzange pinch tongs (pl.)
quetschen to squeeze; to press, to pinch
Quetschung squeeze; bruise (61)
Quibinärcode quibinary code
quietschen to scream, to squeak, to shriek, to creak
Quirlantenne turnstile aerial
Quittung acknowledgment (621.39), (681.3)
Quittungsbetrieb handshaking (681.3)
Quittungskreis acknowledging circuit (621.39)
Quittungsrelais acknowledging relay
Quittungsschalter acknowledging switch, discrepency switch
Quittungssignal acknowledgment signal
Quittungszeichen acknowledgment signal; receipt signal (621.395)
Quotient quotient ratio
Quotientenmeßgerät quotientmeter, ratio meter

R

R [Klemme am Anlasser] A [starter terminal]
R [Widerstand] R [resistance]
Rachenlehre snap gauge
Rad wheel
Rad klemmt sich wheel sticks
Rad um 10 Löcher weiterdrehen to index a wheel 10 holes
Radachse axis of gears (621.8); axle (625)
Radachsenstromkreis axle circuit (656.25)
Raddruck wheel load
Radgestell wheel base
Radkappe wheel guard
Radkeil scotch (block)
Radkranz rim (of a wheel), wheel rim, tyre; tire (US) (625.5)
Radnabe (wheel) hub, nave
Radschutzhaube wheel guard
Radspeiche spoke
Radsplint linchpin

Radstand wheel base, centre distance
Radwiderstandsprüfer rail-wheel bond tester
Radar radar, Radio Detecting And Ranging
Radar zur Erfassung bewegter Ziele moving-target radar
Radar zur Feuerleitung fire-control radar
Radar mit zwei Richtkeulen double-beam radar
Radarabfragefunkfeuer radar beacon, racon
Radarabsorptionsschicht radar paint
Radaranlage großer Reichweite long-range radar
Radaranzeigegerät (range-height) indicator
Radaranzeige mit Festzeichenunterdrückung moving-target indication
Radarbake ramark, raymark
Radarband radar band, K-Band
Radarbedeckung radar coverage
Radarbeleuchtung radar illumination
Radarbildschirm (radar)scope
Radarbildspur radar trace
Radarbildübertragungsstrecke radar link, radar relay
Radarecho radar response
Radareinblendverfahren radar interscan technic
Radarentfernungsmesser range-only radar
Radarerfassungsbereich radar coverage
Radarfeuerleitanlage fire-control radar
Radarflugkontrolldienst radar air traffic service
Radarfrühwarnung early warning (radar)
Radargeschwindigkeitsmeßgerät radar speed meter
radargesteuerter Flugkörper radar-homing missile
Radargleichung radar equation
Radarhöhenmeßgerät radar height finder
Radarimpulslaufzeit radar reflection interval
Radarkontakt radar contact
Radarkontrollbereich radar control area
Radarkuppel radom
Radarlandeverfahren ground-controlled approach, GCA
Radarleitweg radar traffic beam
Radarortungsempfänger radar locator
Radarreflektorboje radar reflector buoy
Radarschirm radar scope
Radarschirmbild radar display
Radarsendebake radar marker, ramark
Radarsichtgerät mit drehender Zeitachse plan position indicator
Radarsichtgerät mit Festzeichenausblendung moving-target indicator
Radarsignaldämpfung radar attenuation
Radarstörecho clutter
Radarstörsender radar jammer
Radarstrahl radar beam
Radartarnung radar camouflage
radarüberwachter ILS-Anflug GCA-monitored ILS approach
Radarwarnempfänger radar detector, radar-warning receiver
Radarzielverfolgung radar tracking
räderlos wheelless
Radial-... radial
Radialbohrmaschine radial-drilling machine
Radialgeschwindigkeit radial speed, radial velocity
Radialkühlung radial system of ventilation
Radialleitung radial transmission feeder
Radialnuten radial slots
Radialstrahl radial (beam)
Radialturbine radial turbine
radial radial
radial gerichtet, radial verlaufend radial
radial speisen to feed radially
radial unterteiltes konzentrisches Kabel separate cable (621.396)
radial zuführen to feed radially
radiale Richtung radial direction
radialer Weg radial path
Radiant radiant
Radiator radiator
Radiatorröhre radiator tube
Radikal radical, radicle (54)
Radio radio (US), wireless
im Radio on the wireless
Radioantenne wireless aerial, receiving antenna
Radioapparat wireless (set), radio, transistor (radio)
Radioapparat mit Plattenspieler radio-grammophone
Radioastronomie radioastronomy
Radio-Ballonsonde radio-sonde balloon (551.5)
Radiobastler amateur
Radioempfänger wireless receiver, radio receiver (US)
Radiofrequenz radio (radiated) frequency
Radiofrequenzraster radio-frequency pattern
Radiogerät radio set, wireless set
Radiohorizont radio horizon
Radioisotop radioisotope
Radiokompaß radio compass, automatic direction finder

Radiolumineszenz radioluminescence
Radiometer radiometer
Radionuklid radionuclide
Radiorecorder radio tape recorder
Radiosonde radio sonde
Radiotechnik wireless engineering, radio engineering
Radioverbindung wireless communication, radio link
radioaktiv radioactive
radioaktiver Abfall radioactive waste, hot waste
radioaktive Abschirmung radiation shield
radioaktive Bestrahlung radioactive irradiation
radioaktive Chemie hot chemistry
radioaktive Halbwertszeit radioactive half-life
radioaktiver Kern radioactive nucleus
radioaktiver Schwebestoff active aerosol
radioaktives Spurenelement tracer (element)
radioaktive Stoffe radioactive materials
radioaktive Strahlen radioactive rays
radioaktive Verseuchung radioactive contamination
radioaktive Werkstoffuntersuchung isotope radiography
radioaktiver Zerfall radioactive decay
radioaktive Zerfallsreihe radioactive series
Radioaktivität radioactivity
Radium radium
Radium-Beryllium Neutronenquelle radium beryllium neutron source
Radiumemanation radium emanation, radon
Radiumfarbe radium paint
radiumhaltig radium bearing
Radiumnadel radium needle
Radiumpräparat radium preparation
Radiumstrahlung radium rays, Becquerel rays (pl.)
Radiumverbindung radium compound
Radius radius (pl.-dii)
Radiusvektor radius vector
Radix radix
Radixschreibweise radix notation, radix representation
Radiziereinrichtung squareroot extraction device (621.316.7)
Radom radom
Radon radon, radium emanation
Raffinadekupfer refined copper
Raffinerie refinery
Rahmen frame (621); loop (621.396); former (621.317)
Rahmenantenne frame aerial, loop aerial; coil aerial (US)
Rahmenkonstruktion framework (624.9)
Rakete rocket
Rakete starten to launch a rocket
Raketenabschußbasis rocket base
Raketenantrieb rocket drive
Raketensteuerung rocket control
Raman-Spektrum Raman spectrum
Rampe ramp
Rampenlicht foot-lights (pl.), proscenium light (79)
Rand rim, brim; edge: border, boundary
Rand rollen to bead, to wire (621.98)
Rand umbiegen to turn an edge over
Randaufhellung edge flare (621.397)
Randaussendung out-of-band radiation
Randbedingung boundary condition
Randbefeuerung boundary light (656.7)
Randfrequenz band-edge frequency
Randintegral contour integral
Randkurve boundary curve
Randlochkarte edge-punched card
Randschicht barrier layer (621.382)
Randwertproblem boundary problem, contour problem
Rändelknopf knurled knob
Rändelmutter knurled nut
Rändelrad thumbwheel
rändeln to knurl, to rim
Rangier-... shunting, marshalling
Rangierbahnhof marshalling yard
Rangierfeld plug board (621.3)
Rangierfunk train shunting radio (service)
Rangierdrahtverbindung cross connection, jumper (621.3)
Rangierlichtsignal signal-repeater light
Rangierschnur patch cord (621.3)
Rangiersignal am Ablaufberg hump signal
Rangiersignalanlage im Führerstand hump engine-cab signal system
Rangierstellwerk shunting tower
rangieren to shunt, to marshal; to switch (US), to patch (621.395.3)
rasseln to rattle
Rasselwecker trembling bell
Raste (foot) rest; stop, notch, catch
Raster raster, grid
Rasterabstand channel spacing (621.396)
Rasterbild half-tone picture
Rasterfeinheit scanning density, fineness of scanning
Rasterleuchte louver troffer, egg-grate troffer
Rastermaß grid dimension
rational rational (51)
rationale Einheiten rationalized units

rationale elektrische Einheiten rational electrical units
rationalisieren to rationalize
Rationalisierung rationalization, economizing (51), (621); work simplification
Ratsche ratchet
Ratschenklinke ratchet stop
Rattern rattle, rattles (621.396.62)
Rattergeräusch rattles (pl.) (621.396.62)
Rattermarken chatter marks
rattern to rattle, to chatter
Rauch smoke
Rauchdichte-Meßgerät smoke-density tester
Rauchgas flue gas, stack gas, chimney gas, furnace gas, waste gas
Rauchgasentstaubungsanlage flue-gas precipitating equipment
Rauchmelder smoke detector, smoke indicator
Rauchverzehrer smoke consumer
rauchen to smoke, to fume, to fumigate
rauchende Säure fuming acid
rauh rough, roughish; crude
rauh werden to gall
rauhe Behandlung rougish use
rauher Betrieb rough usage, arduous duty, crude service, rough service
rauhen to roughen, to gall
Rauhreif hoarfrost, white frost
Rauhreifansatz sleet (621.3)
Rauhreifschutzhaube sleet hood
rauhreifsicher sleetproof
Raum room; region; space
Raum beanspruchen to occupy space
Raum sparen to save space
um Raum zu sparen for compactness
Raum-Zeit space-time (53)
in Raum und Zeit in space and time
Raum zwischen den Elektroden interelectrode space (621.385)
Raum zwischen Gitter und Kathode grid-cathode space
Raumakustik acoustics (of the space)
Raumanteil part by volume
Raumäther ether of space
Raumbedeckung volumetric coverage
Raumbeleuchtung room lighting
Raumdiagonale space diagonal, cube diagonal
Raumdiversity space diversity
Raum-Diversity Empfang space-diversity (reception)
Raumelement element of volume
Raumentladungsgebiet space charge region
Raumfahrttechnik space technique
Raumfahrzeug space craft
Raumgeräusch surrounding noise, background noise, room noise (621.395)
Raumgitter space lattice, atomic structure
Raumheizkörper space heater (621.365)
Rauminhalt volume, cubic capacity, cubic content
Raumklang three-dimensional sound, stereo sound, stereophonic reproduction
Raumladegitter space-charge grid
Raumladung space charge (621.3)
Raumladungsbegrenzung space charge limitation
Raumladungsgebiet space-charge limited current region
Raumladungsgitter space-charge grid
Raumladungskonstante perveance
Raumladungsstrom space current
Raumladungswirkung space-charge effect
Raumladungszone space-charge region
Raumlicht ambient light
Raummaß cubic measure; room dimensions (pl.)
Raumstrahlung space rays, indirect rays (pl.) (621.396)
Raumtemperatur room temperature, ordinary temperature
Raumverhältnisse space conditions
Raumvielfachsystem space-division system (621.395)
Raumwelle sky wave, ionospheric wave, indirect wave (621.396)
Raumwellenfehler ionospheric-path error
Raumwinkel solid angle
raumzentriert body centred (548)
räumlich spatial, in space; cubica
räumlich anordnen to space
räumlich ausbreiten, sich to travel through space
räumlich trennen to space
räumlich unterteilen to pitch
räumlich und zeitlich in space and time
räumliche Änderung space variation
räumliche Lage physical position
räumliche Verteilung spatial distribution, volume distribution
rausgeschleppte Flüssigkeit dragout (621.357)
Rauschen hiss, (rushing) noise, random noise (621.396); snow (621.397); parasitic noise (621.385)
Rauschabstand signal-to-noise ratio
Rauschamplitudenmodulation amplitude modulation with a noise carrier
Rauschbewertungsfilter noise-weighting filter
Rauschfaktor noise factor, noise figure

Rauschfeldstärke radio noise field intensity
Rauschfrequenzmodulation frequency modulation with a noise carrier
Rauschgenerator (standard) noise generator
Rauschklirrmessung intermodulation-noise measurement
Rauschkorrelationsmodulation correlation modulation with a noise carrier
Rauschleistung noise power
Rauschmesser noise meter
Rauschmessung noise measurement
Rauschmodulation noise-carrier modulation
Rauschpegel noise figure
Rauschsender noise generator
Rauschspannung noise voltage, random voltage, fluctuation voltage
Rauschsperre squelch; muting
Rauschtemperatur noise temperature
Rauschträger noise carrier
Rauschunterdrücker denoiser, squelch
Rauschunterdrückung noise quieting, noise suppression
Rauschzahl noise factor, noise figure, noise ratio
rauschen to hiss, to whistle, to rustle
rauscharme Röhre low-noise tube
rauschende Röhre noisy valve
Raute rhomb, rhombus
rautenförmig rhombic, diamond-shaped
Rayleigh-Scheibe Rayleigh disk
Rayongarn rayon (yarn)
RC-Generator, RC-Summer R-C oscillator = resistance-capacitance oscillator, resistance tuned oscillator
RC-Schaltung RC-circuit, resistance-capacitance circuit
RC-Verstärker resistance-capacitance coupled amplifier
Reagens reagent (54)
Reagenz-... test, reagent
Reagenzflasche reagent bottle
Reagenzglas test tube
Reagenzpapier test paper
reagieren auf to respond to, to be responsive to
reagieren mit to react with (54)
Reaktanz reactance
Reaktanzrelais reactance relay
Reaktanzröhre reactance valve
Reaktanzschutz ratio-balance protective system, reactance (distance) protection
Reaktanzspannung reactance voltage, reactive voltage
Reaktion reaction (54)
Reaktion in der festen Phase solid-phase reaction
Reaktionsart mode of reaction
reaktionsfähig reactive (54); responsive (621.3)
Reaktionsgenerator reluctance generator
Reaktionslust reactivity (54)
Reaktionstemperatur temperature of reaction
Reaktionsträgheit inertness
reaktionsunlustig (chemically) inert
Reaktionswärme heat of reaction
Reaktionszeit reaction time; personal equitation
Reaktor reactor, (atomic) pile (539)
Reaktor mit angereichertem Spaltstoff enriched reactor
Reaktor mit flüssiger Brennstoffsuspension fluidized reactor
Reaktor mit Flüssigmetallspaltstoff liquid-metal fuel reactor
Reaktor mit geteiltem Kühlmittelfluß split-flow reactor
Reaktor mit hohem Neutronenfluß high-flux reactor
Reaktor mit langsamen Neutronen thermal reactor, slow reactor
Reaktor mit nasser Spaltstoffsuspension slurry reactor
Reaktor mit organischer Bremssubstanz organic-moderated reactor
Reaktor mit Spektralsteuerung spectral shift reactor
Reaktor mit thermischen Neutronen thermal reactor
Reaktoraufbau reactor construction
Reaktorbehälter reactor vessel
Reaktorgewölbe reactor vault
Reaktorgift nuclear poison
Reaktorgrube reactor pit
Reaktorkern reactor core
Reaktorkuppel reactor sphere
Reaktorleistung reactor output
Reaktorsicherheitshülle reactor containment
Reaktorsteuerung reactor control
Reaktorüberwachung reactor instrumentation; reactor supervision
Reaktorvergiftung pile poisoning
Reaktorzelle cell
Realzeit real time
Realzeitbetrieb real-time operation, real-time processing
realisieren to realize, to put into practice
Rechen-... reckoning
Rechenanlage computer, data processing system

Rechenarbeit arithmetic work
Rechenaufgabe lösen to do a sum
Rechenbefehl arithmetic instruction
Rechenfehler miscalculation
Rechengeschwindigkeit computing speed
Rechengröße operand (681.3)
Rechenmaschine computer, computor, calculating machine
Rechenoperation arithmetic operation
Rechenschieber slide rule
Rechentafel (ready) reckoner, calculator
Rechentechnik calculus, calculating technique
Rechenverfahren method of calculation, method of analysis
Rechenvorschrift computing instruction
Rechenwerk arithmetic logic unit = ALU
Rechenwerkregister arithmetic register
Rechenzentrum computer center
Rechnen reckoning, calculating, computing: arithmetic
rechnen to reckon, to calculate
Rechner computer, calculating machine
Rechner mit fester Wortlänge fixed-length computer
rechnergestützt computer aided
rechnergestütztes Anflugfolgesystem computer assisted approach sequencing
rechnergestützte Konstruktion computer aided design
Rechnerlauf computer run
Rechnerperiode machine cycle
Rechnersprache computer language
rechnerisch mathematical, analytical; by computation
Rechnung calculation; invoice, bill
Rechnungsgang method of calculation, sequence of calculation
rechnungsmäßig according to calculation
Rechteck rectangle, oblong
Rechteckabtastung rectangular scan
Rechteckgenerator square-wave generator
Rechteckimpuls square-topped pulse
Rechteckkurve flat-topped wave (621.3)
Rechteck-Pulsgenerator sine-squared pulse generator
Rechteckrahmen rectangular loop (621.396); rectangular frame (621.314)
Rechteckschleife rectangular loop
Rechteckwelle square wave (621.3); rectangular wave
rechteckig rectangular, oblong, flat-topped
rechteckige Klammern square brackets
Rechte-Hand-Regel right-hand rule
rechter Winkel right angle
im rechten Winkel at right angles (pl.)
rechts right
rechts gelegen right-hand
nach rechts gerichtet rightward
nach rechts verschoben offset to the right
rechtsdrehend clockwise, in clockwise order, dextrorotary (53), (54)
rechtsgängig right-handed
rechtsgängige Schraube right-handed screw
rechtsläufig right-handed; positive-sequence
rechtsläufige Maschine right-handed machine
rechtsläufige Wicklung right-handed winding
rechtsseitig recto
Rechtsdrehung, Rechtslauf right-handed rotation
Rechtsgewinde right-handed thread
Rechtsschweißung rightward welding
Rechtssteuerung right-hand drive (629.113)
rechtweisend true
rechtweisender Kurs true course
rechtweisende Peilung true bearing
rechtwinklig rectangular, right-angled
rechtwinklig abschneiden to square off
rechtwinklig zueinander at right angles (pl.)
rechtwinkliges Dreieck right-angled triangle
rechtwinklige Ecke square corner
rechtwinklige Komponenten resolved parts
rechtwinklige Koordinaten rectangular co-ordinates
recken to stretch
Rectaszension right ascension (52)
Reduktion reduction; deoxidation, reduction (54)
Reduktion auf ein Minimum minimization
Reduktionsfaktor reduction factor
Reduktionsgetriebe reduction gear
Reduktionsmittel reducing agent
Reduktionsmuffe reducing joint (621.315.2)
Reduktionsstück reducing fitting, taper piece (66)
Reduktionsventil reducing valve
Redundanz redundance, redundancy
Redundanzkontrolle redundancy check
Redundanzminderung reduction redundancy
reduzieren to reduce; to deoxidize, to reduce (54)
Reduzierung s. **Reduktion**
Reduzierhülse adapter sleeve
Reduzierventil p. r. valve = pressure reduction valve
Reed-Relais reed relay
reell real (51)
reell machen to rationalize

reelle Achse real axis
reeller Anteil real component
reeller Widerstand ohmic resistance
reelle Zahl real number
Referenz reference
Referenzdiode voltage-reference diode
Referenzspannung reference voltage
Referenzweiß reference white (621.397)
reflektieren to reflect
reflektierend reflecting, reverberatory
reflektierte Wanderwelle reflected wave
Reflektormeter reflectometer
Reflektor reflector (535), (621.32), (621.396); reverberator
Reflektorballon peardrop
Reflektorlampe reflector lamp
Reflektortragring shade carrier ring (621.3)
Reflex reflex
Reflexantenne reflector aerial
Reflexaudionenempfang autoplex reception (621.396.62)
reflexfreier Bildschirm non-glare screen
reflexfreies Glas anti-glare glass
Reflexklystron reflex klystron
Reflexminderung reflectance reduction: glare reduction (535)
Reflexschaltung reflex circuit
Reflexverstärker reflex amplifier
Reflexverstärkung dual amplification (621.396.62)
Reflexwand reflecting wall (621.396)
Reflexion reflection, reflectance
Reflexion am Erdboden ground reflection, earth reflection (621.396)
Reflexionsanode, Reflexionselektrode repeller
Reflexionsdämpfung return loss (621.395)
Reflexionsfähigkeit reflectivity
Reflexionsfaktor return current coefficient (621.39), reflection coefficient
Reflexionsfläche plane of reflection
reflexionsfrei abgeschlossene Leitung terminated line
Reflexionskoeffizient reflection coefficient
Reflexionsmaß eines Flugzeugs aircraft decibel rating
Reflexionsverlust reflection loss
Reflexionsvermögen reflecting power, reflectivity
Reflexionswinkel angle of reflection
Refraktometer refractometer
Refraktor refracting telescope, refractor
Regal, Regalbrett shelf (pl.: -ves)
Regel rule, regulation; standard (s. auch **Regel-...**)
Regel trifft zu a rule applies
gegen eine Regel verstoßen to break a rule
Regelabweichung deviation from standard design, deviation from regulations
Regelbauart standard pattern, standard design
Regel-... regulating, governor, control
Regelabweichung operating differential; control deviation (621.316.7)
regelbar adjustable
regelbare Gegeninduktivität mutual inductometer
regelbarer Generator variable-voltage generator
regelbarer Kondensator variable capacitor
regelbarer Motor adjustable-speed motor, variable-speed motor
Regelbefehl control signal
Regelbereich control range
Regelcharakteristik variable mu characteristic (621.385); regulation characteristic
Regeleinrichtung control device, regulator, governor mechanism, control equipment, controller
Regelfläche area controlled between response curve and horizontal line
Regelgeschwindigkeit control rate
Regelgetriebe variable-speed transmission, final control element
Regelglied controlling unit, positioning element
Regelgröße controlled condition; controlled variable (US), controlled quantity
Regelgüte quality of control, control effectiveness
Regelhub control range
Regelimpuls output variable
Regelkennlinie control characteristic, working characteristic
Regelkreis automatic control system, feedback control system, control loop (US), closed loop
Regelmechanismus governor mechanism, control equipment, control gear
Regelmotor variable-speed motor, varying-speed motor, servomotor
Regelorgan positioning element, controlling element
Regelpilot regulating pilot (621.395.4)
Regelrelais regulating relay
Regelröhre variable mu valve, remote cut-off tube (621.385)
Regelschalter regulating switch, tap-changing switch
Regelspannung control voltage
Regelstab (control) rod (539)

Regelstrecke controlled system, process (621.316.7); regulated line section (621.395)
Regelsystem controlled system, controlled member
Regelsystem mit geschlossener Schleife closed-loop control system
Regelsystem mit Rückführung feedback-control system
Regeltechnik automatic control technology, control engineering
Regelthermostat thermostat regulator
Regeltransformer regulating transformer, variable voltage transformer, tap-changing transformer
Regeltransformator für Hochspannungssteuerung high-tension regulating transformer
Regeltransformator in Sparschaltung auto-connected regulating transformer
Regelventil regulating valve, control valve
Regelverstärker servo amplifier
Regelverzerrung characteristic distortion (621.394)
Regelwicklung regulating winding
Regelwiderstand regulating rheostat, adjustable resistor, variable resistor
Regelzelle regulating cell, end cell (621.355)
Regelzweck control purpose
regelbar adjustable, controllable, variable
regelbarer Antrieb variable drive
regeln to regulate, to control, to govern
regelbares Dämpfungsglied variolosser
geregeltes Netzteil stabilized power supply
geregelte Spannung controlled voltage
Regelung regulation, adjustment
Regelung durch Änderung der Frequenz variable-frequency control
Regelung durch Änderung der Spannung variable-voltage control
Regelung durch Kaskadenschaltung cascade control
Regelung durch Motorgruppierung series-parallel control
Regelung mit Vorhalt rate action
Regelungsgenauigkeit control accuracy
Regelungsüberschreitung overshoot
Regen rain, rainfall
regendicht, regenfest rain-tight, rain-proof
Regenmesser rain gauge, pluviometer; rainfall recorder
Regenstörung rain clutter (621.396.96)
Regenüberschlag wet arcing, wet flashover (621.3)
Regenüberschlagprüfung wet spark-over test
Regenüberschlagspannung wet flashover voltage
Regenüberschlagsweg wet arcing distance
Regeneration regeneration, refreshment
Regenerativ-... regenerative
Regenerativprinzip regenerative principle
Regenerativverfahren regenerative process
regenerieren to regenerate, to rejuvenate
Regenerierung regeneration, rejuvenation (621.385)
Regieraum control cubicle
Regionalsender regional broadcasting station
Register register (681.3)
Registerzeichen register signal
Registrier-... registration, recording
Registriereinrichtung recording equipment, recorder
Registrierinstrument recording instrument, recorder, graphic instrument, grapher
Registrierschrieb record
Registrierstreifen record; chart of a recording instrument
Registriervorgang recording, registering
Registrierwerk register (681.14)
registrieren to record, to register
Registrierung record, recording, registration
Regler governor, regulator, controller
Regler mit Stellgeschwindigkeitszuordnung proportional-speed controller
Regler mit Stellungszuordnung proportional-position action controller
Regler mit Vorhalt rate-action controller
Regler ohne Hilfsenergie self-operated controller, self-acting controller
Reglerantrieb regulator drive
Reglergetriebe governor gear
Reglerhebel governor arm
Reglerkugel governor balls, governor weights
Reglerstellung governor position
regulär regular
reguläre Lichtdurchlässigkeit regular transmission
reguläre Reflexion regular reflection
Regulator governor (621.3)
Regulier-... regulating
Regulierapparat regulating gear
Reguliergewicht governor weight
Regulierschraube adjusting screw
Regulierventil regulating valve
Regulierwicklung regulating winding
Regulierwiderstand regulating resistance: regulating resistor
regulieren to regulate, to govern
regulierte Frequenz regulated frequency

Regulierung regulation
Reib-... frictional, rubbing
Reibahle (tapered) reamer
Reibbelag friction facing
Reibkontakt rubbing contact
Reibradantrieb rim drive (681.84)
Reibradgetriebe friction (wheel) gear, wheel and disc drive
Reibriemen friction belt
Reibsitz friction fit
Reibtrieb friction drive, friction gearing
Reibwirkung rubbing action
reiben to rub; to ream (621.9)
Reibung friction; rub, rubbing
Reibungsbeiwert coefficient of friction
Reibungselektrizität frictional electricity, tribo-electricity
Reibungsfehler frictional error (621.317)
Reibungshöhe frictional head
Reibungskoeffizient s. **Reibungsbeiwert**
Reibungskompensationsspule starting coil (621.317)
Reibungskraft frictional force, frictional grip
Reibungskupplung friction clutch
reibungslos frictionless
reibungsloser Ablauf faultless operation
reibungsloser Lauf smooth running
Reibungsmoment frictional grip
reibungsschlüssig friction-tight
Reibungstrieb friction drive
Reibungsverlust frictional loss
Reibungsvorschub friction feed
Reibungswiderstand frictional resistance, frictional drag
Reibungswinkel angle of friction
Reibungszahl adhesive factor, coefficient of friction
reich an, ... -reich rich in
reichen von ... bis ... to range from ... to ...
reichlich copious, abundant; lavish
reichlich Brennstoff fuel in abundance
Reichweite range, transmission range, coverage, distance getting (621.396); communication range (621.395/6); search capability (621.397)
Reichweite am Erdboden ground range (621.396)
Reichweite über die optische Sicht penetration below optical horizon (621.396)
an Reichweite übertreffen to out-range
in Reichweite within reach
Reif hoar frost, white frost
Reifen tyre (Br.), tire (US) (629.113); hoop (621)
Reihe row; series (51), (621.3); progression, sequence (51)
Reihe bildend serial
der Reihe nach in turn, in order
in Reihe in series (621.3)
in eine Reihe entwickeln to expand in a series (51)
in Reihe schalten, in Reihe zuschalten to join in series (621.3)
Reihenanlage intercommunication telephone (621.395)
Reihenbau series construction
reihenbelastet series-loaded
Reihencode series code
Reihendifferenz common difference (51)
Reihenentwicklung expansion in series
Reihenersatzschaltbild series representation (621.3)
Reihenersatzwiderstand equivalent series resistance
Reihenfeld series field
Reihenfolge sequence
Reihenfolge nach der Güte order of merit
reihenförmige Anordnung row
Reihenkondensator series capacitor
Reihenmotor in-line type engine (621.4); series (-wound) motor (621.3)
Reihenparallelanlasser series-parallel switchgear starter
Reihenparallelschaltung series-parallel connection (621.3)
Reihenparallelschaltung mit Unterbrechung open-circuit transition (621.33)
Reihenparallelwicklung series-parallel winding, multiplex wave winding (621.313)
reihenregulierte Bogenlampe series-wound arc lamp
Reihenresonanz series resonance
Reihenresonanzkreis series (resonant) circuit (621.396)
Reihenschaltstufe series notches (pl.) (621.33)
Reihenschaltung series connection, series combination, cascade (connection)
Reihenschaltungsbogenlampe series-arc lamp
Reihenschaltungssystem series system
reihengeschalteter Kondensator series capacitor
Reihenschlußerregung series excitation
Reihenschlußfeld series field
Reihenschlußkommutatormotor (single-phase) series commutator motor
Reihenschlußlampe series lamp
Reihenschlußmotor series (-wound) motor
Reihenschlußverhalten series characteristic (621.313)

Reihenschlußwicklung series winding, series coil
mit Reihenschlußwicklung serieswound
Reihenspannung series voltage
Reihenuntersuchung mass survey (61)
Reihenwicklung series winding (621.313)
Reihenwiderstand series resistance: series resistor
Rein-... refined, pure
Reinbenzol refined benzene
Reinheitsmagnet purifying magnet (621.397)
Reinmetall pure metal
Reinraum-Technik clean-room technique
rein pure, genuine; absolute (54)
rein darstellen to isolate (54)
rein induktiv purely inductive
rein ohmisch purely resistive
rein ohmscher Widerstand unity power factor impedance, non-inductive impedance
rein sinusförmig simple sinusoidal
rein sinusförmige Kurve a pure sine
rein sinusförmige Spannung simple sinusoidal voltage
reiner Alkohol absolute alcohol
reines Batteriesystem straight storage system (621.32), (621.33)
reiner Dauerstrich pure continuous waves (pl.)
reine Handauslösung non-automatic opening
reine Induktivität pure inductance
reine Kapazität pure capacitance
reines magnetisches Drehfeld pure rotating magnetic field
reine Mathematik pure mathematics
reiner Programmspeicher strict program store
reines Sinusgesetz simple harmonic law
reiner Ton pure note (78)
reine Wechselgröße balanced periodic quantity
Reinheit purity
Reinheitsgrad degree of purity
reinigen to clean, to purity (54); to refine (621); to scavenge (621.43)
Reinigung cleaning, purification; scavenging
Reinigungsanlage purification plant
Reinigungsmaterial cleaning materials (pl.)
Reissmikrophon Reiss microphone, carbon-type microphone, carbon transmitter
Reiß-... drawing, sketching; breaking, rupturing (621), (620.17)
Reißbrett drawing board, sketching board
Reißbrettstift thumb tack
Reißfeder drawing pen, ruling pen
Reißfestigkeit breaking strength, rupturing strength, ultimate strength
Reißlack crackle finish, stress coat (620)
Reißlänge breaking length
Reißlängenprüfmaschine breaking-length tester
Reißlast breaking load
Reißnadel scriber, marking scraper
Reißschiene ruler, T-square
Reißverschluß zip fastener, zipper
Reißzeug drawing set, set of mathematical instruments
Reißzwecke drawing pin
reißen to rupture, to fracture, to break; to tear, to rend
Reitstock tailstock, loose headstock (621.94)
Rekombination recombination
Rekombinationsgeschwindigkeit recombination velocity
Rekombinationskoeffizient recombination rate
rekombinieren, (sich) to recombine
Rekorder recorder, inker (621.394)
Rekristallisation recrystallization
rekristallisieren to recrystallize
Rektazension s. **Rectazension**
Rektifikation, Rektifizierung rectification, redistillation (54), (66)
rektifizieren to rectify, to redistil
rekuperativ recuperative
Rekuperator recuperator
Rekursion recurrence (51)
Rekursionsformel recurrence formula, recurrence relation
Relais relay (621.3)
Relais fällt ab relay drops out
Relais klebt relay sticks
Relais mit Abfallverzögerung slow-release relay
Relais mit Ansprechverzögerung slow-acting relay
Relais mit magnetischem Nebenschluß shunt-field relay
Relais zieht an relay operates, relay responds, relay picks up
Relais mit offener Kontaktbauweise nonsealed relay
Relais mit Schutzrohrkontakten reed relay
durch Relais weiter übertragen to relay
Relais-Ansprechspannung relay pick-up voltage
Relaiseinrichtung relay device
Relaisflattern relay chatter
Relaisfühler relay-type sensor
Relaisfunkstelle radio-relay station
Relaisgeber direct-point repeater, interpolator (621.394)

Relaisklemme relay terminal
Relaiskontakt relay contact
Relaismagnet relay magnet
Relaisprellen relay bouncing
Relaisschreiber relay recorder
Relaisschutz relay protection
Relaissender relay (broadcasting) station
Relaisstation repeating station, repeater (621.39)
Relaisstelle relay station, repeater
Relaisstelle für Richtfunk microwave repeater
Relaissteuerung relay control
Relaiswählsystem all-relay dial system
Relaiswahlvermittlungsstelle all-relay exchange
Relaiswicklung relay winding, relay coil
relativ relative
relativer Abstand burn-up fraction (539)
relative Bezugsdämpfung relative equivalent
relative Einschaltdauer operating factor (621.31); duty cycle
relative Häufigkeit relative abundance (54)
relativer Kippmoment breakdown factor (621.315.1)
relative Lage relative position
relative Luftfeuchtigkeit relative humidity
relative Öffnung lens aperture (77)
relativer Pegel relativ level
relative Schlaglänge lay ratio (621.313)
Relativdrehzahl, Relativgeschwindigkeit relative speed, relative velocity
Relativität relativity
Relativitätstheorie theory of relativity
Relaxation relaxation
Reluktanz (magnetic) reluctance
Reluktanzgenerator reaction alternator
Reluktanzlangsamläufer subsynchronous reluctance motor
remanent remanent, residual
remanente Erregung residual excitation
remanenter Fluß remanent flux
remanente Magnetisierung remanent magnetization, residual magnetization
remanenter Magnetismus remanent magnetism, residual magnetism
remanente Restspannung remanent residual voltage (621.382)
Remanenz magnetic remanence, magnetism, retentivity, residual induction
Remanenzeigenschaft retentivity
Remanenzfeld residual field (621.313)
Remanenzinduktion residual induction, magnetic retentivity
Remanenzrelais retentive-type relay
Remittanz short-circuit reverse transfer admittance (621.382.3)
Reparatur repair, reconditioning
Reparatur an Ort und Stelle field repair
Reparaturarbeit repair work
reparaturbedürftig in need of repair, out of repair
Reparaturwerkstatt repair shop
reparieren to repair, to overhaul
Reproduktion reproduction
reproduzierbar reproducible
Repulsionsmotor repulsion motor (621.313)
Repulsionsmotor mit Doppelbürstensatz repulsion motor with double set of brushes
Repulsionsmotor mit Käfiganker repulsion-induction motor
Repulsionsmotor mit Kurzschlußanker squirrel-cage repulsion motor
Reserve reserve; stand-by, standby; spare
in Reserve stehen to stand by
Reserveader spare conductor
Reserveanlage reserve plant, stand-by equipment
Reservebatterie spare battery; reserve battery, stand-by battery
Reservekanal on-call channel
Reservemaschinen stand-by machines; spare machinery (sing.)
Reservemotor stand-by motor
Reservesammelschiene reserve bus bar, hospital bus bar
Reserveschaltung automatic reserve activation
Reserveschutz back-up protection, reserve protection
Reserveteil spare (part), spare component
Reservezelle emergency cell (621.355)
Reservoir reservoir, cistern
Residuum residuum (pl. -dua)
Residuenberechnung calculus of residua
Resonanz resonance
Resonanz mit der Eigenschwingung period resonance
auf Resonanz abstimmen to tune for resonance
in Resonanz für eine angelegte Frequenz resonant at an applied frequency
in Resonanz sein to resonate
Resonanzabsorption resonance absorption (53)
Resonanzanstieg resonant rise
Resonanzanzeiger resonance indicator
Resonanzbreite width of the resonance curve
Resonanzbremswächter zero-speed resonance plugging switch
Resonanzbuckel resonance hump

Resonanzeinfang resonance absorption; resonance capture (539.17)
dem Resonanzeinfang entgehen to escape resonance
Resonanzfrequenz resonant frequency, resonance frequency
Resonanzfrequenzmesser resonance-frequency meter
Resonanzkreis resonant circuit
Resonanzkurve resonance curve
Resonanzleitung line resonator
Resonanzloch crevasse (621.396)
Resonanzpunkt (self-)resonance point
vom Resonanzpunkt entfernt off resonance
Resonanzrelais tuned relay
resonanzscharf sharply resonant
Resonanzschärfe sharpness of resonance
Resonanzschwingung resonant vibration
Resonanztelefon monotelephone
Resonanzüberhöhung resonant rise
Resonanzverfahren resonance method
Resonanzverstärker tuned amplifier
Resonanzwiderstand dynamic resistance, dynamic impedance, rejector impedance
Resonator resonator (534), (621.396)
Responder responder
Rest rest, remain, remainder, remnant; residue (66), (51)
Restdämpfung net loss; overall loss, insertion loss, structural return loss (US)
Restdämpfung am Pfeifpunkt singing-point equivalent
Restfehlerquote undetected error rate
Restfeld residual field (621.313)
Restgas rest gas, residual gas
Restladung residual charge
Restlichtfernsehen low-light-level television
Restmagnetisierung residual magnetism
Restseitenband vestigal sideband, assymetric sideband
Restseitenbandfilter vestigal-sideband filter
Restseitenbandmodulation vestigal-sideband modulation
Restspannung residual voltage; saturation voltage (621.382.3)
Reststrom residual current, cut-off current; leakage current (621.319.4)
Restwiderstand saturation resistance (621.382.3); end resistance
Restzeit remaining time
resultieren aus to result from
resultierender Fluß resultant flux
resultierende Kraft resultant net force
resultierende Spannung resultant voltage
resultierender Widerstand joint resistance (621.3)
Resultierende resultant, net resultant force
retardieren to retard
retardierendes Potential retarded potential
Retorte retort; distilling furnace, coke oven
Retortenofen retort furnace
Rettungsapparat breather, breathing apparatus
Rettungsboje lifebuoy
Reusenantenne sausage aerial, carge aerial
reversibel reversible
reversible Permeabilität reversible permeability, incremental permeability
Reversier... reversing
Reversierbetrieb reversing operation
Reversierschützensteuerung contactor-type reversing control
revidieren to inspect
Revision inspection
Revolverdrehbank capstan (lathe)
Reynoldssche Zahl Reynold's number
Rezipient bell-jar; receiver
reziprok reciprocal, inverse
reziproke Funktion inverse function
reziprokes Gitter reciprocal lattice
reziproke Potenz inverse power (51)
reziprokes Verhältnis inverse ratio
Reziprozität reciprocity (621.396)
Rhenium rhenium
Rheotron betatron
Rhodium rhodium
rhombisch rhombic, diamond-shaped
Rhombus s. **Raute**
Rhombusantenne rhombic aerial, rhomboid aerial, diamond aerial
rhythmisch rhythmic(al)
Rhythmus rhythm
Richardson-Effekt Richardson effect
Richt... directional, directive
Richtantenne directional aerial, beam aerial, directional array, beam array, directive antenna
Richtantenne mit einseitiger Charakteristik unidirectional aerial
Richtcharakteristik directional characteristic, radiation pattern, directivity pattern
Richtdiagramm directional pattern
Richteigenschaft directivity
mit Richteigenschaft directive
Richtempfang directional reception
Richtempfangsantenne beam-receiving aerial
Richtempfänger unidirectional receiver
Richtfaktor directivity factor

Richtfeder spring control
Richtfunk radio-relay system, microwave radio relay
Richtfunkfeuer directional radio beacon
Richtfunknetz radio-relay network
Richtfunkstation radio-relay station
Richtfunkstrecke radio-relay transmission system, radio link, radio relay, microwave link
Richtkoppler directional coupler
Richtkraft restoring force, versorial force, verticity, directing force
Richtkraftänderung restoring torque gradient
Richtkreisel directional gyro
Richtlautsprecher hoher Leistung loud hailer
Richtmikrophon directional microphone
Richtmoment restoring moment, restoring torque gradient, controlling torque (621.317)
Richtprisma refractor (621.32)
Richtschärfe beam sharpness
Richtsendung beam transmission
Richtstrahl radiated beam (621.396)
Richtstrahlanlage beam system
Richtstrahlantenne s. **Richtantenne**
Richtstrahler directional radiator, directional aerial
Richtstrahlung directive radiation
Richtstrahlungsdichte steradiancy (621.32)
Richtverhältnis einer Richtantenne power ratio of a directional aerial
Richtverkehrsleitung unidirectional circuit (621.395)
Richtverstärkungsfaktor directive gain
Richtwirkung guiding effect, directivity (621.396); rectifying effect (621.314)
Richtwirkungsgrad detector power efficiency
richten to direct, to beam; to guide; to straighten (621)
gerichtet sein nach to point at
gerichtete Strecke vector line
richtig right, true, correct
richtig ausrechnen to work out correctly
richtig bemessen correctly designed
richtig funktionieren to go right
richtig gehen to keep time (681.1)
richtige Absicherung proper fusing
richtige Anpassung correct match (621.396)
richtiges Ansprechen proper operation
richtiges Ergebnis correct result
richtiger Kurs on-course direction
richtiger Wert true value
auf richtiges Maß bringen to true
in die richtige Lage bringen to right
in der richtigen Lage befestigen to fasten into position
richtigstellen to rectify, to correct
Richtung direction
eine Richtung haben to trend
in einer Richtung steuerbar unidirectional
in beide Richtungen steuerbar bidirectional
richtungsabhängig directional
richtungsabhängiges Überstromrelais directional overcurrent relay
richtungsabhängiger Überstromschutz directional overcurrent protection
richtungsgebend directive
Richtungskoppler directional coupler (621.396)
Richtungskosinus direction-cosine
Richtungslautsprecher (großer Leistung) bull horn
Richtungsleitung (ferrite) isolator (621.396.69)
Richtungslogik sign logic
Richtungsrelais directional relay
Richtungsschreiben simplex telegraphy
Richtungsschutz directional-overcurrent protection
Richtungsverkehr one-way communication, simplex operation
Richtungswahlstufe group-selection stage (621.395.3)
Riefe flute, groove
riefen to flute
Riegel latch, bolt
Riegelblech, Riegelschiene latch plate (621.395)
Riegelbolzen interlocking bolt
Riemen belt
Riemen umlegen to shift the belt
Riemenantrieb belt transmission
mit Riemenantrieb kuppeln to belt to gether
Riemenrücklauf slack side of the belt
Riemenschlupf belt slip, creep (621.85)
Riemenscheibe pulley
Riemenspannung tension of the belt
Rieseln trickle, trickling, spraying
Rieselkühler trickling cooler, spray cooler
rieseln to trickle
riffeln to groove, to crimp
Rille groove, grooving; flute; crease
mit einer Rille versehen to recess
Rillenläufer grooved rotor
Rillenkugellager grooved ball bearing
Rillenrad, Rillenscheibe grooved pulley, grooved wheel
Rillenschiene grooved rail, tram rail
Ring ring; sleeve (621)

Ringabschnitt ring segment
Ringanker ring armature (621.313)
Ringantenne circular antenna
Ringanode ring anode
ringförmig ring-shaped, toroidal, annular
ringförmige Abdichtung ring seal
ringförmig angeordneter Leuchtdraht ring filament
ringförmig betriebenes Netz ring-operated system
ringförmiger Körper toroid
Ringkern torus; toroid (core), annular core
Ringkernprüfung Rowland's method (621.318)
Ringkerntransformator toroidal transformer
Ringklammer grip ring
Ringleitung ring main (621.3); loop
Ringmodulator ring-type modulator, rectifier modulator
Ringmolekül ring molecule
Ringmuffe bush
Ringnut annular groove
Ringofen circular kiln
Ringsammelschiene ring bus-bar
Ringschmierung (self-oiling) ring lubrication
Ringschraube ring bolt
Ringsegment ring segment
Ringspeiseleitung ring main (621.3)
Ringspule toroidal coil
Ringübertrager repeating coil (621.395)
Ringwicklung ring winding
Ringzähler ring counter
ringsherum um die Masche around the loop (621.3)
Rinne trough, chute
Rippe rib
mit Rippen versehen to rib
Rippenheizkörper column radiator, ribbed radiator
Riß crack, fissure; break; cleft
Riß in der Korngrenze intercrystalline crack
Rißbildung formation of cracks, cracking
Rißfestigkeit crack resistance (621.315.3)
Ritzel pinion
Ritzel greift ins Rad pinion plays in the wheel
ritzen to scratch
Rm [magnetischer Widerstand] S [magnetic resistance]
Roboter robot
robust rugged
Robustheit ruggedness
Rochellesalz Rochelle salt
Rockwell-Härte Rockwell hardness
Roh-... raw
Rohbaumwolle raw cotton
Roheisen pig iron
Rohglimmer block mica
Rohkupfer black copper
Rohöl crude oil
Rohstoff raw material
roh raw, crude; rough, roughish
rohes Anzeichen, rohe Anzeige rough indication
rohe Behandlung roughish use
rohe Gewalt brute force
roher Kreis rough circle
rohe Messung rough measurement
rohe Peilung coarse bearing
rohe Schätzung rough estimate
rohe Skizze rough sketch
roher Zustand raw state
Rohr reed (58); duct, tube, tubing, pipe (621.64); barrel (623.4)
Rohr-Boden-Potential pipe-to-soil potential
Rohr-Boden-Spannung pipe-to-soil potential difference
Rohre installieren to pipe
Rohranschluß pipe connection
Rohrbündel tube bundle, tube bank
Rohrerder earth rod, ground rod
Rohrfitting pipe fitting
Rohrflansch flange
rohrförmig tubular
Rohrgelenk knuckle (621.64)
Rohrgewinde pipe thread
Rohrheizkörper tubular heater
Rohrknie elbow
Rohrkrümmer pipe bend
Rohrkühler tubular radiator
Rohrlänge length of piping
Rohrleitung pipeline
Rohrleitungsanlage pipe work, tubing, piping
Rohrleitungsarmatur pipefitting
Rohrleitungsmonteur pipefitter
Rohrleitungsnetz, Rohrleitungssystem pipe system, piping
Rohrmast tubular mast
Rohrmuffe pipefitting, pipe joint
Rohrnaht pipe seam
Rohrniet tubular rivet
Rohrpost pneumatic post
Rohrpostanlage pneumatic-tube system
Rohrpostbüchse rabbit
Rohrschelle pipe clip
Rohrschlange coil
Rohrschlitzantenne slotted-cylinder antenna
Rohrschlüssel pipe wrench, twister

Rohrschutz tube protection
Rohrschweißung pipe welding
Rohrverlegung pipe laying; conduit wiring (696)
Rohrwiderstand tubulary-type resistor (621.316.8)
Rohrzange pipe to ngs (pl.), pipe wrench, twister
Röhrchenkondensator miniature (tubular) capacitor
Röhrchensicherung tube fuse
Röhre valve (621.385); tube (621.6) (US), (621.385); pipe (621.6)
Röhre heizen to light a valve (621.385)
Röhre ohne Heizung unlighted valve (621.396)
Röhre mit Geschwindigkeitssteuerung velocity-modulated tube
Röhre mit hochgesetzter Kathode self-biased valve
Röhre mit Mikrophoneffekt microphonic valve
Röhre (auf Emission) prüfen to test a valve, to check a valve
Röhre großer Steilheit high-slope valve
Röhre (vom Gitter aus) steuern to drive a valve (621.385)
Röhre mit Wolframdrahtkathode tungsten-filament valve
zwischen den Röhren liegend intervalve
Röhrenauszieher valve extractor
Röhrenbestückung valve set, valve complement
Röhrenbrumm tube hum
Röhrenelektrometer tube electrometer
Röhrenempfänger valve receiver
Röhrenersatz valve replacement
Röhrenfassung valve holder (621.385); tube socket
Röhrenfuß valve base
Röhrengenerator valve generator, valve oscillator
Röhrengleichrichter valve rectifier, vacuum-tube rectifier, discharge-tube rectifier
Röhrenhals valve neck (621.385)
Röhrenhalterung valve clamp
Röhrenhärte valve hardness (621.385)
Röhrenkappe top cap (621.385)
Röhrenkennlinie valve characteristic
Röhrenkessel tube boiler, tubular boiler, multitubular boiler (621.18); tubular-type tank (621.314.2)
Röhrenkolben valve envelope, bulb
Röhrenkondensator tubular capacitor (621.396)
Röhrenlampe tube lamp, tubular lamp
Röhrenlebensdauer valve life
Röhrenprüfgerät valve tester, valve-test set, valve checker (unit)
Röhrenrauschen valve hiss, Johnson noise
Röhrensatz valve complement, set of valves
Röhrensender valve generator, valve oscillator
Röhrensicherung tube fuse
Röhrensockel valve base, valve holder (621.396); socket (terminal) (621.385)
Röhrensummer s. **Röhrensender**
Röhrenverhalten valve action
Röhrenverlustleistung valve (heat) dissipation
Röhrenverstärker valve amplifier, thermionic amplifier
Röhrenvoltmeter valve voltmeter, thermionic voltmeter
Roll-... roller, rolling
Rollbahn taxiway (351.814)
Rollbahnfeuer taxiway (-marker) light
Rollbewegung rolling motion; roll (629.13)
Rollfeld landing ground, taxiway
Rollfeldüberwachungsradar ASDE, airport surface detection equipment
Rollkreis epicycle
Rollkugel roller ball
Rollreibung rolling friction
Rollschnitt rolling cut (621.397)
Rolltreppe moving staircase; escalator (US)
Rolle roll, roller, pulley (621); (n i c h t t r a g e n d :) reel
Rollenblitzableiter multigap arrester (621.395)
Rollenlager roller bearing
Rollenlaufring roller race
Rollenpunktschweißmaschine roller spot welder
Rollenschaltstück roller-type contact
Rollenschneidmaschine reel-cutting machine (676)
Rollenstromabnehmer pole trolley
Rollentrenner roller switch
Rollenzählwerk recording train, counting train, meter train (621.317)
rollen to roll (621), (625)
rollendes Geräusch roll, rolling noise
rollendes Material rolling stock (625)
gerollter Rand bead (621)
mit gerolltem Rand versehen to bead
rollende Reibung rolling friction
rollender Schnitt wipe (621.397)
Rolls-Royce-Legierung RR-alloy
Röntgen Röntgen, Roentgen; X-ray...
Röntgenanlage, Röntgeneinrichtung X-ray equipment
Röntgenapparat X-ray apparatus
Röntgenaufnahme radiograph, X-ray (picture)

Röntgenaufnahmetechnik radiography
Röntgenbeugungsbild X-ray diffraction pattern
Röntgenbild radiograph, X-ray picture
Röntgenbildverstärker X-ray video amplifier tube
Röntgendosis quantity of X-rays
Röntgendosismesser X-ray dosimeter
Röntgen-Feinstrukturprüfung X-ray crystal analysis, radio-metallography
Röntgenfilm X-ray film
Röntgenkristalllographie X-ray crystallography
Röntgen-Materialprüfung radioexamination of materials
Röntgenologe radiologist
Röntgenphotographie radiography
Röntgenprüfung X-ray examination, radiography
Röntgenröhre X (-ray) tube, Roentgen tube
Röntgenschirm röntgenoscope, fluoroscope
Röntgenschirmtechnik röntgenoscopy
Röntgenspektogramm X-ray spectograph
Röntgenspektrometer X-ray spectrometer
Röntgenspektroskop X-ray spectroscope
Röntgenspektrum X-ray spectrum
Röntgenstrahlen X-rays, Roentgen rays
Röntgenstrahlenbeugung X-ray diffraction
Röntgenstrahlquelle X-ray emitter
Röntgenstrahlung X-radiation; X-rays
Röntgenstreustrahlung scattered X-rays (pl.)
Röntgentherapie X-ray therapy, radiotherapy, röntgentherapy
Röntgentransformator X-ray transformer
Röntgenuntersuchung radioscopy
röntgen to X-ray, to radiograph
röntgenprüfen to X-ray
röntgenographisch radiographic
rosa Rauschen pink noise
Rosetten fittings, ceiling roses (621.32)
Rost grate (621.1); rust (620.1)
Rostansatz, Rostbildung rust formation, rusting
Rost ansetzen to (gather) rust
rostfrei rustless, rust-proof
rostfreier Stahl rustless steel, stainless steel
Rostschutz rust protection
Rostschutzanstrich rust proofing, rust-resisting paint; antirust coating
Rostschutzmasse rust-preventive compound
Rostschutzmittel rust preventive (agent)
rostverhindernd rust-preventive
rosten to rust
Rotbruch, Rotbrüchigkeit hot shortness, red shortness
Rotguß red brass, red metal, A. G. M. = admiralty gun metal
Rotverschiebung red shift (52)
rot red
rotbraun reddish-brown
rotbrüchig hot short
rot färben to redden
rotglühend red-hot
rot-grün-veil [Phasenfarben] red-yellow-blue [phasemarking]
nach Rot umschlagen to turn to red
Rotation rotation; curl (of a vector) (51)
Rotationsellipsoid spheroid (of revolution)
Rotationsenergie angular kinetic energy
Rotationsgeschwindigkeit rotation rate
Rotationskörper figure of revolution, solid of revolution
Rotationsparaboloid paraboloid of revolution
rotatorisch rotational
röten (sich) to redden
rötlich reddish
rötlich-braun reddish-brown
rotieren to rotate, to revolve (**um:** round, about)
rotierendes Feld rotating field
rotierende Funkenstrecke rotary spark gap
rotierender Phasenschieber rotary phase converter, synchronous capacitor
rotierender Umformer rotary converter, revolving commutator
Rotor rotor
Rotor-EMK rotor e. m. f.
rotorgespeiste Maschine inverted machine
Rotorplatten moving plates, rotor vanes (621.319.4)
Rotorplattensatz rotor section
Rotorstern spider
Rotorstrom rotor current
Rotorverbindung rotor interconnection (621.34), (621.8)
Rotorwicklung rotor winding
Routine routine
RS-Flipflop RS flipflop
Rubin ruby
Rubin-Laser ruby laser
rubinrot ruby-red
Ruck jerk, shock
ruckartig jerky
ruckartig bewegen, sich to jerk
Rückansicht rear view
Rückauslösung called-party initiated forward release (621.395)
Rückbewegung back stroke (621.9)
Rückdämpfung front-to-back ratio
Rückdauer fall time (621.316.9), (621.39)

Rückdiffusion back diffusion, rediffusion (542)
Rücken (der Überspannungswelle) tail (of surge wave)
Rückenhalbwertzeit time of half crest on the wave tail
Rückenzeitkonstante tail-time constant
Rückfall reset (621.318)
Rückfalleigenzeit inherent drop-off time
rückfallendes Relais self-reset relay
rückfedernd resilient
Rückfederung resilience, resiliency
Rückfederungszuschlag bending allowance
rückfließend refluent
rückfließender Strom return current
Rückfluß reflux
Rückflußdämpfung (regularity) return loss, structural return loss (US)
Rückfragen call back (621.395)
rückführen to feed back
Rückführung feedback (621.316.7); reset (621)
Rückgang decrease, fall; back stroke, idle stroke (621.9)
Rückgang der Ausgangsleistung fall in output
rückgewinnen to recover
Rückgewinnung recovery
Rückgewinnung des Gleichstromanteils d. c. restoration (621.397)
Rückheizung back heating (621.385)
Rückholmagnet resetting magnet
Rückholseil trolley retriever (621.33)
Rückhördämpfung antisidetone, ATR, sidetone attenuation
Rückhörbezugsdämpfung sidetone-reference equivalent
Rückhub return stroke (621.43)
Rückkehr return, reversion
Rückkehr in den Ausgangszustand recovery
rückkoppeln to couple back, to feed back, to regenerate (621.396.62)
rückgekoppelter Empfänger Armstrong circuit
rückgekoppelter Verstärker feedback amplifier
Rückkopplung (positive) feedback, regenerative feedback, regeneration, reaction (621.396)
als Rückkopplung wirken to be regenerative
Rückkopplungsfaktor feedback factor
Rückkopplungsgleichrichtung regenerative detection
Rückkopplungskreis reaction circuit, feedback circuit
Rückkopplungspfeifen oscillation squeal, singing
Rückkopplungsschaltung feedback network
Rückkopplungsschleife feedback control loop
Rückkopplungssperre singing suppressor
Rückkopplungsweg feed-back path
Rücklauf reversed motion, reflux (66); return trace (621.385); flyback, retrace (621.397); rewinding (681.84)
Rücklaufbremse counter brake
Rücklaufeinrichtung flyback mechanism (621.385)
Rücklaufleistung backward power
Rücklaufspur return trace (621.385)
Rücklauftragrolle return idler
Rücklaufzeit return time (621.385)
rücklaufen to recede (621.9); to recoil
rücklaufende Welle backward wave, reflected wave
rückläufig sein to retrograde, to return
Rückläufigkeit retrogradation
Rückleistung reverse power
Rückleistungsrelais reverse-power relay, power-directional relay
Rückleiter return wire (621.3)
Rückleitung return feeder, return circuit, return wire, return path
Rücklicht tail lamp, tail light (629.113)
rückmelden to back signal
Rückmelder (Drehmelder) response synchro
Rückmeldesignal repeating signal, check-back signal; remote-position indication (621.398)
Rückmischung inverse mixing
rückmodeln to demodulate, to remodulate
Rückmodelung demodulation, remodulation, detection (621.396)
Rückruf ring back (621.395)
rückschalten to step back
Rückschaltklinke backspace pawl
Rückschaltkreis step-back relay
Rückschlag recoil (539.17), (539.18), (623.4); flash (621.791); backfire (621.43)
Rückschlagklappe reverse flow check
Rückschlagventil non-return valve, check valve
rückschließen to work backward (51)
Rückseite reverse (side), back, remote side
rückseitig rear, back
rückseitiger Anschluß rear connection
rückseitige Klemme rear terminal
rückspeisen to give back energy
Rückspeisung feedback
rückspulen to rewind
Rückstand remain, remainder; foots (pl.) (54)
rückstellen to reset, to restore, to zero
Rückstellfeder resetting spring, restoring spring
Rückstellimpuls reset pulse
Rückstellklappe self-restoring indicator
Rückstellkraft restoring force

Rückstellmagnet reset magnet
Rückstellsteilheit restoring torque gradient
Rückstellwicklung reset winding
Rückstellung resetting, reset
Rückstoß recoil (539.17), (539.18), (623.4)
Rückstrahlantenne passive reflector antenna
rückstrahlen to reflect
Rückstrahler reflector
Rückstrahlfehler reradiation error
Rückstrahlfläche redirecting surface (621.32), echoing aerea
Rückstrahlmittel redirecting media (pl.) (621.32)
Rückstrahlquerschnitt echo area (621.396.67)
Rückstrahlunterdrückung radar camouflage
Rückstrahlvermögen reflecting power
Rückstrahlvolumen echoing volume
Rückstrahlung reradiation, reflection
rückstreuen to back-scatter
Rückstreuung back-scattered radiation, back scattering
Rückstrom return current, reverse current
Rückstromauslösung reverse-power tripping, reverse-current release, reverse-current tripping
Rückstromautomat discriminating cut-out
rückstromfreier Gleichrichter perfect rectifier
Rückstromrelais power-reverse relay, reverse-current relay
Rückstromschalter reverse-current circuit breaker, directional circuit breaker, discriminating circuit-breaker
Rückstromschütz reverse-current cut-out, reverse-current contactor
Rückströmung reverse flow
Rücktaste back spacer (681.61); back space (key)
Rücktransformation retransformation (51)
rücktransformieren to retransform (51)
Rücküberlaufanordnung entraide network (621.395)
rückverbinden mit to return to (621.395)
rückverwandeln to retransform (51)
Rückverwandlung retransformation (51)
Rückwand back-plate, rear panel (621); back panel (624)
rückwandern to remigrate
Rückwanderung remigration
rückwärtig backward, rear
rückwärtiger Anschlag backstop
rückwärtige Anschlußleiste rear terminal block
rückwärtiger Mastüberschlag back flashover
rückwärtige Stromstoßgabe revertive impulse dialling
Rückwärts-... back, reverse
Rückwärtsdämpfung reverse attenuation
Rückwärtsdiode backward diode
Rückwärts-Durchlaßspannung reverse conducting voltage, reverse on-state voltage
Rückwärts-Durchlaßwiderstand reverse conducting resistance
Rückwärtsgang reverse gear (621.43); reversed motion
Rückwärtsgang einschalten to go into reverse
Rückwärtskennlinie reverse voltage-current characteristic (621.382.2)
Rückwärtsregelkreis back-bias circuit
Rückwärtsregelspannung amplified back bias, back bias
Rückwärtsrichtung reverse direction (621.382)
Rückwärtsscheinwerfer backing lamp, backing light, back-up lamp, back-up light (629.113)
Rückwärtsspannung reverse voltage (621.382)
Rückwärtsspeisung rear feed (621.396.67)
Rückwärtsstrahlung back-end radiation (621.396.677)
Rückwärtsstrom reverse current (621.382.2)
Rückwärtsverfolgen (eines Anrufs) call tracing (621.395)
Rückwärtsverlustleistung reverse-power dissipation (621.382.2)
Rückwärtswellenoszillator backward wave oscillator
Rückwärtszeichen backward signal (621.395)
Rückwärtszug back pull (621.77)
rückwärts back, backward (s)
rückwärts durchlaufen to recede
rückwärtsfahren to reverse (629.113)
rückwärts gerichtet retrograde
rückwärts leitend reverse conducting
rückwärtsrechnen to work backward (51)
rückwärtssperrend reverse off-state, reverse blocking
rückwärts zählen to count down
Rückwärtswelle backward wave
Rückwärtszipfel back lobe
Rückwatt-... reverse-power
Rückwattauslösung reverse-power tripping
Rückwattrelais reverse-power relay, power-reversal relay
Rückwattschalter reverse-power circuit-breaker
Rückweg return path (621.3)
rückwirken to transfer back (621.396); to react
rückwirken auf to react with
rückwirkend retroactive
Rückwirkung reaction, retroaction, repercussion; back effect

rückwirkungsfrei undirectional (621.53); non-interacting
Rückwirkungsimpedanz motional impedance
Rückwirkungskapazität reflected capacity
Rückwirkungsspannung back voltage
Rückwirkungswiderstand motional resistance, reflected resistance (621.396)
Rückwurfflächen redirecting surfaces (621.32)
Rückwurfmittel redirecting media (pl.) (621.32)
Rückwurfwinkel angle of reflection
Rückzipfelecho back echo (621.396.67)
rückzünden to strike back, to restrike (621.385); to backfire (621.3)
Rückzündung backfire, arcing back (621.316.5)
Rudertelegraph rudder telegraph
Ruf-... calling, ringing (621.395)
Rufanlage staff locator, calling system
Rufanzeiger call indicator (621.395)
Rufbatterie ringing battery
Rufeinrichtung calling device
Ruffortschaltung call forwarding
Rufinduktor magneto ringer, calling magneto
Ruflampe calling lamp, line lamp (621.395)
Ruflampenfeld display panel (621.395)
Rufleitung calling line
Rufmaschine ringer, ringing generator, magneto alternator (621.394)
Ruf- und Signalmaschine ringing and tone generator
Rufnummer subscriber's number
Rufnummernauskunft directory information service
Rufrelais pilot relay (621.395)
Rufschalter ringing key
Rufschnur calling cord
Rufsignal call signal
Rufstellung calling position
Rufstöpsel calling plug
Rufstrom calling current
Rufstrombatterie ringing battery
Rufstromgenerator ringing generator, magneto
Rufstromüberlagerung superposed ringing current
Rufstromversorgung ringing-power supply
Ruftabelau annunciator
Ruftaste ringing key
Rufumformer ringing converter
Rufumgehungsschaltung ringing repeater
Rufumleitung call diversion, call forwarding
Rufumsetzer ringer, decoder (621.395)
Rufverzug post-dialling delay
Rufwecker call bell
Rufweiterleitung call forwarding
Rufweiterleitung bei besetztem Anschluß busy-line transfer
Rufzeichen ringing tone, call signal, call sign (621.395); call letter, call signal (621.396)
Rufzeichenerzeuger ringing interrupter, ringing generator
Rufzustand ringing condition
rufen to call, to ring (621.395)
gerufener Teilnehmer called subscriber
rufender Teilnehmer calling subscriber
Ruhe silence; quietude; rest
in Ruhe at rest
zur Ruhe bringen to bring to rest
Ruheausschlag point of rest (621.317)
Ruheenergie rest energy (539)
Ruhekontakt break contact, normaly closed contact, back contact
Ruhelage, Ruhestellung position of rest, rest position (621.318.5)
Ruhemasse mass of body at rest, rest mass
Ruhepunkt quiescent point
Ruhespannung equilibrium potential
Ruhestellung normal position; off-position (621.316.5)
Ruhestrom static current (621.385); quiescent supply
Ruhestromkreis closed circuit
Ruhestromprinzip holding-current principle
Ruhestromschaltung circuit-opening connection
Ruhewerte steady values
Ruhezeit unused time
Ruhezustand static state, passive state
im Ruhezustand at rest
ruhen to rest
ruhend stationary, static, fixed
ruhende Belastung steady load
ruhendes Bild stationary pattern (621.385)
ruhender Druck steady pressure
ruhende Last dead load, dead weight
ruhendes Programm inoperative program
ruhende Reibung static friction
ruhendes Teil stationary element, static part
ruhiger Gang smooth running
rumpeln to rumble, to flutter
Rumpelspannung wowe, flutter
Rund-... round, circular
Rundanglasung cylindrical seal (621.385)
Runddraht round wire
Rundfeuer flashing (621.313)
Rundfeuerlöscheinrichtung flash suppressor
Rundfunk radio, wireless, broadcasting, radio broadcasting (siehe auch Radio...)

Rundfunk hören to listen in
durch Rundfunk verbreiten to broadcast
Rundfunkansage (radio) announcement
Rundfunkanstalt broadcasting corporation
Rundfunkaufnahmeraum broadcast studio
Rundfunkdienst broadcasting service
Rundfunkempfang broadcast reception
Rundfunkgerät, Rundfunkempfänger radio set, radio receiver
Rundfunkgerätesicherung radio fuse
Rundfunkgesellschaft broadcasting corporation
Rundfunkhändler radio dealer
Rundfunkhörer(in) listener
Rundfunkindustrie radio industry
Rundfunksender broadcasting station, wireless transmitter, radio transmitter, radio station
Rundfunkspule radio coil
Rundfunktechnik radio engineering
Rundfunkteilnehmer listener, viewer
Rundfunktochtersender relay broadcasting station
Rundfunkwerbung radio advertising
Rundgewinde knuckle thread
Rundhohlleiter round wave guide
Rundloch round hole
Rundschreibeinrichtung broadcast unit
Rundschweißung circumferential welding
Rundsichtgerät plan position indicator
Rundsichtdarstellung plan position indication (621.397)
Rundsichtdarstellung mit Mittelpunktaufweitung expanded-centre plan display (621.397)
Rundsichtradar surveillance radar
Rundspulamperemeter repulsion-type ammeter
Rundstange bar of round section
Rundstecker circular connector
Rundsteuersystem centralized multi-service control, centralized ripple system
Rundstrahler isotropic radiator (621.396.67)
Rundstrahlantenne omnidirectional aerial
Rundstrahlbake omni(-directional) radio beacon
Rundstrahlfunkfeuer omnirange beacon
Rundstrahlreflektor omnidirectional lens
Rundstrahlung circular radiation
Rundsuchen azimuth rotate mode
Rundsuchradar air surveillance radar
rund round, tubular, circular sectioned; curved
runde Klammern curved brackets
runde Platte disc, disk
rundes Rohr, runde Röhre circular tube
runde Skala dial
runddrehen to slide (621.94)
rundherum over and over
rundmachen to round
runden to round; to suppress digits (51)
Rundungsfehler round-off error; rounding error (681.3)
Rüschschlauch varnished flex tubing; varnished sleeving (621.396)
Ruß soot, lamp black
rußen to soot; to smoke
Rüstbaum scaffolding pole
Ruthenium ruthenium
Rutherford-Bohrsches Atommodell Rutherford-Bohr model of the atom
Rutil rutile, titania
Rutsche chute; skid
rutschen to slip; to slide; to skid
rutschender Riemen slipping belt
Rydberg-Konstante Rydberg's constant

S

Saalverdunkelungseinrichtung dimming device
Sache thing; cause, matter
der Sache nachgehen to look into the matter
Sachkenntnis (specialized) experience
sachkundig expert
sachkundige Wartung skilled attention, skilled maintenance
Sachverständiger expert
Sachverständigengutachten expert opinion, expert's report
Sachverzeichnis index
Sack sack, pocket
Sackbahnhof dead-end station
Sackloch blind hole
Sackloch mit flachem Boden squared-out hole
Säge saw
Säge schränken to set a saw
Sägeblatt saw blade
Sägezahn sawtooth
Sägezahngenerator sweep generator, sawtooth generator, ramp generator
Sägezahngewinde buttress thread
Sägezahnkurvenform sawtooth wave (621.3)
Sägezahnschwingung sawtooth oscillation
Sägezahn-Signal sawtooth signal
Sägezahnspannung sawtooth voltage (621.396)

Sägenzug stroke of the saw
sägen to saw
Saite string, chord
Saite anstreichen to bow a string
Saite reißt a string snaps
Saitengalvanometer string galvanometer
säkular secular
säkulare Schwankung secular variation
Salizylsäure salicylic acid
Salmiak sal ammoniac; ammonium chloride
Salmiakgeist liquid ammonia, ammonia solution
Salpeter salpetre, nitre; niter (US)
Salpetersäure nitric acid, aqua fortis
salpetrige Säure nitrous acid
Salz salt; sal (54)
Salzgehalt saline content, salinity
Salzlösung brine
Salznebel saltspray
Salzsäure hydrochloric acid, muriatic acid
Salzsprühprobe saltspray test (620.191)
salzig salty, saline, briny
Sammel-... collecting
Sammelanschluß collective line (621.395)
Sammelbehälter collecting container
Sammelektrode collecting electrode, focusing electrode
Sammelgespräch multiple call, bunched call
Sammelkammer collector cell (628.511)
Sammelleitung collecting main, bus line
Sammellinse convex lens
Sammelrufnummer collective number
Sammelschaltung multi destination (circuit) (621.397)
Sammelschiene bus bar; trunk (681.3)
Sammelschienendrosselspule bus reactor
Sammelschienenkuppelfeld bar-coupling panel
Sammelschienenlängstrennschalter bus-bar sectionalizing switch
Sammelschienenschutz bus protection
Sammelschienentrennschalter bus isolator (B); bus-disconnect switch (US)
sammeln to collect, to gather, to accumulate
Sammler storage cell, accumulator
Sammlerbatterie storage battery, secondary battery
Sammlergefäß containing cell (621.35)
Sammlerplattenformierung forming accumulator plates, formation of accumulator plates
Sammlersäure battery acid
Sampling-Verfahren sampling
Samplingoszillograph sampling oscilloscope
Sand sand
Sandguß sand casting (621.7)
Sandpapier sand paper, glass paper, abrasive paper
mit Sandpapier abziehen to sand
Sandstein sandstone
Sandsteinschleifmaschine sandstone wheel
Sandstrahl sand blast
Sandstrahlgebläse sand-blast apparatus, sand-blast machine
Saphir sapphire
SA-Telefon automatic telephone, dialling telephone
Satellit satellite (52)
Satelliten-Erdefunkstelle satellite earth station
Satellitenfunk satellite communication, space radio
Satellitenrechner satellite computer
Sattdampf saturated steam, saturated vapour
satter Erdschluß dead earth (B); dead ground (US)
sättigen to saturate (54), (538), (621.1), (621.385)
gesättigter Dampf saturated vapour, saturated steam
gesättigter Eisenkern saturated-iron core
gesättigte Röhre saturated valve
gesättigte Verbindung saturated compound
gesättigter Zustand saturated state
Sättigung saturation (54), (538), (621.1), (621.385)
in Sättigung übergehen to level-off
Sättigungsamplitude saturation amplitude (621.396.61)
Sättigungsbegrenzung saturation limiting
Sättigungsbereich saturation range
Sättigungscharakteristik saturation curve
mit Sättigungscharakter saturable (621.3)
Sättigungsdichte, Sättigungsfeuchtigkeit saturation density
Sättigungsgrad degree of saturation
Sättigungsinduktion saturation induction, saturation density (538); saturation-flux density
Sättigungspunkt saturation point (536)
Sättigungssiebdrossel swinging choke
Sättigungsspannung saturation voltage
Sättigungsspule saturable inductor
Sättigungsstrom saturation current (621.385)
Sättigungstemperatur saturation temperature (536), (621.385)
Sättigungswert saturation value
Sättigungswiderstand saturation resistance (621.382.3)

Satz period (41); theorem (51); set (621); remnants (pl.) (54), kit
Satzverständlichkeit intelligibility of phrases, sentence articulation
Satz von der Ersatzstromquelle Thevenin's theorem
Satz von Kirchhoff Kirchoff's law
Satz Wechselräder set of change wheels
Satz Werkzeug outfit, tool kit
sauber clean; precise, correct, perfect
sauber einpassen to push-fit
sauber geformt clean-cut
sauber wiedergeben to record faithfully
sauberer Abgleich nice balancing
saubere Anpassung correct match (621.396)
saubere Ausrichtung perfect alignment
sauberer Eingriff exact register
sauberes Geben accurate sending (621.394)
sauberer Kreis perfect circle
sauberer Reaktor cold reactor, clean reactor
sauberer Schnitt clean cut (621.9)
Sauerstoff oxygen
Sauerstoffflasche oxygen cylinder
Sauerstoffgas oxygen gas
sauerstoffhaltig oxygenous
sauerstoffreich rich in oxygen
Sauerstoffschneiden oxy-arc cutting
Sauerstoffverbindungen oxy-compounds
Saug-... suction
Saugdrossel balance coil, interphase reactor (621.314.6); sucking inductor
saugfähig absorbent
Saughöhe suction height
Saugkreis wave trap (621.396.62); sound trap (621.397)
Saugleitung suction trunking; suction duct
Sauglüfter extraction fan
Saugpumpe suction pump, exhaust pump
Saugzug induced draught
Saugzuganlage induced-draught plant
Saugzuggebläse induced-draught fan, (induced-draught) blower
Säule column (72), (621), (621.385); pile, stack (621.3); pillar (624)
Säulenbohrmaschine floor-type drilling machine, column-mounted drilling machine
Säulendarstellung columnar display
Säulenständer pillar support
Säulenstelltransformator column-type adjusting transformer
Säure acid
Säurebestimmung acidmetry
säurebildend acid-forming
Säurebildner acid-forming substance
Säuredichte density of an acid
säurefest acid-proof, acid-resisting
säurefrei acidless
Säuregehalt acidity
Säuremesser acidimeter, battery hydrometer
Säureprüfer cell tester (621.35)
Säurerest acidic portion
Säurewert acid intensity, acidity
Säurezahl acid value
SAW-Ableiter autovalve, variable-resistance arrester
SBA-Verfahren SBA, standard beam approach system
Scatterverbindung scatter link
schaben to scrape, to shave (621.9)
Schaber scraper, scraping tool
Schablone former (621.3); template (621.9); stencil (651); mask (77)
nach Schablone bohren to drill to template
Schablonenstift former pin (621.9)
Schablonenwicklung preformed winding, former-wound coil
Schachbrettmuster check, checkerboard pattern
schachbrettartig gemustert checkered, tesselated
Schacht shaft (621.87), (622), (624); tunnel (622)
Schachtfördergerüst pit-head gear
Schachttür gate
Schachttürkontakt gate switch
Schachttürkontaktüberbrückungsschalter gate by-pass switch
Schachttürverriegelung gate lock
Schaden damage, fault, defect; harm
Schadenersatz compensation
Schadenverhütungsmaßnahme (damage) preventive measure
schadhaft faulty, defective, damaged
schadhafte Isolation faulty insulation
schädlich detrimental, harmful, injurious; noxious
schädliche Gase noxious gases
Schaft arbor, stem (621); shaft (621.9); shank (621.8)
Schaftform shank contour
Schäkel shackle
Schäkelisolator shackle insulator
Schale dish (77); cup; shell
Schalenmanometer cup anemometer
schalenförmig shell-like, cupped, semihollow
Schalenrumpf monocoque
Schall, schall-... sound
schallabsorbierend, schalldämpfend sound absorbant, sound absorptive
Schallabsorption sound absorption

Schallabsorptionsgrad sound-absorbing coefficient
Schallanalysator sound analyser
Schallaufnehmer sound receiver
Schallaufzeichnung sound recording
Schallausbreitung sound propagation
Schallausbreitungsgeschwindigkeit rate of travel of sound
Schallausschlag particle displacement
Schallbrett baffle (board), loudspeaker baffle
Schalldämmung sound insulation
Schalldämmstoff sound absorbant
schalldämpfend sound absorbing, muffling
Schalldämpfer silencer, silencing device (621.4); muffler (681.8)
Schalldämpfung sound absorption
schalldicht sound-proof
Schalldichte sound-energy density
Schalldosismesser sound dosimeter
Schalldruck (effective) sound pressure
mit Schalldruck ansprechendes Mikrophon pressure-actuated microphone
Schalldruckmesser sound-level meter
Schalldruckpegel sound-pressure level
Schalldurchgang transmission of sound
Schallecho sound echo
Schallempfänger sound receiver
Schallfluß volume velocity
Schallgeschwindigkeit sound velocity, sonic speed (534)
Schallgrenze sonic barrier (629.13)
Schallisolationsmaß reduction factor of sound insulation
Schallisolierung sound insulation
Schallkennwiderstand characteristic specific acoustical impedance
Schalleistung acoustic power
Schalleistungsbündelung sound-power concentration
Schallmauer s. **Schallgrenze**
Schallmedium sound-propagation medium
Schallmeß-. . ., Schallmessung sound ranging (623)
Schallortung sound location
Schallpegel sound level
Schallpegeldifferenz sound-level difference
Schallpegelmesser sound level meter
Schallplatte (grammophone) record, disc
Schallplatten-Abspielgerät disc record reproducing equipment, record player
Schallplattenaufnahme disc recording
Schallquelle sound source
schallreflektierend sound-reflecting
Schallreflexion sound reflection
Schallreflexionsfaktor sound-reflection factor
Schallreflexionsgrad sound-reflection coefficient
Schallschlucker sound absorper
Schallschluckgrad sound-absorption coefficient
schallschluckende Wandplatte acoustic(al) tile
Schallschluckmaterial absorbent material
Schallschnelle sound-particle velocity
auf Schallschnelle ansprechendes Mikrophon velocity (-operated) microphone
Schallschutz sound proofing, sound insulation
Schallschwingung sound vibration
Schallsender acoustic radiator, sound generator
Schallstärke sound density, sound energy, flux density
Schallstärkemesser sound meter, volume meter
Schallstrahler acoustic radiator
Schallstrahlung acoustic radiation
Schallstrahlungsdruck acoustic radiation pressure
Schallstrahlungsdruckmesser acoustic radiometer
Schallstrahlungsimpedanz acoustic radiation impedance
Schallstrahlungsreaktanz sound-radiation reactance
Schallstrahlungswiderstand sound-radiation resistance
schalltoter Raum anechoic room, anti-echo room
Schalltransmissionsgrad sound-transmission factor
Schalltrichter acoustic(al) horn, sound horn (621.396.62)
Schallübertragung transmission of sound
Schallwand baffle
Schallwandler acoustical electrical transducer
Schallwelle sound wave
Schalmeiglocke sheep gong (621.395)
Schalt-. . . switching, connecting, breaking (621.3); changing, shifting (629.113)
Schaltalgebra Boolean algebra, switching logic
Schaltanlage switchgear
Schaltanlage mit getrennten Phasen isolated-phase switchgear
Schaltanordnung switching arrangement, scheme of connections
Schaltapparat switching apparatus

Schaltart mode of connection
Schaltbefehl switching command, switching order
Schaltbild wiring diagram, circuit diagram
nach Schaltbild schalten to connect up as in diagram
Schaltbirne suspension push
Schaltdiode switching diode
Schaltdraht jumper wire, hook-up wire
Schaltelement circuit element, circuit constant; switching element (621.316.5)
Schaltfassung switch socket
Schaltfeld switchboard section
Schaltfolge order of connection
Schaltfunktion switching function
Schaltgerät switching gear, switching device
Schaltgerüst frame-type switchboard
Schaltgeschwindigkeit speed of contact operation
Schaltgetriebe gear box, speed box, change-over gear
Schalthaken switch hook
Schalthaus switch house
Schalthebel switch lever, operating lever
Schaltkapazität (stray) wiring capacity, circuit stray
Schaltkasten switch-enclosure, switchboard cell
Schaltkennzeichen signal (621.395.3)
Schaltknebel switch knob
Schaltkontakt switch contact
Schaltkreis circuit
Schaltkreisprüfgerät in-circuit tester
Schaltleiste connecting block
Schaltleistung breaking capacity, rupturing capacity, switching capacity, switched power
Schaltmesser switch knife, switch blade
Schaltpyramide whiffletree switch
Schaltpause circuit blueprint
Schaltphase switching angle
Schaltpilot switching pilot, switching-control pilot (621.395.4)
Schaltplan wiring scheme, connection diagram, circuit diagram
Schaltpult benchboard, desk-type switchboard
Schaltnetz combinatorial circuit
Schaltnetzteil switched-mode power supply = SMPS
Schaltrad ratched wheel
Schaltrichtung forward direction (621.38)
Schaltröhre electronic switch, switching tube
Schaltsäule switch (gear) pillar, switchgear pedestal (621.316)
Schaltscheibe switch wafer (621.396)
Schaltschema wiring scheme, schematic diagram, circuit diagram
Schaltschütz contactor
Schaltsicherung fuse-disconnecting switch
Schaltskizze strapping diagram
Schaltsperre switch-interlocking device
Schaltstange switch hook, switch stick, push rod
für Schaltstangenbetätigung stick operated (621.316.5)
Schaltstation switching station, sectionalizing station
Schaltstecker switch plug
Schaltstelle switching point; switch cabin (621.332)
Schaltstellungsanzeiger switch-position indicator
Schaltstück contact (member), contact segment
Schaltstufe switch step, notch
Schalttafel (panel) switchboard, contactor panel, patchboard, plugboard
Schalttafel mit spannungsführenden Teilen auf der Vorderseite live-front switchboard
Schalttafelanschluß switchboard connection
Schalttafelbalkon switchboard connection
Schalttafelfeld switchboard section, switchboard panel
Schalttafelgerüst switchboard frame, switchboard skeleton
Schalttafelinstrument panel (mounting) meter, switchboard (-type) instrument
Schalttafelrahmen panel framework, switchboard frame
Schalttafelschalter panel switch
Schalttafelwand switchboard panel
Schalttafelzähler switchboard meter
Schaltteilliste parts list
Schaltüberspannung switching surge, switching transient
Schaltuhr time switch, clock relay
Schaltverbindung connection, connecting lead (621.355)
Schaltverlust switch dissipation
Schaltverlustleistung switching-power loss (621.382.3)
Schalt- und Vermittlungsstelle switching center
Schaltvorgang switching breaking (process) (621.3)
Schaltvorrichtung switching device, switching arrangement
Schaltwagen switchgear truck, truck-type switchgear, carriage-type switchgear
Schaltwalze switching drum, tumbler drum

Schaltwarte switch station, control room (621.382)
Schaltweg contact travel
Schaltwerk switching mechanism
Schaltzeichen graphical symbol, diagrammatic symbol
Schaltzeit breaking period, switching time
Schaltzelle switchboard cubicle
schalten to switch (621.316); to wire, to connect (621.315); to shift (the gear), to change the gear (629.113)
schalten parallel zu to switch in parallel with
Schalter switch, breaker, circuit breaker, contact breaker, actuator
Schalter ausklinken to release a switch, to trip a switch
Schalter auslösen to trip a switch
Schalter betätigen to turn a switch, to operate a switch
Schalter mit Deckelverriegelung locked-cover switch
Schalter mit Drehstützer rotating-insulator switch
Schalter mit Druckluftbetätigung pneumatically-operated switch
Schalter mit Einfachunterbrechung single-break switch
Schalter auf Isolierfüßen tropical switch
Schalter mit Kippstützer tilting-insulator switch
Schalter mit Minimalauslösung minimum circuit breaker
Schalter mit rückwärtigem Anschluß back-connected switch
Schalter für Schalttafelverwendung panel switch
Schalter mit Schutzkappe protected switch
Schalter für Seitenbestimmung sense finder (621.396.9)
Schalter mit senkrechter Schaltstrecke vertical-break switch
Schalter umlegen auf to throw a switch to
Schalter mit UMZ-Auslösung definite time-lag circuit breaker
Schalter unter Verschluß locking switch, secret switch, asylum switch
Schalter mit Vielfachunterbrechung multibreak switch
Schalter mit vorderseitigem Anschluß front-connected switch
Schalter mit vorder- und rückseitigem Anschluß front and back connected switch
Schalter mit waagerechter Schaltstrecke horizontal break switch
Schalter mit Zugbetätigung pull switch
Schalter mit Zwischenraum zwische Wand und Grundplatte feet switch
Schalterabdeckblech switch plate
Schalteranschluß switch connection
Schalterauslösung durch Relais indirect tripping of a circuit breaker
Schalterbetätigung circuit-breaker operation
Schalterbetätigungsplatte mit Griff faceplate and handle
Schalterbirne pear switch, pressel switch
Schalterdeckel switch cover
Schalterebene switch deck
Schalterfassung key holder (621.32); switch socket
Schaltergestell switchgear pedestal
Schaltergriff switch knob
Schaltergrundplatte switch base
Schalterisolierung insulator unit (621.316.5)
Schalterkappe switch cover
Schalterkessel switch tank
Schalterklemme switch terminal
Schalterknopf switch knob
Schalteröffnungskraft throw-off force
Schalterprüfung rupturing-capacity test
Schalterreihe row of switches
Schalterscheibe wafer, (rotary) switch deck (621.396)
Schaltersegment section
Schalterstellung switch position, switch setting
Schaltung switching (621.316); circuit (arrangement), (circuit) wiring, network, combination (621.3)
Schaltungsart connection system, mode of connection
Schaltungslehre circuit theory, circuit analysis
Schaltungsschema wiring diagramm
Schaltwerk switching mechanism
Schamotte chamotte, fireclay, refractory clay
scharf sharp, defined, clean-cut, clear-cut; keen; well-focused (77)
scharf abgestimmt sharply tuned
scharf begrenzt sharply defined
scharf gezeichnete Linien clear scribed lines
scharfe Abstimmung sharp tuning
scharfes Bild sharp image, defined image
scharfes Gehör sharp ear, keen hearing
scharfe Grenzfrequenz sharp cut-off (frequency)
scharfe Kante sharp edge, keen edge, narrow edge
scharfe Keule narrow beam (621.396.67)
scharfer Knall sharp report
scharfer Knick sharp turn

scharfer unterer Knick der Kennlinie sharp cut-off (of the characteristic)
scharfe Resonanzkurve narrow resonance curve
scharfer Übergang sharp transition
scharfer Umriß sharp outline
scharfgängig V-threaded (621.8)
scharfgebündelter Strahl pencil beam, sting ray
scharfkantig und auf Maß schneiden to cut true and square
scharfstellen auf to focus on (621.385)
Schärfe sharpness, acuity
Scharfeinstellung focusing (535); sharp tuning (621.396.62)
Scharnier hinge; joint
Scharnierband strap hinge
Scharnierklappe hinged flap
schartig dented, jagged
Schatten shade, shadow
Schattenbild shadow picture, shadow graph (621.386)
Schattenfläche shadow surface
Schattenkeil shadow bar (656.7)
Schattenphotometer shadow photometer, Rumford photometer
Schattenzeigerinstrument shadow-column instrument
Schattenzone shadow region, blind zone (621.371)
schattieren to shade
Schattierung shade, gradation
schätzen to estimate, to take the gauge of
geschätzte Kosten estimated costs
Schätzungsfehler error in estimating
Schaubild graph, chart, diagram
schaubildlich diagrammatic (al)
Schauer shower (55)
Schaufel blade, vane (621.1), (621.2)
Schaufelrad impeller (621.1)
Schauglas sight glass, inspection glass
Schaulampe telltale lamp
Schauloch peep hole, inspection hole
Schaum foam, lather; froth; glair
Schaum bilden to lather
zu Schaum schlagen to beat up
Schaumdielektrikum dielectric foam
Schaumgummi rubber foam
Schaumkeramik foam ceramics (pl.)
Schaumverhütungsmittel antifroth preparation
Schauzeichen visual indicator, drop signal, optical sign
Scheibe disc, disk
Scheibenantenne disc antenna, cheese aerial
scheibenförmig discoid
Scheibenheizung screen demisting (629.113)
Scheibenkondensator disc-type capacitor
Scheibenröhre disc-seal valve
Scheibenspule pancake coil (621.3)
Scheibenwicklung disc-winding, sandwich winding
Scheibenwischer wind-screen wiper (Br.), windshield wiper (US)
Scheiben-Konus-Antenne discone antenna
Scheide-... dividing, separating
Scheideanlage separating plant
Scheidebehälter separating tank, precipitation tank, decanter, separating vessel
scheiden to separate, to precipitate
Scheider separator, precipitator (621.356)
Schein gleam, glare
Scheinwerfer flashlight, searchlight; head lamp (625), (629)
Scheinwerferanleuchtung directional floodlight system
Scheinwerferbeleuchtung spot lighting; flood lighting
Scheinwerferlampe reflector
Scheinwerferleitradar searchlight control radar
Scheinwerferspiegel searchlight reflector
Schein-... apparent (621.3); sham, dummy, mock
Scheinleistung apparent power, voltamperes (pl.)
Scheinleistungsmesser voltampere meter
Scheinleistungszähler trivector, kVA meter
Scheinleitwert admittance
Scheinlot ball-bank indicator (629.13)
Scheinwiderstand impedance, apparent, impedance, apparent resistance
Scheinwiderstand bei festgehaltenem System blocked impedance (534)
Scheinwiderstand bei Normalbetrieb normal impedance (534)
Scheinwiderstandsbrücke impedance bridge
Scheinziel sham target
scheinbar seeming, apparent
scheinbarer Fehlerort resultant fault
scheinbare Höhe der Ionosphärenschichten virtual height of ionospheric layers
scheinbare Lichtstärke apparent intensity
scheinbarer ohmscher Widerstand apparent ohmic resistance
scheinbare Remanenz apparent remanence
scheinbare Schichthöhe apparent layer height (55)
scheinen to shine

Scheitel peak, top (553); crest (535.2); vertex (pl.: -ices) (51)
Scheitelfaktor peak factor, crest factor, amplitude factor
Scheitellinie crest line
Scheitelpunkt vertex (51); zenith (52); turning point
Scheitelspannungsmesser peak voltmeter
Scheitelspeisung vertex feed (621.396.67)
Scheitelsperrspannung crest-working reverse voltage (621.382.2)
Scheitelwert peak, peak value, peak amplitude, crest, maximum
Scheitelwert des Einschaltstromes peak-making current
Scheitelwert einer Kurve tip
Scheitelwert einer Periode peak of a cycle
Scheitelwert der Spannung crest voltage
Scheitelwert der Sperrspannung peak-inverse voltage (621.314.6)
Schellack shellac (varnish)
Schellackanstrich shellac varnish, shellac finish
Schellacklösung shellac solution
Schelle grip (ring), clamp(-ing collar)
Schelleisen conductor clamp, holding clamp (699.83)
Schema scheme, schedule
Schemaschaltbild schematic diagram, wiring diagram
schematisch schematic, diagrammatic(-al), in skeleton form
schematisch darstellen to diagram
schematische Darstellung diagrammatic representation, schematic representation
Schenkel leg (51), (61); limb (621.314.2), (621.318)
Schenkel einer Rhombusantenne leg of a diamond aerial
Schenkel des Stechzirkels leg of a pair of dividers
Schenkelpolgenerator salient-pole generator
Scher-... shearing
Scherbeanspruchung shear (stress)
Scherbolzen shear pin
Scherfestigkeit shear strength
Scherkraft shear force
Scherspannung shear strain, shear stress
Scherversuch shearing test
Scherwirkung shearing effect
scherardisieren to sherardize
Scherardisierung sherardizing
Scherbius-Maschine Scherbius phase advancer
Schere (pair of) scissors; (pair of) shears; shear legs (621.8)
Scherengittertor collapsible gate
Scherenstromabnehmer pantograph (trolley)
Scherentrenner pantograph-type isolator
scheren to shear
Schering-Brücke Schering bridge
Scherung shear; shearing
Schicht layer; film; stratum (55)
mit einer Schicht überziehen to film
Schichtdickenverhältnis metal-distribution ratio (621.357)
Schichtdrehwiderstand variable-film resistor
Schichtfestwiderstand fixed-film resistor
Schichtkathode coated cathode
Schichtwiderstand film resistor, film type resistor, layer-type resistor
schichten to stack
Schiebe-... sliding; shift... (681.3)
Schiebekeil sliding key
Schiebekontakt sliding contact
Schieberegister shift register (681.3)
Schiebeschalter slide switch
Schiebesitz close sliding fit, push fit
Schiebetransformator slide transformer
Schiebetür sliding door
Schiebewiderstand sliding resistor, slider rheostat
schieben to thrust, to push
Schieber slide, slide valve, sliding valve
Schieblehre (sliding) caliper, vernir caliper rule, slide gauge
schief sloping, slanting; out of plumb, inclined; oblique
schief auftreffen auf to fall obliquely on
schief gemessener Abstand slant range
schief gemessene Höhe slant height
schief legen, schief liegen to slant, to bank
schief quetschen to lopside
schiefe Ebene incline, inclined plane
schiefer Kegel oblique cone
Schieflast unbalanced load
Schieflastschutz load-unbalance protection, negativ-sequence protection
schiefwinklig scalene, oblique angled
Schiefer slate, schist (55)
Schieflage tip, bank
schielende Charakteristik squinting lobe (621.396.677)
Schielwinkel squint
Schiene rail (625.1)
Schienenbeeinflussung rail effect (656.25)
Schienenbremse electromagnetic shoe brake
schienengleiche Kreuzung level crossing

Schienenkabel return cable (621.33)
Schienenkontakt rail-current switch; electrical depression bar
Schienenkreuzungssignal highway-crossing signal
Schienenlängsverbinder continuity rail bond
Schienenlasche fish plate (625)
Schienenoberkante rail level
Schienenprofil rail section
Schienenrückleitung track return
Schienenschalter track switch
Schienenseilbahn funicular railway
Schienenstrang track
Schienenstrom rail current
Schienenstrombremse magnetic track braking
Schienenstromkreis pedal circuit
Schienenstromwandler bus bar transformer, bar-type current transformer
Schienenverbinder (track) rail bond, cross-cable bond (621.33)
Schießdübel shoot-in bolt
Schiff ship, vessel
Schiffsantenne shipborne aerial
Schiffsantrieb marine propulsion
Schiffseinfluß calibration error (621.396.9)
Schiffsfunk marine radio
Schiffsfunker ship's radio operator
Schiffsinstallation marine wiring (621.3)
Schiffskreisel gyrostatic stabilizer
Schiffsradar shipborne radar, naval radar
Schiffsender ship transmitter, shipborne transmitter
Schild shield, plate; designation plate, label
Schildlager end-shield bearing
Schimmel mould; mustiness, mildew
Schimmelbeständigkeit fungus resistance
schimmelig mouldy; mildewed, musty
Schimmer shimmer, glimmer, glitter, gleam
schimmern to shimmer, to glimmer, to glitter, to gleam
Schirm umbrella; screen (53), (621.396), (77); shield (538), (621.317/19); shell (621.315.6)
Schirm einbrennen to burn the screen
auf einem Schirm aufleuchten lassen to flash on a screen
Schirmantenne umbrella aerial
Schirmbild pattern, cathode ray, (tube) trace, scope (621.385)
Schirmbilddarstellung display, scan
Schirmbild mit Nullkreisdarstellung open-centre display
Schirmelektrode grading shield (621.315.1)
Schirmfaktor degree of screening
Schirmfläche face area (621.397)
Schirmgenerator umbrella-type alternator (621.313)
Schirmgitter screen grid, shielding grid (621.385)
Schirmgitterblock screen-bypass capacitor (621.385)
Schirmgitterröhre screened pentode, screen-grid valve, screened-plate valve
Schirmgitterverlustleistung screen-grid dissipation
Schirmgitter(vor)widerstand screen-dropping resistor
Schirmisolator petticoat insulator, umbrella insulator
Schirmleiter screening conductor
Schirmring guard ring, grading ring, static end ring (621.315)
Schirmwand shade
schirmen gegen to shield from
geschirmtes Schaltgerät semi-enclosed switchgear
Schirmung shielding
Schlacke slag; scoria (55)
schlaff loose, slack
Schlaffseilschalter slack-rope switch, slack-cable switch (621.876)
Schlag percussion; stroke, bump; impact; (electric) shock (61); lay (621.315)
Schlag bekommen to get a shock (621.3)
Schlagbiegefestigkeit resistance to impact, impact (bend) strength
Schlagbohrmaschine percussion drilling machine
Schlagdämpfer bumper, impact-damping device
schlagfestes Material high impact material
Schlagfestigkeit impact resistance
Schlaghammer pendulum tup
Schlaglänge lay (621.315.1)
Schlaglot brazing spelter
Schlagprobe, Schlagprüfung impact test
Schlagrichtung direction of lay (621.315.1)
Schlagstauchversuch impact-compression test
Schlagwetter fire damp, mine gas
Schlagwetteranzeiger fire-damp indicator, mine-gas indicator
Schlagwetterschutz flame protection, explosion protection
Schlagwetterschutz, erhöhte Sicherheit increased-safety enclosure
Schlagwetterschutzkapselung flame-proof enclosure; explosion-proof enclosure (US)

schlagwettersicher, schlagwettergeschützt explosion-proof, flame-proof, flame-tight, flame-protected (621.313)
schlagwettergeschützter Motor explosion-proof motor (US); flame-proof motor
Schlamm mud, sludge
Schlamm bilden to sludge
Schlammabscheider decanter
Schlammbildung sludging
Schlammpumpe mudpump
Schlammreaktor slurry reactor
Schlämmkreide whiting, French chalk
Schlangenbohrer twisted drill
Schlangenkühler interlaced cooler
Schlankheitsgrad slenderness ratio
Schlauch hose; tube, tubing (629.113), (629.118)
Schlauchanschluß hose tap
Schlauchkupplung hose coupling
Schlauchtrommel hose reel
Schlauchverbindung hose connection
schlecht bad, poor, imperfect
schlecht abgeschirmt poorly screened
schlecht erhalten poorly maintained, in bad repair
schlecht geladen in a low state of charge (621.356)
schlecht isoliert leaky
schlecht repariert in bad repair
schlechter Abgleich, schlechte Angleichung maladjustment
schlechte Arbeit poor workmanship
schlechte Ausrichtung misalignment (621)
schlechte Erdung bad earth, imperfect earth
schlechte Erdungsverhältnisse ground imperfections
schlechter Isolator poor insulator
schlechter Kontakt bad contact, poor contact
schlechter Leiter poor conductor
schlechte Lötstelle rosin joint, dry joint, rosin connection
schlechte Qualität poor quality, inferior quality
in schlechtem Erhaltungszustand in bad repair
Schlechtwetterflugbetrieb all-weather operation
Schlechtwetterkorona foul-weather corona
schleichender Erdschluß earth leakage
Schleif-... slip, slide, collecting (621.3); abrasive, grinding
Schleifarbeit grinding work
Schleifband grinding belt
Schleifdraht slide wire, potentiometer wire (621.317)
Schleifdrahtmeßbrücke slide-wire bridge, meter bridge
Schleifdrahtkontakt slider, sliding contact (621.3)
Schleifmittel abrasive
Schleifring collecting ring, slip ring (621.313)
Schleifringanlasser slip-ring starter
Schleifringdrehstrommotor wound-rotor induction motor
Schleifringisolation slip-ring bush
Schleifringläufer slip-ring rotor; wound rotor (US)
Schleifringleistung slip power (621.313)
Schleifringmotor slip-ring induction motor, wound-rotor induction motor
Schleifringspannung slip-ring voltage
Schleifringspeichen slip-ring spider (sing.)
Schleifringstrom slip-ring current
Schleifriß grinding crack
Schleifschuh skid
Schleifstaub grinding particles (pl.)
Schleife loop (621.396.667), (681.3)
Schleifen schließen sich the loops close
Schleifendipolantenne folded dipole antenna
Schleifenfehlerortsbestimmung Allen's loop test
Schleifengalvanometer torsion-string galvanometer, torsion-loop galvanometer
Schleifenmethode zur Fehlerortsbestimmung Murray's loop test
Schleifenoszillograph moving-coil oscillograph, bifilar oscillograph
Schleifenschaltung loop test (621.315.2)
Schleifenverstärkung closed-loop gain
Schleifenwicklung lap winding, multiple winding, multiple circuit (621.313)
Schleifenwiderstand loop impedance
schleifen to grind
geschliffenes Endmaß ground block
Schleifer slider (621.3); wiper
Schleim slurry
Schlemmkreide carbonate of lime
Schlepp-... trailing
Schleppantenne trailing aerial
Schleppdraht trailing wire
Schleppkabel trailing cable
Schleppkabellokomotive electric cable-reel locomotive
Schleppkanal experiment tank (629.12); water way (532)
Schleppkontakt trailing contact
Schleppwiderstand drag
Schleppzeiger drag hand, follow-up pointer
schleppen to tow, to haul; to drag

schleppen lassen to trail
Schlepper tractor (629.11); tug (629.12)
Schleuder centrifuge; spin dryer, laundry dryer (64)
Schleudergußstück centrifugal casting
schleudern to centrifuge
Schleuse sluice, lock (626)
Schleusenspannung threshold voltage (621.314.63)
Schleusentor sluice gate
Schlickerguß slip casting (666)
Schliere stria (pl.: -ae)
Schlierenbildung striation
schlierig striated
Schließ-... locking; closing
Schließblech locking plate
Schließimpuls make pulse, make impuls
Schließkeil cotter
Schließvorgang closing, locking
Schließvorrichtung closing device, locking device
schließen to shut, to close; to make (621.318.5)
schließen, sich to close
schließen aus to conclude from (51)
geschlossene Maschine enclosed machine
geschlossene Maschine mit Fremdbelüftung enclosed separately-ventilated machine
geschlossener Motor enclosed motor
geschlossener Motor mit Außenbelüftung enclosed fan-cooled motor
geschlossene Nuten totally closed slots
geschlossene Ringwicklung closed-coil armature
geschlossenes Schalttafelgerüst enclosed switchboard
geschlossener Stromkreis closed circuit
geschlossener Umlaufweg complete path (537)
auf dem geschlossenen Umlaufweg around the loop (621.3)
bei geschlossenem Deckel with cover on
in sich geschlossene Figur closed figure
in sich geschlossene Gleichstromwicklung closed-circuit d. c. winding
in sich geschlossener Hohlraum closed cavity
in sich geschlossene Masche closed loop (621.3)
in sich geschlossene Wicklung ring winding, closed-coil winding
Schließungsfunke closing spark, spark at make
Schließer make contact, no-(normally open) contact
Schliff cut, micro-section (669)
Schliffbild micrograph (669)
Schlingenisolator link-type insulator
Schlingerbewegung rolling motion
Schlingerdämpfungseinrichtung roll stabilizer
Schlitten movable carriage (621.396); saddle (621.9); sledge
Schlitz slit; slot
Schlitzanode split anode
Schlitzanodenmagnetron split-anode magnetron
Schlitzantenne slot aerial, pylon antenna
schlitzgespeister Dipol slot-fed dipole
Schlitzkopfschraube slotted-head screw
Schlitzring split ring
Schlitzscheibe slotted washer, slotted disk
Schlitzschienenstromabnehmer plough
Schlitzschienenstromzuführung centre-slot system
Schlitzschienensystem slot system (621.33)
Schlitzschweißung slot weld(ing)
Schlitzstrahler slot antenna
Schlitztransformator split-tube balun
Schlitzverschluß drop shutter (77)
schlitzen to slit, to slot, to split
geschlitzte Unterlegscheibe slotted washer
Schloß lock
Schloßschraube coach screw
Schlosser locksmith, fitter, mechanic
Schlosserarbeit bench work
Schlosserei fitting shop
Schlot funnel, chimney (stack)
schlucken to absorb
Schlucker absorber (534)
Schluckplatte acoustic tile
Schluckstoff sound absorbant
Schluckvermögen absorption capacity
Schlupf slip, slippage
Schlupfdrehzahl slip (speed)
Schlupffrequenz slip (frequency)
Schlupfregler slip regulator
schlüpfen to slip
schlüpfrig slippery
Schluß closing, end; conclusion
Schlußlampe leaving lamp, supervisory lamp (625)
Schlußlicht tail light, rear light (625.2); marker light (656.25)
Schlußlinie closer, closing line
Schlußlinie beim Krafteck equilibrant
Schlußzeichen clearing signal, disconnect signal (621.395); clearback signal, on hook signal, clearing indicator
Schlußzeichen geben to ring off (621.395)
Schlüssel key; spanner (621); cipher, crypto (621.39)
schlüsselfertiges Projekt turnkey project

Schlüsselloch keyhole
Schlüsselmaschine coder, cipher machine, cipher equipment, crypto equipment
Schlüsselschalter key-operated switch
Schlüsselverzeichnis signal book
Schlüsselwort keyword
schlüsseln to code
Schmal-... narrow
Schmalbandempfang narrowband reception
Schmalbandfernsehen low-scan TV, narrow-band television
Schmalbandübertragungssystem thin line transmission system, narrow-band transmission system
Schmalfilm narrow-gauge film
Schmalspurbahn narrow-gauge railway
Schmalspurbahnmotor narrow-gauge motor
Schmalspurstraßenbahn narrow-gauge tram
Schmalsuchen azimuth narrow mode (621.396.96)
Schmalwinkelstreuung narrow-angle diffusion (535)
schmal narrow
schmal machen to narrow
schmales Band narrow band (621.396)
schmales Bündel narrow pencil
schmales Frequenzband narrow frequency range
schmale Resonanzkurve narrow resonance curve
Schmelz-... fusible, smelting
Schmelzbad pool (621.791); molten bath
Schmelzdraht, Schmelzeinsatz fusible element, fusible member, fusible link
Schmelzflußelektrolyse fusion electrolysis
Schmelzleiter fusing conductor
Schmelzofen smelting furnace
Schmelzprozeß melting process
Schmelzpunkt melting point; blowing point (621.316.9)
Schmelzschweißung fusion welding
Schmelzsicherung fusible cut-out, safety fuse, fuse plug
Schmelzstreifen fusible-metal strip, fuse element
Schmelzstrom fusing current, melting current
Schmelztemperatur temperature of fusion, melting point
Schmelztiegel melting crucible, melting pot
Schmelzverlust smelting loss
Schmelzwärme heat (equivalent) of fusion
Schmelzzone fusion zone
Schmelzzustand state of fusion
schmelzbar fusible, smelting
Schmelzbarkeit fusibility
Schmelze melt, pool (66)
schmelzen to melt, to smelt, to fuse; to liquefy
schmelzflüssiges Metall molten metal
Schmetterlingsantenne batwing aerial
Schmetterlings-Drehkondensator butterfly capacitor
Schmied smith, blacksmith
schmiedbar forgeable, malleable
schmiedbarer Guß malleable casting
Schmiede smithy, forge, blacksmith's shop
Schmiedearbeit smithery
Schmiedeeisen W. I. = wrought iron, malleable iron, mild steel
Schmiedegesenk swage block
Schmiedestück forging
schmieden to forge
Schmiege (bevel) protactor (621.9)
schmiegen to osculate (51)
schmiegsam pliable
Schmiegsamkeit pliability
Schmier-... oiling, lubricating
Schmieranlage oiling system, lubricating system
Schmierfett (lubricating) grease
Schmiermittel lubricant, lubricating substance
Schmiernippel (oil gun) nipple, lubricator
Schmieröl lubricating oil
Schmierpresse grease gun
Schmierpumpe lubricating pump
Schmierring oil ring, ring lubricator
Schmierseife soft soap
Schmierstelle oiling point, lubrication point
Schmiersystem lubrication system
Schmiervorrichtung oiling arangement
Schmiere grease
schmieren to oil, to grease, to lubricate; to paste (621.356)
schmierig greasy
Schmierung oiling, lubrication
Schmirgel emery, carborundum, gritstone, abrasive
Schmirgelleinen emery cloth
Schmirgelpapier emery paper, abrasive paper
Schmirgelpulver emery powder
Schmirgelscheibe emery wheel, carborundum wheel
Schmitt-Trigger-Schaltung Schmitt trigger circuit
Schmitzring cylinder bearer (655)
Schmorperle pitting (621.3)
mit Schmorperlen besetzt pitted
Schmorstelle spot of arcing
Schmuck decoration
Schmucklampe decorative lamp

Schmuckleuchte ornamental lantern
Schmutz dirt, mud, smudge
Schmutzabweiser splash apron
Schmutzfang, Schmutzfilter dirt absorber
schmutzig dirty, muddy
Schnallenisolator buckle insulator, shackle insulator
Schnapp-... snap...
Schnappkontakt snap-action contact
Schnappschalter snap (action) switch
Schnappschloß spring lock
Schnappverschluß spring catch, snap lock
Schnarre hummer, buzzer
Schnecke worm (621.83); scroll (72)
Schneckenantrieb worm drive
mit Schneckenantrieb wormdriven
Schneckengetriebe worm gear
Schneckenlinie spiral (line)
Schneckenrad worm pinion, worm wheel, thread worm
Schneckenradübersetzung worm-wheel transmission
Schneckensegment quadrant (621.833)
Schneckenbetrieb, Schneckenuntersetzung worm-reduction gear
Schneelast snow pressure
Schneid-... cutting
Schneidbrennen (torch-)cutting
Schneidbrenner cutting torch, oxygen cutter
Schneideisen thread-cutting die, threading die
Schneidkante cutting edge, tip of the tool, bit (621.9)
Schneidkantenwinkel cutting angle
Schneidmetall cutting metal
Schneidnadel recording cutter (681.84)
Schneidöl cutting oil (621.9)
Schneidplättchen tool bit, tool tip (621.9)
Schneidspäne cuttings
Schneidstichel s. **Schneidnadel**
Schneidvorgang cutting (process)
Schneidwerkzeug cutting tool
schneiden to cut
schneiden, (sich) to intersect
sich schneidende Linien concurrent lines
schneidengelagert supported on a knife edge (621.26)
Schnell-... quick (-acting), rapid, high-speed
Schnellabtastung rapid scan (621.397)
Schnellamtsverkehr demand working (621.395)
Schnellanlauf quick start, rapid starting (621.313)
Schnellantrieb high-speed drive
Schnellauslöser instantaneous trip, instantaneous release, quick-response release
Schnellbahn rapid transit railway
Schnellbereitschaftsaggregat short-break standby set
Schnelldrucker high-speed printer
Schnelleinschalt-... quick-make
Schnellentwickler rapid developer
Schnellerregung high-speed excitation (system)
Schnellgang high speed, rapid travel, rapid motion
Schnelläufermotor high-speed motor
Schnellprüfung rapid checking
Schnellregler quick-acting regulator, high-speed regulator (621.3), (621.316.7); high-speed governor (621.1)
Schnellregelung quick-acting regulation, high-speed regulation
Schnellrelais quick-acting relay, quick-response relay
Schnellschalter fast-action switch, quick-break switch
Schnellschlußauslösung overspeed trip
Schnellschlußeinrichtung quick shut-off device
Schnellschlußschutz overspeed protection
Schnellschlußstab shut-of rod
Schnellschlußventil quick-closing valve
Schnellspannmutter quick nut
Schnellspeicher quick-access store, rapid memory, high-speed storage (681.3)
Schnelltelegraph high-speed telegraph
Schnelltelegraphie high-speed transmission
Schnelltelegraphiesystem high-speed telegraph (transmission) system
Schnelltrennstrecker bail-out plug, quick disconnect plug
Schnellverkehr no-delay service (621.395)
Schnellverkehrsamt toll exchange
Schnellverkehrsleitung toll line
Schnellverkehrsschrank toll board
Schnellversuch quick test
Schnellzug express (train)
schnell rapid, speedy, swift, quick; fast
schnell fahren to speed
schnell laufende Welle high-speed shaft
schnell öffnen to fly open
schnell wirkend quick-acting
schneller Anlauf rapid starting (621.313)
schneller Brutreaktor fast breeder
schnelles Elektron high-speed electron
schneller Reaktor fast reactor
schnelles Teilchen high-speed particle
zu schnell laufen lassen to overspeed, to overrun

Schniewindtband resistance mat, woven sheet (621.36)
Schnitt section, intersection; cut (621.91/.93/.94/.96); die (621.96/.97)
zum Schnitt bringen to intercept (51)
Schnittansicht sectional view
Schnittbandkern C-core, split-tape wound core
Schnittbild sectional view
Schnittebene section plane
Schnittfläche cut face; sectional area
Schnittlinie intersection line
Schnittpunkt junction (625); intercept, intersection (point) (51)
Schnittpunkt der Lote im Dreieck orthocentre
Schnittpunkt der Seitenhalbierenden im Dreieck centroid
Schnittpunkt mit der X-Achse x-intercept
Schnittstelle interface, port (681.3)
Schnittstelle für Fernschreibanschluß und serielle Datenübertragung serial and teletype interface
Schnittverfahren method of sections (531.2)
Schnittzeichnung sectional view, cut-away drawing, sectional representation
Schnüffelventil self-acting valve, sniffing valve
Schnur cord; plumb line
Schnuraufhängung flexible suspension (621.32)
schnurloser Schrank cordless PBX (621.395)
Schnurpaar cord circuit (621.395)
Schnurrolle pulley
Schnurschalter pendant switch, pear switch
Schnurschaltung cord circuit
Schnurverstärker cord-circuit repeater (621.395)
Schnurzug cord gear
schnurren to purr
Schock shock (621.3), (61)
schonende Behandlung gentle manipulation
Schönwetterkorona fair-weather corona
Schoopisierung (zinc) metal spraying
Schornstein funnel, chimney (stack), smoke stack
Schotter crushed rock (624); ballast (625)
Schottky-Diode Schottky barrier diode
Schottky-Effekt Schottky effect
schraffieren to hatch, to shade; (auf Zeichnungen:) to section
Schraffierung, Schraffur hatching
Schraffierungslinie hatch, hatching-line
Schräg-..., schräg inclined, diagonal; sloping; oblique
schräg abfallen to slope
schräger Bürstenhalter trailing brush holder, reaction brush holder
schräge Nuten skewed slots
Schrägaufzug incline (d) hoist
Schrägeinfall oblique incidence (621.396)
Schrägentfernung slant range
Schrägfehler skew error (621.397)
Schräglage tilt, inclined position; bank (629.12)
Schräglauf skew
Schrägschnitt bevel cut (621.9); diagonal section (51)
Schrägsichtradar side-looking radar
Schrägspurverfahren helical recording (681.84)
schrägstellen to tilt, to incline, to skew
Schrägstrich stroke
Schrägverzahnung helical gear(ing)
Schrägwinkelübertragung oblique-incidence transmission
Schrank cupboard; switchboard cabinet, cubicle (621.395)
Schrankbeamtin switchboard operator
Schrankherbeiruf operator recall
Schranklampe switchboard lamp
Schrankstromkreis cord circuit
Schrankvermittlung switchboard exchange
Schranke barrier (621.3); gate (625)
Schraub-... screw, screwed
Schraubdeckel screw cap
Schraubfassung screw socket; screwed-lamp socket
Schraubkappe screw cap
Schraubklammer grip screw
Schraubkontakt screwed contact
Schraublehre screw caliper
Schraubmikrometer micrometer caliper
Schraubsicherung screw-plug (cartridge) fuse
Schraubstecker (plug) adapter
Schraubstock vice
Schraubtrieb screw gearing
Schraube screw; bolt
Schraube anziehen, Schraube einschrauben to drive a screw
Schraube gängig machen to ease a bolt
Schraube klemmt a screw binds
Schraube mit Vierkantkopf square headed bolt
Schraubenbolzen screw bolt, male screw
Schraubendreher screw driver
Schraubeneinstellung screw adjustment, screw setting
Schraubenfeder helical spring
Schraubengang screw pitch
Schraubengewinde (bolt) thread
Schraubenkopf bolt head

Schraubenlinie helical curve, helical line, helix
Schraubenmutter female screw nut
Schraubenschlüssel spanner, wrench, key
Schraubensicherung friction washer, screw locking device
Schraubenventilator propeller fan
Schraubenwindung turn of a screw
schrauben to screw
Schrecksekunde reaction time, personal equation
Schreib-... writing, recording
Schreibdichte recording density
Schreibempfänger recording receiver, recorder
Schreibfeder inkpen, recording pen (621.317)
Schreibfleck scanning spot
Schreibgerät recording instrument, recorder
Schreibgeschwindigkeit recording speed
Schreibkopf recording head, writing head
Schreib-Lesekopf write-read head
Schreib-Lesespeicher random access memory, RAM
Schreibmagnet writing magnet
Schreibmarke cursor
Schreibmaschine typewriter
Schreibstift stylo
Schreibstrahl writing beam (621.396.96)
Schreibstreifen chart (621.317)
Schreibverfahren recording mode
Schreibweise notation (scheme), mode of representation
schreiben to write, to record
schreibendes Fernmeßinstrument tallygraph, recording telemeter
schreibendes Meßinstrument recorder, recording instrument, inkpen recorder
Schrieb record (621.317)
Schritt step; jump (51); signal element (621.39)
Schrittgeschwindigkeit modulation rate, transmission speed
Schritt für Schritt step by step
schritthaltend step-by-step (621.39); on-line (681.3)
schritthaltende Datenverarbeitung realtime operation
schritthaltender Verbindungsaufbau step-by-step operation, stage-by-stage switching
nicht schritthaltender Verbindungsaufbau common-control switching
Schrittmotor stepper motor, stepping motor
Schrittrelais stepping relais
Schrittschaltwähler step-by-step switch, stepping switch
Schrittschaltwerk step-by-step motor, step-by-step mechanism
Schrittspannung surface-voltage gradient
Schrittverzerrung telegraph-signal distortion
Schrittwählersystem step-by-step automatic telephone system
Schrittzähler step counter
schrittweise Annäherung process of trial and error
schrittweise Bewegung step-by-step movement
Schroteffekt shot effect, auto-electronic effect, Johnson noise (621.385)
Schrotrauschen valve noise snow (621.397); shot noise (621.382)
Schrott scrap
schrumpfen to shrink
Schrumpfriß contraction crack
Schrumpflack shrivel varnish
Schrumpfpassung shrink fit
Schrumpfspannung shrinkage stress, contraction strain
Schrumpfung shrinking, contraction
Schub push, (horizontal) thrust; shear force
Schub ausüben to thrust
Schubbeanspruchung shear stress
Schubfestigkeit shear strength, torsional strength
Schubkraft shear force, tangential force
Schublehre (sliding) caliper, vernier caliper rule
Schubmodul to rsional modulus, modulus of rigidity, shear modulus
Schubspannung shear stress, tangential force
Schubtransformator slide transformer
Schubtrenner sliding switch
Schuko = Schutzkontakt
schummern to hatch, to shade
Schuppe scale, flake (621)
Schuppenbildung flaking
Schuppenlochstreifen shedless tape
Schüreisen stoker, stoker's rod
Schütte hopper, chute
Schüttel-... vibration, shaking
Schütteleinrichtung vibrator (621)
Schüttelförderband shaker conveyor
Schüttelprüfung vibration test
Schüttelrost vibrating grate
Schüttelrutsche shaker conveyor
schütteln to shake
schüttelfest vibration-proof
Schutz protection; guard, shield
gegen absichtliche Berührung protection against intentional touch
gegen Berührung mit den Fingern p. against finger touch

gegen Berührung mit Werkzeugen p. against contact with tools
gegen Druckwasser p. against water under pressure
gegen Fremdkörper p. against foreign bodies
gegen großflächige Berührung mit der Hand p. against touch with the flat of the hand
gegen inneren Überdruck p. against internal pressure
gegen Regen p. against rain
gegen schädliche Staubablagerung p. against harmful dust deposits
gegen Schwallwasser p. against water surges
gegen Spritzwasser p. against splash water
gegen Strahlwasser p. against jet water
gegen Tropfwasser p. against dripwater
gegen unbefugten Eingriff tamper proof
gegen zufällige Berührung p. against accidental contact
Schutzabdeckung protective covering
Schutzabstand guard band (621.395)
Schutzanode sacrificial anode (621.35)
Schutzanstrich serving, protective coating (621.315.2)
Schutzanzug overall
Schutzapparat protective apparatus
Schutzart type of protection
Schutzart druckfeste Kapselung flameproof enclosure; explosion-proof enclosure (US)
Schutzart erhöhte Sicherheit increased-safety enclosure
Schutzbereich protection range
Schutzblech protecting plate, shield; mudguard (629.118)
Schutzbrille (safety) goggles (pl.)
Schutzdeckel protection cover
Schutzdrähte screening wires (621.315.1)
Schutzerdung equipment ground, protective earthing, protective ground (CCITT)
Schutzfunkenstrecke protective spark gap
Schutzgas non-oxidizing atmosphere
Schutzgas-Elektrodenschweißung shielded-metal arc welding
Schutzgaskontakt dry reed contact
Schutzgasschweißen (inert-)gas-schielded welding
Schutzgehäuse protective housing, protective casing
Schutzgerät protective device, protective apparatus
Schutzgitter fence; screen grid (621.385)
Schutzhaube, Schutzhülle protective shell
Schutzkanal protection channel; standby channel (621.395)
Schutzkappe protection cap
Schutzkontakt protective contact
Schutzkontaktleiter earthing conductor; grounding contactor (US)
Schutzkontaktsteckdose earthing-contact socket outlet; grounding-type attachment receptacle (US)
Schutzkontaktstecker earthing-contact plug; grounding-type attachment plug (US)
Schutzkorb wire guard
Schutzmantel protective jacket
Schutzmaßnahme protective measure, safeguard
Schutznetz catch net, guard net, cradle guard (621.315.1)
Schutzplatte protecting plate
Schutzplattenordnung baffle system
Schutzrelais protective relay
Schutzring guard ring; guard plate (621.317)
Schutzschalter protective switch, protection circuit breaker; earth-leakage trip
Schutzschicht protective layer, protective sheath
Schutzschild protective shield
Schutzüberzug protective coating (621.315.3)
Schutzumbandlung friction tape (696.6)
Schutzwiderstand protective resistor
Schütz contactor, magnetic switch
Schütz mit Relais automatic tripping contactor
Schützenfeld contactor panel
Schützengruppe contactor bank; unit switch (621.33)
Schützensteuerung contactor control, automatic controller
Schützensteuerwalze contactor controller
Schütz-Sterndreieckschalter contactor-type star-delta starter
schützen to protect, to guard, to safeguard (gegen: against; vor: from)
geschützter Anlasser enclosed starter
geschützte Maschine protected machine (621.3)
geschützter Motor protected motor, semi-enclosed motor
geschützte Sicherung protected fuse
Schwach-. . ., schwach weak, light, lean, feeble
schwach erregt weakly excited
schwach gedämpfter Resonator persistent resonator
schwach gekoppelt low degree of coupling
schwach gekrümmt slightly curved
schwach ionisiert lighty ionized
schwach vergrößerndes Mikroskop low-power microscope

schwaches **Echo** weak echo
schwacher **Empfang** weak signal
schwaches **Feld** feeble field
schwache **Konstruktion** weak design
schwacher **Lichtschein** shimmer
schwacher **Reflex** faint reflection
schwacher **Strom** weak current, faint current
zu schwacher **Strom** undercurrent
schwache **Stromquelle** low-current source
Schwachlast light load
Schwachlastprüfung light-load test (621.317)
Schwachstelle weak spot, fail-soft spot
Schwachstrom light current, low-tension current
Schwachstrombeeinflussung telephone interference (factor) (621.392)
Schwachstromingenieur communication engineer
Schwachstromkabel light-current cable
Schwachstromtechnik communication engineering, light-current engineering
Schwachstromtransformator communication transformer, low-power transformer
schwächen to weaken, to attenuate
Schwaden fume, damp
schwadenbeständig fume-resistant
Schwalbenschwanz dovetail
schwalbenschwanzförmiger Keil dovetail key
Schwalbenschwanznut dovetail keyway
Schwallöten flow soldering
Schwallötstrecke flow-solder line
schwallwasserfest splash-proof
schwallwassergeschützter Motor splash-proof motor
Schwanenhals swan neck, goose neck
schwanken to vary, to fluctuate; to undulate, to oscillate, to pulsate; to swing, to sway...
schwanken um ... als Mittelwert to vary about
schwankender Gleichfluß pulsating flux
Schwankung variation, fluctuation; undulation, oscillation, pulsation; swing, sway
Schwankungen der Anodenspannung plate-supply ripple
Schwankung der Motordrehzahl motor-speed variation
Schwarz-..., schwarz black
Schwarzabhebung black level set-up, lift, pedestal
Schwarzhörer unlicensed listener, deadhead
Schwarzkupfer black copper
Schwarzschulter porch
Schwarzsender unlicensed transmitter (operator)
Schwarzspitze blacker-than-black
Schwarz-Weiß-Bildröhre black-and-white picture tube
Schwarz-Weiß-Fernsehen black-and-white television, monochrome TV
Schwarz-Weiß-Steuerung on-off control
Schwarz-Weiß-Vorlage black-and-white subject
Schwarz-Weiß-Zeichnung black-and-white drawing
Schwarzwert black level
Schwarzwertempfang black recording (621.397)
Schwarzwerthaltung d. c. restoring, black-level control
Schwarzwertklemmung black-level clamping
schwarze Emaille black enamel (621.315.6); black porcelain (666.29)
schwarzer Körper black body (53)
schwarzer Strahler black body, full radiator (53)
schwarze Tusche India ink
schwarzlackiert black enamelled (621.315.6)
Schwärze black
Schwebebahn suspension railway
Schwebespannung floating voltage
Schwebeteilchen suspended matter
schweben to float
schwebend suspended, floating
Schwebung beat
Schwebungsempfang beat reception, heterodyne reception
Schwebungsfrequenz b. f. = beat frequency
Schwebungsmeßbrücke heterodyne bridge
Schwebungsnull zero beat
Schwebungssummer b. f. o. = beat frequency oscillator
Schwebungston beat note, difference tone
Schwefel sulphur
Schwefelblüte flowers (pl.) of sulphur
Schwefelkies iron pyrite
Schwefelkohlenstoff carbon disulphide
Schwefelsäure sulphuric acid
Schwefelwasserstoff sulphuretted hydrogen, hydrogen sulphide
schweflig sulphurous
schweflige Säure sulphurous acid
Schweigekegel cone of silence, cone of nulls (621.396)
Schweiß-... welding
Schweißaggregat welding unit
Schweißbogen welding arc
Schweißbrenner welding burner, welding torch, blow pipe
Schweißdraht welding rod

Schweißelektrode welding electrode
Schweißflamme welding flame
Schweißgenerator welding generator
Schweißgut deposit
Schweißkabel welding lead
Schweißkonstruktion welded work
Schweißkopf welding head
Schweißlage weld layer
Schweißlichtbogen welding arc
Schweißmaschine welding machine, welder
Schweißmaschinenmaul welder throat
Schweißmaterial weld
Schweißmetall filler metal
Schweißmittel welding flux
Schweißnaht welding seam, welded seam
Schweißpaste welding flux
Schweißperle globule, bead
Schweißprofil welded steel section
Schweißprüfung welding test
Schweißrahmen welded frame
Schweißraupe welding fillet
Schweißspannung (welding) arc voltage
Schweißstab welding rod
Schweißstrom welding current
Schweißstromkreis welding circuit
Schweißtakter welding timer
Schweißtransformator welding transformer
Schweißumformer welding converter
Schweißumspanner welding transformer
Schweißverbindung welded joint
Schweißvorgang welding (process)
Schweißvorrichtung welding jig
Schweißwerkstatt welding shop
Schweißzuleitung welding lead
Schweiße weld metal, filling metal, weld (puddle)
schweißen to weld
Schweißer welder, welding operator
schwelen to smoulder; to carbonize at low temperature
Schwell-... threshold
Schwellfeldstärke threshold strength
Schwellfestigkeit fatigue limit
Schwellwert threshold, threshold level, threshold strength
Schwellwertlogik threshold logic
Schwelle threshold (53), (69); sleeper (625.1)
Schwelle der Hörempfindung threshold of audibility
Schwellenenergie threshold energy (539)
Schwellenfeuer threshold (-marker) light (351.814)
Schwellenhochleistungsfeuer high-intensity threshold light
Schwellenkennzeichnung runway-threshold marking
Schwellenspannung threshold voltage
Schwellung swell, swelling
Schwemmsand silt, drift sand
Schwenk-... swing; tumbler
Schwenkarm swing panel, swinging bracket (621.317)
Schwenkrad tumbler gear (621.83)
Schwenkrahmen hinged frame
Schwenktransformator phase-shifting transformer
Schwenkvorrichtung tumbler device, swivel
schwenken to swirl, to swivel; to shift (621.3); to change in direction (621.396.67)
schwenkbar hinged, swivelling, swinging
Schwer-... heavy
Schwerflüchtigkeit low volatility
Schwerhörigenapparat hearing aid
Schwerhörigkeit hard hearing, defective hearing
Schwerkraft gravity, gravitation
der Schwerkraft folgen to gravitate
Schwerkraftrückstellung gravity control (621.317)
Schwerlastanlauf heavy starting
Schwerlastbetrieb heavy-duty service
Schweröl heavy oil, heavy-gauge naphtha
Schwerpunkt C. G = centre of gravity; point of balance, centroid
Schwerspat heavy spar, baryte
Schwerwasserreaktor heavy-water reactor
schwer heavy; difficult, hard; serious
schwer bearbeitbar difficult to machine, hard to work
schwer lasten auf to weigh upon
schwer löslich sparingly soluble
schwer schmelzbar infusible
schwer zugänglich hard-to-get-at
schwere Beanspruchung severe strain
schwere Belastung heavy loading conditions (pl.), heavy load
schwerer Betrieb arduous duty
schweres Ion heavy ion, large ion
schwerer Kern heavy nucleus
schwere Masse heavy mass
schwerer Unfall serious accident
schweres Wasser heavy water
schwerer Wasserstoff heavy hydrogen, deuterium
schwerflüchtig difficult to volatilize
schwergängig slugish
schwerhörig hard hearing
schwerkraftartig pseudo-gravitational

schwerwiegend serious, weightful
Schwere gravity
Schwerefeld gravitational field
schwierig difficult
schwierige Zugänglichkeit difficulty of access
Schwierigkeit difficulty; trouble
Schwimmbadreaktor swimming-pool reactor
Schwimmer float (621.43)
Schwimmermanometer float-pressure gauge
Schwimmerschalter float switch
Schwinden fading (621.396)
schwinden to fade (away) (621.396); to shrink (621.74)
Schwinderscheinung fading effect
Schwindmaß shrinkage dimension
Schwing-... vibrating, oscillating
Schwingaudion oscillating detector, ultra-audion oscillator, oscillion
Schwingungsdämpfung antihunt (circuit) (621.316.7)
schwingfähig resonant, oscillatory
schwingfähiger Kreis oscillatory circuit
Schwingfrequenz oscillation frequency
Schwinggleichrichter vibrating-reed rectifier, tuned-reed rectifier
Schwinghebel rocking lever
Schwinghebelanordnung rod and lever mechanism
Schwingimpuls vibratory impulse
Schwingkondensator-Elektrometer vibrating-reed electrometer
Schwingkreis resonant circuit, oscillating circuit, tank, tuned circuit
Schwingkreis des Steuersenders master-oscillator circuit
Schwingkreisdämpfung damping
Schwingkreiskopplung tuned-anode coupling (621.396.64)
Schwingkreisspule oscillator coil
Schwingkreisstrom tank-circuit current
Schwingkristall oscillating crystal
Schwingleistung vibrating power; oscillatory power (621.37)
Schwingmetall rubber-metal vibration damper
Schwingneigung tendency to oscillation
Schwingquarz piezoelectric oscillator, quartz resonator, (quartz) crystal
Schwingspule oscillator coil; moving coil (534)
Schwingtopf closed coaxial resonator, cavity resonator
Schwingzunge reed (534); vibrating reed (621.317)
schwingen to oscillate; to swing, to vibrate
schwingen um to vibrate about
schwingendes Feld oscillating field
schwingender Kreis oscillating circuit
schwingende Leitung resonant line
Schwinger resonator, oscillator, vibrator
Schwingung vibration, oscillation, swing
Schwingungen erzeugen to excite oscillations
Schwingungen dämpfen to damp oscillations
Schwingungen unterdrücken to suppress oscillations
in Schwingungen versetzen to set in vibration (sing.)
Schwingungsanregung excitation of oscillations
Schwingungsbauch internode, antinode
Schwingungsbeanspruchung vibratory stresses (pl.), alternating stress
durch Schwingungsbeanspruchung zu Bruch gehen to break by vibration
schwingungsdämpfend vibration absorbing
Schwingungsdämpfer vibration damper, vibration isolator, antifatigue device, sourdene (621.315.1), (621.395.72)
Schwingungsdämpfung absorption of vibrations
Schwingungsdauer oscillating period, periodic time, time of swing
Schwingungseigenschaft vibrational characteristic
Schwingungserregung excitation of oscillations
schwingungsfest vibration-proof
Schwingungsform mode of motion (534)
schwingungsfrei nonoscillatory, nonresonating; vibration free
Schwingungsgeber oscillator, sine-wave generator
Schwingungsisolation vibration insulation
Schwingungsknoten node, nodal point
Schwingungsmesser vibration meter, vibrograph
Schwingungsordnung mode of oscillations
schwingungsschluckend vibration absorbing
Schwingungstyp s. **Schwingungsordnung**
Schwund fading (621.396); shrinkage (621.7); loss of weight
nach Schwund wiederkommen to fade in (621.396)
Schwundausgleich automatic volume control, automatic gain control
Schwundbereich fading margin
schwundfreie Nahzone primary service area (621.396)
schwundmindernde Antenne antifading aerial
Schwundregelung automatic volume control

Schwundregelung nach rückwärts back-bias circuit
Schwundregelungseinrichtung anti-fading device (621.396.62)
Schwundriß shrinkage crack
Schwungrad flywheel
Schwungradanlasser inertia starter (621.4)
Schwungradantrieb flywheel drive
Schwungradgenerator flywheel-type alternator
Schwungradkreis tank circuit (621.396.61/.62)
Schwungradstrom (tank circuit) circulating current (621.396)
Schwungradsynchronisierung flywheel synchronisation
Scott-Schaltung Scott connection
Scott-Umspanner Scott (connected) transformer
Sechseck hexagon
Sechseckschaltung hexagon connection
Sechseckspannung hexagon voltage
sechseckig hexagonal
Sechskantkopf hexagon head, hex head
Sechskantmutter hexagon nut
Sechskantprofil hexagon (section)
Sechskantschraube hexagon screw
Sechskantsteckschlüssel hexagonal socket wrench
Sechsphasenstrom six-phase current
Sechsphasenumspanner six-phase transformer
sechsphasig six-phase, hexaphase
Sechspolmischröhre hexode mixer (tube)
Sechspolröhre hexode, six-electrode valve
Sedimentierung sedimentation, settling
Seebeck-Effekt Seebeck effect, thermo-electric effect
See ... sea ..., ocean ..., marine ...
Seefunk marine radio communication
Seefunkdienst maritime service
Seegangsecho sea return, wave clutter (621.396.96)
Seekabel submarine cable, underwater cable
Seekabeltelegraphie submarine telegraph system
Seemeile sea mile, nautical mile (s. Anh.)
Seenavigationsfunkdienst maritime radio navigation service
Segment segment
Segmenttankerscheibe segmental core disc (621.313)
Segmentantenne pillbox antenna
Segmentkopf cup-head (621.8)
Segmentschnecke quadrant (621.8)
Segmentteilung segment pitch
Seh-... visual, optic
Sehfähigkeit eyesight; vision
Sehschärfe visual acuity
Sehweite visual distance (535)
Sehwinkel optic angle
Sehne chord (51)
Sehnenwicklung chord winding, short-pitch winding (621.313)
Sehnungsfaktor pitch factor, winding factor, chording factor (621.313)
Seide silk
Seidenisolation silk insulation
Seidenpapier tissue paper
Seidenumspinnung silk covering, silk braiding
seiden silken
seidenumsponnen silk-covered
seidenumsponnener Draht silk-covered wire
Seil cable; rope (66); fall (621.8)
Seilantrieb cable drive, rope drive
mit Seilantrieb cable-driven, rope-driven
Seilbahn ropeway, cable railway
Seilbremse rope brake
Seildurchhang sag
Seileck link polygon
Seilförderung rope haulage
Seilführung rope guide
Seilkern rope core, cable core
Seilklemme cable clamp
Seilscheibe rope pulley, sheave
Seilschlinge tie-wire (621.315.6)
Seilschwebebahn overhead-rope railway, aerial ropeway
Seilsteuerung rope control
Seilsteuerung für Fahrstühle rope lift control
Seiltransmission rope transmission
Seiltrieb rope drive, cable drive
Seiltrommel rope drum, hoisting drum, winch (drum)
Seilverbinder rope connector, tension joint
Seilverbindung rope connection
Seilwinde (hoisting) winch
Seilzug tow line
mit Seilzug funicular
Seilzugförderung funicular traction
seismisch seismic
seismische Welle seismic wave
Seismograph seismograph
Seite side (auch 51); page (655)
Seite einer Schleife leg of a loop (621.317)
Seitenabtastbereich azimuth scan coverage (621.396.96)
Seitenansicht side elevation, end elevation, side view, lateral view
Seitenband sideband (621.396)

Seitenbandanteil sideband component
Seitenbandfrequenz sideband frequency
Seitenbandleistung side(band) energy
Seitenbandrauschen sideband noise
Seitenbandrest vestigial sideband (621.396)
Seitenbandtelefonie sideband telephony
Seitenbandunterdrückung sideband suppression
Seitenbestimmung sense finding, sense adjustment (621.396.9)
Seitendruck lateral pressure
Seitendrucker page-at-time printer
Seitenfläche lateral surface
Seitenhalbierende median
Seitenhalterung steady span (621.33)
Seitenkanal side channel (621.39)
Seitenplatte side panel
Seitenriß s. **Seitenansicht**
Seitenschalter sense finder (621.396.9)
Seitensichtradar side-looking radar
Seitenschneider side-cutting pliers
Seitenschub side thrust, outthrust
Seitenverhältnis aspect ratio, squareness ratio
Seitenwand side wall, side panel
Seitenwind side wind, cross wind
Seitenwinkel azimuth angle
Seitenwinkelanzeiger cursor, azimuth strobe (629.13)
Seitenzipfel side lobe (621.396.67)
seitlich lateral
seitlich abschnappen to whip sidewards
seitlich verschieben to displace
seitlich versetzt out of line
seitlicher Anschluß side lead, side connection
seitliches Ausweichen side bending
seitliche Bewegungsfreiheit lateral working space (621.315)
seitlicher Druck s. **Seitendruck**
seitlicher Stromabnehmer side-running trolley
seitliche Zuführung side lead
Sektor sector
Sektoraustastung zone blanking (621.396.96)
Sektorfunkfeuer sector-scanning beacon
Sektorhorn sectoral horn (521.396.67)
Sektorinstrument sector-pattern instrument
Sektorkabel sector cable, "D" cable
Sekundär-... secondary
Sekundäranschluß secondary terminal
Sekundärauslöser shunt trip (coil)
Sekundärbatterie secondary battery
Sekundärelektron secondary electron
Sekundärelektronenemission secondary-electron emission
Sekundärelektronenverstärker secondary-electron multiplier, photomultiplier tube
Sekundärelement secondary cell
Sekundäremission secondary emission, impact emission (621.385)
Sekundärgruppe supergroup (621.395.4)
Sekundärgruppendurchschaltefilter through-supergroup filter
Sekundärgruppenlage supergroup allocation
Sekundärgruppenumsetzer supergroup modulator
Sekundärleistung output
Sekundärradar secondary radar, secondary-surveillance radar
Sekundärspannung secondary voltage
Sekundärstrahler secondary radiator, passive aerial; parasitic aerial
Sekundärstrahlung secondary radiation, parasitic radiation
Sekundärstrom secondary current
Sekundärwicklung secondary (winding) (621.314.2)
sekundär..., secondary
sekundär abtrennen to open-circuit
sekundäre Belastung secondary load
sekundäre Nennleistung rated output
sekundäre Nennspannung rated secondary voltage
sekundärer Nennstrom rated secondary current
sekundäre Röntgenstrahlung secondary X-rays (pl.)
sekundäre Transformatorwicklung transformer secondary
Sekunde second
Sekundenzeichen second mark
Sekurit safety glass
Selbstabgleich self-adjustment, self-balancing (method), autobalance
selbstabgleichend self-adjusting, self-balancing
selbstabgleichende Brücke self-balancing bridge
selbstabgleichender Spannungsteiler self-balancing potentiometer
selbstabgleichender Widerstandsgeber potentiometric recorder (621.317)
selbständig self-sustained; self-contained; independent
selbständige Entladung self-sustained discharge (621.385)
selbständiges Gerät self-contained set
selbständige Kettenreaktion self-maintaining nuclear chain-reaction
Selbstanlasser automatic starter (621.313)

Selbstanlauf auto restart
selbstanlaufender Synchronmotor self-starting synchronous motor, autosynchronous motor
selbstanlaufender Umformer self-starting rotary converter
Selbstanschlußamt automatic (telephone) exchange
Selbstanschlußsystem dial-telephone system, automatic telephone system
Selbstaufheizung internal heating
Selbstauslöseeinrichtung automatic-tripping system
Selbstauslösung automatic release, automatic tripping
selbstbegrenzend self-limiting
selbstbremsende Kettenreaktion self-limiting chain reaction
selbsteinsetzender Sender self-starting oscillator
Selbstentladung self discharge; self depreciation (621.35)
Selbstentlade-Zeitkonstante self-discharge time constant
Selbstentzündung spontaneous combustion, spontaneous ignition
selbsterregt self-excited
selbsterregter Generator self-excited generator
selbsterregter Mischkreis autodyne circuit
selbsterregte Schwingung self(-exciting) oscillation
selbsterregter Sender self-oscillator
Selbsterregung self-excitation, building-up (621.313)
Selbstfahreraufzug self-service lift, automatic push-button lift (621.8)
Selbsthaltekontakt self-holding contact, lock-in contact
selbsthaltend self-sustaining
Selbsthalteschaltung self-holding circuit
selbstheilend self-healing, self-sealing
Selbstinduktion, Selbstinduktivität self-induction, self-inductance
Selbstinduktionsspannung voltage of self-induction
Selbstkontrolle automatic checking (681.3)
selbstkorrigierend error correcting
Selbstkühlung self-cooling, self-ventilation, induced ventilation (621.313), (621.314)
selbstgekühlt self-cooled, self-ventilated
selbstgekühlter Motor self-cooled motor
selbstleuchtend self-luminous
selbstlöschender Fehler transient fault
selbstlöschendes Zählrohr self-quenched counter tube
selbstprüfend selfchecking
selbstprüfender Code error-detecting code (681.3)
Selbstprüfung built-in check, machine check
selbstregelnd self-regulating
selbstregelnder Umspanner self-regulating transformer, automatic variable-voltage ratio transformer
Selbstregelung self-regulation, inherent regulation
Selbstschalter automatic circuit breaker, automatic cut-out
selbstschreibend self-recording
selbstschwingender Mast tower radiator
selbstsperrend self-locking
Selbststeuereinrichtung autopilot
Selbststeuerung self-control, automatic control, autocontrol; automatic pilot, gyropilot, robot pilot (629.13)
selbsttätig self-acting, automatic
selbsttätige Bereicheinstellung autoranging
selbsttätige Fehleranzeige automatic fault signalling
selbsttätige Gebührenermittlung automatic message accounting
selbsttätiges Kraftwerk autoplant, automatic power station
selbsttätiges Ladeaggregat automatic generating plant
selbsttätiger Plattenwechsler record changer
selbsttätige Rückstellung automatic reset
selbsttätige Rufgabe automatic ringing (621.395)
selbsttätiges Schlußzeichen automatic ring-off
selbsttätige Spannungsregelung automatic voltage regulation
selbsttätiges Unterwerk automatic substation
selbsttätiges Verkehrssignal automatic traffic signal
selbsttätige Wiedereinschaltung automatic reclosing
selbsttätige Zeichengebung automatic signalling (621.395)
selbsttätige Zugbeeinflussung automatic train stop (656.25)
selbsttragend self-supporting
Selbstwahl direct dialling
Selbstwähltelefonanlage automatic telephone system
Selbstzündung self-ignition
Selektion selectivity (621.396.6); selection
Selektionskurve selectivity curve, selectivity characteristic (621.396.62)
selektiv (frequency) selective

selektiver Anruf selective ringing, selective calling
selektive Eigenschaft selective property
selektiver Empfänger selective receiver
selektiver Pegelmesser selective level meter
selektive Reflexion selective reflection
selektiver Schwund selective fading (621.396)
selektiver Spannungsmesser heterodyne voltmeter
selektive Strahlung selective radiation
Selektivität selectivity, selectance, selective action
Selektivruf selective call, selective ringing
Selektivschutz selective protection (621.31)
Selen selenium
Selengleichrichter selenium rectifier
Selenwiderstand selenium resistor (cell)
Selenzelle selenium cell
Selsynempfänger, Selsyngeber Selsyn transformer
Selsynverfahren Selsyn system
seltene Erden rare earths, lanthanides
Semaphor semaphore (signal)
Sende-... transmitting
Sendeantenne transmitting aerial
Sendeart class of emission
Sendebezugsdämpfung transmission reference equivalent (621.395)
Sendebereich coverage (of a transmitter); transmission range
Sendeeinrichtung transmitting station, transmitting installation (621.394)
Sende-Empfangsgerät transceiver
Sende-Empfang-Relais disabling relay (621.396)
Sende-Empfangs-Umschalter t. r. switch = transmitter-receiver switch, transmit-receive switch
Sende-Empfangs-Verriegelung t. r. box = transmitter-receiver box
Sende-Empfangs-Weiche duplexer
Sendefrequenz transmitted frequency, transmitter frequency
Sendefrequenzband frequency band of emission
Sendegerät transmitting set
Sendegeschwindigkeit speed of transmission
Sendeimpuls main puls, main bang (621.396.96)
Sendekanal transmitting channel (621.395), (621.396)
Sendeleistung transmitting power
Sendepause no-transmission interval, unmodulated interval
Senderaum studio (621.396), (621.397)
Senderelais transmitting relay
Senderöhre transmitting valve, power valve
Senderöhre mit Anodenkühlung cooled-anode transmitting valve
Senderöhre mit Wasserkühlung water-cooled transmitting valve
Sendeseite generator end, input end, sending end
Sendestation transmitting station (621.396); transmitting terminal (621.394.6)
Sendeverstärker transmitter amplifier
Sendeverzerrung transmit distortion
Sendewelle outgoing wave, transmitted wave
Sendezentrale transmitter-control station
senden to broadcast, to transmit, to put a transmitter on the air (621.396); to send, to ship, to mail
sendende Station station on the air
Sender transmitter, transmitting station, sender (621.396), (621.397)
senderfremde Kombinationsschwingungen products of mutual intermodulations
Sender abfragen to interrogate a transmitter (621.396)
Sender hochfahren to bring the transmitter up to full load
Sender mit Anodenschwingkreis tank-circuit oscillator (621.396.61)
Sender ausstrahlen lassen to put a transmitter on the air
Sender mit Richtstrahler beam-transmitting station
Sender mit weichem Schwingungseinsatz self-starting oscillator
Senderabstand spacing between transmitters
Sender-Empfänger transceiver; transmitter-receiver; walkie-talkie (portable)
Senderendröhre transmitter output valve
Sendernetz broadcasting network
Senderreichweite range of a transmitter
Sendersperröhre antitransmit-receive tube
Sendersperrschalter antitransmit-receive switch
Senderweiche transmitter-combining filter
Sendung transmission, emission; programme
Sendung beenden to go off the air
Sendung beginnen to go on the air
sengen to singe
Senke drain
senken to sink, to countersink (621.95)
Senkblei plumb bob
Senkkopfniet countersunk rivet, flush rivet

senkrecht vertical, plumb, perpendicular, normal, square
senkrecht abschneiden to square (621.9)
senkrecht auf ... square to ..., square with
senkrecht fallen to fall plumb
senkrecht in der Phase in quadrature
senkrecht polarisiert vertically polarized
senkrechter Auflagerdruck normal reaction
senkrechter Druck normal pressure, normal stress
senkrechter Einfall vertical incidence (621.396)
senkrechter Schnitt vertical section
senkrechte Schweißung nach oben upward welding
senkrechte Seite upright side (51)
senkrechter Zylinder right cylinder (51)
Senkrechte vertical, perpendicular line
Senkrechte ziehen to drop a perpendicular
Senkrechtstart VTOL, vertical take-off and landing
sensibilisieren to sensitize (77)
Sensibilisierung sensitization
Sensor sensor
Sensoreinrichtung sensory equipment (621.316.7)
Sensortastenfeld touchpanel keyboard
Sensortastensteuerung sensor control, touch-button control
Separation separation
Separationsgrad efficiency of separation
Separator separator
separieren to separate
Sequenz-Farbfernseh-Verfahren sequential colour television
sequenzieller Betrieb one-at-a-time mode (681.3)
Sequenziellrechner consecutive computer
Sequenziellrechner mit Sprunglauf arbitrary-sequence computer
Serie series, run
in Serie schalten to connect in series
Serienarbeit repetition job
Serienblindwiderstand series reactance
Seriendrucker serial printer, one-at-a-time printer
Serienfabrikation series production, mass production
Serieninduktivität series inductance
Serienmodell standard modell, standard design
Serien-Parallel-Wandler serial-parallel converter, staticizor
Serienparallelwicklung multiplex wave winding
Serienreife production stage
Serienschalter two-circuit single interruption-switch (696)
Serienschaltung series connection, cascade connection
Serienübertragung serial transmission
Serienverbindung polling call (621.395)
Serienwiderstand series resistance, series resistor
seriell serial
serielle Abtastung serial scanning
serieller Zugriff serial access
Serpentin serpintine (55)
Servomechanismus servomechanism
Servomotor servo motor, pilot motor, power cylinder
Servoverstärker servo amplifier
Setzeingang triggering input
setzen, sich to settle, to subside
Sextant sextant, heliometer
S-förmig S-shaped
S-förmige Schweifung ogee
Shapingmaschine shaping machine
Sheddach shed roof, saw roof, northlight roof
sicher safe, secure; sure; proof; fast
sicher befestigen to secure, to fasten
sicher gegen ...... -proof
sichere Abbindung secure bonding (621.396)
Sicherheit safety; security (623)
Sicherheitsabstand safety clearance; margin of safety; interference guards bands (pl.) (621.39)
Sicherheitsauslöser safety trip
Sicherheitsfaktor safety factor
Sicherheitsfilm safety base (film)
Sicherheitsfunkdienst safety (radio) service
Sicherheitsfunkenstrecke safety gap
Sicherheitsglas safety glass, triplex glass
Sicherheitsgrad factor of safety, coefficient of safety, level of safety
Sicherheitsgrenze safe limit
Sicherheitsgrubenlampe miner's safety lamp, Davy lamp
Sicherheitsgurt, Sicherheitsgürtel safety belt
Sicherheitsmaßnahme safeguard
Sicherheitsrutschkupplung safety slipping clutch
Sicherheitsschalter safety breaker
Sicherheitsschloß safety lock, Yale lock
Sicherheitssprengstoff mit flüssigem Sauerstoff liquid oxygen explosive
Sicherheits-Unbedenklichkeitsbescheinigung clearance, security clearance (35)
Sicherheitsventil safety valve, emergency valve, pressure relief

Sicherheitsvorrichtung safety appliance, safety device
Sicherheitsvorschriften safety rules, safety regulations; safety code (sing.)
sichern gegen to safeguard against, to protect from
Sicherung fuse (621.3); protector (621.395)
Sicherung auswechseln to replace a fuse
Sicherung mit Blasspule magnetic blow-out fuse
Sicherung durchbrennen to blow a fuse
Sicherung geht durch a fuse blows
Sicherung mit Signalgabe alarm fuse
Sicherung des Spannungspfades potential fuse
Sicherung überbrücken to bridge a fuse
Sicherung mit Unterbrechungsmelder indicating fuse
Sicherungausfalltrenner dropout-fuse cut-out
Sicherungsautomat (miniature automatique) circuit breaker
Sicherungsbemessung fuse rating
Sicherungsdraht fuse wire
Sicherungseinrichtungen fuse gear (sing.)
Sicherungseinsatz fuse link
Sicherungselement fuse element, fuse unit, fuse block
Sicherungsfüllung fuse filler
Sicherungsgrundplatte fuse support, fuse base
Sicherungshalter fuse carrier, fuse holder
Sicherungskasten fuse box, cutout box
Sicherungskontakt fuse contact
Sicherungskörper fuse holder
Sicherungslasttrenner fuse load-break switch
Sicherungsnennstrom rated fuse current
Sicherungspatrone fuse cartridge, enclosed fuse
Sicherungsrohr fuse tube
Sicherungsschalter fuse switch
Sicherungssockel fuse base, fuse socket; protector springs (pl.) (621.395)
Sicherungstafel fuse board, fuse panel
Sicherungsträger fuse holder
Sicherungszange fuse puller
Sicherungs-... safety..., guard
Sicherungsring guard ring, lockwasher
Sicherungsstift safety pin
Sicherungsunterlegscheibe lock washer, friction washer
Sicht sight; visibility; view
Sichtanflug visual contact approach
Sichtanzeige visible indication
mit Sichtanzeige visual(-type)
Sichtanzeiger visual detector
Sichtbarkeit visibility
Sichtbehinderung obstruction of line-of-sight path
Sichtbeobachtung visual observation
Sichtgerät display terminal, display console (681.3); visual display unit (621.397); scope
Sichtlinienwinkel angle of sight
Sichtspeicherröhre viewing storage tube
Sichtverbindung line-of-sight connection
Sichtverhältnisse seeing (sing.), visibility conditions
Sichtweite sight; line of sight
in Sichtweite within sight
sichtbar visible, visual
sichtbare Anzeige visual indication
sichtbares Bild visual image
sichtbare Erscheinung visible phenomenon
sichtbare Strahlung visible radiation, photic radiation
sichtbare Veränderung visible change
sicken to bead
Sickerbecken oozing basin
Sickergrube trickle pool
Sieb filter (621.3); sieve, strainer, screen (66)
Siebblech perforated plate, perforated metal screen
Siebdrossel filter choke, filter reactor
Siebdruck screen print
Siebeinrichtung sifting device
Siebgewebe screen wire
Siebkette wave filter (621.392)
Siebkondensator filter capacitor
Sieböffnung sieve spacing
Siebschaltung filter circuit, filter network, frequency filter
Siebschaltung mit Querkondensator im Eingang capacitor input filter
Siebspule filter coil, filter reactor
Siebtrennschärfe filter discrimination
Siebwirkung filtering action
sieben to filter (621.3); to sieve, to sift, to riddle, to screen (66), (622)
Siebeneck heptagon
Siebenpolröhre heptode
Siedehitze boiling heat
Siedepunkt boiling point
Siedepunktserhöhung elevation of the boiling point
Siedewasserreaktor boiling-water reactor, water-boiler reactor
Siemens (-Leitwerteinheit) reciprocal ohm = mho = 1/ohm
Siemens-Doppel-T-Anker Siemens H-armature
Siemens-Drehspulfernschreiber Siemens moving-coil type telewriter

Siemens-Einheit (des Widerstandes) = 0, 9407 Ohm Siemens unit, mercury unit
Siemens-Hell-Schreiber Siemens-Hell teleprinter, Hell printer
Siemens-Herdverfahren, Siemens-Martin-Prozeß open-hearth process
Siemens-Martin-Stahl Siemens-Martin steel, open-hearth steel
Siemens-Rekorder, Siemens-Schnellschreiber high-speed recorder
Sigmareaktor sigma pile
Signal signal
Signal geben to signal
Signalanlage signalling system
Signalbrücke gantry (625.1)
Signaldiode signal diode
Signaleinstellungsrückmelder lever indication (656.25)
Signalflügel signal arm
Signalflügelkontakt (signal) arm contact, signal-circuit controller
Signalgeber signal transmitter
Signalgemisch composite picture signal
Signal/Geräuschabstand signal-to-noise ratio
Signalgeschwindigkeit signal velocity
Signalglocke signalling bell
Signalhebel signal lever
Signalhorn bugle
Signalinversion signal inversion (621.395)
Signallampe signal lamp, indication lamp, pilot lamp
Signallampenrelais signal-lamp relay
Signalleine signalling rope
Signalleitung signalling line, signalling lead
Signalmast signal post, semaphore (625.1)
Signalmelder signal proving (625)
Signalmischgerät sync mixer (621.397)
Signalpegel signal level
Signal-Rauschverhältnis signal-to-noise ratio
Signalscheinwerfer signal light, traffic-control projector (656.7)
Signalspannung (signal) voltage
Signalstellmotor signal motor (656.25)
Signalstellung signal indication (656.25)
Signalstellungsanzeige im Führerstand cab signal (656.25)
Signal-Telefonhäuschen signal box (625)
Signalumformer signal converter
Signalverfolger signal tracer
Signalweg signal path
Signalwegenetz signal network (621.395.3)
signalisieren to signal
Signalisierung außerhalb des Sprachbandes outband signalling (621.395)
Signalisierung innerhalb des Sprachbandes inband signalling
Signalisierungskanal signalling channel
signifikante Ziffer significant digit
Sikkativ siccative
Silbenverständlichkeit intelligibility of syllables, syllable articulation
Silbenverständlichkeitsprobe articulation test
Silber silver
Silber aufbrennen auf ... to fire silver on ...
Silberamalgam silver amalgam
mit Silberauflage silver-plated, silver-faced
Silberbronze aluminium powder
Silber-Chlorid-Element silver-chloride cell
Silberfolie silver foil
Silberglätte (silver) litharge
Silberkabel shielded-aerial connection (621.396)
Silberkontakt silver contact
Silberniet silver contact (rivet)
Silberpapier silver paper, tin-foil
Silberplattierung silver cladding
Silberstahl silver steel
Silbervoltameter silver voltameter
Silberzinkakkumulator silver-zinc-accumulator, silver-storage battery
Silbermann-Kabel graded cable
Silikat silicate
Silikon silicone
Silikon-Glasseiden-Hartfasergewebe silicon-fiberglass laminate
Silikongummi silicone rubber
Silikonöl silicone oil
Silizium silicon
Siliziumdioxid silica
Siliziumgleichrichter silicon rectifier
Siliziumkarbid silicon carbide
Siliziumnitrid-Passivierung silicon-nitride passivation
Silizium-Planartransistor silicon planar transistor
Siliziumthyristor silicon-controlled rectifier = SCR
Siliziumtransistor silicon transistor
Silumin silicon alloy
Simplexbetrieb simplex operation, single operation
Simplextelegraphie simplex (telegraph) system
Simulator simulator
Simulierer simulator program
simultan simultaneous
Simultanbetrieb concurrent working
Simultan-Farbfernseh-Verfahren simultaneous color television

sin = sinus sin = sine
sin²-Impuls sine-squared pulse
1/sin²-Antenne cosecant squared aerial
singen to sing
singende Bogenlampe Duddell arc
singender Lichtbogen singing arc
singulär singular (51)
Singularität singularity
Sing. = Hyperbelsinus sinh. = hyperbolic sine
sinken to sink, to decline, to fall
sinken auf to fall back to
Sinn sense
Sinnbild symbol
Sinter sinter
Sinterkarbid sintered carbide
Sinterkohle sintered coal
Sinterlager sintered bearing
Sintermagnet sintered magnet
Sintermetall sintered metal
Sinter-Tantalkondensator capacitor tantalum sintered anode
sintern to sinter
Sinus, sinus sine
Sinusbussole sine galvanometer
Sinusform sine-wave form
sinusförmig sine-wave, sinusoidal; sinoidal (US)
sinusförmig veränderliches Feld sinusoidal field
sinusförmige Änderung sine-law variation
sinusförmige Größe sinusoidal quantity, simple harmonic quantity
sinusförmiger Verlauf sine-law variation
sinusförmige Zeitablenkung sinusoidal sweep (621.385)
Sinusgesetz sine law
Sinushalbwelle half-sinusoid
Sinuskurve sine curve
Sinusleistung siehe Nennausgangsleistung
Sinussatz sine rule
Sinusschwingung harmonic oscillation, sinusoidal oscillation
Sinusspannung sine-wave voltage
Sinusstrom sine-wave current
Sinuswelle sine wave, sine curve
Sinuswellengenerator sine-wave generator, sinusoid generator
Sirene siren
Sisalhanf sisal (hemp)
Sitz fit; seat (621.6)
Skala, Skalen-... scale, dial
Skala auf dem Registrierstreifen chart scale of a recording instrument
Skala mit unterdrücktem Nullpunkt set-up scale, suppressed scale (621.317)
mit einer Skala versehen to scale, to graduate
über die Skala hinaus ausschlagen to go off scale
Skalenablesung scale reading
Skalenbeleuchtung dial illumination, scale illumination
Skaleneinstellung dial setting
Skaleneinteilung graduation
mit Skaleneinteilung versehen to graduate, to index
Skalenendwert maximum scale value
Skaleninstrument direct-reading instrument
Skalenkorrektur scale correction
Skalenlampe instrument lamp (621.317); dial lamp (621.396.62)
Skalenring graduated ring, index collar
Skalenscheibe dial (621.317); scale plate (621.396.62)
Skalenseil dial cord
Skalenstrich hairline (621.396.62)
Skalenteil scale division
Skalenübereinstimmung dial tracking
Skalenzeiger dial pointer
Skalar, skalar scalar (51)
skalares Feld scalar field
skalare Funktion scalar function
skalare Größe scalar (quantity)
skalares Potentialfeld scalar potential field
skalares Produkt scalar product, inner product, dot product
Skalen-... s. unter **Skala**
Skandium scandium
Skelettbau framed building
Skineffekt skin effect, Heavyside effect (621.3)
Skizze sketch, draught; scratch
nach Skizze arbeiten to work from scratch
sklavische Nachahmung Chinese copy
Skleroskophärte scleroscope hardness
Smaragd emerald
S.-M.-Stahl = Siemens-Martin-Stahl s. dort
Sn Sn (54); T (US) (669.18)
Sockel socket (621), (621.3); mount (621.396); base (621)
Sockelanschluß socket terminal
Sockelfeder socket receptacle (621.385)
Sockelgeschoß basement
Sockelimpuls pedestal
Sockelkontakt socket receptacle (621.385)
Sockelschalter socket switch
Sockelstift contact pin (621.385)
Sockeltransformator adapter transformer (621.315.6)

Soda soda
Soffitte soffit, tubular lamp, festoon lamp
Soffittenbeleuchtung strip lighting
Soffittenfassung tubular lamp holder
Soffittenlampe tubular lamp
Sofortbereitschaftsaggregat ready-reserve generating set
Sofortruf immediate ring
Sofortsperre immediate busy
Sofortverkehr demand working (621.395)
Software software
Software-Entwicklung software design, software engineering
softwarekompatibel software-compatible
Sog wind suction
Sohlplatte, Sohlenplatte sole plate (621.369)
Solar-..., solar solar
Solarbatterie solar battery
Solargenerator solar generator
Solarkonstante solar constant
Solarstörung solar noise (621.396)
Solarstromversorgung solar power supply
Solarzelle solar cell
Solenoid solenoid
solide solid, firm, stout
solide Bauart solid make, stout construction
solide Konstruktion solid design
Solleistung rated output
Sollmaß design size, intended size
Sollwert rated value, desired-value, set point, reference value
Sollwerteinsteller set-point adjustment
Sollzahl target number
Sollziffer target (figure)
Sonar-Boje sonobuyo
Sonde probe (61), (621.396); sound (61)
Sonder-... special, particular, exceptional
Sonderausrüstung special equipment
Sonderdienst special service
Sonderfunkdienst special radio service
Sondergebiet special branch, special field
Sondergenehmigung special licence
Sonderkonstruktion special design
Sondermotor special duty motor, non-standard motor
Sondernetz functional network
Sondertarif exceptional rate
Sonderzeichen additional character (621.394); special character (681.3)
sonderbar geformt oddly shaped
sondieren to probe, to sound
Sonne sun (siehe auch unter Solar...)
Sonnenenergie sun power, solar power
Sonnenfleck sunspot
Sonnenkraftwerk solar power station
Sonnenlicht sunlight
Sonnenstrahlung solar rays (pl.), solar radition
Sonnentätigkeit solar activity
Sonnenzelle solar cell
Sonnemann-Kabel "SO" cable
sorbieren to sorb
Sorte kind, sort; species, class; grade, quality
Source source (621.382.3)
Source-Schaltung common source
Spachtelmasse daub
spachteln to spaddle, to daub
Spalt, Spalte slit, slot, cleft, cleavage, gap
Spalt-... cleavable; fissible (519.17)
Spaltausbeute fission yield
Spaltbruchstücke fission fragments
Spaltfähigkeit fissionability (539)
Spaltglimmer sheet mica
Spaltmagnetron slot-type magnetron
Spaltneutronen fission neutrons
Spaltpol shaded pole, shielded pole (621.313.3)
Spaltpolinstrument shaded-pole instrument
Spaltpolmotor shaded-pole motor, shielded-pole motor, split-pole motor
Spaltpolumformer split-pole rotary converter
Spaltprodukt fission product
Spaltstoff fissible matter, nuclear fuel
Spaltstoffaufarbeitung fuel reconditioning, fuel regeneration, fuel reprocessing
Spaltstoffelement fuel element
spaltbar fissile, fissionable
spaltbares Atom fissile atom; fissionable atom (US)
spaltbarer Kern fissible nucleus
spaltbare Kernart fissible nuclid
spaltbares Material fissible matter
Spaltbarkeit fissionability (539)
Spalte (einer Tabelle:) column
Spaltenbezeichnung column designation
Spaltenüberschrift column heading
spalten to cleave, to split
Spaltung splitting, cleavage; fission (539)
Spaltung mit langsamen Neutronen slow-neutron fission, thermal-neutron fission
Spaltung mit schnellen Neutronen fast (neutron) fission
Span chip, sliver, shaving
spanabhebend bearbeiten to machine
spanabhebende Bearbeitung machining, metal-removing work, metal cutting
spanlose Formung non-cutting shaping
Spanquerschnitt chip section

Spann-... clamping, tightening, setting
Spannanker pole guy, guy anchor
Spannband brace, taut strip, taut band
Spannbandaufhängung taut-band suspension
Spannbandgalvanometer torsion-string galvanometer
Spannbolzen tie bolt, hold-down bolt
Spannbügelschalter taut-tape attachment
Spanndraht guy wire, bracing wire, bridle (621.33)
Spanndrahtanker guy anchor
Spanndrahtgalvanometer marine galvanometer
Spanndrahtschwelle dead man, anchor log (621.315)
Spanneisen tommy bar, clamp
Spannhaken tenter hook
Spannkraft work-holding power (621.881)
Spannmutter clamping nut
Spannrahmen tenter frame (67)
Spannriemen tightening belt
Spannring expanding collar, setscrewing
Spannrolle tightening pulley, tension pulley
Spannschiene slide rail (621.313)
Spannschraube hold-down screw, locking bolt, clamping bolt; tensioning bolt
Spannvorrichtung tensioning device, clamping device, fixture (621.9)
Spannweite span length, (width of) span (621.315.1)
Spannzwinge clamp
spannen to span, to tie, to tighten, to clamp; to tension; to chuck (621.9)
gespannte Saite tense string, stretched chord
gespanntes Seil tense cord, tense rope, tense cable
Spannung tension, voltage, potential (difference), pressure (621.3); stress (620.18); (deformierend:) strain
Spannung abnehmen to extract a voltage
unter Spannung stehend energized, alive to earth, live, hot
Spannung am Leitungsanfang sending-end voltage
Spannung am Leitungsende received voltage, receiving-end voltage
Spannung an eine Wicklung legen to put a voltage on a winding
Spannung an einen Widerstand anlegen to put a voltage across a resistance
Spannung anlegen, Spannung aufdrücken to impress a voltage, to apply a potential, (**zwischen:** across), (**an**: to)
Spannung auf der Leitung line voltage
Spannung aufnehmen to absorb voltage, to consume voltage
Spannung bei Ausgleichsvorgängen transient voltage
Spannung bei Belastung on-load voltage
Spannung bei Leerlauf no-load voltage, open-circuit voltage
Spannung einführen to inject a voltage
Spannung Eins unit voltage
Spannung erhöhen to boost, to increase the voltage, to raise the voltage
Spannung fällt ab voltage drops, potential falls
Spannung fällt ab an ... voltage is lost in ...
Spannung gegen Erde voltage to ground
Spannung gegen den Sternpunkt voltage to neutral
Spannung in Durchlaßrichtung forward voltage
Spannung innerhalb der Elastizitätsgrenze elastic stress
Spannung Null zero voltage
Spannung in Sperrichtung reverse voltage
Spannung des stationären Endzustandes final voltage
Spannung steigt voltage rises
Spannung treibt ein Elektron voltage urges an electron
Spannung verbrauchen to consume voltage, to absorb voltage
Spannung zwischen voltage from ... to ..., voltage across ...
Spannung zwischen den Außenleitern outer-conductor voltage
Spannung zwischen Rohr und Boden pipe-ground potential (621.64)
Spannungsabfall potential drop, fall in potential, voltage dip, voltage drop
Spannungsabfall in der Elektrode electrode drop (621.791)
Spannungsabfall bei Lastabwurf regulation "up"
Spannungsabfall auf der Leitung line drop
Spannungsabfall in der Röhre tube drop (621.314.6)
Spannungsabfall von 10 Volt verursachen to drop 10 volts
spannungsabgestimmt voltage-tuned
spannungsabhängig voltage-responsive, voltage-sensitive, voltage-dependent
spannungsabhängiger Widerstand voltage-sensitive resistor; non-linear resistor (621.316.9)

Spannungsänderung voltage variation
Spannungsanisotropie magnetic allotropy (538)
Spannungsanstieg voltage rise, voltage increase
 Spannungsanstieg bei kapazitiver Last impedance rise
 Spannungsanstieg bei Lastabwurf voltage regulation (621.313)
 Spannungsanstiegsgeschwindigkeit rate of voltage rise
... mit Spannungsanwurf voltage restraint (621.318)
Spannungsanzeiger voltage indicator
Spannungsausfall loss of voltage, voltage breakdown
Spannungsauslösung shunt-trip
Spannungsbauch antinode, voltage antinode
Spannungsbereich voltage range
spannungsbetätigt voltage-operated, voltage-controlled
Spannungsdämpfung voltage attenuation
Spannungs-Dehnungs-Diagramm stress-strain diagram, load-extension diagram
Spannungsdreieck voltage triangle
Spannungseinführung voltage injection
Spannungseinheit unit (of) voltage
Spannungseisen voltage magnet, shunt stator, potential element (621.317)
Spannungsermittlung durch elektrische Widerstandsmessung ikeda test
Spannungsfall s. **Spannungsabfall**
Spannungsfehler voltage error, ratio error (621.317)
Spannungsfeld field of stress
Spannungsfestigkeit electric strength
spannungsfrei dead (621.3); stress free (621.79)
 spannungsfrei machen to isolate (621.3); to stress-relief, to relief the stress (621.79)
 Spannungsfreiglühen stress-relief heat-treatment
spannungsführend live, alive, hot; (gegen Erde:) alive to earth
 spannungführender Draht live wire
 spannungführender Leiter live conductor
 spannungführender Teil live part
spannungsgespeist voltage fed
spannungsgesteuerter Oszillator voltage-controlled oscillator
Spannungsgleichgewicht voltage balance
Spannungsgleichhalter voltage regulator, regulated voltage supply unit, voltage stabilizer
Spannungsgradient voltage gradient
Spannungskern shunt stator, potential element (621.317)
Spannungsknoten voltage note
Spannungskomparator voltage comparator
Spannungskonstanthalter voltage regulator, voltage stabilizer
spannungslos dead, off-circuit, de-energized (621.3)
Spannungsmeßbereichumschalter voltage-range switch
Spannungsmesser voltmeter
Spannungsnormal voltage standard
Spannungsnulldurchgang zero voltage
Spannungspegel voltage level
Spannungspfad voltage circuit, potential circuit (621.317)
Spannungsprobe high-voltage test
Spannungsprüfer charge indicator, cell tester (621.355); potential tester, voltage detector (621.315); voltage detector, voltage indicator
Spannungsprüfung bei Regen wet flashover test
Spannungsprüfung mit stufenweiser Spannungssteigerung step method of voltage testing
Spannungsquelle power source
Spannungsreferenzdiode voltage-reference diode
Spannungsregelung voltage control, voltage regulation
 Spannungsregelung durch Bürstenverschiebung brush-shifting control
Spannungsregler voltage regulator
Spannungsreihe contact series, electrochemical series
Spannungsrelais voltage relay
Spannungsresonanz series resonance (621.3)
 Spannungsresonanzkreis resonant circuit, acceptor circuit
Spannungsrichtungsrelais voltage directional relay
Spannungsrichtverhältnis detector voltage efficiency (621.382.2)
Spannungsrißbildung stress cracking
Spannungsrückgangsrelais fall-in-voltage relay, undervoltage relay (621.313)
Spannungsrückkehr voltage recovery
Spannungsrückkopplung voltage feedback
Spannungssättigung voltage saturation
Spannungsschwankung voltage fluctuation, voltage swing, variation in voltage
Spannungssicherheit voltage rating
Spannungssonde voltage probe (621.396)
Spannungsspitze voltage peak, peak voltage, spike

Spannungsspule voltage coil, potential coil (621.317)
Spannungsstabilisierung voltage regulation, voltage stabilization
Spannungsstabilisator voltage stabilizer, constant-voltage regulator
Spannungsstabilisatordiode voltage-regulator diode
Spannungssteuerung voltage control
Spannungsstoß surge
Spannungsstufe voltage range, voltage step
Spannungsstufenregler step voltage regulator
Spannungssucher voltage indicator
Spannungsteiler potential divider, potentiometer
Spannungsteiler für Kompensationsapparat volt box (621.317)
Spannungsteilerwiderstand voltage-dividing resistor
in Spannungsteilerschaltung potentiometer fashion
Spannungsteilung voltage division
Spannungstrichter area of risk, earth-resistance area
Spannungsüberhöhung voltage magnification (621.396); stress concentration
Spannungsübersetzung voltage transfer (621.396); voltage ratio
Spannungsumschaltung voltage commutation
Spannungsumschaltung unter Last on-load tap-changing
Spannungsumsetzung (= Ausgleichsspannung) degree of voltage rectification
Spannungsunsymmetrieschutz voltage phase-balance protection
Spannungsverdoppler voltage doubler
Spannungsverdopplungsschaltung voltage-doubler circuit, cascade-voltage doubler
Spannungverdreifacher voltage tripler
Spannungsvergütung strainhardening
Spannungsverstärker voltage amplifier
Spannungsverstärkung voltage amplification; voltage gain (621.375)
Spannungsverteilung standing-wave pattern (621.317), (621.396); distribution of potentials (621.31)
Spannungsverteilungsdiagramm stress-variation diagram
Spannungsvervierfacher voltage quadrupler
Spannungswaage absolute electrometer
Spannungswandler voltage transformer, potential transformer
Spannungswelle voltage surge
Spannungszeiger potential indicator, voltgauge
Spannungszuführungspunkt point of voltage application
Spannungszusammenbruch voltage dip, breakdown of voltage
Spannungszustand state of strain, conditions (pl.) of stress
Spar-... restrictive
Sparschaltung restrictive circuit (621.3); two-circuit method (621.317); economy circuit
Spartransformator auto-transformer
Spat spar
Spätzündung retarded ignition
Speckglanz greasy lustre
Speckstein soapstone, steatite
Speiche spoke, arm (621); radius (61)
Speichengriff star handle
Speichenrad capstan wheel (621.9); spider
Speicher boiler, reservoir (621.36); store, storehouse; loft, attic (728); memory; storage (US), store (681.3); water heater (64)
Speicher mit sequentiellem (seriellem) Zugriff sequential-access storage, serial storage
Speicher mit wahlfreiem Zugriff random access storage, random access store, RAM
Speicher ohne Zerstörung des Speicherinhaltes beim Herauslesen der Information nondestructive readout memory
Speicheradressenregister memory address register
Speicherausdruck dump
Speicherbefehl store instruction
Speicherdichte bit density
Speicherelement storage element
Speicherflasche storage bottle
Speicherheizung thermal storage heating
Speicherkapazität memory capacity
Speicherkern memory core
Speicherkraftwerk storage power station, power station with reservoir
Speicherkreis storage circuit
Speicheroszilloskop storage oscilloscope
Speicherplatte storage target (621.385)
Speicherplatz storage location (681.3)
Speicherregister memory data register
Speicherröhre storage tube
Speicherschutz memory protection
Speichervermittlung message storage exchange
Speichervermögen storage capacity, output capacity (621.355); stored energy (538.22)
Speicherwirkung storage action

Speicherzeit carrier storage time (621.382.3); retention time (621.396.96)
Speicherzelle (storage) location, (storage) cell
Speicherzugriff memory access
Speicherzuweisung memory allocation
Speicherzyklus memory cycle
speichern to store, to hoard
gespeicherte Verbindung delayed call (621.395)
speicherprogrammiert stored program controled
Speicherung storage
Speise-... feeder, feeding, supply
Speiseanschluß feeder ear (621.33)
Speisebrücke feeding bridge
Speiseleitung feeder, feeding line, supply main (621.3); feedline (621.396.67); feeding pipe (621.1)
Speiseleitung eines Strahlernetzes radial feeder
Speiseleitungsverteiler feeder-distribution centre (696.6)
Speisekabel feeder cable
Speisepumpe feed pump, feed engine
Speisepunkt feeding point (621.3); feed point (621.396.67)
Speiseschalter feeder switch
Speisespannung supply voltage
Speisestromdämpfung feeding loss
Speisestromkreis feeder circuit
speisen to feed, to energize, to supply (621.1), (621.3), (621.396.67)
gespeist werden durch to feed through (621.3)
Speisung feed, feeding; excitation
Speisung einer Antenne excitation of an aerial
Speisung eine Wicklung excitation of a winding
Spektral-... spectral, spectroscopic
Spektralanalyse spectral analysis, spectroscopic analysis
Spektralanteil der Strahlungsdichte spectral radiant intensity
Spektralfarbe prismatic colour
Spektrallinie (spectral) line (535)
Spektralphotometer spectrophotometer
Spektralverkämmung frequency interlacing (621.397)
Spektralwert spectral-distribution coefficient
spektral spectral
spektrale Auflösung spectral resolution
spektrale Aufnahmecharakteristik pickup spectral characteristic (621.397)
spektrale Empfindlichkeit spectral sensitivity
spektrale Empfindlichkeitsverteilung current-wavelength characteristic (621.383)
spektraler Farbanteil, spektrale Farbdichte excitation purity
spektrale Verteilung der Ausstrahlung spectral emissivity
Spektrometer spectrometer
Spektrophotometer spectro photometer
Spektroskop spectroscope
Spektroskopie spectroscopy
Spektrum spectrum (pl.: -tra), prism
Spektrumanalysator spectrum analyzer
Spektrum n-ter Ordnung spectrum of the nth order
Spengler tinsmith
Sperr-... blocking, locking, barring
Sperrbelastung inverse load
Sperrbereich attenuation region; blocking state region (621.382.2), (621.392); inverse voltage range (621.314.6); cut-off range (621.385), (621.382); attenuation band, rejection band
Sperrdrossel suppression choke
Sperreingang inhibiting input (621.374)
Sperrfilter band-elimination filter
Sperrgatter inhibition gate
Sperrgebiet cut-off (621.385); restricted area
ins Sperrgebiet hinein past cut-off
Sperrgitterröhre barrier-grid tube
Sperrhaken pawl, detent
Sperrhebel locking lever
Sperrholz plywood, three-ply
Sperrimpuls disable pulse; inhibit pulse (681.3)
Sperrkennlinie blocking-state characteristic, reverse voltage-current characteristic (621.382.2)
Sperrklinke pawl, detent; ratchet
Sperrkreis rejector circuit, antiresonant circuit, wave trap (621.3)
Sperrmauer dam
Sperrmuffe stop joint (621.315.2); isolating joint (621.39)
Sperrelais lockout relay
Sperrichtung reverse direction, blocking direction (621.314.6)
in Sperrichtung betrieben reversed biased
Sperröhre ATR-tube
Sperroszillator blocking oscillator
Sperrschalter lockout switch
Sperrschicht depletion layer, barrier layer, junction
Sperrschichtberührungsspannung punch-through voltage, penetration voltage, reach-through voltage (621.382)

Sperrschicht-Feldeffekttransistor junction gate field-effect transistor, junction FET
Sperrschichtgleichrichter barrier rectifier, depletion layer rectifier, junction rectifier
Sperrschichtkapazität depletion-layer capacitance, barrier capacitance
Sperrschicht-Photoeffekt photovoltaic effect
Sperrschichtphotozelle photovoltaic cell, rectifier cell
Sperrschichttemperatur junction temperature (621.382)
Sperrschichtzelle blocking-layer cell
Sperrschwinger blocking oscillator
Sperrspannung inverse voltage, reverse voltage, off-state voltage, blocking-state voltage (621.314.5) (621.382.2); de-ionization potential
Sperrspannungsscheitelwert crest working reverse voltage
Sperrspitzenspannung inverse peak voltage (621.314.6)
Sperrstellung bind position, blocked state (621.8)
Sperrstrom reverse current, back current, reverse leakage, off-state current, blocking current
Sperrtaste locking key (621.395)
Sperrtopf cup, choke joint (621.396)
Sperrtopfantenne quarter-wave shirt dipole, sleeve dipole
Sperrträgheit recovery effect
Sperrverlustleistung blocking-state power loss
Sperrverzögerungsstrom reverse recovery current
Sperrverzögerungszeit reverse recovery time
Sperrverzugsladung recovered charge
Sperrwiderstand off-state resistance, blocking resistance, reverse (d. c.) resistance
Sperrzeichen blocking signal
Sperrzustand off-state, blocking state
Sperrzustand in Rückwärtsrichtung reverse blocking state, reverse off-state (621.382)
Sperre choke (621.3); barr(i)age, dam (626); baffle (533)
sperren to lock, to block, to bar; to bias off (621.385); to reject, to discriminate (621.3)
Spezial-... special, particular
Spezialausführung special design, modified version
Spezialerfahrung specialized experience
Spezialkonstruktion special design
Spezialmotor special purpose motor
spezialisieren, sich to specialize (**auf**: in)
Spezialist specialist
Spezies species
Spezifikation specification, statement of particulars
spezifiziert detailed, itemized
spezifisch specific
spezifischer Abbrand specific burn-up, fuel irradiation level (539)
spezifische Belastung specific utilization coefficient (621.313); unit load
spezifische Dämpfung attenuation constant
spezifische Dehnung elongation per unit force
spezifische Dichte specific density
spezifische Drehung specific rotation
spezifischer Druck specific pressure
spezifisches Gewicht specific weight, specific gravity, density
spezifische Konstanten primary constants
spezifische Ladung des Elektrons electron-charge mass ratio
spezifischer magnetischer Widerstand specific magnetic resistance, reluctivity
spezifischer Schallwiderstand specific acoustical impedance
spezifischer Strömungswiderstand der Längeneinheit flow resistivity per unit length
spezifische Verlustleistung specific power loss
spezifisches Volumen specific volume
spezifische Wärme specific heat, thermal capacity
spezifischer Wechselstromwiderstand effective resistivity
spezifischer Widerstand (volume) resistivity, insulativity
Sphäre sphere
sphärisch spherical
sphärische Aberration spherical aberration
sphärisches Dreieck spherical triangle
sphärische Geometrie spherical geometry
sphärische Linse spherical lens
sphärische Trigonometrie spherical trigonometry
Spiegel mirror, glass; reflector (621.396)
Spiegelantenne reflector antenna
Spiegelbild (mirror) image
Spiegeleisen spiegeleisen
Spiegelfrequenz image frequency (621.396)
Spiegelfrequenzdämpfung image rejection
Spiegelfrequenzfestigkeit image response, image-interference ratio
Spiegelfrequenzselektion signal-to-image ratio
Spiegelfrequenzsicherheit image-frequency rejection

Spiegelfrequenzunterdrückung image suppression
Spiegelgalvanometer mirror galvanometer, reflecting galvanometer
Spiegelglas plate glass
Spiegelinstrument reflecting instrument
Spiegelmagnetometer mirror magnetometer
Spiegeloptik, mit catoptric
Spiegelreflexionskoeffizient specular reflection factor
Spiegelscheinwerfer reflector floodlight, (specular) reflector beacon (351)
spiegeln to mirror, to reflect
spiegeln, sich to be reflected
spiegelnde Reflexion mirror reflection
Spiegelung reflection, reflexion
Spiel (erwünschtes:) clearance, (unerwünschtes:) play (621)
Spielart variety, variation
Spielraum margin, free space; clearance
Spin spin (53)
Spindel spindle, screw (jack) (621)
Spindelöl spindle oil
Spindelpresse screw press
Spinne spider (59), (534.86)
Spiral-... spiral
Spiralabtastung spiral scan
Spiralantenne spiral antenna
Spiralbohrer twist drill
Spiralfeder spiral spring
Spiralkegelrad spiral bevel gear
Spiralnut spiral groove
Spiralschleifdraht Wheatstone's rheostat
spiralverzahntes Rad helical gear
Spiralwindung spiral turn
Spiralzähne spiral teeth
Spiralzahnrad spiral gear
Spirale spiral (line)
spiralig spiral
spiralig aufwinden to spiralize
spiralig winden, sich to spiral
Spiritus spirit
Spirituslack spirit varnish
spitz pointed (51), (53); sharp (621); peaked; acute (51)
spitz auslaufen to taper off
spitz zulaufen to taper
spitz zulaufend tapered
spitze Kurvenform peaked wave (621.3)
spitze Nadel sharp needle
spitze Resonanzkurve pointed resonance curve
spitzer Winkel acute angle
spitzwinklig acute angled
spitzwinklige Kreuzung diamond crossing
Spitze point; top, peak; tip; vertex (pl.: vertices)
Spitze abfahren to take the swings (621.311)
Spitze erreichen to top
Spitze der Hysteresisschleife tip of the hysteresis loop
Spitze kappen to top
Spitze eines Mastes point of a pole
Spitze des Schweißstabs electrode tip
Spitze des Steckerstifts tip of the plug pin
auf der Spitze (stehend) on end
mit einer Spitze versehen to tip
Spitzenanzeiger peak indicator
Spitzenbegrenzer peak clipper
Spitzendiode point-contact diode
Spitzendrucker demand printer
Spitzenelektrode point electrode (621.3)
Spitzengleichrichter peak detector
Spitzenkontakt point contact
Spitzenlager pivot bearing
Spitzenlast maximum demand, peak load
Spitzenleistung maximum capacity, maximum power demand; peak output (power), peak envelope power (621.396.61)
Spitzenreibung friction of the pivot (621.317)
Spitzenschreiber maxigraph
Spitzenspannung peak voltage, needle-point voltage
Spitzenspannungsmesser peak voltmeter, peak-reading voltmeter
Spitzensperrspannung inverse peak voltage (621.314.6); peak off-state voltage
Spitzenstrom peak current
Spitzentransistor point-contact transistor
Spitzenwert peak, peak value, peak amplitude
Spitzenwert der Spannung in Durchlaßrichtung peak-forward voltage
Spitzenzähler peak meter, maximum demand meter
Spitzenzähler mit Zeitverzögerung lagged-demand meter (621.317); Geiger counter (539)
Spitzenzählung intermittent integration (621.317)
Spitzzange long-nose plier
spitzen to point, to sharp
spleißen to splice
gespleißte Verbindung spliced joint
Spleißstelle (cable) splice
Spleißverbindung spliced joint
Splint split pin, cotter pin, linchpin
Splitter splinter, sliver
splittern to splinter, to sliver

spontane Kettenreaktion spontaneous nuclear reaction
Sporn spur
Sprach-... voice, speech, vocal, speaking
Sprachaudiometer speech audiometer
Sprachband voice band, speech range (621.395)
Sprachfrequenz speech frequency, voice frequency, talking to ne (534); telephone frequency
Sprachfrequenzbereich voice-frequency range
sprachgesteuerter Pegelregler vogad = voice-operated gain adjusting device
sprachgesteuertes Relais voice-operated relay (621.395)
Sprachentschlüsselung speech decoding
Sprach-IC speech synthesizer
Sprachinverter speech inverter
Sprachlabor language laboratory
Sprachmodulation voice modulation
sprachmodulierte Wellen voice-modulated waves
Sprachpegel speech level
Sprachrohr speaking tube, speaking trumpet
Sprachschutz speech protection (621.395)
Sprachschutzschaltung guard circuit
Sprachsicherheit speech security
Sprachsimulator voder = voice-operation demonstrator
Sprachverschlüsselung speech scrambling, speech inverting, ciphony
Sprachsperre vodas = voice-operated device, anti-sing (621.395)
Sprachspitze voice peak
Sprachverständigung voice communication
Sprachverständlichkeit speech intelligibility
Sprachsynthesizer speech synthesizer
Sprachverwürfler speech scrambler
Sprache speech, voice (53); language
spratzen to sizzle
Sprech-... talking, voice
Sprechbatterie talking battery, microphone battery (621.395)
Sprechfilm talking film, talking picture, sound film
Sprechfunk radio telephony
Sprechkanal telephone channel, voice channel
Sprechkapsel transmission capsule, telephone transmitter
Sprechkopf sound head
Sprechkreis speech circuit, voice grade circuit
Sprechleistung voice power, speech power, acoustical speech power
Sprechorgan s. **Sprachorgan**
Sprechprobe talking test
Sprechrelais voice-operated relay (621.395)
Sprechrufzeichen voice-call sign
Sprechsperre s. **Sprachsperre**
Sprechspitze voice peak
Sprechstelle telephone station
Sprechstellung speaking position (621.395)
Sprechstrom voice current, speech signal
Sprechstromkreis telephone circuit
Sprechtaste push-to-talk button, push-to-talk key (621.39); speaking key, microphone switch (621.395)
Sprechverkehr intercommunication
Sprechvermögen faculty of speech
Sprechwand (loud)speaker baffle
Sprechwechselstrom audio current (621.395)
Sprechweg speech path
Spreizring expanding collet
Spreizstütze straddle
Spreng-... detonating, explosive, blasting
Sprengniet explosive rivet
Sprengring snap ring, split ring, circlip
Sprengstoff explosive, blasting agent, blasting compound
sprengen to blow up, to explode, to blast
Springschreiber start-stop printer (621.394)
sprinkeln to sprinkle
Sprinkler sprinkler
Sprinkleranlage sprinkler plant
Sprinklerdüse sprinkler nozzle
Spritz-... spray
Spritzanlage spraying plant
Spritzdüse spray tube, atomizer
Spritzflasche sprinkling bottle
Spritzguß pressure die-casting; injection moulding (66)
Spritzlack spraying varnish
Spritzlackierung spray (lacquer painting)
Spritzpistole spray gun
Spritzwasser splash water
spritzwassergeschützt weatherproof, splashproof (621.3)
spritzwassergeschütztes Schaltgerät weatherproof switchgear
spritzen to squirt, to spray, to splash
gespritzter Leuchtdraht squirted filament
spröde brittle
spröde machen to embrittle
Sprödigkeit brittleness
Sprosse sprocket, spoke
sprudeln to effervesce, to squirt
Sprüh-... spray (621); corona (621.31)
Sprühätzvorrichtung spray etcher

Sprühbüschel (corona) tuft (discharge)
Sprühelektrode active electrode (628); discharge electrode
Sprühentladung corona (discharge)
Sprühkühler spray cooler
Sprühnebel spray
Sprühregen small rain
Sprühschirm corona shield
Sprühschutz corona protection
mit Sprühschutz versehen corona protected
Sprühstörung precipitation static (interference) (621.391)
Sprühstrom corona (discharge) current
Sprühverfahren spraying process
Sprühverlust corona loss
sprühen to spray (621); to fizz (621.396); to form corona (621.31)
Sprung jump (51); crack, fissure (621.315.62); hop (621.396)
Sprungantwort transient response
Sprungbefehl branch instruction, jump instruction, control-transfer instruction (681.3)
Sprungentfernung jump-over distance, skip distance (621.396)
Sprungfunktion discontinuous function, impulse function, Heaviside unit function
Sprungkontakt snap-action contact
Sprungsignal bar signal (621.397)
Sprungspannung initial inverse voltage
Sprungverzerrung transient distortion
sprungweise in a series of jumps
Sprungwelle voltage surge
Sprungwellenprobe surge-pressure test
Sprungzone silent zone, skip zone
Spule coil, solenoid, inductor (621.3); bobbin (621.315); spool, reel (621)
Spule unter Strom setzen to energize a coil
Spule mit Verlusten resistive coil, resistive inductor
Spule wickeln to wind a coil
90°-Spule retarder (621.317)
auf Spulen aufwickeln to assemble on reels
mit einer Spule verkettet sein to link a coil
spulenbelastet coil-loaded (621.392)
Spulendraht coil wire
Spulenfeld coil section, loading section (621.392)
Spulenfluß flux linking a coil
Spulenkasten spool, loading case
Spulenkern inductor core, core (of coil)
Spulenkern für induktive Abstimmung slug
Spulenkopf overhang, end winding
Spulenkopfstreuung leakage around overhang of coils, end-winding leakage (621.313)
Spulenkörper coil form, coil former, bobbin
Spulenlack coil varnish
Spulenmuffe loading pot
Spulenpunkt loading point
Spulenrahmen former (621.3)
Spulenrelais solenoid relay
Spulensatz bank of coils (621.396)
Spulenschalter coil switch (621.396)
Spulenseite coil side
Spulensonde magnetic proofplane
Spulentrommel spool
Spulenwange spool head, spool end
Spulenwickelmaschine coil-winding machine
Spulenwicklung coil winding, concentrated winding (621.313)
Spulenwiderstand coil resistance; swamping resistance (621.317)
spulen to reel, to coil
spülen to scavenge (621.43)
Spülmaschine dishwasher
spunden to groove-and-tongue
Spur track, trace; gauge (625.1)
Spur bei Härteprüfung indentation
Spurbreite track width (681.84)
Spurelement track element; spot (681.84)
Spurkranz wheel flange, wheel rim (625.2)
Spurlager step bearing, step box
Spurplanstellwerk track-interlock system
Spurweite rail gauge, (width of the) track
spuren to track
Stab rod, bar
Stabantenne rod aerial, flagpole
Stabausdehnungsthermostat stem-type thermostat
Stabbatterie dry pile
Staberder ground rod (US); earthing rod
Stabgitter rod lattice
Stabgleichrichter rod-type rectifier
Stabisolator rod-type insulator, stick insulator
Stabkräfte member forces
Stablampe electric to rch
Stabmagnet bar magnet
Stab(strom)wandler bar-type (current) transformer
Stabwalzwerk bar mill
Stabwicklung bar winding (621.313)
stäbchenförmig prismatic
stabil stable, stout; non-radioactive (539)
stabil gebaut substantially made
stabiles Fundament stable foundation
stabiles Gleichgewicht stable equilibrium

stabiler Gleichgewichtspunkt stable equilibrium point
stabile Konstruktion stout design
stabile Lage stable position
Stabilisationsröhre VR tube = voltage-regulating tube, voltage stabilizer
Stabilisator stabilizer (621.385)
Stabilisatordiode voltage-regulator diode
stabilisieren to stabilize
stabilisieren bei to keep stationar at
stabilisiertes Netzanschlußgerät regulated power supply
stabilisierte Stromquelle regulated voltage source
Stabilisierschaltung base-line stabilizer, stabilizer circuit (621.397)
Stabilisierung stabilization
Stabilisierungskreisel gyrostat
Stabilisierungswiderstand stabilizing resistance
Stabilität stability
Stabilitätsbedingung stability condition, condition of equilibrium
Stabilitätsgrenzleistung stability limit, power limit (621.313)
Stabilitätskriterium criterion for stability
Stabilovolt stabilovolt, glow-tube regulator
Stadt town
Stadtbahn urban railway, metropolitan railway
Stadtwerke municipal works
staffeln to stagger
Staffel staggered arrangement; progressive grading (621.395)
Staffelung grading (621.395); stagger, staggered arangement
Stahl steel
Stahl mit hohem Kohlenstoffgehalt high-carbon steel
Stahl mit niedrigem Kohlenstoffgehalt low-carbon steel
Stahl mit Schneidauflage tipped tool
Stahlakkumulator alkaline accumulator, steel accumulator
Stahlaluminium steel-cored aluminium
Stahlaluminiumseil aluminium-steel conductor, aluminium-cable steel-reinforced
Stahlband steel band, steel ribbon, steel tape
Stahlbandbewehrung steelband armouring, steel-tape armouring
stahlbandbewehrt steel-tape armoured
Stahlbau steel construction, structural steelwork
Stahlblech sheet steel, steel sheet
stahlblechgekapselt sheet-steel clad, metal clad
Stahlbürste steel-wire brush
Stahldraht steel wire
stahldrahtbewehrtes Kabel steel-wire armoured cable
Stahleinlage steel reinforcement
Stahlguß(stück) steel casting
Stahlkern steel core
Stahlkupferdraht copper weld, copper-clad steel-wire cable
Stahlmast steel tower, tubular steel pole
Stahlpanzerrohr steel conduit, screwed conduit (621.3)
Stahlrohr steel conduit, tubular steel
Stahlrohrleitung plain steel conduit
Stahlrohrrahmen tubular steel frame
Stahlröhre metal valve (621.385)
Stahlschmiedestück steel forging
Stahlseele steel core
Stahlseil steel rope
Stahlspäne steel wool (sing.)
Stahlstäbchenamperemeter surge-crest ammeter
Stahlstäbchenmessung magnetic-link record
Stahlwellblech corrugated steel
Stahlwolle steel wool
Staku copper-clad steel
Stakudraht copper-clad iron alloy wire, copperweld (steel conductor) (621.315.1)
Stakuseil aluminium-steel conductor (621.315)
Stamm distribution line (621.396.67)
Stammkreis physical circuit (621.394/.395)
Stammleitung side circuit
stampfen to ram, to stamp, to tamp; to pitch (629.21)
Stand level; stand, state
Stand der Technik state-of-the-art
Standhöhe level
Standleitung permanent circuit, dedicated line
Standlinie fiducial line; position line (621.398.96)
Standmaß glass measure, graduated flask, gauge glass
Standort site, locality, position
Standortidentifizierungssignal positional signal, identification signal
Standrohr (syphon) standpipe, gauge glass
Standverbindung point-to-point transmission, point-to-point connection; dedicated line (621.395)
Standard standard
Standardabweichung standard deviation (51)

Standardatmosphäre standard radio atmosphere (621.396)
Standardausführung standard design
Standardbrechung standard refraction
Ständer stand, standard (621); stator, stationary armature (621.313); (s. auch **Stator**)
Ständeranker stationary armature
Ständerblech stator lamination
ständergespeist stator fed
ständergespeister Mehrphasen-Nebenschlußmotor doubly fed, polyphase-shunt commutator motor
Ständerjoch stator yoke
Ständerrahmen standard
Ständerspannung stator voltage
Ständerspule stator coil
Ständerstab stator conductor, armature conductor
Ständerstreuung stator leakage
Ständerstrom stator current
Ständerwicklung stator winding
ständig wiederholend, sich constantly recurring, repetitive
Stange stick, rod, bar
von der Stange arbeiten to work from the bar (621.9)
Stangendicke bar diameter
Stangenfuß butt
Stangenstromabnehmer pole trolley
Stanniol tinfoil
Stanze punch-press, punching machine
stanzen to punch, to stamp, to pierce
Stanzrest chad (621.394.6)
Stapelbetrieb batch processing (681.3)
Stapeleinrichtung stacker
Stapelspeicher stack, LIFO = Last In - First Out
Stapelverarbeitung batch processing
Stapelzeiger stackpointer
stapeln to stack, to stock, to pile
Stapler stacker, piler
stark strong, powerful, intense, heavy
stark beanspruchen to work hard
stark belastet heavily loaded
stark rauschende Röhre noisy valve
starke Abnutzung heavy wear
starke Beanspruchung hard usage, intense stress
starke Belastung heavy load
starke Bestrahlung strong irradiation
starker Druck intense pressure
starke Hitze intense heat
starke Lampe intense light
starke Säure strong acid
starke Strahlung strong radiation, intense radiation
starker Strom heavy current, intense current, strong current
starker Verschleiß heavy wear
zu stark angespannt overstrung
zu stark bemessen to over-design
Stärke thickness; diameter; strength; intensity (621.3); starch (621.35), (66)
Stärke des Feldes strength of the field
Stärke eines Stromes strength of a current
Starkstrom heavy current, power current
Starkstromanlage heavy current installation, power installation
Starkstrombeeinflussung telephone interference (factor) (621.392)
Starkstrombeleuchtung power lighting
Starkstromfunkstörung radioinductive interference, man-made interference
Starkstromgeräusch power induction, induced noise (US) (621.395)
Starkstrominstallateur wiring contractor
Starkstromkabel (electric) power cable
Starkstromkondensator heavy current capacitor, power capacitor
Starkstromkreis heavy-current circuit, power circuit
Starkstromleitung power line
Starkstromnetz power mains, heavy-current system, power grid, electric mains
Starkstromrelais heavy-duty relais
Starkstromseekabel submarine power cable
Starkstromtechnik heavy-current engineering, electrical power engineering
starr rigid, stiff, solid
starr eingespannt direction fixed (531)
starr erden to earth solidly, to ground solidly
starr geerdetes Netz solidly earthed system, solidly grounded system
starr kuppeln to couple rigidly (621.1)
starre Aufhängung rigid suspension (621.33)
starre Einspannung der Enden defined-end restraint (62)
starrer Körper rigid body
starre Regel positive rule, hard and fest rule
starre Vorschrift rigid regulation
Starrheit rigidity
Start start; take-off (629.13)
Startbahn runway (629.13)
Startbahnfeuer runway (-marker) light
Startbahnhochleistungsfeuer high-intensity runway light
Startbahnunterflurfeuer runway flush-marker light

Starthebel, Startknopf starter push-button
Startleistung take-off power
Start-Stop start-stop
Start-Stop-Schreiber start-stop printer
Start-Stop-Telegraphie start-stop telegraphy
Start-Stop-Verzerrungsgrad start-stop distortion
Starter starter (621)
Starterbatterie starter battery
Statik statics (pl.)
Station station (625), (621.3); ward (61)
Station im Sendebetrieb station on the air
Stationskennungsgerät station-identification device
Stationskennzeichen (station) signature, identification (621.396)
Stationszeichen signature tune, identification signal (621.396)
stationär stationary
stationär halten bei to keep stationary at, to keep steady at
stationärer Ausschlag point of rest, stationary deflection (621.317)
stationäre Endspannung final voltage
stationärer Endzustand steady state (condition)
in stationärem Endzustand for steady conditions
stationäres Feld stationary field
stationärer Sperrstrom resistive reverse current (621.382.2)
stationieren to station
statisch static
statische Aufladung static electrification, static charge
statische Auslösung definite-time (overcurrent-) trip (621.316.5)
statische Berechnung static calculation
statische Elektrizität static electricity
statisches Elektrodenpotential (ohne Strombelastung) static electrode potential
statischer Frequenzwandler static frequency changer
statische Kennlinie static characteristic (621.385)
statische Kippleistung steady-state stability limit (621.311)
statischer Kondensator static capacitor
statische Ladung static charge
statischer Regler proportional position action controller
statische Reibung static friction
statischer Speicher static memory
statische Stabilität steady-state stability
statischer Umformer static converter, static inverter
statisches Voltmeter electrostatic voltmeter
statischer Zustand static state, steady state, state of rest
Statistik statistics (pl.)
statistisch statistic(al)
statistischer Fehler random error
statistisches Material statistica, information; statistical data (pl.)
statistische Qualitätskontrolle statistical quality control
statistisches Verfahren statistical method
statistisch verteiltes Rauschen white noise
statistische Verteilung statistical distribution
Stativ (floor) stand, tripod
Stativbein telescopic leg
Stator stator (621.313), (621.319.4); carcase (621.313); (s. auch **Ständer**)
Statoranker stationary armature
Statorblech stator lamination
Stator-EMK stator e. m. f.
Statorgehäuse, Statorgestell stator frame, carcase (621.313)
Statorplatte stator plate, stator vane (621.319.4)
Statorspule stator coil, stationary coil
Statorstreureaktanz Poitier reactance
Statorstrom stator current
Statorwickung stator winding, stationary armature
Statuswortregister status word register
Staub dust
Staubablagerung dust accumulation
Staubabscheider dust precipitator, air cleaner
staubdicht dust-proof
Staubelektrisierung dust electrification
Staubexplosion dust explosion
Staubfang dust catcher
Staubfilter dust filter
staubförmiger Brennstoff pulverized fuel
Staubkörnchen speck of dust
Staubsammelkammer collector cell
Staubsauger vacuum cleaner, electric sweeper; dust exhauster
Staubteilchen dust particle
Staubwolke dust cloud
staubig dusty
stauchen to upset; to compress
Stauchgetriebe upset gear (621.791)
Stauchung compression strain, compressive strain
Staudamm, Staumauer dam
Staudruck ram pressure (532), (533)

Staufferbüchse grease cup
Staufferfett cup grease, consistent grease
Steatit steatite, soapstone
stechender Geruch pungency, poignant odour
Stechuhr time clock, watchman's clock
Stechzirkel (pair of) dividers (pl.)
Steck-... plug-in
  Steckbaugruppe plug-in-type subassembly
  Steckdose socket (outlet), wall socket, receptacle outlet, convenience outlet, mains socket, female socket
    Steckdose auf Grundplatte mounting receptacle
    an die Steckdose anschließen to plug into connection
  Steckeinheit plug-in package, plug-in unit
  Steckfassung lamp jack, contact cap
  Steckkarte plug-in printed-circuit board
  Steckschlüssel socket wrench boxspanner
  Stecksockelrelais plug-in relay
  Steckspule plug-in coil
  Stecktafel pegboard
  Stecktransformator plug-in transformer
  Steckverbinder plug-and-socket connector, plug connector, socket joint
  Steckverbindung plug-in connection
  Steckverteilertafel cross-connecting board
  Steckvorrichtung plug-in device, plug and socket, coupler
stecken to plug
  steckbar pluggable
    steckbarer Baustein plug-in module
    steckbarer Einschub plug-in tray
  gesteckter Stöpsel plug in position (621.317)
Stecker plug, connector, male plug
  Stecker in die Klinke stecken to slip a plug into the jack
  Steckeranschluß plug connection, plug-in outlet
    mit Steckeranschluß plug-connected
    Steckeranschlußstelle receptacle outlet
  Steckerbuchse plug socket
  Steckerleiste plug receptacle, frame connector
  Steckerloch plug hole, pin-jack
  Steckerschalter plug switch
  Steckersicherung plug fuse
  Steckerstift contact pin, pin plug, male plug, prong
  Steckerstrippe plug-terminated cord
  Steckerteil male and/or female connector
Stefan-Boltzmannsches Strahlungsgesetz Stefan-Boltzmann law of radiation
Steghohlleiter ridge waveguide
Stegleitung flat underplaster wire, flat webbed wire
stehen to stand
  stehen frei von ... to stand clear from ...
  stehenbleiben to stop
  stehendes Bild stationary pattern
  stehende Welle standing wave, stationary wave (621.392)
Stehlampe standard (lamp), floor standard
Stehwellenverhältnis standing wave ratio (621.396)
steif stiff, stout, rigid
  steifer Draht stout wire
  steife Pappe stiff cardboard
Steife torque per unit elongation
Steifigkeit stiffness, rigidity
Steigeisen pole climber, climbing iron
Steigleitung rising main
Steigen der Stromstärke rise of current
steigen auf to rise to, to build up to
  steigende Charakteristik rising characteristic
  steigende Potenz rising power (51)
  steigende Spannung rising voltage
  steigende Tendenz upward trend, upward tendency
steigern to augment, to increase; to enhance
  steigern, sich to increase, to build up, to run up
Steigerung increase, augmentation; enhancement
Steigung grade, (rising) gradient; inclination, incline (625); lead, pitch (621.8); riser (69)
  Steigungsmesser inclinometer
  Steigungswinkel angle of twist, angle of lead (621.9)
    Steigungswinkel der Wendel helix angle
steil steep
  steil abfallen to drop steeply
  steil ansteigen to rise steeply
  steiler Anstieg high rate of rise
  steile Flanke steeply-falling side (621.396)
  steiles Gefälle steep gradient, rapid slope
  steiler Konus steep taper
  steile Röhre high-slope valve (621.385)
  steile Wellenfront, steile Wellenstirn sharp wave front, steep front of the wave
Steilheit mutual conductance, (grid plate) transconductance, electrode admittance, rate of amplitude change (621.385); inclination, slope (625)
  Steilheit der beim Ausgleichsvorgang wiederkehrenden Spannung transient recovery voltage rate (621.316.5)

Steilheit des Spannungsanstiegs rate of voltage rise
Steilheit der Wellenfront steepness of the wave front
Steilheitsrelais rate of change relay
Stein stone
Steinbau stone construction, stone structure
Steinbettung stone bedding
Steinbrecher stone breaker, stone crusher, (stone) crushing machine
Steinkohle hard coal, mineral coal
Steinkohlenteer coal tar, gas tar
Steinlager jewel, jewelled bearing (621.317), (681)
Steinöl petroleum
Steinpfanne cup jewel
Steinsalz rock salt, sodium chloride
Steinsalzfenster sodium-chloride window
Steinschraube rawlbolt, anchor bolt
Stell-... set, adjusting, operating
Stellantrieb positioner, actuator, servomotor
Stellbefehl control signal (621.316.7)
Stellbereich operating range, to tal range of controller output, maximum possible range
Steller positioner, final control element
Stellgeschwindigkeit adjusting velocity, speed of adjustment, floating speed
Stellglied actuator, regulating unit, positioning element, positioner, servo actuator
Stellgrad ratio of actual steady-state controller, output to maximum, controller output
Stellgröße regulated quantity, actuating variable
Stellhebel adjusting lever, control lever
Stellkeil adjusting wedge
Stellkraft operating force
Stellkurbel operating crank
Stellmoment operating torque
Stellmotor power cylinder, servo motor
Stellmutter set screw, adjusting nut
Stellort control point
Stellring set collar, setting ring
Stellschlitten brush slide
Stellschraube set screw, adjusting screw
Stellstrom control current
Stellwerk interlocking plant, switch-box, signal-box, interlocking apparatus (625.1); positioner, correcting unit (621.316.7)
Stellwerkswärter leverman (625.1)
Stellzeit actuating time, floating speed
Stellage stillage
Stellar-... stellar
Stelle position; place, spot; digit, cipher (51)
Stellen hinter dem Komma decimal points
Stellen weglassen to suppress digits
auf n Stellen genau berechnen to work to n places of decimals
Stellenwertschreibung base notation, positional notation (51), (681.3)
Stellenwertverschiebung cyclic shift
Stellenzahl number of digits, number of decimals (51)
stellen to put; to set, to adjust, to position
stellendes Regelsystem positioning control system
Stellit stellit
Stellung position, place
Stellungsanzeiger position indicator; vane type relay
Stellungsgeber remote-position indicator
Stellungswechsel switch-over, make or break (621.316)
Stempel prop (622); stamp (651); impress, imprint (656)
Stereo..., stereo..., stereophonic...
Stereo-Aufnahme- und Wiedergabegerät stereo recorder
Stereo-Empfang stereophonic reception
Stereogerät stereo console
Stereo-Mischpult stereo mixer
Stereo-Multiplexsignal stereo-broadcasting signal
Stereo-Verstärker stereo amplifier
Stereowiedergabe stereophonic reproduction
Stereophonie stereophonic sound system
stereophonisch stereophonic
stereophonischer Effekt stereophonic effect
stereophoner Fernsehton stereophonic TV sound
steril sterile
Sterilisator sterilizer
Stern star (52), (621.3); asterisk (51)
Sterndreieck star-delta
Sterndreieckanlasser star-delta starter
Sterndreieckanlauf star-delta starting
Sterndreieckschalter star-delta switch, "Y"-delta star
Sterndreiecktransformation star-delta transformation
Sterndreieck-Umschalter star-delta throw-over switch
Sternenlicht starlight
sternförmig asteroid
Sternmotor radial engine (621.43)
Sternnachführungsgerät astrotracker, star-tracking system

Sternnetz star-shaped network, star-type network
Sternpunkt star point, neutral point, centre point (621.3)
Sternpunktsanode zero-point anode
Sternpunkterdung neutral ground
Sternpunktklemme neutral terminal
Sternpunktleiter neutral conductor, marked conductor, neutral wire
Sternschaltung star connection, "Y"-connection
sterngeschaltet star-connected, "Y"connected
sterngeschaltete Wicklung "Y"-winding
Sternschreiber P. P. I. = plan position indicator (621.397)
Sternspannung star voltage, phase voltage
Sternsystem radial configuration
Sternviererkabel star quad cable
Sternwarte observatory (52)
Sternchen asterisk (51)
mit einem Sternchen versehen to asterisk
Stethoskop stethoscope
stetig steady, continuous; smooth (621.3)
stetig regelbar continuously adjustable
stetig übergehen in to blend into
stetig veränderlich infinitely variable, continuously variable
stetige Funktion continuous function
stetige Regelung, stetige Steuerung continuous control
stetiges Regelverfahren controlled continuous process
Stetigbahnsteuerung continuous-path control
Steuer rudder, helm (629.12), (621.13); tax (3)
Steuer-... control, controlling
Steueranschluß gate terminal (621.314.63), control-current terminal
Steuerbefehl control order, control signal
Steuerbefehlsspeicher control register
Steuerbus control bus
Steuerdifferentialgeber synchro control differential transmitter
Steuerdifferentialempfänger synchro control differential receiver, synchro control differential transformer
Steuerdrehmelder control synchro (621.316.7)
Steuerdruck control load
Steuereinrichtung control mechanism, controlling device, control unit
Steuerelektrode control electrode (621.385)
Steuerempfänger synchro control receiver, synchro transformer
Steuerfähigkeit controllability
Steuerflüssigkeit controlling agent
Steuerfolie intersheath (621.315.2)
Steuerfrequenz pilot frequency, driver frequency (621.395)
Steuerfunke trigger spark
Steuerfunktion wiedererhalten to resume control
Steuergeber control transmitter, synchro control transmitter
Steuergerät control gear, controller, control device
Steuergestänge, Steuergetriebe control mechanism, control gear
Steuergitter control grid, signal grid (621.385); buncher grid, buncher electrode
Steuergröße controlling quantity, actuating variable
Steuerhebel control lever
Steuerimpuls drive pulse; control pulse
Steuerkabel control cable
Steuerkanal control channel
Steuerkennlinie transfer characteristic; control characteristic (621.316.7)
Steuerkette open-loop control (621.316.7)
Steuerknüppel joystick
Steuerkreis master-oscillator circuit, driver, driving circuit
Steuerkurs (aircraft) heading
Steuerleistung control power, driving power (621.396)
Steuerleitung control line, control lead
Steuernocken manoeuvring cam
Steuerpegel signal level (621.396)
Steuerprogramm executive program (681.3)
Steuerquittungsschalter control-discrepancy switch
Steuerpult control desk
Steuerquarz piezoelectric frequency stabilizer, control crystal, oscillator crystal
Steuerquittungsschalter control-discrepancy switch
Steuerregister control register
Steuersäule control column (621.1)
Steuerschalter (cam-shaft) controller; control switch, sequence switch (621.395)
Steuerschalttafel control board, control panel
Steuerschaltung control circuit
Steuerschaltwalze pilot controller
Steuerschema control scheme
Steuerseil control cable
Steuersender master oscillator, exciter
Steuersender-Kraftverstärker m. o. p. a. = master-oscillator power amplifier
Steuersignal control signal (681.3)

Steuersonde control probe (539.17)
Steuerspannung control voltage; gate voltage (621.314.63), signal voltage (621.39)
Steuerspannungskennlinie total characteristic (621.385)
Steuerspur control track
Steuerstrom control current
Steuerstromkreis control circuit
Steuerstufe exciter stage
Steuersystem control system
Steuersystem ohne Rückkopplung open-loop control system
Steuertafel control board, control panel
Steuerventil control valve, pilot valve
Steuerverfahren control system, control scheme, control method
Steuerverlustleistung gate power loss (621.314.63)
Steuerwalze (master) controller, barrel controller, drum controller, control drum
Steuerwelle cam-shaft
Steuerwerk control unit (681.3)
Steuerwicklung control winding (621.318)
Steuerwirkung control action, guiding effect
Steuerwirkung ausüben auf to exercise control over
Steuerzeichen control character (681.3)
steuerbar controllable
steuerbare Einkristall-Gleichrichterzelle silicon controlled rectifier = SCR
steuerbare Spannungsquelle variable-voltage source
steuern to control, to guide, to monitor; to time
gesteuerte Funkenstrecke timed-spark gap
gesteuerte Rakete guided rocket, guided missile
gesteuerter (modulationsabhängiger) Träger controlled carrier
Steuerung control (mechanism); guiding
Steuerung von Arbeitsabläufen process control
Steuerungsablauf control sequence
Steuerungskommando control signal
Steuerungsprogramm executive routine (681.3)
Stia-Zähler mercury meter
Stibitzcode excess-three code (681.3)
Stich stitch; engraved plate, engraving
Stich eines Bogens rise of an arch
in Stich schalten to tee (621.3)
Stichbogen segmental arch
Stichflamme narrow flame, thin flame
Stichleitung dead-end feeder, stub (621.3); tee (621.315)
Stichmaß feeler (gauge) (620.1)
Stichprobe random test; representative specimen, sample
Stichprobenentnahme sampling, drawing of samples
Stichprobenkontrolle sampling inspection
Stichprobenprüfung random test, sampling test
Stichsäge piercing saw
Stichspeiseleitung teed feeder
Stichel turning chisel
Stickoxid nitric oxide
Stickoxidul nitrous oxide
Stickstoff nitrogen, azote
Stickstoffbindung fixation of nitrogen
stickstoffhaltig nitrogenous
Stickstoffkühlung inertair cooling
Stickstoff-Lichtbogenofen Birkeland and Eyde furnace
Stickstoffperoxid nitrogen peroxide
Stickstoffverbindungen nitrocompounds
Stiel handle, shaft, shank
Stielstrahler dielectric aerial
Stift pin, stud; sprig
Stiftfassung prong grip (621.385)
Stiftkeil sprig
Stiftkontakt pin contact
Stiftrad cog wheel
Stiftschraube stud (-bolt) (621.882)
Stiftsockel plug-type base, pin-type socket (621.385)
Stiftzahn sprocket
Stil style
Stilb stilb
still still, silent, quiet
stillegen to shut down, to close down
stillsetzen to bring to rest
stillstehend at rest, stationary, fixed; non-operative
stillstehender Anker fixed armature (621.313)
stillstehendes Bild stationary pattern (621.385)
stillstehende Wicklung stationary winding
Stillstand standstill, rest, stoppage, zero r. p. m. (621.3); stagnation
zum Stillstand bringen to bring to a stop, to bring to a stand (still)
zum Stillstand kommen to come to a stop, to come to a stand (still)
Stillstandswächter zero-speed monitor
Stillstandszeit period of rest, downtime
Stillstandszeit in Reserve standby period
Stimme voice
Stimmbänder vocal chords

Stimmgabel tuning fork
stimmgabelgesteuerte Uhr tuning-fork clock
Stimmgabelsender mechanical master oscillator, tuning-fork oscillator, fork oscillator, maintained tuning fork
Stimmgabelschwebung fork beat
Stimmumfang gamut, register
stimmen to tune (78); to hold good (51)
gestimmt sein auf to be unison with
Stirn front; end grain (67)
Stirn der Welle wave front
Stirnansprechstoßspannung front-of-wave impulse sparkover-voltage
Stirnfläche frontal area, (top) face, end plane
Stirnlager end-journal bearing
Stirnlampe cap lamp
Stirnrad spur gear
Stirnradverzahnung spur gearing
Stirnseite front side, face
Stirnstoß front joint, flange edge joint
Stirnstreuung brow leakage (621.313)
Stirntrieb spur gearing
Stirnverbindungen overhang (sing.) (621.313)
Stirnwand front wall, face; facade (72)
Stirnwandtür end door (625)
stochastisches Rauschen stochastic noise
Stockwerk floor, storey, story
Stockwerksanzeiger position indicator, lift-car annunciator (621.3)
Stockwerkslichtrufanzeiger lift-car flash signal-device
Stockwerksruf waiting-passenger indicator (621.876)
Stockwerksrufanzeiger lift-car annunciator
Stockwerkschalter floor switch
Stoff matter, substance, material; "stuff"; cloth, tissue, fabric (67)
Stoffübergang mass transfer
Stopfbuchse, Stopfbüchse (stuffing) gland, stuffing-box
Stopfbuchspackung stuffing-box packing, gland packing
Stopfen plug, stopper; wad
stopfen to plug, to stop up; to tamp (66)
Stoppbefehl breakpoint instruction (681.3)
Stoppuhr stopwatch
Stöpsel plug (621.3)
Stöpsel stecken to plug
Stöpsel ziehen to unplug
Stöpselkasten resistance box
Stöpselloch plug hole, pin jack
Stöpselpatrone plug cartridge
Stöpselring sleeve (621.395)
Stöpselschalter plug cut-out
Stöpselsicherung plug(-type) fuse
Stöpselumschalter plug commutator
Stöpselverbindung plug connection
Stöpselwiderstand (plug-type) resistance box (621.317)
stöpseln to plug, to patch
Stör-... interfering, parasitic, disturbing, spurious
Störabstand signal-to-noise ratio, noise margin
Störausstrahlung spurious radiation
Störaustastschaltung interference blanker, interference eliminator
Störbeeinflussung telephone interference (factor) (621.392)
Störbefreiungsschaltung static eliminator (621.396.62)
Störbegrenzer noise limiter
Störecho clutter (629.396.96)
Störecho-Unterdrückung anticlutter (circuit)
Störeindruck annoyance (534)
Störfeldstärke noise-field intensity, field strength of the unwanted signal
Störfeldabstand radio field-to-noise ratio
Störfestigkeit interference rejection
Störfilter noise filter
Störgeräusch (disturbing) noise, static noise (621.396.62)
Störgröße disturbance (621.316.7)
Störmodulation spurious modulation
Störpegel static noise, background noise, noise level
Störpegelzählung background count
Störreflektor parasitic reflector
Störresonanzen unwanted responses
Störschutz mains filter, noise suppression, interference elimination
Störschutzantenne anti-static aerial
Störschutzfilter interference filter, interference trap
Störschutzkondensator (anti-)interference condenser
Störschwingung parasitic oscillation
Störsender jamming transmitter; jammer, interfering transmitter
Störsenderunterdrückung anti-jamming
Störsicherheit noise immunity
Störsignal spurious signal, interfering signal
Störspannung disturbing voltage, noise voltage, interference voltage (621.385); noise level (621.396)
Störspannungsabstand noise margin (621.374)
Störsperre noise filter (621.396.62)

Störstelle imperfection, impurity (621.382)
Störstellenhalbleiter extrinsic semiconductor
Störstellenkompensation impurity compensation (621.382)
Störstellenleitung extrinsic conduction (621.382)
Störstellen-Übergang junction
Störstrom disturbing current (621.382); interference current
Störunterdrückung noise suppression
Störwirkung disturbing effect, spurious effect
Störzeichen interfering signal
Störzipfel parasitic lobe (621.396.67)
Storchschnabel pantograph
stören to disturb; to jam (621.396); to interfere with
störend beeinflussen to interfere with
störender Einfang parasitic capture (539)
gestörter Abzweig faulty branch (621.3)
gestörter Stromkreis faulty circuit
Störung trouble, fault (621.3); interference (621.395); disturbance; break down, malfunction, failure
Störung beseitigen to clear the trouble; debugging (681.3)
Störung durch Ausstrahlung transmission jamming (623)
Störung durch Kreuzmodulation cross-talk interference
Störungsannahme fault recording (621.395)
Störungsdienst fault-complaint service
störungsfrei trouble free
Störungsquelle source of trouble, source of interference
Störungsquelle lokalisieren to isolate the source of trouble
störungssicher fail safe
störungssichere Logikschaltung high-noise immunity logic
Störungsstelle maintenance section, trouble department, trouble desk (621.395/.6); position of fault
Störungssuche trouble shooting, interference location, fault location
Störungssucher trouble shooter (621.3), (621.396)
Störungstableau fault-indicator panel
Störungswegschaltung rapid auto reclosing (621.316)
Stoß impact, percussion, push, blow (621); shock (534), (621.3); concussion (534); impulse, surge (621.3)
Stoßanregung shock excitation (534)
stoßartig impulsive
Stoßausschlag throw (621.317)
Stoßbeanspruchung impact conditions (pl.)
Stoßbelastung shock load; surge load (621.3)
Stoßdämpfer dashpot, shock absorber (621.113)
Stoßdämpfung mismatch loss, reflection loss (621.395)
stoßdämpfende Aufhängung shock-absorbing suspension, shock-absorbing mounting
stoßerregt shock excited
stoßerregter Sender shock-excitation (-type) transmitter
Stoßerregung shock excitation, high-speed excitation system (621.313.), (621.396); superexcitation
Stoßerregungswiderstand forcing resistor
stoßfest shockproof, high-impact
Stoßfestigkeit impact strength, resistance to impact
stoßfrei non-reflecting (621.39)
stoßfreier Abschluß matched load
stoßfreie Abschlußwiderstände image impedances (621.392)
stoßfreie Umschaltung bumpless transfer (621)
Stoßfrequenz collision frequency; surge frequency (621.3)
Stoßfunktion impulse function
Stoßgenerator surge generator, lightning generator
Stoßgesetze laws of impact
Stoßionisation ionization by collision, ionization by impact
Stoßkraft impetus
Stoßkurzschlußstrom maximum asymmetric short-circuit current
Stoßkurzschlußstromverhältnis initial short-circuit ratio
Stoßkurzschlußreaktion transient reactance
Stoßlast s. **Stoßbelastung**
Stoßleistungsgenerator surge-power generator
Stoßmagnetisierung flash magnetization
Stoßpegel impulse level
Stoßprozeß collision process (537)
Stoßprüfanlage lightning generator (set)
Stoßprüfung flash test, surge (pressure) test (621.315.6)
Stoß-Rückwärtsverlustleistung surge reverse power dissipation (621.382.2)
Stoßschweißung percussion welding
Stoßschwelle joint sleeper (625)

Stoßspannung surge voltage, impulse voltage, transient voltage; non-repetitive peak reverse voltage (621.382)
Stoßspannungsanlage surge-test station, surge-voltage generator
Stoßspannungsbeanspruchung surge-voltage stress
Stoßspannungsfestigkeit impulse strength
Stoßspannungsgenerator surge-voltage generator
Stoßspannungsschutz surge protection
Stoßspitzenspannung non repetitive peak reverse voltage (621.314.63), (621.382.2)
Stoßspitzensperrspannung peak-transient reverse voltage
Stoßstelle interface, junction
Stoßstelle im Magnetkreis magnetic joint
Stoßstrom surge current; rush of current
Stoßstrom-Grenzwert maximum rated surge-forward current (621.314.63), (621.382.2)
Stoßüberschlag impulse flashover
Stoßüberschlagsprüfung impulse spark-over test
Stoßüberschlagsverzögerung time to impulse flashover
Stoßverhalten impulse response
Stoßvorgang transient
stoßweise by jerks, intermittent
stoßweise Belastung impact load
Stoßwelle surge (621.3); shock wave
Stoßwirkung impact effect
stoßen to push, to kick, to strike
stoßendes Teilchen striking particle, impinging particle
gestoßenes Partikel struck particle
Stößel kicker
stottern to stutter, to stammer; to sputter (621.4)
straff tight
straff spannen to tighten
Strahl ray, beam, shaft (535), (537), (539); jet (532)
Strahl aussenden to throw a beam (621.396.67)
Strahlen aussenden to emit rays
Strahl sperren to gate (621.397)
Strahlablenker, Strahlabweiser jet deflector (621.2)
strahlangetrieben jet-propelled
Strahlantrieb jet propulsion
Strahlauftastung beam trigger
Strahlaustastung beam blanking
Strahlauswanderung wandering (621.396)
Strahlbreite spread (621.396.67)
Strahlbündelungsröhre beam-power tube
Strahldüse jet
Strahlenbündel bundle of rays, pencil (535); beam (621.396)
Strahlenerder buried radials (pl.); crow-foot earthing; radial counterpoise
Strahlenerzeugungssystem electron gun
Strahlenfalle beam trap, ray-locking device
strahlenförmig radial
strahlenförmig verlaufen to radiate
strahlenförmig betriebenes Netz radially-operated system
Strahlengefährdung radiation hazard
Strahlenkegel conical beam (621.396)
Strahlenschaden radiation damage
Strahlenschutz biological shielding (539); X-ray protection (621.385)
Strahlenschutzröhre metal X-ray tube, self-protected X-ray tube (621.386)
Strahlenschutz (wand) biological shield (539)
Strahlführung beam control (527)
Strahlgeschwindigkeit scan speed (621.397)
Strahlkörper radiant body, radiator
Strahlofen radiant heater, radiant-heat fire
Strahlreiter beam rider (623)
Strahlrohr beam hole (539)
Strahlröhre beam tube, power tube
Strahlspannung electron beam potential
Strahlsperrspannung blanking voltage
Strahlsperrung gating (621.397)
Strahlstärke beam power (621.396.67)
Strahlsteuerung ray control (621.385)
Strahlstrom beam current, cathode-ray current
Strahlstromkennlinie beam-current characteristic (621.385)
Strahltriebwerk J. P. = jet propulsion
Strahlverdichtung beam compression
strahlen to radiate, to beam, to flare
strahlendes Atom radiating atom
strahlender Körper radiant body
strahlender Punkt radiant point
Strahler radiant, radiator (621.396)
Strahlerzeile row of radiators (621.396.677)
Strahlung radiation; rays (pl.)
mit Strahlung verbunden radiative
Strahlungsausbeute radiant efficiency
Strahlungsdetektor radiation detector
Strahlungsdiagramm radiation pattern (621.396.677); beam-shaping characteristic, directional pattern
Strahlungsdichte radiant (energy) density, power flux density, power per unit area; emittance

Strahlungsdosis radiation dose (539.16/.18), (621.386)
Strahlungsdurchlässigkeit radiant transmittance
Strahlungseinfang radiative capture (539)
strahlungserregt, strahlungsgekoppelt parasitically exited
strahlungserregte Antenne parasitic aerial
strahlungserregter Teil parasitic element
Strahlungsfaktor radiation coefficient
Strahlungsfeld radiated field, radiation field (621.396)
Strahlungsfläche radiating surface
Strahlungsfluß radiant flux
Strahlungsfrequenz radiation frequency
Strahlungsgleichgewicht radiative equilibrium
Strahlungsheizkörper radiant heater (621.365)
Strahlungsheizung radiation heating; panel heating
Strahlungshöhe radiation height
Strahlungsintensität radiant intensity
Strahlungskrankheit radiation sickness (539)
Strahlungsleistung radiated power
Strahlungsmesser radiometer, radiation counter
Strahlungsmessung radiometry; radiation measurement
Strahlungspyrometer radiation pyrometer
Strahlungsquant corpuscle of radiation, radiation quantum
Strahlungsquelle radiation source, radiant
Strahlungsstabilität radiation stability (539)
Strahlungstrockner radiation drier
Strahlungsverlust radiation loss
Strahlungswärme radiant heat
Strahlungswiderstand radiation resistance (621.396.67)
Strahlungswinkel angle of radiation, wave angle
Strahlungswirkungsgrad radiation efficiency (621.396.67)
stramm tight
stramm anziehen to tighten, to lace tight, to stretch tight
stramm sitzen to be a tight fit
stramm sitzend tight-fitting, close-fitting
strammer Sitz tight fit
strammen to tauten
Strang phase (winding) (621.31); track (625); skein (677)
Strangklemme line terminal
Strangpresse extrusion machine
Strangspannung star voltage, "Y" voltage, voltage to neutral
Strangstrom phase current
Strangwicklung phase winding
Straße street, road
Straßenbahn tramway, streetcar (line), trolley line, tram(-car)
Straßenbahnleitung tram mains, trolley line
Straßenbahnmast tram standard
Straßenbahnmotor tram motor
Straßenbahnoberleitung tram mains, overhead line
Straßenbahnschiene tram rail
Straßenbahnwagen tramcar; streetcar (US)
Straßenbeleuchtung street illumination, street lighting
Straßenfahrzeug road vehicle
Straßenverkehr road traffic
Stratosphäre stratosphere
Strebbeleuchtung coal-face lighting (622)
Strebe strut, support, stay
streckbar ductile, stretchable
Streckbarkeit ductility, stretchability
Strecke length, distance; line section (625); trunk line (621.396.67)
Strecke freiblocken to clear a section
Streckenabschnitt (track) section, block
Streckenabschnitt freiblocken to open a block
Streckenapparat portable telephone set (625)
Streckendämpfung path attenuation (621.396); loss per line, loss per section (621.395)
Streckenentblockung section clearing
Streckenisolator section insulator (621.33)
Streckenmarkierungsfunkfeuer en route marker beacon (351)
Streckenprofil railway profile (625.1)
Streckenrundsichtradar RSR, en-route surveillance radar
Streckenschalter sectionalizing switch (621.33); track switch (625)
Streckenschutz pilot protection
Streckensteuerung point to point control, straight-cut control (621.9)
Streckentrenner mit Luftisolation air-section break (621.33)
strecken to stretch, to extend
Streckgrenze yield point, elastic limit
Strecklänge gauge length (620)
Streckmetall expanded metal
Streckung stretch (ing)
Streich-... striking
Streichfläche striking surface
Streichmaß strike measure
Streichmasse coating

streichen to strike; to trowel (693); to paint, to wash, to coat (698)
Streifzünder graze fuse
Streifen strip, stripe; streak; tape; lane (621.397)
Streifenaufwickler tape winder
Streifendoppler tape reproducer
Streifendrucker tape printer
Streifenidentifizierung lane identification
Streifenleitertechnik strip-line technique, micro-strip technique
Streifenleser tape reader
Streifenlocher reperforator, tape perforator, paper tape reperforator, tape punch set
Streifenschreiber tape printer (621.394); strip-chart recorder
Streifensicherung plate fuse, metal-strip fuse, open fuse
streifen to streak; to graze
streifender Einfall grazing incidence, glancing incidence
streng strict, rigid, severe, rigorous
strenger mathematischer Beweis rigorous mathematical proof
strenge Prüfung severe test
strenge Rechnung rigorous analysis
strenge Regel rigid rule, hard-and-fast rule
strenge Vorschrift strict regulation
Streu-... scattered, diffusing, stray; leakage (621.31)
Streuausbreitung scatter propagation
Streubereich range (51), (519.2)
Streublindwiderstand leakage reactance
Streuecho scatter echo
Streufaktor leakage coefficient, circle coefficient (621.313.3)
Streufeld stray field, fringing of a field
Streufeldmaschine diverter-pole generator
Streufeldtransformator leakage-reactance transformer
Streuflächen scattering surfaces (621.32)
Streufluß stray flux, leakage flux
Streuinduktivität leakage inductance
Streukapazität stray capacitance; stray capacity, strays (pl.)
Streukopplung stray coupling
Streuleitwert leakage permeance
Streulicht stray light, scattered light
Streupol diverter pole
Streureaktanz quadrature-axis subtransient reactance, leakage reactance
Streuscheibe scattering screen (535)
Streuspalt leakage gap
Streuspannung reactance voltage
Streustrahlung scattered radiation, secondary radiation
Streustrahlungsübertragung scatter propagation
Streustrom stray current, parasitic current, leak current
Streustromableitung electric drainage
Streustromkorrosion electrolysis of underground structures, anodic corrosion, stray-current corrosion
Streutransformator leakage transformer
Streuvermögen diffusing power
Streuwandler high-leakage current transformer
Streuwelle scattered wave
Streuwirkung diffusing effect, scattering action
streuen to stray, to scatter
streuend diffusive, scattering
gestreutes Licht scattered light, diffuse light
gestreute Strahlung (back-) scattered radiation
Streuung diffusion, scatter (ing), leakage (538); variance, standard deviation (51)
Strich stretch, line; dash (621.395); prime (51); point (538)
Strich (marke) hairline (535)
x Strich = x′ x prime = x′
Strichcode bar code
Strichfahren holding load (621.311)
Strichgitter graticule (535)
Strichmuster fringe
Strichpeilung cursor-line bearing
strichpunktierte Linie dash-dotted line
Strichstärke line width
stricheln to dot
„Strippe" flex, flexible, string
Strippenbruch dislead
Strippenwiderstand lead resistance
Stroboskop stroboscope, stroboscopic device
stroboskopisch stroboscopic(al)
stroboskopisches Verfahren stroboscopic method
stroposkopische Zählerscheibe stroboscopic meter disc
Strohhalmelektrometer moving dielectric voltmeter
Strom current (621.3); stream (55)
Strom abführen von ... to convey current from ...
Strom abgeben an ... to contribute current to ..., to deliver current to
Strom abnehmen to collect current, to consume current (621.31)
Strom aufnehmen to consume current, to draw current

Strom durchlassen to pass current
Strom in Durchlaßrichtung forward current
Strom durchschicken durch ... to pass a current through ...
Strom eilt gegenüber der Spannung um 30° nach current lags the voltage by 30 °
Strom Eins unit current
Strom einschalten to make a current, to turn on the current
Strom ein- und ausschalten to make and break the current
Strom in Durchlaßrichtung forward current
Strom im (stationären) Endzustand final current
Strom entnehmen aus ... to draw current from ...
Stromerzeugen to generate current
Strom fließt ab von ... a current leaves ...
Strom fließt zu a current approaches
Strom führen to carry current
Strom glatt durchlassen to pass current freely
Strom gleichbleibender Richtung unidirectional current
Strom hindurchtreiben to drive current
Strom liefern to deliver current, to supply current
Strom nimmt zu current rises
Strom setzt aus current cuts off
Strom in Sperrichtung inverse current, reverse current
Strom verbrauchen to consume current
Strom mit wechselndem Vorzeichen bidirectional current
Strom zuführen zu ... to convey current to ...
stromabhängig current-operated
  stromabhängige Verluste load losses, impedance losses (621.314.2)
Stromabnahme decrease of current
Stromabnehmer collector, commutator, brush (621.3); trolley (standard), pantograph (625); (current) consumer (621.316)
  Stromabnehmerbürste current-collecting brush
  Stromabnehmerdrehstuhl swivelling trolley head
  Stromabnehmerfuß trolley base
  Stromabnehmergabel trolley harp
  Stromabnehmer-Kippvorrichtung trolley reverser
  Stromabnehmerkopf trolley head
  Stromabnehmerrolle trolley wheel
  Stromabnehmerrollenlager trolley bush
  Stromabnehmerstange trolley boom
Stromabschaltung power cut
Stromänderungsgeschwindigkeit rate of current change
Stromanforderung current demand
Stromanstieg rise of current, build-up of current
  bei Stromanstieg on build-up
  Stromanstiegrelais rate of change relay, surge relay
  Stromanstiegsschutz rate of change protection
  Stromanstiegssteuerung rise-in-current control (621.791)
Stromaufnahme current input
Stromausbreitung current distribution
Stromausfall cutting out of the circuit, interruption of power supply
Stromausgleichsbatterie floating trickle battery
Strombahn path of current
Strombauch current loop, current antinode
Strombedarf current demand, current drain, current requirement
Strombegrenzer current limiter
Strombegrenzung current limitation
  Strombegrenzungsdrossel current-limiting reactor, protective reactance coil
  Strombegrenzungsschalter throttle switch
  Strombegrenzungswiderstand load-limiting resistor, current-limiting resistor
Strombelag ampere-turns per cm
Strombelastbarkeit load-carrying capacity, current-carrying capacity
Strombelastung current loading, power load
Strombereich current range
strombetätigt current operated
Stromdichte current density
Stromdifferentialrelais balance-current relay, balance relay
Stromdifferentialschutz Beard protective system
Stromdurchgang current passage, current flow
  Stromdurchgang sperren to block the current flow
stromdurchlässig conductive
Stromeinheit unit (of) current
Stromeintritt current intake
Stromeisen current element, series stator, current magnet (621.317)
Stromentnahme current drain, consumption of current
Stromerzeuger power generator
Stromerzeugung generation of electricity
Stromfaden current streamer
Stromfehler (current) ratio error
Stromfluß current flow

Stromflußwinkel angle of (current) flow (621.314.6)
stromführend current carrying, live
stromführender Teil current-carrying part
Stromführungszeit on-period (621.314)
stromgespeist current fed (621.396.67)
stromgesteuert current operated, current-controlled
Stromimpuls pulse of current
Stromkern series stator, current magnet
Stromkorrosion anodic corrosion
Stromkosten power expenses
Stromkreis circuit, electric circuit
Stromkreis ausschalten to break a circuit, to open a circuit
Stromkreis einschalten to close a circuit, to make a circuit
Stromkreis schließen to close a circuit, to make a circuit
durch einen Stromkreis hindurchtreten (lassen) to thread through a circuit
... in den Stromkreis einschalten to throw ... into circuit
Stromkreisdämpfung circuit attenuation
Stromkreisparameter circuit parameter
Stromkurve current waveform
Stromlaufplan circuit diagram
Stromleiter (electric) conductor
Stromlinie streamline
Stromlinienform streamline shape
stromlinienförmig streamline, streamlined
stromlinienförmig machen to streamline
stromlos currentless, at-zero current, dead
stromloses Schalten no-load switching
Strommeßbereich current range
Strommesser amperemeter, ammeter
Strommesser mit Nebenwiderstand shunted meter
Strommodulation current modulation
Strommoment radiation factor, metre-ampere (621.396.67)
Stromnetz power-supply system, network
Stromnulldurchgang zero current
Strompfad current circuit (621.317); current path
Strompegel current level
Stromquelle current source, power source
Stromquelle treibt einen Strom durch ... a source sends a current via ...
Stromregelröhre ballast lamp, barretter
Stromregler current regulator
Stromresonanz parallel resonance, anti-resonance
Stromresonanzkreis anti-resonant circuit
Stromrichter static converter; rectifier (621.314.6)
Stromrichtung current direction
Stromrichtungsrelais power directional relay
Stromrichtungsumkehr reversal of the current
Stromrückgang decrease of current
Stromrückgangsrelais undercurrent relay
Stromrückkopplung current feedback
Stromschiene contact rail, live rail, third rail, conductor line (621.33)
Stromschienenisolator third-rail insulator
Stromschienenlokomotive third-rail-electric locomotive
Stromschienensystem third-rail system, conductor-rail system
Stromschienenverbinder conductor-rail bond
Stromschleife current loop
Stromschluß circuit closing
Stromschritt mark, marking signal (element)
Stromschwankung current undulation, current fluctuation
Stromsenke current sink
Stromspannung voltage
Stromspannungskennlinie voltage-current characteristic
Stromspitze peak current
Stromspule current coil, current element (621.317)
Stromstärke current intensity, current strength, amperage
Stromsteilheit rise of on-state current
Stromstoß current rush, current (im)pulse, current surge
stromstoßgeschaltete Uhr impulse clock
Stromstoßrelais impulse relay, pulse-operated relay
Stromstoßschalter impulse switch
Stromtor thyratron, controlled rectifier, gasfilled three-electrode valve
Stromüberlastung current overload
Stromübersetzungsverhältnis current ratio
stromunabhängige Auslösung definite-time (overcurrent) trip
stromunabhängige Zeitverzögerung definite time lag
Stromunterbrecher contact breaker, current interrupter
Stromunterbrechung current interruption
Stromverbrauch power consumption, current consumption
Stromverbraucher current consumer, user of electric power

Stromverdrängung skin effect, Heaviside effect; current displacement
Stromversorgung current supply, power supply, electric supply, power pack, powerplant
Stromversorgung einschalten to turn on the power supply
Stromversorgungsgeräusch hum, battery supply circuit noise (US)
Stromversorgungskabel power cable
Stromverstärkung current amplification, current gain (621.396.64)
Stromverstärkungsfaktor short-circuit forward current transfer ratio (621.314.33)
Stromverteilung power distribution (621.315); current distribution (621.385)
Stromverteilungskasten power distribution box
Stromwaage ampere balance, Thomson balance, electrodynamic balance
Stromwächter automatic controller
Stromwandler current transformer, series transformer
Stromwandler mit eigenerregter Zusatzwicklung auto-compounded current transformer (621.317)
Stromwandlerauslösung transformer overcurrent tripping
Stromwärme Joulean dissipation
Stromwärmeverluste ohmic losses; copper losses
Stromweg current path
Stromwelle surge of current
Stromwender commutator, reversing key (621.316); commutator, collector (621.313)
Stromwendung reversal of the current, commutation
Stromzeiger current vector, current phaser
Stromzuführung power supply leads
Stromzunahme increase of current
strömen to flow; to stream (55)
strömendes Medium fluid
Strömung current, flow, stream
Strömung aufrechterhalten to keep flowing (532), (532.33)
Strömungsanzeiger flow indicator (532), (532.33)
Strömungsdruck stream thrust (532), (532.33)
Strömungslogik fluid logic
Strömungsmesser flow meter (532), (532.33)
Strontium strontium
Strowgerwähler Strowger selector
Strowger-Wählsystem Strowger automatic telephone system
Struktur structure
strukturierte Anweisung structured statement (681.3)
strukturierte logische Schaltung structured array
strukturierte Programmierung structured programming
Studio studio (621.396), (621.397)
Studium study; studies (pl.)
Stück piece
aus Stücken zusammensetzen to piece up, to piece together
in Stücke springen to fly to bits, to break into pieces
Stückgewicht weight per piece
Stückliste list of components, item list, parts list
Stückprüfung routine inspection, routine test
stückeln to piece
Stufe step (621.3), (629.12), (694); stage (621.396); stair, stage (694)
Stufenfunktion step function
Stufenkopplung interstage coupling (621.396.64)
stufenlos smouth, stepless (621.3)
stufenlose Drehzahlregelung stepless-speed control, continuous-speed control
stufenloser Drehzahlbereich continuous-speed range
stufenlos regelbarer Motor infinitely variable-speed motor
Stufenmarkengeber step-strobe marker (621.397)
Stufenmodulation double modulation
Stufenrad stepped gear
Stufenregelung ratio control (621.314.2)
Stufenschalter step switch, multiple contact switch (621.316.7); tap changer (621.314.2)
Stufenschalter für spannungslose Schaltung off-voltage tap changer
Stufenschalter für Lastschaltung on-load tap changer, load tap changer
in Stufenschaltung in sequence
Stufenverstärkung gain of the stage
stufenweise S/S = step by step
stufenweise Belastungssteigerung step stress (620.1)
stufenweise Integration step-by-step integration
stufenweise vorwachsende Blitzentladung stepped leader
Stufenwicklung tap winding (621.314.2)
Stufenziehen rate growth (621.382)
stufen to grade

Stufung gradation, stepping
stumm silent, quiet, mute
stumme Entladung mute (lightning) discharge
Stummabstimmung silent tuning, quiet tuning
Stummfilm silent film
Stumpf frustrum (51); stump, butt (621)
Stumpf-Abbrennschweißung flash-butt welding
Stumpfgleis dead-end siding
Stumpfschweißung butt weld; butt welding
Stumpfstoß butt joint
stumpf obtuse (51); dull (535), (621.9); blunt
stumpf abschneiden to truncate
stumpf ansetzen an... to butt to...
stumpfes Ende stock
stumpfe Oberfläche dull surface
stumpfe Schneide dull edge
stumpfer Stoß butt joint
stumpfes Werzeug dull tool
stumpfer Winkel obtuse angle, blunt angle
stumpfschweißen to butt weld, to buttweld
Stunde hour
Stundenlast one-hour rating; one-hour load
Stundenleistung hourly output
Stundenzähler hours counter
stündlich hourly, per hour
Stütz-... supporting
Stützdraht supporting wire
Stützgerüst supporting frame
Stützgewinde buttress thread
Stützisolator pin-type insulator; (ohne Stütze) post insulator
Stützmauer retaining wall, sustaining wall, breast wall, supporting wall
Stützpfeiler supporting pillar, buttress
Stützpunkt base, support
Stützrelais latch relay
Stützweite span (621.315.1)
Stütze support, strut, buttress
stützen to support, to strut, to buttress; to drive (621.396)
Stützer pin-type insulator, line post (insulator)
subharmonisch subharmonic
subjektiv subjective
subjektiv wahrgenommener Ton subjective tone
subjektive Lautstärke equivalent loudness
Subkadmiumneutronen subcadmium neutrons (539)
sublimieren to sublime
submikroskopisch submicroscopic
Substanz substance
Substitution substitution
Substrat substrate
subtil subtile
Subtilität subtilety
Subtrahend subtrahend
subtrahieren von to subtract from
Subtraktion subtraction
subtransient subtransient
suchen nach to search for, to look for
Such... search
Suchanlage staff-finding system, staff-locating system
Suchempfänger scanning receiver, search receiver
Sucher (view-) finder (77)
Suchgerät locator, detector; small object detector
Suchlicht spot light (629)
Suchradar search radar
Suchroutine diagnostic routine
Suchspule search coil
Suchstellung choice
Suchwähleranschlußsystem Lorimer automatic telephone system
Suchzeit searching time (621.396.96); access time (681.3)
Suchzyklus search cycle
südweisend south-seeking
Sulfat sulphate
...-sulfat vitriolate of...
sulfatieren to sulphate (621.356)
Sulfit sulphite
summarisch summary
Summation summation
Summationswerk register
Summe sum, total amount
Summe über alles grand total
Summenausdruck summation term
Summenerfassung bulk registration (621.395)
Summenfrequenz sum frequency
Summen-Häufigkeitsfunktion cumulative distributive function
Summenprobe sum check, summation check (681.3)
Summenrechnung bulk billing (621.395)
Summenrelais to talizing relay
Summensignal composite signal
Summenstromschutz self-balancing protective system
Summenstromwandler summation transformer, balance transformer
Summenwirkung combined effect
Summenzähler summator, summatory meter
Summenzählwerk totalizer, counter register
summen to hum, to buzz, to purr
Summer hummer, buzzer (643); oscillator, signal generator, tone generator (621.39)

Summer mit Gitter- und Anodenabstimmung TP-TG = tuned plate, tuned grid (oscillator)
Summer mit veränderlicher Frequenz variable-frequency oscillator
Summerzusatz tone oscillator (621.396.61)
summieren to sum up, to add
sich summieren zu to add up to, to sum up to
summierendes Meßgerät summation instrument
summierendes Wattmeter integrating wattmeter
Summiereinrichtung summator
Summierglied summing network
Summierung addition, summing up, summation
Summierverstärker summing amplifier
Summierwerk register, to talizer
Sumpf swamp, sump, marsh
Superhet (empfänger) superheterodyne receiver, superhet
Superikonoskop supericonoscope
Superorthikon image orthicon
Superoxid peroxide
Superposition superposition
Supplement supplement (51)
supraleitend super-conductive
supraleitender Zustand supraconductive state
Supraleiter superconductor, supraconductor
Supraleitfähigkeit superconductivity, supraconductivity
Supraleitungsspeicher cyrogenic storage
surren to purr
Suszeptibilität susceptibility
Swanfassung bayonet cap (621.32)
Swan-Kleinsockellampe s. b. c. lamp = small bayonet cap lamp
Syenit syenite
Symbol symbol
Symbol einer logischen Schaltung logic symbol
Symbolsprache symbolic language
symbolisch symbolic
symbolischer Befehl pseudo instruction
symbolischer Code symbolic code
symbolische Darstellung symbolic representation
symbolischer Operationsbefehl mnemonic operation code
symbolisches Rechenverfahren complex quantity notation (621.3)
symbolische Schreibweise symbolic notation (621.3)
symbolisches Verfahren symbolic method (621.3)
in symbolischer Darstellung symbolically
Symmetrie symmetry, impedance unbalance (621.392)
Symmetrieachse axis of symmetry
Symmetriedämpfung unbalance attenuation
Symmetrierglied line balance converter, balun, balance converter
Symmetrierleitung balun, pawsey stub, balancing line
Symmetriertopf bazooka, line balance converter
Symmetrier- und Transformationsleitung balancing and matching line
Symmetrierübertrager balun, bazooka, balance converter, balance-to-unbalance transformer
Symmetrierverstärker paraphase amplifier (621.396)
symmetrieren to balance (621.39)
Symmetrierung symmetrization; balancing
symmetrisch symmetrical, balanced
symmetrisch gegen Erde balanced to ground
symmetrische Belastung symmetrical load, balanced loading
symmetrische Doppelleitung balanced twin, twin lead
symmetrisches Drehstromsystem symmetrical three-phase system
symmetrischer Eingang balanced input
symmetrische Last s. **symmetrische Belastung**
symmetrisches Leitungssystem balanced-line system
symmetrische Lichtverteilung symmetrical light distribution
symmetrisches Mehrphasenspannungssystem symmetrical polyphase set of potential differences, balanced polyphase system
symmetrischer Sägezahn back-to-back saw tooth (621.3)
symmetrischer Transistor bidirectional transistor
symmetrischer Übertrager push-pull transformer
symmetrische Übertragungsleitung balanced transmission line
symmetrische Vierdrahtleitung four-wire balanced line
symmetrischer Vierpol balanced four-terminal network, symmetrical transducer, symmetrical quadripole
Synchron-... synchronous
Synchron-Asynchron-Umformer synchronous-induction motor generator
Synchronbetrieb synchronous operation
Synchronfunkenstrecke synchronous spark gap

Synchrongenerator synchronous generator, synchronous alternator
Synchronisation synchronization, synchronizing
Synchronisationsfehler jitter (621.315)
Synchronisationszeichen sync character (621.394.5)
Synchronisator synchronizer
Synchronlauf synchronized rotation, synchronized operation
Synchronmaschine synchronous machine
Synchronmotor synchronous motor
Synchronmotor mit asynchronem Anlauf self-starting synchronous motor, induction-type synchronous motor
Synchronschaltung synchronization (621.31); synchro mesh (629.113)
Synchronsignal synchronizing signal, sync pulse (621.397)
Synchronsteuerung synchronization control
Synchrontelegraphiesystem synchronous telegraph system
Synchronuhr synchronous clock
Synchronuhrmotor synchronous time motor
Synchronwert synchronizing-signal level
synchron synchronous
synchron bleiben mit to keep in step with
synchron laufen to run in synchronism
synchron mit in step with
synchrone Drehzahl synchronous speed, pole-speed ratio
synchroner Einankerumformer binary converter
synchrone Impedanz synchronous impedance
synchrones Moment synchronous torque, running torque
synchroner Phasenschieber synchronous capacitor, synchronous-phase modifier
synchrone Reaktanz synchronous reactance
synchrone Streureaktanz quadrature-axis synchronous reactance
Synchronisier-... synchronizing
Synchronisiereinrichtung synchronizing equipment
Synchronisierimpuls synchronizing pulse, timing pulse, sync pulse
Synchronisierlampe synchronizing lamp
Synchronisierleitung synchronizing circuit
Synchronisierschalter paralleling switch
Synchronisiervoltmeter synchronizing voltmeter
Synchronisierwandler synchronizer, synchronizing transformer
Synchronisierwelle timing shaft
synchronisieren to synchronize
synchronisierende Kraft synchronizing force, synchronous torque
synchronisierendes Moment synchronizing moment, synchronizing torque, pull-in torque
synchronisierter Asynchronmotor synchronized induction motor, synchronous induction motor, synduct motor, Synaut motor
Synchronisierung synchronization, phasing-in, parallelizing, phasing (621.397)
Synchronisierungsdrossel synchronizing reactor
Synchronisierungsimpuls timing pulse
Synchronisierungsrelais synchronizing relay
Synchronisierungsschalter parallelizing switch
Synchronisierungsschiene synchronizing bus bar
Synchronismus synchronism
aus dem Synchronismus fallen to fall out of synchronism, to fall out of step
in Synchronismus bleiben mit to keep in step with
in Synchronismus kommen to fall in step
Synchronoskop synchronoscope, rotary synchronizer, synchronism indicator
Synthese synthesis (pl.: -eses), synthetic process
synthetisch synthetic(al)
synthetischer Gummi synthetic rubber
synthetisches Isoliermaterial synthetic insulation material
synthetisieren to synthesize, to build up (54)
System system; movement, moving system (621.317)
System von Interferenzlinien interference fringe system
Systemabweichung system deviation
Systemdämpfung system loss, total transmission loss, basic loss
systemeigene Wahl out-of-band signalling (621.395)
Systementwicklung system engineering
Systemfeder hairspring (621.317)
Systemkapazität effective capacity (621.385)
Systemwirkungsgrad system efficiency
systematisch systematic(al)
Szintillation scintillation (53)
Szintillationsschirm spinthariscope (539.16)
Szintillationszähler scintillation counter
szintillieren to scintillate (53)

T

T [Periodendauer] T [length of period]
t [Temperatur] t [temperature]
t [Zeit] t [time]
tabellarisch tabular
Tabelle table, schedule, chart
  Tabelle aufstellen to draw up a table, to compile a table
  Tabelle mit Materialkonstanten property table
  Tabellendifferenz tabular difference
  Tabellenkopf head (of a table), column heading
  Tabellensammlung set of tables
  Tabellenwerk table work
  Tabellenwert tabular value
tabellieren to tabulate
Tabelliermaschine tabulator
Tabellierung tabulation
Tablette tablet; wafer (621.382)
tablettieren to cake
tabulieren to tabulate
Tachodynamo tachogenerator
Tachograph tachograph
Tachomaschine tachometer generator
Tachometer tachometer, speedometer, revolution indicator
tachometrisch tachometric
Tachosignalgeber tacho-pulse generator
Tafel table (51); cake; panel (621.3)
  Tafeldifferenz tabular difference
  Tafelwert tabular value
Tag day
  bei Tage in the daytime
  über Tage overground
  unter Tage underground
  Tagesanfuhr, Tagesbedarf one day's supply
  Tagesausstoß, Tagesleistung daily output
  Tageslicht daylight
    Tageslichtlampe daylight lamp, sunlight lamp
täglich daily, diurnal
  täglicher Bereich diurnal range
  tägliche Schwankung diurnal variation
  tägliche Energieabgabe daily demand
Takt measure (78); stroke (621); clock, cycle (681.3)
  Taktfeuer rhytmic light (35)
  Taktfrequenz clock rate, clock frequency
  Taktgeber clock, clock pulse generator, master clock, timer; impulse generator
    Taktgebersignale scanning agents (621.396.96); timing pulses
  Taktgenerator timing generator, clock generator
  Taktpuls clock pulse, strobe (681.3)
  taktunabhängige Arbeitsweise asynchronous operation
  Taktzyklus clock cycle
Tal valley (91); trough (621.3)
  Talpunkt valley point
  Talspannung valley-point voltage
Talk talc(um)
Tandem tandem
tang = tangens tan = tangent
Tangente tangent
  Tangentenbussole tangent galvanometer
Tangential-. . ., tangential tangential
  tangentiale Beanspruchung tangential stress
  tangentiale Feldstärke tangential stress
  tangentialer Schub tangential thrust
  Tangentialdruck tangential pressure
  Tangentialkraft tangential force
  Tangentialstrahl tangent ray
tangieren to touch, to be tangent to
Tank tank
  Tankkreis tank circuit
Tannenbaumantenne broadside array, billboard array, pinetree aerial
Tannenbaumkristall dentrite
Tannenbaumverteiler tag-end connector-plug
Tantal tantalum
  Tantal-Elektrolytkondensator tantalytic capacitor, tantalum-electrolytic capacitor
  Tantalfestkondensator capacitor solid tantal, tantalum capacitor
  Tantalfolien-Kondensator tantalum-foil capacitor
  Tantallampe tantalum lamp
  Tantal-Sinder-Anode porous tantalum anode
T-Antenne T-aerial
Tarif tariff, rate
  Tarifschaltuhr multi-rate tariff switch
  Tarifzähler special-tariff meter
Tasche bag, pocket (621)
  Taschendosimeter pocket chamber
  Taschenfunkgerät pocketphone
  Taschenlampe torch light, pocket lamp, flash lamp
    Taschenlampenbatterie pocket-lamp battery
    Taschenlampenbirne pocket-lamp bulb
    Taschenlampengehäuse pocket-lamp case
  Taschenmeßgerät pocket meter
  Taschenvoltmeter pocket voltmeter
Tast . . . key . . ., keying
  Tastatur keyboard
    Tastaturcodierer keyboard encoder

Tastaturwahl keyboard dialling
Tastfilter keying filter
Tastfühler feeler arm; sensor
Tastgerät keyer (621.394)
Tastklick key-click
Tastkopf probe (621.317)
Tastschalter momentary-contact switch, dip switch
Tastverhältnis duty factor (621.394)
Tastwahl push-button selection, key pulsing, key sending (621.395)
Taste key
Taste drücken to depress a key
Taste einklinken to latch a key
Taste ist offen key is up
Taste ist zu key is down
Tastenfeld keyboard
Tastenfolge key sequence
Tastenkontrollampe buttons pilot lamp
Tastenlocher keypunch
Tastenrechenmaschine digital computer
Tastensatz key set, key-board, bank of keys
Tastenwahl push-button dialling
Tastenwähler key sender (621.395)
tasten to key (621.394), (621.396)
Taster calipers, callipers (pl.), push-button switch; feeler (621.8)
Tastung keying
tätig busy, active
Tätigkeit activity, action
in Tätigkeit setzen to put in(to) operation
Tätigkeitsbericht record
tatsächlich effective, actual, real; positive
tatsächlich aufgeladen positively charged
tatsächliche Leistung actual power (621.3)
tatsächliche Leistungsfähigkeit effective capacity
Tatzenlagerung bar suspension, axle suspension (621.3)
Tatzlagermotor axle-hung motor, nose-suspended motor (621.33)
Tau dew (55); rope, cable (67)
Taupunkt dew point
Taupunktshygrometer dew-point hygrometer
Taupunktspsychrometer wet-and-dry bulb thermometer
taub deaf (61); dead (622)
taub machen to deafen
taubes Gestein dead rock
Tauch-... diving; dipping; immersion
Tauchanker(magnet) plunger-type magnet
Tauchankerrelais plunger-type relay
Tauchapparat diving apparatus
Tauchbad dipping bath
Tauchfähigkeit submersibility
Tauchgalvanisierung dip plating
Tauchkernrelais solenoid relay
Tauchkessel dipper (621.319)
Tauchkolben plunger (621.6), (621.396), (66)
Tauchkolbenpumpe plunger pump
Tauchlack dipping varnish
Tauchlötung dip soldering
Tauchmotor submersible motor
Tauchprobe immersion test
Tauchprozeß dipping process, immersion
Tauchsieder immersion heater
Tauchspulmikrophon moving-coil microphone, electrodynamic microphone
Tauchspulenprinzip moving-coil principle, electrodynamic principle
Tauchverzinkung hot-dip galvanizing
tauchverzinnen to (hot) dip
Tauchverzinnung hot tinning
Tauchvorgang dipping
Tauchvorrichtung immersion device
Tauchzündung striking by dipping electrode (621.314.6)
tauchen to immerse, to dip; to dive
getauchte Elektrode dipped electrode, coated electrode (621.79)
tauen to thaw; to melt
tauiger Belag bloom
Täuschung durch Ausstrahlung transmission deception (623)
Tausendstel thousandth, point 00 ones
Tausendwähler thousands selector
Taxi taxi(meter)
Taylorsche Reihe Taylor's series
Technik engineering, technology; technics (pl.); technique
Techniker technician; engineer; technologist
technisch technical; engineering, technological
technisch durcharbeiten to engineer
technische Abteilung engineering department
technische Angaben technical data
technische Anleitung technical manual
technische Bedingungen specifications
technischer Beirat advisory council
technische Chemie applied chemistry
technische Einheiten m. k. s. units
technische Einzelheit technical detail, technicality
technische Elektronik electronic engineering
Technische Hochschule institute of technology, technical university; engineering college
technischer Laie non-technical person

technisches Maßsystem engineer's system, m. k. s.-system
technische Mathematik engineering mathematics
technische Physik applied physics
technische Röntgenuntersuchung industrial radiography
technische Salzsäure muriatic acid
technisches Stadium engineering stage
technische Überwachung technical inspection, technical supervision
technische Voraussetzungen technical premises
technisches Wissen technical knowledge
technische Zeitschrift technical magazine
Technologie technology
technologisch technological
Teer tar
Teerderivat tar derivate
Teil part, portion, component (portion), section, element, member
Teil mit nicht definiertem Potential disconnected part, floating part
Teilabschaltung partial disconnection
Teilamt satellite exchange
Teilansicht partial view
Teilausfall partial failure
Teilbild field, frame (621.397)
Teilbildflimmern line flicker
Teiler divider
Teilfeld partial field
Teilgalvanisierung partial plating
Teilkreis pitch circle (621)
Teilkreisdurchmesser P. C. D. = pitch circle diameter (621.8)
Teilkurzschluß partial short
Teilkurzschlußstrom component fault current
Teilleistung partial power
Teilleiterkabel split-conductor cable
Teilmaschine ruling machine (535)
Teilnehmer customer (33); subscriber (621.39)
Teilnehmer rufen to call up a subscriber
Teilnehmeranschluß subscriber's station, subscriber's line
Teilnehmerbetrieb time sharing (operation) (681.3)
Teilnehmer Hauptanschluß main telephone station
Teilnehmerkennung subscriber verification
Teilnehmerklasse class of service
Teilnehmerklinke subscriber's jack, answering jack, local jack (621.395)
Teilnehmerleitung subscriber's line, party line, exchange line, local loop, station line
Teilnehmerruf subscriber ringing signal
Teilnehmerrufnummer subscriber's number
Teilnehmerschaltung line circuit
Teilnehmerschleife subscriber's loop
Teilnehmerseite drop side (of an equipment)
Teilnehmervermittlung PBX = private branch exchange
Teilnehmerwahlstufe subscriber's stage
Teilparabolspiegel cut paraboloid
Teilsperre partial barring
Teilstreckentechnik message switching; store and forward (621.394)
Teilstreckentechnik mit paketweiser Übertragung packet switching (621.394)
Teilsumme sub-total
Teilstrich graduation (mark), scale division
Teilstrom component current
Teilvermittlungsstelle satellite exchange
Teilwert item
Teilwicklung partial winding
teilbar divisible
Teilbarkeit divisibility
Teilchen particle, corpuscle
aus Teilchen bestehend corpuscular
Teilchenbeschleuniger particle accelerator
Teilcheneinfang particle capture
Teilcheneinfang mit Ausstrahlung radiative capture (539)
Teilchenflußdichte particle-flux density (539)
Teilchengeschwindigkeit particle velocity
Teilchenwolke swarm of particles
Teilchenzähler particle counter
teilen, (sich) to divide; to split
teilen durch to divide by (51)
geteiltes Gehäuse split frame (621.313); split case
Teiler divider; measure; attenuator (621.317)
Teilertastkopf attenuator probe (621.317)
Teilung division (61); pitch (621.8); splitting, fission (539)
Teilungsring index collar
T-Eisen T-iron, T-section (pl.: tees)
tektonisch tectonic
Telefon telephone, phone (s. auch **Fernsprech-...**)
am Telefon on the phone
Telefon mit Hufeisenmagnet bipolar receiver
Telefonamt telephone exchange (office)
Telefonanlage telephone installation
Telefonanruf telephone call, ring
Telefon-Anrufbeantworter- und Aufnahmegerät automatic answering and recording equipment
Telefonanschluß subscriber's line

Telefonapparat telephone set, telephone apparatus, telephone equipment
Telefondraht telephone wire
Telefongebühren telephone fees, (telephone) charges
Telefongespräch call, talk on the phone
Telefon-Hausanschlußleitung telephone drop
Telefonhörer telephone receiver
Telefonkabel telephone cable
Telefonkabine telephone booth, telephone box
Telefonkanal telephone channel
Telefonklingel station ringer, telephone bell
Telefonleitung telephone line, telephone wire
Telefonleitung anzapfen to tap a telephone line
Telefon-Nebenanschluß telephone extension
Telefonnummer call number
Telefonrelais telephone relay
Telefonschnur flex, flexible (cord)
Telefonteilnehmer subscriber, party
Telefonvermittlung telephone exchange, telephone switch board
Telefonzelle siehe **Telefonkabine**
Telefonzentrale telephone exchange
Telefonie telephony
Telefonie mit Trägerunterdrückung quiescent-carrier telephony
auf Telefonie on (the) phone (621.396)
telefonieren to telephone, to phone, to ring up, to call
telefonisch telephonic; on the phone
telefonische Nachricht telephonic message
Telefonist(in) telephonist, telephone operator
Telegramm telegram, cable, wire
Telegraph telegraph
Telegraphenalphabet telegraph code
Telegraphenamt telegraph office
Telegraphenapparat telegraph receiver, telegraph transmitter
Telegraphenbeamter telegraphist, telegraph operator
Telegraphenleitung telegraph line, telegraph circuit
Telegraphenmast telegraph pole
Telegraphenrelais telegraph relay, incrementer relay
Telegraphenschlüssel cipher code
Telegraphentaste back contact key
Telegraphie telegraphy, record communication
Telegraphie-Alphabet telegraphy code
Telegraphieempfang telegraphy reception
Telegraphieempfänger telegraphy receiver
Telegraphiefunk radiotelegraphy
Telegraphiegeräusch telegraph noise (621.394); thump
Telegraphie-Phantomkreis phantom circuit
Telegraphie-Rundschreibeinrichtung telegraph-broadcast unit
Telegraphiesender telegraphy transmitter
Telegraphie-Speichersystem telegraph message file and retrieval system
Telegraphie-Speichervermittlung computer controlled message switching system
Telegraphieumsetzer telegraph converter
Telegraphieverkehr telegraph transmission
Telegraphieverstärker telegraph repeater
Telegraphieverzerung telegraph distortion
Telegraphiezeichen code signal, telegraph character
telegraphieren to telegraph, to wire, to flash
Telegraphiergeschwindigkeit telegraph-transmission speed
Telegraphierschritt telegraph-signal element
Telegraphiersystem system of telegraphy
telegraphisch telegraphic
telegraphische Verbindung telegraph(ic) connection
Telegraphist telegraphist, telegraph operator
Telemetrie telemetry
Telephon telephone (s. **Telefon**)
Teleskop telescope
Teleskopmast telescopic mast, extension mast
Teleskopfeder telescopic spring
Teleskoprohr telescopic tube
Telexnetz teleprinter network
Tellerröhre door-knob tube
Tellur tellurium
Temp tephigram (551.5)
Temperatur temperature, heat level
Temperatur nach Grad Celsius temperature on Cent scale
Temperatur um 10 K erhöhen to raise the temperature through 10 K
Temperatur fällt temperature falls, temperature drops
Temperatur unter Null sub-zero temperature
Temperatur am heißesten Punkt hot-spot temperature
Temperatur steigt temperature rises
auf Temperatur kommen to run hot
Temperaturabfall temperature drop
Temperaturänderungsgeschwindigkeit rate of change of temperature
Temperaturanstieg rise in temperature
Temperaturausgleich temperature balance; equalization of temperature
Temperaturbereich temperature range

Temperaturbestimmung thermometry
Temperaturdämpfungsausgleich temperature-dependent attenuation balance
Temperatureinfluß influence of temperature
Temperaturfühler pyrometer probe, thermometer probe, temperature sensor, thermal sensor, temperature detector
Temperaturfühler mit Schutzrohr thermometer tube
Temperaturgeber temperature transmitter
Temperaturgefälle temperature gradient
Temperaturgleichgewicht temperature balance
Temperaturinversion temperature inversion
Temperaturkoeffizient des Widerstandes resistance temperature coefficient
Temperaturkonstanz temperature stability
temperaturkonstanter Schnitt A. T. cut (537.228)
Temperaturmessung temperature measurement
Temperaturprogramm time-temperature program
Temperaturregelung temperature control, temperature regulation
Temperaturregler thermostatic control, thermostat(ic) regulator, heat governor
Temperaturrückgang drop in temperature
Temperaturschalter thermostat, thermostatic switch, temperature-actuated switch
Temperaturschreiber temperature recorder, thermograph
Temperaturschwankung temperature variation
Temperaturspannung voltage equivalent of thermal energy
Temperatursteuerung temperature control
Temperaturstrahler thermal radiator
Temperaturwechselprüfung temperature-cycle test
Temperatursturz thermal shock
Temperatursturzprobe alternate immersion test
Temperaturunterschied difference in temperature
Temperaturwelligkeit temperature ripple
Temperaturzunahme temperature rise
Temperaturzustand thermal condition
Temperguß malleable cast iron
tempern to temper (621.7); to cook (621.314.6)
Tempo rate, measure, tempo
Tempo 100 20 w. p. m. = 20 words per minute (621.394)
mit einem Tempo von 10 je Stunde at a rate of 10 per hour
Temposteigerung increase of velocity, rate increase
temporär temporary
Tendenz tendency (**zu**: to, **nach**: towards)
einer Tendenz entgegenwirken to offset a tendency, to counteract a tendency
Tensor tensor (51)
Tensorrechnung tensorial calculus
Tera- = 10^{12} trega
Teraohm tregohm
Term term
Termin term, fixed date
Terminplan (time) schedule, time table; production schedule
Terminal terminal
Terminologie terminology
ternär ternary
Terpentin turpentine, "turps"
Terrain terrain
Terrakotta terra cotta
terrestrisch terrestrial
terrestrisches Kraftwerk terrestrial power station
terrestrische Wärme terrestrial heat
tertiär tertiary
Tertiärgruppe mastergroup (621.395.4)
Tertiärgruppendurchschaltefilter through-mastergroup filter
Tertiärwicklung tertiary winding, stabilizing winding
Tesla-Transformator Tesla transformer
Test check, test
Testbild TV-test chart, test pattern
Testprogramm test program
Tetrachlorkohlenstoff carbon tetrachloride
Tetraeder tetrahedon
Tetrode tetrode, four-electrode valve, quadrode
TE-Welle H-wave, transverse electric wave (621.396)
Teufenzeiger depth indicator
Text text
Text in offener Sprache plain language text
Text-Editor text editor
Textverarbeitung word processing
Textil-... textile, fabric
Textilband fabric tape
Textilfaser textile fibre
Textilsieb cloth filter
Textur texture
Texturblech grain-oriented sheet, oriented lamination (621.314)
TF = Trägerfrequenzsystem (siehe dort)
TF-System für Kleinkoaxialpaare small diameter coaxial cable CF-system
T-Form T-shape, Tee-shape
T-Form haben to be the shape of a Tee

T-förmig T-shaped, Tee-shaped
T-Glied T-section, T-network (621.392)
Thallofidzelle thalofide cell
Theodolit theodolite
theoretisch theoretical
theoretische Physik mathematical physics
Theorie theory; theoretics (pl.)
Theorie aufstellen to advance a theory
thermionische Energieumwandlung thermionic-energy conversion
thermisch thermal
thermisch bedingte Abweichung thermal drift
thermisch bedingtes Weglaufen thermal runaway
thermisch behandeln to anneal
thermische Aufzeichnung thermal writing
thermischer Auslöser thermal trip
thermische Auslösung thermal release; thermal tripping (621.316.7)
thermische Behandlung heat treatment
thermische Belastung thermal loading
thermische Bewegung thermal agitation
thermische Blinkschalteinrichtung thermal flasher
thermischer Brüter thermal breeder, thermal reactor
thermischer Durchbruch thermal breakdown, equivalent breakdown (621.382)
thermischer Einfluß thermal influence, influence of temperature
thermisches elektrisches Meßgerät electrothermic instrument
thermische Elektronenemission thermionic (electron) emission
thermischer Elektronenstrom thermionic current
thermische Energie heat energy
thermische Grenzleistung heating limit
thermische Instabilität thermal runaway (621.317), (621.39); thermal breakdown (621.382)
thermische Ionen thermions, thermal (-velocity) ions
thermische Kathode thermionic cathode
thermisches Kraftwerk thermal power plant
thermische Leitfähigkeit thermal conductivity
thermisches Neutron thermal (-velocity) neutron
thermisches Ohm thermal ohm
thermisches Rauschen thermal noise, resistance noise
thermischer Reaktor thermal reactor
thermischer Reduzierkoeffizient thermal derating factor (621.382)
thermisches Relais thermostat relay, temperature relay
thermische Restspannung zero field thermal residual voltage (621.382)
thermisch spaltbar fissile
thermische Spaltung thermal fission (539)
thermische Überlastbarkeit thermal overload capacity
thermisches Überlastrelais overload temperature relay
thermische Umsatzgleichung thermochemical equation
thermische Unruhe thermal agitation
thermischer Wandler thermionic converter
thermischer Widerstand thermal resistance
thermischer Wirkungsgrad thermal efficiency
thermischer Wirkungsquerschnitt thermal cross section
Thermistor thermistor
Thermit thermite
Thermitschweißung thermite welding
Thermo-..., thermo-... thermo-..., thermal
Thermoauslöser thermal circuit breaker, thermal cut-out
Thermoblinkrelais thermal flashing relay
Thermochemie thermochemistry
thermochemisch thermochemical
Thermodiffusion thermal diffusion
Thermodrucker thermal printer
Thermodynamik thermodynamics (pl.)
thermodynamisch thermodynamic(al)
thermodynamischer Wirkungsgrad thermodynamic efficiency
thermoelektrisch thermoelectric
thermoelektrischer Effekt Seebeeck effect, thermoelectric effect
thermoelektrische Kettenschaltung cascaded thermoelectric device
thermoelektrischer Kühlblock thermoelectric cooler
thermoelektrischer Pyrometer thermoelectric pyrometer, thermocouple pyrometer
thermoelektrische Spannungsreihe thermoelectric series
thermoelektrisches Thermometer thermoelectric thermometer, thermel
Thermoelektrizität thermoelectricity
Thermoelement thermocouple, thermoelectric junction, temperature detector
Thermoelementschenkel thermoelectric element
Thermo-EMK thermoelectromotive force, thermo e. m. (621.36)
Thermo-Galvanometer thermogalvanometer

Thermo-Meßinstrument thermoinstrument
Thermometer thermometer
Thermometer zeigt 70° C thermometer reads 70 ° C
Thermometereichung thermometer calibration
Thermometergrad thermometer point
Thermometerhütte thermometer screen (55)
Thermometerrohr thermometer stem
Thermometerskala thermometer scale, thermometer graduation
Thermometerstutzen thermometer pocket
thermonukleare Reaktion thermonuclear reaction
Thermophon thermophone
Thermoplast thermoplastic (material)
thermoplastisch thermoplastic
thermoplastisches Material thermoplastic (material)
Thermorelais thermal relay
Thermosäule thermoelectric generator, thermopile, thermoelectric pile
Thermoschalter temperature switch, thermal delay switch, thermal cut-out, temperature-control device
Thermospannung thermo e. m. f., thermoelectric voltage
Thermostat thermostat; constant-temperature oven
Thermostatregler thermostatic controller
Thermostatrelais thermostat relay
thermostatisch thermostatic
Thermoumformer thermo-converter, thermal converter
Thermoumformer-Strommesser thermocouple ammeter
Thermoumformer-Voltmeter thermo-voltmeter
Thomsonsche Brücke Thomson bridge, double bridge
Thomson-Effekt Thomson effect
Thomsonsche Schwingungsgleichung Kelvin formula
Thomson-Zähler Thomson meter
thorieren to thoriate
thorierte Kathode thoriated filament (621.385)
thoriertes Wolfram thoriated tungsten
thorierte Wolframkathode thoriated tungsten cathode
Thorium thorium
Thoriumoxid, Thoroxid thoria
Thury-Anlage Thury transmission system
Thury-Regler Thury regulator
Thury-Schaltung Thury transmission system, series-distribution system
Thyratron thyratron, gas triode, gas-filled three-electrode valve, gas-tube switch
Thyristor thyristor, silicon controlled rectifier, SCR
Thyristordiode diode thyristor
Thyristorgleichrichter silicon-controlled rectifier circuit
Thyristorregler thyristor controller
Thyristortriode triode thyristor
Thyrit thyrite
THz = 10^{12} Hz
Ticker ticker, telegraph sounder (621.396); chopper (621.3)
Tief-... deep, underground
Tiefbaß-Lautsprecher subwoofer
Tiefbau underground mining (622); underground construction (624)
Tiefbrandelektrode deep-penetration electrode
Tiefkühlung low-temperature cooling; deep freezing (621.56)
Tiefkühlschrank deep freezer
Tiefpaß low-pass filter (621.395)
Tief- und Hochpaß low-and high-pass filter
Tiefpaß in Pi-Schaltung shunt-condenser input filter
Tiefpaß in T-Schaltung series-inductance input filter, choke-input filter
Tiefpegelwahl low-level selection
Tiefseekabel deep-sea cable
Tiefstrahler high-bay reflector, concentrating reflector
Tieftemperaturanlage cryogenic equipment
Tieftemperaturtechnik cryogenic technique
Tieftemperaturspeicher cryogenic memory
Tieftonbereich bass
Tieftonblende bass control (621.396.62)
Tieftonlautsprecher low-frequency speaker, woofer, boomer, bass loudspeaker
Tiefziehen deep drawing
Tiefziehprobe deep-drawing test
tief deep; low-pitched (78)
tief gegliedert, tief gestaffelt deep in array, deep-braced
tiefer Ton low(-pitched) note, low pitch
tiefgeätzt deep-etched
tiefziehen to deep draw
Tiefe depth
Tiefe messen to fathom
Tiefenanzeiger position indicator (622); depth indicator, fathometer (626)
Tiefenlotung depth sounding
Tiefenwirkung penetration (536)

Tiegel crucible
Tiegelofen crucible furnace, Canley furnace, Helberger furnace
Tiegelstahl crucible steel
tierisch animal
tierische Elektrizität animal electricity
Tintenschreiber ink writer, recording pen (621.317)
Tintenstrahldrucker ink-jet printer
tippen to typ(write) (651); to touch
Tipp-Probe busy test (621.395)
Tipptastenfeld touchpanel keyboard
Tirrill-Regler Tirrill regulator, automatic regulator (621.3)
Tisch table, desk; stage, base plate (621.9)
Tischapparat desk-type telephone
Tischbohrmaschine bench drill
Tischlampe table lamp, table standard
Tischrechner desk calculator
Tischtelefon s. **Tischapparat**
Titan titanium
titrieren to titrate
T-Muffe T joint, branch joint
TM-Welle transverse magnetic wave, E-wave
Tochtergerät, Tochterinstrument repeater, slave
Tochter-Flip-Flop slave flip-flop
Tochterkern daughter nucleus
Tochtersender slave transmitter, slave station (621.397); relay station (621.396)
Tochtervermittlungsstelle slave exchange
Tod death
Todesstrahlen death rays
Todzeit dead time
tödlich, todbringend fatal, lethal, deadly, deathly
tödliche Dosis lethal dose
tödlicher Unfall fatal accident, fatality
Toepler-Influenzmaschine Toepler-Holtz machine
Toleranz tolerance
Toleranzanzeiger tolerance indicator
Toleranzdosis tolerance dose, maximum-permissible dose, maximum-permissible exposure (US) (621.386)
Toleranzgrenze margin, tolerance limit, tolerance level
Toleranzlehre go and not-go gauge
Toleranzmeßbrücke limit bridge
tolerieren to tolerate
Toluol toluol, toluene
Ton sound, tone (534); note, ring (78); clay, potter's earth (66)
Tonabnehmer (sound) pickup, (phono) pickup
Tonabnehmeranschluß pickup (connection) socket
Tonabnehmereinsatz pickup cartridge
Tonabnehmersystem phono cartridge
Tonanhebung accentuation
Tonanschlußleitung sound-program circuit (621.396.9)
Tonarm tone arm, pickup arm
Tonaufnahme sound recording; (sound) record
Tonaufnahmegerät audio recorder, sound recorder
Tonaustauschleitung sound-program exchange circuit (621.396.9)
Tonband magnetic tape, recording tape
Tonbandbreite sound bandwith
Tonbandgerät tape recorder
Ton-Bild-Bandgerät sound-image tape recorder
Tonblende tone control (621.396.62)
Tonblende zur Höhenabschneidung treble-cut control
Tonfalle sound trap
Tonfilm sound film
Tonfilmprojektor sound film projector
Tonfilm-Vorführapparat sound film projector
tonfrequent audio frequency, voice-frequency; audible
tonfrequent modulierte Wellen tone-modulated waves
tonfrequente Prüfspannung test tone
tonfrequenter Schwebungston audio-frequency beat note
Tonfrequenz a. f. = audio frequency, voice frequency, sonic frequency
Tonfrequenzabstimmung note tuning
Tonfrequenzbereich voice-frequency range, audio range
Tonfrequenzfernwahl v. f. dialling = voice-frequency dialling (621.395)
Tonfrequenzfilter audio filter
Tonfrequenzgenerator tone generator, low-frequency oscillator, audio-frequency oscillator
Tonfrequenzgleichrichter sound discriminator (621.397)
Tonfrequenzkanal audio channel
Tonfrequenzmeßsender (audio frequency) signal source (621.395)
Tonfrequenzoszillator siehe **Tonfrequenzgenerator**
Tonfrequenzruf voice-frequency ringing

Tonfrequenzrufumsetzer voice-frequency relay set; v. f. signalling transmitter
Tonfrequenzsieb audio filter
Tonfrequenzspannungsquelle tone source
Tonfrequenzspektrum audible spectrum
Tonfrequenzsperre audio-frequency filter, audio-frequency rejector (circuit)
Tonfrequenzteil audio-frequency section
Tonfrequenztelefone voice-frequency telephony
Tonfrequenztelegraphie voice-frequency telegraphy
Tonfrequenzübertrager audio-frequency transformer
Tonfrequenzverstärker audio amplifier
Tonfrequenzwiedergabe sound reproduction
Tonfrequenzwiedergabetreue audio fidelity
Tonfunk sound radio, radio (transmission), sound broadcasting
Tonfunkentelegraphie Lepel system of wireless telegraphy, singing-spark telegraphy
Tongenerator tone generator, signal generator, audio oscillator
Tonhöhe pitch (of the note)
Tonkanal sound channel, audio channel, voice channel (621.396), (621.397)
Tonkontrolleinrichtung sound-program control equipment
Tonkopf sound head
Tonleiter (musical) scale
Tonleitung sound-program circuit
Tonleitungsschaltstelle sound-program switching point, sound-program center
Tonminimum sound minimum (621.317); critical silence
Tonmischung sound mixing; dubbing (621.396.9)
tonmoduliert sound modulated
Tonrad phonic wheel
tonrichtig tone-compensated (621.396.62)
Tonröhre earthenware pipe
Tonruf tone ring
Tonschrieb sound record (ing)
Tonsender aural transmitter, sound transmitter
Tonsignal sound signal
Tonsignal-Rhythmus tone cadence
Tonsperrkreis audio trap, sound trap
Tonspur phonogram (77); groove (681.84)
Tonträger sound carrier (frequency), audio carrier; sound recording medium
Tonträgersperre sound (-carrier) trap
Tontreppe sound-carrier step
Tonüberwachung monitoring telephone
Tonwahl inband signalling (621.395)
Tonwert density step
Tonwertblende tone-control aperture
Tonwertskala tone wedge
Tonwiedergabe sound reproduction
tönen to sound, to sing (78); to tinge, to tint (535)
tönender Funke singing spark, musical spark
tönender Lichtbogen singing arc, singing spark
tönende Telegrafie interrupted continuous waves telegraphy
Tonnage tonnage
Tonne barrel (67); ton (621.317)
Tonnenverzeichnung barrel distortion (621.397)
Tonnenwicklung barrel winding (621.313)
Tönung tinge, tint, tone, shade
Topas topaz
Topfkern pot core (621.396) **Topfkernspule** pot-type coil
Topfkreis cavity resonator, tank circuit
Toplicht toplight (629.13)
Topographie topography
Tor gate (621.374)
Torfernsprecher doormen converser
Torimpuls gate puls
Torschaltung gate, gate circuit
Torsteuerung gating
Tornado tornado, cyclone
Tornisterfunkgerät manpack radio
Toroid toroid, doughnut
toroidförmig toroidal, doughnutshaped
Toroidspule toroidal coil, single ring coil
Torpedo torpedo
Torpedofisch electric cat fish
Torpedorohr torpedo tube
Torsion torsion, twist (ing)
auf Torsion beanspruchen to twist
Torsionsbeanspruchung torsional strain
Torsionsdehnung torsional strain
Torsionsenergie shear-strain energy
Torsionsfeder torsion bar
Torsionsfestigkeit torsional strength
Torsionsgalvanometer torsion galvanometer
Torsionskraft torsional force
Torsionsmeßgerät torsion meter
Torsionsmodul modulus of rigidity
Torsionsmoment torsional moment, twisting moment, torque
Torsionsprüfung torsion test
Torsionsschwingung torsional vibration
Torsionsspannung torsional stress
Torsionsversuch torsion test
Torsionswaage torsion balance
Torus torus
tot dead

toter Gang lost motion (621); backlash (621.8); dead travel (621.9)
tote (ruhende) Last dead load
totes (anorganisches) Material dead matter
toter Punkt dead point, dead-centre
tote Zone silent zone, skip zone, zone of silence
total total
totale Finsternis total eclipse
totale Reflexion total reflection
Totalausfall catastrophic failure, blackout failure, breakdown
Totmannknopf dead-man's handle
Totmannknopffunktion dead-man's feature
Totpunkt dead point, dead-centre
Totpunktstellung dead-centre position
Totzeit dead time; down time, outage time
Totzone dead zone
Touren revolutions
auf Touren bringen, auf Touren kommen to run up to speed, to get up to speed
Tourenzahl (number of) revolutions
T-Profil T-section
Trabant satellite (52)
Tracer-Chemie tracer chemistry
Tracer-Element tracer element
Trag-... supporting, carrying
Tragarm support arm, bracket
Tragbalken (supporting) beam, joist, transom
tragbar portable; bearable, permissible
tragbare Batterie portable battery
tragbares Gerät portable appliance
Tragbarkeit portability; bearability
Tragbrett tray (621.356)
Tragdraht catenary wire, supporting wire, span wire, bearer wire (621.3); messenger (621.33), (621.39)
Tragdrahtanordnung catenary system (621.33)
Tragfähigkeit (load) carrying capacity, bearing capacity, maximum load
Tragkabel supporting cable, suspension wire
Tragkette suspension chain (621.315.6)
Tragkettenzusatzgewicht suspension insulator weights (pl.)
Tragkonsole mounting bracket
Tragkonstruktion supports (pl.), supporting structure
Tragkraft portative force, carrying capacity; lifting capacity, hoisting capacity (621.8); tractive power (621.318.2)
Traglager rotor bearing
Tragmast supporting tower, suspension tower
Tragöse eye bolt, ring bolt
Tragplatte bearing plate
Tragrolle roller
Tragschelle carrying clamp
Tragseil bearer cable, supporting cable, messenger (621.33); messenger, supporting strang (621.39)
Tragseil-Luftkabel catenary-wire, aerial cable
Tragturm supporting tower (624)
träge inert (54); sluggish (621.317)
träge Masse inertial mass
träges Meßinstrument sluggish measuring instrument
träge Sicherung delay (action) fuse, slow fuse (621.3)
tragen to support, to carry, to bear
tragender Bauteil structural element, constructional member
tragende Wand load-bearing wall, main wall; supporting wall
Träger carrier (54), (539), (621.396); girder, beam, joist (624)
Träger unterdrücken to suppress the carrier (621.396)
Träger wieder zusetzen to resupply the carrier (621.396)
Trägeranreicherung enhancement mode (621.382)
Trägerbeweglichkeit carrier mobility (621.382)
Trägerdiffusion carrier diffusion
Trägererzeugung mit Hilfsfrequenz pilot synchronizing (621.396)
trägerfrequente Nachrichtenübertragung carrier transmission
Trägerfrequenz carrier frequency
Trägerfrequenzbespulung carrier vading
Trägerfrequenz-Grundleitung multichannel circuit
Trägerfrequenz-Kanal carrier channel
Trägerfrequenzleitung carrier line
Trägerfrequenz-System carrier system
Trägerfrequenztelefonie carrier (wave) telephony
Trägerfrequenztelefonie auf Hochspannungsleitungen power-line carrier telephony
Trägerfrequenztelegraphie carrier telegraphy
Trägerfrequenzübertragung carrier transmission
Trägerfrequenzverschiebung carrier-frequency shift
Trägerfrequenzverstärker carrier repeater
Träger-Geräusch-Abstand carrier-to-noise ratio

Trägerinjektion carrier injection
Trägerinterferenz carrier interference
Trägerlawine electron avalanche; ion avalanche
Trägerleistung carrier power
Trägermaterial base material (621.382)
Trägerpegel carrier level (621.39)
Trägerplättchen substrate (621.382)
Trägerrest carrier leak, residual carrier (621.39)
Trägerrost girder grillage
Trägerschwebung carrier beat (621.395.4)
Trägerspannung carrier (frequency) voltage
Trägerstaueffekt hole storage (621.382)
Trägerstaueffektbeschaltung surge-absorbing capacitors (621.314.6)
Trägersteuerung carrier control
Trägerunterdrückung carrier suppression
Trägerverarmung depletion mode 8621.382)
Trägerverwischung carrier dispersal
Trägerwelle carrier (wave) (621.396)
Trägerzusatzeinrichtung speech restorer (621.396)
Trägheit inertia
trägheitsarm fast in operation, low inertia
trägheitsfrei instantaneous
trägheitsfreie Steuerung instantaneous control
Trägheitskraft inertia force
Trägheitsmoment moment of inertia, inertia factor
Trägheitsnavigation inertial navigation
Trägheitsradius radius of gyration
Trajektorie trajectory
Traktrix curve of pursuit
tränken to impregnate, to soak
Tränkerder impregnated earth, impregnated ground
Tränkmasse impregnation compound
Tränkung impregnation steeping
Transadmittanz transadmittance
Transatlantikkabel (trans)atlantic cable
Transduktor magnetic amplifier, transductor
Transfer, Transferierung transfer
Transferbefehl transfer instruction (681.3)
Transfergeschwindigkeit bit rate (681.3); data-transfer rate (621.39)
Transferwert transfer value
transferieren to transfer
Transfiguration transfiguration
transfigurieren to transfigure
Transformation transformation (51), (54), (621.3)
Transformationsglied impedance-matching device
Transformator transformer (621.3) (s. auch **Umformer**)
Transformator aufnehmen to dismantle a transformer
Transformator 1:1 one-to-one transformer
Transformator mit Anblaskühlung air-blast transformer, forced air-cooled transformer
Transformator zur Erzeugung eines künstlichen Nullpunktes neutralator
Transformator mit Gleichrichter transrector
Transformator mit Ölumlaufkühlung forced oil-cooled oil-immersed transformer
Transformator in Scott-Schaltung Scott (connected) transformer
Transformator mit beweglicher Sekundärwicklung moving-coil transformer
Transformator mit Stickstoffüllung inertair transformer
Transformator mit Wasserkühlung water-cooled transformer
vom Transformator abgegebene Leistung transformer output
vom Transformator aufgenommene Leistung transformer input
Transformatorenausfall transformer failure, transformer breakdown, transformer fault
Transformatorbetrieb transformer operation (621.314); self-rectifying operation (621.386), (621.396.61)
Transformatorenblech transformer lamination, transformer sheet
Transformatorenfeld transformer branch, transformer bay
Transformatorenjoch transformer yoke
Transformatorenkern transformer core
Transformatorenkessel transformer tank
Transformatorenklemme transformer terminal
Transformatorenkonstruktion transformer design, transformer construction
Transformatorenkreis transformer circuit
Transformatorenkupplung transformer coupling
Transformatorenleistung transformer rating
Transformatorenleistungsabgabe transformer output
Transformatorenöl transformer oil
Transformatorensatz bank of transformers, transformer bank
Transformatorenschalter transformer switch
Transformatorenschutz transformer protection
Transformatorspannung transformer voltage
Transformatorspule transformer bobbin, transformer coil
Transformatorenstation transformer station

Transformatorenstrom transformer current
Transformatorentype transformer type
Transformatorverluste transformer dissipation (sing.), transformer losses
Transformatorenwicklung transformer winding
Transformatorwirkung transformer action, transformer effect
Transformatorenzelle transformer cell, transformer vault
transformatorisch transformer
transformatorisch gespeist transformer-fed
transformatorische Kopplung transformer coupling
transformatorische Rückkopplung Meissner circuit
transformieren to transform (51), (621.3)
transformierte Funktion transformed, transform
transformierter Widerstand coupled impedance
Transformierte transform (51)
transiente Kurzschlußzeitkonstante direct-axis transient short-circuit time constant
transienter Wärmewiderstand transient thermal impedance (621.382)
Transimpedanz transimpendance
Transistor transistor (621.396.64)
Transistor-Ersatzschaltung transistor equivalent circuit
Transistorprüfgerät transistor measuring equipment
Transistorradio transistor (radio)
Transistorschalter transistor switch
Transistorschaltung transistor circuit, semiconductor circuit
Transistor-Tedrode tetrode transistor
transistoriert transistorized
Transitfrequenz transition frequency (621.382.3)
transitorisch transitory, transient
Transitron transitron
Transitübertragung transit transmission (621.395)
Transitvermittlungsstelle transit exchange (621.395)
Translation translation
Translationsbewegung translatory motion
Translationsfläche plane of translation
Transmission drive, shafting, overhead motion
Transmissionswelle line-shaft, main shaft
Transmittanz (small-signal) short-circuit forward transfer admittance
transparent transparent
transparenter Bildschirmvorsatz control overlay (621.396.96)
Transparentabdeckung perspex cover
Transparentauflage overlay
Transparentpapier transparent paper, tracing paper
Transparenz transparence, transparency
Transport transport(ation), conveyance, transfer, carriage, haulage
auf dem Transport nach in transit to
Transportanlage conveyance plant, conveyor system
Transportband belt conveyor
Transportkasten carrying case
Transportklinke feed pawl (621.394.6)
Transportkratze scraper conveyer
Transportmittel means of transport[ation (US)]
Transportrinne conveyor trough
Transportschnecke conveyor screw, screw conveyor
transportabel transportable, conveyable; portable
transportieren to transport, to transfer, to carry, to haul
Transpositionsfehler transposition error
Transuran transuranic element
Transversale transversal line
Transversalaufzeichnung transverse magnetization (691.84)
Transversalwelle transverse wave
transzendent transcendent(al) (51)
transzendente Funktion transcendental function
transzendente Zahl transcendental number
Trapez trapezium (51)
Trapezgewinde tetragonal thread, acme thread
Trapezimpuls trapezoidal pulse
Trapezverzerrung keystone distortion (621.397)
Trap trap (621.382)
Trasse trace (transmission) route (621.315)
Trassierung line location, route layout
Traufe gutter, dripstone
träufeln to trickle
Traverse suspension bracket, crossarm (621.315)
Travertin travertine
treffen auf to hit, to strike
getroffenes Teilchen struck particle
Treffpunkt point of impact, point of collision
Treib-... driving
Treibgas fuel gas
Treibkeil driving wedge
Treibrad driver (wheel)
Treibriemen (driving) belt, leatherbelt
Treibrolle capstan
Treibscheibe driving pulley
Treibseil driving rope

Treibsitz driving fit
treiben to drive, to propel, to move
treibende Kraft moving force
Treiber driver (621.396)
Treiberleistung (grid) driving power
Treiberröhre driver valve
Treiberstufe driver stage, driving stage
Treibertransformator driver transformer
Treidellokomotive to wing engine
Trend trend
Trenn-... separating, separative; isolating; dividing
Trennblech separator plate
Trennerdungsschalter combination isolating/earthing switch; combination disconnecting/grounding switch (US)
Trennfaktor separation factor
Trennfeld separating field (539)
Trennfunkenstrecke isolating gap
Trennklinke break jack
Trennkontakte isolating contacts
Trennmuffe insulating joint, section box (621.315.2)
Trennprozeß separative process
Trennrelais cut-off relay
Trennröhre buffer valve (621.396)
Trennschalter isolating switch, disconnecting switch, disconnecting link
Trennschalter mit Schaltstangenbetätigung hook-stick disconnecting switch (621.316.5), (621.395)
Trennschärfe selectivity (621.396.62)
Trennschicht stripping compound (621.35)
Trennsicherung isolating fuse, (fusible) cutout (621.315.6)
Trennsieb separating mesh, separating sieve
Trennstelle disconnection; testing point (621.395)
Trennstrich dividing line
Trennstück disconnecting link (US); isolating link
Trennstufe separator buffer, buffer stage (621.396.64)
Trenntransformator disconnecting transformer
Trennverbinder disconnector
Trennverstärker buffer amplifier
Trennwand partition (wall), dividing wall, barrier; diaphragm
Trennwandplatte partition slab
Trennweiche diplexer
Trennwiderstand isolating resistor
Trennzyklotron calutron
trennbar separable
Trennbarkeit separability
trennen to separate; to disconnect, to isolate (621.3); to clear (621.395)
trennend separating, separative
trennfähig separable
getrennt anordnen to space apart
getrennt anordnen um 180° to space 180° apart
getrennt betätigt independently operated
getrennt werden to be cut off (621.395)
getrennte Einheit separate unit
Trennung separation; disconnection
Trennung auf chemischem Wege chemical separation
Trennung der Variablen separation of the variables
Trennungsgüte efficiency of separation
Trennungslinie dividing line
Trennungsstrich dash
Treppe stairs (pl.), stairway
Treppenbeleuchtung stair lighting
Treppenfunktion step function
Treppenhausschalter landing switch
Treppenspannung staircase voltage
Treppenwicklung stepped winding, split winding
Tretantrieb treadle operation
Tretkontakt foot (-operated) contact
Tri = Trinitrotoluol trotyl = trinitrotoluene
Triac triac, bidirectional triode thyristor
Triangulierung triangulation
Trichloräthylen trichlorethylene
Trichter funnel, hopper; horn (534.86), (621.396)
Trichterhals throat of a horn (534.86)
Trichterlautsprecher horn (loud) speaker
Trichteröffnung mouth of the horn (621.396)
Trickaufnahme trick sequence
Trickmischgerät studio mixer
Trieb drive, gear
Triebachse driving axle, live axle (625)
Triebgestell motor bogie (625)
Triebkraft (prime) mover
Triebmoment deflecting torque, deflecting force (621.317)
Triebrad drive wheel, driving wheel
Triebscheibe driving pulley
Triebstange side rod (621.33)
Triebsystem motor element (621.395.62); driving system (621.317)
Triebwagen electric motor car, motor coach (621.33); accumulator car; self-propelled carriage (625)
Triebwagenzug motor-coach train
Triebwerk driving mechanism (621.317);

motion work (681.11); propulsion unit (629.13)
Triebzapfen driving axle
Triftraum drift space
Triftröhre velocity-modulation tube
Trigger trigger
Triggerröhre trigger tube, trigatron
Trigonometrie trig = trigonometry
trigonometrisch trigonometrical
trigonometrische Funktion trigonometrical function
trigonometrischer Punkt triangulation station
trigonometrische Reihe trigonometrical series
triklin triclinic
Trillion trillion (Br), quintillion (US)
Trimmen trimming (621.396), (622), (629.12), 629.13); tracking (621.396.62)
Trimmkondensator trimmer, trimming condenser, padding capacitor, padding condenser (621.396)
trimmen to trim; to track
Trimmer trimmer, trimming condenser, padder
Trinom, Trinomialausdruck trinomial (51)
Triode triode, three-element tube, three-electrode valve
Triodenschaltung triode circuit, triode connection
Tripel, Triplett triplet (535)
Tripelspiegel triplane
Tritt step, tread
außer Tritt fallen to fall out of step, to fall out of synchronism, to pull out (621.313)
in Tritt fallen, in Tritt ziehen to pull into synchronism, to pull into step (621.313), (621.396.61)
Trittgrenze 150° setting (621.314.6)
Trocken-... dry(-type)
Trockenanlage drying plant
Trockenbatterie dry battery, dry pile, Leclanché battery
Trockenelektrolyt dry-electrolytic capacitor
Trockenelement dry cell, Leclanché cell
Trockengleichrichter dry-contact rectifier, metal oxide rectifier, metallic rectifier, semi-conductor rectifier
Trockengut stock (621.369)
Trockenkammer drying chamber
Trockenmittel desicant
Trockenrasierer electric shave, electric razor
Trockenschleuder centrifugal drier, spin drier
Trockenschrank exsiccator, desiccator, drying cabinet
Trockenstrahler radiant heater, heat lamp (621.369)
Trockentransformator dry-type transformer, air-cooled transformer
Trockenüberschlagspannung dry flashover voltage
Trockenwandler dry-type instrument transformer
trocken dry
trockener Dampf dry steam
trockene Destillation dry distillation, destructive distillation
trockenes Klima dry climate
trockene Lötstelle dry joint
trockene Luft dry atmosphere
trockene Überschlagspannung dry flashover voltage
trocknen to dry, to desiccate, to dehydrate
Trolitul trolitul
Trolleybus trolley bus
Trommel drum (534), (621), (621.3); bobbin (621.315)
Trommel des Magnetscheiders magnetic pulley
Trommelanker drum armature (621.313)
Trommelfilter revolving filter
Trommelinduktor, Trommelläufer non-salient pole rotor (621.313)
Trommelkabel trailing cable
Trommelläufer drum rotor, non-salient pole rotor
Trommelofen drum furnace
Trommelschreiber drum-chart recorder
Trommelsieb cylindrical sieve, cylindrical screen
Trommelspeicher drum storage
Trommelwicklung drum winding (621.313)
Trompete trumpet, bugle
Tropen tropics
Tropenausführung tropical design
tropensicher tropicalized, tropic-proof
Tropf-... drain, dripping
Tropfelektrode dripping electrode
Tropffilter percolator
Tropfnässe dripping wetness
Tropfwasser drip(-water)
tropfwasserdicht drip-proof
tropfwassergeschützte Leuchte rain-proof lighting fitting
tropfwassergeschützte Maschine drip-proof machine
Tropfwasserschutz protection against dripping water
Tröpfchenspur cloud track (53)
tröpfeln to trickle; to drip, to dribble
Tropfen drop
tropfen to drop, to drip

Tropopause tropopause
Troposphäre troposphere
troposphärische Überhorizontverbindung troposcatter transmission
troposphärische Reflexion tropospheric reflection
troposphärische Streuung tropospheric scattering
troposphärischer Wellenleiter tropospheric radio duct
trübe dim, turbid; opaque, milky
Trübung turbidity; milkiness; opalescence; antenna effect (621.396); rain clutter (621.396.96)
Trust trust
T-Schaltung T-network
T-Schiene T-rail
T-Schlitz T-slot, Tee-slot
T-Schutz blooming (535)
mit T-Schutz versehen to bloom (535)
T-Schweißung Tee-weld
TS-Diagramm temperature-entropy diagram
T-Stück pipe-Tee, branch-T
T-Träger T-section
Tübbing tubbing
Tübbingring tubbing ring
Tübbingsegment tubbing segment
Tubus tube (535)
tuckern to motorboat (621.396.62)
tünchen to limewash, to whitewash
Tunnel tunnel; canyon (539)
Tunneldiode tunnel diode
Tunneleffekt tunnel effect
tunneln to tunnel
Tunnelofen tunnel furnace, tunnel kiln
Tunnelvorgang tunnel action (621.382)
Tür door, gate
Türglocke doorbell
Türkontakt car-door electric contact (629); door contact (696)
Türlautsprecher door loudspeaker
Türöffner door opener
Türschalter door switch
Türsprechanlage housedoor intercom, doorman converser
Türverriegelung gate interlock (621.34), (621.8)
Turbine turbine
Turbo-... turbo
Turboabscheider cyclone separator
Turbodynamo turbo dynamo
turboelektrische Lokomotive turbine-electric locomotive
Turbogebläse turbo blower, turbo compressor
Turbogenerator turbo-generator, turbo-alternator, turbo-dynamo
Turbogenerator mit Getriebe geared turbo-generator
Turbokompressor turbo compressor, rotary compressor
turbulenter Bereich turbulent region
Turbulenz turbulence
Turm tower
Turmförderung tower winding
Turmalin tourmaline
Turnstile-Antenne batwing antenna, turnstile
Tusche drawing ink, Indian ink
Tuschezeichnung ink tracing
tuten to toot, to hoot
T-Verzweigung T-junction
Type type, design, class; letter (655)
Typendrucktelegraph (type) printing telegraph
Typenhebel type lever (681.6)
Typenlegalisierung type approval
Typenleistung design merit; rated output
Typenprüfung type-acceptance test
Typenrad daisy wheel, type wheel
Typenraddrucker (daisy) wheel printer (621.394.6), (681.3)
typenrecht, typengetreu true to type
Typenschild rating plate; name plate (US)
typisch für typical of
typisieren to typify
Typisierung typification
Typus type, mode

U

U [Spannung] U [potential difference]
U-Bahn(tunnel) tube; subway (US)
überaperiodische Dämpfung overdamping
überaperiodisch gedämpft overdamped
Überbau superstructure
überbeanspruchen to overburden, to overload
Überbeanspruchung overburden, overload, overwork
überbelasten to overload (621.3)
Überbelastung overload
überbelichten to over-expose (77)
Überbelichtung over-exposure, over-exposition
überbemessen to over-design, to over-dimension, to overrate
Überbestrahlung over-exposure to radiation
überbewerten to overrate, to overvalue

Überblendung cross fade (654.19); lap dissolving (621.397)
überbrücken to bypass, to shunt (621.3); to bridge (621.3), (624); to span (624)
überbrückter Kondensator shunted condenser
Überbrückung bypass, bridging, shunting (621.3); jumper wire (621.395)
Überbrückungsdraht jump wire
Überbrückungskondensator bypass condenser, bridging condenser, parallel capacitor, shunt capacitor
Überbrückungsschalter bypass switch; field-breaking switch, field-discharge switch (621.314)
Überbrückungsschaltung bypass circuit, shunt circuit
Überbrückungsschwingkreis resonant shunt
Überbrückungswiderstand field-breaking resistor, field-discharge resistor (621.318); shunt resistor
überdachte Anlage covered plant
überdämpft overdamped
Überdämpfung overdamping
überdauern to outlast
überdecken to cover; to obscure, to swamp (621.396)
überdehnen to overstretch
überdimensionieren to oversize
überdrehen to overspeed, to "overrev" (621.4); to overwind (681)
Überdrehzahlschutz overspeed protection
Überdruck overpressure; overload
Überdruckventil overload valve
übereinander anordnen to superpose; to stack
übereinstimmen to concur, to harmonize; to coincide
Übereinstimmung concordance, harmony, conformity
übererregen to overexcite
übererregter Synchronmotor over-excited synchronous motor
Übererregung over-excitation
Überfallruf bank-raid alarm
Überfangglas flashed glass
überfettet superfatted
Überfluß an prodigality of
überführen in to convert into
überführen nach to transport to, to transfer to
Überführung transport, transference (**nach**: to); conversion (**in**: into); viaduct
Überführung in den gasförmigen Zustand gasification
Überführungsdraht bridle wire
Überführungszahl transport number (621.35)

Übergang transition; junction (621.382)
Übergangsbogen transition curve (625.1)
Übergangsfrequenz transition frequency; crossover frequency (654.19)
Übergangsgebiet transition range (621.37)
Übergangskonus flared section (621.396.67)
Übergangspunkt transition point
Übergangsradius blending radius
Übergangsstadium transition interval
Übergangsstecker plug adapter
Übergangsvorgang transient
Übergangswahrscheinlichkeit transition probability
Übergangswiderstand contact resistance
Übergangszeit transit time; transition time (621.37)
Übergangszone transition zone; junction (621.382)
Überhang overhang, overhung portion
überhängen to overhang
überhärten to overcure
überhitzen to overheat; to superheat
überhitzter Dampf superheated steam
Überhitzer (steam) superheater
Überhitzung superheating, excess heating
überholen to overhaul, to recondition; to pass, to overtake, to outrun
Überholung preventive maintenance, overhaul(ing), reconditioning; passing
Überholungsgleis loop line, passing loop (625.1)
Überhorizontübertragung beyond-the-horizon transmission, forward scatter transmission (621.372)
Überhorizontverbindung over the horizon radio system, transhorizon link
Überhorizontverbindung für ionosphärische Ausbreitung ionospheric scatter equipment
überkapptes Kabelende sealed cable end
überkippen to tilt over
überkochen to boil over
Überkommutierung overcommutation
überkompoundiert overcompounded (621.313)
Überkompoundierung overcompounding (621.313)
Überkopfschweißung overhead welding
überkreuz multiplizieren to multiply across
überkritisch supercritical
überkritisch koppeln to overcouple
überkritische Dämpfung overdamping
überkritische Drehzahl speed above the critical
überkritischer Reaktor supercritical reactor
überladen to overcharge (621.356); to supercharge (621.4)

Überladung overcharge (621.356); supercharge (621.4)
Überlagerer heterodyne oscillator, local oscillator
überlagern, (sich) to (super)heterodyne (621.396)
überlagern über to superpose on, to superimpose on, to heterodyne with
überlagerte Belastung superimposed load (624)
überlagertes Feld superimposed field
überlagerter Gleichstromanteil standing d. c. component (621.397)
überlagerter Gleichstromruf superposed ringing
überlagertes Netz overlay network (621.395)
Überlagerung superposition, superimposition, interaction, mixing (621.396)
Überlagerungsempfang superheterodyne reception, supradyne reception, heterodyne reception, beat reception
Überlagerungsempfang mit höherer Überlagerungsfrequenz infradyne reception
Überlagerungsempfänger superheterodyne receiver, superhet
Überlagerungsfrequenz heterodyne frequency
Überlagerungsfrequenzmesser heterodyne frequency meter
Überlagerungsheulen fringe howl, self whistle (621.396)
Überlagerungspfeifen heterodyne whistle, heterodyne interference, self whistle
Überlagerungsoszillator beat frequency oscillator, heterodyne oscillator, local oscillator
Überlagerungsprinzip principle of superposition
Überlagerungssender heterodyne oscillator, local oscillator
Überlagerungsstufe mixer (stage), converter (stage) (621.396)
Überlagerungstelegraphie super (im)posed telegraphy, super-audio telegraph system, carrier telegraphy
Überlagerungstelegraphieempfang heterodyne code reception
Überlagerungsverfahren heterodyne method
Überlagerungswellenmesser heterodyne wavemeter, oscillator wavemeter
Überlandleitung long-distance (transmission) line
Überlandzentrale rural power station, district power station, rural utility
Überlänge surplus length
überlappen, (sich) to overlap
überlappt geschweißtes Rohr lap-welded pipe
überlappt wickeln to lap
überlappte Spule lap coil
überlappter Stoß lap joint (621.7)
Überlappung overlap, lapping
Überlast overload, superload, unsafe load
Überlastregelung overload control
Überlastschalter overload circuit breaker
Überlastschutz overload protection
Überlaststrom overcurrent
Überlaststrom bei Rückzündung oder Kurzschluß fault electrode current, surge electrode current (621.382)
Überlastbarkeit overload capacity
überlasten to overload
Überlauf overflow
Überlaufanzeiger overflow indicator
Überlaufdaten excessive data, overflow data (681.3)
Überlaufverkehr overflow traffic (621.395)
überlaufen to overflow, to flow over
Überleitung cutover (621.395)
übermäßig beanspruchen to stress unduly, to overstress
Übermikroskop ultramicroscope, electron microscope
übermitteln to convey; to transmit
Übermittlung transmission
Übermodulation overmodulation
übermodulieren to overmodulate
überneutralisieren to overneutralize
überprüfen to test, to check
Überprüfung test, check
Überregelung overshooting
Überreichweite transhorizon propagation, over-the-horizon reception
übersättigen to supersaturate
Übersättigung supersaturation
Überschall ultrasound (s. a. **Ultraschall)**
Überschallfrequenz supersonic frequency
Überschallgeschwindigkeit supersonic speed
Überschallkanal supersonic tunnel (533)
überschalten nach to switch over to
Überschaltwiderstand preventive resistance (621.316.5); transition resistance
überschiebbarer Kabelstrommeßwandler slip-over current transformer
überschießen to overshoot
überschießende Tausendstel odd thousandth
Überschlag flashover, sparkover, arcover (621.3)
Überschlagprüfung flashover test
Überschlagspannung arcover voltage, arcover potential, sparkover voltage

Überschlagspannung zwischen Spitzen needle-point voltage
Überschlagstoßspannung impulse-flashover voltage
Überschlagsverzögerung impulse inertia
Überschlagwechselspannung power-frequency flashover voltage
Überschlagsweg spark length
überschlagen to flashover, to sparkover, to arcover (621.3)
überschlagende Flamme down draught (621.78)
überschneiden, sich to cross, to overlap
Überschneidungspunkt , crossover point
überschreiben to overwrite
überschreiten to exceed
Überschreitung exceeding, excess
Überschuß an excess of
Überschußblindverbrauchszähler excess reactive power meter
Überschußelektron excess electron (621.382)
Überschußelektronenleitung (excess) electron conduction
Überschußhalbleiter n-type semiconductor
Überschußleistung dump power
Überschußleitung n-type conduction, excess conduction, electron conduction
Überschwingen ringing, overshoot
überschüssig redundant, supernatant
übersetzen to gear (621.8); to transform (621.3); to translate (41)
Übersetzer compiler (681.3); translator
Übersetzer für höhere Programmiersprachen high level compiler (681.3)
Übersetzung gear ratio, transmission ratio, speed ratio (621.83); transformation (621.314); translation (41)
Übersetzungsfehler ratio error (621.317); misinterpretation (41)
Übersetzungsgetriebe transmission gear, step-up gear, speed transformer
Übersetzungsverhältnis ratio of transformation, turns ratio (621.314.2); (s. auch **Übersetzung**)
Übersetzungsverhältnis Eins unity turn ratio
übersichern to overfuse
Übersichtsplan general plan, block diagram
Übersichtsschaltplan basic circuit diagram, block diagram
Überspannung excess voltage, overvoltage; voltage surge, surge voltage
Überspannungsableiter surge arrester, surge dissipator, surge diverter; lightning arrester, lightning diverter (US)
Überspannungsableiter mit spannungsabhängigem Widerstand autovalve (arrester) (621.316.9)
Überspannungsauslösung overvoltage tripping
Überspannungsbegrenzer surge-voltage protector
Überspannungsrelais overvoltage relay, maximum voltage relay
Überspannungsschutz overvoltage protection, excess-voltage protection; transient protection, surge protection
Überspannungsschutzkondensator surge-protection capacitor
Überspannungswelle surge, transient
Überspannungswanderwelle travelling overvoltage
Überspannungszählwerk surge recorder
Übersprechen cross talk
Übersprechfaktor cross-talk factor
Übersprechpegel cross-talk level
Übersprechstörung cross-talk trouble, cross-talk interference
Überstandard-Brechung superrefraction
übersteuern to overload, to overamplify, to overdrive (621.396)
Übersteuerung overloading, overamplification
überstrahlen to outshine, to overshadow
Überstrahlung blooming (621.397)
überstreichen to traverse; to recoat, to repaint
Überstrom overcurrent, excess current, overload; rush current; overload forward current (621.314.5)
Überstromabschaltung non-re-entrant current limiting
Überstromauslöser overload release, overload trip, overcurrent trip
Überstromauslösung overcurrent circuit breaking
Überstromautomat overload trip
Überstromrelais overcurrent relay, overload relay
Überstromschalter overcurrent circuit breaker
Überstrom-Rückstromschalter overcurrent and reverse current circuit breaker
Überstromschnellauslöser instantaneous overcurrent trip
Überstromschutz overcurrent protection, overload protection
Überstromspule overcurrent coil
Überstromstoß rush current
Übertemperatur raised temperature, excess temperature, overtemperature, temperature rise
Übertrag carry (681.3)

Übertragsbit carry bit
übertragbar transmissible (621.396)
übertragen to transmit (621.396)
übertragener Widerstand reflected resistance
Übertrager audio-frequency transformer, telephone transformer
Übertrager mit Eisenkern iron-core transformer
Übertragerkopplung transformer coupling
übertragergekoppelt transformer-coupled
Übertragung transmission, communication (621.392), (621.5), (621.6)
Übertragung durch Mehrfachreflexion multihop transmission
Übertragungsanlage sound system (621.375)
Übertragungsart type of transmission
Übertragungsband communication band
Übertragungsbelag propagation constant (621.392)
Übertragungsbereich transmission range, transmission region (621.392); frequency response (654.19); communication range (621.395), (621.6)
Übertragungsdämpfung transmission loss (621.395); path attenuation (621.396)
Übertragungsexponent gamma, contrast gradient (621.397)
Übertragungsfaktor transmission coefficient, steady-state gain (621.316.7)
Übertragungsgeschwindigkeit transmission speed, transmission rate, transfer rate
Übertragungsgewinn transmission gain
Übertragungsgrenzleistung power handling capacity (621.315)
Übertragungsgüte fidelity (534.86); (621.396.62); transmission performance, transmission quality (621.395)
Übertragungskennlinie transfer characteristic (621.39), (621.374)
Übertragungskonstante propagation constant (621.39)
Übertragungskurve transfer curve
Übertragungsleitung transmission line
Übertragungsmaß transfer constant, effective transmission factor
Übertragungsmeßgerät transmission test set
Übertragungsnetz transmission network, transmission system, grid system
Übertragungspegel transmission level (621.395)
Übertragungsunterbrechung outage (621.395)
Übertragungsverfahren transmission mode
Übertragungsverlust transmission loss, loss in transmission
Übertragungsverzug transfer lag (62); transmission lag (621.39)
Übertragungsweg transmission path
Übertragungswinkel transmission angle
Übertragungswirkungsgrad transmission efficiency
Überverbrauch excess consumption
Überverbrauchszähler excess meter
überwachen to supervise, to superintend, to inspect, to control, to monitor
Überwachung supervision, superintendence, inspection, control, monitoring
Überwachungsantenne monitor aerial
Überwachungsarbeit supervisory work
Überwachungsbeamter supervisor, superintendent
Überwachungsgerät monitor, checking equipment, supervisory device
Überwachungskanal supervisory channel
Überwachungskontakt controlling contact
Überwachungslampe pilot lamp, telltale lamp, supervisory lamp
Überwachungsprogramm monitor program; tracing program (681.3)
Überwachungspult supervisory control desk
Überwachungsradar surveillance radar
Überwachungstisch monitor console
Überwachungszeichen control mark
Überweisungsvermittlung toll centre (621.395)
Überwurfmutter cap nut, screw cap, bond nut
überzähliger Zahn hunting to oth
überziehen mit to coat with
Überzug coating, serving; covering, sheath
üblich customary, common, normal, standard
übliche Auslegung, übliche Bemessung normal rating
übliche Form standard shape
übliche Frequenz normal frequency, standard frequency
U-Bogen U-bend
U-Boot submarine
U-Boot orten to locate a submarine
U-Boots-Anzeige submarine detection
U-Bügel U-bolt
Übung practice, exercise; training
Übungsgerät practice-type set (621.396)
U-Eisen channel (iron)
U-förmig U-shaped
Uhr clock; watch
Uhr aufziehen to wind up a clock
Uhr mit elektrischem Aufzug electrically wound clock
Uhr mit Selbstaufzug self-winding clock
Uhrfeder main spring

Uhrwerk clockwork, movement
Uhrwerksreserve spring reserve
Uhrzeiger hand of a clock
gegen den Uhrzeigersinn counter-clockwise, against the sun
im Uhrzeigersinn clockwise
UKW = Ultra-Kurzwellen VHF = very high frequency
UKW-Band v. h. f. band = very high frequency band
UKW-Drehfunkfeuer VHF-omnidirectional Range, VOR
UKW-Empfang v. h. f. reception
UKW-Empfangsteil tuner
UKW-Frequenz v. h. f. = very high frequency
UKW-Sprechfunk VHF-mobile radio; citizen-radio
Ulbrichtsche Kugel sphere photometer
Ultrakurzwelle = UKW
ultramarin ultramarine
Ultrarot, ultrarot infrared
ultra-rot-empfindliche Platte, Ultrarotplatte infrared plate (77)
Ultrarotempfänger infrared detector
Ultrarotortung infrared location
Ultrarotstrahlung infrared radiation
Ultraschall ultrasonics (pl.), ultra sound (s. auch **Überschall)**
Ultraschallfrequenz supersonic frequency, supra-acoustic frequency, ultra-audiofrequency, ultra-sound frequency
Ultraschall-Laufzeitkette supersonic delay line
ultraschallmodulierte Wellen super-audio-frequency modulated waves, ultrasonic modulated waves
Ultraschallortung supersonic sounding
Ultraschall-Reinigungsgerät ultrasonic cleaning tank
Ultraschallschweißen ultrasonic brazing
Ultraschallschwinger ultrasonic generator
Ultraschall-Wandstärkenmesser supersonic gauge
Ultrastrahlung ultragamma radiation
Ultraviolett-..., ultraviolett ultraviolet (s. auch UV)
Ultraviolettempfindlichkeit ultraviolet sensitivity, ultraviolet response
Ultraviolettlampe ultraviolet lamp, quartz lamp
Ultraviolettlicht ultraviolet light
Ultraviolettmikroskop ultraviolet microscope
Ultraviolettphotographie ultraviolet photography
Ultraviolettstrahlung ultraviolet rays (pl.)
umbandeln to tape, to serve, to wrap (621.3)
umbandeltes Kabel served cable
Umbandelung taping, serving
umbiegen to bend over
umbördeln to peen over, to bend over, to crimp, to bead
umdrehen (sich) to turn round; to rotate, to revolve
Umdr. = Umdrehung rev. = revolution; rotation, turn
Umdr./min = U/min = Umdrehungen pro Minute r. p. m. = revolutions per minute
Umdrehungsparaboloid paraboloid of revolution
Umdrehungszahl number of revolutions per minute, speed
Umdrehungszähler revolution counter, revolution indicator, gyrograph
Umfang range, compass, sphere; extent; bulk, volume; periphery, perimeter, circumference
Umfang des Fertigungsprogramms range of products
Umfang überschreiten to be beyond the scope
Umfang überstreichen to sweep around the periphery
Umfangsgeschwindigkeit s. f. p. m. = surface feet per minute, peripheral speed, rim speed
Umfangskraft tangential force
Umfangslänge perimeter
Umfangsschweißung circumferential welding
Umflechtung concentric(al) braiding
umflochtene Elektrode braided electrode
umformen to transform, to convert (621.3); to re-arrange (51)
Umformer power converter, converter; dynamotor (US); transducer (621.316.7)
Umformerlokomotive motor-generator locomotive
Umformersatz motor-generator set
Umformerwerk converting plant, converter station (621.3)
Umformung conversion (621.3); manipulation, transformation (51)
Umformung einer Gleichung transformation of an equation
Umformung von Energie conversion of energy
Umgebung environs (pl.), environment; surroundings (pl.)
Umgebung des Arbeitspunktes operating region (621.385)
umgebungsbedingt environment-induced
umgebungsbedingte Beanspruchung environmental stress

Umgebungsbedingungen environmental conditions, ambient conditions
Umgebungslärm surrounding noise, ambient noise
Umgebungstemperatur ambient temperature
Umgehungsschiene by-pass busbar (621.316)
umgruppieren to regroup, to rearrange
Umgußmaterial recast material
umhüllen to sheathe, to cover, to coat
umhüllte Elektrode sheathed electrode, covered electrode, coated electrode
umhüllte Leitung sheathed wire, weatherproof wire
Umhüllung sheating, cover, coating, casing
Umkehr reversal, reversion; return
Umkehr der Drehrichtung reversal of rotation
Umkehr der Leistungsrichtung power reversal
Umkehranlasser reversing starter
Umkehrbogen return bend
Umkehrgetriebe reversing gear
Umkehrmotor reversing motor, reversible motor, bidirectional motor
Umkehrpunkt point of inflection (621.316.7)
Umkehrschalter reversing switch
Umkehrschütz reversing contactor
Umkehrstufe phase-reversing stage
Umkehrverbindung revertive call (621.395)
Umkehrverfahren reversing process (655)
umkehrbar reversible
umkehrbarer chemischer Vorgang reversible reaction
umkehrbares Element reversible cell
Umkehrbarkeit reversibility
Umkehrbarkeit der Wirkungsweise reversibility of function
umkehren to reverse, to invert; to return
umgekehrt inverted, reciprocal, reversed, converse
umgekehrt proportional inversely proportional
umgekehrter Konus reversed taper
umgekehrte Lage inverted position
umgekehrte L-Antenne inverted "L"-aerial
umgekehrte Lichtmodulation negative light modulation
umgekehrte Polarität inverse polarity
umgekehrte Proportionalität inverse proportion
umgekehrter Prozeß reciprocal process
umgekehrte Richtung reverse direction
umgekehrtes Verhältnis inverse ratio
in umgekehrter Reihenfolge in reverse order
umkippen to tilt over, to overturn, to upset
umkleiden to sheathe
Umkleidung sheathing
umklemmen to interchange the leads at the terminals, to reconnect
umklöppeln to braid
Umklöppelung braiding
umkonstruieren to redesign, to modify
Umkreis circum-circle; circumference, zone
im Umkreis von within a radius of
umladen to reload; to charge in opposite direction, to charge inversely
Umlauf circulation, rotation, circuitation
Umlauf gegen Uhrzeigersinn anticlockwise rotation
Umlauf im Uhrzeigersinn clockwise rotation
auf einem Umlauf round a mesh
Umlaufbahn orbit, orbital path (52), (539)
Umlaufdämpfung feedback loss
Umlaufgeschwindigkeit speed of circulation
Umlaufintegral line integral round a circuit, contour integral
Umlaufintegralbildung circuitation
Umlaufkühlung closed-circuit cooling
Umlaufluftkühlung closed-circuit air cooling
Umlaufpeiler rotating aerial direction-finder
Umlaufsinn sense of rotation
Umlaufspeicher circulating memory, circulating store, cyclic store (681.3)
Umlaufsweg path of circulation; path of integration
Umlaufszeit period of rotation (52)
umlaufen to revolve, to rotate
umlaufende Kontaktscheibe revolving contact-maker
umlaufender Leitstrahlpeiler Robinson direction finder
umlaufende Maschinen moving machinery, rotating machinery (sing.)
umlaufendes Polsystem rotating field magnet
umlaufender Rahmenpeiler rotating loop direction finder
umlaufender Stromwender revolving commutator
umlaufende Trommel revolving drum (621.317)
umlaufender Vektor rotating vector, phasor, sinor
Umlegen call transfer (621.395)
umleiten to deflect, to bypass
Umlenkplatte baffle plate
Umlenkrolle angle sheave, guide pulley, idler pulley
umlöten to resolder
Umlötsicherung heat coil (fuse)
Umluft recirculated air, circulating air
Ummagnetisierung magnetic(al) reversal

ummanteln to sheathe
Ummantelung sheathing
umnieten to peen over, to crimp, to clinch
umnumierieren to renumber
umordnen to rearrange
Umpoleffekt turnover (621.317)
umpolen to change poles, to change the polarity
Umpolung reversal of polarity
umpumpen to circulate; to pump over (**von... nach...** from... to ...)
Umrandung rim
umrechenbar convertible
umrechnen to manipulate (51); to translate (621.395)
umrechnen auf to convert to, to reduce to
Umrechner director (621.395.34)
Umrechnung conversion, reduction
Umrechnungsfaktor reduction factor, conversion factor
Umrechnungstafel conversion table
Umrechnungszahl conversion figure
Umrechnungszahl von mittlerer horizontaler auf mittlere sphärische Lichtstärke spherical reduction factor
Umrechnungszahl vom el.-magn. zum el.-st. c. g. s.-System unitary ratio
Umrichter (static) frequency converter, (static) frequency changer
Umriß contour, outline (51)
Umrißanleuchtung outline lighting
Umrißzeichnung outline drawing
Umroller rewinder
umrüsten to refit
umschalten to throw over, to switch over, to cutover, to change over (621.3); to shift (629.113)
Umschalter throw-over switch, change-over switch, double-throw switch, reversing switch, reversing key
Umschaltautomatik automatic change-over device
Umschalteinrichtung change-over circuit; shifting device (629.113)
Umschalthebel change lever
Umschaltkontakt throw-over contact, change-over contact
Umschaltkontakt mit Unterbrechung two-way breake-before-make contact
Umschaltmoment switching torque (621.313)
Umschaltstromverhältnis transient-current ratio (621.382.3)
Umschaltzeichen escape character (621.394)
Umschaltung change-over, reconnection; switch over; shifting
Umschlagszeit change-over time, transit time (621.318)
umschließen to encircle, to enclose, to retain
umschlossener Raum enclosed volume
Umschlingungswinkel angle of contact
umschmelzen to remelt
Umschmelzmetall remelt (metal)
umsetzen in to turn into; to translate (621.395); to transpose (621.394); to convert (621.314)
Umsetzer rebroadcasting transmitter; transposer, converter (621.316.7)
Umsetzung conversion, transformation, translation
umspannen to transform (621.3)
Umspanner transformer (s. auch **Transformator**)
Umspanner mit Anblaskühlung air-blast transformer
Umspanner mit Wasserkühlung water-cooled transformer
Umspannerbauart transformer type
Umspannerrelais transformer relay
Umspannerwicklung transformer winding, transformer bobbin
Umspannerzelle transformer cell, transformer vault
Umspannung transformation (621.3)
Umspannwerk transformer station, transformer plant
umspinnen to braid
umsponnen covered (621.3)
umstellen von... auf... to change over from... to...
umsteuerbar reversible
Umsteuerbarkeit reversibility
umsteuern to reverse
Umsteuerung reversing, reversal
Umtastpeiler switched-beam direction finder
Umwälzeinrichtung agitator (621.35)
Umwälzgeschwindigkeit speed of circulation
Umwälzheizung plenum heating
Umwälzung agitation (621.35)
umwandelbar transformable (51), (54), (621.3); transmutable (539); convertible (621.3)
umwandeln to transform; to transmute; to convert, to transfigure
Umwandlung transformation; transmutation; conversion, transfiguration; transition (54)
Umwandlung von zu... transition from... to...
Umwandlungspunkt change point, transition point
Umwandlungstemperatur transition temperature, critical temperature

Umwandlungszeit transition period, transition interval
Umwegecho multiple-reflection echo
Umwelt environment
Umweltprüfraum environmental chamber
Umwerter director, translator (621.395)
UMZ = unabhängig-Maximal-Zeit definite time (-lag)
UMZ-Auslösung independent timelag circuit breaking
UMZ-Relais independent time element, constant time element, definite time element
umzeichnen to replot
unabgeglichen unbalanced (621.317)
unabgeschirmt unshielded (621.3); unscreened (621.396)
unabgeschirmte Doppelleitung open-wire twin
unabgestimmt untuned
unabhängig von independent of; irrespective of
unabhängiges Überstromzeitrelais definite time-lag overcurrent relay
unabhängig-Maximal-Zeit s. **UMZ**
unabhängige Veränderliche independent variable
unabhängige Zeitverzögerung definite time-lag, fixed time-lag
unangreifbar durch impervious to, unattacked by
unaufgeladener Zustand uncharged state (537), (621.355)
unauflöslich irresolvable
unausführbar impracticable
unausgewuchtet unbalanced, out of balance
unbearbeitbar nonmachinable
unbearbeitet unfinished
unbearbeiteter Zustand rough state, raw state
unbedingter Sprung unconditional jump (681.3)
unbegrenzt unlimited, unbounded
unbekannt unknown
unbekannte Eigenschaft unknown property
unbekannte Größe unknown quantity
Unbekannte unknown (51)
Unbekannte enthalten to involve a variable
unbelastet unstressed (621); unloaded (621.3)
unbelastete Antenne unloaded aerial
unbemannter Flugkörper RPV = remotely piloted vehicle, drone
unbemanntes Flugzeug pilotless aircraft
unbemannte Station automatic station, unattented station
unbenutzt herumliegen to lie idle, to lie waste
unbenutzt liegen to lie by
unberechenbar incalculable, incomputable
unbesetzter Höhenschritt dead level (621.395)
unbeschädigt intact
unbespultes Kabel unloaded cable
unbestimmt undetermined, indefinite, non-discript, vague
Unbestimmtheitsrelation uncertainty principle
unbewaffnetes Auge naked eye
unbeweglich immovable, immobile
unbeweglich machen to immobilize
unbewehrtes Kabel unarmoured cable
unbewiesen unverified
unbiegsam inflexible
unbrauchbar unserviceable, unusable
unbrennbar incombustible, noninflammable
unbrennbarer Film safety (cellulose acetate) film
unbrennbares Öl pyranol
Unbunt-Bereich achromatic locus (535)
undefinierbar undefinable, nondescript
undicht leaky, untight
undicht werden to spring a leak
Undichtigkeit leak, leakage
Undichtigkeiten aufweisen to leak
Undichtigkeitsverlust loss by leakage, leakage loss
UND-Schaltung AND-circuit, AND-element (621.374)
UND-Verknüpfung AND-operation
Undulator undulator (621.394)
undurchdringlich für impenetrable to
undurchlässig für impermeable to, impervious to
undurchlässig für Röntgenstrahlen radiopaque
undurchsichtig non-transparent, opaque
Undurchsichtigkeit opacity
uneben uneven, bumpy
unecht spurious; improper (51)
unechter Bruch improper fraction
unedles Metall base metal
unelastisch inelastic
unempfindlich gegen immune from (61); insensitive to, non-responsive to
unempfindlicher Film slow film, low-speed film
unempfindliches Instrument low-sensitivity instrument
unendlich infinite, unending
unendlich klein infinitesimal, next to nothing
unendliche Reihe unending series (51)
der unendlich ferne Punkt INF. = infinity
in unendlicher Folge endless(ly)
Unendliches, das Unendliche infinity
unentzifferbar indecipherable
unentzündlich uninflammable
unerforscht unexplored
unerwünscht spurious, undesired
unerwünschte Ausstrahlung spurious emission

unerwünschte Nebenresonanz spurious resonance
unerwünschter Nebenschluß sneak path
Unfall accident
Unfallschutzmaßnahme safety precaution, protective measure
Unfallschutzvorrichtung safety device
Unfallverhütungsvorschrift safety rule
unformatiert unformatted
ungedämpft undamped, unattenuated
ungedämpfte Feldstärke unabsorbed fieldstrength
ungedämpfte Schwingung undamped oscillation, sustained oscillation
ungedämpfte Welle c. w. = continuous wave, undamped wave, unattenuated wave
ungedämpfte sprachmodulierte Wellen speech-modulated continuous waves
ungeerdet unearthed, ungrounded
ungeerdete Anlage unearthed system, ungrounded system, insulated supply system
ungefährer Kreis rough circle
ungefedert unsprung
ungefederte Massen unsprung weight (sing.) (625.2), (629.113)
ungehärteter Stahl soft steel
ungeladen uncharged (537), (621.355)
ungeladener Zustand uncharged state
ungelenkte Rakete free rocket
ungelöschter Kalk quick lime
ungenau inexact, inaccurate, out of truth
Ungenauigkeit inexactitude, inaccuracy
ungenützt ineffective
ungenutzte Zeit ineffective time
ungeprüft ex plant
ungerade odd (51)
ungerade Ordnungszahl odd order
ungerade Zahl odd number
ungeradzahlig odd(-numbered)
ungeradzahlige Oberwelle odd(-numbered) harmonic
ungerichtet non-directional, omnidirectional
ungerichtete Antenne nondirectional antenna
ungerichtet betriebene Leitung bothway trunk (621.395)
ungesättigt unsaturated
ungesättigtes Normalelement unsaturated standard cell
ungesättigte Verbindung unsaturated compound
ungespeister Reflektor beam reflector, parasitic reflector (621.396.677)
ungeschirmt unshielded (621.3)
ungeschützte Maschine open machine
ungeschützte Schaltgeräte open switchgear
ungesteuert uncontrolled
ungesteuerte Gleichrichterröhre hot-cathode gas diode
ungewöhnlich unconventional, exceptional
ungewollte Kopplung stray coupling, undesired coupling
ungleiche Vorzeichen unlike signs
ungleichmäßig uneven, non-uniform
ungleichnamige Pole unlike poles
ungleichnamige Pole ziehen sich an unlike poles attract
ungünstigster Fall worst case (620.1)
unhörbar inaudible, sub-audio
unhörbar tiefe Frequenz sub-audio frequency
Unhörbarkeit inaudibility
Unijunction-Transistor unijunction transistor
unipolar unipolar, homopolar
unipolarer Impuls single-polarity pulse
unipolarer Leiter unidirectional conductor
Unipolargenerator homopolar alternator, acyclic generator
Unipolarmaschine unipolar machine, acyclic machine
Unipolartransistor unipolar transistor
unisolierter Draht uninsulated wire, bare wire
Universal-. . ., universal universal, G. P. = general-purpose
Universalgelenk universal joint
Universalgleichrichter multi-purpose rectifier
Universalmaschine universal machine, general-purpose machine
Universalmaterial all-round material
Universalschalter general-use switch
universell universal, multi-purpose . . .
universelle Verwendbarkeit universality
unklassifiziert non-graded, unclassified
unkontrollierbar uncontrollable
unkritisch uncritical (621.396)
unlackiert unvarnished
unlegiert unalloyed, plain
unlösbar insolvable, irresolvable (51); undetachable
unlösbare Aufgabe problem defying solution
unlöschbarer Speicher nonerasable storage
unlöslich insoluble, indissoluble (54)
unmagnetisch unmagnetized, non-magnetic
unmeßbar unmeasurable, immeasurable
unmeßbar klein below measurable limits
unmittelbar immediate, direct
unmittelbare Adressierung immediate addressing
unmittelbare Aufhängung direct suspension system

unmittelbare Auslöung direct tripping (621.316.5)
unmittelbare Regelung direct control
unmittelbarer Zugriff immediate access
unmoduliert unmodulated
unmodulierter Betrieb mit Anodenwechselspannung unmodulated continuous wave
unmodulierter Träger unmodulated carrier, resting frequency
unmodulierte Welle unmodulated wave, continuous wave
unmoduliertes Zeichen unmodulated signal
unpraktisch impractical
unqualifiziert unqualified
unregelmäßig irregular
unregelmäßig geformt irregular shaped, oddly shaped
unregelmäßiges Rauschen random noise (621.396)
unregelmäßige Verzerrung fortuitous distortion
Unregelmäßigkeit irregularity
unrein unclean, impure
Unreinheit uncleanlines, impurity
Unruhe agitation (621.35); trouble balance (wheel) (681.11); flutter
Unruhfeder hairspring (681.11)
unruhiges Bild unsteady image (621.397)
unruhiger Lauf uneven running
unrund out of roundness, out of truth
unrund laufen to run out of truth
unrund quetschen to squeeze out of concentricity
unrunder Kollektor uneven commutator
unsaubere Bearbeitung bad finish
unsauberer Kreis imperfect circle, rough circle
unscharf unsharp, diffuse, smearedout (535)
unscharfes Bild diffuse image
Unschärfe blurring (621.397); lack of focus
unschraffiert unshaded
unselbständige Glimmentladung abnormal glow
unsichtbar invisible
unstabil unstable
unstarr nonrigid
unstetig discontinuous; in a series of jumps
unstetige Funktion discontinuous function
Unstetigkeit discontinuity
Unstetigkeitsfläche surface of discontinuity
Unstimmigkeit discrepancy
Unsymmetrie unsymmetry, asymmetry, unbalance; impedance unbalance (621.392)
Unsymmetriedämpfung unbalance attenuation
Unsymmetriefehler asymmetrical effect, lack-of-symmetry error (621.396.9)
Unsymmetrielast out-of-balance load, asymmetrical load
Unsymmetrierelais phase-balance relay, phase-reversal relay
Unsymmetrieschutz phase-reversal protection, phase-balance protection
Unsymmetriespannung asymmetric(al) voltage
Unsymmetriestrom out-of-balance current
unsymmetrisch unsymmetric(al), asymmetric(al), unbalanced (621.3)
unsymmetrische Belastung out-of balance load, unbalanced load (621.3)
unsymmetrische Leitung unbalanced line
unsymmetrisches Spannungsdreieck unbalanced voltages (pl.)
unsymmetrischer Verstärker single-ended amplifier
untätig inactive, idle
untauglich unfit
untauglich machen to disable
unteilbar indivisible
Unteilbarkeit indivisibility
unten below, at the bottom
unten an, unten auf at the foot of
nach unten gerichteter Lichtstrom downward flux
nach unten wirkend downward acting
nach unten wirkender Druck downward pressure
unter under, below; bottom, lower
untere Grenze low limit, bottom limit
untere Grenzfrequenz lower cut-off frequency
unterer Heizwert low calorific value, net heat value
unterer hemisphärischer Lichtstrom lower hemispherical flux
untere Hörgrenze lower audible limit
unterer Index suffix, subscript
unterer Isolationspegel minimum flashover voltage
unterer Knick der Kennlinie cut-off (621.385)
unter Last under load, on load
unter Null below zero, sub zero
unter pari below pari
unteres Seitenband lower side band
unter Spannung stehen to be live, to be alive
unterer Totpunkt bottom dead centre
unter Wasser befindlich submerged
Unteransicht bottom view
unteraperiodisch underdamped
unteraperiodische Dämpfung periodic damping, underdamping

Unterband lower band
Unterbau foundation, substructure
unterbelasten to underrun
unterbelichten to underexpose
unterbrechen to interrupt, to break (621.3)
unterbrechende Umschaltung break before make
unterbrochen discontinuous
unterbrochenes Feuer occulting beacon (656.7)
Unterbrecher interrupter, breaker
Unterbrecherkontakt interrupter contact, breaker point (621.4)
Unterbrechung interruption, (circuit-)break; discontinuity; interrupt (681.3)
Unterbrechung der Übertragung durch Regen rainfall outage
Unterbrechungsanforderung interrupt request (681.3)
unterbrechungsfreier Umschalter transfer switch, make-before-break switch
unterbrechungsfreie Stromversorgung power fail protection
Unterbrechungsfunke break spark, circuit-opening spark
Unterbrechungsklinke break jack
unterbrechungsloser Umschaltekontakt make-before-break contact
unterbrechungslos schalten make before break
Unterbrechungsstöpsel infinity plug (621.317)
unterbrochene Wendel space winding (621.32)
Unterdeterminante minor determinant
Unterdruck diminished pressure, (partial) vacuum
Unterdruckbremse vacuum servo brake
Unterdruckwächter minimum-pressure governor
unterdrücken to suppress
unterdrückter Nullpunkt inferred zero, suppressed zero (621.317)
unterdrückter Träger suppressed carrier
Unterdrückung suppression
Unterdrückung der Kreuzmodulationsprodukte suppression of cross modulation
Unterdrückung der Raumwelle suppression of the sky wave
Unterdrückung von Störsignalen suppression of spurious signals
Unterdrückungsgrad von Versorgungsspannungsänderungen supply voltage rejection
Unterflansch bottom flange, lower flange
Unterflurfeuer flush marker light, flush runwaylight
Unterfrequenz subharmonic, submultible; underfrequency (621.313)
Unterfrequenzschutz underfrequency protection
Untergehen im Rauschen loss in the noise
Untergewicht underweight, short weight
Untergröße undersize
Untergrund underground, subsoil
Untergrundbahn underground, subway, tube (railway)
Untergruppe subassembly, subdivision
unterhalb below, inf. = infra
unterhalb der Resonanz below resonance
unterhalten to maintain
Unterhaltung maintenance, upkeep
Unterhaltungsarbeiten upkeep work, maintenance work (sing.)
Unterhaltungskosten maintenance cost (sing.)
unterheizen to underheat (621.385)
unterirdisch underground, buried (in the ground), subterraneous, subsurface
unterirdisches Kabel burried cable
unterirdisches Rohr underground pipe
unterirdischer Transformator subway-type transformer
unterirdischer Verstärker buried repeater
unterirdisches Werk subsurface work
unterklemmen to secure under terminals (621.3)
Unterkommutierung undercommutation
unterkritisch subcritical (539)
unterkritische Dämpfung underdamping
unterkritisch gedämpft underdamped
unterkritischer Reaktor subcritical reactor
unterkühlen to supercool
Unterlage bolster
mit Unterlage versehen to bolster
Unterlagen information (sing.), (pertinent) data
Unterlagerungsmodelung subcarrier modulation
Unterlagerungstelegraphie subaudio telegraphy
Unterlagstück packing block
Unterlast underload
Unterlastschaltung on-load switching
Unterlastung derating (621.396.67)
Unterlegscheibe (lock)washer
Untermaß undersize
mit Untermaß undersized, below size
Unternormal substandard, secondary standard (621.32)
unternormal subnormal
unternormal schlecht below the mark
unterordnen to subordinate
Unterprogramm subroutine (681.3)
Unterputzinstallation buried wiring, concealed wiring

Unterputzschalter recessed switch (696)
Unterputzschalter flush-mounted switch, recessed switch (696)
unterschätzen to underrate, to underestimate
unterscheidbar discernible
unterscheiden to discern, to differentiate, to discriminate
sich der Art nach unterscheiden to differ in kind
unterscheidend distinctive
Unterschied difference, distinction
Unterschied dem Grade nach difference of degree
Unterschied machen to differentiate
zum Unterschied von distinct from
Unterschirmisolator petticoat insulator
unterschlächtig undershot (621.2)
unterschnitten undercut
unterschreiten to underpass, to be lower than
Unterschriftenleser signature reader
Unterseil belance rope (621.34)
Unterseite underside, lower side
untersetzen to gear down (621.8); to step down (621.3), (621.8)
untersetzter Motor back-geared motor
Untersetzung reduction unit; step-down ratio
Untersetzungsgetriebe reduction gear, back-gearing, step-down gear
Untersetzungsschaltung counting-down device
untersichern to underfuse
Unterspannung low voltage, undervoltage, minimum voltage; low tension, low pressure
Unterspannungsauslösung undervoltage circuit-breaking, undervoltage tripping
Unterspannungsklemme low-voltage terminal
Unterspannungsrelais undervoltage relay, low-voltage relay, minimum-voltage relay
Unterspannungsschutz (phase) undervoltage protection
Unterspannungsschutzsystem undervoltage protection
Unterspannungsspule novolt coil, undervoltage coil (621.316); l.t.w. = low tension winding (621.319)
Unterspannungsverteilung secondary power distribution
Unterspannungswicklung low-tension winding (= l.t.w.)
Unterstandard-Brechnung subrefraction (621.371)
Unterstation substation, distributing centre
Unterstromrelais undercurrent relay
unterstützen to support
Unterstützungspunkt support (point), (lever) fulcrum
untersuchen to test, to check, to examine; to investigate
Untersuchung test, examination; investigation
zur Untersuchung vorliegend on trial
Untersuchungsobjekt test specimen
Untersumme subtotal
untersynchron subsynchronous
Untersystem slave system, subsystem
Unterteil lower part, bottom part, base
unterteilen to subdivide, to sectionalize
unterteilte Dämpferwicklung open amortiseur winding (621.313)
unterteilte Primärwicklung split primary
unterteilte Sammelschienen sectionalized bus bars
unterteilte Tafel sectional panel
Unterteilung subdivision; partition
Untertemperatur deficient temperature
Unterton subharmonic (tone)
Untervermittlung subexchange
Unterwasser... underwater...
Unterwasserantenne submerged aerial
Unterwasserbeleuchtung underwater floodlighting
Unterwasserhorchgerät hydrophone, underwater sound detector
Unterwasserkabel submarine cable, underwater cable
Unterwassermotor submersible motor
Unterwasserschallempfänger hydrophone
Unterwasser-Schallortung underwater sound ranging
Unterwasserschallsender submarine sound projector
Unterwasserschallsignal submarine signal
Unterwasser-Schneidbrenner submersible cutting-torch
Unterwasserverstärker submarine repeater, submerged repeater
Unterwerk substation, switch-cabin (621.3)
Unterwind forced draft, forced draught (621.18)
untrennbar inseparable, non-separable
untrennbar verbunden mit inherent in
untrennbarer Teil integral part
unüberbrückt unbypassed (621.3)
ununterbrochen uninterrupted, permanent
ununterbrochene Entladung permanent discharge (621.353), (621.356)
ununterscheidbar indiscernible, irrecucible
ununterteilbar irresolvable
unveränderlich unvarying, unvariable
Unveränderlichkeit unvariability

unverbrannt unburnt
unverbrennbar incombustible
unverdrillt untwisted
unverdünnt undiluted; neat (54)
unverhältnismäßig disproportionate, out of proportion
unverlierbare Mutter captive nut
unverpackt verladen to load in bulk
unversehrt intact
unverständlich unintelligible
  unverständliches Nebensprechen babble, unintelligible cross talk, inverted cross talk (621.395)
unverwechselbar non-interchangeable
unverzerrt undistorted
  unverzerrte Ausgangsleistung undistorted power output (621.396.62), (621.396.64)
  unverzerrte Sprechleistung undistorted signal power (621.396.62)
  unverzerrte Welle undistorted wave
unverzögert instantaneous
  unverzögerte Auslösung instantaneous trip
  unverzögerte automatische Verstärkungsregelung instananeous automatic gain control
  unverzögerte Sicherung instantaneous fuse (621.3)
unvollkommener Erdschluß partial earth (fault)
unvollständige Lichtstreuung incomplete diffusion
unvollständige Wahl uncomplete dialling (621.395)
unwägbar imponderable
unwirtschaftlich uneconomical, inefficient
Unwucht unbalanced mass, flutter
unzerstörbar indestructible, indelible
unzugänglich inaccessible, uncomeatable
unzulässige Überlastung unsafe load
unzusammenhängend incoherent
unzuverlässig unreliable
U-Profil U-section, channel
Uran uranium
  Uran-Atom uranium atom
  Uranbombe uranium bomb
  Uranbrenner uranium pile
  Uranerz uranium ore
  Uranhexafluorid uranium hexafluoride
  Uran-Isotop uranium isotope
  Uranklumpen uranium lump
  Urankraftwerk uranium power plant, uranium power station
  Uranlagerstätte uranium deposit
  Uranoxid uranium oxide
  Uranpecherz pitchblende
  Uranreaktor uranium reactor, uranium pile
  uranreich rich in uranium
  Uranspaltung uranium fission
  Uranstab uranium bar, uranium rod
  Uranstrahlung uranium rays (pl.)
  Uranverbindung uranium compound
Ureichkreis master telephone, transmission reference system (621.395)
Urgestein primary rock (55)
Urkilogramm standard kilogram (me)
Urlader bootstrap loader, initial loader program (681.3)
Urmaß standard measure
Urmeter standard meter
Urnormal standard
U-Rohr U-tube, syphon (standpipe)
  U-Rohrbarometer syphon barometer
Urspannung impressed voltage, impressed e. m. f.
Ursprung (point of) origin
  Ursprungsdaten source data (681.3)
  Ursprungssprache source language
  Ursprungsvermittlungsstelle originating exchange (621.395)
ursprünglich original
  ursprüngliches Feld parent field
  ursprüngliche Genauigkeit built-in accuracy
U-Träger U-girder, U-section, channel iron
UV ultraviolet (s. auch **Ultraviolett**)
  UV-Dosismesser ultraviolet light meter, UV-dosimeter
  UV-durchlässiges Glas vita glass
  UV-Lampe ultraviolet lamp
  UV-Licht ultraviolet light
  UV-löschbares PROM ultraviolet erasable PROM
  UV-Strahlung ultraviolet rays
  UV-Therapie ultraviolet therapy, actinotherapy
Uviol-Lampe uviol lamp

V

V = Volt V = volt
vagabundieren to stray
  vagabundierender Strom vagabond current, stray (earth) current
Vakuum vacuum
  im Vakuum, unter Vakuum in vacuo
  Vakuumableiter vacuum protector, vacuum arrester
  Vakuumbremse vacuum brake

Vakuumdichtung vacuum seal
Vakuumfaktor gas ratio
Vakuumgüte tube hardness (621.385)
Vakuumimprägnierung vacuum impregnation
Vakuumkolben evacuated envelope
Vakuumlampe vacuum lamp
Vakuummesser vacuum gauge
Vakuumofen vacuum oven
Vakuumpackung vacuum pack
Vakuumpumpe vacuum pump (621.64); exhaust pump, suction pump (621.1)
Vakuumröhre vacuum valve, vacuum tube
Vakuumschalter vacuum switch
Vakuumschlauch vacuum tubing
Vakuum-Thermoelement vacuum thermocouple, vacuum junction
Vakuumtopf vacuum pan
Vakuumwellenlänge free-space wavelength (53)
Valenz valence, valency
Valenzband valence band
Valenzbindung valency bond, valence link
Valenzelektron valence electron, outer-shell electron
Valenzzustand atomic orbit
Vanadium vanadium
van de Graaff-Generator van de Graaff machine
van der Waalsches Gesetz van der Wall's law
V-Antenne V-aerial, V-antenna
Varactor Diode varactor diode
Variation variation
Variationsverfahren variational method, variational calculus
Variometer variometer (mutual) inductometer, variable inductor, continously adjustable inductor
Varistor varistor
VDE siehe **Verband Deutscher** Elektrotechniker
VDE-Vorschriften VDE Standards, VDE Regulations, VDE Rules
V-Doppekette semi-strain insulator
Vektor vector (line)
Vektor der Leistungsdichte Pointing's vector
Vektoranalysis vector(ial) analysis
Vektordarstellung vector representation
Vektorfeld vector field
Vektorfunktion vector function
Vektorgröße vector quantity
Vektorpotentialfeld vector potential field
Vektorprodukt vector product, outer product, cross product
Vektorschreibweise vector notation
Vektorskop vectorscope
Vektorsumme vector sum, vectorial sum
vektoriell vectorial
vektoriell addieren to add vectorially
vektorielle Differenz vectorial difference
vektorielle Größe vector quantity
Ventil valve; rectifier (621.3) (siehe auch Gleichrichter)
Ventil auslösen to trip a valve
Ventilableiter valve-type arrester
Ventilhub stroke of the valve
Ventilröhre valve tube
Ventilwirkung valve effect
Ventilation ventilation, windage (621.313)
Ventilator ventilator, (ventilating) fan, (air) blower
Ventilatorflügel fan blade
Ventilatorgebläse centrifugal fan, (fan) blower
Ventilatorgehäuse fan casing
ventilieren to ventilate
Venturi-Rohr Venturi tube
veralten to become obsolete, to go out of date
veraltete Maschinenanlage obsolete machinery, out-of-date machinery
veränderlich variable, variant
veränderliche Drehzahl variable speed
veränderliches Feld variable field
veränderliche Frequenz variable frequency
veränderliche Gegeninduktivität mutual inductor variometer
veränderliche Induktivität variable inductance, adjustable inductor
veränderlicher Kondensator variable capacitor, adjustable capacitor
veränderliche Steilheit variable mu (621.385)
veränderliches Übersetzungsverhältnis variable ratio
veränderlicher Widerstand variable resistance; variable resistor
Veränderliche variable (51)
Veränderlichkeit variability
verändern to alter, to change
sich verändern to vary
verankern to anchor (629.12); to stay, to guy (621.315)
Verankerung anchorage (629.12); anchor tie, stay(ing) (621.315)
veranschlagen to estimate
verarbeitbare Leistung power handling capacity
verarbeiten to make up, to work up, to handle; to machine
verarmtes Material depleted material (539)
Verarmung depletion (621.382)
Verarmungsbetrieb depletion mode operation
Verarmungs-IG-FET depletion type field-effect transistor

Verarmungs-Isolierschicht-Feldeffekttransistor depletion type field-effect transistor
veraschen to incinerate, to calcine
Veraschung incineration, calcination
verästelt tree-like, forked
verästelter Blitz forked lightning
Verband bandage (61); binding (624); association, federation (33)
Verbandkasten ambulance box, first-aid box
Verband Deutscher Elektrotechniker Association of German Electrical Engineers
verbessern to improve, to upgrade; to correct, to rectify
Verbesserung improvement, advance; correction
verbesserungsfähig perfectible, improvable
verbeulen to bruise
verbiegen to bend, to deform, to force out of straight
sich verbiegen to bend, to deflect
verbinden to join, to link; to connect, to interconnect (621.3); to tie together
verbinden mit to put through to, to connect with (621.395)
Verbinder tension joint (621.315.1); splicing ear (621); bond (621.33)
Verbinderprüfung joint testing; bond test(ing)
Verbindung junction, connection (621.3); compound (54); interconnection (621.39)
Verbindung aufbauen to build up a connection, to set up a connection (621.395)
Verbindung eingehen mit to combine with (54)
Verbindung haben to intercommunicate (629.13); to be in contact (**mit:** with)
Verbindung lösen to unmake; to disconnect (621.39)
in Verbindung mit in combination with, in connection with
in Verbindung stehen to communicate, to intercommunicate
Verbindungsaufbau mit direkter Wählereinstellung step-by-step switching
Verbindungsaufbau mit Rücksprung rerouting
Verbindungsaufbau, nicht schritthaltend common-control switching
Verbindungsaufbau, schritthaltend stage-by-stage switching
Verbindungsdose junction box, joint box, access fitting (621.315.6)
Verbindungsglied link
Verbindungskasten junction box (621.395)
Verbindungslasche connection plate
Verbindungsleitung connecting wire, interconnection, connection (line), tie line
Verbindungslinie line of communication (621.39); connecting line (51), (621)
verbindungsprogrammierte Regelung wired-program control
Verbindungspunkt junction (point)
Verbindungsschiene connection bar (621.13)
Verbindungsschnur connection cord, answering cord, calling cord (621.395)
Verbindungsstecker bridging connector
Verbindungsstelle junction point, juncture
Verbindungsstöpsel connection plug
Verbindungsstück jointing fitting, linkage
Verbindungswärme heat of formation of a compound
verblassen to fade
verbleien to lead (-cover)
verbleit lead coated
verblocken to interlock
Verblockung interlocking
Verblockungseinrichtung interlocking device
verborgen latent
verborgener Fehler latent defect
verbotenes Band forbidden band, energy gap (621.382)
Verbrauch consumption
verbrauchen to consume, to spend
verbrauchtes Bad spent solution
verbrauchter Elektrolyt foul electrolyte (621.357)
verbrauchte Luft vitiated air
verbrauchte Säure spent acid
Verbraucher (power) consumer, user, customer; load (621.31)
Verbraucheranlage consumer's installation, utilization equipment
Verbraucherseite load end, output end, demand side; sink (621.392)
verbreiten to propagate, to spread; to broadcast
verbrennbar combustible
verbrennen to burn
Verbrennung burn(ing), combustion
Verbrennung unterhalten to support combustion
Verbrennungskammer combustor, combustion chamber
Verbrennungskraftmaschine internal combustion engine
Verbrennungsraum combustion chamber
Verbund-... compound (621.313.2); interconnected (621.316)
Verbundbetrieb interconnection of power systems (621.316)
Verbundgenerator compound(ed) generator
Verbundlokomotive compound locomotive

Verbundmaschine compound engine (621.1); compound machine (621.3)
Verbundnetz interconnected system, grid system
Verbundröhre multi-unit tube, multi-system valve, duplex tube
Verbundsystem interconnected system, grid system (621.3)
Verbundwicklung compound winding (621.313.2)
verchromen to chromium-plate
Verchromung chromium-plating
verdämmen to tamp (62)
Verdämmung tamper
Verdämmungsmaterial tamping (material)
verdampfen to vaporize, to evaporate
verdampfend evaporative
Verdampfer vaporizer, vaporizing plant, evaporator (621.1)
Verdampferrohr generating tube (621.1)
Verdampferspeisepumpe evaporator feed pump
Verdampfung vaporization, evaporation
Verdampfungsgeschwindigkeit rate of evaporation
Verdampfungskühlung vapour cooling
Verdampfungsleistung evaporative power (621.1)
Verdampfungspunkt vaporizing point, evaporation point
Verdampfungstemperatur vaporizing temperature
Verdampfungswärme heat (equivalent) of vaporization
Verdampfungswirkungsgrad vaporization efficiency
Verdampfungszahl steam rising coefficient
verdecken to mask (534.1), (621.396); to obscure, to swamp (621.396)
verdeckte Nummerierung closed numbering (621.395.3)
Verdeckung der Signale durch Rauschen masking of the signal by noise
verderben to spoil, to vitiate, to deteriorate
verdichten to compress; to compact, to swage; to pack (681.3)
sich verdichten to thicken, to condense
Verdichtung compression; thickening, condensation
Verdichtungsgrad compression ratio
verdoppeln to double
Verdoppler doubler
Verdopplerschaltung doubler circuit, voltage-doubling circuit
Verdopplung doubling
Verdopplungsstufe doubling stage
verdrahten to wire (621.3)
Verdrahtung (circuit) wiring; cabling (621.39)
Verdrahtungsplan wiring diagram, circuit diagram
Verdrahtungswiderstand wiring resistance
verdrallen to twist (around)
verdrängen to displace
Verdrängung displacement (629.12), (53)
verdrehen to twist
Verdrehung twist, torsion, torsional strain; angular displacement (53)
Verdrehungsbeanspruchung torsional strain
Verdrehungsfestigkeit torsional strength
Verdrehungsmeßgerät torsion meter
Verdrehungsprüfung torsion test
Verdrehungswinkel angle of twist, distortion angle
Verdreifacher tripler
Verdreifacherschaltung tripler circuit
verdrillen to twist (up), to intertwist, to twin together (621.3); to transpose (621.315.1), (621.392)
verdrillte Doppelleitung twisted pair (621.3)
verdrillte Litze litz, twisted flex, twisted flexible cord
Verdrillung transposition (621.315.1), (621.392); twisting
Verdrillungsisolator transposition insulator
Verdrillungsschema pole diagram
verdrosselter Anschluß impedance bond
verdübeln to dowel
verdunkeln to dim, to obscure; to eclipse; to black-out (35)
Verdunkelung dimming, obscuration; eclipse; blackout
Verdunkelungshebel dimmer lever (621.3)
Verdunkelungsregler dimmer
Verdunkelungstransformator dimmer transformer
Verdunkelungswiderstand dimming resistor, rheostat dimmer
verdünnen to thin; to dilute; to attenuate; to rarefy
verdünntes Gas rarefied gas
verdünnte Lösung dilute solution
verdünnte Luft attenuated air
verdünnte Salzsäure dilute hydrochloric acid
Verdünnungsmittel diluent
verdunsten to vaporize, to evaporate
Verdunster vaporizer
Verdunstung vaporization, evaporation

Verdunstungswärme heat (equivalent) of vaporization
veredeln to improve, to refine, to enriche, to process
vereinheitlichen to unify, to standardize
vereinigen to join, to bond
vereisen to ice, to congeal
vereist ice-covered
Vereisung icing
verengen, (sich) to contract, to constrict
verengen to narrow
Verengung narrowing, necking
Veresterung esterification
Verfahren process, method, technique, procedure
Verfahren ausarbeiten to evolve a process; to develop a method
Verfahrenstechnik process industries (pl.), design engineering
verfestigen to work-harden
sich verfestigen to stiffen
verfinstern to darken, to obscure, to eclipse
Verflacherschaltung graduator (621.394)
verflechten to plait, to interlace
verflüchtigen to volatilize
verflüssigen to liquefy; to condense
verflüssigtes Gas liquefied gas
Verflüssigung liquefaction; condensation
verfolgen to track
verfolgen nach Zielrückstrahlung skin tracking
Verfolgungskurve curve of pursuit
Verfolgungsradar tracking radar
verformen to strain; to deform, to distort, to contort
Verfügbarkeit availability
Verfügbarkeitszeitraum up-time
verfugen to joint (693)
Verfügungsfrequenz assigned frequency
vergällen to denature, to methylate
vergären to ferment
vergasen to gasify
vergeuden to waste
Vergießen potting
vergießen to shed, to spill; to seal in
vergossene Batterie sealed-in battery
Vergußmasse (potting) compound
vergiften to poison, to intoxicate
Vergiftung poisoning, intoxication (54), (61), (539)
Vergiftung eines Katalysators poisoning of a catalyst
Vergiftung eines Kernreaktors poisoning of an atomic pile
vergipsen to plaster
verglasen to vitrify; to glaze
Vergleich comparison, correlation
Vergleich des Zeichenverlaufs signal comparison
Vergleichsgröße comparison quantity
als Vergleichsgröße dienend fiducial
Vergleichslogik servo logic (621.317)
Vergleichsmessung comparative measurement
Vergleichsnormal standard of comparison
Vergleichsschaltung comparator circuit (621.37); differential connection (621.316.9)
Vergleichssignal reference signal
Vergleichsspannung comparison voltage, reference voltage
Vergleichswiderstand comparison resistance
Vergleichszähler substandard meter (621.317)
Vergleichszweige proportional arms (621.3)
vergleichen mit to compare with, to compare against, to compare to
vergleichende Gegenüberstellung comparison table
vergolden to gild
Vergoldung gilding; gold plating (621.35)
vergrößern to expand, to augment; to magnify (535); to enlarge (77), (535)
vergrößerter Bildausschnitt expanded scope
vergrößerter Maßstab enlarged scale
Vergrößerung augmentation; magnification (535); enlargement (77), (535)
Vergrößerungsapparat enlarger (77); magnifier (535)
Vergrößerungsglas magnifier, magnifying glass
Verguß seal, sealing
Vergußmasse sealing compound, insulating compound (621.3)
Vergußmasse auf Ölbasis oilbase type compound
vergüten to heat-treat (Stahl); to age harden (Leichtmetall)
Verhalten (gegenüber äußeren Einflüssen) response
Verhalten gegenüber Hochfrequenz high frequency response
Verhalten gegen Stoßlast impulse response
Verhältnis ratio, proportion; rate
Verhältnis Breite zu Höhe width to height ratio
Verhältnis Länge zu Dicke width to thickness ratio
verhältnisgleich zu proportional to
Verhältnisgrößen ratio of two quantities
verhütten to smelt
verjüngen, sich to taper (621)
Verkauf sale

Verkaufsautomat vending machine
Verkaufsbüro sales office
Verkaufspreis selling price
verkaufen to sell
Verkehr traffic (625); intercommunication (621.39)
in Verkehr stehen to intercommunicate (**mit:** with)
Verkehrsampel traffic control light, traffic light
Verkehrsart type of operation (621.39)
Verkehrsausgleich traffic balancing (621.39)
Verkehrsbelastung traffic load
Verkehrsdichte traffic density
Verkehrsdisziplin circuit discipline (621.39)
Verkehrseinrichtungen transport facilities
Verkehrsgüte grade of service, traffic quality (621.395)
Verkehrsleistung eines Leitungsbündels traffic capacity
Verkehrsmenge traffic volume, traffic amount (621.395)
Verkehrsproblem traffic problem
Verkehrsregelung traffic control
verkehrsreiche Zeit rush hour (sing.)
verkehrsschwache Zeit slack hours (pl.)
Verkehrssignal traffic (control) signal
Verkehrssignalanlage traffic control equipment, traffic signalling equipment, semaphoric traffic control system
Verkehrsstauung traffic jam
Verkehrsstörung traffic interruption
Verkehrswert traffic intensity, traffic flow (621.395)
Verkehrswesen traffic engineering
verkeilen to wedge, to scotch
verketten to interlock (621); to interlink (621.3)
verkettet sein mit to link with
verkettete Spannung mesh voltage, delta voltage, voltage between phases
Verkettung daisy chain (621.394.5), (681.3)
verkitten to seal, to cement together (535)
verkleiden to cover, to case
verkleinern to reduce, to scale down (51)
verkleinerter Maßstab reduced scale
Verkleinerung reduction
Verkleinerungsarbeit reduction work
Verklinkung release gear (621.316.5)
verknoten to knot, to intertwine
verknüpfen mit to link up with
verknüpfter Scheinwiderstand mutual impedance
Verknüpfung relationship; logical operation (621.374)
Verknüpfungsglied logic element, logical element (621.374)
Verknüpfungsschaltung logical circuit
verkohlen to char, to carbonize
verkoken to coke, to carbonize
verkorken to cork, to stopper
verkupfern to copper-plate
verkupferte Kohle coppered carbon
verkürzen to shorten, to foreshorten
verkürzte Lebensdauerprüfung accelerated life test
Verkürzung foreshortening (621.396.67)
verladen to load, to ship, to entrain
Verladerampe loading ramp
Verladestraße loading road
verlagern to displace
verlagern einer Verbindung to rearrange (621.395)
verlängern to prolong, to enlongate, to produce (51)
verlängertes Rotationsellipsoid oblong spheroid, prolate spheroid
Verlängerung prolongation; elongation, lengthening; extension
Verlängerungsleitung extension line; artificial line (621.372.5)
Verlängerungsrohr lengthening pipe, extension pipe
Verlängerungsschnur extension cord, extension flex
Verlängerungsstück lengthening piece, lengthening bar
verlangsamen to slow down
Verlangsamer moderator (54), (539)
verlaufen to run, to wander (621.395)
verlegen to lay, to wire, to install; to misplace
verlegtes Kabel laid cable
Verlegung laying, wiring, installation
Verlegung über Putz surface wiring
Verlegung unter Putz buried wiring, concealed wiring
Verlegung in Rohr conduit wiring
Verlegungsart system of wiring
Verlegungsschiff cable ship
verleimen to glue
verlitzen to strand
verlöten to seal, to solder together; to braze together
Verlust loss; drop; loss probability
Verlust durch Isolationsfehler leakage (loss)
verlustarm low-loss
verlustarmer Isolator low-loss insulator (621.396)
verlustarme Leitung low-loss line

verlustbehaftet wasteful, dissipative, imperfect
verlustbehaftetes Dielektrikum imperfect dielectric
verlustbehafteter Kondensator imperfect capacitor
Verlustbestimmung aus Kühlluftmenge und -erwärmung air-calorimeter test (621.313)
Verlustenergie wasted energy, energy loss
Verlustfaktor loss factor; dielectric loss factor, dielectric loss index (621.316.5)
Verlustfaktormesser loss factor meter
Verlustleistung power dissipation (621.385); power loss (621.3); leakage power (621.39)
verlustlos, verlustfrei loss-free, no-loss, dissipationless, wattless (621.3)
verlustloses Dielektrikum no-loss dielectric
verlustlose Induktivität pure inductance
verlustloser Isolator no-loss insulator
velustlose Kapazität pure capacitance
verlustloser Kondensator no-loss condenser, perfect condenser
verlustlose Leitung line without losses
verlustloses Nachbild basic network (621.395)
Verlustmessung wattmeter method (621.318)
verlustreich wasteful
Verluststrom leakage current
Verlustverkehr lost traffic (621.395)
Verlustwahrscheinlichkeit loss probability (621.395)
Verlustwiderstand loss resistance, equivalent series resistance
Verlustwinkel (dielectric) loss angle, (dielectric) phase difference
Verlustziffer figure of loss, grade of service (621.396); specific core loss (621.314)
vermaschtes Netz mesh network, meshed network
vermaschter Regelkreis multi-loop control system
Vermaschung (inter)meshing
vermauern to wall up
vermessen to survey, to measure
Vermessung survey(ing)
Vermessungsingenieur surveyor
Vermessungsinstrumente surveying instruments
vermindern to diminish, to lessen, to reduce
vermindern um to reduce by
verminderter Druck diminished pressure
verminderter Träger reduced carrier
Verminderung diminution, reduction, weakening
Verminderung der Abstimmschärfe flattening of the tuning
Verminderung der Lastspitze durch Nichtzusammenfallen load diversity
vermischen to mix (up), to blend
sich vermischen to intermingle
vermitteln to connect, to put through (621.395)
Vermittlung exchange, central office (621.395)
Vermittlung eines Gesprächs connection of a call
Vermittlungsamt exchange, central office, switching center
Vermittlungsapparat operator's telephone set
Vermittlungsbeamter/-in operator (621.395)
Vermittlungsgüte switching quality
Vermittlungsklinke connecting jack (621.394)
Vermittlungspult attendant's cabinet, operators board, operators position
Vermittlungssaal operating room
Vermittlungsschnur connecting cord
Vermittlungsschrank telephone switchboard
Vermittlungsschrank eines Handamtes manual switchboard
Vermittlungsstelle exchange (equipment), office, switching center
Vermittlungssystem switching system
Vermittlungstaste talking key
vermittlungstechnische Information control information
vermittlungstechnische Peripherie peripheral units
vermuffen to joint (621.3)
Vermuffung sleeve joint, (cable) jointing (621.315.2)
vernebeln to fog
vernichten to annihilate (539)
Vernichtung annihilation (539)
Vernichtungsstrahlung annihilation radiation
vernickeln to nickel-plate
Vernickelung nickel-plating
verpacken to pack up, to wrap up; to case; to crate; to box
Verpackung packing, wrapping; casing; crating, boxing
Verpackungskosten packing expenses
Verpackungsmaterial packing material
verpfuschen to make a mug of
verquer cross
verquetschen to squeeze out of concentricity, to burr, to crumple
verquetscht skewed, lopsided
verreiben to rub down
verriegeln to lock, to interlock, to latch, to bar
verriegelbarer Schalter lockable switch
Verriegelung locking mechanism, interlocking
Verriegelungseinrichtung interlocking device

Verriegelungshebel locking lever
Verriegelungskeil locking wedge
Verriegelungskontakt interlocking contact
Verriegelungsmechanismus interlocking mechanism
Verriegelungsrelais interlocking relay
Verriegelungsschalter interlocking switch
Verriegelungsschutz interlocking contactor
Verriegelungsspannung gate voltage
Verriegelungsstellung lock position
Verriegelungsvorgang intelocking
Verringerung reduction; decrease, diminution
Verringerung der Abmessungen gain in compactness
verrippen to rib
verrosten to rust
verrotten to rot
verrutschen to lurch
versagen to fail
Versager failure
Versand shipping, shipment, forwarding
Versandkappe shipping seal (621.315)
Versandvorschrift shipping instruction, forwarding instruction
Versatzbetrieb offset (621.396)
verschachteln to interleave; to interlace; to nest (681.3)
verschachteltes Arbeiten interlace operation
verschachteltes Programm nested program
verschalen to board, to encase, to enclose
verschalte Schalttafel enclosed switchboard
verschärfen to sharpen
Verschärfung sharpening (621.396)
verschieben to displace (531); to shift (537) (681.3); to relocate (681.3)
um 90° verschoben 90 ° out of phase, in quadrature (621.3)
Verschieben (der Bildschirmanzeige) scrolling
Verschiebung displacement, shift, shifting (537)
90°-Verschiebung phase quadrature (621.3)
Verschiebung in der Phase shift in phase
Verschiebungsdichte dielectric flux density
Verschiebungsfluß displacement flux, dielectric flux
Verschiebungsoperator transfer operator (51)
Verschiebungsspannung lumped voltage, cut-off bias potential, cut-off grid potential (621.385)
Verschiebungsstrom displacement current
Verschiebungsvolumen volume displacement
Verschiebungszahl permittivity, inductive capacity
verschieden different (**von**: from, **hinsichtlich**: in)
verschiedene Einzelteile sundries
verschiedenfarbig differently coloured
verschlammen to sludge, to silt
verschlechtern to deteriorate
Verschlechterung deterioration
verschleiern, sich to fog (77)
Verschleiß wear, abrasion
verschleißfest wear-resisting, abrasion-resistant
Verschleißfestigkeit wear reistivity, resistance to wear, abrasion resistance
Verschleißteil high-mortality part, part subject to wear
verschleißen to wear (out)
verschließen to shut, to lock
Verschluß lock, locking (device), closure; shutter (77)
unter Verschluß locked-in
Verschlußdeckel cover plate
Verschlußeinrichtung locking device
Verschlußkappe sealing cap, cup plug
Verschlußkeil keypiece (621.313); locking key
Verschlußklappe shutter
Verschlußpflock (für Luftloch) vent peg
Verschlußpfropfen stopper
Verschlußschalter locking switch, asylum switch
Verschlußschraube locking screw
Verschlußstöpsel inspection plug (621.355)
Verschlußtafel interlocking chart (625.1)
verschlüsseln to cipher, to code, to encode, to encrypt
verschlüsselter Sprechverkehr ciphony
Verschlüsselungseinrichtung, Verschlüsselungsmaschine coder, cipher equipment, cipher machine
verschmelzen to melt (together); to fuse (together), to merge, to amalgamate
Verschmelzung fusion, fusing; merger, amalgamation
verschmieren to smear, to blur
verschmiert smeared-out
verschmutzen to contaminate
verschränken to interlace; to skew (621.313); to transpose
verschränkte Stabwicklung cable-and-bar winding
verschränkte Wicklung twisted winding, transposed winding
verschrauben to screw, to bolt
Verschraubung screw connection, screw joint
verschwelen to carbonize
Verschwelung low-temperature carbonization
verschwinden to disappear, to fade, to vanish; to drop out of sight

verschwommen nebulous, diffuse
verschwommenes Bild diffuse image
verseifen to saponify
verseilen to strand
verseilter Draht stranded wire
verseiltes Kabel stranded cable, rope-lay cable
Verseilung twisting, stranding
versengen to singe
versenken to sink, to countersink, to bury
versenkt C'sk = countersunk
versenkter Einbau flush mounting
versenkter Schalter sunk switch (696.6)
Versenkung sinking
versetzen to offset, to stagger; to displace, to shift
versetzt anordnen to stagger
versetzte Bürsten staggered brushes (621.313)
versetzte Impulsfolgefrequenz staggered pulse repetition frequency
versetzter Raster interleared pattern
mit versetzter Achse out of line
Versetzung dislocation (621.382)
versickern to ooze away, to sink in
versiegeln to seal up
versilbern to silver
versilbert silver-plated
versilberter Draht silver-coated wire
versilberter Leiter silver surfaced conductor
versorgen mit to supply with
Versorgung supply; coverage (654.19)
Versorgung aus dem Netz mains supply
Versorgungsbereich service aera
Versorgungsbetrieb utility service, public utility
Versorgungsgebiet supply region, supply district
Versorgungskabel service cable
Versorgungsleitung supply line, distribution trunk line (621.3); pipeline
Versorgungslücke coverage gap
Versorgungsnetz mains, supply (mains)
Versorgungsspannung supply voltage
Versorgungssystem power supply system
Versorgungsunternehmen supply authority, utility (company); power supply company
verspannen to brace, to stay, to guy
Verspannung (rope) bracing
Verspannungsdraht guy wire, stay wire, bracing wire
Verspannungstau stay
versprengen to sprinkle
verspröden to embrittle
Verständigung intercommunication (621.395), (621.6)
Verständigungsprobe talking test (621.395)
verständlich understandable, intelligible (621.395)
verständliches Nebensprechen intelligible crosstalk, uninverted crosstalk (US)
Verständlichkeit articulation, intelligibility, audibility
Verständlichkeitsversuch articulation test, listening test, intelligibility test
verstärken to amplify, to magnify, to boost (621.39); to reinforce, to strengthen (624), (621); to intensify (67)
Verstärker amplifier (621.39); telephone repeater (621.395)
Verstärker steuern to drive an amplifier
Verstärker verarbeitet eine Spannung von ... amplifier handles a voltage of...
Verstärkerabstand repeater spacing
Verstärkeramt repeater station
Verstärkerdurchlaßbereich amplifier frequency response
Verstärkerfeld repeater section
Verstärkerfolie (salt) intensifying screen, (salt) intensifying leaf (621.386)
Verstärkergestell repeater bay
Verstärkermaschine rotary amplifier, rotating regulator
Verstärkerröhre amplifier tube
Verstärkersaal repeater room, amplifier room
Verstärkerstation repeater station
Verstärkerstufe amplification stage
Verstärkerwicklung amplifying winding
Verstärkung amplification, magnification, gain (621.396); intensification (77); reinforcement (624); transmission gain (621.395)
Verstärkung der Stufe gain of the stage
Verstärkung je Stufe gain per stage
Verstärkungsabfall loss of gain
Verstärkungseinlage reinforcement ply (624)
Verstärkungseinstellung gain setting
Verstärkungsfaktor amplification factor, amplification constant (621.385), (621.396); speed factor (539)
Verstärkungskurve gain characteristic, amplification curve
Verstärkungsmaß measure of amplification, gain
Verstärkungsregelung gain control, gain controlling
Verstärkungsrippe reinforcing rib
Verstärkungssicherheit gain margin, amplification margin (621.375)
Verstärkungsüberhöhung peaking
Verstärkungsvermögen amplifying ability

Verstärkungswarze strengthening boss
verstäuben to sputter
versteifen to stiffen; to strut
Versteifung stiffening, stiffener; strut (624)
Versteifungsplatte gusset plate
Versteifungsrippe rib stiffener
versteilen to peak, to steepen
Versteilerstufe peaker (621.397)
verstellbar adjustable, controllable
verstellbares Endmaß adjustable parallel blocks (pl.)
verstellbarer Schraubenschlüssel monkey (wrench)
verstellbare Winkellehre slide level
verstellen to adjust, to shift
Verstellmotor correcting motor, adjusting motor
Verstellschraube setting screw
Verstellstativ adjustable stand
verstiften to cotter-pin
verstimmen to mistune, to de-tune; to misalign
Verstimmungsmeßverfahren detuning measuring method
verstreben to brace, to strut, to stay
Verstrebung brace, strut, strutting, stay
verstreuen to scatter
Versuch test, experiment
Versuche anstellen to conduct experiments
Versuche durchführen to carry out tests
Versuchsanlage model plant, pilot plant
Versuchsaufbau breadbord
Versuchsbedingungen conditions of experiment, test conditions
Versuchsdauer time rating (621.313)
Versuchsergebnis test result
Versuchsgruppe test team
Versuchslaboratorium research laboratory
Versuchsmodell test model
Versuchsreaktor experimental reactor
Versuchsreihe series of experiments, test series
versuchsweise tentative experimental; on approval
Versuchszweck experimental purpose; test objective
verstümmeln to mutilate
vertauschen to exchange; to interchange
Vertauschung interchange, exchange
Vertauschung der Klemmenanschlüsse interchange of leads of the terminals
verteilen to distribute
verteilte Induktivität distributed inductance
verteilte Kapazität distributed capacitance
verteilte Wicklung distributed winding
Verteiler distributor, distribution frame, cable-terminal box, splitter
Verteilerdose, Verteilerkasten distributing box, distribution box, terminal box, junction box, link box
Verteilergestell distribution frame (621.395)
Verteilersammelschienen submains, distribution mains
Verteilerschalter section switch
Verteilerschalttafel distribution board
Verteilerstation distribution centre; load center (US)
Verteilerstreifen branch strip
Verteilertafel distribution panel, branch switchboard
Verteilung distribution
Verteilungsfunktion distributive function
Verteilungsleitung distribution feeder, subfeeder
Verteilungsnetz distribution network, distributing system
Verteilungspunkt subcentre, distribution point
vertiefen to deepen (621)
vertieft sunk
Vertiefung deepening
Vertiefung in einem Werkstück sinking in a workpiece
vertikal vertical
vertikale Abtastung vertical sweep
vertikale Antennenzeile vertical stacked array
vertikaler Bildfang vertical hold control
vertikale Bündelung vertical directivity
vertikale Polarisation vertical polarization
vertikale Projektion vertical projection
vertikaler Strich vertical line, down rule
vertikale Verschiebung vertical displacement
Vertikale vertical, plumb fine
Vertikalablenkung field sweep, vertical deflection (621.397)
Vertikal-Ablenktransformator vertical-scan transformer
Vertikalantenne Marconi antenna, vertical antenna
Vertikal-Austastlücke field-blanking interval
Vertikalbildlageregler vertical control (621.397)
Vertikaldiagramm vertical polar diagram (621.396.67)
Vertikalfräsmaschine vertical milling machine
Vertikalfrequenz field frequency, frame frequency (US)
Vertikalfrequenzteiler field divider
Vertikalintensität vertical intensity (538.7)
Vertikalrücklauf frame fly back

Vertikalschnitt vertical section
Vertikalsynchronsignal field synchronizing signal, field syne pulse, vertical synchronizing pulses
Vertragsinstallateur installation contractor, wiring contractor
verunreinigen to contaminate
Verunreinigung impurity, contamination
Verunreinigung der Luft atmospheric(al) pollution
Verunreinigungselement impurity element (621.382)
verursachen to cause, to evoke
vervielfachen to multiply
Vervielfacher (electron) multiplier (621.385)
Vervielfacher Photozelle photo-multiplier cell
Vervielfältigungsapparat duplicator, copying machine
vervierfachen to quadruple
vervollständigen to complete
Verwackeln jitter (621.397)
verwalten to administer; to manage; to supervise
Verwaltung administration, management
verwandeln in to convert (in)to
verwandte Wissenschaften allied sciences
verweben to interweave
Verweilzeit lingering period (539)
verwerfen to reject, to refuse
verwerfen, sich to warp
verwickeln to entangle
verwickelt complex, intricate, complicated
verwirklichen to realize, to accomplish
Verwirrungszone confusion region (621.397); mush area (654.19)
verwittern to weather, to disintegrate, to decompose
verwittert weather-worn
Verwitterung disintegration (process), discomposition, weathering
Verwitterungsprüfung weathering test
verworren diffuse; entangled
verwürfeln to hat, to scramble
Verwürflungsgerät scrambler
Verzeichnis list, index, catalogue
Verzeichnis der Abkürzungen key of abbreviations
in ein Verzeichnis aufnehmen to schedule; to enter in(to) a list
Verzeichnung distortion (621.397)
verzerren to distort, to contort
verzerrte Kurvenform distorted waveshape
verzerrte Welle distorted wave
Verzerrer, Verzerrerschaltung distortion circuit
Verzerrung distortion, contortion
Verzerrung der Kurvenform wave distortion
Verzerrungsfaktor distortion factor
verzerrungsfrei distortionless
verzerrungsfreie Leitung distortionless line (621.395)
Verzerrungsfreiheit accuracy of signal repetition, signal fidelity, absence of distortion
Verzerrungsleistung distortion power, harmonic power
Verzerrungsmesser distortion measuring set
verziehen, sich to warp, to distort
verzinken to zinc, to galvanize
verzinkter Draht galvanized wire
verzinktes Eisenblech galvanized sheet iron
verzinnen to tin(-plate)
vezinnter Kupferdraht tinned copper wire
Verzinnung tin coating, tin plating
Verzögerer moderator (539)
verzögern to decelerate (51); to retard (51); (621.3); to delay (621.3); to moderate (539)
verzögertes Ansprechen delay action
verzögerte Auslösung time-lag circuit breaking, delayed tripping, time-lag tripping
verzögerte Bremsung delayed application of the brake (625)
verzögerte Regelspannung delayed bias (621.396)
vezögerte Schwundregelung d. a. v. c. = delayed automatic volume control
verzögerter Unterspannungsschutz time undervoltage protection
verzögerte Vorspannung delay bias (621.396)
verzögerte waagerechte Zeitablenkung R-scan (621.385)
Verzögerung delay, deceleration, lag
verzögerungsfrei instantaneous
Verzögerungsleitung delay line (621.394)
Verzögerungsmaß delay constant
Verzögerungsrelais time-delay relay
Verzögerungsschaltung delay circuit, time-lag circuit, retardation network
Verzögerungsspannung delay bias
Verzögerungsstellung retard position
Verzögerungszeit time delay, delay time; recovery time (621.382)
Verzonen zoning
Verzoner zoner (621.395.3)
verzweigen, sich to branch; to furcate, to bifurcate, to take two paths (621.3)
Verzweigung junction, branching, bifurcation; branch (681.3)
Verzweigungsbefehl transfer command, branch instruction

Verzweigungskabel branching cable
Verzweigungsmuffe multiple joint, dividing box, bifurcating box, trifurcating box (621.315.2)
Verzweigungspunkt branching point, junction (point)
einem Verzweigungspunkt zufließen to flow into a junction
verzwillingen to twin (54)
V-Form V-formation (629.13)
V-Form-Winkel dihedral angle
v. H. = vom Hundert p. c. = per cent
Vibration vibration
Vibrationsgalvanometer vibration galvanometer
Vibrationsmikrophonie vibration microphonism
vibrieren to vibrate; to quaver (534)
vibrierend vibrant; quavering
Vickers-Härtezahl V. P. N. = Vickers pyramid number
Video . . . video
Videoaufzeichnung video recording
Videobandgerät video tape recorder = VTR
Videofrequenz video frequency
Videokopf video head
Video-Koppelpunkt video crosspoint connector
Video-Koppelsystem video crossbar system
Video-Meßgenerator video frequency generator, video test signal generator
Video-Recorder VTR = video tape recorder
Video-Regel- und Regenerierverstärker video control and regenerating amplifier
Video-Signal composite picture signal, video signal
Videoverstärker video amplifier
Vidikon vidicon (camera tube)
Vidikon-Filmabtaster vidicon telecine
vieladrig multi-cable, multi-core
vieladriges Kabel multi-core cable, multiconductor cable
vieladrige Leitung multi-core line
Vieldrahtantenne multiple-wire aerial, multi-wire aerial
Vieleck polygon
vieleckig polygonal, many-sided
Vielfach- . . ., vielfach multiple
Vielfachdose multiple jack, multicontact jack
Vielfachfeld multiple switchboard
Vielfachfunkenstrecke multigap arrester
Vielfachinstrument multirange instrument, multimeter, circuit analyzer
Vielfachkabel bank cable (621.395)
Vielfachklinke s. **Vielfachdose**
Vielfachmeßgerät multirange meter, multimeter
Vielfach-Pralltechnik multiple bounce technique
Vielfachschaltung multiple connection, subscriber's multiple (621.39)
Vielfachsteckdose multi-contact socket
Vielfachsystem multiple system (621.395)
Vielfachverbindungen bank wires, multiple connection (621.395)
Vielfachzugriff multiple access
vielfältig complex, multiple
vielfarbig many-coloured, polychromatic
Vielflächner polyhedron
Vielganggetriebe multichange gear
Vielkammermagnetron multicavity magnetron
Vielkanaltelefonie multichannel telephony
vielpaariges Kabel multiway cable
vielphasig polyphase
vielpolig multicontact, multipole
vielpoliger Schalter multipole switch
vielpolige Steckdose multicontact socket
vielpoliger Stecker multicontact plug, multipin plug
Vielschicht . . . multilayer
Vielschichtkondensator multilayer capacitor
Vielschichtleiterplatte multilayer (printed) board
Vielschlitz-Magnetron multisegment magnetron
vielseitig many-sided; versatile
vielseitige Verwendbarkeit versatility
vielstufig multistage
vielstufige Turbine multistage turbine
vielstufiger Verstärker multistage amplifier
vieradriges Kabel four-core cable
Vierdraht . . . four-wire
Vierdrahtschaltung four-wire circuit
Vierdrahtverstärker four-wire system amplifier, four-wire repeater
Viereck quadrangle, tetragon
Viereckschaltung O network
Viereckwähler Strowger selector
viereckig square, quadrangular
viereckiger Querschnitt square section
Vierelektrodenröhre four-electrode valve
Vierer quad (621.394), (62.5)
Viererbündel quadruble bundle, quadruplex-bundle conductor
Viererkabel quad cable, quadded cable
Vierereitung phantom circuit
vierfach quadruple, quad
Vierfach-Operationsverstärker quad-op-amp

vierfach polumschaltbarer Motor four-speed pole-changing motor
Vierflächner tetrahedron
Vierkanal-Duplex-Telegraphie four channel duplex telegraph system
Vierkanal-Übertragung four channel transmission; quadrophonics (654.19)
Vierkant-... square-section
Vierkantdraht square-section wire
Vierkantfeder square-section spring
Vierkantstange square bar, rectangular bar
Vierkantwelle square shaft
Vierkursfunkfeuer four-course beacon
Vierkursfunkfeuer mit Sicht- und Höranzeige visual-aural radio range
Vierleiterkabel four-core cable
Vierleiternetz for-wire systeme, secondary neutral grid
Vierphasenanlage, Vierphasensystem four-phase system
Vierpol quadripole, two-port network, four-terminal network
Vierpol in H-Schaltung H netword
Vierpol in L-Schaltung L network
Vierpol in Polygonschaltung O network
Vierpol in P-Schaltung P network
Vierpol in T-Schaltung T network
Vierpoldämpfung image attenuation
Vierpoldämpfungsmaß image-attenuation coefficient
Vierpolersatzschaltung four-pole equivalent circuit
Vierpolröhre mit Pentodeneigenschaften beam (power) tube
Vierpolübertragungsmaß image transfer coefficient, image transfer constant (US)
Vierpolwinkelmaß image-phase change coefficient, image-phase constant (US)
vierpolig quadripolar, four-pole
vierpolige Maschine four-pole machine
Vierschichtgleichrichter N-P-N-P device
vierseitig four sided, quadrilateral
Vierstellenzählwerk standard register (621.317)
vierstellig four-place
Viertakter, Viertaktmotor four-stroke engine
Viertonbildtelegraphie Alexanderson system of phototelegraphy
Viertel fourth, quarter
Vierteldrehung quarter turn
Viertelkreis quarter circle, quadrant (51)
Viertelperiode one-fourth cycle
Viertelwelle quarter wave
Viertelwellenlänge quarter wavelength
Viertelwellenübertragung quarter-wave transmission
vierteln to quarter
... vierten Grades biquadratic
vierwertig tetravalent (54)
Vinylchlorid vinylchloride
virtuell virtual
virtuelle Adresse virtual address
virtuelles Bild virtual image
virtuelle Kathode virtual cathode
virtueller Speicher virtual memory
virtuelle Verschiebung virtual displacement
Visier sight (623.4)
Visiereinrichtung sighting device
Visiergerät diopter
Visierkompaß azimuth compass
visieren to sight
viskos viscous
Viskose viscose
Viskoseseide viscose silk
Viskosität viscosity
Viskositätsmeßgerät viscosimeter
visuell visual
visuelle Anzeige visual display
visueller Störabstand subjective S/N-ratio
visueller Wirkungsgrad light efficiency
Vitriol vitrio, copperas
V-Kerbe notch
V-Nut Vee-groove (621.791)
Vocoder vocoder
Vogelperspektive bird's-eye view
Voll-... full, full-scale, full-sized
Volladdierer full adder
Vollalarm actual warning (35)
Vollamtsberechtigung direct outward dialling, DOD
Vollanalyse complete analysis (54)
Vollast full load
Vollastleistung full-load output
Vollaststrom full-load current
Vollausschlag full-scale deflection, maximum deflection (621.317)
Vollbahn standard gauge railway
Vollbahngleis standard gauge track
Vollbetrieb full-scale service, full swing
Volldampf full speed
Vollduplex full duplex
Vollgas full throttle (629.113)
Vollsperre total barring (621.395)
Vollweggleichrichter full-wave rectifier
Vollweggleichrichtung full-wave rectification
voll full

voll-amtsberechtigte Nebenstelle unrestricted extension
voll aufgeladen fully charged (621.3)
voll ausgesteuert controlled to full value, driven to full output
voll ausschlagen to deflect full-scale
voll ausschreiben to extend (51)
voll besetzt full up
voll in Betrieb sein to work full swing
voll durchmoduliert completely modulated
voll kreisförmig solid circular
volles Tageslicht full daylight
auf vollen Touren laufen to run at high speed
in vollem Umfange to the full
vollautomatische Unterstation fully automatic substation
vollelektrisch all-electric
vollelektrische Signalanlage all-electric signalling
vollelektrische Verblockung all-electric interlocking
vollelektronisch all-electronic
vollgummi-isoliert tough rubber sheathed (621.3)
vollgummi-isoliertes Kabel C. T. S. = cab tyre sheath cable
vollimprägniert fully impregnated
vollwandiges Rohr continuous tube
völlig full, complete, total, perfect; quite
völlig abgeschirmt perfectly screened
völlig abschrauben to unscrew right off
völlig durchdringen to pervade
völlig flach dead flat
völlige Auslöschung complete cancellation
völlige Austauschbarkeit pin-for-pin replacement
bei völliger Dunkelheit in total darkness
vollständig total, complete (s. auch **völlig**)
vollständig von Zug entlasten to remove all pull off...
vollständige Aussteuerung complete modulation
vollständige Einzelheiten full details, complete data
vollständiges Gleichgewicht complete equilibrium
vollständiger Kreis perfect circle
vollständiger Nichtleiter ideal dielectric
vollständige Streuung perfect diffusion (535)
vollständiger Umlauf complete rotation (621.3)
vollständige Unterlagen full details, complete data
vollständige Verbrennung perfect combustion, complete combustion
Vollständigkeit completeness; degree of perfection
Vollständigkeitsprüfung parity check (681.3)
Volt volt
Voltampere volt-ampere
Volt-Ampere-Stundenzähler volt-amperehour meter
Voltmeter voltmeter
Voltmeterumschalter voltmeter switch
Voltameter voltameter
Volta-Effekt Volta effect
Voltasche Säule voltaic pile, galvanic pile
Volta-Spannung voltaic electricity
Volumen volume
Volumen Eins unit volume
Volumabnahme contraction in volume
Volumenausdehnung volumetric expansion
Volumbestimmung volumetric measurement
Volumelement element of volume
Volumgewicht weight of volume
Volumprozente percentage (sing.) by volume
Volumverhältnis ratio by volume
volumetrisch volumetric(al)
voneinander abhängig interdependent
voneinander abweichen to vary
voneinander trennen to isolate
sich voneinander entfernen to fly apart
VOR VHF omnidirectional range
vor (im Zuge einer Schaltung) ahead of
vorangehen to precede
vorangehende Stufe preceding stage
Voranmeldegespräch person-to-person call
Voranschlag forecast, estimation; budget
Voranstrich prewash
Voraussage prognostication; forecast (551.509)
voraussagen to prognosticate; to forecast
voraussichtliche Lebensdauer life expectancy, life expectation
Vorbehandlung pre-treatment
vorbeibewegen an, sich to move past
vorbeilassen to let pass
vorbeirutschen an to shave past
vorbeischwenken an, vorbeischwingen an to swing past
vorbelasten to preload, to bias
Vorbelastung pre-load, initial load, bias, load resistor (621.314.6.)
vorbereiten to prepare
vorbereitend einstellen to preset
vorbereitende Einstellung presetting
Vorbereitung preparation
als Vorbereitung für in preparation for, preparatory of
vorbeugen to prevent, to obviate

vorbeugende Maßnahme preventive measure
vorbeugende Unterhaltung preventive maintenance
Vorderansicht front view, front elevation
vordere Austastschulter front porch (621.397)
vordere Polkante leading pole horn, leading pole tip
Vorderfläche front face
Vorderflanke leading edge
Vordergrundprogramm foreground program (681.3)
Vorderkante leading edge (621.3), (629.13); lip (621.9)
Vorderrad front wheel
Vorderradantrieb front-wheel drive
Vorderseite face front, front face, front side
vorderseitig front, forward
vorderseitig belegen mit to face with (693)
vorderseitiger Anschluß front connection (621.3)
Vordrossel series inductance
voreilen gegen to lead (621.3)
voreilend gegen in advance of, in front of (621.3)
voreilende Phase leading phase
Voreilung lead (621.3)
Voreilungswinkel angle of lead (**gegen**: over) (621.3)
Voreinflugzeichen outer marker (beacon)
voreinstellen to preset, to pre-adjust
Voreinstellung presetting, pre-adjustment
Vorentladung predischarge (621.3)
vorerwärmen to preheat
Vorerwärmung preheat, preheating
Vorfeuerung forehearth
vorführen to project, to feature (77); to demonstrate, to present
Vorführung (ocular) demonstration, performance, presentation
Vorführungsgerät projector (77)
Vorfunkenstrecke series gap (621.3); auxiliary spark-gap (629.13)
Vorgang event, phenomenon (pl.: -na), process
vorgegebene Randbedingungen predetermined boundary conditions
Vorgelege gearing, transmission gear; back gear
vorgeschaltet ahead of
vorgeschaltete Sicherung line-side fuse
vorgeschalteter Widerstand series resistor
vorgesetzte Kennung prefix (621.395)
Vorgruppe 3-channel pregroup (621.395.4)
Vorhaltekurs constant bearing course (527)
Vorhalt rate action, derivative action (621.316.7)
Vorhaltzeit derivative-action time; rate time (US)
vorhanden sein to exist, to be on hand, to prevail
vorhandener Zustand prevailing condition
vorheizen preheating
Vorheizzeit preheating time
vorher einstellen s. **voreinstellen**
vorhergehende Stufe preceding stage (621.396.62), (621.64)
vorherrschen to prevail, to predominate
Vorimprägnierung preimpregnation
Vorimpuls preknock pulse, pretrigger impulse (621.396.96)
Vorionisationelektrode keep-alive electrode
Vorkalkulation cost estimation, precalculation
Vorkondensator series capacitor (621.317)
Vorlast preload, initial load
Vorlauf winding (681.84)
vorlaufende Welle foreward wave
vorläufig provisional, preparatory, temporary; for the present
vorläufiger Bericht preliminary report, interim report
vorläufige Installation temporary wiring
vorläufige Konstruktion provisional design
Vorlicht ambient light (621.397)
Vormagnetisierung magnetic bias
Vormagnetisierungsfeld polarizing field
Vormagnetisierungsfeldstärke polarizing magnetic force
Vormagnetisierungsgleichstrom bias d.c.
Vormagnetisierungsinduktion polarized induction
Vormerken booking (621.395)
Vormerkgespräch delayed call
vorn ahead, in front
vorn abschließen mit to front by
nach vorn schieben to push forward
nach vorn schleudern to project
Vorrang priority
Vorrangbezeichnung precedence designation
Vorranggespräch precedence call
Vorrangverarbeitung priority processing (621.39), (681.3)
Vorrat stock, provision, supply, reserve
Vorratsader spare core, spare wire (621.395)
Vorratsbehälter storage bin, reservoir
Vorratsbehälter-Kathode dispenser cathode with reservoir
Vorratskathode dispenser cathode
Vorratslager magazine, storage dump
Vorrichtung device, appliance, gadget, contrivance

Vorrichtung zum 90°-Abgleich power-factor adjusting device (621.317)
Vorrücksignal proceed signal (625)
Vorschaltdiode series diode
Vorschaltgerät für Niederspannungsleuchtstofflampe ballast for low voltage fluorescent lamp
vorschalten to connect in series
vorgeschaltet sein to precede, to be in series with (621.3)
vorgeschaltete Induktivität series inductance
vorgeschalteter Widerstand series resistor
vorschreiben to prescribe, to direct, to specify
vorgeschriebener Grenzwert specified limit
vorgeschriebener Weg specified path (51)
Vorschrift regulation, rule, specification, provision
Vorschriften beachten to honour regulations, to adhere to rules, to comply with regulations
den Vorschriften entsprechen to be in accordance with regulations, conform to specifications
in Vorschriften festlegen to codify
Vorschriftenbuch für Elektrotechnik electrical code
vorschriftsmäßig according to regulations
vorschriftsmäßig schutzgeerdet effectively grounded
Vorschub travelling movement feedrate (621.9); (in-)feed, feeding (621.9); carriage (621.394.6)
Vorschubgeschwindigkeit feed rate
Vorschubgetriebekasten feed box
Vorschubspindel, Vorschubwelle feed spindle
Vorselektion preselectivity (621.396.62)
Vorsicht caution
Vorsicht! Hochspannung - Lebensgefahr Danger High Voltage, Caution High Voltage
Vorsichtsmaßnahme precautionary measure
Vorsignal preliminary signal, distance signal (625)
Vorspannbatterie biasing battery
vorspannen to bias
vorgespannter Gleichrichter biased rectifier
vorgespannt in Sperrichtung reversed biased
Vorspannung bias (voltage) (621.396)
ohne Vorspannung unbiased
Vorspannungswiderstand biasing resistor
vorspringen to project, to protrude
vorspringender Teil projection
Vorsprung protrusion, projection; start (**vor**: over)
Vorsprung gewinnen vor to get ahead of
Vorstadium preliminary (stage)
Vorstand board of management
Vorstecker linchpin
Vorsteckring guard ring
Vorsteckstift cotter pin, locking pin, linchpin
vorstehen s. **vorspringen**
vorstehende Glimmerisolation high mica (621.313)
vorstehender Griff protruding grip
vorstehender Punkt salient point, high spot
Vorstufe preceding stage, prestage (621.396.62), (621.396.64)
Vorstufen-Modulation low power modulation
Vortrieb propulsion
vorübergehend transient, transitory, temporal, temporary
Voruntersuchung preliminary research (53)
Vorvakuum(druck) forepressure
Vorverbrennung precombustion
Vorverschiebung der Bürsten forward shift of brushes
Vorverstärker preamplifier, low-level amplifier
Vorversuch exploratory test
Vorverzerrung accentuation, pre-emphasis
Vorwähler subscriber's rotary line switch, preselector (621.395)
Vorwärmer economizer, preheater
Vorwärmzeit warm-up time
Vorwarnung early warning, alert signal (35)
Vorwarn-Radar longrange radar device, early warning device
Vorwärts-... forward
Vorwärtsauslösung forward release
Vorwärtsbewegung forward motion
Vorwärtsgang forward speed
Vorwärtsregler forward-acting regulator
Vorwärtsrichtung forward direction (621.382)
Vorwärts-zu-Rückwärtsverhältnis front-to-back ratio
Vorwärtsstreuung forward scattering (621.396)
Vorwärtsverlustleistung forward power loss
Vorwärtsverweis forward reference
Vorwärtszeichen forward signal (621.395)
vorwärts zählen to count up
Vorwiderstand series resistor (voltage), dropping resistance, ballast resistance; voltage multiplier (621.317)
Vorwiderstandslampe ballast bulb
Vorzahl characteristic
Vorzeichen sign
Vorzeichen der Spannung voltage sign
Vorzeichen wechseln to alternate in polarity
Vorzeichenumkehr polarity reversal, polarity inversion; reversal of sign
Vorzugsrichtung preferred direction (54)
Voutenbeleuchtung cornice lighting

VU-Meter vu-meter (621.375)
Vulkanfiber hard fibre, vulcanized fibre
vulkanisieren to vulcanize, to cure
V-Winkel dihedral angle (629.13)

W

W = Watt w = watt
Waage (weighing) scales (pl.), balance; Scales (pl.), Libra (52)
in die Waage bringen to level
in die Waage richten to bed down level
Waagebalken weigh beam, beam of a balance
waagerecht horizontal, level
waagerecht halten to keep level
waagerechte Zeitablenkung A-scan, horizontal scan (621.397)
Waagerechte horizontal line
Waagschale scale; loading tray
Wabenspule honeycomb coil (621.396)
Wachs wax
Wachsbildung wax formation (621.315.2)
Wachsdraht waxed wire, paraffined wire, wax-coated string
Wachsform wax mould (621.357)
Wachspapier waxed paper
wachsen to wax
gewachste Strippe wax(-coated) string
Wachstum growth; increase
Wachstumsgeschwindigkeit rate of growth; creep rate (66)
Wachstumskennzahl grouth figure (621.395)
Wachstumsvorgang process of growth
Wächter watchman
Wächterkontrollanlage watchman control system
Wächterkontrolluhr watchman's clock
Wackelei jitter (621.397)
Wackelburst alternating burst (621.397)
Wackelkontakt loose contact, loose connection, slack joint
wackelig jittery, to ttery, slack, shaky
wackeln to totter, to waggle, to rock
wägbar ponderable
Wägbarkeit ponderability
wägen to weigh
Wagen car, carriage; wag(g)on (625); carriage (621.394.6)
Wagen mit Einzelachsantrieb individual geardriven axle carriage
Wagenachse (car) axle
Wagenführer wattman, driver
Wagengestell car frame
Wagenheber (lifting) jack
Wagenpark rolling stock
Wagenrücklauf carriage return (621.394.6)
Wagnerscher Hammer hammer, hammer break, hammer interrupter, trempler
Wagnersche Hilfsbrücke, Wagnersche Hilfserde Wagner earth, Wagner earthing device
Wägung weighing (experiment) (54)
Wahl selection, dialling
Wahlabrufzeichen begin-to-end signal
Wahlaufforderung proceed to dial
Wahlaufnahme dial-signal receiver
Wahlende end of selection
Wahlschalter selector switch
Wahlspeicher dialling register
Wahlstufe selection stage
Wahlumsetzer dial converter
Wähl-... selective; automatic (621.39)
Wählamt automatic (telephone) exchange, dial central office
Wählapparat dial telephone set
Wählende end of dialling
Wählendezeichen end-of-selection signal (621.395)
Wählfernsprechverkehr automatic telephony, dial telephone communication
Wählimpuls selective impulse, dial impulse (621.395)
Wählimpulsverstärker pulse repeater
Wählnetzgebiet automatic area (621.395)
Wählschalter selector, selecting switch
Wählscheibe (calling) dial
Wählsystem dial telephone system, automatic (dial) system
Wähltastatur keyboard
Wähltelefon autophone
Wähltelefonanlage automatic telephone system
Wählton dial tone, proceed-to-send tone
Wähltonverzug pre-dialing delay
Wählumschalter intercommunication switch
Wählvermittlung dial exchange, dial switching system
Wählvermittlungsstelle automatic exchange (621.395)
Wählzeichen pulsing signal
wählen to choose; to select; to dial (621.395)
Wähler selector (621.395)
Wähler heben to raise a selector
Wählerarm selector switch arm
Wählergestell selector frame

Wählerkontaktarm wiper
Wählerrelais stepping relay
Wählersaal switch-room, auto-room, selector room
Wählerscheibe calling dial
wahlfreier Zugriff random access, direct access
wahllos at random, in random fashion
wahr true
wahre Induktivität true inductance
wahre Remanenz true remanence
wahrer Wert true value
Wahrheitstabelle truth table (621.374)
wahrnehmbar intelligible, noticeable, discernible
wahrnehmen to observe, to notice, to perceive, to realize
Wahrnehmung observation, apprehension; sensation
wahrscheinlicher Fehler probable error
Wahrscheinlichkeit probability
Wahrscheinlichkeit, dem Resonanzeinfang zu entgehen resonance escape probability (539)
Wahrscheinlichkeitsrechnung probability calculus
Walz-... rolling, rolled
Walzblech rolled plate
Walzglas rolled glass
Walzschweißung sealed assembly rolling
Walzstahl rolled steel
Walzstich rolling operation, rolling pass
Walzvorgang rolling
Walzwerk rolling mill
Walzwerkantrieb rolling-mill drive
Wälz-... rolling
Wälzachse axis of rolling (531)
Wälzbahn centrode, rolling curve (531)
Wälzbewegung rolling motion
Wälzebene rolling plane
Wälzfläche rolling surface
Wälzgerade rolling line
Wälzgetriebe rolling gearing
Wälzkreis rolling circle
Wälzlager anti-friction bearing, rolling-contact bearing
Walze roll, roller (621); platen (651)
Walzendrehbank roll turner's lathe
Walzendrucker on-the-fly printer (681.3)
Walzenelektrisiermaschine Lemstrom machine
Walzenschalter drum controller, drum switch, barrel controller
Walzenstraße rolling train
Wand wall; partition
Wandapparat wall-type telephone, wall telephone set
Wandbefestigung wallmounting
Wandbelag wall coat (621.385)
Wanddurchführung wall-entrance bushing (621.315.6)
Wandfassung batten-lamp holder
Wandheizung panel heating
Wandkonsole wall bracket
Wandleuchte wall fitting
Wandmontage wall mounting
Wandplatte wall board, partition slab; tile
Wandschalter surface switch (696.6)
Wandstärke wall thickness
Wandsteckdose socket outlet
Wandstecker wall plug
Wandtelefon s. **Wandapparat**
Wandverteilergestell wall-mounted distribution rack
Wanderfeldröhre traveling-wave tube (US), travelling-wave valve (Br)
Wanderfeld-Magnetfeldröhre traveling-wave magnetron
Wanderfläche migration area (539)
wandern to travel, to move (621)
Wanderstecker wander plug (621.317)
Wandertransformator movable transformer
Wanderungsgeschwindigkeit ionic speed, ionic velocity (621.335); migration speed (539)
Wanderungsrichtung direction of travel
Wanderwelle traveling wave, surge, transient wave (621.3)
Wanderwellenantenne progressive-wave antenna, traveling-wave antenna
Wanderwellenschutz surge protection, transient protective device
Wandler transducer (621.316.7); instrument transformer (621.317); converter (681.3)
Wandlerempfindlichkeit transducer sensibility
Wandlerzähler transformer-connected meter
Wandlung change, transformation
Wanne tub, vat, trough, tank
in Wannen anordnen to louver (621.32)
Wannenfutter cell lining (669.71)
Ward-Leonard-Regelung Ward Leonard speed control
Warenzeichen trade mark
warm warm, hot
warm aufziehen to shrink on
warm einwickeln to muffle
warm werden to run hot
warme Lötstelle warm junction (621.36)
warmbrüchig hot short
warmhärtend thermosetting, thermoplastic
warmhärtender Kunststoff thermoplastic, thermosetting plastic
warmpressen to stamp (621.6)

warmrissig hot short
warmvergütet heat-treated
warmwalzen to hot roll
Warmwalzwerk hot-rolling mill
Wärme warmth, heat
Wärme abführen, Wärme abgeben to give up heat, to dissipate heat, to disburse heat
Wärme aufnehmen to absorb heat
Wärme ausstrahlen to radiate heat
Wärme entziehen to draw heat from
Wärme erzeugen to generate heat
Wärme leiten to conduct heat
Wärme speichern to store heat
auf Wärme ansprechend heat responsive; thermophile
Wärmeabgabe thermal dissipation, heat development, heat dissipation
Wärmeabgabevermögen heat-dissipating ability
Wärmeableiter heat sink
Wärmeanzeiger thermoscope
Wärmeäquivalent heat equivalent
Wärmeaufnahme heat absorption, gain of heat
Wärmeaustausch heat transfer, heat exchange
wärmebehandeln to heat-treat
Wärmebehandlung heat-treatment
einer Wärmebehandlung unterziehen to subject to heat-treatment, to heat-treat
wärmebeständig heat-resistant, heat-resisting
Wärmebeständigkeit heat resistance, thermal resistivity; thermal stability
Wärmebewegung thermal agitation, thermal motion
Wärmebilanz heat balance
Wärmebildgerät thermal imaging (night sight) device (623)
Wärmeblitz heat flash
wärmedämmend heat insulating
Wärmedämmung heat insulation
Wärmedehnung heat expansion, thermal expansion
Wärmedehnungszahl coefficient of thermal expansion
Wärmedurchgang passage of heat
Wärmedurchgangswiderstand thermal resistance
wärmedurchlässig diathermal
Wärmedurchschlag high-temperature breakdown (621.3)
Wärmeeigenschaft thermal property
Wärmeeinfluß influence of heat
Wärmeeinheit unit of heat, thermal unit, specific heat unit
Wärmeenergie heat energy
Wärmeentwicklung heat evolution, heat development, heat production
Wärmeerzeugung heat generation
wärmefest heat-proof
Wärmeinhalt heat content, to tal heat
Wärmeisolation, Wärmeisolierung heat insulation, thermal insulation
wärmeisolieren to heat-insulate
wärmeisolierend heat-insulating, thermo-insulating
Wärmekapazität heat capacity, thermal capacitance, calorific capacity
Wärmekontakt thermal contact
Wärmekonvektion heat convection
wärme- und korrosionsfester Ni-Stahl Ni-resist
Wärme-Kraft-Kopplung combined power and heating system
Wärmekraftmaschine heat engine, combustion engine
Wärmekraftwerk thermal power plant
Wärmeleistung thermal power
wärmeleitend heat conductive
Wärmeleiter heat conductor
Wärmeleitfähigkeit heat conductivity, thermal conductivity
Wärmeleitrohr heat pipe
Wärmeleitung heat conduction, thermal conduction
Wärmeleitungswiderstand thermal resistance
Wärmemenge heat quantity
Wärmemessung heat measurement, thermal mesaurement
Wärmeplatte electric hot plate
Wärmeprüfung high-temperature test, thermal test
Wärmepumpe heat pump
Wärmequelle source of heat
Wärmerauschen thermal noise, thermal agitation, resistance noise
Wärmeregler heat governor
Wärmerückgewinnung heat recovery
Wärmerückzündung hot arc back
Wärmeschock thermal shock
Wärmeschrank thermostat, thermostatic chamber; high temperature room
Wärmeschutz heat protection, heat-insulating jacket
Wärmespannung thermal stress
Wärmespeicher heat store
wärmespeichernd heat-storing
Wärmespeicherung heat storage
Wärmestauung accumulation of heat
wärmestrahlend heat-radiating
Wärmestrahlung heat radiation, radiant heat

Wärmetauscher heat exchanger, calorifier
Wärmeübertragung heat transfer, heat transmission, heat exchange
Wärmeverbrauch heat consumption
Wärmeverlust loss of heat
Wärmeversorgung supply of heat
Wärmewächter heat controller
Wärmewelle heat wave
Wärmewiderstand thermal resistance
Wärmewirkung heating effect, calorific effect
Wärmewirtschaft heat economy, fuel economy
Wärmezähler heat meter
Wärmezeitkonstante thermal time constant
Wärmezufuhr supply of heat, heat supply
Wärmezustand thermal condition
wärmen to warm
Warn-... alarm, warning, signalling
Warnanzeige attention display
Warnapparat signalling apparatus, warning device
Warneinrichtung alarm device
Warnempfänger alarm receiver, warning receiver
Warngerät predictor, alarm device, warning device
Warnglocke alarm bell, signalling bell
Warnlampe warning light
Warnrelais alarm relay
Warnschalter alarm switch
Warnsignal danger signal, warning signal
Warntableau warning panel
warnen vor to warn of, to caution against
Warnung warning, alarm, danger signal
Warnung betätigen to sound an alarm
Warnungsschild warning plate, notice board
Wartbarkeit maintainability
Warte control panel; control room
Wartebelastung queue length, waiting traffic (621.395)
Wartepunkt holding point (351.814.33)
Warteschlange queue
Warteschleife holding pattern, holding track (351.814.33); queuing system
Wartesystem delay system (621.395)
Warteton hold-on tone
Wartewahrscheinlichkeit probability of delay (621.395)
Wartezeit stand-by period, waiting-on time; delay (621.395); latency time (681.3)
Wartezeit-Theorie queuing theory
warten to wait; to maintain, to upkeep
warten auf Freiwerden to park on busy, to camp on busy
wartendes Programm waiting program
Wärter attendant, watchman (681.17); signalman (625)
Wartung (preventive) maintenance, upkeep, service, attendance
Wartungsanleitung maintenance instructions
wartungsfrei maintenance-free
Wartungspersonal maintenance staff
Wartungsprogramm maintenance schedule; service routine (681.3)
Wartungsvorschrift maintenance instruction
Wäscheschleuder spin drier, laundry drier, centrifugal drying machine
Wäschetrockner washer dryer
Waschkessel wash boiler
Waschmaschine washing machine
Wasser water
Wasser abgeben, Wasser aufnehmen to water
Wasser entziehen to dewater
mit Wasser extrahieren to lixiviate
mit Wasser versorgen to (supply with) water
Wasserabgabe dehydration
Wasserablauf water waste; waste-water duct, drain
Wasseranstauung slack-water
wasseranziehend hygroscopic
Wasseraufnahme water absorption
Wasseraufnahmefähigkeit hygroscopicity, water capacity
Wasserbedarf water demand
Wasserbildung formation of water (54)
Wasserbremse water brake
Wasserdampf water vapour, aqueous vapour, steam
Wasserdampfreaktor water-boiler reactor
wasserdicht watertight, waterproof
wasserdicht gekapselte Maschine watertight machine
wasserdicht machen to waterproof
wasserdicht sein to hold water
wasserdichte Imprägnierung waterproof serving, waterproofing
wasserdichte Kabelstopfbuchse watertight gland
wasserdichte Leuchte watertight fitting; waterproof luminaire (US)
Wasserdichtigkeit watertightness
Wasserdruck water pressure, hydraulic pressure, hydrostatic pressure
wasserdurchlässig pervious to water
wasserdurchlässige Schicht pervious stratum
Wassereinbruch rush of water
wasserentziehend desiccating
Wasserentziehung dewatering, desiccation
Wasserfall waterfall

Wasserfalle moisture trap
wasserfrei free of water, anhydrous
Wassergas (blue) water gas
Wassergasgenerator water-gas generator
Wassergehalt water content
wassergekühlt water-cooled
wassergekühlter Motor water-cooled engine (621.4)
Wasserglas water glass, soluble glass (54)
Wasserhahn water tap, faucet
wasserhaltig watery, hydrous, aqueous
Wasserhaltung drainage, draining
Wasserhaltungsmaschine draining machine
Wasserhaltungspumpe draining pump, water-raising pump
wässerig-homogener Reaktor aqueous homogeneous reactor
Wasserinhalt water volume
Wasserkraft water power, white coal, hydraulic power
Wasserkraftgenerator hydroelectric generating set, waterwheel generator
Wasserkraftmaschine hydroelectric machine
Wasserkraftwerk water-power plant, hydro-electric generating station, hydro-electric power station
Wasserkreislauf water circulation
Wasserkühler water cooler
Wasserkühlung water cooling
wassergekühlter Reaktor water-cooled reactor
Wasserkunde hydrography
Wasserleitung water conduit; water supply, water main
Wassermantel water jacket
mit einem Wassermantel water-jacketed
Wassermenge eines Stroms flow of a stream
Wassermotor hydraulic engine
Wasserpumpe water (supply) pump, hydraulic pump
Wasserrechen water screen (621.3)
Wassersäule water column, head of water
Wassersäulenmaschine water-pressure engine
Wasserschwall rush of water
Wasserspiegel water level
Wasserstand water level, head (of water)
Wasserstandsanzeiger water gauge, water-level indicator
Wasserstandsrohr water-level gauge
Wasserstoff hydrogen
Wasserstoffelektrode hydrogen electrode (621.35)
wasserstoffgekühlt hydrogen-cooled
Wasserstoffionenkonzentration hydrogen ion concentration
Wasserstoffzähler gas voltameter, gas coulometer
Wasserstrahl water jet
Wasserstrahlerder water-jet arrester
Wasserstrahlpumpe water-jet pump
Wassertarif water rate
Wasserturbine water turbine
Wasseruhr water meter
Wasserumlauf water circulation
Wasserversorgung water supply
Wasserwerk water works (pl.), pumping plant
Wasserwert water equivalent, calorific capacity
Wasserwiderstand drag; water resistor (621.3)
Wasserwiderstandsableiter watercolumn arrester
Wasserzufuhr water supply
wässerig watery, hydrous, aqueous
wässeriges Gel hydrogel
wässerige Lösung hydrous solution, aqueous solution
wässeriges Sol hydrosol
Watt watt
wattlos wattless
wattloser Strom wattles current, idle current
Wattmeter wattmeter
Wattstunde watt-hour
Wattstundenkapazität watt-hour capacity
Wattstunden-Motorzähler motor-type watt-hour meter
Wattstundenzähler watt-hour meter
Watte cotton (wool)
W. E. = Wärmeeinheit Joule (s. Anh.)
weben to weave
gewebter Filz woven felt (677)
Weber = elektromagnetische Stromeinheit E. M. U. = electromagnetic unit
Wechsel change; alternation (621.3); bill of exchange, raft (332.1)
Wechselanteil pulsating component, periodic component, alternating component, a. c. component
Wechselbeanspruchung alternating stress
Wechselbeziehung reciprocal relationship, interrelation
Wechselbiegeversuch reversed bending test
Wechselbrenner variable head welding to rch (621.791)
Wechsel-EMK alternating e. m. f.
Wechselfeld pulsating field, alternating field
Wechselfeldhysteresis static hysteresis

Wechselgröße alternating quantity, periodic quantity
Wechselgröße mit überlagertem Gleichanteil pulsating quantity
Wechselklima change climates
Wechselkomponente varying component, alternating component, pulsating component
Wechsellicht fluctuating light
Wechsellichtempfindlichkeit dynamic sensitivity (621.385)
Wechselmagnetisierung alternating magnetization, linear hysteresis
Wechselpolgenerator heteropolar generator
Wechselpotentialdifferenz oscillating potential difference
wechselrichten to invert (621.314.5)
Wechselrichter d. c. / a. c. converter, inverter (621.314.5); vibrator (621.39)
Wechselrichter mit Wiedergleichrichtung C. M. B. converter, d. c. static converter
Wechselrichtung inversion (621.314.5)
Wechselschalter two-way switch, double-throw switch, throw-over switch
Wechselschaltung two-way wiring
Wechselschlagbiegeversuch alternating impact bending test
Wechselschließer transfer contact, change-over contact
Wechselschreiben half-duplex operation (621.394)
wechselseitig reciprocal, mutual
Wechselspannung a. c. voltage, alternating voltage, alternating potential difference
Wechselspannung mit überlagerter Gleichspannung pulsating voltage, pulsating potential difference
Wechselspannungsprüfung a. c. test
Wechselsprechanlage intercommunication equipment one way
Wechselsprechen simplex voice communication, half-duplex operation
Wechselstrom a. c. = alternating current
Wechselstrom mit gleich verlaufenden Halbwellen symmetrical a. c.
Wechselstromanteil varying component, alternating component, ripple effect
bei Wechselstrom anzeigen to indicate on a. c. (621.317)
Wechselstrombetrieb a. c. operation (621.386), (621.396.61)
mit Wechselstrombetrieb a. c. operated
Wechselstrombremse a. c. magnetic brake
Wechselstrombrücke a. c. bridge
wechselstromgeheizte Röhre a. c. heated tube (filament)
Wechselstromgenerator a. c. generator
Wechselstromgenerator mit konzentrierter Wicklung monotooth alternator (621.313)
Wechselstrom-Gleichstrom-Umformer inverted rotary converter, Auvert-Ferrand permutator, a. c./d. c. converter (621.314.5)
Wechselstrom-Hubmagnet alternating electromagnet
Wechelstrominduktivität incremental inductance
Wechselstromkabel a. c. cable
Wechselstrom-Kommutatormotor a. c. commutator motor
Wechselstrom-Kompensationsapparat a. c. potentiometer
Wechselstromkomponente a. c. component
Wechselstromkreis a. c. circuit
Wechselstromlehre basic alternating current theory
Wechselstromleistung a. c. power
Wechselstrom-Lichtbogenschweißung a. c. arc welding
Wechselstrommagnetverstärker magnetic amplifier with a. c. output
Wechselstrommaschinen a. c. machinery (sing.)
Wechselstrommotor a. c. motor, single-phase motor
Wechselstromnetz a. c. mains network
Wechselstromnetzanschluß a. c. operation, a. c. supply
Wechselstromnetzmodell a. c. network anaylser
Wechelstrompermeabilität incremental permeability
Wechselstromprüfung a. c. test
Wechselstrom-Reihenschlußmotor series-conduction motor
Wechselstromseite a. c. terminals (pl.)
Wechselstromstehspannung power-frequency withstand voltage
Wechselstromtelegraphie VF telegraphy, voice-frequency carrier telegraphy
Wechselstromtherapie sinusoidal therapy
Wechselstromtrockenüberschlagsprüfung power-frequency dry-flashover test
Wechselstrom-Verstärkungsfaktor a. c. current transfer ratio
Wechselstromwecker magneto bell
Wechselstromwiderstand a. c. resistance, impedance
Wechselstromzähler a. c. meter

Wechseltauchmaschine für Korrosionsprüfung alternate immersion corrosion machine
Wechseltauchversuch alternate immersion test
Wechselverkehr simplex operation
Wechselverkehrsverbindung half-duplex circuit, one-way reversible operation
Wechselwirkung interaction
in Wechselwirkung stehen to interact
Wechselwirkungswiderstand interaction loss
wechseln to alternate, to pulsate (621.3); to vary
wechselnde Belastung variable load
wechselnde Betriebsart varying duty
wechselnde Betriebsbedingungen variable service conditions
wechselnde Farbfolge reversing color sequence, RCS (621.397)
wechselnde Drehzahl varying speed
Wechsler change over contact
Weckdienst wake-up service
Wecker (station) ringer, call bell (621.395); alarm clock (681.11)
Weg way; path (51), (53), (621.3); route (621.396); road (625)
Weg des geringsten Widerstandes line of least resistance
auf einem Wege wirken to act through a space
vom Wege ablenken to pull off the course
wegabhängige Steuerung displacement-sensitive control; follow-up control (629.12)
Wegabschnitt section of a path
Wegebesetztton congestion tone, trunk-busy tone
Weglänge path length
Wegesuchen path finding, path search
Wegesuchabschnitt path-finding section (621.395)
Wegmeßgeber position sensor, position transducer
Wegregelung distance control, travel control, position control
Wegschreiber travel recorder
Wegunterschied path difference
Wegweiser signpost, direction indicator board
weg away, off
wegbrechen to break away
wegbrennen to burn away
wegdrosseln to choke out
wegfangen to capture
wegfeilen to file off
wegfließender Strom outgoing current
wegfressen, wegkorrodieren to eat away, to fret
wegkompensieren to cancel
wegschleifen to grind off
wegschmelzen to melt away
wegschneiden to cut away, to trim away
wegstimmen to tune out
Wegfang parasitic capture (539)
Wehnelt-Elektrode modulator electrode
Wehnelt-Unterbrecher Wehnelt interrupter
Weich-. . ., weich soft; mild (669)
weiche Aufhängung flexible suspension
weich geglühter Draht soft-annealed wire
weichmagnetisch soft magnetic
weiche Röntgenröhre gas (filled) X-ray tube
weiche Röntgenstrahlung soft X-rays (pl.)
weicher Stahl mild steel, soft steel
weiche Strahlung soft radiation, low-energy radiation
weiches Wasser soft water (621.18)
Weicheisen soft iron, mild steel, ductile iron
Weicheisenabschirmung soft iron shield
Weicheisenamperemeter moving-iron ammeter
Weicheisendraht mild steel wire
Weicheiseninstrument soft iron instrument, moving-iron instrument
Weicheisenkern soft iron core
Weicheisennadel induced needle (621.317)
Weicheisenstück soft iron mass
Weicheisenvoltmeter moving-iron voltmeter
weichglühen to normalize (669)
Weichkupfer annealed copper
Weichlot (soft) solder
weichlöten to (soft-)solder
Weichlötung soft soldering
Weichtasten soft keying
Weiche points (pl.), switch (625); bifurcation (621.39)
Weiche umlegen to throw over a point
Weichenantrieb point machine
Weichenriegelkontakt lock-circuit controller
Weichenschloß points lock
Weichensignal points indicator, switch signal
Weichenstellmaschine lock-and-switch movement
Weichenstellungs-Kontrollschalter switch-machine point detector
Weichenstellungssignal switch signal
Weichenstellvorrichtung switch machine
Weichenstellwerk switch machine lever
Weichenüberbrückung continuity rail bond
Weichenverriegelung points lock
weiß white
weiße Kohle = Wasserkraft white coal
weißes Rauschen white noise, fluctuation noise, random noise
weißes Roheisen white iron

weißbinden to whitewash
weißglühend white hot, incandescent
Weißblech tinned iron, tinplate
Weißblechbüchse tin; can (US)
Weißglut incandescence
zur Weißglut bringen to incandesce
Weißmetall white metal, bearing metal, babbitt metal, pewter
Weißpause black-and-white print
Weißpegel white level (621.397)
Weißscher Bezirk domain (538)
Weißspitze whiter-than-white, peak white (621.397)
Weißwert (peak) white level (621.397)
weit wide, far; vast, broad
weit vom Vorschriftsmäßigen entfernt wide off the mark
bei weitem am größten much the largest
bei weitem nicht lang genug not half long enough
weitabständig widely spaced
weitgehende Näherung close approximation
weitmaschig wide-mesh, open-mesh
weitmaschiges Netzwerk open mesh
weitreichend far-reaching
weiter reichen als to outrange
weiterlaufen to progress (621.396)
weiterleiten to propagate, to transmit
weiterschicken to send forth
Weit... long-distance, long haule
Weitsprechverkehr long-distance telephony
Weitverkehr long-distance communication, long-range communication
Weitverkehrsgespräch long-distance call
Weitverkehrsnavigation long-range navigation
Weitverkehrs-Navigationsverfahren Loran = Long Range Navigation
Weitverkehrstelegraphie long-distance telegraphy
Weitwinkel wide-field (77), (535)
Weitwinkelobjektiv wide-field objective
Weitwinkelradar panoramic radar
Weitwinkelstreuung wide-angle diffusion (535)
Weite width; breadth; distance
Weiterruf periodic ring
Wellblech corrugated sheet (-iron), corrugated sheeting
Welle wave (621.3); arbor, shaft, journal (621)
Welle absetzen to neck down a shaft (621)
Welle beugen um... to bend waves round...
Welle verlängern to extend a shaft (621)
mit Wellen kuppeln to shaft (621)
wellenartig wave-like
Wellenausbreitung wave propagation, travel of (radio) waves
Wellenband wave band (621.396)
Wellenbauch antinode
Wellenbereich wave range, wave band
Wellenberg (wave) crest
Wellenbewegung wave motion, undulation
Wellenbock Vee-block (621)
Wellenbrechung wave diffraction
Wellendämpfung image attenuation (621.39)
Wellenfilter wave-guide filter
Wellenform wave shape
wellenförmig wave-like, undulating
Wellenfront wave front, surge front
wellengerader Kondensator straight-line wavelength capacitor
Wellengeschwindigkeit wave velocity
wellengewickelt wave-wound (621.313)
Wellengleichung wave-motion equation (51)
Wellenkeil key, sunk key, wheel key, wedge, spline (621)
Wellenknoten node
Wellenkopf wave head
Wellenlager shaft bearing, journal bearing (621)
Wellenlagerbock shaft-bearing bracket (621)
Wellenlänge wavelength
in der Wellenlänge einstellbar adjustable for wavelength
Wellenlängenbereich wave band
Wellenlast journal load (621)
Wellenleiter wave guide, duct
Wellenleiter-Ausbreitungstyp trapped mode
Wellenleiterdicke duct width
Wellenlinie waved rule, wavy line
Wellenmechanik wave mechanics (pl.) (53)
wellenmechanisch wave-mechanical
Wellenmesser wavemeter
Wellenneigung wave tilt
Wellennut spline, shaft slot (621)
Wellenpaket wave package
Wellenplan frequency allocation
Wellenring shaft collar (621)
Wellenrücken wave tail, surge tail
Wellenschalter wave-change switch, range switch, tuning switch, band switch
Wellensieb wave filter
Wellenspannung ripple (voltage), surge voltage
Wellenstirn wave front, wave head
Wellenstrahlung wave radiation
Wellenstrahlung von... radiation of waves from...

Wellenstrom surge current, undulatory current, ripple current
Wellenstumpf shaft butt, shaft extension, stub shaft (621)
Wellental trough (621.3)
Wellentunnel propeller-shaft tunnel (629)
Wellentyp mode (of oscillation)
Wellenumwandlungen in Hohlkabeln wave-mode conversion
Wellenwicklung wave winding, spiral winding, multiplex winding, wave connection (621.313)
Wellenwiderstand wave impedance, surge impedance, characteristic impedance, iterative impedance
Wellenwiderstand des (leeren) Raumes intrinsic impedance (of free space)
Wellenzapfen journal neck (621)
Wellenzug wave train, train of waves (621.396); jig (538)
Wellenzuteilung wave allocation
wellen to wave; to corrugate
sich wellen to warp
gewellte Linie waved line, waved rule
gewelltes Flammrohr corrugated flue tube
wellig wavy, warped, creasy
Welligkeit ripple; standing wave ratio (621.396.6)
Welligkeitsfaktor ripple factor (621.314.5); voltage standing wave ratio (621.396.6)
Wellmantel corrugated sheath
Wellrohrhohlleiter axially corrugated guide
Weltraum space
Weltraumfunkdienst space-radio service
Wendel helix (pl.: -ices); coil, spira
Wendelantenne helical antenna, helix antenna
Wendelfeder helical spring
Wendelhohlkabel helix wave guide
Wendelleitung helical transmission line
wendelförmig helical, helicoid
wendeln to coil, to loop
gewendelte Handapparateschnur coiled handset cord
gewendelter Leuchtdraht helical filament, looped (carbon) filament, looped (tungsten) filament
Wendepol commutating pole, auxiliary pole, reversing pole, interpole
Wendepolmotor interpole motor
Wendepolnebenschluß auxiliary pole shunting
Wendepolwicklung commutating pole winding, interpole winding
Wendepunkt turning point; interflection point (51)
Wendeschalter reversing switch
Wendezeiger turn indicator (629.13)
Wende- und Lagezeiger turn and bank indicator
Wendung turn
wenig little
wenig ergiebige Stromquelle low-current source
wenig Masse besitzen to be light in weight
mit zu wenig ... deficient in ...
weniger Rauschen less noise
Werg oakum
Werk work; works (pl.), plant, factory; establishment
ab Werk ex works
Werkbahn works railway
Werkbank bench, workbench, vice bench
Werkdirektor works manager
werkgerecht workmanlike
Werkmannsarbeit workmanship
Werksprüfung bench test, factory test
Werkstatt (machine) shop, workroom, workshop
Werkstattleuchte factory fitting (621.32)
Werkstattnormal working standard
Werkstattoleranz shop tolerance
Werkstattzeichnung working drawing
Werkstoff (work) material
Werkstoffehler defect in material
Werkstoffprüfung testing of materials
Werkstück piece, work (piece)
Werkzeug tool, instrument, implement
Werkzeug mit Widiaschneide tipped tool
mit Werkzeugen bestücken to tool
Werkzeugbesteck tool kit
Werkzeugkasten tool box, tool chest, tool kit
Werkzeugmaschine machine tool
Werkzeugmaschinensteuerung machine tool control
Werkzeugmaschinensteuerung mit Computer computerized numerical control = CNC
Werkzeugschlosser tool maker
Werkzeugstahl tool steel
Wert value
Werte anzeigen to read values
Werte am Arbeitspunkt steady values (621.385)
Wert Eins unity
Wert des Vorwiderstandes series resistance
Werte zuerteilen to give values to
Wertanalyse value engineering analysis
Wertarbeit craftsmanship, precision work
wertdiskretes Signal n-level signal
Werteglättung data smoothing (51)
wertkontinuierliches Signal level-continous signal

Wertetabelle truth table (621.374)
Wertminderung depreciation
Wertigkeit valence, valency
Western-Electric-Wählsystem Western Electric automatic telephone system
Weston-Element Weston (normal) cell
Wetter weather (55); air (622)
dem Wetter ausgesetzt weatherbeaten
Wetter- und Kollisionsbordradar cloud and collision warning system
Wetterbedingungen climatic conditions
wetterbeständig weather resistant
Wettereinfluß influence of weather
wetterfest weather-proof
wetterfeste Leitung weather-proof insulated wire
wetzen to whet
Wetzstein whetstone
WEZ = Westeuropäische Zeit Greenwich time
Wh = Wattstunde wh = watthour
Wheatstone-Brücke Wheatstone bridge, resistance bridge
Wheatstonescher Schnelltelegraph Wheatstone's telegraph
Whitworth-Gewinde Whit. = Whitworth (thread)
Wichte specific density, specific gravity
Wickel convolution; condenser element, section (621.314.4)
Wickelangaben winding data, winding specifications
Wickelautomat automatic winding machine
Wickeldorn mandrel
Wickeleinrichtung winding device
Wickelhärte (der Spulen) firmness of cops
Wickelkasten former
Wickelkern mandel (621.7), (621.9); core (621.3)
Wickelkondensator paper capacitor
Wickelkopf overhang of coils, end turns
Wickelkopfkappe end bell (621.313)
Wickelkopfstreuung leakage around overhang of coils, flank dispersion
Wickelkörper former (621.3)
Wickellage serving
Wickelmaschine winding machine (621.3)
Wickelquerschnitt winding cros section
Wickelraum (winding) space (621.31)
Wickelrolle reel drum
Wickelschablone former (621.3)
Wickelschritt pitch
Wickelschritt auf der kommutatorabgewandten Seite back pitch (621.313)
Wickelschritt auf der Kommutatorseite front pitch
Wickelverbindung wire wrap (connection)
wickeln to wind, to coil, to reel, to roll
gewickelter Läufer wound rotor
Wicklung winding (621.3)
Wicklung einfädeln to thread a winding
Wicklung speisen to energize a winding
Wicklung mit verkürztem Nutenschritt fractional pitch winding
Wicklung mit zweietagigem Wickelkopf skew-coil winding
Wicklungsabdeckung winding shield (621.313)
Wicklungsabschnitt section (621.3)
Wicklungsachse direct axis
Wicklungsbandage binding band
Wicklungsdurchschlag winding puncture, winding breakdown
Windungserwärmung winding-temperature rise
Wicklungsfaktor slot factor breadth coefficient, winding factor, copper factor
Wicklungsfläche winding area
Wicklungsinduktivität winding inductance
Wicklungskapazität winding capacitance; (Wicklung gegen Wicklung:) interwinding capacity
Wicklungskopf end winding, end turn
Wicklungsquerschnitt winding cross-section
Wicklungsrichtung direction of winding
Wicklungsschaltbild winding diagram
Wicklungsschritt, Wicklungsteilung winding pitch
Wicklungssinn sense of winding
Wicklungswiderstand winding resistance
Widerhall echo, repercussion
widerhallen to echo, to resound, to reverberate
Widerlager springer, shoulder, bolster
widerspiegeln to reflect, to mirror, to image
Widerstand resistance: (Widerstandskörper:) resistor; rheostat
Widerstand an (durchlaufende) Leitung anschließen to join a resistance across a line
Widerstand Eins unit resistance
Widerstand einschalten to throw in resistance
Widerstand bei festgehaltener Membran damped impedance (534)
Widerstand für die gegenläufige Komponente negative-sequence resistance
Widerstand mit Induktivität inductive resistor
Widerstand im Wasserbad immersed rheostat
Widerstand der Zuleitung lead resistance, wiring resistance
über einen Widerstand geerdet resistance-grounded

Widerstandabgriff rheostat slider
Widerstandsanlasser starter resistor, rheostatic starter
Widerstandsanpassung impedance match (ing)
widerstandsbehaftet resistive
widerstandsbehafteter Draht resistive wire
widerstandsbehaftete Spule resistive coil, resistive inductor
Widerstandsbeiwert drag coefficient (629.12), (629.13); resistivity (621.3)
Widerstandsbelastung resistive load (ing)
Widerstandsbremse dynamic brake (621.33)
Widerstandsbremsschalter rheostatic braking controller
Widerstandsbremsung resistance braking, rheostatic braking
Widerstandsdraht resistance wire, resistive wire, resistor wire
Widerstandsdreieck impedance triangle, impedance diagram
Widerstandseigenschaft resistivity; resistive property
Widerstandselement resistance element, elemental resistance
Widerstandserwärmung resistance heating
widerstandsfähig tough, sturdy, resistant
widerstandsfähig gegen resistant to
Widerstandsfähigkeit stability, strength, ruggedness
Widerstandsfähigkeit gegen resistance to
Widerstandsfolie resistive film
Widerstandsfunktion impedance function
Widerstandsgehäuse resistor housing
Widerstandsgewebe woven sheet, resistance mat
Widerstands-Hammerschweißung resistance percussive welding
Widerstandshartlötung resistance brazing
Widerstandsheizung resistance heating
Widerstandskasten resistance box (621.317)
Widerstandskopplung resistive coupling
widerstandsgekoppelt resistance-coupled
widerstandsgekoppelter Verstärker resistance-capacitance-coupled amplifier
Widerstandskörper resistor (621.3); (in Strömung:) baffle
Widerstandslampe ballast tube
Widerstandslegierung resistance alloy
widerstandslos resistanceless, non resistive, perfectly conductive
Widerstandsmaterial high resistive material
Widerstandsmeßbrücke resistance bridge
Widerstandsmeßgerät ohmmeter
Widerstandsmeßstreifen strain gauge
Widerstandsmikrophon rheomicrophone
Widerstandsmoment section modulus (539.384)
Widerstandsofen resistance furnace, resistor furnace
Widerstandsoperator impedance operator, vector impedance
Widerstandspunktschweißmaschine resistance spot welder
Widerstandspyrometer resistance pyrometer
Widerstandsrauschen resistance noise, thermal noise, Johnson noise
Widerstandsregler rheostatic controller
Widerstands-Restspannung zero-field resistitive residual voltage (621.382)
Widerstandsrohr ballast tube
Widerstandsrolle resistance coil
Widerstandssatz resistance frame, resistor bank
Widerstandsscheibe resistive disk
Widerstandsschicht resistive film
Widerstandsschweißung resistance welding
Widerstandsschweißgerät resistance welder
Widerstands-Spannungsteiler resistance voltage divider
Widerstandsspule resistance coil
Widerstandsstufe resistance notch, step of a resistance
Widerstands-Stumpfschweißung resistance butt welding
Widerstandsthermometer resistance thermometer
Widerstandstoleranz tolerance on resistance
Widerstandstransformation impedance transformation
Widerstands-Transistorlogik resistor-transistor logic
Widerstandsverlust rheostatic loss
Widerstandsverstärker resistance-capacitance-coupled amplifier
Widerstandszweig resistance branch
widerstehen to resist, to withstand
widerstehendes Moment moment of resistance
Wiedemann-Frantzsche Zahl Wiedemann-Frantz constant
wieder absinken auf to fall back to
Wiederanlauf restart
wieder anlaufen lassen to restart
Wiederanruf timed recall
Wiederaufbau re-establishment, reconstruction
Wiederaufbau des Flusses restoration of flux
wieder aufbauen to rebuild, to reconstruct, to re-establish
wieder aufbereiten to recondition; to reprocess (539)

Wiederauffinden von Daten data retrieval
wieder aufladen to recharge (621.3)
Wiedereinfang recapture
wieder einfangen to recapture
wieder einschalten to reconnect, to reclose
Wiedereinschaltung reconnection
Wiedereinschaltung der Beleuchtung re-illumination
Wiedereinschaltrelais reclosing relay
Wiedereinschaltzeit reclosing time
wieder einschreiben to rewrite (681.3)
Wiedererfassungszeit reacquisition time
wiedererhitzen to reheat
Wiedererhitzung reheat
wiedererstarren to resolidify
Wiedergabe reproduction; play back (681.84)
Wiedergabe in natürlicher Größe full-sized reproduction
Wiedergabeverstärker playback amplifier
Wiedergabegüte quality of reproduction
Wiedergabetreue fidelity of reproduction
wieder in Gang setzen to restart
wiedergeben to reproduce; to delineate
wieder gefrieren to regelate, to refreeze
wiedergewinnen to recover
Wiedergewinnung recovery
wiederherstellen to restore, to readjust
Wiederherstellung restoration, readjustment
wiederholen, sich to recur, to be repetitive, to rerun (681.3)
Wiederholung, Wiederkehr recurrence, repetition
Wiederholungsfrequenz recurrence rate, repetition rate, repetitive frequency
wiederkehrend repetitious, repetitive, periodical
wiederkehrende Spannung (transient) recovery voltage (621.316.5)
wiederkehrende Spannung der Betriebsfrequenz normal-frequency recovery voltage
wiederkommen (nach Schwund) to fade in (621.396)
wieder auf Null bringen to restore to zero, to reset
wieder steuerfähig werden to regain control (621.385)
wieder zünden to re-ignite, to restrike
wiegen to weigh
gewogenes Mittel effective average (519)
Wiensche Brücke Wien bridge
Wiensches Strahlungsgesetz Wien radiation law
Wiensches Verschiebungsgesetz Wien displacement law
wilde Kopplung stray coupling
wilde Schwingung parasitic oscillation, spurious oscillation (621.396)
wild verlegt hay wire (621.3)
Willemit willemite
Wilsonsche Nebelkammer Wilson cloud chamber
Wimshurst-Maschine Wimshurst machine
Wind wind (55)
Winddruck windage, wind pressure (621.315.1)
Windkanal wind tunnel (533)
Windkraft wind power, boreal energy
Windkraftanlage wind power plant
Windkraftgenerator wind driven generator
Windlast wind load
Windmesser wind gauge, anemometer
Windrichtung wind direction, direction of wind
Windrichtungsanzeiger wind indicator
Windrichtungsfeuer lighted wind tee, lighted tetrahedron (656.7)
Windschutz (scheibe) windshield
Windsog wind suction
Windstärke wind force
Windstille calm
Windstoß gust (of wind)
Windwiderstand wind resistance
Winde hoist, windlass, spill
winden to wind, to hoist
sich windend sinuous, meandering
Windung (lap) turn (621.3)
mit einer Windung verkettet sein to link a turn
Windungsisolation interturn insulation
Windungsschluß fault between turns, interturn short-circuit
Windungsschluß-Schutzsystem mid-point protective system
Windungsverhältnis turn(s) ratio (621.314.2)
Windungsverhältnis Eins unity turn(s) ratio
Windungszahl number of turns, winding turns
Windungszähler turns count
Winkel angle (51); square (744)
Winkel bilden mit to be at an angle with, to make an angle with
Winkel einschließen to contain an angle
Winkel (eingeschlossen) zwischen ... angle adjoined by ... and ...
Winkel gegen eine Fläche angle about a plane
Winkel zwischen zwei Linien angle contained by two lines
Winkel an der Spitze point angle, apex angle, vertical angle
Winkel von ... umschlingen to embrace an angle of ...
um einen Winkel drehen to move through an angle, to swing through an angle

Winkelabhängigkeit angular dependence
Winkelabstand angular distance, angular spacing
winkelarm low-inductance
winkelarmer Brückenzweigwiderstand low-inductance ratio box
winkelarmer Stöpselwiderstand low-inductance resistance box
Winkelauflösung angular resolution
Winkelbereich angular spread
Winkelbeschleunigung angular acceleration
Winkelbewegung angular movement
Winkelbock angle bracket
Winkeldämpfung angular loss
Winkelentkopplung angular decoupling (621.396.67)
Winkelfehler phase-angle error, residual error, impurity (621.317)
winkelfrei purely resistive (621.317)
winkelfreier Kondensator perfect condenser
winkelfreier Widerstand pure resistance, non-inductive resistance
Winkelfunktion circular function, bigonometric function
Winkelgeschwindigkeit angular velocity, angle velocity
Winkelgrad degree
Winkelhalbierende bisector
Winkelhebel cranked lever, bell-crank lever, toggle
Winkelhebellager bell-crank hanger
Winkellage orientation
Winkelmast angle pole, angle tower (621.315)
Winkelmesser protractor
Winkelmessung measurement of angles
Winkelmodulation angle modulation
Winkelreflektorantenne corner-reflektor antenna
Winkelspiegel optical square
Winkelstecker-M angle jack
Winkelstecker-W angle plug
Winkelstück angle adapter
Winkelsupport swivel slide (621.94)
Winkeltrieb angular drive
Winker (direction) indicator (629.113)
winklig angular
Winksignal wig-wag signal
Winter-Eichberg-Motor Latour-Eichberg motor
winzig faint, minute, negligible
winziger Fehler negligible error
winziger Strom faint current
Wippbewegung seesaw motion
wippen to seesaw
Wippenschalter rocker-dolly switch, flip-flop switch
Wirbel whirl (532.5); eddy, vortex (532), (533); eddy (538)
Wirbelablösung vortex shedding
Wirbelbremse water-brake (621)
Wirbelfeld rotational field
wirbelfreies Feld lamellar field, non-cyclical field, irrotational field
Wirbelstrom eddy current, circulating current
Wirbelstrombremse eddy-current brake
Wirbelstromdämpferscheibe copper damper
Wirbelstromdämpfung eddy-current damping
Wirbelstromkanone electric gun
Wirbelstromläufer squirrel cage rotor
Wirbelstrommotor deep-bar squirrel-cage motor
Wirbelstromscheibe Arago's disc
Wirbelstromverlust eddy-current loss
Wirbelstromweg eddy-current path
Wirbelsturm tornado
Wirbelwind whirlwind
wirbeln to whirl, to swirl; to eddy
Wirk-... active, co-phase
Wirk- und Blindkompensator coordinate potentiometer
Wirkdämpfung effective attenuation, transducer loss (US)
Wirkkomponente co-phase component, active component, resistive component
Wirklast active power (load)
Wirkleistung real power, active power, true power, wattful power, true watts (pl.)
Wirkleistung der gegenläufigen Komponente negative-sequence active power
Wirkleistung des mitläufigen Systems positive-sequence active power
Wirkleistung der symmetrischen Komponente der k-ten Folge k-sequence active power
Wirkleistungsmesser wattmeter
Wirkleitwert conductance
Wirkspannung wattfull voltage, useful voltage, in-phase voltage
Wirkstrom wattful current, useful current, in-phase current
Wirkwiderstand (effective) resistance, true resistance
Wirkwiderstands-Distanzrelais resistance relay
wirklich real, true
wirkliches Übersetzungsverhältnis true ratio (621.317)
wirksam effective, efficient, operative

wirksame Anpassung active balance
wirksam werden to take effect
wirksame Höhe effective height
wirksame Kapazität effective capacitance
wirksame Kühlung efficient cooling
wirksame Reflexionsfläche effective scattering cross-section (621.397)
wirksamer Widerstand effective resistance (621.396)
Wirksamkeit efficacy, effectiveness, efficiency
Wirkung effect, action, operation, efficiency
Wirkung beeinträchtigen to lessen the efficienty
Wirkungsbereich sphere of influence
Wirkungsgrad (energy) efficiency (621.355)
Wirkungsgrad der Umformung conversion efficiency
wirkungsmäßig functional
Wirkungsquerschnitt cross section (539)
Wirkungsquerschnitt für Einfang capture cross section (539.17)
Wirkungsweise function, efficiency, principle of action
wirr diffuse (535); entangled; confuse(d)
wirtschaftlich economic(al)
wirtschaftliche Betriebsführung economy in operation, economic operation
wirtschaftliche Fertigung productive efficiency, economy in production
Wirtschaftlichkeit (operating) economy
Wirtschaftlichkeit des Betriebes factory efficiency
wischen to wipe
Wischer wiper
Wischermeldung fleeting alarm
Wischkontakt, Wischrelais wipe contact, flick contactor, impulse contact, momentary contact, passing contact
Wismut bismuth
wispern to whisper
Wissen (schaft) science
Wissenschaftler scientist
wissenschaftlich scientific
wissenschaftliche Schreibweise scientific notation
Witterung atmospheric(al) conditions (pl.), weather
der Witterung aussetzen to weather
witterungsabhängig dependent on atmospheric conditions
Witterungsbedingungen atmospheric conditions, climatic conditions
Witterungseinfluß atmospheric influence
wobbeln to wobble, to warble
Wobbelfrequenz sweep rate, sweep frequency
Wobbelhub sweep width
Wobbelsender wobbulator, sweep generator
Woche week
Wochenbericht weekly report
Wochenproduktion weekly output
wöchentlich weekly; a week
wöchentliche Arbeitszeit working week
wohltemperiert equally tempered (534)
Wollastondraht Wollaston wire
Wolfram tungsten
Wolframblech plate tungsten
Wolframbogenlampe tungsten arc lamp, pointolite lamp
Wolframdrahtlampe tungsten filament lamp
Wolframdrahtröhre tungsten filament valve
Wolframheizfaden tungsten filament
Wolframkarbid tungsten carbide
Wolframkathode tungsten emitter
Wolframkathodenröhre bright emitter valve
Wolframlampe tungsten lamp
Wolframleuchtfaden tungsten filament (621.32)
Wolke cloud
Wolkenhöhenmesser cloud ceilometer
Wolkenuntergrenze cloud base, ceiling
Wolle wool
Wollfaden woollen thread
wollig woollen, woolly
Wommelsdorf-Maschine Wommelsdorf machine
Wort word
Wortabstand interword space
Wortlänge word length
wortorganisierter Speicher word-organized storage
Wortverständlichkeit word articulation, word intelligibility
WT-Telegraphie voice frequency (carrier) telegraphy, audio frequency telegraphy, VF-telegraphy
WT-Verstärker VF-amplifier (621.394)
Wucht velocity energy; vehemence, impetus; force
wuchtig vehement, impetuous; forceful
Wulst swell(ing), bulge, upset, bulb
Wulst bilden to bulge, to bulb
wulstig bulbous
wunder Punkt in der Konstruktion weakness in design
Wünschelrute dowsing rod, divinig rod
Wünschelrutengänger dowser, water finder
Würfel cube (51); die (pl.: dice)
Würfelantenne cubical antenna
würfelartig, würfelförmig cubical, cubiform
Würfelform cube shape

Würfelwiderstand insulativity, volume resistivity (621.3)
Würgesitz wringing fit
Würgeverbindung solderless wrap, wire wrap
Würgeverbindung mit Hülse twisted sleeve joint
Wurzel root (51); (58); (61); radix (pl.: -ices) (51)
Wurzel ziehen to root
Wurzelzeichen radical (sign) (51)
Wurzelziehen evolution (51)

X

X [Blindwiderstand] X [reactance]
x-drähtiges Seil x-strand cable
Xenon xenon
Xenon-Hochdrucklampe high-pressure xenon lamp
Xerographischer Drucker xerographic printer
X-Nut double Vee groove (621.791)
XP-Gespräch personal call
X-Schnitt X-cut, Curie cut (537)
X-Strahlung X-rays (pl.)
X-Wachs cable wax
Xylol xylole, xylene
X-Y-Scheiber XY recorder, plotter

Y

Y [Leitwertoperator] Y [quadrature component of admittance]
Yagi-Antenne Yagi array
Yaleschloß Yale lock
Y-Schnitt 30° cut (537)
Yttrium yttrium
Yttrium-Eisen-Granat YIG, yttrium iron garnet

Z

Z [Widerstandsoperator Scheinwiderstand] Z [impedance (operator)]
Z-Markierungfunkfeuer Z-marker beacon
Zacke prong
Zackenschrift variable-area (sound) track
Zacken dent, pip (621.397)
zackig dented
zackig einreißen, zacken to dent
zäh tough; sluggish
zähe Flüssigkeit sluggish fluid, viscous fluid
zähes Gefüge tough temper
zäher Stahl tough steel
zähflüssig sluggish, viscous, sticky
Zahl number, figure
Zahl eins unity
Zahl der Wicklungen number of windings
Zahl der Windungen number of turns
an Zahl übertreffen to outnumber
in Zahlen in figures
Zahlenangabe numerical data, figures
Zahlenbeispiel numerical example
Zahlendarstellung number representation
Zahlenfaktor numerical factor
Zahlencode cipher code
Zahlenkonstante numerical constant
zahlenmäßig numerical, in figures
zahlenmäßig ermitteln to evaluate
sich zahlenmäßig belaufen auf to figure
Zahlenreihe row of figures, series of figures
Zahlensystem scale of notation
Zahlentripel number triplet
Zahlenverhältnis ratio, rate
Zahlenwert numerical value, numerical constant
Zahlenwerte zuerteilen to allot numerical values
Zähl-... counting, metering
Zähleinrichtung counter, counting device
Zähleinsatz start of charging (621.395)
Zählgeschwindigkeit counting rate
Zählimpuls counting pulse; metering pulse (621.395)
Zählkreis counting circuit
Zählmagnet magnetic counter
Zählrelais stepping relay, counting relay
Zählrohr Geiger counter, counting tube (53)
Zählrohrbereich counter range
Zähltakt clock frequency (681.3); metering rate (621.395)
Zähltaste meter key, register key (621.395)
Zählwerk register, counter, registering mechanism, counting mechanism
Zählwerkablesung register reading
Zählwerkskonstante register ratio
Zählwerksreibung counter friction
Zählwerksskala register face
zählen to count, to enumerate, to register
zählendes Meßinstrument registering instrument

Zähler counter (539), (681.3); meter, energy meter, integrating meter (621.3); numerator (51)
Zähler plombieren to seal a meter
Zählerabgleich meter calibration
Zählerableser meter reader
Zähleranzeige meter indication
Zählereicher meter tester
Zählereichwerkstatt standardizing room
Zählereigenverbrauch meter loss
Zählerfehler meter error (621.317)
Zählergestell meter rack (621.396)
Zählerlagerstein meter jewel
Zählerplatz meter mounting board
Zählerprüfamt standardizing laboratory, meter testing station
Zählerprüfung meter calibration, meter testing
Zählerprüfung beim Abnehmer installation test of a meter
Zählerprüfung in der Werkstatt laboratory test of a meter
Zählerrevision inspection of a meter
Zählerschaltuhr meter change-over clock, time switch
Zählerscheibe meter disc (621.317)
Zählerstand meter indication
Zählerstandsdrucker printometer
Zählertafel meter panel
Zählerwicklung meter winding
Zählung registering, count (ing)
Zahn tooth (p.: teeth) (61), (621); cog (621.2)
Zahnbreite width of tooth, tooth breadth
Zahnbreite/Nutbreite tooth ratio
Zahnbruch tooth breakage
Zahnform tooth contour
Zahnfrequenz tooth pulsation frequency (621.313)
Zahnfuß root of the tooth (621.313)
Zahnklinkenstange ratchet bar
Zahnkopf tooth tip
Zahnkopfbreite width of tooth at top
Zahnkopfhöhe (über Teilkreis) addendum
Zahnkopfstreuung tooth top leakage (621.313)
Zahnkranz toothed rim, spur wheel
Zahnkupplung tooth clutch
Zahnprofil tooth contour
Zahnrad toothed wheel, rack wheel, spur gear; cog-wheel (621.21)
Zahnradbahn rack railway, cog-wheel railway
Zahnradfolge gear train
Zahnradfräser gear cutter
Zahnradgetriebe (toothed-wheel) gear
Zahnradlokomotive cog locomotive
Zahnradpumpe geared pump
Zahnradsatz gear train
Zahnradtrieb gearing, gear drive
Zahnradübersetzung gear transmission, gearing
Zahnsättigung tooth saturation (621.313)
Zahnspitze tooth tip (621.313)
Zahnstange toothed bar, rack
Zahnstangenritzel rack pinion
Zahnstangentrieb rack and pinion device
Zahnteilung pitch of teeth
Zahnwurzel root (of the tooth) (621.8)
Zähnung indentation, serration
Zambonische Säule Zamboni's dry pile
Zange tongs, pliers (pl.); electrode holder (621.791)
Zangenmeßgerät clamp-on meter
Zangenstromwandler split-care type transformer instrument, clip-on current transformer, hook-on current transformer, clamp-on current trans-former
Zapfen gudgeon peg, pin, stud
Zapfenlager pin bearing (621.317)
Zapfenschraube pivot screw
zapfen to tap
Zaponlack lacquer
Zarge rim (68)
zarte Linie delicate line
Zaun fence
ZB = Zentralbatterie common battery, universal battery
ZB-Amt, ZB-Vermittlung common battery central office
ZB-Telegraphieanlage central battery telegraph system
Z-Diode z-diode, voltage-reference diode
Zeche pit, colliery, coal-mine
Zedernholz cedar (wood)
Zedernholzöl cedar wood oil
Zehneck decagon
zehneckig decagonal
Zehner (number of) ten
Zehnernummerung decimal numeration
Zehnerpotenz order of magnitude
Zehner-Zählröhre tens-counting tube
Zehntel tenth; point ones
Zehntelgramm decigramme
Zehntelsekundenschalter tenthsecond timer
Zeichen sign; signal (621.396); character, symbol (51), (621.394), (681.3)
Zeichen aufnehmen to receive a signal
Zeichen fallen ein a signal is received (621.396)
Zeichendauer character duration

Zeichendrucker character printer, character-at-a-time printer, serial printer
Zeichenerkennung character recognition, pattern detection
Zeichenelement signal component, signal element
Zeichenfolge string
Zeichengenerator character generator
Zeichengeschwindigkeit operating speed
Zeichenlesen character sensing
Zeichenleser character reader
zeichenparallel parallel by character
Zeichenpause spacing wave, space
Zeichenpegel signal level (621.396)
Zeichenpunkt dot
Zeichenruf interrupted ringing
Zeichenschreibröhre character-storage tube
Zeichenspannung signal voltage (621.396)
Zeichenstrich dash
Zeichenübertragung letter transmission (621.394)
Zeichenumsetzer translator (621.394)
Zeichenumsetzrelais translating relay
Zeichenunterdrückung blanking (681.3)
Zeichenverzerrung telegraph signal distortion
Zeichenvorrat character set, character repertoire
Zeichen-... drawing, tracing
Zeichenbrett drawing board
Zeichenbüro drawing office
Zeichendreieck triangle, set square
Zeichenebene plane of the paper
Zeichengerät drawing device; plotter (681.3)
Zeichenmaschine drafting machine (US); drawing machine
Zeichenmaßstab scale rule
Zeichenpapier drawing paper
Zeichenparallel parallel by character
Zeichenschärfe definition (535)
Zeichentusche drawing-ink
Zeichenutensilien drawing materials
zeichnen to trace, to draft; to design
Zeichner draftsman (US); draughtsman, plotter
zeichnerisch darstellen to depict
Zeichnung drawing, draught; diagram
Zeichnungsentwurf draft
Zeichnungsleser drawing reader
Zeigefinger index
zeigen to show; to display, to manifest
sich zeigen to manifest oneself, to show
zeigen nach to point to (wards)
Zeiger index (pl.: -ices), needle, pointer, indicator (621.317); time vector, time amplitude, rotating vector, sinor, phasor (621.3), (538)
Zeiger der Scheinleistung vector power
Zeigerausschlag meter deflection, deflection (of the pointer)
Zeigerdarstellung vector representation
Zeigerdiagramm (rotating) vector diagram
Zeigerdifferenz vectorial difference
Zeigerdrehknopf pointer knob
Zeigerinstrument pointer instrument
Zeigerschreibinstrument direct acting recording instrument
Zeigerschreibweise vector notation
Zeigersumme vector sum, vectorial sum
Zeile line, row; scanning line, picture strip (621.397)
Zeile einer Tabelle row of a table
Zeilenablenkung scanning (621.397)
Zeilenablenkspule scanning coil
Zeilenabstand line-to-line spacing
Zeilenabtastung rectilinear scanning
Zeilendauer line duration
Zeilendrucker line printer, line-at-a-time printer
Zeilendurchlauf line slip
Zeilenfolge line sequential (621.397)
Zeilenfortschaltung line feed
Zeilenfrequenz line frequency, strip frequency (621.397)
Zeilenraster raster
Zeilenkippschaltung line scan circuit
Zeilenoffset line offset
Zeilenpaarigkeit line pairing
Zeilenrauschen clamper noise
Zeilenreißen tearing (621.397)
Zeilenrücklauf line flyback
Zeilensprungverfahren interlaced scanning
Zeilenspule scanning coil (621.397)
Zeilenstreifigkeit clamper noise
Zeilentransformator line transformer, flyback transformer
in der Zeilenrichtung laufen to scan
Zeilenversatz periodical line displacement
Zeilenvorschub line feed
zeilenweise Abtastung line by line scanning
zeilenweise Abtastung ohne Zeilensprung sequential scanning
Zeilenwobbelung spot wobble
Zeilenzugriff random access (681.3)
Z-Eisen Z-iron, Z-section
Zeit time
Zeit abstoppen to time
Zeit Eins unit time
Zeit Null zero time

Zeit bis zur Erreichung des Gleichgewichts equilibrium time
Zeitablenkung time base, sweep (621.385)
Zeitablenkgenerator sweep generator, scanning generator
Zeitablenkschaltung sweep circuit
Zeitablenkspannung sweep voltage
Zeitabschnitt time interval, period
Zeitachse time axis, time base
Zeitansage time announcement
Zeitbasis time basis, timebase
Zeitbasis-Generator time base generator, sweep generator
Zeitbasisverschiebung sweep displacement
Zeitbemessung timing
Zeitdauer duration, length of time, period
zeitdiskretes Signal time-detined signal
Zeiteinheit unit (of) time
Zeitelement, Zeitglied timer, constant-time element, time limit attachment
Zeitfehlerausgleich time-equalization circuit
Zeitgatter time gate
Zeitgeber timer, master clock, timing circuit (681.3); cycle timer (621.316.7)
Zeitgleichung equation of time (52)
Zeithaftstelle trap (621.382)
Zeitimpulszählung time-pulse metering
Zeitintervallmesser time interval meter
Zeitkonstante time constant, time factor
Zeitkonstante der Auferregung exciter build-up time constant
zeitkontinuierliches Signal time-continous signal
Zeitkreis sweep circuit
Zeitlastprüfung time-loading test
Zeitlupe slow motion (camera)
Zeitlupenaufnahme slow-motion sequence
Zeitmarke time trace
Zeitmarkengeber marker, marking generator (621.317)
Zeitmarkenschreiber chronograph
Zeitmarkierungs-... rangemarking... (621.397)
Zeitmaßstab time scale
Zeitmessung time metering
Zeitmultiplex time-division multiplex
Zeitmultiplexbetrieb time-sharing operation (621.395)
Zeitnormal time standard
Zeitnullinie time base (621.385)
Zeitplan time table, schedule
Zeitplangeber program transmitter
Zeitpunkt instant (of time)
Zeitraffer quick-motion apparatus, time accelerator
Zeitrafferaufnahme quick-motion picture
Zeitraum time interval, period
Zeitrelais time (delay) relay
Zeitschalter timer, time switch, time-lag switch
Zeitscheibe time slice, timeslot (681.3)
Zeitschreiber chronograph
Zeitsignal time signal
Zeitskala time scale
Zeitspanne time interval
Zeitspannung sweep voltage
Zeitstaffelung time grading (621.316)
Zeittaktgebermaschine time-pulse sender
Zeitvektor time vector, plane vector, revolving vector (538.55); time amplitude (621.3)
Zeitverzögerung time lag, time delay
Zeitvielfachsystem time-division (multiplex) system
Zeitwaage watch rate recorder
Zeit-Weg-Steuerung time-travel control, time-distance control
Zeitzeichen time signal
Zeitzeichengeber chronopher
zeitlich temporal, temporary
zeitlich definierter Impuls timed pulse (621.397)
zeitlich falsch anbringen to mistime
zeitlich festlegen to time
zeitlich gut gewählt well-timed
zeitlich konstant steady
zeitlich linear linear in time
zeitlich unbegrenzt beyond limited periods
zeitlich veränderlich time-variable
zeitlich veränderlich sein to vary with the time
zeitlich veränderliche Funktion time function (51)
zeitlicher Ablauf time history; chronological order
zeitliche Ableitung time derivative
zeitlicher Abstand time interval
zeitliche Änderungsgeschwindigkeit time-rate of change
zeitliche Begrenzung time limitation
zeitliche Folge chronological order
zeitliche Kapazitätsveränderung capacity drift with time
zeitliche Konstanz time stability, steadiness, permanence of value, permanency (621.317)
zeitlicher Mittelwert mean value with respect to time
zeitliche Nachwirkung time yield

zeitliches Programm time schedule
zeitliche Verwacklung time jitter (621.397)
zeitliche Verzögerung time delay
zeitlicher Zyklus time cycle
zeitlinear linear with time
zeitweilig, zeitweise temporary, intermittent; at times
zeitweiliger Fehler intermittend defect
zeitweise Abschaltung partial disconnection
Zelle cell, accumulator (621.3); cubicle (621.316)
Zelle mit Photokathode photoemissive cell
in eine Zelle hineinladen to put into a cell
Zellenanlage cell structure (621.316)
Zelleneinbau cell mounting
Zellenschalter multiple-contact switch, accumulator switch, regulating switch (621.335)
Zellenspannung cell voltage
Zellentür cell door (621.316)
Zellenverschluß inspection plug (621.355)
Zellglas plastic film
Zellophan cellophane
Zellstoff cellulose
Zellstoffbrei chemical pulp
Zellstoffwatte cellulose wadding
Zellteilung fission
Zelluloid celluloid, xylonite
Zellulose cellulose
Zelluloselack cellulose varnish
mit Zelluloselack lackieren to cellulose
Zelluloselackierung cellulose spraying
Zement cement; concrete
Zement bindet ab cement sets
mit Zement verkleiden to cement
Zementanstrich cement covering
Zementfaserplatte cement fibre slab
zementieren to cement
Zenerdiode Zener diode, avalanche diode, breakdown diode
Zenerdurchbruch Zener breakdown (621.382)
Zenerspannung Zener voltage
Zenit zenith
zenital zenithal
Zentesimalsystem centesimal system
Zentigramm centigramme
Zentiliter centilitre
Zentimeter centimetre
zentimeterweise by inches
Zentimeterwellen centimetric waves, SHF, super high frequency
Zentner cwt. = hundredweight (s. Anh.)
Zentral-... zentral central
zentral zusammenfassen to centralize
zentraler Antrieb central drive
zentraler Bordrechner main computer, main airborne computer
zentrale Lebensdauer median life
Zentralbatterie central battery, universal battery, common battery
Zentralbatteriesystem central battery signalling
Zentraleinheit central processing unit = CPU
Zentralheizung central heating
Zentrallast central load
Zentralprojektion radial projection
Zentralrechner central data processor, central computer
Zentralschmierung central lubrication
Zentralspeicher central memory
Zentralsteuerung common control
Zentraluhrenanlage electrical time-distribution system
Zentralwert median (519.2)
Zentrale central station, power station (621.313)
zentralisieren to centralize
Zentralisierung centralization
zentrieren to centre
Zentrierung centering
Zentrifugal-..., zentrifugal centrifugal
Zentrifugalanlasser centrifugal starting switch
Zentrifugalkraft centrifugal force
Zentrifugalkupplung centrifugal coupling, centrifugal clutch
Zentrifugalpumpe centrifugal pump
Zentrifugalregler centrifugal governor
Zentrifugalschalter centrifugal switch
Zentrifuge centrifuge, separator
zentrifugieren to centrifuge, to separate
Zentripetalkraft centripetal force
Zentriwinkel polar angle
Zentrum centre
Zeppelin-Antenne Zeppelin aerial
zerbrechen to break to pieces, to rupture, to fracture
zerbrechlich breakable, fragile
zerdrücken to crush
Zerfall disintegration (process), decomposition, decay
Zerfallsenergie disintegration energy
Zerfallskonstante transformation constant, decay constant, disintegration constant (539.17)
Zerfallsprodukt disintegration product, decomposition product, decay product, fission product
Zerfallsprodukte metabolons (539.16)
Zerfallsprozeß disintegration process, process of decay

zerfallen to disintegrate, to decompose, to decay
zerfallen in to fall into, to separate into
zerfasern to tease, to shred, to defibrate
zerfressen to corrode
Zerhacker chopper, vibrator
Zerhacker mit Wiedergleichrichtung synchronous type vibrator
Zerhackerpatrone vibrator unit
Zerhackerwirkung chopper action
zerkleiner to grade, to disintegrate; to break, to crush
Zerknall explosion, burst
zerknallen to explode, to burst
zerlegen in to break apart into; to resolve into (51)
Zerlegungsdreieck resolution triangle (531.2)
zermahlen to grind, to crush
zermalmen to pound
zerquetschen to mash, to smash, to crush
zerreiben to rub down
Zerreißen rupture
Zerreißdiagramm stress-strain diagram
Zerreißfestigkeit rupturing strength, ultimate strength
Zerreißmaschine testing machine
Zerreißprüfung test to destruction
Zerreißstab tension test bar, steel specimen
Zerreißversuch tension test, tensile test
zerreißen to rupture, to rend, to dismember, to tear (to pieces)
zersetzen, sich s. **zerfallen**
Zersetzung s. **Zerfall**
Zersetzungsspannung decompositionpotential, minimum potential (621.385)
Zersetzungstemperatur decomposition level
zersprühen to spray
zerschellen to fall to pieces
zerschmettern to shatter
zerstampfen to pound
zerstäuben to sputter, to pulverize, to atomize; to spray
zerstören to destroy, to demolish
zerstörendes Lesen destructive readout
Zerstörung destruction, demolition
bis zur Zerstörung prüfen to test to destruction
zerstörungsfrei non-destructive
zerstörungsfreies Lesen non-destructive readout
zerstörungsfreie Werkstoffprüfung non-destructive testing of materials
zerstreuen to scatter, to dissipate, to disperse
Zerstreuungslinse concave lens
zertrümmern s. **zerstören**
ZF = Zwischenfrequenz i. f. = intermediate frequency (s. auch dort)
ZF-Empfang supersonic heterodyne reception
ZF-Generator i. f. oscillator
ZF-Gleichrichter second detector, i. f. detector (621.396.62)
ZF-Röhre i. f. valve
ZF-Sicherheit i. f. protection
ZF-Spannung i. f. voltage
ZF-Stufe i. f. stage
ZF-Transformation i. f. t. = intermediate frequency transformation
ZF-Verstärker i. f. amplifier
Z-förmig Z-shaped
Zickzack zig-zag
Zickzack-Fahrdrahtaufhängung inclinded catenary suspension
Zickzacknietung zig-zag riveting
Zickzack-Punktschweißung staggered spot-welding
Zickzackschaltung zig-zag connection, interconnected star system (621.314.2)
Zickzackstreufluß zig-zag flux
Zickzackstreuung zig-zag air-gap leakage, belt dispersion (621.313)
Zickzackweg zig-zag path
Ziegel(stein) brick
Zieh-... drawing, pulling
Ziehbereich tuning range, wavelength coverage (621.396.61)
Zieheisen, Ziehklinge scraper, spoke shave
Ziehfähigkeit ductility
Ziehfeder drawing pen, ruling pen
Ziehfett pulling grease (621.315.2)
Ziehimpuls trigger pulse (621.397)
Ziehspan scraping
Ziehwerkzeug drawing die, scraping tool
ziehen to draw, to pull
gezogener Stahl drawn steel
gezogener Stöpsel plug out of box (621.317)
Ziel target, mark; aim, object, objective
Ziel der Untersuchung objective of the investigation
Ziel verfehlen to miss a target
Zielbereich destination area (621.395)
Zieldeckungskurs line-of-sight course
Zieldeckungsverfahren three-point system
Zielerfassung target acquisition
Zielerfassungsradar acquisition radar
Zielflug homing (flight), radio homing
Zielfluggerät homing device
Zielfunkfeuer homing beacon, NDB = nondirectional beacon
Zielkennzeichnung destination code marking

Zieltaster automatic dialer
Zielsuchgerät target seeking equipment
Zielverfolgung tracking
Zielverfolgungsradar target tracking radar
Zielvermittlungsstelle terminating exchange, destination exchange (621.395)
Zielweg track
Zielwegdarstellung track display
zielen nach to aim at
Ziffer figure, cipher, digit, number, numeral
Zifer mit höchstem Stellenwert most significant digit
Ziffer mit niedrigstem Stellenwert least significant digit
Zifferblatt dial, face
Ziffernanzeige digital display
ziffernmäßig in figures, numerical, digital
Ziffernrechner digital computer
Ziffernröhre numerical-readout tube, Nixie tube
Ziffernschaltung figure shift (621.394)
Zifferntaste figures key (621.394)
Zimmer room, chamber
Zimmerantenne indoor aerial
Zimmertemperatur room temperature
Zink zinc (54); spelter (621)
Zinkblech zinc plate
Zink-Blei-Sammler zinc-lead accumulator
Zinkblende zinc blende
Zinkdach zinc roof(ing)
Zinkdraht zinc wire
Zinkelektrode zinc electrode
Zinkerz zinc ore
Zinkoxid zinc white
Zink-Silber-Element chloride of silver cell
Zinksilikat willemite, electrical calamine
Zinkspritzguß zinc alloy die-casting
Zinksulfat zinc sulphate, white copperas
Zinksulphid zinc sulphide
Zinkvitriol white vitriol, white copperas
Zinkweiß zinc white
zinken to dovetail (68)
Zinn tin (54); older (621.71)
Zinnchlorid stannic chloride
Zinnfolie tin foil, silver-paper
zinnplattiert tin-plate
Zinntropfen bead of solder
Zipfel lobe, ear (621.396.67)
Zirkel pair of compasses, dividers (pl.)
mit dem Zirkel abstechen to step off
Zirkelschenkel leg of the compass
Zirkonium zirkonium
Zirkonfenster zirkonium window (621.386)
Zirkonlampe zirkonium lamp
zirkular polarisiert circular polarized
Zirkulation circulation
Zirkulator circulator
zirkulieren to circulate
Zischen hiss, fizz, whiz
Zischstörung precipitation static (621.396)
Zisterne cistern; storage tank (539)
Zitronensäure citric acid, salt of lemon
Zitteraal gymnotus electricus, electric eel
zitterig jittery
zittern to tremble; to quake
Zitterspannung dither voltage
Z-Markierungsfeuer zero marker
Zn Zn (54)
Zoll in. = inch (pl.: ins. = inches); duty, tariff; customs (pl.)
1 Zoll Maulweite 1 in. A/C = 1 inch across
1 Zoll Wassersäule 1 in. water
Zollstock yardstick, folding rule
zölliges Brett inch board
zonal zonal
Zone zone, area; region (621.382)
Zonenamt zone centre (621.395)
Zonengrenze zonal boundary
Zonenfolge sequence of regions
Zonenübergang junction, transition region (321.382)
gezogener Zonenübergang grown junction (621.382)
Zoner zoner (621.395.34)
zu- und abschalten to turn on and off (621.3)
Zubehör accessories, attachments (pl.)
Zubehörteil accessory, attachment, fitment, item
zubinden to tie up, to lace up
Zubringerband feed conveyor; transfer line
Zubringergleis side track, siding
Zubringerteilgruppe grading group (621.395)
Zubringerübertragung outside broadcast transmission
Zubringerverkehr shuttle traffic
Zucken des Zeigers dip (621.317)
zudecken to cover (in), to block out
Zufall chance, accident
durch Zufall by chance, by accident
Zufallsanordnung random order
Zufallsausfälle random failures
Zufallsauswahl random sampling
Zufallseinfluß chance cause
Zufallsergebnis by chance result
Zufallsgenerator random generator
Zufallsgesetz law of probability
Zufallsgröße random quantity
Zufallsstichprobe random sample

Zufallsstichprobenuntersuchung random sampling
Zufallsstreubereich random interval (620.1)
Zufallstreffer chance hit
Zufallszahl random number
zufällig accidental, incidental, contingent, casual, occasional; by chance, per chance
zufällig auf etwas stoßen to come across a thing
zufällige Berührung accidental contact
zufälliger Fehler random error
zufällige Möglichkeit contingency, chance, occasion
zufließen zu to enter (621.3); to flow to(wards)
Zufluß inflow, influx, onflow
Zuflußleitung feeding pipe (621.64)
zufügen to add
Zufuhr supply, feed
Zufuhr der Betriebsspannung steady supply (621.385)
Zufuhrkanal feeding channel
Zufuhrstrecke supply line
zuführen to supply, to feed, to inject (621.3)
zugeführte Spannung injected voltage
Zuführung lead(-in); injection; feeder (621.315)
Zuführungsdraht, Zuführungsleitung lead wire, power supply line
Zuführungsrohr feed pipe
Zug stress, pull, tension (621); traction (621.3), (538); lifting tackle (621.8); train (625); groove (623); draught (621.1)
auf Zug beanspruchen to stress
auf Zug beanspruchtes Bauelement tension member
Zuganker tie(-bar)
Zugbalken tie beam
Zugbeanspruchung stress, pull, tensile load, tensile train
Zugbeeinflussungsanlage mit festen Kontrollpunkten intermittent train control
Zugbeeinflussungshebel automatic train stop
Zugbeeinflussungsort indication point (656.25)
Zugbeleuchtung train lighting
Zugbeleuchtungsdynamo train-lighting dynamo
Zugbeleuchtungslampe traction lamp
zugbetätigtes Lichtsignalrelais approach-lighting relay (621.33)
zugbetätigte Verblockung electric approach locking
Zugbremsung train deceleration
Zugdehnung tensile strain
Zugdraht tie-wire
Zug-Druck-Dauerprüfung tension-compression fatigue test
Zugentlastung stress relief
Zugentlastungsschelle cord grip
Zugfeder dran spring
Zugfestigkeit U. T. S. = (ultimate) tensile strength
Zuggewicht train tonnage (656); tensioning bob
Zuggeschwindigkeit train speed
Zugglied tensile member, tension member
Zughaken tow hook (625)
Zugklammer cable grip (621.315.2)
Zugkraft tractive force (621.33), (538); tractive power (621.318)
Zuglast tensile load; train tonnage (65)
Zugleistung tractive power; hauling capacity
Zugleitung train line
Zugmaschine traction engine, tractor
Zugmeldekreis approach circuit (65)
Zugpendel rise-and-fall pendant (621.32); cord adjuster (621.315.6)
Zugregler draught regulator, damper
Zugregelung damper control
Zugschalter pull switch
Zugseil sash cord; hoisting cable, tracker wire
Zugseite tension side (539)
Zugseite eines Riemens tight side of a belt
Zugspannung tension, tensile stress
Zugstab, Zugstange tie bar, tension member, tension rod
Zugstärke strength of draught (621)
Zugversuch tensile test
Zugang access, approach, ingress
Zugangskennung access code
Zugangszahl access code (621.395.3)
Zugangsöffnung access opening
zugänglich come-at-able, accessible
zugeben to add
zugehörig appertaining, corresponding
zugehöriger Empfänger associated receiver
zugehöriges Profil corresponding profile
zugehöriges Rad mating gear
zugehöriger Teil complement
Zugriff access
Zugriff in stochastischer Reihenfolge random access
Zugriffsberechtigung access authorization
Zugriffszeit access time
zugriffszeitfreie Speicherung zero-access storage
Zuhaltung tumbler, ward (68)
zuhören to listen
Zuhörer listener

Zuhörerschaft audience
zukorken to cork, to stopper
zulassen to admit, to license; to permit
zugelassenes Maximum safe maximum
zulässig admissible, safe, permissible, recognized
zulässiges Abmaß permissible deviation
zulässige Abweichung tolerance, permissible derivation
zulässige Belastung safe carrying capacity, safe load; maximum permissible load (621.3)
zulässige Frequenzabweichung frequency tolerance (621.396); recognized frequence deviation
zulässige Grenze permissible limit
zulässige Höchstdrehzahl maximum safe speed
zulässiger Höchststrom safe carrying capacity, maximum permissible current (621.313), (696.6)
zulässige Höchsttemperatur maximum safe temperature
zulässiger Höchstwert safe maximum
zulässige Kontaktbelastung contact rating, permissible contact duty
zulässige Schaltleistung admissible switching power
zulässige Spannung voltage rating (621.3); safe working stress
zulässige Sperrspannung inverse voltage rating
zulässige Strombelastung safe carrying capacity
zulässige Überlast recognized overload
zulässige Verzerrung recognized distortion, admitted distortion
Zuleitung terminal wire, supply line, lead
Zuleitungen zu den Klemmen vertauschen to interchange the leads to the terminals
Zuleitungsdraht supply line wire, lead wire
Zuleitungsinduktivität lead inductance
Zuleitungskabel service cable, cable feeder, supply cable
Zuleitungsschnur flexible cable, "flex"
„Zuleitungsstrippe" "flex"
Zuleitungswiderstand lead resistance
zulöten to seal up
Zunahme increase (**an**: in)
Zünd-... igniting; detonating, initiatory
Zündanlage ignition system (621.4)
Zündanode starting anode (621.314.5)
Zündeinstellung ignition timing (adjuster)
Zündelektrode ignition electrode
zündfähig ignitable
Zündfähigkeit ignitability, detonating ability
Zündflamme ignition flame
Zündfunke igniting spark, ignition spark
Zündfunkengleichrichter pilot spark rectifier
Zündgrenzen limits of inflammability, explosive limits
Zündkabel ignition wire (621.4); shot-firing cable (66)
Zündkennlinie control characteristic, ignition characteristic
Zündkerze spark plug, sparking plug, ignition plug
Zündkreis blasting circuit, firing circuit
Zündleitung ignition cable
Zündmagnet (ignition) magneto
Zündmaschine magneto (621.4); firing apparatus, shot-firing generator (66)
Zündmittel igniting agent, means of ignition
Zündneigung proneness to ignition
Zündprobe, Zündprüfung ignition test
Zündpunkt igniting point
Zündschalter ignition switch
Zündschlüssel ignition key
Zündspannung gate-trigger voltage, ignition voltage, firing potential (621.385)
Zündspannung am Gitter critical grid voltage
Zündstörungen ignition noise (sing.) (621.396)
Zündstrom ignition current; gate trigger current (621.382)
Zündtemperatur ignition temperature, igniting point
Zündtransformator ignition transformer
Zündübertrager (gain) booster (623.4)
Zündverzug ignition delay; gate controlled delay time (621.314.63)
Zündzeit gate controlled turn-on time (621.314.63)
zünden to ignite (621.3), (621.4); to detonate, to prime; to fuse
gezündet alight (621.385), (537)
Zündung igniting, ignition, primning; firing, detonation; arc-through
Zündungseinstellung timing of ignition
Zündungsvorgang ignition process
Zunge tongue (61); reed (681)
Zungenfrequenzmesser reed frequency meter, vibrating reed instrument
Zungenlautsprecher reed loudspeaker
zuordnen to collate (681.3); to allocate
Zuordner (route) translator (621.395)
Zuordnungstabelle cross reference table
zupfropfen to wad; to stopper
zurechtschneiden to trim
zurück back, backward(s)

zurückbleiben gegen to lag (621.3); to be behind
zurückbleiben hinter to lag (621.3); to stay behind; to fall short of
zurückbleibende Magnetisierung residual magnetization, remanence
zurückblocken to open a block (625)
zurückfallen to fall back; to reset (621.318.5)
zurückfallen auf to fall back to, to jump back to
zurückführen auf to reduce to (51); to trace back to, to feed back to
zurückkommen auf to fall back to, to return to
zurückrechnen to work backward
zurückrufen to ring back (621.395)
zurücksetzen to set back, to reverse; to reduce in price
zurückschalten auf to switch back to, to return to
zurückschlagen to strike back
zurückschwingen to swing back
zurücksetzen to reset (681.3)
zurückspulen to rewind
zurückstrahlen to reradiate, to reflect
zurückverfolgen to retrace
zurückweichen to recede
zurückweisen to reject
Zurückweisung bei der Abnahme rejection
zurückwiegen to reweigh
zurückziehen to draw back, to withdraw, to retract
zusammen together
zusammen ergeben to total to, to sum up to
zusammenballen, sich to conglomerate, to agglomerate
zusammenbauen to assemble
zusammenbrechen to break down, to give way, to collapse
Zusammenbruch (einer Rechneroperation) crash (681.3)
zusammendrehen to twist (up)
zusammendrücken to compress, to tamp
zusammenfallen to coincide (**mit**: with); to break down, to collapse
zusammenfalten to fold (up)
zusammenfassen to collect (51); to resume, to wind up
zusammenfrieren to regelate
zusammenfügen to fit togehter, to join, to connect
Zusammenhalt cohesion, cohesiveness
zusammenhalten to hold together, to cohere
zusammenhaltend cohesive
Zusammenhang conjunction, connection
im Zusammenhang mit in conjunction with
in diesem Zusammenhang in this connection
zusammenhanglos incoherent
zusammenhängen to be coherent
zusammenhängend coherent; continuous
zusammenkuppeln to couple up
zusammenlegbar collapsible
zusammenpassen to fit to gether
zusammenrechnen to reckon up
zusammenschalten to tie together (621.31); to compound, to connect (621.3)
zusammenschieben (lassen), sich to telescope
zusammenschmelzen to fuse to gether
zusammensetzen to compose, to compound, to synthesize (621.397)
sich zusammensetzen aus to be compounded of, to consist of
Zusammensetzung composition, combination, assembly
zusammenstellen to compose, to collect
Zusammenstellung combination; aggregate
Zusammenstoß impact, collision, crash
zusammenstoßen to collide
zusammenstoßen mit to knock against
zusammenstückeln to piece up
zusammenstürzen to collapse
Zusammentreffen convergence, coincidence; meeting
zusammentreffen to converge (51) (**bei**: at); to meet; to coincide (**mit**: with)
zusammenwirken to act together, to interact
zusammenziehen, sich to contract
Zusammenziehung contraction
Zusatz supplement, attachment, addition; option
Zusatzagens addition agent
Zusatzanode extra anode (621.314.6)
Zusatzausrüstung extra equipment; extras (pl.)
Zusatzbatterie booster battery
Zusatzbeanspruchung additional stress
Zusatzdämpfung obstruction loss, additional loss
Zusatzgebühr extra charge, additional fee
Zusatzgitter extra grid (621.385)
Zusatzlast additional load, superimposed load
Zusatzleiter extra conductor
Zusatzluft make-up air, additional air flow
Zusatzmaschine booster (621.3)
Zusatzmetall filler metal (621.79)
Zusatzregister extension register
Zusatzspannung series voltage, boost
Zusatztransformator series transformer, booster transformer

Zusatzverlust extra loss, stray loss (621.313)
Zusatzwecker extension bell (621.395)
Zusatzwiderstand extra resistance, spurious resistance
Zusatzzahn hunting to oth
zuschalten to add, to switch on (621.31)
zu- und abschalten to turn on and off, to switch on and off, to make and break
zuschließen to lock (up)
zuschmelzen to seal (up)
zuschneiden to true
zuschrauben to screw up
zuspitzen to thin, to taper, to peak (621.39)
zugespitzter Bruch pointed fracture
zusprechen to phonogram
zugesprochenes Telegramm phonogram
Zustand state; condition, conditions (pl.); stage; status (681.3)
Zustand des stabilen Gleichgewichts condition of stable equilibrium, state of equilibrium
Zustand der Ionosphäre ionospheric conditions (pl.)
Zustand des Leermitlaufens windmilling conditions (pl.) (629.13)
Zustand des Verfalls dilapidation
im Zustande der Verflüssigung liquescent
Zustandsänderung change of state
Zustandsbit status bit
Zustandsdiagramm constitution diagram
Zustandsgleichung equation of state
Zustandsschaubild constitution diagram, phase diagram
zustöpseln to stopper
Zustrom inflow, influx
zuströmen to flow in
zuteilen to (pre-)assign, to allot
zugeteilte Frequenz assigned frequency
Zuteilung pre-assignment (621.396)
Zutritt access, admission
zuverlässig reliable, sound, positive
zuverlässige Maschine sound machine, reliable machine
Zuverlässigkeit reliability, dependability
Zuverlässigkeit im Betrieb service reliability, dependability in operation
Zuverlässigkeitsprüfung reliability test
zuweisen to allocate
zugewiesener Frequenzbereich allocated band
Zwang force, constraint
Zwangsauslösung forced release
zwangsbelüftet, zwangsventiliert force-ventilated, forced-air cooled
Zwangsbelüftung forced ventilation
Zwangslaufverfahren compelled signalling (621.395.3)
zwangsschlüssig positive
zwangsschlüssige Sicherungsscheibe positive lock washer
Zwangsumlaufkessel circulation boiler, pressurized boiler
zwängen to force, to urge
Zwecke tack
Zwei . . ., zwei two, twin, double
Zweiadresscode two-address code
zweiadriges Kabel twin-wire cable, two-core cable
zweiatomig diatomic
Zwei-aus-Fünf-code two-out-of-five code
Zweibahn-Kommutator two-track commutator
Zweibandtelefonie two-band telephony
Zweibein bipod
Zweibereichschalter two-position range selector
zweidimensional two-dimensional
Zweidraht- . . . two-wire
Zweidrahtbrückenverstärker 21 type repeater (621.396)
Zweidrahtfernsprechleitung two-wire telephone line
Zweidrahtgabelverstärker 22 type repeater (621.395)
Zweidraht-Gegensprechverkehr simultaneous two-wire conversation
Zweidrahtkreis two-wire telephone circuit
Zweidrahtleitung two-wire line
Zweidrahtschaltung two-wire telephone circuit
Zweidrahtsystem two-wire system (621.395)
Zweidrahtverstärker two-wire system amplifier, two-wire repeater, two-way repeater
Zweielektrodenröhre two-electrode valve; two-element tube (US)
Zweieranschluß two-party line
Zweierbündel two-conductor bundle, duplex-bundle conductor
zweietagige Wicklung two-range winding
zweifach . . . double, duplex
Zweifachkanal-Frequenzumtastung double-frequency shift keying
Zweifach-Konusantenne biconical horn-antenna
zweifach polumschaltbarer Motor two-speed pole-changing motor
Zweifach-Richtantenne bidirectional antenna
Zweifachunterbrechung double break

Zweifarbträgerverfahren two subcarrier system (621.397)
Zweifrequenzlokomotive dual-frequency locomotive
Zweiganggetriebe two-speed gear
Zweigitterröhre dual-grid tube, double-grid tube
zweigleisige Strecke double-track line
Zweikammerklystron two-resonator klystron
Zweikreisschaltung (zur Zählerprüfung) two-circuit method (for meter testing)
zweilagige Wicklung double-layer winding
Zweileiter-... two-wire
Zweileiteranlage two-wire system (621.3)
Zweileiterkabel twin cable, duplex cable, "D" cable
Zweileiternetz two-wire system
Zweileiterzähler two-wire meter
zweimal baumwollisoliert double-cotton covered
zweimotorig twin-engined
Zweiphasen-... two-phase, quarter-phase
Zweiphasen-Dreileitersystem two-phase three-wire system
Zweiphasen-Fünfleiteranlage two-phase five-wire system
Zweiphasengenerator two-phase generator, two-phase alternator
Zweiphasenmotor two-phase motor
Zweiphasenstrom two-phase current
Zweiphasen-Vierleitersystem two-phase four-wire system
Zweiphasenwicklung two-phase winding
Zweiphasenzähler two-phase meter
zweiphasig two-phase, biphase, quarter-phase
zweiphasige Wirkung two-phase winding
Zweipol dipole; two-pole (circuit element), two-terminal network (621.392)
Zweipol-... two-pole, two-terminal
Zweipolröhre diode, two-electrode valve
zweipolig two-pole, two-terminal, double-pole, bipolar
zweipoliger Anker two-pole armature
zweipolige Installation two-conductor wiring
zweipoliger Motor bipolar motor
zweipoliger Schalter two-pole switch, double-pole switch
zweipolige Steckdose two-pole socket
zweipoliger Stecker two-pole plug, two-pin plug
zweipoliger Umschalter double-pole double-throw switch
Zweipunktregler two-limits regulator, two-point stabilizer; two-step controller, two-position controller (US)
Zweipunktsteuerung two-limits control, bistable control, black-white control, flicker control
Zweipunktverhalten on-off action, two-step controller action
zweireihig double-row
zweireihiges Kugellager double-row ball bearing
Zweirichtungs-Transistor bi-directional transistor
Zweirichtungszähler reversible counter
Zweischlitzmagnetron split-anode magnetron
zweischneidig cutting both ways
Zweiseitenband Übertragung double-sideband transmission
zweiseitig gerichtet bidirectional
zweiseitiger Anschluß dual service, plural service
zweiseitige Richtantenne bidirectional antenna, bilateral antenna
zweistellig two-digit
Zweistellungsregelung two-position action
Zweistofflegierung binary alloy
Zweistrahlkathodenstrahlröhre dual-beam cathode-ray tube
Zweistrahloszillograph double-beam oscilloscope, dual-beam oscilloscope
Zweistufen-..., zweistufig two-stage, two-step
zweistufiges Relais two-step relay
zweistufiges Verfahren double-stage process
zweistufige Verstärkung two-stage amplification
Zweitakt two-stroke (cycle)
Zweitaktmotor, Zweitakter two-stroke engine
Zweiwattmeterschaltung two-watt-meter method
Zweiweggleichrichter full-wave rectifier
Zweiwegverstärker forked repeater
zweiwertig divalent (54); double valued (51)
Zweiwicklungsrelais two-element relay, double-winding relay
Zweiwicklungs-Synchrongenerator double-winding synchronous generator
Zweizweckröhre two-purpose valve
zweizweigige Wellenwicklung two-circuit winding (621.313)
Zweig branch; department
Zweiggesellschaft subsidiary company, affiliated company
Zweigleitung branch line

Zweigstelle subbranch
Zweigstrom branch current
zweit second
zweiter Gleichrichter second detector (621.396.62)
zweite Güte(klasse) second class
zweiter Hauptsatz second law of thermodynamics
zweite Hitze reheat
zweite Ordnung second order
Zweitluft secondary air
Zweitwicklung secondary (winding) (621.314.2)
Zwergempfänger midget set
Zwergfassung midget base, miniature base
Zwerglampe midget lamp, miniature lamp, pygmy lamp
Zwickel gusset
Zwickzange pincers (pl.)
Zwielicht twillight, gloaming
Zwillinge Twins, Gemini (52)
Zwillinge bilden to twin (54)
Zwillingsbildung twinning, twin formation (54)
Zwillingsklinke jack pair
Zwillingskontakt bifurcated contact, twin contact
Zwillingskristalle twinned crystals
zwingen to force
zwingend vorschreiben to exact
zwischen den Elektroden interelectrode
Zwischenbasisschaltung grounded-base circuit
Zwischenbild intermediate image (535)
Zwischenbild-Kameraröhre image camera tube (621.397)
Zwischendose pull box (621.315.6)
Zwischenfrequenz (siehe auch **ZF**) intermediate frequency
Zwischenfrequenzgenerator i. f. oscillator = intermediate frequency oscillator
Zwischenfrequenzspannung i. f. voltage
Zwischenfrequenzteil i. f. section
Zwischenfrequenzübertrager i. f. transformer
Zwischenfrequenzverstärker i. f. amplifier
zwischenfügen to fit between
Zwischenfutter extension socket (621.9)
Zwischengetriebe intermediate gearing
Zwischenglied intermediary, intermediate link, connecting link
Zwischenglühung intermediate annealing
Zwischenkabel intermediate cable
zwischenkeilen to wedge between
Zwischenkern compound nucleus, intermediate nucleus (539)
Zwischenklemme intermediate terminal
Zwischenkonus flared section (621.396.67)
Zwischenlage interlacing
zwischenmolekular intramolecular
Zwischenrad idler
Zwischenraum gap, separation, interspace, clearance
Zwischenrechnung manipulation (51)
zwischenschalten to tap into, to insert between (621.3)
Zwischenschaltung insertion between (621.3)
Zwischenschaltstellung transition stop
Zwischenschicht interface, intervening layer, intermediate layer
Zwischenschichten einlegen to interleave (621.3)
zwischenschichten to sandwich
Zwischensignal intermediate signal (625.1)
Zwischensockel valve adapter (621.385)
Zwischenspeicher temporary storage, register
Zwischenstation repeater, repeating station, intermediate station (621.394), (621.396)
Zwischenstellenumschalter interthrough switch
Zwischenstufe interstage
Zwischenstufenmodulation amplification of modulated high frequency (621.396.61)
Zwischenstück connecting piece, intermediary
Zwischenübertrager interstage transformer, intervalve transformer (621.396.64)
Zwischenübertragung mit Eisenkern iron-cored intervalve transformer
Zwischenverstärker intermediate repeater
Zwischenverteiler intermediate distribution frame
Zwischenvorgelege intermediate gearing
Zwischenwahlzeit interdigital interval (621.395)
Zwischenwelle countershaft, jack shaft (621.335)
Zwischenzeilenflimmern interline flicker
Zwischenzeilenverfahren interlacing (621.397)
Zwischenzeit interval, intervening time, meantime
zwitschern to cheep
zyklisch cyclic(al)
zyklische Blockprüfung cyclic redundancy check
zyklischer Code cyclic code
zyklische Vertauschung cyclic interchange
Zykloide cycloid
Zyklotron cyclotron
Zyklus cycle
Zyklusimpuls commutator pulse
Zykluszählung cycle count
Zykluszeit cycle time

Zylinder cylinder (51), (621)
Zylinderkoordinaten cylindrical polars
Zylinderlinse cylindrical lens
Zylinder-Spalt-Magnetron hole-and-slot anode magnetron
Zylinderspule solenoid
Zylindersymmetrie cylindrical symmetry
zylindrisch cylindrical

Part II
English-German

A

A = ampere A = Ampere
Å A.E. (Ångström-Einheit = 10^{-8} cm)
abampere elektromagnetische c.g.s.-Stromeinheit
to abandon a call einhängen, ein Gespräch beenden (621.396)
A-battery Heizbatterie
abbreviated dialling Kurzwahl
abbreviated number Kurzrufnummer
abcoulomb elektromagnetische c.g.s.-Ladungseinheit
aberration Aberration
abfarad elektromagnetische c.g.s.-Kapazitätseinheit
abhenry elektromagnetische c.g.s.-Einheit der Induktivität
abmho elektromagnetische c.g.s.-Einheit des Leitwertes
abnormal anomal, abnormal
abnormal glow unselbständige Glimmentladung
abnormally polarized waves anomal polarisierte Wellen
abohm elektromagnetische c.g.s.-Widerstandseinheit
abort (Programm-)Abbruch (681.3)
above über, hinaus
above ground oberirdisch
above ground level über Erde
above the rated value über den Nennwert hinaus
above sea level über N.N.
to abrade abschleifen, abschaben, abreiben, abschürfen
to abrase abreiben, abschürfen
abrasion Abrieb, Abnutzung, Abschürfung
abrasion resistance Abriebfestigkeit, Kratzfestigkeit, Verschleißfestigkeit
abrasion resistant verschleißfest
abrasive Schleif-..., schleifend; Schmirgel
abrasive paper Sandpapier, Schleifpapier
abrasive wheel Schleifscheibe, Schmirgelscheibe
abscissa (pl.: -ae) Abszisse
absent-subscriber service Abwesenheitsdienst (621.395.3)
absolute absolut, vollständig
absolute address absolute Adresse
absolute alcohol absoluter Alkohol, reiner Alkohol
absolute altimeter absolut anzeigender Höhenmesser
absolute block absoluter Block, unbedingtes Halt (625)
absolute delay Festverzögerung (621.397)
absolute efficiency idealer Wirkungsgrad
absolute electrical units absolute elektrische Einheiten
absolute electrometer Spannungswaage
absolute level absoluter Leistungspegel
absolute limiting value absoluter Grenzwert
absolute loader Absolutlader (681.3)
absolute maximum rating Grenzwert, maximaler Betriebswert
absolute peak Jahresspitze (621.316)
absolute permissive block absoluter permissiver Block (625)
absolute sensitivity Grenzempfindlichkeit, absolute Empfindlichkeit
absolute system of units absolutes Einheiten-System
absolute value Absolutwert
absolute zero (of temperature) absoluter Nullpunkt (der Temperatur)
to absorb absorbieren, schlucken, aufnehmen
to absorb energy from Energie entziehen
to absorb moisture Feuchtigkeit aufnehmen
to absorb power Leistung aufnehmen, Leistung verbrauchen
to absorb voltage Spannung aufnehmen, Spannung verbrauchen
absorbed dose absorbierte Dosis (621.384)
absorbency Saugfähigkeit
absorbent saugend, saugfähig
absorbent material Schallschluckmaterial
absorbent paper Löschpapier, Fließpapier
absorber Absorptionsstoff, Schlucker, Dämpfer, Absorber (534); Absorptionsmasse (54); Blindantenne (621.396)
absorber valve Absorptionsröhre
absorptance Absorptionskoeffizient
absorption Absorption (53), (54), (621.396); Zusatzdämpfung (621.396)
absorption area Absorptionsfläche
absorption bands Absorptionsbanden
absorption capacity Schluckvermögen, Absorptionsvermögen
absorption cartridge Filterzelle (621.314.2)
absorption circuit Saugkreis, Absorptionskreis
absorption coefficient Absorptionskoeffizient
absorption coil Nullpunktdrossel
absorption control Absorptionsmodulation (621.396); Absorptionssteuerung (539)
absorption cross section Absorptionsfläche (621.396); Absorptionsquerschnitt
absorption current Nachwirkungsstrom (537)

absorption dynamometer Leistungsbremse (621)
absorption factor Absorptionskoeffizient (535)
absorption frequency meter Absorptions-Frequenzmesser, Resonanzfrequenzmesser
absorption line Absorptionslinie (535)
absorption loss Dämpfungsmaß; Absorptionsverlust
absorption modulation Absorptionsmodulation
absorption of moisture Feuchtigkeitsaufnahme
absorption power meter Absorptionsleistungsmesser
absorption spectrum Absorptionsspektrum (536), (537)
absorption trap Absorptions-Sperrkreis
absorption-type refrigerator Absorptionskühlschrank
absorption of vibrations Absorption von Schwingungen, Schwingungsdämpfung
absorption wavemeter Absorptionswellenmesser (621.396)
absorptive absorbierend
absorptive power Absorptionsvermögen
abstatampere elektrostatische c.g.s.-Einheit des Stromes
abstatcoulomb elektrostatische c.g.s.-Einheit der Ladung
abstatfarad elektrostatische c.g.s.-Einheit der Kapazität, cm
abstatvolt elektrostatische c.g.s.-Einheit der Spannung
to abstract ab-, ent-, ausziehen, absondern
to abstract energy from Energie entziehen aus
abundance Häufigkeit (z.B. des Vorkommens)
abundance determination Häufigkeitsbestimmung, Gehaltsbestimmung
abundance ratio Isotopenhäufigkeitsverhältnis
abvolt elektromagnetische c.g.s.-Einheit der Spannung
a.c. = alternating current Wechselstrom
a.c. capacitor Starkstromkondensator
a.c. commutatormachine Wechselstromkommutatormaschine
a.c. commutator motor Wechselstromkommutatormotor
a.c. component Wechselstromkomponente
a.c. current transfer ratio Wechselstrom-Verstärkungsfaktor (621.375)
a.c. dump Netzausfall
a.c. magnetic biasing Wechselstrom-Vormagnetisierung, Wechselstrom-Löschung
a.c. magnetic brake Wechselstrombremse
a.c. mains network Wechselstromnetz
a.c. motor Wechselstrommotor
a.c. operated mit Wechselstrombetrieb
a.c. operation Wechselstrombetrieb, Wechselstromnetzanschluß
a.c. plate resistance innerer Widerstand (einer Röhre) (621.385)
a.c. potentiometer Wechselstrom-Kompensationsapparat
a.c. transmission Wechselstromübertragung, gleichstromfreie Übertragung (621.397)
a.c. voltage Wechselspannung
a.c. winding Wechselstromwicklung
to accelerate beschleunigen
to accelerate subsequently nachbeschleunigen
accelerated life test geraffte Lebensdauerprüfung
accelerated reliability test verkürzte Zuverlässigkeitsprüfung
accelerated torque Beschleunigungsmoment
accelerated voltage Beschleunigungsspannung (621.385)
accelerating grid Nachbeschleunigungsanode (621.385), (621.397)
accelerating loss Anfahrverlust, Beschleunigungsverlust (621.33)
accelerating potential Beschleunigungsspannung (621.385)
accelerating power Beschleunigungsleistung (621.33)
acceleration Beschleunigung
acceleration along the path Bahnbeschleunigung (53)
acceleration due to gravity Schwerebeschleunigung, Erdbeschleunigung
accelerator Beschleuniger, Beschleunigungselektrode
accelerator anode Beschleunigungsanode (621.385)
accelerometer Beschleunigungsmesser
to accentuate hervorheben, anheben
to accentuate contrast Kontrast verstärken (621.396)
to accentuate harmonics Oberwellen herausheben
accentuation Vorverzerrung, Preemphasis, Tonanhebung
accentuator Anhebeschaltung, Preemphasisschaltung
acceptable zulässig, annehmbar
acceptable quality level = AQL zulässige Fehlerrate
acceptance Abnahme
acceptance certificate Abnahmeprotokoll

acceptance official Abnahmebeamter
acceptance specifications Abnahmebedingungen
acceptance test Abnahmeprüfung
acceptor Akzeptor
acceptor circuit Spannungsresonanzkreis, Saugkreis
acceptor impurity Akzeptorverunreinigung
access Zugang, Zutritt, Zugriff
access authorization Zugriffsberechtigung
access code Zugangskennung
access fitting Verbindungsdose, Abzweigdose (621.315.6)
access method Zugriffsart
access opening Zugangsöffnung, Bedienungsöffnung
access time Zugriffszeit, Speicherzugriffzeit
accessibility Erreichbarkeit (554.5); Zugänglichkeit
accessories (pl.) Zubehör
accessory Zubehör-..., Hilfs-...
accessory charge Nebengebühr, Nebenkosten (pl.) (621.317)
accessory fitment Zubehörteil
accidental zufällig, unabsichtlich
accidental contact zufällige Berührung
accidental ground Erdschluß
accompanying audio channel Tonträgerkanal
accounting period Ablesezeitraum (621.317)
to accumulate sammeln, sich sammeln, sich ansammeln
accumulating register Addierwerk
accumulation Ansammlung
accumulation of heat Wärmestauung
accumulative sich anhäufend
accumulator Akkumulator, Sammler; Akkumulator (681.3)
accumulator battery Akkumulatorenbatterie
accumulator car Batteriefahrzeug, Triebwagen
accumulator cell Akkumulatorenzelle
accumulator charger Batterielader, Batterieladeeinrichtung
accumulator charging Batterieladung
accumulator container Akkumulatorengefäß, Batteriekasten
accumulator grid (plate) Gitterplatte (621.35)
accumulator insulator Gestell-Isolator (621.35)
accumulator plate Akkumulatorenplatte
accumulator lamp Akkulampe, Batterielampe
accumulator-lead plate Akkumulatorbleiplatte
accumulator locomotive Akkumulatorenlokomotive
accumulator switch Zellenschalter (621.35)
accuracy Genauigkeit, Fehlergrenzen
accuracy of adjustment Abgleichgenauigkeit, Einstellgenauigkeit
accuracy of measurement Meßgenauigkeit (621.317)
accuracy rating Fehlergrenzen-Klassifizierung
accuracy of reading Ablesegenauigkeit
accuracy of signal repetition Verzerrungsfreiheit
accurate genau
accurate sending sauberes Geben (621.394), (621.6)
acetate Acetat
acetate wire azetatfolien-isolierter Draht
acetylene Azetylen
achromatic achromatisch
achromatism Achromatismus
ACIA = asynchronous communications interface adapter Interface für asynchronen Datenaustausch
acid Säure
acid-proof säurefest
acid-resistant, acid-resisting säurebeständig, säurefest
acid-resisting paint säurefeste Farbe
acidimeter Säuremesser
acidity Säurewert, Säuregehalt
acidless säurefrei
acknowledge positive Rückmeldung
to acknowledge zustimmen (625); bestätigen, quittieren (621.39)
acknowledger Quittungsschalter (656.26)
acknowledging Zustimmung (625)
acknowledging circuit Quittungskreis (621.39)
acknowledging contactor Quittungsschalter (656.25)
acknowledging relay Quittungsrelais
acknowledging switch Quittungsschalter (621.39)
acknowledgment signal Quittungszeichen
acme thread Trapezgewinde
acorn Eichel
acorn(-type) tube, acorn valve Eichelröhre, Knopfröhre
acoustic(al) akustisch
acoustic absorptivity akustisches Absorptionsvermögen
acoustic coupler Akustikkoppler
acoustic delay line akustische Verzögerungsleitung
acoustical electrical transducer akustischer Wandler, Schallwandler
acoustic feedback akustische Rückkopplung
acoustic horn Schalltrichter

acoustic impedance akustischer Widerstandsoperator, akustischer Scheinwiderstand
acoustic lining akustische Verkleidung
acoustic memory akustischer Speicher
acoustical quality Hörsamkeit
acoustical quantity akustische Größe
acoustic radiator Schallstrahler
acoustic radiometer Schallstrahlungsdruckmesser
acoustic reactance akustischer Blindwiderstand
acoustic resistance akustischer Wirkwiderstand
acoustic shock reducer Knackschutz (621.395), (621.396)
acoustic sounding Echolotung
acoustical speech power Sprechleistung
acoustic tile schallschluckende Wandplatte, Schluckplatte
acoustic vibration akustische Schwingung
acoustics (pl.) Akustik, (Raum-)Akustik
acoustics of the space Raumakustik
acquisition Zielerfassung
acquisition radar Zielerfassungsradar
acquisition and tracking radar Erfassungs- und Verfolgungsradar
actinic aktinisch
action Wirkungsweise; Aktion, Tätigkeit, Wirkung
to be in action in Betrieb sein, betrieben werden
action current Aktionsstrom
action and reaction Kraft und Gegenkraft
to activate aktivieren, in Tätigkeit setzen
activated carbon Aktivkohle
activation Aktivierung; Anregung (54)
activation cross section Aktivierungsquerschnitt
activation energy Aktivierungsenergie
active „heiß" (539); aktiv, arbeitend, belastet
active aerial Primärstrahler, Erreger, aktiver Strahler
active balance wirksame Anpassung
active balance return loss Echodämpfung, aktive Fehlerdämpfung
active component Wirkkomponente (537); aktives Bauelement
active current Wirkstrom
active deposit aktiver Niederschlag
active electric network aktiver Stromkreis, elektrisches Netzwerk
active electrode Sprühelektrode (628.511)
active fuse Betriebssicherung
active filter aktives Filter
active homing Eigenzielsteuerung
active level aktiver Pegel
active material Emissionssubstanz
active paste aktive chemische Masse (621.355)
active power Wirkleistung
active power load Wirklast
active program ablauffähiges Programm
active repair time Instandsetzungszeit
active return loss Echodämpfung (US)
active transducer Energiewandler mit Energiequelle
active transition aktiver Pegelübergang
active volt-amperes Wirkleistung
active voltage Wirkspannung
active waste radioaktiver Abfall
active wires (pl.) aktive Leiterlänge
activity Aktivität
actual wirklich, tatsächlich
actual adress echte Adresse
actual frequency Istfrequenz, Arbeitsfrequenz
actual level absoluter Pegel
actual power tatsächliche Leistung, Wirkleistung (621.3)
actual resistance Echtwiderstand
actual value Istwert
actual warning Vollalarm (35)
to actuate betätigen, in Gang setzen
actuate message Ausgabebefehl
actuating Betätigung
actuating arm Antriebshebel, Betätigungshebel
actuating current Ansprechstrom, Betätigungsstrom
actuating signal Betätigungssignal
actuating switch Betätigungsschalter
actuating time Stellzeit
actuating variable Stellgröße
actuating voltage Betätigungsspannung
actuation Betätigung
actuation by boot Fußbetätigung
actuator Schalter, Stellantrieb; Stellglied (621.316.7)
acyclic azyklisch
acyclic generator Unipolargenerator
acyclic machine Unipolarmaschine, Nöggerathmaschine
to adapt anpassen, (zweckentsprechend) abwandeln
adaptation (zweckbestimmte) Abwandlung, Anpassung
adapter Paßstück, Zwischenteil, Adapter, Anpassungsglied
adapter plug Schraubstecker
adapter sleeve Reduzierhülse

adapter transformer Sockeltransformator (621.315.6)
Adcock direction finder Adcock-Peiler (621.396)
A/D converter A/D-Wandler, Analog Digital Wandler
to add zuschalten (621.3); auftragen (621.791); addieren (51)
to add vectorially vertoriell addieren
added material aufgeschweißtes Material, Schweiße
adder Addierwerk im Rechenwerk; Addierer (681.3)
adder circuit Addierschaltung
add instruction Additionsbefehl
additional zusätzlich, Zusatz...
additional character Sonderzeichen
additional load Zusatzlast
additional loss Zusatzdämpfung
additional stress Zusatzbeanspruchung
additive colour mixing additives Farbsystem
additive mixing additive Mischung
address Adresse (681.3)
address bus Adreßbus
address counter Befehlszähler
address part Adressenteil, Adreßraum
address register Adreßregister
addressing Adressierung
addressing mode Adressierungsart
ADF = automatic direction finder automatischer Peiler
ADF approach Anflug mit automatischer Peilung, ADF-Anflug
to adhere to festhalten an, festhaften an
to adhere to directions Anweisungen beachten
to adhere to rules Vorschriften einhalten, Regeln beachten
adhesion Adhäsion (53); Haftvermögen, Haftung (621.35)
adhesive Klebemittel
adhesive klebend, klebrig
adhesive factor Reibungszahl
adhesive strength Haftfestigkeit, Klebkraft
adhesive tape Klebeband, Klebestreifen
adhesive wax Klebwachs
adhesiveness Klebfähigkeit, Haftfähigkeit
adiabatic adiabatisch
adiabatic compression adiabatische Kompression
adiabatic expansion adiabatische Expansion
adiabatic line Adiabate
adjacent anliegend (513); anstoßend, benachbart
adjacent carrier trap Nachbarbildfalle
adjacent channel Nachbarkanal (621.935.4)
adjacent-channel interference Nebensprechen aus dem Nachbarkanal
adjacent-channel selectivity Nachbarkanalselektion
adjacent turns benachbarte Windungen, Nachbarwindungen
adjacent video carrier trap Bildträgersperre
adjoining anliegend (513)
to adjust abgleichen, einstellen, justieren, einrichten
to adjust a meter for a scale ein Instrument auf einen Meßbereich umschalten
adjustable justierbar, einstellbar, abgleichbar
adjustable capacitor veränderlicher Kondensator
adjustable inductor veränderliche Induktivität
adjustable parallel blocks Beilagekeile, verstellbare Beilage
adjustable resistance Abgleichwiderstand
adjustable resistor veränderbarer Widerstand, einstellbarer Widerstand; Regelwiderstand
adjustable spanner Verstellschlüssel, Engländer
adjustable speed einstellbare Drehzahl
adjustable speed motor drehzahlregelbarer Motor
adjustable varying-speed motor drehzahlregelbarer Motor
adjustable varying-voltage control Drehzahlregelung mit Kompoundcharakteristik (621.313)
adjustable voltage control Drehzahlregelung mit Nebenschlußcharakteristik (621.313)
adjustable voltage regulator einstellbarer Spannungsregler
adjustable for wavelength in der Wellenlänge einstellbar
adjustable wrench Verstellschlüssel, Patentschlüssel, Engländer
adjusting Stell-..., Einstell-...
adjusting device Einstellvorrichtung
adjusting knob Einstellknopf
adjusting lever Stellhebel
adjusting motor Stellmotor
adjusting nut Einstellmutter, Stellmutter
adjusting screw Einstellschraube, Stellschraube, Regulierschraube, Justierschraube
adjusting wedge Stellkeil
adjustment Einstellung, Abgleich
adjustment area Einstellbereich
adjustment error Abgleichfehler
adjustment range Einstellbereich
admiralty Admiralität

admiralty unit (of capacitance) 1/900 Mikrofarad = 1000 cm
admissible zulässig
admissible switching power zulässige Kontaktbelastung, zulässige Schaltleistung
admission Zutritt, Einlaß, Einströmung
admittance Leitwert, Scheinleitwert, Leitwertoperator
admittance chart Leitwertdiagramm
admittance meter Admittanz-Meßgerät
to adsorb adsorbieren
adsorbed molecules adsorbierte Moleküle
adsorption Adsorption
adsorptive Adsorptions-...
adsorptive power Adsorptionsfähigkeit
advance Voreilung
in advance of voreilend gegen
advance element Differenzierglied, Vorhalteglied (Regelung)
advance wire Konstantandraht
to advance vorschieben, anstellen, vorstellen
to advance the brushes die Bürsten vorverschieben
to advance a step um einen Schritt vorschieben
to advance the volume control die Lautstärke aufdrehen, den Lautstärkeregler aufdrehen
advanced ignition Frühzündung
advertising literature Prospektmaterial
to aerate lüften, belüften
aerial Antenne (siehe auch **antenna**)
aerial array Antennengruppe, Antennenfeld
aerial base Antennenfuß
aerial bay Antennenfeld
aerial center Antennenschwerpunkt
aerial circuit Antennenkreis
aerial coil Antennenspule
aerial coupling Antennenkopplung
aerial coupling circuit Antennenkopplungskreis
aerial current Antennenstrom
aerial curtain Antennenwand
aerial diagram Antennendiagramm
aerial-directivity pattern Antennenrichtdiagramm
aerial discharger Antennen-Ableiter
aerial effect Hochantenneneffekt (621.396.9)
aerial efficiency Antennenwirkungsgrad
aerial element Antennenelement
aerial feed Erreger (621.396.67)
aerial-feed impedance Antenneneingangswiderstand
aerial filter Antennenweiche
aerial-following control Antennennachlaufsteuerung
aerial gain Antennengewinn
aerial insulator Antennenisolator
aerial mount Antennenuntersatz
aerial pattern Antennencharakteristik
aerial pedestal Antennenfuß, Antennenuntersatz
aerial power Antennenleistung
aerial-radiation resistance Antennenstrahlungswiderstand
aerial reflector Antennenreflektor
aerial resistance Antennenwiderstand
aerial spacing Antennenabstand
aerial support Antennenträger
aerial switch Antennenschalter
aerial system Antennensystem
aerial tuning Antennenabstimmung
aerial-tuning capacitor Antennenabstimmkondensator
aerial-tuning inductance Antennenabstimmspule
aerial-tuning unit Antennen-Abstimmglied, Antennen-Anpassungsglied
aerial wire Antennendraht, Antennenlitze
aerial Luft-...
aerial cable Luftkabel
aerial cable line Luftkabelleitung, oberirdische Kabelleitung
aerial cable way Seilschwebebahn
aerial conductor Freileitung
aerial navigation Luftnavigation, Flugnavigation
aerial railroad Schwebebahn
aerial ropeway Seilbahn, Drahtseilbahn
aerodrome Flugplatz, Flughafen
aerodrome-surface movement indicator Flugfeldüberwachungsradar
aerodynamic aerodynamisch
aerodynamics (pl.) Aerodynamik
aeronautic aeronautisch
aerophar (ungerichtetes) Funkfeuer
aeroplane Flugzeug, Aeroplan
aeroplane effect Flugzeugeffekt
aerosol Aerosol
aerostatic aerostatisch
a.f. = audio frequency Tonfrequenz, Niederfrequenz (siehe auch **audio frequency**)
A.F.C. = automatic frequency control automatische Scharfabstimmung, Abstimmautomatik
affinity Affinität, chemische Verwandtschaft
affix Affix, Index, Anhängezeichen
afflux velocity Anströmgeschwindigkeit
afire in Flammen, brennend [nicht attributiv gebraucht]

after acceleration Nachbeschleunigung
after-effect Nachwirkung
afterglow Nachleuchten, Nachleucht-...
to afterglow nachleuchten
after office hour service Nachtdienst
A.G.C. = automatic gain control Schwundregelung, selbsttätige Verstärkungsregelung
age Alter, Alterung
to age altern, künstlich altern, alt werden
  to age-harden vergüten (Leichtmetall)
ageing Alterung
  ageing process Alterungsverfahren
  ageing test Alterungsprüfung
agent Agens
agglomeration Anhäufung (548)
to aggravate gravieren, erschweren
aggregate Aggregat
aggregate Gesamt-...
  aggregate area Gesamtfläche
aggressive angreifend (54)
agitation Umwälzung (621.35)
agitator Umwälzeinrichtung (621.35)
A.G.M. = admiralty gun metal Rotguß
agricultural landwirtschaftlich
  agricultural service transformer Landwirtschaftstransformator, Sonderreihentransformator
ah = ampere hour Ah = Amperestunde
ahead of vor (im Zuge einer Schaltung)
aid to air navigation Flugnavigationshilfe
to aid helfen, in gleicher Richtung wirken
  to aid a field ein Feld verstärken
  aided tracking halbautomatische Verfolgung (623)
AI = artificial intelligence künstliche Intelligenz
Aiken Code Aiken-Code
air Luft; Wetter (62)
  air admission Luftzutritt
  air-to-air-missile Luft-Luft-Flugkörper
  air blast Luftstrom
    air-blast circuit breaker Luftschalter, Preßluftschalter
    air-blast transformer Transformator mit Anblasekühlung, fremdbelüfteter Transformator
  air blower Gebläse, Ventilator
  air break Luftrennstrecke
    air-break contactor Luftschütz
    air-break switch Luftschalter
    air-break switchgear Luftschalter
  air bubble Luftblase
  air-calorimeter test Verlustbestimmung aus Kühlluftmenge und -erwärmung (621.313)
air capacitor Luftkondensator
air cell Luftsauerstoffzelle (621.35)
  air-cell battery Luftsauerstoffbatterie
air cleaner Luftfilter, Luftreinigungsfilter
air condenser Luftkondensator (621.3); luftgekühlter Kondensator (621.1)
air-conditioned klimatisiert
air conditioning Klimatisierung, Klima-...
air-cooled luftgekühlt
  air-cooled condenser luftgekühlter Kondensator
  air-cooled transformer luftgekühlter Transformator
air cooling Luftkühlung
air-core eisenlos
  air-core choke Luftdrosselspule, eisenlose Drosselspule
  air-core transformer Lufttransformator
air-cored coil Luftspule
  air-cored reactor Luftspule
air current Luftstrom, Luftströmung
air-data computer Flugdatenrechner, Luftwerterechner
air-dehumidifying device Luftentfeuchter
air-dielectric variable condenser Luftdrehkondensator
air-driven mit Preßluftantrieb
air duct Kühlkanal, Kühlschlitz (621.313)
air escape Luftaustritt
air-filled luftgefüllt
air flow Luftströmung
air friction Luftreibung
air gap Luftspalt
  air-gap flux Luftspaltfluß
  air-gap flux density Luftspaltinduktion
  air-gap line geradliniger Teil der Leerlaufkennlinie (621.313)
air-ground... Bord-Boden...
air hose Luftschlauch, Preßluftschlauch
air humidity Luftfeuchtigkeit
air insulation Luftisolation
air intake Luftansaugung, Luftaufnahme
air-interception radar Jägerleitradar
air line Freileitung (621.394)
air lock Luftschleuse
air mass Luftmasse (55)
air moistening Luftbefeuchtung
air operated pneumatisch
air power Druckluft = Druckluftversorgung
air pressure Luftdruck
  air pressure gauge Luftdruckmesser
air-pressure Preßluft-...
air resistance Luftwiderstand

air section break Streckentrenner mit Luftisolation (621.33)
air space Luftspalt, Zwischenraum, Luftraum
air space cable luftisoliertes Kabel
air spaced mit Luftisolation, mit Luftzwischenraum
air speed indicator Fahrtmesser
air starter Preßluftanlasser (621.4)
air-to surface missile Luft-Boden Flugkörper
air surveillance Luftraumüberwachung
air surveillance radar Rundsuchradar
air switch Luftschalter
air terminal Auffangvorrichtung, Auffangstange (69); Flughafen-Abfertigung (656.7)
air tight luftdicht
air-traffic control Flugsicherung, Flugverkehrskontrolle
air-type trimmer Lufttrimmer
air wires Luftdrähte (621.396)
airborn fliegend, Bord...
airborne intercept radar Bordabfangradar
airborne radio equipment Bordfunkanlage
aircraft Flugzeug, Luftfahrzeug
aircraft altitude Flughöhe
aircraft-control system Flugzeugsteuerung
aircraft-decibel rating Reflexionswert eines Flugzeugs
aircraft establishment Flugerprobungsstelle
aircraft heading Steuerkurs
aircraft-radio regulations Flugfunkvollzugsordnung
airheater Lufterhitzer
airport Flughafen, Flugplatz
airport beacon Flughafenleuchtfeuer, Flugplatzkennfeuer
airport lighting Flughafenbefeuerung
airport surface detection equipment Rollfeldüberwachungsradar
airport-surveillance radar Flughafenüberwachungsradar
airway Luftverkehrsweg
airway beacon Flugsicherungsbake
airway-lighting installation Befeuerung
aislometer Hochspannungselektrometer
alarm Alarm, Warnung
alarm bell Warnglocke, Meldeklingel
alarm device Warneinrichtung
alarm display Alarmanzeige
alarm on failure Ausfallalarm
alarm relay Warnrelais
alarm system Alarmeinrichtung, Warneinrichtung
alarm switch Warnschalter
Alexanderson alternator Alexanderson-Generator
Alexanderson system of phototelegraphy Viertonbildtelegraphie
A.L.F. = absorption limiting frequency Dämpfungsfrequenzgrenze
algebra Algebra
algebraic(al) algebraisch
ALGOL = algorithmic language ALGOL
algorithm Algorithmus
alight brennend, in Brand; gezündet (Gasentladung) [nicht attributiv gebraucht]
to align ausrichten, auf Gleichlauf abstimmen, abgleichen
alignment Ausrichtung, gegenseitige Lage; Gleichlauf (621.396); Einpegelung, Abgleichung
alignment pin Paß-Stift, Führungsstift
alignment of tuned circuits Gleichlaufabstimmung
to aline ausrichten, fluchten
alive scharf (623); spannungsführend (621.3)
alive circuit eingeschalteter Stromkreis, spannungsführender Kreis
alive to earth spannungsführend (gegen Erde), auf Spannung, unter Spannung stehend
alkali Alkali, Lauge
alkali metals Alkalimetalle
alkaline alkalisch, alkalin, laugenhaft
alkaline accumulator alkalischer Sammler, NC-Akkumulator (621.35)
alkaline cell alkalischer Sammler, Edison-Akkumulator
alkaline photo-cell Alkaliphotozelle
alkaline-type battery alkalischer Sammler, NC-Sammler, Edison-Akkumulator
all alle, ganz
all-climate proof klimafest
all-electric vollelektrisch
all-electric interlocking vollelektrische Verblockung
all-electric signalling vollelektrische Signalanlage
all-electronic vollelektronisch
all in eingeschaltet
all-insulated switch berührungsschutzsicher isolierter Schalter
all-mains Allstrom-...
all-mains radio Allstromradio
all-mains set Allstromgerät
all metal Ganzmetall-...
all-pass network Allpaß
all-purpose Allzweck-...
all-relay dial system Relaiswählsystem

all-relay exchange Relaiswählvermittlung
all-round Allzweck-..., überall brauchbar
all-round looking device Panoramaeinrichtung (621.397)
all-round material Universalmaterial
all-up weight Fluggewicht
all-watt motor phasenkompensierter Asynchronmotor
all-wave receiver Allwellenempfänger
all-weather landing Allwetterlandung
all-weather operation Schlechtwetterflugbetrieb
all-welded ganz geschweißt
Allan cell Elektrolyseur (621.357)
Allen screw Inbus-Schraube
Allen's loop test Schleifenfehlerortsbestimmung
alligator clip Krokodilklemme
to allocate zuweisen, zuordnen
allocated band zugewiesener Frequenzbereich
to allot zuteilen, erteilen
to allot numerical values Zahlenwerte zuerteilen
allotter Vorwähler, Anrufordner (621.394)
allotropic(al) allotropisch
allowable tension zulässige Beanspruchung
allowance Berücksichtigung, Zuschlag, Zugabe (bei Maßen), Verschnitt; Abmaß
alloy Legierung, Mischung
alloy steel legierter Stahl
alloy technique Legierungsverfahren, Legierungstechnik
to alloy legieren
alloyed junction legierter Übergang, legierter Zonenübergang (621.382)
alloyed transistor Legierungstransistor
alloying Legierungs-...
alloying elements Legierungsbestandteile
alloying inclusion Beilegierung
Allström relay photoelektrisches Relais
along axis in Achsrichtung
along a line längs einer Leitung
alpha Alpha
alphanumeric alphanumerisch
alpha particle Alphateilchen
alpha radiation Alphastrahlung
alphabetic aphabetisch
to alternate abwechseln, sich abwechseln, abwechselnd eintreten
to alternate in polarity das Vorzeichen wechseln, abwechselnd verschiedenes Vorzeichen haben
alternate abwechselnd, alternativ, gleichwertig
alternate channel Ausweichkanal
alternate frequency Ausweichfrequenz
alternate immersion corrosion machine Wechseltauchmaschine für Korrosionsprüfung
alternate immersion test Wechseltauchversuch, Temperatursturzprobe
alternating wechselnd, Wechsel-...
alternating burst alternierender Burst, Wackelburst
alternating circuit Wechselstromkreis
alternating component Wechselstromanteil, Wechselstromkomponente
alternating current = a.c. Wechselstrom
alternating current arc welding Wechselstrom-Lichtbogenschweißung
alternating current balancer Ausgleichstransformator (621.316.1)
alternating current bridge Wechselstrombrücke
alternating current cable Wechselstromkabel
alternating current distribution Wechselstromverteilung
alternating current generator Wechselstromgenerator
alternating current hum Netzbrumm (621.396.64)
alternating current machinery Wechselstrommaschinen (pl.)
alternating current meter Wechselstromzähler
alternating current motor Wechselstrommotor
alternating current resistance innerer Widerstand (621.385); Wechselstromwiderstand
alternating current test Wechselstromprüfung
alternating current theory Wechselstromlehre
alternating current transmission Wechselstromübertragung
alternating current winding Wechselstromwicklung
alternating impact bending test Wechselschlagbiegeversuch
alternating magnetization Wechselmagnetisierung
alternating potential difference Wechselspannungsdifferenz
alternating quantity Wechselgröße
alternating stress Wechselbeanspruchung, Schwingungsbeanspruchung
alternating voltage Wechselspannung
alternation Halbwelle, Wechsel (621.3); abwechselnde Betätigung

alternative frequency Ausweichfrequenz
alternative route Ausweichleitweg (621.395)
alternative routing Ausweichleitwegzuteilung
alternator Wechselstromgenerator
alternator transmitter Maschinensender
altimeter Höhenmesser
altitude Höhe
altitude meter Höhenmesser
altitude resistance Höhenfestigkeit
ALU = arithmetic and logical unit
alum = aluminium Alu = Aluminium
alumina Tonerde; Aluminiumoxid
aluminium Aluminium
aluminium alloy Aluminium-Legierung
aluminium-cable steel-reinforced Stahlaluminiumseil
aluminium can Aluminiumbecher (621.396.64)
aluminium castings (pl.) Aluminiumguß
aluminium conductor Aluminiumleiter (621.315.1)
aluminium-electrolytic capacitor Aluminium-Elektrolytkondensator
aluminium furnace Aluminiumofen
aluminium rectifier Elektrolytgleichrichter
aluminium-sheathed cable Aluminiummantelkabel
aluminium solder Aluminiumlot
aluminium-steel cable Stahlaluminiumseil
aluminium-steel conductor Stahlaluminiumleiter, Stalu-Seil (621.315.1)
aluminium wire Aluminiumdraht
aluminized screen Bildschirm mit aufgedampfter Aluminiumhaut
aluminum (US) Aluminium
a.m. = amplitude modulation Amplitudenmodulation
amalgam Amalgam
to amalgamate amalgamieren; verschmelzen
amalgamated amalgamiert; fusioniert
amalgamation Amalgamierung; Fusionierung, Fusion
amateur Amateur, Bastler
amateur radio Amateurfunkdienst
amateur transmitter Amateursender
amber Bernstein
ambiguity Mehrdeutigkeit
ambient umgebend
ambient light Raumlicht, Nebenlicht; Vorlicht (621.397)
ambient noise Raumgeräusch
ambient temperature Umgebungstemperatur
amenable zugänglich
amenable to calculation berechenbar
American Wire Gauge = A.W.G. Amerikanische Drahtlehre
amine Amin
ammeter Amperemeter, Strommesser
ammonia Ammoniak, NH_3
ammonia solution Salmiakgeist
ammonium Ammonium
ammonium chloride Ammoniumchlorid, Salmiak
amorphous amorph
amorphous graphite amorpher Graphit
amortisseur Dämpferwicklung (621.313); Dämpferrahmen (621.317)
amortisseur winding Dämpferwicklung (621.313)
amp = ampere A = Amp = Ampere
amperage Stromstärke
ampere Ampere
ampere balance Stromwaage
ampere bars per cm Strombelag, AW/cm
ampere conductors Ampere-Stäbe (621.313)
ampere-conductors per inch Strombelag
ampere hour Amperestunde
ampere-hour capacity Amperestunden-Kapazität
ampere-hour efficiency Gütegrad (621.355)
ampere-hour meter Amperestundenzähler
Ampere's law Biot-Savartsches Gesetz
ampere-meter Amperemeter, Strommesser
Ampere's principle (Gesetz der Anziehung und Abstoßung zwischen stromführenden Leitern)
Ampere's rule Ampersche Schwimmerregel
Ampere's table Amperescher Rahmen
Ampere's theory of magnetization Molekularstromtheorie
ampere-turn Amperewindung (AW)
ampere-turns per cm Strombelag
ampere wires Ampere-Stäbe (621.313)
amperite tube Eisenwasserstoffwiderstand
amphoteric amphoter
amphoteric electrolyte amphoterer Elektrolyt
amplidyne Querfeldverstärkermaschine
amplidyne circuit Amplidynschaltung
amplifiable verstärkbar, zu verstärken
amplification Verstärkung (621.396.64)
amplification constant Verstärkungsfaktor (= Kehrwert des Durchgriffs); Verstärkung (621.396.64)
amplification curve Verstärkungsgang
amplification factor Verstärkung, Verstärkungsfaktor
amplification margin Verstärkungssicherheit (gegen Selbsterregung) (621.375)

amplification of modulated high frequency Zwischenstufenmodulation (621.396.61)
amplification stage Verstärkerstufe
amplifier Verstärker (621.396.64)
amplifier-frequency response Verstärkerdurchlaßbereich, Frequenzgang des Verstärkers
amplifier tube Verstärkerröhre
to amplify verstärken (621.396.64)
amplified back bias Rückwärtsregelspannung
amplifying Verstärker-..., Verstärkungs-...
amplifying ability Verstärkungsvermögen, Verstärkereigenschaft
amplifying chain Verstärkerfolge, Verstärkerkette
amplifying winding Verstärkerwicklung
amplitude Amplitude, Betrag, Wert, Größtwert, Scheitelwert
amplitude control Amplitudenregelung
amplitude discriminator Amplituden-Selektor
amplitude distortion nichtlineare Verzerrung, Amplitudenverzerrung
amplitude factor Scheitelfaktor
amplitude-frequency response Frequenzabhängigkeit
amplitude increase Amplitudenanstieg
amplitude keying Amplitudentastung
amplitude limiter Amplitudenbegrenzer
amplitude-modulated amplitudenmoduliert
amplitude modulation Amplitudenmodulation
amplitude resonance Amplitudenresonanz
amplitude response Amplitudenabhängigkeit, Amplitudengang
amplitude separator Amplitudentrennstufe
amplitude swing Amplitudenhub
amplitude-time value Amplitudenzeitwert
ampmeter = ammeter Amperemeter, Strommesser
analogic(al) analog
analog computer Analogrechner
analog-frequency conversion Analog-Frequenz-Umsetzung
analog-digital converter Analog-Digital-Umsetzer, AD-Wandler
analog-digital conversion Analog-Digital-Umsetzung
analog-measuring instrument Analog-Meßgerät
analog signal analoges Signal, Analogsignal
analogous entsprechend, analog, ähnlich (**to:** zu)
to analyse analysieren, berechnen, mathematisch behandeln
analysis Analysis, Rechnung (51); Analyse (54)
analytic(al) analytisch, rechnerisch
analytical function analytische Funktion
analyser Analysator (53)
anaphoresis Anaphorese
anchor Anker
anchor bolt Ankerbolzen (621); Steinschraube
anchor clamp Abspannklemme (621.315)
anchor escapement Ankergang, Echappement (681.11)
anchor gap Antennenschutzfunkenstrecke (621.396.67)
anchor log Spanndrahtschwelle, Abspannpflock (621.315)
anchor pole Abspannmast (621.315.1)
anchor tower Abspannmast (621.315.1)
to anchor verankern
anchorage Verankerung
ancillary untergeordnet, Hilfs-...
ancillary equipment Hilfsgeräte (pl.), Hilfseinrichtung
AND circuit UND-Schaltung
AND operation UND-Verknüpfung
Anderson's bridge Anderson-Brücke
anemometer Anemometer, Windmesser
aneroid Aneroid-..., Dosen-...
aneroid barometer Dosenbarometer, Aneroidbarometer
angle Winkel, Ecke (51)
angle about a plane Winkel gegen eine Fläche
angle adjoined by zwischen ... eingeschlossener Winkel
to be at an angle to einen Winkel bilden mit
angle adapter Winkelstück
angle of approach Anflugwinkel
angle-of-approach light Anflugwinkelfeuer
angle bar Winkeleisen, L-Eisen
angle of arrival Einfallswinkel
angle of beam Öffnungswinkel
angle-bevel gear Winkel(kegel)rad
angle bracket Winkelbock, Winkelkonsole
angle of contact Umschlingungswinkel
angle contained by two lines Winkel zwischen zwei Linien
angle of current flow Stromflußwinkel (621.314.6)
angle cutting Diagonalschnitt, Gehrungsschnitt
angle of departure Abstrahlwinkel
angle of depression negativer (Höhen-)Winkel, Neigungswinkel
angle of elevation Höhenwinkel, Erhebung, Elevation, Erhebungswinkel
angle of flare Öffnungswinkel, Konuswinkel (621.396)
angle of flow Stromflußwinkel (621.314.6)
angle of friction Reibungswinkel

angle of incidence Auftreffwinkel, Einfallswinkel
angle iron Winkeleisen, L-Eisen
angle jack Winkelstecker
angle of lag Nacheilungswinkel
angle of lead (over) Voreilungswinkel (gegen) (621.3); Vorstellwinkel; Steigungswinkel (621.9)
angle modulation Phasen(winkel)modulation
angle of pitch Anstellwinkel
angle plug Winkelstecker
angle pole Winkelmast (621.315)
angle pulley Ablenkrolle
angle of radiation Strahlungswinkel, Abstrahlwinkel
angle of reflection Reflexionswinkel, Rückwurfwinkel, Ausfallwinkel
angle of refraction Brechungswinkel
angle of repose Böschungswinkel
angle section Winkeleisen, L-Eisen
angle sheave Umlenkrolle
angle of sight Sichtlinienwinkel
angle of slope Böschungswinkel
angle tower Winkelmast (621.315)
angle velocity Winkelgeschwindigkeit
Ångstrom Unit Ångström-Einheit = 10^{-8} cm
angular kantig, eckig, Winkel-..., winkelförmig
angular acceleration Winkelbeschleunigung
angular advance of the brushes Bürstenverschiebung (nach vorn)
angular aperture Öffnung, Apertur (535)
angular decoupling Winkelentkopplung (621.396.67)
angular dependence Winkelabhängigkeit
angular displacement Verdrehung
angular displacement of a polyphase transformer Phasenverschiebung zwischen korrespondierenden Klemmen eines Transformators
angular distance Winkelabstand
angular drive Winkeltrieb
angular frequency Kreisfrequenz
angular kinetic energy Rotationsenergie
angular loss Winkeldämpfung
angular momentum Drehimpuls
angular movement Drehbewegung, Winkelbewegung
angular phase difference Phasenwinkel
angular resolution Winkelauflösung
angular spacing Winkelabstand
angular spread Winkelbereich
angular variation Phasenschwankung, Phasenschub
angular velocity Winkelgeschwindigkeit
angular width Keulenhalbwertsbreite (621.396.67)
anhydride Anhydrid
anhydrous wasserfrei
aniline Anilin
animal tierisch
animal electricity tierische Elektrizität
anion Anion
anionic anionaktiv
anisotropic anisotrop
anisotropic conductivity anisotrope Leitfähigkeit
anisotropic field Anisotropiefeldstärke
anisotropy Anisotropie
to anneal glühen, thermisch behandeln, anlassen, ausglühen
annealed copper Weichkupfer
annealing Glühung, Glühbehandlung
annealing furnace Glühofen
to annihilate vernichten, in Nichts verwandeln (539)
annihilation Vernichtung (539)
annihilation radiation Vernichtungsstrahlung
announcement device Ansagegerät
announcer Ansager (654.19)
annoyance Belästigung, Störeindruck (534)
annual jährlich
annual efficiency Jahreswirkungsgrad
annual load curve Jahresbelastungskurve
annual mean Jahresmittel
annual variation Jahresgang, jährlicher Gang
annular ringförmig, Ring-...
annular gearing Innenverzahnung
annular groove Ringnut
annunciator Ruftableau (621.3); Fallklappentafel, Signaltafel
annunciator board Anzeigetafel
anode Anode
anode a.c. conductance Kehrwert des inneren Widerstandes (621.385)
anode a.c. resistance innerer Widerstand (621.385)
anode-balancing coil Anodenausgleichsdrossel (621.314.5)
anode battery Anodenbatterie
anode bend Anodenknick, Krümmung der Anodenkennlinie
anode-bend detection Anodengleichrichtung
anode-bend rectification Anodengleichrichtung (621.396)
anode-breakdown voltage Anodenzündspannung
anode butt Anodenstummel (621.365)
anode cap Anodenkappe

anode characteristic Anodenkennlinie
anode circuit Anodenkreis
anode cleaning anodische Reinigung (621.35)
anode conductance Kehrwert des inneren Widerstandes
anode converter Gleichrichterröhre (621.396.64)
anode-current characteristic Anodenstromkennlinie (621.385)
anode dark space Anodendunkelraum
anode d.c. power-input Anodengleichstromleistung
anode d.c. resistance Anodengleichstromwiderstand
anode detector Anodengleichrichter
anode-differential resistance Innenwiderstand einer Röhre
anode dissipation Anodenverlustleistung (621.385)
anode drop Anodenfall (537)
anode efficiency anodische Stromausbeute, Anodenwirkungsgrad
anode glow anodisches Glimmlicht
anode impedance innerer Widerstand (621.385)
anode layer anodische Schicht
anode modulation Anodenmodulation
anode modulator Anodenmodulationsröhre (621.396.61)
anode mud Anodenschlamm (621.35)
anode pickling anodische Beizung
anode rays Kanalstrahlen, Ionenstrahlen
anode-ray current Kanalstrahlstrom, Ionenstrom
anode rectification Anodengleichrichtung (621.396.6)
anode resistance Anodenwiderstand
anode resistor Widerstand im Anodenkreis (621.396.6)
anode ring Anodenring
anode segment Anodensegment (Magnetron)
anode shell Anodenmantel
anode shield Anodenabschirmung
anode slime Anodenschlamm (621.35)
anode sputtering Anodenzerstäubung
anode-supply filter circuit Anodensiebschaltung, Anodensiebkreis
anode-tapping point Anodenabgriff
anode-valve cap Anodenkappe
anode voltage Anodenspannung
anode-voltage drop Anodenspannungsabfall
anodic anodisch
anodic corrosion Stromkorrosion, anodische Korrosion
anodic oxydation anodische Oxidation
anodic passivation anodische Passivierung
anodic polarization anodische Polarisation (621.35)
anodic production of solids anodische Erzeugung von Schutzschichten
anodization Eloxierung
to anodize eloxieren
anomalous anomal, abnorm
anomalous charging current anomaler Ladestrom
anomalous conduction anomale Leitfähigkeit
anomalous conduction current anomaler Leitungsstrom
anomalous discharge current anomaler Entladestrom
anomalous dispersion anomale Dispersion
anomalous displacement dielektrische Nachwirkung
anomalous magnetization magnetische Nachwirkung
anomalous propagation Überreichweite, anomale Ausbreitung
anomaly Anomalie
anotron Glimmgleichrichterröhre
anouncement Ansage
AN radio range AN-Richtfunkfeuer
ANSI = American National Standards Institute
to answer abfragen; melden
answer-back code Namengeberzeichen (621.394)
answer back device Namengeber
to answer a call einen Anruf beantworten
answer lamp Abfrage- und Schlußlampe
answer signal Beginnzeichen; Meldesignal
answering Abfrage
answering cord Verbindungsschnur (621.395)
answering delay Meldeverzug
answering jack Abfrageklinke
answering key Abfrageschalter
antenna (US) Antenne (s. auch **aerial**)
antenna array Antennenanordnung, Antennensystem, Antennengruppe
antenna assembly Antennenanlage
antenna base Antennensockel, Antennenfuß
antenna core Ferrit-Antenne
antenna coupler Antennen-Koppler, Antennenankopplung
antenna effect Antenneneffekt, Trübung
antenna gain Antennengewinn
antenna-radiation pattern Antennenrichtdiagramm
antenna selector Antennenwahlschalter
Anthygron served lead-sheathed wire Antigronbleimantelleitung

Anthygron weatherproof wire Anthygronleitung
anti-... Gegen-..., Anti-...
anti-capacitance kapazitätsarm
anti-cathode Antikathode
anti-clockwise gegen den Uhrzeigersinn
anti-clockwise rotation Umlauf gegen den Uhrzeigersinn
anticlutter facility Seegang-Entstörung, Enttrübung (621.396.96)
anticlutter gain control Nahechodämpfungsschaltung (621.397)
anticollision radar Kollisions-Warnradar
anti-creep device Leerlaufhemmung (621.317)
anti-dazzle lamp Blendschutzlampe
anti-dazzling Blendschutz-...
anti-fading aerial schwundmindernde Antenne
anti-fading device Schwundregelungseinrichtung (621.396.62)
anti-fatigue device Schwingungsdämpfer (621.315.1)
anti-freeze transformer Frostschutztransformator
anti-friction reibungsmindernd
anti-friction bearing Wälzlager
anti-friction metal Lagermetall
anti-froth preparation Schaumverhütungsmittel
anti-glare Blendschutz-...
anti-glare glass blendfreies Glas, reflexfreies Glas
anti-glare spectacles (pl.) Blendschutzbrille
antihunt (circuit) Schwingungsdämpfung (621.316.7)
anti-inductive Induktionsschutz-...
anti-interference choke Entstördrossel
anti-interference condenser Störschutzkondensator, Entstörkondensator (621.396)
antijamming Störungsunterdrückung, Maßnahmen gegen Störungen
antimicrophonic klingsicher (621.396)
antinode Bauch, Schwingungsbauch
antinoise microphone geräuscharmes Mikrophon
antiparallel coupling Antiparallelschaltung
antiparticle Antiteilchen
antiresonance Stromresonanz
antiresonant circuit Sperrkreis, Stromresonanzkreis
antirust coating Rostschutzanstrich
anti-secondary grid Bremsgitter (621.396)
antisidetone Rückhördämpfung (621.395)
antispray film Decköl (621.355)
antistatic aerial Störschutzantenne
antistatic cloth Antistatiktuch
antitransmit-receive switch Sendersperrschalter (621.396.96)
antitransmit-receive tube = ATR Sendersperröhre
antivibration klingsicher (621.396); erschütterungsfrei
antivibration device Schwingungsdämpfer (621.315.1)
antivibration lamp-holder erschütterungssichere Fassung
antidyne circuit gegengekoppelter Sender (621.396.61)
antihunt circuit Dämpfungsschaltung im Regelkreis
antilogarithm Numerus
antimony Antimon
"A" operator A-Beamter, A-Beamtin (621.395)
apartment-house aerial Gemeinschaftsantenne
aperiodic aperiodisch, aperiodisch gedämpft
aperiodic aerial aperiodische Antenne
aperiodic circuit aperiodischer Kreis (621.396)
aperiodic damping aperiodische Dämpfung
aperiodic instrument aperiodisch gedämpftes Instrument
aperture Öffnung, Loch; Blende, Apertur; Fenster (53)
aperture mask Schattenmaske, Lochmaske
apex (pl.: **apices**) Spitze (des Kegels), Ecke (des Dreiecks)
apex angle Winkel an der Spitze, Öffnungswinkel
A-pole A-Mast
apostilb Apostilb
apparatus Apparat
apparatus-location diagram Apparateverteilungsplan
apparent anscheinend, scheinbar, Schein-...
apparent candlepower scheinbare Lichtstärke (Lichtstärke einer äquivalenten Punktlichtquelle)
apparent impedance Scheinwiderstand
apparent inductance scheinbare Induktivität
apparent ohmic resistance scheinbarer ohmscher Widerstand
apparent power Scheinleistung (621.3)
apparent remanence scheinbare Remanenz
apparent resistance Scheinwiderstand
apparent voltamps (pl.) Scheinleistung
apparent watts (pl.) Scheinleistung
apple-shaped diagram Nierencharakteristik (621.396)
apple tube Rücksteuerröhre (621.397.67)

Appleton-layer F-Schicht
appliance Einrichtung, Gerät, Vorrichtung; elektrisches Haushaltsgerät
appliance branch circuit Gerätestromkreis
appliance connector Gerätesteckdose
appliance inlet Gerätestecker
appliances ... -geräte
application Anwendung, Anlegen, Aufbringen
application connector Anwenderstecker
application magnet Bremsanlegemagnet (65)
application program Anwenderprogramm
application of varnish Lackanstrich, Lackauftrag
application of voltage Anlegen einer Spannung
applicator electrode Elektrode
to apply zutreffen, gelten; anlegen (**to:** an); auflegen (**to:** auf); anwenden (**to:** auf)
to apply the brake Bremse anziehen, Bremse betätigen
applied energy aufgewandte Energie
applied physics technische Physik, angewandte Physik
applied potential angelegte Spannung
to apply a potential eine Spannung anlegen (**across:** zwischen)
a rule applies eine Regel trifft zu
to apply a signal to the grid Gitterwechselspannung anlegen, Modulation aufs Gitter geben
the voltage applied angelegte Spannung
appr. = approximately ungefähr
appreciable merklich, wahrnehmbar
appreciable-sized von merklicher Größe
approach Annäherung, Näherung (51); Weg (**to:** zu); Anflug (656.7) (**to:** auf); Zugang, Zufahrt
approach circuit Zugmeldekreis (625)
approach-control radar Anflugüberwachungsradar
approach lane Anflugschneise
approach light Anflugrichtungsfeuer (656.7)
approach-lighting relay zugbetätigtes Lichtsignalrelais (621.33)
approach path Anflugweg
approach sequence Anflugfolge
to approach sich nähern; annähern (51); anfliegen (656.7)
to approach and recede hin- und hergehen, sich hin- und herbewegen
to approximate annähern (51); sich nähern
approximate ungefähr, annähernd, angenähert
approximate formula Näherungsformel
approximate value Näherungswert
approximation Näherung (51); Annäherung
apron floodlight Vorplatzflutlichtleuchte (656.7)
aqua ammonia Ammoniakwasser
aqua fortis Salpetersäure
aqua regia Königswasser
aquarium reactor Bassin-Reaktor, Schwimmbadreaktor
aqueous wässerig
aqueous electrolysis wässerige Elektrolyse
aqueous-homogeneous reactor wässerig-homogener Reaktor
aqueous solution wässerige Lösung
aqueous vapour Wasserdampf
Arago's disc Wirbelstromscheibe
araldite Araldit
arbitration Zuweisung (681.3)
arbitrary willkürlich, frei wählbar, beliebig
arbitrary-sequence computer Sequentiellrechner mit Sprunglauf
arbor Welle, Schaft, Wellenstumpf
arc Bogen, Kreisbogen (51); Lichtbogen (621.3)
′ of arc Bogenminute
arcback Rückzündung (621.3)
arc-blow-out Blasmagnet
arc brazing Lichtbogenlötung
arc buffle Ablenkplatte (621.314.6)
arc chamber Löschkammer
arc chute Lichtbogenkammer (621.316.5)
arc of a circle Kreisbogen
arc column Lichtbogensäule, Plasma
arc cutting Lichtbogenschneiden (621.791)
arc deflector Blasmagnet (621.325); Lichtbogenschutzkammer (621.316.5)
arc discharge Lichtbogenentladung
arc drop Lichtbogenspannungsabfall
arc-drop voltage Lichtbogenspannung
arc electrode Lichtbogenfußpunkt
arc failure Überschlag
arc flame Flammenbogen
arc furnace Lichtbogenofen
arc lamp Bogenlampe
arc-lamp carbon Lichtbogenkohle, Bogenlampenkohle
arc light Bogenlampe
arc lighting Bogenlampenbeleuchtung, Bogenlicht
arc over Überschlag (621.3)
arc-over voltage Überschlagsspannung
arc quenching Funkenlöschung
arc shield Lichtbogenschutzkammer (621.316.5)
arc spectrum Bogenspektrum
arc stream Plasma, positive Säule, Lichtbogenplasma (537)
arc-stream voltage Plasmaspannung

arc suppressor Lichtbogenlöscheinrichtung, Erdschlußlöscheinrichtung (621.316.9)
arc through Zündung, Durch-Zündung
arc transmitter Lichtbogensender, Poulsensender
arc voltage Lichtbogenspannung, Bogenspannung, Schweißspannung
arc welding Lichtbogenschweißen, Lichtbogenschweißung
arc-welding converter Schweißumformer
arc-welding generator Schweißgenerator
to arc Lichtbogen bilden
to arc over überschlagen (621.3)
architecture (of a microprocessor) Architektur (eines Mikroprozessors)
arcing Lichtbogenbildung, Lichtbogenüberschlag
arcing back Rückzündung (621.316.5)
arcing contact Abreißkontakt (621.316.5)
arcing ground Erdschlußlichtbogen
arcing horn Lichtbogenschutzhorn (621.315.6)
arcing ring Lichtbogenschutzring (621.315.6)
arcing time Abschaltdauer, Löschzeit (621.316.9); Lichtbogenbrenndauer (621.316.5)
arcing tips Abbrennkontakte (621.316.5)
arcing voltage Lichtbogenspannung
arduous hart, schwer
arduous duty rauher Betrieb, schwerer Betrieb
area Zone, Fläche, Querschnittsfläche, Gebiet
area of contact Berührungsfläche
area enclosed between response curve and horizontal line Regelfäche (621.316.7)
area of risk Gefahrenzone; Spannungstrichter (621.3)
area of the triangle Fläche des Dreiecks
areal flächenmäßig
areal extension Flächenausdehnung
Argand burner Argandbrenner
argon Argon
argon-arc welding Argonarc-Schweißung
Argonaut(-research) reactor Argonaut Reaktor
argument Argument (51); Beweis, Argument, Beweisführung
argumentation Beweisführung, Argumentation
arithmetic(al) arithmetisch
arithmetic instruction Rechenbefehl, arithmetischer Befehl
arithmetic logic unit = ALU Arithmetisch-Logische Einheit, Rechenwerk
arithmetic means arithmetisches Mittel, arithmetischer Mittelwert
arithmetic operation Rechenoperation
arithmetic progression arithmetische Reihe
arithmetic register Rechenwerkregister
arithmetic unit Rechenwerk
arithmetic work Rechenarbeit
arm Arm; Speiche; Schenkel; Zweig, Brückenzweig (621.3); Waffe; Signalflügel (625)
arm contact Signalflügelkontakt (625)
arm of the moment Hebelarm des Moments
arm opposite to Brückenzweig gegenüber... (621.317)
armature Anker (621.3), (538)
armature band Ankerbandage
armature bar Ankerstab
armature circuit Ankerstromkreis (621.313), (621.317)
armature clearance Pimpelluft, Stegluft
armature coil Ankerspule
armature conductor Ankerleiter, Ankerstab, Stab
armature control Hauptstromregelung (621.34)
armature core Ankerblechpaket, nackter Anker, Ankerkern, Ankereisen
armature-core disk Ankerblech
armature-core plate Ankerblechpaket
armature current Ankerstrom
armature duct Ankerkühlschlitz
armature end connection Ankerfahne
armature-end connections Stirnverbindungen (621.313)
armature end plate Ankerdruckplatte
armature flux Ankerkraftschluß
armature gap Ankerabstand, Luftspalt
armature head Ankerendplatte (621.313)
armature-iron loss Ankereisenverlust
armature key Ankerwellenkeil; Nutenverschlußkeil (621.313)
armature lamination Ankerblech
armature of a magnet Magnetanker
armature punchings Ankerblechschnitte (621.313)
armature pusher Ankerbügel
armature quill Nabe, Ankernabe (621.313)
armature reaction Ankerrückwirkung
armature-reaction excited machine Querfeldmaschine (621.313)
armature repound Ankerprellung
armature resistance Ankerwiderstand
armature shaft Ankerwelle
armature sleeve Ankerhülse, Nabenhülse (621.313)
armature slot Ankernut
armature-slot ripple Ankernutwellen
armature spider Polkreuz, Polstern,

Ankerspeichenstern, Nabenstern, Ankernabenstern (621.313)
armature spindle Ankerachse (621.317)
armature stamping Ankerblechschnitt, Ankerblech
armature stop Ankeranschlag
armature stray flux Ankerstreufluß
armature stroke Ankerhub
armature stud Pimbel
armature travel Ankerhub
armature voltage Ankerspannung
armature winder Ankerwickelmaschine
armature winding Ankerwicklung
armature wire Ankerdraht (621.313)
armour Armierung, Bewehrung (621.315.2)
armour clamp Bewehrungsschelle
to armour armieren, bewehren (621.315.2)
amoured cable armiertes Kabel, Panzerkabel
armoured hose Panzerschlauch, bewehrtes Kabel
armoured switchgear gußgekapselte Schaltanlage
armoured wire bewehrter Draht
armouring Bewehrung, Armierung (621.315.2)
Armstrong circuit rückgekoppelter Empfänger (621.396.62)
Armstrong-Orling detector Kapillardetektor
aromatic aromatische Verbindung
aromatic compound aromatische Verbindung
aromatic hydrocarbon aromatischer Kohlenwasserstoff
Aron clock-meter Aronscher Pendelzähler
Aron meter Aronzähler
around um ... herum, ringsherum
around the loop auf dem geschlossenen Umlaufweg, ringsherum um die Masche (621.3)
around-the-world signal Erdumlaufzeichen (621.396)
to arrange anordnen, aufbauen, gestalten
to arrange plugs in position Stöpsel stecken
array Anordnung, Gruppe, Feld (621.396); Tabelle (51); Vielfachschaltung (621.382)
array factor Gruppenfaktor (621.396.6)
array network Gruppenschaltung (621.392)
arrest Arretierung
arrester Ableiter, Überspannungsableiter
arrester gap Ableiterfunkenstrecke
arrester relay Blitzschutzrelais
arrival Ankunft, Eintreffen
arrival curve Kabelendstrom, Kurvenform des Empfangszeichens, Empfangszeichen (621.394)
arrow Pfeil, Richtungspfeil
arrow head Pfeilspitze; Pfeilrichtung
arrow-head direction Pfeilrichtung
arsenic Arsen
articulated voice-coil arrangement Doppelspullautsprecher
articulation Verständlichkeit
articulation reference Ersatzdämpfung
articulation test Verständlichkeitsversuch, Silbenverständlichkeitsprobe
artificial künstlich, Kunst-...
artificial ageing künstliche Alterung
artificial aerial Kunstantenne, künstliche Antenne
artificial circuit künstliche Leitung
artificial ear künstliches Ohr
artificial fibre Kunstfaser
artificial horizon künstlicher Horizont
artificial light Kunstlicht
artificial line Kunstleitung, Leitungsnachbildung
artificial load Belastungswiderstand (621.313)
artificial neutral point künstlicher Nullpunkt, künstlicher Sternpunkt
artificial nuclear disintegration künstlicher Kernzerfall
artificial resin Kunstharz
artificial silk Kunstseide
artificial sunlight künstliches Sonnenlicht
artificial white (black) künstliches Weiß (Schwarz)
artwork generation (Fertigungs-)Unterlagenerstellung
ASA = American Standard Association
asbestos Asbest
asbestos board Asbestpappe
asbestos-cement plate Asbestzementplatte
asbestos-covered cable Asbestmantelkabel
asbestos-covered conductor asbestisolierter Leiter
asbestos-covered wire asbestisolierter Draht
asbestos fibre Asbestfaser
asbestos gasket Asbestdichtung
asbestos millboard Asbestpappe
asbestos packing Asbestliderung, Asbestpackung, Asbestdichtung
asbestos ribbon Asbestband
asbestos rope Asbestfaden
asbestos slate (board) Asbestschieferplatte
asbestos tape Asbestband
A.S.C. = automatic selectivity control automatische Bandbreiteneinstellung
ASCII = American Standard Code for Information Interchange [International

genormter 7-Bit-Code für den Datenaustausch]
A-scan waagerechte Zeitablenkung (621.397)
A.S.D.I.C. = Allied Submarine Detection Investigation Committee, Asdic Unterwasserschallortung
askew schief, windschief
aspect Aspekt; Verhältnis; Betrachtungsweise
aspect ratio Bildverhältnis, Seitenverhältnis (621.397)
asphalt Asphalt
asphalt cardboard Teerpappe
asphalt coating Asphaltanstrich
asphalt paper Teerpapier
ASR = airport surveillance radar Rundsicht-Radar
to assemble zusammenbauen, montieren; assemblieren (681.3)
to assemble parts montieren
to assemble on reels aufspulen, auf Spulen wickeln
assembler Assembler, Programmumsetzer, Assemblierer (681.3)
assembler directive Assembleranweisung
assembling shop Montagehalle
assembly Montage, Zusammenbau; Aufbau, Anordnung; Baugruppe (621); Satz
assembly (of apparatus) Apparatesatz, Gerätesatz
assembly drawing Montagezeichnung
assembly instructions Montageanleitung
assembly jig Baugerüst, Montagegerüst, Montagelehre, Montagevorrichtung
assembly line Fließband, Montageband
to assign zuschreiben, zuerteilen, beilegen; abstellen, übereignen
assigned frequency zugeteilte Frequenz, Kennfrequenz, Verfügungsfrequenz
to assign values to Werte zuerteilen, Werte zuschreiben
assistance traffic handvermittelter Verkehr
to associate zuordnen, verknüpfen; zusammen vorkommen mit
associated electric circuits angeschlossene Stromkreise, mitangeschlossene Stromkreise
associated matter Begleitsubstanz
associated receiver zugehöriger Empfänger
Association cable Normkabel
associative storage Assoziativspeicher
astable multivibrator astabiler Multivibrator, freilaufender Multivibrator
astable trigger circuit astabile Kippschaltung, astabiler Multivibrator
astatic(al) astatisch
astatic control astatische Regelung
astatic galvanometer astatisches Galvanometer
astatic needle astatische Magnetnadel
astatic wattmeter astatisches Wattmeter
asterisk Sternchen, Stern (51)
to asterisk mit einem Sternchen versehen, besternen
astigmatism Astigmatismus
astigmatism control Punktschärferegler
astronavigation Astro-Navigation
astrotracker Sternnachführungsgerät
astronomy Astronomie
"A"-switchboard A-Schrank (621.395)
asylum switch Schalter unter Verschluß
asymmetric(al) unsymmetrisch, asymmetrisch
asymmetrical cell Gleichrichterzelle
asymmetrical conduction asymmetrische Leitfähigkeit, unipolare Leitfähigkeit
asymmetric conductor unipolarer Leiter
asymmetrical effect Unsymmetriefehler (621.396.9)
asymmetric sideband transmission Restseitenbandübertragung
asymmetric voltage Unsymmetriespannung
asymmetry Unsymmetrie
asymptote Asymptote
asymptotic(al) asymptotisch
asymptotic(al) representation asymptotische Darstellung
asymptotic voltage Dauerdurchschlagspannung
asynchronous asynchron, Asynchron-...
asynchronous communications interface adapter asynchron arbeitende Übertragungs-Schnittstellenschaltung
asynchronous condenser asynchroner Phasenschieber
asynchronous generator Asynchrongenerator
asynchronous machine Asynchronmaschine
asynchronous motor Asynchronmotor
asynchronous operation Asynchronverfahren (621.394.6); asynchrone Arbeitsweise, taktunabhängige Arbeitsweise (681.3)
asynchronous starting asynchroner Anlauf, asynchrones Anlassen
asychronous three-phase motor Drehstromasynchronmotor
asynchronous transmission asynchrone Übertragung
A.T. = ampere turns AW = Amperewindungen
A.T. cut temperaturkonstanter Schnitt (537.228)
atlantic atlantisch
atlantic cable Atlantikkabel, Transatlantikkabel

atmosphere Luftraum, Atmosphäre
atmospheric(al) atmosphärisch, Luft-...
atmospheric absorption Absorption in der Atmosphäre (621.396)
atmospheric conditions Witterungsbedingungen, Witterung
atmospheric contaminant atmosphärischer Einfluß
atmospheric disturbances atmosphärische Störungen, atmosphärischer Störpegel
atmospheric duct Führungsschicht
atmospheric electricity Luftelektrizität
atmospheric noise atmosphärischer Störpegel, atmosphärisches Rauschen, atmosphärische Störung
atmospheric oxygen Luftsauerstoff
atmospheric pollution Verunreinigung der Luft
atmospheric pressure Luftdruck; normaler Druck, Atmosphärendruck
atmospherics atmosphärische Störungen, luftelektrische Störungen, atmosphärischer Störpegel (621.396)
atom Atom
atom bomb Atombombe
atomic atomar, Atom-...
atomic battery Atombatterie
atomic binding energy Atombindungsenergie, atomare Bindungsenergie
atomic bomb Atombombe
atomic bond Atombindung
atomic clock Atomuhr
atomic disintegration Atomzerfall
atomic energy Atomkraft, Atomenergie
atomic frequency standard atomares Frequenznormal
atomic fusion Atomfusion, Kernverschmelzung
atomic heat Atomwärme
atomic hydrogen welding Arcatomschweißung
atomic lattice Atomgitter
atomic linkage Atombindung
atomic mass Atomgewicht
atomic nucleus Atomkern
atomic number Kernladungszahl, Ordnungszahl, Atomzahl
atomic orbit Valenzzustand
atomic physics (pl.) Atomphysik
atomic pile Atombrenner, Atomkraftbatterie, Atommeiler
atomic power Kernenergie, Atomenergie
atomic power plant Atomkraftwerk, Kernkraftwerk
atomic radius Atomradius
atomic shell Atomhülle, Atomschale
atomic structure Atombau, Atomaufbau; Raumgitter
atomic torso Atomrumpf
atomic transformation Atomumwandlung
atomic valence Atomwertigkeit
atomic volume Atomvolumen
atomic weight Atomgewicht
atomistic atomistisch
atomization Feinzerstäubung, Atomisierung
to atomize atomisieren, feinst verteilen, feinzerstäuben
atonic nicht-abgestimmt (nur 531)
atonic interrupter nicht-abgestimmter Unterbrecher
ATR tube = anti-transmit-receive tube Gegensperröhre, Nullode, Sendersperröhre
to attach to anschließen an, hängen an, anhängen an, befestigen an, befestigen auf, anheften an
attached to angebaut an, befestigt an
attachable aufsteckbar
attachment Zusatz, Zusatzvorrichtung, Befestigung, Anschluß (**to:** an); Bindung (**to:** an, mit)
attachment screw Befestigungsschraube, Klemmschraube
attachments (pl.) Zubehör, Zubehörteile
to attain erreichen
to attain speed auf Drehzahl kommen
attendance Hörerschaft, Auditorium; Wartung
attendant Wärter, Maschinenwärter; Bedienungsperson in Nebenstellenanlagen
attendant's cabinet Vermittlungspult, Abfragepult
attended station bemannte Station
attention Aufsicht, Wartung
attention display Warnanzeige
to attenuate dämpfen, schwächen; verdünnen
attenuated air verdünnte Luft
attenuating band Dämpfungsbereich, Sperrbereich
attenuation Dämpfung, Abschwächung; Dämpfungsmaß (621.392)
attenuation coefficient Dämpfungsmaß
attenuation compensator Dämpfungsentzerrer (621.395)
attenuation constant Dämpfungskonstante
attenuation distortion Dämpfungsverzerrung
attenuation equalizer Dämpfungsentzerrer (621.395)
attenuation factor Dämpfungsfaktor
attenuation length Gesamtdämpfung (621.395)
attenuation network Dämpfungsvierpol
attenuation plan Dämpfungsplan (621.395)

attenuation ratio Dämpfungsmaß
attenuation region Sperrbereich (621.392)
attenuation unit Dämpfungslied
attenuator Dämpfungsregler; Dämpfungsglied (621.392)
attenuator probe Teilertastkopf (621.317)
attenuator set Dämpfungsregler, Dämpfungskette
atto 10^{-18}
to attract anziehen
attracted disc electrometer Plattenelektrometer
to attract moisture Feuchtigkeit anziehen
attraction Anziehung
attraction-type ammeter Flachspul-Amperemeter
attractive anziehend, Anziehungs-...
attractive force Anziehungskraft
at-zero current stromlos
Å.U. = Ångstrom Unit Å.E. = Ångström-Einheit = 10^{-8} cm
audibility Hörbarkeit, Verständlichkeit
audible hörbar
audible busy signal Besetztzeichen (621.395.34)
audible frequency Tonfrequenz, Hörfrequenz
audible harmonic hörbarer Oberton
audible indication akustische Anzeige
audible response hörbarer Empfang, Hörempfang (621.396)
audible ringing signal Rufzeichen, Freiton (621.395.34)
audible sensation Höreindruck, Hörempfindung
audible signal akustische Anzeige
audible sound level Geräuschpegel
audio Tonfrequenz-..., Hör-...
audio amplifier Niederfrequenzverstärker, Tonfrequenzverstärker
audio carrier Tonträger
audio channel Tonkanal
audio current Sprechwechselstrom (621.395); Modulation (621.396)
audio fidelity Tonfrequenzwiedergabetreue
audio filter Tonfrequenzfilter, Tonfrequenzsieb
audio frequency = AF Tonfrequenz, Niederfrequenz, NF
audio-frequency amplification Tonfrequenzverstärkung
audio-frequency beat note tonfrequenter Schwebungston, hörbarer Schwebungston
audio-frequency circuit NF-Tonleitung (621.397)
audio-frequency driving power Modulationsleistung
audio-frequency oscillator Niederfrequenzgenerator, Tongenerator
audio-frequency rediffusion niederfrequente Drahtfunkverteilung, Drahtfunk
audio-frequency telegraphy Tonfrequenztelegraphie, W.T. = Wechselstromtelegraphie
audio-frequency transformer Übertrager, Tonfrequenzübertrager, Niederfrequenzübertrager
audio-frequency wire broadcasting niederfrequenter Drahtfunk
audio generator NF-Generator
audio monitor Tonüberwachung
audio oscillator Tongenerator, Tonfrequenzgenerator
audio recorder Tonaufnahmegerät
audio signal NF-Signal
audio-signal generator Niederfrequenzgenerator
audio trap Tonsperrkreis
audio-trunk cable Niederfrequenz-Fernkabel
audiogram Audiogramm, Hörbarkeitsdiagramm
audiometer Otometer, Audiometer
audion Audion
auditorium Auditorium, Hörsaal
auditory Gehör-..., hörmäßig
auditory perspective plastisches Hören
auditory sensation Hörempfindung
aura Gasleuchten (621.385)
aural das Ohr betreffend
aural direction-finder Hörempfangspeiler
aural setting Hörabgleich
aural transmitter Tonsender
aural-type device akustisches Gerät, Gerät mit Höranzeige, Hörgerät
aurora Polarlicht
austenite Austenit
Austin formula Austinsche Ausbreitungsformel (621.396)
authorized frequency genehmigte Frequenz
auto selbsttätig, auto-..., Auto-...
autoalarm receiver Autoalarmempfänger
autobalance Selbstabgleich
auto-capacitive coupling kapazitiv-galvanische Kopplung
auto-catalytic autokalytisch
auto-compounded current transformer Strommeßwandler eigenerregter Zusatzwicklung (621.317)
auto-connected regulating transformer Regeltransformator in Sparschaltung
autocontrol Selbststeuerung
autocorrection Eigenkorrektur

autodial Druckknopfwähler (621.395.34)
autogalvanic corrosion elektrolytische Korrosion
autoheterodyne reception Überlagerungsempfang
auto-inductive coupling induktivgalvanische Kopplung
autodyne circuit Schwingaudionschaltung
autodyne reception Überlagungsempfang
autogenous Autogen-..., autogen
autogenous welding Autogenschweißung
automanual (telephone) exchange halbautomatisches Amt
automatic Automat (621.9)
automatic automatisch, selbsttätig
automatic acceleration automatische Anfahrsteuerung (621.33)
automatic accessment of message fees automatische Gebührenerfassung
automatic aiming automatische Nachführung (621.397)
automatic alarm automatische Warnanlage
automatic alarm receiver automatischer Alarmempfänger
automatic amplitude control Schwundausgleich
automatic answering and recording equipment telefon. Anrufbeantworter und Aufnahmegerät
automatic approach coupling automatische Anflugaufschaltung
automatic arc welding automatische Lichtbogenschweißung
automatic back bias automatische Rückwärtsregelung
automatic black-level control automatische Schwarzwerthaltung
automatic block system automatisches Blocksystem
automatic call booking automatischer Rückruf
automatic calling automatischer Ruf
automatic carriage automatischer Vorschub (621.394.6)
automatic checking Selbstkontrolle (681.3)
automatic circuit-breaker automatischer Leistungs-Schalter, Selbstschalter
automatic circuit breaking automatische Abschaltung
automatic closing automatische Einschaltung
automatic coding automatische Programmierung
automatic combustion control automatische Kesselsteuerung
automatic control automatische Regelung
automatic control technology Regeltechnik
automatic controller Wächter, Stromwächter; Schütz; automatischer Regler
automatic cut-out Selbstschalter, Automat (621.3)
automatic (dial) system Wähltelefonsystem
automatic dialer Zieltaster
automatic direction finder Radiokompaß, automatischer Peiler
automatic dispatching device automatisches Abfahrtssignal (656)
automatic exchange Selbstanschlußamt, Wählamt, Wählvermittlungsstelle
automatic fault signalling selbsttätige Fehleranzeige
automatic feeder Selbstanlegeeinrichtung (655)
automatic field suppressor automatische Feldschwächungseinrichtung (621.316.9)
automatic flight controller automatische Flugüberwachungs- und Steueranlage, Selbststeuergerät
automatic floor-stop operation automatische Stockwerkschaltung (621.876)
automatic following automatische Zielverfolgung; automatische Nachführung
automatic frequency control automatische Scharfabstimmung, automatische Frequenzregelung
automatic gain control Schwundausgleich, Fadingregelung, automatische Verstärkungsregelung
automatic gate lock automatische Schachtverriegelung (621.876)
automatic generating plant selbsttätiges Ladeaggregat
automatic interrupter Wagnerscher Hammer
automatic level regulation automatische Pegelregelung
automatic load response Bedarfsautomatik (621.311)
automatic lock-on automatische Aufschaltung (621.396.96)
automatic message accounting selbsttätige Gebührenermittlung (621.395)
automatic message switching system automatische Telegraphie-Speichervermittlung
automatic monitor automatische Frequenzsteuerung (621.396.61)
automatic opening automatische Abschaltung
automatic phase control automatische Phasenregelung
automatic pilot automatische Kurssteuereinrichtung, Autopilot (629.13)

automatic power station automatisches Kraftwerk
automatic printing machine Kopierautomat
automatic programming automatisches Programmieren (518.5)
automatic punch Stanzautomat
automatic railway signalling automatische Bahnsignalanlage
automatic reclosing automatische Wiederzuschaltung; Kurzschlußfortschaltung, Kurzunterbrechung (621.316)
automatic record changer automatischer Plattenwechsler
automatic regulation automatische Regelung
automatic regulator automatischer Regler, Schnellregler, Tirillregler (621.3)
automatic release Selbstauslösung
automatic reserve activation Reserveschaltung
automatic reset automatische Rückführung, automatische Rückstellung; automatische Nullstellung
automatic ringing selbsttätige Rufgabe (621.395)
automatic reversible battery booster automatischer Batteriespannungsregler
automatic routiner automatische Prüfeinrichtung (621.395)
automatic scanning receiver automatischer Suchempfänger
automatic scanning unit Abfrageautomatik (621.398)
automatic sequencing Programmsteuerung
automatic selectivity control automatische Bandbreiteneinstellung
automatic short circuiter automatischer Kurzschließer (621.313)
automatic short-circuit protection automatische Kurzschlußsicherung, kurzschlußsicher
automatic shutoff automatische Abschaltung
automatic shutter automatischer Verschluß
automatic signalling device automatische Signalanlage (656.25); automatische Meldeanlage
automatic starter Selbstanlasser (621.313)
automatic starting control automatische Anlaßsteuerung
automatic starting resistor Selbstanlasser (621.313)
automatic stop automatische Abschaltung
automatic substation selbsttätiges Unterwerk, automatisches Unterwerk, bedienungsloses Unterwerk
automatic synchronizer automatische Synchronisiereinrichtung
automatic telephone exchange Selbstanschlußtelefonamt, Wählamt
automatic telephone system Wähltelefonanlage, Selbstwählsystem
automatic tracking automatische Zielverfolgung, automatische Nachführung
automatic traffic signal selbsttätiges Verkehrssignal
automatic train control automatische Zugsteuerung (656.25)
automatic train control application automatische Zugbremsung (656.25)
automatic train stop Zugbeeinflussungshebel
automatic transfer equipment automatische Umschalteinrichtung
automatic transmission Maschinentelegraphie (621.394)
automatic tripping contactor Schütz mit Relais
automatic tripping device Selbstauslöseeinrichtung, automatische Unterbrechung, automatische Auslösung
automatic variable voltage ratio transformer automatischer Regelumspanner, selbstregelnder Umspanner
automatic voltage regulation selbsttätige Spannungsregelung
automatic volume control Schwundregelung, Fadingregulierung, automatische Lautstärkereglung
automatic volume expansion automatische Dynamikregelung
automatic welding Maschinenschweißung, Automatenschweißung
automatic winding machine Wickelautomat
automaton (pl.: **-ta**) Automat
automation Automation, Automatisierung
automobile Automobil, Auto, Kraftwagen
automobile body Karosserie; „Masse“ (621.3)
automobile lighting Autolichtanlage
automobile starter Anlasser
automotive Auto-...
autonomous navigation bodenunabhängige Navigation
autophone SA-Telefon, Wähltelefon, Selbstanschlußtelefon
autopilot Autopilot, Selbststeuereinrichtung
autoplex reception Reflexaudionempfang (621.396.62)
auto restart Selbstanlauf
autoscrutineer Gleichlaufselbststeuerung (621.394)
autosyn Drehmelder
auto-synchronous motor selbstanlaufender Synchronmotor

autotransformer Autotransformator, Spartransformator, Einwicklungstransformator
autovalve SAW-Ableiter; Überspannungsableiter mit spannungsabhängigem Widerstand (621.316.9)
Auvert-Ferrand permutator Wechselstrom-Gleichstrom-Umformer (621.314.5)
auxiliary Eigenbedarfsanlage
auxiliary Hilfs-...
auxiliary anode Hilfsanode
auxiliary button Hilfstaste
auxiliary coil Hilfsspule
auxiliary contact Hilfskontakt
auxiliary electrode Hilfselektrode
auxiliary machine Hilfsmaschine
auxiliary motor Hilfsmotor
auxiliary phase Hilfsphase, Anlaufwicklung
auxiliary pole Hilfspol, Wendepol
auxiliary pole shunting Wendepolnebenschluß
auxiliary position light Hilfspositionslampe (629.13)
auxiliary program Hilfsprogramm (681.3)
auxiliary relay Hilfsrelais
auxiliary spark gap Hilfsfunkenstrecke; Vorfunkenstrecke (629.113)
auxiliary stabilizing series winding Hilfsreihenschlußwicklung
auxiliary storage Hilfsspeicher
auxiliary switch Hilfsschalter
auxiliary transformer Hilfstransformator
auxiliary winding Hilfswicklung, Hilfsphase
available verfügbar, vorhanden, zur Verfügung
available machine time nutzbare Maschinenzeit
available power maximal abgebbare Leistung, verfügbare Leistung
available width Nutzbreite
availability Erreichbarkeit (621.395); Verfügbarkeit
avalanche Lawine (537)
avalanche breakdown Lawinendurchbruch
avalanche rectifier Gleichrichter mit reversibler Durchbruchscharakteristik, Lawineneffektgleichrichter
avalanche voltage Lawinendurchbruchsspannung
a.v.c. = automatic volume control automatische Schwundregelung
avdp. = avoirdupois [Gewichtssystem]
average Mittelwert, mittlerer Wert
to average mitteln; im Mittel betragen
averaging recorder Spitzenverbrauchszähler (621.317.8)
average Mittel-..., mittler, Durchschnitts-...
average candle power mittlere sphärische Lichtstärke
average current arithmetischer Strommittelwert
average delay mittlere Wartezeit (65)
average density mittleres spezifisches Gewicht
average life mittlere Lebensdauer
average load Durchschnittslast, Durchschnittsbelastung, mittlere Last
average noise factor Band-Rauschzahl
average output Durchschnittsausstoß, mittlere Ausgangsleistung
average picture-signal level Bildsignalmittelwert
average rate of change of the output voltage mittlere Flankensteilheit der Ausgangsspannung
average traffic flow mittlere Verkehrsdichte (65)
average transfer rate mittlere Übertragungsgeschwindigkeit
average value Mittelwert, arithmetischer Mittelwert (621.3)
average voltage arithmetischer Spannungsmittelwert
aviation Luftfahrt
avionics Flugelektronik
Avogadro's number Avogadro'sche Zahl
avoirdupois Avoirdupois (Gewichtsmaßsystem)
A.W.G. = American Wire Gauge
"A"-wire a-Ader
axial axial
axial ball bearing Axialkugellager
axial ducts axiale Lüftungsschlitze (621.313)
axial flow Axialfluß
axial-flow fan Axialverdichter, Axialgebläse
axial flow pump Axialpumpe
axial leads axiale Anschlußdrähte, axiale Anschlüsse
axial pitch axiale Teilung, Axialteilung (621.8)
axial pressure Längsdruck, Axialschub
axial section Axialschnitt, Längsschnitt
axial system of ventilation Axialkühlung
axial thrust Axialschub
axial-thrust bearing Axialdrucklager
axial ventilation Axialbelüftung
axially corrugated guide Wellrohrhohlleiter
axis (pl.: **axes**) Achse
axis of abscissae Abszissenachse
across axis quer zur Achse
axis angle Achsenwinkel (621.8)

axis of commutation Kommutierungsachse (621.313)
axis of gears Radachse (621.8)
axis of imaginaries imaginäre Achse
axis of reference Bezugsachse; Koordinatenachse
axis of rolling Wälzachse (531)
axis of rotation Drehachse
axis of symmetry Symmetrieachse
axle Achse, Wagenachse; Drehachse (621.317)
axle base Achsstand
axle bearing Achslager
axle box Achsbüchse (625)
axle circuit Radachsenstromkreis
axle counter Achszähler
axle generator Achsdynamo, Lichtmaschine (652)
axle-generator pole changer Achsdynamo-Umschalter
axle-generator regulator Achsdynamoregler (65)
axle-hung motor Motor mit Achsaufhängung, Tatzlagermotor (621.33)
axle journal Achsschenkel
axle load Achslast, Achsdruck
axle mounted auf der Achse sitzend
axle pressure Achsdruck (652.2)
axle suspension Achsaufhängung, Tatzenlagerung (621.33)
Ayrton-Perry winding induktionsfreie Wicklung
azimuth Azimut
azimuth angle Seitenwinkel (656.7)
azimuth compass Visierkompaß
azimuth narrow mode Schmalsuchen
azimuth rotate mode Rundsuchen
azimuth scan coverage Seitenabtastbereich
azimuth strobe Seitenwinkelzeiger
azimuth wide mode Breitsuchen
azimuthal azimutal
azote Stickstoff

B

B.A. = British Association
B.A.thread Verbands-Normgewinde
BABS = beam approach beacon system Leitstrahlanflug-Funkfeuersystem
babbitt Lagermetall; Bleihammer
babbitt hammer Bleihammer
babbitt metal Weißmetall
babble unverständliches Nebensprechen, Babbeln
back Rücken, Rückseite
back-to-back connection Kreuzschaltung (621.314.6)
back-to-back devices antiparallel geschaltete Bauteile
back-to-back saw-tooth symmetrischer Sägezahn (621.3)
back of a generator Kupplungsseite eines Generators, Generator-AS
back of a motor Antriebsseite eines Motors, Motor-AS
back of a panel Schalttafelrückseite
back rückseitig, hintenliegend, Rücken-..., Rück-...
back ampere turns Gegenamperewindungen
back bias Gegenspannung, Rückwärtsregelspannung
back bias circuit Schwundregelung nach rückwärts, Rückwärtsregelkreis
back connected switch Schalter mit rückwärtigem Anschluß
back contact Ruhekontakt
back contact key Telegraphentaste, Morsetaste
back current Rückstrom, Sperrstrom
back diffusion Rückdiffusion
back echo Rückzipfelecho
back effect Rückwirkung
back electromotive force gegenelektromotorische Kraft, Gegen-EMK
back-end radiation Rückwärtsstrahlung (621.396.67)
backfire Rückzündung (621.3); Rückschlag (bei Frühzündung) (621.4)
back firing = backfire
back flashover rückwärtiger (Mast-)Überschlag
back focal length Brennweite
back gear Vorgelege
back-geared motor untersetzter Motor, Getriebemotor
back gearing Untersetzungsgetriebe
background noise Hintergrundrauschen
background radiation Hintergrundstrahlung, Grundstrahlung
back heating Rückheizung (621.385)
back light Kontrollichtöffnung; Rücklicht
back lobe Rückwärtszipfel
back panel Rückwand
back pitch Wickelschritt auf der kommutatorabgewandten Seite (621.313)

back plate Auflagescheibe, Stützscheibe, Rückwand
back-plate electrode Gegenelektrode
back porch hintere Austastschulter (621.397)
back potential Gegenspannung
back pressure Gegendruck, Druck auf der Saugseite
back pull Rückwärtszug, Bremszug (621.7)
back resistance Sperrwiderstand
back-scattered radiation Streustrahlung, gestreute Strahlung, Rückstreuung
back scattering Rückstreuung
back shunt Spitzkehre
back-shunt keying Antennenkreistastung
backspacer Rücktaste (681)
backspacing Bandrücksetzung (681.3)
back stay Abspanndraht
back stroke Rückhub, Rückgang, Rückbewegung
backtracking [Suchalgorithmus zur Lösung eines Problems durch „trial and error"]
back voltage Rückwirkspannung; Sperrspannung
to back unterstützen, hinterlegen, rückwärts versehen mit
to backfire rückzünden (621.3)
to back-off hinterschneiden, hinterschleifen
to back-out sich zurückziehen aus, zurückgehen (Preßstempel)
backed saw Metallblattsäge
to back-scatter rückstreuen (53)
to back-signal rückmelden
backup Datensicherung (681.3)
backup copy Sicherungskopie
backup lamp, backup light Rückwärtsscheinwerfer (629.113)
backup memory backup Speicher, Bereitschaftsspeicher
backup protection Reserveschutz
backup system Bereitschaftssystem (681.3)
backing lamp, backing light Rückwärtsscheinwerfer (629.113)
backing material Hinterlage, Grundmetall
backing pump Vorpumpe
backing store Zusatzspeicher, Hilfsspeicher, Ergänzungsspeicher
backing strip Schweißunterlage
backbone Hochspannungsverteilungsschiene
background Hintergrund, Grund-...
background brightness Grundhelligkeit (621.397)
background count Störpegelzählung (539)
background hiss Grundgeräusch
background illumination mittlere Helligkeit (621.397)
background noise Grundgeräusch, Störpegel, Raumgeräusch, Nebengeräusch
background processing Hintergrundverarbeitung (681.3)
backlash Spiel, Lose, toter Gang; Flankenspiel (621.8)
backlog of orders Auftragspolster, Auftragsüberhang
backstop rückwärtiger Anschlag
backward rückständig; rückwärtig; rückwärts
backward-acting regulator Rückwärtsregler
backward diode Rückwärtsdiode
backward lead, backward shift (of brushes) Bürstenverschiebung gegen den Drehsinn
backward-looking radar Rückwärtsradar
backward power Rücklaufleistung
backward signal Rückwärtszeichen (621.395)
backward travelling gegenläufig (621.394)
backward wave rücklaufende Welle, Rückwärtswelle
backward wave oscillator Rückwärtswellenoszillator
backwardation Konventionalstrafe (bei Lieferverzug)
bad schlecht, minderwertig
bad contact schlechter Kontakt
bad earth schlechte Erde (621.3)
in bad order in schlechtem Zustand
bad spot Bandfehlstelle (681.3)
baffle Sperre; Widerstandskörper (532); Schallbrett (621.395.62); Schallwand
baffle board Schallbrett
baffle plate Ablenkblech, Umlenkplatte (621); Glockenboden, Kolonnenboden (66); Hohlleiterblende
baffle system Puffersystem, Schutzplattenanordnung
bag Tasche, Beutel
bagpipes Dudeltonstören
bag type Beuteltype (621.352)
baggy beutelig, ausgebeult
bail-out plug Schnelltrennstecker
to bake backen; ausheizen (621.38); brennen (666)
to bake out formieren (621.314.6)
to bakelise bakelisieren
bakelised paper Hartpapier
bakelite Bakelit
bakelite-bonded wheel Schleifscheibe mit Kunstharzbindung
bakelite derivative Bakelitprodukt
bakelite paper Bakelitpapier

bakelite varnish Bakelitlack
baking coal backende Kohle, Backkohle
baking varnish ofentrockener Lack, Einbrennlack
BAL = business application language
balance Waage; Gleichgewicht, Gleichgewichtslage; Abgleichzustand (621.317); Unruhe (681); Abgleich, Symmetrierung
balance bob Ausgleichsgewicht, Gegengewicht
balance coil Ausgleichsdrossel, Saugdrossel
balance control Frequenzgangausgleich (621.395.62)
balance converter Symmetrierglied
balance indicator Nullinstrument (621.317)
out of balance unausgewuchtet (621); unabgeglichen (621.317)
balance point Abgleichpunkt (621.317)
balance-return loss Fehlerdämpfung
balance rope Unterseil (621.34), (622)
balance-to-unbalance transformer Symmetrieübertrager
balance weight Gegengewicht, Ausgleichsgewicht; Balanciergewicht
balance wheel Unruhe (681)
balance winding Ausgleichswicklung
to balance abwägen; ins Gleichgewichtbringen, abgleichen (621.317); auswuchten, wuchten (621.0); kompensieren (621.3); symmetrieren (621.39)
to balance out ausbalancieren, ausgleichen, abwägen
balanced ausgewogen, abgeglichen; symmetriert, symmetrisch
balanced amplifier Gegentaktverstärker
balanced-armature loud speaker Magnetlautsprecher
balanced-bridge transition Ausgangsbrückenschaltung
balanced capacitor Ausgleichskondensator, Entkopplungskondensator
balanced circuit symmetrische Schaltung
balanced condition eingeregelter Zustand
balanced currents Gegentakt-Ströme
balanced-current relay Stromdifferentialrelais
balanced-flow servo-motor Stellmotor mit Ölausgleich
balanced four-terminal network symmetrischer Vierpol
balanced to ground erdsymmetrisch, symmetrisch gegen Erde
balanced input symmetrischer Eingang
balanced-line system symmetrisches Leitungssystem
balanced load symmetrische Last
balanced loading symmetrische Belastung
balanced modulator Modulator mit unterdrücktem Träger, Gegentaktmodulator (621.396.61)
balanced periodic quantity reine Wechselgröße
balanced-polyphase system balanciertes Mehrphasensystem
balanced relay Differentialrelais
balanced three-phase system balanciertes Drehstromsystem
balanced three-wire system balanciertes Dreileitersystem
balanced transmission line symmetrische Übertragungsleitung
balanced twin symmetrische Doppelleitung
balancer Ausgleichsmaschine (621.3)
balancer booster Ausgleichsregelsatz (621.316.1)
balancer transformer Ausgleichsdrosselspule (621.316.1)
balancing Abgleichs-..., Ausgleichs-...
balancing aerial Kompensationsantenne
balancing capacitance Gegengewicht (621.396)
balancing circuit Ausgleichsschaltung
balancing gear Ausgleichsgetriebe
balancing machine Auswuchtmaschine
balancing and matching line Symmetrier- und Transformationsleitung
balancing motor Abgleichmotor (621.317)
balancing network Leitungsnachbildung (621.395)
balancing potentiometer Abgleichpotentiometer
balancing resistor Ausgleichswiderstand
balancing rings Ausgleichsverbindungen (621.313)
balancing speed Gleichgewichtsgeschwindigkeit (625)
ball Kugel
ball-bank indicator Kugelneigungsmesser, Scheinlot
ball bearing Kugellager
ball and bucket viscosimeter Kugelviskosimeter
ball cage Kugelkäfig
ball electrode Kugelelektrode
ball eye Aufhängeöse (621.315.1)
ball handle Kugelgriff
ball hardness Kugeldruckhärte
ball lightning Kugelblitz
ball mill Kugelmühle
ball race Laufring (621.822)
ball and socket Kugelgelenk

ball-thrust washer Axialkugellager, Axialdrucklaufring
to ball ballen, sich ballen
ballast Bettung (625.1)
ballast bulb Vorwiderstand, Vorwiderstandslampe
ballast leakage Bettungsableitung
ballast for low voltage fluorescent lamp Vorschaltgerät für Niederspannungsleuchtstofflampe
ballast resistance Ballastwiderstand, Lastwiderstand
ballast tube Widerstandslampe, Eisen-Wasserstoff-Widerstand, Stromregelröhre
ballistic ballistisch
ballistic factor Dämpfungsverhältnis (621.317)
ballistic galvanometer ballistisches Galvanometer
balljoint Kugelgelenk
balloon reflector Ballonreflektor
balun Symmetrierleitung, Symmetrierübertrager
banana jack Bananensteckerbuchse
banana pin Bananenstecker
banana plug Bananenstecker
band Band; Bande; (Schleif-)Band, Bereich, Bandage (621.313); Bandbreite (621.39)
band-armoured cable bandeisenbewehrtes Kabel
band articulation Bandverständlichkeit
band brake Bandbremse
band-center indication Mittenanzeige
band chain Gelenkkette, Bandkette
band-edge energy Energieband (621.382)
band-edge frequency Randfrequenz
band-elimination filter Bandsperre, Sperrfilter (621.39)
band filter Bandfilter
band of frequencies Frequenzband
band limit Bandgrenze
band of losses Bandagenverluste (621.313)
bandpass Bandpaß
bandpass amplifier bandfiltergekoppelter Verstärker
bandpass characteristic Bandfiltereigenschaft, Bandfilterkennlinie
bandpass filter Bandpaß, Bandfilter
bandpass flatness Bandfilterwelligkeit
bandpass response Durchlaßkurve, Bandfilterkennlinie
bandpass tuning Bandfilterabstimmung (621.396.62)
band-rejection filter Bandsperre (621.39)
band spectrum Bandenspektrum
bandspread Bandspreizung
band-stop filter Bandsperre
band switch Wellenschalter (621.396)
bandwidth Bandbreite, Kanalbreite (621.39)
bandwidth control Bandbreitenregler
bang-bang control Zweipunktsteuerung, Zweipunktregelung
bank Satz, Schaltung, Gruppe, Bank; Böschung; Querneigung, Schaltbahn (621.395)
bank cable Vielfachkabel (621.395)
bank of capacitors Kondensatorbatterie
bank of coils Spulensatz (621.396)
bank indicator Querneigungsmesser
bank-raid alarm Überfallruf
bank of transformers Transformatorensatz
bank winding Stufenwicklung
bank-and-wiper switch Hebdrehwähler
bank wires Vielfachverbindungen (621.395)
bank-wound stufengewickelt, kapazitätsarm
to bank in die Kurve legen, schief liegen, sich schief legen, sich neigen
banked track geneigte Bahn
banked winding kapazitätsarme Wicklung (621.317)
B.A. Ohm (B. A. Unit = British Association Unit) (Einheit des Widerstandes = 0,9886 Ohm)
bar Stange, Stab, Vollstange; Stab (621.313); Balken (621.397)
bar chart Balkendiagramm
bar code Strichcode, Balkencode
bar-coupling panel Sammelschienen-Kuppelfeld
bar diameter Stangenstärke, Stangendicke
bar form Stangenform
bar generator Balkengenerator (621.397)
bar graph Balkendiagramm; Bandanzeige
bar of hexagonal section Sechskantstange
bar magnet Stabmagnet
off the bar von der Stange (621.94)
bar of round section Rundstange
bar signal Sprungsignal (621.397)
bar suspension Tatzenlagerung (621.33)
bar-type current transformer Schienenstromwandler, Durchsteckstromwandler, Stabstromwandler
bar winding Stabwicklung (621.313)
bar-wound armature Anker mit Stabwicklung
bar and yoke method Jochverfahren (538.22)
to bar sperren; absperren, aussperren; verriegeln; unberücksichtigt lassen, absehen von
bar quer (51) (z. B. $\bar{X}$ lies: X bar = X quer)
barb Widerhaken

bare frei, unbedeckt, bloß; los, bar; blank, nicht isoliert (621.3), (621.1)
bare conductor blanker Leiter
bare electrode blanke Elektrode (621.791)
down to the bare metal bis aufs blanke Metall
barely extinguished eben erloschen, eben aus
barium Barium
Barkhausen jump Barkhausen-Sprung (538.2)
Barkhausen-Kurz oscillator Barkhausen-Kurz-Generator
Barlow's wheel Barlowsches Rad
barometric barometrisch
barometric effect Luftdruckeinfluß
barometric pressure Luftdruck
barrage Sperre, Damm; Sperrfeuer (623.4)
barrage jamming Breitbandstörung
barrel Faß; Lauf; Rohr (623.4); Muffe, Drehbankspindel (621.94)
barrel controller Kontroller, Steuerwalze, Walzenschalter
barrel distortion Tonnenverzeichnung
barrel winding Faßwicklung, Korbwicklung (621.313)
barreter Baretter, Stromregelröhre
barrier Trennwand, Barriere, Schranke (621.3); Sperrschicht (621.382)
barrier capacitance Sperrschichtkapazität
barrier-film rectifier Sperrschichtgleichrichter
barrier frequency Abdeckfrequenz, Grenzfrequenz
barrier-grid tube Sperrgitterröhre
barrier layer Sperrschicht, Grenzschicht (621.382)
barrier-layer capacitor Sperrschicht Kondensator
barrier-layer effect Halbleitereffekt (621.314.6)
barrier-layer photoelectric cell Sperrschichtphotozelle
barring motor Durchdrehmotor
baryte Baryt, Schwerspat
baryte paper Barytpapier
basalt Basalt
BAS-signal Bildaustastsynchronsignal
base Basis, Fundament, unteres Ende; Stützpunkt (623); Grundzahl, Basis (51); Grundlinie (51), (52); Grundfläche (51); Grundplatte, Fundament (621); Röhrensockel (621.3); Grundstoff (54); Grundmetall (621.35); lange Kathede (51); Basis; Trägermaterial (621.382)
at base am Fußpunkt
base address Basisadresse
base band Basisband (u)
base contact Basisanschluß
base current Basisstrom
base frequency Grundfrequenz, Basisfrequenz (621.395.4)
base line Grundlinie, Nullinie
base-line stabilizer Nullinienstabilisiereinrichtung, Stabilisierschaltung (621.397)
base load Grundlast, Grundbelastung, Lastminimum (621.31)
base-loaded antenna Antenne mit Verlängerungsspule (im Fußpunkt), fußpunktbelastete Antenne
base material Trägermaterial (621.382)
base notation Stellenwertschreibung (681.3)
base number Grundzahl; Basiszahl (681.3)
base plate Tisch (621.95); Grundplatte
base region Basiszone (621.382)
base register Indexregister, Basisadreßregister (681.3)
base resistance Basiswiderstand
base speed Normaldrehzahl, Grunddrehzahl
base station Feststation, Feste Landfunkstelle
base terminal Basisanschluß (621.382)
base voltage Basisspannung
base unedel, minderwertig
base metal unedles Metall
basement Kellergeschoß, Sockelgeschoß
BASIC = Beginners All-purpose Symbolic Instruction Code BASIC
basic grundlegend, Grund-...; basisch, laugenartig
basic alternating current theory Wechselstromlehre
basic circuit Grundschaltung
basic circuit diagram Grundschaltbild, Übersichtsschaltplan
basic electricity Grundlagen (pl.) der Elektrotechik
basic equation Grundgleichung, Ausgangsgleichung
basic equipment Grundausrüstung
basic frequency Grundschwingung
basic group Grund-Primärgruppe (621.395.4)
basic knowledge Grundwissen, Grundlagenkenntnisse (pl.)
basic loss Systemdämpfung
basic machine cycle Grundzyklus
basic mastergroup Grund-Tertiärgruppe (621.395.4)
basic network Nachbildung, Leitungsnachbildung, verlustloses Nachbild (621.395)
basic noise Grundgeräusch

basic reference atmosphere Standard Atmosphäre
basic supergroup Grund-Sekundärgruppe (621.395.4)
basic super-master group Grund-Quartärgruppe (621.395.4)
basic transmission loss Grundübertragungsdämpfung
basic value Ausgangswert
basicity basisches Verhalten, Basizität
basing point Ausgangspunkt
basis Basis, Grundlage; Basis (Elektrode)
basis for measurement Meßbasis
on a . . . basis über . . . als Basis
basket Korb
basket coil Korbspule (621.396)
basket-wave coil Korbspule (621.396)
basket winding Korbwicklung, Faßwicklung (621.396)
bass Baß, Tieftonbereich
bass boost Baßanhebung
bass compensation Baßanhebung (621.396.62)
bass control Tieftonblende, Klangfarbenregler (621.396.62)
bass loudspeaker Tieftonlautsprecher
batch Stoß, Ladung; Charge; Eintrag, Einsatz
batch mixer Knetmischer (666.3)
batch processing Stapelbetrieb (681.3) [automatisches Abarbeiten von Betriebssystem-Befehlen]
batch production Serienfabrikation
bath Bad
bath voltage Badspannung (621.35)
bathing-tub diagram Badewannenkurve (620.1)
batten Latte, Leiste
batten lampholder Wandfassung (621.315.6)
battens Lampenleiste (621.32)
to batter verhauen, verkeilen
battered thread verdrücktes Gewinde
battery Batterie (621.3), (623.4)
battery of accumulators Akkumulatorenbatterie
battery acid Batteriesäure, Sammlersäure
battery backup Netzausfallschutz durch Batterie
battery bell Gleichstromwecker
battery booster Ausgleichslademaschine, Batteriezusatzmaschine (621.355)
battery box Batteriegehäuse, Batteriekasten
battery call Batterieruf, Batterieanruf
battery capacity Batterie-Kapazität, Batteriespeichervermögen
battery cell Zelle
battery charger Batterieladegerät, Ladegleichrichter, Batterielader; Akkumulatorladegerät
battery charging Batterieladung
battery-charging rectifier Batterieladegleichrichter
battery clip Batterieklemme
battery of condensers Kondensatorenbatterie
battery container Batteriekasten, Batteriegehäuse
battery dialling Batteriewahl (621.395.34)
battery discharge Batterieentladung
battery drain Batteriestromverbrauch
battery-driven vehicle Batteriefahrzeug, Triebwagen
battery eliminator Direktspeisegerät, Netzspeisegerät
battery gauge Batterieprüfer (621.317)
battery hydrometer Säuremesser
battery ignition Batteriezündung
battery lamp Akku-Lampe
battery meter Batteriezähler
battery-operated batteriegespeist, mit Batteriebetrieb
battery operation Batteriebetrieb
battery radio Batterieempfänger
battery rack Batteriegestell
battery rail car Akkumulatorentriebwagen
battery regulating switch Zellenschalter
battery ringing Batterieruf
battery set Batteriegerät, Batterie-Empfänger
battery supply Batterie-Stromversorgung
battery supply circuit noise Stromversorgungsgeräusch, Netzbrumm
battery switch Zellenschalter
battery tester Batterieprüfer
battery traction Batterieantrieb (652), (629.1)
batwing antenna Superturnstilantenne, Schmetterlingsantenne
Baud Baud
Baudot multiplex printing telegraph Baudot-Drucktelegraph
Baudot telegraph system Baudot-System
baulk Balken
B.A. Unit s. **B.A. Ohm**
Baur's constant 1 mm Durchschlagfestigkeit
bauxite Bauxit
bay Abschnitt (621.396.677); Kammer, Bucht; Gebäudeabschnitt (zwischen zwei Stützen), Hallenschiff; Gestell (621.39)
bay lamp Gestellampe
bayonet Bajonett
bayonet cap Swan-Sockel, Bajonettsockel (621.32)

bayonet holder Swan-Fassung, Bajonettfassung (621.32)
bayonet joint Bajonettverbindung
Bazooka Symmetriertopf (621.396)
B battery Anodenbatterie
B.B.C. = British Broadcasting Corporation Britische Rundfunk-Gesellschaft
BCD = binary coded decimal BCD-Code
BCI = Broadcast interference Rundfunkstörung
b.c. lamp = bayonet cap lamp Lampe mit Swan-Fassung
BDOS = basic disk operating system
B.D.S.T. = British double summer time Britische doppelte Sommerzeit
beacon Bake, Funkbake; Funkfeuer, Leuchtfeuer
beacon tracking Verfolgen nach Zielausstrahlung (623)
bead Perle, (Isolier-)Perle; Schweißraupe; eingerollter Rand (621); Korn (623.44)
bead lightning Perlschnurblitz
bead of solder Lötperle, Zinntropfen
bead test Perlenprüfung (54)
to bead sicken, bördeln, einrollen, Rand rollen, mit gerolltem Rand versehen
bead support Abstandshalterung durch Scheiben (621.315.2)
beaded transmission line Koaxialleitung mit Abstandshaltern (621.315.2)
beading Durchführungsperle, Einschmelzperle
beading glass Einschmelzglas (621.385)
Bead protective system Stromdifferentialschutz
beam Balken, Träger, Tragbalken, Schwelle; Waagebalken; Leitstrahl, Keule, Strahl, Elektronenstrahl, Richtstrahl, Leitstrahl (621.396.67); Hilfsbalken, Querträger (69); Strahl (53)
beam aerial Richtantenne
beam angle Keulenbreite, Keulenwinkel, Strahlwinkel, Öffnungswinkel (621.396.677)
beam approach beacon system BABS-System, Leitstrahlanflug-Funkfeuersystem
beam array Richtstrahlantenne, Richtantenne
beam axis Keulenachse
beam of a balance Waagebalken
beam-bending magnet Ablenkmagnet
beam blanking Strahlaustastung
beam compression Strahlverdichtung
beam control Strahlführung
beam current Strahlstrom (621.385)
beam current characteristic Strahlstromkennlinie (621.385)
beam-forming plate Abweiseranode, Bündelungselektrode
beam of ions Ionenstrahl
beam-lead technology Beam-Lead-Technik
beam of light Lichtstrahl, Lichtbüschel
beam-lobe switching Keulenumtastung (621.396)
beam-power amplifier tube Leistungsverstärkerröhre
beam-power tetrode Endtetrode mit Elektronenbündelung
beam of protons Protonenstrahl
beam-receiving aerial Richtempfangsantenne
beam reflector ungespeister Reflektor (621.396.677)
beam rider (Leit-)Strahlreiter (623)
beam-rider guidance Leitstrahllenkung
beam-shaping characteristic Strahlungscharakteristik, Antennencharakteristik
beam sharpness Richtschärfe
beam spot Leuchtfleck, Fleck (621.385)
beam system Richtstrahlanlage
beam switching Keulenumtastung (621.397)
beam transmission Richtsendung
beam-transmitting station Sender mit Richtstrahler
beam trap Strahlfalle (621.385)
beam trigger Helligkeitssteuerung, Strahlauftastung
beam power Strahlstärke (621.396.67)
beam (power) tube Strahlröhre, Bündelungsröhre, Vierpolröhre mit Pentodeneigenschaften
beam width Keulenbreite, Strahlbreite, Öffnungswinkel
beam width at half power Halbwertsbreite
to beam richten, bündeln (621.396.677)
beans (pl.) Kohlengrus
to bear against, on anliegen gegen, sich abstützen auf
bearer-cable Tragseil (621.33)
bearer wire Tragdraht
bearing Lager (621); Einfluß; Peilung, Winkelpeilung; tragende Fläche, Auflager (621); Kurs (621.396.93)
bearing angle Peilwinkel, Azimut-Peilung
bearing bracket Lagerkonsole (621.313)
bearing-bracket assembly Lager mit Lagerbock
bearing bush Lagerbuchse
bearing capacity Tragfähigkeit
bearing correction Funkbeschickung, Peiler-Korrekturkurve
bearing cup Lagerschale
bearing error Azimutfehler
bearing evaluation Peilauswertung

bearing friction Lagerreibung
bearing indication Kursanzeige
bearing metal Lagermetall, Weißmetall
bearing pedestal Lagerbock
bearing pin Tragstift
bearing plate Tragplatte
bearing pressure Lagerdruck
bearing resolution Azimut-Auflösung, Seitenwinkelauflösung
bearing shell Lagerschale
bearing sleeve Lagerschalen (pl.)
bearing stress Auflagerdruck
bearing support Lagerbock
bearing surface tragende Fläche, Auflagefläche
bearing washer Lagerring, Lagerscheibe
beat Schwebung (volle Periode), Takt
beat frequency Schwebungsfrequenz
beat-frequency oscillator Schwebungssummer, Überlagerungsoszillator
beat method Nullfrequenzton-Methode (621.317)
beat note Schwebungston
beat reception Schwebungsempfang, Überlagerungsempfang
beater Schlagwolf; Holländer
beating oscillator Schwebungssummer
beaver-tail beam Biberschwanzkeule
Beck arc lamp Beck-Lichtbogenlampe
Becquerel rays (pl.) Radiumstrahlung
bed Bett (621.9); Bohrtisch (621.9); Lagerstätte, Schicht (55)
bed casting Gußbett (621.94)
bed-plate Grundplatte
bed rock gewachsener Fels
bed slide Führungsbett, Führungsschiene
to bed betten, lagern
to bed brushes Bürsten einschleifen (621.3)
to bed down level auswiegen, in die Waage richten
bedding Auflage
bedding of jute Juteumspinnung
beech Buche, Buchenholz
bega (Vorsatz) 10^9, Giga
begin of tape Bandanfangsmarke
begin-to-end signal Wahlabrufzeichen
beginning of conversation Gesprächsbeginn (621.395)
begohm 10^9 Ohm
behind hinter, nach
to be behind nacheilen gegen, zurückbleiben gegen
bel Bel
to belch ausstoßen, auswerfen
to belch smoke Rauch ausstoßen
Belisha beacon Fußgänger-Verkehrszeichen, Ampel
bell Glocke; Klingel
bell button Klingelknopf
bell crank Winkelhebel
bell-crank lever Winkelhebel
bell glass Glasglocke
bell jar Glasglocke, Rezipient
bell pull Klingelzug
bell push Klingelknopf
bell-shaped pulse glockenförmiger Impuls (621.397)
bell transformer Klingeltransformator
bell wire Klingeldraht
Bell receiver, Bell telephone Bell-Telefon
Bellini-Tosi direction finder Kreuzrahmenpeiler
bellow the mark unternormal, schlecht
bellow measurable limits unmeßbar klein
belt Gürtel (623); Riemen (621)
belt conveyor Förderband, Bandförderer
belt dispersion Zickzack-Streuung, doppelt verkettete Streuung (621.313)
belt drive Riementrieb
belt furnace Durchlaufofen
belt pull Riemenzug
belt sander Bandschleifmaschine
belt slip Riemenschlupf
belt transmission Riementrieb, Riemenantrieb
to belt gürten, mit Gürtel versehen
to belt together mit Riemenantrieb kuppeln
belted cable Gürtelkabel
belting Riementrieb
bench Bank; Werkbank; Steilküste; Labortisch
bench drill Tischbohrmaschine
benchmark Küstenecho, Küstenmarke, Festzeichen (621.397)
benchmark program Bewertungsprogramm (681.3)
bench test set up Prüfaufbau
bench work Schlosserarbeit
benchboard Schaltpult, Steuerpult (US)
bend Biegung, Bogen, Krümmer, Knie (621), (66); Kurvenstück (625)
bend strength Biegefestigkeit
bend test Biegeversuch
bend-test number Biegezahl
to bend biegen, verbiegen, beugen
to bend edges at right angles abkanten
to bend in einbördeln
to bend over umbiegen, umbördeln
to bend a wire over on itself einen Draht umknicken, einen Draht zusammenfalten

to bend up aufbiegen
to bend to and fro hin- und herbiegen
to bend waves Wellen beugen
to bend waves round ... Wellen umbeugen
bending Biegung, Abkröpfung
bending allowance Rückfederungszuschlag
bending coil (Elektronenstrahl-)Krümmungsspule
bending moment Biegemoment
bending-moment diagram Momentdiagramm
bending strain Biegebeanspruchung, Biegung
bending strength Biegefestigkeit
bending stress Biegespannung, Biegungsspannung
bentonite Bentonit, Fullererde
benzene Benzol
benzine Benzin
berry transformer Drehtransformator, Drehregler
BESA = British Engineering Standard Association (entspr. DNA = Deutscher Normen-Ausschuß)
Bessel function Besselsche Funktion
beta disintegration Betazerfall
beta particle Beta-Teilchen
beta ray Betastrahl
beta transformation Beta-Umwandlung
betatron Betatron, Elektronenschleuder, Rheotron
bevel Abschrägung, Facette
bevel cut Schrägschnitt, Gehrungsschnitt
bevel edge Facette
bevel gear Kegelrad, Kegeltrieb
to bevel abschrägen, facettieren
bevelled poles abgeschrägte Pole (621.313)
Beverage aerial Beverage-Antenne, Langdrahtantenne
beyond jenseits, über ... hinaus
beyond cut-off ins Sperrgebiet (621.385)
beyond-the-horizon transmission Überhorizontübertragung
beyond the limit über den Grenzwert hinaus
beyond limited periods zeitlich unbegrenzt
b.f. = beat frequency Schwebungsfrequenz
b.f.o. = beat-frequency oscillator Schwebungssummer
B-H curve B-H-Kurve, Magnetisierungskurve
B-H loop Hysteresisschleife
B.H.N. = Brinell hardness number Brinellhärte
b.h.p. = brake horse power Brems-PS, Bremsleistung
bias Vorbelastung (621); Vorspannung (621.38); Vormagnetisierung
bias battery Gitterbatterie, Gittervorspannbatterie
bias d.c. Vormagnetisierungs(gleich)strom
bias distortion einseitige Verzerrung (621.394)
bias resistance Kathodenwiderstand, Kathodenvorwiderstand
bias resistor Vorspannungswiderstand
bias voltage Vorspannung, Gittervorspannung
to bias vorspannen; vorbelasten
to bias back die Vorspannung erhöhen
to bias off sperren, negativ vorspannen bis zur Stromlosigkeit
biased bell polarisierter Wecker
biased differential protective system Längsdifferentialschutz (621.316.9)
biased rectifier vorgespannter Gleichrichter
biased telephone ringer Wechselstromwecker
biasing battery Vorspannbatterie
biasing resistor Vorspannungswiderstand
biasing transformer Differentialschutzwandler
biasing voltage Vorspannung
bi-band telephony Zweiband-Trägerfrequenztelefonie
bichromate Bichromat
bichromate cell Chromsäureelement
biconical antenna Zweifach-Konusantenne, Doppelkonusantenne
bidirectional bidirektional, zweiseitig gerichtet, doppelseitig gerichtet
bidirectional antenna nach zwei Seiten gerichtete Antenne, zweiseitige Richtantenne
bidirectional current Strom mit wechselndem Vorzeichen
bidirectional transistor symmetrischer Transistor
bidirectional triode thyristor Triac, bidirektionaler Transistor
bidirectional motor Umkehrmotor
bifilar bifilar
bifilar oscillograph Schleifenoszillograph
bifilar suspension bifilare Aufhängung
bifilar winding bifilare Wicklung
bifocal bifokal, mit doppeltem Brennpunkt, mit zwei Brennweiten
to bifurcate gabeln, sich gabeln, sich verzweigen
bifurcated contact Zwillingskontakt, geteilter Kontakt, Gabelkontakt
bifurcating box Gabelmuffe
bilateral zweiseitig, beidseitig
bilateral antenna zweiseitige Richtantenne
bilateral characteristic Achtercharakteristik (621.396.6)
billboard array Tannenbaumantenne, Antennenwand, Querstrahler

billi nano = 10^{-9}
billi condenser Wickelkondensator
billion (Br) Billion, 10^{12}
billion (US) Milliarde, 10^{9}
billohm nano-Ohm = 10^{-9} Ohm
bimetal Bimetall
bimetal relay Bimetallrelais
bimetal strip Bimetallstreifen
bimetallic Bimetall-...
bimetallic release Bimetallauslöser
bimetallic trip Bimetallauslöser
bimetallic wire Manteldraht
bin Fach (659)
binary binär, Zweistoff-...
binary alloy Zweistofflegierung
binary character Binärzeichen
binary circuit Binärschaltung
binary cell binäres Speicherelement
binary code Binärkode
binary-coded binär verschlüsselt
binary-coded decimal Binärkode für Dezimalziffern
binary converter asynchroner Einankerumformer (621.314); Binärwandler
binary counter Binärzähler
binary-to-decimal conversion Binär-Dezimal-Umwandlung
binary digit Binärzeichen, bit
binary digid system Binärsystem
binary divider Binärteiler
binary electrical quantity binäre elektrische Größe
binary inverter Binärinverter
binary notation binäre Zahlendarstellung
binary number system binäres Zahlensystem
binary operation Binäroperation
binary point Binärkomma
binaural binaural, beidohrig
binaural effect Binauraleffekt, Stereoeffekt
to bind binden, festbinden
bind position Verriegelungsstellung, Sperrstellung
binder Bindemittel, Bindelack; Mähbinder, Binder (63); Binder (69)
binding Fressen (621); abbinden, Binde-..., Bindung
binding agent Bindemittel
binding band Wicklungsbandage, Bandage (621.313)
binding clip Befestigungsklemme
binding energy Bindungsenergie (54), (53)
binding post Anschlußschraube, Lochklemme (621.317); Anschlußklemme, Apparateklemme (621.395); Lötstützpunkt
binding screw Anschlußschraube
binding substance Bindemittel
binding wire Bindedraht; Bandagendraht
binocular binokular
binomial Binom
binomial binomial
binomial antenna array Richtantenne mit Nebenzipfelreduzierung, Binominalantenne
binomial theorem binomischer Lehrsatz
biological shield Strahlenschutz(-wand)
biophysics Biophysik
Biot-Savart law Biot-Savartsches Gesetz
biphase zweiphasig
bipod Zweibein
bipolar dipolartig, Dipol-... (53); zweipolig, bipolar (621.313)
bipolar motor doppelpoliger Motor, zweipoliger Motor
bipolar receiver Telefon mit Hufeisenmagnet
bipolar transistor bipolarer Transistor
bipolar winding zweipolige Wicklung
biprism Doppelprisma
biquadratic biquadratisch, ... vierten Grades
birdnesting Düppelklumpenbildung (621.396.96)
bird's-eye view Vogelperspektive
Birkeland and Eyde furnace Stickstoff-Lichtbogenofen
Birmingham Wire Gauge = B.W.G. Birmingham-Drahtlehre
biscuit Bisquitporzellan
bisectional horn antenna Hornantenne mit zwei Seitenwänden
bisector Winkelhalbierende
bismuth Wismut
bistable bistabil
bistable circuit bistabile Schaltung
bistable trigger circuit bistabiler Multivibrator, bistabile Kippschaltung
bit Schneide, Schneidkante; Schneidplättchen; Kloben, Kolben (621.791.3); Schneidwerkzeug
bits and pieces Kleinteile
bit Bit, Binärzeichen
bit density Bitdichte
bit error rate Bitfehlerquote, Bitfehlerhäufigkeit, Bitfehlerrate
bit location Bitstelle
bit-parallel bitparallel
bit rate Bitfolge, Bitfrequenz, Bitgeschwindigkeit (681.3); Transfergeschwindigkeit
bit-slice microprocessor Bit-Slice-Mikroprozessor [Mikroprozessoren beliebiger Wortlänge, deren Komponenten

aus 2- oder 4-Bitelementen (slices = Scheiben) aufgebaut werden]
bitumen Bitumen, Asphalt
bitumen cable Massekabel
bituminous bituminös
bituminous compound Asphaltmasse
black schwarz
blacker-than-black Schwarzspitze (621.397)
black body schwarzer Körper (53); schwarzer Strahler
black bolt (unbearbeitete) schwarze Schraube, Stellmacherschraube
black box „schwarzer Kasten"
black copper Schwarzkupfer
black enamelled schwarzlackiert
black failure Totalausfall
black lead Graphit
black level Schwarzwert (621.397)
black-level clamping Schwarzwertklemmung
black-level control Schwarzwerthaltung
blackout Verdunkelung
blackout failure Totalausfall
black punch Locheisen
black recording Schwarzwertempfang
black scope Dunkelschirm
black signal Schwarzwert (621.397)
blacksmith Schmied, Grobschmied
black-and-white Schwarz-Weiß-...
black-and-white control Zweipunktsteuerung
black-and-white-drawing Schwarz-Weiß-Zeichnung
black-and-white picture tube Schwarz-Weiß-Bildröhre
black-and-white print Weißpause
black-and-white subject Schwarz-Weiß-Vorlage, Schwarz-Weiß-Bild (-vorlage)
black-and-white television Schwarz-Weiß-Fernsehen
blade Blatt (Propeller, Feile, Spaten, Ruder, Säge, Stichmaß, Anreißwinkel); Schaufel (Turbine, Wasserrad, Raddampfer); Klinge; Messer (621.316)
blade angle Anstellwinkel
blade-contact connector Messerleiste, Messerkontaktstecker
blade guide Messerführung (621.316.5)
blade latch Messerverriegelung
blank Rohling; Platine; Leerzeichen, Leerstelle (681.3)
blank instruction Leerbefehl (681.3)
blank line Leerzeile
blank panel Leerfeld, Blindfeld
to blank vorschneiden, Blechschnitte stanzen
blanked picture signal BA-Signal, Bildsignal mit Austastung
to blank out austasten, dunkeltasten (621.397)
blanket Brutzone (539)
blanketing Überstrahlung
blanketing frequency Abdeckfrequenz
blanking Austasten, Dunkeltasten (621.397); Zeichenunterdrückung (681.3)
blanking interval Austastlücke
blanking level Austastpegel, Dunkelsteuerwert
blanking pulse Austastimpuls, Dunkelsteuerimpuls
blanking signal Austastsignal (621.397)
blanking voltage Strahlsperrspannung, Austastspannung
blanking zone Austastbereich, Dunkelsteuerbereich
blast Preßluftstrom; Druckwelle, Sprengladung; Wind (66)
blast contact Konvektionsschalterkontakt; Löschkammerkontakt
blast effect Druckwirkung
blasting Spreng-...
blasting agent Sprengstoff
blasting circuit Zündleitung, Zündkabel
blasting machine Zündmaschine (66)
bleb Blase (66)
to bleed anzapfen
to bleed off the charge (elektrische) Ladung abführen
bleeder Schutzwiderstand, Ableitwiderstand, Entladewiderstand
bleeder resistance Belastungswiderstand
bleeder resistor Ableitwiderstand, Entladewiderstand
blemish Fehlerstelle, Schönheitsfehler
to blend mischen, verschneiden
to blend into (stetig) übergehen in
to blend a line into a curve eine Linie glatt in eine Kurve übergehen lassen
blending radius Abrundungsradius, Übergangsradius
blende Blende (548); Zinkblende
blind blind; Blindplatte
blind approach Blindanflug
blind current Blindstrom
blind duplex teletypewriter communication Gegenschreiben ohne Mitlesemöglichkeit
blind flying Blindflug, Blindfliegen
blind hole Sackloch
blind landing Blindlandung
blind-landing system Blindlandeanlage
blind power Blindleistung

blind region tote Zone (621.396)
blind riser verlorener Kopf (621.7)
blind spots (pl.) tote Zone (621.396)
blind zone Schattenzone (621.396)
to blink blinken
blinker light Flackerlicht; Flackerfeuer (656.7)
blip Pips (621.1); Echozeichen (621.397)
blister Blase, Auswuchs
blister formation Blasenbildung
to blister blättern, abblättern; blasig auftreiben, Blasen werfen
blob Lötperle, Schweißperle, Lichtbogenperle
block Block, Klotz; Häuserblock; Blockstrecke (65); Wortgruppe, Block (621.394), (681.3)
block condenser Block-Kondensator
block contact Klotzkontakt
block diagram Blockschaltbild, Ablaufplan, Flußdiagramm, Übersichtsplan
block-error rate Blockfehlerquote
block gauge Endmaß
block mica Rohglimmer, Naturglimmer
block and pulley Flaschenzug
block of sal ammoniac Salmiakstein
block schematic diagram Blockschaltbild
block section Blockabschnitt, Blockstrecke, Block (625)
block signal Blocksignal
block system Blocksystem
block tin Barrenzinn
block transfer Block-Übertragung (681.3)
to block blockieren, sperren
to block the current flow Stromdurchgang sperren
to block off abriegeln, absperren
to block out abdecken, zudecken, überdecken
blocked-grid keying Gittertastung, Gittersperrspannungstastung
blocked impedance Scheinwiderstand bei festgehaltenem System (534)
blocking Sperre, Sperrung; Durchstoßen (621.385); Blockierung, Zustopfen
blocking Block-..., Sperr-...
blocking capacitor Blockkondensator, Block
blocking layer Sperrschicht, Doppelschicht
blocking oscillator Sperrschwinger, Blockschwinger
blocking signal Sperrzeichen, Sperrsignal
blocking-signal station Blockstation (625)
blocking state Sperrzustand
to bloom mit T-Schutz versehen, vergüten (535)
blooming T-Schutz, Vergütung (535); Leuchtfleck-Überhellung, Leuchtfleckaufweitung, Überstrahlung
blossom Düppelblüte (623)
to blot out ausstreichen, unleserlich machen
blow Blas-..., Löt-...
blow hole Gasblase, Gußblase, Innenlunker
blow lamp Lötlampe
blow-off valve Sicherheitsventil, Ausblaseventil, Dampfabblaseventil
blow-out coil Blasspule (621.32); Löschspule (621.316.9)
blow-out fuse Blassicherung
blow pipe Schweißbrenner, Lötrohr
blow torch Lötlampe
to blow blasen; durchbrenen (621.3)
to blow a fuse eine Sicherung durchbrennen
a fuse blows eine Sicherung geht durch
blown fuse durchgegangene Sicherung
blow out proof kurzschlußsicher (621.382)
blower Lüfter (621.313); Lüfter, Gebläse, Ventilator; Blaskopf
blower fan Gebläse
blowing Abschmelzen, Durchgehen (621.3); Blasvorgang
blowing current Abschmelzstrom (621.316.9)
blowing point Schmelzpunkt (621.316.9)
blue blau
blue copperas Kupfervitriol
blue dip Amalgamierungsbad (621.357)
blue glow Gasleuchten (621.385)
blueprint Blaupause
blue vitriol Kupfervitriol
blunt stumpf
blunt angle stumpfer Winkel
to blur verwischen, verschmieren
blurring Unschärfe (621.397); Fleckigkeit
B.O.A. = British Optical Association Britische Optische Gesellschaft
board Brett, Diele, Dielenbrett; Karton, Kartonpappe; Platine, Leiterplatte
board machine Rundsieb (67)
board of management Vorstand, Geschäftsleitung
Board of Trade Handelsamt, Handelsministerium, Gewerbeamt
Board of Trade ohm gesetzliches Ohm
Board of Trade panel Prüftafel (656.25)
Board of Trade regulations Sicherheitsvorschriften
Board of Trade standards beglaubigte Normalien
Board of Trade unit Kilowattstunde
bobbin Spule (621.314.2); Spulenkörper; Trommel
Bode diagram Bode-Diagramm
bodiless körperlos, masselos

body Körper; Karosserie (629.113); „Masse“ (621.3)
body of the bearing Lagerkörper
body-centred raumzentriert (548)
body colour Grundierung
body screw Halsschraube
bogie Fahrgestell, Drehgestell; Karren (62)
bogie frame Drehgestellrahmen
bogie furnace Durchlaufofen
bogie pin Drehzapfen
to boil kochen
boiler Kocher; Erhitzer; Kessel, Dampfkessel
boiler feeding pump Kesselspeisepumpe
boiling Sieden, Kochvorgang
boiling Koch-..., Siede-...
boiling heat Siedehitze
boiling plate Kochplatte (621.369)
boiling point Kochpunkt, Siedepunkt
boiling-water reactor Siedewasserreaktor
bolometer Bolometer
bolt Bolzen, Schraube, Schraubenbolzen, Riegel
bolt head Schraubenkopf
bolt hole Bolzenloch, Schraubenloch
to bolt anschrauben, verschrauben, verriegeln
Boltzmann's constant Boltzmannsche Konstante
Boltzmann-Stefan law Stefan-Boltzmannsches Gesetz
bomb Bombe (621.1), (623)
bomb calorimeter Heizwertbombe, Kalorimeterbombe
bomb-release computer Ausklinkrechner (623)
to bombard bombardieren (621.385)
bombardment Bombardement, Bombardierung (621.385)
bond Bindung, Valenzbindung (54); Bindung (621); Schienenverbinder (621.33); Bindeglied; Bindemittel
bond nut Überwurfmutter
to bond binden; vereinigen, durchverbinden, abbinden, bonden (621.3), (621.396)
to bond rubber to metal Gummi auf Metall fest aufbringen
to bond together elektrisch leitend verbinden; abbinden (621.396)
bonding Abbindung (621.396)
bonding Binde-...
bonding agent Bindemittel (666)
bonding clip Erdungsschelle, Beidrahtschelle, Verbindungsklemme, Abbindeklemme
bonding medium Klebstoff
bonding ring Erdungsschelle, Erdungsring
bonding wire Abbindedraht (621.396)
book Buch
book capacitor Quetschkondensator
book of instructions Bedienungsanweisung
book-keeping machine Buchhaltungsmaschine
booking Vormerken (621.395)
booking of call Gesprächsanmeldung
booklet Druckschrift, Broschüre
Boolean algebra Boolesche Algebra
Boolean gate Boolsches Schaltglied
to boom dröhnen (534)
boomer Tieftonlautsprecher
boost Zusatzspannung
boost charge Nachladung (621.355)
boost pressure Zusatzdruck, Ladedruck
to boost anheben, nachschieben; Druck zusetzen; Spannung erhöhen
booster Zusatzmaschine (621.3); Zündübertrager, Übertragersatz (623)
booster Zusatz-..., Nachhilfs-...
booster battery Zusatzbatterie
booster pump Zusatzpumpe
booster transformer Zusatztransformator, Zusatzumspanner
bootstrap [Ladeprogramm, das den Beginn des Rechnerbetriebs einleitet]
bootstrap circuit [Anodenbasisschaltung mit Gitter-Kathodeneingang]
bootstrap loader Urlader (681.3)
"B" operator B-Beamter, B-Beamtin
boracic acid Borsäure
borax bead testing Lötperlenprobe
border line Grenzlinie
border-line case Grenzfall
bore Bohrung; lichte Weite, Durchmesser; Bohrloch (622)
bore axis Seelenachse, Bohrungsachse
boreal Wind-...
boreal energy Windkraft
boreal-power station Windkraftwerk
boring machine Bohrwerk
boring-machine spindle Bohrspindel, Bohrwerkspindel
borrow Übertrag
boss Buckel, Knauf, Warze; Wellenende, Wellenstumpf
to boss mit einem Knauf versehen
BOT = begin of tape Bandanfangsmarke
B.O.T. = Board of Trade Handelsamt, Handelsministerium, Gewerbeamt
B.O.T. Unit kWh, Kilowattstunde
bothway trunk ungerichtet betriebene Leitung (621.395)
bottom Boden
bottom-dead centre unterer Totpunkt
bottom edge unterer Rand, Unterkante
bottom face Grundfläche, Bodenfläche

bottom limit untere Grenze
bottom view Unteransicht, Ansicht von unten
Boucherot circuit Boucherot-Schaltung
Boucherot motor Boucherot-Motor, Motor mit Stromverdrängungsläufer, Doppelwicklungs-Asynchronmotor
boulders (pl.) Geschiebe, Geröll
to bounce prellen (621.318.5); springen, zurückspringen, abprallen
bouncing plötzliche Änderung der Bildhelligkeit, Überlagerung von Stoßspannungen (621.397); Prellen (621.318)
bound Rückprall, Rücksprung; Grenze, Bereichsgrenze (681.3)
bounds of possibility Grenzen der Möglichkeit
boundary Grenze; Grenzschicht (621.382)
boundary curve Randkurve
boundary friction Grenzreibung
boundary layer Grenzschicht
boundary lights (pl.) Randbefeuerung (656.7)
boundary line Grenzlinie
boundary lubrication Grenzschmierung
boundary plane Grenzfläche
boundary surface Grenzfläche (51), (53)
bounding Grenz-...
bounding surface Grenzfläche (51)
boundless grenzenlos, unermeßlich
Bourdon tube Bourdonrohr
bow Bug (629); Bogen, Geigenbogen; Bügel
bow collector Bügelstromabnehmer
bow compasses (pl.) Nullenzirkel, Stechzirkel
bow-like bügelartig
bow saw Bügelsäge
bow strip Bügelbelag (621.33)
bow trolley Bügelstromabnehmer (621.33)
Bowden control Bowdenzug
Bowden-Thomson protective system Pfannkuch-Schutz
box Kasten, Gehäuse; Büchse, Kiste, Schachtel; Führerstand (625); Muffe, Kabelmuffe (621.3)
box-frame motor Straßenbahnmotor (mit Kastengehäuse)
box spanner Steckschlüssel
box-type brush-holder Kastenbürstenhalter
box-type frame Kastenrahmen (621.313)
box-type plates Kastenplatten (621.355)
box-wood Buchsbaumholz
to box in Kästen packen, einpacken
boxed verpackt; in Originalpackung
boxed contacts Löschkammerkontakte (621.316.5)
brace Spannband, Band; Klammer; Strebe
bracing Verspannung, Verstrebung
bracing wire Spanndraht; Drahtverspannung
bracket Konsole; Lagerbock, Wandstütze; Halter, Gestell, Ständer; Klammer (621)
bracket arm Fahrleitungsausleger (621.33)
bracket arm suspension Auslegeraufhängung (621.33)
bracket pole Auslegermast
bracket type supporting structure Auslegermast (621.33)
brackets Klammern (51)
brackets closed Klammer zu
brackets open Klammer auf
in brackets in Klammern
brad Flachnagel, Hakennagel
bradawl Spitzbohrer
Bragstad converter Kaskadenumformer
braid Bespinnung, Beflechtung; Geflecht; Zopf
to braid bespinnen, beflechten, umklöppeln
braided electrode umflochtene Elektrode
braided wire geflochtene Litze; besponnener Draht
braiding Bespinnung, Beflechtung, Umklöppelung
brake Bremse
brake band Bremsband
brake block Bremsklotz, Bremsbacke
brake collar Bremsband, Bremsschelle
brake control device Bremswächter
brake disk Bremsscheibe (621.317)
brake drum Bremstrommel
brake dynamometer Bremswaage, Pronyscher Zaum
brake handle Bremshebel, Bremsgriff
brake horse power Bremsleistung, Brems-PS
brake jet Bremsstrahl, Bremsdüse (621.2)
brake lever Bremshebel
brake lining Bremsbelag
brake linkage Bremsgestänge
brake magnet Bremsmagnet
brake mechanism Bremseinrichtung, Bremsvorrichtung
brake motor Bremsmotor
brake pipe Bremsleitung
brake shoe Bremsschuh
brake test Bremsstandprüfung, Bremsprobe
to brake bremsen, abbremsen
braking Bremsung, Brems-...
braking controller Bremsschalter (621.33)
braking current Bremsstrom
braking device Bremseinrichtung
braking distance Bremsweg, Bremsstrecke
braking effect Bremswirkung
braking force Bremskraft
braking magnet Bremsmagnet (621.317)
braking method Bremsverfahren

breaking by short-circuiting the armature Ankerkurzschlußbremsung
braking system Bremssystem
branch Zweig; Abzweig (625), (621.3); Verzweigung, Stromkreis (621.3); Teil; Ast; Abteilung; Verzweigung, Sprung (681.3)
branch address Verzweigungsadresse
branch circuit angeschlossener Stromkreis, Anschluß, Nebenstromkreis
branch-circuit distribution centre Hauptverteilertafel
branch conductor Abzweigleitung
branch connection Anzapfung, Anschluß (621.3)
branch current Zweigstrom, Teilstrom
branch exchange Nebenstellenanlage
branch instruction Sprungbefehl, branch instruction (681.3)
branch joint Abzweigmuffe, T-Muffe (621.315.2); Abzweigpunkt; Abzweigklemme
branch line Zweigleitung, Abzweig (621.3); Nebengleis, Nebenstrecke, Abzweigung, Anschlußbahn (625)
branch-off Abzweigung, Abzweigungspunkt
branch strip Verteilerstreifen
branch switch Abzweigschalter
branch switchboard Verteilertafel (621.315.6)
branch-T T-Stück, Abzweig (621.3), (666)
to branch (-off) abzweigen, sich gabeln, sich verzweigen
branching Verzweigung, Verästelung; Abzweigung (681.3)
branching filter Frequenzweiche
branching jack Mithörklinke (621.395)
branching point Verzweigungspunkt, Abzweigpunkt
Branly coherer Feilspankohärer
brass Messing; Lagerschale (auch wenn nicht aus Messing)
brass block Messingklotz (621.317)
brass rivet Messingniet
brass wire Messingdraht
brassy messingen, messingähnlich
Braun tube Braunsches Rohr
to braze hartlöten
to braze on auflöten
brazen messingen
brazing spelter Hartlot, Schlaglot
breadboard layout Brettschaltung, Brettaufbau, Versuchsaufbau
breadth Breite, Weite
breadth coefficient Wicklungsfaktor, Sehnungsfaktor (621.313)
breadth of range Bereichbreite
break Bruch, Riß (620.1); Knick (51); Sprung (51); Unterbrechung (621.3); Bruchstelle (621); Pause
breakaway cable Abreißkabel
breakaway plug Abreißstecker
breakaway receptacle Abreißbuchse
break before make unterbrechend, öffnen bevor schließen
break contact Öffner, Ruhekontakt
break distance Kontaktabstand (621.316.5)
breakdown Störung, Ausfall
breakdown region Durchbruchsbereich
break guard Bremswächter
break impulse Abschaltspannung, Abschaltüberspannung, Öffnungsimpuls
break jack Abschaltklinke; Anodenschutzbuchse (621.396.62); Trennklinke
break period Öffnungsdauer
break-make ratio Impuls-Pausen-Verhältnis
breakpoint Programmstop, Haltepunkt (681.3)
breakpoint instruction Stoppbefehl
break pulse Öffnungsstromstoß
break spark Abschaltfunke, Unterbrechungsfunke
breakup constituents Bruchstücke
break time Ausschaltzeit
to break brechen, reißen, zu Bruch gehen (620.1); brechen, zerkleinern; unterbrechen, abschalten (621.3); brechen, abbauen (622)
to break apart auseinanderbrechen
to breay away entgleisen; abbrechen, wegbrechen; abreißen
to break a circuit einen Stromkreis unterbrechen
to break down in Einzelteile zerlegen, auseinandernehmen, abmontieren, abbauen; zusammenbrechen, Panne kriegen, fehlerhaft werden, betriebsunfähig werden; ungültig werden (51); abreißen (535)
to break down a fault einen Fehler festbrennen (621.315.2)
to break open aufbrechen
to break a rule gegen eine Regel verstoßen
to break through durchstoßen (629.13)
to break up abwracken, verschrotten, auseinandernehmen; auseinanderbrechen
to break up into unterteilen in
to break up into lengths der Länge nach unterteilen
to break by vibration durch Schwingungsbeanspruchung zu Bruch gehen
breakable coupling Brechkupplung
breakage Bruch

breakdown Zusammenbruch, Bruch (621); Panne; Durchschlag (621.3); Kippen (621.313), Ausfall; Durchbruch (621.382); Aufschlüsselung (33)
breakdown diode Z-Diode
breakdown factor relatives Kippmoment (621.313.3)
breakdown of insulation Isolationsdurchbruch
breakdown rating Durchschlagsfestigkeit (537.5)
breakdown torque Kippmoment, abgebremstes Drehmoment (621.313)
breakdown voltage Durchschlagsspannung, Durchbruchspannung (681.3)
breaker Schalter (621.3); Brecher, Steinbrecher
breaker controller Anlaßschalter (621.313)
breaker point Unterbrecherkontakt (621.43)
breaker starter Anlaßschalter (621.313)
break-in Einlaufen, Einlaufvorgang
break-in wear Einlaufverschleiß, Abnutzung bei Einlaufen
breaking Abschaltvorgang; Abschalten; Schalt-...; Brechvorgang, Zerkleinerung
breaking arc Abschaltlichtbogen
breaking capacity Abschaltleistung, Schaltleistung
breaking current Abschaltstrom (621.316.5)
breaking load Bruchlast, Bruchbelastung, Reißlast
breaking period Abschaltzeit
breaking spark Abreißfunke
breaking strain Bruchfestigkeit
breaking strength Bruchfestigkeit, Reißfestigkeit, Knickfestigkeit
breaking stress Bruchspannung
breakover current Kippstrom (621.382), (621.314.63)
breakover point Kippunkt
breast transmitter Brust-Mikrophon
to breathe atmen; belüften
breather Atemgerät, Rettungsapparat
breather plug Entlüfungsschraube
breathing vent Atmungsöffnung, Belüftungsöffnung
to breed crystals Kristalle züchten
breeder reactor Brutreaktor, Brüter
breeding Brüten (539)
breeding gain Brutgewinn
breeding processing efficiency Brutausbeute
breeding ratio Brutverhältnis
breezy grusig
Brewster angle Brewsterscher Winkel, Pseudo-Winkel
brick Ziegel, Ziegelstein; Baustein
brick building method Baustein-Bauweise
bridge Brücke (621.1), (621.317), (629); Kurzschlußbrücke (621.3)
in bridge im Nebenschluß, in Parallelschaltung (621.395)
bridge assembly Brückenschaltung, Brückenanordnung
bridge balance Brückenabgleich
bridge circuit Brückenschaltung
bridge-connected im Nebenschluß
bridge connected rectifier Gleichrichterbrücke, Gleichrichter in Brückenschaltung
bridge-duplex system Brücken-Duplexschaltung
bridge fuse Trennsicherung (621.315.6)
bridge measurement Brückenmessung
bridge rectifier circuit Gleichrichter in Brückenschaltung
bridge section Brückenglied (621.317)
bridge transition Reihen-Parallelschaltung (621.33)
to bridge überbrücken (621.3); überspannen, überbrücken (624.2/8)
to bridge across etwas überbrücken
to bridge a distance eine Entfernung überbrücken
to bridge over überbrücken
bridging capacitor Überbrückungskondensator
bridging connector Verbindungsstecker
bridging telephone Nebenhörer
bridle Spanndraht (621.33)
bridle wire Überführungsdraht
bright hell, blank
bright drawn blank gezogen
bright electro-plating Glanzplattieren, Glanzgalvanisieren (621.35)
to brighten glänzen, blankmachen, aufhellen; auftasten (621.397)
brightness Helligkeit, Leuchtdichte
brightness control Helligkeitssteuerung, Helligkeitsregelung
brightness of image Bildhelligkeit
brightness ratio Helligkeitsverhältnis
brilliance Leuchtkraft, Glanz, Kontrast (621.397)
brilliance modulation Helligkeitssteuerung, Helligkeitsmodulation
brilliancy Leuchtkraft, Leuchtdichte
brilliant glänzend, hell, leuchtend, blendend, brillant (77)
brimstone Schwefel
brine Sole, Salzlösung
Brinell-hardness Brinellhärte
Brinell-hardness number Brinellhärte, Brinellzahl

to bring mit sich bringen, heranbringen, bringen
to bring down herabsetzen, herunterbringen
to bring down to a diagram in ein Diagramm übertragen
to bring into engagement in Eingriff bringen, einkuppeln
to bring into focus fokussieren
to bring into force in Kraft setzen
to bring into phase in Phase bringen
to bring a line to eine Leitung anschließen
to bring out nach außen führen
to bring out to zum Anschluß an ...bringen, durchführen nach ...
to bring to rest zur Ruhe bringen, stillsetzen
to bring to a stand zum Stillstand bringen
to bring to a stop zum Stillstand bringen, anhalten, abstoppen
briny salzig, Sol ...
briquette Brikett, Preßstein, Preßkohle
British Association Ohm British Association Einheit = 0,9866 Ohm
British Association Unit = British Association Ohm
British Standard candle Britische Normalkerze = 1,02 int. Kerzen
British Standard Wire Gauge Britische Norm-Drahtlehre
brittle spröde, brüchig, mürbe
brittleness Brüchigkeit, Sprödigkeit
broad breit
broad-area rectifier Großflächengleichrichter
broad-band Breitband- ..., breitbandig (621.396)
broad-band antenna Breitbandantenne
broad-band carrier system Breitband-TF-System
broad-band reception Breitbandempfang
broad-band receiver Breitbandempfänger
broad face of a rectangle lange Seite eines Rechtecks
broad fault location Fehlereingrenzung
broad flat resonance curve breite flache Resonanzkurve
broad gauge Breitspur (625.1)
broad pulse Impuls von doppelter Horizontalfrequenz (621.397)
broad resonance curve breite Resonanzkurve
broadbanding Breitbandauslegung (621.396)
broadcast Rundfunk- ...
broadcast corporation Rundfunkgesellschaft
broadcast reception Rundfunkempfang
broadcast studio Rundfunkaufnahmeraum, Rundfunkstudio
broadcast unit Rundschreibeeinrichtung (621.394)
to broadcast senden, durch Rundfunk verbreiten
broadcasting Rundfunk
broadcasting corporation Rundfunkgesellschaft, Rundfunkanstalt
broadcasting network Sendernetz (654.19)
broadcasting service Rundfunkdienst
broadcasting station Rundfunksender
broadside Breitseite (623); in Querstrahlanordnung (621.396.67)
broadside aerial Querstrahlantenne, Querstrahler
broadside array Querstrahler, Querstrahlgruppe, Tannenbaumantenne
broadside curtain of aerials Querstrahlantennenwand
broadside directional aerial Querstrahlerantenne, Querstrahlrichtantenne
broadside to querab von, querab zu
Broca galvanometer astatisches Nadelgalvanometer
broker's call Makelverbindung (621.395)
bronze Bronze
bronze conductor Bronzeseil, Bronzeleiter (621.315.1)
bronze wire Bronzedraht
brown braun
brown coal Braunkohle
brown iron ore Brauneisenstein, Brauneisenerz
brown leakage Stirnstreuung (621.313)
Brown and Sharpe Wire Gauge Brown und Sharpe Drahtlehre
bruise Beule, Quetschung (61), Quetschstelle, Druckstelle (621)
to bruise verbeulen, verquetschen
brush Bürste; Pinsel; Bürste (621.3)
brush arm Bürstenstift, Bürstenbolzen (621.313)
brush box Bürstenhalterkasten, Bürstenhalter (621.313)
brush contact Bürstenkontakt
brush current Bürstenstrom
brush discharge Büschelentladung
brush-friction loss Bürstenreibungsverlust
brush gear Bürstengestell
brush heel ablaufende Bürstenkante
brush holder Bürstenhalter
brush-holder arm Bürstenstift
brush-holder stud Bürstenhalterstift
brush lead Bürstenverschiebung
brush lifter Bürstenabhebeeinrichtung
brush motor Kommutatormotor
brush pressure Bürstendruck

brush resistance Bürstenwiderstand
brush ring Bürstenring
brush rocker Bürstengestell, Bürstenbrücke
brush-rocker ring Bürstenhalterring
brush shift Bürstenverschiebung
brush-shifting control Spannungsregelung durch Bürstenverschiebung
brush-shifting device Bürstenversteller
brush slide Stellschlitten
brush sparking Perlfeuer
brush spindle Bürstenstift, Bürstenbolzen
brush spring Bürstenfeder, Bürstenandruckfeder (621.313)
brush stud Bürstenbolzen, Bürstenstift
brush switch Bürstenschalter
brush yoke Bürstengestell
to brush bürsten, pinseln; streifen
brute-force jamming Massivstörung (623)
B.S.B. = British Standard Beam NP = Normalprofil
B-scan Sternschreiberbild (621.397)
B. S. F. = British Standard Fine Britisches Normal-Feingewinde
B. S. P. = British Standard Pipe Britisches Normal-Rohrgewinde
B. S. S. = British Standard Specification Britische Norm-Vorschrift
B. S. T. = British Summer Time Britische Sommerzeit
B. S. W. = British Standard Whitworth Britisches Normal-Whitworthgewinde
B. and S. Wire Gauge = Brown and Sharpe Wire Gauge Brown und Sharpe Drahtlehre
"B" switchboard B-Vermittlungsschrank (621.395)
B. T. U. = Board-of-Trade Unit Kilowattstunde
B. Th. U. = British Thermal Unit Britische Wärmeeinheit = 1,055 kJ
bubble Blase
bubble memory (Magnet-)Blasenspeicher
to bubble in Blasenform aufsteigen, blubbern; Blasen bilden
bubbling Blasenbildung
Buchholz protection Buchholzschutz
Buchholz protective device Buchholzschutz
to buck out aufheben, auslöschen
to buck a rivet einen Niet setzen, einen Niet schlagen
bucket Eimer; Baggergefäß, Baggereimer; Kübel
bucket-brigade devices = BBD Eimerkettenschaltung
bucketful Eimer (als Hohlmaß)
buckle Schnalle
buckle insulator Schnallenisolator, Isolierschnalle, Fahrleitungs-, Abspannisolator
to buckle ausbeulen, sich beulen, beulen, Beulen bilden; ausbuckeln; knicken, ausknicken
buckling Knickung; Krummwerden, Verwerfen, Ausbeulen, Ausbuckeln
buckling load Knickbeanspruchung
buckling strength Knickfestigkeit
buckling stress Knickspannung
buckling test Knickversuch
to buff schwabbeln
buffer Puffer; Prellbock; Trennstufe (621.396.6); Pufferspeicher (681.3)
buffer amplifier Trennverstärker
buffer battery Pufferbatterie
buffer element Pufferglied
buffer machine Ilgner-Maschine
buffer resistance Ballastwiderstand
buffer salt Puffersalz (621.35)
buffer solution Pufferlösung
buffer spring Pufferfeder
buffer stage Trennstufe (621.396.64)
buffer store Pufferspeicher (681.3)
buffer valve Trennröhre (621.396)
to buffer puffern
buffers Puffermittel (621.35)
buffing Schwabbeln
bug halbautomatische Handtaste, Bug (621.394); Störung, Fehler (681.3)
bugle Horn, Signalhorn, Trompete
build-up Aufbau Anstieg
build-up of current Stromanstieg
build-up factor Aufbaufaktor (539)
build-up period Anschwingzeit
build-up of self excitation Aufbau der Selbsterregung, Auferregung
build-up time Einschwingzeit, Anstiegszeit
on build-up beim Stromanstieg
building-up (of oscillation) Aufschaukeln (von Schwingungen)
to build bauen, erbauen, aufbauen; errichten, machen
to build-in einbauen
to build up anwachsen; aufbauen; synthetisieren (54); auftragschweißen
to build up a connection eine Verbindung aufbauen, durchverbinden
to build up a field ein Feld aufbauen
building Gebäude, Bauwerk, Bau, Konstruktion
building block Baustein, Bauelement
building block system Baukastensystem
building out... Ausgleich..., Abgleich...

building-out capacitor Ausgleichkondensator, Abgleichkondensator
building-out circuit Ausgleichschaltung
building-out network Ergänzungsnetzwerk (621.395)
building-out section Anpassungsglied (621.39)
building up Auftragen (621.35), (621.7); Selbsterregung, Anspringen, Anfachung (621.313)
built-in Einbau-..., eingebaut
built-in accuracy ursprüngliche Genauigkeit
built-in aerial eingebaute Antenne (621.396.62)
built-in check eingebaute Prüfeinrichtung, Selbstprüfung
to built-out abgleichen (621.395)
built-up Aufbau-..., zusammengesetzt
built-up curve Aufbaukurve; Anstieg, Kurvenanstieg
built-up rotor Scheibenrotor (621.313)
bulb Lampe, Glühlampe, Birne, Glühbirne; Kolben, Lampenkolben, Röhrenkolben; Gleichrichterkolben; Wulst, Knolle
bulb-holder Lampenfassung, Fassung
bulb temperature Kolbentemperatur
to bulb einen Wulst bilden
bulbed kolbenförmig, knollig
bulbous wulstig, aufgebaucht
bulge Ausbuchtung, Wölbung; Beule, Wulst
bulge side gewölbte Seite
to bulge Beulen werfen, Wulst bilden
to bulge (out) sich ausbeulen
bulk Umfang; wesentlicher Teil, Hauptteil, Masse; Platzbedarf, Raumbedarf, Gesamtvolumen
bulk billing Summenrechnung (621.395)
bulk modulus Kehrwert der Kompressibilität (539)
bull horn Richtungslautsprecher
bumpless transfer stoßfreie Umschaltung (621)
bunch Bund, Bündel; Paket, Packen
bunch of cables Kabelbündel
bunch of electrons Elektronenpaket
bunch of pairs Leitungsbündel (621.394)
to bunch bündeln (621.396.67); paketieren
to bunch electrons Elektronen paketieren
bunched cables (pl.) Kabelbündel
bunched call Sammelgespräch
bunched conductors (pl.) Leiterbündel
buncher Modulationsgitter, Modulationsfeld (621.385); Modulationssystem (Klystron)
buncher electrode Geschwindigkeitsmodulationsgitter, Steuergitter (Klystron)
buncher grid Steuergitter (Klystron)
bunching Bündelung; Phasenfokussierung (621.385)
bunching strip Parallelklinkenstreifen
bundle Bündel
bundle of rays Strahlenbündel
to bundle bündeln, in ein Bündel zusammenfassen
bung Pflock, Pflöckchen; Stopfen
Bunsen Bunsenbrenner
Bunsen burner Bunsenbrenner
Bunsen cell Bunsen-Element
Bunsen-photometer Bunsen-Photometer
buoyancy Auftrieb
Bureau of Standards Eichamt, Bundesanstalt (für Normen)
burden Last (621); Bürde, Last (621.317)
burglar alarm Einbruchsicherung, Diebstahlsicherung
buried eingegraben, unterirdisch
buried aerial Erdantenne
buried cable Erdkabel
buried layer vergrabene Schicht (621.382)
buried radials (pl.) Strahlenerder
buried repeater unterirdischer Verstärker, Kabelverstärker
buried wiring Unterputzinstallation
to burn brennen (621.1), (621.3); verbrennen; brennbar sein
to burn away wegbrennen
to burn coal Kohle verfeuern
to burn in einbrennen (620.1)
burn-in test Einbrenntest
to burn off abbrennen
to burn out ausbrennen (621.3); ausbrennen (621.791); durchbrennen (621.3)
burned-out lamp durchgebrannte Lampe
to burn to a red heat auf Rotglut erhitzen
to burn the screen den Schirm einbrennen (621.397)
burner Brenner
burner nipple Brennerdüse (621.791)
burning Brennen, Brennvorgang; Verbrennung; Mikrophongeräusch
burning on breaking Ausschaltabbrand (621.316.5)
burning glass Brennglas
burning-off Abbrand
burning period Brennzeit, Brenndauer
burn mark Brandfleck
burn-off Abbrand (621.311)

burnt-out ausgebrannt
burnt-out bulb durchgebrannte Glühlampe
burnt-out valve ausgebrannte Röhre
burn-up Abbrand (539)
burn-up fraction relativer Abbrand (539)
burr Delle, Vertiefung; Grat
to burr verdrücken, verquetschen
burst Schwall, Ausbruch; Bruch, Zerknall; Burst, Farbsynchronimpuls (621.397)
to burst bersten, platzen, zerknallen, explodieren; sich entladen, sich ergießen
to burst into flame in Flammen aufgehen, sich entzünden
bursting limit Berstdruck
to bury eingraben, versenken, hineinversenken (s. auch **buried**)
bus Autobus, Bus, Omnibus; Sammelleitung, Schiene (621.316); Bus (681.3)
bus-bar Sammelschiene
bus-bar sectionalizing switch Sammelschienen-Längstrennschalter
bus-bar transformer Schienenwandler, Schienenstromwandler
bus-disconnect switch Sammelschienentrennschalter
bus enable [Bus-Freigabesignal]
bus extender Buserweiterung
bus isolator Sammelschienentrennschalter
bus line Sammelleitung
bus-line coupler Leitungskupplung (625)
bus protection Sammelschienenschutz
bus reactor Sammelschienen-Drosselspule
bus system Bussystem (681.3)
bus way Leitungskanal
bush Lagerbüchse, Lagerschale, Futterschale, Lagerfutter; Hülse, Ringmuffe, Buchse; Ring, Distanzring
bushing Buchse; Durchführung (621.3)
bushing clamp Durchführungsklemme
bushing current transformer Durchführungswandler, Durchführungs-Stromwandler
bushing insulator Durchführungsisolator
busy besetzt (621.395)
busy back Besetztzeichen (621.395)
busy condition Besetztzustand
busy flash signal Besetztzeichen (621.395)
busy hour Hauptverkehrsstunde (621.395)
busy lamp Besetztlampe
busy line besetzte Leitung (US)
busy-line transfer Rufweiterleitung bei besetztem Anschluß
busy test Besetzt-Prüfung, Tipp-Probe (621.395)
busy time Belegungszeit (621.395)
busy tone Besetztzeichen, Besetztton (621.395)
butt stumpfes Ende, dickes Ende, Stangenfuß
butt joint Stumpfstoß, Stumpfschweißung; stumpfer Stoß
butt-laced joint stumpf genähter Stoß (621.8)
butt weld Stumpfschweißung
to butt up to stumpf ansetzen an, stumpf anschweißen an
to buttweld stumpfschweißen
buttwelded joint Stumpfschweißung
butterfly nut Flügelmutter
butterfly connection Froschbein-Verbindung (621.313)
butterfly valve Drosselklappe
button Knopf
button control Druckknopfsteuerung
button pilot lamp Tastenkontrollampe
button switch Druckknopfschalter
buttress Strebepfeiler, Stütze, Stützpfeiler
buttress thread Stützgewinde, Sägezahngewinde
buyo Boje
to buzz summen, brummen, schnarren
buzz stick Meßstange (621.315.1), (621)
buzzer Summer, Schnarre
B.W.G. = Birmingham Wire Gauge Birmingham-Drahtlehre
B-wire b-Ader
by durch, von, bei
by rail auf dem Schienenwege, mit der Eisenbahn, per Bahn
by road auf der Straße, auf dem Straßenwege
by law (örtliche) Vorschrift, gesetzlich
bypass Umgehung, Umleitung; Umführung; Nebenschluß-..., Neben-..., Parallel-..., Überbrückungs-... (621.3); Überbrückung; Nebenweg
bypass busbar Umgehungsschiene (621.316)
bypass capacitor Überbrückungskondensator, Ableitkondensator
bypass discharge path Nebenentladungsweg
bypass switch Überbrückungsschalter
to by-pass ableiten, umleiten, überbrücken
by-product Nebenprodukt, Abfallprodukt
by-product circuit Telegraphie-Phantomkreis
byte Byte (681.3)
byte serial byteseriell

C

C = coulomb C = Coulomb
cab Führerstand (625.282)
cab signal Signalstellungsanzeige im Führerstand (656.25)
cabin-address equipment Kabinenansagegerät
cabinet Kasten, Gehäuse; Geräteschrank; Kabinett
cable Seil, Drahtseil, Tau (621.1); Kabel, Leitung (621.2); Telegramm; als Längenmaß = 100 fathoms = 182,9 m
cable armour (ing) Kabelbewehrung
cable bandage Kabelschutzbinde
cable and bar winding Wicklung mit Gitterstab-Oberleiter, verschränkteStabwicklung
cable bond Mantelverbinder (621.315.2)
cable box Kabelmuffe
cable chamber Kabelschacht
cable clamp Kabelschelle, Seilklemme
cable conductor Kabelader
cable conduit Kabelkanal
cable connection Kabelverbindung
cable core Kabelseele, Kabelader
cable coupling Kabelverbindung, Kabelkupplung
cable current transformer Kabelstrommeßwandler
cable distribution Kabelaufteilung
cable-distribution box Kabelverzweiger
cable drive Seiltrieb, Seilantrieb
cable drum Seilscheibe, Kabeltrommel
cable duct Kabelkanal
cable entry Kabeleinführung
cable equipment Kabelzubehör
cable-extension arm einseitige Querträger für Luftkabel
cable eye Kabelöse
cable fault Kabelfehler
cable filler Kabelbeilauf
cable fitting Kabelverbinder, Kabelmuffe
cable form ausgeformtes Kabelende, Kabelbaum
cable gear Kabelgeschirr (621.315.2)
cable gland Anschlußstutzen
cable grip Kabelklemme, Zugklammer (621.315.2)
cable harness Kabelbaum
cable joint Kabelmuffe
cable jointing Kabelverbindung, Kabelvermuffung
cable-junction box Kabelanschlußkasten
cable laying Kabelverlegung
cable lug Kabel(löt)öse
cable paper Kabelpapier, Isolierpapier
cable P. F. = cable power factor Kabelverlustwinkel
cable pole line auf Masten verlegte Luftkabelleitung
cable post Kabelmast, Kabelaufführung
cable power factor Kabelverlustwinkel
cable rack Kabelgestell
cable railway Seilbahn, Drahtseilbahn
cable relay Heberrelais (621.394)
cable running Kabelführung, Leitungsführung
cable section Kabelabschnitt
cable sheath Kabelmantel
cable-sheath bond Mantelverbindung
cable ship Kabelschiff, Verlegungsschiff
cable sling Kabelaufhänger
cable socket Kabelschuh
cable splice Spleißstelle, Kabelmuffe
cable stripper Kabelmesser, Gerät zum Abisolieren von Kabel
cable subway begehbarer Kabelkanal
cable-support bracket Kabelhalterung
cable suspender Kabelaufhänger
cable tank Imprägniertank
cable television Kabelfernsehen
cable terminal Kabelklemme, Kabelanschluß
cable-terminal box Verteiler, Kabelschrank
cable termination Kabelabschluß
cable tree Kabelbaum
cable trench Kabelgraben
cable troughing Kabelkanal
cable tunnel Kabeltunnel, Kabelkanal
cable wax X-Wachs
cabling Verkabelung; Kabellegung
cabling harness Kabelbaum
cabling plan Kabelführungsplan
"cab-tyre" cable Gummischlauchleitung
"cab-tyre" sheathing Gummischlauchmantel
Cache memory Cache-Speicher
CAD = computer aided design
cadger Ölkännchen
cadmium Kadmium, Cadmium
cadmium cell Normalelement; Kadmiumzelle
cadmium-control rod Kadmium-Regelstab
cadmium copper Kadmiumbronze
cadmium electrode Kadmium-Normalelektrode
cadmium-nickel storage battery Nickel-Kadmium-Akkumulator
cadmium-plated kadmiumplattiert, kadmiert
cadmium plating Kadmiumplattierung, Kadmierung

cadmium screen Kadmiumschirm
cadmium-selenide cell Kadmium-Selen-Zelle
cadmium-vapour lamp Kadmiumdampflampe
caesium Caesium
caesium cell Caesium-Photozelle
cage Käfig; Förderkorb, Aufzugkorb; Gehäuse
cage aerial Reusenantenne
cage housing Schutzkappe, Gitterabdeckung
cage rotor Kurzschlußläufer, Käfigläufer
cage-type negative plate Kastenplatte (621.355)
to cake brikettieren; tablettieren; backen, zusammenkleben
to calcine kalzinieren, brennen, veraschen; entwässern, rösten
calcite Kalkspat
calcium Kalzium
to calculate berechnen
calculating machine Rechenmaschine, Rechengerät
calculation Berechnung, Rechnung
calculator Rechner, Berechner; Rechentafel, Nomogramm
calculator chip Rechnerbaustein
calculator mass memory system Rechner mit Massenspeicher
calculus Rechentechnik; Differentialrechnung, Differentialkalkül
to calibrate eichen, kalibrieren; einmessen
calibrated attenuator Eichteiler (621.317)
calibrating terminal Prüfklemme, Potentialklemme
calibration Eichung, Kalibrierung
calibration accuracy Eichgenauigkeit
calibration chart Eichdiagramm, Eichkurvenblatt
calibration curve Eichkurve
calibration error Eichfehler (621.85); Schiffseinfluß (621.396.9)
calibration mark Eichmarke
calibration point Eichpunkt
calibration scale Eichskala, Einstellskala (621.318.5)
calibration voltage Eichspannung (621.317)
calibrator Eichgerät
calibre Schieblehre; Kaliber (623.4)
caliper Schublehre, Schieblehre
calipers Taster (sing.)
call Telefongespräch, Gespräch, Anruf, Abruf, Ruf (621.39); Aufruf eines Programms (681.3)
call attempt Belegungsversuch (621.395)
call back Rückfragen, Rückruf
call bell Rufwecker
call board Aufruftafel
call device Rufeinrichtung
call distribution Anrufverteilung
call diversion Rufumleitung
call fee Gesprächsgebühr, Gebühr
call finder Anrufsucher
call forwarding Ruffortschaltung, Rufumleitung (621.395)
call-forwarding to attendant Abwurf auf den Bedienplatz
call-in Aufruf
call-indicator Rufanzeiger, Anrufklappe (621.395); Fallklappentableau (643.6)
call letters Rufzeichen
call meter Gesprächszähler, Anrufzähler
call mix Belegungsspektrum
call number Telefonnummer
call routing Leitweglenkung (621.395)
call sharing Anrufteilung
call sign Rufzeichen, Rufname
call signal Rufsignal, Rufzeichen (621.395)
call-tracing circuit Fangschaltung
call transfer Besuchsschaltung; Umlegen (621.395)
call waiting Anklopfen (621.395)
to call anrufen, telefonisch anrufen; rufen
to call up a subscriber einen Teilnehmer anrufen
called-party (initiated forward) release Auslösen durch den gerufenen Teilnehmer, Rückauslösung
called subscriber angerufener Teilnehmer
Callender recording bridge selbstabgleichende schreibende Schleifdrahtbrücke
calling Anruf, Anrufvorgang; Ruf-...
calling cord Verbindungsschnur (621.395)
calling current Rufstrom
calling device Rufeinrichtung (621.394)
calling dial Wählscheibe, Nummernscheibe
calling frequency Ruffrequenz, Anruffrequenz
calling lamp Anruflampe (621.395)
calling line anrufende Leitung; Rufleitung
calling-party release Auslösen durch den rufenden Teilnehmer
calling plug Rufstöpsel
calling position Rufstellung (621.395)
calling rate Gesprächsdichte
calling relay Anrufrelais
calling subscriber rufender Teilnehmer
calling system Rufanlage
calliper s. **caliper**
calomel Kalomel
calomel electrode Kalomelelektrode
calomel half cell Kalomelelektrode
calorie Kalorie (Wärmeeinheit)
calorific Wärme-..., Heiz-...

calorific capacity Wärmekapazität, Wasserwert
calorific effect Wärmewirkung, Heizwirkung; Heizwert
calorific power Heizleistung
calorific value Heizwert
calorifier Wärmetauscher
calorimetric kalorimetrisch
Calrod heating element Bakkerrohr
calutron = California University Cyclotron Kalutron, Trennzyklotron
calx Verbrennungsprodukt, Oxid
CAM = computer aided manufactoring
cam Nocken, Daumen
cam contact Nockenkontakt
cam control Nockensteuerung
cam controller Nockenfahrschalter
cam drum Nockenscheibe
cam gear Nockenrad; Nockentrieb
cam groove Kurvenschlitz, Führungsschlitz, Kurvenscheibe
cam-limit switch Nockenendschalter
cam-operated master controller Nockenfahrschalter
cam roller Nockenrolle, Nockenring
cam shaft Nockenwelle, Steuerwelle
cam-shaft controller Steuerschalter, Kontroller
cam-type control switch Nockensteuerschalter, Befehlsnockenschalter
cam wheel Nockenrad
camber Wölbung
cambric Baumwollgewebe, Cambrik
camera tube Bildaufnahmeröhre (621.397)
to camp on busy warten auf Freiwerden (621.395)
camshaft Nockenwelle
can Abschirmbecher (621.395); Abschirmung (621.396); Kanne, Kanister; Büchse, Dose; Hülse, Kartusche (539.17); Gehäuse
Canada balsam Kanadabalsam
canal (offener) Kanal
canal rays Kanalstrahlen
to cancel aufheben, streichen, ausstreichen; kürzen (51); wegnehmen, auslöschen, wegkompensieren (621.3)
to cancel out each other sich gegenseitig auslöschen, sich aufheben
cancelling key Löschtaste
cancelling voltage drops sich aufhebende Spannungsabfälle
cancellation Aufhebung, Streichung; Kürzung (51); Auslöschung
candelabrum Kandelaber
candle Kerze, Wachskerze; Kerze
candle foot s. **foot candle**
candle lamp Kerzenlampe
candle metre s. **metre-candle**
candle power Lichtstärke
cane Rohr
canister Kanister, Blechbüchse
Canley furnace Tiegelofen
cannon Kanone; Glaswalze, Glaszylinder (66)
canopy switch Führerstandautomat (621.33)
caoutchouc Kautschuk, Rohgummi
cap Kapsel; Deckel, Kappe (621.315.6)
cap of a bearing Lagerdeckel
cap lamp Stirnlampe
cap nut Überwurfmutter
cap and pin type insulator Stützisolator
to cap mit einer Kappe versehen, überkappen; zubleiben (621.315.2); sockeln (621.32)
capped end zugebleites Ende (621.315.2)
capacitance Kapazität, kapazitiver Widerstand, kapazitiver Blindwiderstand (621.3)
capacitance bridge Kapazitätsmeßbrücke
capacitance coupling kapazitive Kopplung
capacitance current Ladestrom, kapazitiver Strom
capacitance diode Kapazitätsdiode
capacitance divider kapazitiver Teiler
capacitance earth Gegengewicht, kapazitive Erdung
capacitance meter Kapazitätsmesser
capacitance standard Kapazitätsnormal
capacitance per unit length Kapazitätsbelag
capacitive kapazitiv
capacitive coupling kapazitive Kopplung
capacitive load kapazitive Last
capacitive reactance kapazitiver Blindwiderstand
capacitive storage welding Kondensatorschweißung
capacitor Kondensator (siehe auch **condenser)**
capacitor ceramic Keramikkondensator
capacitor-coated foil Kondensator mit metallisierter Lackfolie
capacitor coil Kondensatorwickel
capacitor electrolytic Elektrolytkondensator
capacitor foil and paper Metallfolienpapierkondensator
capacitor-input filter π-Tiefpaß, Siebschaltung mit Querkondensator im Eingang
capacitor metallized paper Metallpapierkondensator
capacitor metallized plastic Kondensator mit metallisierter Kunststoffolie
capacitor microphone Kondensatormikrophon
capacitor motor Kondensator-Motor (621.313)

capacitor polystyrene Polystyrolkondensator
capacitor silvered mica Glimmerkondensator
capacitor solid tantal Tantalfestkondensator, Tantalkondensator mit festem Elektrolyten
capacitor-start motor Motor mit kapazitiver Hilfsphase, Motor mit Kondensatoranlauf
capacitor store Kondensatorspeicher
capacitor tantalum foil Tantalfolienkondensator
capacitor tantalum sintered anode Sinter-Tantalkondensator
capacitor transformer kapazitiver Spannungswandler
capacity Leistungsfähigkeit, Speicherfähigkeit, Fassungsvermögen; Förderleistung, Produktionsvermögen; Leistung; Kapazität (621.3); Meßbereich (621.317)
capacity drift with time zeitliche Kapazitätsveränderung
capacity effect Hochantenneneffekt (621.396.67)
capacity factor Belastungsfaktor, Ausnutzungsfaktor
capacity to ground Erdkapazität, Kapazität gegen Erde
capacity lag Nacheilung durch Trägheit
capacity-metering device Kapazitätsmesser
capacity ratio Kapazitätsverhältnis [C_{min}/C_{max}]
capacity top kapazitive Kopflast, kapazitives Dach (621.396.67)
capillarity Kapillarität
capillary Kapillare
capillary attraction Kapillarkraft, Kapillaranziehung
capillary detector Kapillardetektor
capillary electrometer Kapillarelektrometer
capillary lamp Quecksilberhochdrucklampe
capillary tube Kapillarröhrchen, Kapillare
capping Abdeckleiste
capstan Spill, Gangspill; Revolverdrehbank, Spindel; Treibrolle
capstan lathe Revolverdrehbank
capstan wheel Speichenrad
captive nut unverlierbare Mutter
capture Einfang (539)
capture cross section Wirkungsquerschnitt für Einfang, Einfangquerschnitt (539)
capture gamma radiation Einfang-Gammastrahlung (539)
to capture einfangen, abfangen (539)
captured electron eingefangenes Elektron
car Auto, Automobil; Wagen, Gondel (629); Fahrkorb (621.3); Förderwagen (622)
car aerial Autoantenne
car body Karosserie, Autokarosserie, Chassis
car-door electric contact Türkontakt (621.34), (621.876)
car frame Wagengestell, Chassis
car-gate electric contact Türkontakt (621.34), (621.876)
car shed Wagenhalle
car sidelamp Scheinwerfer, Autolampe
car-switch (lift) control Fahrkorbschaltung, Kabinensteuerung, Innensteuerung (621.876)
car-switch operation Kabinenschaltung (621.876)
carbide Karbid
carbide tip Hartmetallauflage
carbide tool Hartmetallwerkzeug
carbohydrate Kohlenwasserstoff, Kohlenwasserstoffverbindung
to carbolize mit Phenol behandeln
carbon Kohle = Kohlenstoff; Kohle; Bogenlampenkohle, Kohlepapier
carbon arc Kohlelichtbogen
carbon-arc lamp Kohlebogenlampe
carbon-arc welding Kohlelichtbogenschweißung
carbon brush Kohlenbürste (621.3)
carbon-combustion cell Brennstoffelement
carbon (consuming) cell Brennstoffelement
carbon contact Kohlekontakt (621.316)
carbon dioxide Kohlendioxid, Kohlensäure
carbon disulphide Schwefelkohlenstoff
carbon electrode Kohlenelektrode
carbon-filament lamp Kohlefadenlampe
carbon-film resistor Kohleschichtwiderstand
carbon granule Kohlekörnchen
carbon lamp Kohlefadenlampe
carbon microphone Kohlemikrophon, Reiß-Mikrophon
carbon moderator Graphitbremse (539)
carbon monoxide Kohlenoxid
carbon-pile regulator Kohledruckregler
carbon-plate rheostat Massewiderstand
carbon potentiometer Masse-Potentiometer
carbon resistor Kohlewiderstand
carbon rheostat Kohledruckregler
carbon rod Kohlestab, Kohlestift
carbon-rod resistor Karbowidwiderstand, Karbowid
carbon-stick resistor Karbowidwiderstand, Karbowid
carbon tetrachloride Tetrachlorkohlentstoff
carbon-tetrachloride fuse Concordia-Sicherung

carbon transmitter Kohlemikrophon, Reiß-Mikrophon
carbon-type microphone Kohlemikrophon, Reiß-Mikrophon
carborundum Karborundum
carcase, carcass Skelett; Karkasse (629.113); Polgestell, Statorgestell, Stator (621.313)
card Karte, Billett; Kartonpapier, Karton, Pappe; Karde (67)
card aligner Kartenanschlag
card bed Kartenhalter, Kartenmagazin
card code Lochkartencode
card controlled lochkartengesteuert
card feed Kartentransport, Kartenzuführung
card index Kartei
card-leading edge Kartenvorderkante
card puncher Kartenlocher
card reader Kartenleser
card reproducer Kartendoppler
card stacker Kartenablage
card-to-tape converter kartengesteuerter Streifenlocher
card-trailing edge Kartenhinterkante
cardboard Karton, Pappe
Cardew voltmeter Hitzdrahtvoltmeter
cardinal Kardinal-..., Haupt-...
cardinal number Kardinalzahl, Grundzahl
cardinal point Haupthimmelsrichtung
cardiod, cardioid Herzkurve, Kardioide
cardioid diagram Nierencharakteristik
cardioid pattern Kardioidendiagramm
cardioid reception Kardioiden-Diagramm
cardiod, cardioid herzförmig
Carey-Foster bridge Carey-Foster-Brücke (621.317)
carnauba wax Carnaubawachs
carriage Transport, Abfuhr, Anfuhr; Beförderung; Wagen, Wagengestell; Laufkatze, Laufwerk (621.8); Lafette (623.4); (Wagen-)Vorschub (621.394.6)
carriage return Wagenrücklauf (621.394.6)
carriage-type switchgear Schaltwagen
carrier Träger, Trägerwelle (621.396); Träger (539), (54); Flugzeugträger (623)
carrier beat Trägerschwebung
carrier channel Trägerfrequenzkanal
carrier control Trägersteuerung, Hapug-Modulation
carrier coupler Kopplungskondensator (621.315), (621.396)
carrier-current telephone communication Trägerfrequenztelefonie
carrier of current Elektrizitätsträger
carrier diffusion Trägerdiffusion (621.382)
carrier dispersal Trägerverwischung
carrier frequency Trägerfrequenz
carrier-frequency shift Trägerfrequenzverschiebung
carrier-frequency shifting Trägerfrequenzumtastung
carrier-frequency voltage Trägerspannung
carrier-frequency wire broadcasting hochfrequenter Drahtfunk
carrier injection Trägerinjektion (621.382)
carrier interference Trägerinterferenz
carrier leak Trägerrest
carrier level Trägerpegel
carrier line TF-Leitung
carrier loading TF-Bespulung
carrier mobility Trägerbeweglichkeit
carrier to noise ratio Träger/Geräusch-Abstand
carrier offset Offsetbetrieb (621.396)
carrier power Trägerleistung
carrier repeater Trägerfrequenzverstärker
carrier-storage time Speicherzeit, Entladeverzug (621.382)
carrier suppression Trägerunterdrückung
carrier system Trägerfrequenzsystem
carrier telegraphy Trägerfrequenztelegraphie, Überlagerungstelegraphie
carrier telephony Trägerfrequenztelefonie
carrier telephony along lines leitungsgerichtete Telefonie
carrier-terminal equipment Trägerfrequenzgerät, Trägerfrequenzendstelle
carrier transmission trägerfrequente Übertragung
carrier voltage Trägerspannung, H. F.-Spannung
carrier wave Trägerwelle
carrier-wave telephony Trägerfrequenztelefonie
carry Übertrag (681.3)
carry bit Übertragsbit
to carry tragen, befördern, transportieren; führen, mit sich führen, übertragen (681.3)
to carry current Strom führen
to carry down to a diagram in ein Diagramm eintragen
to carry out tests Versuche durchführen
carrying Trag-...
carrying cable Tragseil
carrying case Transportkasten
carrying capacity Nutzlast, Tragfähigkeit; Strombelastbarkeit, Belastbarkeit; Platzzahl (625)
carrying clamp Tragschelle

carrying current Grenzstrom
Cartesian kartesisch
Cartesian coordinates kartesische Koordinaten
Cartesian grid kartesisches Koordinatensystem
cartridge Patrone (621.3); Kassette; Tonabnehmer (681.84)
cartridge drive Kassettenlaufwerk
cartridge fuse Patronensicherung (621.3)
cartridge playback Kassetten-Abspielgerät
cartridge-tape recorder Kassetten-Tonbandgerät
cartridge-type brush holder Bürstenhalterpatrone
cascade Kaskade, Reihenschaltung, Kettenschaltung
cascade amplifier Mehrstufenverstärker, Kaskadenverstärker
cascade connection Kaskadenschaltung, Kettenschaltung
cascade control Kaskadenregelung (621.313)
cascade converter Kaskadenumformer
cascade motor Kaskadenmotor
cascade starter Kaskadenanlasser
cascade thermoelectric device thermoelektrische Kettenschaltung
cascade transformer Kaskadenwandler
cascade voltage doubler Spannungsverdoppelungsschaltung
cascode amplifier Kaskodeverstärker, Kaskadenverstärker
case Gehäuse, Büchse, Kasten; Fall, Vorkommnis
in case of emergency im Notfall
case capacitance Restkapazität, Querkapazität
case hardening Einsatzhärtung
to case überfangen, überziehen, umkleiden
cased cord Litzenleiter mit Hanfseele
cased glass überfangenes Glas
cash register Registrierkasse
casing Gehäuse; Umhüllung, Verrohrung (622)
Cassegrain antenna Cassegrainantenne
cassette Kassette
cassette recorder Kassettenrecorder
cast gegossen, Guß-...
cast aluminium Gußaluminium
cast brass Messingguß
cast iron Gußeisen
cast-iron casing Gußgehäuse
cast resin Gießharz
cast steel Gußstahl
castable gießbar
casting Guß, Gußstück
casting alloy Gußlegierung
casting mould Gußform
castor oil Rizinusöl
catalysis Katalyse
catalyst (material) Katalysator, Katalysatormasse
catalytic katalytisch
cataphoresis Kataphorese
catastrophic failure Totalausfall
catch Arretierung; Mitnehmer, Raste
catch net Fangnetz, Schutznetz (621.315.1)
to catch fangen, auffangen
catcher Auskoppelungsglied, Auskoppelungsfeld (Klystron)
catcher grid Auskoppelungsgitter (Klystron)
catcher space Auskoppelraum
catenary Kettenlinie
catenary curve Kettenlinie
catenary hanger Tragdraht, Hängedraht
catenary suspension Fahrdrahtaufhängung
catenary system Tragdrahtanordnung (621.33)
catenary wire Tragdraht
catenary-wire aerial cable Tragseil-Luftkabel
cathamplifier Kathodenverstärker
catharometer Hitzdrahtanemometer
cathetron Gleichrichter mit Außengittersteuerung
cathion Kation
cathode Kathode
cathode area Kathodenoberfläche, Kathodenfläche
cathode arrester Kathodenfallableiter
cathode-basis connection Kathodenbasisschaltung (621.396.62)
cathode bias Kathodenvorspannung, gehobene Kathode
cathode-bias resistor Kathodenwiderstand, Kathodenvorwiderstand
cathode border Kathodensaum
cathode-bypass capacitor Kathodenblock
cathode characteristic Emissionskennlinie
cathode cleaning kathodische Reinigung
cathode-coupled durch den Kathodenwiderstand gekoppelt, kathodengekoppelt
cathode current Kathodenstrom
cathode dark space Kathodendunkelraum, kathodischer Dunkelraum
cathode desintegration Kathodenabbau, Kathodenzerstäubung
cathode drop Kathodenfall
cathode fall Kathodenfall
cathode filament Glühkathode
cathode follower Kathodenverstärker, Anodenbasisschaltung, Kathodenfolger

cathode glow kathodisches Glimmlicht
cathode heating Kathodenheizung
cathode-heating power (Kathoden-)Heizleistung
cathode-heating time Kathoden-Anheizzeit (621.385)
cathode layer Kathodenschicht
cathode oscillograph Kathodenstrahloszillograph
cathode pickling kathodische Beizung
cathode ray Kathodenstrahl
cathode-ray current Kathodenstrahlstrom, Strahlstrom
cathode-ray direction finder Kathodenstrahlpeiler
cathode-ray indicator magisches Auge
cathode-ray instrument Kathodenstrahloszillograph
cathode-ray oscillograph Kathodenstrahloszillograph (mit Fotografiereinrichtung)
cathode-ray oscilloscope Kathodenstrahloszilloskop, Oszilloskop, Elektronenstrahloszilloskop
cathode-ray particle Kathodenstrahlteilchen, Elektron
cathode-ray pencil Kathodenstrahlbündel
cathode-ray spot Kathodenstrahlfleck
cathode-ray storage tube Kathodenstrahlspeicherröhre
cathode-ray television tube Fernsehröhre
cathode-ray tube Braunsches Rohr, Kathodenstrahlröhre
cathode-ray tube display Kathodenstrahl-Schirmbilddarstellung
cathode-ray tube store Kathodenstrahlspeicherröhre
cathode-ray tube trace Schirmbild
cathode-ray tuning indicator magisches Auge, Abstimmanzeigeröhre
cathode resistor Kathoden(vor)widerstand
cathode sleeve Kathodenröhrchen (621.385)
cathode spot Kathodenfleck
cathode sputtering Kathodenzerstäubung
cathode stream Kathodenstrom, Emissionsstrom
cathode terminal Kathodenanschluß, Kathodenklemme
cathodic kathodisch
cathodic polarization kathodische Polarisation
cathodic protection kathodischer Schutz, Cumberlandverfahren, galvanischer Korrosionsschutz
cathodophone Ionisationsmikrophon
catholyte kathodischer Elektrolyt
catkin-type valve Senderöhre (bei der die Anode einen Teil der vakuumdichten Hülle bildet)
cation Kation
catoptric katoptrisch, mit Spiegeloptik
cat's whisker Detektordrähtchen
C.A.T.V. = community antenna television
Cauchy's integral theorem Cauchyscher Integralsatz
caustic ätzend, beißend, kaustisch
caustic . . . . . . hydroxid
caustic alkali metals Ätzalkalien, Ätzalkalimetalle
caustic lime Ätzkalk
caustic potash Ätzkali
caustic soda Ätznatron
caustic-soda cell Alkalielement
Caution High Voltage Vorsicht! Hochspannung – Lebensgefahr
cautionary bell Warnklingel
cavern-type power station Kavernenkraftwerk
cavitation Kavitation, Hohlraumbildung
cavity Hohlraum, Höhlung; Lunker, Groblunker; Hohlraumresonator
cavity resonator Hohlraumresonator, Topfkreis (621.396)
Cb Nb, Niobium
C.B. = central battery Z. B. = Zentralbatterie
CPU = central processing unit
c.c. = cubic centimetre ccm = Kubikzentimeter
c.c. lamp cap = central contact lamp cap (s. dort)
c.c. = continuous current Gleichstrom
c.c. voltage Gleichspannung
C-core Schnittbandkern
CCS = continous commercial service Dauerbetrieb
ccw = counterclockwise
CD = compact disc
CD-digital-player Compact-Disc-Spieler
cedar Zeder, Zedernholz
cedar wool oil Zedernholzöl
ceiling Decke; Steighöhe, Gipfelhöhe
ceiling of amplification Maximalverstärkung
ceiling plate Deckenrosette
ceiling-reflecting lamp indirekte Leuchte
ceiling rose Deckenrosette
ceiling switch Deckenschalter
ceiling voltage Höchstspannung, höchst erreichbare Spannung, maximaler Spannungswert
celestial Himmels-. . ., himmlisch
celestial guidance astronomische Führung (623.4)
celestial navigation astronomische Navigation

cell Zelle; Zelle, Element; Speicherzelle (681.3)
cell cavity Ofenwanne (66)
cell connector Schaltverbindung (621.355)
cell constant innerer Widerstand (eines Elementes) bei spezifischem Widerstand 1
cell door Zellentür (621.316)
cell mounting Zelleneinbau
cell structure Zellenanlage (621.316)
cell tester Säureprüfer; Spannungsprüfer (621.355)
cellophane Zellophan
cellular netzförmig, Netz-...
celluloid Zelluloid
cellulose Zellulose, Zellstoff; Zelluloselack
cellulose spraying Zelluloselackierung, Zelluloselack-Spritzanlage
cellulose varnish Zelluloselack
cellulose wadding Zellulosewatte, Zellstoffwatte
to cellulose mit Zelluloselack lackieren
Celsius Celsius
cement Zement; Klebstoff
cement-fibre slab Zementfaserplatte
cement-floor board Zementdiele
to cement zementieren, mit Zement verkleiden; kleben
to cement on ankleben, aufkleben; verkitten (535)
cemented carbide Hartkarbid, Schneidmetall
center (US) Zentrum, Mittelpunkt (s. a. **centre**)
center of the band Bandmitte (621.396)
center frequency Mittenfrequenz
center-loaded antenna mittenbelastete Antenne
center of rotation Drehpunkt
center tap Mittelabgriff
to center about seinen Mittelpunkt haben bei
centerfed mittengespeist, mit Mittelspeisung (621.396.67)
centering Zentrierung
centering control Bildstandsregelung (621.397)
centesimal Zentesimal-...
centesimal system Zentesimalsystem
centigradation Hunderterteilung
Centigrade Celsiusgrad
Centrigrade Heat Unit = 1,892 kJ
centigramme Zentigramm
centilitre Zentiliter
centimetre Zentimeter
centimetre cube Kubikzentimeter
centimetre square Quadratzentimeter
centimetric Zentimeter-...
centimetric radar SHF-Radar
centimetric waves Zentimeterwellen
central zentral, Zentral-..., in der Mitte liegend, Mittel-...
central axis Hauptachse, Mittelachse
central battery Zentralbatterie
central-battery exchange ZB-Vermittlung
central-battery signalling Zentralbatterie-Rufsystem
central-battery telegraph system ZB-Telegraphieanlage
central conductor Mittelleiter
central-contact (lamp) cap Steckfassung mit Mittelkontakt
central data processor Zentralrechner
central deflection Mitteldurchbiegung
central drive Mittelantrieb; zentraler Antrieb
central load Zentrallast, Mittenlast, Last im Mittelpunkt
central lobe Mittellappen (621.396.67)
central lubrication Zentralschmierung
central memory Zentralspeicher
central office Hauptamt, Fernsprechvermittlung (621.395)
central office toll dialing Fernwahlvermittlung
central plane Mittelebene
central point Mittelpunkt
central portion Mittelteil
central position Mittelstellung
central processing unit Zentraleinheit (681.3)
central station Kraftwerk, Zentrale (621.311)
central unit Zentraleinheit
centralization Zentralisierung
to centralize zentral zusammenfassen, zentralisieren
centralized control Zentralsteuerung
centralized multi-service control Rundsteuersystem
centralized ripple system Rundsteuersystem
centre, center (US) Zentrum, Mittelpunkt; Körnerspitze (621.9); Kern, Innerstes; mittler, Mittel-...
centre of attraction Anziehungszentrum
centre of bore Seelenachse
centre to centre Mitte-Mitte
centre-to-centre distance Mittelpunktabstand
centre conductor Mittelleiter, Innenleiter
centre core Mittelkern, Kernstopfen (621.396)
centre of curvature Krümmungsmittelpunkt
centre distance Achsenabstand, Achsabstand, Radstand
centre of distribution Lastschwerpunkt (621.316)

centre frequency Mittelfrequenz, Mittenfrequenz
centre of gravity Schwerpunkt
centre of gyration Drehpunkt, Drehachse
centre line Mittellinie (51); Anfluggrundlinie (656.7)
out of centre exzentrisch, außermittig
centre pin Mittelstift
centre point Mittelpunkt; Sternpunkt (621.3)
centre-pole suspension Mittelmastaufhängung (625.6)
centre of a range Mittelpunkt eines Bereiches, Bandmitte
centre section Mittelteil
centre-slot system Schlitzschienenstromzuführung
centre of span Mitte der Spannweite, Mitte Spannweite
centre of suspension Aufhängepunkt
centre tap Mittelanzapfung
centre-tapped mit Mittelanzapfung, mit Mittelabgriff (621.3)
centre-zero instrument Instrument mit Nullpunkt in der Mitte
to centre zentrieren; Nullpunkt einstellen, in Mittelstellung bringen
to centre the cathode spot Kathodenfleck in Mittelstellung bringen
to centrefeed im Mittelpunkt speisen (621.396), (621.397)
centreline Mittellinie
centres (pl.) Mittelpunktsabstand
to be at 10 ft. centres 10 Fuß Mittelpunktsabstand haben
centrifugal Zentrifugal-..., Schleuder-...
centrifugal casting Schleuderguß, Schleudergußstück
centrifugal clutch Fliehkraftkupplung, Zentrifugalkupplung
centrifugal coupling Zentrifugalkupplung
centrifugal drying machine Trockenschleuder, Wäscheschleuder
centrifugal fan Kreisellüfter, Ventilator, Kreiselgebläse
centrifugal force Fliehkraft, Zentrifugalkraft
centrifugal governor Fliehkraftregler
centrifugal starter Kreiselanlasser, Fliehkraftanlasser
centrifugal starting switch Zentrifugalanlasser
centrifugation Zentrifugieren, Schleudern
centrifuge Zentrifuge, Schleuder
centripetal mittelpunktstrebig, Zentripetal-...
centripetal force Zentripetalkraft
centrode Wälzbahn (531.1)
centroid Schwerpunkt; Schnittpunkt der Seitenhalbierenden im Dreieck
ceramic keramisch
ceramic bead keramische Abstandshalter (621.315.2)
ceramic capacitor keramischer Kondensator
ceramic fuel keramischer Heizstoff (539.17)
ceramic insulating material keramisches Isoliermaterial
ceramic insulation keramische Isolation
ceramic insulator keramischer Isolator
ceramic package keramisches Gehäuse
ceramic tube Keramikröhre
ceramics (pl.) Keramik
cer-DIP package CERDIP-Gehäuse
ceresin Ceresin
cermet resistor Metallglasurfestwiderstand
cesium = Caesium
c.ft. = cubic foot Kubikfuß = 28,32 Liter
c.ft. per min Kubikfuß pro Minute = 28,32 l/min
cg. = centigram Zentrigramm
C.G. = centre of gravity Schwerpunkt
C.G.S. system cgs-System
chad Stanzrest (621.394.6)
chadless (punched) tape Lochstreifen ohne Abfall
chafe Abschürfung, Schürfwunde
to chafe rubbeln, schuffeln, schürfen
chaff Düppel (623)
chaff burst Düppelstoß
chafing dish Wärmeplatte
chain Kette; Feldmeßkette [Längenmaß von 66 ft = 20,117 m]
chain block Kettenflaschenzug
chain call Kettengespräch
chain conveyor Kettenförderer, Förderkette
chain-dotted strichpunktiert
chain fission yield Isobarenausbeute (539)
chain insulator Kettenisolator, Hängekette
chain link Kettenglied
chain printer Kettendrucker
chain pulley Kettenrad, Kettenrolle
chain reaction Kettenreaktion
chain sprocket Kettenrad
chain of television lines Fernsehleitungskette
chain winding Zweischicht-Korbwicklung (621.313)
chalk Kreide
challenger Abfragesender
chamber Kammer
chamber kiln Kammerofen
chamfer Fase, Abschrägung, Kantenbruch
to chamfer abfasen, fasen, Kanten brechen

chamotte Schamotte
chandelier Kronleuchter
change Änderung, Wechsel, Verwandlung
change climates Wechselklima
change gears Wechselräder
change lever Umschalthebel, Schalthebel
change-over Umstellung; Umschaltung; Übergang
change-over contact Wechsler, Umschaltekontakt, Wechselschließer
change-over switch Umschalter
change-over time Umschlagszeit
change point Umwandlungspunkt, Haltepunkt
change in pole combinations Polumschaltung
change-pole motor polumschaltbarer Motor
change of rotation Drehrichtungsumkehr
change of section Querschnittsänderung
change of speed Drehzahländerung
change-speed motor Regelmotor, drehzahlregelbarer Motor; Motor mit mehreren Drehzahlstufen
change of state Zustandsänderung
to change ändern, sich ändern, verändern
to change in direction schwenken (621.396.677)
to change-over umschalten
changed-number anouncement Ansage einer neuen Rufnummer
channel Kanal, Nachrichtenkanal, Informationskanal; Einlauf (621.74); U-Eisen, U-Profil
channel capacity Kanalkapazität
channel coding Kanalcodierung (621.397)
channel converter Kanalumsetzer
channel division Kanalbreite (621.396)
channel iron U-Eisen
channel modulator Kanalumsetzer
channel pattern Kanalraster
channel resistance Kanalwiderstand (621.382)
channel selector Kanalwähler (621.396); Spurwähler (681.3), (681.84)
channel spacing Kanalabstand, Kanalraster, Rasterabstand (621.395), (621.396)
channel switch Programmschalter, Kanalwahlschalter (621.397)
channel translation Kanal-Umsetzung (621.395.4)
channel translator Kanal-Umsetzer
channel working Mehrfachausnutzung (621.394)
to channel in einfädeln (351.814)
to channel to hinleiten nach (621.3)
to channelize Kanal bilden (621.395)
to char verkohlen
character Zeichen, Schriftzeichen
character crowding Zeichenraffung
character duration Zeichendauer
character generator Zeichengenerator
character printer Zeichendrucker
character reader Zeichenleser, Klarschriftleser
character recognition Zeichenerkennung
character repertoire Zeichenvorrat
character sensing Zeichenlesen
character set Zeichenvorrat
character-storage tube Zeichenspeicherröhre (621.397)
character-at-a-time-printer Einzelzeichendrucker
characteristic Charakteristik; Kennlinie; Numerus, Vorzahl; Merkmal, Kennzeichen
characteristic charakteristisch, kennzeichnend (**of**: für)
characteristic angular phase difference Phasenfolgewinkel (der symmetrischen Komponente)
characteristic curve Kennlinie, Charakteristik
characteristic distortion charakteristische Verzerrung, Einschwingverzerrung (621.395)
characteristic frequency Eigenfrequenz
characteristic impedance Wellenwiderstand
characteristic line (of a vector field) Feldlinie
characteristic radiation charakteristische Strahlung (621.386)
characteristic spectra charakteristische Röntgenspektren
characteristic spread Normalverteilung
characteristic surface (of a thermionic valve) Kennlinienfläche (621.385)
characteristic wave impedance Feldwellenwiderstand
charcoal Holzkohle
charge Ladung; Charge; Einsatz, Beschickung, Füllung; Gebühr
charge carrier Ladungsträger, Träger (621.382)
charge-carrier diffusion Ladungsträgerdiffusion
charge-carrier injection Ladungsträgerinjektion
charge-carrier mobility Ladungsträgerbeweglichkeit
charge-coupled devices ladungsgekoppelte Bauelemente
charge indicator Spannungsprüfer (621.35); Gebührenanzeiger (621.395)
charge of magnetism magnetische Ladung
charge for maximum demand Maximumtarif
charge-resistance furnace Elektroofen mit direkter Beheizung des Heizgutes
charge transfer Ladungsaustausch, Ladungstransport

to charge aufladen; chargieren, beschicken, einsetzen; berechnen, in Rechnung stellen; beauftragen
to charge a battery eine Batterie laden
to charge up sich aufladen
charging Lade-..., Ladevorgang; Beschickung; Gebührenerfassung (621.395)
charging capacitor Ladekondensator, Glättungskondensator
charging circuit Ladekreis, Ladeschaltung
charging current Ladestrom
charging facilities Beschickungseinrichtungen
charging generator Ladegenerator, Ladeaggregat
charging rate Ladestromstärke
charging rectifier Ladegleichrichter
charging resistor Ladewiderstand
charging set Ladegerät
charging voltage Ladespannung
chart Diagramm; Karte = Seekarte; graphische Darstellung, Schaubild; Schreibstreifen (621.317); Tabelle
chart-comparison unit Karteneinblendung (623)
chart diagram Nomogramm
chart paper Diagrammstreifen
chart-recording instrument Registrierinstrument
chart of a recording instrument Registrierstreifen
chart scale Skala auf dem Registrierstreifen
chartered electrical engineer Mitglied des I. E. E.
chassis Fahrgestell, Chassis; Chassis (621.396.64)
chassis ground Masse
to chatter rattern, flattern (621.313), (621.9); prellen (621.318.5)
chatter marks Rattermarken
check Kontrolle; Kontrollstelle; Kontrollmarke; Prüfung; Nachprüfung, Test, Überprüfung; Probe (51); Festhalten, Anhalten, Behinderung
check analysis Kontrollanalyse
check-back signal Rückmeldesignal
check bit Prüfbit
check digit Prüfzeichen
check mark Kontrollmarke
check measurement Kontrollmessung, Nachmessung
check nut Gegenmutter
checkout Test (681.3)
check point Kontrollstelle, Kontrolle
check rail Leitschiene
check reading Kontrollablesung
check receiver Kontrollempfänger (621.396)
check room Prüfraum, Prüfstand
check routine Prüfprogramm (681.3)
check sample Probestück, Prüfmuster
check test Kontrollversuch
check valve Absperrventil; Rückschlagventil
to check kontrollieren; nachprüfen; Probe machen; kollationieren (655); in Grenzen halten
to check out austesten
to check for specified limits auf Maßhaltigkeit prüfen, auf Einhaltung der vorgeschriebenen Toleranzen prüfen
to check for truth auf Genauigkeit prüfen
checking routine Prüfprogramm
checkered schachbrettartig gemustert
checkered plate Riffelblech
cheek of the vice Schraubstockbacke
to cheep zwitschern
cheese aerial Scheibenantenne, Segmentantenne
cheese-headed screw Zylinderkopfschraube, Flachkopfschraube
chemical Chemikalie; chemisch
chemical action chemische Reaktion, chemische Einwirkung
chemical activity chemische Aktivität
chemical analysis chemische Analyse
chemical balance chemische Waage, Analysenwaage
chemical compound chemische Verbindung
chemical constitution chemischer Aufbau, chemische Zusammensetzung
chemical deposition chemische Beschichtung, galvanische Beschichtung
chemical energy chemische Energie
chemical plant chemische Fabrik
chemical property chemische Eigenschaft
chemical reaction chemische Reaktion
chemical separation chemische Trennung, Trennung auf chemischem Wege
chemical symbol chemisches Zeichen
chemically inert chemisch inert, chemisch träge, reaktionsunlustig
chemically pure chemisch rein
chemistry Chemie
chestnut coal Nußkohle
Chicago rail bond Nietschienenverbinder (621.33)
chief engineer Oberingenieur
to chill abkühlen, abschrecken
china Porzellan
Chinese copy sklavische Nachahmung
chip Span, Schnitzel; Halbleiterkristall, Halbleiterkristallscheibe [auch mit integrierter Schaltung], Halbleiterbaustein, Chip (621.382)

chip enable Chip-Freigabe (681.3)
chip select Chip-Auswahl (681.3)
chip slices Prozessorelemente (681.3)
to chip ausbrechen, röckeln; behauen, meißeln
chipping Ausbröckeln, Schartenbildung
chippings Späne (621.9); Schnitzel
Chireix-Mesny array Chireix-Antenne
Chireix modulator Chireix-Modulator
choice Suchstellung (621.395)
choke Drossel, Sperre; Drossel, Drosselspule
choke-capacitance coupling LC-Kopplung
choke coil Drosselspule
choke control Drosselmodulation
choke coupling Drosselkopplung
choke flange Drosselflansch (621.37)
choke-input filter Sieb aus T-Gliedern, Tiefpaß in T-Schaltung
choke joint Drosselstoß (621.391); Sperrtopf (621.396); induktive Hohlleiter-Kopplung
choke modulation Drosselmodulation, Heising-Modulation
to choke sperren, drosseln; verstopfen, zusetzen
to choke out wegdrosseln
to choke up verstopfen, zusetzen
choker Drossel, Drosselspule
choking Drossel-...
choking coil Drosselspule
choking effect Drosselwirkung
to chop hacken, zerkleinern
chopper Ticker, Zerhacker, Meßzerhacker (621.3)
chopper amplifier Chopper-Verstärker
chopper-bar controller Fallbügelregler
chopper-bar recorder Fallbügelschreiber, Punktschreiber
chopping Amplitudenbegrenzung (621.396)
chopping action Zerhackerwirkung
chord Saite; Sehne (51); Tiefe, Profiltiefe (533)
chord winding Sehnenwicklung, gesehnte Wicklung (621.313)
chording factor Sehnungsfaktor
chroma Farbwert
chroma information Farbinformation (621.397)
chromatic chromatisch (534), (535)
chromatic aberration chromatische Aberration
chromaticity Farbart
chromaticity diagram Normfarbtafel
chromatics (pl.) Farbenlehre
chromatron Einstrahlröhre mit Farbsteuergitter
chrome Chrom, chromfarbig; Chrom-...
chrome-nickel Chromnickel-...
chrome-nickel wire Chromnickeldraht
chromic acid Chromsäure
chrominance Chrominanz, Farbdifferenz
chrominance demodulator Farbdemodulator
chrominance modulator Farbmodulator
chrominance signal Farbartsignal
chrominance subcarrier Farbträger
chromium Chrom
chromium-plated verchromt
chromium plating Verchromung
chromosphere Chromosphäre (52)
chronograph Chronograph, Zeitschreiber, Zeitmarkenschreiber
chronometer Chronometer
chronopher Zeitzeichengeber
C. H. U. = Centigrade Heat Unit (s. dort)
chuck Futter, Spannfutter, Bohrfutter (621.9)
chuck jaw Spannbacke, Futterbacke
chug Tuckern, Getucker (621.4)
chute Rutsche, Schütte, Rinne, Gleitrinne
chute feed Füllrutsche
cifax Faksimileverschlüsselung
cinder Schlacke
cine Kino-...
cine camera Kinokamera
cinema Kinematograph, Kino, Kinotheater
cinema projector Kinoprojektor
cinematograph Kinematograph, Kino
cinematographie kinematographisch
cinematographic projector Kinoprojektor
cipher Ziffer; Null; Stelle, Stellenzahl; Chiffre, Schlüssel
cipher code Telegraphiercode, Zahlencode
cipher machine Schlüsselgerät, Schlüsselmaschine
to cipher chiffrieren, verschlüsseln
ciphony verschlüsselter Sprechverkehr, Sprachverschlüsselung
circle Kreis; Kreis, Umkreis, Zirkel, Bereich; Ronde (66)
circle coefficient Streufaktor (621.313.3)
circle diagram Kreisdiagramm (621.3)
to circle kreisen; sich auf Kreisbahnen bewegen
circulating current Kreisstrom
circlip Sprengring
circuit Stromkreis, Kreis; Kreislauf; Schaltung
circuit analyzer Vielfachmeßinstrument
circuit arrangement Schaltung, Schaltanordnung
circuit blueprint Schaltpause
circuit board Platine, gedruckte Schaltung
circuit board artwork Entwurf des Leiterbahnbildes, Platinenentwicklung
circuit break Leitungsbruch, Unterbrechung
circuit breaker Schalter, Ausschalter, Sicherungsautomat

circuit-breaker mechanism Schaltmechanismus, Schaltergetriebe
circuit-breaker mounting Schaltereinbau
circuit breaking Abschaltung, Abschaltvorgang
circuit-breaking arc Ausschaltlichtbogen
circuit closing Einschaltung, Stromschluß, Einschaltvorgang
circuit-closing connection Arbeitsstromkreis
circuit-commuted recovery time Freiwerdezeit (621.314.63)
circuit constant Kreisgröße, Schaltelement, Konstante des Stromkreises (621.3)
circuit controller Kontroller, Schaltwalze, Walzenschalter
circuit diagram Schaltbild, Stromlaufplan, Verdrahtungsplan
circuit element Schaltelement
circuit-holding time Belegungsdauer
circuit lay-out card Kabelplan
circuit noise Leitungsrauschen, Leitungsgeräusch (621.395)
circuit opening Abschaltung eines Kreises
circuit-opening connection Ruhestromschaltung
circuit-opening spark Unterbrechungsfunke
circuit parameter Schaltelement, Stromkreisparameter
circuit simulation Nachbildung (eines Netzes oder Stromkreises)
circuit stability Frequenzkonstanz (621.396)
circuit stray Schaltkapazität
circuit switching Durchschaltetechnik (621.395)
circuit tester Schaltungsprüfer, Leitungsprüfer
circuit theory Schaltungslehre
circuit voltage Leiterspannung
circuit wiring Schaltung, Verdrahtung
circuital magnetization Längsmagnetisierung
circuital vector field Wirbelfeld
circuitation Umlaufintegralbildung
circular kreisförmig, rund, kreisrund, Rund-...
circular antenna Kreisantenne, Ringantenne
circular arc Kreisbogen
circular conductor kreisförmiger Leiter, runder Leiter
circular connector Rundstecker
circular cone Kreiskegel
circular frequency Kreisfrequenz
circular function Kreisfunktion, Winkelfunktion
circular magnetization Rundmagnetisierung
circular mil Querschnitt eines Drahtes von $^1/_{1000}$ Zoll Durchmesser, als Flächeneinheit = $0{,}785 \cdot 10^{-6}$ Quadratzoll = $0{,}506 \cdot 10^{-3}$ mm^2
circular motion Kreisbewegung
circular polarization kreisförmige Polarisation
circular polarized zirkularpolarisiert
circular radiation Rundstrahlung
circular saw Kreissäge
circular scan Kreisabtastung (621.396.96)
circular section Kreisquerschnitt
circular in section mit rundem Querschnitt, mit Kreisquerschnitt
circular sectioned mit Kreisquerschnitt, rund
circular sweep kreisförmige Zeitablenkung (621.385)
circular trace Kreisbahn, Kreisspur
circular tube rundes Rohr, runde Röhre
circularly polarized kreisförmig polarisiert
circularity Kreisförmigkeit, Kreisform
to circulate zirkulieren, umlaufen; umpumpen, umlaufen lassen; herumschicken
to circulate about umlaufen um, sich drehen um
circulating umlaufend, Umlauf-...
circulating air Umluft
circulating current Wirbelstrom, Schwungradstrom, Schwingkreisstrom
circulating-oil lubrication Ölumlaufschmierung
circulating reactor Kreislaufreaktor
circulating store Umlaufspeicher (681.3)
circulation Zirkulation, Umlauf, Kreislauf
circulation boiler Zwangsumlaufkessel
circulator Zirkulator (621.372)
circum-... um..., rund...
circum centre Mittelpunkt des Umkreises
circum circle Umkreis (51)
circumference Umfang
around the circumference auf dem Umfang
circumferential Umfangs-...
circumferential key Ringkeil
circumferential welding Rundschweißung, Umfangsschweißung
to circumvent a patent ein Patent umgehen
cistern Zisterne; Wasserbehälter, Reservoir, Hochbehälter
cistern wagon Kesselwagen
citizen-band radio telephone = CB radio Jedermann-Funkgerät, CB-Funksprechgerät
civil zivil
civil engineering Bauwesen, Bauingenieurwesen
cl. = centilitre cl = Zentiliter
C.L. = centreline Mittellinie

clack Klappen, Klapp
clad bedeckt, umkleidet; plattiert; Brennelementhülle (539)
clad metal plattiertes Metall, plattiertes Material
clad steel plattierter Stahl
cladding Brennelementhülle (539)
claim Anspruch, Forderung; Beanstandung; Patentanspruch
clamp Klammer, Zwinge, Klemme; Spannzwinge, Spannfutterklemme
clamp ammeter Dietze-Anleger, Zangenamperemeter (621.317)
clamp bushing Klemmhülse
clamp circuit Klemmschaltung
clamp-on meter Zangenmeßgerät
clamp voltage Klemmspannung
to clamp klemmen, festklemmen, klammern, spannen
clamped amplifier Klemmverstärker
clamped connection Klemmverbindung
clamper noise Zeilenrauschen, Zeilenstreifigkeit (621.397)
clamping Klammern, Spannvorgang; Pegelhaltung (durch Klemmschaltung)
clamping appliances Spannvorrichtungen
clamping bolt Spannschraube, Klemmschraube
clamping circuit Klemmschaltung (621.397)
clamping collar Klemmring, Schelle
clamping device Spannvorrichtung
clamping distortion Klemmverzerrung
clamping nut Klemmschraube; Spannmutter
clamping pulse Klemmimpuls
clamping rings Spannringe (621.313.2)
clamping screw Klemmschraube
clang Gongton, Klingen
clang of a bell Glockenton
clank Klirren
clapper Klöppel
Clark cell Clark-Element
class Klasse, Sorte, Type; Klasse, Schulraum; Klasse, Wagenklasse
class of accuracy Güteklasse
class of emission Sendeart
class A amplifier A-Verstärker
class AB amplifier AB-Verstärker
class A insulation Isolationsklasse A
class A operation A-Betrieb (entspr. AB, B, C)
class B amplification B-Verstärkung (621.396.64)
class B modulation B-Modulation
class B output stage Endstufe in B-Betrieb
class H insulation = high-temperature insulation Hochtemperaturisolation, Heißisolation
class O insulation Trockenisolation
classification Klassifikation, Klassifizierung, Klasseneinteilung, Ordnung; Klassierung, Sortierung
classification lamp, classification light Kennlampe (625.2)
to classify klassifizieren
clatter Klappern, Geklapper
to clatter klappen
claw Klaue
claw-pole stationary-field generator Klauenpolgenerator
to clean säubern, reinigen
to clean up gases Gase absorbieren
clean sauber, rein
clean-cut sauber geformt, gut ausgepreßt (621.9); scharf geschnitten (655); sauber, scharf (621.394), (621.396.3)
clean reactor kalter Reaktor, sauberer Reaktor
clean-room technique Rein-Raumtechnik
cleaning Säuberung, Reinigung
to cleanse reinigen; (chemisch) reinigen
to clear reinigen, klären, säubern; (einen Fehler) abschalten (621.3); trennen, auslösen (621.395); ohne Berührung vorbeikommen an; aufräumen; löschen (681.3)
to clear a complex fraction den Nenner eines komplexen Bruches reell machen
to clear one another sich gegenseitig nicht berühren
to clear a section eine Strecke freiblocken, einen Streckenabschnitt entblocken
to clear the trouble die Störung beseitigen
clear klar, selbstverständlich
All clear Entwarnung (35)
in the clear im Lichten, in der lichten Weite
clear-back signal Schlußzeichen (621.395)
clear-forward signal Auslösezeichen (621.395)
clear of gas gasfrei
clear glass Klarglas
clear ring klarer Ton, reiner Ton
clear scribed line scharf gezeichnete Linie
clear space leerer Raum, freier Raum
clear text Klartext
clear width lichte Weite, lichte Breite
clearance Abstand, Zwischenraum, Spiel, Spielraum, Luft; freie Höhe, lichte Höhe; Räumung, Reinigung; Abrechnung, Freigabe (351.814.33); Sicherheits-Unbedenklichkeitsbescheinigung (623)

clearance fit Laufsitz
clearance limit Freigabegrenze (351.814)
clearing Schlußzeichengabe (621.394); Löschen (681.3)
clearing circuit Blockstromkreis (656.25)
clearing indicator Schlußzeichen(-Anzeige)
clearing key Schlußtaste (621.395); Löschtaste (681.3)
clearing lamp Schlußlampe (621.395)
clearing relay Schlußrelais (621.395)
clearing of screen Löschen des Bildschirms
clearing signal Schlußzeichen, Abläuten (621.395)
clearing time Abschaltzeit (621.316.5); Abschmelzzeit (621.316.9)
cleat Dübel
cleat wiring Dübelverlegung
to cleat auf Dübeln verlegen (621.3)
cleavable spaltbar
cleavage Spalte, Spaltung
cleavage plane Spaltfläche (548)
to cleave spalten
cleft Spalte, Riß, Sprung
to clench verklammern, umnieten (s. **to clinch**)
clevis U-Schelle
click Klicken, Knack, Klick
click absorber Gehörschutz (621.395)
click control Knackprobe (621.395)
click method Knackprüfung
click suppressor Knackschutz
to click klicken, knacken
climate Klima
climatic klimatisch
climatic conditions Wetterbedingungen
climax Gipfelpunkt, Scheitel
climbing irons Klettereisen
climbing space Kletterraum (621.315)
to clinch festklammern, umnieten (vgl. **to clench**)
to clinch over a rivet einen Nietkopf schlagen, einen Niet setzen
clip Klammer, Klemme; Büroklammer, Heftklammer, Briefklammer
clip-on transformer Anlegewandler, Zangenwandler
to clip (in der Amplitude) beschneiden, begrenzen (621.396); festklammern, klammern, zusammenheften; beschneiden, ausschneiden
to clip off abschneiden, wegschneiden
clipper (Schwellwert-)Begrenzer, Amplitudenbegrenzer, Abkapper, Clipper (621.396); Karabinerhaken
clipper stage Abtrennstufe
clipping Begrenzung (621.396)
clock Uhr; Taktgeber, Takt (681.3)
clock diagram Kreisdiagramm (538)
clock frequency Taktfrequenz; Zähltakt (681.3)
clock gauge Meßuhr
clock generator Taktgeber, Taktgenerator
clock meter Uhrzähler, Aronzähler, Pendelzähler
clock pulse Taktpuls
clock radio Uhrenradio
clock rate Taktfrequenz
clock relay Schaltuhr
to clock Kontrolluhr stechen
to clock in, (on) Eingang stechen
to clock off, (out) Ausgang stechen
clockspring core Spiralkern, Wickelkern
clockwise, in clockwise order im Uhrzeigersinn, rechtsdrehend
clockwork Uhrwerk
Clophen Clophen
clophen-immersed transformer Clophen-Transformator
to close einschalten; schließen, abschließen; einrücken; sich schließen
to close a circuit einen Stromkreis einschalten, einen Stromkreis schließen
to close down schließen, stillegen
to close in from sich nähern von
to close in on sich schließen um, sich zusammenziehen um
to close up into übergehen in
closed geschlossen
closed block geblockte Blockstrecke
closed cavity geschlossener Hohlraum, abgeschlossener Hohlraum
closed circuit geschlossener Stromkreis
closed-circuit condition eingeschalteter Zustand
closed-circuit cooling Kreislaufkühlung, Umlaufkühlung
closed-circuit direct current winding in sich geschlossene Gleichstromwicklung
closed-circuit system Ruhestromanlage
closed-circuit television, CCTV Industrielles Fernsehen, Kabelfernsehen
closed-circuit ventilation system Umlaufkühlung
closed-circuit voltage Spannung bei Belastung
closed coaxial resonator Schwingtopf
closed-coil armature Anker mit in sichgeschlossener Wicklung, geschlossene Ringwicklung, Gleichstromwicklung
closed-coil armature winding = closed coil winding

closed-coil winding in sich geschlossene Wicklung, Grammescher Ring
closed-core transformer Transformator mit geschlossenem Kern
closed electric circuit geschlossener elektrischer Stromkreis
closed-end line am Ende kurzgeschlossene Leitung
closed figure in sich geschlossene Figur
closed loop geschlossene Masche, geschlossene Schleife (621.3); Regelkreis (621.316.7)
closed-loop control system Regelsystem mit geschlossener Schleife
closed-loop gain dynamische Schleifenverstärkung
closed-loop series street lithting system Reihenlampenbeleuchtung
closed magnetic circuit geschlossener magnetischer Kreis
closed numbering verdeckte Numerierung (621.395.3)
closed slot geschlossene Nute (621.313)
closed stub durch Kurzschlußbügel abgeschlossen
closed subscriber group geschlossene Teilnehmergruppe (621.395)
close eng, nahe
close analysis genaue Analyse
close approach gute Annäherung
close approximation gute Näherung
close coupling enge Kopplung, feste Kopplung
close electrode spacing geringer Elektrodenabstand
close fit Feinsitz
close-fitting eng anliegend, genau passend, stramm sitzend
close-grained feinkörnig, feinfaserig
close lid dicht schließender Deckel
close measurement genaue Messung
close-range fire control Nahbereich-Feuerleitradar
close sliding fit Schiebesitz
close spacing geringer Abstand
close tolerance enge Toleranz
closer Schlußlinie
closing Schließung, Schließvorgang, Einschaltung; Schluß
closing coil Einschaltspule (621.316.5)
closing device Schließvorrichtung, Einschaltvorrichtung
closing line Schlußlinie
closing relay Einschaltrelais
closing spark Schließungsfunke
closure Abschluß, Abdichtung, Verschluß
closure check Schlußprüfung = Prüfung auf Schluß
cloth Stoff, Gewebe
cloth screen Tuchfilter, Textilsieb
cloud Wolke; Hauch (auf einer Oberfläche)
cloud base Wolkenuntergrenze
cloud ceilometer Wolkenhöhenmesser
cloud chamber Nebelkammer
cloud and collision warning radar Wetter-und Kollisionsbordradar
cloverleaf antenna Kleeblattantenne
cluster Nest (55)
clutch Kupplung; Klaue
clutch disk Kupplungsscheibe
clutch facing Kupplungsbelag
clutch fork Kupplungsgabel
clutch lining Kupplungsbelag
to clutch packen; kuppeln
clutter Störflecke, Störecho (621.397)
clutter diagramm Festzeichenbild, Festzielbild (621.396.96)
clutter gating Festzeichenaustastung, Festzielaustastung
C.M.B. converter Umrichter, Wechselrichter mit Wiedergleichrichtung
CMOS = Complementary Metal Oxide Semiconductor CMOS
CNC = Computerized Numerical Control numerische (Werkzeug-)Maschinensteuerung mit Computer
coal Kohle
coal dust Kohlenstaub
coal face Kohlenstoß, Streb (622)
coal-face lighting Strebbeleuchtung (622)
to coalesce verschmelzen, ineinander übergehen
coalescence Verschmelzung
coarse grob
coarse adjustment Grobeinstellung
coarse bearing rohe Peilung
coarse control Grobsteuerung
coarse crushing Grobzerkleinerung
coarse-grained grobkörnig; grobfaserig
coarse grinding Grobmahlung, Grobzerkleinerung
coarse-pitch thread grobgängiges Gewinde
coarse setting Grobeinstellung
coarse thread Grobgewinde
coastal refraction Küstenbrechung, Küsteneffekt
coast station Küstenfunkstelle
coat Mantel; Auftrag, Anstrich; Verputz, Verkleidung, Beschichtung
to coat überziehen, bedecken, ummanteln, beschichten, bestreichen, streichen (**with:** mit)

coated cathode pastierte Kathode, Schichtkathode
coated electrode umhüllte Elektrode (261.791)
coating Anstrich, Bedeckung, Umhüllung; Streichmasse; Belag
coating machine Beschichtungseinrichtung
coating of paint Farbanstrich
coaxial koaxial, konzentrisch
coaxial cable konzentrisches Kabel, Koaxialkabel
coaxial cable fit konzentrische Muffe
coaxial connector konzentrischer Stecker
coaxial line koaxiales Kabel, konzentrisches Kabel, konzentrische Leitung
coaxial relay Koaxialrelais
coaxial waveguide junctions Koaxial-Hohlleiter-Übergang
cobald Kobalt
COBOL = Common Bussiness Oriented Language
co-channel operation Gleichkanalbetrieb (621.397.9)
co-channel transmitter Gleichkanalsender
cock Hahn (623), (696)
cockpit Führerraum, Pilotenraum (629.13)
code Gesetzbuch, Vorschriftenbuch; Code, Schlüssel; Bezeichnungsweise; Kennzahl, Kennung (621.395.3)
code beacon Kennungsbake, Kennungsfeuer (656.7)
code converter Code-Umsetzer, Codewandler
code garbling Codeverwirrung
code detector Code-Erkennung
code element Code-Element
code light Kennfeuer, Blinkfeuer (656.7)
code receiver Telegraphieempfänger
code recognition Code-Erkennung
code regulations Normvorschriften
code ringing Morseruf (621.395)
code selector Rufnummerwähler, Amtswähler (621.395)
code signal Telegraphiezeichen, Morsezeichen
code-signal reception Telegraphieempfang, Morseempfang
code symbol mathematisches Zeichen
code transparent transmission codetransparente Übertragung
code translation Codeumsetzung
code wire (US) zugelassene (Norm-)Leitung (696)
to code schlüsseln, verschlüsseln, chiffrieren, codieren
coded passive reflector antenna getastete Rückstrahlantenne
coded plug Programmstecker
codec = coder-decoder
coder Schlüsselmaschine, Verschlüsselungseinrichtung, Chiffriermaschine, Codiergerät
to codify in Vorschriften festlegen, vorschreiben
coding Codierung
coding tube Codierungsröhre (51)
coefficient Koeffizient, Beizahl, Beiwert, Kennzahl
coefficient of amplification Verstärkungsfaktor
coefficient of capacitance Kapazität, Kapazitätswert
coefficient of coupling Kopplungskoeffizient, Kopplungsfaktor
coefficient of cubical expansion Ausdehnungskoeffizient
coefficient of diffusion Diffusionskoeffizient
coefficient of discharge Ausflußkoeffizient
coefficients of electrostatic induction Induktionskoeffizienten
coefficient of expansion Ausdehnungskoeffizient
coefficient of friction Reibungskoeffizient, Reibungswert, Reibungszahl
coefficient of harmonic distortion Klirrfaktor
coefficient of leakage Streukoeffizient
coefficient of linear expansion linearer Ausdehnungskoeffizient
coefficient of magnetization Suszeptibilität
coefficient of mutual induction Koeffizient der gegenseitigen Induktion, Gegeninduktivität
coefficients of potential Potentialkoeffizienten
coefficient of reflection Reflexionskoeffizient
coefficient of safety Sicherheitsgrad
coefficient of self-induction Selbstinduktionskoeffizient, Induktivität
coefficient of thermal expansion Wärmedehnungskoeffizient, Wärmedehnungszahl
coefficient of transmission Durchlaßfaktor (621.392); Übertragungsfaktor
coefficient of transparency Durchlässigkeit, Durchsichtigkeit (535)
coefficient of utilization Ausnutzungsfaktor, Beleuchtungswirkungsgrad (621.32)
coercimeter Koerzitivkraftmeßgerät
coercive Koerzitiv-..., koerzitiv
coercive field Koerzitivfeldstärke
coercive force Koerzitivkraft
coercive intensity Koerzitivkraft
coercivity Koerzitivkraft, koerzitive Eigenschaft
to coffer wasserdicht machen, abdichten
cog Zahnstift, Zahn

cog wheel Stiftrad, Zahnrad, Kammrad
cog-wheel railway Zahnradbahn
to cohere zusammenhalten, haften, kohärieren
coherent kohärent
coherent moving target indication kohärente Festzeichenunterdrückung
coherent pulse radar Pulskohärenzradar
coherent scattering kohärente Streuung (539)
coherer Kohärer
coherer effect Kohärerwirkung
cohesion Kohäsion, Zusammenhalt
cohesive zusammenhaltend, Kohäsions-...
cohesive force Kohäsionskraft
cohesiveness Zusammenhalt
coil Spule (621.3); Schlange, Rohrschlange; Wendel
coil aerial Rahmenantenne (US)
coil-and-capacitor circuit LC-Schaltung
coil constant Spulengüte
coil-driven loud speaker Tauchspulenlautsprecher
coil form Spulenkörper
coil ignition Spulenzündung, Abreißzündung
coil-loaded spulenbelastet, pupinisiert, bespult
coil loading Pupinisierung, Spulenbelastung, induktive Belastung (621.392)
coil section Spulenfeld (621.39)
coil side Spulenseite
coil switch Spulenschalter, Wellenumschalter (621.396)
coil varnish Spulenlack, Isolierlack
coil winding Spulenwicklung
coil-winding machine Spulenwickelmaschine
to coil wendeln, wickeln, spulen
to coil up aufspulen, aufhaspeln, aufwickeln
coiled coil Doppelwendel
coiled-coil filament Doppelwendel (621.32)
coiled-coil lamp Doppelwendellampe
coin Münze
coinbox telephone for local calls Ortsmünzfernsprecher
coinbox telephone for long distance calls Fernwählmünzfernsprecher
coin collector (set) Münzfernsprecher
coincidence Koinzidenz, Zusammentreffen, Gleichzeitigkeit
coincidence circuit Koinzidenzschaltung
coincidence counter Koinzidenzzähler
coincidence element Äquivalenzglied
coincidence gate Koinzidenz-Tor
coincidence range finder Mischbildentfernungsmeßgerät, Koinzidenzentfernungsgerät
coincident koinzident; zusammentreffend, zusammenfallend
coincident-current memory Koinzidenzspeicher
cold Kälte
cold-producing machinery Kältemaschinen (pl.)
cold trap Kühlfalle, Ausfrierfalle
cold kalt
cold cathode kalte Kathode, Kaltkathode
cold-cathode tube Kaltkathodenröhre
cold-crushing strength Kaltdruckfestigkeit (66)
cold-curing synthetic raisin (kalt) aushärtendes Kunstharz
cold-drawn kaltgezogen
cold-drawn wire kaltgezogener Draht
cold electronic discharge Dunkelentladung
cold emitter valve Kaltkathodenröhre
cold reactor kalter Reaktor
cold-strained wire kaltgereckter Draht
colinear kolinear, auf der gleichen Geraden liegend
colinear array Antennenzeile (621.396.677)
to collapse zusammenschrumpfen, zusammenbrechen, zusammenfallen (**to:** auf)
collapsible zusammenlegbar, zusammenklappbar, Klapp-..., Falt-...
collapsible-type undercarriage einziehbares Fahrwerk (629.13)
to collate ordnen, zuordnen, mischen (681.3)
to collect sammeln, auffangen, zusammentragen, zusammenfassen (51)
to collect current Strom abnehmen (621.313)
to collect electrons Elektronen einfangen, Elektronen übernehmen
collecting Sammel-...
collecting container Sammelbehälter
collecting electrode Sammelelektrode, Auffangelektrode; passive Elektrode (621.35)
collecting main Sammelleitung
collecting ring Schleifring
collection Sammlung, Einsammlung, Einziehung
collection of tables Tabellenwerk, Tafelsammlung
collective number Sammelrufnummer
collector Kollektor (621.313), (621.382), (621.385 Klystron); Kollektor (-Elektrode) (621.396.64); Stromabnehmer (625)
collector-base cut-off current Kollektor-Basis-Restrom
collector-base voltage Kollektor-Basis-Spannung
collector current Kollektorstrom
collector cut-off current Kollektorreststrom

collector depletion layer capacitance Kollektorsperrschichtkapazität
collector junction Kollektorsperrschicht, Kollektor-Basis-Zonenübergang
collector plate Auffangelektrode, Auffanganode (621.385Klystron)
collector rings Schleifringe
collector series resistance Kollektorbahnwiderstand
collector shoe Kontaktschuh (621.33)
collector strip Bügelbelag, Abnehmerbügel (621.33)
collector terminal Kollektor-Anschluß
to collide kollidieren, zusammenstoßen, stoßen
to collimate bündeln
collimation mast Justiermast
collimator Kollimator
collinear auf einer Linie liegend
collinear antenna array lineare Dipolgruppe, Antennenzeile
collision Zusammenstoß, Kollision, Stoß
collision avoidance Kollisionsschutz
collision frequency Stoßfrequenz
collision process Stoßprozeß (537)
collision warning Kollisionswarnung
collodion Kollodium
collodion film Kollodiumhäutchen
colloid Kolloid
colloid rectifier Kolloidgleichrichter
coolloid(al) kolloidal
colloidal filament gespritzter Faden (621.32)
colloidal particles kolloidale Teilchen
colophony Kolophonium
coloration Färbung (534), (535)
colorific farberzeugend, farbgebend, färbend
colorimeter Kolorimeter
colour Farbe
colour-bar generator Farbbalkengenerator
colour blind farbenblind
colour burst Farbsynchronsignal
colour code Farbkennzeichnung, Farbcode
colour-coded mit Farbkennzeichnung
colour decoder Farbdecoder
colour encoder Farbcoder
colour gamut Farbbereich
colour killer Farbabschalter
colour light signal Lichtsignal, Tageslichtsignal (65)
colour mask Farbmaske
colour match Farbanpassung
colour-matrix unit Farbmatrixschaltung
colour mixture Farbmischung
colour picture Farbbild
colour picture screen Farbschirm (621.397)
colour-picture signal Farbbildsignal
colour-picture transmission Farbbildübertragung
colour-picture tube Farbbildröhre
colour position light signal mehrbegriffiges Tageslichtsignal (65)
colour-purity error Farbverfälschung
colour range Farbbereich
colour registration Farbverfälschung
colour rendering Farbwiedergabe
colour saturation Farbsättigung
colour sensation Farbempfindung, Farbeindruck
colour slight Farbdia
colour-stimulus specification Farbvalenz
colour substance Farbsubstanz, Farbkörper, Pigment
colour television Farbfernsehen
colour-television camera Farbfernsehkamera
colour temperature Farbtemperatur
colour triad Farbtripel
colour triangle Farbdreieck
colour-video signal Farbbildsignalgemisch
colourimetry Farbmetrik
colourless farblos
colourplexer Farbcoder
Colpitt's circuit kapazitive Dreipunktschaltung (621.396.61)
column Säule; Pfeiler (72); (Tabellen-)Spalte (51); Ständer (621); Säule (35); Kolonne (66)
column heading Tabellenkopf, Spaltenkopf, Spaltenüberschrift
column radiator Rippenheizkörper
column-type adjusting transformator Säulenstelltransformator
column of water Wassersäule
COMAL = common algorithmic language
comb Spitzenabnehmer (621.319); Kamm
comb collector Spitzenabnehmer (621.319)
comb lightning arrester Spitzenableiter
comb-shaped pole Kammpol
combination Zusammensetzung, Zusammenstellung, Kombination; Schaltung, Schaltanordnung (621.3); Verbindung (54)
combination frequency Kombinationsfrequenz
combination pliers (pl.) Kombinationszange
combinational logic kombinatorische Logikschaltung
combinatorial analysis Kombinationsrechnung
combinatorial circuit Schaltnetz, Verknüpfungsschaltung, kombinatorische Schaltung
to combine kombinieren, zusammenziehen; (in

sich) vereinigen; verbinden, sich verbinden (54); sich vereinigen
to combine in parallel parallelschalten
to combine together (sich) zusammensetzen, (sich) verbinden
to combine with (sich) verbinden mit, reagieren mit, eine Verbindung eingehen mit
combined gemeinsam, zusammengefaßt; gebunden (54)
combined carbon gebundener Kohlenstoff
combined distribution frame kombiniertes Verteilergestell
combined effect Gesamtwirkung, resultierende Wirkung
combined power Scheinleistung
combined power and heating system Kopplung von Wärme- und Kraftversorgung, Wärme-Kraft-Kopplung
combining weight Verbindungsgewicht
combustible Brennstoff
combustible brennbar, verbrennbar
to come kommen
come-at-able zugänglich
to come into action ansprechen, in Betrieb kommen
to come into effect in Kraft treten
to come off abgehen, sich ablösen, abblättern
to come into operation in Betrieb kommen
to come to rest außer Betrieb kommen, zum Stillstand kommen
to come short nicht befriedigen, nicht genügen, hinter den Ansprüchenzurückbleiben
to come to a stop anhalten, zum Stillstand kommen
command Befehl, Kommando (621.316.7), (621.398)
command hot loop Führungsdirektleitung (623)
command language Befehlssprache (681.3)
command variable Führungsgröße (621.316.7)
commensurable kommensurabel
commercial öffentlich; Werbesendung (654.19)
commercial item handelsüblicher Artikel
commercial network öffentliches Fernmeldenetz
commercial radio network öffentliches Rundfunknetz
to commit to writing schriftlich niederlegen
commodity Verbrauchsgut
common üblich, gemeinsam, gebräuchlich, allgemein gebräuchlich
common antenna Gemeinschaftsantenne
common base (configuration) Basisschaltung (621.382)
common battery Zentralbatterie = Z.B.
common battery central office Z.B.-Amt, Z.B.-Vermittlung (621.395)
common battery system Z.B.-System
common battery telephone set Z.B.-Apparat
common channel interference Gleichkanalstörung (621.396.9)
common chord Dreiklang, Akkord
common collector Kollektorschaltung
common control Zentralsteuerung
common-control switching nicht schritthaltender Verbindungsaufbau
common difference Reihendifferenz (51)
common drain Drain-Schaltung
common emitter Emitterschaltung
common factor gemeinsamer Teiler
common frequency working Gleichwellenbetrieb, Gleichwellenfunk
common gate Gateschaltung
common hardening steel Flußstahl
common leg gemeinsamer Pol (621.392)
common limb gemeinsamer Schenkel, Mittelschenkel
common logarithm Briggscher Logarithmus
common measure gemeinsamer Teiler
common mode Gleichtakt...
common mode input Gleichtakteingang
common mode interference Gleichtaktstörung
common mode rejection ratio Gleichtaktunterdrückung
common mode signal Gleichtakt-Signal
common mode voltage amplification Gleichtakt-Spannungsverstärkung
common multiplier gemeinsamer Teiler, gemeinsamer Faktor
common normal Berührungsnormale
common point gemeinsamer Anschlußpunkt (621.3)
common point of concurrency gemeinsamer Schnittpunkt
common return gemeinsame Rückleitung
common screw gewöhnliche Holzschraube
common source Source-Schaltung
common trunk gemeinsame Abnehmerleitung
communal aerial Gemeinschaftsantenne
to communicate kommunizieren, in Verbindung stehen; mitteilen, Nachrichten übermitteln
communication Mitteilung, Veröffentlichung, Nachricht; Verbindung (621,1), (621,39); Nachrichtenwesen
communication band Übertragungsband, Sendebandbreite, Kanalbreite (621.396)
communication cable Nachrichtenkabel

communication channel Nachrichtenkanal
communication circuit Nachrichtenkreis, Fernmeldeverbindung
communication engineer Nachrichteningenieur, Fernmeldetechniker, Schwachstromingenieur
communication line Fernmeldeverbindung, Fernmeldeleitung
communication range Nachrichtenreichweite, Übertragungsbereich; Reichweite (621.395), (621.396)
communication satellite Fernmeldesatellit
communication theory Kommunikationslehre, Kybernetik
communications tower Fernmeldeturm
communication transformer Übertrager
community antenna system Gemeinschaftsantennen-Anlage
to commutate kommutieren
commutating Kommutierungs-..., Wende-...
commutating angle Kommutierungswinkel (621.345.5)
commutating field Kommutierungsfeld, Hilfspolfeld
commutating machine Kommutatormaschine, Kollektormaschine
commutating period Kommutierungszeit
commutating pole Wendepol
commutating pole winding Wendepolwicklung
commutating windings (pl.) Kommutierungswicklung, Hilfswicklung, Wendepolwicklung
commutating zone Kommutierungszone
commutation Kommutierung
commutation field Wendefeld, Kommutierungsfeld
commutation machine Kommutatormaschine, Kollektormaschine
commutation margin Löschwinkel, Kommutierungswinkel (621.314.5)
commutation period Kommutierungszeit, Kommutierungsdauer
commutation pole Wendepol
commutation windings (pl.) Wendepolwicklung
commutative kommutativ (51)
commutator Kommutator, Stromwender, Kollektor
commutator armature Kommutatoranker, Anker mit Kommutator
commutator bar Kollektorlamelle, Kollektorsegment
commutator bar assembly einbaufertiger Kollektorkörper
commutator brush Kommutatorbürste, Kollektorbürste
commutator connector Kollektorfahne
commutator grinder Kollektorschleifmaschine
commutator hub Kollektorhülse
commutator insulating rings (pl.) Kollektorringisolation
commutator-insulating strips Segmentisolation (sing.) (621.313)
commutator loss Stromwenderverluste, Kollektorverluste (pl.)
commutator lug Kollektorfahne
commutator mica Kollektorglimmer
commutator motor Kollektormotor, Kommutatormotor
commutator pulse Zyklusimpuls
commutator rectifier mechanischer Gleichrichter
commutator ring Kollektorring
commutator ripple Kollektortöne (pl.)
commutator riser Kommutatorfahne
commutator section Lamelle, Kollektorlamelle
commutator segment Kollektorlamelle
commutator shell Kollektornabe, Kommutatornabe
commutator-shell insulation Kollektorisolierhülse
commutator sleeve Kollektorhülse, Kommutatorhülse
commutator tag Kommutatorfahne, Kollektorfahne
commutator tube Kollektorhülse, Kommutatorhülse
compact dicht, fest, gedrängt, kompakt
compact disc Compact-Platte (681.84)
to compact verdichten
compandor Dynamikregler, Kompander (621.395), (621.396)
companion Begleit-...
companion (matter) Begleitsubstanz
companion sideband das andere Seitenband
comparative verhältnismäßig, Vergleichs-...
comparative measurement Vergleichsmessung
comparator Komparator
comparator circuit Vergleichsschaltung
comparison Vergleich, Gegenüberstellung, Vergleichs-...
comparison resistance Vergleichswiderstand
comparison voltage Vergleichsspannung, Referenzspannung
compass Kompaß; Leitfaden; Umfang
compass bearing Kompaßpeilung
compass card Kompaßscheibe
compass direction Himmelsrichtung

compass error Kompaßfehler
compass needle Kompaßnadel
compass saw Stichsäge
compasses (pl.) Zirkel
compatibility Kompatibilität, Verträglichkeit
compelled signalling Zwangslaufverfahren (621.395.3)
to compensate for . . . kompensieren, . . . ausgleichen, abgleichen
compensated alternator kompoundierter Wechselstromgenerator
compensated induction motor kompensierter Asynchronmotor
compensated motor Motor mit Kompensationswicklung
compensated repulsion motor kompensierter Repulsionsmotor
compensated semiconductor Kompensationshalbleiter
compensated series motor Wechselstromhauptschlußmotor mit Kompensationswicklung
compensating Ausgleichs- . . ., Kompensations . . .
compensating coil Ausgleichsspule
compensating diaphragm Ausgleichsmembran
compensating field Kompensationsfeld, Ausgleichsfeld
compensating pole Wendepol (621.313.2)
compensating resistor Ausgleichswiderstand
compensating weight Gegengewicht, Ausgleichsgewicht
compensating winding Kompensationswicklung
compensation Kompensation; Ausgleich
compensation charge Erhaltungsladung, Dauerladung (621.355)
compensation of power factor Leistungsfaktorverbesserung, Blindstromkompensation
compensating voltage Ausgleichsspannung, Kompensationsspannung
compensation voltmeter Kompensationsspannungsmesser
compensator . . . ausgleicher; Spartransformator (621.314.2); Neutralisierungs-, Entzerrer (621.396)
compensator balancer Ausgleichtransformator (621.316.1)
compensator starter Transformatoranlasser (621.313.3)
compensator transformer Spartransformator (621.313)
compensatory leads Ausgleichsleitungen (621.317)
compiler Compiler, Befehlsumsetzer, Sprachübersetzerprogramm (681.3)
compiler-level languages Compilersprachen
complement Komplement (51)
complementary komplementär, sich ergänzend
complementary colour Komplementärfarbe
complementary MOS = CMOS komplementäre MOS-Technik, CMOS
complementary transistors komplementäre Transistoren
to complete fertigstellen, vervollständigen, komplettieren
to complete a circuit einen Stromkreis schließen
complete vollständig
complete analysis Vollanalyse (54); eingehende Untersuchung (51)
complete cancellation völlige Auslöschung
complete combustion vollständige Verbrennung
complete cycle Kreisprozeß (536)
complete diffusion vollständige Streuung (535), 621.32)
complete equilibrium vollständiges Gleichgewicht
complete modulation 100-prozentige Modulation; vollständige Aussteuerung (681.84)
complete path geschlossener Umlaufweg (537)
complete rotation vollständiger Umlauf (621.3)
completely mesh-connected circuit vollständige Polygonschaltung
completely modulated volldurchmoduliert
complex kompliziert, vielfältig; komplex; Mehrstoff- . . ., Komplex- . . .
complex harmonic electromotive force Zeiger der EMK
complex harmonic potential difference Spannungszeiger
complex harmonic quantity sinusförmige Größe in komplexer Schreibweise
complex harmonic voltage Spannungszeiger
complex impedance Widerstansoperator
complex integration Integration im Komplexen
complex notation komplexe Darstellung
complex number komplexe Zahl
complex plane komplexe Ebene
complex quantity komplexe Größe, komplexe Zahl
complex quantity notation komplexe Schreibweise, symbolisches Rechenverfahren (621.3)

complex quantity representation komplexe Darstellung, symbolisches Rechenverfahren
complex sinusoidal current Stromzeiger
complex sinusoidal quantity sinusförmige Größe in komplexer Schreibweise
complex variable komplexe Veränderliche
complexity Kompliziertheit
compliance Gegebenheit; Nachgiebigkeit (531); Nadelnachgiebigkeit (681.48)
compliance of a mechanical element Nachgiebigkeit (Kehrwert der Federkonstante)
compole = commutating pole Wendepol
component Teil, Bestandteil, Maschinenteil; Komponente; Baugruppe, Gruppe von Einzelteilen, Bauelement
component current Teilstrom
component force Kraftkomponente
component layout Bauplan (621.396)
component part Bestandteil, Zubehörteil
component portion Teil, Bestandteil
component velocity Geschwindigkeitskomponente
composite zusammengesetzt
composite attenuation Betriebsdämpfung
composite cable gemischtadriges Kabel, kombiniertes Kabel
composite colour Farbbildsignalgemisch
composite picture signal (Fernseh-)Signalgemisch, Video-Signal, BAS-Signal (621.397)
composite plate galvanisch aufgebrachte Legierungsschicht
composite signal Summensignal, Frequenzgemisch
composited circuit mehrfach ausgenützte Leitung (621.394), (621.395)
composition Zusammensetzung; Satz (655); Komposition (78)
composition-type resistor Massewiderstand (621.3)
compound Verbindung (54); Masse, Vergußmasse, Kabelmasse (621.3)
compound alternator kompoundierter Wechselstromgenerator
compound brush Metallkohlebürste (621.313)
compound-catenary suspension Mehrdrahtaufhängung (621.33)
compound coil Kompensationsspule (621.317)
compound-connected transistors Darlington Schaltung
compound dynamo Kompoundgenerator (621.313)
compound engine Verbundmaschine (621.3)
compound excitation Kompounderregung (621.313)
compound-filled switchgear masse-isolierte Schaltanlage
compound filling Ausgußmasse
compound generator Kompoundgenerator, Doppelschlußgenerator
compound locomotive Verbundlokomotive
compound magnet Magazinmagnet
compound motor Kompoundmotor, Doppelschlußmotor
compound nucleus Zwischenkern (539)
compound signal Mehrfrequenzzeichen (621.395)
compound strip Bimetallstreifen
compund winding Verbundwicklung, Kompoundwicklung (621.313)
compound-wound kompoundiert, Kompound-..., mit Kompoundwicklung (621.3)
compound-wound dynamo Kompoundgenerator
compound-wound generator Kompoundgenerator
compound-wound motor Kompoundmotor, Motor mit Kompoundwicklung
to compound zusammensetzen; kompoundieren (621.3); imprägnieren, mit Masse füllen (621.3)
compounding Kompoundierung (621.313)
compounding coil Kompoundwicklung (621.313)
to compress komprimieren, zusammendrücken, verdichten; (durch Druck) vergüten, druckvergüten; drücken, auf Druck beanspruchen
compressed zusammengedrückt, unter Druckbeanspruchung, unter Druck stehend; Druck-..., Preß-...
compressed air Druckluft, Preßluft
compressed-air capacitor Druckgaskondensator, Druckluftkondensator
compressed-air dielectric Druckluftisolation
compressed-air lamp Druckluft-Sicherheitslampe
compressed-air line Druckluftleitung
compressed gas Preßgas
compressed-gas condenser Preßgaskondensator
compressed-iron core Massekern
compressed scale gedrängte Skala (621.317)
compressible kompressibel, zusammendrückbar

compression Druck; Kompression, Verdichtung, Zusammendrückung
compression cable Druckkabel
compression fitting Preßverbinder
compression gauge Druckmesser, Manometer
compression ratio Kompressionsverhältnis, Verdichtungsgrad, Kompression (621.4)
compression side Druckseite (621.5)
compression spring Druckfeder
compression strain Druckdehnung, Kompression, Stauchung
compression strength Druckfestigkeit
compression stress Druckbeanspruchung
compression test Stauchversuch
compression type refrigerator Kompressionskühlschrank
compressive drückend, Druck-...
compressive load Drucklast, Druckbelastung, Druckbeanspruchung
compressive strain Druckbeanspruchung; Druckdehnung = Kompression, Stauchung, Zusammendrückung
compressive strength Druckfestigkeit
compressive stress Druckbeanspruchung, Druckspannung
Compton effect Compton-Effekt
Compton scattering Compton-Streuung
computable berechenbar
to compute at berechnen auf, berechnen zu
to compute by berechnen aus, berechnen nach, ausrechnen
computer, computor Rechner, Rechengerät, Computer, EDV-Anlage
computer aided rechnergestützt
computer aided design rechnergestützte Konstruktion
computer-assisted approach sequencing rechnergestütztes Anflugfolgesystem
computer-controlled message switching system Telegraphie Speichervermittlung
computer-oriented language Programmiersprache
computer oriented language programming Programmierung in Maschinensprache
computer operation Programmablauf
computer process Rechenprozeß
computer process control rechnergesteuerter Prozeßablauf, Prozeßleittechnik, Prozeßsteuerung
computer run Rechnerlauf
computer science Informatik
computerized numerical control = CNC numerische Steuerung mit Computer
computing Rechen...
computing centre Rechenzentrum
computing instruction Maschinenbefehl (681.3); Rechenvorschrift
computing machine Rechenmaschine
computing speed Rechengeschwindigkeit
concatenated motor Kaskadenmotor
concatenation connection Kettenschaltung, Kaskadenschaltung (621.313)
concave konkav, hohl
concave cathode Hohlkathode
concave lens Konkavlinse, Zerstreuungslinse
concave mirror Hohlspiegel
concave washer Hohlscheibe
concealed wiring Unterputzinstallation
concentrate Konzentrat
to concentrate (sich) konzentrieren (**upon:** auf); anreichern, konzentrieren (66); zusammenhalten, zusammenfassen
to concentrate into zusammenfassen zu (51)
concentrated konzentriert (54), (621.3)
concentrated inductance konzentrierte Induktivität
concentrated load konzentrierte Last
concentrated winding konzentrierte Wicklung, Spulenwicklung
concentrating reflector Tiefstrahler
concentration Konzentration, Konzentrierung, Anreicherung; Anhäufung; Bündelung
concentration cell Konzentrationselement
concentration line Sammelleitung (621.395)
concentration stage Konzentrationsstufe (621.395)
concentrator Konzentrator
concentric(al) konzentrisch
concentric(al) braiding Umflechtung, Beklöppelung
concentric(al) cable konzentrisches Kabel
concentric circles konzentrische Kreise
concentric(al) conductor line konzentrisches Kabel
concentric-lay cable mehrfachkonzentrisches Kabel; mehrlagiges Seil
concentric line konzentrische Leitung, konzentrisches Kabel
concentric plug and socket konzentrische Steckverbindung
concentric winding Wicklung mit Spulen ungleicher Weite (aber gleicher Achse) (621.313); konzentrische Wicklung (621.314.2)
concentric wiring einpoliger Rohrdraht
conclusive schlüssig, beweiskräftig
concordant gleichsinnig, gleichlaufend
concrete Beton, Zement

to concrete betonieren, mit Beton auskleiden, mit Beton verkleiden
concreted steel work Stahlbau mit Betonverkleidung, Stahlskelettbau
concurrence Gleichzeitigkeit, Parallelität
concurrent gleichzeitig, parallel gehend; mit gemeinsamem Schnittpunkt
to be concurrent sich schneiden
concurrent lines sich schneidende Linien
concurrent working Simultanbetrieb
concussion Erschütterung, Stoß
condensable kondensierbar
condensation Kondensation
condensation pump Diffusionspumpe
to condense (sich) kondensieren; sich verdichten, dichter werden
condenser Kondensator (siehe auch **capacitor**) (621.1), (621.3); Kondensor (535); Vorlage, Kühlvorlage (66)
condenser aerial Netzantenne
condenser bank Kondensatorengruppe
condenser bushing Kondensatordurchführung, Nagelsche Klemme
condenser case Kondensatorbecher; Kondensatorwanne (621.396)
condenser coil Kühlschlange
condenser constant Kondensatorgüte (621.3)
condenser coupler kapazitiver Spannungsteiler
condenser electroscope Kondensatorelektroskop
condenser loudspeaker Konsensatorlautsprecher
condenser microphone Kondensatormikrophon
condenser pick-up kapazitiver Tonabnehmer
condenser receiver Kondensatortelefon
condenser transmitter Kondensatormikrophon
condenser-type bushing Kondensatordurchführung, Nagelsche Klemme
condenser-type terminal Kondensatordurchführung, Nagelsche Klemme
condensible kondensierbar
condensing Kondensations-..., Kondensor-...
condensing lens Kondensor, Kondensorlinse
condensing steam engine Kondensationsdampfmaschine
condensive mit kapazitiven Eigenschaften
condensive circuit kapazitiver Stromkreis
condensive load kapazitive Last; kapazitive Belastung, voreilende Last
condensive reactance kapazitiver Blindwiderstand
condition Bedingung; Zustand; Verhältnis
out of condition nicht gebrauchsfähig
condition of balance Gleichgewichtsbedingung, Abgleichbedingung (621.317)
condition of equilibrium Gleichgewichtsbedingung; Gleichgewichtszustand
conditions of experiment Versuchsbedingungen
condition of stable equilibrium stabiles Gleichgewicht, Zustand des stabilen Gleichgewichts
conditions of stress Spannungszustand (sing.)
to condition konditionieren (67); klimatisieren (628)
conditioned air klimatisierte Luft
conditional jump bedinger Sprung
conditional transfer of control bedingter Sprung (681.3)
to conduct ausführen, durchführen; leiten
to conduct away abführen, ableiten
to conduct experiments Versuche anstellen, Versuche durchführen
to conduct heavily kräftig leiten, gut leiten
conductance Leitwert, Wirkleitwert, Ableitung
conducting leitend, leitfähig
conducting current Durchlaßstrom
conducting material leitendes Material
conducting path leitende Verbindung, leitender Weg
conducting period Durchlaßzeit (621.314.6)
conducting salt Leitsalz (621.35)
conducting state Durchlaßzustand (621.382)
conducting state characteristic Durchlaßkennlinie
conducting state power loss Durchlaßverlustleistung
conducting state region Durchlaßbereich
conduction Leitung, Stromleitung, Stromdurchgang
conduction band Leitungsband
conduction current Leitungsstrom
conduction direction Durchlaßrichtung (621.314.6)
conduction electron Leitungselektron
conduction period Durchlaßzeit (621.385)
conductive leitend, stromdurchlässig
conductive connection leitende Verbindung
conductive coupling galvanische Kopplung
conductive discharge Widerstandsentladung
conductivity Leitfähigkeit, Leitvermögen, Leitung
conductivity bridge Kohlrausch-Brücke
conductivity meter Leitfähigkeitsmesser

conductivity modulation Leitfähigkeitsmodulation
conductivity test Fehlerbestimmung aus dem Spannungsabfall
conductor Leiter, Stromleiter, Leitung
conductor bundle Leitungsbündel
conductor loading Seilbeanspruchung, Leiterbeanspruchung
conductor rail Stromschiene
conductor-rail anchor Stromschienenfestpunkt (621.33)
conductor-rail bond Stromschienenverbinder
conductor-rail insulator Stromschienenisolator
conductor-rail system (of electric traction) Stromschienensystem
conductor of the second class Leiter zweiter Klasse
conductor size Leiterdicke
conduit Leitungsrohr, Schutzrohr, Verrohrung; Kabelkanal (aus Formsteinen)
conduit box Dose (621.315.6)
conduit connection Leitungsanschluß (696)
conduit fittings (pl.) Rohrleitungszubehör (696)
conduit run Mehrfachrohrkanal
conduit system unterirdische Stromzuführung (621.33)
conduit wiring Rohrverlegung, Verlegung in Rohr
cone Konus, Kegel
cone antenna Konusantenne, Kegelantenne
cone friction clutch Kegelreibungskupplung
cone-headed rivet Kegelkopfniet
cone loudspeaker Konuslautsprecher
cone of nulls Nullkegel (621.396.93)
cone of silence Nullkegel, Schweigekegel (621.396.93)
cone speaker Konuslautsprecher
conference call Konferenzgespräch
conference circuit Konferenzschaltung
configuration Konfiguration, Anordnung, Beschaffenheit
to confine begrenzen
confined discharge arrester gekapselter Blitzschutz
confined space begrenzter Raum, abgeschlossener Raum
confined space wave Hohlrohrwelle
confirmatory test Kontrollversuch
confocal konfokal
confocal coordinates elliptische Koordination
conformal konform
conformal representation konforme Abbildung
conform to specifications den Vorschriften entsprechen
confusion region Verwirrungszone (621.397)
to congeal gefrieren, erstarren; gelieren, vereisen
congealing point Gefrierpunkt, Erstarrungspunkt
congelation Gefriervorgang; Gelierung
congelation point Gefrierpunkt, Erstarrungspunkt, Stockpunkt
congested gedrängt, verstopft
congested traffic dichter Verkehr
congestion Blockierung (621.395)
congestion tone Wegebesetztton
congruence Kongruenz
congruent kongruent
congruent triangle kongruentes Dreieck
conic(al) konisch, kegelförmig
conical beam Strahlenkegel (621.386)
conical horn Konustrichter
conical horn antenna Hornstrahler, Trichterstrahler
conic section Kegelschnitt (51)
conical scan konische Abtastung
conics (pl.) Lehre von den Kegelschnitten
coniform kegelförmig
conjugate Konjugierte (51); konjugierter Widerstand (621.3)
conjugate of a complex harmonic voltage konjugierter Spannungszeiger
conjugate konjugiert
conjugate-attenuation konjugiert komplexe Dämpfung
conjugate branches entkoppelte Zweige
conjugate complex konjugiert komplex
conjugate complex attenuation coefficient konjugiert komplexes Dämpfungsmaß
conjugate complex number konjugiert komplexe Zahlen
conjugate impedances konjugierte Widerstandsoperatoren, konjugiert komplexer Widerstand
conjugate transfer coefficient konjugiert komplexes Übertragungsmaß
conjunction Zusammenhang; Konjunktion (52)
to connect anschließen, verbinden, schalten; kuppeln, zusammenfügen; vermitteln (621.395)
to connect in cascade in Kaskade schalten
to connect to earth an Erde legen, erden
to connect in parallel parallel schalten
to connect in series in Reihe schalten, hintereinanderschalten, in Serie schalten
to connect through durchverbinden

to connect up anschließen, anschalten (**to:** an)
to connect up as in diagram nach Schaltbild schalten
to connect with in Eingriff bringen mit; durchverbinden, verbinden mit, anschließen an (621.395)
connected load Anschlußwert
connected in mesh in Dreieck geschaltet, dreieckgeschaltet
connected in parallel im Nebenschluß
connected position Betriebsstellung (bei metallgekapselter Schaltanlage)
connecting Anschluß-...
connecting block Anschlußleiste, Schaltleiste, Klemmleiste
connecting flange Anschlußflansch
connecting line Anschlußgleis (625); Anschlußleitung (621.3)
connecting link Anschlußstück, Bindeglied, Zwischenglied
connecting matrix Koppelvielfach (621.395)
connecting network Koppelanordnung
connecting piece Zwischenstück, Anschlußstück
connecting plug Verbindungs-Stöpsel
connecting point Anschlußpunkt
connecting rod Schubstange, Pleuelstange
connecting screw Anschlußschraube
connecting terminal Anschlußklemme
connecting wire Anschlußdraht, Anschlußleitung, Verbindungsleitung, Verbindungsdraht
connection Verbindung, Anschluß (621.3)
connection bar Verbindungsschiene (621.3)
connection board Anschlußplatte
connection box Anschlußdose, Anschlußkasten
connection cable Anschlußkabel
connection of a call Vermittlung eines Gesprächs
connection diagram Schaltplan, Leitungsplan
connection panel Anschlußfeld
connection plate Verbindungslasche, Knotenblech
connection strap Verbindungslasche
connection stub Anschlußstutzen
connection system Schaltungsart, Schaltart
connector Stecker, Gerätestecker, Steckverbinder
connexion = connection
consecutive aufeinanderfolgend, fortlaufend
consecutive computer Sequentiellrechner
consecutive phases aufeinanderfolgende Phasen
consequent konsequent, folgerichtig
consequent pole unbewickelter Pol
consequent pole generator (Generator, bei dem nur jeder zweite Pol bewickelt ist)
consequent pole motor (Motor, bei dem nur jeder zweite Pol bewickelt ist)
consequent pole winding polumschaltbare Wicklung
conservation Erhaltung, Einsparung
conservation of energy Erhaltung der Energie
conservation of momentum Impulserhaltung
conservative estimate vorsichtige Schätzung
consistent (miteinander) verträglich, vereinbar; zusammenhängend, fortdauernd; einheitlich; konsistent
consistent grease Staufferfett
Consol beacon Consol Drehfunkfeuer
console Bedienungspult, Konsole
console command processor Konsolen-Befehls-Verarbeiter
consonance Konsonanz, Einklang, Harmonie
constancy Konstanz
constant Konstante (51); Schaltelement, Konstante des Stromkreises (621.3)
constant of entropy equation Boltzmannsche Konstante
constant of integration Integrationskonstante
constants of the line Leitungskonstanten
constants primary spezifische Konstanten
constant konstant (**in:** hinsichtlich); ständig
constant bearing course Vorhaltekurs (621.396.93)
constant current konstanter Strom, Dauerstrom
constant current charge Ladung mit konstantem Strom (621.355)
constant current dynamo Konstantstromgenerator
constant current equivalent circuit Kurzschlußersatzschaltbild (621.385)
constant current generator Konstantstromgenerator, Generator mit eingeprägtem Strom
constant current motor Thurymotor, Konstantstrommotor
constant current regulation Konstantstromregelung
constant current street lighting system Straßenbeleuchtung mit konstantem Strom
constant current system Thurysystem
constant current transformer Boucherot-Transformator
constant current welding Konstantstromschweißung

constant frequency drive Frequenzsteuerung (621.396.61)
constant load konstante Last, Dauerbelastung
constant luminance principle Prinzip der konstanten Leuchtdichte
constant in magnitude der Góße nach konstant
constant-potential acceleration Gleichspannungs-Beschleuniger
constant-potential dynamo Konstantspannungsgenerator
constant-potential generator Konstantspannungsgenerator
constant-potential transformer Transformator mit konstanter Primärspannung
constant power Dauerleistung, konstante Leistung
constant-power generator Konstantleistungsgenerator
constant-power welding Konstantleistungsschweißung
constant pressure konstanter Druck
constant-speed motor Motor mit konstanter Drehzahl
constant-temperature oven Thermostat
constant-time element UMZ-Relais, Zeitelement
constant time lag stromunabhängige Verzögerung
constant voltage Dauerspannung, konstante Spannung
constant-voltage charge Ladung mit konstanter Spannung (621.335)
constant-voltage dynamo Konstantspannungsmaschine
constant-voltage equivalent circuit Leerlaufersatzschaltbild
constant-voltage generator Konstantspannungsgeber
constant-voltage regulator Spannungskonstanthalter, Spannungsregler
constant-voltage welding Konstantspannungsschweißung
at constant volume bei konstantem Volumen
constantan Konstantan
constituent Bestandteil, Schaltelement
constituent zugehörig, Bestand-...
constituent crystal kristallinischer Bestandteil
constituent matter Bestandteil; Materie
constituent particle Bauteilchen (539.15)
constituent substance Bestandteil
constitution Aufbau, Konstitution; Verfassung
constitution diagram Zustandsschaubild, Zustandsdiagramm
constitutional Zustands-...
constitutional formula Konstitutionsformel, Strukturformel (54)
to constrict verengen, einziehen
constriction Verengung
to construct konstruieren (nur 51); bauen, aufbauen
construction Konstruktion (nur 51); Bau, Aufbau
under construction in Bau, im Bau
construction supervision Bauaufsicht
construction weight Baugewicht
constructional Bau-..., Konstruktions-...
constructional detail Konstruktionsdetail
constructional feature Konstruktionsmerkmal
constructional member Bauelement, tragender Bauteil
constructional sections Bauteile, Konstruktionselemente
constructional work Baukonstruktion
consulting engineer beratender Ingenieur
to consume verbrauchen, aufnehmen
to consume current Strom aufnehmen, Strom verbrauchen
to consume power Leistung verbrauchen, Leistung abnehmen
to consume voltage Spannung verbrauchen, Spannung aufnehmen
consumer Verbraucher, Abnehmer
consumer appliance Konsumgüter (pl.)
consumer's installation Hausinstallation, Verbraucheranlage, Abnehmeranlage
consumer's terminals Verbraucherklemmen, Übergabeklemmen
consumption Verbrauch
consumption of current Stromverbrauch, Stromentnahme
consumption of energy Energieverbrauch
consumption of fuel Brennstoffverbrauch
contact Kontakt, Verbindung; Berührung; Kontaktkopie, Kontaktabzug (77)
contact area Kontaktfläche, Berührungsfläche
contact arm Kontaktarm
contact-bank assembly Kontaktfeld
contact blade Kontaktmesser
contact bounce Kontaktprellung
contact breaker Stromunterbrecher, Schalter
contact brush Kontaktbürste
contact burn Kontaktabbrand
contact chatter Kontaktprellen
contact clip Kontaktwange (621.316.5)
contact conductor Kranleitung (621.34), (621.8); Kontaktelektrode (621.357)
contact current-closing rating Nenneinschaltstrom
contact disk Kontaktscheibe

contact drop Kontaktspannungsabfall
contact duty Kontaktbelastung
contact electricity Kontaktspannung, Voltapotential
contact electrode Elektrode, Stromzuführung (66)
contact e. m. f. Kontaktspannung, Voltapotential
contact finger Kontaktfinger, Kontaktstift
contact float Kontaktspiel
contact follow Kontaktmitgang, Kontaktmitnahme
contact force Kontaktdruck, Kontaktkraft
contact-interrupting rating Nennabschaltleistung
contact jaw Kontaktbacke
contact lever Kontakthebel, Taste
contact line Berührungslinie
contact maker Kontaktgeber, kontaktgebendes Glied
contact material Kontaktmaterial
contact member Schaltstück
contact microphone Körperschallmikrophon, Kontaktmikrophon
contact noise Kontaktrauschen, Kontaktgeräusch; Kratzgeräusche (621.395)
contact piece Kontaktstück
contact pin Kontaktstift, Sockelstift (621.385)
contact pitting Kontaktabbrand
contact plate Kontaktstück, Kontaktklotz
contact plating Tauchgalvanisierung
contact point Berührungspunkt, Auflagepunkt; Kontaktpunkt
contact position Kontaktstellung
contact potential Kontaktpotential
contact pressure Kontaktdruck
contact-pressure movement Durchdruck (621.316.5)
contact rail Kontaktschiene, Stromschiene
contact-rail anchor Stromschienenfestpunkt
contact rating Kontaktbemessung, zulässige Kontaktbelastung
contact resistance Kontaktwiderstand, Übergangswiderstand
contact roller Kontaktroller
contact screw Kontaktschraube
contact segment Kontaktsegment, Schaltstück
contact separation Kontakttrennung; Kontaktabstand
contact series Spannungsreihe
contact shoe Stromschienenschuh, Stromschuh
contact spring Kontaktfeder, Andrückfeder
contact stress Flächenpressung
contact stud Kontaktstift, Kontaktklotz
contact surface Berührungsfläche, Kontaktfläche (621.3)
contact time Kontaktdauer
contact travel Kontaktweg, Schaltweg
contact voltage Kontaktspannung; Berührungsspannung
contact-wear allowance zulässige Kontaktabnutzung
contact wire Fahrdraht, Fahrleitung (621.33)
contactor Schütz, Schaltschütz
contactor control Schützensteuerung
contactor controller Schützensteuerwalze
contactor panel Schalttafel, Schützenfeld
contactor starter Anlaßschütz
contactor-type reversing control Reversierschützensteuerung
contactor-type star-delta starter Schütz-Sterndreieckschalter
contactor-type starting control Anlaßschützensteuerung
to contain enthalten; sich (ohne Rest) teilen lassen durch... (51)
to contain an angle einen Winkel einschließen
container Behälter
containing ... -haltig
containing cell Akkumulatorenglas, Sammlergefäß
to contaminate verunreinigen; verseuchen (539)
contamination Verunreinigung, Verschmutzung; Verseuchung (539)
content Gehalt, Inhalt, Fassungsraum
contents (pl.) Inhalt, Inhaltsverzeichnis; Innenausrüstung, Inneneinrichtung
contents gauge Flaschendruckmanometer
to continue fortsetzen, weiterführen, verlängern
to continue in force in Kraft bleiben
continuity Kontinuität, Zusammenhang; Durchgang (621.3)
continuity cable bond Muffenverbinder (621.315.2)
continuity check Durchgangsprüfung, Kontinuitätsprüfung
continuity checker Leitungsprüfer
continuity fitting Rohrverbinder (696)
continuity rail bond Schienenlängsverbinder, Weichenüberbrückung
continuity test Prüfung auf Durchgang
continuous laufend, durchlaufend, fortlaufend; dauernd, stetig, Dauer..., kontinuierlich, ununterbrochen
to be continuous durchlaufen, entlanglaufen, anhalten
continuous belt furnace Durchlaufofen

continuous colour sequence = CCS kontinuierliche Farbfolge
continuous control stetige Steuerung, stetige Regelung
continuous controller action kontinuierliche Reglerwirkung
continuous current Gleichstrom (siehe auch **d.c.**)
continuous current generator Gleichstromgenerator
continuous current locomotive Gleichstromlokomotive
continuous current meter Gleichstromzähler, Gleichstrommeßinstrument
continuous current motor Gleichstrommotor
continuous current plant Gleichstromanlage
continuous current traction Gleichstrombahnbetrieb
continuous current transformer Gleichstrommotorgenerator
continuous current voltage Gleichstromspannung
continuous-curve distance-time protection Distanzschutz mit steiler Auslösekennlinie
continuous duty DB = Dauerbetrieb
continous duty with intermittend loading Dauerbetrieb mit Aussetzbelastung
continous duty with short-time loading Dauerbetrieb mit Kurzzeitbelastung
continuous electrode Söderberg-Elektrode, selbstbackende Elektrode
continuous earth Dauererdschluß
continuous form Endlosformular
continuous forward current Dauergleichstrom (621.382.2)
continuous forward off-state voltage Gleichsperrspannung in Vorwärtsrichtung
continuous function stetige Funktion
continuous integration Dauerzählung (621.317)
continuous load Dauerbelastung, Dauerlast
continuous loading gleichmäßige Bespulung, gleichmäßige Pupinisierung
continuous loop recorder Magnetbandgerät mit endlosem Band
continuous operation Dauerbetrieb, Dauerbelastung (621.3)
continuous output Dauerleistung
continuous path control Bahnsteuerung, kontinuierliche Bahnsteuerung (621.396.93), (621.9)
continuous progressive code einschrittiger Code
continuous rating Nenndauerlast, DB-Last
continuous reverse voltage Gleichsperrspannung in Rückwärtsrichtung, Rückwärtsgleichspannung (621.382.2)
continuous rheostat stetig regelbarer Widerstand
continuous running Dauerlauf, Dauerbetrieb
continuous signal Dauerkennzeichen (621.395)
continuous spectrum kontinuierliches Spektrum
continuous speed control stufenlose Drehzahlregelung
continuous test Dauerbelastungsprobe (621.35)
continuous tube massives Rohr, vollwandiges Rohr
continuous use Dauerbenutzung
continuous wave Dauerstrich, ungedämpfte Welle (621.396)
continuous-wave radar pulsfreies Rückstrahlverfahren, Dauerstrichradar
continuous weld durchlaufende Naht, durchlaufende Schweißnaht
continuous winding durchlaufende Wicklung, eingängige Wicklung
continuously adjustable stetig veränderlich, stetig regelbar
continuously adjustable inductor Variometer
continuously adjustable resistor stetig regelbarer Widerstand
continuously integrating meter Zähler für durchgehende Zählung
continuum Kontinuum
to contort verzerren, verformen
contortion Verzerrung, Verformung
contour Kontur, Umriß
contour integral Umlaufintegral
contour line Höhenschichtlinie, Schichtlinie
to contract (sich) kontrahieren, (sich) verengen, (sich) zusammenziehen; (sich) zuziehen
contractible zusammenziehbar
contraction Kontraktion, Zusammenziehung, Zusammendrückung
contraction crack Schrumpfriß
contraction strain Schrumpfspannung
contraction in volume Volumabnahme, Volumkontraktion
contractor Vertragsinstallateur; Dynamik-Presser (621.396.61)
contraption Klapperatismus
contra-rotating field gegenläufiges Drehfeld
contrast Kontrast
contrast amplifier Dynamik-Verstärker (621.396.62)
contrast gradient Übertragungsexponent (621.397)
contrast photometer Kontrastphotometer

contrast range Maximal-Kontrast
contrasty kontrastreich, hart (77)
to contribute beitragen, liefern; übergeben, abgeben
to contribute current to Strom abgeben an
to contribute energy Energie abgeben, Energie liefern, Energie zuführen
contrivance Vorrichtung, „Dreh"
control Steuerung; Regelung, Betätigung, Bedienung; Bedienungsknopf; Kontrolle, Aufsicht, Beaufsichtigung
control accuracy Regelungsgenauigkeit
control action Steuerwirkung
control board Steuertafel, Steuerschalttafel
control bus Steuerbus; Betätigungsschiene
control cable Steuerseil, Bowdenzug; Steuerkabel, Betätigungskabel
control channel Steuerkanal
control character Steuerzeichen (681.3)
control characteristic Zündkennlinie; Steuerkennlinie, Regelkennlinie (621.316.7)
control circuit Steuerstromkreis, Steuerschaltung
control-circuit transformer Steuerkreisumspanner
control column Steuersäule (629.13)
control console Bedienungspult, Bedienungstisch
control counter Befehlszähler, Befehlzählregister (681.3)
control crystal Steuerquarz
control cubide Regieraum (654.19)
control current Steuerstrom
control cut-out switch Steuerstromunterbrecher
control desk Steuerpult, Schaltpult, Bedienungspult, Übcrwachungspult
control deviation Regelabweichung, Regelfehler
control device Regelgerät
control-differential transmitter Differentialgeber, Steuerdifferentialgeber
control-discrepancy switch Steuerquittungsschalter
control electrode Steuerelektrode (621.385)
control element Regelelement, Steuerorgan
control engineering Regeltechnik
control equipment Regeleinrichtung
control error Regelfehler
control exchange Leitvermittlung (621.395)
control factor Regelfaktor
control gate Schütz (626)
control gear Steuergetriebe, Steuergerät, Steuergestänge
control generator Steuergenerator
control grid Steuergitter (621.385)
control-grid-plate transconductance Steilheit (621.385)
control by guide beam Leitstrahlführung
control handle Stellknebel
control-impulse store Kommandospeicher (621.398)
control initiation Kommandogabe
control input Führungsgröße (621.316.7)
control lead Steuerleitung
control lever Steuerhebel, Stellhebel
control light Kontrollampe
control limit switch Grenzstromsteuerschalter
control line Steuerleitung, Betätigungsleitung
control loop Regelkreis
control magnet Steuermagnet
control mark Überwachungszeichen
control mechanism Steuergestänge, Steuereinrichtung, Steuerung; Regelmechanismus
control order Steuerbefehl
control overlay transparenter Bildschirmvorsatz (621.396.96)
control panel Steuerschalttafel, Steuertafel, Warte
control point istwert, Regelpunkt
control-point transmitter istwertgeber
control power Steuerleistung
control probe Steuersonde (539)
control range Regelbereich, Steuerbereich
control rate Regelgeschwindigkeit
control register Befehlsregister, Steuerregister (681.3)
control relay Steuerrelais
control rod Regelstab, Regelstange (539)
control room Bedienungswarte
control scheme Steuerschema, Steuerverfahren
control set Regelsatz
control sheet Meldebogen, Kontrollzettel
control signal Stellbefehl, Regelbefehl (621.316.7); Kommando (621.398); Steuersignal (681.3)
control sequence Steuerungsablauf, Befehlsfolge (681.3)
control switch Steuerschalter, Betätigungsschalter
control synchro Steuerdrehmelder
control system Steuerverfahren
control tower Kontrollturm, Flugdienstleitung
control-transfer instruction Sprungbefehl (681.3)
control transmitter Steuergeber
control unit Steuereinrichtung, Steuergerät; Steuereinheit; Steuerwerk (681.3)

control valve Steuerventil, Regelventil
control voltage Steuerspannung, Betätigungsspannung; Regelspannung
control winding Steuerwicklung (621.318)
control wiring Steuerleitungen (pl.)
to control steuern; regeln; überwachen, kontrollieren; aussteuern, ansteuern
controlled to half value halb ausgesteuert
controllability Steuerfähigkeit
controllable steuerbar, regelbar
controlled gesteuert, geregelt
controlled avalanche rectifier Lawinen-Gleichrichterdiode mit eingegrenztem Durchbruchsbereich
controlled carrier (modulationsabhängig) gesteuerter Träger
controlled condition Regelgröße
controlled continuous process stetiges Regelverfahren
controlled mercury-arc rectifier gesteuerter Quecksilberdampfgleichrichter
controlled quantity Regelgröße
controlled rectifier gesteuerter Gleichrichter, gittergesteuerter Gleichrichter
controlled speed axle generator drehzahlgesteuerter Achsdynamo (625)
controlled system Regelsystem
controlled transmitter Sender mit Steuerstufe
controlled variable geregelte Größe
controller Kontroller, Steuerschalter, Steuerwalze; Regler; Kontrolleur, Prüfer; Steuergerät, Regeleinrichtung
controller construction diagram Fahrschalter-Schaltbild
controller mechanism Steuereinrichtung
controller notch Schaltstufe
controller output Stellgröße
controller position Kontrollerstellung, Schaltwalzenstellung
controller regulator automatischer Anfahrschalter
controlling Steuer-..., Regel-...
controlling agent Steuerflüssigkeit
controlling contact Überwachungskontakt
controlling device Steuereinrichtung
controlling element Regelorgan
controlling mechanism Steuereinrichtung
controlling quantity Steuergröße, Stellgröße
controlling section Steuerabschnitt
controlling torque Richtmoment (621.317)
controlling unit Regelglied
convection Konvektion
convection circuit breaker Konvektionsschalter
convection cooling Konvektionskühlung
convection current Konvektionsstrom, Leitungsstrom
convection stream elektrischer Wind, Ionenwind
convective Konvektions-...
convective discharge Entladung durch Konvektionsströme
convector Konvektionsofen (621.3)
convector circuit breaker Konvektorschalter
convenience outlet Steckdose
convenience receptacle Steckdose, Gerätesteckdose
to converge konvergieren (51); zusammenlaufen
to converge at zusammentreffen bei, sich schneiden bei
to converge upon konvergieren nach
convergence, convergency Konvergenz (51)
convergence electrode Konvergenzelektrode
convergent konvergent
conversation counter Gesprächszähler
conversation time Gesprächsdauer
conversational mode Dialogbetrieb (621.395), (681.3)
converse piezo-electric effect reziproker piezoelektrischer Effekt
conversion Umformung, Umsetzung, Konvertierung, Umwandlung; Umrechnung; Umwicklung (621.313); Mischung (621.396)
conversion efficiency Wirkungsgrad der Umformung
conversion of energy Energieumwandlung
conversion figure Umrechnungszahl
conversion gain Mischverstärkung
conversion loss Mischdämpfung
conversion mutual conductance Mischsteilheit
conversion table Umrechnungstafel
conversion transconductance Mischsteilheit
to convert umformen, umsetzen, umwandeln (**into**: in, zu); verwandeln, überführen, (**into**: in); umschalten; umrechnen (**to**: auf)
converter Umformer (621.313); Bessemerbirne, Konverter (66); Mischstufe, Überlagerer, Umsetzer (621.396); Wandler (681.3)
converter fuel element Konverterbrennelement
converter reactor Konversionsreaktor
converter tube Mischröhre
converter valve Mischröhre
converting plant Umformerwerk (621.3)
converting station Umformerstation, Umformerwerk
convex konvex
convex lens Konvexlinse, Sammellinse
to convey befördern, tragen; übertragen, vermitteln, übermitteln

to convey current from Strom abführen von
to convey current to Strom zuführen zu
to convey movement to Bewegung übertragen auf, etwas in Bewegung setzen
conveyance Beförderung, Transport
conveyor Förderer, Förderanlage, Beförderungseinrichtung; Transportband
conveyor belt Förderband
conveyor furnace Förderbandofen
conveyor screw Transportschnecke, Förderschnecke
conveyor trough Transportrinne
convolution Wickel, Windung
to cook kochen; tempern (621.314.6)
cooker Kocher, Kochgerät, Küchenherd
cooking appliance Kocher, Kochgerät
to cool kühlen, abkühlen
to cool down abkühlen, herunterkühlen (**to:** auf)
to cool off abkühlen
cooled anode valve Senderöhre mit Rippenkühlung; Senderröhre mit Anodenkühlung
coolant Kühlmittel
coolant pump Kühlpumpe
cooler Kühler (621)
cooler element Kühlelement
Coolidge tube Coolidge-Röhre (621.386)
Coolidge radiator tube Coolidge-Röhre mit Rippenkühlung
cooling Kühlung, Kühl-...
cooling chamber Kühlraum
cooling collar Kühlflansch
cooling fin Kühlrippe
cooling jacket Kühlmantel
cooling oil Kühlöl
cooling period Abkühlzeit, Auskühlzeit
cooling surface Kühlfläche, abkühlende Oberfläche
cooling tank Kühltank
cooling tower Kühlturm, Kühlkolonne
cooling tunnel Kühlofen
cooling vane Kühlfahne
cooling velocity Kühlgeschwindigkeit, Abkühlungsgeschwindigkeit
cooling vessel Kühltank
cooling water Kühlwasser
Cooper-Hewitt lamp Quecksilberdampflampe
Cooper-Hewitt rectifier Quecksilberdampfgleichrichter
co-ordinate Koordinate; Koordinierte
co-ordinate geometry analytische Geometrie
co-ordinate grid reference system Quadratnetzverfahren mit Bezugskoordinaten
to co-ordinate zuordnen, beiordnen
co-ordinated insulation abgestufte Isolation
co-ordinate beigeordnet, koordiniert
co-ordinate potentiometer Wirk- und Blindkompensator
copaline (Isoliermasse für H.F.-Kabel)
cophasal gleichphasig
co-phase gleichphasig
co-phase component Wirkkomponente, gleichphasige Komponente
coplanar in der gleichen Ebene liegend, koplanar
copper Kupfer
copper arsenide Arsenkupfer
copper-bearing kupferhaltig
copper brush Kupferbürste
copper can kupferner Abschirmbecher, kupferner Abschirmtopf
copper clad kupferummantelt, mit Kupferauflage
copper-clad iron alloy wire Kupfermanteldraht (621.385)
copper-clad laminate kupferkaschierte Leiterplatte
copper-clad steel Kupfermantelstahl, Staku
copper-clad steel conductor Stahlkupferleiter (621.315.1)
copper-clad wire Kupfermanteldraht (621.385); Stakudraht (621.315)
copper conductor Kupferleiter
copper contact Kupferkontakt
copper-cored carbon Kupferdochtkohle
copper damper Wirbelstromdämpferscheibe
copper-eutectic solder eutektisches Kupferlot
copper factor Füllfaktor (621.313)
copper loss Kupferverlust (621.3)
copper-oxide rectifier Kupferoxidulgleichrichter
copper-strip ground conductor Banderder
copper sulphate Kupfersulfat
copper tape Kupferband
copper tube Kupferrohr
copper voltameter Kupfervoltameter
copperweld (steel conductor) Stahlkupferdraht; Stakudraht
copper welding Kupferschweißung
copper wire Kupferdraht
to copper-plate verkupfern
copper plating galvanische Verkupferung
copper kupfern
copperas Vitriol, Kupfervitriol
coppery kupferig
copying machine Kopiergerät

cord Schnur, Kabel, Verbindungskabel; Strick, (dünnes) Seil; Kordstoff; (Holzmaß = 128 Kubikfuß = 3,63 cbm)
cord adjuster Zugpendel
cord circuit Schnurschaltung, Schrankstromkreis
cord-circuit repeater Schnurverstärker
cord gear Schnurzug
cord grip Fassungsnippel, Leitungsschelle, Zugentlastungsschelle
cord shortener Schnurbehälter (621.32)
cordless board schnurloser Schrank (621.395)
core Kern, Gußkern (621.7); Kernhaus; Kern (621.314.2); Seele (621.791), (67); Kern (52), (53); Leiterseil, Litze, Ader (621.3); Spaltzone (539)
core array Kernmatrix
core balance Quotientenwerk (621.317); Differentialschutz (621.316.9)
core-balance protective system Fehlerstromschutzsystem, Erdstromunsymmetrieschutz
core conductor Leiterseele, Mittelleiter
core cross section Kernquerschnitt (621.3)
core diameter Kerndurchmesser
core flux Kernfluß, Hauptfluß (621.314.2)
core lamination Kernblech
core loss Eisenverlust (621.313)
core matrix Kernmatrix
core memory Kernspeicher (51)
core of network Kernbereich (621.395)
core plate Kernblech (621.314.2)
core-plated punches papierisoliertes Dynamoblech
core screening Aderabschirmung
core store Kernspeicher (681.3)
core-type induction furnace Induktionsofen, Niederfrequenzofen
core-type transformer Kerntransformator
to core Kerne einlegen (621.7); mit einem Kern versehen (621.3)
cored mit Kern
cored electrode umhüllte Elektrode (621.791)
cored hole eingegossenes Loch
cored solder Lötdraht (mit Flußmittel)
coreless kernlos, ohne Kern
coreless armature Ankerwicklung ohne Eisenkern
coreless-type induction furnace kernloser Induktionsofen, HF-Induktionsofen
cork Kork
corkscrew Korkenzieher
corkscrew rule Korkenzieherregel
corner Ecke, Eckpunkt (51)
corner bracing Eckversteifung, Eckstrebe
corner of a bridge Brückeneckpunkt (621.317)
corner frequency Eckfrequenz
corner iron Winkeleisen
corner reflector antenna Winkelreflektorantenne
corner rounding Eckenabrundung
cornice Gesims, Simsband
cornice lighting Voutenbeleuchtung
corona Korona-..., Glimm-..., Sprüh-...
corona attenuation Koronadämpfung
corona discharge Koronaentladung, Glimmentladung; Sprühverlust
corona (discharge) current Koronastrom
corona loss Koronaverlust
corona protected mit Sprühschutz versehen
corona protection Sprühschutz
corona tuft discharge Sprühbüschel
coronograph Koronograph
corposant St. Elms-Feuer
corpuscle Teilchen, Körperchen, Korpuskel
corpuscle of radiation Strahlungsquant, Photon
corpuscular korpuskular, aus Teilchen bestehend
copuscular radiation Korpuskularstrahlung
to correct korrigieren, verbessern, berichtigen; ausgleichen; Korrektur lesen
to correct for Korrekturen anbringen für
to correct readings Ablesungen korrigieren
to correct barometer readings den abgelesenen Barometerstand auf NN reduzieren
correct richtig
correct match richtige Anpassung, saubere Anpassung (621.396)
correctly designed richtig bemessen
correcting element Stellglied (621.316.7)
correcting motor Stellmotor (621.316.7)
correcting network Ausgleichsschaltung (621.3); Entzerrerschaltung (621.39)
correction Korrektur, Verbesserung, Berichtigung
correction curve Korrekturkurve; Funkbeschickungskurve (621.396.93)
correction of error Fehlerausgleich
correction factor Korrekturfaktor
correction term Korrektionsglied
correction time Ausregelzeit
corrective berichtigend, Korrektur-...
corrective apparatus Phasenschieber (621.3)
corrective maintenance instandsetzende Unterhaltung
corrective network Ausgleichsschaltung (621.395)

corrector Entzerrer (621.39)
corrector circuit Entzerrerschaltung
correlate Korrelierte (51); Korrelat
to correlate korrelieren, aufeinander abstimmen; verknüpft sein, in Beziehung stehen (**to:** zu)
correlation Vergleich, Beziehung, Korrelation
correlation analysis Korrelationsrechnung
correlation coefficient Korrelationskoffizient (51)
to corrode korrodieren, rosten
corrodible korrodierbar
corrosion Korrosion
corrosion fatigue Korrosionsermüdung
corrosion medium Korrosionsmittel
corrosion-proof korrosionsbeständig, korrosionsfest
corrosion protect Korrosionsschutz
corrosion-resistant korrosionsfest, korrosionsbeständig
corrosive korrodierend, Korrosions-...
corrosive agent Korrosionsmittel, angreifende Substanz
corrosive sublimate Quecksilbersublimat
corrugated gewellt, Well-...
corrugated cardboard Wellpappe
corrugated flue tube Wellrohr, gewelltes Flammrohr
corrugated sheet Wellblech
corrugated tubing Wellschlauch
corrugation Wellung, Welle; Rippe
corundum Korund, Korundum
cos = cosine cos = Cosinus
cosec = cosecant $1/_{\sin}$ = Cosecans
cosecant Cosecans
cosecant squared aerial $1/_{\sin}{}^2$-Antenne, Cosec-Quadrat-Antenne
cosech = hyperbolic cosecant $1/_{\sin}$ = Hyperbelosecans
cosh = hyperbolic cosine cosh = Hyperbelcosinus
cosine Cosinus
cosine directivity Cosinus-Charakteristik (621.396)
cosine law Lambertsches Cosinusgesetz
cosine rule Cosinussatz (51)
cosmic kosmisch
cosmic radiation kosmische Strahlung
cosmic rays kosmische Strahlen; kosmische Strahlung (sing.)
Cosmotron Kosmotron
cost estimation Kostenschätzung
cot = cotangent cot = Cotangens
cotangent Cotangens
coth = hyperbolic cotangent coth = Hyperbelcotangens
cotter Schließkeil, Keil
cotter pin Vorsteckstift, Splint
cotter way Keilschlitz, Keilnut
to cotter-pin verstiften
cotton Baumwolle; Watte (54)
cotton-braided covering Baumwollumklöppelung
cotton-covered baumwollisoliert, baumwollumsponnen
cotton-covered wire baumwollisolierter Draht
cotton covering Baumwollumspinnung
cotton waste Putzwolle
cotton wrapping Baumwollumspinnung
coulisse Kulisse (621)
coulomb Coulomb
Coulomb repulsion Coulomb(sche) Abstoßung
coulombmeter Amperestundenzähler, Ah-Zähler
coumaric resin Kumaronharz
count Zählung
to count zählen; Bedeutung haben, Wert besitzen
to count down rückwärts zählen
to count up vorwärts zählen
counter Zähler, Zählwerk; Zählrohr
counter-frequency meter Digitalfrequenzmesser, Frequenzzähler
counter friction Zählwerksreibung (621.317)
counter range Anlaufgebiet (539)
counter registration Zählwerksanzeige (621.317)
counter Gegen-...
counter admittance Gegenadmittanz (621.392)
counter balance Gegengewicht
counter brake Gegenbremse, Rücklaufbremse
counter cell Gegenzelle, gegengeschaltete Regelzelle (621.355)
counterclockwise gegen den Uhrzeigersinn, links herum
counter-compound winding Gegenkompoundwicklung (621.313)
counter contact Gegenschaltstück
counter-current Gegenstrom-...
counter-current flow Gegenstrom
countercurrently im Gegenstrom
counter-direction drive gegenläufiger Antrieb (621.316.5)
counter electromotive force gegenelektromotorische Kraft
counter e.m.f. Gegen-EMK
counterflow Gegenstrom
counter impedance Gegenimpedanz (621.392)

counterpoise Gegengewicht (621.396)
counter pressure Gegendruck
countersunk(-head) rivet Senkkopfniet
countervoltage Gegenspannung
counterweight Gegengewicht
to counterbalance aufwiegen; auswiegen; auswuchten; entgegenwirken
to counterconnect gegenschalten
to counterpoise auswiegen, austarieren, durch Gegengewichte ausbalancieren
to countersink versenken, senken
counting Zähl-...
counting circuit Impulszählschaltung
counting-down device Frequenzteilerschaltung, Untersetzungsschaltung
counting mechanism Zählwerk
counting pulse Zählimpuls
counting rate Zählgeschwindigkeit
counting relay Zählrelais
counting train Rollenzählwerk
counting tube Zählröhre
country code Landeskennzahl (621.395)
couple Kräftepaar; Paar; Zugpaar (625)
to couple kuppeln (**through:** durch, vermittels); koppeln (621.396)
to couple out energy Leistung auskoppeln
to couple from auskoppeln aus (621.396)
to couple rigidly starr kuppeln (621.1)
to couple up ankuppeln, zusammenkuppeln
to couple up with ankuppeln an
coupled gekuppelt; gekoppelt, verknüpft (621.396)
coupled cicuit angekoppelter Kreis
coupled impedance transformierter Widerstand, angekoppelter Widerstand
coupled resistance eingekoppelter Widerstand
coupled switch gekuppelter Schalter
coupler Steckvorrichtung, Kupplung, Gerätesteckvorrichtung (621.3)
coupler plug Gerätestecker, Kupplungsstecker
coupler socket Gerätesteckdose, Kupplungsdose
couplet Paar, Antennenpaar (621.396)
coupling Kupplung (625); Kopplung (621.396); Schlauchkupplung
coupling capacitor Kopplungskondensator, Koppelkondensator
coupling coefficient Kopplungskoeffizient, Kopplungsgrad, Kopplungsfaktor
coupling condenser Gitterblock (621.396.64); Koppelkondensator
coupling effect Koppelwirkung
coupling element Kopplungsglied, Kopplungselement
coupling end Kupplungsseite, Antriebsseite
coupling factor Kopplungsfaktor
coupling impedance Kopplungswiderstand (621.392)
coupling loop Koppelschleife
coupling rod Kuppelstange
coupling shaft Kupplungswelle, Kuppelwelle
coupling transformer Kopplungstransformator
course Kurs; Richtung
course bearing Kurspeilung
course calculator Kursrechner
course deviation Kursabweichung
course-deviation indicator Kursanzeiger
course-indicating radio beacon Leitstrahlfunkbake
course-laid endwise Läuferschicht
course of life Lebensdauer, Lebenszeit
course light Streckenfeuer (351.814.33)
course-line computer Kursrechner
cover Kappe, Deckel, Abdeckung, Umschlag, Haube
cover glass Deckglas (535)
cover plate Abdeckplatte, Deckel; Lasche, Decklasche (621.8)
cover plate and handle Schalterbetätigungsplatte
cover strip Deckstreifen, Deckleiste
with cover on bei geschlossenem Deckel
to cover decken, verkleiden, zudecken, umhüllen; umfassen, überstreichen; ausreichen für
to cover a range einen Bereich umfassen, einen Bereich überstreichen
coverage Bedeckung, Überdeckung; Inhalt, Gegenstand; Reichweite, Erfassungsbereich (621.396.96); Empfangsbereich (621.396); Auffaßreichweite (621.397)
coverage area Bedeckungsbereich
coverage of a radar set Auffaßbereich eines Funkmeßgerätes
coverage of reception Reichweite des Empfangs
covered umsponnen, isoliert (621.3); bedeckt, umhüllt
covered electrode umhüllte Elektrode (621.791); umhüllter Schweißstab
covered plant überdeckte Anlage
covered space zurückgelegter Weg
covered wire isolierter Draht
covering Abdeckung, Hülle, Verkleidung
cowling Abdeckung, Schornsteinaufsatz
c.p. = candle power Lichtstärke
CPM = critical path method

CPM = control program for microprocessors [spezielles Betriebssystem für Mikroprozessoren]
CPS = characters per second
CPU = central processing unit
crackle Knistern
crackle finish Reißlack
cradle Wiege; Motorwiege (621.313); Gabel (621.395.6)
cradle dynamometer Pendelgenerator
cradle guard Schutznetz (621.315.1)
cradle relay Kammelrelais
cradle switch Gabelumschalter, Hakenumschalter
cramp Krampe, Klammer
to cramp verklammern, klammern
crane Kran
crane controller Kransteuerschalter
crane driver Kranführer
crane hook Kranhaken
crane magnet Hubmagnet
crane motor Kranmotor
crane rating DAB = Dauer-Aussetzbetrieb (621.313)
crank Kurbel; Winkelhebel
crank axle Kurbelwelle, Triebachse (625)
crank case Kurbelgehäuse, Kurbelwanne
crank bearing Kurbellager
crank cheek Kurbelwange
crank drive Kurbeltrieb, Kurbelantrieb
crank driven mit Kurbelantrieb
crank gear Kurbeltrieb, Kurbelgetriebe
crank handle Handkurbel
crank inductor Kurbelinduktor
crank pin Kurbelzapfen
crank-pin box Pleuelkopf
crank plate Kurbelscheibe
crank shaft Kurbelwelle
to crank kröpfen; ankurbeln
cranked gekröpft, Kurbel-...
cranked handle Handkurbel
cranked lever Winkelhebel
crash Absturz, Zusammenstoß; Programmabsturz, Zusammenbruch (681.3)
crash recorder Flugschreiber (629.13)
crate Kiste, Lattenverschlag, Lattenkiste
to crate mit Latten verschlagen; in Kisten verpacken
crawling Lauf mit Schleichdrehzahl
crawling speed Schleichdrehzahl, Kriechgang (621.313)
creak Knarren, Quietschen
to crease knittern, brechen
creasy faltig, wellig
creep Kriechen, elastische Nachwirkung (620.1); Schlupf (621.8); Kriechen (66)
creep curve Kriechkurve
creep flashover Gleitfunkenüberschlag
creep limit Dauerstandfestigkeit
creep rate Kriechgeschwindigkeit, Wachstumsgeschwindigkeit (66)
creep recovery Kriecherholung
creep strain Kriechdehnung
creep strength, creep stress Dauerstandfestigkeit
to creep kriechen
creepage Hochkriechen (621.357)
creepage current Kriechstrom
creepage distance Kriechweg
creepage proof kriechstromfest
creeping Kriechvorgang, Kriechen (539); Spannungsvortrieb (621.317)
creeping discharge Gleitentladung
creeping distance Kriechweg (621.3)
crest oberer Rand, Gipfel, Spitze; Höchstwert, Scheitelwert; Wellenberg; Gewindespitze
crest amplitude höchste Spitze, größter Scheitelwert
crest factor Scheitelfaktor
crest line Scheitellinie
crest value Scheitelwert
crest voltage Scheitelwert der Spannung
crest voltmeter Scheitelspannungsmesser, Scheitelspannungsvoltmeter
crest working forward voltage Scheitelspannung in Vorwärtsrichtung
crest working reverse voltage Scheitelsperrspannung
crevasse (scharfe) Einsattelung, Resonanzloch (621.396)
crevice Spalte, Spalt, Hohlraum
to crimp umnieten, umbördeln, umkrempen; prägen, riffeln, wellen
criss-cross gekreuzt, kreuz und quer
to criss-cross kreuzschraffieren
criterion (pl.: **-ria**) Kriterium
criterion for the stability Stabilitätsbedingung, Stabilitätskriterium
critical kritisch
critical angle Grenzwinkel
critical coupling kritische Kopplung
critical damping aperiodische Dämpfung
critical frequency kritische Frequenz, Grenzfrequenz (621.396)
critical gradient Durchschlagfestigkeit
critical grid current Gitterzündstrom (621.385)
critical grid voltage kritische Gitterspannung, Gitterzündspannung (621.385)

critical magnitude kritische Größe
critical mass kritische Masse
critical path method Netzplantechnik
critical pressure kritischer Druck
critical rate of rise of off-state voltage kritische Spannungssteilheit (621.314.63)
critical rate-of-rise of on-state current kritische Stromsteilheit
critical reactor kritischer Reaktor
critical resistance Grenzwiderstand
critical silence Tonminimum
critical span Grenzspannweite, kritische Spannweite
critical state kritischer Zustand
critical temperature kritische Temperatur
critical velocity kritische Geschwindigkeit
critical voltage Anfangsspannung (621.315.6)
critical-voltage parabola Kennlinie einer Magnetfeldröhre
critically damped kritisch gedämpft, aperiodisch gedämpft, im aperiodischen Grenzzustand
CRO = cathode ray oscillograph K.O. = Kathodenstrahloszillograph
crocodile clip Krokodilklemme
Crooke's dark space Crookescher Dunkelraum
cross Kreuz; Kreuzung; Kreuzstück
cross-ampere turns Querfeldamperewindungen (621.313)
cross arm Traverse (621.315.1)
cross-assembler Cross-Assembler [Programmübersetzer zur Übersetzung eines Assemblerprogramms des Rechners A in die Maschinensprache des Rechners B]
cross-band transponder Antworteinrichtung mit Frequenzumsetzung
crossbar Querbalken, Querstange; Riegelbalken
crossbar concentrator Schalterkonzentrator
crossbar distributor Kreuzschienenverteiler
crossbar selector Koordinatorschalter (621.395)
crossbar switch Crossbar-Wähler
cross bearing Kreuzpeilung
cross bracing Verstrebung, Kreuzversteifung, Windverband
cross-breaking strength Biegefestigkeit
cross-checking Gegenprobe
cross-coil instrument Kreuzspulinstrument
cross compiler [übersetzt ein Programm in den Maschinencode eines anderen Rechners]
cross-connecting board Steckverteilertafel, Rangierfeld
cross connection Kreuzschaltung
cross-connection field Auskreuzfeld
cross coupling gegenseitige Störung
cross direction Querrichtung
cross fade Überblendung
cross feed Querkopplung, Übersprechen
cross field Querfeld, quergerichtetes Feld
cross-field dynamo Querfelddynamo
cross flux Querfeld (621.313)
cross guide Kreuzkoppler
cross hairs (pl.) Fadenkreuz (535)
cross-hairs diaphragm Fadenkreuzblende
cross handle Kreuzgriff
cross hatching Kreuzschraffur
cross-induction field Ankerquerfeld
cross-jet circuit breaker Querstromschalter (621.315.5)
cross level Querlageanzeiger (629.3); Libelle
cross modulation Kreuzmodulation, nichtlineares Nebensprechen (621.396)
cross multiplication Überkreuz-Multiplizieren
cross neutralization Gegentaktneutralisation
crossover Überschneidungspunkt, Kreuzungspunkt
crossover frequency Übergangsfrequenz
crossover network Frequenzweiche
crossover point Überschneidungspunkt, Kreuzungspunkt
crosspoint Kreuzungspunkt; Koppelelement (621.39)
crosspoint matrix Koppelfeld
cross pointer Kreuzzeiger
cross product vektorielles Produkt, äußeres Produkt, Vektorprodukt
cross reference table Zuordnungstabelle
cross section Querschitt; Wirkungsquerschnitt (539)
cross section of nuclear reaction Wirkungsquerschnitt
cross-sectional Querschnitts...
cross-sectional deformation Querschnittsänderung
cross-shaded querschraffiert, schattiert
cross span Querverspannung
cross-span suspension Spanndrahtaufhängung (621.33)
cross talk Übersprechen, Nebensprechen
cross-talk attenuation Nebensprechdämpfung
cross-talk coupling Nebensprechkopplung
cross-talk factor Übersprechfaktor
cross-talk interference Störung durch Kreuzmodulation (621.396); Übersprechstörung (621.395)
cross-talk isolation Übersprechdämpfung

cross-talk measurement Übersprechmessung, Messung der Übersprechdämpfung (621.395)
cross-talk trouble Übersprechstörung (621.395); Kreuzmodulation (621.396)
cross wind Seitenwind
cross wires (pl.) Fadenkreuz
cross-wise kreuzweise
cross-wise movement Querbewegung
to cross kreuzen, überqueren; schneiden, kreuzen; sich überschneiden; verschränken (621.8)
to cross one another sich überschneiden, sich kreuzen
crossed coil instrument Kreuzspulinstrument, Kreuzspulgerät (621.317)
crossed Nicols gekreuzte Nicols
to cross-hatch kreuzschraffieren, überblenden (621.397)
cross widrig, verquer; Quer-..., Kreuz-...
crossing Kreuzungspunkt, Kreuzung; Herzstück (625)
crossing bell Kreuzungsläutewerk
crossing line Kreuzungslinie
crossing loop Kreuzungsgleis (625)
crossing point Kreuzungspunkt
crowbar Brecheisen
crowbar SCR (schneller) Abschalthyristor
crow-foot earthing Strahlenerder
crown Krone; Scheitel
crown-rail bond Nietschienenverbinder
Crown glass Crownglas
CR-test (of cables) = copper resistance test Widerstandsprüfung (621.315.2)
CRT = Cathode Ray Tube Braunsches Rohr, Kathodenstrahlröhre
crucible Tiegel, Schmelztiegel
crude roh; ungereinigt; rauh
crude metal Rohmetall
crude oil Rohöl
crude service rauher Betrieb
to crumple verknautschen, verquetschen
to crush zerquetschen, zerkleinern, zerbrechen; zerquetscht werden, brechen
crushing Mahlvorgang, Zerkleinerung
crushing strength Druckfestigkeit (66)
crushing stress Druckbeanspruchung
cryogenic equipment Tieftemperaturanlage
cryogenic memory Tieftemperaturspeicher, Supraleitungsspeicher
cryogenic technique Tieftemperaturtechnik
cryolite Kryolith
cryotron Kryotron
crypto ... Krypto..., Schlüssel...
crypto equipment Schlüsselgerät, Chiffriergerät
cryptodyne reception Reflexempfang
crystal Kristall (54); Quarz (621.396)
crystal analysis camera Kristallanalysenkamera
crystal control Quarzsteuerung (621.396.61)
crystal-controlled quarzgesteuert
crystal cut Quarzschnitt
crystal detector Kristalldetektor, Kristallgleichrichter, Halbleitergleichrichter
crystal diode Kristalldiode, Halbleiterdiode
crystal drive Quarzsteuerung (621.396.61)
crystal-driven quarzgesteuert
crystal element Kristalldetektor (621.396.62)
crystal filter Quarzfilter, Kristallfilter
crystal frequency Quarzfrequenz
crystal holder Quarzhalterung
crystal lattice Kristallgitter
crystal microphone Kristallmikrophon, piezoelektrisches Mikrophon
crystal mixer Kristalldiodenmischstufe
crystal mount Detektorfassung
crystal mounting Quarzhalterung, Quarzfassung, Detektorfassung
crystal oscillator Quarzsender, Quarzoszillator
crystal oven Quarz-Thermostat (621.396)
crystal pickup Kristalltonabnehmer
crystal puller Kristallziehvorrichtung
crystal receiver Detektorempfänger
crystal rectifier Kristalldetektor
crystal set Detektorapparat, Detektorempfänger
crystal speaker Kristall-Lautsprecher
crystal spots Einschlüsse (621.357)
crystal temperature Kristalltemperatur (621.382)
crystal wafer Kristallscheibe
crystalline kristallin
crystalline form Kristallform; kristalline Form
crystalline growth Kristallwachstum
crystallization Kristallisierung, Kristallisation
crystallite Kristallit
to crystallize kristallisieren
to crystallize out auskristallisieren
crystallizing temperature Kristallisationspunkt
crystallography Kristallographie
C-scan Azimut-Höhenbild im Funkmeßgerät
C'sk = countersunk versenkt
c. t. = cross talk (s. dort)
c. t. unit = cross-talk unit (s. dort)
C.T.S. = cab tyre sheath cable vollgumiisoliertes Kabel, Gummischlauchleitung
cu. = cubic Kubik-...

cu. ft. Kubikfuß = 1728 Kubikzoll = 28,32 Liter
cu. ft/min. Kubikfuß pro Minute = 28,32 l/min
cu. in. Kubikzoll = 16,387 cm^3
cub. = cubic Kubik-...
cube Würfel, Kubus; dritte Potenz
cube diagonal Raumdiagonale
cube root dritte Wurzel
cube shape Würfelform
to cube in die dritte Potenz erheben
cubic kubisch, würfelförmig; Kubik-...
cubic capacity Fassungsvermögen, Rauminhalt
cubic content Rauminhalt
cubic equation kubische Gleichung
cubic foot Kubikfuß (s. Anhang)
cubic inch Kubikzoll (s. Anhang)
cubic measure Raummaß
cubic metre Kubikmeter
cubical räumlich; wüfelförmig, würfelig
cubical antenna Würfelantenne
cubical expansion kubische Dehnung
cubicle Zelle, Schaltzelle, Schrank (621.3)
cubiform würfelförmig, würfelartig
cue Unterprogrammaufruf (681.3)
cuff Manschette (621)
to culminate gipfeln, kulminieren, einen Höhepunkt erreichen; durch den Meridian gehen (52)
Cumberland protection kathodischer Schutz
cumulative kumulativ, sich häufend, fortschreitend, nach und nach erfolgend
cumulative compound winding Kompoundwicklung, mitkompoundierende Wicklung
cumulative error integrierender Fehler
cumulative grid rectification Audiongleichrichtung
cumulative ionization lawinenartige Ionisation
cumulative process Kettenreaktion
cup Behälter, Tasse, Glocke; Tasse voll; Kalotte; Sperrtopf (621.396); Becher (621.35)
cup anemometer Schalenanemometer, Schalenkreuz
cup grease Staufferfett
cup head Segmentkopf (621.8)
cup jewel Steinlager, Lagerstein (621.317)
cup-shaped tassenförmig, Hütchen-..., Napf-..., napfförmig
cup-shaped condenser Hütchenkondensator
cup tool Setzhammer (621.8)
to cup hohl drücken, tiefen; aushöhlen
cupped schalenförmig, hohl, ausgehöhlt, napfförmig
cupped edge hochgebogener Rand
cupola furnace Kupolofen
cupric Cupri-... (zweiwertige Kupferverbindung)
cupro-nickel alloy Kupfernickel-Legierung
cuprous Cupro-... (einwertige Kupferverbindung)
cuprous oxide Kupferoxidul
cuprox rectifier Kupferoxidulgleichrichter
to cure kurieren; konservieren, haltbar machen; aushärten, vulkanisieren; fermentieren
to cure bakelite Bakelit aushärten
Curie Curie (539)
Curie constant Curie-Konstante
Curie cut X-Schnitt (537)
Curie effect Curieeffekt
Curie's law Curiesches Gesetz
Curie point Curiepunkt
curing Abbinden, Abbindungsvorgang; Alterung, Aushärtung; Fermentierung, Fermentation
curl Rolle, Locke; Kantenverrundung; Rotation (51)
curl of a vector Rotation eines Vektors (51)
current Strom (621.3); Strömung; Luftzug; gegenwärtig
current amplification Stromverstärkung (621.395)
current antinode Strombauch
current balance Amperewaage
current-carrying stromführend
current-carrying capacity Strombelastbarkeit
current-carrying conductor stromführender Leiter
current-carrying rating Nennstrom
current circuit Strompfad (621.317)
current coil Stromspule (621.317)
current-collecting brush Stromabnehmerbürste
current collector (US) Stromabnehmer (621.33)
current consumer Stromabnehmer, Stromverbraucher, Konsument
current-controlled stromgesteuert
current cuts off Strom setzt aus
current density Stromdichte
current demand Strombedarf, Stromanforderung
current direction Stromrichtung
current directional relay Richtungsrelais, Stromrichtungsrelais
current distribution Stromverteilung (621.385)
current drain Stromentnahme, Strombedarf
current efficiency Gütergrad (621.355)
current element Stromeisen (621.317)
current-fed stromgespeist (621.396.67)

current feedback Stromrückkopplung
current fluctuation Stromschwankung
current flow Stromfluß, Stromdurchgang
current gain Stromverstärkung
current of hydrogen Wasserstoffstrom
current impulse Stromstoß
current indicator Strommesser, Stromzeiger
current input Stromaufnahme
current inrush Belastungsstoß
current instruction register aktuelles Befehlsregister
current intake Stromeintritt
current intensity Stromstärke
current interruption Stromunterbrechung
current-leakage path Ableitweg, Kriechweg
current leaves... Strom fließt ab von...
current level absoluter Strompegel
currentless stromlos
current limitation Strombegrenzung
current limiter Strombegrenzer (621.317)
current-limiting inductor Strombegrenzungsdrossel, Kurzschlußdrosselspule
current-limiting reactor Strombegrenzungsdrossel
current-limiting resistor Strombegrenzungswiderstand
current loading Strombelastung
current loop Stromschleife; Strombauch
current magnet Stromeisen (621.317)
current margin Zeichenstrom (621.394)
current modulation Strommodulation
current-operated strombetätigt, stromgesteuert, stromabhängig
current-operated earth-leakage circuit breaker with overcurrent protection Fehlerstrom-Schutzschaltung mit Überstromauslösung
current overload Stromüberlastung
current passage Stromdurchgang
current path Strompfad
current-phase balance protection Stromdifferentialschutz
current pulse Stromstoß, Impuls
current range Strommeßbereich
current rating Nennstromstärke
current regulator Stromregler (621.791)
current relay Stromrelais
current requirement Strombedarf
current response entstehender Strom
current rush Stromstoß
current sheet Flächenstrom
current sink Stromsenke, Belastung
current source Stromquelle, Spannungsquelle
current square meter Stromquadratmeßgerät, Effektivstrommesser
current streamer Stromfaden
current strength Stromstärke
current supply Stromversorgung
current surge Stromwelle
current tap Mehrfachstecker, Anschlußwürfel
current transformer Stromwandler
current-type telemeter strombetätigtes Fernmeßgerät
current undulation Stromschwankung
current vector Stromzeiger
current waveform Stromkurve, Kurvenform des Stromes
current-wavelength characteristic spektrale Empfindlichkeitsverteilung (621.383)
current weigher Stromwaage
current-weighting factor Frequenzbewertungsfaktor eines Stromes
curriculum Lehrplan, Fachrichtung
cursor Läufer (681); Blinker, Leuchtmarke, Cursor
cursor home Cursor-Anfangsstelle
cursor-line bearing Strichpeilung
curtain Vorhang
curtain array of aerials Antennenwand (621.396)
curvature Krümmung
curve Kurve, krumme Linie; Linienzug (auch gebrochen)
curve of cyclic magnetization Magnetisierungsschleife
curve displays humps Kurve weist Buckel auf
curve follower Kurvenleser (621.316.7)
curve of horizontal distribution horizontales Polardiagramm (621.32)
curve of light distribution Polardiagramm (der Lichtverteilung) (621.32)
curve of pursuit Hundekurve, Traktrix, Verfolgungskurve
curve sheet Kurvenblatt
curve trace Kurvenverlauf
curve tracer Kurvenzeichner; Doppelspiegelmagnetometer
curve of vertical distribution vertikales Polardiagramm (621.32)
to curve krümmen, abbiegen, kurven
to curve an electron around along a path ein Elektron auf eine gekrümmte Bahn zwingen
curved gekrümmt, gebogen, krumm
curved brackets runde Klammern
curved line krumme Linie, Kurve
curved rail Bogenschiene
curved shears (pl.) Kurvenschere

curvilinear krummlinig
curvilinear system of co-ordinates nichtlineares Koordinatensystem (mit nichtlinear geteilten Achsen); krummliniges (z. B. elliptisches) Koordinatensystem
cushion Kissen, Polster; Puffer
cushion mounting Einbaupuffer, gepolsterter Einbau
customer Verbraucher, Abnehmer; Teilnehmer (621.395), Kunde
cut Schnitt, Schnittfuge; Einschnitt, Abschnitt; Schliff (66); Span, Spanstärke, Spantiefe (621.9); Schnittwunde (61)
cut allowance Verschnittzugabe, Zuschlag für Verschnitt, Verschnitt
cut-away drawing Schnittzeichnung
cut face Schnittfläche, Anschnittfläche
cut-in Eingriffstellung (65)
cut-in loop Zugbeeinflussungsschleife (65)
cutoff Einsatzpunkt, unterer Knick der Kennlinie (621.385); Grenz-...
beyond cutoff ins Sperrgebiet (621.385)
cutoff bias (Gitter-)Sperrspannung
cutoff characteristic Begrenzungseigenschaft
cutoff current Reststrom; Ausschaltspitzenstrom (621.316.9)
cutoff frequency Grenzfrequenz; Abdeckfrequenz, kritische Frequenz (621.371)
cutoff grid potential Gittersperrspannung
cutoff limiting Begrenzung durch Gitterstromeinsatz
cutoff region Sperrbereich (621.382)
cutoff relay Löschrelais, Trennrelais (621.394)
cutoff voltage Entladegrenzspannung (621.385); Abschnürspannung (621.382)
past cutoff ins Sperrgebiet hinein
cutout Ausschalter, Auslöser, Automat; Ausschnitt
cutout base Sicherungssockel
cutout board Verteilertafel, Sicherungstafel
cutout box Sicherungskasten (621.315.6)
cutout loop Zugbeeinflussungsschleife
cutover Überleitung (621.395), Umschaltung
cut paraboloid Teilparabolspiegel
30° cut Y-Schnitt (537.228)
cut-through Durchschaltung
cut and try Hin- und Herprobieren, Probieren
cut-and-try process Probiermethode, Iterationsverfahren
to cut schneiden, ausschneiden
to cut across lines of force Kraftlinien schneiden
to cut down beschneiden, heruntersetzen
to cut glass Glas schneiden; Glas schleifen
to cut to length auf Länge schneiden, ablängen
to cut lines of force Kraftlinien schneiden
to cut metal Metall sägen
to cut off abstellen, abschalten
to cut off the steam Dampf abstellen
to cut through durchschalten; durchverbinden
to cut true and square scharfkantig und auf Maß schneiden
to cut and try probieren, ausprobieren
cutter feed Hohlraum-Schlitz-Speisung (621.396.677)
cutting Schneidvorgang, Schneiden; Schneidbrennen
cuttings Schneidspäne, Drehspäne
cutting Schneid-...
cutting angle Schneidwinkel, Schneidkantenwinkel
cutting edge Schneidkante, Schneide
cutting metal Schneidmetall, Hartmetall
cutting oil Bohröl, Schneidöl (621.9)
cutting speed Schnittgeschwindigkeit (621.9)
cutting tool Schneidwerkzeug, Drehstahl
cutting torch Schneidbrenner
cutting waste Schnittverlust, Verschnitt
cutting-in Aufschalten (621.395)
C-voltage Gitterspannung
c.w. = continuous wave Dauerstrich, ungedämpfte Welle
C-wire Meßleitung, Prüfdraht, Prüfader
cwt. = hundredweight Zentner (1 cwt. = 112 lbs = 50,8 kg)
cyan Cyan
cyanide Cyandid
cybernetics Kybernetik
cycle Periode (51), (53), (621.3); Zyklus, Arbeitstakt (621), (681.3)
cycle count Zykluszählung; Schleifenzählung (681.3)
cycle of magnetization Magnetisierungsschleife
cycle of operation Arbeitszyklus
cycle rate counter Periodenzähler
cycles per second Hertz, Perioden pro Sekunde
cycle time Belegungszeit, Zykluszeit (681.3)
to cycle periodisch verlaufen, sich periodisch abspielen; zyklisch durchlaufen; pendeln
to cycle-shift zyklisch vertauschen
cyclic(al) periodisch
cyclic code zyklischer Code
cyclic interchange zyklische Vertauschung

cyclic irregularity Ungleichförmigkeit (621.311)
cyclic magnetization Ummagnetisierung
cyclic redundancy check = CRC zyklische Blockprüfung
cyclic shift Stellenwertverschiebung (681.3)
cyclic store Umlaufspeicher (681.3)
cyclic variation periodische Veränderung
cyclic with periodisch wiederkehrend mit
cyclical field Wirbelfeld
cyclicly magnetized condition eingeschwungener Zustand der Ummagnetisierung
cyclogram Lissajous-Figur
cycloid Zykloide
cyclometer Kilometerzähler (629)
cyclometer dial Rollenzählwerk (621.317)
cyclone separator Zyklon, Turboabscheider
cyclotron Zyklotron
cylinder Zylinder (51), (621)
cylinder bearer Schmitzring (655)
cylinder-electrical machine Walzen-Elektrisiermaschine
cylinder function Zylinderfunktion
cylinder lining Zylinderbüchse
cylindrical zylindrisch
cylindrical lens Zylinderlinse
cylindrical parabola Zylinderparaboloid
cylindrical polars Zylinderkoordinaten
cylindrical rotor Trommelläufer (621.313)
cylindrical seal Rundanglasung (621.385)
cylindrical symmetry Zylindersymmetrie
cylindrical winding Zylinderwicklung
cypher Null (im Dezimalbruch) s. **cipher**

D

dab Tupfen, Farbtupfen, Farbklecks
dag = decagram Dekagramm
daily täglich
daily keying element Tagesschlüssel (623)
daisy chain Verkettung (621.394.5), (681.3)
daisy-wheel printer Typenraddrucker
dal. = decaliter Dekaliter
dam. = decametre Dekameter
dam Damm, Staudamm, Staumauer
to dam abdämmen, stauen, dämmen, eindämmen
damage Schaden, Beschädigung, Zerstörung; Entschädigung
damp Dunst, Schwaden
damp-proof feuchtigkeitsicher
to damp anfeuchten, dämpfen; dämpfen (534)
to damp oscillations Schwingungen dämpfen
damp feucht
damp atmosphere feuchte Luft
damp course Isolierschicht (693)
damp-proof wiring Feuchtrauminstallation
damp room feuchter Raum, Feuchtraum
damp weather feuchtes Wetter
damped gedämpft
damped harmonic system gedämpfter harmonischer Schwinger
damped impedance Widerstand bei festgehaltener Membran (534.86)
damped oscillation gedämpfte Schwingung
damped sinusoidal quantity gedämpfte Sinusschwingung
damped vibration gedämpfte Schwingung
damped wave gedämpfte Welle
damper Rauchklappe, Rauchschieber, Zugregler; Dämpfer (534)
damper winding Dämpferwicklung (621.313)
damping Dämpfung (621.317)
damping coil Dämpferspule
damping coefficient Abklingkonstante, Dämpfungskonstante
damping constant Dämpfungskonstante (d = 1/Q) (621.396)
damping factor Dämpfungsfaktor, logarithmisches Dekrement
damping grid Dämpferspule, Dämpferkäfig
damping magnet Dämpfermagnet
damping power Dämpfungsfähigkeit, Dämpfungsvermögen
damping ring Dämpferring
damping time Beruhigungszeit (621.317)
dampint torque Dämpfungsmoment (621.317)
damping winding Dämpferwicklung
danger Gefahr (**to:** für)
danger coefficient Gefährdungsschwelle, Gefahrschwelle
danger of combustion Brandgefahr
Danger High Voltage Vorsicht! Hochspannung-Lebensgefahr
danger to life Lebensgefahr
danger signal Warnsignal, Warnung
Daniell cell Daniell-Element
daraf 1/Farad = Ω/s
dark dunkel
dark current Dunkelstrom (621.383)
dark discharge Dunkelentladung
dark field Dunkelfeld (535)
dark heat dunkle Glut

dark-ray radiator Dunkelstrahler
dark red heat dunkle Rotglut
dark resistance Dunkelwiderstand
darkroom lamp Dunkelkammerlampe
dark space Dunkelraum (621.385)
darktrace tube Dunkelschriftröhre, Blauschriftröhre
Darlington power transistors Darlington-Schaltung, Darlington-Leistungstransistoren
d'Arsonval current d'Arsonval-Strom (615.84)
d'Arsonval galvanometer Drehspulgalvanometer
d'Arsonval instrument d'Arsonval-Instrument, Drehspulinstrument
dash Morsestrich, Strich; Trennungsstrich, Gedankenstrich
dash-dotted line strichpunktierte Linie
dashed line gestrichelte Linie
dashboard Instrumentenbrett, Armaturenbrett (629.113)
dashpot Stoßdämpfer, Dämpfungszylinder
DAT = digital audio tape
data . . . Daten. . .
data acquisition Datenerfassung
data bank Datenbank
data block Datenblock
data bus Datenbus
data carrier Datenträger
data collection Datenerfassung
data communication Datenübertragung
data compaction Datenverdichtung
data compression Datenverdichtung
data display Anzeige, Sichtgerät
data-entry terminal Dateneingabegerät, Dateneingabeterminal
data file Datei
data flow Datenfluß
data flowchart Datenflußplan, Flußdiagramm
data format Datenformat
data input Dateneingabe
data link Datenübertragungsverbindung
data logger Datendrucker
data medium Datenträger
data movement Datentransfer
data output Datenausgabe
data preparation Datenvorbereitung
data processing Datenverarbeitung
data-processing system Datenverarbeitungsanlage, Rechenanlage, Rechner
data-processing terminal equipment Datenendeinrichtung
data protection Datenschutz
data recording Datenaufzeichnung
data reduction Datenreduzierung
data representation Datendarstellung
data retrieval Wiederauffinden von Informationen
data security Datensicherung
data service Datendienst
data shift Datenverschiebung
data sink Datenempfänger, Datensenke
data smoothing Werteglättung
data source Datenquelle
data storage Datenspeicherung
data terminal Daten-Ein- und Ausgabegerät
data-transfer rate Datentransfergeschwindigkeit
data transmission Datenübertragung
data word (Daten-)Wort
datel service Dateldienste
datum (pl.: **-ta)**Wert, Zahlenangabe, Unterlage, Angabe; vorgegebene Tatsache
daub Spachtelmasse
to daub spachteln
daughter nucleus Tochterkern (539)
d.a.v.c. = delayed automatic volume control verzögerte Schwundregelung
davit-arm pole Peitschenmast
daylight Tageslicht
daylight lamp Tageslichtlampe
db = decibel db = decibel
d.b. = double break doppelte Unterbrechung (621.316.5)
d.c. = direct current Gleichstrom
d.c./a.c. converter Wechselrichter
d.c. to a.c. ratio Ausgangsgleichstrom/Eingangswechselstrom (621.316.5)
d.c. amplifier Gleichspannungsverstärker
d.c. arc Gleichstromlichtbogen
d.c. arc welding Gleichstromlichtbogenschweißung
d.c. arc-welding converter Gleichstrom-Schweißumformer
d.c. arc-welding generator Gleichstrom-Schweißgenerator
d.c. balancer Ausgleichsmaschine (621.316.1)
d.c. damp diode Nullpegeldiode
d.c. commutating machine Gleichstromkommutatormaschine
d.c. component Gleichstromkomponente, Gleichstromanteil
d.c. compound-wound generator Gleichstromdoppelschlußmotor
d.c. conductance Gleichstromleitwert
d.c. coupled direkt gekoppelt, galvanisch gekoppelt

d.c. current transformer Gleichstromwandler (621.317)
d.c. distribution Gleichstromverteilung
d.c. dynamo Gleichstromgenerator
d.c. form factor Mittelwert/Effektivwert (621.314.5)
d.c. generator Gleichstromgenerator
d.c. leak current Leck-Gleichstrom
d.c. machine Gleichstrommaschine
d.c. mains Gleichstromnetz
d.c. meter Gleichstromzähler
d.c. motor Gleichstrommotor
d.c. neutral grid Mittelleitersystem, Gleichstrom-Drehleiternetz
d.c. operation Gleichstrombetrieb
d.c. plant Gleichstromanlage
d.c. plate resistance (innerer) Gleichstromwiderstand (621.385)
d.c. potential Gleichspannung
d.c. power Gleichstromleistung
d.c. receiver Gleichstromempfänger
d.c. resistance Gleichstromwiderstand
d.c. restoration Rückgewinnung des Gleichstromanteils (621.397)
d.c. restorer Nullpunktfesthalteschaltung
d.c. restorer diode Nullpegeldiode
d.c. restoring Schwarzwerthaltung (621.397)
d.c. series-wound generator Gleichstromreihenschlußgenerator
d.c. series-wound motor Gleichstrom-Hauptschlußmotor
d.c. shunt-wound generator Gleichstromnebenschlußgenerator
d.c. static converter Wechselrichter mit Wiedergleichrichtung, Gleichstrom-Gleichstrom-Umformer, Gleichstromwandler
d.c. telegraphy Gleichstromtelegraphie
d.c. three-wire system Gleichstrom-Dreileiternetz
d.c. traction Gleichstrombahnbetrieb
d.c. transformer Gleichstromwandler
d.c. transmission Gleichstromübertragung, Gleichstromkopplung (621.397)
d.c. voltage Gleichspannung
d.c. winding Gleichstromwicklung
under d.c. conditions bei Gleichstrom
D-cable Sektorkabel, Zweileiter-Sektorkabel
d.c.c. = double cotton covered
deactivation abschalten, außer Betrieb setzen, in den Ruhezustand zurückführen
dead tot; ganz, völlig; spannungslos (621.3)
dead accurate haargenau
dead-beat aperiodisch gedämpft, aperiodisch
dead centre Totpunkt
dead-centre position Totpunktstellung
dead coil tote Spule (621.313)
dead conductor nichtspannungsführender Leiter, spannungsloser Leiter
dead earth (satter) Erdschluß, vollkommener Erdschluß
dead end Sackgasse, Sackbahnhof
dead-end clamp Abspannklemme
dead-end effect Nebenresonanz (621.396)
dead-end insulator (US) Abspannisolator
dead-end pedestel (for ground wire) Abspannbock (für Erdseil)
dead-end portal Abspannportal
dead-end siding Stumpfgleis
dead-end switch Wellenschalter mit Kurzschlußkontakten
dead-end thimble Abspannkausche
dead-end tower Abspanmast, Endmast
dead-ended feeder Stichleitung
dead-ended line Ausläuferleitung
dead ending Abspannung
dead-front-type switchboard (Schalttafel ohne spannungsführende Teile auf der Vorderseite)
dead ground satter Erdschluß
d.c./a.c. converter Wechselrichter
dead load tote (ruhende) Last
dead man Spanndrahtschwelle, Abspannpfahl (621.315)
deadman's feature Totmannfunktion (625)
deadman's handle Totmannknopf
dead matter totes (anorganisches) Material
dead point toter Punkt, Totpunkt
dead reckoning Kursrechnung
dead-reckoning device Kursrechner
dead-reckoning navigation Koppel-Navigation
dead segment blinde Lamelle (621.313)
"dead short" „satter Kurzer", Vollkurzschluß
dead space Schwebungslücke; tote Zone (621.371)
deaed spot Empfangslücke, tote Zone (621.371)
dead studio nachhallarmes Studio
dead surface matte Oberfläche
dead time Todzeit
dead travel toter Gang
dead weight Eigengewicht; ruhende Last, Belastungsgewicht, totes Gewicht
deaeration Entlüftung
deaeration reservoir Entlüftungskammer, Entlüftungsbehälter
deaf taub
deaf aid Hörapparat

deafener Schwingungsdämpfer (621.315.1)
deafening ohrenbetäubend
to deal algebraically with algebraisch behandeln
death rays Todesstrahlen
deathnium center Haftstelle, Fangstelle (621.382)
deattenuation Entdämpfung (621.395)
to debounce entprellen
debris Trümmer, Schutt, Gesteinsschutt, Trümmerschutt; Abfälle
to debug Fehler beseitigen
debug program Fehlersuchprogramm
debugger [Unterstützungsprogramm zur Fehlersuche]
debugging Fehlerbeseitigung (681.3)
to debur entgraten
deca-... deka-...
decade Dekade, Jahrzehnt; Dekade (621.317)
decade box Dekadenwiderstand (621.317)
decade dial Dekadenkurbel, Kurbeldekade (621.317)
decade-resistance box Dekadenwiderstand, Kurbelwiderstand, Dekadenwiderstandskasten (621.317)
decadic counter dekadischer Zähler
decagram(me) Dekagramm
decal Abziehbild; Schiebemarke (621.317)
decalitre Dekaliter
decametre Dekameter
decametric waves Kurzwellen
decanter Scheidebehälter, Schlammabscheider
decay Abfall; Zerfall
decay characteristic Nachleuchtcharakteristik
decay coefficient logarithmisches Dekrement
decay constant Zerfallskonstante
decay factor logarithmisches Dekrement
decay product Zerfallsprodukt
decay time Abklingzeit
to decay abfallen, abklingen, ausschwingen; zersetzen; sich zersetzen, zerfallen, verwittern
decaying conduction current anomaler Leitungsstrom
Decca Navigator System Decca Navigations-System
to decelerate verzögern
decelerating electrode Bremselektrode
deceleration Verzögerung, Bremsverzögerung, Auslauf
decentralization Aderverlagerung (621.315.2)
decentralized dezentralisiert
deception Täuschung (623)
decibel Dezibel, dB
decibel meter Pegelmesser, Dämpfungsmeßgerät
decigram(me) Dezigramm, Zehntelgramm
decilitre Deziliter
decimal Dezimalbruch
decimal Dezimal-...
decimal balance Dezimalwaage
decimal-to-binary conversion Dezimal-Binär-Umwandlung
decimal classification Dezimal-Klassifikation
decimal coinage Dezimalwährung
decimal digit Dezimalziffer
decimal fraction Dezimalbruch
decimal notation Dezimalsystem
decimal number Dezimalzahl
decimal numeration Dezimalzählung; Zehner-Nummerung
decimal place Dezimalstelle
decimal point Komma (im Dezimalbruch), Dezimalkomma
decimal points Dezimalstellen, Stellen hinter dem Komma
decimal scale Dezimalsystem, Dezimalskala
decimal system Dezimalsystem
to decimalize in einen Dezimalbruch verwandeln
decimetre Dezimeter
decimetric waves Dezimeterwellen
decinormal zehntelnormal (54)
to decipher dechiffrieren, entschlüsseln
deciphering, decipherment Entschlüsselung, Dechiffrierung
decision box Entscheidungssymbol
decision element Verknüpfungsglied (621.374)
decision level Entscheidungsschwellwert
deck Deck (629.12); Fahrbahn (624.2); Stockwerk, Stockwerksboden (622); Kartenstapel, Kartensatz (681.3)
declination Deklination (52), (538.7)
decline Abnahme; Baisse; Abfall, Abfallen
to decline abfallen, abnehmen, sinken, nachlassen
to declutch entkuppeln, auskuppeln
to decode entschlüsseln, dechiffrieren
decoder Rufumsetzer (621.395); Dekodierer
decoding Dekodieren
decoding matrix Dekodiermatrix
to decohere entfritten
to decompose verwittern; (sich) zersetzen, zerfallen
decomposition Zerfall, Zersetzung, Verwitterung
decomposition potential Zersetzungsspannung
decomposition voltage Zersetzungsspannung
decompounded motor Gegenkompoundmotor
decompounding Abwärtskompoundierung
decompounding winding Gegenkompoundwicklung
to decompress entspannen (626)

to decontaminate entgiften (623); entseuchen (539)
decontamination Entgiftung (623); Entseuchung, Dekontamination (539)
to decorate dekorieren, verzieren
decorative lamp Schmucklampe, Ornamentlampe
decorative lighting Schmuckbeleuchtung
to decouple entkoppeln, neutralisieren (621.396)
decoupling network Entkopplungsschaltung
decoy Scheinanlage, Bold; Lockziel, Täuschziel (623)
decrease Abnahme, Rückgang
decrease of current Stromabnahme, Stromrückgang
decrease of output Leistungsrückgang
decrease in weight Gewichtsabnahme
decrement Dekrement, Verminderung, Abnahme
to decrement [schrittweises Herunterzählen um jeweils einen Schritt] (681.3)
decremeter Dämpfungsmesser
dedicated line Standverbindung (621.395)
to dedust entstauben, abstauben
Dee D-Elektrode (539.17)
de-emphasis Deemphasis
to de-energize abschalten
de-energized stromlos, spannungslos, abgeschaltet
deep tief
in deep array tief gestaffelt, tief gegliedert
deep-bar squirrel-cage motor Wirbelstrommotor
deep colour satte Farbe, tiefe Farbe
deep drawing Tiefziehen
deep-drawing test Tiefungsversuch, Tiefziehprobe
deep-etched tiefgeätzt
deap sea cable Tiefseekabel
to deep-draw tiefziehen
defect Schaden, Defekt, Fehler, Mangel; Fehlstelle (621.382)
defect electron Defekt-Elektron
defect in material Werkstoffehler
defective schadhaft, fehlerhaft, mangelhaft
defective hearing Schwerhörigkeit, Hörschwäche
defective workmanship Bearbeitungsfehler, schlechte Arbeit
defibrator Zerfaserer (67)
to defibrate zerfasern
deficiency rate Ausfallrate
deficient in mit Mangel an, mit zu wenig..., ...arm
deficient in electrons mit Elektronenmangel, elektronenarm, positiv geladen
deficient in temperature zu kalt
deficient temperature Untertemperatur
to define definieren, erklären
defined image klares Bild, scharfes Bild
defined end restraint starre Einspannung der Enden (620)
defining slit Auflösespalt (535)
definite genau festliegend
definite area bestimmte Fläche
definite integral bestimmtes Integral
definite time UMZ = unabhängige Maximal-Zeit (621.317)
definite time element unabhängiges Zeitrelais
definite time lag (strom-)unabhängige Zeitverzögerung
definite time-lag circuit breaker Schalter mit UMZ-Auslösung
definite time-lag relay UMZ-Relais, unabhängiges Maximalzeitrelais
definite time-lag tripping begrenzt verzögerte Auslösung
definite time limit stromunabhängige Verzögerungszeit
definite-time overcurrent trip statische Auslösung, stromunabhängige Auslösung (621.316.5)
definition Definition; Zeichenschärfe, Bildschärfe (535); Fehlerkennbarkeit (621.386); Auflösung (621.397)
by definition definitionsgemäß
to deflect ableiten, umlenken, abführen, auslenken; (sich) verbiegen; beugen (535)
deflecting Ablenk-..., Ablenkung, Ausschlag, Auslenkung
deflecting force Triebkraft, Triebmoment (621.317)
deflecting plate Ablenkplatte, Ablenkanode (621.385)
deflecting torque Triebmoment (621.317)
deflection Ablenkung (621.397); Umlenkung; Ausschlag (621.317); Abbiegung, Durchbiegung
deflection amplifier Ablenkverstärker
deflection circuit Ablenkschaltung (621.397)
deflection coefficient Durchbiegungskoeffizient
deflection coil Ablenkspule (621.385), (621.397)
deflection control Ablenksteuerung (621.397)
deflection defocusing Fleckunschärfe (621.397)
deflection factor (Kehrwert der Empfindlichkeit) (621.385)
deflection indicator Durchbiegungsmesser

deflection means Ablenkmittel
deflection plane Ablenkebene (621.397)
deflection plates Ablenkplatten
deflection potentiometer anzeigende Kompensationsschaltung
deflection sensitivity Ablenkempfindlichkeit (621.385)
deflection yoke Ablenkspule, Ablenkspulenjoch
deflector Abweiser, Ablenkblech
deflector electrode Ablenkplatte (621.385)
deflector plate Ableitblech, Abweiseblech
deflexion = deflection (s. dort)
deflexional method Ausschlagsverfahren
defocusing Entfokussieren
de Forest system Audioempfang (621.396)
to deform verformen, deformieren
deformation Deformierung, Verformung, Formänderung
deformation of the lattice Gitterverformung
to defrost auftauen, abtauen
defruiter Gerät zur Unterdrückung nichtsynchroner Antworten, Defruiter (621.396.96)
to defy solution sich nicht lösen lassen
deg. = degree Grad
to degas entgasen
degasification Entgasung
to degauss entmagnetisieren (629.12)
degaussing belt Entmagnetisierungskabel, Entmagnetisierungsschleife, MES
degaussing cable Entmagnetisierungskabel
degaussing equipment MES, Entmagnetisierungseinrichtung (629.12)
degeneracy Entartung (51), (53)
to degenerate gegenkoppeln; degenerieren, entarten
degenerate degeneriert, entartet
degeneration Entartung (53); Gegenkopplung (621.396.64)
degenerative gegengekoppelt, Gegenkopplungs-...
degenerative feedback Gegenkopplung
degenerative voltage Gegenkoppelspannung
degradation Güteverlust
degradation failure Driftausfall (620.1)
to degrease entfetten
degree Grad, Winkelgrad, Altgrad; Grad, Ausmaß; Faktor; Diplom
by degrees gradweise, nach und nach, allmählich
degree Centigrade Celsiusgrad, Grad Celsius
degree of coupling Kopplungsgrad
degree of current rectification Gleichrichtungsfaktor
degree of dissociation Dissoziationsgrad
degree of freedom Freiheitsgrad
degree of hardness Härtegrad (s. Anhang)
degree of humidity Feuchtigkeitsgrad
degree of latitude Breitengrad
degree of longitude Längengrad
degree mark Gradstrich
degree of modulation Modulationsgrad
degree of perfection Vollständigkeit, Güte
degree of purity Reinheitsgrad, Feingehalt (669)
degree of saturation Sättigungsgrad
degree of screening Schirmfaktor
degree of uniformity Gleichförmigkeitsgrad
degree of voltage rectification Spannungsumsetzung = Ausgangsgleichspannung/Eingangswechselspannung (621.314.5)
to dehydrate entwässern, entfeuchten, trocknen, dehydrieren
dehydration Entwässerung, Entfeuchtung, Wasserabgabe, Trocknung, Dehydrierung
to de-ice enteisen
de-icer Enteiser, Enteisungseinrichtung
deion circuit-breaker Deion-Schalter, Deionisationsschalter
deionization Entionisierung
deionization potential Sperrspannung, Entionisierungsspannung (621.385)
deionization rate Entionisierungsgeschwindigkeit
deionization time Entionisierungszeit
to deionize entionisieren
deionizing Entionisierung
deionizing potential Löschspannung (621.385)
del Nabla-Operator
delay Verzögerung
delay action verzögertes Ansprechen
delay-action circuit breaker Schalter mit verzögerter Auslösung
delay-action fuse träge Sicherung
delay bias verzögerte Vorspannung, Verzögerungsspannung, verzögerte Regelspannung (621.396)
delay circuit Verzögerungsschaltung
delay constant Verzögerungsmaß
delay difference Laufzeitunterschied
delay distortion Laufzeitverzerrung, Phasenverzerrung
delay element Zeitglied; Verzögerungsglied
delay equalizer Phasenausgleichkette, Laufzeitentzerrer

delay fuse träge Sicherung (621.3)
delay line Verzögerungsleitung (621.394)
delay-line memory Laufzeitspeicher
delay system Wartesystem (621.395)
delay time Verzögerung, Verzögerungszeit, Abklingzeitkonstante (621.397); Abfallverzögerung
delay vs. frequency Laufzeitgang
to delay verzögern, aufschieben
delayed application verzögerte Bremsung (656)
delayed automatic volume control verzögerte Schwundregelung
delayed call Vormerkgespräch
delayed pulse generator Impulsverzögerungsgenerator
delayed test Kapazitätsprüfung nach Lagerung (621.35)
to delete löschen (fehlerhafter Zeichen) (681.3)
delicate empfindlich, fein zart
delicate balance empfindliche Waage
delicate line dünne Linie, zarte Linie
delicate test empfindliche Prüfung
delicate work feine Arbeit, heikle Arbeit
to delineate wiedergeben, aufzeichnen
to deliquesce flüssig werden, sich verflüssigen
deliquescent im Zustand des Flüssigwerdens, schmelzend
to deliver current Strom liefern, Strom abgeben
to deliver power Leistung abgeben
delivery Lieferung, Ablieferung, Auslieferung, Abgabe
delivery of energy Energieabgabe, Energielieferung
delivery of gas Gaslieferung, Gasabgabe
Dellinger fade-out Mögel-Dellinger-Effekt
Delon rectifier Delon-Gleichrichter; mechanischer Gleichrichter
delta Delta; Dreieck (621.3)
delta clock Deltatakt (681.3)
delta connected dreieckgeschaltet
delta connection Dreieckschaltung
delta function Deltafunktion
delta rays langsame Betastrahlen
delta voltage Dreieckspannung
delta winding Dreieckswicklung
deltoid dreieckförmig
demagnetization Entmagnetisierung
demagnetization by continued reversals Entmagnetisierung durch Ummagnetisierung
demagnetization curve Entmagnetisierungskurve
to demagnetize entmagnetisieren
demagnetizing turns Gegenamperewindungen
demand Bedarf, Leistungsbedarf (621.31); Anforderung, Nachfrage
demand factor Leistungsspitze/Anschlußwert, Gleichzeitigkeitsfaktor
demand indicator Spitzenzeiger (621.317)
demand of an installation mittlere Anschlußleistung
demand limiter Strombegrenzer (621.317)
demand meter Spitzenzähler
demand printer Spitzendrucker, Prinotometer (621.317)
demand of a system mittlere Netzleistung
to demodulate rückmodeln, demodulieren
demodulation Demodulation
demodulator Demodulator
demonstrative geometry darstellende Geometrie
demountable abnehmbar, ausbaubar
to denature denaturieren, vergällen
dendrite Dendrit
denoiser Rauschunterdrücker
denominator Nenner
dense dicht; gut gedeckt
dense texture dichtes Gefüge
density Dichte, spezifisches Gewicht; Schwärzung (77)
density of an acid Säuredichte
density of charge Ladungsdichte
density of flux changes Flußwechseldichte
density modulation Dichtemodulation
density of steam Dampfdichte
dent Zahn, Zacken
to dent zacken, zackig einreißen
dentrite Tannenbaumkristall
deoxidation Reduktion, Reduzierung, Desoxidierung
department Abteilung
dependent on abhängig von
dependent variable abhängige Variable (51)
depth indicator Teufenzeiger
to depict wiedergeben, (zeichnerisch) darstellen
to depict graphically graphisch darstellen
to de-plate anodisch behandeln
depleted ausgebraucht, erschöpft; verarmt, abgereichert (539)
depleted acid matte Säure
depleted solution ausgebrauchte Lösung (77)
depletion Verarmung
depletion layer Sperrschicht (621.382)
depletion-layer capacitance Sperrschichtkapazität
depletion-layer rectifier Sperrschichtgleichrichter
depletion mode Trägerverarmung

depletion-mode operation Verarmungsbetrieb
depletion-type field-effect transistor Verarmungs-Isolierschicht-Feldeffekttransistor, Verarmungs-IGFET
depolarization Depolarisation
depolarization loss Polarisationsentkopplung
depolarizing mix Depolarisator
depolarizer Depolarisator, Depolarisierungsmittel
depolymerization Depolymerisation
deposit Niederschlag, Ablagerung, Bodensatz (54); Ablagerung, Lagerstätte, Lager (55); Auftrag, Schweißgut (621.791)
deposit of flue dust Flugascheablagerung
to deposit ablagern, niederschlagen, abscheiden, niederschmelzen; umspeichern (681.3)
deposited metal niedergeschlagenes Metall
depositing Ablagerungsvorgang, Absetzen
deposition Ablagerung, Lagerstätte; Aufbringung, Beschichtung
deposition of copper galvanische Verkupferung
to depress herunterdrücken, herabdrücken
to depress a key eine Taste drücken
to depress a pushbutton einen Druckknopf betätigen, auf einen Druckknopf drücken
depression Druckminderung; Depression; Tiefdruckgebiet, Depression (551.5)
depression bar Fühlschiene (625)
depression of the freezing point Gefrierpunktserniedrigung
Deprez-d'Arsonval galvanometer Drehspulinstrument
depth Tiefe; Dicke, Wandstärke; Teufe (622)
depth indicator Teufenanzeiger (622)
depth of penetration Eindringtiefe (54), (621.3)
depth sounding Tiefenlotung
depth of water Wasserstand, Wassertiefe
to derail entgleisen, zur Entgleisung bringen
derating Lastminderung; Unterlastung (621.396.69)
Déri-motor Dérimotor
derivation Ableitung, Abzweigung
derivative Folgeerscheinung, Folgeprodukt, Derivat (54); Differential, Ableitung, Abgeleitete (51)
derivative action D-Verhalten, Vorhalt (621.316.7)
derivative-action time Vorhaltzeit
derivative controller D-Regler
derivative corrector Differenzierentzerrer (621.397)
derivative equalizer Differenzierentzerrer, Differentialentzerrer
n'th derivative n-te Ableitung
to derive ableiten, herleiten (**from:** von); differenzieren, ableiten (51)
derived circuit Überbrückungsschaltung
derived units abgeleitete Einheiten
to descale entrosten
to describe beschreiben
to describe a circle einen Kreis beschreiben
to describe in einbeschreiben
description Beschreibung; Bauart
descriptive beschreibend, kennzeichnend
descriptive legend erläuternde Aufschrift, Erläuterung, Beschriftung
descriptive literature Beschreibung
to desiccate trocknen, austrocknen
desiccating wasserentziehend
desiccation Trocknung
desiccator Trockenschrank, Exsikkator; Trockner
design Entwurf, Konstruktion; Ausführung, Typ
design criterion Entwurfsrichtlinie
design-drawing paper Entwurfpapier
design formula Konstruktionsregel
design merit Konstruktionsgüte; Typenleistung
design principles Konstruktionsgrundlagen
design size Sollmaß
design stress zulässige Belastung, zulässige Beanspruchung
design voltage Konstruktionsspannung, Nennspannung
to design konstruieren, entwerfen; ausführen; zeichnen
designation plate Bezeichnungsschild
designed rating Nennbeanspruchung, Nennlast
designer Konstrukteur, Designer, Formgestalter
to desilver entsilbern
desired value Sollwert, Aufgabesollwert
desk Pult, Schreibpult; Schreibtisch
desk calculator Tischrechner
deskstand telephone set Tischtelefon, Tischapparat
desk switchboard Schaltpult
desktop publishing [Textverarbeitung und Druckvorbereitung]
desk-type telephone Tischapparat, Tischtelefon
Dessauer connection Dessauerschaltung
destination area Zielbereich (621.395)
destination-code marking Zielkennzeichnung (621.395)
destination exchange Zielvermittlungsstelle
destruction Zerstörung, Abbau, Vernichtung; Abbau (54)
destructive zerstörend

destructive distillation trockene Destillation
destructive reading löschendes Lesen, Auslesen mit Zerstörung des Speicherinhalts (681.3)
destructor Müllofen, Abfallverbrennungsanlage
detachable abnehmbar, lösbar
detachable key switch Abzweigschalter mit Aufsteckgriff
detachable link ausbaubares Glied, lösbarer Teil
detachment loser Teil
detail Detail, Einzelheit
detail contrast Feinkontrast (621.397)
to detail detaillieren, spezifizieren
detailed drawing Detailzeichnung
to detect demodulieren (621.39); erfassen (621.396.96); feststellen, entdecken, ermitteln
detection Feststellung, Aufspürung; Erfassung, Ermittlung (54); Demodulation, Rückmodelung, Gleichrichtung (621.396)
detection range Erfassungsreichweite (621.396.96)
detector Detektor; Gleichrichter (621.396); Nullinstrument (621.317)
detector branch Nullzweig (621.317)
detector circuit Detektorkreis
detector contact Überwachungskontakt (625)
detector-power efficiency Richtwirkungsgrad, Richtleistungswirkungsgrad (621.382.2)
detector tube Gleichrichterröhre, Demodulatorröhre
detector-voltage efficiency Spannungsrichtverhältnis (621.382.2)
detent Sperrklinke
detent mechanismus Rastwerk
detergent Reinigungsmittel, Entfettungsmittel
determinant Determinante
detonation Detonation, Sprengung; Zündung; Explosion
to detune verstimmen (621.396)
detuning Verstimmung (621.396)
detuning measuring method Verstimmungsmeßverfahren (621.317)
deuterium Deuterium, schwerer Wasserstoff
deuteron Deuteron
to develop (sich) entwickeln; entwickeln (77); sich entwickeln lassen (77)
development Entwicklung; Entwicklung (77)
development of gas Gasentwicklung
development of a surface Abwicklung einer Fläche
development system Entwicklungssystem (681.3)
deviable ablenkbar
to deviate ableiten; abweichen
deviation Abweichung; Ablenkung; Deviation (538); Frequenzhub (621.396)
deviation of clock Gang (681)
deviation factor of a wave prozentuale Abweichung von der Sinusform
deviation meter Hubmeßgerät
deviation from the rated value Regelabweichung, Abweichung vom Nennwert
deviation ratio Frequenzhub, Hubverhältnis (621.396)
deviation from reference value Abweichung vom Sollwert
device Einrichtung, Gerät, Vorrichtung, Apparat
dew Tau
dew point Taupunkt
dew-point hygrometer Taupunkthygrometer
Dewar flask Dewarsche Flasche
to dewater entwässern, Wasser entziehen
dewatering Entwässerung, Wasserentziehung
dewatering centrifuge Trockenschleuder
dewatering tower Trockenkolonne
dextrorotary rechtsdrehend (53), (54)
D.F. = direction finding Peilung
D.F. equipment Peilgerät, Peilausrüstung
D.F. loop Peilrahmen
D.F receiver Peilempfänger
3-D-hearing plastisches Hören, 3-D-Hören
dia. = diameter Durchm. = Durchmesser
diacathode rays Kanalstrahlen
diagnostic routine Fehlersuchprogramm, Diagnoseprogramm (681.3)
diagonal Diagonale, diagonal
diagram Diagramm, Schaubild, Zeichnung, graphische Darstellung, Kurvenblatt
diagram panel (optische) Meldetafel (625)
diagram of reception Richtcharakteristik
to diagram graphisch darstellen, schematisch darstellen
diagrammatic(al) schematisch, in Kurvenform dargestellt, schaubildlich
diagramatic display Betriebsschaubild
diagrammatic(al) representation schematische Darstellung
diagrammatic(al) symbol Schaltzeichen
dial (runde) Skala, Skalenscheibe; Zifferblatt; Kompaßscheibe; Wählerscheibe, Nummernscheibe (621.395)
dial central office Wählamt, Wählvermittlung, automatisches Amt
dial cord Skalenseil
dial converter Wahlumsetzer
dial-decade box Dekadenkurbelwiderstand

dial-decade condenser Dekaden-Kurbelkondensator
dial exchange Wählvermittlung
dial illumination Skalenbeleuchtung
dial impulse Wählimpuls
dial indicator Meßuhr
dial instrument Anzeigeinstrument
dial lamp Skalenlampe (621.396.62)
dial-pattern bridge Doppel-Brücke (621.317)
dial pointer Skalenzeiger (621.317)
dial-pulse springs Nummernschalter-Impulskontakt
dial pulsing Nummernschalterwahl
dial reading Skalenablesung
dial-signal receiver Wahlaufnahme (621.395)
dial setting Skaleneinstellung
dial switch Nummernschalter (621.395)
dial-switching system Wählvermittlungssystem
dial-telegraph system Zeigertelegraphie
dial-telephone set Wählapparat, Wähltelefon
dial-telephone system Wählsystem, Selbstanschlußsystem
dial tone Amtsfreizeichen
dial train Zählwerksgetriebe (621.317)
to dial wählen (621.395)
dialling Wahl
dialling register Wahlspeicher
dialling tone Amtszeichen (621.395)
diamagnetic diamagnetisch
diamagnetism Diamagnetismus
diameter Durchmesser; Stärke, Dicke
diametral Durchmesser-...
diametral pitch 1/Modul, pi/Zahnteilung (621.833)
diametral voltage Durchmesserspannung (621.313)
diametral winding Durchmesserwicklung
diamond Diamant; Rhombus
diamond antenna Rhombusantenne
diamond bit Diamantbohrkrone
diamond crossing spitzwinklige Kreuzung
diamond-shaped rhombisch, rautenförmig
diamond-type coil Ankerformspule (621.313.2)
diaphragm Diaphragma, Trennwand; Membran; Membran (621.395); Blende (535)
diaphragm current osmotischer Strom
diaphragm gauge Aneroidbarometer
diaphragmless microphone membranloses Mikrophon
diathermal Diathermie-..., wärmedurchlässig
diathermal heating Diathermie, Diathermiebehandlung
diathermic coagulation Hochfrequenz-Koagulierung
diathermic surgery HF-Chirurgie
diathermy Diathermie
diatomic zwei-atomig
dichromate element Dichromatelement
dichromatic zweifarbig
dictaphone Diktaphon, Diktiergerät
dictating machine Diktiermaschine
die (pl.: **dies**) Schnitt, Preßwerkzeug (621.9); Kokille (621.7); Prägestempel; Ziehstein; Schneidbacke (621.9); Sockelplatte, Fußleiste (72)
die casting Kokillenguß
die head Schneidkopf (621.9); (Nieten-Setzkopf)
die steel Matrizenstahl, Werkzeugstahl
to die away (down, out) abklingen, vergehen, ausschwingen
dielectric Dielektrikum
dielectric dielektrisch, isolierend
dielectric absorption Nachwirkungsverlust, dielektrische Absorption
dielektric aerial Stielstrahler; dielektrische Antenne
dielectric-coated waveguide dielektrisch beschichtetes Hohlkabel
dielectric coefficient Dielektrizitätskonstante
dielectric constant Dielektrizitätskonstante
dielectric current dielektrischer Strom
dielectric displacement dielektrische Verschiebung
dielectric flux density dielektrische Flußdichte, elektrische Induktion, Verschiebung
dielectric guide dielektrischer Leiter
dielectric heating HF-Heizung, HF-Erwärmung, dielektrische Erwärmung, kapazitive Erwärmung
dielectric hysteresis dielektrische Nachwirkung
dielectric hysteresis loss dielektrischer Hysteresisverlust
dielectric loss dielektrischer Verlust
dielectric loss angle Verlustwinkel
dielectric loss factor Verlustfaktor ($\varepsilon \cdot \tan \delta$)
dielectric loss index Verlustfaktor ($\varepsilon \cdot \tan \delta$)
dielectric parameter-measuring set Dielektrikum Meßgerät
dielectric phase difference Verlustwinkel
dielectric polarization dielektrische Polarisation
dielectric property isolierende Eigenschaft, dielektrische Eigenschaft
dielectric resistance Isolationswiderstand
dielectric rigidity Durchschlagsfestigkeit
dielectric rod dielektrischer Draht

dielectric strain elektrische Beanspruchung
dielectric strength (dielektrische) Durchschlagsfestigkeit
dielectric stress elektrische Feldstärke, elektrische Beanspruchung
dielectric susceptibility elektrische Suszeptibilität
dielectric test Spannungsprüfung
dielectric transmission line dielektrische Übertragungsleitung
dielectric viscosity dielektrische Nachwirkung
dielectric wave guide dielektrischer Wellenleiter
dielectric wire dielektrischer Leiter
Diesel Diesel, Dieselmotor
Diesel-driven generating set Dieselaggregat
Diesel-electric drive Dieselelektrischer Antrieb
Diesel-electric locomotive Dieselelektrische Lokomotive
difference Unterschied, Differenz
difference of degree Unterschied dem Grade nach
difference frequency Differenzfrequenz
difference in frequency Abweichung in der Frequenz
difference in phase Phasenunterschied
difference of potential Potentialdifferenz, Spannung
difference in pressure Druckdifferenz, Druckunterschied
difference tone Schwebungston, Differenzton
different verschieden, unterschiedlich (**from:** von), (**in:** hinsichtlich)
differential Abweichung, Regelabweichung; Differential (51)
differential differentiell, Differential-...
differential amplifier Differentialverstärker, Differenzverstärker
differential analyzer Differentialanalysator
differential arc lamp Differentialbogenlampe
differential barometer Differentialbarometer
differential calculus Differentialrechnung, Differentialkalkül
differential capacitor Differenzial-(Dreh-)Kondensator
differential compounding Gegenkompoundierung
differential compound winding Gegenkompoundwicklung
differential connection Vergleichsschaltung (621.316.9)
differential current relay Differenzstromrelais
differential duplex system Differentialgegenverkehrssystem (621.394)
differential electromagnet Differential-Magnet
differential equation Differentialgleichung
differential excitation Differentialerregung, Gegenverbunderregung
differential fading selektiver Schwund
differential gain differentiellerVerstärkungsfehler
differential galvanometer Differential-Galvanometer
differential gear Differentialgetriebe, Differential
differential generator Differentialgeber, Drehmelder (621.316.7)
differential input Differentialeingang (621.375)
differential iron tester Differential-Epsteinapparat
differential leakage doppelt verkettete Streuung (621.313)
differential manometer Differentialmanometer
differential microphone Differentialmikrophon
differential mode voltage amplification Differenz-Spannungsverstärkung (621.375)
differential modulation Konsekutivmodulation
differential motor Differentialempfänger, Drehmelder (621.316.7)
differential operator Differentialoperator
differential permeability differentielle Permeabilität
differential phase differentieller Phasenfehler
differential pressure gauge Differentialmanometer
differential protection Differentialschutz
differential pulley Differentialflaschenzug
differential relay Differential-Relais
differential repeater Differentialempfänger, Drehmelder (621.316.7)
differential resistance differentieller Widerstand
differential series winding Gegenreihenschlußwicklung
differential shunt winding Gegennebenschlußwicklung
differential susceptibility differentielle Suszeptibilität
differential synchro Differentialdrehmelder (621.316.7)
differential transformer Differentialtransformator
differential transmitter Differentialgeber
differential winding Differentialwicklung
differentially wound motor Kompoundmotor

to differentiate differenzieren (51); unterscheiden, differenzieren, einen Unterschied machen
differentiating circuit Differenzierschaltung, Differenzierkreis, differenzierendes Netzwerk
differentiation Differentiation, Differenzierung, Ableitung (51); Unterscheidung
differentiator Differenzierschaltung, Differenzierungseinrichtung, Differentiator
difficulty Schwierigkeit
difficulty of access schwierige Zugänglichkeit, Unzugänglichkeit
to diffract beugen (535)
diffraction Beugung (535)
diffractive beugend
diffractive edge Beugungskante
diffractive grating Beugungsgitter
diffusate Diffusionsprodukt
to diffuse diffundieren (**to:** nach)
diffused junction diffundierter Übergang, diffundierter Zonenübergang
diffused transistor diffundierter Transistor
diffuse verwirrt; diffus, unscharf; wirr
diffuse illumination diffuse Beleuchtung
diffuse image verschwommenes, unscharfes Bild
diffuse reflecting power diffuses Reflexionsvermögen
diffuse reflection diffuse Reflexion
diffuse reflection factor Koeffizient der diffusen Reflexion
diffuse transmission diffuse Durchlässigkeit (535)
diffuse transmission factor Koeffizient der diffusen Durchlässigkeit
diffused lighting indirekte Beleuchtung; diffuses Licht
diffuser Diffusor
diffusing Streu-...
diffusing effect Lichtstreuung, Streuwirkung
diffusing media lichtstreuende Stoffe
diffusing power Streuvermögen
diffusing screen Mattscheibe
diffusing surface lichtstreuende Fläche
diffusion Diffusion, Streuung; Verteilung
diffusion furnace Diffusionsofen
diffusion length Diffusionslänge (629.382)
diffusion potential Diffusionsspannung (621.382)
diffusive streuend, diffundierend, Diffusions-...
diffusive flow Diffusionsströmung
digit Ziffer, Stelle (51)
digit delay Ein-Bit-Verzögerungsglied
digit emitter Impulsgeber
n-digits figure n-stellige Zahl
digital Digital...
digital-to-analog converter Digital-Analogwandler
digital audio tape Digitaltonband
digital circuit Digitalschaltung, logische Schaltung
digital computer Digitalrechner
digital control digitale Prozeßsteuerung
digital data conversion enquipment Digitalumsetzer
digital display Digitalanzeige, Digitalbildschirmdarstellung
digital electrical quantity digitale elektrische Größe
digital indication Digitalanzeige
digital measuring instrument Digital-Meßgerät, digitalanzeigendes Meßinstrument
digital meter Digitalmeßgerät
digital readout digitale Anzeige
digital signal digitales Signal
digital transducer Digitalumsetzer
digital voltmeter Digitalvoltmeter
to digitize digitalisieren, digital umsetzen
digitizer A/D-Umsetzer
dihedral angle V-Winkel, V-Formwinkel (629.13)
DIL = dual in line
to dilate (sich) dehnen
dilatibility Dehnbarkeit
dilatation Dehnung
dilatometer Dilatometer, Dehnungsmesser
to dilute verdünnen (**to:** auf)
dilute verdünnt
dilute solution verdünnte Lösung
diluent Verdünnungsmittel
dilution Verdünnung
to dim abblenden
dimmed light gedämpftes Licht
dim light Abblendlicht
dimension Dimension, Abmessung; Dimension (51)
dimension saw Formatsäge (67)
dimensional dimensional; Abmessungs-...
dimensional elastic range elastischer Bereich (620.1)
dimensioning rules Bemessungsregeln
dimensionless dimensionslos
to diminish verkleinern, vermindern (62); abklingen (621.37)
diminished pressure verminderter Druck, Unterdruck
diminution Verkleinerung, Abnahme (**of:** von), (**in:** an)

dimmer Verdunkelungsregler, Abblendregler
dimmer control Helligkeitssteuerung
dimmer switch Abblendschalter
dimmer transformer Verdunkelungstransformator
dimming Verdunkelung
dimming device Saalverdunkelungseinrichtung
dimming resistance Verdunkelungswiderstand (621.32)
dimming resistor Verdunkelungswiderstand (621.32)
dimming switch Verdunkelungsschalter (621.32)
diode Diode, Zweipolröhre
diode detection Diodengleichrichtung
diode detector Diodengleichrichter
diode rectification Diodengleichrichtung
diode thyristor Thyristordiode
diode-transistor logic Dioden-Transistor-Logik, DTL-Technik
dionic water tester Leitfähigkeitsmesser (für Kesselspeisewasser)
diopter Diopter, Visiergerät
dioptric dioptrisch
dioxide Dioxid
DIP = dual in-line package Dual-In-Line-Gehäuse, DIP
DIP-connector DIL-Steckverbinder, DIL-Stecker
dip Durchhang; Einsattelung (621.396); Inklination, magnetische Inklination (538); plötzlicher Ausschlagrückgang, Zucken (des Zeigers) (621.317)
dip plating Tauchgalvanisierung
dip soldering Tauchlötung
dip switch Tastschalter
to dip tauchschwingen (629); tauchen, eintauchen, tauchfärben, tauchverzinnen; untergehen (52); sich senken (625)
dipping bath Tauchbad
dipping movement Nickbewegung (629.11)
dipping needle Inklinatorium, Inklinationsnadel
dipping process Tauchprozeß
dipping varnish Tauchlack
diphase (rare) Zweiphasen-...
diplex operation Diplexbetrieb
diplex wireless reception Doppelwellenempfang
diplexer Diplexer, Trennweiche (621.396)
dipolar doppelpolig
dipole Dipol (53), (621.396)
dipole aerial Dipolantenne
dipole array Dipolanordnung, Dipolwand
dipole moment Dipolmoment
dipole oscillator Dipol-Oszillator
dipped electrode getauchte Elektrode
dipper Tauchkessel
dipper relay Tauchkontaktrelais
direct direkt, unmittelbar
direct access Direktzugriff
direct-acting recording instrument Zeigerschreibinstrument; Tintenschreiber
direct addressing direkte Adressierung
direct-arc furnace Héroult-Ofen
direct axis magnetische Längsachse, Polachse, Längsfeldachse (621.313)
direct-axis component of armature current Längsfeldkomponente des Statorstromes
direct-axis component of mmf Längsdurchflutung (621.313)
direct-axis subtransient open-circuit time constant Subtransientzeitkonstante des Längsfeldes bei Leerlauf
direct-axis subtransient reactance Subtransientreaktanz des Längsfeldes; Anfangsreaktanz
direct-axis subtransient short-circuit time constant Subtransientzeitkonstante des Längsfeldes bei Kurzschluß
direct-axis synchronous reactance synchrone Längsfeldreaktanz, Synchronreaktanz
direct-axis transient open-circuit time constant Transientzeitkonstante des Längsfeldes bei Leerlauf
direct-axis transient reactance Transientreaktanz des Längsfeldes
direct-axis transient short-circuit time constant Transientzeitkonstante des Längsfeldes bei Kurzschluß, transiente Kurzschlußzeitkonstante
direct-axis voltage Längsfeldspannung, Längsfeld-EMK, Haupt-EMK
direct capacitance (US) Teilkapazität (C_{mn})
direct-connected excitation Eigenerregung (621.313)
direct control system direkt gesteuertes System (621.395)
direct coupled galvanisch gekoppelt (621.3)
direct-coupled exciter direkt gekuppelte Erregermaschine
direct-coupled generator direkt gekuppelter Generator
direct-coupled transistor logic = DTCL direkt gekoppelte Transistorlogik
direct coupling galvanische Kopplung
direct current Gleichstrom (Zusammensetzungen s. unter **d.c.**)
direct data capture direkte Datenerfassung

direct dialing Selbstwahl
direct distance dialing Fernwahl (621.395)
direct drive direkter Antrieb; Direktantrieb (681.84)
direct-drive motor Achsmotor
direct electromotive force Gleichstrom-EMK
direct exchange line Amtsleitung
direct feeder direkte Speiseleitung
direct(ly) heated direkt geheizt (621.385)
direct-heated cathode direkt geheizte Kathode
direct heating direkte Heizung
direct illumination direkte Beleuchtung
direct key selector Direktrufeinrichtung (621.395)
direct lighting direkte Beleuchtung
direct (telephone) line Teilnehmerleitung (621.395)
direct on-line starting direktes Anlassen
direct manual control direkte Handbetätigung (621.316.5)
direct memory access direkter Speicherzugriff
direct outwart dialling Vollamtsberechtigung
direct operation direkte Betätigung
direct pick-up receiver Ballempfänger
direct-point repeater Relaisgeber (621.394)
direct potential difference Gleichstrompotentialdifferenz, Gleichspannung
in direct proportion direkt proportional
direct ray Bodenwelle (621.396)
direct reading mit Skalenablesung, direkte Ablesung, anzeigend (621.317)
direct-reading direction finder direktanzeigender Peiler (621.396)
direct-reading instrument Skaleninstrument, direkt anzeigendes Instrument
direct-relation telemeter Proportional-Fernmeßgerät
direct resistance coupled amplifier Gleichstromverstärker, direktgekoppelter Verstärker
direct route Direktweg (621.395)
direct suspension direkte Aufhängung (621.33)
direct-suspension hanger Fahrdrahtaufhänger
direct-suspension system unmittelbare Aufhängung
direct trip direkte Auslösung (621.316.5)
direct voltage Gleichspannung
direct wave direkte Welle, Bodenwelle
direction Richtung; Anweisung, Verfügung, Richtlinie; Direktion, Vorstand; Leitung; Führung
direction cosine Richtungskosinus
direction of current Stromrichtung
direction finder Peiler, Funkpeiler (621.396)
direction-finding Peilung, Peil-...
direction-finding aerial Peilantenne
direktion-finding aid Peilhilfe
direction-finding beam Peilstrahl
direction-finding loop Peilrahmen
direction fixed starr eingespannt (531)
direction indicator Blinklicht, Blinker, Winker (629)
direction of lay Schlagrichtung (621.315.1)
direction of maximum radiation Hauptstrahlrichtung
direction of polarization Polarisationsrichtung
direction of propagation Laufrichtung, Fortpflanzungsrichtung
direction of rotation Drehrichtung
direction switch Fahrtrichtungsschalter
direction of travel Ausbreitungsrichtung (621.395), 621.396)
directional gerichtet, Richtungs-..., Richt-...
directional aerial Richtantenne, Richtstrahlantenne
directional characteristic Richtcharakteristik; Richteigenschaft
directional coupler Richtungskoppler (621.396)
directional gyro Richtkreisel, Lagekreisel
directional microphone Richtmikrophon
directional overcurrent protection Richtungsschutz
directional pattern Richtdiagramm, Strahlungsdiagramm
directional radiator Richtstrahler
directional radio beacon Richtfunkfeuer
directional receiver Richtempfänger
directional reception Richtempfang
directional relay Rückstromrelais, Richtungsrelais
directional tripping magnet Rückstromauslöser
directive richtunggebend; mit Richteigenschaft, Richt-...
directive antenna Richtantenne
directive efficiency Richtfaktor
directive force Richtkraft (538)
directive gain Richtverstärkungsfaktor
directive radiation Richtstrahlung
directive radio beacon Richtstrahlbake
directivity Richtwirkung, Richteigenschaft, Antennengewinn in Strahlungsrichtung
directivity angle Hauptstrahlwinkel (in der Vertikalebene)
director Direktor, Leitantenne,

Führungsantenne (621.396);
Nummernspeicher; Umwerter (621.395)
directory (Datei-)Verzeichnis
directory information service Rufnummernauskunft
directrix Leitlinie
dirigible Zeppelin, lenkbares Luftschiff
"dis" Leitungsbruch
to disable untauglich machen, außer Betrieb setzen, lahmlegen, sperren
disable pulse Sperrimpuls
disabling relay Sende-Empfang-Relais (621.396)
disacidification Entsäuerung
disappearing filament pyrometer Glühfadenpyrometer
to disassemble demontieren, auseinandernehmen
disassembler Disassembler (681.3) [Rückübersetzungsprogramm von Maschinencode in Assembler]
to disburse heat Wärme abgeben
disc Scheibe, (runde) Platte; Teller (s. auch **disk**)
disc antenna Scheibenantenne
disc armature Scheibenanker
disc brake Wirbelstrombremse (621.3)
disc capacitor Plattenkondensator
disc commutator Flachkommutator, Scheibenkollektor
disc discharger rotierende Funkenstrecke (621.396)
disc drive Plattenlaufwerk (681.3)
disc file Plattendatei
disc insulator Kappenisolator
disc operating system Plattenbetriebssystem
disc recording Schallplattenaufnahme, Schallplattenschneiden
disc record reproducing equipment Schallplatten-Abspielgerät
disc-seal valve Scheibenröhre
disc spacer Abstandscheibe
disc storage Plattenspeicher
disc-type capacitor Scheibenkondensator
disc valve Tellerventil
disc winding Scheibenspulenwicklung
discharge Entladung (621.3); Ausguß, Liefermenge, Schüttmenge (621.9); Ausströmen, Ausströmung
discharge channel Entladungskanal (621.3)
discharge current Entladestrom; Ableiterstrom (621.316.9)
discharge cock Ablaßhahn
discharge electrode Niederschlagselektrode, Sprühelektrode; Abscheidungselektrode (628)
discharge-inception voltage Ionisationsansprechspannung
discharge key Entladeschalter
discharge lamp Entladungslampe
discharge path Entladungsweg, Entladungsbahn
discharge rate Entladegeschwindigkeit
discharge rays (pl.) ionisierende Strahlung
discharge resistance Überbrückungswiderstand
discharge spark Entladungsfunke
discharge surge Entladungsstoß
discharge tube Entladungsröhre
discharge-tube rectifier Röhrengleichrichter
discharge velocity Ausflußgeschwindigkeit
discharge voltage Entladespannung
to discharge entladen, abführen; ablassen; ausfließen; sich entladen; entlasten; entlassen; abschießen, abfeuern
to discharge a battery eine Batterie entladen
to discharge a condenser einen Kondensator entladen
discharging current Entladestrom
discharger Parallelfunkenstrecke (621.315); Widerstandszünder (662.2)
discoid scheibenförmig
to discolo(u)r verfärben; sich verfärben
discoloration Verfärbung, Mißfärbung
discone antenna Scheiben-Konus-Antenne, Diskon-Antenne
to disconnect abschalten, ausschalten, abklemmen (621.3); trennen (621.39); abtrennen, entkuppeln (625)
disconnect signal Schlußzeichen
disconnected abgetrennt, isoliert (621.3); abgetrennt, entkuppelt
disconnected grid freies Gitter, offenes Gitter
disconnected part Teil mit nichtdefiniertem Potential
disconnecting Ausschaltung, Abschaltung, Abschaltvorgang
disconnecting contact Abschaltkontakt
disconnecting lamp Schlußlampe (621.395)
disconnecting link Auslöseglied; Trennstück; Trennschalter
disconnecting switch Trennschalter
disconnecting transformer Trenntransformator
disconnection Abschaltung, Ausschaltung (621.3); Trennung, Trennstelle
disconnector Trennverbinder
disconnexion (s. **disconnection**)
to discontinue unterbrechen, nicht fortsetzen
discontinuity Unstetigkeit
discontinuous diskontinuierlich, unstetig, unterbrochen

discontinuous function Sprungfunktion; unstetige Funktion
discontinuous integration Spitzenzählung (621.317)
discontinuous operation Aussetzbetrieb
discontinuous running AB = Aussetzbetrieb
discontinuously integrating meter Spitzenzähler
to discriminate unterscheiden, unterschiedlich behandeln, diskriminieren
to discriminate against ausscheiden; sperren (621.392); benachteiligen
to discriminate in favour of bevorzugen; durchlassen (621.392)
to be discriminated against ausgesiebt werden, abgesperrt werden (621.392)
discriminating Rückstrom-..., Selektiv-...
discriminating circuit-breaker Rückstromschalter
discriminating cut-out Rückstromautomat
discriminating protective system Selektivschutzsystem
discriminating relay Rückstromrelais
discrimination Unterscheidung; Auflösung (621.396.96)
discriminator Doppeltarifzähler, Blocktarifzähler (621.317); Diskriminator (621.396.96)
disembodied körperlos, massefrei
to disengage auskuppeln, ausrücken; losmachen; ausklinken; freimachen (54)
to disengage from losmachen von; sich ablösen von
disengageable ausklinkbar
disengaged nicht besetzt, frei (621.395); nicht im Eingriff (621.8)
disengaged line freie Leitung
disengaged selector freier Wähler
disengagement Freisetzung, Freimachen, Freiwerden (54)
dish Teller, Schale, Schüssel
dish antenna Parabolantenne
dish reflector konkaver Reflektor, Parabolreflektor
dishwasher Spülmaschine (64)
to dish tiefen, tellerförmig eindrücken
disintegrable zersetzbar, zersetzungsfähig
to disintegrate zersetzen; zerfallen, verwittern; zerkleinern
disintegration Zersetzung, Zerfall, Verwitterung; Zerkleinerung
disintegration constant Zerfallskonstante
disintegration energy Zerfallsenergie
disintegration of filament Fadenzerstäubung (621.395)
disintegrator Desintegrator, Schlagwolf; Zerkleinerer
disjunction Disjunktion
disk Scheibe (s. **disc**); Magnetplatte (681.3)
disk drive Plattenlaufwerk
disk file Plattendatei
diskette floppy disk, diskette
diskette operating system Disketten-(Betriebs)System
disk storage Plattenspeicher
"dis-lead" Strippenbruch, Leitungsunterbrechung (621.3)
to dislocate auseinanderzerren, (gewaltsam) trennen
dislocation (gewaltsame) Trennung; Versetzung (621.382)
to dislodge hinauswerfen; freimachen
to dislodge electrons Elektronen befreien
to dismantle öffnen, abbauen, demontieren, abmontieren
to dismantle a transformer einen Transformator aufnehmen
dismantling Abbau, Demontage
to dismount demontieren
disordering Fehlstellenerzeugung (621.382)
dispensation Nachlieferung (621.385)
to dispense with überflüssig machen, weglassen
dispenser cathode with reservoir Vorratsbehälterkathode
dispersable dispergierbar
dispersal Zerstreuung, Dispersion
to disperse fein verteilen, dispergieren; zerstreuen
dispersion Dispersion, Zerstreuung; Vereinzelung, Auftrennung, Aufteilung
dispersion coefficient Streukoeffizient (621.313)
to displace verlagern; verdrängen; (seitlich) verschieben
displacement Verdrängung, Verschiebung; Verlagerung; Verwerfung (55); Verdrängung, Deplacement (629.12); Auslenkung (534); Distanzadresse (681.3)
displacement current Verschiebungsstrom
displacement-sensitive control wegabhängige Steuerung
display Ausstellung, Schau; Anzeige, Darstellung, Wiedergabe; Schirmbilddarstellung, Sichtgerät (621.396.96)
display console Bildschirmkonsole, Sichtgerät
display register Anzeigeregister
display scope Bildschirm
display statement Ausgabebefehl

display terminal Sichtgerät, Bildschirm (-Ein)-Ausgabegerät
display unit Sichtgerät (621.397); Anzeigeeinheit
to display darstellen, zeigen, zur Schau stellen, zur Schau tragen; entfalten, aufweisen
disproportionate unverhältnismäßig
disruptive zerreißend; plötzlich, abrupt; brisant (662.2)
disruptive breakdown Durchschlag (621.3)
disruptive discharge Funkenentladung, plötzlicher Durchschlag
disruptive electric field strength Durchbruchsfeldstärke
disruptive explosive brisanter Sprengstoff
disruptive spark Durchschlagfunken
disruptive strength Durchschlagfestigkeit (621.3)
disruptive voltage Durchschlagspannung, Durchbruchspannung
to dissipate abführen, zerstreuen; verbrauchen, (in Wärme) verwandeln
to dissipate energy Energie verbrauchen, Energie als Wärme abgeben
to dissipate energy into heat Energie in Wärme verwandeln
to dissipate heat Wärme abführen, Wärme abgeben
dissipation Zerstreuung, Abgabe; Verlustleistung (621.385)
dissipation factor dielektrischer Verlustfaktor
dissipationless verlustlos (621.3)
dissipative verlustbehaftet, dämpfend (621.3)
dissipator Ableiter (621.316.9)
to dissociate (sich) dissoziieren
dissociated component Dissoziationsprodukt
dissociation Dissoziation
dissolubility Löslichkeit
dissoluble löslich
to dissolve lösen, auflösen (54)
to dissolve off herauslösen, herunterlösen
dissolvent Lösungsmittel
distance Abstand, Entfernung
distance apart gegenseitiger Abstand
distance control Fernsteuerung (621.398); Wegregelung (621.316.7)
distance getting Reichweite (621.396)
distance-marking light Entfernungsfeuer
distance-measuring equipment Entfernungsmeßanlage, DME (621.396.93)
distance occupied by Abstand zwischen
distance piece Abstandsstück
distance protection Distanzschutz
distance relay Distanzrelais
distance signal Vorsignal (625)
distant entfernt, weit weg
distant field Fernfeld (621.371)
distant station Gegenstelle
to distil destillieren
to distil over überdestillieren
distilled water destilliertes Wasser
distillation Destillation
distinctive unterscheidend, Unterscheidungs-..., Kenn-..., unterschiedlich
distinctive feature Unterscheidungsmerkmal
distinctive mark Kennmarke, Kennung
to distort verzerren, verformen
distorted waves (pl.) nichtsinusförmige Kurvenform, verzerrte Kurvenform
distortion Verzerrung (535), (621.3); Verformung, Verziehen (621); Verzeichnung (621.397)
distortion bridge Klirrfaktormeßbrücke
distortion circuit Verzerrerschaltung, Verzerrer
distortion factor Klirrfaktor, Verzerrungsfaktor
distortion-factor meter Klirrfaktormeßgerät
distortion of field Feldverzerrung
distortion-measuring set Verzerrungsmesser
distortion power Verzerrungsleistung
distortionless verzerrungsfrei
distortionless line verzerrungsfreie Leitung (621.395)
distress signal Notruf, Notsignal
distress traffic Notverkehr
to distribute verteilen
distributed amplifier Leitungsverstärker, Kettenverstärker
distributed capacity verteilte Kapazität
distributed capacity due to wiring Schaltkapazität
distributed floodlight system Flutlichtanleuchtung
distributed inductance verteilte Induktivität
distributed system dezentrales System
distributed winding verteilte Wicklung (621.313)
distributing Verteilungs-...
distributing box Verteilerdose, Verteilerkasten
distributing centre Verteilungsstation, Unterstation
distribution coefficient Spektralwert
distributing frame Verteilergestell, Verteiler
distributing mains Hauptverteilungsleitung
distributing network Verteilungsnetz
distributing pillar Verteilersäule (621.315)
distributing point Verteilerpunkt (621.315)
distributing system Verteilungsnetz

distribution Verteilung; Ausgabe
distribution board Verteilerschalttafel
distribution box Verteilerkasten, Verteilungsdose
distribution centre Verteilungspunkt, Hauptverteilung (696)
distribution of electricity Stromverteilung (621.316)
distribution factor Sehnungsfaktor (621.313)
distribution feeder Verteilungsleitung (696.6)
distribution frame Verteilergestell, Verteiler (621.39)
distribution fuse board Verteiler-Sicherungstafel
distribution line Stamm (621.396.67); Verteilerleitung
distribution main Verteilungssammelschiene
distribution network Verteilungsnetz
distribution panel Verteilertafel
distribution pillar Verteilersäule (621.315)
distribution of power Energieverteilung, Kraftverteilung
distribution switchboard Verteilungsschalttafel
distribution transformer Netztransformator
distribution trunk line Versorungsleitung
distributive distributiv (51)
distributor Verteiler
distributor duct Kabelkanal (aus Formsteinen)
to distune verstimmen
distuning Verstimmung
to disturb stören
disturbance Störung
disturbing störend, Stör-...
disturbing current Störstrom
disturbing effect Störwirkung
disturbing noise Störgeräusch
disturbing voltage Störspannung
to disuse nicht mehr benutzen, die Benutzung einstellen
dither voltage Zitterspannung
diurnal täglich, Tages-...
diurnal range täglicher Bereich, tägliche Schwankung
diurnal variation täglicher Gang
div div (51)
divalent zweiwertig
to diverge auseinanderlaufen, divergieren, auseinandergehen
divergence Divergenz (51); Divergenz, Öffnungswinkel; Apertur (535)
divergent auseinanderlaufend, divergierend; kritisch (539.17)
diversity Vielfalt
diversity factor Gleichzeitigkeitsfaktor (621.31)
diversity reception Mehrfachempfang, Diversity-Empfang
diverter regelbarer Nebenwiderstand, Ableitwiderstand; Überspannungsableiter (621.316.9)
diverter-pole generator Streufeldgenerator
diverter-type field rheostat Feldüberbrückungswiderstand
divide line Wasserstand, Östand
to divide teilen, dividieren (51); einteilen, teilen; sich teilen
to divide by teilen durch; sich ohne Rest teilen lassen durch (51)
to divide-by-two circuit Halbierschaltung
to divide equally sich gleichmäßig verteilen, sich gleichmäßig aufteilen
to divide into ohne Rest aufgehen in (51); sich teilen in
to divide out ausdividieren
divided verteilt, geteilt
divided winding Wicklung mit parallelen Zweigen
dividend Dividend (51)
divider Teiler, Divisor (51)
dividers (pl.) Zirkel, Stechzirkel
dividing Teil-..., Trenn-..., Scheide-...
dividing box Verzweigungsmuffe (621.315.2)
dividing line Trennstrich, Trennungslinie
dividing wall Trennwand
divisibility Teilbarkeit
divisible teilbar, einteilbar
division Teilung, Division (51); Teilung, Trennung; Teilstrich (621.317); Abteilung, Gruppe
division of the current Stromverteilung (621.385)
divisor Divisor (51)
d.k. test Kapazitätsprüfung (621.315.2)
dl. = decilitre Deziliter
D-layer D-Schicht
DMA = direct memory access
DME = distance measuring equipment
DNC = direct numerical control
to do tun, handeln, verrichten
do-nothing instruction Leerbefehl (681.3)
to do work Arbeit verrichten (**on:** an)
to do work on a body einem Körper Energie zuführen
document-mark reader Markierungsleser
document sorter Belegsortierer
dog Mitnehmer
Doherty transmitter Doherty-Sender
dolby Dolby
Dolezalek electrometer Dolezalek-Elektrometer

dolly Preßstempel, fahrbarer Untersatz
domain Elementarbereich, Weißscher Bezirk (538)
dome Haube; Kuppel; Gewölbe
dome cover gewölbter Deckel
dome-shaped gewölbt
domed gewölbt
domed head screw Linsenkopfschraube
domestic häuslich, Haushalts-...
domestic appliance Haushaltsgerät
domestic application Haushaltsgebrauch, Haushaltsgerät
domestic consumer Haushaltsverbraucher
domestic demand Haushaltsbedarf
domestic facilities Haushaltsmaschinen
domestic installation Hausinstallation, Haushaltsinstallation
domestic refrigerator Haushaltskühlschrank
domestic telephone Hausfernsprecher
domestic wiring Hausinstallation
dominant dominant, vorherrschend, beherrschend, ausschlaggebend
dominant mode Haupttypus, Hauptform (621.396)
dominant wave Grundwelle, Hauptwelle (621.37)
donkey engine Hilfsmaschine
donkey pump Hilfspumpe
donor Donator (621.382)
door Tür, Tor
doorbell Türglocke
door contact Türkontakt
door interlock Türverriegelung
door knob Türknopf
door knob tube Tellerröhre, Knopfröhre
door loudspeaker Türlautsprecher
doormen converser Türsprechanlage
door opener Türöffner
door switch Türschalter
dopant Dotiermittel
to dope impfen, dotieren, dopen
Doppler averaged cross section Doppler gemittelter Wirkungsquerschnitt (529)
Doppler frequency Dopplerfrequenz
Doppler Navigation Doppler-Navigation
Doppler's principle Dopplersches Prinzip
DOS = disk operating system
dosage meter Dosiermesser, Dosismeßgerät
dose Dosis
dose-rate meter Intensimeter (589)
to dose dosieren
dosimeter s. dosage meter
dot Punkt, Morsepunkt; Zeichenpunkt; Punkt (als Zeichen der Differentiation nach der Zeit)
dot-cycle Periodendauer der Punktfolge, Schrittperiode (621.394)
dot frequency Punktfrequenz
dot interlace Punktsprungverfahren
dot-matrix printer Punktmatrixdrucker
dot product skalares Produkt, inneres Produkt
dot sequential Punktfolge
to dot punktieren, stricheln
dotted line punktierte Linie
to double doppeln, verdoppeln
to double-Vee groove aus-ixen (621.791)
double doppelt, zweifach, Doppel-...
double-acting doppelt wirkend
double-acting steam engine doppelt wirkende Dampfmaschine
double armature Doppelanker
double-arm signal zweiflügeliges Signal (625)
double-beam oscilloscope Doppelstrahl-Oszillograph, Zweistrahloszillograph
double-beam radar Radar mit zwei Richtkeulen
double bond Doppelbindung (54)
double break Zweifachunterbrechung
double-break circuit-breaker Schalter mit doppelter Unterbrechung
double-break switch zweipoliger Schalter
double bridge Thomsonsche Brücke
double-button microphone Differentialmikrophon
double-catenary suspension Doppeltragdrahtaufhängung (621.33)
double coating zweimaliger Anstrich
double-commutator motor Doppelkommutatormotor
double contact wires (pl.) Doppelfahrdraht, Doppelfahrleitung
double convex lens bikonvexe Linse
double-cotton covered zweimal baumwollisoliert = BB
double-covered doppelt gelascht (621.884); doppelt isoliert (621.3)
double-current generator Doppelstromgenerator
double-current sounder Doppelstrom-Ticker
double-current system Doppelstrom-System (621.394)
double-current working Doppelstrombetrieb
double deck motor Asynchronmotor mit Anlaßund Betriebswicklung
double delta connection Doppeldreieckschaltung
double diode Duo-Diode, Doppel-Zweipolröhre

double edged zweischneidig
double-end control doppelseitiger Führerstand (621.33)
double-ended drive doppelseitiger Antrieb, beidseitiger Antrieb
double filament lamp Doppelfadenlampe
double flanged butt weld Bördelschweißung
double frequency oscillations zweiwellige Schwingungen
double frequency shift keying Zweikanalfrequenzumtastung (621.394)
double helical gear Stirnrad mit Winkelverzahnung, Pfeilrad
double image Echobild (621.397)
double-layer winding zweilagige Wicklung (621.313)
double-linkage leakage doppelt verkettete Streuung
double modulation Stufenmodulation
double modulator Modulator mit Trägerunterdrückung
double pentode Doppel-Fünfpolröhre
double phantom circuit Achterschaltung (621.39)
double-point information Doppelmeldung (621.398)
double-pole zweipolig
  double-pole circuit-breaker zweipoliger Schalter
  double-pole cut-out doppelpoliger Ausschalter, doppelpoliges Schütz
  double-pole double-throw switch zweipoliger Umschalter
  double-pole fuse zweipolige Sicherung
  double-pole switch zweipoliger Schalter
  double-pole two-way switch zweipoliger Umschalter
double-pressure-... Zweidruck-...
double reception doppelwelliger Empfang
double reduction drive doppeltes Untersetzungsgetriebe
double refraction Doppelbrechung
double-row zweireihig
  double-row ball bearing zweireihiges Kugellager
double screen Mehrschichtenbildschirm
double-sideband transmission Zweiseitenbandübertragung
double-sided beidseitig
double-silk covered wire doppeltseidenisolierter Draht
double spiral coil bifilare Wicklung
double squirrel-cage induction motor Doppelkäfig-Induktionsmotor
double strain yoke Abstandshalter (621.315.1)
double strip Bimetallstreifen
double subscripts Doppelindizes
double tariff Doppeltarif (621.317.8)
  double-tariff meter Doppeltarifzähler
double-throw circuit-breaker Umschalter
double-throw switch Umschalter, Wechselschalter
double transmission zweiwellige Ausstrahlung (621.396.4)
double-trolley system Obus-Fahrdrahtanlage, Doppelfahrdrahtanlage
double Vee groove Doppelkeilnut, X-Nut (621.791)
double-way rectifier Zweiweggleichrichter
double-winding synchronous generator Zweiwicklungs-Synchrongenerator
double-wire system zweipolige Installation
double-yoke method Doppeljochmethode (528)
double zig-zag connection Doppel-Zickzack-Schaltung
doubly-fed motor doppeltgespeister Motor, Repulsionsmotor
  doubly-fed polyphase shunt commutator motor ständergespeister Mehrphasen-Nebenschlußmotor
doubly re-entrant winding zweigängige Wicklung
doubler Verdoppler
  doubler circuit Verdopplerschaltung
  doubler stage Verdopplungsstufe
doublet Dipol
  doublet aerial Dipolantenne
doubling Dopplung (621.7); Verdopplung
doughnut Toroid
  doughnut-shaped toroidförmig
dovetail Schwalbenschwanz
  dovetail joint Schwalbenschwanzverbindung
  dovetail key schwalbenschwanzförmiger Keil
    doventail keyway Schwalbenschwanznut
  dovetail teeth eingesetzte Zähne, Prismen (621.313)
to dovetail mit Schwalbenschwanz ineinanderfügen, zinken (68)
dovetailing Verzinkung (68)
dowel Dübel
to dowel verdübeln
down herunter, herab, hinunter, hinab
  down conductor Ableitung (699.83)
  down grade Gefälle
  down lead Antennenableitung
    down-lead cable Niederführungskabel
  down the line längs der Leitung (621.3)
  down rule vertikaler Strich, Längsstrich

downtime Ausfallzeit, Stillstand
downhand welding Abwärtsschweißung
downward nach unten zu
downward-acting nach unten wirkend, Druck-...
downward-acting lever Druckhebel
downward flux nach unten gerichteter Lichtstrom
doz. = dozen Dtz. = Dutzend
d.p. = double pole zweipolig (s. dort)
dr. = drachm, dram (s. dort)
drachm Drachme (Gewichtseinheit vorzugsweise im Apotheker-System; 1 Drachme = 1/8 Unze = 3,89 g)
draft Entwurf, Zeichnungsentwurf
draft copy Entwurfsexemplar
drafting machine Zeichenmaschine
draftsman Zeichner, Detailkonstrukteur
drag Luftwiderstand, Wasserwiderstand, Schleppwiderstand; Hemmschuh, Bremsschuh
drag coefficient Widerstandsbeiwert (c_w)
drag-in mitgeschleppte Flüssigkeit (621.357)
drag hand Schleppzeiger
drag-out rausgeschleppte Flüssigkeit (621.357)
drag rope Schlepptau
to drag nachschleppen, schleppen, hemmen
dragging brake anliegende Bremse
drain Drain, Senke; Absaugelektrode (621.382.3); Drainage; Kanalisationsrohr, Abfluß, Ablauf; Abzugsgraben
drain current Drainstrom (621.382)
drain cut-off current Drainreststrom (621.382)
drain region Drainzone (621.382)
drain-source voltage Drain-Source-Spannung (621.382)
drain terminal Drain-Anschluß (621.382)
to drain entwässern, drainieren; abziehen, entleeren, ablaufen lassen
to drain away abführen, ableiten (621.3)
to drain away the charge Ladung abführen
to drain off abziehen
drainage Drainage, Entwässerung; Ablauf, Abfluß; Wasserhaltung (622)
drainage coils Differential-Erdungsdrosseln
drainage of oil Ölverlust, Ölaustritt (621.315.2)
draining Drainierung, Entwässerung
draining machine Wasserhaltungsmaschine, Schöpfmaschine
DRAM = dynamic RAM
draught Zug, Zugluft, Luftstrom; Zug, Schornsteinzug; Entwurf, Skizze, Zeichnung
draughtsman Zeichner
draw Entnahme, Abzug
draw-bar pull Kraft am Zughaken (621.33)
draw-in box Einziehschacht (621.315.2)
draw-out equipment unit ausziehbarer Geräteblock, Einschub
draw spring Zugfeder
draw-out switchgear Schaltwagen, Schaltschlitten, ausfahrbares Schaltgerät
draw-off tackle Abziehvorrichtung
draw tongs (pl.) Seilspannklemme, Froschklemme
draw vice Seilspannklemme
to draw ziehen, recken; zeichnen, aufzeichnen; einsaugen, nachsaugen (621.7); aussetzen (66); ziehen, abheben, abziehen
to draw an arc einen Lichtbogen ziehen
to draw a circle einen Kreis schlagen (**around**: um)
to draw current from Strom entnehmen aus
to draw in einzeichnen; einziehen, schmaler werden
to draw a load current Belastungsstrom entnehmen
to draw off abzapfen, abziehen (54); dekantieren
to draw out herausziehen
to draw a parallel eine Parallele ziehen
to draw power Leistung entziehen, Leistung entnehmen
to draw up regulations Vorschriften aufstellen
to draw up a table eine Tabelle aufstellen
to draw wire Draht ziehen
drawing Zeichnung, Plan
drawing board Reißbrett, Zeichenbrett
drawing die Ziehstein; Ziehwerkzeug
drawing engine Fördermaschine
drawing ink Zeichentusche, Tusche
drawing material Zeichenmaterial
drawing paper Zeichenpapier
drawing pen Ziehfeder, Reißfeder
drawing pin Reißzwecke
drawing reader Zeichnungsleser
drawing rope Förderseil
drawing set Reißzeug
drawn gezogen
drawn steel gezogener Stahl
drawn wire filament gezogener Glühdraht
dredger Bagger; Baggerschiff
D-region D-Region, D-Schicht
dressing plant Aufbereitungsanlage (622)
to dribble tröpfeln
drier Trockner
drift Abtrift; Frequenzwanderung (621.396); Drift (621.382); Nullpunktverschiebung
drift field Driftfeld

drift mobility Driftbeweglichkeit
drift space Triftraum
drift stabilization Driftstabilisierung, Driftausgleich
drift velocity Driftgeschwindigkeit
drill Bohrer
drill bit Bohrerschneide
drill chuck Bohrfutter
drill gauge Bohrlehre
drill gun Handbohrmaschine
drill jig Bohrlehre, Bohrvorrichtung
drill stem Bohrerschaft
to drill bohren (621.8)
to drill out ausbohren
to drill to template nach Schablone bohren
drilling Bohren, Bohrvorgang; Bohr-...
drilling machine Bohrmaschine
drillings Bohrspäne; Bohrmehl (sing.)
drip Tropfwasser
drip-proof tropfwasserdicht, tropfwassergeschützt
drip-proof motor tropfwassergeschützter Motor
drip-proof switchgear tropfwassergeschützte Schaltgeräte (pl.)
drip-tight tropfwasserdicht
drip tray Tropfschale, Tropfenfänger
to drip tröpfeln, abtropfen
dripping Tropfen, Tröpfeln; ausgelassenes Fett, auszulassendes Fett
dripping cooler Rieselkühler
dripping electrode Tropfelektrode
drippings (pl.) **of solution** abtropfende Lösung
drive Antrieb, Trieb; Transmission; Laufwerk; Ansteuerung (621.39)
drive belt Antriebsriemen
drive current Ansteuerungsstrom
drive end Antriebsseite
drive mechanism Laufwerk
drive pulse Ansteuerungsimpuls, Steuerimpuls (621.374)
drive side Antriebsseite
drive wheel Förderscheibe
to drive treiben, antreiben; fahren, lenken (629.113); mitziehen, stützen (621.396)
to drive an amplifier einen Verstärker speisen, auf einen Verstärker arbeiten, einen Verstärker steuern
to drive current Strom (hindurch-)treiben
to drive a grid ein Gitter speisen, ein Gitter beaufschlagen, ein Gitter steuern
to drive the grid below cut-off Gitter ins stromlose Gebiet aussteuern, durch Gitterspannung sperren
to drive a screw eine Schraube anziehen, eine Schraube einschrauben
to drive a valve eine Röhre steuern (621.385)
driven angetrieben, betrieben
driven aerial gespeiste Antenne, Primärantenne, Erreger
driven element gespeistes Antennen-Element
driven to full output voll ausgesteuert
driven oscillator geführter (gestützter) Generator (621.396)
driven shaft angetriebene Welle, Abtriebswelle
driver Treiber, Steuerteil (621.396); Mitnehmer, Treibrad (621); Fahrer, Lokomotivführer
driver's cab Führerstand (625)
driver pinion Antriebsritzel
driver stage Treiberstufe (621.396)
driver transformer Treibertransformator (621.396)
driver valve Treiberröhre (621.396)
driving Antrieb
driving axle Triebachse (625)
driving belt Treibriemen
driving circuit Steuerkreis
driving disc Antriebsscheibe
driving engine Antriebsmaschine
driving force Antriebskraft
driving gear Antrieb; Antriebsräder (pl.)
driving mechanism Antriebsmechanismus, Triebwerk
driving pin Mitnehmerstift
driving position Führerstand
driving power Antriebsleistung (621); Steuerleistung, Treiberleistung (621.396)
driving pulley Triebscheibe, Treibscheibe, Antriebsscheibe
driving rod Schubstange
driving rope Antriebsseil, Treibseil
driving shaft Antriebswelle, Triebwelle
driving sit Treibsitz
driving speed Antriebsdrehzahl
driving stage Treiberstufe
driving torque Triebmoment, Antriebsmoment
driving wedge Treibkeil
driving wheel Triebrad, Antriebsrad, Treibrad
drone unbemannter Flugkörper
drop Tropfen; Abfall, Fallen; Verlust; Spannungsabfall (621.3); Schauzeichen, Fallklappe (621.395)
drop annunciator Fallklappentableau, Fallklappentafel
drop away Abfallwert, Abfallstrom, Abfallspannung (621.318)
drop of barometer fallender Luftdruck,

Absinken des Luftdrucks, Fallen des Barometers
drop dead HALT vollständiger Halt, unwiederruflicher Halt (681.3)
drop forging Gesenkschmieden
drop in Störsignal
drop in horsepower Leistungsabfall
drop indicator panel Klappenschrank
drop of potential Spannungsabfall
drop-in winding Träufelwicklung (621.313)
drop-out Abfallen (621.318); Aussetzfehler, Signalausfall
drop-out current Abfallstrom (621.318)
drop-out fuse cutout Sicherungsausfalltrenner
drop-out voltage Abfallspannung
drop repeater Abzweigverstärker (621.395)
drop shutter Schlitzverschluß (77); Fallklappe (621.318)
drop side of an equipment Teilnehmerseite, Vermittlungsseite eines Gerätes
drop signal Fallklappe, Schauzeichen
drop stampings Gesenkschmiedestücke
drop switchboard Klappenschrank
drop in temperature Temperaturrückgang
drop of voltage Spannungsabfall
drop weight Fallgewicht
to drop fallen, abfallen; kleiner werden, abnehmen; hinfallen lassen, hinwerfen; tropfen, tröpfeln, ableiten
to drop-forge gesenkschmieden, im Gesenk schmieden
to drop off abfallen, nachlassen
to drop a perpendicular ein Lot fällen, eine Senkrechte ziehen (51)
to drop steeply steil abfallen
to drop 10 volts einen Spannungsabfall von 10 Volt verursachen, einen Spannungsabfall von 10 Volt aufnehmen
dropper Aufhänger (621.33)
dropping Abwurf, Abwurf-...
dropping loop Fallbügel
dropping resistance Vorwiderstand
dropping resistor Vorwiderstand
to drown ertränken, überschwemmen; ertrinken, ersaufen
d.r. test Isolationsprüfung (621.315.2)
drum Trommel (534), (621); Seiltrommel, Kesseltrommel; Faß, Metallfaß; Schaft, Säulenschaft (72)
drum apparatus Magnetscheider
drum armature Trommelanker (621.313); Doppel-T-Anker
drum-chart recorder Trommelschreiber
drum controller Walzenschalter, Kontroller; Steuerwalze
drum furnace Trommelofen
drum motor Bandrollenmotor, Trommelläufer
drum starter Walzenanlasser
drum storage Trommelspeicher
drum switch Walzenschalter
drum washer Waschtrommel
drum winding Trommelwicklung (621.313)
Drumm accumulator Nickel-Zink-Akkumulator
dry trocken
dry atmosphere trockene Luft
dry battery Trockenbatterie
dry bulb trockenes Thermometer (55)
dry cell Trockenelement
dry-contact(-type) rectifier Trockengleichrichter
dry-core cable Papierkabel
dry-disc rectifier Trockengleichrichter
dry-electrolytic capacitor Trockenelektrolyt, trockener Elektrolytkondensator, Trockenelektrolytkondensator
dry flashover voltage trockene Überschlagspannung, Trockenüberschlagspannung
dry ice Trockeneis, Kohlensäureschnee
dry joint trockene Lötstelle, kalte Lötstelle
dry pile Trockenbatterie, Stabbatterie
dry plate Trockenplatte (77)
dry-plate rectifier Trockenplattengleichrichter, Trockengleichrichter
dry-reed relay Reed-Relais, Schutzgasrelais
dry running Blindversuch
dry spark-over test trockene Überschlagsprüfung, Trockenüberschlagsprüfung
dry-type Trocken-...
dry-type instrument transformer Trockenwandler (621.317)
drying Trockenvorgang, Trocknung
drying cabinet Trockenschrank
drying drum Trockentrommel, Trommeltrockner
drying-out insulation Isolationstrocknung (621.313)
drying oven Trockenofen
drying tower Trockenturm
Drysdale potentiometer Wechselstrom-Kompensator
d.s.c. = double silk covered SS = zweimal seideisoliert
DTP = desktop publishing
D-type flip-flop D-Flip-Flop, Auffangflipflop

dual Doppel-..., zweifach, dual
dual amplification Reflex-Verstärkung (621.396.62)
dual beam CRT Zweistrahlkathodenstrahlröhre
dual control Doppelsteuerung
dual-frequency locomotive Zweifrequenzlokomotive
dual-grid tube Doppelgitterröhre, Zweigitterröhre
dual-in-line package = DIP Dual-in-Line Gehäuse
dual operation Doppelsteuerung (621.876)
dual purpose Zweizweck-...
dual service Doppelversorgung, zweiseitige Versorgung
dubbing Tonmischung (621.396.9)
duplexer Antennenweiche, Duplexer
duct Rohr, Leitungsrohr, Kabelschutzrohr (621.395); Leitung; Leitungsschacht, Schacht; Kanal; Umhüllung, Führung, Wellenleiter, Leitschicht (621.371)
duct bank Mehrfachkanal, Kabelkanal
duct sealing Kanalverguß, Kanalverschluß (621.315.2)
duct-ventilated motor Motor mit Durchzuglüftung
duct width Wellenleiterdicke (621.371)
ducter Widerstandsmeßgerät (für kleine Widerstände)
ductile duktil, verformbar, bildsam; ziehbar, ziehfähig; geschmeidig
ductile iron Weicheisen
ductility Duktilität, Verformbarkeit; Ziehfähigkeit; Geschmeidigkeit; Streckbarkeit, Bildsamkeit
Duddell arc singende Bogenlampe
Duddell oscillograph Schleifen-Oszillograph
Duddell thermo-ammeter Duddell-Galvanometer
Duddell thermo-galvanometer Duddell-Galvanometer
duffer ELOGM-Peilschiff (623)
dull matt, schwach, dunkel, stumpf
dull heat dunkle Glut
dull lustre matter Glanz, Mattglanz
dull red heat dunkle Rotglut
dull tool stumpfes Werkzeug
dumb aerial Antennennachbildung
dummy Blind-..., Schein-..., Attrappe; Füllsignal, Leersignal (681.3)
dummy aerial Blindantenne, Kunstantenne, Phantomantenne
dummy load künstliche Antenne, Kunstlast
dummy road Blindstrecke (622)
dump Abladeplatz, Lagerplatz; Speicherauszug, Speicherausdruck (681.3)
dump power Überschußleistung
to dump abladen, auf Lager geben; abschalten
duo-decimal conversion Duodezimalcodeumsetzung
duplex Duplex-..., Doppel-...
duplex artificial lined Vierpol-Leitungsnachbildung
duplex bundle conductor Zweierbündel
duplex cable Zweileiterkabel
duplex catenary suspension Doppelfahrdrahtaufhängung (621.33)
duplex circuit Duplexverbindung, Gegenverkehrsverbindung
duplex diode Duodiode, Doppeldiode, Doppelzweipolröhre
duplex operation Duplexbetrieb, Gegenverkehr, Gegensprechen
duplex protective system Stromrichtungsschutz mit Hilfsader
duplex switchboard Duplexschrank
duplex telegraphy Duplex-Telegraphie
duplex tube Verbundröhre
duplex voice communication Gegensprecher
duplex winding Serien-Parallelwicklung
to duplex duplex schalten (621.394)
duplexer Sende-Empfangsweiche, Antennenweiche, Duplexer
duplicate Duplikat, Abschrift
duplicate feeder Zusatzspeiseleitung
duplicate lines (pl.) Doppelleitung
duplicate service doppelseitige Versorgung
duplicator Vervielfältigungsapparat
duralumin Duralumin
duraluminium Duraluminium
duration Dauer, Zeitdauer
duration of combustion Brenndauer, Verbrennungszeit
duration of hold Haltezeit (621.316.7)
duration of a pulse Impulsdauer (621.397)
duration of short-circuit Kurzschlußdauer
dusk (Abend-)Dämmerung
dusk belt Dämmerungsgürtel
dust Staub
dust accumulation Staubansammlung, Staubablagerung
dust catcher Staubfang
dust cloud Staubwolke
dust coal Staubkohle
dust collection Staubabscheidung, Entstaubung
dust core Eisenpulverkern (621.396)

dust-cored coil Eisenkernspule, Eisenpulverkernspule
dust cover Staubdeckel (621)
dust electrification Staubelektrisierung
dust explosion Staubexplosion
dust figures Lichtenbergsche Figuren
dust filter Staubfilter
dust particle Staubteilchen
dust precipitation Staubabscheidung
dust precipitator Staubabscheider
dust-proof staubischer, staubgeschützt, staubdicht
dust-proof machine staubdichte Maschine
dust-proof switchgear staubdicht gekapselte Schaltgeräte (pl.)
dust settler Staubabscheider
dust-tight staubdicht
dust-tight machine staubdichte Maschine
to dust stauben, einstauben, einstäuben, pudern; abstauben
duty Betriebsart (621.313); betriebliche Anforderung; Dienst
duty classification Betriebsarten (pl.); Schalthäufigkeit (621.318)
duty cycle Arbeitszyklus, Betriebsspiel; Schalterprüfbedingungen (pl.); Tastverhältnis (621.394); relative Einschaltdauer; Nutzungsfaktor
duty-cycle factor Typenfaktor; Nutzungsfaktor (621.313)
duty factor Tastverhältnis
duty of a machine Betriebsart (621.313)
DVOR = Doppler very high frequency omnidirectional radio range DVOR
dwell period Verweildauer
dwt. = pennyweight (Gewichtseinheit im Troy-System)
dye Farbstoff, Farbe (677)
dying-down time Ausschwingzeit
dynamic dynamisch
dynamic balancing dynamische Auswuchtung
dynamic brake Widerstandsbremse (621.33)
dynamic braking elektrische Bremsung
dynamic characteristic dynamische Charakteristik, Arbeitskennlinie (621.396)
dynamic complementary MOS dynamische CMOS-Technik
dynamic condenser Synchronphasenschieber
dynamic electrode potential Elektrodenpotential bei Stromfluß
dynamic equilibrium dynamisches Gleichgewicht
dynamic flow chart Datenflußplan
dynamic impedance Resonanzwiderstand
dynamic loudspeaker dynamischer Lautsprecher
dynamic memory dynamischer Speicher
dynamic plate resistance innerer Widerstand (621.385)
dynamic pressure dynamischer Druck
dynamic range Aussteuerbereich
dynamic resistance Resonanzwiderstand (621.396)
dynamic response Frequenzverhalten
dynamic scattering mode dynamische Streuung (539)
dynamic sensitivity Wechsellichtempfindlichkeit (621.385)
dynamic slope Arbeitssteilheit
dynamic state dynamischer Zustand
dynamic storage dynamischer Speicher
dynamic stresses (pl.) dynamische Beanspruchung
dynamically balanced dynamisch ausgewuchtet
dynamics Dynamik
dynamo Dynamo, Dynamomaschine, Generator
dynamo-electric dynamoelektrisch
dynamo-electric generator Dynamomaschine, Dynamo
dynamo sheet Dynamoblech
dynamometer Dynamometer, Leistungsbremse, Leistungsmesser
dynamometer car Meßwagen (652.1)
dynamometer-type instrument dynamometrisches Meßinstrument (651.317)
dynamometer-type wattmeter dynamometrisches Wattmeter
dynamometric(al) dynamometrisch
dynamometric(al) brake Leistungsbremse
dynamomotor (US) Umformer
dynatron Dynatron (621.396)
dynatron oscillations Dynatronschwingungen
dyne dyne, dyn (c.g.s.-Einheit der Kraft = 1 cmg/s^2 = 10^{-6} N)
dynode Dynode, Sekundäremissionskathode
dwell time (Kontakt-)Ruhezeit

E

e (charge of the electron) e (Ladung des Elektrons)
E = earth connection E = Erdklemme
ear Ohr; Lappen, Zipfel (621.396.677); Öse

earcap Hörermuschel
earphone Hörer, Kopfhörer; Ohrhörer
earphones Doppelkopfhörer
earpiece Kopfhörer; Handapparat (621.395)
ear-response characteristic Ohrkurve
early früh, frühzeitig
early failures Frühausfälle
early stages of a set Eingangsstufen eines Geräts (621.396)
early warning Frühwarnung, Fernwarnung (621.397)
early-warning radar Radarfrühwarnung
EAROM = electrically alterable ROM elektrisch veränderbarer Festwertspeicher
earth (s. auch **ground**) Erde, Erdboden; Erde, Erdball; Erde, Erdung (621.3); Masse (621.39)
earth circuit Erdstromkreis
earth coil Erdinduktorspule
earth-conduction telegraphy Erdtelegraphie
earth conductivity Bodenleitfähigkeit
earth conductor Erdleitung, Erdseil
earth connection Erdverbindung, Erdanschluß, Erdung
earth-connection point Erdungspunkt, Erdungsanschluß
earth-continuity conductor Erdschutzleiter
earth current Erdschlußstrom, Erdstrom
earth-current equalizer Erdstromausgleicher
earth curvature Erdkrümmung
earth detector Erdschlußzeiger
earth's electric field luftelektrisches Feld
earth electrode Erder, Erdelektrode
earth fault Erdschluß
earth-fault reactor Gestelldrossel (621.313)
earth filling Erdfüllung
earth inductor Erdinduktor (638.7)
earth-inductor compass magnetelektrischer Kompaß
earth lead Erdleitung
earth leakage Erdschluß, schleichender Erdschluß
earth-leakage current Erdschlußstrom
earth leakage indicator Erdschlußprüfer, Erdschlußanzeiger
earth-leakage monitor Erdschlußwächter
earth-leakage relay Erdschlußrelais
earth-leakage trip Erdschlußautomat; Heinisch-Riedl-Schutzschalter, Schutzschalter, Fehlerspannungs-Schutzschalter
earth magnetism Erdmagnetismus
earth mat Gegengewicht
earth pitch Erdpech
earth plate Erdungsblech, Erdungsplatte, Plattenerder
earth potential Erdpotential
earth potential of radio-frequency hochfrequenzmäßiges Erdpotential
earth pressure Erddruck, Erdbeanspruchung
earth protective sheath geerdete Abschirmung
earth's radius Erdradius
earth resistance Erdungswiderstand, Erdausbreitungswiderstand (621.396)
earth-resistance area Spannungstrichter (621.316.9)
earth resistance meter Erdungsmesser
earth return Erdrückleitung
earth rod Erdungsrohr, Rohrerder, Staberder
earth sheath Erdungsmantel
earth shield geerdete Schirmwicklung
earth's surface Erdoberfläche
earth station Erdefunkstelle
earth system Erdnetz, Gegengewicht
earth terminal Erdklemme, Erdungsklemme
earth wire Erdungsdraht, Erdleitung
to earth erden
to earth direct direkt erden, starr erden; an Masse legen
earthed (s. auch **grounded**) geerdet
earthed circuit geerdeter Stromkreis
earthed conductor geerdeter Leiter
earthed neutral geerdeter Sternpunkt, Nullpunkt
earthed neutral conductor geerdeter Mittelleiter, Nulleiter
earthed neutral system geerdetes Netz, Netz mit geerdetem Mittelpunkt
earthed point Erdungspunkt
earthed pole geerdeter Pol, geerdeter Außenleiter
earthed switch Gehäuse-Erdungsschalter
earthed system geerdetes System (621.316)
earthed terminal geerdete Klemme, Erdklemme
earthed wire Erdseil (621.315); geerdeter Draht
earthed wiring einpolig geerdete Installation
earthing (s. auch **grounding**) Erdung, Erdanschluß; Erdverbindung; Masseanschluß; Erdschluß
earthing auto-transformer Erdungsdrossel (621.316.9)
earthing cable bond Mantelerdung
earthing clip Erdungsklammer, Erdungsklemme
earthing coil Erdungsspule
earthing conductor Erdleiter, Schutzkontaktleiter

earthing connection Erdverbindung; Erdungspunkt
earthing-contact plug Schutzkontaktstecker
earthing-contact socket outlet Schutzkontaktsteckdose
earthing core Erdungsader (621.395)
earthing electrode Erder
earthing grip Erdschelle, Erdanschlußschelle
earthing inductor Erdungsdrossel
earthing isolator Erdungstrennschalter
earthing mat Netzerder
earthing plug Erdungsstecker, Erdungsstöpsel
earthing reactor Erdungsdrossel
earthing resistance Erdungswiderstand, Erdausbreitungswiderstand
earthing rod Staberder
earthing screw Erdungsschraube
earthing strip Erdungsband, Banderder
earthing switch Erdungsschalter, Erdtrennschalter
earthing terminal Erdklemme, Erdungsklemme
earthy erdig; geerdet, auf Erdpotential (621.3)
ease Erleichterung, Leichtigkeit
ease of installing leichter Einbau
ease of operation leichte Bedienbarkeit
to ease erleichtern, leichter machen; langsamer fahren (629.12)
to ease a bolt eine Schraube gängig machen
to ease out leicht auslösen, vorsichtig auslösen
easy leicht, einfach
easy-to-get-at leicht zugänglich
easy-to-melt leicht schmelzbar
easy-to-operate leicht bedienbar, leicht zu bedienen
to eat away wegfressen, korrodieren
ebonite Hartgummi, Ebonit
ebullient kochend, aufsprudelnd
ebullition Hochkochen, Aufsprudeln, Sieden mit Blasenentwiccklung, Aufwallen
eccentric Exzenter; exzentrisch, außermittig
eccentric poles exzentrisch gebohrte Polschuhe
eccentric rod Exzenterstange
eccentric shaft Exzenterwelle
eccentric stud Exzenterstutzen, Exzenterstift
eccentricity Exzentrizität; Achsabstand, Abweichung von der Mittellinie
Eccles-Jordan circuit Eccles-Jordan-Schaltung, bistabiler Multivibrator
Eccles-Jordan trigger Flip-Flop Auslöser, Eccles-Jordan-Trigger
ECCM = electronic counter-counter measures (623)
echelon Staffelung, Echelon-...
echelon grating Echelongitter, Staffelgitter
echo Echo
echo area Rückstrahlquerschnitt (621.396.67)
echo attenuation Echodämpfung
echo depth sounding Echolotung
echo elimination Echounterdrückung, Echosperrung (621.395)
echo eliminator Echosperre (621.395)
echo ranging Echoortung, aktive Unterwasserortung (623)
echo room Hallraum
echo signal Echozeichen (621.397)
echo sounder Echolot
echo strength Echostärke (621.397)
echo suppressor Echosperre (621.39)
echo-transmission time Echolaufzeit
to echo hallen, widerhallen, Echo geben
echoing area Rückstrahlfläche
echoing volume Rückstrahlvolumen (621.397)
ECL = emitter coupled logic emittergekoppelte Logikschaltung
ECM = electronic countermeasures (623)
ECMA = European Computer Manufactoring Association
econimical load Regellast
economic life wirtschaftliche Lebensdauer
economizer Rauchgasvorwärmer, Ekonomiser, Speisewasservorwärmer, Vorwärmer
economy Ersparnis, Wirtschaftlichkeit
economy circuit Sparschaltung
economy coil Spartransformator (621.3)
economy resistance Haltewiderstand (621.318); Vorwiderstand der Haltespule
economy of space Raumersparnis
e.c. speed indicator = eddy current speed indicator Wirbelstromtachometer
eddy Wirbel
eddy current Wirbelstrom
eddy current brake Wirbelstrombremse
eddy current losses Wirbelstromverluste
to eddy wirbeln
edge Kante, Rand; Schneide
edge block Eckstück, Abstandsblock (621.313), (621.314.2)
edge connector Federleiste
edge flare Randaufhellung (621.397)
on edge hochkant, auf der Kante stehend
edge punched cards Randlochkarten
edge runner Kollergang
edge steepness Flankensteilheit
edge tool Schneidewerkzeug
edge winding Hochkantwicklung (621.313)
edge-wise hochkant
edge-wise wound hochkant gewickelt

edgewise instrument Profilinstrument
Edison battery Edison-Batterie, alkalische Batterie, NC-Batterie
Edison cell Edison-Element, Edison-Sammler, NC-Sammler
Edison distribution system Dreileiter-Verteilungsnetz
Edison effect Edison-Effekt (621.32)
Edison lamp Edison-Lampe, Glühlampe
Edison screw cap Edison-Gewinde (621.32)
Edison screw holder Edison-Fassung (621.32)
Edison thread Edison-Gewinde
editor Herausgeber, Schriftleiter; Editor; Textverarbeitungsprogramm (681.3)
EDP = electronic data processing elektronische Datenverarbeitung
to educe ableiten; herausholen (54)
EEPROM = electrically erasable PROM
EEROM = electrically erasable ROM
eidophor tube Eidophorröhre
effect Wirkung, Einwirkung, Einfluß, Effekt, Leistung
effect of external magnetic field magnetischer Fremdfeldeinfluß
to effect bewirken, herbeiführen
effective wirksam, effektiv; praktisch; tatsächlich; Effektiv-... (621.317)
effective a.c. resistance Wechselstrom-Wirkwiderstand, Echtwiderstand
effective area Antennenwirkfläche, wirksame Antennenfläche
effective aerial height effektive Antennenhöhe
effective amplification Wirkverstärkung
effective attenuation Wirkdämpfung
effective attenuation of a termination circuit Gabeldämpfung (621.396)
effective average gewogenes Mittel (519.2)
effective cadmium cut-off effektive Kadmium-Schwellenenergie
effective capacitance wirksame Kapazität (621.3); Systemkapazität (621.385)
effective capacity Nutzleistung, tatsächliche Leistungsfähigkeit
effective conductance Wechselstromwirkleitwert
effective conductivity Wechselstromleitfähigkeit
effective current Effektivwert des Stromes
effective height wirksame Höhe, effektive Höhe (621.396.67)
effective induction area of the control current loop induktive Restfläche des Steuerstromkreises eines Halleffekt-Bauelementes
effective induction area of the output loop induktive Restfläche des Hallstromkreises
effective input capacitance dynamische Gitterkapazität
effective power effektive Leistung, Nutzleistung
effective pressure Arbeitsdruck
effective radiated power effektive Strahlungsleistung
effective radius of the earth äquivalenter Erdradius
effective range Nutzmeßbereich, Meßbereich (621.317)
effective resistance Echtwiderstand, Wirkwiderstand (621.3)
effective resistivity spezifischer Wechselstromwiderstand
effective scattering cross-section wirksame Reflexionsfläche (= Querschnitt einer gleich stark reflektierenden Kugel) (621.397)
effective transmission factor Betriebsübertragungsmaß
effective value Effektivwert, quadratischer Mittelwert
effective voltage Effektivspannung
to effervesce perlen, sprudeln
efficiency Wirkungsweise; Erfolg; Wirkungsgrad; Ausnutzung; visueller Nutzeffekt (535)
efficiency depreciation Abnahme der Lichtausbeute (621.32)
efficiency engineer Zeitstudien-Ingenieur
efflorescence Ausblühen
efflorescent ausblühend, auskristallisierend
effluent Ausfluß, Ausflußmenge
effluent ausfließend, abfließend, austretend
effluve Büschelentladung
effluvium Ausdünstung
efflux Ausfluß, Ausströmung
effusion Erguß; Ausströmung (von Gasen)
egg-grate troffer Rasterleuchte
egg-insulator Eiisolator
e.h.f. = extremely high frequency EHF, Millimeterwellen
e.h.p. = electrical horse-power elektrisches PS = 746 Watt
e.h.t. = extra high tension Höchstspannung
e.h.t.-stack Hochspannungsstabgleichrichter
eigenvalue Eigenwert (51)
eight-element dipol array Achterfeld
eight segment magnetron Achtschlitzmagnetron
eight segment plate Achtschlitzanode

Eindhoven galvanometer Saitengalvanometer
to eject ausstoßen, auswerfen (621.9); hinauswerfen, hinausschleudern
to eject electrons Elektronen emittieren
ejection Auswerfen, Ausstoßen (621.9)
ejector Auswerfer, Ausstoßer (621.9); Ejektor, Dampfstrahlpumpe (621.1)
to elapse ablaufen, verfließen
elastance (Kehrwert der Kapazität)
elastic Gummiband
elastic elastisch
elastic chuck Klemmfutter
elastic collision elastischer Stoß
elastic contraction elastische Kontraktion
elastic limit Elastizitätsgrenze, Streckgrenze
elastic modulus Elastizitätsmodul
elastic strain elastische Dehnung, elastische Beanspruchung
elastic stress Spannung innerhalb der Elastizitätsgrenze
elastic stretch elastische Dehnung
elasticity Elastizität; Federkraft, Federung
E-layer E-Schicht (621.396)
elbow Ellenbogen (61); Krümmer, Knie, Rohrknie; Leitungsbogen
electret Elektret
electric (s. auch **electrical**) elektrisch
electric approach locking zubetätigte Verblockung (621.33)
electric arc Lichtbogen, elektrischer Lichtbogen
electric arc furnace elektrischer Lichtbogenofen
electric arc welding elektrische Lichtbogenschweißung
electric automatic push-button lift Fahrstuhl mit automatischer Druckknopfsteuerung, Selbstfahrer (621.876)
electric automatic self-service lift Fahrstuhl mit Selbststeuerung, Selbstfahrer (621.876)
electric axis Elektrische Achse (539)
electric battery elektrische Batterie
electric bell elektrische Klingel
electric blanket Heizdecke
electric boiler Elektrokessel, Boiler
electric brake elektrische Bremse
electric braking elektrische Bremsung
electric brazing Elektro-Hartlötung
electric bulb elektrische Lampe, Glühbirne
elektric cable reel locomotive Schleppkabellokomotive
electric capacity altimeter kapazitiver Höhenmesser (629.13)
electric carpet Heizteppich
electric cat-fish Zitterwels
electric charge elektrische Ladung
electric circuit elektrischer Stromkreis
electric circuit analysis Wechselstromlehre
electric clock elektrische Uhr
electric CO indicator elektrischer CO-Messer
electric CO recorder CO-Gehaltschreiber
electric coal cutter elektrische Schrämmaschine
electric coffee grinder elektrische Kaffemühle
electric conductor elektrischer Leiter, Stromleiter
electric connection elektrischer Anschluß
electric controller elektrischer Kontroller, Schaltwalze
electric control-trail car Anhänger mit Führerstand
electric cooker Elektroherd
electric coulomb force Coulombsche Anziehungskraft, Coulomb-Kraft
electric coupler elektrische Wagenkupplung (621.33)
electric coupler plug Kupplungsstecker
electric coupler socket Kupplungsdose
electric crab-reel mine locomotive elektrische Grubenlokomotive mit Spillwinde
electric crane Elektrokran
electric current elektrischer Strom
electric depth finder elektrisches Echolot
electric detonator elektrischer Zünder (622)
electric discharge elektrische Entladung
electric door operator elektrischer Türöffner
electric double layer elektrische Doppelschicht (681.382)
electric doublet elektrischer Dipol
electric drainage Streustromableitung
electric drive elektrischer Antrieb
electric dumbwaiter (US) elektrischer Küchenaufzug, Speisenaufzug; elektrischer Gepäckaufzug
electric eel Zitteraal
electric equipment elektrische Ausrüstung; Elektrogeräte (pl.)
electric eye elektrisches Auge, Photozelle
electric fence elektrischer Weidezaun
electric field elektrisches Feld
electric field strength elektrische Feldstärke
electric fire elektrischer Ofen, elektrischer Strahlofen, Heizsonne
electric fire bowl Heizsonnenreflektor
electric fish Zitterfisch
electric flash signal elektrisches Lichtsignal
electric flux elektrischer Fluß
electric force elektrische Feldstärke

electric furnace Elektroofen
electric fuse elektrischer Zünder (623.4), (622)
electric gas lighter elektrischer Gasanzünder
electric glue pot elektrisch beheizter Leimtopf, elektrischer Leimkocher
electric gramophone elektrischer Plattenspieler
electric grinder Elektroschleifer, Motorschleifstein, Schleifbock
electric gun Strahlsystem
electric haulage Elektrokarrentransport
electric heat Elektrowärme
electric heating elektrische Heizung, Elektroheizung
electric hoist Elektrozug
electric horn Elektrische Hupe
electric hotplate elektrische Kochplatte, Heizplatte, Wärmeplatte
electric igniter elektrischer Zünder (622)
electric ignition elektrische Zündung (621.4)
electric impulse clock elektrische Nebenuhr mit Stromstoßsteuerung
electric incline railway elektrischer Bremsberg
electric indication lock(ing) elektrische Verriegelung (646.25)
electric induction elektrische Induktion
electric industry Elektroindustrie
electric installation elektrische Anlage, Leitungsanlage
electric intensity elektrische Feldstärke
electric interlocking machine elektrisches Stellwerk
electric iron elektrisches Bügeleisen
electric jar Leydener Flasche, Kleistsche Flasche
electric lamp elektrische Lampe, elektrische Leuchte, elektrischer Beleuchtungskörper
electric lift elektrischer Fahrstuhl
  electric lift controller Fahrstuhlsteuerschalter
electric light cable Beleuchtungskabel
electric lighting elektrische Beleuchtung
electric line elektrische Leitung
electric lock(ing) elektrische Verriegelung
electric locomotive Elektrolokomotive, E-Lok
electric mains elektrisches Netz, Starkstromnetz
electric metal hochwertiges HF-Eisen, HF-Eisen
electric milker elektrische Melkmaschine
electric moment elektrisches Moment
electric motor Elektromotor
  electric motor car Triebwagen (621.33)
electric network Netzwerk, elektrische Schaltung, Netz
electric operation elektrische Betätigung
electric oscillation elektrische Schwingung
electric phonograph recorder elektrischer Plattenschneider, Aufnahmegerät
electric plant elektrische Anlage, Elektrizitätswerk
electric polarization elektrische Polarisation
electric potential elektrisches Potential
  electric potential difference elektrische Potentialdifferenz
electric power elektrische Leistung
  electric power cable Starkstromkabel
  electric power substation elektrisches Unterwerk
electric pressure elektrische Spannung
electric propulsion of ships elektrischer Schiffsantrieb
electric pulley block Elektrozug (621.861)
electric pyrometer elektrisches Pyrometer
electric radiator elektrischer Strahlofen, Heizstrahler
electric railway elektrische Bahn
electric range Elektroherd
electric razor elektrischer Rasierapparat, Trockenrasierer
electric repeater Spiegelfeld (625.1)
electric route locking elektrische Fahrstraßenverriegelung
electric safety lamp elektrische Grubenlampe
electric section locking elektrische Verblockung
electric shaver Elektrorasierer, Trockenrasierer
electric shock elektrischer Schlag, Schlag (61)
electric sign Leuchtreklameschild
  electric sign advertising Lichtreklame
electric signal box elektrisches Stellwerk
electric sleep Elektronarkose
electric soldering elektrisches Löten
electric spark elektrischer Funken
electric strength Spannungsfestigkeit
electric stress elektrische Beanspruchung, elektrische Feldstärke
electric supply Elektrizitätsversorgung, Stromversorgung; elektrischer Anschluß
  electric supply lines elektrische Versorgungsleitungen
  electric supply meter Elektrizitätszähler
electric surface density Oberflächenladungsdichte
electric surface resistance Oberflächenwiderstand
electric sweeper (elektrischer) Staubsauger

electric switch-lever lock elektrische Weichenstellverriegelung
electric switch lock elektrische Weichenverriegelung
electric switching locomotive elektrische Rangierlokomotive
electric tension elektrische Spannung
electric test elektrische Prüfung
electric time locking elektrische Zeitsperre (65)
electric toaster (elektrischer) Brotröster
electric tools Elektrowerkzeuge
electric torch elektrische Taschenlampe
electric tower car elektrischer Turmwagen (621.33)
electric traction elektrischer Bahnbetrieb, elektrische Zugförderung
electric traffic locking elektrische Fahrstraßenverriegelung
electric trailer Straßenbahnanhänger
electric train elektrischer Zug
electric train line elektrische Zugleitung
electric transducer elektrische Übertragungsschaltung, Vierpol
electric transmission elektrische Kraftübertragung
electric truck Elektrokarren
electric vacuum cleaner elektrischer Staubsauger
electric ventilator elektrischer Ventilator
electric voltage elektrische Spannung
electric water heater (elektrischer) Heißwasserspeicher
electric welder Elektroschweißer
electric welding Elektroschweißung
electric wire Leitungsdraht
electric wiring elektrische Leitungen (pl.)Lichtnetz, Elektro-Installation
electrical (s. auch **electric**) elektrisch
electrical accident elektrischer Unfall
electrical apparatus elektrischer Apparat
electrical appliances Elektrogeräte
electrical axis elektrische Achse (537.228)
electrical calamine Zinksilikat
electrical circuit elektrischer Stromkreis
electrical code Vorschriftenbuch für Elektrotechnik
electrical communication elektrische Nachrichtenübertragung, Fernmeldewesen
electrical conductivity elektrische Leitfähigkeit
electrical contact elektrischer Kontakt
electrical contractor Elektro-Unternehmer, Vertragsinstallateur
electrical control elektrische Steuerung
electrical degree elektrischer Grad
electrical depression bar Schienenkontakt
electrical discharge elektrische Entladung
electrical endosmose elektrische Endosmose
electrical energy elektrische Energie
electrical engineer Elektro-Ingenieur
electrical engineering Elektrotechnik
electrical equipment elektrische Ausrüstung
electrical exposure elektrische Belichtung (77)
electrical fence Elektro-(Weide-)zaun
electrical fire Elektrobrand, durch Elektrizität verursachter Brand
electrical flux Verschiebungsfluß, dialektrischer Fluß, elektrischer Fluß
electrical generating capacity Kraftwerksleistung, installierte Erzeugerleistung
electrical goods Elektroartikel
electrical ignition system elektrische Zündanlage (621.43)
electrical installation elektrische Installation, Lichtnetz
electrical instrument elektrisches Meßinstrument, Instrument, Gerät
electrical insulating material elektrisches Isolationsmaterial
electrical interlocking elektrische Verriegelung
electrical length elektrische Länge (621.396.67)
electrical load elektrische Last, elektrische Belastung
electrical machine Elektromaschine, elektrische Maschine
electrical machinery elektrische Maschinenanlage, Elektromaschinen (pl.)
electrical measurement elektrische Messung
electrical measurements (pl.) elektrische Meßtechnik
electrical metal tubing J-Rohr (696.6)
electrical plant elektrische Anlage
electrical power elektrische Leistung
electrical precipitation elektrische Staubabscheidung
electrical property elektrische Eigenschaft
electrical quantity elektrische Größe
electrical recording elektrische Registrierung
electrical reproduction elektrische Wiedergabe
electrical reset elektrische Rückstellung
electrical resistance elektrischer Widerstand
elektrical resistivity spezifischer Widerstand
electrical scan elektrische Abtastung
electrical screw thread Stahlpanzerrohrgewinde
electrical signal elektrisches Signal
electrical signalling elektrisches Signalwesen
electrical welding Elektroschweißung

electrical work elektrische Arbeit
electrically alterable ROM = EAROM elektrisch veränderbarer Festwertspeicher
electrically conducting elektrisch leitend
electrically erasable ROM = EEROM elektrisch löschbares ROM
electrically operated brake elektrisch betätigte Bremse
electrically trip-free... mit Freiauslösung
electrically wound clock Uhr mit elektrischem Aufzug
electrician Elektriker, Elektromonteur
electricity Elektrizität
electricity meter Elektrizitätszähler
electricity supply Stromversorgung, Elektrizitätsversorgung
electricity works (pl.) Elektrizitätswerk
electrification Elektrifizierung; Elektrisierung; Aufladung (537.2)
to electrify elektrifizieren, auf elektrischen Betrieb umstellen; elektrisieren, elektrisch aufladen
electrified track elektrifizierte Strecke, elektrifiziertes Gleis
electrization Elektrisierung
to electrize elektrisieren
electroacoustic(al) elektro-akustisch
electroacoustic transducer elektroakustischer Wandler
electroadhesion Johnsen-Rahbek-Effekt
electroadhesion telephone Johnsen-Rahbek-Telefon
electroanalysis Elektrolyse
electrobiology Bioelektrizität
electrocapillarity Elektrokapillarität
electrocapillarity microphone Kapillar-Mikrophon
electrocardiogram Elektrokardiogramm
electrocardiograph Elektrokardiograph
electrocautery Kauterisation
electrochemical elektrochemisch
electrochemical condenser Elektrolyt-Kondensator
electrochemical equivalent elektrochemisches Äquivalent
electrochemical passivation elektrochemische Passivisierung
electrochemical series galvanische Spannungsreihe
electrochemical valve Elektrolytgleichrichter
electrochemistry Elektrochemie
electrochord elektrisches Klavier
electroconductive elektrisch leitend
electrode Elektrode
electrode admittance Steilheit (621.385)
electrode boiler Elektrodenkessel
electrode cable detector Kabelfehlersuchgerät (621.315.2)
electrode capacity Elektrodenkapazität
electrode characteristic Strom-Spannungskennlinie bei konstanten Gitterspannungen (621.385)
electrode conductance Gittersteilheit (Wirkkomponente bei konstantem Potential aller anderen Elektroden)
electrode current Elektrodenstrom
electrode dissipation Elektrodenverlust (-leistung) (621.385); Gitterverlustleistung
electrode drop Spannungsabfall in der Elektrode (621.791)
electrode economizer Elektrodenkragen (621.357)
electrode holder Zange (621.791)
electrode impedance innerer Widerstand (621.385)
electrode plate Elektrodenplatte
electrode potential Elektrodenpotential; elektrochemisches Potential
electrode radiator Kühlflügel (621.396.69)
electrode resistance Elektrodenwiderstand
electrode system Elektrodensystem
electrode tip Elektrodenspitze, Spitze des Schweißstabes
electrode voltage Elektrodenpotential (gegen Kathode)
electrodeless discharge elektrodenlose Entladung
electrodeposition elektrolytische Abscheidung
electrodessication elektrische Verschorfung, Verödung
electrodissolution elektrolytische Auflösung; galvanische Zersetzung, anodische Korrosion
electrodynamic elektrodynamisch
electrodynamic balance Stromwaage
electrodynamic force elektrodynamische Kraft
electrodynamic headphone dynamischer Kopfhörer, Tauchspulenkopfhörer
electrodynamic instrument elektrodynamisches Instrument
electrodynamic loudspeaker elektrodynamischer Lautsprecher, Tauchspulenlautsprecher
electrodynamics (pl.), **electrodynamism** Elektrodynamik
electrodynamometer Dynamometer
electroengraving elektrolytische Ätzung
electroerosion Elektroerosion
electrofacing Elektro-Plattierung
electrofilter Elektrofilter

electroforming Galvanoplastik
electrogalvanizing galvanische Verzinkung
electrogilding galvanische Vergoldung
electrographic ink leitfähige Tinte
electrohydraulic elektrohydraulisch
electrokinetics Elektrodynamik
electrolier elektrische Krone (621.32)
electrolimit gauge elektrische Grenzlehre
electrology Elektrotherapie
electrolyser Elektrolyseur
to electrolyse elektrolysieren
electrolysis Elektrolyse
electrolysis of underground structures Streustromelektrolyse, Streustromkorrosion
electrolyte Elktrolyt
electrolytic elektrolytisch
electrolytic arrester Elektrolytableiter
electrolytic bath Elektrolysebad, elektrolytisches Bad
electrolytic capacitor Elektrolytkondensator
electrolytic cell elektrolytisches Element, elektrolytische Zelle
electrolytic cleaning elektrolytische Reinigung
electrolytic condenser Elektrolytkondensator
electrolytic conductor elektrolytischer Leiter, Leiter zweiter Klasse
electrolytic convection Polarisationsstrom
electrolytic copper Elektrolytkupfer
electrolytic corrosion elektrolytische Korrosion, Stromkorrosion
electrolytic detector Polprüfpapier
electrolytic dissociation elektrolytische Dissoziation
electrolytic extraction elektrolytische Raffination
electrolytic furnace Elektrolyt-Ofen, Elektrolyseofen
electrolytic instrument Elektrolyt-Zähler
electrolytic interrupter Wehnelt-Unterbrecher
electrolytic ionization elektrolytische Dissoziation
electrolytic iron Elektrolyteisen
electrolytic lightning arrester Elektrolyt-Ableiter (621.316.9)
electrolytic meter Elektrolyt-Zähler
electrolytic oxidation Eloxierung
electrolytic parting of metals elektrolytische Trennung von Metallen
electrolytic pickling galvanische Beizung
electrolytic potential Elektrodenpotential
electrolytic rectifier Elektrolyt-Gleichrichter
electrolytic reduction elektrolytische Reduktion
electrolytic refining elektrolytische Raffination
electrolytic resistance elektrolytischer Widerstand
electrolytic solution voltage Lösungsspannung
electrolytic surge arrester Elektrolyt-Ableiter (621.316.9)
electrolytic valve Elektrolyt-Gleichrichter
electrolytic water tester Speisewasserprüfer (621.317)
electrolytic wire bar Elektrolyt-Barren
electromagnet Elektromagnet
electromagnetic elektromagnetisch
electromagnetic brake elektromagnetische Bremse
electromagnetic compatibility elektromagnetische Verträglichkeit
electromagnetic component magnetische Komponente (535.13)
electromagnetic coupling elektromagnetische Kupplung
electromagnetic crack detector magnetischer Rißprüfer, Flutunggerät
electromagnetic damping Wirbelstromdämpfung
electromagnetic deflection elektromagnetische Ablenkung
electromagnetic delay line elektromagnetische Verzögerungsleitung
electromagnetic emission EM, elektromagnetische Strahlung
electromagnetic field elektromagnetisches Feld
electromagnetic fluid clutch magnetische Flüssigkeitskupplung
electromagnetic focusing elektromagnetische Fokussierung
electromagnetic horn Hornstrahler (621.396.67)
electromagnetic induction elektromagnetische Induktion
electromagnetic instrument Dreheiseninstrument
electromagnetic lens magnetische Linse (535)
electromagnetic locking elektrische Verriegelung, elektrische Verblockung
electromagnetic loudspeaker elektromagnetischer Lautsprecher
electromagnetic percussive welding magnetische Stoßschweißung, induktive Stoßschweißung
electromagnetic radiation elektromagnetische Strahlung
electromagnetic release elektromagnetische Auslösung, elektrische Auslösung
electromagnetic relay elektromagnetisches Relais

electromagnetic retroaction induktive Rückkopplung
electromagnetic shoe brake Schienenbremse
electromagnetic spectrum elektromagnetisches Spektrum
electromagnetic switch Schalter mit Magnetantrieb
electromagnetic theory of light elektromagnetische Lichttheorie
electromagnetic units elektromagnetische Einheiten
electromagnetic vibration elektromagnetische Schwingung
electromagnetic wave elektromagnetische Welle
electromagnetically operated switch Schalter mit Magnetantrieb
electromagnetics, electromagnetism Elektromagnetismus
electromatic traffic signalling selbsttätige Lichtsignalverkehrsregelung
electromechanical elektromechanisch
electromechanical interlocking machine elektrisches Stellwerk
electromedical elektromedizinisch
electrometals furnace Induktionsofen, Rinnenofen
electrometallization galvanische Metallisierung
electrometallurgy Elektrometallurgie
electrometer Elektrometer
electrometer tube (US) Elektrometerröhre
electrometer valve (Br.) Elektrometerröhre
electrometric analysis elektrolytische Analyse, Polarisationsanalyse
electromobile Elektromobil
electromotive elektromotorisch
electromotive force elektromotorische Kraft, EMK
electromotive intensity Feldstärke, Potentialgradient
electromotive series galvanische Spannungsreihe
electromotor Elektromotor
electron Elektron (53), (66)
electron accelerator Elektronenbeschleuniger
electron avalanche Elektronenlawine, Trägerlawine
electron beam Elektronenstrahl
electron-beam potential Strahlspannung
electron-beam valve Elektronenstrahlröhre
electron bombardment Elektronenbombardement
electron camera Ikonoskop
electron charge-mass ratio spezifische Ladung des Elektrons
electron concentration Elektronenkonzentration, Elektronendichte
electron conduction Elektronenleitung, N-leitung, Überschußleitung
electron-coupled elektronengekoppelt (621.396)
electron-coupled frequency changer elektronengekoppelter Frequenzumsetzer
electron current Elektronenstrom
electron deceleration Elektronenbremsung
electron diffraction Elektronenbeugung
electron discharge elektronische Entladung
electron emission Elektronenemission
electron-emissive elektronenemittierend
electron emitter Elektronenquelle
electron-excess region Elektronenüberschußzone
electron flow Elektronenstrom, Elektronenfluß
electron gas Elektronengas, Elektronenplasma
electron gun Elektronenquelle, Kathode; Strahlerzeugungssystem, Elektronenkanone
electron image elektronisches Bild
electron-image tube Bildwandler
electron jet Elektronenstrahl
electron-jet instrument Elektronenstrahlgerät
electron lens Elektronenlinse
electron microscope Elektronenmikroskop
electron motion Elektronenbewegung
electron multiplier Sekundärelektronenvervielfacher
electron number Ordnungszahl (54)
electron optics (pl.) Elektronen-Optik
electron oscillator Laufzeitgenerator
electron-oscillator detector Klystron, Laufzeitempfängerröhre
electron plasma Elektronenplasma
electron particles Elektronen, (negative) Elementarteilchen
electron-positron pair Elektron-Positron-Paar
electron-power tube Senderöhre, Leistungsröhre
electron ray Elektronenstrahl
electron-ray indicator-tube magisches Auge; Elektronenstrahlröhre
electron scanning elektronische Abtastung
electron scattering Elektronenstreuung
electron sheath Elektronenwolke
electron shell Elektronenhülle, Elektronenschale
electron spin Elektronendrall, Elektronenspin

electron stream Elektronenstrom
electron telescope Infrarot-Bildwandler
electron theory Elektronentheorie
electron trajectory Elektronenbahn
electron-transit time Elektronenlaufzeit, Laufzeit
electron transition Elektronenübergang
electron trap Elektronenfehlstelle
electron tube Elektronenröhre
electron valve Elektronenröhre
electron volt Elektronenvolt, eVolt
electron waves Materiewellen des Elektrons
electronegative elektro-negativ
electronic elektronisch, Elektronen-...
electronic charge Elementarladung, Elektronenladung
electronic computer Elektronenrechner
electronic data processing = EDP elektronische Datenverarbeitung, EDV
electronic device elektronisches Bauteil, elektronisches Gerät
electronic director elektronischer Umrechner (621.395)
electronic discharge Elektronische Entladung
electronic engineering (technische) Elektronik; HF-Technik
electronic flash Elektronenblitz
electronic grade silicon [Silizium hoher Reinheit als Ausgangsmaterial für Halbleiterschaltungen]
electronic instrument elektronisches Gerät, elektronisches Meßgerät
electronic lens Elektronenlinse
electronic microscope Elektronenmikroskop
electronic oscillations Laufzeitschwingungen
electronic oscillator Laufzeitgenerator
electronic pacemaker elektronischer Herzschrittmacher
electronic path Elektronenbahn
electronic pen elektronischer Stift, Lichtgriffel
electronic piano Klavier mit elektrischer Tonabnahme
electronic rectifier Röhrengleichrichter
electronic relay elektronisches Relais, Stromtor, Thyratron
electronic scanning elektronische Abtastung
electronic switch Schaltröhre, elektronischer Schalter
electronic test pattern elektronisches Testbild
electronic timer elektronische Zeitschaltung
electronic tube Elektronenröhre
electronic tuning elektronische Abstimmung
electronic voltmeter elektronisches Voltmeter
electronic warfare elektronische Kampfführung (623)
electronics (pl.) Elektronik
electro-optics Elektrooptik
electro-osmosis Elektroosmose
electropathy Elektrotherapie
electrophore, electrophorus Elektrophor
electrophoresis Elektrophorese
electrophorus Elektrophor
to electroplate galvanisieren
electroplating Galvanisierung, galvanische Metallabscheidung
electroplating bath Galvanisierungsbad, galvanisches Bad
electroplating plant Galvanisieranstalt
electroplating rectifier Galvanik-Gleichrichter
electroplating vat Galvanisiertrog
electropneumatic brake elektrische Luftdruckbremse
electropneumatic contactor Druckluftschütz mit Magnetsteuerung
electropneumatic control Druckluftsteuerung mit elektrischer Betätigung
electropneumatic controller elektrischer Steuerschalter mit Druckluftbetätigung
electropneumatic interlocking machine Druckluftstellwerk mit elektrischer Betätigung
electropneumatic rock-drill Gesteinsbohrer
electropneumatic valve elektrisch betätigtes Druckluftventil
electropositive elektropositiv
electroprecipitation Elektro-Phorese
electroscope Elektroskop
electroscopic powder Schwefelblüte und Mennige
electro-slag remelting furnace Elektro-Schlacke-Umschmelzofen
electrostatic elektrostatisch
electrostatic adhesion elektrostatische Adhäsion
electrostatic attraction elektrostatische Anziehung
electrostatic charge elektrostatische Ladung
electrostatic cohesion elektrostatische Haftung, elektrische Klebkraft
electrostatic component elektrische Komponente (535.13)
electrostatic coupling elektrostatische (kapazitive) Kopplung
electrostatic deflection Thermorelais, elektrostatische Ablenkung
electrostatic field elektrostatisches Feld
electrostatic field strength elektrische Feldstärke
electrostatic filter elektrostatisches Filter

electrostatic flux dielektrischer Fluß
electrostatic flux density dielektrische Verschiebung
electrostatic focusing elektrostatische Strahlkonzentrierung, elektrostatische Bündelung
electrostatic force elektrostatische Kraft
electrostatic generator Influenzmaschine, elektrostatischer Generator
electrostatic induction dielektrische Verschiebung
electrostatic instrument elektrostatisches Instrument
electrostatic Kerr effect Kerr-Effekt
electrostatic lens elektrische Linse
electrostatic loudspeaker elektrostatischer Lautsprecher, Kondensatorlautsprecher
electrostatic machine Influenzmaschine, Elektrisiermaschine
electrostatic microphone Kondensatormikrophon
electrostatic optical stress elektrische Doppelbrechnung
electrostatic precipitation Elektro-Staubabscheidung
electrostatic precipitator Elektrofilter
electrostatic printer elektrostatischer Drucker
electrostatic refraction Brechung der elektrischen Feldlinien
electrostatic retroaction kapazitive Rückkopplung
electrostatic shield statische Abschirmung, kapazitiver Schirm
electrostatic speaker statischer Lautsprecher
electrostatic stress elektrostatische Beanspruchung, elektrische Feldstärke
electrostatic telephone statisches Telefon, Kondensatortelefon
electrostatic tube of force elektrische Kraftröhre
electrostatic unit elektrostatische Einheit
electrostatic voltmeter (elektro-)statisches Voltmeter
electrostatic wattmeter elektrostatisches Wattmeter
electrostatics (pl.) Elektrostatik
electrostriction Elektrostriktion
electrosurgery Elektro-Chirugie
electrotechnics, electrotechnology Elektrotechnik
electrotellurograph Erdstrommeßgerät
electrotherapeutics (pl.), **electrotherapy** Elektrotherapie
electrothermal elektrothermisch
electrothermal efficiency elektrothermischer Wirkungsgrad
electrothermal instrument Hitzdrahtinstrument
electrothermal recording elektrothermische Aufzeichnung
electrothermal relay Thermorelais
electrothermal release (elektro)thermische Auslösung
electrothermancy, electrothermy Thermoelektrizität
electrothermics (pl.) Elektrowärme
electrothermic instrument thermisches elektrisches Meßgerät
electrotonus Elektrotonus
electrotype Druckstock, Galvano
to electrotype Druckstock auf galvanischem Wege herstellen
elctrotypograph elektrische Setzmaschine
electrotyping Galvanoplastik
electrovalent link elektronische Bindung (54)
electrovital bio-elektrisch
element Teil, Abschnitt; Element, Grundlage, Grundelement; Bauteil, Bauelement
element carrier Heizkörperträger
element of length Längenelement
element of volume Raumelement, Volumenelement
elements of... Grundzüge der...
elemental elementar, auf ein Element bezüglich
elementary elementar, einfach
elementary charge Elementarladung
elementary combinatorial circuit elementare Verknüpfungsschaltung
elementary dipole Elementardipol (621.396.67)
elementary length Längenelement
elementary particle Elementarteilchen (539)
elementary wave Elementarwelle
to elevate hochschwenken; in der Höhe verstellen; hochheben
elevated duct hochliegender Wellenleiter (621.372)
elevated temperature erhöhte Temperatur, höhere Temperatur
elevation Riß, Ansicht, Aufriß; Seitenansicht (US); Höhenlage; Erhöhung
elevation angle Höhenwinkel, Erhebungswinkel
elevation of the boiling point Siedepunkterhöhung
elevation-position indicator E.P.I. Höhenwinkel-Entfernungssichtgerät
elevation rod Auffangstange (69)
elevator Elevator; Höhenruder (629.13); Fahrstuhl, Lift, Aufzug (US); (s.**lift**)

ellipse Ellipse
ellipsoid Ellipsoid
elliptic(al) elliptisch
elliptical function elliptische Funktion
elliptic polarization elliptische Polarisation
elliptically polarized wave elliptisch polarisierte Welle
to elongate verlängern, elongieren, dehnen
elongated hole Langloch
elongation Dehnung, Verlängerung
elongation on break Bruchdehnung
elongation on fracture Bruchdehnung
elongation per unit force spezifische Dehnung, Federsteife
Elverson oscilloscope Lichtblitzstroboskop
to emanate ausströmen, emanieren
emanation Ausfluß, Ausfluß, Ausströmung; Emanation (54)
to embed betten, einbetten, einlagern, einhüllen
embedded panel heating Tieftemperatur-Strahlungsheizung
embedded pilot Erdseilkabelader
embedded temperature detector eingebautes Kontrollthermometer
to emboss prägen, drücken
embossment Prägung
to embrace umfassen, umschlingen (51)
to embrittle verspröden, spröde machen
emerald Smaragd
emergency (case) Notfall, Notstand, Not-...
emergency brake Notbremse
emergency call Notruf
emergency exchange Ausweichvermittlung
emergency frequency Notfrequenz
emergency light Notlicht, Notlampe
emergency lighting Notbeleuchtung, Panikbeleuchtung
emergency measure Notmaßnahme
emergency power supply Notstromversorgung
emergency release push Notdruckknopf, Notauslöser
emergency rod Abstellstab (539)
emergency service Notbetrieb, Notdienst
emergency shutdown Notabschaltung
emergency stop Notschalter (621.34), (621.876)
emergency switch Notschalter
emergency valve Sicherheitsventil, Notventil
emery Schmirgel
emery cloth Schmirgelleinen
emery paper Schmirgelpapier
emery powder Schmirgelpulver
emery wheel Schmirgelscheibe
e.m.f. = electromotive force EMK = elektromotorische Kraft
emission Emission, Austritt
emission characteristic Emissionskennlinie
emission efficiency Elektronenausbeute
emission of electrons Elektronenemission
emission of gas Ausströmen von Gas, Gasaustritt
emission of ions Ionenmission
emission noise Emissionsrauschen
emission saturation echte Sättigung, Kathodensättigung (621.385)
emission spectrum Emissionsspektrum
emission velocity Austrittsgeschwindigkeit
emissive Austritts-..., emittierend
emissive material emittierendes Material
emissive power Emissionsvermögen
emissivity Ausstrahlungsvermögen; Emissionsvermögen
to emit ausstrahlen, aussenden, emittieren
emitron camera Ikonoskop (621.397)
emittance Strahlungsdichte
emitter Emitter
emitter-base voltage Emitter-Basis-Spannung
emitter-base cut-off current Emitter-Basis-Reststrom
emitter bias Emittervorspannung
emitter-coupled logic = ECL emittergekoppelte Logik
emitter current Emitterstrom
emitter depletion layer Emitter-Sperrschicht
emitter follower Verstärker in Emitterschaltung, Emitterfolger
emitter-injection efficiency Emitterwirkungsgrad
emitter junction Emitter-Sperrschicht, Emitter-Basis-Zonenübergang
emitter region Emitterzone
emitter series resistance Emitterbahnwiderstand
emitter terminal Emitteranschluß, Glühkathode (621.385); (Emitter-)Elektrode (621.396.64)
emitting surface Emissionsfläche
empirical empirisch, Erfahrungs-...
empirical data Erfahrungszahlen
empirical rule Erfahrungssatz, Erfahrungsregel
empty leer
empty space leerer Raum
E.M.U. = electromagnetic unit elektromagnetische Einheit; elektromagnetische Stromeinheit (= Weber)
emulator Emulator [Hardware-Nachbildung]
emulsification Emulgierung
to emulsify emulgieren

emulsion Emulsion
emulsive emulgierend
enable... Freigabe...
enable pulse Freigabepuls
enamel Email, Emaille; Lack
enamel-insulated wire Lackdraht, Emaildraht
enamel mark Lackmarke, Lackzeichen, Lackstrich
enamel paint Isolierlack
enamel rheostat emaillierter Widerstand
to enamel emaillieren; lackisoliert (621.3)
enamelled plate Emailleschild
enamelled wire Lackdraht
encapsulated gekapselt
encapsulation Kapselung, Becher, Gehäuse
to encase in einhüllen in; einbauen in; verkleiden mit, ummanteln mit
encased capacitor Becherkondensator
to encipher chiffrieren
to encircle umschließen, umschlingen, umfassen, einrahmen
to enclose einschließen; einlegen, beilegen; abschirmen (621.3); kapseln
enclosed bus gekapselte Schiene
enclosed fan-cooled motor geschlossener Motor mit Außenbelüftung
enclosed fuse Sicherungspatrone, Patronensicherung (621.3)
enclosed fuse cut-out gekapselte Sicherung
enclosed machine geschlossene Maschine (621.3)
enclosed separately ventilated machine geschlossene Maschine mit Fremdbelüftung
enclosed shape Innenausschnitt
enclosed starter gekapselter Anlasser
enclosed switchboard geschlossenes Schalttafelgerüst, verschalte Schalttafel
enclosed ventilated motor gekapselter Motor mit Durchzuglüftung
enclosed volume umschlossener Raum
to encode verschlüsseln, codieren
encoder Codierer
encoding Codieren
to encrypt verschlüsseln
end Ende
end approach Auflaufschuh (621.33)
end bell Wickelkopfkappe (621.313); Motorschild, Endkappe
end-of-block signal Blockendesignal
end cap Abschlußkappe
end cell Regelzelle (621.335)
end connection Stirnverbindung (621.313.1)
end contact Endkontakt
end of conversation Gesprächsende (621.395)
end correction Korrektur für endliche Länge
end door Stirnwandtür (625.2)
end effect Einfluß der Enden
end elevation Seitenansicht, Seitenriß
end to end aneinandergereiht
end-to-end signalling durchgehende Signalisierung (621.395)
end of file Ende der Datei
end-fire array Längsstrahlerreihe (621.396.67)
end fixture Einspannung der Enden
end frame Lagerschild
end grain Stirn (67)
end-journal bearing Stirnlager
end leakage Stirnstreuung (621.313)
end on mit dem Ende vorneweg
end-on directional aerial Längsstrahlrichtantenne, Längsstrahler, Yaghi-Antenne
end plane Endfläche, Stirnfläche
end plate Endplatte, Druckplatte (621.313)
end play (axiales) Spiel, toter Gang
end position Endstellung
end pressure (axialer Schub)
end restraint Einspannung der Enden
end ring Kommutatorring
end-of-selection signal Wahlendzeichen (621.395)
end shield Lagerschild (621.313)
end-shield assembly komplettes Lagerschild (mit Lager)
end-shield bearing Schildlager
end stop Endanschlag, Anschlag
end of tape Bandendmarke (681.3)
end thrust (axialer) Schub
end turns Wickelkopf
end winding Spulenkopf
end-winding leakage Spulenkopfstreuung
to end enden, beenden; ablängen
ending Ende, Schluß; Sprechende (621.39)
ending fitting Abspannklemme
endless endlos, in unendlicher Folge
endless belt endloser Riemen, in sich zurücklaufendes Förderband, endloses Band
endless winding in sich geschlossene Wicklung
endodyne Überlagerungs-Schwebungsempfänger (621.396)
endodyne reception Überlagerungs-Schwebungsempfang
endothermic endotherm
endurance Ausdauer
endurance limit Dauerfestigkeit, Ermüdungsgrenze
endurance tensile strength Dauerzugfestigkeit

endurance test Dauerversuch, Dauerfestigkeitsprüfung
endwise in der Längsrichtung
energetic energisch, heftig
to energize speisen, Energie liefern an; erregen (621.318); aktivieren
to energize a coil eine Spule unter Strom setzen
to energize a winding eine Wicklung speisen
energized unter Spannung stehend, eingeschaltet, erregt
energizing power Anzugserregung
energy Energie, Energieinhalt, Arbeitswert
energy amplification Leistungsverstärkung (621.396.64)
energy band Energieband (621.382)
energy of burning Verbrennungswärme
energy component Wirkkomponente (621.3)
energy consumption Energieaufnahme
energy converter Energiewandler
energy dissipation Energieverlust
energy dissipation during turn-off time Ausschaltverlustenergie (621.314.63)
energy dissipation during turn-on time Einschaltverlustenergie (621.314.63)
energy efficiency Wirkungsgrad, energetischer Wirkungsgrad (621.355)
energy flow Energiefluß, Energiestrom
energy-flux density Energieflußdichte (539)
energy gap Bandabstand (621.382)
energy intake Energieaufnahme, Leistungsaufnahme
energy level Energieniveau
energy losses Energieverluste
energy meter Zähler
energy packet Energiepaket
energy-rich energiereich
energy spectrum Energiespektrum, Geschwindigkeitsverteilung (539)
energy supply Energielieferung
energy term Energieterm, Energieniveau
energy transfer Energietransport, Energieübertragung
energy yield Energieausbeute
to engage kuppeln, in Eingriff bringen, einrücken; engagieren, einstellen, anstellen; einstecken (621.3)
to engage with in Eingriff sein mit, in Berührung sein mit, zur Berührung kommen mit
to engage with a contact einen Kontakt berühren
engaged besetzt; im Eingriff (621.8)
engaged line besetzte Leitung
engaged selector besetzter Wähler (621.395)
engaged signal Besetztzeichen
engaged test Besetzt-Prüfung
engaged tone Besetzt-Zeichen (621.395)
engagement Eingriff; Einkuppeln, Einrücken
engaging Eingriffs-..., Kupplungs-...
engaging lever Kupplungshebel
engaging link Kupplungsglied
engine Motor, (Kraft-)Maschine, Maschine (nicht 621.31!)
engine bearing Motorenlager
engine block Motorenblock
engine driven mit Motorantrieb
engine governor Maschinenregler
engine house Krafthaus, Maschinenhalle
engine operator Maschinenwärter, Maschinist
engine relay Geschwindigkeitsüberwachungsrelais (656.25)
engine-room telegraph Maschinentelegraph
engineer Ingenieur, Techniker
engineer fitter Motorenschlosser
engineer's system technisches Maßsystem
engineering Ingenieurwesen, Technik
engineering channel Dienstkanal (621.39)
engineering college Ingenieurschule
engineering department Technische Abteilung
engineering industry Maschinenindustrie
engineering mathematics technische Mathematik
engineering stage technisches Stadium
to engrave gravieren, stechen
engraving Gravierung
engraving machine Graviermaschine
engraving needle Graviernadel
to enhance anreichern
enhancement mode Trägeranreicherung (621.382)
enhancement-mode operation Anreicherungsbetrieb
enhancement-type field-effect transistor Anreicherungs-Isolierschicht-Feldeffekttransistor, Anreicherungs-IG-FET
to enlarge größer werden, sich aufweiten; vergrößern (535), (77); vergrößerungsfähig sein, sich vergrößern lassen (77)
enlarged scale vergrößerter Maßstab
enlargement Vergrößerung (535), (77); Erweiterung, Aufweitung
enquiry Anforderung einer Antwort (681.3)
to enrich anreichern
enriched material angereichertes Material (539)
enriched reactor angereicherter Reaktor, Reaktor mit angereichertem Spaltstoff
enrichment Anreicherung

enrichment factor Anreicherungsfaktor
en route auf Strecke
en-route marker beacon Streckenmarkierungsfunkfeuer
en-route surveillance radar Streckenrundsichtradar
ensign Abzeichen, Marke
to enter zufließen zu, eintreten in (621.3); eingehen in (51)
to enter a holder at an angle of... im Halter unter einem Winkel von... sitzen
to enter into eingreifen in
entering edge auflaufende Bürstenkante, Vorderkante (621.313)
entering guide Einsteckführung, Einführung, Führungsstift
entraide network Rücküberlaufanordnung (621.395)
entrance ell (Haus-)Einführungspfeife (696.6)
entropy Entropie
entry Anfang, Anfang eines Abzweiges; Eintritt, Eingang; Zugang; Einführung, Hineinstecken
entry point Einsprungstelle (681.3)
to enumerate zählen
to envelop einhüllen; einwickeln
envelope Mantel, Hülle; Kolben (621.385); Hüllkurve (51)
envelope delay Gruppenlaufzeit
envelope-delay distortion Gruppenlaufzeitverzerrung
envelope of gas Gashülle
environmental chamber Umweltprüfraum
environmental stress umgebungsbedingte Beanspruchung
environment induced umgebungsbedingt
EOT = end of tape Bandendmarke
epicadmium neutrons Epikadmiumneutronen (539)
epicycle Epizykel, Aufkreis, Rollkreis
epicyclic gearing Planetengetriebe
epicycloid Epizykloide
epidiascope Epidiaskop
epitaxial transistor Epitaxial-Transistor, epitaktischer Transistor
epithermal neutrons epithermische Neutronen (539)
epithermal reactor epithermischer Reaktor
epoxy glass Epoxyd-Glashartgewebe
epoxy resin Epoxyd-Gießharz
EPROM = erasable programmable ROM lösch- und programmierbarer Festwertspeicher
EPROM-programmer EPROM-Brenner, Eprommer
Epstein hystersis tester Epstein-Apparat
equable temperature ausgeglichene Temperatur
to equal gleich sein; gleich werden; gleichsetzen
equals ist gleich, ergibt
equal gleich, gleichartig; ebenbürtig
equal angle gleichschenkliges L-Eisen, gleichschenkliges Winkeleisen
equal-area web flächentreues Abbildungsnetz
equal-energy white Gleichenergieweiß
at equal spacing, equally spaced in gleichem Abstand
equal tempered wohltemperiert (534)
equality Gleichheit; Gleichsetzung; Äquivalenz
equality circuit Äquivalenzschaltung
equality sign Gleichheitszeichen
equalization Entzerrung; Glättung
to equalize gleichmachen, egalisieren, abgleichen, entzerren
equalizer Ausgleicher, Entzerrer, Ausgleichsschaltung
equalizer ring Ausgleichsring (621.313)
equalizer switch Ausgleichsschalter
equalizing Ausgleichs-...
equalizing bar Ausgleichsschiene (621.3)
equalizing cable Ausgleichskabel
equalizing charge Ausgleichsladung (621.355)
equalizing connection Ausgleichsverbindung (621.313)
equalizing current Ausgleichsstrom
equalizing network Entzerrerschaltung
equalizing pulse Ausgleichsimpuls (621.397)
to equate gleichsetzen
equation Gleichung
equation for balance Abgleichbedingung (621.317)
equation of higher degree Gleichung höheren Grades
equation of moments Momentengleichung
equation solver Polynomrechner
equation of state Zustandsgleichung
equation of time Zeitgleichung (52)
equatorial plane Mittelebene (senkrecht zur Achse)
equiangular gleichwinklig
equidistant äquidistant, gleichweit entfernt, in gleichem Abstand
equilateral gleichseitig
equilibrant Schlußlinie (beim Krafteck)
to equilibrate ausbalancieren, ins Gleichgewicht bringen
equilibration Ausbalancierung, Gewichtsausgleich
equilibrium (pl.: **-ia**) Gleichgewicht
equilibrium concentration Gleichgewichtskonzentration

equilibrium electrode potential Gleichgewichts-Elektrodenspannung
equilibrium potential Ruhepotential
equilibrium reaction potential Mindestzersetzungsspannung
equilibrium time Anlaufzeit, Einlaufzeit, Zeit bis zur Erreichung des Gleichgewichtes
equimultiple mit gemeinsamem Teiler
to equip ausrüsten
equiphase gleichphasig
equiphase surface Phasenfläche, Fläche gleicher Phase (535)
equipment Ausrüstung, Einrichtung, Vorrichtung, Gerät, Gerätesatz, Apparat
equipment box Gerätekasten (621.315.6)
equipment clearance line Profil, Bahnprofil, Wagenprofil (625)
equipment ground Körpererdung, Gehäuseerdung, Schutzerdung
equipment internal climate Geräteinnenraumklima
equipment in parallel operation aktive(s) Reserve(-gerät)
equipment practice Einschubsystem, Geräteaufbausystem
equipotential Äquipotential-...
equipotential connection Ausgleichsverbindung, Ausgleichsleiter (621.313)
equipotential line Niveaulinie, Äquipotentiallinie
equipotential method Phantomschaltung
equipotential space Äquipotentialfläche
equipotential surface Niveaufläche, Äquipotentialfläche
equisignal line Linie gleicher Feldstärke, Leitstrahl (621.396)
equisignal radio beacon Leitstrahlbake, Funkstrahlbake
equisignal radio range zone Leitstrahlsektor (621.396.93)
equivalent Gegenwert
equivalent articulation loss Ersatzdämpfung
equivalent of heat Wärmeäquivalent
equivalent gleichwertig, äquivalent; Ersatz-...
equivalent circuit Ersatzschaltung, Ersatzstromkreis, Ersatzschaltbild
equivalent conductivity Äquivalentleitfähigkeit
equivalent disturbing voltage äquivalente Störspannung
equivalent earth radius äquivalenter Erdradius
equivalent input voltage/current drift äquivalente Eingangsdrift (621.395)
equivalent isotropically radiated power äquivalente isotrope Strahlungsleistung
equivalent junction temperature Ersatzsperrschichttemperatur
equivalent ladder network Ersatzkettenleiter
equivalent logarithmic increment mittleres logarithmisches Inkrement
equivalent loudness subjektive Lautstärke
equivalent network Ersatzschaltung
equivalent periodic line Ersatzkettenleiter
equivalent reactance wirksamer Blindwiderstand
equivalent resistance Ersatzwiderstand
equivalent resistivity Äquivalent-Widerstand (621.35)
equivalent sine wave sinusförmige Welle mit gleichem Effektivwert
equivalent thermal network thermische Ersatzschaltung
equivalent uniform line homogene Ersatzleitung
equivalent weight Äquivalenz-Gewicht
to erase löschen
erasable storage löschbarer Speicher
erasable programmable ROM = EPROM lösch- und programmierbarer Festspeicher
erase button Löschtaste (681.84)
erase character Löschzeichen
erasing Löschen
erasing head Löschkopf
to erect bauen, aufstellen, montieren; errichten (51)
to erect an ordinate eine Ordinate auftragen
to erect a perpendicular line ein Lot fällen
erection Errichtung, Bau, Aufbau, Montage, Aufstellung
erection crane Montagekran, Baukran
erection sag Montagedurchhang (621.315.1)
erection stresses innere Spannungen, innere Kräfte, Einbauspannungen
erg Erg
to erode erodieren, auswaschen
erosion Erosion, Auswaschung
erosive erodierend, Erosions-...
erratic erratisch, regellos; umherirrend; wechselnd, unregelmäßig, streuend
erroneous fehlerhaft, mit Fehlern behaftet
erroneous bit Fehlerbit
error Fehler, Irrtum, Beobachtungsfehler; Anzeigefehler
error of approximation Approximationsfehler
error in bearing Peilwinkelfehler
error burst Fehlerhäufung
error correcting selbstkorrigierend

error correction Fehlerkorrektur
error curve Fehlerkurve, Funkbeschickung (621.396)
error detection Fehlererkennung
error-detection code Fehlererkennungscode, Fehlersuchcode
error in estimating Schätzungsfehler
error free fehlerfrei
error indicator Fehleranzeige
error in pointing Einstellfehler, Anzeigefehler
error-proof fehlersicher
error rate Fehlerrate, Fehlerhäufigkeit
error in reading Ablesefehler
to be in error by fehlerhaft sein um
erythemeter Erythem-Dosismeter
escals Nuten mit einer Polteilung Abstand
escalator Rolltreppe
escape Entweichen; Ausweg; Feuerleiter
escape character Umschaltzeichen
escape hatch Ausstiegklappe, Fluchtklappe, Notausstieg (629.13)
to escape entweichen, entrinnen, entgehen, entkommen; ablaufen, auslaufen
to escape resonance dem Resonanzeinfang entgehen (539)
escapement Echappement, Hemmwerk, Ankergang (681)
escutcheon Bezeichnungsschild
E.s. lamp cap Edison-Gewindesockel
essential wichtig, wesentlich
essential connections (pl.) Grundschaltbild (621.3)
establishment Einrichtung, Errichtung; Institut, Werk, Unternehmen
ester Ester
ester number Veresterungszahl
esterification Schätzung, Abschätzung, Voranschlag; Wertschätzung
E.S.U. = Electrostatic Unit E.S.E. = elektrostatische (c.g.s.-)Einheit
to etch ätzen, anätzen (669); ätzen (655)
etched circuit geätzte Schaltung
etchand Ätzlösung
ethane Äthan
ether Äther (53), (54)
ether drift Ätherwind
ether of space Raumäther, Weltäther
etheric telegraphy Funktelegraphie
etherophone Theremin-Instrument
ethylene Äthylen
E.T.U. = Electrical Trades Union Elektriker-Gewerkschaft
eutectic Eutektikum
eutectic eutektisch
eutectic alloy eutektische Legierung
eutectic mixture Eutektikum, eutektisches Gemisch
to evacuate evakuieren (621.315.2); leerpumpen; herausholen, herausziehen; räumen
evacuated envelope Vakuumkolben
evacuated space luftleerer Raum
evacuation Evakuierung (621.315.2); Auspumpung
evacuation process Pumpvorgang
to evaluate auswerten, berechnen, zahlenmäßig ermitteln
evaluation Auswertung, Berechnung, Zahlenrechnung
to evaporate verdampfen, wegdampfen, verdunsten; aufdampfen
to evaporate off wegdampfen
evaporation Verdampfung, Wegdampfen, Verdunstung; Eindampfung
evaporation point Verdampfungspunkt
evaporative verdampfend, Verdampfungs-...
evaporative power Verdampfungsleistung, Dampfleistung (621.18)
evaporator Verdampfer (621.18)
even eben, glatt, gleichmäßig; geradzahlig, gerade (51)
even-numbered geradzahlig
even-order von geradzahliger Ordnung
even-order harmonics geradzahlige Oberwellen
even parity gerade Parität (681.3)
even wear gleichmäßige Abnutzung
evenly distributed gleichmäßig verteilt
evenly spaced in gleichmäßigen Abständen
event counter Ereignis-Zähler (681.3)
event recorder Meßwertschreiber, Störungsschreiber
evident klar ersichtlich, evident
to evoke hervorrufen, erzeugen, verursachen
to evoke a voltage drop einen Spannungsabfall hervorrufen
evolute Evolvente
evolute connection Evolventenstirnverbindung
evolution Entwicklung, Entfaltung; Wurzelziehen (51)
evolution of gas Gasentwicklung
evolution of heat Wärmeentwicklung
to evolve entwickeln, herausarbeiten
to evolve a process ein Verfahren ausarbeiten
E-wave E-Welle, TM-Welle (621.396)
Ewing hysteresis tester Hysteresisschleifenschreiber
exact genau, exakt
exact balance genauer Abgleich

exact register sauberer Eingriff; gute Passung
exact sciences exakte Wissenschaften
exact solution genaue Lösung (51)
exalted carrier reception Empfang mit Trägeranhebung
examination Prüfung, Untersuchung, Examen
examination of metals Metallprüfung
examination work Prüfungsarbeit
to examine prüfen, untersuchen, examinieren
to examine a line Leitung prüfen (621.395)
to excavate ausgraben; baggern (an Land); ausschachten
excavator Löffelbagger, Bagger
exceptional ungewöhnlich, Ausnahme-..., ausnehmend, außerordentlich
exceptional rate Sondertarif, Ausnahmetarif (625)
excess Überschreitung; Ausschreitung, Exzeß; Übermaß, übrige Länge, übrige Menge; Überschuß (**of:** an)
excess air Luftüberschuß (621.18)
excess conduction Überschußleitung
excess consumption Mehrverbrauch, Überverbrauch
excess current Überstrom
excess electron Überschuß-Elektron
excess-electron conduction Überschußelektronenleitung
excess heating Überhitzung
in excess im Überschuß, im Übermaß; darüber hinaus
in excess of... über... hinaus
excess meter Überverbrauchszähler
excess pressure Überspannung, zu hohe Spannung (621.316)
excess-reactive power meter Überschußblindverbrauchszähler
excess temperature Übertemperatur
excess-three-code Drei-Exzeß-Code, Stibitzcode (681.3)
excess and total meter Gesamt- und Überverbrauchszähler
excess voltage Überspannung (621.313)
excess-voltage preventer Überspannungsschutz (621.313)
excess-voltage protection Überspannungsschutz (621.313)
excess weight Übergewicht
excessive data Überlaufdaten
exchange Amt, Vermittlungsstelle, Vst. (621.395); Tausch, Austausch
exchange area Anschlußbereich (621.395)
exchange battery Amtsbatterie
exchange line Teilnehmerleitung
exchange number Amtsnummer
exchange reaction Austauschreaktion (64)
in exchange for im Austausch gegen
to exchange wechseln, austauschen, ersetzen (**by:** durch), (**for:** für)
exchangeability Austauschbarkeit
excitability Erregbarkeit, Anregbarkeit (534); (621.396)
excitable erregbar, anregungsfähig
excitant Elektrolyt (621.35)
excitation Anregung, Erregung, Speisung
excitation of an aerial Speisung einer Antenne, Anregung einer Antenne
excitation anode Erregeranode (621.314.5)
excitation band Anregungsband (539)
excitation energy Anregungsenergie (539)
excitation field Erregerfeld
excitation flux Erregerfluß
excitation loss Erregerverlust
excitation of oscillations Schwingungsanregung, Schwingungserregung
excitation purity spektrale Farbdichte, spektraler Farbanteil
excitation voltage Erregerspannung
excitation of a winding Speisung einer Wicklung
excitative erregend, anregend
excitative doublet Erregerdipol
to excite anregen, erregen, erzeugen
to excite a dynamo einen Generator erregen
to excite oscillations Schwingungen erzeugen
excited atom angeregtes Atom
excitement Erregung
exciter Erregermaschine, Steuersender; aktive Antenne, Erreger, aktiver Strahler (621.396)
exciter anode Erregeranode (621.314.6)
exciter build-up time constant Zeitkonstante der Auferregung (621.313)
exciter circuit Erregerkreis (621.314.6)
exciter-field rheostat Erregerfeldwiderstand, Feldregler
exciter resistance Feldregler, Erregerwiderstand (621.313)
exciter response Änderungsgeschwindigkeit der Erregerspannung (621.313)
exciter set Erregersatz
exciter stage Steuerstufe
exciter system Anregungssystem (621.396)
exciting Erreger-..., erregend
exciting ampturns Erreger-AW
exciting circuit Erregerkreis
exciting coil Erregerspule
exciting current Erregerstrom, Magnetisierungsstrom

exciting field Erregerfeld
exciting generator Erregermaschine
exciting voltage Erregerspannung (621.313); Gitterspannung (621.314.6)
exciting winding Erregerwicklung
excitron Excitron
exclusion Ausschluß, Ausschließung; Unterdrückung
exclusion of air Luftabschluß
exclusive exchange line Amtsleitung
exclusive OR circuit Exklusiv-ODER-Gatter
excursion Ablenkung, Auslenkung; Ausschlag
execution time Ausführungszeit (681.3)
executive Exekutive; leitender Angestellter
executive cycle Ausführungszyklus (681.3)
executive instruction Ausführungsbefehl (681.3)
executive program Steuerprogramm, Betriebsprogramm (681.3)
to exercise üben, ausüben
to exercise control over Steuerwirkung ausüben auf
to exfoliate abblättern, (sich) schuppen, abschilfern
exfoliation Abschuppung, Abschilferung
exhaust Auspuff, Auslaß
exhaust air Abluft
exhaust box Auspufftopf
exhaust pump Vakuumpumpe, Saugpumpe
to exhaust erschöpfen; auspumpen, evakuieren
exhausted erschöpft, verbraucht, ausgebraucht (54)
exhausted battery erschöpfte, entladene Batterie, ausgebrauchte Batterie
exhauster Exhaustor, Absauger, Absaugeinrichtung
exhaustion Erschöpfung; Verbrauchtsein
exhaustion of steam Dampfausströmung
exit Ausgang; Ausgabebefehl
exosmose Exosmose
exothermal exotherm
exothermic(al) exotherm
exothermic nuclear reaction exotherme Kernreaktion
to expand expandieren, (sich) ausdehnen, (sich) dehnen; ausmultiplizieren, ausschreiben (51); vergrößern
to expand in a series in eine Reihe entwickeln (51)
to expand with heat sich bei Erwärmung ausdehnen
expanded-centre plan display Rundsichtdarstellung mit Mittelpunktaufweitung
expanded metal Streckmetall
expanded scope vergrößerter Bildausschnitt
expander Dehner, Expander
expander circuit Dehnerschaltung, Expanderschaltung (621.395); Erweiterungsschaltung (621.374)
expanding Dehn-..., Spreiz-...
expanding collet Spannring, Spreizring
expansibility Dehnbarkeit, Ausdehnungsfähigkeit
expansible dehnbar
expansion Expansion, Dehnung, Ausdehnung, Wärmedehnung, Aufweitung, Erweiterung
expansion circuit-breaker Expansionsschalter
expansion coupling Dehnungskupplung
expansion crack Dehnungsriß (69)
expansion joint Ausdehnungsglied, Kompensationsrohr, Kompensationsbogen; Dehnungsfuge (693)
expansion match Zwischenglas = hinsichtlich der Wärmedehnung angepaßtes Glas (621.38)
expansion in series Reihenentwicklung
expansion stage Expansionsstufe (621.395.3)
expansion switch Expansionsschalter
expansion tank Ausdehnungsgefäß (69), (621.314.2)
expansion vessel Ausdehnungsgefäß (621.314.2)
expansive ausgedehnt, weitläufig, Dehn-..., Expansions-...
expansive force Expansionskraft
to expel austreiben, herausdrücken
to expel the volatiles from ... die flüchtigen Bestandteile abtreiben aus ... (54)
to expend verbrauchen, verzehren, ausgeben, verausgaben; ausweiten
to expend energy Energie verbrauchen
expenditure of energy Energieverbrauch, Energieaufwand
experiment Experiment, Versuch, Untersuchung
experiment tank Schleppkanal (629.12)
to experiment experimentieren
experimental experimentell
experimental confirmation experimentelle Bestätigung
experimental findings (pl.) experimenteller Befund
experimental probe Prüfsonde, Meßsonde
experimental proof experimenteller Nachweis, experimenteller Beweis
experimental reactor Versuchsreaktor
experimental run Probelauf, Probebetrieb
experimentation Experimentieren
experimenter Experimentator

expert Fachmann, Sachverständiger; sachverständig, fachmännisch
expert engineer Fachingenieur
expert opinion Sachverständigengutachten
expert worker Facharbeiter
explicit explizit, eingehend, detailliert
to explode explodieren; sprengen
exploded view Ansicht der Einzelteile, Ansicht in auseinandergenommenem Zustand, Explosionsdarstellung, aufgelöste Darstellung
exploder Zündübertrager, Übertrager (623.4); Sprengkapsel, Zündhütchen
exploratory aufklärend, vorfühlend, Probe-...
exploratory experiment Stichversuch
exploratory test Vorversuch, Stichversuch
exploratory work Forschungsarbeit
to explore aufklären, erforschen; schürfen
exploring brush Prüfbürste (621.313)
exploring coil Prüfspule, Suchspule
to explore a surface eine Fläche abtasten
explosion Explosion, Zerknall
explosion limit Berstdruck
explosion-pot circuit breaker Löschkammerschalter
explosion-proof schlagwettergeschützt (621.313); druckfest (621.316.5); explosionsgeschützt
explosion-proof enclosure Schlagwetterschutzkapselung [A]
explosion-proof motor schlagwettergeschützter Motor
explosion-proof transformer schlagwettergeschützter Umspanner
explosion protection Schlagwetterschutz
explosion of a steam boiler Dampfkesselexplosion
explosion-tested schlagwettergeprüft
explosion-type engine Otto-Motor
explosion vent Explosionsschutz (621.314.2)
explosive Sprengstoff; explodierend, explosiv, explosibel, explosionsartig, Spreng-...
explosive gas explosionsfähiges Gas, zündfähiges Gemisch
explosive limits Explosionsgrenzen, Zündgrenzen
explosive rivet Sprengniet
explosiveness Explosionsfähigkeit
exponent Exponent
exponential Exponentialfunktion
exponential exponentiell, Exponential-...
exponential character exponentieller Verlauf, Exponential-Charakter
exponential function Exponentialfunktion
exponential horn Exponentialtrichter (534); Exponentialhornstrahler (621.396.67)
exponential law Exponentialgesetz
exponential notation Exponentialschreibweise
exponential pile Exponentialbatterie (539)
exponential quantity exponentiell verlaufende Größe
exponential sweep exponentielle Zeitablenkung (621.397)
exponentially damped sinusoidal quantity exponentiell gedämpfte sinusförmige Größe
to expose belichten (77)
exposed freiliegend, offen, offenliegend, zugänglich; belichtet, bestrahlt (77)
to be exposed freiliegen, der Berührung zugänglich sein
exposed wiring offene Leitungsverlegung, Leitungsverlegung auf der Wand
exposed work freiliegend, offen (621.3)
exposure Belichtung, Aufnahme (77)
exposure to light Belichtung
exposure meter Belichtungsmesser
exposure time Belichtungszeit
expressible in terms of ... angebbar in ...
expression Ausdruck; Formelausdruck
expulsion Austreibung, Vertreibung
expulsion fuse (unit) Löschrohrsicherung
expulsion tube Löschrohr
expulsion-type (lightning) arrester Löschrohrableiter
to exsiccate austrocknen
exsiccation Austrocknung
exsiccator Trockner, Trockenschrank, Exsikkator
to extent dehnen, strecken, ausdehnen, erweitern; voll ausschreiben (51); übertragen (65); entfalten (35)
to extend over sich erstrecken über
to extend the range den Meßbereich erweitern (621.317)
to extend a shaft eine Welle verlängern
to extend to reichen bis, sich erstrecken bis; verlängern bis (51)
extended ausgedehnt
extended antenna ausgefahrene Antenne
extended frequency range breites Frequenzband
extended period längere Zeit
extension Dehnung, Verlängerung; Nebenanschluß, Nebenstelle (621.395)
extension bell Zusatzwecker
extension cord Verlängerungsschnur
extension handle Griffverlängerung, Ansatzgriff, verlängerter Griff

extension line Nebenanschlußleitung (621.395)
extension register Zusatzregister
extension set Nebenanschluß (621.395)
extension station Nebenanschluß
extensional mode Längsschwingungen
extensive ausgedehnt, umfangreich, ausgiebig, weitreichend, extensiv
extensive reflector Breitstrahlreflektor
extensometer Dehnungsmesser
exterior äußerlich, Außen-...
exterior lighting Außenbeleuchtung
exterior view Außenansicht
external äußerlich, äußer, Außen-...
external armature alternator Innenpolmaschine (621.313)
external characteristic äußere Charakteristik (621.313)
external circuit äußerer Stromkreis
external conductor äußerer Leiter, Mantelleiter, leitender Mantel
external dimensions Außenabmessungen
external excitation Fremderregung
external fan Außenlüfter
external field generator Außenpolgenerator
external gas-pressure cable Gasaußendruckkabel
external magnetic field äußeres magnetisches Feld, magnetisches Fremdfeld
external memory Außenspeicher, externer Speicher
external modulation Fremdmodulation
external pole generator Außenpolgenerator
external remanent residual voltage äußere remanente Restspannung (eines Halleffektbauelementes)
external resistance äußerer Widerstand (621.3)
external rotor motor Außenläufermotor
external self-inductance äußere Selbstinduktion
external thread Außengewinde
external voltage Fremdspannung (621.395)
externally operable von außen zu betätigen
extinction Erlöschen, Extinktion
extinction of lamps Ausschalten von Lampen
extinction of light Ausbleiben des Lichts, Stromausfall
extinction potential Löschspannung (621.385)
to extinguish löschen, auslöschen; ausschalten (621.32)
extinguisher Feuerlöscher
extinguishing Lösch-...
extinguishing medium Löschmittel
extra zusätzlich, Zusatz-...
extra anode Zusatzanode (621.314.6)
extra charge Zusatzgebühr; Zusatzkosten (pl.)
extra current Ausgleichstrom, freier Strom, Extrastrom
extra equipment Zusatzausrüstung
extra grid Zusatzgitter (621.385)
extra-high-pressure mercury-vapour lamp Quecksilberdampfhöchstdrucklampe
extra high tension Höchstspannung (261.315)
extra-high voltage line Höchstspannungsleitung
extra loss Zusatzverlust (621.313)
extra-low voltage lighting Kleinstspannungsbeleuchtung
extra resistance Zusatzwiderstand
extra-terrestrial noise kosmisches Rauschen
to extract herausziehen, herauslösen, ausziehen, extrahieren (54); entnehmen, abnehmen (621.3); ziehen, ausziehen (51)
to extract a root eine Wurzel ziehen
to extract a voltage eine Spannung abnehmen
extractable ausziehbar
extraction Extraktion; Herstellung; Gewinnung
extraction fan Sauglüfter
extractor Auszieher (621.9); Ausziehvorrichtung
extraneous äußer, von außen kommend, außen angebracht
extraneous field äußeres Fremdfeld
extraneous voltage Fremdspannung
extranuclear außerhalb des Kerns befindlich, Hüllen-... (639)
extranuclear electron Hüllenelektron
extranuclear structure Hüllenaufbau
extraordinary außerordentlich
extraordinary ray außerordentlicher Strahl (535)
extraordinary wave außerordentliche Welle
to extrapolate extrapolieren (**towards:** nach)
extrapolated boundary extrapolierte Reaktorbegrenzung (539)
extrapolation Extrapolation, Extrapolierung
extreme Extremwert, Extremum (pl.: -ma), Höchstwert
extremely high frequency EHF, Millimeterwellen
extrinsic base resistance Basiswiderstand
extrinsic conduction Störstellenleitung (621.382)
extrinsic resistance Bahnwiderstand (621.382)
extrinsic semiconductor Störstellenhalbleiter
to extrude herauspressen, strangpressen
extruded bar Preßstrang, Profilstange
extrusion Strangpressen, Herausdrücken
extrusion dye Preßdüse, Strangpreßform
extrusion machine Strangpresse
to exude ausschwitzen

eye Auge; Öse, Öhr
eyebolt Ösenschraube, Tragöse
eyeestimate visuelle Abschätzung
eyelens Okular
eyelet Öse, Durchziehloch; Einschmelzöse
eyepiece Okular
eye protection Augenschutz
eye protective glass Augenschutzglas
eye protective glasses (pl.) Schutzbrille
eyestrain Augenermüdung
eyeletting Ösenverbindung

F

F. = Fahrenheit F = Fahrenheit
f. = farad F = Farad
f. = feet (pl.) Fuß (1 Fuß = 0,305 m)
fabric Gewebe
fabric lining Textilfutter, textiler Belag
Fabroil Tenax, Tenacit
face Stirnfläche, Fläche, Vorderseite; Zifferblatt
face area Schirmfläche (621.397)
face-centred flächenzentriert (54)
face chuck Planfutter (621.9)
face diagonal Flächendiagonale
face to face aufeinandergelegt, flach geschichtet
face grinder Flächenschleifmaschine
face of a lens Linsenfläche, Linsenvorderfläche
face plate Planscheibe (621.9)
face-plate controller Flachbahnsteuerschalter, Flachbahnkontroller
face-plate and handle Schalterbetätigungsplatte mit Griff
face-plate step switch Flachbahnstufenschalter
face-plate starter Flachbahnanlasser (621.3)
face-shear mode Flächenscherungsschwingung
face up mit der Fläche nach oben
to face gegenüberstellen; gegenüberstehen, sich gegenübersehen; plandrehen (621.9)
to face with verkleiden mit, (vorderseitig) belegen mit
facet Facette
facility Einrichtung; Erleichterung; Hilfseinrichtung, Hilfsmittel
facing Belag; Verschalung; Flächen-..., Außen-...
facing plates sich gegenüberstehende Platten
facing point Spitzweiche (625)
facsimile Faksimile, Reproduktion
facsimile recorder Faksimileschreiber
facsimile telegraph Bildtelegraph, Faksimiletelegraph
facsimile transmission Bildübertragung
factor Faktor (51); Faktor, Einflußgröße (53)
factor of assurance Prüfspannungsverhältnis (621.315.2)
factor of merit (Stromkonstante eines Galvanometers bei 1 Ohm Widerstand und 10 s Schwingungsdauer)
factor of proportionality Proportionalitätsfaktor
factor of safety Sicherheitsgrad
factorial n = n! n Fakultät = n!
to factorize in Faktoren zerlegen
factory Fabrik, Fabrikanlage, Werk, Betrieb
factory fitting Werkstattleuchte (621.32)
factory length allocation Gruppierung der Fertigungslängen von Kabeln
factory test Werksprüfung
to fade schwinden (621.396); verblassen
to fade in (nach Schwund) wiederkommen (621.396)
to fade out schwinden (621.396); ausblassen; ohnmächtig werden (61)
fade unit Schwundregler (621.396)
fading Schwinden, Schwund, Fading (621.396); Ausblassen, Verblassen
fading effect Lautstärkeschwankung
fading margin Schwundbereich
fading trouble Schwundstörung
fadometer Farbfestigkeitsprüfer
to faggot paketieren, bündeln
Fahr. = Fahrenheit F = Fahrenheit
fan-in Belastungsfaktor des Eingangs (621.374)
fail-safe system störungssicheres System
failure Fehler (621.3); Ausfall
failure criteria Ausfallkriterien
failure frequency Ausfallhäufigkeit
failure rate Ausfallrate
faint schwach, geringfügig, leise
faint current schwacher Strom, winziger Strom
faint line dünne Linie
faint noise leises Geräusch, schwaches Geräusch
faint reflection schwacher Reflex
fair-weather corona Schönwetterkorona
fall Abfall, Fallen, Fall; Heruntergehen, Abwärtsgehen; Senkung, Abfall; Wasserfall; Seil (des Flaschenzuges)
fall of current Fallen des Stromes, Abnahme der Stromstärke

fall in output Leistungsrückgang, Rückgang der Fabrikation
fall pendant Zugpendel (621.32)
fall of potential Spannungsabfall
fall in pressure Druckabfall, Druckrückgang
fall of pressure Druckverlust
fall time Abfallzeit (621.382), (621.318); Rückendauer, Abfallzeit (621.316.9), (621.397); Flanken-Abfallzeit (621.374)
fall-in-voltage relay Spannungsrückgangsrelais
to fall fallen, abfallen, heruntergehen, sinken, sich senken, abnehmen; umstürzen, einstürzen
to fall along sich abspielen auf
to fall away abfallen
to fall back to sinken auf, abfallen auf, wieder absinken auf, zurückfallen in
to fall into disuse nicht mehr benutzt werden, außer Gebrauch kommen
to fall down zu Bruch gehen
to fall freely frei fallen
to fall into zerfallen in; münden (91)
to fall into groups in Gruppen zerfallen
to fall into a notch einklinken, sich einklinken
to fall obliquely on schief auftreffen auf, schief einfallen auf
to fall off abfallen, absinken
to fall out of step außer Tritt fallen
to fall out of synchronism aus dem Synchronismus fallen, außer Tritt fallen (621.313)
to fall over a distance eine Strecke durchfallen (531)
to fall plumb senkrecht fallen
to fall short of zurückbleiben hinter, nicht entsprechen
to fall in step in Synchronismus kommen
faller Vorreiber
fallibility Fehleranfälligkeit
fallible fehleranfällig
falling Abfall, Abfallen
falling characteristic fallende Charakteristik
falling-off in speed Drehzahlabfall
falling weight test Fallgewichtsprüfung (662)
false falsch
false alarm Fehlalarm
false bottom Doppelboden (629.12); Blindboden (693)
family of characteristics Kennlinienfeld
family of curves Kurvenschar
fan Ventilator, Lüfter; Fächer; Lüfterrad
fan aerial Fächerantenne (621.396.67)
fan beam Fächerkeule
fan blade Gebläseflügel, Ventilatorflügel
fan casing Ventilatorgehäuse
fan-cooled mit Anblasekühlung
fan-cooled motor Motor mit Eigenbelüftung
fan cooling Anblaskühlung
fan light Oberlicht
fan-like fächerähnlich
fan marker beacon Fächerfunkfeuer
fan-ventilated motor gekapselter Motor mit Eigenbelüftung
to fan fächeln, entfachen
fan-in Belastungsfaktor des Eingangs (621.374), (681.3)
fan-out Belastungsfaktor des Ausgangs (621.374)
fan-fold paper Endlosformular
fanned-beam antenna Fächerstrahlantenne
fanwise fächerartig
far weit, weit entfernt
far-end cross-talk Gegennebensprechen, Fern-Nebensprechen
far station Gegenstelle
farad Farad
faraday (Einheit der Elektrizitätsmenge = 96 490 Coulomb)
Faraday cage Faraday-Käfig, Abschirmkäfig
Faraday cell Faraday-Kammer (535)
Faraday's dark space Faradayscher Dunkelraum
Faraday's ring Faradayscher Ring
Faraday tubes elektriche Feldröhren
faradic current Faraday-Strom (615.84)
faradism, faradization Faradisierung (615.84)
faradmeter Kapazitätsmesser
fast schnell; fest, sicher, zuverlässig
fast-action switch Schnellschalter
fast (neutron) fission Spaltung mit schnellen Neutronen
fast in operation trägheitsarm
fast pulley Festscheibe (621.8)
fast reactor schneller Reaktor
to fasten befestigen, festmachen, anschrauben, anbinden, anschlagen (**to:** an), festkeilen
to fasten into position in der richtigen Lage befestigen
fastener Befestigung, Befestigungsstück
fastening Befestigung
fastening bolt Befestigungsbolzen
fastening screw Befestigungsschraube
fastness Lichtechtheit, Beständigkeit
fatal nachteilig; tödlich
fatal accident tödlicher Unfall
fathom Faden (Längenmaß; 1 fathom = 2 yard = 6 Fuß = 1,829 m)

to fathom loten, Tiefe messen
fathometer Echolot
fatigue Ermüdung (61), (620.1), (621.385)
fatigue allowance Ermüdungszuschlag
fatigue bending Dauerbiegung
fatigue failure Ermüdungsbruch, Dauerbruch
fatigue fracture Dauerbruch, Ermüdungsbruch
fatigue limit Ermüdungsgrenze, Dauerfestigkeit
fatigue strength Ermüdungsfestigkeit, Dauerfestigkeit
fatigue test Dauerfestigkeitsprüfung, Dauerversuch
fatigue testing machine Dauerprüfmaschine
fault Fehler (621.3), Schaden
to be at fault schuld sein; fehlerhaft sein, defekt sein
fault bus Erdungsschiene; K-Leiter (621.316.9)
fault current Fehlerstrom, Körperschlußstrom
fault diagnosis Fehlerdiagnose
fault-electrode current Überlaststrom bei Rückzündung oder Kurzschluß (621.382)
fault finding Fehlersuche
fault-to-ground current Erdschlußstrom
fault indication Fehleranzeige
faultless fehlerlos
fault localization Fehlerortsbestimmung, Fehlereingrenzung
fault recording Störungsannahme (621.395); Fehleraufschrieb
fault resistance Fehlerwiderstand
fault signalling network Pfannkuch-Kabelschutz
fault time Ausfallzeit
fault-voltage circuit breaker Fehlerspannungsschutzschalter
faulted line fehlerbehaftete Leitung
faulty fehlerhaft, schadhaft, fehlerbehaftet
faulty branch gestörter Abzweig (621.3)
faulty circuit gestörter Stromkreis
faulty insulation schadhafte Isolation
faulty selection Falschwahl (621.395)
faulty transformer beschädigter Transformator, schadhafter Umspanner
Faure accumulator Gitterplattensammler, Pasteplatten-Akkumulator
Faure plate Faure-Platte
fcp. = foolscap Foolscap [$13^1/_2'' \times 17''$]
FCS = fire control system
feasibility study Durchführbarkeitsstudie
feather Feder (59); Feder, Befestigungskeil (621)
to feather fiedern
feathery fiederförmig
feature Merkmal, Kennzeichen, Eigenschaft
fed gespeist (s. **to feed**)
feeble schwach, gering
feeble field schwaches Feld
feeble voltage geringe Spannung
feed Speisung; Vorschub (621.9); Beschickung (669.162.2)
feedback Rückkopplung, Rückführung
feedback amplifier rückgekoppelter Verstärker
feedback around Rückkopplung über
feedback circuit Rückkopplungskreis
feedback control loop Rückkopplungsschleife, geschlossener Regelkreis
feedback control system Regelkreis, Regelsystem mit Rückführung
feedback factor Rückkopplungsfaktor
feedback loss Umlaufdämpfung, Rückkopplungsdämpfung
feedback network Rückkopplungsschaltung
feedback path Rückkopplungsweg
feed conveyor Aufgabeband, Zubringerband
feed forward Mitkopplung
feed mechanism Vorschubwerk (621.325)
feed part Transportklinke
feed point Speisepunkt (621.396.67)
feed-in point Einspeisepunkt
feed system Speiseleitung (621.396.67)
feed tank Speisewasserbehälter (621.2)
feed-through capacitor Durchführungskondensator
feed-through input Durchschleifeingang (621.397)
feed-trip stop Endauslöser, Endanschlag, Begrenzer (621.9)
to feed speisen, zuführen (621.3); speisen (621.396.67); beschicken (669.162.2); vorschieben (621.9); unterhalten, nähren, beschicken (66)
to feed-back rückkoppeln, rückführen
to feed-in einspeisen
to feed radially radial zuführen (621)
to feed through gespeist werden durch (621.3); durchführen
feeder Speiseleitung (621.3); Einguß (621.74); Aufgabevorrichtung, Beschickungsvorrichtung; Zufuhr, Zufuhrkanal, Zufuhrrinne; Nebenfluß (91)
feeder box Abgangszelle (621.316.1)
feeder bus-bar Leitungssammelschiene
feeder cable Speisekabel
feeder chute Aufgaberutsche, Speiserutsche (621.1)

feeder circuit Speisestromkreis
feeder distribution centre Speiseleitungsverteiler (696.6)
feeder ear Speiseanschluß (621.33)
feeder panel Speiseleitungsfeld, Leitungsfeld
feeder pillar Speisepunktsäule
feeder protection Speiseleitungsschutz
feeder reactor Leitungsdrossel
feeder switch Speiseleitungsschalter, Speiseschalter
feeder voltmeter Leitungsvoltmeter
feeding Beschickung, Speisung; Aufgabe, Anlegen; Füllen; Füllung; Vorschub; Unterhaltung, Erhaltung
feeding cable Speisekabel
feeding funnel Aufgabetrichter
feeding line Speiseleitung (621.3)
feeding loss Speisestromdämpfung
feeding magazine Fülltrichter (621.9)
feeding pipe Zuflußleitung, Speiseleitung (621.1)
feeding point Speisepunkt (621.3)
feeding screw Förderschnecke
feedline Speiseleitung, Energieleitung (621.396.67)
feel Gefühl, Fingerspitzengefühl, Griff
feel of the control Gefühl für die Steuerung
to feel abtasten, fühlen
feeler Fühler, Taster; Stichmaß (620.1)
feeler gauge Innenlehre, Stichmaß, Dickenlehre
feeler switch Prüfschalter, Fehlerprüfschalter (621.315.1)
feet (pl.) Füße; Fuß
n feet on edge (mit) n Fuß Kantenlänge
feet switch (Schalter mit Zwischenraum zwischen Wand und Grundplatte)
fel(d)spar Feldspat
felt Filz, Dachpappe
felt packing Filzdichtung
felt wad Filzpfropfen
felt-wick lubricator Dochtöler (621.82)
female weiblich
female die Matrize
female screw Schraubenmutter
female socket Steckdose
female thread Muttergewinde, Innengewinde
fence Hecke, Zaun, Einzäunung; Gitter, Schutzgitter, Schutzverkleidung; Schutzblech (621.9)
to fence verkleiden, vergittern, schutzverkleiden
Fermi level Fermi-Niveau
Ferranti effect Ferranti-Effekt (621.315)
Ferranti-Hawkins protective system Kabelerdstromdifferentialschutz
Ferraris instrument Ferraris-Instrument, Drehfeldinstrument
Ferraris motor Ferrarismotor, Drehfeldmotor
ferreous eisenhaltig
ferric Ferri-... [dreiwertige Eisenverbindung]
ferric induction Magnetisierung
ferric oxide Eisen-III-Oxid
ferrite Ferrit
ferrite core Ferritkern
ferrite isolator Ferrit-Richtungsleitung (621.395.69)
ferrite ring core Ferrit-Ringkern
ferrite rod aerial Ferritantenne
ferritic ferritisch
ferrocart Ferrocart
ferro concrete Eisenbeton
ferro-concrete pole Eisenbetonmast
ferrodynamic ferrodynamisch
ferroelectric material ferroelektrisches Material
ferromanganese Manganeisen [12% Mn]
ferromagnetic ferromagnetisch
ferromagnetism Ferromagnetismus
ferrometer Ferrometer, Eisenprüfgerät
ferrous Ferro-... (zweiwertige Eisenverbindung); Eisen-...
ferrous coated tape Magnetband
ferrous metals Eisenmetalle
ferrous oxide Eisenoxidul
ferrule Quetschhülse, Verbinder
ferry (Widerstandslegierung ähnlich Konstantan)
fertile Brütermaterial (539)
festoon Girlandenschmuck
festoon drier Hängetrockner (77)
festoon lamp Soffittenlampe
festoon lighting Schmuckbeleuchtung; Lampenketten (pl.)
FET = field effect transistor Feldeffekttransistor
fetch Abruf (681.3)
fetch cycle Fetchzyklus, Holzyklus
fibre Faser, Fiber
across fibre quer zur Faser
fibre board Leichtbauplatte
fibre core Hanfseele
fibre link Glasfaserverbindung
fibre optic Faseroptik, Glasfaseroptik, Fiberoptik; Lichtleitertechnik, Lichtwellentechnik
fibre slab Faserplatte, Leichtbauplatte
fibre wall board Hartfaserplatte
fibre washer Fiberunterlegscheibe
fibril Fibrile

fibrillation Zermahlung, Zerfaserung (67)
fibrous faserig, faserähnlich, Faser-..., Textil-...
fibrous material Faserstoff
fibrous structure Faserstruktur
fictitious vorgetäuscht, erfunden, gedacht
fictitious power (geometrische Differenz aus Schein- und Wirkleistung)
fid Schmitz, Keilchen
fidelity Übertragungsgüte, getreue Wiedergabe (534), (621.396.62)
fiducial als Vergleichsgröße dienend, Vergeichs-..., Basis-...
fiducial line Basislinie (52)
fiducial value Basiswert (621.317)
fiduciary anvertraut, auf Vertauensbasis
fiduciary value Basiswert, Ausgangswert
field Feld (53), (621), (623), (63); Revier (622); Halbbild (621.397)
fields (pl.) Erregersystem, Feld (621.313)
field of activity Arbeitsgebiet
field ampere-turns Gegenamperewindungen; Ankerrückwirkung (sing.) (621.313)
field blanking interval Vertikal-Austastlücke
field breaking resistor Überbrückungswiderstand (621.318)
field breaking switch Überbrückungsschalter (621.318)
field break-up switch Feldtrenner
field cable Feldkabel
field calibration Eichung im Feld fortschreitender Wellen (534)
field change Feldänderung
field coil Feldspule
field configuration Feldbild
field control Feldregelung, Feldschwächung (621.313)
field current Feldstrom, Erregerstrom
field-current regulator Erregerstromregler
field density Felddichte
field direction Feldrichtung
field discharge Feldentladung
field discharge resistor Überbrückungswiderstand (621.318)
field discharge switch Überbrückungsschalter (621.318)
field distribution Feldverteilung
field diverter rheostat Feldüberbrückungswiderstand (621.313)
field duration Halbbilddauer
field-effect transistor Feldeffekt-Transistor
field-enhanced emission Feldemission
field excitation Felderregung
field of force Kraftfeld
field frequency Vertikalfrequenz (621.397)
field intensity Feldstärke
field intensity meter Feldstärkemesser
field iron loss Ständereisenverlust (621.313)
field losses Erregungsverluste (621.313)
field magnet Feldmagnet
field magnet-winding Feldwicklung, Erregerwicklung
field obstruction Feldstörung (621.396)
field offset Frequenzversatz (621.397)
field pattern Feldbild
field pole Pol, Induktorpol, Feldpol (621.313)
field-programmable logic arrays = FPLAs vom Anwender programmierbare Logikanordnungen
field regulator Feldregler
field repair Reparatur an Ort und Stelle
field rheostat Feldregler, Feldregelwiderstand
field sensitivity Feld-Übertragungsfaktor (621.396.67)
field sequentitial Halbbildfolge
field spider Polrad, Polstern
field spool Spulenkasten, Feldspulenkasten (621.313)
field strength Feldstärke
fieldstrength measuring set Feldstärkemeßgerät (621.396)
field strength of the unwanted signal Störfeldstärke
field of stress Spannungsfeld
field suppressor Feldschwächungseinrichtung
field sweep Vertikal-Ablenkung (621.397)
field sync pulse Vertikal-Synchronsignal
field system Erregerfeld; Pole (pl.)
field telephone Feldtelefon
field tube Feldröhre, Kraftröhre
field of view, field of vision Gesichtsfeld, Blickfeld
field winding Feldwicklung
field winding coil Feldspule
field wire Feldkabel
FIFO = first-in, first-out
fifth voltage range Spannungsbereich über 660 Volt
fig. = figure Fig. = Figur, Abb. = Abbildung
figure Abbildung, Figur, Zahl, Ziffer, Zahlzeichen
to be a figure eine Form haben
figure-eight pattern Achterfigur, Achtermuster, Achterkurve, Doppel-Cosinus-Charakteristik
figure-of-eight-reception Achterdiagramm, Doppelkreisdiagramm
figure-eight trolley wire Profilfahrdraht
figures key Zifferntaste (621.394)

figure of loss Verlustziffer
figure of merit Empfindlichkeit, Gütezahl (621.317)
figure of revolution Rotationskörper
figure shift Ziffernschaltung (621.394.6)
to figure figürlich darstellen, sich zahlenmäßig belaufen auf
filament Draht, Drähtchen; Kathode, Heizfaden (621.385); Glühfaden (621.326)
filament battery Heizbatterie
filament bombardment Kathodenbombardement
filament cathode direkt geheizte Kathode
filament current Heizstrom (621.385)
filament efficiency Heizmaß (621.385)
filament electrometer Fadenelektrometer, Saitenelektrometer
filament-heating transformer Heiztransformator (621.396)
filament lamp Glühlampe, Drahtlampe
filament power supply Heizleistungsquelle, Heizspannungsquelle
filament resistance Heizwiderstand (621.385); Fadenwiderstand (621.32)
filament supply Heizbatterie, Heizstromversorgung
filament transformer Heiztransformator (621.385)
filament-type cathode direkt geheizte Kathode
filament-type tube direkt geheizte Röhre
filament voltage Heizspannung
file Datenbank, Datei
file protection Dateischutz
file structure Dateiorganisation
to file einsortieren, einordnen, ablegen, speichern
filings Feilspäne; Feilicht (sing.)
filings coherer Feilspankohärer
to fill füllen, anfüllen; ausfüllen; beschweren (67)
to fill to capacity ganz füllen
to fill in increments stufenweise füllen
to fill out (up), ausfüllen, auffüllen
fill character Füllzeichen
fill piece Füllstück, Füllansatz
filler Füllmasse, Füllmaterial (621.791); Füllstoff (67); Beilauf (621.315.2); Zuschlag (66); Füllgas (621.32)
filler metal Zusatzmetall, Auftragmetall (621.791)
filler valve Füllventil
fillet Schweißraupe; Hohlkehle; Abrundung; Leiste
fillet gauge Abrundungslehre
fillet weld Kehlschweißung
filling Füllung; Beilauf (621.315.2); Füllstoff (67)
filling-level meter Füllstandsanzeiger
filling material Vergußmasse, Kabelvergußmasse, Kabelmasse
filling medium Füllmittel
filling metal Schweiße (621.791)
film Film, Filmband, Filmstreifen (77); Film (77); Film, dünne Schicht
film badge Filmdosimeter, Filmplakette (539)
film projector Filmprojektor, Kinoprojektor
film resistor Schichtwiderstand
film scanner Filmabtaster
filter Filter, Sieb, Siebschaltung (621.3)
filter attenuation band Filtersperrbereich (621.395)
filter capacitor Glättungskondensator, Siebkondensator
filter circuit Siebschaltung, Glättungsschaltung
filter choke Siebdrossel
filter coil Filterspule, Siebspule
filter crystal Filterquarz
filter network Siebschaltung
filter-paper Filtrierpapier
filter passband Durchlaßbandbreite (621.39)
filter reactor Filterspule, Siebspule, Siebdrossel
filter stop band Bandsperre (621.39); Filtersperrbereich (621.395)
filter tissue (feines) Filtergewebe, Seihtuch
filter transmission band Durchlaßfrequenzband (621.39)
to filter filtrieren, filtern, seihen; filtern, sieben, glätten (621.3)
to filter out aussieben, absieben (621.3)
filtering Filterung; Filter-..., Sieb-...
filtering action Filterwirkung, Siebwirkung
filtrate Filtrat
to filtrate filtrieren
filtration Filtration, Filtrierung, Filterung
fin Flosse, Seitenflosse; (Heizkörper-)Rippe
final endgültig, endlich; End-...
final approach Endanflug (351.814.33)
final control element Regelorgan, Steller, Stellglied, Regelglied
final current Strom im (stationären) Endzustand, Endstrom
final form endgültige Gestalt
final limit switch Endausschalter (621.876)
final modulation Direktmodulation
final position Endstellung (621.3167)
final point Endpunkt

final result Endergebnis, Endresultat
final selector Leitungswähler
final stage Endstufe
final temperature Endtemperatur
final terminal-stopping device Endausschalter (621.876)
final voltage (stationäre) Endspannung
finality endgültiger Zustand, endgültige Form; Endausführung
to finalize endgültig festlegen
finder Suchfernrohr, Leitfernrohr; Sucher (77)
finder switch Anrufsucher
finding Ergebnis, Befund, Urteil; Anrufsuchen (621.395)
to fine feinen, klären, raffinieren; dünner machen
fine fein; dünn; schön
fine adjustment Feineinstellung, Feinabgleich
fine control Feinsteuerung
fine crack Haarriß
fine-grained feinkörnig
fine line dünne Linie, dünner Strich
fine-measuring instrument Feinmeßinstrument
fine mechanics Feinmechanik
fine-mesh feinmaschig
fine migration Feinwanderung
fine-regulating knob Feinstellknopf
fine-regulating rod Feldregelstab (539)
fine setting Feineinstellung, Feintrieb
fine strand dünndrähtige Litze
fine structure Feinstruktur
fine thread Feingewinde
fine tuning Feinabstimmung
fine wire dünner Draht, feiner Draht, Feindraht
fine-wire dünndrähtig, feindrähtig
fine-wire coil dünndrähtige Spule
fine-wire fuse dünndrähtige Sicherung, Feinsicherung
finely divided fein verteilt
finely subdivided fein unterteilt
fineness Feinheit
fineness of scanning Rasterfeinheit
finery Stahlofen, Frischofen; Stahlveredlung
finger Finger
finger-push switch Druckknopfschalter
finger stop Fingeranschlag
finger-type contact Fingerkontakt
finish Oberfläche, Oberflächenzustand; Art der Oberflächengüte, Glätte; Deckanstrich, Anstrich, Lackierung
finish machining Feinbearbeitung (621.9)
to finish fein bearbeiten, glätten, schlichten; anstreichen, lackieren, abschließen, beenden
finished article Fertigware
finished product Fertigprodukt, Fertigware
finished state bearbeiteter Zustand
finishing Lackierung, Anstrich; Ausrüstung (67)
finishing coat Deckanstrich, Anstrich
finishing operation letzter Arbeitsgang
finishing rate Entladestrom
finsen Erythemdosis
fir Tannenholz, Tanne
fire Feuer; Feuer, Brand, Feuersbrunst
fire alarm Feuermelder
fire-alarm box Feuermelder
fire ball Kugelblitz; Meteor
fire clay Schamotte, feuerfester Ton
fire control Feuerleitung (623)
fire control radar Feuerleitradar
fire-damp indicator Schlagwetteranzeiger
fire detector Feuermelder
fire direction Feuerleitung (623)
fire-direction device Feuerleitgerät (623)
fire door feuersichere Tür, Feuertür
fire engine Feuerspritze, Feuerwehrspritze, Motorspritze
fire extinguisher Feuerlöscher
fire foam Löschschaum
fire hazard Brandgefahr
on fire in Brand, brennend
fire point Löschwache, Feuerwehrstützpunkt (35); Brennpunkt, Zündpunkt (66)
fire-proof feuerfest, feuersicher
fire-proof curtain eiserner Vorhang
fire resistance Feuerbeständigkeit
fire resistant feuerbeständig
fire resisting feuerhemmend
fire resistive feuerbeständig
fire-retarding feuerhemmend
fire risk Feuersgefahr
firespread Flugfeuer, Funkenflug
to fire in Brand setzen, anzünden, anbrennen; zünden; heizen, anheizen, einbrennen
to fire silver on ... Silber auf ... aufbrennen
to fire up anheizen
firing Brennmaterial, Feuerung; Heizung; Abschießen, Abfeuern; Zündung
firing apparatus Zündmaschine (622)
firing cable Zündleitung
firing circuit Zündkreis
firing key Zündschlüssel, Zündschalter
firing point Zündspannung, Zündpunkt (621.385)
firing potential Zündspannung (621.385)
firing temperature Brenntemperatur (666)
firing time Anheizzeit (64); Brenndauer, Brennzeit (66)

firkin Fäßchen (1 Firkin = 40,9 Liter)
firm fest, solide
firm insulating material festes Isoliermaterial
firm material festes Material, fester Stoff
firmness of cops Wickelhärte (der Spulen)
firmware [in Festwertspeichern gespeicherte Software]
first erste (-r, -s)
first aid erste Hilfe
first choice erste Wahl, beste Sorte
first-class erstklassig, erstrangig
first-class conductor Leiter 1. Klasse, metallischer Leiter
first-degree algebraic equations (pl.) lineares Gleichungssystem
first detector H. F.-Gleichrichter, Modulationsstufe, Mischstufe (621.396.62)
first law of thermodynamics erster Hauptsatz (der Wärmelehre)
first point of contact Auflaufpunkt
first selector erster Wähler (621.395)
first voltage range Spannungsbereich unter 30 Volt
to fish durchfädeln, einfädeln
fish wire Einziehdraht
fishbone aerial anfangsgespeister Längsstrahler, Fischgrätenantenne
fissible spaltbar, spaltfähig (539)
fissible matter für Kernspaltung brauchbares Material, Atombrennstoff (539)
fissible nucleus spaltbarer Kern
fissible nuclide spaltbare Kernart
fissile spaltbar
fissile atom spaltbares Atom
fission Spaltung, Zellteilung; Kernspaltung (539)
fission fragments Spaltbruchstücke
fission neutrons Saltneutronen
fission product Spaltprodukt
fission yield Spaltausbeute
fissionability Spaltbarkeit (539)
fissionable spaltbar
fissionable material Kernspaltstoff
fissure Spalt, Spalte, Bruch, Riß
fit Sitz, Passung; Anfall, Schlag (61)
to be a good fit gut passen, gut sitzen
a good fit on gut passend auf
to fit passen; anpassen, montieren
to fit in einpassen; einfügen, hineinfügen, einmontieren
to fit in with (sich) einfügen in, (sich) anpassen an
to fit into passen in, hineinpassen; einpassen in, einsetzen in, einmontieren
to fit on anprobieren, anpassen, aufprobieren, aufpassen
to fit out ausrüsten, ausstaffieren
to fit together zusammenfügen, zusammenpassen
to fit with versehen mit, ausrüsten mit
fit geeignet, passend; trainiert; reif, tauglich (**for:** für, zu)
fitments Paßteile, Zubehörteile
fitter Monteur, Schlosser, Montageschlosser
fitting Armaturteil, Paßstück
fitting shop Schlosserei, Montagewerkstatt
fitting stand Montagebock
fittings Armaturen, Beschläge, Rosetten (621.32)
five fünf
five-electrode valve Fünfpolröhre
five-heat switch Kochplattenumschalter
five-place logarithm fünfstelliger Logarithmus
five-point jack fünfpolige Klinke
five-track punched tape Fünfkanal-Lochstreifen
five-wire system Fünfleitersystem
five-wire two-phase system Fünfleiter-Zweiphasensystem
to fix festlegen, festsetzen; binden (54); fixieren (77); bestimmen (51); festmachen, anmachen, befestigen (621)
to fix up with ausrüsten mit
fixation Fixierung (77); Bindung (54)
fixation of nitrogen Stickstoffbindung
fixative Fixativ
fixed fest, festgehalten, unveränderlich; fest vorgegeben, gegeben
fixed across ... quer zu ...
fixed aerial direction-finder Festantennenpeiler
fixed armature stillstehender Anker (621.313)
fixed bias feste Vorspannung (621.396)
fixed capacitor Festkondensator
fixed carbon gebundener Kohlenstoff (66)
fixed carbon-film resistor Kohleschichtfestwiderstand
fixed charge Grundgebühr
fixed disk Festplatte
fixed-echo elimination Festzeichenunterdrückung (621.396.96)
fixed film resistor Schichtfestwiderstand
fixed frequency transmitter Festfrequenzsender
fixed grid bias feste Gittervorspannung
fixed handle circuit-breaker Schalter ohne Freiauslösung
fixed installation fest verlegte Leitungen (pl.)

fixed location point festes Widerlager, Auflager
fixed metal-film resistor Metallschicht-Festwiderstand
fixed metal oxide film resistor Metalloxidschicht-Festwiderstand
fixed oxygen gebundener Sauerstoff
fixed plates feste Platten (621.396); fester Plattensatz, Stator (sing.)
fixed point Festpunkt, Fixpunkt; Festkomma (681.3)
fixed radio service fester Funkdienst
fixed resistance Festwiderstand, fester Widerstand
fixed resistor Festwiderstand, fester Widerstand
fixed service conditions feste Betriebsbedingungen
fixed station feste Funkstelle
fixed store Festspeicher
fixed time lag unabhängige Zeitverzögerung
fixed-tuned fest abgestimmt
fixed voltage regulator Festspannungsregler
fixed wire-wound resistor Drahtfestwiderstand
fixing Befestigung; Fixierung; Festsetzung
fixing bath Fixierbad
fixing screw Befestigungsschraube
fixture Vorrichtung, Spannvorrichtung (621.9)
fixture stud Leuchtenhalter (696)
to fizz zischen, sprühen
to fizzle leise zischen
to flag kennzeichnen, markieren (621.395), (681.3)
flagged call gekennzeichnete Verbindung (621.395)
flag Marke, Kennzeichen, Flagge
flag bit Kennzeichenbit
flagpole Stab, Stange, Stabantenne
flake Flocke, Schuppe
flakes of graphite Graphitflocken (669)
flaking Schuppenbildung, Abschuppung
flame Flamme; Schweißlampe, Lichtbogen
flame cap Aureole
flame carbon Effektkohle
flame coloration Flammenfärbung (54)
flame-cored carbon Effektdochtkohle
flame lamp Kerzenlampe (621.32)
flame-proof explosionssicher, explosionsgeschützt, schlagwettergeschützt, schlagwettersicher
flame-proof enclosure Schlagwetterschutzkapselung [B]; druckfeste Kapselung
flame-proof machine schlagwettergeschützte Maschine
flame-proof material schlagwettergeschütztes Material
flame-proof motor schlagwettergeschützter Motor
flame-proof switch schlagwettergeschützter Schalter
flame-proof transformer schlagwettergeschützter Umspanner
flame protection Schlagwetterschutz
flame retarding feuerhemmend
flame-tight schlagwettergeschützt
flame-tight switchgear schlagwettergeschützte Schalteinrichtung
flame-tight transformer schlagwettergeschützter Umspanner
to flame flammen
flaming arc Effektlichtbogen, Effektbogen, Flammbogen
flange Flansch, Rohrflansch, Profilflansch
flange breadth Flanschbreite
flange coupling Scheibenkupplung, Kupplungsflansch
flange-edge joint Stirnstoßverbindung
flange joint Fanschverbindung
flange motor Flanschmotor
flange-protected machine schlagwettersichere Maschine mit Breitflansch
flange protection Breitflansch-Schlagwettersicherung
flange width Flanschbreite
to flange mit Flansch versehen, umkrempen, krempen
flanged mit Flansch (versehen)
flanged joint Flanschverbindung
flanged to angeflanscht an
flank Flanke
flank dispersion Wickelkopfstreuung
flap Klappe; Steuerklappe; Umschlagklappe
flap attenuator Dämpfungsschwert (621.37)
flap door Klapptür
flap-down handle umklappbarer Henkel, Klappgriff
flare Fackel; flackernder Brand; Auskragung (629.12); Aufweitung (621.396.67); trichterförmige Erweiterung
flare angle Öffnungswinkel (621.396.67)
flare path befeuerte Landebahn (629.13)
to flare (sich) aufweiten, (sich) konisch erweitern; flackern, strahlen, flackernd brennen
flared section Zwischenkonus, Übergangskonus (621.396.67)

flary verschleiert (77)
flash Lichtblitz, Aufblitzen; Schlag, plötzlicher Stoß; Preßgrat (621.9)
flash arc Rückzündung (621.314.6)
flashback Rückschlag, Rückzündung (621.314.6)
flash boiler Flammenrohrkessel
flash-butt welding Abbrenn-Stumpfschweißung
flash guard Lichtbogenkammer
flash lamp Blitzlichtlampe, Taschenlampe
flash-lamp bulb Taschenlampenbirne
flash of light Lichtblitz
flashlight Blitzlicht; Scheinwerfer, Suchlicht
flashlight battery Taschenlampenbatterie
flash magnetization Stoßmagnetisierung, Impulsmagnetisierung
flashover Überschlag (621.3)
flashover distance Durchschlagsentfernung
flashover potential Überschlagspannung
flashover test Überschlagprüfung
flashover voltage Überschlagsspannung
flash plate dünner Überzug (621.357)
flash point Entflammungspunkt, Entflammungstemperatur
flash-to-report Laufzeit, Schallaufzeit
flash signal Lichtsignal
flash suppressor Rundfeuerlöscheinrichtung (621.313)
flash test Stoßprüfung (621.315.6)
flash-tight schwallwasserdicht
flash tube Blitzröhre, Entladungsröhre
flash welding Abbrennschweißung; Abschmelzschweißung
to flash blitzen, aufblitzen, blinken; telegraphieren; aufblitzen lassen, abschießen, feuern; überschlagen
to flash a getter ein Getter abschießen (621.385)
to flash on a screen auf einem Schirm aufleuchten lassen
to flashover überschlagen (621.3)
flashed glass Überfangglas
flasher Blitzschalter, Blinkeinrichtung
flashing Abdeckung, Überfangung, Abdeckblech, Abweisblech (69); Rundfeuer
flashing Flamm-...
flashing beacon Blitzfeuer (656.7)
flashing of filament Fadenformierung (621.326)
flashing indication Blinkanzeige
flashing light Blitzlichtbake, Blinkfeuer (656.7)
flashing period Abbrennzeit (621.791); Brenndauer (623)
flashing point Flammpunkt
flask Flasche (54); Formkasten (621.7); Korbflasche
flat Mietwohnung, Etagenwohnung; Abflachung, Fläche; Einbrandstelle, Kollektorbrandstelle (621.313)
flat flach, eben; breitbandig (621.396.62.64); flau (77); schal (66); zu tief, einen halben Ton tiefer (53)
flat aerial Antennenwand, „Matratze" (621.396.67)
flat-bar copper Flachkupfer
flat-bar iron Flacheisen
flat colour flaue Farbe
flat compounding vollständige Kompoundierung
flat curve flache Kurve
flat earth ebene Erdoberfläche (621.396)
flat edgewise pattern instrument Flachprofilgerät
flat ground ebener Boden
flat-head screw Flachkopfschraube
flat-headed rivet Flachkopfniet
flat maximum flaches Maximum
flat pack Flatpack
flat pliers (pl.) Flachzange
flat resonance curve flache Resonanzkurve
flat response breitbandige Frequenzkurve
flat section bar Flachstange, Flacheisen
flat sketch ebene Skizze
flat spring Blattfeder
flat surface ebene Fläche, glatte Oberfläche
flat-surface milling Planfräsen
flat top flacher Kopf (einer Kurve), flaches (Kurven-)Dach
flat-top antenna horizontale Langdrahtantenne
flat-top structure Flachdach(bau) (621.396.67)
flat-topped flach, rechteckig, oben flach
flat-topped wave Rechteckkurve (621.3)
flat-tuning Grobabstimmung
flat twin cable Flachkabel (696)
flat-type relay Flachrelais
flat-type spring Blattfeder
flat unterplaster wire Stegleitung
flat variation langsame Änderung
flat webbed wire Stegleitung
to flatten abflachen, flachmachen; flachdrücken
flattening of the tuning Verminderung der Abstimmschärfe
flaw Sprung, Schönheitsfehler; Fehler (621.7)

flaw detector Rißprüfer
flaw discrimination Fehlererkennbarkeit (621.386)
F-layer F-Schicht
fleeting alarm Kurzimpulsmeldung, Wischermeldung
Fleming's rule Handregel (621.3)
flewelling circuit Pendelrückkopplungsschaltung
flex Anschlußschnur, bewegliche Leitung, Telefonschnur; „Strippe"
to flex biegen (Schmiegsamkeitsprüfung)
flexibility Biegsamkeit; Anpassungsfähigkeit
flexible Anschlußschnur; bewegliche Leitung, Telefonschnur; Litze
flexible biegsam, anpassungsfähig
flexible cable Zuleitungsschnur, Anschlußschnur; Litzenkabel
flexible conductor biegsame Leitung, Litzenleitung
flexible cord Anschlußschnur, bewegliche Leitung, Telefonschnur
flexible drive biegsame Welle
flexible lead Geräteanschlußschnur, Anschlußlitze
flexible metal conduit Metallschlauch (69)
flexible metal tube Metallschlauch, biegsames Rohr, Wellrohr
flexible mounting elastische Aufhängung
flexible numbering freizügige Rufnummernzuteilung
flexible pigtail Anschlußlitze (621.313)
flexible shaft biegsame Welle
flexible suspension Schnuraufhängung, weiche Aufhängung
flexible tube biegsames Rohr
flexible varnished tubing Lackschlauch
flexible wire Anschlußlitze, biegsame Litze
flexible wiring Installation mit biegsamen Leitungen
flexile biegefähig, biegsam, flexil
flexural mode Biegeschwingung
flexural strength Biegefestigkeit
flexural vibration Biegeschwingung
flexure Biegung, Krümmung, Knickung; Krümmung (51)
to flick „schnipsen"
to flick back zurücklaufen, zurückflitzen
flick contactor Wischkontakt, Wischrelais
flicker Flimmern (535); Flackern
flicker control Zweipunktsteuerung
flicker effect Funkeleffekt (621.385)
flicker light Blinklicht, Flackerfeuer
flicker photometer Flimmerphotometer
to flicker flimmern
flight Flug...
flight-base line Anfluggrundlinie
flight control Flugleitung
flight control system Flugregler
flight data recorder Flugdatenschreiber
flight-path computer Flugwegrechner
flight simulator Flugsimulator
flintglass Flintglas
flip-chip transistor Flip-chip Transistor
flip-flap switch Wippenschalter
flip-flop circuit bistabile Kippschaltung, Flip-Flop (621.2)
flip-flop storage Flip-Flop Speicher
float Schwimmer, Floß; Schaufel (621.2)
float loading Pufferung
float pressure gauge Schwimmermanometer
float(s) (pl.) Rampenlicht
float switch Schwimmerschalter
to float schwimmen, aufsteigen, hochsteigen (durch Auftrieb); auf freiem Potential liegen, im Potential nicht definiert sein
to float charge puffern (621.355)
floatage Schwimmsel
floating Flotation, Schwimmaufbereitung
floating ohne definiertes Potential, erdfrei (621.3); schwebend; Schwimm-...
floating action integrierende Regelung; astatisches Verhalten (621.316.7)
floating battery Pufferbatterie
floating carrier system Hapugmodulation (621.396)
floating grid offenes Gitter, freies Gitter (621.385)
floating input erdfreier Eingang
floating light Feuerschiff, Leuchtboje
floating neutral nichtgeerdeter Nulleiter, nichtgeerdeter Mittelleiter
floating point Gleitkomma (681.3)
floating solder bath Lötbad
floating speed Stellgeschwindigkeit (621.316.7)
floating-trickle battery Stromausgleichsbatterie
floating voltage Schwebespannung; Leerlaufgleichspannung
flock Fussel, Flocke
flocks (pl.) flockige Masse
flood discharge Höchstwassermenge
floodlight Flutlichtleuchte, Flutleuchte
floodlight system Flutlichtanleuchtung
floodlighting Anstrahlung
floodlit angestrahlt, mit Scheinwerfern beleuchtet, angeleuchtet
floor Fußboden tragende Decke; Fußboden, Dielung; Stockwerk; Brückenfahrbahn

floor contact Fußbodenschalter, Bodenkontakt
floor insulator Deckendurchführung (621.315.6)
floor polisher Bohner, Bohnermaschine
floorspace Bodenfläche, Grundfläche
floor standard Stehlampe
floor switch Stockwerkschalter
floor-type Säulen-..., Ständer-...
floor-type drilling machine Säulenbohrmaschine
to flop kippen, schwanken
to flop over umschwenken
floppy disk Diskette, Floppy-Disk
floppy-disk drive (Disketten-)Laufwerk
flotsam Schwimmsel
flow Fluß, Strömung, Erguß
flow breaks down Strömung reißt ab (533)
flowchart Flußdiagramm, Ablaufplan, Befehlsschema
flow control Durchflußregelung
flow diagram Flußdiagramm
flow heater Durchflußerhitzer
flow indicator Strömungsanzeiger
flow meter Strömungsmesser, Durchflußmesser
flow property Fließeigenschaft
flow relay Strömungsrelais
flow resistivity spezifischer Strömungswiderstand der Längeneinheit (535)
flow and return circuit Hin- und Rückleitung
flow sheet Bearbeitungskarte, Arbeitsablaufkarte
flow soldering Schwallöten
flow soldering line Schwallötstrecke
flow of a stream Wassermenge eines Stromes, Wasserführung
to flow fließen, strömen
to flow in at einströmen bei, eintreten bei
to flow into a junction einem Verzweigungspunkt zufließen
to flow through durchfließen
to flow to waste ablaufen, weglaufen
flower of sulphur Schwefelblüte
to fluctuate schwanken, sich ändern
fluctuating light Wechsellicht
fluctuation Schwankung, Änderung
fluctuation of power Leistungsschwankung
fluctuation of voltage Spannungsschwankung
flue Esse, Rauchrohr, Rauchkanal, Fuchs; Flammrohr; Ofenzug, Kesselzug, Fussel
flue gas Rauchgas, Abgas; Rauchgase (pl.)
flue-gas precipitating equipment Rauchgasentstaubungsanlage
flue-gas testing Rauchgasprüfung, Rauchgasanalyse
fluff Faserabfall; Flaum; Fusseln (pl.)
fluffy fusselig, flaumig
fluid Flüssigkeit; strömendes Medium
fluid detector Flüssigkeitsmelder
fluid friction Flüssigkeitsreibung
fluid level Flüssigkeitsstand
fluid level detector Flüssigkeitsstand-Melder
fluid logic Strömungslogik
fluid substance Flüssigkeit
fluidic fluidisch
fluidized reactor Reaktor mit flüssiger Brennstoffsuspension, Flüssigspaltstoff-Reaktor
fluor = fluorspar Flußspat
to fluoresce fluoreszieren
fluorescence Fluoreszenz
fluorescence screen Leuchtschirm, Fluoreszenzschirm
fluorescence spot Leuchtfleck
fluorescent fluoreszierend, Floreszenz-...
fluorescent lamp Leuchtstofflampe, Leuchtstoffröhre
fluorescent light Fluoreszenzlicht
fluorescent screen Leuchtschirm, Fluoreszenzschirm
fluorescent X-rays (pl.) charakteristische Röntgenstrahlung
fluorine Fluor
fluoroscope Röntgenschirm, Fluoreszenzschirm
fluoroscopic screen Leuchtschirm, Fluoreszenzschirm
fluoroscopy Leuchtschirmbetrachtung (621.386)
fluorspar Flußspat
flush eben, flach, nicht vorstehend, bündig
flush-butt welding Stumpf-Abbrennschweißung
flush with a face aus einer Ebene nicht vorstehend, in einer Ebene liegend
flush-marker light Unterflurfeuer
flush-mounted switch Unterputzschalter
flush mounting meter Einbauinstrument (621.317)
flush-mounting switch Einbauschalter
flush pattern Einbautype
flush-pattern instrument Einbauinstrument, Instrument für versenkten Einbau
flush rivet Senkkopfniet
flush-runway light Unterflurfeuer
flute Rille, Nut, Riefe; Spiralnut (621.9); Flöte (534)
to flute mit Nuten versehen, nuten, riefen

fluting Längsnutung, Riefung; Kannellierung (72)
flutter Unruhe, Rumpelspannung, Unwucht, Flattern
flutter and wow Gleichlaufschwankung (681.84)
flux Fluß (53); Flußmittel, Lötpaste, Schweißpaste
flux ball Kugelspule (538)
flux coefficient (Produkt aus Scheitelfaktor und Formfaktor für die Verteilung des Polflusses) (621.313)
flux-cutting Kraftlinienschneiden
flux density Flußdichte, Induktion
flux line Flußlinie, Kraftlinie
flux linkage Flußverkettung
flux linking a coil Spulenfluß, der mit einer Spule verkettete Fluß
flux paste Lötpaste; Schweißpaste
flux sensitivity Flußempfindlichkeit (eines Hallgenerators)
flux shunting Feldschwächung (621.313)
a flux is taken out of a face ein Fluß tritt aus einer Fläche aus
flux tuft Flußbüschel, Flußbündel (538)
to flux schmelzen, niederschmelzen; mit Flußmitteln bestreichen
fluxmeter Fluxmeter
fly Flug-..., Schwung-...
fly ash Flugasche
fly ball Schwungkugel
fly-ball governor Schwungkugelregler, Zentrifugalregler
fly wheel Schwungrad
fly-wheel synchronisation Schwungradsynchronisation
fly-wheel type alternator Schwungradgenerator
to fly fliegen; fliehen
to fly apart auseinanderfliegen; sich voneinander entfernen; sich abstoßen, sich fliehen
to fly off auslösen (621.3)
to fly open aufspringen, sich öffnen (621.3); aufsprengen, gewaltsam öffnen, schnell öffnen, aufstoßen
flyback Fleck-Rücklauf (621.385)
flyback mechanism Rücklaufeinrichtung
flyback power supply Rücklaufspannungsgewinnung
flyback time Fleckrücklaufzeit
flyback transformer Zeilentransformator
flying Flug-..., fliegend
flying bomb Flugkörper
flying lead fliegende Leitung, Anschlußleitung
flying spot scanning Lichtpunktabtastung
f. m. = frequency modulation FM = Frequenzmodulation
foam Schaum
foam ceramics (pl.) Schaumkeramik
focal fokal
focal length Brennweite
focal point Brennpunkt
to focalize fokussieren, in den Brennpunkt stellen
focus Brennpunkt; Brennfleck, Fleck (621.385)
focus coil Fokussierungsspule (621.385)
to focus fokussieren, in den Brennpunkt stellen; scharfstellen, punktscharf stellen; sammeln, konzentrieren
to focus on sammeln auf, scharfstellen auf, bündeln auf
focusing Fokussierung, Scharfeinstellung; Fokussierungs-...
focusing coil Fokussierungsspule (621.385)
focusing control Bildschärferegler
focusing device Fokussierungseinrichtung
focusing electrode Sammelelektrode, Bündelungselektrode
focusing lamp Projektionslampe (621.32)
focusing magnet Abbildungsmagnet
focusing screen Mattscheibe (77)
fog (dicker) Nebel; Schleier (77)
fog horn Nebelhorn
fog light Nebelscheinwerfer
fog suction plant Absauganlage (621.357)
to fog vernebeln, einnebeln; schleiern, sich verschleiern (77)
foil Folie, Blatt
foil-mica capacitor Glimmerblock
fold Knick, Kniff, Falte; Einsattelung
to fold falten, knicken, zusammenfalten; abkanten (621.7)
folded doublet Faltdipol
folded dipole antenna Schleifendipolantenne, Faltdipol
... fold ...-fach, ...-fältig
folder Faltblatt, Faltprospekt, geknicktes Papierblatt; Ordner, Hefter
folding Faltung, Kniffung; Bruch, Kniff
folding machine Abkantmaschine (621.9)
folding rule Zollstock
follow current nachfolgender Betriebsstrom (621.316.9)
to follow folgen, nachfolgen, verfolgen; aufeinanderfolgen
to follow through mitgehen, bis ans Ende gehen

follow-on tool Folgewerkzeug
follow-up control wegabhängige Steuerung (629.12)
follow-up pointer Schleppzeiger
follower Abtaststift, Führungsstift (621.91); Nachlaufwerk; Verstärker (621.39)
font Ölbehälter
foot (pl.: **feet**) Fuß (61); Fuß (unteres Ende); Fußpunkt (51); Fuß (als Längenmaß) = 12 Zoll = 30,5 cm
footcandle Fußkerze (Lichteinheit entsprechend – aber nicht gleich – Lux) (s. Anh.)
footcandle meter Beleuchtungsmesser
foot control Fußschaltung
footlambert 1 lumen/sq. ft.
foot lever Fußhebel
foot-operated mit Fußbetätigung, fußbetätigt
foot of the perpendicular Fußpunkt des Lotes
foot-pound Fußpfund (Arbeitseinheit) = 1 Fuß · 1 Pfund = 0,1383 mkg
foot poundal (Arbeitseinheit im technischen Maßsystem) = 412 400 erg = 0,04124 J
foot rest Trittleiste, Fußraste, Raste
foot run lfd. Fuß = laufende Fuß, Fuß, laufende Länge
foot switch Fußschalter, Pedalschalter
foot treadle Fußhebel, Pedal
footage laufende Länge
footfall Fußtritte (pl.), Fußschall
footings Gründungen (621.315.1)
foots (pl.) Bodensatz, Rückstand (54)
forbidden band verbotenes Band (621.382)
force Kraft; Wucht, Gewalt, Zwang
forces of adhesion Adhäsionskräfte
force of attraction Anziehungskraft
by force gewaltsam, zwangsweise
force diagram Krafteck
force-displacement characteristic Kraft-Weg-Diagramm
force factor elektroakustischer Übertragungsgrad
force-feed lubrication Druckschmierung
force fit Preßsitz
force polygone Kräftepolygon, Krafteck
force pump Druckpumpe
force of repulsion Abstoßungskraft, abstoßende Kraft
force scale Kräftemaßstab
force on unit area Kraft pro Flächeneinheit
force per unit elongation Steife
force-ventilated zwangsventiliert, zwangsbelüftet
to force zwingen, zwängen, drücken, pressen; erzwingen, durchsetzen; forcieren; übertreiben; aufdrücken
to force in eindrücken
to force into hineinpressen, hineindrücken
to force on aufpressen
to force out herausdrücken, heraustreiben
to force out of straight (gewaltsam) verbiegen, krümmen
to force through hindurchtreiben durch
forced erzwungen, Druck-..., Zwangs-...
forced-air cooled fremdbelüftet, zwangsbelüftet
forced-air cooled oil-immersed transformer Öltransformator mit Anblasung, OF-Transformator
forced-air cooling Fremdbelüftung, Anblasekühlung, Druckluftkühlung
forced commutation erzwungene Kommutierung, Zwangskommutierung
forced current erzwungener Strom
forced draft erzwungener Luftstrom, Unterwind (621.1); forcierte Luftkühlung (621.396)
forced-draft fan Frischluftgebläse
forced (feed) lubrication Druckschmierung
forced-oil cooling Ölumlaufkühlung
forced-oil cooled oil-immersed transformer Transformator mit Ölumlaufkühlung, OSA-Transformator
forced oscillations erzwungene Schwingungen
forced release Freischalten (621.395); Zwangsauslösung
forced ventilation Zwangsbelüftung, Fremdbelüftung
forced vibration erzwungene Schwingung
forceps (pl.) Pinzette (61)
forcing pump Druckpumpe
forcing-off device Abdrückvorrichtung
forcing resistor Stoßerregungswiderstand (magn. vorerregte Spule)
fore Vorschiff (629.12)
fore-and-aft level Längstrimmzeiger
foreground program Vordergrundprogramm
forehearth Vorfeuerung
foreign fremd; ausländisch, auswärtig
foreign atom Fremdatom
foreign exchange line Ausnahmeanschluß (621.395)
foreign material Fremdmaterial
foreign matter Fremdstoff (54)
foreman Meister, Rottenführer, Obermonteur
forepart Vorderteil

forepressure Vorvakuum, Vorvakuumdruck; Druck auf der Druckseite (621.63)
forerunner Vorläufer, Vortype
to foreshorten (perspektivisch) verkürzen
foreshortening (perspektivische) Verkürzung; Verkürzung (621.396.67)
forestaller Quittungsschalter (656.25)
forestalling switch Quittungsschalter (625)
forestalling whistle Quittungspfeifsignal (625)
foreward(s) vorwärts, vorwärts...
foreward motion Vorwärtsbewegung, Hingang
foreward wave vorlaufende Welle, hinlaufende Welle (621.3)
forge Schmiede; Hütte, Hüttenwerk
forging Schmieden, Schmiedevorgang; Schmiedestück
forging alloy Knetlegierung (699.71)
fork Gabel, Straßengabel, Gabelung
fork beat Stimmgabelschwebung
fork connection Zickzackschaltung
fork oscillator Stimmgabeloszillator
forked gegabelt
forked lightning verästelter Blitz
forked repeater Zweiwegverstärker
forked terminal Kabelschuh
form Form; Spulenkörper (621.3); Formular
form of construction Bauweise
form cutter Profilfräser
form factor Formfaktor (621.3)
form-factor indicator Formfaktormeßbrücke
form-factor meter Formfaktormeßbrücke
to form bilden, formen, formieren (621.385); drücken, formpressen, profildrehen (621.9)
to form an image ein Bild entwerfen, ein Bild erzeugen
to form a plate eine Platte formieren (621.355)
to form to shape auf Maß formen, Form geben
formed coil Formspule
formaldehyde Formaldehyd
formalin Formalin
formation Bildung, Herstellung, Aufbau; Formation (35), (55)
formation of blow holes Blasenbildung (621.791)
formation of cracks Rißbildung
formation of fog Nebelbildung
formation of steam Dampfentwicklung
formation voltage Formierungsspannung (621.313)
formation of water Wasserbildung, Wasserentwicklung, Wasseraustritt (54)
formatter Formatierer (681.3)
former Spulenkörper, Wickelkörper (621.3); Vorlage, Schablone; Wickelschablone (621.3); Biegeschablone (621.9); Drehschablone (621.9); Rahmen, Drehspulrahmen (621.317); Abbiegeblock (621.313)
former pin Abtaststift, Schablonenstift (621.9)
former-wound coil Formspule, Ankerformspule (621.313)
formic acid Ameisensäure
forming Formierung (621.355), (621.385)
forming of accumulator plates Sammlerplatten-Formierung
forming process Formierungsvorgang
formula Formel (51), (54)
formulae of balance Abgleichbedingungen (621.317)
formula holds good Formel gilt
FORTRAN = FORmula TRANslation FORTRAN (technisch wissenschaftliche Programmiersprache)
fortuitous distortion unregelmäßige Verzerrung
forward vorderseitig, vorwärts, Vorder-...
forward-acting regulator Vorwärtsregler
forward bias Durchlaßspannung, in Durchlaßrichtung angelegte Spannung
forward blocking state Vorwärts-Sperrzustand (621.382), (621.314.63)
forward characteristic Durchlaßkennlinie (621.314.5)
forward conducting state Vorwärts-Durchlaßzustand (621.382), (621.314.63)
forward current Vorwärtsstrom, Strom in Durchlaßrichtung
forward direction Durchlaßrichtung (621.314.6); positive Richtung (537)
forward lead of the brushes Bürstenvorverstellung (621.313)
forward motion Vorwärtsbewegung
forward power loss Durchlaßverlust (621.314.5)
forward recovery time Durchlaßverzögerungszeit, Durchlaßverzug (621.314.63)
forward reference Vorwärtsverweis
forward release Vorwärtsauslösung
forward-scatter transmission Überhorizont-Richtfunk, Vorwärtsstreuausbreitung
forward scattering Vorwärtsstreuung
forward shift (of brushes) Vorverschiebung (der Bürsten)
forward signal Anregungssignal (621.316.7); Vorwärtszeichen (621.395)
forward slope resistance Vorwärts-Ersatzwiderstand (621.382.2)
forward speed Vorwärtsgang

forward sweep Fleckhinlauf, Hinlauf (621.385)
forward traverse Arbeitshub, Arbeitsweg (621.9)
forward voltage Vorwärtsspannung, Durchlaßspannung, Spannung in Durchlaßrichtung (621.382.2)
forward voltage-current characteristic Vorwärtskennlinie (621.382.2)
F. O. W. = optimum working frequency optimale Betriebsfrequenz
foto flashlight Photo-Blitzgerät
Foucault current Foucault-Strom, Wirbelstrom
Foucault's pendulum Foucaultsches Pendel
to foul (versehentlich) falsch treffen, beschädigen; sich festsetzen, sich verstopfen, sich zustopfen
foul faul, übel, eklig
foul electrolyte verbrauchter Elektrolyt (621.357)
foul-weather corona Schlechtwetterkorona
to found gründen (624); gründen, stiften, begründen; gießen (621.74)
foundation Fundament, Gründung (624); Baugrund (624); Unterbau (625); Stiftung; Grundlagen (pl.)
foundation bold Ankerschraube, Fundamentschraube
foundation pressure Bodendruck, Bodenpressung
to founder nachgeben, nachrutschen, einstürzen (624)
founding Gießen, Guß
foundry Gießerei
foundry ingot Gußbarren, Rohrmassel
four channel duplex telegraph system Vierkanal-Duplex-Telegraphie
four-core cable Vierleiterkabel, vieradriges Kabel
four-course beacon Vierkursfunkfeuer
four-electrode valve Vierpolröhre, Vierelektrodenröhre, Tetrode
four-frequency diplex telegraphy Doppeldiplextelegraphie
four-phase system Vierphasensystem, Vierphasenanlage
four-place vierstellig
four-pole Vierpol; vierpolig
four-pole machine vierpolige Maschine
four-sided vierseitig, Vierseit-... (51)
four-speed pole changing motor vierfach polumschaltbarer Motor
four-stroke Viertakt-...
four-stroke cycle Viertaktprozeß
four-stroke engine Viertaktmotor, Viertakter
four-terminal network Vierpol
four-valve receiver Vierröhrenempfänger
four-wire Vierdraht-...
four-wire balanced line symmetrische Vierdrahtleitung
four-wire-circuit Vierdrahtschaltung
four-wire repeater Vierdrahtverstärker (621.395)
four-wire system amplifier Vierdrahtverstärker
Fourier integral Fourier-Integral
Fourier series Fouriersche Reihe
Fourier transform Fourier-Transformierte
fourth vierter, Viertel
fourth rail vierte Schiene (621.33)
fourth voltage range Spannungsbereich 250 ... 660 Volt
fox message Fox-Code (621.394.6)
fraction Teil, Bruchteil, Anteil; Bruch (51); Fraktion (54)
fraction line Bruchstrich
fractional anteilig; Bruch-..., bruchzahlig
fractional deviation Abweichung in Prozent
fractional distillation fraktionierte Destillation (54)
fractional horsepower motor Kleinmotor
fractional line Bruchstrich
fractional number Bruchziffer
fractional pitch winding Wicklung mit verkürztem Wicklungsschritt
fractionation Fraktionierung
fractioned distillation fraktionierte Destillation
fracture Bruch, Zerbrechen, Zerreißen; Bruch (61)
fracture by fatigue Ermüdungsbruch, Dauerbruch
fracture plane Bruchfläche
fracture structure Bruchgefüge
to fracture brechen, reißen, zerbrechen, zerreißen
fractured piece Bruchstück, abgebrochenes Stück
fragmentation of nucleus Kernzertrümmerung
Frahm frequency meter Frahmscher Zungenfrequenzmesser
frame Rahmen, Gestell; Gestell (621.3); Stabwerk (624); Türstock (69); Bild (621.397), (77); Bandsprosse (681.3)
frame aerial Rahmenantenne
frame coil Bildablenkspule
frame connector Steckerleiste
frame-cooled mantelgekühlt
frame duration Bilddauer (621.397)
frame flyback Vertikalrücklauf

frame frequency Taktfrequenz; Vertikalfrequenz; Bildfrequenz, Bildwechselfrequenz (621.397)
frame-grounding circuit Gehäuseerdung
frame mounting Gestelleinbau (621.316.5)
frame pulse Bildimpuls (621.397)
frame slip Bilddurchlauf (621.397)
frame-type switchboard Schaltgerüst, Schaltpult
to frame formen, gestalten; formulieren, in Worte fassen; einrahmen, einfassen
framend Fachwerk-...
framed building Skelettbau
framework Gitterwerk, Rahmenkonstruktion (624); Traggerüst, Gestellrahmen
framing Bildlageeinstellung (621.397)
Franklin aerial Franklin Antenne
Frauenhofer's lines Frauenhofersche Linien
to fray durchschreiben, durchscheuern, abwetzen, verschleißen, ausfransen
to free freimachen, ablösen, loslösen, lösen
to free from reinigen von, befreien von
free frei
free airway freier Luftzutritt
free charge freie Ladung
free cutting steel Automatenstahl (621.9)
free cyanide überschüssiges Cyankali (621.357)
free electron freies Elektron
free fatty acids freie Fettsäuren
free of ground erdfrei
free-handle circuit-breaker Schalter mit Freiauslösung
free headroom lichte Höhe, Durchfahrthöhe
free-moving freibeweglich
free oscillation freie Schwingung
free path freie Weglänge
free programmable frei programmierbar
free rocket ungelenkte Rakete
free routing freie Leitwegzuteilung (621.395)
free-running freilaufend, nicht angetrieben, lose, selbstgeführt (621.314.5)
free-running multivibrator astabiler Multivibrator, frei schwingender Multivibrator
free-running speed Gleichgewichtsgeschwindigkeit (621.33)
free space freier Raum, leerer Raum; Spielraum; Freiraum...
free-space attenuation Freiraumdämpfung
free-space propagation Freiraumausbreitung
free-space wavelength Luftwellenlänge (621.396); Vakuumwellenlänge (53)
free surface freie Oberfläche
free vibration freie Schwingung
free waves freie Wellen
free wheel Freilaufrad, Freilauf (629)
free wheeling clutch Freilaufkupplung
freedom from ...-freiheit, ...-losigkeit
freedom from afterglow Nachleuchtfreiheit
freedom from harmonics Oberwellenfreiheit
freedom from secondary lobes Nebenzipfelfreiheit (621.396.677)
freedom of movement Bewegungsfreiheit
to freeze gefrieren, erstarren, einfrieren; einfrieren lassen, blockieren; festhalten; verbindlich festlegen
to freeze readings Anzeige(n) festhalten
freezer Kühltruhe, Kühleinrichtung
freezing Gefriervorgang, Einfrieren, Erstarrung; Festlegung
freezing of the fillet Erstarren der Schweißraupe
freezing point Gefrierpunkt, Erstarrungspunkt
freight locomotive Güterzuglokomotive
French chalk Schlämmkreide
freon Frigen [$CCl_2 F_2$]
frequency Frequenz, Häufigkeit; Frequenz (621.3)
frequency accuracy Frequenzgenauigkeit
frequency allocation Frequenzverteilung, Frequenzzuweisung
frequency allotment plan Frequenzverteilungsplan
frequency analysis Frequenzanalyse
frequency analyzer Frequenzanalysator
frequency assignment Frequenzzuteilung
frequency band Frequenzband
frequency band of emission Sendeband, Sendefrequenzband (621.396)
frequency changer Frequenzwandler, Frequenzumsetzer
frequency characteristic Frequenzgang, Frequenzcharakteristik
frequency compensation Frequenzgangentzerrung
frequency control Frequenzsteuerung
frequency converter Frequenzwandler, Umrichter
frequency coverage Frequenzbereich
frequency demultiplier Frequenzteiler
frequency departure Frequenzabweichung, Frequenzauswanderung
frequency dependence Frequenzabhängigkeit
frequency deviation Frequenzabweichung; Frequenzhub
frequency discriminator Frequenzdiskriminator

frequency distortion lineare Verzerrung, Dämpfungsverzerrung
frequency distribution Häufigkeitsverteilung (51)
frequency diversity Frequenzdiversität
frequency divider Frequenzteiler
frequency division Frequenzteilung
frequency-division multiplex Frequenzmultiplex
frequency doubler Frequenzverdoppler
frequency doubling Frequenzverdopplung
frequency drift Frequenzwanderung, Frequenzablage
frequency error Frequenzeinfluß (621.317)
frequency filter Sieb, Filter (621.392); Siebschaltung
frequency fluctuation Frequenzschwankung
frequency indicator Frequenzmesser, Frequenzanzeiger
frequency inselective breitbandig aperiodisch
frequency interlace Frequenzverschachtelung
frequency inversion Frequenzumkehrung
frequency-measuring bridge Frequenzmeßbrücke
frequency meter Frequenzmesser
frequency-modulated frequenzmoduliert
frequency modulation Frequenzmodulation
frequency multiplication Frequenzvervielfachung
frequency multiplier Frequenzvervielfacher
frequency of oscillation Schwingfrequenz
frequency prediction Frequenzvorhersage
frequency pulling Frequenzziehen
frequency range Frequenzbereich, Frequenzband
frequency rating Nennfrequenz
frequency-reducing system Frequenzteilerschaltung
frequency regulator Frequenzregler
frequency relay Frequenzrelais
frequency response Frequenzgang, Frequenzabhängigkeit, Übertragungsbereich
frequency-response locus Ortskurve des Frequenzgangs
frequency scanning elektronische Abtastung, Wobbeln, Frequenzwobbelung
frequency selective frequenzabhängig, selektiv
frequency selective filter Frequenzweiche
frequency sensitivity Trennschärfe, Selektivität
frequency shift Frequenzumsetzung (621.394)
frequency shift keying Frequenzumtastung
frequency spectrum Frequenzspektrum
frequency stability Frequenzkonstanz, Frequenzstabilität
frequency standard Frequenznormal
frequency swing Frequenzauslenkung, Frequenzhub, Hub
frequency synthesizer dekadischer Frequenzgenerator
frequency tolerance Frequenztoleranz, zulässige Frequenzabweichung (621.396)
frequency translation Frequenzumsetzung (621.395.4)
frequency tripler Frequenzverdreifacher
frequency tripling Frequenzverdreifachung
frequency-type telemeter Fernmesser mit Frequenzumsetzung, Impulsfrequenz-Fernmesser
frequency of unity current transfer ratio Einsfrequenz, Frequenz bei Stromverstärkung (621.395)
frequency used in the test Meßfrequenz
Fresnel zone Fresnel Zone
to fret wegfressen
fretsaw Laubsäge
friability Bröckligkeit
friable bröcklig, zerbrechlich
friction Reibung
friction belt Reibriemen
friction clutch Reibungskupplung
friction disk Friktionsscheibe
friction drive Reibungstrieb, Reibtrieb
friction facing Reibbelag, Kupplungsbelag
friction feed Reibungsvorschub
friction fit Reibsitz
friction gear Reibradgetriebe
friction of the pivot Spitzenreibung (621.317)
friction tape Klebeband
friction-tight reibungsschlüssig
friction washer Sicherungsunterlegscheibe, Schraubensicherung
frictional Reibungs-..., Reib-...
frictional drag Reibungswiderstand (533)
frictional electricity Reibungselektrizität
frictional error Reibungsfehler (621.317)
frictional force Reibungskraft
frictional grip Reibungskraft
frictional head Druckhöhe zur Überwindung des Reibungswiderstandes, Reibungshöhe
frictional loss Reibungsverlust
frictional resistance Reibungswiderstand
frictionless reibungslos, reibungsfrei
fringe Franse, Strichmuster
fringe aerea Interferenzgebiet (654.19)
fringe of a field Rand eines Feldes
fringe howl Einsatzpfeifen, Überlagerungsheulen (621.396)

frings of the operating range Grenze des Empfangsbereiches (621.396)
to fringe ausfransen
fringing Ausfransung, faseriger Rand; Streuflußausbildung (621.313)
fringing coefficient (Verhältnis der effektiven zur geometrischen Polfläche) (621.313)
fringing of a field Streufeld, Randfeld, Randwirkung
to frit fritten
to frizzle brutzeln
frog Frosch (59); Herzstück, Kreuzungsherz (625.1)
frog-leg winding Schleifenwicklung mit zusätzlicher Wellenwicklung als Ausgleichswicklung
FROM = Fusible ROM Festwertspeicher mit Schmelzbrücken
front Front; Front, Frontseite, Vorderseite, Vorderwand; Vorder-..., Nichtantriebsseite (621.313)
front and back connected switch Schalter mit vorder- und rückseitigem Anschluß
front-to back ratio Vorwärts-zu-Rückwärts-Verhältnis, Rückdämpfung
front-connected switch Schalter mit vorderseitigem Anschluß
front connection vorderseitiger Anschluß (621.3)
front contact Arbeitskontakt
front cord Verbindungsschnur
front cover Kappe (621.317)
front elevation Aufriß, Vorderansicht
front face Vorderfläche, Vorderseite
front feed Frontspeisung (621.396.67)
in front of voreilend gegen (621.3); vor, räumlich vor (621)
front joint Stirnstoß
front loading system Frontlader (681.84)
front panel Vorderwand, Frontplatte
front pitch kollektorseitiger Wicklungsschritt
front porch vordere Austastschulter (621.397)
front side Vorderseite, Stirnseite
front view Vorderansicht, Ansicht von vorn
front of wave Wellenfront
front-of-wave impulse sparkover-voltage Stirnansprechstoßspannung
to front blicken auf
to front by vorn abschließen mit, vorderseitig versehen mit
frontal Stirn-...
frontal area Stirnfläche
frost Reif, Frost
frost crack Frostriß (69)
frost-proof frostgeschützt, frostsicher
to frost mattieren
frosted lamp mattierte Lampe, Opalglaslampe
frosting Mattierung
froth Schaum
frothy schaumig
frustrum Stumpf
frustum of a cone Kegelstumpf
frying Brutzeln (Mikrophongeräusch) (621.305.61)
FSK = frequency shift keying
ft. = foot oder **feet** Fuß (Längenmaß) = 12 Zoll = 0,305 m
ft. lb. = foot pound Fußpfund = 0,138 mkg
ft. super = Quadratfuß
fuchsin Fuchsin
fuel Brennstoff, Brennmaterial, Heizmaterial; Kraftstoff, Treibstoff
fuel air-lock Brennelementschleuse
fuel burn out Brennelementschaden
fuel cell Brennstoffelement
fuel consumption Brennstoffverbrauch, Kraftstoffverbrauch
fuel economy Brennstoffwirtschaft; Kraftstoffwirkungsgrad; Wärmewirtschaft
fuel element Brennelement (539)
fuel element cave Brennelementgruft
fuel-fired brennstoffbeheizt
fuel flowmeter Kraftstoffverbrauchsmesser
fuel reconditioning Spaltstoffaufarbeitung
fuel reprocessing Brennstoffaufarbeitung, Spaltstoffaufarbeitung
fuel rod Brennstab (539)
fuel storage Brennstofflagerung, Brennstoffspeicherung
fuel supply Brennstoffversorgung, Kraftstoffversorgung
fulcrum Unterstützungspunkt, Drehpunkt
fulgurite Blitzröhre, Fulgurit (55)
full voll, vollständig
full adder Volladdierer
full details vollständige Einzelheiten, vollständige Unterlagen
full duplex Vollduplex
full excitation Vollerregung, Höchsterregung
full line ausgezogene Linie
full-load Vollast-...
full-load characteristic Vollastcharakteristik (621.313)
full-load current Vollaststrom
full-load output Vollastleistung, Vollast
full-load short circuit ratio Nennkurzschlußverhältnis

full magnetic controller Steuerschalter mit elektromagnetischer Betätigung
full-page ganzseitig
full parallel notch Fahrstellung in Parallelschaltung (621.33)
full parallel position Fahrstellung in Parallelschaltung (621.33)
full pitch winding Durchmesserwicklung
full radiator schwarzer Strahler
full-scale Voll-...
full-scale deflection voller Anschlag, Vollausschlag (621.317)
full-scale service Vollbetrieb
full series notch Fahrstellung in Reihenschaltung (621.33)
full series position Fahrstellung in Reihenschaltung (621.33)
full-sized in natürlicher Größe, Voll-..., Normal-...
full-sized reproduction Wiedergabe in natürlicher Größe
full speed Volldampf, mit voller Kraft, Höchstgeschwindigkeit
in full swing in vollem Gang
full-throttle Vollgas (629.113)
full-wave Vollwellen-...
full-wave bridge circuit Doppelwegbrückenschaltung, Zweiweg-(Gleichrichter-)Schaltung
full-wave dipole Grenzwellendipol
full-wave rectification Vollweggleichrichtung, Zweiweggleichrichtung
full-wave rectifier Vollweggleichrichter, Zweiweggleichrichter
to full beschweren (67), (77)
Fuller cell Bichromat-Element
fuller board Preßspan
fuller's earth Fullererde, Bleicherde
to fuller verstemmen
fully völlig, vollständig
fully charged voll aufgeladen (621.3)
fully enclosed gekapselt, ganz gekapselt
fully enclosed switchgear ganz gekapselte Schaltgeräte (pl.)
fully formed character printer Ganzzeichen-Drucker
to fulminate blitzen, donnern; verpuffen, explodieren
fultograph galvanischer Bildtelegraph
fume Dunst, Schwaden
fume cupboard Abzug (54)
fume-resistant schwadenbeständig
to fume rauchen, Rauch abgeben, Dunstabgeben
fuming acid rauchende Säure
fumivorous rauchverzehrend
function Funktion (51); Funktion, Tätigkeit, Arbeitsweise, Wirkungsweise; Funktion, Dienstobliegenheit
function generator Funktionsgenerator (621.397)
function key Funktionstaste
function matrix (for a sequential circuit) Arbeitsmatrix (einer Folgeschaltung) (621.374)
function table Arbeitstabelle (621.374)
to function funktionieren, arbeiten, wirken
functional funktional (51); funktionell, wirkungsmäßig
functional diagram Betriebsschaubild
functional key Auslösetaste, Betätigungstaste (51), (65)
functional network Sondernetz
fundamental fundamental, grundsätzlich, Grund-..., Grundlagen-..., Grundwellen-..., grundwellig (621.3)
fundamental active power Grundwellenwirkleistung
fundamental component Grundwellenanteil, Grundwellenkomponente
fundamental crystal Grundschwingungsquarz
fundamental frequency Grundfrequenz
fundamental frequency component Grundwellenanteil
fundamental hum Grundwellenbrumm, 50-Hz-Brumm
fundamental k-sequence reactive power Grundwellenblindleistung des k-Systems
fundamental mode Hauptschwingbereich, Grundschwingbereich
fundamental negative-sequence reactive power Grundwellenblindleistung der gegenläufigen Komponente
fundamental note Grundton
fundamental oscillation Grundschwingung
fundamental particle Elementarteilchen (539)
fundamental positive-sequence reactive power Grundwellenblindleistung der mitläufigen Komponente
fundamental research Grundlagenforschung
fundamental unit Grundeinheit (53)
fundamental wave Grundwelle
fundamental wavelength Grundwellenlänge, Eigenwellenlänge
fundamental zero-sequence Grundwellenblindleistung derNullkomponente
fungicide schwammtötend, pilzverhindernd
fungus (pl.: -gi) Pilz, Holzschwamm

fungus resistance Schimmelbeständigkeit
funicular mit Seilzug, Seil-...
funicular railway Schienenseilbahn
funicular traction Seilzugförderung
funnel Trichter; Schornstein (625), (629.12); Schornstein, Esse, Kamin, Schlot
fur Pelz; (pelzförmiger) Ansatz, Belag, Kesselstein, Kruste (621.1)
to fur Kesselstein entfernen
fur. = furlong (s. dort)
to furbish polieren, glätten
to furbish up aufpolieren, ausbessern, putzen
to furcate gabeln, sich verzweigen
furcate gegabelt
furcation Gabelung
furlong Furlong (Längenmaß) = 1/8 Meile =201,2 m)
furnace Feuerung, Ofenraum, Brennraum; (Industrie-)Ofen
fuse Sicherung (621.3); Zünder (623)
fuse base Sicherungssockel
fuse block Sicherungselement
fuse blows Sicherung geht durch
fuse board Sicherungstafel
fuse box Sicherungskasten
fuse carrier Sicherungshalter
fuse cartridge Sicherungspatrone
fuse clip Einsteckkontakt (621.316.9)
fuse contact Sicherungskontakt
fuse cut-out Abschmelzsicherung, Schmelzsicherung
fuse element Sicherungselement, Schmelzstreifen, Schmelzdraht
fuse filler Sicherungsfüllung, Löschmittelfüllung
fuse holder Sicherungskörper, Sicherungsträger, Sicherungshalter
fuse link Sicherungseinsatz, Schmelzeinsatz
fuse load-break switch Sicherungslasttrenner
fuse member Schmelzeinsatz
fuse panel Sicherungstafel
fuse plug Schmelzeinsatz
fuse puller Sicherungszange
fuse rating Sicherungsbemessung, Absicherung
fuse socket Sicherungssockel
fuse support Sicherungsgrundplatte
fuse switch Sicherungsschalter
fuse tongs (pl.) Sicherungszange, Hochspannungszange
fuse tube Sicherungsrohr (621.3)
fuse unit Sicherungselement, Sicherung
fuse wire Sicherungsdraht, Schmelzdraht
fuse with enclosed fuse element geschlossene Sicherung
to fuse schmelzen, abschmelzen, durchbrennen (621.3); zünden (623.4)
to fuse in einschweißen (621.791)
to fuse together zusammenschmelzen
fused electrolyte Schmelzflußelektrolyt
fused salt Salzschmelze
fusegear Sicherungseinrichtungen (pl.)
fusibility Schmelzbarkeit
fusible schmelzbar, Schmelz-...
fusible arc-welding electrode Abschmelzelektrode
fusible cut-out Schmelzsicherung, Abschmelzsicherung, Trennsicherung
fusible element Schmelzelement, Schmelzeinsatz
fusible metal strip Schmelzstreifen
fusible read only memory Festwertspeicher mit Schmelzbrücken
fusible wire Schmelzdraht
fusing Durchbrennen, Abschmelzen, Abschmelzvorgang
fusing conductor Schmelzleiter
fusing current Abschmelzstrom, Ansprechstrom
fusing point Abschmelztemperatur
fusing time Abschmelzzeit
fusion Durchbrennen, Abschmelzen (621.3); Schmelzen, Schmelzvorgang (66)
fusion electrolysis Schmelzflußelektrolyse
fusion point Schmelzpunkt
fusion welding Schmelzschweißung
fusion zone Schmelzzone
fuze (US) = **fuse** (s. dort)

G

G (transconductance) S (Steilheit)
G = gauge Lehre (als Dickenangabe bei Drähten und Blechen)
g = gram (me) g = Gramm
gabble Gebabbel, unverständliche Sprechgeräusche (pl.)
to gabble babbeln
gadget Vorrichtung, Mechanismus, Hilfseinrichtung; „Dingsda"; Paßstück
gage (US) = **gauge** (s. dort)
gain Gewinn, Zuwachs, Nutzen, Erweiterung; Gewinn (621.396); Verstärkung, Verstärkungsfaktor (621.39)

gain booster Zündübertrager, Übertragersatz (623)
gain characteristic Verstärkungskurve
gain in compactness Einsparung an Raum, Verringerung der Abmessungen (pl.)
gain control Verstärkungsregelung
gain controlling Verstärkungsregelung, verstärkungsregelnd
gain of heat Wärmeaufnahme
gain margin Vertärkungssicherheit (gegenSelbsterregung), Verstärkungsgrenze (621.375)
gain setting Verstärkungseinstellung, eingestellte Verstärkung
gain of the stage Stufenverstärkung, Verstärkung der Stufe
gain per stage Verstärkung je Stufe
to gain gewinnen, erlangen
to gain electrons Elektronen gewinnen
to gain energy Energie aufnehmen
galena Bleiglanz
galena detector Bleisulfiddetektor, Bleiglanzdetektor
Gall chain Gallsche Kette
to gall rauhen, feinrissig werden, rauh werden, sich aufrauhen (621.94)
gallium Gallium
gallium-arsenide laser = Ga-As laser Gallium-Arsenid-Laser
gallon Gallone (Hohlmaß; 1 Imp. Gallone (Br.)= 4,546 Liter; 1 US-Gallone = 3, 782 Liter)
gallows (pl.) Galgen
galvanic galvanisch
galvanic battery Primärbatterie, galvanische Batterie
galvanic cell Primärelement, galvanisches Element
galvanic current galvanischer Strom, Gleichstrom
galvanic electricity galvanische Elektrizität
galvanic pile Voltasche Säule
galvanism Galvanismus
galvanist Galvaniseur
galvanization Feuerverzinkung; Galvanisierung (61)
to galvanize feuerverzinken; galvanisieren (61)
galvanized verzinkt
galvanized iron verzinktes Eisen
galvanized sheet iron verzinktes Eisenblech
galvanized wire verzinkter Draht
galvanograph Elektrotypie (655)
galvanometer Galvanometer
galvanometer constant Galvanometer-Konstante
galvanometer resistance Galvanometer-Nebenwiderstand
galvonometer shunt Galvanometer-Nebenwiderstand
galvanoplastic Galvanoplastik
galvanoscope Galvanoskop
galvano-therapy Galvano-Therapie
galvo = galvanometer Galvanometer
gamma Gamma; Übertragungsexponent (621.396.96)
gamma corrector Gradationsentzerrung (621.397)
gamma decay Gammazerfall (539)
gamma radiation Gammastrahlung (539)
gamma quantum Gammaquant
gamut Stimmumfang, Tonbereich
gang Arbeitsgruppe, Kolonne; Mannschaft, Belegschaft
gang condenser Mehrfachkondensator, Mehrgangkondensator
gang control Gleichlaufsteuerung (621.396)
gang switch Gruppenschalter, Paketschalter, Schalter mit mehreren Ebenen
to gang gleichlaufabstimmen, auf Gleichlaufabstimmen
to gang together kuppeln, mechanisch kuppeln, auf Gleichlauf bringen
ganged circuits mechanisch gekoppelte Schwingkreise
ganged tuning Einknopfabstimmung
gantry Signalgerüst, Signalbrücke (625.1); Portalgerüst, Portal; Gerüst, Bockgerüst
gap Spalt, Zwischenraum, Fuge; Maulweite (621.9); Luftspalt (621.3); Nullstelle (621.396.77)
gap arrester Funkenableiter, Hörnerableiter (621.3)
gap-breakdown voltage Ansprechspannung (621.316.9)
gap clearance Spaltbreite, Luftspaltbreite, Luftspalt
gap coding Pausenkodierung
gap filler Fernsehfüllsender
gap flux Luftspaltfluß
gap width Spaltweite, Spaltbreite
gap reluctance (magnetischer) Luftspaltwiderstand
Garton lightning arrester Bendmannschutz
gas Gas
gas accumulation Gasansammlung
gas alarm Grubengassignalapparat, Gasmelder

gas-amplification factor innerer Verstärkungsfaktor
gas bag Gaszelle, Gasbehälter, Ballonett (629.13), (629.113)
gas blast Preßluftstrom
  gas-blast switch Druckluftschalter, Preßgasschalter
gas bleb Gasblase, Blase (in festen Körpern)
gas bubble Gasblase (in Flüssigkeiten)
gas cell Gaselement, Gasbeladungselement, Belüftungselement
gas-charged gashaltig, gasbeladen
gas constant Gaskonstante
gas contents (pl.) Gasgehalt
gas-cooled reactor gasgekühlter Reaktor
gas coulombmeter Wasserstoffvoltameter, Knallgasvoltameter, Wasserstoffzähler
gas cylinder Gasflasche, Stahlflasche
gas density Gasdichte
gas detector Schlagwetterprüfgerät, Gasanzeiger
gas discharge Gasentladung (621.3)
  gas-discharge display Gasentladungsanzeige
  gas-discharge lamp Gasentladungslampe
  gas-discharge tube Gasentladungsröhre
gas-electric car benzinelektrischer Wagen
gas-electric drive benzinelektrischer Antrieb
gas-electric generating set benzinelektrischer Generator
gas engine Gasmotor
gas envelope Gashülle, Gasmantel (621.791)
gas explosion Gasexplosion
gas-filled gasgefüllt
  gas-filled cable Druckgaskabel
  gas-filled cavity gasgefüllter Hohlraum
  gas-filled lamp gasgefüllte Lampe
  gas-filled rectifier gasgefüllte Gleichrichterröhre
  gas-filled three-electrode valve Gastriode, Stromtor, Thyratron
  gas-filled tube gasgefüllte Röhre
  gas-filled tube rectifier Gasentladungsgleichrichter
gas flow Gasströmung
gas focusing Gaskonzentrierung (621.385)
gas-free gasfrei
gas formation Gasentwicklung
gas groove Gasabflußrinne (621.357)
gas heating Gasheizung
gas leakage Gasaustritt
gas lighter Gasanzünder
gas piping Gasrohrnetz, Gasnetz
gas pocket Gasblase, Gaspore (66)
gas pressure Gasdruck
  gas-pressure cable Druckgaskabel
  gas-pressure indicator Gasdruckmesser
gas-proof gasdicht
gas ratio Vakuumfaktor
gas seal Gasabschluß (621.314.2)
gas sentinel Gaswächter, Gasprüfgerät
gas shield Gasmantel (621.791)
  gas-shielded welding Schutzgasschweißen
gas thread Gasgewinde
gas tight gasdicht
gas triode Gastriode, gasgefüllte Triode, Stromtor, Thyratron
gas tube Gasentladungsrohr, gasgefüllte Röhre
  gas-tube switch Thyratron, Stromtor
gas turbine Gasturbine
gas-turbo-electric locomotive gasturboelektrische Lokomotive
gas voltameter Wasserstoffzähler
gas X-ray tube gasgefüllte Röntgenröhre, weiche Röntgenröhre
gaseous gasförmig
  gaseous compound gasförmige Verbindung
  gaseous diffusion Gasdiffusion
  gaseous discharge Gasentladung (621.3)
  gaseous matter gasförmige Materie
  gaseous rectifier tube gasgefüllte Gleichrichterröhre
  gaseous state gasförmiger Zustand
  gaseous void gaserfüllter Hohlraum
to gasify vergasen, in den gasförmigen Zustandüberführen
gasiform gasförmig
gassing accumulator kochender Akkumulator
gasket Dichtung, Dichtungsscheibe
  gasket board Dichtungspappe, Dichtungsmaterial; Klingerit
  gasket material Dichtungsmaterial
gasoline (US), (engl. zuweilen **gasolene**) Treibstoff, Kraftstoff
  gasoline-electric locomotive benzinelektrische Lokomotive
  gasoline engine Benzinmotor
    gasoline-engine driven mit Benzinantrieb, mit Benzinmotorantrieb
gassing Gasen (621.355)
gassy gasführend, gashaltig
  gassy nebula Gasnebel (52)
gate Tor, Eingang; Einlauf (621.7), (621.2); Gatter (621.382); logische (Verknüpfungs-)Schaltung
  gate apparatus Leitapparat (621.2)
  gate by-pass switch Schachttürkontakt-Überbrückungsschalter

gate circuit Torschaltung (621.374); Auftastschaltung (621.397)
gate contact Türkontakt (621.876)
gate-controlled delay time Zündverzug (621.314.63)
gate-controlled rise time Durchschaltzeit
gate-controlled turn-off time Löschzeit
gate-controlled turn-on time Zündzeit
gate current Gate-Strom, Steuerstrom (621.314.63)
gate cut-off current Gate-Reststrom
gate interlock Türverriegelung (621.34), 621.876)
gate latch Türriegel, Türverschluß
gate-lock Schachttürverriegelung
gate-non-trigger current höchster nichtzündender Steuerstrom
gate power loss Steuerverlustleistung
gate pulse Torimpuls (621.374); Auftastimpuls (621.397)
gate switch Schachttürkontakt
gate terminal Steueranschluß (621.382)
gate trigger current Zündstrom (621.382)
gate turn-off voltage Mindestabschaltspannung
gate voltage Steuerspannung (621.374); Verriegelungsspannung, Sperrspannung (621.397)
to gate austasten, ausblenden; Strahl sperren (621.397)
gateway Toreinfahrt, Einfahrt
gateway trunk circuit Durchgangsfernleitung (621.394)
to gather rust Rost ansetzen
to gather together zusammenfassen, ordnen (51)
gating Austastung, Strahlsperrung, Impulssperrung (621.397); Torsteuerung (621.374)
gating time Öffnungszeit
gauge Maß, Maßstab, Lehre, Normalmaß; Kontrollgerät, Lehre, Meßuhr; Spur (625.1); Wandstärke, Dicke, Blechdicke, Drahtstärke, Drahtdurchmesser
gauge block Endmaß
gauge glass Standmaß, Mensur (54); Standrohr, Wasserstand, Ölstand (621)
to gauge lehren, messen, abmessen, ausmessen; eichen, abschätzen
to gauge the wind Wind messen, Windstärkemessen
gauging Meß-...
gauging face Meßfläche
gauging head Meßkopf
Gaussian distributian Gauß'sche Verteilung, Normalverteilung
Gaussian units cgs-Einheiten
gaussmeter Gaußmeter, magnetischer Induktionsmesser
gauze Gaze
gauze brushes Gazebürsten, Kupferbürsten (621.313)
gauze filter Filtergaze, Gazefilter, Drahtsieb
gauze-protected machine durch Gazeabdeckung schlagwettersichere Maschine
G. C. A. = ground controlled approach system bodengesteuertes Anflugsystem
GCA-monitored ILS approach radarüberwachter ILS-Anflug
G-clamp Schraubzwinge
G. C. M. = greatest common measure größter gemeinsamer Teiler
gear Rad, Zahnrad, Getrieberad; Getriebe, Mechanismus, Trieb
gear-box oil Getriebeöl
gear cutter Zahnradfräser
in gear eingekuppelt
gear drive Zahntrieb
gear lever Schalthebel, Kupplungshebel
out of gear ausgekuppelt, außer Betrieb
gear ratio Übersetzungsverhältnis, Übersetzung (621.8)
gear reducer Untersetzungsgetriebe
gear train Zahnradfolge, Zahnradsatz, Rädersatz
gear transmission Zahnradübersetzung, Zahnradübertragung
to gear anschirren; einkuppeln
to gear down herunterübersetzen, untersetzen
to gear into eingreifen in, kämmen in
to gear together mechanisch kuppeln; eingreifen
to gear up hinaufübersetzen, übersetzen
geared Getriebe-...
geared locomotive Getriebelokomotive (621.33)
geared motor Getriebemotor (621.3)
geared pump Zahnradpumpe
gearing Getriebe, Vorgelege, Zahnradantrieb, Verzahnung, Zahnradübersetzung
gearless motor Achsmotor (621.33)
GEE Hyperbelnavigation
Gehrcke oscilloscope Glimmlichtoszillograph
Geiger-Müller-counter Geiger-Müller-Zähler
Geier-Müller-tube-amplifier Geiger-Müller-Zählrohr
Geissler discharge Geißler-Entladung
Geissler tube Geißlersche Röhre

gel Gel
general allgemein, generell, General-..., Gemein-...
general expenses allgemeine Ausgaben, Gemeinkosten
general field floodlight system Flutlichtplatzbeleuchtung
general lighting Allgemeinbeleuchtung
general management Geschäftsleitung, Direktion
general overhaul Grundüberholung
general plan Übersichtsplan
general-purpose Allzweck-..., Universal-...
general-purpose machine Universalmaschine
general register Mehrzweckregister
general-use switch Universalschalter
general view Gesamtansicht, Gesamtbild
to generalize verallgemeinern
generalized sinusoidal quantity modulierte Sinusgröße
to generate erzeugen
to generate current Strom erzeugen
to generate electrical energy elektrische Energie erzeugen
to generate heat Wärme erzeugen
to generate power Leistung erzeugen
to generate steam Dampf erzeugen
generating Erzeugung
generating capacity Leistung, installierte Leistung
generating electrostatic voltmeter rotierendes elektrostatisches Voltmeter, Feldmühle
generating plant Kraftwerk; Kraftanlagen (pl.)
generating set Generatorensatz, Maschinensatz (621.31)
generating station Kraftwerk
generating station auxiliaries Kraftwerkhilfsmaschinen
generating station auxiliary power Eigenbedarfsleistung (621.311)
generating station capacity installierte Kraftwerksleistung
generating station efficiency Kraftwerkswirkungsgrad
generating tube Verdampferrohr (621.18)
generating voltmeter Rotationsvoltmeter
generation Generation (57); Erzeugung
generation of electricity Stromerzeugung, Elektrizitätserzeugung
generation of heat Wärmeerzeugung, Wärmeentwicklung
generation of steam Dampferzeugung
generator Generator, Stromerzeuger (621.3); Lichtmaschine (629.113); Autogengenerator, Entwicklungsbehälter (66)
generator bus bar Generator-Sammelschiene
generator end Sendeseite, Eingang (621.394), (621.396)
generator field control Leonardsteuerung
generator output Generatorleistung
generator protection Generatorschutz
generator voltage regulator Maschinenspannungsregler
generator winding Generatorwicklung
gentle vorsichtig, zart, sanft, leicht
gentle heating vorsichtige Erwärmung
gentle manipulation schonende Behandlung
geocentric geozentrisch
geo-electric geo-elektrisch
geographic pole geographischer Pol
geomagnetic erdmagnetisch
geomagnetic surveying geomagnetische Erzsuche
geometric(al) geometrisch
geometrical distortion Geometriefehler (621.397)
geometric locus geometrischer Ort, Ortskurve (538)
geometric(al) mean geometrisches Mittel
geometrical point geometrischer Punkt
geometrical progression geometrische Reihe
geometrician Geometer
geometry Geometrie; geometrische Anordnung
geophysic(al) geophysikalisch
geophysics (pl.) Geophysik
geothermal power station geothermisches Kraftwerk
geothermic gradient geothermische Tiefenstufe
geothermic power station Erdwärmekraftwerk
German deutsch
German silver Neusilber
German text Fraktur
germanium Germanium
germanium diode Germaniumdiode
G. E. S. cap = Goliath Edison screw cap Goliath-Edisongewinde
to get at herankommen, drankommen an
get-at-able zugänglich
to get matted sich verfilzen, durcheinandergeraten
to get 120 as the sum 120 als Summeherausbekommen (51)
to get under way in Gang kommen, anfangen
getter Getter (621.385)
ghost Echobild, Geist, Geisterbild (621.397)
ghost echo Geisterecho
ghost working Phantom-Betrieb (621.394)

giant riesig, Riesen-...
giant crane Riesenkran
giant reflector Riesenspiegel
gilbert Gilbert
gill Lappen; Lamelle (58); Gill (Hohlmaß = 1/4 pint = 142 cm^3)
gimbal Kardan
gimbals (pl.) kardanische Aufhängung
gimbal axis Kardanachse
gimbal bearing Kardanlager
gimbal ring Kardanring, Kardanrahmen
gimbal kardanisch
gimbally mounted kardanisch gelagert, kardanisch aufgehängt
Giorgi system Giorgi-Maßsystem, MKSA-System
girder Binder, Träger, Balkenträger; Unterzug
girder-type frame genietetes Gehäuse (621.313)
to give geben
to give back zurückgeben, zurückliefern
to give back energy Energie zurückliefern, rückspeisen
to give energy to Energie übertragen auf, Energie abgeben an
to give forth hergeben, liefern, ausgeben, abgeben
to give up heat to Wärme abgeben an
to give off light Licht abgeben, Lichtausstrahlen
to give out herausgeben, hergeben, abgeben, loslassen, aussenden; ermatten, nachlassen
to give out a note einen Ton von sich geben, einen Ton abgeben
to give out power Leistung abgeben
to give place to Platz machen für, das Feldräumen für, das Feld überlassen an
to give rise to hervorbringen, erzeugen
to give values to Werte zuerteilen, Zahleneinsetzen für
to give way nachgeben, zusammenbrechen
glance (flüchtiger) Blick; Glanz (55)
glance coal Glanzkohle, Anthrazit
to glance (flüchtig) hinschauen; streifen, oberflächlich berühren
glancing incidence streifender Einfall (535)
gland Drüse (61); Buchse, Stopfbuchse, Stopfbüchse, Wellendurchführung; Eichel
gland packing Stopfbüchsenpackung
glandiform eichelförmig
glare Glanz, Blendung
glareless blendungsfrei, nichtblendend
glareless lighting blendungsfreie Beleuchtung
glass Glas; Glas, Trinkglas; Glas, Fernglas; Barometer, Wetterglas; Spiegel
glasses (pl.) Brille
glass bead Glasperle
glass bulb Glaskolben
glass electrode Glaselektrode
glass-enclosed fuse Röhrchensicherung (621.396)
glass envelope Glashülle, Glaskolben
glass-envelope tube Glasröhre (621.385)
glass fibre Glasfaser
glass half cell Glaselektrode
glass hard glashart
glass jar Glasbehälter, (Akkumulator-)Glas
glass measure Standmaß, Mensur
glass-to-metal seal Einschmelzung; Anglasung, Glasdurchführung
glass paper Glaspapier, Sandpapier
glass slide Objektträger (535)
glass-tube fuse Glasrohrsicherung
glass vessel Glasgefäß, Becherglas
glass wool Glaswolle
glassivation Glasierung
glassy glasig, glasartig
glassy semiconductor amorpher Halbleiter
Glauber's salt Glaubersalz
glaze Glasur
to glaze verglasen; glasieren, mit Glasurversehen, glasig werden; glänzen = glänzende Oberfläche geben, satinieren
glazed tile Fliese, Kachel
glazing Verglasung; Satinierung (67), (77)
glazing sheet Glänzblech, Satinierblech (77)
to gleam glimmern, schimmern
glider Segelflugzeug
glide slope Gleitweg (621.396.93)
glide-slope transmitter Gleitwegsender (621.396.93)
glide tone stetig veränderter Ton
gliding angle Gleitwinkel (629.13)
glim lamp (rare) Glimmlampe
to glimmer blinken
glimpse frequency Blitzfrequenz
to glint glitzern, flimmern
to glisten glitzern, funkeln
glitches Störspannungsspitzen
to glitter glitzern
gloaming Zwielicht, Abenddämmerung
globe Globus; Kugel, Lampenkugel
globe lightning Kugelblitz
globe photometer Kugelphotometer, Ulbrichtsche Kugel
globe-strain insulator Eierisolator
globular kugelförmig, kugelig
globular discharge geschichtete Entladung
globular lightning Kugelblitz

globule Kügelchen; Schweißperle
gloss Glanz; Politur
to gloss glänzen; glänzend machen
glossimeter Glanzmeßgerät
glow Glut, Glimmen, Glühen; Glimmentladung (621.385)
glow discharge Glimmentladung
glow-discharge anemometer Glimmstrom-Anemometer
glow-discharge lamp Glimmröhre
glow-discharge microphone Glimmlichtmikrophon, Kathodophon
glow lamp Glimmlampe, Glimmröhre
glow regulator Stabilisator, Glimmröhrenstabilisator, Stabilovolt
glow-starter switch Glimmzünder
glow-switching diode Glimmschaltdiode
glow tube Glimmröhre
to glow glimmen, glühen
glower Glühstab, Nernststäbchen
glue Leim
to glue leimen, verleimen, anleimen
gluey leimig, klebrig
gluing Verleimung
gluing process Verleimung, Leimverfahren, Leimvorgang
gluten Kleber (63)
glycerin(e) Glyzerin
gm = gram(me) g = Gramm
G. M. T. = Greenwich mean time Mittlere Greenwich-Zeit
go Gang, Versuch
on the go in Gang, in Betrieb
to be on the go laufen
to go gehen, laufen
to go bad schlecht werden
go-between Mittelding
to go by sich behandeln lassen nach, sich lösen lassen nach
to go into conduction leitend werden, Strom durchlassen, beginnen Strom durchzulassen
to go far weit reichen, lange vorhalten
to go through a hysteresis loop eine Hysteresisschleife durchlaufen
to go on-hook einhängen (621.395)
to go negative negativ werden, ins Negative gehen
go - no go Gut-Schlecht (621.9)
go and not-go gauge Toleranzlehre
to go off losgehen, explodieren
to go off-hook den Handapparat aushängen (621.395)
to go off scale über die Skala hinausgehen, über die Skala hinausausschlagen
to go into operation in Betrieb gehen
to go out ausgehen, erlöschen
to go to pieces in Stücke gehen, in Stücke brechen
to go positive positiv werden, ins Positive gehen
to go into production in Fertigung gehen; dieFabrikation aufnehmen
to go into reverse rückwärts gehen, den Rückwärtsgang einschalten
to go right richtig funktionieren
to go round ausreichen, „rundgehen"; umhergehen, um ... herum gehen
to go to waste in den Abfall kommen, wegfallen; weglaufen
to go with gleichlaufen mit
gone fuse durchgebrannte Sicherung
goggles (pl.) Schutzbrille
gold Gold
gold beater's skin Goldschlägerhaut
gold bounded diode Golddraht-Diode
gold leaf Schaumgold, Goldblättchen
gold-leaf electroscope Goldblattelektroskop
gold plating Vergoldung
Goldschmidt alternator Goldschmidt-Maschine
Goliath Edison screw holder Edison-Goliathfassung
gong Klinger, Glocke, Gong
goniometer Goniometer
goniometry Goniometrie, Vermessungskunde
good gut
good-class hochwertig, gut
good fit guter Sitz
good heat kräftige Hitze
good quality gute Qualität, hochwertig
good-quality material hochwertiges Material, gutes Material
a good 10 yards reichliche 10 Yards, gut 10 Yards
goods locomotive Güterzuglokomotive
goose neck Schwanenhals
Gott's method kapazitive Kabelfehlerortsmessung
to govern beherrschen, festlegen, bestimmen (51); regieren, beherrschen; regeln, steuern (621)
governor Regler, Regulator; Drehzahlregler
governor arm Reglerhebel; Bügel
governor mechanism Regelmechanismus, Regelapparatur
governor position Reglerstellung
governor response Ansprechen des Reglers
G. P. O. = General Post Office Postzentralamt
gr. = grammes g = Gramm

grad = gradient grad = gradient (51)
to gradate abstufen, abschattieren
gradation Stufung, Abstufung; Gradation (77); Schattierung
gradation distortion Gradationsfehler
grade Grad, Sorte, Klasse, Güteklasse, Qualität; Steigung, Neigung, Gefälle, Neugrad, Gon (514)
grade of coal Kohlensorte, Kohlenkörnung
grade of hardness Härtegrad
grade of service Verkehrsgüte, Dienstgüte (621.395)
to grade abstufen, stufen; zerkleinern; klassieren, klassifizieren, (nach Güte oder Größe) sortieren
graded-base transmitter Drifttransistor
graded cable Silbermann-Kabel; Kabel mit abgestufter Isolation
graded insulation abgestufte Isolation
gradient Gradient (51); (53); Steigung, Gefälle
gradient fiber Gradientenfaser
grading Klassierung; Sortierung, Klassifizierung; Mischung (621.395)
grading coefficient Stufenfaktor (621.316.7)
grading group Zubringerteilgruppe (621.395)
grading ring Schirmring (621.315.1)
grading rule Klassifizierungsvorschrift
grading screen Sortiersieb, Klassiersieb
grading shield Schirmelektrode (621.315.1)
gradual steigend, schrittweise, fortschreitend, allmählich
gradual cut-off schleichender Einsatz (621.385)
graduate Akademiker; Standmaß, Mensur (54)
to graduate teilen, mit einer Skala versehen, graduieren; abstufen
to graduate in terms of einteilen nach
graduated abgestuft, graduiert
graduated flask Meßflasche, Standmaß
graduated ring Gradring, Skalenring
graduation Abstufung; Gradeinteilung, Skaleneinteilung; Skalenteil, Teilstrich
graduation line Teilstrich
graduation mark Teilstrich, Markierungsstrich
graduator Verflacherschaltung (621.394)
to graft einsetzen, hineinsetzen, transplantieren, aufpfropfen; (mit Schnur) bewickeln
grain (Korn (63), (77); körniges Pulver; Korn, Kristallkorn (53); Faser, Faserrichtung (66), (67); Gran (Gewichtseinheit = 64, 8 mg)
against the grain gegen den Strich, gegen die Faserrichtung
grain alcohol Äthylalkohol
grain boundary Korngrenze
grain-oriented transformer sheet kornorientiertes Transformatorblech
grain size Korngröße
to grain körnen; masern (674)
grained metal (Metall-)Granalien (pl.)
grainfree, grainless kornfrei
grainy körnig, faserig
gram Gramm
gram-equivalend Grammäquivalent
gram-molecule Mol
Gramme ring Grammescher Ring
gramme (veraltet), siehe **gram**
gramophone Grammophon
gramophone motor Grammophonmotor, Plattenspielmotor
gramophone needle Grammophonnadel
gramophone pick-up Tonabnehmer, Tonarm, Pickup
granular körnig
to granulate körnen, granulieren
granulated carbon Kohlegrieß
granulated metal Granalien (pl.), gekörntes Metall (66)
granulation Granulation, Körnung, Granulierung
granule Körnchen, Korn
graph graphische Darstellung, Kurvenblatt, Diagramm; Charakteristik; Schaubild
graph paper Millimeterpapier
grapher Registriergerät
graphic graphisch; anschaulich
graphic board Grafikkarte (681.3)
graphic construction graphische Konstruktion
graphic display graphische Anzeige
graphic formula Konstitutionsformel (54)
graphic instrument Registriergerät
graphic panel Betriebsschaubild
graphic plotter Kurvenschreiber
graphic recorder Kurvenschreiber, Linienschreiber
graphic representation graphische Darstellung
by graphics graphisch, auf graphischem Wege, nach einem graphischen Verfahren
graphite Graphit
graphite brush Graphitbürste, Kohlebürste
graphite coating Graphitbelag
graphite crucible Graphittiegel
graphite electrode Graphitelektrode
graphite-moderated (uranium) pile graphitgebremster Uranbrenner, Atombrenner mit Graphit als Bremssubstanz
graphite-moderated reactor graphitgebremster Reaktor, graphit-moderierter Reaktor

graphite resistance Kohlewiderstand, Kohleschichtwiderstand
to graphite graphitieren
graphitic graphitisch
to graphitize graphitieren
graphitized filament graphitierter Glühfaden
grate Gitter, Rost
graticule Strichgitter (535)
grating Gitterwerk (72); Gitter (535)
grating constant Gitterkonstante (535)
grating and dot generator Bildmustergenerator (621.397)
grating space Gitterkonstante (535)
Graetz rectifier Gleichrichter in Graetzschaltung
to grave eingraben, gravieren
gravel Kies, Schotter
graving chisel Grabstichel, Graviernadel
to gravitate der Schwerkraft folgen, gravitieren, der Schwerkraft unterliegen
gravitation Gravitation, Schwerkraft, Schwerewirkung
gravitational Schwere-...
gravitational constant Erdbeschleunigung, Gravitationskonstante
gravitational field Schwerefeld
gravity Schwere, Schwerkraft, Schwerebeschleunigung, Erdbeschleunigung
gravity casting (einfacher) Guß, Formguß
gravity control Schwerkraftrückstellung (621.317)
gravity switch Hakenschalter (621.395.6)
graybody grauer Körper, Graustrahler
Gray code Gray-Code
gray scale Graukeil
to graze streifen, leicht berühren
graze fuse Streifzünder, Aufschlagzünder, hochempfindlicher Zünder
grazing incidence streifender Einfall
grease Fett, Schmiere, Schmiermittel
grease cup Staufferbüchse, Fettbüchse
grease-free fettfrei
grease-lubricated fettgeschmiert
to grease fetten, schmieren, einschmieren, mit Fett schmieren
greasy fettig, schmierig
greasy lustre Speckglanz
great groß
great circle Großkreis, größter Kugelkreis
great-circle path Großkreiskurs
great-circle route Großkreiskurs
greatest common divisor größter gemeinsamer Teiler (51)
greatest common measure größter gemeinsamer Teiler
Greek fire griechisches Feuer
green grün
green copperas grüner Vitriol, Eisenvitriol
green sand mould grüne Gußform, Naßgußform (621.7)
Greenwich time Greenwich-Zeit, WEZ = Westeuropäische Zeit
grey cardboard Graupappe
grey iron Grauguß
grey iron castings Graugußteile
grey scale generator Graustufengenerator
to grey mattieren (77); grau machen
grid Gitter (621.385); Netz, Leitungsnetz, Kabelnetz (621.315); Rohrnetz, Ferngasnetz (621.6); Koordinatennetz (51)
grid a. c. Gitterwechselstrom
grid battery Gitterbatterie
grid bias Gittervorspannung
grid-bias battery Gitterbatterie, Gittervorspannbatterie
grid-bias connection Gitterbasisschaltung
grid cap Gitterkappe
grid capacitor Gitterkondensator, Gitterblock
grid-cathode space Raum zwischen Gitter und Kathode
grid-cathode tube capacity Gitter-Kathodenkapazität
grid choke Gitterdrossel
grid circuit Gitterkreis, Gitterzweig
grid condenser Gitterkondensator, Gitterblock
grid conductance Gitterableitung
grid control Gittersteuerung
grid controlled gittergesteuert
grid current Gitterstrom
grid current characteristic Gitterstromkennlinie
grid-current damping Gitterstromdämpfung
grid-current modulated gitterstrommoduliert
grid-current modulation Gitterstrommodulation
grid detector Gittergleichrichter, Audion
grid dimension Rastermaß
grid-dip meter Grid-dip-Meter, Asorptionsfrequenzmesser
grid-driving power Gittersteuerleistung
grid emission Gitteremission
grid-glow relay Glimmrelais
grid-glow tube Glimmrelais, gittergesteuerte Glimmröhre
grid keying Gittertastung (621.385), (621.396)
grid leak Gitterableitung, Gitterableitwiderstand
grid-leak detection Audiongleichrichtung
grid-leak detector Audion

grid-leak rectification Audiongleichrichtung
grid-leak resistance Gitterableitwiderstand
grid limiting Gitterstrombegrenzung
grid modulation Gittermodulation
grid plate Gitterplatte (621.355)
grid-plate transconductance Steilheit (621.385)
grid-plate tube capacity (innere) Gitteranodenkapazität
grid potential Gitterpotential, Gittervorspannung
grid potentiometer Gitterpotentiometer
grid rectifier Gittergleichrichter
grid resistance Gitterwiderstand
grid resistor Gittervorwiderstand, Gitterdämpfungswiderstand
grid suppressor Gitterwiderstand, Gitterdämpfungswiderstand
grid sweep Gitterwechselspannung, Gitterspannungsschwankung, Gitteramplitude
grid system Fernleitungsnetz, Verbundnetz
grid tank Gitterschwingkreis
grid technique Gittertechnik, Bleiblendenanordnung, Bleiblende (621.386)
grid voltage Gitterspannung
grid-voltage/grid-current curve Gitterstromkennlinie
grid-voltage supply Gitterspannungsquelle
grid-voltage swing Gitterspannungsschwankung, Gitteramplitude
grid wire Gitterdraht (621.385)
to grind mahlen, zermahlen, schleifen
to grind coal Kohle mahlen
to grind a lens eine Linse schleifen
grinder Mühle, Quetsche
grinders (pl.) Brodelstörung, Rauschen (621.396)
grinding Schleifen, Mahlen; Schleifvorgang, Mahlvorgang
grinding belt Schleifband
grinding crack Schleifriß
grinding fineness Mahlfeinheit
grinding particles (pl.) Schleifstaub
grindstone Schleifstein
grip Klammer, Schelle, Nippel; Umklammerung, Griff; Klammereisen, Greifer
grip ring Schelle, Ringklammer
grip screw Schraubklammer
to grip fassen, erfassen, packen; einspannen, umklammern, umfassen, umgreifen; greifen, halten
grit Schlacke, Asche, Flugasche; Kies, Grus
gritstone Schmirgel
gritty kiesig, sandig (621.74)
groove Nut (674); Ringnut, Rille, Laufrille; Nut (621.313); Zug (623); Rille, Tonspur (681.8)
to groove nuten, rillen, mit einer Nut versehen, riffeln; kannelieren (72)
grooved gerillt, genutet
grooved armature Anker mit Nuten
grooved ball bearing Rollenkugellager
grooved pulley Rillenscheibe, Seilscheibe, Nutrolle
grooved rail Rillenschiene
grooved rotor Rillenläufer
grooved trolley wire Profilfahrdraht (621.33)
grooved wheel Rillenrad, Rillenscheibe
grooved wire Profilfahrdraht (621.33)
grooving Riffelung, Nutung; Rille
gross Brutto-...
gross calorific value oberer Heizwert
gross heat value oberer Heizwert
gross length Bruttolänge
gross weight Bruttogewicht
ground (s. auch **earth**) Erde, Erdboden, Erdoberfläche, Boden, Grund, Erde, Erdung, Erdschluß, Masse (621.3); Bodensatz (54)
ground absorption Dämpfung der Bodenwelle, Bodendämpfung (621.396)
ground aerial Erdantenne
ground-to-air communication Boden-Bord-Verbindung
ground-to-air radiotelephony Boden-Bord-Sprechverkehr
ground back scatter Bodenrückstreuung
ground-based duct Oberflächenwellenleiter
ground-base traffic control bodengesteuerte Verkehrskontrolle (351)
ground bus Erdungssammelleitung
ground-cable bond Kabelerdungsschelle
ground clamp Erdungsklemme, Erdklemme
ground clutter unerwünschte Bodenechos
ground coating Grundanstrich
ground controlled approach-GCA Radarlandeverfahren
ground controlled interception bodengesteuerte Jägerleitung
ground damp Bodenfeuchtigkeit
ground dependent navigation bodenabhängige Navigation
ground echo Festzeichen (621.397)
ground effect Bodeneinfluß (621.396)
ground fault Erdschluß
ground-fault current Erdschlußstrom
ground-fault monitor Erdschlußwächter
ground imperfections schlechte Erdungsverhältnisse

ground indication Erdschlußanzeige
ground indicator Erdschlußanzeiger
ground installation Bodenanlage
ground level Boden, Erdboden, Bodennähe
above ground level über Erde
ground lug Erdklemme, Erdanschluß
ground marker Bodenmarker (656.7)
ground outlet Schuko-Steckdose, Schutzkontaktsteckdose
ground plan Grundriß
ground-plane antenna Viertelwellenantenne mit Gegengewicht
ground protection Erdschlußschutz
ground radar Bodenradar
ground range horizontale Reichweite (621.396); Reichweite am Erdboden
ground reflexion Reflexion am Erdboden, Bodenreflexion (621.396)
ground relay Erdschlußrelais
ground resistance Erdungswiderstand
ground return Erdrückleitung (621.315), (621.394); Bodenecho, Bodenzeichen (621.396.96)
ground rod Erdungsstange
ground station Bodenstation (621.396)
ground strap Erdungslasche
ground system Erdungsanlage (621.396.67)
ground terminal Erdklemme (621.315)
ground transformer Erdungstransformator, Erdungswandler, Erdungsdrossel
ground water Grundwasser
ground-water level Grundwasserspiegel
ground wave Bodenwelle (621.396)
to ground (s. auch **to earth**) erden, an Masse legen (621.3)
grounded-base circuit Basisschaltung
grounded capacitance Betriebskapazität eines Leiters bei Erdung aller übrigen
grounded-cathode circuit Kathodenbasisschaltung
grounded plate circuit Anodenbasisschaltung
ground geschliffen (s. auch **to grind**)
ground block geschliffenes Endmaß
ground glass Mattscheibe, Mattglas
ground-glass screen Mattscheibe (35)
ground-in eingeschliffen
grounding (s. auch **earthing**) Erden; Erdungs...
grounding clamp Erdungsklemme, Erdklemme
grounding conductor Erdleiter; Schutzkontaktleiter
grounding screw Erdungsschraube
grounding switch Erdungsschalter, Erdtrennschalter
grounding transformer Erdungstransformator, Erdungswandler
grounding-type attachement receptacle Schutzkontakt-Steckdose
grounding-type attachement plug Schutzkontaktstecker
groundwork Fundament, Grundlage
group Gruppe (51); Primärgruppe (621.395.4)
group aerial Antennengruppe
group delay Gruppenlaufzeit
group-delay distortion Gruppenlaufzeitverzerrung
group directional characteristic Gruppenrichtcharakteristik
group drive Gruppenantrieb (621.9)
group driving Gruppenantrieb (621.9)
group frequency Funkenfolgefrequenz, Impulsfrequenz (621.396); Gruppenfrequenz (621.395.4)
group link Primärgruppenverbindung
group modulator Primärgruppenumsetzer (621.395.4)
group operation gemeinsamer Antrieb (621.316.5)
group-reference pilot Gruppenpilot
group retardation Gruppenlaufzeitverzögerung
group-selection stage Richtungswahlstufe (621.395.3)
group selector Gruppenwähler
group translation Gruppenumsetzung
group translator Gruppenumsetzer (621.395.4)
group (of trunks) Leitungsbündel; Bündel (621.395)
group velocity Gruppengeschwindigkeit (621.392)
grouping Gruppierung, Eingruppierung, Ordnung, Anordnung; Reihen-Parallel-Schaltung (621.3)
grown junction gezogener Zonenübergang, gezogener p-n-Übergang (621.382)
growth Wachstum; Aufwachsen (621.382)
growth figure Wachstumskennzahl (621.395)
grub screw Madenschraube, Gewindestift
grumble Grummeln, Brummen
to grumble brummen
to guarantee garantieren, sicherstellen, zusichern
guard Wache, Schutzvorrichtung, Schutzblech, Schutznetz, Schutzkorb; Funkbereitschaft mit sendebereitem Sender
guard band Schutzbereich, Schutzabstand (621.395)
guard circuit Sprachschutzschaltung
guard cradle Schutznetz (621.315.1)

guard net Schutznetz (621.315.1)
guard plate Schutzring (621.317)
guard ring Schutzring, Schirmring (621.3); Vorsteckring, Sicherungsring
guard wire Schutzdraht, Schutznetz
to guard against schützen vor, schützen gegen, abdecken gegen
guarded berührungsgeschützt (621.3); abgedeckt
gudgeon Bolzen, Drehzapfen, Zapfen
gudgeon pin Kolbenbolzen
gudgeon-pin hole Kolbenbolzenloch
guesswork Überschlag, Schätzung
guidance Anhaltspunkt, Anleitung; Führung, Führungsschiene; Lenkung (623)
guidance radar Leitradar
guidance system Fernlenksystem; Leitsystem, Lenksystem (623)
guide Hinweis, Richtlinie; Richtschnur; Führung (621); Hohlrohrleiter (621.396)
guide bar Führungsschiene, Führungsstange
guide beam Leitstrahl
guide block Führungsstück
guide cable Leitkabel, Fahrrinnenkabel (527)
guide cylinder Führungszylinder
guide fillet Führungsleiste
guide piece Führungsstück
guide pillar Führungssäule (621.97)
guide pin Führungsstift
guide pip Führungsspitze, Führungspips
guide plate Führungsplatte
guide pulley Leitrolle, Umlenkrolle
guide rail Führungsschiene
guide roller Führungsrolle
guide slot Führungsschlitz
guide stem Führungsschaft
guide strip Führungsstreifen, Führungsband
to guide führen, leiten, richten; lenken, steuern, fernsteuern
guided missile gesteuerter Flugkörper, gesteuerte Rakete
guided rocket gesteuerte Rakete, Lenkrakete
guided wave leitungsgerichtete Welle; geführte Welle (621.396)
guiding effect Richtwirkung, Steuerwirkung
Guildinus theorem Guldinsche Regel
guillotine Guillotine; Stanze, Schnittstanze; Schere, Papierschneidemaschine
to gum gummieren, kleben, verkleben; verharzen, klebrig werden
to gum to ankleben an
gummy klebrig
gun Geschütz, Kanone; Gewehr (623); Elektronenquelle, Elektronenkanone (621.385)
gun current gesamter Strahlstrom
gun dicrector Feuerleiteinrichtung, Kommandogerät
gun-laying Feuerleitung, Feuerleit-...
gun-laying radar Feuerleitradar
gun metal Bronze
Gunn effect Gunn-Effekt
gunshot Schußweite, Schußentfernung
gusset Knotenblech, Zwickel
gusset plate Knotenblech, Versteifungsplatte
guttapercha Guttapercha
guy Abspannung, Halterung
guy anchor Spanndrahtanker, Abspannpflock, Abspannanker (621.315)
guy rope Abspannseil, Halteseil
guy(-shackle) insulator Pardunenisolator
guy wire Abspanndraht, Spanndraht, Pardune
to guy abspannen, anseilen
gymnotus electricus Zitteraal
gypsum Gips
to gyrate kreiseln, kreisen
gyration Kreiseln, Kreiselbewegung, Kreisbewegung, Drehung
gyratory kreisend, Kreisel-...
gyratory motion Kreiselbewegung
gyro ... Kreisel ...
gyrocompass Kreiselkompaß
gyrofrequency Gyrofrequenz
gyromagnetic effect gyromagnetischer Effekt, Einstein-de-Haas-Effekt
gyropilot Kreiselpilot
gyroscope Kreisel, Kreiseleinrichtung
gyrostabilized plattform kreiselstabilisierte Plattform
gyrostatic compass Kreiselkompaß
gyrostatic stabilizer Schiffskreisel

H

h = henry H = Henry
h = hour h = Std. = Stunde
hack Picke; Einschnitt, Kerbe
hack saw Metallsäge, Metallbügelsäge
hack-saw blade Metallsägeblatt
haematite Hämatit
haft Heft, Messerheft
hair Haar; Haare (pl.)
hair line Haarstrich; Faden (535.8)
hair lines (pl.) Fadenkreuz (535.8)

hairline Strich, Strichmarke (535); Skalenstrich (621.396.2); Längsriefe (621.9), (66)
hairsieve Haarsieb
hairspring Unruhfeder (681); Systemfeder, Rückstellfeder (621.317)
halation Lichthof, Lichthofbildung (77)
half (pl. **-ves**) Hälfte
half halb
half adder Halbaddierer
half cell Meßelektrode, Normalelektrode (621.35)
half-closed slot halbgeschlossene Nut
half-coiled winding Wicklung mit einer Windung je Polpaar
half cycle Halbwelle
half-deflection method Halbwertverfahren
half-duplex circuit Wechselverkehrsverbindung, Halbduplexverbindung
half-duplex operation Wechselschreiben (621.394); Wechselsprechen
half-finished halbfertig
half-finished products Halbfabrikate; Halbzeug (sing.)
half-hard halbhart
half-hard brass halbhartes Messing
half life Halbwertszeit (539)
half-line pulse Impuls von doppelter Horizontal-Frequenz (621.397)
half maximum value Halbwert
half normal bend 135°-Bogen (621.315.6)
half period Halbwelle; Halbperiode, Halbwertszeit (539)
half period average value Halbwellenmittelwert (621.3)
half nut Halbmutter
half open halboffen
half-power angle Halbwertwinkel
half-power beam width Halbwertsbreite
half-power point Halbwertspunkt (621.396), (621.396.67)
half-round halbrund
half-round bar Halbrundstange
half section halber Querschnitt, Halbquerschnitt
half-silvered halbversilbert, halbdurchlässig versilbert
half-sinusoid Sinushalbwelle
half-sinusoidal Halbwellen-..., halbwellenförmig
half size halbe Größe, in halber Größe, Maßstab 1:2
half-tone Halbton, Raster-...
half-tone distortion Graduationsverzerrung (621.397)
half-tone picture Halbtonbild, Rasterbild, Punktrasterbild
half-value layer Halbwertsdicke (621.386)
half-value thickness Halbwertsdicke (621.386)
half-watt lamp Halbwattlampe
half-wave Halbwellen-...
half-wave cylinder dipole antenna Halbwellenrohrdipolantenne
half-wave differential protection Halbwellendifferentialschutz
half-wave dipole Halbwellendipol
half-wave doublet Halbwellenantenne, Hertzscher Dipol
half-wave frame Halbwellenschleife (621.396)
half-wave-length aerial Halbwellenantenne, Halbwellendipol
half-waver radiator Halbwellendipol (621.396.67)
half-wave rectification Halbwellengleichrichtung, Einweggleichrichtung
half-wave rectifier Halbwellengleichrichter, Einweggleichrichter
half-wave transmission Halbwellenübertragung (621.315)
halide Halogensalz
Hall effect Hall-Effekt
Hall angle Hallwinkel
Hall constant Hallkonstante
Hall generator Hallgenerator
Hall mobility Hall-Beweglichkeit
Hall plate Hallplättchen
Hall probe Hallsonde
Hall residual voltage for zero control current Restspannung bei Steuerstrom Null (eines Halleffekt-Bauelementes)
halogen Halogen
to halve halbieren; anblatten (69)
halving Überblattung (69)
hammer Hammer, Klöppel; Hammerwerk; Wagnerscher Hammer (621.3)
hammer blow Hammerschlag, Schlag mit dem Hammer
hammer break Hammerunterbrecher, Wagnerscher Hammer
hammer-effect varnish Hammerschlaglack
hammer-finished gehämmert
hammer interrupter Hammerunterbrecher, Wagnerscher Hammer
hammer-shaped hammerförmig

Hamming Code Hamming Code [Code mit Fehlererkennung]
to hamper behindern, erschweren
hand Hand; Arbeitskraft; Handschrift, Unterschrift; Handbreite, Handbreite (als Längenmaß = 4 Zoll); Zeiger (681.11); Drehsinn (621.313); Seite, Richtung
hand-actuated handbetätigt
hand actuating Handbetätigung
hand adjustment Handeinstellung
hand of a clock Uhrzeiger
hand drill Handbohrmaschine
hand-driven handbetätigt, handbetrieben
hands-free telephone Freisprechtelephon
hand geared mit Handantrieb
hand generator Kurbelinduktor
handicraft Handfertigkeit, Handwerk
handie-talkie Handfunksprecher
handiness of operation bequeme Bedienung, handliche Bedienbarkeit
hand jack Handwinde
hand-keyed handgetastet
hand lamp Handlampe
hand lever Handhebel, Betätigungshebel
handmade handgefertigt, handgearbeitet
hand motion Handbetrieb, Handantrieb
hand nut Flügelmutter
hand-operated handbetätigt
hand-operated exchange Handamt, Handvermittlung (621.395)
hand-operated regulator Handregler
handoperation Handbetätigung, Handarbeit, Arbeitsgang von Hand
hand punch Handlocher
hand rail Handleiste
hand receiver Handapparat (621.395)
hand regulation Handregelung
hand reset ... mit Handrückstellung
hand restoring ... mit Handrückstellung
handset Handapparat
hand setting Handeinstellung
handshaking Quittungsbetrieb (621.394.5)
hand shears (pl.) Blechschere
hand tool Handwerkszeug
hand-type microphone Handmikrophon
hand vice Feilkloben, Handkloben, Klöbchen
handwheel Handrad
hand winding Handwicklung (621.313)
handle Griff, Handgriff, Stiel, Heft, Handhabe; Klinke (694)
handle bar Lenkstange (629)
handle type fuse Griffsicherung
to handle handhaben, behandeln, verarbeiten
to handle power Leistung verarbeiten
to handle a voltage eine Spannung verarbeiten
handler [Routine zur Steuerung eines peripheren Gerätes]
handling Handhabung, Behandlung
handling device Betätigungseinrichtung
handling instruction Behandlungsanweisung, Bearbeitungsanweisung
handling time Bearbeitungszeit
to hang hängen, aushängen, aufhängen
to hang on am Apparat bleiben, in der Leitung bleiben (621.395)
to hang together zusammenhängen
to hang up einhängen (621.395)
hanger Aufhänger, Aufhängeeinrichtung
hanging lengths (pl.) Lichtleiste (621.32)
hangover time Nachwirkzeit
hangup (Selbst-)Blockierung des Arbeitsablaufes (681.3)
Hankel function Hankelsche Funktion
hard hart, schwer
hard alloy harte Legierung, Hartmetall
hard blow harter Schlag, trockener Schlag
hardcopy Kopie, Ausdruck, Hardcopy
hard disk Festplatte
hard drawn hart gezogen
hard facing Auftragsschweißung
hard-and-fast rule strenge Regel, starre Vorschrift
hard fibre Vulkanfiber
hard-frozen festgefroren
hard-gas circuit breaker Hartgasschalter
hard-glass fibre Hartglasgewebe
hard-hearing schwerhörig, harthörig
hard keying Harttastung
hard lead Hartblei
hard paper Hartpapier
hard porcelaine Hartporzellan
hard rubber Hartgummi
hard-sectored hartsektoriert
hard solder Hartlot
hart-soldering Hartlöten, Hartlötung
hard-store enamel Einbrennlack
hard superconductor harter Supraleiter
hard-to-get-at schwer zugänglich
hard tube Hochvakuumröhre
hard-tube rectifier Hochvakuumgleichrichter
hard-type cathode ray tube Hochvakuum-Kathodenstrahlröhre
hard usage starke Beanspruchung
hard valve Hochvakuumröhre; Hochvakuum-..., mit Hochvakuumröhren
hard water hartes Wasser (54), (621.18)

hard wear starke Verschleißbeanspruchung, starker Verschleiß
hard-wearing verschleißfest
hard wiring festverlegte Installation, Drahtleitung, Drahtverbindung
hard-wired logic festverdrahtete logische Schaltung
hard-wood fibre slab Hartfaserplatte
hard X-rays (pl.) harte Röntgenstrahlung
to harden härten, hart machen; hart werden
hardenable härtbar
hardener härtender Bestandteil, Härter; Härtungsbad (77)
hardness Härte (54), (628), (66), (535), (621.386); Höhe des Vakuums (621.385)
hardness scale Härteskala
hardness test Härteprobe, Härteprüfung
to hard solder hart löten
hardware Eisenwaren, Eisenteile, Kleinteile (pl.); Hardware, Maschinenausstattung (681.3)
hardwood Hartholz
H-armature Doppel-T-Anker (621.3)
harmful interference starke Störungen
harmonic Harmonische, Oberton, Oberwelle
harmonic absorber Saugkreis
harmonic analyzer Oberwellen-Analysator
harmonic antenna Oberwellenantenne
harmonic component Oberwellenkomponente, Harmonische
harmonic detector Klirrfaktormeßbrücke
harmonic distortion Klirrdämpfung
harmonic-distortion coefficient Klirrfaktor
harmonic filter Oberwellenfilter, Oberwellensieb
harmonic generator Oberwellenerzeuger, Frequenzvervielfacher
harmonic interference Oberwellenstörung
harmonic-mode crystal Oberschwingungsquartz
harmonic power Oberwellenleistung, Verzerrungsleistung
harmonic stopper Oberwellensiebschaltung
harmonic suppressor Oberwellensieb
harmonic tone Oberton
harmonic harmonisch
harmonic analyser harmonischer Analysator
harmonic distortion nichtlineare Verzerrung, Klirrfaktor
harmonic motion harmonische Bewegung, Sinusschwingung
harmonic progression harmonische Reihe (51)
harmonic selective ringing abgestimmter Tonfrequenzruf
harmonic series Harmoniegesetz (534)
to harmonize harmonisieren; zueinander passen; zueinander passend machen
harness Kabelbaum
Hartley circuit Dreipunktschaltung mit induktiver Spannungsteilung
to hat verwürfeln
hatch Luke, Tür; (untere) Türhälfte; Einstieg, Ausstieg (629.13); Schraffierungslinie, Schraffurlinie
to hatch schraffieren, schummern
hatching Schraffur, Schraffierung
to haul fördern, transportieren, schleppen
haulage Förderung; Abholung; Transport; Güterbeförderung
haulage installation Förderanlage (622)
haulage mining locomotive Förderlokomotive, Grubenlokomotive
hauling Förderung (622)
hauling capacity Zugleistung (625.1)
hauling device Fördereinrichtung
hauling machine Fördermaschine
hauling rope Förderseil
hay-wire wild verlegt (621.3)
hazard Gefahr, Gefährdung, Gefahrenquelle
hazard lighting Hindernisbefeuerung, Flugwarnbefeuerung
hazardous risikobehaftet, gefährlich
haze Dunst (55)
hazy dunstig (55)
hazy pane beschlagene Scheibe
H-cable H-Kabel, Höchstädterkabel, gürtelloses Kabel
H. C. F. = highest common factor größter gemeinsamer Teiler
HDDR = high density digital recording HDDR-Verfahren
HDLC = high level data link control
head Kopf; Kopf, Kopfteil; Tabellenkopf; Druckhöhe, Wasserstand; Zug, Saugzug, Zugstärke (621.18); Überschrift
head band Kopfbügel, Bügel (621.395)
head-and-chest set Brustsprechsatz, Brustgeschirr
head circuit Kopfhörerkreis
head of department Abteilungsleiter
head drum Bandführungstrommel (681.84)
head end Empfangsstelle (621.396.67)
head-end system Zugbeleuchtung von einer Lok-Dynamo
head extremity Kopfende
head gap Magnetkopfspalt; Kopfabstand
head gear Fördergerüst, Förderturm (622)
head of a hammer Hammerkopf

headlamp Scheinwerfer (625), (629); Kopflampe (61); in die Flügel eingebaute Radarantenne
headless screw Madenschraube
headlight Scheinwerfer
head of the mast Mastkopf
head-on collision zentraler Stoß
headphone Kopfhörer
head pressure Bandandruck (681.84)
head race Oberwasserkanal, Oberwassereinlauf
head receiver Kopfhörer
head room lichte Höhe, Bauhöhe
headset Kopfhörer, Doppelkopfhörer, Kopfhörergarnitur
head stack Mehrspurkopf
head-switching circuit Kopfumschalter (681.84)
head-to-tape contact Bandkontakt (681.84)
head-tip projection Kopfüberstand (681.84)
head-up-display Blickfeldanzeige, Blickfelddarstellungsgerät (629.13)
head of water Wassersäule, Wasserstand, Druckhöhe
head wheel Kopfrad (681.84)

to head vorstehen; obenanstehen, drüberstehen; überschreiben, mit Kopf versehen; einen Kopf anstauchen

header Kopfstück, Bodenplatte; Vorsatz; Binder (69); Obertrommel (621.1)

heading Überschrift, Titel, Oberbegriff (655); Steuerkurs (351.814)
- **heading of a table** Tabellenkopf, Tabellenüberschrift

to hear hören; abhören, überhören

hearer Hörer

hearing Gehör; Hörweite, Rufweite; (mündliche) Verhandlung
- **hearing aid, hearing appliance** Schwerhörigenapparat, Hörgerät, Hörhilfe

heart Herz (61)
- **heart-shaped** herzförmig
- **heart-shape reception** Kardioiden-Diagramm

hearth Herd, Feuerstelle; Gestell (66)

heat Wärme; Hitze; Hitzeeinwirkung
- **heat absorbing** Wärme aufnehmend
- **heat absorption** Wärmeaufnahme
- **heat balance** Wärmebilanz
- **heat capacity** Wärmekapazität
- **heat coil** Heizspule, Heizwicklung; Lötsicherung (621.395)
- **heat conduction** Wärmeleitung
- **heat conductive** wärmeleitend
- **heat conductivity** Wärmeleitfähigkeit
- **heat conductor** Wärmeleiter
- **heat consumption** Wärmeverbrauch
- **heat content** Wärmeinhalt
- **heat controller** Wärmewächter
- **heat convection** Wärmekonvektion, Wärmetransport
- **heat cycle** Erwärmungszyklus
- **heat development** Wärmeentwicklung, Wärmeabgabe
- **heat disbursing** wärmeabgebend
- **heat dissipation** Wärmeableitung
- **heat-dissipating ability** Wärmeabgabevermögen
- **heat emissivity** spezifische Wärmeabgabe
- **heat energy** Wärmeenergie, thermische Energie
- **heat engine** Wärmekraftmaschine
- **heat equivalent** Wärmeäquivalent
  - **heat equivalent of fusion** Schmelzwärme
  - **heat equivalent of internal work** innere Wärme
  - **heat equivalent of vaporization** Verdampfungswärme
- **heat of evaporation** Verdampfungswärme
- **heat evolution** Wärmeentwicklung
- **heat exchanger** Wärmetauscher
- **heat expansion** Wärmedehnung
- **heat flash** Wärmeblitz
- **heat of formation** Bildungswärme
  - **heat of formation of a compound** Verbindungswärme
- **heat of fusion** Schmelzwärme
- **heat generation** Wärmeerzeugung
- **heat governor** Wärmeregler, Temperaturregler, Thermostat
- **heat insulating** wärmedämmend, wärmeisolierend
  - **heat-insulating jacket** Wärmeschutz, Isoliermantel
- **heat insulation** Wärmeisolation, Wärmedämmung, Wärmeisolierung
- **heat level** Temperatur
- **heat of the liquid** Flüssigkeitswärme
- **heat measurement** Wärmemessung
- **heat meter** Wärmezähler
- **heat pipe** Wärmeleitrohr
- **heat production** Wärmeerzeugung, Wärmeentwicklung
- **heat-proof** wärmefest, hitzebeständig; wärmesicher
- **heat protection** Wärmeschutz
- **heat pump** Wärmepumpe
- **heat quantity** Wärmemenge
- **heat-radiating** wärmestrahlend

heat radiation Wärmestrahlung
heat rays Wärmestrahlen; Wärmestrahlung (sing.)
heat of reaction Reaktionswärme
heat recovery Wärmerückgewinnung
heat release Wärmeentwicklung
heat resistance Wärmebeständigkeit
heat-resistant wärmebeständig
heat-resisting wärmebeständig, hitzefest
heat run Erwärmungsversuch
heat sink Wärmeleiter, Kühlkörper
heat of solution Lösungswärme
heat storage Wärmespeicherung
heat-storing wärmespeichernd
heat supply Wärmezufuhr
heat transfer Wärmeübertragung, Wärmeaustausch
heat transmission Wärmeübergang, Wärmeübertragung
heat treatment Wärmebehandlung, Vergütung, thermische Behandlung
heat value Heizwert
heat wave Hitzewelle, Wärmewelle
to heat heizen; erhitzen, erwärmen
to heat bond heißverleimen, warmverleimen
to heat treat wärmebehandeln, vergüten
to heat up anheizen, aufwärmen, aufheizen
heater Heizkörper, Heizelement, Heizapparat; Heizkörper, Heizfaden (621.385)
heater current Heizstrom (621.385)
heater element Heizelement, Heizkörper
heater-type cathode indirekt geheizte Kathode
heater-type tube indirekt geheizte Röhre
heater voltage Heizspannung (621.385)
heater winding Heizwicklung (am Transformator) (621.396)
heating Heizung, Aufheizung, Erwärmung; Raumheizung
heating coefficient Wärmeabgabeziffer (621.313)
heating coil Heizwicklung, Heizspirale (621.3); Heizschlange (66)
heating compartment Heizfach, Heizraum
heating conductor Heizleiter (621.3)
heating crack Warmriß, Erwärmungsriß
heating current Heizstrom
heating cycle Erwärmungszyklus; Heizzyklus
heating effect Wärmewirkung, Heizwirkung
heating element Heizkörper, Heizelement
heating limit thermische Grenzleistung
heating platen Heizelement, Heizstreifen
heating power Heizleistung
heating resistor Heizwiderstand
heating surface Heizfläche
heating test Erwärmungsversuch
heating time Anheizzeit, Aufheizzeit; Anwärmzeit (621.385)
heating unit Heizelement (621.365)
heating-up period Aufheizzeit
heating value Heizwert
heating voltage Heizspannung
heating wire Heizdraht (621.365)
heavily . . . stark-. . ., hoch-. . .
heavily loaded hochbelastet, hochbeansprucht
heavily stocked mit reichem Lagerbestand
heavily stressed hochbeansprucht
Heaviside effect Stromverdrängung, Hautwirkung, Skineffekt
Heaviside expansion theorem Heavisidesche Regel
Heaviside layer Heaviside-Schicht
Heaviside operational calculus Heaviside Operatoren-Rechnung
Heaviside unit function Sprungfunktion, Heaviside-Einheitsspannungsstoß
Heaviside units (pl.) Heavisidesches Maßsystem
heavy schwer, massebehaftet, massebelastet
heavy circuit hochbelasteter Stromkreis
heavy current starker Strom; Starkstrom, Starkstrom-. . .
heavy current circuit Starkstromkreis, Starkstromnetz
heavy-current component Bauelement der Starkstromtechnik
heavy current engineering Starkstromtechnik
heavy current installation Starkstromanlage
heavy current switch Hochstromschalter
heavy current system Starkstromnetz
heavy duty hochbelastet, hochbelastbar, Schwerlast-. . .
heavy duty relay Starkstromrelais
heavy-gauge dick
heavy-gauge sheet dickes Blech
heavy-gauge wire dicker Draht
heavy-grade naphta Schweröl
heavy hydrogen schwerer Wasserstoff
heavy line dicker Strich
heavy load schwere Last, hohe Belastung
heavy loading conditions (pl.) schwere Belastung, hohe Beanspruchung
heavy mass schwere Masse
heavy nucleus schwerer Kern
heavy oil Schweröl
heavy oil engine Schwerölmotor
heavy spar Schwerspat
heavy water schweres Wasser
heavy-water reactor Schwerwassereaktor

heavy wear starke Abnutzung
heavy wire dicker Draht
hecto ... Hekto-..., 100
hectog. = hectogram(me) Hektogramm, 100 g
hectograph Hektographenapparat
hectol. = hectolitre hl = Hektoliter, 100 Liter
hectom. = hectometre Hektometer, 100 Meter
heel Hacken, Absatz (685)
to heel überholen, krängen (629.12); behacken, mit Absätzen versehen (685)
heeling error Rollfehler, Mißweisung durch Rollen (527)
Hefner candle Hefner-Kerze
Hefner lamp Hefnerlampe
Hefner unit Hefner-Kerze
height Höhe
height of fall Fallhöhe
height of flight Flughöhe
height gauge Höhenreißer (621.9)
height indicator Höhenmesser
height-position indicator Entfernungs-Höhensichtgerät
height of suction Saughöhe
Heising modulation Heising-Modulation
Helberger furnace Tiegelofen
helical schraubenförmig, gewendelt
helical antenna Wendelantenne
helical curve Schraubenlinie
helical gear spiralverzahntes Rad; Schrägverzahnung
helical gearing Spiralverzahnung, Schraubverzahnung
helical line Schraubenlinie
helical recording Schrägspurverfahren (681.84)
helical spring Wendelfeder, Schraubenfeder
helical transmission line Wendelleitung
helicoid schraubenförmig, wendelförmig
heliograph Heliograph, Sonnentelegraph; Lichttelegraph
heliometer Sextant
helios Leuchtdichte (621.32)
helioscope Helioskop
helium Helium
helium-cooled reactor heliumgekühlter Reaktor
helium-neon gas Maser Helium-Neon-Gas-Maser
helix (pl. **-ices**) Wendel, Schraubenlinie
helix angle Steigungswinkel der Wendel
helix antenne Wendelantenne
helix of wire Drahtwendel
Hellesen cell Trockenelement
Hell-printer Siemens-Hell-Schreiber
Helmholtz coil arrangement Goniometer
Helmholtz coil phase shifting circuit Drehfeld-Phasenschieber
helve Heft, Griff
hematite Hämatit, Roteisenstein
hemispherical halbkugelig, halbkugelförmig
hemitropic field (Feldsystem, bei dem nur jeder zweite Pol eine Erregerwicklung hat) (621.313)
hemp Hanf
hemp-fixing clamp (for insulators) Isolatorenaufhanfkluppe
hemp packing Hanfpackung
hemp rope Hanfseil
hempen core Hanfseele
henry (h) Henry = H
henrymeter Induktivitätsmesser
hepatic gas Schwefelwasserstoff
heptode Heptode, Siebenpolröhre
Heraeus lamp Hanauer Quarzlampe, Höhensonne
hermetic hermetisch
hermitic seal dichter Verschluß
hermitically sealed dicht abgeschlossen, hermetisch abgeschlossen
Heroult furnace Heroult-Ofen
Hertz aerial Hertzsche Antenne, Hertzscher Dipol
Hertz oscillator Hertzscher Dipol
Hertzian Hertz-..., Hertzsche
Hertzian telegraphy drahtlose Telegraphie, Funkentelegraphie
Hertzian waves Hertzsche Wellen, elektromagnetische Wellen
heterodyne Überlagerungs-...
heterodyne bridge Schwebungsmeßbrücke
heterodyne code reception Überlagerungs-Telegraphieempfang
heterodyne filter 9-kHz-Sperre
heterodyne frequency Überlagerungsfrequenz
heterodyne frequency meter Überlagerungsfrequenzmesser
heterodyne interference Überlagerungspfeifen
heterodyne method Überlagerungsverfahren
heterodyne oscillator Überlagerungssender, Oszillator
heterodyne reception Überlagerungsempfang
heterodyne voltmeter selektiver Spannungsmesser
heterodyne wavemeter Überlagerungswellenmesser
heterodyne whistle Überlagerungspeifen
to heterodyne überlagern
to heterodyne between überlagern mit
heterogeneous heterogen
heteropolar Wechselpol-

heteropolar generator Wechselpol-Generator
heterostatic method Nadelschaltung (621.31)
Heurtley magnifier Bolometerkopf-Verstärker
Heusler alloy Heuslersche Legierung
Hewlett insulator Hewlett-Isolator
hexadecimal hexadezimal
hexagon Sechseck; Sechskantprofil
hexagon bar Sechskantstange
hexagon connection Sechseckschaltung (621.3)
hexagon head Sechskantkopf
hexagon nut Sechskantmutter
hexagon screw Sechskantschraube
hexagon voltage Sechseckspannung
hexagonal sechseckig
hexagonal socket wrench Sechskant-Steckschlüssel
hexane Hexan
hexaphase sechsphasig, Sechsphasen-...
hexavalent sechswertig (54)
hexode Hexode, Sechspolröhre
hexode mixer Mischhexode, Sechspolmischröhre
hexode mixer tube Mischhexode, Sechspolmischröhre
h. f. = high frequency HF = Hochfrequenz (s. dort)
hg. = hectogram(me) Hektrogramm = 100 Gramm
Hg-delay line Quecksilber-Verzögerungsleitung
hhd. = hogshead (s. dort)
hiatus Lücke, Kluft, Spalt, Unterbrechung
hicap Kondensator mit hoher D. K.
hickey Lampenträger
hickory Hickory, Hickoryholz
Hicks hydrometer Perlensäureprüfer (621.355)
hidden conductor unter Putz verlegter Leiter
hierarchical network hierarchisches Netz
high hoch
high-activity waste hochaktiver Abfall
high band frequenzhöheres Band, Oberband
high barometer hoher Barometerstand, hoher Luftdruck
high-bay reflector Tiefstrahler
high-breaking-capacity fuse = HBC fuse Hochleistungssicherung
high-capacity Hochleistungs-...
high-capacity electrical motor Großelektromotor
high-capacity engine Großmotor
high-capacity machine Hochleistungsmaschine
high-carbon steel Stahl mit hohem Kohlenstoffgehalt
high-class fit Edelsitz, Edelpassung
high definition große Auflösung (621.397)
high-definition television Fernsehen mit hoher Zeilenzahl
high-density paper dichtes Papier (621.315.2)
high-duty Hochleistungs-..., Qualitäts-...
high-duty steel Hochleistungsstahl
high-duty tool Hochleistungswerkzeug
high efficiency hoher Wirkungsgrad
high-efficiency engine Hochleistungsmaschine, Hochleistungsmotor; Qualitätsmaschine
high energy reaction hochenergetischer Prozeß
high fidelity gute Tonqualität, hochwertige Tonwiedergabe, Hi-Fi, Heimstudio...
high-fidelity set Gerät mit guter Tonwiedergabe, Hi-Fi-Gerät
high-flux reactor Hochflußreaktor, Reaktor mit hohem Neutronenfluß
high frequency Hochfrequenz; Kurzwellenbereich (siehe auch **radio**)
high-frequency hochfrequent
high-frequency alternating current hochfrequenter Wechselstrom
high-frequency alternator Hochfrequenzgenerator, Hochfrequenzmaschine
high-frequency amplification Hochfrequenzverstärkung
high-frequency carrier telegraphy HF-Trägertelegraphie
high-frequency engineering Hochfrequenztechnik
high-frequency heating Hochfrequenzheizung
high-frequency (induction) furnace Hochfrequenz-Induktionsofen
high-frequency resistance Hochfrequenzwiderstand
high-frequency response Hochfrequenzverhalten, Verhalten gegenüber Hochfrequenz
high-frequency speaker Hochtonlautsprecher
high-frequency transformer Hochfrequenztransformator, Hochfrequenzübertrager
high-frequency treatment HF-Behandlung, Diathermie
high-frequency voltage Hochfrequenzspannung
high gain hohe Verstärkung
high-gain amplifier hochverstärkender Verstärker
high gear schneller Gang

high-grade hochwertig, Qualitäts-..., Hochleistungs-...
high-grade alloy hochlegiert
high-grade fidelity gute Klangtreue, hohe Wiedergabegüte
high-grade nuclear fuel hochgradiger Spaltstoff, hochangereicherter Spaltstoff
high heat value oberer Heizwert
high impact material stoßfestes Material, schlagfestes Material
high impedance hoher Scheinwiderstand
high-impedance mit hohem Scheinwiderstand, hochohmig
high-intensity runway light Startbahnhochleistungsfeuer
high-intensity threshold light Schwellenhochleistungsfeuer
high latitude hohe Breite (91)
high-leakage current transformer Streuwandler
high level hohes Niveau
high-level language höhere Programmiersprache
high-level modulation Endstufenmodulation
high-level pile Hochleistungsbatterie, Hochleistungsbrenner (539)
high-level selection Hochpegelwahl (621.395)
high-level tank Hochbehälter
high light Licht, leuchtender Punkt (621.397.7)
high lights Glanzstellen (621.357)
high limit obere Grenze
high mica vorstehender Glimmer (621.313)
high moisture content hoher Feuchtigkeitsgehalt
high-molecular-weight substance hochmolekularer Stoff
high mortality part Verschleißteil
high nickel content hoher Ni-Gehalt
high nickel content material hochnickelhaltiges Material
high-noise immunity logic HNIL störsichere Logikschaltung
high-note response Hochtonwiedergabe
high-numbered mit hoher Nummer
high oil pressure cable Druckkabel
highpass filter Hochpaß
high percentage hoher Prozentsatz
high-percentage hochprozentig
high-permeability hohe Permeabilität
high-permeability material hochpermeables Material
high pitch hoher Ton
high-pitched hoch (53)
high-pitched note hoher Ton
high polish glänzende Politur
high-polish hochglanzpoliert
high power hohe Leistung, große Leistung
high-power Hochleistungs-..., Leistungs...
high-power choke modulator Antennenmodulationsdrossel
high-power engine Hochleistungsmotor
high-power loudspeaker Hochleistungslautsprecher
high-power modulation Endstufenmodulation
high-power motor Hochleistungsmotor (621.3)
high-power plant Großkraftwerk
high-power reactor Hochleistungsreaktor
high-power signal generator Leistungsmeßsender
high-power station Großkraftwerk
high-power transistor Leistungstransistor
high-power transmitter Großsender
high-power tube Endröhre, Leistungsverstärkerröhre
high-power microscope hochauflösendes Mikroskop
high pressure hoher Druck, Hochdruck (55)
high-pressure compressor Hochdruckkompressor
high-pressure mercury vapour lamp Quecksilberhochdrucklampe
high-pressure xenon lamp Xenon-Hochdrucklampe
high rate of rise steiler Anstieg
high range oberer Bereich
high resistance hoher Widerstand
high resistive hochohmig
high resistive material Widerstandsmaterial
high resolution hohes Auflösungsvermögen, hohe Auflösung (535)
high-resolution microscope hochauflösendes Mikroskop
high-resolution plate Feinkornplatte (77)
high-rupturing-capacity fuse = HRC fuse Hochleistungssicherung
high-side voltage Oberspannung (621.314.2)
high slope große Steilheit (621.385)
high-slope valve steile Röhre, Röhre großer Steilheit (621.385)
high speed hohe Geschwindigkeit; hohe Drehzahl
high-speed circuit-breaker Schnellschalter
high-speed drive Schnellantrieb
high-speed excitation system Schnellerregung, Stoßerregung
high-speed film hochempfindlicher Film

high-speed motor Schnelläufermotor
high-speed passenger service Personenschnellverkehr
high-speed printer Schnelldrucker
high-speed recorder Schnellschreiber
high-speed steel Schnellstahl (621.9)
high-speed storage Schnellspeicher
high-speed telegraph Schnelltelegraph
high-speed telegraph transmission system Schnelltelegraphie-System
high-speed whine Hochheulen; hoher Heulton
high spot vorstehender Punkt
high sulfur hoher Schwefelgehalt
high temperature hohe Temperatur
high tension Hochspannung; Oberspannung (siehe auch **high voltage**)
high-tension battery Hochspannungsbatterie, Anodenbatterie
high-tension busbars Hochspannungssammelschienen
high-tension bushing Hochspannungsdurchführung
high-tension cable Hochspannungskabel
high-tension connection Oberspannungsanschluß
high-tension d. c. railway Hochspannungsgleichstrombahn
high-tension detonator Hochspannungszünder, Spaltzünder (66)
high-tension fuse Hochspannungssicherung
high-tension ignition Abreißzündung (621.43)
high-tension insulator Hochspannungsisolator
high-tension line Hochspannungsleitung
high-tension magneto Hochspannungszündmaschine (621)
high-tension measurement Hochspannungsmessung
high-tension motor Hochspannungsmotor
high-tension overhead line Hochspannungsfreileitung
high-tension power Hochspannungsleistung
high-tension power plant Hochspannungsanlage
high-tension regulating transformer Regeltransformator für Hochspannungssteuerung
high-tension side Oberspannungsseite, HV-Seite
high-tension supply Hochspannungsanschluß
high-tension switch Hochspannungsschalter
high-tension switchgear Hochspannungsschalteinrichtung
high-tension terminal Oberspannungsklemme, HV-Anschluß
high-tension transformer Hochspannungstransformator
high-tension transmission Hochspannungskraftübertragung
high-tension transmission line Hochspannungsleitung
high-tension winding Hochspannungswicklung, Oberspannungswicklung
high-usage route Direktweg (621.395)
high vacuum Hochvakuum
high-vacuum pump Hochvakuumpumpe
high-vacuum thermionic rectifier Hochvakuumgleichrichter
high-vacuum tube Hochvakuumröhre
high voltage Hochspannung, hohe Spannung, HV = Hochvolt (siehe auch **high tension**)
high-voltage bus bar Hochvoltsammelschiene
high-voltage bushing Hochvoltdurchführung
high-voltage cable Hochspannungskabel
high-voltage line Hochspannungsleitung
high-voltage substation Hochvoltstation
high-voltage technique Hochspannungstechnik
high-voltage test Hochspannungsprüfung, Spannungsprobe
high-voltage winding Hochspannungswicklung, Hochvoltwicklung, HV-Wicklung
high water Hochwasser
high wattage hohe Leistung
high-wattage lamp Hochleistungslampe, Großlampe
higher harmonics höhere Harmonische
higher nickel steel mehr Nickel enthaltender Stahl
highest reading höchste Ablesung, höchste Einstellung
highly inflammable leicht entzündlich
highly ionized stark ionisiert
highway Verkehrsstraße, Hauptstraße (625); Multiplexleitung (621.39)
highway crossing bell Lokomotivläutewerk
highway crossing signal Schienenkreuzungssignal
hinder part rückwärtiger Teil
hinder wall Hinterwand
hindrance Behinderung

hinge Scharnier, Gelenk, Angel; Drehachse, Drehpunkt
hinge clip Drehpunktkontakt, Drehkontaktwange (621.316.55)
hinge line Angel; Drehachse
hinge point Drehpunkt, Aufhängepunkt
to hinge mit Scharnier anlenken; drehbar aufhängen
hinged armature Klappanker
hinged door Klapptür
hinged flap Anhängefahne, Ausklappbogen; Scharnierklappe
hinged frame Schwenkrahmen
hinged lid Klappdeckel
hinged louvre Klappjalousie
hipernik s. **hypernik**
H-iron Doppel T-Eisen
hiss Zischen, Zischton; Rauschen (621.396)
to hiss zischen
hit Schlag; Treffer (623.14)
to hit treffen, treffen auf, auftreffen auf, aufschlagen auf, aufstoßen auf, schlagen, schlagen auf
to hit against anstoßen an, anstreifen an, anschlagen an
Hittorf dark space Kathodendunkelraum
Hittorf tube Hittorfsche Röhre
hi = hectolitre hl = Hektoliter = 100 Liter
hm = hectometre hm = Hektometer = 100 Meter
H-network Vierpol in H-Schaltung
to hoard horten, Vorräte sammeln, speichern
to hoard energy Energie speichern
hoarfrost Rauhreif
hoarse heiser, rauh
Hoechstaedter cable Höchstädter Kabel, H-Kabel
hogshead (Faß von etwa 200 l)
hoist (winch) Winde, Aufzug, Hebezeug, Elevator
to hoist winden, aufwinden, hochziehen, fördern, hochziehen
hoisting Hochziehen, Förderung
hoisting cable Zugseil, Aufzugseil
hoisting cage Förderkorb
hoisting drum Seiltrommel
hoisting engine Fördermaschine
hoisting pulley Förderscheibe
hoisting rope Förderseil, Aufzugseil
hoisting shaft Treibschacht
hoisting winch Seilwinde, Haspel
hoistway Fahrschacht, Fahrstuhlschacht
hoistway-door electric contact Fahrschachttürkontakt
hoistway-door interlock Fahrschachttürverriegelung
hold Halt; Laderaum (629.12)
hold control Regelung der Kippfrequenz
hold excitation Halteerregung
hold lamp Besetztlampe
hold-on contact Haltekontakt
hold-on tone Warteton (621.395)
hold range Haltebereich
hold up gebundene Menge, Einsatz, Füllung
to hold halten, festhalten; fassen, halten; gelten, gültig sein; stimmen; enthalten, aufnehmen, fassen
to hold constant konstant halten
to hold course Kurs halten, Kurs einhalten
to hold down festhalten, spannen
hold-down bolt Spannbolzen, Spannschraube
hold-down screw Halteschraube, Spannschraube
to hold fast festbremsen (621.313)
to hold frequency Frequenz fahren, Frequenz halten (621.311)
to hold (good) gelten, gültig sein, stimmen
to hold the image das Bild (durch Nachleuchten) festhalten (621.397)
to hold load Last fahren, Grundlast fahren, Strich fahren (621.311)
to hold true with gelten für
to hold up hochhalten, tragen, Tragkraft haben
to hold water wasserdicht sein, dicht sein; hieb- und stichfest sein
holder Halter, Halterung, Fassung
holdfast Klammer, Mauerklammer
holding Halte-...
holding appliance Haltevorrichtung, Halteeinrichtung
holding coil Haltespule, Haltewicklung
holding contact Haltekontakt
holding current Haltestrom
holding-current principle Ruhestromprinzip
holding device Haltevorrichtung, Halteeinrichtung
holding-down bolt Spannschraube, Halteschraube
holding-down plate Andrückplatte
holding-down screw Andrückschraube
holding line Halteleitung
holding magnet Haltemagnet
holding pattern Warteschleife (656.7)
holding point Wartepunkt (656.7)
holding ratio Halteverhältnis
holding relay Halterelais
holding time Belegungszeit (621.396)

holding voltage Haltespannung
holding wire Prüfdraht (621.315.2)
hole Loch, Öffnung; Defekt-Elektron, Elektronenmangelstelle (621.382)
hole basis system Einheitsbohrungsystem
hole conduction Defektleitung, P-leitung, Löcherleitung, Elektronenmangelleitung
hole connection Durchgangsverbindung
hole current Löcherstrom
hole mobility Löcherbeweglichkeit (621.382)
hole-and-slot anode Zylinder-Spalt-Magnetron
hole storage Trägerstaueffekt (621.382)
hollow hohl
hollow brick Hohlziegel, Hohlstein
hollow circular rohrförmig
hollow conductor Hohlrohrleitung, Hohlrohr (621.396); Hohlseil (621.315)
hollow conduit Hohlrohrleitung (621.396)
hollow cone Hohlkegel
hollow die Hohlform
hollow drill Hohlbohrer
hollow fibre Hohlfaser
hollow girder Kastenträger
hollow member Hohlkörper
hollow pipe Hohlrohr, Hohlrohrleitung (621.396)
hollow-pipe wave Hohlrohrwelle
hollow screw Inbusschraube
hollow-shaft motor drive Hohlwellenantrieb
hollow-space resonator Hohlraumresonator
hollow sphere Hohlkugel
hollow waveguide Hohlwellenleiter, Hohlraumleitung (621.396)
hologram Hologramm
home Heim, Wohnung
home computer Heim-Computer
home flight Zielflug
home movie Heimkino
Home Office fuse Sicherung nach den Vorschriften des Home Office
Home Office switch Schalter nach den Vorschriften des Home Office
home position Ausgangsstellung
home signal Einfahrtsignal (625.1)
home study Heimstudium
home study tuition Fernunterricht
to home zielfliegen, zielsuchen; nach Hause fliegen
homing beacon Zielflugfunkfeuer, Zielflugbake
homing device Zielfluggerät, Zielflugeinrichtung
homodyne reception Empfangsanlage mit nachträglichem Trägerzusatz
homogeneity Homogenität
homogeneity of X-rays Homogenität einer Röntgenstrahlung
homogeneous homogen
homogeneous equation homogene Gleichung
homogeneous glass gewöhnliches Glas (621.32)
homogeneous grading homogene Mischung
homogeneous reactor homogener Reaktor
homologous homolog
homologue Homologon, Homolog
homology Homologie
homopolar Gleichpol-..., gleichpolig
homopolar alternator Gleichpolgenerator, Gleichpolmaschine (621.313)
homopolar generator Gleichpolmaschine (621.313)
to hone honen, feinschleifen, abziehen
honeycomb coil Wabenspule (621.396)
to honour regulations Vorschriften beachten
hood Aufsatz, Abdeckung, Haube
to hood mit einer Haube abdecken, mit einem Aufsatz versehen
hook Haken, Kranhaken
hook flash Gabelschlag
hook-on transformer Zangenwandler, Anlegewandler
hook operation Schaltstangenbetätigung
hook spanner Hakenschlüssel
hook stick Schaltstange
hook-stick disconnecting switch Trennschalter mit Schaltstangenbetätigung (621.395), (621.316.5)
hook (-type) switch Hakenumschalter (621.395)
hook-up wire Schaltdraht
to hook haken, einhaken, anhaken
to hook in einhaken, einhängen
to hook out aushaken, aushängen
Hooke's law Hookesches Gesetz
hoop Reifen
hoop-drop relay Fallbügelrelais
hoot Tuten, Hupen
to hoot tuten, hupen
hop Funkfeld, Sprung
hopper Schütt-Trichter, Schütte, Trichter; Kartenmagazin (681.3)
hopper (bottom) car Trichterwagen, Kübelwagen
horizon Horizont
horizon limitation Begrenzung durch optische Sicht, Begrenzung durch die Erdkrümmung
horizontal horizontal, waagerecht
horizontal break switch Schalter mit waagerechter Schaltstrecke

horizontal component horizontale Komponente
horizontal control Längsregler, Längsregelung (621.316.7)
horizontal deflection Horizontalablenkung
horizontal directivity horizontale Richtwirkung, horizontale Bündelung
horizontal engine liegende Maschine
horizontal intensity Horizontalintensität (538)
horizontal line Waagerechte
horizontal pattern Horizontal-Strahlungsdiagramm
horizontal plane Horizontalebene, waagerechte Ebene
horizontal polarized horizontal polarisiert
horizontal rule Querlinie
horizontal section Horizontalabschnitt
horizontal synchronizing Horizontalsynchronisation, Zeilensynchronisation
horizontal thrust Horizontaldruck, Schub
horn Horn, Hornsubstanz; Horn, Hupe (629.13); Horn (681); Horntrichter, Hornstrahler (621.396.67)
horn antenna Hornstrahler
horn arrester Hörnerableiter
horn-break fuse Hörnersicherung
horn-break switch Hörnerschalter
horn gap Hörnerfunkenstrecke
horn-gap switch Hörnerschalter
horn lightning arrester Hörnerblitzableiter
horn-like hornähnlich
horn loudspeaker Trichterlautsprecher
horn-parabolic antenna Hornparabolantenne
horn radiator Hornstrahler (621.396.67)
horn speaker Trichterlautsprecher
horse-power Pferdestärke (als Leistungseinheit = 550 ft. lbs./sec. = 1,014 PS)
horse-power meter Leistungsmesser, Leistungswaage
horse-shoe Hufeisen
horseshoe filament Haarnadel-Glühfaden
horseshoe magnet Hufeisenmagnet
horseshoe-shaped hufeisenförmig
hose Schlauch
hose connection Schlauchverbindung
hose-proof machine spritzwassergeschützte Maschine
hose reel Schlauchtrommel
hose tap Schlauchanschluß, Schlauchanschlußstelle
hospital Krankenhaus
hospital bus-bar Reserve-Sammelschiene, Hilfssammelschiene
hospital switch Notstromumschalter
hot heiß; spannungsführend
hot arc back Wärmerückzündung
hot atom hochangeregtes Atom
hot-cathode Glühkathode
hot-cathode gas diode ungesteuerte Gleichrichterröhre, Gasdiode (mit Glühkathode)
hot-cathode mercury-vapour rectifier Glühkathoden-Quecksilberdampfgleichrichter
hot-cathode tube Glühkathodenröhre
hot-cathode X-ray tube Glühkathoden-Röntgenröhre
hot-dip galvanizing Tauchverzinkung
hot dipping Tauchverzinnung, Feuerverzinnung
hot galvanizing Feuerverzinkung
hot line Direktverbindung, Heißer Draht
hot pad Heizkissen
hot plate Kochplatte
hot pressing Warmpressen
hot reserve einschaltbereite Reserve
hot-rolled warmgewalzt
hot rolling Warmwalzen
hot short warmbrüchig, rotbrüchig, warmrissig
hot shortness Warmbrüchigkeit, Rotbrüchigkeit; Warmbruch, Rotbruch
hot-spot temperature Temperatur am heißesten Punkt
hot-strip instrument Hitzbandinstrument
hot tinning Tauchverzinnung
hot waste radioaktiver Abfall
hot wire Hitzdraht (621.317)
hot-wire altitude meter Hitzdrahthöhenmesser (629.13)
hot-wire ammeter Hitzdrahtamperemeter
hot-wire anemometer Hitzdraht-Anemometer
hot-wire cautery Hitzdrahtausbrennung, Elektrokauterisation, Glühkaustik (61)
hot-wire expansion instrument Hitzdrahtinstrument (621.317)
hot-wire magnifier Bolometerkopfverstärker
hot-wire oscillograph Hitzdrahtoszillograph
hot-wire telephone Hitzdrahttelefon
hot-wire voltmeter Hitzdrahtvoltmeter, Hitzdrahtspannungsmesser
hot-wire wattmeter Hitzdrahtwattmeter, Hitzdrahtleistungsmesser
hour Stunde
hour of service Betriebsstunde, Brennstunde (621.385)
hours counter Betriebsstundenzähler

hours hand Stundenzeiger, kleiner Zeiger
hours meter Zeitzähler
hourly stündlich
hourly rating Stundenleistung
house Haus
house-connection line Hausanschlußleitung, Hausanschluß
housedoor intercom Türsprechanlage
house-keeping data Eigenbetriebsdaten, Anlagen- und Systemwerte
house service Hausanschlußleitung, Hausanschluß
house-service meter Haushaltszähler (621.317)
house wiring hausinstallation (621.3)
to house einbauen (in ein Gehäuse); enthalten, aufnehmen
household Haushalt
household appliances Haushaltsgeräte
household demand Haushaltsbedarf
household facilities Haushaltsgeräte
housing Wohnung, Unterbringung; Gehäuse; Ständer (621.9)
Howard protective system Gestellschlußschutz
howl Heulen; Pfeifen (621.396)
howl back Rückheulen
to howl heulen; pfeifen (621.396)
h. p. = horse power (s. dort)
H. P. = high pressure (HD = Hochdruck-...)
"H" pole Doppelmast
h. t. = high tension (s. dort)
H T L = high-threshold-logic
hub Nabe
hub shell Hohlnabe
hue Färbung, Farbton
Hughes printing telegraph Hughes-Schreiber, Hughes-Apparat
hull Rumpf (629.12/13)
hull gauge Wandstärke, Kesselwandstärke
hum Brummen; Brumm, Netzbrumm, Stromversorgungsgeräusch (621.396)
hum-bucking coil Entbrummerspule, Brummentstörungsspule
hum level Brummpegel
hum pick-up Brummeinstreuung
hum voltage Brummspannung
to hum brummen, summen
human menschlich
the human element der menschliche Einfluß, Bedienungseinfluß
human engineering Ergonomie
human operation Handbedienung, Handbetrieb
humid feucht
humid atmosphere feuchte Luft
to humidify anfeuchten
humidity Feuchtigkeit, Luftfeuchtigkeit
hummer Summer, Schnarre
humming Brumm
hump Buckel; Ablaufberg (656.2)
hump-engine cab-signal system Rangiersignalanlage im Führerstand (65)
hundreds selector Hunderterwähler
hundredweight Zentner (1 hundredweight = 112 lbs. = 50,8 kg)
to hunt pendeln (621.311), (621.316); aufprüfen, freie Leitung suchen (621.395); suchen, nachlaufen (621.396)
to hunt to steady state einpendeln
hunting Pendeln (621.313.3); Anrufsuchen, Freiwahl (621.395); Nachlaufen
hunting tooth überzähliger Zahn, Zusatzzahn (621.83)
Hurwitz relation Hurwitz-Bedingung
hutch Gehäuse, Umhüllung
h. v. = high voltage (s. dort)
H.-wave H-Welle, TE-Welle (621.396)
hybrid Hybrid-..., doppelartig; Gabelschaltung (621.395)
hybrid balance Gabelabgleich
hybrid circuit Hybrid-Schaltung
hybrid coil Gabelspule, Gabeldrossel (621.395); Brückenübertrager (621.395)
hybrid computer Hybrid-Rechner, Analog-Digital-Rechner
hybrid coupler Drei-dB-Koppler
hybrid integrated circuit hybride integrierte Schaltung
hybrid junction Verzweigung
hybrid repeater Gabelverstärker (621.395)
hybrid set Gabel, Gabelschaltung (621.395)
hybrid termination (circuit) Gabelschaltung
hybrid transformer Gabelübertrager
hydraulic hydraulisch
hydraulic pressure Wasserdruck, Öldruck, Flüssigkeitsdruck
hydraulic pump Druckpumpe
hydraulics Hydraulik
hydro-... Hydro-..., Wasser-...
hydrocarbon Kohlenwasserstoff
hydrocarbon compound Kohlenwasserstoffverbindung
hydrochloric acid Salzsäure
hydrocyanic acid Cyanwasserstoff, Blausäure
hydrodynamic hydrodynamisch
hydrodynamics (pl.) Hydrodynamik
hydro-electric hydroelektrisch, Wasserkraft-...

hydro-electric generating set Wasserkraftgenerator
hydro-electric generating station Wasserkraftwerk
hydro-electric machine Wasserkraftmaschine
hydro-electric plant Wasserkraftwerk
hydro-electric power station Wasserkraftwerk
hydrofluoric acid Flußsäure
hydrogel (wäßriges) Gel
hydrogen Wasserstoff
hydrogen bond Wasserstoffbrücke (54)
hydrogen chloride Salzsäure
hydrogen cooling Wasserstoffkühlung (621.313)
hydrogen electrode Wasserstoffelektrode (621.35)
hydrogen embrittlement Wasserstoff-Versprödung
hydrogen gas thermometer Wasserstoffthermometer
hydrogen ion concentration Wasserstoffionenkonzentration, pH-Wert
hydrogen-like atom wasserstoffähnlichesAtom
hydrogen peroxyde Wasserstoffsuperoxid
hydrogen sulphide Schwefelwasserstoff
hydrogenation Hydrierung
hydrogenous wasserstoffhaltig
hydrometallurgy Lösungsmetallurgie
hydrometer Areometer
hydrophone Hydrophon, Unterwasserschallempfänger
hydroplane Gleitboot; Tiefensteuerfläche, Tiefenruder (623)
hydrostatic hydrostatisch
hydrostatic pressure Wasserdruck, hydrostatischer Druck
hydrous wäßrig
hydrous solution wäßrige Lösung
... hydroxide ... lauge, ... hydroxid
hydroxyl Hydroxyl
hydroxyl group Hydroxyl-Gruppe, OH-Gruppe
hygrometer Hygrometer, Feuchtemesser
hygroscope Hygroskop, Feuchteanzeiger
hygroscopic hygroskopisch, wasseranziehend
hygroscopicity Hygroskopie, Wasseranziehungskraft, Wasseraufnahme
hyperbola Hyperbel
hyperbolic hyperbolisch
hyperbolic cosine Hyperbelcosinus
hyperbolic cotangent Hyperbelcotangens
hyperbolic function Hyperbelfunktion
hyperbolic horn Hyperbelhornstrahler
hyperbolic logarithm natürlicherLogarithmus
hyperbolic navigation Hyperbel-Navigation
hyperbolic pattern Hyperbelfeld (621.396.93)
hyperbolic sine Hyperbelsinus
hyperbolic tangent Hyperbeltangens
hyperboloid Hyperboloid
hyper frequency Höchstfrequenz
hypernik Hypernik [Eisen-Nickel-Legierung]
hypo = hyposulphite of sodium Fixiersalz
hypotenuse Hypotenuse
hypothesis (pl. **-ses**) Hypothese
hypsogram Pegeldiagramm
hysteresis Hysteresis, Hysterese
hysteresis coefficient Hysteresiskoeffizient (der Steinmetzformel)
hysteresis curve Hysteresiskurve, Hysteresisschleife
hysteresis distortion Hystereseverzerrung
hysteresis heat Hysteresiswärme
hysteresis index Hysteresis-Exponent
hysteresis loop Hysteresisschleife, Hystereseschleife
hysteresis loss Hysteresisverlust
hysteretic Hysterese-..., Hysteresis ...
hysteretic constant Hysteresiskonstante, Steinmetzkonstante
hysteretic index Hysteresisverlust, Ummagnetisierungsverlust

I

I-beam I-Träger, I-Profil, Doppel-T-Träger
IC = integrated circuit integrierte Schaltung
IC socket IC-Fassung
IC-photomasking IC-Fotomaskierung
IC-surface passivation Oberflächenpassivierung (621.382)
IC-tester IC-Prüfgerät
ice Eis
ice-covered eisbedeckt, vereist
ice loading Eisbelastung, Eislast (621.315.1), (621.396.67)
ice tray Eisschale, Gefrierschale (64)
to ice vereisen
Iceland spar Kalkspat
icing Vereisung, Eislastbildung
iconoscope Ikonoskop
I.C.W. = interrupted continuous waves Telegraphie tönend

ID = internal diameter Innendurchmesser, innerer Durchmesser, lichte Weite
ideal ideal
ideal dielectric ideales Dielektrikum, vollständiger Nichtleiter
identification Markierung; Auffindung; Kennung, Kennzeichnung; Identifizierung
identification badge Firmenabzeichen
identification colour Kennfarbe
idenfification equipment Erkennungsgerät
identification friend-foe Freund-Feind-Identifizierung (623)
identification mark Identifizierungszeichen (621.395)
identification plate Firmenschild
identification sign(al) Erkennungssignal, Erkennungszeichen
identification sleeve Bezeichnungshülse
to identify kennzeichnen, kennen; identifizieren, herauskennen
identity Identität, Kennung
identiy keyer Kennungsgeber
idiostatic idiostatisch
idiostatic method idiostatische Schaltung (621.317)
idle leerlaufend, unbelastet; müßig, nutzlos, unproduktiv; Blind... (621.3)
idle component Blindkomponente (538)
idle condition Freizustand (621.395)
idle current Blindstrom, wattloser Strom
idle-current meter Blindstrommesser, Blindstromamperemeter
idle-current wattmeter Blindleistungsmesser
idle mode [Leerlauf]
idle motion Leerlauf
idle pulley Leerscheibe, Losscheibe
idle-running speed Leerlaufdrehzahl
idle time Leerlaufzeit, Wartezeit
idle wire inaktiver Draht
to idle leerlaufen
idling speed Leerlaufdrehzahl
idler Leerlaufrolle, (leer mitlaufende, nicht angetriebene) Tragrolle, Stützrolle
idometer Idometer
IEC = International Electrotechnical Commission
IEC-bus IEC-Bus
IEEE = Institute of Electrical and Electronic Engineers (entspr. VDE)
i.f. = intermediate frequency ZF = Zwischenfrequenz
i.f. amplifier ZF-Verstärker
i.f. oscillator Zwischenfrequenz-Generator, ZF-Oszillator
i.f. protection ZF-Sicherheit
i.f. stage ZF-Stufe
i.f.t. = intermediate frequency transformation ZF-Umsetzung
i.f. valve ZF-Röhre
I.F.F. = identification of friend or foe Kennungsabfrage, Kenngerät, Freund-Feind-Kennung
IG-FET = insulated gate FET Feldeffekt-Transistor mit isolierter Gate-Elektrode
ignitability Entzündlichkeit, Zündfähigkeit
ignitable zündfähig, entzündlich, entflammbar, brennbar
to ignite zünden, entzünden, anzünden, sich entzünden, anbrennen
igniting Zündung, Zündvorgang, Zünd-...
igniting agent Zündmittel
igniting device Zündvorrichtung, Zündapparat
igniting point Zündpunkt, Zündtemperatur
igniting spark Zündfunke
ignition Zündung (621.4), 621.385); Entzündung
ignition characteristic Zündkennlinie
ignition current Zündstrom
ignition electrode Zündelektrode
ignition flame Zündflamme
ignition noise Zündstörungen (pl.) (621.396)
ignition plug Zündkerze (621.4)
ignition rating Nennleistung für Abreißzündung (621.355)
ignition spark Zündfunke (621.43)
ignition switch Zündschlüssel (629.113)
ignition system Zündanlage (621.4)
ignition temperature Entzündungstemperatur, Zündtemperatur
ignition test Zündprüfung, Zündprobe
ignition time Zündzeitpunkt
ignition transformer Zündtransformator
ignition voltage Zündspannung (621.385)
ignition wire Zündkabel
ignitron Ignitron, Gleichrichter mit Zündstift
ignitron welding timer Ignitronschweißtakter
to ignore vernächlässigen, nicht berücksichtigen, unberücksichtigt lassen
ignore character Auslassungszeichen (681.3)
ignored conductor Leiter mit schwebendem Potential, freier Leiter
i.h.p. = indicated horse-power indizierte Leistung
Ikeda test Spannungsermittlung durch elektrische Widerstandsmessung
ikonophone Fernsprechanlage mit Gegenfernsehen
Ilgner wheel Ilgner-Rad

illuminance Beleuchtungsstärke
illuminant Lichtquelle (53); Leuchtöl, Leuchtgas, Leuchtbetriebsstoff
illuminant c Normlichtart c
to illuminate erleuchten, beleuchten, illuminieren
illuminated dial Leuchtskala (621.317)
illuminating apparatus Beleuchtungseinrichtung (535)
illuminating gas Leuchtgas
illumination Beleuchtung, Beleuchtungsstärke, Illumination
illumination control Dämmerungsschalter
illumination darkening Saalverdunkelung
illumination measurement Beleuchtungsmessung
illumination photometer Beleuchtungsmesser
Illuminator Beleuchtungsradar (623)
illuminometer Beleuchtungsmesser
to illustrate bebildern, bildlich darstellen
ILS = Instrument Landing System ILS, Instrumentenlandesystem (321.396.93)
image Bild, Spiegelbild
image aerial Bildantenne
image attenuation Vierpoldämpfung, Wellendämpfung
image attenuation coefficient Vierpoldämpfungsmaß
image camera tube Zwischenbild-Kameraröhre
image clearness Bildschärfe
image converter Bildwandlerröhre
image dissector Bildzerleger, Farnsworth-Röhre
image frequency Spiegelfrequenz (621.396)
image-frequency rejection Spiegelselektion, Spiegelfrequenzsicherheit
image impedance stoßfreier Abschlußwiderstand, Kennwiderstand (621.392)
image-intensifier tube Bildverstärkerröhre
image interference Spiegelfrequenzstörung
image-interference ratio Spiegelfrequenzsicherheit
image modulation Bildmodulation
image orthicon Orthicon, Superorthikon
image-phase constant Phasenmaß bei stoßfreiem Abschluß, Vierpolwinkelmaß
image rejection Spiegelfrequenzdämpfung
image response Spiegelfrequenzfestigkeit
image sensor Bildsensor
image suppression Spiegelfrequenzunterdrückung
image-transfer coefficient Vierpolübertragungsmaß
to image widerspiegeln, Bild entwerfen
imaginary imaginärer Anteil, Imaginärteil; imaginäre Zahl
imaginary imaginär, gedacht
imaginary component Imaginärteil
imaginary cut gedachter Schnitt
imaginary figure imaginäre Zahl
imaginary fulcrum gedachter Drehpunkt
imaginary part imaginärer Teil
imaginary term imaginärer Anteil
to imbed einbeten
to imbibe einsaugen, aufnehmen
imbibition Einsaugung, Flüssigkeitsaufnahme
to imbricate schuppig überlappen
imbricated winding Faßwicklung, Korbwicklung
immeasurable unmeßbar
immediate unmittelbar, direkt, jetzt vorliegend
immediate access unmittelbarer Zugriff
immediate addressing unmittebare Adressierung
immediate busy Sofortsperre (621.395)
immediate ring Sofortruf
to immerse eintauchen, tauchen
immersed liquid quenched fuse Ölsicherung
immersed in oil unter Öl
immersed rheostat Widerstand im Wasserbad
immersible Unterwasser-...
immersion Tauchung, Tauchprozeß, Immersion
immersion device Tauchvorrichtung
immersion heater Tauchsieder
immersion test Tauchprobe, Tauchversuch, Korrosionsversuch
immiscible nicht mischbar
immobile unbeweglich
to immobilize unbeweglich machen, die Bewegungsfreiheit nehmen
immovable unbeweglich
to immure einmauern
impact Aufprall, Auftreffen, Aufschlagen, Zusammenstoß
impact bend(ing) strength Schlagbiegefestigkeit
impact-compression test Schlagstauchversuch
impact conditions (pl.) Stoßbeanspruchung
impact-damping device Schlagdämpfer
impact effect Stoßwirkung
impact emission Sekundäremission (621.385)
impact force Anschlagkraft
impact load Stoßlast, stoßweise Belastung
impact printer [Drucker mit mechanischem Anschlag]
impact of radiation Einfall von Strahlung

impact resistance Schlagfestigkeit
impact strength Schlagbiegefestigkeit
impact test Schlagprobe, Schlagprüfung
impact velocity Auftreffgeschwindigkeit
to impact auftreffen, aufprallen
impalpable unfühlbar, gewichtslos
to impart zuerteilen, erteilen, geben, übertragen, verleihen
impassable unpassierbar
impedance Scheinwiderstand, Wechselstromwiderstand, Impedanz
impedance bond Drosselverbinder, verdrosselter Anschluß
impedance bridge Scheinwiderstandsbrücke, Impedanzmeßbrücke
impedance coil Drosselspule, Drossel
impedance compensator Entzerrer (621.395)
impedance coupled drosselgekoppelt
impedance coupling L.-C.-Kopplung, Impedanzkopplung
impedance diagram Widerstandsdreieck
impedance drop Wechseltrom-Spannungsabfall
impedance factor (Kehrwert des Leistungsfaktors)
impedance function Widerstandsfunktion
impedance losses stromabhängige Verluste (621.314.2)
impedance match Widerstandsanpassung, Anpassungswiderstand
impedance matching Widerstandsanpassung, Impedanzanpassung
impedance-matching device Transformationsglied
impedance mismatch Impedanzfehlanpassung
impedance operator Widerstandsoperator
impedance relationship Widerstandsbeziehung
impedance relay Impedanzrelais, Distanzrelais
impedance rise Spannungsanstieg bei kapazitiver Last
impedance transformation Widerstandstransformation
impedance transformer Anpassungsübertrager
impedance triangle Widerstandsdreieck
impedance unbalance Unsymmetrie (621.395)
impedance-unbalance measuring set Fehlerdämpfungsmeßbrücke
impedance voltage Gesamtspannungsabfall; Nennkurzschlußspannung
impedance voltage of a transformer Kurzschlußspannung eines Transformators
to impede behindern
impeder Drossel, Begrenzungsdrossel
to impel antreiben
impellent Antriebsmittel
impeller Propeller, Schaufelrad
impenetrable to undurchdringlich für
imperfect unvollständig, schlecht, unsauber, ungenau
imperfect dielectric verlustbehaftetes Dielektrikum
imperfect earth schlechte Erdung
imperfect magnetic circuit nichtgeschlossener magnetischer Kreis
imperfection Störstelle, Fehlordnung (621.382)
impermeability Undurchlässigkeit
impermeable to undurchlässig für
impervious to undurchlässig für, unangreifbar durch
imperviousness Undurchlässigkeit
impetuous wuchtig, heftig
impetus Wucht, Stoßkraft, Impuls
to impinge auftreffen, auffallen
to impinge in eintauchen in
to impinge on auffallen auf
impingement Auftreffen, Aufschlag
implement Hilfsgerät, Hilfsmittel, Werkzeug, Gerät
implementation [erstmalige Inbetriebnahme eines Programms]
implosion Implosion (621.385)
impliedly implizite
to imply voraussetzen, annehmen, in sich schließen, enthalten, beinhalten
implied addressing implizierte Adressierung
imponderable unwägbar
to impose auferlegen, aufbringen, erzwingen
to impose a limitation eine Grenze setzen
imposed boundary conditions erzwungene Randbedingungen
to impoverish verarmen
impoverishment Verarmung
impracticable unausführbar; unzweckmäßig, unhandlich
impractical unpraktisch, unzweckmäßig
impregnant Imprägnierungsmittel
to impregnate tränken, imprägnieren
impregnated earth Tränkerder
impregnated insulating material imprägniertes Isoliermaterial
impregnating material Imprägnierung, Imprägnierungsmaterial
impregnating tank Imprägnierbottich
impregnation Tränkung, Imprägnierung
imprägnation compound Tränkmasse, Imprägnierungsmasse
impregnation liquid Tränkflüssigkeit
impress Abbild, Stempel, Prägung, Prägestempel

to impress aufdrücken, prägen; beeindrucken
to impress a voltage eine Spannung aufdrücken, eine Spannung anlegen
impressed eingeprägt
impressed electromotive force Urspannung, eingeprägte elektromotorische Kraft
impressed e. m. f. eingeprägte E. M. K., Urspannung
impressed force eingeprägte Kraft
impressed forces äußere Kräfte
impressed voltage eingeprägte Spannung, Urspannung
imprint Stempel, Prägestempel
to imprint stempeln, einprägen
improper ungenügend, unzureichend, schlecht; falsch, unpassend, unziemlich
improper fraction unechter Bruch
improper number irrationale Zahl
improved safety erhöhte Sicherheit
improvised behelfsmäßig
impulse Impuls, Antrieb, Stoß, Anstoß
impulse action Impulswahl (621.395)
impulse circuit Stoßkreis (621.315.6); Impulskreis (621.395)
impulse-circuit breaker Ölstromschalter
impulse counter Impulszähler, Impulszählwerk
impulse distortion Impulsverzerrung
impulse-driven clock impulsgesteuerte Uhr, Uhr mit Stromstoßfortsteuerung
impulse excitation Stoßerregung
impulse flashover Stoßüberschlag
impulse flash-over voltage Überschlagstoßspannung
impulse flashover voltage-time characteristic Stoßspannungscharakteristik
impulse frequency Impulsfrequenz, Impulsfolgefrequenz
impulse function Stoßfunktion; Sprungfunktion
impulse generator Impulsgenerator; Stoßgenerator; Taktgeber
impulse inertia Überschlagverzögerung, Durchschlagverzögerung
impulse level Stoßspannungsfestigkeit; Stoßpegel
impulse machine Impulsgeber, Stromstoßgeber
impulse meter Impulszähler
impulse noise Impulsrauschen
impulse period Impulsabstand
impulse radiation Bremsstrahlung (321.386)
impulse ratio Impulsverhältnis (621.397)
impulse relay Stromstoßrelais
impulse repeater Impulsübertrager (621.395.34)
impulse response Stoßverhalten, Verhalten gegen Stoßspannung; Verhalten gegen Stoßlast
impulse separator Impulsausleseschaltung (621.396)
impulse spark-over test Stoßüberschlagsprüfung
impulse strength Stoßspannungsfestigkeit
impulse summator Impulszähler (621.317)
impulse switch Stromstoßschalter
impulse train Impulsgruppe, Impulszug
impulse-type telemeter Impulsfernmeßgerät
impulse voltage Stoßspannung
impulsive impulsiv; stoßartig; antreibend
impurity Unreinheit, Verunreinigung, Beimengung; Fehlwinkel, Winkelfehler (621.317); Fremdatom, Störstelle (621.382)
impurity compensation Störstellenkompensation
impurity-diffusion technique Diffusionsverfahren, Diffusionstechnik
impurity element Verunreinigungselement (621.382)
in. = inch (pl. **ins. = inches**) (s. dort)
1 in. diameter 1 Zoll im Durchmesser
1 in. water 1 Zoll WS = 1 Zoll Wassersäule
inaccessibility Unzugänglichkeit
inaccessible unzugänglich
inaccuracy Ungenauigkeit
inaccurate ungenau, unsauber
inactive inaktiv, nicht aktiv, unwirksam
inaudibility Unhörbarkeit
inaudible unhörbar
inband signalling Signalisierung im Sprachband; Tonwahl (621.395)
incalculability Unberechenbarkeit
incalculable unberechenbar
to incandesce weißglühend werden; weißglühend machen, zur Weißglut bringen
incandescence Weißglut
incandescence lamp Glühlampe
incandescent glühend, weißglühend, Glüh-...
incandescent arc lamp Mischlichtlampe
incandescent column Leuchtsäule, Leuchtplasma
incandescent lamp Glühlampe, Glühbirne
incandescent light Glühlampe, Glühlicht
incandescent lighting Glühlampenbeleuchtung
in-carbonization Inkohlung
incedent einfallend (535)
incendiary Brand-..., Feuer-...
incentive wage Akkordlohn
incentive worker Akkordarbeiter

in-centre Mittelpunkt des Inkreises, Mittelpunkt des einbeschriebenen Kreises
inception Konzeption; Auslösung, Anfang
inceptive beginnend, anfangend
inch Zoll (Längenmaß = 2,54 cm)
inch board zölliges Brett, Zollbrett
by inches zollweise
inching control Langsamsteuerung (621.97)
inching speed Kriechgang
inching starter Kriechanlasser
to incide einfallen (535)
incidence Einfallen, Eintreffen, Auftreffen, Auffallen (535)
incidence wire Diagonaldraht, Verspannung (629.13)
incident einfallend, auffallend (**upon:** auf); eintretend (**in:** in)
incident wave einlaufende Welle, ankommende Welle
to incinerate veraschen
in-circle Inkreis; einbeschriebener Kreis
in-circuit emulator = ICE In-Circuit Emulator, Testadapter
in-circuit tester Schaltkreisprüfgerät
to incise einschneiden, einkerben
incision Einschnitt, Einkerbung
inclinable kippbar, schrägstellbar
inclinable press Kippresse
inclination Neigung; Neigung, Steigung; Inklination (538)
inclinatorium Inklinationsnadel (538.7)
incline schiefe Ebene; Ablaufbahn, Bremsberg; Neigung, Steigung, Schräge
to incline (sich) neigen, schief stellen, schräg stellen
inclined schräg, schiefgestellt, geneigt, Schräg-...
to be inclined at an angle of ... to um ...schief stehen gegen ...
inclined catenary suspension Schrägdrahtaufhängung (621.33)
inclined hoist Schrägaufzug (63)
inclined plane schiefe Ebene, Rampe
to inclose abschließen, abschirmen (621.396)
to include einschließen, (mit) berücksichtigen; enthalten; einbauen, mit einbauen; zum Einbau vorsehen
included angle eingeschlossener Winkel (51)
inclusion Berücksichtigung; Einschluß (55); Gehalt, Bestandteil (54)
inclusion in a table Aufnahme in eine Tabelle
inclusive OR ODER-Verknüpfung, inklusives ODER
incoherence Zusammenhanglosigkeit, Inkohärenz
incoherent zusammenhanglos, unzusammenhängend, inkohärent
incombustibility Unbrennbarkeit, Unverbrennbarkeit
incombustible unbrennbar, unverbrennlich
incoming ankommend, eintreffend
incoming call ankommendes Gespräch, ankommender Ruf (621.395)
incoming exchange line ankommende Amtsverbindung
incoming feeder ankommende Speiseleitung, Einspeisung
incoming track Ankunftsgleis, Einfahrtsgleis
incoming trunk ankommende Teilnehmerleitung
incommensurability inkommensurable Eigenschaft
incompatibility Unverträglichkeit, Unvereinbarkeit
incompatible unverträglich, unvereinbar
to be mutually incompatible sich gegenseitig ausschließen
incomplete unvollständig
incomplete diffusion unvollständige Lichtstreuung
incompressibility Inkompressibilität, Unzusammendrückbarkeit
incompressible zusammendrückbar, inkompressibel
incomputable unberechenbar
inconclusive nicht schlüssig
incondensable nicht kondensierbar
inconsequential bedeutungslos, ohne Folgen
inconsiderable unbeachtlich, unbedeutend
inconstancy Inkonstanz
inconstant inkonstant
to incorporate einbauen, einsetzen, einverleiben; innewohnen
to be incorporated in zugehören zu, innewohnen
incorporation Einbau
incorrodible korrosionsfest, nicht korrodierend
increase Zunahme, Zuwachs, Steigerung, Wachsen
increase of current Stromzunahme
6 db increase 6 db Zunahme, 6 db Steigerung
increase in moisture content Feuchtigkeitsaufnahme
increase in momentum Impulszuwachs
increase on Steigerung gegenüber
increase of output Leistungssteigerung
increase in pressure Drucksteigerung

increase in velocity Geschwindigkeitszunahme
increase in weight Gewichtszunahme
to increase steigern, erhöhen; zunehmen, sich steigern, wachsen
to increase as zunehmen mit, ebenso steigen wie
to increase by electro-deposition elektrolytisch verstärken, galvanisch verstärken
increased-safety enclosure Schlagwetterschutz mit erhöhter Sicherheit, Schutzart erhöhte Sicherheit
increment Zuwachs, Stufe; Erhöhung (681.3)
increment iron losses Oberwelleneisenverluste
to increment inkrementieren [schrittweise um 1 erhöhen] (681.3)
incremental zusätzlich, anwachsend, zunehmend, stufenweise
incremental computer digitaler Integrator
incremental hysteresis loss differentieller Hysteresisverlust, Oberwelleneisenverluste
incremental inductance Wechselstrominduktivität; Induktivität für überlagerten (kleinen) Wechselstrom
incremental induction differentielle Induktion
incremental magnetizing force differentielle magnetische Feldstärke (Scheitelwert)
incremental permeability differentielle Permeabilität
incrementer (relay) Telegraphenrelais
indecipherable unentzifferbar, nicht leserlich
indefinite unbestimmt, unbeschränkt; unendlich, unendlich lange
indefinitely beliebig lange
indelible unzerstört, waschecht
indelible pencil Kopierstift
indent Anforderung; Kerbe
to indent einschneiden, hineinbeißen; zähnen; eindrücken (655); Doppel ausfertigen
indented joint Verzahnung (693)
indentation Vertiefung, Eindruck, Spur (Härteprüfung); Zähnung
independent of unabhängig von
independent drive (Betrieb mit) Steuerstufe (621.396.61)
independent feeder direkte Speiseleitung
independent sideband transmission Übertragung mit unabhängigen Seitenbändern, ISB-Übertragung
independent time element UMZ-Relais, Zeitrelais
independent time-lag circuit breaking UMZ-Auslösung
independent trip Hilfsstromauslösung
independent unit Grundeinheit (53)
independent variable unabhängige Variable
independently operated getrennt betätigt
indestructible unzerstörbar
indeterminate unbestimmt
index Zeigefinger (61); Zeiger (621.317); (pl.**-ices**) Index; Exponent (51); Anzeichen (**to:** für); (pl. **-exes**) Liste, Preisliste, Verzeichnis, Inhaltsverzeichnis, Sachverzeichnis, Register
index card Karteikarte
index collar Skalenring, Teilungsring
index hole Indexloch
index notation Potenzschreibweise, Potenzform (51)
index of operation Arbeitsmodul (621.394.6)
index of refraction Brechungsindex
index register Indexregister (681.3)
index stripe Indexstreifen
to index mit einem Index versehen, indizieren; mit einer Skaleneinteilung versehen, teilen
to index a wheel 10 holes ein Rad um 10 Löcher weiter drehen
indexed addressing indizierte Adressierung
indexing tone Kennfrequenz
India ink chinesische Tusche
India rubber cable Gummikabel
in-dialling Durchwahl
to indicate anzeigen, bezeichnen, angeben; indizieren (621.1)
to indicate on a. c. bei Wechselstrom anzeigen (621.317)
indicated horse-power indizierte Leistung
indicating Anzeige-...
indicating control switch Quittungssteuerschalter
indicating device Anzeigeeinrichtung
indicating fuse Sicherung mit Unterbrechungsmeldung
indicating instrument Anzeigeinstrument
indicating lamp Meldelampe, Anzeigelampe, Signallampe
indicating range Anzeigebereich
indication Anzeige; Anzeichen, Hinweis; Indikation (61)
indication error Anzeigefehler
indication point Zugbeeinflussungsort (656.25)
indicative of hinweisend auf
indicator Anzeiger, Hinweis; Indikator (54); (621.1), (621.4); Winker (629.113); Sichtgerät (621.306), (621.397); Anzeigegerät (621.317); Zeiger
indicator board Anzeigetafel
indicator lamp Anzeigelampe, Meldelampe
indicator light Kontrollampe

indicator panel Anzeigetafel
indicator tube Anzeigeröhre
indicial response Pulsamplitudengang, Übergangsfunktion
indirect indirekt
indirect addressing indirekte Adressierung
indirect arc furnace Lichtbogenofen mit indirekter Heizung
indirect-control system indirekt gesteuertes System (621.395.3)
indirect fittings indirekte Leuchten (621.329)
indirect heating indirekte Heizung (621.385)
indirect illumination indirekte Beleuchtung
indirect lighting indirekte Beleuchtung
indirect mode Indirektverfahren (527)
indirect operation indirekte Betätigung
indirect rays (pl.) Raumstrahlung (621.396)
indirect tripping of a circuit breaker indirekte Auslösung eines Schalters, Schalterauslösung durch (Zwischen-)Relais
indirect wave Raumwelle
indirectly heated cathode indirekt geheizte Kathode
indirectly heated walve indirekt geheizte Röhre
indiscriminate wahllos, regellos, wild
indissolubility Unlöslichkeit
indissoluble unlöslich
indistinct undeutlich, verwaschen
indium Indium
individual indivuduell, einzeln, Einzel-...
individual axle drive Einzelachsantrieb
individual consumption Einzelverbrauch, Kleinverbrauch
individual distortion individueller Verzerrungsgrad
individual drive Einzelantrieb (621.9)
individual gear-driven axle einzeln angetriebene Achse
individual gear-driven axle carriage Wagen mit Einzelachsantrieb
individual item Einzelposition
individual line Teilnehmer(anschluß)leitung, Einzelanschluß (621.395)
individual load Einzellast (621.3)
individual trunk individuelle Abnehmerleitung (621.395)
individual workplace lighting Arbeitsplatzbeleuchtung
indivisibility Unteilbarkeit
indivisible unteilbar
indoor Innenraum-..., Innen-..., Haus-...
indoor aerial Innenraumantenne, Zimmerantenne
indoor climates Innenraumklima
indoor insulator Innenraumisolator
indoor transformer Innenraum-Transformator, Innenraum-Umspanner
indoor wiring Inneninstallation
indrive end Abtriebsseite
to induce induzieren (53); induzieren, erzeugen, hervorrufen
induced charge induzierte Ladung
induced control current induzierte Steuerspannung (eines Halleffektbauelementes)
induced current induzierter Strom
induced draught Saugzug
induced draught fan Saugzuggebläse
induced electricity induzierte elektrische Ladung
induced electromotive force induzierteelektromotorische Kraft
induced magnetization Magnetisierung
induced needle Weicheisennadel (621.317)
induced noise Starkstromgeräusche
induced nuclear disintegration induzierter Kernzerfall
induced radioactivity induzierte Radioaktivität, künstliche Radioaktivität
induced ventilation künstliche Belüftung, Selbstkühlung (621.313), (621.314)
induced voltage Mantelspannung (621.315.2); Induktionsspannung
inductance Induktivität, Spule
inductance bridge Induktivitätsmeßbrücke
inductance coil Drosselspule, Drossel
inductance coupling induktive Kopplung
inductance meter Induktivitätsmesser
inductance standard Induktivitätsnormal
inductance per unit length Induktivitätsbelag
inductance-voltage drop induktiver Spannungsabfall
induction Induktion
induction brazing Induktionslötung
induction coil Induktionsspule, Funkeninduktor
induction compass Erdinduktorkompaß, magnetelektrischer Kompaß
induction frequency converter asynchroner Frequenzumformer, Drehfeldfrequenzwandler
induction furnace Induktionsofen
induction generator Asynchrongenerator
induction hardening Induktionshärtung
induction heating Induktionsheizung; induktive Erwärmung
induction instrument Drehfeldinstrument

induction lamp Hochfrequenz-Glimmlampe, elektrodenlose Glimmlampe
induction loudspeaker dynamischer Lautsprecher
induction machine Induktionsmaschine, Drehfeldmaschine
induction meter Induktionszähler, Ferrariszähler
induction motor Induktionsmotor, Asynchronmotor
induction motor-generator asynchroner Motorgenerator
induction motor meter Induktionszähler, Ferrariszähler
induction phase shifter Drehtransformator
induction pick-up coil Induktions(aufnehmer)-Spule
induction regulator Drehregler, Drehtransformator
induction relay Ferraris-Relais
induction resistance welding induktive Widerstandsschweißung
induction starting asynchrones Anlassen, asynchroner Anlauf
induction-type synchronous motor Synchronmotor mit asynchronem Anlauf
induction-type instrument Ferrarisinstrument
induction voltage regulator Drehregler, Drehtransformator
induction voltmeter Ferraris-Voltmeter
induction wattmeter Drehfeldwattmeter, Drehfeldleistungsmesser
inductionless induktionsfrei
inductive induktiv
inductive action Induktionswirkung
inductive branch induktiver Zweig
inductive capacity Verschiebungszahl, Dielektrizitätskonstante, D. K.
inductive circuit induktiver Stromkreis
inductive coupling induktive Kupplung
inductive current induktiver Strom
inductive drop induktiver Spannungsabfall
inductive impedance Scheinwiderstand mit induktivem Anteil, induktiver Scheinwiderstand
inductive interference induktive Störbeeinflussung
inductive leak Drosselerdung (621.394)
inductive load induktive Last, induktive Belastung
inductive loading induktive Belastung
inductive pick-off Induktionsfühler
inductive reactance induktiver Blindwiderstand
inductive resistance induktiver Widerstand
inductive resistor Widerstand mit Induktivität
inductive rise induktiver Spannungsanstieg (bei kapazitiver Last)
inductively coupled induktiv gekoppelt
inductivity Induktivität
inductometer Variometer
inductor Spule; Induktor; Läufer (621.313)
inductor alternator Gleichpolmaschine (621.313)
inductor circuit Läuferstromkreis, Induktorstromkreis
inductor coil Induktorspule
inductor core Spulenkern, Induktorballen
inductor current Induktorstrom, Erregerstrom, Feldstrom
inductor loudspeaker magnetischer Lautsprecher
inductor slot Induktornut, Läufernut
inductor-type synchronous generator Reluktanzgenerator
inductor voltage Induktorspannung, Läuferspannung, Feldspannung
inductor winding Induktorwicklung, Läuferwicklung
industrial industriell, Industrie-...
industrial frequency Netzfrequenz, normale Frequenz
industrial radiography technische Röntgenuntersuchung
industry Industrie
ineffective unwirksam
ineffective time ungenutzte Zeit
inefficiency Leerlauf, Mangel an Leistungsfähigkeit; Unwirschaftlichkeit
inefficient unwirtschaftlich, unwirksam, nicht leistungsfähig
inelastic unelastisch
inelasticity Mangel an Elastizität
inert träge, inert, reaktionsunlustig; inaktiv
inert cell Feldelement (621.352)
inert constituent träge Komponente, träger Bestandteil (54)
inert gas inertes Gas
inert-gas equipment Schutzgasgeräte (pl.)
inertair cooling Stickstoffkühlung
inertair transformer Transformator mit Stickstoffkühlung
inerteen Clophen (621.315.6)
inertia (Massen-)Trägheit, Beharrungsvermögen
inertia constant Zeitkonstante des Auslaufs bei Nennlast (621.313), Vollastauslaufzeit (621.313) (Energieinhalt der rotierenden Massen/Nennleistung)

inertia factor Trägheitsmoment
inertia force Trägheitskraft, Massenkraft
inertia of needle Anzeigeträgheit (621.317)
inertia starter Schwungradanlasser, Fliehkraftanlasser (621.4)
inertial mass träge Masse
inertial navigation Trägheitsnavigation, bodenunabhängige Navigation
inertness Reaktionsträgheit, Mangel an Reaktionsfähigkeit
inexactitude Ungenauigkeit
inexpert Laie, Nichtfachmann
inextensible nicht dehnbar, nicht dehnungsfähig
inextinguishable nicht löschbar, unlöschbar
inextricable unlösbar
INF. = infinity das Unendliche, der unendlich ferne Punkt
inf. = infra unterhalb, infra (siehe dort)
infeed Vorschub, Eingabe
to infer from von ... aus schließen auf, folgern aus
inferred zero unterdrückter Nullpunkt (621.317)
inference Schluß, Folgerung
inferior unterlegen; kleiner (**to:** als); minderwertig, schlecht, unterdurchschnittlich
inferior fuel minderwertiger Brennstoff
inferior quality schlechte Qualität
to infiltrate infiltrieren (in fein verteiltem Zustand) einführen; einsickern (lassen)
infinite unendlich
infinite regulation kontinuierliche Regelung, stufenlose Regelung
infinitely variable stetig veränderlich
infinitely variable speed motor feinregelbarer Motor, stufenlos regelbarer Motor
infinitesimal infinitesimal, unendlich klein
infinitesimal calculus Infinitesimalrechnung
infinity Unendlichkeit, Unendliches
infinity plug Unterbrechungsstöpsel (621.317)
to inflame (sich) entzünden, anstecken
inflammability Entzündlichkeit, Entflammbarkeit
inflammable brennbar, entzündlich, feuergefährlich
inflammable gas brennbares Gas
inflammable matter feuergefährliches Material
to inflate aufblasen, aufblähen; aufpumpen (629.113)
inflation Inflation; Aufpumpen (629.113)
to inflect biegen, beugen
inflection Biegung, Beugung; Einbuchtung
inflection point Wendepunkt (51)
inflexibility Mangel an Biegsamkeit
inflexible unbiegsam
inflexion Kurvenknick, Knickpunkt
inflow Zufluß, Zustrom
influence Einfluß (621.317); Influenz
influence of heat Wärmeeinfluß
influence of moisture Feuchtigkeitseinfluß
influence of temperature Temperatureinfluß
influx Zufluß, Zufuhr, Zustrom
information Unterrichtsmaterial; Information, Material; Unterlagen (pl.); Nachricht (621.39); Information (681.3)
information bit Informationsbit
information channel Informationskanal
information density Informationsdichte
information drain Nachrichtensinke
information path Nachrichtenpfad
information processing Informationsverarbeitung
information sink Nachrichtensinke
information source Nachrichtenquelle
information storage Nachrichtenspeicherung
information storage medium Informationsträger
information theory Informationstheorie (621.39)
information transfer Informationsübertragung
informative lehrreich, aufschlußreich; informativ
infra ... Infra ..., Unter ...
infra-acoustic frequency Infraschallfrequenz, unhörbar tiefe Frequenz
infra-audible frequency Infraschallfrequenz
infradyne reception Überlagerungsempfang mit höherer Überlagerungsfrequenz
infrasonic frequency Infraschallfrequenz
infrasound frequency Infraschallfrequenz
infrared Infrarot, Ultrarot
infrared converter Infrarotwandler, Infrarotgerät
infrared microscope Infrarot-Mikroskop, Bildwandler-Mikroskop
infrared plate Ultrarotplatte, ultrarotempfindliche Platte (77)
infrared radiation ultrarote Strahlung, Ultrarotstrahlung
infrared therapy Infrarottherapie
infusible nicht schmelzbar, schwer schmelzbar
ingate Einlaufkanal, Einlauf (621.7)
in gear eingekuppelt
ingot Barren, Block, Ingot
ingot steel Flußstahl, Flußeisen
ingress Eindringen; Zugang
inharmonic nicht harmonisch (51), (621.3)
inherent anhaltend, innewohnend
inherent in untrennbar verbunden mit

inherent drop-off time Rückfalleigenzeit
inherent error mitgeschleppter Fehler
inherent forward current transfer ratio inhärentes Gleichstromverhältnis (621.382.3)
inherent losses Eigenverluste
inherent noise Egenrauschen (621.396.62)
inherent property natürliche Eigenschaft
inherent random noise Eigenrauschen
inherent regulation Selbstregelung; Spannungsanstieg bei Abwurf der Nennlast (621.313)
inherent resistance Eigenwiderstand, innerer Widerstand
inherent response time Eigenzeit
inherent self-excitation Läufererregung
to inhibit blockieren, bremsen, sperren
inhibiting input Sperreingang
inhibiting signal Blockierungssignal
inhibition gate Sperrgatter
inhibitor Inhibitor, Bremssubstanz (621.357)
inhomogeneity Inhomogenität
inhomogeneous inhomogen
initial einleitend, Initial-..., Erst-..., Anfangs-...
initial approach Anfangsanflug
initial condition Anfangsbedingungen, Beginnzustand
initial equation Ausgangsgleichung
initial expenses Anlagekosten, Anschaffungskosten
initial heating Anheizung
initial instruction Ladeprogramm (681.3)
initial inverse voltage Sprungspannung
initial load Vorbelastung, Vorlast
initial loader program Urlader
initial permeability Anfangspermeabilität
initial pressure Anfangsdruck
initial program load Ladeprogramm, Urlader
initial reactance Anfangsreaktanz
initial short-circuit ratio Stoßkurzschlußstromverhältnis
initial state Ausgangszustand
initial value Anfangswert
initial velocity Anfangsgeschwindigkeit
initial voltage Anfangsspannung
to initiate einleiten, anregen, auslösen, anstoßen
initiating contact Anregekontakt
initiation, initialization Einleitung, Anstoß, Anfang; Auslösung
initiation time Anwurfzeit (629)
initiator Zünder
initiatory einleitend, zündend, Zünd-...
to inject einspritzen, injizieren; eingeben, einführen; einführen, anlegen (621.3)
to inject on the grid dem Gitter zuführen (621.385)
to inject a voltage eine Spannung einführen, eine Spannung zuführen, eine Spannung anlegen
to inject with tränken mit
injected pile geimpfter Reaktor (539)
injected voltage zugeführte Spannung, Fremdspannung
injection Injektion; Einführung, Zuführung (621.3); Einspritzung (621.4); Ladungsträgerinjektion (621.382)
injection engine Einspritzmotor
injection grid Mischgitter
injection laser Injektionslaser
injection modulator Hilfsgittermodulator
injection moulding Spritzguß
ink Tinte; Tusche; Farbe, Druckfarbe
ink-jet printer Tintenstrahldrucker
ink ribbon Farbband
ink tracing Tuschzeichnung
ink writer Tintenschreiber (621.317)
to ink in (mit Tusche) ausziehen
inker Rekorder, Tintenschreiber, Farbschreiber (621.394)
inking mechanism Tintenschreiber (621.3)
inlay Inlay-Verfahren (621.397)
inlet Einlaß (621.1), (621.4); (Leitungs-)Einführung
in-line in einer Linie, ausgerichtet, hintereinander, gerade; hintereinander geschaltet
in-line construction gleichachsige Anordnung
in-line type engine Reihenmotor (621.43)
iner inner, Innen-...
inner conductor Innenleiter (621.315.2)
inner core Innenschicht (67)
inner layer Mittelschicht, Innenschicht (67)
inner lining Innenfutter, Buchse (621)
inner main Mittelleiter
inner marker Platzeinflugzeichen
inner pole generator Innenpolgenerator
inner pole type alternator Innenpolmaschine (621.313)
inner product inneres Produkt, skalares Produkt (51)
inner self-inductance innere Induktivität
inner-shell electron inneres Elektron, Elektron der inneren Schalen
inoperative außer Betrieb, stillgelegt; nicht betriebsfähig; unwirksam
inoperative program ruhendes Programm
inorganic anorganisch
inorganic chemistry anorganische Chemie

inorganic matter anorganische Substanz; anorganische Bestandteile (pl.)
in phase gleichphasig; in Phase; Gleichtakt...
in-phase amplifier Gleichtaktverstärker
in-phase component Wirkkomponente, gleichphasige Komponente (538.55)
in-phase current gleichphasiger Strom, Wirkstrom
in-phase voltage gleichphasige Spannung, Wirkspannung
input Eingangsleistung; Eingang, Eingangsseite; Eingabe (681.3)
input admittance Eingangsleitwert
input block Eingabeblock (681.3)
input-buffer register Eingabepufferregister
input capacitance Eingangskapazität
input capacitance-short output Kurzschluß-Eingangskapazität (621.382.3)
input capacitor Eingangskondensator
input coupling capacitor Eingangskoppelblock
input current range Arbeitsbereich des Eingangsstromes (621.375)
input device Eingabegerät
input end Eingangsklemmen (pl.), Eingangsseite, Sendeseite, Eingang
input equipment Eingabegerät
input impedance Eingangswiderstand
input loading factor Eingangslastfaktor (621.374)
input noise voltage Eingangsrauschspannung
input offset current Eingangsfehlstrom, Eingangsoffset-Strom (621.375)
input-output device Ein-/Ausgabegerät (681.3)
input power Eingangsleistung
input refered to potential potentialbezogener Eingang
input resistor Eingangsimpedanz
input routine Eingabeprogramm
input stage Eingangsstufe, erste Stufe (621.396.64)
input terminal Eingangsklemme
input transformer Eingangstransformator
input variable Eingangsgröße (beim Regler)
input voltage range Arbeitsbereich der Eingangsspannung (621.375)
inquiry Abfrage
inquiry station Abfragestation (681.3)
inquiry terminal Abfrageterminal
in radius Radius des einbeschriebenen Kreises, Radius des Inkreises
inrush (plötzliches) Einströmen; Einbruch
inrush-current Einschaltstrom
to inscribe einbeschreiben (51); einbauen; mit Inschrift versehen
inscription Aufschrift, Inschrift
insecure unsicher, nicht sicher befestigt, nicht fest, locker
insensible unmerklich
insensitive unempfindlich (**to:** gegen)
insensitivity Unempfindlichkeit (**to:** gegen)
inseparability Untrennbarkeit
inseparable untrennbar, nicht abtrennbar
insert Einlage
to insert einschalten, einfügen, einsetzen
to insert a flux into a face einen Fluß in eine Fläche eintreten lassen
inserted ceiling Zwischendecke
insertion Einschaltung, Einfügung; Einbau, Einlegen (621.9)
insertion between Zwischenschaltung (621.3); Einfügung zwischen
insertion gain Einfügungsgewinn (621.395); Einfügungsverstärkung (621.375)
insertion loss Einfügungsdämpfung, Restdämpfung (621.395)
insertion wall Einschubwand (693)
inset Einsatzbild, Bild im Text; Nebenkarte; Einfügung, Zwischenschaltung
inset diagram Schaubild im Text
inside inner, Innen-...
inside conductor Innenleiter (621.315.2), (621.396)
inside diameter Innendurchmesser, lichteWeite
inside dimension Innenmaß
inside face Innenfläche, innere Fläche
inside measurements Innenabmessungen, Innenmaße
inside radius Innenradius, innerer Radius
inside surface innere Oberfläche
insolubility Unlöslichkeit (54)
insolubility face Mischungslücke (66)
insoluble unlöslich (54)
to inspect beaufsichtigen, überwachen; prüfen, inspizieren, revidieren
inspection Besichtigung, Nachschau, Revision, Inspektion, Prüfung
by inspection aus der Anschauung, anschaulich klar, beim (bloßen) Ansehen, auf den ersten Blick
inspection cover Schaulochdeckel
inspection department Prüffeld
inspection glass Schauglas
inspection hole Schauloch
inspection lamp Handlampe; Faßlampe, Kessellampe
inspection of a meter Zählerrevision
inspection pit Montagegrube, Revisionsgrube

inspection plug Zellenverschluß, Verschlußstöpsel (621.355)
inspection room Prüfraum, Kontrollraum
inspection torch Faßlampe, Kessellampe; Handlampe
instability Instabilität, Labilität
instable instabil, labil
to install (auch **to instal**) installieren, einbauen, verlegen, einrichten
installed capacity installierte Leistung
installation Installation, Einbau, Verlegung, Einrichtung, Aufbau
installation contractor Vertragsinstallateur, Installationsunternehmer
installation equipment Installationsgerät, Installationswerkzeug
installation instructions Montageanleitung
installation kit Einbausatz
installation material Installationsmaterial
installation site Aufbauplatz, Aufstellungsort
installation test of a meter Zählerprüfung beim Abnehmer
to instance aufführen, angeben, (als Beispiel) aufzählen
instantaneous augenblicklich, Augenblicks-..., momentan, Moment-...; verzögerungsfrei, trägheitsfrei, unverzögert
instantaneous automatic gain control unverzögerte automatische Verstärkungsregelung
instantaneous centre of movement Wälzpunkt
instantaneous connection Momentschaltung
instantaneous control trägheitsfreie Steuerung
instantaneous current Augenblickswert des Stromes
instantaneous fuse Momentzünder (662); unverzögerte Sicherung, flinke Sicherung (621.3)
instantaneous overcurrent trip Überstrom-Schnellauslöser
instantaneous picture Momentbild, Momentaufnahme (77)
instantaneous tripping unverzögerte Auslösung, sofortige Auslösung
instantaneous relay Momentrelais
instantaneous value Augenblickswert
to instil eintropfen, tropfenweise einfüllen
institute Institut; Vereinigung
institute of technology Technische Hochschule
to institute einrichten
instruction Befehl (681.3); Unterricht, Instruktion, Anleitung, Anweisung, Unterweisung, Handbuch
instruction address Befehlsadresse (681.3)
instruction address register Befehlsadreßregister, Befehlszähler
instruction book Bedienungsanweisung, Gebrauchsanweisung
instruction code Befehlscode (681.3)
instruction counter Befehlszählregister
instruction cycles Befehlszyklen
instruction format Befehlsaufbau
instruction list Befehlsliste
instruction manual (Bedienungs-)Anleitung
instruction plate Bedienungsschild
instruction register Befehlsregister
instruction sequence Befehlsfolge
instruction set Befehlsvorrat
instruction slip Kennblatt
instruction time Befehlszeit
instructions (pl.) **for use** Gebrauchsanweisung
instruction word Befehlswort
instrument Instrument, Gerät; Werkzeug, Meßgerät
instrument approach Instrumenten-Anflugverfahren
instrument base Instrumentenfuß
instrument control Rückstellung, Rückstellkraft, Richtmoment (621.317)
instrument correction Instrumentenkorrektur
instrument flight Instrumentenflug, Flug nach den Instrumentenflugregeln
instrument flight rules = IFR Instrumentenflugregeln
instrument housing Instrumentengehäuse
instrument lamp Skalenlampe, Skalenbeleuchtungslampe
instrument landing system Instrumentenlandesystem (621.396.93)
instrument with locking device Fallbügelinstrument
instrument multiplier äußerer Vorwiderstand (621.317)
instrument navigation Instrumenten-Navigation
instrument panel Armaturenbrett, Instrumentenbrett
instrument proper Meßwerk (621.317)
instrument shop Instrumentenraum
instrument shunt Nebenwiderstand (621.317)
instrument switch Instrumentenschalter, Meßbereichsumschalter
instrument transformer Meßwandler
instrumental werkzeugähnlich, nützlich als Werkzeug
instrumental error Gerätefehler
instrumentation instrumentelle Ausrüstung, Instrumentenbestand

insulance Isolationswiderstand
insulant Isolierstoff
insulant isolierend
insulant material Isoliermaterial
to insulate isolieren (621.1/3)
insulated isoliert, Isolier-...
insulated bolt Isolierklemme
insulated bushing Isolierbuchse, Durchführungsisolator
insulated cable isoliertes Kabel, isolierte Leitung
insulated circuit ungeerdeter Stromkreis
insulated clip isolierte Leitungsklemme (696)
insulated component isolierter Teil, isolierter Bestandteil
insulated conductor isolierter Leiter
insulated from earth gegen Erde isoliert
insulated-gate filed-effect transistor Feldeffekttransistor mit isolierter Steuerelektrode, Isolierschicht-Feldeffekttransistor, IG-FET
insulated-gate thin-film field-effect transistor Dünnschicht-Feldeffekttransistor, Dünnfilm-Transistor
insulated hook isolierter Leitungsträger (696)
insulated middle wire isolierter Mittelleiter, nicht geerdeter Mittelleiter
insulated neutral (conductor) isolierter Mitteilleiter
insulated part isolierter Teil
insulated rail joint Isolierstoß (625)
insulated return system doppelpolige Speisung (621.313)
insulated sleeving Isolierschlauch
insulated supply system ungeerdetes System (621.31)
insulated support Stützisolator (621.315.6)
insulated terminal Isolierklemme
insulated turnbuckle isoliertes Spannschloß
insulated winding isolierte Wicklung
insulated wire isolierter Draht
insulating isolierend, Isolier-...
insulating beads Isolierperlen
insulating boots Gummistiefel
insulating bush Isolierbuchse
insulating cardboard Isolierpappe
insulating coating Isolieranstrich
insulating compound Isoliermasse, Kabelmasse, Vergußmasse
insulating coupling Isolierkupplung
insulating course Isolierschicht, Isolierlage (69)
insulating gloves Isolierhandschuhe, Gummihandschuhe
insulating guide Isolierrohr, Isolierfutter, isolierende Führung
insulating hose Isolierschlauch
insulating jacket Isoliermantel
insulating joint Isoliermuffe (621.315.2)
insulating lacquer Isolierlack
insulating layer Isolierschicht
insulating liquid Isolierflüssigkeit
insulating mat Isoliermatte
insulating material Isoliermaterial, Isolierstoff, Isolierung; Dämmstoff (69)
insulating medium Isoliermittel, Isolierstoff
insulating oil Isolieröl
insulating pipe Isolierrohr
insulating plastic Isolierkunststoff
insulating power Isoliervermögen, Isolierfähigkeit
insulating property isolierende Eigenschaft
insulating resistance Isolationswiderstand
insulating screen Isolierschirm, Isolierwand, isolierende Zwischenwand
insulating sleeve Isolierhülse
insulating stand isolierender Untersatz, Isolierunterlage
insulating substance Isolierstoff
insulating support Isolierstütze, Stützisolator
insulating tape Isolierband
insulating tube Isolierrohr
insulating varnish Isolierlack
insulation Isolation, Isolierung
insulation board Dämmplatte (693)
insulation check Isolationsprüfung
insulation defect Isolationsschaden
insulation failure Isolationsfehler
insulation felt Isolierfilz
insulation indicator Isolationsprüfer
insulation resistance Isolationswiderstand
insulation sleeve Isolierumhüllung
insulation test Isolationsprüfung
insulation tester Isolationsprüfer
insulativity Isolierfähigkeit; spezifischer Widerstand, Würfelwiderstand
insulator Isolator, Isoliermaterial
insulator-arcing horn Isolatorschutzhorn (621.315.6)
insulator-arcing ring Isolatorenschutzring (621.315.6)
insulator-arcing shield Isolatorschutzschild (621.315.6)
insulator arcover Isolator-Überschlag
insulator cap Isolatorkappe
insulator-grading shield Lichtbogenschutzring (gleichzeitig zur Verbesserung der Spannungsverteilung)

insulator losses Isolationsverluste (621.396.67)
insulator pin Isolatorenstütze
insulator string Isolatorenkette
insulator suspension Isolatorenaufhängung
insulator unit Schalterisolierung (621.316.5)
intact unversehrt, unbeschädigt, intakt
intake Auszugöffnung, Eintritt
intangible (sinnlich) nicht wahrnehmbar
int. comb. = internal combustion Verbrennungskraft-...
integer ganze Zahl, ganzzahlig, ganz
integral integral
integral equation Integralgleichung
integral sign Integralzeichen
integral ganzzahlig, als Ganzes, vollständig; eingebaut, angebaut
integral action I-Verhalten (621.316.7)
integral action time Nachstellzeit (621.316.7)
integral control I-Regelung
integral and derivative control ID-Regelung
integral horsepower motor Motor von einigen kW
integral part untrennbarer Teil, integrierender Bestandteil
integral-proportional action IP-Regelung, IP-Verhalten
integral-proportional derivative action IPD-Regelung, IPD-Verhalten
integral slot winding Ganzlochwicklung
integral submultiple ganzzahliger Bruchteil
integral value ganzzahliger Wert; Gesamtwert
integrand Integrand
integrant integrierender Bestandteil
to integrate integrieren
to integrate by parts partiell integrieren
integrated circuit integrierte Schaltung (siehe IC)
integrated-demand meter Spitzenzähler
integrated-injection logic I^2L-Logik
integrated network integriertes Netz
integrating integrierend, Integrier-...
integrating circuit Integrationsschaltung, Integrierkreis
integrating device Integriereinrichtung; Zählwerk
integrating instrument integrierendes Instrument, Zähler
integrating mechanism Integrierwerk; Zählwerk
integrating meter Zähler
integrating network Integrator, Integriereinrichtung
integrating photometer Mittelwertsphotometer
integrating sound level meter Schallpegelmesser mit Mittelungseinrichtung
integrating wattmeter summierendes Wattmeter, Zähler
integration Integration
integration method Integrationsverfahren
integration by parts partielle Integration
integrator Integrator, Integriereinrichtung, Integrationsschaltung
integrator circuit Integrierschaltung
integrator mechanism Integrierwerk
intelligence Nachricht (621.396); Nachrichteninhalt; Intelligenz; Nachrichtendienst (327.8); Information
intelligence bearing die Nachricht enthaltend
intelligibility Verständlichkeit (621.395)
intelligibility of phrases Satzverständlichkeit
intelligibility test Gehörprüfung, Verständlichkeitsprüfung
intelligible verständlich, wahrnehmbar (621.395)
intelligible crosstalk verständliches Nebensprechen
intended size Sollmaß
intense stark, intensiv, dicht, hell, kräftig
intense colour satte Farbe, kräftige Farbe
intense current starker Strom, kräftiger Strom
intense field starkes Feld, kräftiges Feld
intense heat intensive Hitze, starke Hitze
intense illumination helles Licht, starkes Licht, helle Beleuchtung
intense light helles Licht; starke Lampe
intense pressure starker Druck
intense stress starke Beanspruchung
intense voltage hohe Spannung
intensification Verstärkung (77); Helltastung (621.397)
to intensify verstärken (77); helltasten (621.397)
intensifier electrode Nachbeschleunigungselektrode
intensifier tube Lichtverstärkerröhre
intensifying leaf Verstärkerfolie (621.386)
intensifying ring Nachbeschleunigungselektrode (621.385)
intensifying screen Verstärkerfolie (621.386)
intensimeter Dosismesser, Dosismeßgerät
intensity Stärke, Intensität, Helligkeit (621.397)
intensity of activation Aktivierungsenergie
intensity brightening Aufleuchten, Auftasten (621.397)
intensity fading Schwund, Fading
intensity level Intensitätspegel
intensity of loading Beanspruchungshöhe, Beanspruchung
intensity of magnetization Magnetisierung

intensity modulation Helligkeitssteuerung, Helltastung (621.397)
intensity of sound Lautstärke
intensive intensiv, kräftig
intensive arc lamp Intensiv-Bogenlampe
intensive cooling intensive Kühlung, kräftige Kühlung
intensive heating kräftiges Erhitzen
intensive reflector Tiefstrahlreflektor
intentional absichtlich
interaction gegenseitige Beeinflussung, Wechselwirkung; Zusammenwirken, Überlagerung, Rückwirkung; Austauschvorgang
interaction impedance Kopplungswiderstand (621.372.5)
interaction loss Wechselwirkungsdämpfung
interactive mode Dialogbetrieb (681.3)
interatomic distance Atomabstand
interbase region Basiszwischenzone
to interblend (miteinander) mischen (54)
to intercalate interpolieren
intercalation Interpolation, Interpolierung
intercarrier noise Intercarrierbrumm
intercarrier system Differenzträgerverfahren
intercept Schnittpunkt; Abschnitt
intercept service Bescheiddienst, Hinweisdienst (621.395)
to intercept zum Schnitt bringen (51); herausschneiden; auffangen, abhören, mithören; abfangen (623)
intercepted angle umfaßter Winkel
intercepted message abgefangene Nachricht, mitgehörte Nachricht
to intercept and proceed the track einfädeln (351.814)
interception Abhördienst (621.396); Zusammentreffen, Sichfinden (629.13); Abfangung, Unterfangung (693)
interchange Austausch, Austauschvorgang, Vertauschung
interchange current Ausgleichsstrom
interchange of energy Energieaustausch
interchange of instruction Erfahrungsaustausch, Gedankenaustausch
interchange of leads to the terminals Vertauschung der Klemmenanschlüsse
to interchange austauschen, auswechseln
to interchange electrons Elektronen austauschen
to interchange energy Energie austauschen
to interchange the leads to the terminals die Zuleitungen zu den Klemmen vertauschen, umklemmen
to interchange position die Lage wechseln, abwechselnd verschiedene Lagen einnehmen, sich in der Lage abwechseln
interchangeability Austauschbarkeit, Auswechselbarkeit
interchangeable austauschbar, auswechselbar
interchangeable components Austauschbauteile
interchangeable contact piece auswechselbarer Kontakt, Abtrennkontakt
interchanger Austauscher, Tauscher
interchanging parts Austauschbauteile
intercity television link Fernsehfernleitung
intercom. = intercommunication Eigenverständigung (629.13); Bordsprechanlage
intercom. system Wechselsprechanlage, Gegensprechanlage
to intercommunicate in Verkehr stehen, in Verbindung stehen, Verbindung haben, Verkehr haben (621.396)
intercommunication Eigenverständigung, Eigenverständigungsanlage (629.13); Verständigung, Sprechverkehr, Verkehr (621.396)
intercommunication equipment one way Wechselsprechanlage
intercommunication equipment two way Gegensprechanlage
intercommunication switch Wählschalter, Wählumschalter
intercommunication telephone Reihenanlage, Haustelefonanlage
to interconnect verbinden
interconnected Verbund... (621.316)
interconnected damper winding geschlossene Dämpferwicklung
interconnected star system Zickzackschaltung (621.314.2)
interconnected system Verbundnetz
interconnectable probe Durchgangsmeßkopf
interconnectable wattmeter Durchgangsleistungsmeser
interconnector Ausgleichsleitung (621.315)
interconnecting Verbindungs-...
interconnecting feeder Verbindungsleitung (621.316); Kupplungsleitung, Kuppelleitung
interconnecting main Ausgleichsleitung (621.315)
interconnection Netzkupplung, Kupplungsleitung, Verbindungsleitung; Verbindung (621.39)
interconnection of power system Parallelbetrieb von Netzen, Verbundbetrieb (621.316)

interconnection tie Kuppelleitung, Kupplungsleitung
intercooler Zwischenkühler (621.63)
intercrystalline interkristallin, Korngrenz-...
intercrystalline corrosion Korngrenzkorrosion, interkristalline Korrosion
intercrystalline crack Riß in der Korngrenze
intercrystalline cracking interkristalline Rißbildung
interpedendance gegenseitige Abhängigkeit
interdigital interval Zwischenwahlzeit (621.385)
interdigital magnetron Doppelkamm-Magnetron
interelectrode ... zwischen den Elektroden
interelectrode capacitance Elektrodenkapazität (621.385)
interelectrode space Raum zwischen den Elektroden, Entladungsraum (621.385)
interface Grenzschicht, Grenzfläche, Zwischenschicht, Stoßstelle; Schnittstelle, Interface (621.39), (681.3)
interface region Grenzgebiet
to interfere with (störend) beeinflussen, stören, beeinträchtigen, einwirken auf; interferieren mit (535)
interference Störung, Beeinflussung (621.396); Interferenz (535); Übergreifen
interference blanker Störaustastschaltung
interference caused by the commutator Kollektorstörungen (pl.) (621.396)
interference condenser Störschutzkondensator
interference elimination Störschutz, Entstörung
interference fading Interferenz-Schwund
interference filter Störschutzfilter
interference fit Haftsitz
interference fringe Interferenzstreifen
interference fringe system System von Interferenzlinien
interference-guard band Schutz-Frequenzband, Sicherheitsabstand (621.39)
interference pattern Interferenzmuster, Interferenzbild
interference photometer Interferenzphotometer
interference preventer Krachtöter (621.396)
interference rejection Störfestigkeit
interference suppressor Krachtöter (621.396)
interference trap Störschutzfilter
interference voltage Störspannung
interfering störend, Stör-...
interfering signal Störzeichen, Störsender, Störsignal (621.396)
interfering transmitter Störsender
interferometer Interferometer
interferric space Luftspalt
intergranular Korngrenz-..., Zwischenkorn-...
interior inner, Innen-...
interior communication Bordsprechanlage
interior conduit I-Rohr, Installationsrohr
interior lighting Innenbeleuchtung
interior view Innenansicht
interior wiring Innen-Installation
interior-wiring system ground Installationserdung
interlace operation verschachteltes Arbeiten
interlacing Zwischenlage (621); Zwischenzeilenverfahren (621.397)
to interlace ineinanderfügen, verflechten
interlaced cooler Schlangenkühler
interlaced scanning Zeilensprungverfahren (621.397)
to interleave Zwischenschichten einlegen (621.3); durchschießen (655); verschachteln
interline flicker Zwischenzeilenflimmern (621.397)
to interlink verketten (621.3); kuppeln (621.316)
interlinking Verbundbetrieb (621.316)
interlock Verriegelung
to interlock (gegenseitig) verriegeln; verblocken; verketten, sich verhaken
interlocking Verriegelung, Verblockung; Verriegelungsvorgang
interlocking apparatus Stellwerk (625)
interlocking bolt Riegelbolzen
interlocking chart Verschlußtafel (625)
interlocking contact Verriegelungskontakt
interlocking contactor Verriegelungsschütz
interlocking device Verriegelung, Verblockung; Verriegelungseinrichtung, Verblockungseinrichtung
interlocking equi-signal system Leitstrahlverfahren
interlocking mechanism Verriegelungsmechanismus
interlocking plant Stellwerk (625); Verriegelungsanlage, Verblockungsanlage
interlocking relay Verriegelungsrelais, Blockierungsrelais
interlocking signals Blocksignale (656.25)
interlocking switch Verriegelungsschalter
intermediary Zwischenglied, Zwischenstück; Mittler, Vermittler
intermediate Zwischenprodukt
intermediate Zwischen-..., Mittel-...
intermediate annealing Zwischenglühung
intermediate cable Zwischenkabel
intermediate circuit Zwischenkreis
intermediate-distribution frame Zwischenverteilergestell (621.39)

intermediate frequency Zwischenfrequenz
intermediate-frequency amplifier Zwischenfrequenzverstärker
intermediate-frequency section Zwischenfrequenzteil
intermediate-frequency transformer Zwischenfrequenzübertrager
intermediate gearing Zwischengetriebe, Zwischenvorgelege
intermediate high voltage Mittelspannung
intermediate image Zwischenbild (535)
intermediate layer Zwischenschicht
intermediate link Zwischenglied
intermediate nucleus Zwischenkern (539)
intermediate pinion Zwischenritzel, Zwischenrad
intermediate-pressure turbine Mitteldruckturbine
intermediate repeater Zwischenverstärker, Leitungsverstärker
intermediate signal Zwischensignal (625.1)
intermediate station Zwischenstation (621.394)
intermediate storage Zwischenspeicher
intermediate terminal Zwischenklemme
intermedial waves Grenzwellen (50-200 m)
intermediate wiring Wechselschaltung, Mehrfachschaltung (621.3)
intermetallic compound halbleitende Verbindung
intermeshing Vermaschung
to intermingle mischen, sich vermischen, durcheinanderkommen, durcheinandergeraten
intermittent aussetzender Fehler, aussetzender Erdschluß (621.3)
intermittent intermittierend, zeitweilig aussetzend
intermittent-contact recorder Fallbügelschreiber
intermittent current intermittierender Strom
intermittent defect zeitweiser Fehler, aussetzender Fehler
intermittent discharge aussetzende Entladung
intermittent duty Aussetzbetrieb
intermittent earth intermittierender Erdschluß, aussetzender Erdschluß
intermittent fault aussetzender Erdschluß, intermittierender Fehler
intermittent integration intermittierende Zählung, Spitzenzählung
intermittent light Blinkfeuer (656.7)
intermittent load aussetzende Belastung, Aussetzlast
intermittent operation Blockschwingerbetrieb (621.396); aussetzender Betrieb
intermittent rating Belastung bei Aussetzbetrieb
intermittent running Aussetzbetrieb (AB)
intermittent service Aussetzbetrieb (621.313)
intermittent-service area Bodenwellenrandzone (621.396)
intermittent test Kapazitätsprüfung bei Programmentladung (621.35)
intermittent train control Zugbeeinflussungsanlage mit festen Kontrollpunkten
intermittent trouble aussetzender Fehler, intermittierender Fehler, aussetzende Störung
to intermix miteinander mischen
intermodulation Kreuzmodulation
intermodulation distortion Intermodulationsverzerrung
intermodulation-noise measurement Rauschklirrmessung
internal innerlich, inner, Innen-...
internal capacitance Eigenkapazität
internal characteristic innere Charakteristik (621.313)
internal circuit innere Schaltung; Inlandsverbindung (621.395)
internal combustion engine Verbrennungskraftmaschine
internal communication Bordsprechanlage
internal conductor Innenleiter
internal drift field inneres Driftfeld
internal equivalent temperature innere Ersatztemperatur (621.382)
internal feedback innere Rückkopplung
internal field alternator Innenpolmaschine (621.313)
internal frosting Innenmattierung (621.326)
internal impendance innerer Widerstand (621.385)
internal line impedance innerer (Wechselstrom-)Widerstand
internal modulation Eigenmodulation, eingebaute Modulation
internal resistance Innenwiderstand, innerer Widerstand; Ausgangswiderstand (eines Gerätes), Ausgangswirkwiderstand
internal self-inductance innere Selbstinduktion
internal stresses innere Spannungen
internal thread Innengewinde
internal voltage EMK, innere Spannung
internal wiring and assembly diagram Bauschaltplan
internally frosted innenmattiert
international international

international candle internationale Kerze (veraltet)
international exchange Auslandsvermittlungsstelle
international standard internationale Norm
international telephone line grenzüberschreitender Leitungsabschnitt
international units internationale Einheiten
internode Schwingungsbauch, Bauch
interoffice mail Hauspost
interoffice trunk Ortsverbindungsleitung, OVL (621.395)
interhase-connecting pipes Phasenkuppelrohre (621.316.5)
interphase reactor Saugdrossel, Ausgleichsdrossel (621.314.5)
interphone Haustelefon; Bordsprechanlage
interphone system Haustelefonanlage
interplane zwischen den Flächen
interpolation Interpolation
interpolator Relaisgeber (621.396)
interpole Wendepol (621.313)
interpole motor Wendepolmotor
interpole winding Wendepolwicklung
interposition trunk Platzverbindungsleitung, Übergabeleitung
interpreter Dolmetscher; Zuordner; Interpretierer; Zuordnungsprogramm (681.3)
to interrelate (miteinander) verknüpfen
interrelation Verknüpfung, Wechselbeziehung
to interrogate abfragen (621.39)
interrogation command Abfrage (621.398)
interrogation pulse Abfrageimpuls
interrogation-suppressed time delay Gesamtlaufzeit eines Transponders
interrogator Abfragesender, Abfrageeinrichtung (621.396)
interrogator responder Interrogator-Responsor
interrupt Unterbrechung, Interrupt (681.3)
interrupt logic Abfragelogik
interrupt handler [spezielles Ein-Ausgabeprogramm]
interrupt request Unterbrechungsanforderung
to interrupt unterbrechen, abschalten
interrupted continuous waves telegraphiemodulierte ungedämpte Wellen
interrupted current unterbrochener Gleichstrom
interrupted ringing Zeichenruf
interrupted spot welding Mehrfachstoß-Punktschweißung
interrupter Unterbrecher
interrupting Unterbrecher-..., Abschalt-...
interrupting capacity Ausschaltleistung
interrupting contact Unterbrecherkontakt, Ausschaltkontakt
interrupting device Abschalteinrichtung
interrupting power Abschaltleistung
interrupting rating Abschaltleistung (621.316.9)
interrupting time Abschaltzeit (621.316.5)
interruption Unterbrechung, Abreißen
interruption arc Abreißbogen
interruption of an arc Abreißen eines Lichtbogens
interruption current Abschaltstrom
interruption of current Stromunterbrechung
interruption of discharge Abreißen einer Entladung
interruption of power supply Stromausfall
interruptive Abschalt-...
interruptive capacity Abschaltleistung (621.316.5)
to intersect (sich) schneiden
intersecting Einschneiden (52)
intersection Schnitt, Schnittpunkt, Schnittlinie
intersection line Schnittlinie
intersection point Schnittpunkt
intersheath Steuerfolie (621.315.2)
interspace Zwischenraum, dazwischen liegender Raum
intersperse dispergieren, fein verteilen, einstreuen
interstage Zwischenstufe
interstage coupling Stufenkopplung (621.396.64)
interstage transformer Zwischenübertrager
interstage-tuning coupling Bandfilterkopplung (zwischen den Stufen)
interstellar interstellar
interstellar noise interstellares Rauschen, kosmisches Rauschen (621.396)
interstice Zwischenraum, Hohlraum, Lücke
interstitial im Zwischenraum liegend
intertangled verfilzt, verschlungen
interthrough switch Zwischenstellenumschalter (621.395)
intertoll dialing Fernwahl
intertoll trunk Fernleitung
intertrack bond Gleisverbinder
intertripping Mitnahmeschaltung (621.318.5)
interturn insulation Windungsisolation
interturn short circuit Windungsschluß
to intertwine verschlingen, verknoten
to intertwist (miteinander) verdrillen
interval Intervall, Zwischenraum, (zeitlicher) Abstand, Zeitspanne, Pause

to intervene dazwischenliegen, dazwischentreten, intervenieren
intervening layer Zwischenschicht
to interweave verweben, verschachteln; ineinanderwickeln (621.314.2); verschränken (621.313)
interwinding capacity Kapazität, Wicklung gegen Wicklung (621.314.2); Wicklungskapazität
interword space Wortabstand
intimate innig, eng
intimate contact innige Berührung
intimate mixture innige Mischung
to intoxicate vergiften
intoxication Vergiftung
intraband telegraphy Einlagerungstelegraphie
intramolecular intermolekular, zwischenmolekular
intrinsic angeboren, innewohnend
intrinsic brightness Leuchtdichte (535)
intrinsic brilliancy Helligkeit, Kontrast
intrinsic conduction Eigenleitung
intrinsic impedance (of free space) Wellenwiderstand des leeren Raumes
intrinsic induction Magnetisierung
intrinsic mobility Eigenbewegung
intrinsic permeability Suszeptibilität
intrinsic semiconductor Eigenhalbleiter
intrinsical innerlich, wesentlich, von Haus aus
intrinsically safe von Haus aus sicher, von sich aus ungefährlich, eigensicher
to introduce einschalten, zwischenschalten; einfügen, einführen
introduction Einführung; Einleitung; Hineinstecken, Einschaltung
intrusion Aufschalten (621.395)
invalid ungültig
to invalidate ungültig machen
invar Invar
invariancy Invarianz
inventory Inventar; Bestandsverzeichnis, Inventarverzeichnis, Lagerliste
inverse Umkehrung, Gegenteil
inverse reziprok; Gegen-...
inverse anode voltage Sperrspannung (621.314.5)
inverse current Sperrstrom
inverse direction of operation inverser Betrieb
inverse feedback Gegenkopplung, negative Rückkopplung
inverse load Sperrbelastung
inverse mixing Rückmischung
inverse neutral telegraph transmission Einfachstrom-Arbeitsstrombetrieb
inverse peak voltage Sperrspitzenspannung, Spitzensperrspannung (621.314.6)
inverse power reziproke Potenz (51)
inverse power factor (Kehrwert des Leistungsfaktors)
inverse power series fallende Potenzreihe
inverse powers fallende Potenzen
inverse proportion umgekehrte Proportionalität
inverse ratio umgekehrtes Verhältnis, reziprokes Verhältnis
inverse speed motor Motor mit Hauptstromcharakteristik
inverse time element AMZ-Glied
inverse time-lag relay AMZ-Relais, abhängiges Maximal-Zeit-Relais
inverse time lag tripping abhängig verzögerte Auslösung
inverse time lag with definite minimum begrenzt abhängige Verzögerung
inverse time limit AMZ-... (Abhängig-Maximal-Zeit)
inverse time-overcurrent protection abhängiger Überstromzeitschutz
inverse-time-overcurrent tripping dynamische Überstromauslösung (621.316.5)
inverse voltage Sperrspannung (621.314.6)
inverse voltage rating zulässige Sperrspannung (621.314.6)
inversely as umgekehrt proportional zu
inversely proportional umgekehrt proportional
inversion Umkehrung; Inversion (621.3); Wechselrichtung (621.314.5)
inversion density Inversionsdichte, Intrinsicdichte (621.382)
to invert umdrehen, umkehren; wechselrichten (621.314.5)
inverting amplifier invertierender Verstärker
inverted umgekehrt, umgedreht
inverted amplifier Gegentaktverstärker in Gitter-Basisschalter
inverted brush contact fester Bürstenkontakt (621.316.5)
inverted crosstalk unverständliches Nebensprechen
inverted "L" aerial umgekehrte L-Antenne
inverted line circuit Entzerrerschaltung
inverted machine rotorgespeiste Maschine
inverted position umgekehrte Lage
inverted rectifier Wechselrichter
inverted repulsion motor Repulsionsmotor mit kurzgeschlossenem Feld
inverted rotary converter Wechselstrom-Gleichstrom-Umformer

inverted speech invertierte Sprache
inverted tube Bremsfeldröhre (621.396)
inverted vacuum tube Kathodenverstärkerröhre
inverter Wechselrichter, Umformer (621.314.5); Inverter (681.3)
investigator Forschungsarbeiter, Forschungsingenieur
involute Evolvente
involute connection Evolventenstirnverbindung
involute gearing Evolventenverzahnung
involution Potenzierung
to involve enthalten, beinhalten
to involve a phase angle einen Phasenwinkel haben
to involve one variable eine Unbekannte enthalten
inward dialling Durchwahl
iodine Jod
ion Ion
ion activity wirksame Ionenkonzentration
ion beam Ionenstrahl, Kanalstrahl
ion bombardment Ionenbombardement
ion burn Ionenfleck (621.385)
ion concentration Ionenkonzentration
ion exchange Ionenaustausch (54)
ion implantation process Ionenimplantations-Prozeß
ion migration Ionenwanderung
ion spot Ionenfleck
ion trap Ionenfalle (621.397)
ion yield Ionenausbeute
ionic Ionen-...
ionic breeze Ionenwind
ionic centrifuge Ionenschleuder (539)
ionic conduction Ionenleitung
ionic current Ionenstrom
ionic focusing Gasfokussierung, Gaskonzentration
ionic jet Ionenstrahl
ionic mobility Ionenbeweglichkeit
ionic speed Ionengeschwindigkeit, Wanderungsgeschwindigkeit
ionic valve Ionenrohr, gasgefüllle Röhre, Gasentladungsröhre
ionic velocity Ionengeschwindigkeit, Wanderungsgeschwindigkeit
ionic wind elektrischer Wind, Ionenwind
ionic-wind voltmeter Ionenwindvoltmeter
ionics Ionenlehre
ionization Ionisation
ionization chamber Ionisationskammer
ionization by collision Stoßionisation, Stoßionisierung
ionization current Ionisationsstrom
ionization gauge Ionisationsmanometer
ionization by impact Stoßionisation
ionization potential Ionisierungsspannung
ionization rate Ionisationsgeschwindigkeit
to ionize ionisieren
ionized ionisiert
ionized gas ionisiertes Gas
ionized-gas detector Glimmlichtdetektor
ionizid layer ionisierte Schicht
ionizing potential Ionisierungsspannung
ionogen Ionenbildner
ionometer Röntgen-Dosismesser
ionosphere Ionosphäre
ionosphere layer Ionosphärenschicht
ionospheric Ionosphären..., ionosphärisch
dionospheric conditions (pl.) Zustand in der Ionosphäre
ionospheric cross modulation ionosphärische Kreuzmodulation, Luxemburgeffekt
ionospheric forecast Funkwettervorhersage
ionospheric focusing ionosphärische Fokussierung
ionospheric layer Ionosphärenschicht
ionospheric path Ausbreitungsweg in der Ionosphäre
ionospheric-path error Raumwellenfehler
ionospheric scatter propagation ionosphärische Streuausbreitung, ionosphärischer Überhorizont-Richtfunk
ionospheric shell Ionosphärenhülle
ionospheric sounding Ionosphärenlotung
ionospheric wave Raumwelle
IPD-action controller IPD-Regler
ipot induktives Potentiometer, Spulenpotentiometer
i. r. drop ohmscher Spannungsabfall
iridescent irisierend, schillernd
iridium Iridium
iris Iris
iris diaphragm Irisblende
i^2r loss Kupferverluste (pl.)
iron Eisen; Bügeleisen
iron alloy Eisenlegierung
iron-bound regulations eiserne Vorschriften
iron carbide Eisenkarbid
iron casting Eisenguß, Gußstück
iron-clad eisengekapselt
iron-clad case Eisenkapselung, Gußkapselung
iron-clad switchgear gußgekapseltes Schaltgerät
iron-clad wattmeter eisengeschlossenes Wattmeter

iron core Eisenkern
iron-core modulation Drosselmodulation
iron-core reactor Eisendrosselspule
iron-cored mit Eisenkern
iron-cored coil Eisenkernspule
iron-cored inductor Eisenkernspule
iron-cored intervalve transformer Zwischenübertrager mit Eisenkern (621.396.64)
iron-cored transformer Eisenkerntransformator, Eisentransformator, Übertrager mit Eisenkern
iron-cored wattmeter eisengeschlossenes Wattmeter
iron curve Magnetisierungskurve des Eisens
iron-cutting saw Eisensäge, Kaltsäge
iron dust Eisenpulver (621.396)
iron-dust core Eisenpulverkern (621.396)
iron-dust slug Pulverkernstopfen, Eisenpulverkern (621.396)
iron-dust tuning slug Abstimmkern aus Eisenpulver
iron filings Eisenfeilspäne; Eisenfeilicht (sing.)
iron glance Eisenglanz
iron-hydrogen resistance Eisen-Wasserstoff-Widerstand
iron loss Eisenverlust (621.3)
iron metals Eisenmetalle
iron-nickel storage battery Nickel-Eisen-Akkumulator
iron ore Eisenerz
iron punching Blechschnitt
iron pyrite Eisenpyrit, Schwefelkies
iron specimen Eisenprobe
iron stamping Blechschnitt
iron-strip earth conductor Banderder
iron vane Eisenfahne, Eisenblättchen (621.317)
iron wire Eisendraht
to iron bügeln, plätten
to iron out ausbügeln, glätten, einebnen
ironing effect glättende Wirkung
ironless ohne Eisenkern
ironless circuit eisenloser Kreis
ironstone Eisenerz, Eisenstein
irradiancy Bestrahlungsstärke
to irradiate bestrahlen, bescheinen, einstrahlen
irradiation Einstrahlung, Bestrahlung
irradiation channel Bestrahlungskanal (539)
irrational irrationale Zahl, irrational
irreducible ununterschreitbar
irregular unregelmäßig
irregular-shaped unregelmäßig geformt
irregularity Unregelmäßigkeit
irregularity factor Ungleichförmigkeitsgrad
irreparable unreparierbar
irresolvable unlösbar, unauflöslich; ununterteilbar
irresponsive to nicht ansprechend auf
irreversibility Nichtumkehrbarkeit
irreversible nicht umkehrbar, irreversibel
irreversible process nicht umkehrbarer Prozeß
irreversible transformation irreversible Umwandlung
irrodational field wirbelfreies Feld
I-scan polare Abtastung, Polarbild (621.397)
ISDN = integrated system digital network
I-section Doppel-T-Profil, I-Profil
isinglass Fischleim
island effect Inselbildung (621.385)
Island furnace Stickstoffofen mit rotierender Elektrode
Isle of Man connection Zickzackschaltung
ISO = International Organization for Standardization
isocandle ... gleicher Lichtstärke
isocandle diagram Lichtstromdiagramm
isochromatic gleichfarbig
isochronism Gleichlauf
isochronous isochron
isochronous circuits abgestimmte Kreise
isochronous distortion Isochron-Verzerrungsgrad (621.394)
isoclinal isoklin, gleich geneigt; Isoklinal-... (548)
isocline Isokline, Linie gleicher Neigung (51); Isokline (538.7)
isodynamic line Linie gleicher Horizontalfeldstärke (52)
isolantite Isolierpreßmaterial
to isolate isolieren, voneinander trennen, abtrennen; rein darstellen (54); aussondern, extrahieren; freischalten (621.395)
isolated conductor Einzelleiter
isolated crane freistehender Kran
isolated by elevation oberhalb des Handbereichs liegend
isolated generating plant Blockstation
isolated neutral isolierter Sternpunkt
isolated neutral system isoliertes Netz
isolated phase switchgear Schaltanlage mit getrennten Phasen
isolated section abgetrennter Teil
isolating Trenn-..., Isolier-...
isolating air gap Lufttrennstrecke
isolating contacts Trennkontakte
isolating fuse Trennsicherung

isolating gap Trennfunkenstrecke
isolating joint Isoliermuffe
isolating link Trennstück
isolating resistor Trennwiderstand, Entkopplungswiderstand
isolating switch Trennschalter
isolating transformer Isoliertransformator
isolating valve Absperrventil
isolation Abtrennung, Absperrung, Ausschaltung
isolator Trennschalter
isolit Lackpappe (Handelsname)
isolux ... gleicher Beleuchtungsstärke
isolux diagram Lichtverteilungsdiagramm
isomorphous gleichgestaltig, isomorph
isoperm Isoperm
isosceles gleichschenklig
isotope Isotop
isotope battery Isotopenbatterie
isotope radiography radioaktive Werkstoffuntersuchung
isotopic abundance Isotopenhäufigkeit
isotron Isotron, Laufzeit-Isotopentrenner
isotropic isotrop
isotropic body isotroper Körper
isotropic conductivity isotrope Leitfähigkeit
isotropic dielectric isotroper Isolierstoff
isotropic radiator isotroper Strahler, Kugelstrahler (621.396.67)
to issue ausgeben, herausgeben; herauskommen, heraustreten (**trough, from:** aus)
to issue a patent ein Patent erteilen
item Position, Punkt, Posten; Teilwert, Bestandteil, Bauteil, Zubehörteil, Teil
item-list Stückliste
to itemize aufteilen, positionsweise angeben, spezifizieren
to iterate wiederholen, iterieren (51)
iterated fission expectation asymptotische Spalterwartung (539)
iteration Iteration (51)
iterative attenuation (US) Kettendämpfung
iterative impedance Wellenwiderstand, Kettenwiderstand
iterative phase constant Kettenwinkelmaß
iterative propagation constant Kettenübertragungsmaß
iterative transfer constant Kettenübertragungsmaß

J

J = Joule Joule
Jablochkoff candle Jablochkoff-Kerze (veraltet)
jack Klinke, Buchse (621.395); Wagenheber, Heber, Winde, Hebebock
jack shaft Zwischenwelle (621.335)
jack strip Klinkenstreifen, Klinkenreihe
to jack up aufbocken, hochbocken
jacket Mantel (621), Gehäuse
to jacket mit einem Mantel versehen, ummanteln
jacketed pan Doppelmantelkessel
jag scharfe Kante
jagged scharfkantig
jalousie Jalousie
jam nut Kontermutter
to jam (sich) zustopfen (621.396); (sich)verstopfen, sich zusetzen; (sich) klemmen, (sich) festklemmen (621.9); einzwängen, hineinzwängen
to jam against sich verfangen in, sich stauen gegen
to jam in sich festklemmen in
to jam a receiver einen Empfänger zustopfen
jammer Störsender (623)
jamming Stauung; Verklemmung, Zustopfen; vorsätzliche Störung (23)
jamming transmitter Störsender (621.396)
jar Dose, (flaschenartiger) Behälter; Stoß, Schlag, Erschütterung
jaw Kiefer (61); Backe (621)
jaw crusher Backenbrecher
jaw lining Backenfutter, Backenfütterung; Bleibacken (pl.)
jaw-type terminal Maulklemme
jerk Ruck, Zuck, Rucken
to jerk rucken, sich ruckartig bewegen, anrucken
jerky sprungartig, ruckartig
jet Düse, Strahldüse, Strahlrohr; Strahl; Jet, Diamentenkohle
jet deflector Strahlabweiser, Strahlablenker, Ablenkblech (621.2)
jet nozzle Strahlrohr
jet propulsion Strahlantrieb, Düsenantrieb
jet of steam Dampfstrahl
jet-wave rectifier Quecksilberstrahl-Gleichrichter
jewel Edelstein, Juwel; Lagerstein, Steinlager (621.317), (681)
jewel(led) bearing Steinlager, Diamantlager
jib crane Auslegerkran
JFET = Junction FET Sperrschicht-Feldeffekt-Transistor

jig gedämpfter Wellenzug (621.396); Vorrichtung, Bohrlehre; Halterung (621.317)
jig saw Laubsägemaschine
to jig aufspannen, in eine Vorrichtung einlegen, spannen, einspannen; waschen, separieren (66)
jigger Antennen-Transformator
jigging Vorrichtungsbau
jingle Klappern, Schlüsselklappern
to jingle klimpern, klappern
jitter Wackelei, Verwackeln (621.397); Flimmern, Flackern, Flattern, Zittern; Synchronisationsfehler
jittery zitterig, wackelig
J-K flip-flop J-K Flip-Flop
job Arbeit; Arbeitsplatz; Stück, Werkstück; Auftrag, Job (681.3)
job step Jobschritt, Befehl an das Betriebssystem (681.3)
jogging Klapperschaltung (621.34)
joggle Verzapfung, Falz
to joggle verzapfen, verzahnen
Johnson noise Wärmerauschen, thermisches Rauschen
join Verbindungslinie
to join vereinigen, zusammenfügen, verbinden; sich anschließen
to join capacitively kapazitiv zusammenschalten
to join in parallel parallel anschließen
to join a resistance across a line einen Widerstand an eine (durchlaufende) Leitung anschließen
to join in series in Reihe schalten, in Reihe anschließen
to join to anschließen an (621.3); befestigen an (621)
to join together zusammengarnieren (666)
joining piece Ansatzstück
joiner Tischler, Möbeltischler, Schreiner; Fügemaschine (67)
joint Gelenk (61), (621); Anschluß, Verbindung, Scharnier (621), (621.3); Fuge (69); Knoten, Knotenpunkt (624); Schienenstoß (625)
joint box Verbindungsdose
joint load Knotenpunktlast, Knotenkraft (624)
joint sleeper Stoßschwelle (625)
joint testing Verbinderprüfung
joint vereinigt, gemeinsam
joint resistance resultierender Widerstand, Muffenwiderstand (621.3)
to joint Kanten hobeln; fügen (67); fugen, verfugen (69); vermuffen (621.3)
jointed gemufft; mit Gelenk verbunden
jointed coupling Gelenkkupplung
jointing Vermuffung (621.315.2); Verbindung, Zusammenschaltung
jointing fitting Verbindungsstück
to jolt hopsen; rütteln
jolting device Rütteleinrichtung, Rütteltisch
Joule heating effect Joulesche Wärmeentwicklung
Joule's law Joulesches Gesetz
Joulean dissipation Joulesche Wärme, Stromwärme
journal Wellenzapfen (621)
journal bearing Achslager
journal load Wellenlast, Lagerkraft, Auflagedruck, Lagerdruck, Querkraft
joystick Steuerknüppel
jump Abschnitt, Schritt (51); Sprung (51) (681.3)
jump instruction Sprungbefehl
jump-over distance Sprungentfernung (621.396)
jump spark Überschlagsfunken
jump weld stumpfe T-Naht (621.791)
jump wire Überbrückungsdraht
to jump hüpfen, springen, überspringen
to jump across überspringen, springen über
to jump back zurückfallen (**to:** auf)
jumper fliegender Anschluß, Anschlußstrippe, Rangierdraht, Kurzschlußbügel
jumper receptacle Kupplungsdose
jumper wire Kurzschlußbügel (621.396)
junction Verbindungspunkt, Verzweigungspunkt, Verbindung, Verzweigung, Knotenpunkt (621.3); Schnittpunkt, Kreuzungspunkt (51); Anschlußpunkt, Stoßstelle (621.3); Lötstelle (621.396); Sperrschicht, Zonenübergang, Übergang, Grenzschicht, Störstellenübergang (621.382)
junction box Anschlußdose, Verteilerdose, Abzweigdose
junction circuit Amtsverbindung (621.395)
junction diode Flächendiode
junction FET Sperrschicht-Feldeffekttransistor
junction-gate field-effect transistor Feldeffekttransistor mit PN-Übergang, Sperrschicht-Feldeffekttransistor
junction key Nachbaramtstaste (621.395)
junction line Amtsverbindung (621.395)
junction plate Knotenblech
junction point Verzweigungspunkt; Abzweigpunkt, Verbindungsstelle, Verbindungspunkt
junction rectifier Sperrschichtgleichrichter
junction selector Amtswähler

junction stage Durchschaltebene (621.395.3)
junction surface aktive Fläche (621.382)
junction temperature Sperrschichttemperatur
junction transistor Flächentransistor
juncture Verbindungsstelle, Fuge, Naht
jute Jute
jute-insulated cable Kabel mit Juteisolation
jute thread Jutefaden
to juxtapose nebeneinander setzen, nebeneinander stellen, nebeneinander legen
juxtaposition Umsetzung, Umstellung

K

kaolin Kaolin, Porzellanerde
Kaplan turbine Kaplanturbine
Kapp coefficient (Quotient aus Fluß × Windungszahl und EMK) (621.313)
kapp line (obs) (Einheit des magnetischen Flusses = 6000 M)
Karolus cell Karoluszelle
Karolus system of phototelegraphy Karolussystem für Bildtelegraphie
kathode = cathode (s. dort)
kc = kc.p.s. kHz
keen scharf, tüchtig
keen edge scharfe Kante, scharfe Schneide
keen hearing scharfes Gehör, scharfes Ohr
keen tool scharfes Werkzeug, scharfer Stahl
to keep halten, behalten, festhalten, sich halten; aufbewahren
to keep alight brennend halten; gezündet halten (621.585)
to keep its charge seine Ladung beibehalten
to keep cool kühl halten, kühl aufbewahren
to keep flowing am Fließen halten, in Fluß halten; die Strömung aufrechterhalten
to keep level waagerecht halten; den Pegel halten
to keep pointed at nach ... ausgerichtet halten, anbinden (621.396.67)
to keep the same length gleich lang halten
to keep the same weight gleich schwer halten, gleich schwer bleiben
to keep in service in Betrieb halten
to keep stationary festhalten, stabilisieren, stationär halten (**at:** bei)
to keep in step with synchron bleiben mit
to keep time richtig gehen (681)
of keeping qualities haltbar, beständig
keep-alive electrode Vorionisationselektrode
keeper Anker, Eisenschluß
K-electron capture K-Einfang, Einfang eines K-Schalen-Elektrons
kell factor Kellfaktor (621.397)
Kellog switch Kellogschalter
Kelvin balance Kelvinsche Stromwaage
Kelvin bridge Thomson-Brücke, Doppelbrücke
Kelvin electrometer Quadrant-Elektrometer
Kelvin formula Thomsonsche Schwingungsgleichung
Kelvin quadrant electrometer Kelvin-Quadratelektrometer
Kennelly-Heaviside layer Ionosphäre, Heaviside-Kennelly-Schicht
kenotron Hochvakuumgleichrichterrohr
kerf Schnittspalt, Fuge (621.791), (621.9); Schnittfläche (67)
kerosene Leuchtöl
Kerr's constant Kerrkonstante
Kerr cell Kerrzelle
key Schlüssel, Schraubenschlüssel; Taste (621.394), (681.3), (621.317); Keil, Wellenkeil (621); Kippschalter (621.3)
key to abbreviations Verzeichnis der Abkürzungen
keyboard Tastatur, Tastenfeld; Wähltastur (621.395)
keyboard dialling Tastaturwahl
keyboard encoder Tastaturcodierer
keyboard perforator Handlocher (621.394)
key click Tastklick (621.394)
key controlled getastet, telegraphiemoduliert
key holder Schalterfassung (621.32)
keyhole Schlüsselloch
key is down Taste ist zu, Taste ist gedrückt (621.394)
key-operated switch Schlüsselschalter
keypiece Nutenkeil, Verschlußkeil (621.313)
key pulsing Tastwahl
keypunch Handlocher, Tastenlocher
key sender Tastenwähler (621.395)
key sequence Tastenfolge
key to signs Zeichenerklärung, Zeichenerläuterung
key switch Fassungsschalter
key is up Taste ist offen, Taste ist oben (621.394)
key-verifying unit Lochkartenprüfgerät
keyword Schlüsselwort
to key aufkeilen, festkeilen, (mit Keil) befestigen (621); tasten (621.394), (621.396); fixieren, festlegen
keyed to fest verbunden mit

keyer Impulszentrale, Tastgerät
keying Tastung (621.394)
keying tube Taströhre
keying filter Tastfilter
to keypunch ablochen (681.3)
keystone distortion Trapezverzerrung
kg. cal. = kilogram calorie kcal = Kilokalorie
kick Stoß
to kick stoßen, schlagen
kicking coil Drosselspule, Drossel
kicker Stößel, Ausstoßer
kieselguhr Kieselgur
Kilburn Scott furnace Drehstrom-Stickstoffofen
kilderkin Faß (als Hohlmaß verschieden je nach Inhalt)
killer circuit Austastschaltung (621.397)
kiln Brennofen; Kalkofen
kiln-dried ofentrocken
kiloamp Kiloampere, kA
kilobyte = 2^{10} Byte = 1024 Byte = 1 Kilobyte
kilocycles (per second) Kilohertz, kHz
kilogram(me) Kilogramm
kilohertz Kilohertz
kiloline (Einheit des magnetischen Flusses = 1000 M)
kilolitre Kiloliter
kilometre Kilometer
kilometric waves Langwellen
kiloohm Kiloohm
kilovar Blindkilowatt, Blind-kVA
kilovarhour Blindkilowattstunde
kilovolt Kilovolt
kilovoltampere Kilovoltampere
kilovoltampere-hour meter kVA-Zähler
kilowatt Kilowattstunde
kilowatt-hour meter Kilowattstunden-Zähler
kinematic kinematisch
kinematic viscosity kinematische Zähigkeit
kinescope Fernsehprojektionsröhre
kinetic kinetisch
kinetic energy kinetische Energie
kinetic theory of gases kinetische Gastheorie
king point Einfahrtsweiche, Hauptweiche (625)
kink Knick, „Katzenkopf", Klinken
to kink sich vertörnen, „Katzenköpfe bilden"
kipp oscillations Kippschwingungen
Kipp's apparatus Kippscher Apparat
Kirchhoff's laws Kirchhoffsche Gesetze
kit Kasten, Satz, Bausatz
kitchen range Küchenherd
Kjellin furnace Induktionsrinnenofen
klaxon Hupe
klydonograph Klydonograph
klystron Klystron, Laufzeitröhre
klystron oscillator Klystronsender
to knead kneten
kneading Knet-. . ., Kneten, Knetvorgang
kneading mill Kneter, Knetmaschine, Knetmühle
knee Knie (61); Knie, Kniestück, Krümmer (621.6); Auslegertisch (621.9)
to knick knicken, sich verschlingen, Knicke bekommen
knife (pl. **-ves**) Messer; Abstechstahl (621.9)
knife-blade connector Messersteckverbinder
knife blade contact Messerkontakt
knife edge Messerschneide; Lagerschneide, Schneidenlager (681)
knife edge diffraction Kantenbeugung
knife switch Messerschalter, Hebelschalter
knife tool Schlichtstahl; Abstechstahl
knob Knopf, Drehknopf, Einstellknopf
knock Schlag; Klopfen (621.4)
to knock against zusammenstoßen mit, kollidieren, abbrechen
to knock out herausschlagen, herausstoßen
to knock out electrons Elektronen auslösen
knock-out bar Ausstoßstift (621.9)
knocker Klopfer, Türklopfer
knockout Ausbrechöffnung (696)
knot Knoten; Ast (674); Knoten = Seemeile/Stunde (1 knot = 6080 ft/h)
Knowles cell Hydrolyseur
knuckle Knöchel (61); Rohrgelenk (621.6); Puffer, Mittelpuffer (625)
knuckle action Kniehebelwirkung
knuckle joint Gelenkverbindung
knuckle thread Rundgewinde
to knurl kordeln, rändeln
knurled knob Rändelknopf
knurled nut Rändelmutter, Kordelmutter
knurled ring Rändelring, Kordelring
Koepsel permeameter Köpsel-Apparat
Kohlrausch bridge Kohlrausch-Brücke
Krämer control Drehzahlregelung mit Hintermotor
Krarup cable Krarup-Kabel
krypton Krypton
k-sequence k-System
k-sequence active power Wirkleistung der symmetrischen Komponente der k-ten Folge
kV = kilovolt kV = Kilovolt
kVA = kilovoltampere kVA = Kilovoltampere
kVA meter Scheinleistungszähler
kVAr = kilovoltampere reactive kVAr = Kilovar
kW = kilowatt kW = Kilowatt
kWh = kilowatthour kWh = Kilowattstunde

kWh meter kWh-Zähler
kWhrs. = kilowatthours kWh = Kilowattstunden
kybernetics Kybernetik

L

lab. = laboratory Lab. = Labor = Laboratorium
label Anhängezettel, Anhängeschild, Etikett, Anhänger; Aufklebezettel; Bezeichnungsschild; Merkmal (681.3)
to label mit Anhängern versehen, mit Aufklebezetteln versehen, bekleben; bezeichnen, benennen
labelled instructions Etikett mit Gebrauchsanweisung, Anhänger mit Bedienungsanweisung
lability (rare) Labilität
to labor (US) mühsam laufen, schwer laufen (s. auch **to labour**)
laboratory Laboratorium
laboratory desk Labortisch, Laboratoriumstisch
laboratory equipment Laboratoriumsgerät, Laboratoriumseinrichtung
laboratory test of a meter Zählerprüfung in der Werkstatt
laboratory tool Laboratoriumshilfsmittel, Laboratoriumswerkzeug, Laborgerät
labour Arbeit; Arbeitskräfte (pl.)
labour-saving arbeitssparend
to labour (schwer) arbeiten, „schuften", sich mühen, sich quälen; schwer laufen
labourer (ungelernter) Arbeiter
labyrinth Labyrinth
labyrinth packing Labyrinth-Packung
labyrinth protection Labyrinth-Explosionsschutz
labyrinth seal Labyrinthdichtung
to lace klöppeln, flechten
to lace on aufwickeln auf
to lace wire on a reel Draht auf eine Rolle wickeln
to lace through durchflechten, durchfädeln
to lace tight anziehen, stramm holen
to lace up zubinden
lacing Verschnürung
lack Fehlen, Mangel
lack of Fehlen von, ...-mangel
lack of focus Unschärfe
lack of symmetry error Unsymmetriefehler (621.396.9)
to lack fehlen
Lacour converter Kaskadenumformer
lacquer Lack, Zaponlack, Farblack; Bindemittel, Klebmittel; farbloser Lack (681); blonder Lack (681)
lacquer film Lackfilm
to lacquer lackieren, blond lackieren (681)
lacquered wire Lackdraht
ladder network Kettenleiter, Kettenschaltung (621.3)
to lade beladen, belasten, befrachten (**with:** mit)
laevo-rotary (optisch) linksdrehend (535), (54)
lag Nacheilung, Verzögerung; Schlag (67)
lag circuit Verzögerungsschaltung
lag in phase Nacheilung (621.3)
to lag nacheilen gegen
to lag behind nacheilen gegen (621.3)
to lag by nacheilen um
current lags the voltage by 30° Strom eilt gegenüber der Spannung um 30° nach
lagged-demand meter Spitzenzähler mit Zeitverzögerung
lagging load nacheilende Last, induktive Last
lagging power factor nacheilender Leistungsfaktor, induktiver Leistungsfaktor
laid verlegt (s. **to lay**)
laid cable verlegtes Kabel
Lambert Lambert
Lambert's law Lambertsches Gesetz (55)
lamellar field wirbelfreies Feld
lamellar magnetism, lamellar magnetization Quermagnetisierung
lamine (pl. **-nae**) Schicht, Schichte (55)
laminar laminar
laminar flow Laminarströmung
laminate Preßstoff
to laminate lamellieren; blechen, aus Blechen aufbauen
laminated lamellenförmig, lamelliert
laminated block board Tischlerplatte mit Stäbcheneinlage
laminated brush lamellierte Bürste (621.3)
laminated conductor unterteilter Leiter, Röblstab
laminated contact Bürstenkontakt (621.316.5)
laminated core Blechkern
laminated form of structure Lamellenstruktur
laminated magnet lamellierter Magnet
laminated pole lamellierter Pol, geblechter Pol
laminated pole-shoe lamellierter Polschuh
laminated rotor core Läuferblechpaket
laminated yoke Blechjoch, lamelliertes Joch

laminating wood Schichtholz
lamination Lamelle, lamelliertes Blech, Blechschnitt; Dynamoblech, Transformatorenblech
lamination insulation Blechisolation
laminator Beschichtungseinrichtung
lamp Glühlampe, Lampe
lamp base Lampensockel
lamp bulb Lampenkolben, Kolben; Birne, Lampe
lamp cap Lampenabdeckung, Lampenschirm; Lampensockel
lamp cord Fassungslitze; Aufhängelitze
lamp-holder Lampenhalter; Lampenfassung, Fassung
lamp holder plug Fassungsstecker
lamp light Lampenlicht
lamp post Laternenpfahl
lamp receptacle Lichtsteckdose
lamp regulator Laderegler; (621.33)
lamp resistance Lampenwiderstand
lamp shade Lampenschirm
lamp socket Lampenfassung
lamp stand (64) Lampenfuß
lamp synchronoscope Synchronisierlampen (pl.)
lamp wire connector Lüsterklemme
Lancashire boiler Flammrohrkessel, Zweiflammrohr-Kessel
lanceolate lanzettförmig
land Land; Land, Grund, Boden; Hinterschneidungsfläche (621.9)
land-based an Land, am Boden, ortsfest
land-base station feste Landfunkstelle, ortsfeste Funkstelle
land-mobile station bewegliche Landfunkstelle
land return Bodenecho
to land landen, an Land gehen; landen, ausladen, ausschiffen
landing Landung, Ausschiffung, Ausladung; Treppenabsatz, Podest
landing-area floodlight Landeflächenbeleuchtung (656.7)
landing-call push Fahrstuhl-Stockwerksdruckknopf
landing flares Landepfadbefeuerung
landing floodlight Landungsscheinwerfer
landing-direction light Landerichtungsfeuer (656.7)
landing ground Landeplatz, Flugplatz, Rollfeld
landing light Landeleuchte, Landescheinwerfer; Landebahnleuchte
landing rue Landebahn, Landeweg
landing switch Treppenhausschalter
landing wire Spanndraht (zur Aufnahme von Gewichtskräften) (629.13)
landmark beacon Hindernisbefeuerung (656.7)
lane identification Streifenidentifizierung, Streifenkennung (621.396.96)
Langevin ion Langevin Ion
langvage editor Sprachen-Editor
language laboratory Sprachlabor
lantern laterne; Handlampe
lantern slide Lichtbild, Diapositiv
lap Überlappung; Läppen, Läppvorgang; Windung, Runde, „einmal rum"
lap coil überlappte Spule
lap dissolving Überblendung (621.397)
lap joint überlappter Stoß, Lappnaht; Laschenverbindung (621.791)
lap-seam weld überlappte Schweißung
lap-welded pipe überlappt geschweißtes Rohr
lap winding Schleifenwicklung (621.313)
to lap überlappen, überlappt wickeln; umrändern; läppen
lapel Rockaufschlag
lapel microphone Knopflochmikrophon
Laplace's equation Laplacesche Gleichung
Laplace transform Laplace-Transformierte
Laplace transformation Laplace-Transformation
lapped überlappt, geläppt
lapse Lapsus, (geringfügiger) Fehler; Verstreichen, Vergehen
lapse of time Zeitverlauf, Ablauf der Zeit
lard oil tierisches Öl, tierisches Fett
large groß, weit
large-area contrast Grobkontrast (621.397)
large-area plate Großflächenplatte
large-area radiator Flächenstrahler
large grain grobes Korn, Grobkorn
large-grained grobkörnig
large ion schweres Ion
large power motor Motor mit mehr als 1 PS
large scale Groß-..., in großem Maßstabe
large-scale integration Großintegration
large-scale production Großserienfertigung
large signal Großsignal
large size große Abmessungen (pl.)
large volume große Menge, große Lautstärke (621.396), großes Volumen
in large in natürlicher Größe
laryngaphone Kehlkopfmikrophon
laser (light amplifiction by stimulated emission of radiation) Laser
laser(-beam) printer Laser(-strahl) Drucker
laser scanner Laser-Lesegerät (681.3)

to lash (fest) anbinden
to lash on to festbinden an
last Last (Hohlmaß)
last-choice rout Letztweg (621.395.3)
last in, first out = LIFO Kellerspeicher
last-party release Auslösen durch die zuletzt auflegenden Teilnehmer
lat. = latitude Br. = Breite
latch Klinke, Drücker, Riegel; Verriegelung, Verklinkung; Auffang-Flipflop, Latch
latch key Türdrücker, Drücker; Haustürschlüssel
latch plate Riegelschiene, Riegelblech (621.395)
latch relay Stützrelais
to latch verriegeln, feststellen, einklinken
to latch a pushbutton einen Druckknopf einschnappen lassen; eine Taste feststellen
latching current Einraststrom (621.314.63)
latency time Wartezeit (681.3)
latent latent, verborgen
latent defect verborgener Fehler, nicht offenkundiger Fehler
latent heat latente Wärme
latent heat of evaporation Verdampfungswärme
latent heat of fusion Schmelzwärme
latent heat of the steam Verdampfungswärme
lateral seitlich, Seiten-...
lateral area Mantelfläche (514)
lateral axis Querachse (629.13)
lateral blower Querstromlüfter
lateral conductor seitliche Ableitung (621.315.1)
lateral deviation error Peilfehler durch Großkreisausbrechung
lateral pressure seitlicher Druck, Seitendruck
lateral surface Seitenfläche, Mantelfläche
lateral view Seitenansicht
lateral working space seitliche Bewegungsfreiheit (621.315)
laterally correct drawing seitenrichtige Zeichnung
lath Latte, Leiste
lathe Drehbank; Furnierschäler (67)
Latour-Winter-Eichberg motor Winter-Eichberg-Motor
lattice Gitter, Gitterwerk (72); Gitter (54); Hyperbelschar (621.396.9)
lattice circuit Kettenleiter
lattice coil Gitterspule (621.396)
lattice-connection Gitterstirnverbindungen (pl.)
lattice defect Gitterfehler (54)
lattice deformation Gitterstörstelle (54)
lattice imperfection Gitterfehler (621.382)
lattice mast Gittermast (621.315)
lattice network Kettenleiter, Differentialfilter
lattice parameter Gitterkonstante (548)
lattice pitch Gitterabstand
lattice section Kreuzglied
lattice winding Faßwicklung, Korbwicklung
Laundry drier Wäscheschleuder, Wäschetrockner
lava insulation $Mg_3H_2Si_4O_{12}$
law Gesetz
law of conservation of energy Energiesatz, Energieprinzip
law of conservation of momentum Impulssatz
law of equilibrium Gleichgewichtsbedingungen
laws of impact Stoßgesetze
law of mass action Massenwirkungsgesetz
law of mass conservation Massenerhaltungsgesetz, Gesetz von der Massenkonstanz
law of momentum Impulssatz
law of Poiseuille Poiseuillesches Gesetz
law of refraction Brechungsgesetz
Lawrence tube Gitterablenkröhre, Einstrahlröhre mit Farbsteuergitter (621.397)
lay Schlaglänge (621.315.1)
lay ratio relative Schlaglänge (621.315.1)
to lay legen, hinlegen; auflegen, auftragen
to lay a cable ein Kabel legen, ein Kabel verlegen
to lay a coat of paint einen Farbanstrich auftragen
to lay off aufzeichnen, anzeichnen, anreißen (621.9)
to lay out entwerfen, anordnen, auslegen
to lay a pipeline eine Rohrleitung legen, eine Rohrleitung verlegen
to lay a rope ein Seil schlagen
to lay underground in die Erde verlegen
layer Lage, Schicht
layer of earth Erdschicht, Erdlage
layer-type resistor Schichtwiderstand
layer of wire Drahtlage
laying Verlegen, Verlegung
layout Entwurf, Planung, Auslegung; Anordnung
to layout entwerfen, auslegen
lay-up Stapel (674)
lb = libra = pound Pfund (**1 lb** = 453,6 g)
lb. ft. = footpound Fußpfund (**1 lb. ft.** = 1356 Nm)

lb./sq.in. = pound per sqare inch Pfund/Quadratzoll (= 6,9 kN/m^2)
LCD = liquid crystal display Flüssigkristallanzeige
L.C.M. = lowest common multiple = least common multiple (kleinster) Generalnenner
lead Zuführung, Zuleitung (621.3); Führung (621); Steigung (621.8) (bei n-gängigen Schrauben ist **lead = n · pitch**); Voreilung (621.3)
lead in Einführung, Anschlußleitung
lead-in wire Einführung, Einführungsdraht, Durchführungsdraht
lead inductance Zuleitungsinduktivität, Induktivität der Zuführungen
lead polarity Polarität der Klemmen
lead resistance Widerstand der Zuleitungen, Zuleitungswiderstand, „Strippenwiderstand"
lead and return Hin- und Rückleitung
lead-through bushing Durchführung
lead wire Zuführungsleitung, Zuführungsdraht
lead Blei
lead accumulator Blei-Akkumulator, Bleisammler
lead acid battery Bleisammlerbatterie, Bleisammler
lead cable Bleikabel
lead cell Bleiakkumulator, „Bleiakku"
lead-coated verbleit
lead-covered mit Bleimantel
lead-covered cable Bleikabel, Kabel mit Bleimantel
lead covering Bleimantel, Bleiummantelung
lead dioxide Bleisuperoxid
lead hammer Bleihammer
lead jointing Bleimuffe
lead lining Bleibelag, Bleifutter, Bleiauskleidung
lead mold Bleiform (621.357)
lead pipe Bleirohr
lead plate Bleiplatte, Bleiblech
lead rubber Bleigummi
lead seal Bleidichtung
lead sheath Bleimantel, Bleihülle
lead sheathed cable Bleimantelkabel, Bleikabel
lead sheathing Bleimantel, Bleiumhüllung
lead tree Bleibaum (621.35)
to lead führen, anführen, leiten; voreilen gegen (621.3)
to lead by an angle of ... um ... voreilen gegen
to lead in einführen
leading-in cable Einführungskabel, Hausanschlußkabel
leading-in tube Einführung, Einführungsisolator
leading-in wire Einführungsleitung, Anschlußleitung
leaden bleiern
leader Führer; Vorlauf (67)
leader cable Leitkabel, Fahrrinnenkabel (527)
leading voreilend (621.3)
leading current voreilender Strom
leading edge Vorderkante (621.3); Vorderflanke (621.374)
leading phase voreilende Phase
leading pole horn vordere Polkante
leading pole tip vordere Polkante
leading power factor kapazitiver Leistungsfaktor
leading term Anfangsglied (51)
leaf (pl. **-ves**) Blatt
leaf of a gauge Blatt einer Dickenlehre
leaf electrometer Blattelektrometer
leaf spring Blattfeder
leaf of a spring Federblatt
leaflet Datenblatt
leak Leck, Leckstelle; Undichtigkeit (621.9); Ableitung, Isolationsfehler (621.3)
leak coil Überbrückungsspule
leak conductance Ableitung
leak detector Lecksucher
leak resistance Ableitungswiderstand
to leak leck sein; leck werden, Undichtigkeiten aufweisen
to leak away abfließen (621.3); sickern, wegsickern
to leak off abfließen (621.3), (621.64)
leakage Undichtigkeit, Leckstelle, Loch (621.6); Isolationsfehler, Ableitung (621.3); Verlust durch Undichtigkeit; Verlust durch Isolationsfehler; Streuung (53); Neutronenfluß (539)
leakage airgap Streuspalt (53)
leakage around overhang of coils Spulenkopfstreuung (621.313)
leakage capacitance Ableitkapazität
leakage coefficient Streufaktor
leakage current Fehlerstrom, Kriechstrom, Ableitungsstrom, Isolationsstrom, Verluststrom; Reststrom (621.374)
leakage detector Isolationsmesser
leakage distance Kriechweg, Kriechstrecke
leakage factor Kopplungsfaktor (Statorfluß/Nutzfluß)
leakage flux Streufluß
leakage inductance Streuinduktivität
leakage path Kriechstrecke

leakage power Verlustleistung
leakage-protective system Fehlerstromschutz
leakage reactance Streublindwiderstand, Streureaktanz
leakage-reactance transformer Streufeldtransformator
leakage relay Erdschlußrelais
leakage resistance Ableitungswiderstand
leakage transformer Streutransformator
leakance Ableitung (Kehrwert des Isolationswiderstandes)
leakance loss Ableitungsdämpfung
leakance per unit length Ableitungsbelag
leaky undicht; schlecht isoliert
leaky grid detector Gittergleichrichter, Audion-Schaltung
leaky-waveguide antenna Hohlleiterschlitzantenne
lean mager
lean gas Schwachgas
leased telephone circuit geschaltete Fernsprechleitung, Mietleitung
least common multiple Generalnenner
least significant bit Bit mit dem niedrigsten Stellenwert
leather Leder
leather belt Lederriemen, Ledergurt; Treibriemen
leather washer Lederscheibe, Lederdichtungsring
to leave lassen, verlassen, austreten
leaving current austretender Strom, wegfließender Strom
leaving edge ablaufende Kante
leaving lamp Schlußlampe (65)
Lecher circuit Lecherleitung
Lecher wires Lecherleitung (sing.), Lecherdrähte
Leclanché battery Trockenbatterie, Beutelelement
Leclanché cell Trockenelemente, Leclanché-Element
lecture bench Experimentiertisch
LED = light emitting diode Leuchtdiode, Lumineszenzdiode
ledge Ionisationsstufe (621.371)
leeway Abtrift
left(-hand) links, links gelegen
left-hand drawing linke Zeichnung
left-hand drive Linksantrieb; Linkssteuerung
left-hand machine linksläufige Maschine
left-hand motion Bewegung nach links
left-hand part linker Teil
left-hand rule Linke-Hand-Regel (621.3)
left-hand side linke Seite
left-hand view Ansicht von links
left-handed linksgängig, linksdrehend
left-handed rotation Linksdrehung (535)
left-handed screw Linksschraube, linksgängige Schraube
left-handed thread linksgängiges Gewinde, Linksgewinde
left-to-right von links nach rechts gerichtet
left-ward nach links gerichtet
left-ward welding Linksschweißung
leg Bein, Schenkel (61), (621)
leg of a bridge Brückenzweig (621.317)
leg of a circuit Außenleiter, Phasenleiter (621.315)
leg of the compass Zirkelschenkel
leg of a diamond aerial Schenkel einer Rhombusantenne
leg of a pair of dividers Schenkel eines Stechzirkels
leg of a loop Seite einer Schleife (621.317)
legal gesetzlich
legal ohm gesetzliches Ohm
legal standard wire gauge Britische Normen-Drahtlehre = **British S. W. G.**
to legalize gesetzlich festlegen, legalisieren
legalized unit gesetzliche Einheit
legend Beschriftung; Aufdruck, Inschrift; Erläuterung
Lemstrom machine Walzenelektrisiermaschine
Lenard rays Lenard-Strahlen
Lenard tube Lenard-Röhre
Lenard window Lenardfenster
length Länge, Abschnitt; (Normal-)Länge
length gauge Längenmesser, Längenlehre
length movement Längsbewegung
length of piping Rohrstück, Rohrlänge, Ende Rohr
length of stroke Hub, Hubweg, Hublänge
length of time Zeitdauer
lengthening Verlängerung
lengthening bar Verlängerung, Verlängerungsstück
lengthening piece Verlängerungsstück
lengthening pipe Verlängerungsrohr
lengthways in Längsrichtung
lengthwise längs, in Längsrichtung
lens Linse (535)
lens antenna Linsenantenne
lens aperture (relative) Öffnung (77)
lens system Linsensystem
Lenz's law Lenzsche Regel
Lepel system of wireless telegraphy Tonfunkentelegraphie

lesion Verletzung, Beschädigung, Läsur
less weniger; vermindert um; ohne
a bolt less the nut eine Schraube ohne die (dazugehörige) Mutter
less noise weniger Rauschen
less than weniger als, kleiner als (51)
to lessen vermindern
to lessen the efficiency die Wirkung beeinträchtigen; den Wirkungsgrad herabsetzen
to let lassen, zulassen, erlauben
to let into a surface in eine Oberfläche einlegen
to let off steam Dampf ablassen
to let through durchlassen (621.3)
lethal letal, tödlich, todbringend
lethal dose tödliche Dosis
letter Brief; Buchstabe; Type
letter-culling equipment Briefsortieranlage (65)
letter-facing equipment Briefaufstellanlage (65)
letters patent Patent, Patentschrift
letter transmission Zeichenübertragung (621.394)
letter shift Buchstabenschaltung (621.394.6)
to letter beschriften; mit Bezeichnungen versehen, bezeichnen
to letter up mit Buchstaben bezeichnen
lettering Beschriftung, Buchstabenbezeichnung
level Spiegel; Niveau, Stand, Standhöhe, Pegel; Pegel (621.395), (621.6)
level adjustment Einpegeln
level-continous signal wertkontinuierliches Signal
level controller Pegelregler (921.395), (621.6)
level conversion Pegelumsetzung
level crossing schienengleiche Kreuzung (625)
level of illumination Bildhelligkeit (621.397)
level indicator Flüssigkeitsstandanzeiger
level of language Sprachebene
level meter Pegelmesser
level raiser Verstärker (621.394)
level recorder Pegelschreiber
level-recorder chart Pegeldiagramm
level of safety Sicherheitsgrad
level tracer Pegelbildgerät (621.397)
level water Grundwasser
to level einpegeln; auswiegen, in die Waage bringen; nivellieren; Planieren, einebnen
to level off flach werden, flach auslaufen, nicht mehr weiter steigen (oder fallen), in Sättigung übergehen
to level up horizontieren
level eben, waagerecht
level compounding vollständige Kompoundierung
level track ebenes Gleis
level with gleich hoch wie, abschneidend mit
levelling Nivellement; Planierung
levelling pad Horizontierungsfläche
levelling rod Nivellierlatte
lever Hebel
lever arm Hebelarm
lever control Hebelsteuerung
lever drive Hebelantrieb
lever fulcrum Hebeldrehpunkt, Unterstützungspunkt
lever indication Signalstellungs-Rückmelder (656)
lever load Hebelbelastung
lever machine elektrisches Stellwerk (625)
lever motor Stellgliedantrieb
lever-operated mit Hebelantrieb
lever operation Hebelantrieb, Hebelbetätigung
lever press Hebelpresse
lever-type starter Hebelanlasser (621.313)
to lever hebeln
leverage Hebelübersetzung; Hebelanordnung
leverman Stellwerkswärter (625)
Leyden battery Batterie Leidener Flaschen
Leyden jar Leidener Flasche, Kleistsche Flasche
l. f. = low frequency (siehe dort) NF = Niederfrequenz
l. f. amplifier NF-Verstärker
i. f. voltage NF-Spannung
to liberate befreien; freimachen, entwickeln (54)
to liberate electrons Elektronen freimachen
to be liberated frei werden (54)
liberation Befreiung; Freimachung
liberation from Freisetzung aus (54)
liberation of electrons from a surface Auslösung von Elektronen aus einer Oberfläche
librarian [Programm zur Verwaltung von Bibliotheken (681.3)]
library Befehlsbibliothek (681.3)
licence, license Lizenz, Zulassung; Erlaubnis
to license, to licence lizenzieren, zulassen, genehmigen, konzessionieren
licensee Lizenzträger; Konzessionsinhaber
Lichtenberg figures Lichtenbergfiguren
to lie liegen, ruhen
to lie above sea level über dem Meeresspiegel liegen
life (pl.: **-ves**) Leben, Lebensdauer
life expectancy Lebenserwartung, voraussichtliche Lebensdauer

life expectation voraussichtliche Lebensdauer
life load bewegte Last, bewegliche Last
life period Lebensdauer (539)
life-raft radio Funkgerät für Rettungsfloß
life size Originalgröße, Lebensgröße
life test Lebensdauerversuch; Brennprobe (621.326)
life time Lebensdauer
life utility Brauchbarkeitsdauer
lifeboat radio Funkgerät für Rettungsboote
lifebuoy Rettungsboje
LIFO = Last In, First Out Kellerspeicher, Stack, Stapelspeicher
lift Auftrieb (532), (533); Aufzug, Fahrstuhl, Lift; Schräglage, Trimmwinkel (629.12), (621.13); Daumen, Nockendaumen (621); Schwarzabhebung (621397)
lift-car annunciator Stockwerksanzeiger
lift-car flash-signal device Stockwerkslichtrufanzeiger
lift control Fahrstuhlsteuerung
lift controller Fahrstuhlsteuerschalter
lift electric travelling cable Kabinenkabel
lift machine Fahrstuhlantriebsmaschine
lift-potential switch Fahrstuhlnullspannungsschalter
lift shaft Fahrstuhlschacht
lift slack-cable-switch Fahrstuhlschlaffseilschalter
lift wire Spanndraht (zur Aufnahme von Auftriebskräften) (629.13)
to lift heben, anheben, abheben
to lift the brushes die Bürsten abheben (621.313)
to lift from abheben von
to lift the handset den Hörer aushängen (621.395)
to lift into hineinheben, einheben
to lift off abheben
to lift out herausheben, ausheben
lifter Arretierungshebel (621.317)
lifting Hebe-...
lifting bridge Hebebrücke, Hubbrücke
lifting capacity Tragfähigkeit, Hubkraft (621.318)
lifting device Hebezeug
lifting lug Aufhängebügel
lifting magnet Hubmagnet, Lastmagnet
lifting tackle Hebezeug, Flaschenzug, Zug (621.8)
light Licht; Lichtquelle, Licht; Streichholz
light barrier Lichtschranke
light beam Lichtstrahl
light cable Beleuchtungskabel, Lichtkabel
light-centre length Lichtpunktabstand (621.326)
light corpuscle Lichtkorpuskel
light efficiency visueller Wirkungsgrad
light-emmitting diode lichtemittierende Diode, Leuchtdiode, Lumineszenzdiode
light flash Lichtblitz
light-gathering power Lichtstärke (77)
light-ground illumination Hellfeldbeleuchtung (535)
light intensity Lichtstärke
light loss Lichtverlust
light meter Photometer, Lichtmesser; Beleuchtungsmesser
light pen Lichtgriffel, Lichtstift
light pencil Lichtbündel, Lichtstrahl (535)
light and power Licht und Kraft
light-proof lichtdicht; lichtecht
light rays Lichtstrahlen; Lichtstrahlung (sing.)
light relay lichtelektrisches Relais
light-resisting lichtbeständig, lichtecht
light-restricting lichthemmend
light screen Lichtschirm
light-sensitive lichtempfindlich
light-sensitive cell lichtempfindliche Zelle, Photozelle
light-signalling installation Lichtrufanlage
light socket Lichtsteckdose
light source Lichtquelle
light-tight lichtdicht
light-trapped lichtdicht abgeschlossen
light wave Lichtwelle
light year Lichtjahr (52)
to light leuchten; anzünden, zünden; befeuern (351.814)
to light a discharge lamp eine Entladungslampe zünden
to light a fire ein Feuer anzünden, ein Feuer anmachen, anheizen
to light a lamp eine Lampe einschalten (621.3)
to light a tube eine Röhre heizen (621.385)
to light up aufleuchten; zünden
lighted tetrahedron Windrichtungsfeuer (656.7)
lighted wind indicator Licht-Windanzeiger (656.7)
light leicht, schwach; matt, blaß
light aeroplane Kleinflugzeug, Leichtflugzeug
light alloy Leichtmetallegierung
light alloys Leichtmetalle
light ashes Flugasche (sing.)
light current Schwachstrom
light-current cable Schwachstromkabel

light-current engineering Schwachstromtechnik
light engineering Feinmechanik
light load Schwachlast
light load test Schwachlastprüfung (621.317)
light metal Leichtmetall
light nucleus leichter Kern
light railway Kleinbahn, Feldbahn
light stroke leichter Anschlag (681)
light-weight leicht, Leicht-...
light-weight design Leichtbauweise
light in weight mit wenig Masse
to be light in weight wenig Masse besitzen
lightly ionized schwach ionisiert
to lighten erleuchten, aufleuchten; blitzen; erleichtern, leichter machen
lighter Feuerzeug; Feueranzünder
lighthouse Leuchtturm
lighting Beleuchtung
lighting branch (circuit) Lichtstromkreis
lighting circuit Lichtstromkreis, Lichtnetz
lighting current Lichtstrom
lighting density Leuchtdichte
lighting fitting Leuchte, Beleuchtungskörper, Beleuchtungsarmatur
lighting fittings Beleuchtungsarmaturen; Beleuchtungszubehör (sing.)
lighting installation Beleuchtungsanlage, Lichtanlage
lighting load Lichtlast
lighting meter Lichtzähler, Haushaltszähler, Lichtstromzähler
lighting outlet Brennstelle (696)
lighting peak Lichtspitze
lighting power Leuchtkraft
lighting switch Lichtschalter
lighting transformer Lichttransformator
lightness Leichtigkeit, Masselosigkeit, geringe Masse
lightning Blitz
lightning arrester Blitzableiter, Überspannungsableiter (621.3)
lightning conductor Blitzableiter
lightning current Blitzstrom
lightning discharge Blitzentladung
lightning diverter Blitzableiter, Überspannungsableiter (621.315)
lightning flash Blitzschlag
lightning generator Stoßgenerator, Stoßprüfanlage
lightning guard Grobschutz, Grobblitzschutz (621.394)
lightning protection Blitzschutz
lightning protector Blitzschutz, Blitzableiter
lightning rod Blitzauffangstange
lightning speed Blitzgeschwindigkeit
lightning stroke Blitzschlag, Blitzeinschlag
lightning surge Blitzüberspannung
lightning switch Blitzschutzschalter, Antennenschalter
lightship Feuerschiff
lightspot Lichtfleck
lignite Lignit, Braunkohle
like gleich, gleichartig, gleichnamig, gleichgerichtet
like currents gleichgerichtete Ströme
like pair Doppelleitung mit gleichsinnigen Strömen
like poles gleichnamige Pole
like poles repel gleichnamige Pole stoßen sich ab
likelihood Wahrscheinlichkeit
likehood of errors Fehlermöglichkeit
limb Schenkel (621.314.2), (621.318); Flügel (72)
lime Kalk (54); Linde (58); Zitrone (58)
lime spot cathode Glühkathode mit Emmissionsfleck
limit Grenze; Grenzwert, Grenze (51)
limit of audibility Hörgrenze
limit bridge Toleranzmeßbrücke
limit dimension Grenzmaß
limit of elasticity Elastizitätsgrenze
limit frequency Eckfrequenz
limit indicator Strombegrenzer (621.317)
limits of inflammability Grenzen des Zündbereiches, Zündgrenzen
limit of measurement Meßgrenze
limit of proportionality Proportionalitätsgrenze
limit of sensitivity Empfindlichkeitsgrenze
limit sensor Grenzwertmelder
limit speed Grenzdrehzahl
limit switch Endschalter, Endlagenschalter
limit of travel Endanschlag; Ausschlagsbegrenzung
to limit begrenzen; limitieren
limitation Begrenzung, Einschränkung
limitation of deflection größte zulässige Durchbiegung; Ausschlagsbegrenzung (621.317)
limited begrenzt, beschränkt
limited by space charge raumladungsbegrenzt (621.395)
limiter Begrenzer; Amplitudenbegrenzer (621.396)
limiter stage Begrenzerstufe (621.396)
limiting Begrenzung, Grenz-...
limiting action Begrenzung, Begrenzerwirkung

limiting case Grenzfall
limiting circuit Begrenzerschaltung
limiting creep stress Dauerstandfestigkeit
limiting current Grenzstrom (621.316.5)
limiting frequency Grenzfrequenz
limiting friction Grenzreibung
limiting overload characteristic Grenzstromkennlinie
limiting reactor Begrenzungsdrossel, Kurzschlußdrossel
limiting sensitivity Grenzempfindlichkeit (621.396.62)
limiting speed Grenzgeschwindigkeit
limiting stop Begrenzungsanschlag
limiting surface Grenzfläche
limiting value Grenzwert
limiting value of mean on-state current Dauergrenzstrom
linchpin Vorsteckstift, Vorstecker, Splint, Radsplint
line Linie (51); Leitung (621.3); Spektrallinie (535); Strecke, Eisenbahnlinie, Autobuslinie (625); Zeile (621.397), (51); Äquator, Linie (91); Linie (Längenmaß = $^1/_{12}$ Zoll = 2, 12 mm); Montageband
line of action Angriffslinie (einer Kraft); Eingriffslinie (621.833)
line amplifier Leitungsverstärker, Linienverstärker
line balance Leitungsentzerrer (621.395); Leitungssymmetrie
line-balance converter Symmetriertopf, Symmetrierglied (621.396)
line-blanking interval Horizontal-Austastlücke
line breaker Fahrleitungsschalter (621.33)
line broadcasting Drahtfunkt
line-busy tone Anschlußbesetzton
line characteristic Wellenwiderstand, Charakteristik (621.39)
line circuit Freileitungskreis (621.394); Teilnehmerschaltung (621.395)
line circuit-breaker Leitungsschalter
line concentrator Leitungskonzentrator, Leitungsdruckschalter (621.395)
line conductor Leitungsseil, Leiter (621.315.1)
line configuration Leitungsanordnung
line contact Linienberührung
line of contact Berührungslinie; Eingriffslinie (621.833)
line-control unit Kanalgerät (621.398)
line current Netzstrom, Gesamtstrom
line disconnection Leitungsabschaltung
line with dissipation Leitung mit Verlusten
line driver Leitungsverstärker, Leitungstreiber
line drop Spannungs(ab)fall auf der Leitung
line duration Zeilendauer
line entry Einspeisung
line equalization Leitungsausgleich
line feed Zeilenfortschaltung, Zeilenvorschub
line filter Filter aus abgestimmten Leitungen
line finder Anrufsucher (621.395)
line flicker Teilbildflimmern
line flyback Zeilenrücklauf
line of force Kraftlinie
lines of force proceed from a pole Kraftlinien treten aus einem Pol aus
line frequency Zeilenfrequenz, Horizontalfrequenz (621.397); Netzfrequenz (621.31)
line-to-ground fault Erdschluß
line grouping Linienanordnung
in line ausgerichtet
line insulator Freileitungsisolator
line integral Linienintegral
line integral round a circuit Umlaufintegral
line lamp Ruflampe (621.395)
line of least resistance Weg des geringsten Widerstandes
line-by-line scanning zeilenweise Abtastung
line-to-line spacing Zeilenabstand
line lock-out Abwurf auf Teilnehmerschaltung
line loss Leitungsverlust
lines of magnetic flux magnetische Flußlinien
lines of magnetic force magnetische Kraftlinien
line of magnetism magnetische Feldlinie, Kraftlinie
line noise Leitungsgeräusch
line offset Zeilenoffset
out of line seitlich versetzt, mit versetzter Achse
line pairing Zeilenpaarigkeit
line of position Peilstandlinie
line post (insulator) Stützer
line of pressure Kraftrichtung, Druckrichtung
line printer Zeilendrucker
line production (Fließ-)Bandarbeit, Fließbandfertigung
line pulsing Laufzeitkettentastung
line recorder Linienschreiber (621.317)
line relay Leitungsrelais, Anrufrelais (621.395)
line for release Abwurfleitung
line resistance Leitungswiderstand
line resonator Leitungsschwingkreis, Resonanzleitung (621.396)
line-scan circuit Zeilenkippschaltung
line-scan transformer Horizontalablenk-Transformator

line scratches Kratzgeräusche (621.395)
line selector Leitungswähler, Leitungssucher (621.395)
line sequential Zeilenfolge
line shaft Transmissionswelle
line-of-sight Sichtweite, optische Sicht
line-of-sight connection Sichtverbindung
line of sight course Zieldeckungskurs, Dreipunktkurs, Sichtlinienkurs (527)
line simulator künstliche Leitung, Leitungsnachbildung
line-side fuse vorgeschaltete Sicherung
line slip Zeilendurchlauf
line stretcher Posaune (621.317)
line spectrum Linienspektrum
line sweep Horizontalablenkung
line switch Leitungsschalter, Netzschalter
line-synchronizing signal Horizontal-Synchronsignal
line system Leitungsnetz
line technique Leitungstechnik (621.397)
line telegraphy Drahttelegraphie
line telephony Drahttelefonie
line television Drahtfernsehverbindung
line terminal Strangklemme
line of thrust Drucklinie
line-at-a-time printer Zeilendrucker
line transformer Zeilentransformator (621.397)
line transmission Leitungsübertragung
line of travel Bahn, Weg
up and down the line hin und her auf der Leitung (621.3)
line of a vector Feldlinie
line verification Anschlußkennung (621.396)
line visibility sichtbare Zeilenstruktur (621.397)
line voltage Spannung auf der Leitung, Leiterspannung, Netzspannung
line-voltage variation Netzspannungsschwankung
line wire Leitungsdraht
to line füttern, ausfüttern, auskleiden, ausmauern, belegen
to line out entwerfen, aufreißen
to line up ausrichten (621); abgleichen, trimmen (621.3); einfädeln (351.814)
linear linear
linear accelerator Linearbeschleuniger
linear actuator Antriebseinrichtung für Schubbewegung
linear amplification lineare Verstärkung (621.396.64)
linear amplifier Linearverstärker
linear antenna array lineare Dipolgruppe
linear capacity law kapazitätsgerade Charakteristik (621.396)
linear circuit Analogschaltkreis
linear coefficient of expansion linearer Koeffizient der Wärmedehnung
linear combination linearer Ausdruck (51); lineare Superposition, lineare Überlagerung (51), (621.3)
linear connected system linear verknüpftes System
linear crosstalk lineares Nebensprechen
linear decrement lineare Amplitudenabnahme
linear detector linearer Gleichrichter
linear distance Entfernung, Abstand
linear distortion lineare Verzerrung
linear electric constant Leitungsbelag
linear electron accelerator Elektronenlinearbeschleuniger
linear expansion lineare Wärmedehnung
linear function lineare Funktion
linear hysteresis Ummagnetisierung, Wechselmagnetisierung
linear integrated circuit analoger IC
linear modulation lineare Modulation
linear rectifier linearer Gleichrichter
linear scanning zeilenweise Abtastung
linear sweep lineare Zeitablenkung (621.385)
linear with time zeitlinear, zeitlich linear
linearly polarized wave linear polarisierte Welle
linearity Linearität
linearity curve Modulationskennlinie (621.396.61)
linearity error Linearitätsfehler
lineman Elektromonteur (621.315)
linen Leine, Leinen-...
linen ribbon Leinenband
to linger verweilen, nachleuchten (621.397)
lingering period Verweilzeit
lining Futter, Belag, Auskleidung
lining of a bearing Lagerbuchse, Lagerfutter
link Glied, Bindeglied; Zugstange, Lasche; Verbindungslinie, Kuppelstange (625); (alsLängenmaß s. Anh.); Unterprogrammrücksprung (681.3)
link-by-link abschnittsweise (621.395)
link coil Koppelspule (621.396)
link-polygon Seileck
link-type insulator Hewlett-Insolator, Schlingenisolator
to link verbinden; verketten mit; sich verketten mit
to link a coil mit einer Spule verkettet sein
to link a flux mit einem Fluß verkettet sein

to link a winding mit einer Wicklung verkettet sein
to link with umschlingen, verkettet sein mit
to link up with verknüpfen mit
to link a yoke ein Joch umschlingen
to be linked with verkettet sein mit
linked switches verkuppelte Schalter
linking switch Kuppelschalter
linkage Kupplung, Verbindungsstück; Verkettung (621.39)
linkage gear Kuppelgestänge
linker [Verbinder von object moduls zu einem lauffähigen Programm]
lino Linol
linoleum Linoleum
linseed oil Leinöl
lip Tülle (666.5); Schneidkante (621.95); Vorderkante (621.9)
liquation Seigerung
liquefaction Verflüssigung
to liquefy verflüssigen, schmelzen
liquefied gas verflüssigtes Gas
liquescent im Zustande der Verflüssigung; schmelzend
liquid Flüssigkeit
liquid-fuse unit Flüssigkeitssicherung, Ölsicherung
liquid-junction potential Potentialsprung zwischen zwei Elektrolyten, Kontaktpotential zwischen Flüssigkeiten
liquid-level control Flüssigkeitsstandregelung
liquid-metal fuel reactor Reaktor mit Flüssigmetallspaltstoff
liquid pressure Flüssigkeitsdruck
liquid-quenched cut-out Ölsicherung
liquid resistance Flüssigkeitswiderstand
liquid seeks its level Flüssigkeit stellt sich waagerecht ein
liquid starter Flüssigkeitsanlasser
liquid flüssig
liquid air flüssige Luft
liquid-air trap Kühlfalle mit flüssiger Luft
liquid ammonia Ammoniakwasser, Salmiakgeist
liquid crystal Flüssigkristall
liquid-crystal display = LCD Flüssigkristallanzeige
liquid fuel flüssiger Brennstoff, flüssiger Stoff
liquid gas Flüssiggas
liquid insulation flüssige Isolation, Isolierflüssigkeit
liquid matter Flüssigkeit, flüssiger Stoff
liquid oxygen flüssiger Sauerstoff
LISP = list processsing
Lissajou's figure Lissajou-Figur
list Liste; Verzeichnis; Schlagseite (629.1)
list of components Stückliste
to list in eine Liste eintragen; Schlagseite haben; Schlagseite bekommen (629.12)
to listen zuhören; horchen
to listen in abhören, mithören
listener Hörer (621.396)
listener echo Mitecho, mitläufiges Echo
listening Hören, Horchen
listening key Mithörtaste
listening test Verständlichkeitsversuch
listing Auflistung, Ausdruck (681.3)
litharge Bleiglätte, Glätte
lithium Lithium
litmus Lackmus
litre Liter
littelfuse Kleinsicherung, Röhrchensicherung
little load wenig Last
litz HF-Litze, verdrillte Litze
litz wire Litzendraht
litz-wound mit HF-Litze gewickelt
litzendraht Litzendraht, HF-Litze
litzendraht wire Litzendraht, HF-Litze
live lebendig; spannungsführend, „heiß“ (621.3)
live axle Triebachse (625)
live broadcast Direktübertragung (654.19)
live cable test cap Bleikappe (Zubleiung) für unter Spannung stehendes Kabel
live conductor spannungsführender Leiter
live-front switchboard Schalttafel mit spannungsführenden Teilen auf der Vorderseite
live load bewegliche Last
live part spannungsführender Teil
live rail Stromschiene
live weight bewegliche Masse, bewegliche Last
live wire spannungsführender Draht, Außenleiter
L network L-Schaltung, Vierpol in L-Schaltung
load Last, Belastung, Bürde; Ladung; Ausgangswiderstand (621.396)
load-bearing capacity Tragfähigkeit
load-bearing wall tragende Wand
load-break switch Lasttrenner (621.315)
load-carrying capacity Strombelastung, Belastbarkeit
load center Belastungsschwerpunkt
load circuit Lastkreis, Belastungskreis
load current Arbeitsstrom, Ausgangsstrom
load curve Belastungskurve
load-dispatching plant Lastverteileranlage
load diversity Gleichzeitigkeitsfaktor (621.31)
load divider Lastverteiler

load-duration curve geordnete Belastungskurve
load end Empfangsseite, Ausgangsseite, Verbraucherseite, Ausgang, Ende
load-extension diagram Spannungs-Dehnungs-Diagramm
load factor Lastvielfaches (629.13); Belastungsfaktor
load-factor rating Nennleistung für Aussetzbetrieb
load-frequency control of interconnected systems Netzkennlinienregelung
load impedance Belastungswiderstand, Belastungsscheinwiderstand
load instruction Ladebefehl (681.3)
load-levelling relay Lastbegrenzungsrelais
load limit Belastungsgrenze
load-limiting relay Lastbegrenzungsrelais
load-limiting resistor Strombegrenzungswiderstand
load limitation Leistungsbegrenzung
load line Belastungskennlinie (621.396)
load losses stromabhängige Verluste (621.314.2)
load matching Lastanpassung
on load unter Last, bei Belastung
load peak Lastspitze, Belastungsspitze
load power factor Leistungsfaktor der Belastung
load-ratio control transformer Regeltransformator (Unterlast-Regelung)
load regulator Leistungsregler, Belastungsregler
load rejection Lastabwurf
load-rejection test Lastabschaltversuch
load resistance Belastungswiderstand
load resistor Belastungswiderstand, Vorbelastung, Grundbelastung (621.314.6)
load sharing Belastungsteilung (621.395)
load shedding Lastabwurf
load-shifting resistor Lastverschiebungswiderstand
load test Belastungsprobe
load-unbalanced protection Schieflastschutz
load variation Lastschwankung, Belastungsschwankung
load voltage Belastungsspannung, Spannung bei Last
to load laden, beladen, belasten, beschicken
to load a battery eine Batterie belasten
to load a cable ein Kabel pupinisieren (621.395)
to load an oven einen Ofen beschicken
to load plastics Kunststoffe füllen
loaded antenna belastete Antenne
loaded circuit pupinisierter Kreis; belasteter Kreis
loaded line beschaltete Leitung, belastete Leitung; pupinisierte Leitung, bespulte Leitung
loaded region (of a reactor) aktiver Teil (eines Reaktors)
loader Lader (681.3)
loading Ladung, Belastung; Verladung
loading beld Ladeband
loading capacity Ladefähigkeit
loading case Spulenkasten
loading coil Pupinspule (621.395); Verlängerungsspule (621.396.67)
loading-coil spacing Spulenabstand
loading conditions (pl.) Belastungszustand
loading point Spulenpunkt
loading pot Spulenmuffe
loading ramp Laderampe
loading section Spulenfeld (621.392)
loading tray Waagschale, Gewichtsschale
loadline Wasserlinie (beladen)
loadstone Magneteisenstein
lobe Ohrläppchen, Lappen (61); Lappen, Keule, Zipfel, Schleife (621.396.67)
lobe frequency Keulenumtastfrequenz (621.396.67)
lobe pattern Antennendiagramm
lobe switching Keulenumtastung, Keulenschwenkung
local örtlich, lokal
local battery Ortsbatterie = OB
local battery exchange OB-Vermittlung, OB-Amt
local battery telephone set OB-Telefonapparat, OB-Apparat
local-cable Ortskabel
local-carrier reception Empfang mit Trägerzuführung
local circuit Ortsleitung
local-communications network Ortsnetz
local corrosion örtliche Korrosion, elektrolytische Korrosion
local exchange Ortsamt
local fitting Arbeitsplatzleuchte
local-hot spot örtliche Überhitzungsstelle
local jack Teilnehmerabfrageklinke
local-junction circuit Ortsverbindungsleitung
local-line network Oberleitungsnetz
local loop Teilnehmerleitung, Teilnehmerschleife
local office Ortsamt, Ortshauptamt
local oscillations (pl.) Oszillatorfrequenz

local oscillator Überlagerungssender, Eigenoszillator (621.696.62)
local tariff Ortstarif
local-telephone area Ortsnetzbereich
local-television link Fernsehortsverbindung
to localize lokalisieren, örtlich begrenzen
localized circuit quasistationärer Stromkreis
localized heating örtliche Überhitzung (621.313)
localizer beacon Landekurssender (621.396.32)
to locate fixieren, festlegen, führen (621.9); orten (629.12), (621.13); anbringen, einbauen
to locate a plane ein Flugzeug orten
to locate by the sense of smell mit der Nase orten
locating Sitz-..., Paß-...
locating edge Anschlagleiste
locating face Sitzfläche, Paßfläche
locating point Auflagepunkt, Paßpunkt
locating mark Paßmarke
locating rest Fixierungsring, Fixierungsanschlag
location Örtlichkeit, Gegend; Räumlichkeit, Ort (auf einer Leitung); Ortung, Ortsbestimmung; Speicherzelle (681.3)
location counter Befehlszähler
location diagram Verteilungsplan, Anordnungsplan
location of faults Fehlerortsbestimmung, Fehlersuche
location finding Ortung
location hole Befestigungsloch, Fixierungsloch, Paßloch
location with regard to Lage zu ..., Lage gegenüber ...
location of the trouble Ermittlung der Schadenstelle, Auffindung der Störungsstelle
lock Schloß, Türschloß; Verschluß, Gewehrschloß; Schleuse (626); Klemme
lock-circuit controller Weichenriegelkontakt
lock-in amplifier Lock-in Verstärker, phasenstarrer Verstärker
locknut Feststellmutter; Kontermutter; Klemmring
lock-out relay Sperrelais
lock-out switch Sperrschalter
lock position Verriegelungsstellung
lock washer Sicherungs-Unterlegscheibe
to lock verschließen, zuschließen, verriegeln, sperren, blockieren
to lock in place einrasten
to lock out abwerfen (621.395)
to lock to ketten an, binden an
to lock up zuschließen
locked cover switch Schalter mit Deckelverriegelung
locked-in unter Verschluß
locked-in stress innere Spannung, eingefrorene Spannung
locked-joint tube gefalztes Rohr
locked-kilovolt-amperes Festbremsleistung
locked oscillator Mitnahmeoszillator
locked-rotor torque Drehmoment bei festgebremstem Rotor
lock-and-switch movement (US) Weichenstellmaschine
lockable switch verriegelbarer Schalter
locking Verschluß, Verriegelung
locking bolt Riegelschraube; Spannschraube, Feststellschraube; Verschlußschraube
locking device Verschluß, Verschlußeinrichtung; Arretierung
locking lamp holder diebstahlsichere Fassung
locking lever Sperrhebel, Feststellhebel; Riegelhebel
locking pin Vorsteckstift; Prisonstift, Arretierstift
locking plate Schließblech
locking ring Klemmring, Feststellring
locking screw Klemmschraube, Feststellschraube
locking switch Verschlußschalter, Schalter unter Verschluß
locking wedge Verriegelungskeil
locksmith Schlosser
lockwasher Unterlegscheibe, Sicherungsring
locomotion Bewegung, Ortsveränderung, Verkehr
locomotive Lokomotive, „Lok"
locomotive engine Lokomotive
locomotive engineer Lokomotivführer
locomotive wheel Lokomotivrad, Triebrad
locus (pl. **-ci**) Ort, geometrischer Ort (51); Ortskurve (621.3)
locus circle Ortskreis
locus of points Ortskurve, geometrischer Ort
lodestone Magneteisenstein
Lodge coherer Metalldetektor, Spitzendetektor
Lodge-Muirhead coherer Quecksilberdetektor
to lodge unterbringen (**in:** in)
log Holzklotz, Rohstamm; Betriebsbuch, Logbuch, Log; Log (629.12)
to log protokollieren (681.3)
log = logarithm log = lg = Logarithmus
log e = natural logarithm ln = natürlicher Logarithmus
log plotting logarithmisches Diagramm
logarithm Logarithmus (pl.: -men)

logarithmic logarithmisch
logarithmic amplifier logarithmischer Verstärker
logarithmic chart logarithmische Darstellung
logarithmic decrement logarithmisches Dekrement
logarithmic horn logarithmischer Trichter(-lautsprecher)
logarithmic paper Logarithmenpapier, logarithmisches Papier
logarithmically spaced logarithmisch gestuft, in logarithmischer Stufung
logic logisch; Schaltalgebra; Logik
logic array logische Mehrfachschaltung
logic comparator Logikkomparator
logic element Verknüpfungsglied
logic function Schaltfunktion
logic level logischer Zustand
logic network logische Schaltung (681.14)
logic operation logische Verknüpfung
logic probe Logiktastkopf
logic shift logische Verschiebung
logic symbol Symbol einer logischen Schaltung
logical logisch, funktionell
logical add logische Addition
logical circuit logische Schaltung, Verknüpfungsglied (681.14)
logical decision logische Entscheidung
logical element Verknüpfungsglied
logical high logisch hoch
logical instruction logischer Befehl
logical operation Verknüpfung, logische Operation
logistics Logistik
long lang
long-bar work Arbeit von der Stange (621.9)
long chord winding übersehnte Wicklung
long distance weite Entfernung, großeEntfernung
long-distance call Ferngespräch (621.395)
long-distance center Fernvermittlungsstelle
long-distance communication Weitverkehr
long-distance dialling Fernwahl
long-distance line Fernleitung, Überlandleitung
long-distance network Fernnetz
long-distance switching Fernvermittlung
long-distance telegraphy Weitverkehrstelegraphie
long-distance telephony Weitsprechverkehr
long-distance traffic Fernverkehr
long-distance transmission Überlandleitung, Fernleitung
long division Division durch Zahl größer als 12
long handled mit langem Stiel
long haul Weitverkehr
long hundred 120 Stück
4 in. long 4 Zoll lang
long life lange Lebensdauer
long line Fernleitung
long-lived langlebig
long-lived fission product langlebiges Spaltprodukt
long measure Längenmaß
long-nose plier Spitzzange
long-path bearing Großkreispeilung
long persistence lange Nachleuchtzeit (621.397)
long-pull magnet Kegelankermagnet
long range große Reichweite, große Entfernung
long-range communication Weitverkehr
long-range navigation Weitverkehrsnavigation
long-range radar device Vorwarnradar, Weitbereichsradargerät
long-tailed pair kathodengekoppelte Gegentaktstufe
long-term test Langzeitprüfung
long-time Dauer-...
long waves Langwellen, lange Wellen (über 3000 m)
long-wave coil Langwellenspule
long-wave range Langwellenbereich
long-wave receiver Langwellenempfänger
long wearing abnutzungsfest, abriebfest
long-wearing quality lange Haltbarkeit
long-wire aerial Langdrahtantenne
longevity Langlebensdauer
longeron Holm, Längsholm, Längsträger (629.13)
longitude Länge (91)
longitudinal längs-...
longitudinal electromotive force Längs-EKM
longitudinal magnetization magnetische Längsaufzeichnung
longitudinal recording Längsspurverfahren (681.84)
longitudinal resonance Längsresonanz
longitudinal rib Längsrippe
longitudinal seam Längsnaht, Längsschweißnaht
longitudinal section Längsschnitt
longitudinal-section wave Längsschnittwelle (621.396)
longitudinal stress Längsbeanspruchung
longitudinal voltage Längsspannung (621.3)

longitudinal wave Longitudinalwelle
longitudinal welding Längsschweißen, Längsschweißung
loom Webstuhl
loop Schleife, Rahmen; Programmschleife (681.3)
loop aerial Rahmenantenne
around the loop auf dem geschlossenen Umlaufweg, ringsherum um die Masche (621.3)
loop circuit Schienenstromschleife
loop current Schleifenstrom (656.25)
loop dialling Schleifenwahl (621.394)
loop feeder Hausanschlußleitung, Hausanschluß (621.315.1)
loop gain Schleifenverstärkung
loop galvanometer Schleifengalvanometer
loop hole Schlupfloch
loop impedance Schleifenwiderstand
loop service Einschleifung (621.316)
loopstick antenna Ferrit(stab)antenne
loop test Schleifenschaltung (621.315.2)
loop of wire Drahtschleife
to loop schlingen, wendeln; Looping fliegen
looped filament gewendelter Draht, gewendelter Leuchtdraht
to loop in einschleifen
looping-in Einschleifen (621.316)
to loose lösen
loose lose, schlaff; ungenau, schludrig
loose cap loser Lagerdeckel
loose collar Stellring
loose connection Wackelkontakt, lose Verbindung
loose cord lose Strippe
loose coupling lose Kupplung (621.396)
loose friction fit loser Reibsitz
loose headstock Reitstock (621.9)
loose jaw Einlegebacke, Backenfutter
loose pully Losscheibe, lose Scheibe
to loosen lockern, lösen; sich lockern, locker werden, sich lösen
to lop kappen, abschneiden, wegschneiden
to lopside schief quetschen, seitlich beschneiden
lopsided schief getrimmt (629.12); verquetscht
Loran = Long Range Navigation Weitverkehrs-Navigationsverfahren, Hyperbel-Navigationsverfahren
Lorimer automatic telephone system Suchwähler-Selbstanschlußsystem
lorry Plattenwagen, Rollwagen; Lastkraftwagen, Lkw
to lose verlieren; nachgehen (681)
to lose control die Steuerwirkung verlieren, die Gewalt verlieren (**of:** über)
to lose electrons Elektronen abgeben
to lose time Verspätung bekommen (625)
loss Verlust; Dämpfung (621.395)
loss angle Verlustwinkel
loss of charge Ladungsverlust
loss coefficient (Kehrwert des Wirkungsgrades minus Eins = $1/(\eta - 1)$ (621.313)
loss of definition Einbuße an Zeichenschärfe
losses of energy Energieverluste
loss factor Verlustfaktor; Ausschußfaktor
loss-factor meter Verlustfaktormesser
loss-free verlustfrei, verlustlos
loss-free dielectric verlustloses Dielektrikum
loss-free line verlustfreie Leitung
loss of gain Verstärkungsabfall
loss of gas Gasverlust
loss of heat Wärmeverlust
loss by leakage Undichtigkeitsverlust, Ableitungsverlust
loss per line Streckendämpfung
loss in the noise Untergehen im Rauschen
loss of pressure Druckverlust
loss resistance Verlustwiderstand (621.396)
loss in transmission Übertragungsverlust
loss of voltage Ausbleiben der Spannung, Spannungsausfall
loss of weight Gewichtsverlust, Schwund
loss by windage Luftreibungsverluste, Ventilationsverluste (pl.) (621.313)
losser tube (US) Absorptionsrohr (621.396)
lossless line verlustfreie Leitung
lossy line Leitung mit hoher Dämpfung
lost motion toter Gang, Leerlauf
lost traffic Verlustverkehr (621.395)
lot Anteil, Posten, Menge, Los; Parzelle; Sendung, Charge
loud laut
loud hailer Richtlautsprecher hoher Leistung
loudness (level) Lautstärke, Lautheit, Lautstärkepegel
loudness-volume equivalent Bezugsdämpfung
loudspeaker Lautsprecher
loudspeaker baffle Schallwand, Sprechwand, Einbauwand
louver = louvre Jalousiewand
louver troffer Rasterleuchte
louver-type aperture Öffnung mit Jalousiewand
louver-type slot Jalousieschlitz
low tief, niedrig
low-angle fading antenna Nahschwundantenne

low-angle radiation Ausstrahlung unter kleinem Erhebungswinkel
lower audible limit untere Hörgrenze
low band, lower band Unterband
low barometer niedriger Barometerstand, niedriger Luftdruck
low capacity niedrige Kapazität; kapazitätsarm
low-capacity wiring kapazitätsarme Verdrahtung
low-carbon mit niedrigem C-Gehalt, kohlenstoffarm
low-carbon steel kohlenstoffarmer Stahl
low-current fuse niedrige Sicherung
low-current source wenig ergiebige Stromquelle, schwache Stromquelle
low cut-off frequency untere Grenzfrequenz
low-definition television Fernsehen mit geringer Auflösung (Zeilenzahl unter 200)
low degree of coupling schwach gekoppelt
low-density paper poröses Papier (621.315.6)
low-flux reactor Niederflußreaktor
low frequency Niederfrequenz
low-frequency amplification Niederfrequenzverstärkung
low-frequency amplifier Niederfrequenzverstärker
low-frequency filter Niederfrequenzsieb, Netzfilter
low-frequency furnace Niederfrequenzofen
low-frequency impedance-corrector Tiefenentzerrer für Nachbildung
low-frequency resistance Niederfrequenz-Widerstand
low-frequency speaker Tieftonlautsprecher
low-frequency transformer Niederfrequenztransformator
low-gear langsamer Gang
low-grade minderwertig
low-grade coal minderwertige Kohle
low heat value unterer Heizwert
lower hemispherical flux unterer hemisphärischer Lichtstrom
low-and-high-pass filter Tief- und Hochpaß
low impedance niedriger Widerstand
low-impedance niederohmig
low-impedance path niederohmiger Weg
low-inductance induktionsarm
low-inductance ratio box winkelarmer Brückenzweigwiderstand
low-inductance resistance box winkelarmer Widerstand, winkelarmer Stöpselwiderstand
low level niedriges Niveau, niedriger Pegel
low-level amplifier Vorverstärker, Eingangsverstärker
low-level design Flachbauweise (621.316.5)
low-level language niedrige Programmiersprache
low-level selection Tiefpegelwahl (621.395)
low light level television Restlichtfernsehen, Nachtfernsehen
low limit untere Grenze
low loss niedriger Verlust
low-loss verlustarm
low-loss insulator verlustarmer Isolator (621.396)
low-loss line verlustarme Leitung
low-noise tube rauscharme Röhre
low-note tiefer Ton
low-numbered mit niedriger Nummer
low-pass filter Tiefpaß (621.395)
low-pass network Tiefpaßkettenleiter
low permeability kleine Permeabilität
low-permeability path Weg mit niedriger Permeabilität
low pitch tiefer Ton, niedrige Tonhöhe
low potential niedrige Spannung
low power niedrige Leistung
low-power beacon Niederleistungsfunkfeuer
low-power microphone Mikrophon mit kleiner Ausgangsleistung
low-power microscope schwach vergrößerndes Mikroskop
low-power modulation Vorstufenmodulation
low-power reactor Niederleistungsreaktor
low-power Schottky TTL Low-Power Schottky TTL, TTL-Schaltung mit Schottky-Dioden
low-powered mit geringer Leistung
low pressure niedriger Druck, Niederdruck
low-pressure Niederdruck-...
low-pressure mercury-vapour lamp Quecksilberdampfniederdrucklampe
lowest reading niedrigste Ablesung, niedrigste Einstellung
low range unterer Bereich
low resistive niederohmig
low-scan TV Schmalbandfernsehen
lower side Grundlinie (51); Unterseite
lower sideband unteres Seitenband
lower sound leiser Ton
low speed geringe Geschwindigkeit, niedrige Drehzahl
low-speed film unempfindlicher Film
low state of charge schlechter Ladezustand (621.356)

in a low state of charge schlecht geladen, fast entladen (621.356)
low temperature tiefe Temperatur
low-temperature brittleness Kaltbrüchigkeit
low-temperature cooling Tiefkühlung
low-temperature heater Tieftemperaturheizkörper
low-temperature trap Kühlfalle, Ausfrierfalle
low tension (siehe auch **low voltage**) niedrige Spannung; Niederspannung, Niederspannungs-...
low-tension cable Niederspannungskabel
low-tension detonator Niederspannungszünder, Brückenzünder (662.2)
low-tension ignition Abreißzündung (621.43)
low-tension insulator Niederspannungsisolator
low-tension line Niederspannungsleitung
low-tension magneto Niederspannungs-Zündmaschine
low-tension power plant Niederspannungskraftanlage
low-tension regulating transformer Regeltransformator für Niederspannungssteuerung
low-tension switch Niederspannungsschalter
low-tension switch-gear Niederspannungsschaltgeräte (pl.)
low-tension terminal Niederspannungsklemme, Unterspannungsklemme
low-tension winding Unterspannungswicklung, Niederspannungswicklung
low tide Niedrigwasser, Ebbe
lowest useful frequency niedrigste brauchbare Weitverkehrsfrequenz
low volatility Schwerflüchtigkeit
low-volatility liquid schwerflüchtige Flüssigkeit
low voltage (siehe auch **low tension**) niedrige Spannung, Niederspannung
low-voltage bus Niederspannungsverteilungsschiene
low-voltage distribution Niederspannungsverteilung
low-voltage insulator Niederspannungsisolator
low-voltage motor Kleinspannungsmotor
low-voltage relay Unterspannungsrelais
low-voltage release Unterspannungsauslösung
low-voltage supply system Niederspannungsversorgungsnetz
low-voltage system Niederspannungsanlage (unter 750 V Betriebsspannung)
low-voltage terminal Unterspannungsklemme
low-voltage winding Unterspannungswicklung, Niederspannungswicklung, NV-Wicklung, Niedervoltwicklung
to lower herabsetzen, senken; herunterlassen, absenken; kleiner werden, absacken, sinken
LP = low pressure ND = Niederdruck
L-section L-Glied (621.392)
LSB = least significant bit Bit mit dem niedrigsten Stellenwert
L. S. I. = large scale integration integrierte Halbleiter-Großschaltung, Großintegration
LSD = least significant digit
L. T. = low tension N. Sp. = Niederspannung
l.t.w. = low tension winding
Lu = Lutecium Lutetium
lubricant Schmiermittel
to lubricate schmieren, ölen, fetten
lubricating schmierend, Schmier-...
lubricating apparatus Schmierapparat
lubricating grease Fett, Abschmierfett, Staufferfett
lubricating groove Schmiernut
lubricating oil Schmieröl
lubricating pad Schmierpolster
lubricating substance Schmiermittel, Schmierstoff
lubricating system Schmiervorrichtung, Schmiersystem
lubrication Schmierung, Schmieren
lubricator Schmiernippel, Schmiertopf, Öler
lucid klar, durchsichtig
L.U.F. = lowest useful frequency niedrigste brauchbare Frequenz, LUF
lug Öse, Warze (621); Fahne, Öse, Lötfahne, Lötöse (621.3); Lageröse, Lagerböckchen
to lug schmieren (621.9)
aluminium lugs when turned Aluminium schmiert beim Drehen
lumen Lumen
luminare (US) Leuchte (621.32)
luminance Helligkeit; Leuchtdichte (621.397)
luminance signal Leuchtdichtesignal, Helligkeitssignal, Luminanzsignal
luminator crystal Leuchtquarz
luminescence Lumineszenz, Leuchten
luminescence lamp Leuchtstofflampe
luminescent material Luminophor

luminescent screen Leuchtschirm
luminosity Leuchtkraft; Lichtstärke (52); Lichtstärke, Helligkeit (535)
  luminosity factor (Anteil des sichtbaren Lichts an der Gesamtstrahlung)
luminous leuchtend, Leucht-...
  luminous (annunciator) board Leuchtschrifttafel
  luminous annunciator switch Leuchtbildschalter
  luminous arc lamp Gasentladungslampe
  luminous-band instrument Lichtbandinstrument
  luminous call system Lichtrufanlage
  luminous ceiling Leuchtdecke
  lumonous (circuit) diagram Leuchtschaltbild
  luminous crystal Leuchtquarz
  luminous discharge (leuchtende) Gasentladung
    luminous-discharge tube Gasentladungsrohr, Leuchtrohrreklameschild
  luminous efficiency Lichtausbeute
  luminous flux Lichtstrom
  luminous-gas lamp Gasentladungslampe
  luminous glow Glimmentladung
  luminous intensity Lichtstärke
  luminous paint Leuchtfarbe, Leuchtfarbenanstrich
  luminous-pointer galvanometer Lichtmarkengalvanometer
  luminous sensitivity Lichtempfindlichkeit (621.385)
  luminous spot Lichtfleck
  luminous substance Leuchtsubstanz, Leuchtmasse
Lummer-Brodhun photometer Lummer-Brodhun-Photometer
lump Klumpen, Stück, Brocken; konzentriertes Schaltelement (621.3)
  lump capacitance konzentrierte Kapazität
  lump inductance konzentrierte Induktivität
to lump zusammenfassen, konzentriert anordnen
lumped konzentriert
  lumped capacitor konzentrierte Kapazität
  lumped characteristic Kennlinie der Steuerspannung (621.385)
  lumped circuit Ersatzschaltung
  lumped constants konzentrierte Schaltelemente
  lumped-constant transmission line Kettenleiter
  lumped elements konzentrierte Schaltelemente
  lumped network quasistationärer Stromkreis
  lumped voltage Verschiebungsspannung
lune Sichel (52)
lurch Verrutschen
lustre Glanz
lux Lux
  luxmeter Beleuchtungsmesser
Luxemburg effect ionosphärische Kreuzmodulation, Luxemburgeffekt
L.W.L. = low water line Hauptwasserlinie (629.12)
lye Lauge

M

M = meg(a)-... M = Meg(a)-...
m = metre m = Meter
mA mA = milliampere
machinable (spanabhebend) bearbeitbar, zerspanbar
machine Maschine, Arbeitsmaschine
  machine addresse Maschinenadresse
  machine bed Maschinenfundament, Maschinenrost
  machine check Selbstprüfung, automatische Prüfung
  machine-code instruction Maschinenbefehl
  machine cycle Maschinenzyklus, Rechnerperiode
  machine drawing Maschinenzeichnung; Maschinenzeichnen
  machine finished maschinenbearbeitet
  machine language Maschinensprache
  machine program Maschinenprogramm
  machine ringing Generator-Rufanlage
  machine readable maschinenlesbar
  machine ruler Liniiermaschine
  machine screw Maschinenschraube
  machine shop Maschinenraum, mechanische Werkstatt, Werkstatt; Dreherei
  machine-spoilt work time maschinenbedingte Ausfallzeit
  machine-telegraph system Maschinentelegraphie
  machine time Maschinenzeit (621.9)
  machine tool Werkzeugmaschine
    machine-tool control Werkzeugmaschinensteuerung
  machine torque Maschinenmoment
  machine winding Maschinenwicklung
to machine (spanabhebend) bearbeiten, mechanisch bearbeiten
  to machine easily sich leicht bearbeiten lassen
  machined face bearbeitete Fläche
machinery Maschinerie; maschinelle Einrichtung, Maschinenpark

machinery allowance Bearbeitungszugabe (621.9)
machining (spanabhebende) Bearbeitung, Zerspanung
machining operation Bearbeitungsgang
machining shop mechanische Werkstatt, Dreherei
machinist Maschinist, Maschinenwärter
macro... Makro...
macroassembler Makroassembler
macro instruction Makrobefehl
magazine Lager, Magazin, Vorratslager; Magazin (623); Kassette
magic magisch, zauberhaft
magic eye magisches Auge (621.385), (621.396)
magic lantern laterna magica
magnet Magnet
magnet coil Magnetspule
magnet core Magnetkern, Kern
magnet frame Magnetgestell (621.313)
magnet shell Magnetglocke (621.317)
magnet steel Magnetstahl
magnet-type loudspeaker Magnetlautsprecher
magnet wheel Polrad (621.313)
magnet winding Magnetwicklung
magnet wire Spulendraht (621.313)
magnet yoke Magnetjoch
magnetic magnetisch, Magnet-...
magnetic adhesion magnetische Klebkraft
magnetic allotropy Spannungsanisotropie (538)
magnetic alloy Magnetlegierung
magnetic amplifier magnetischer Verstärker, Transduktor
magnetic amplifier with a. c. output Wechselstrommagnetverstärker
magnetic armature loudspeaker magnetischer Lautsprecher
magnetic attraction magnetische Anziehung
magnetic axis magnetische Achse (538)
magnetic balance Magnetwaage
magnetic bearing Magnetkompaß-Peilung, mißweisender Kurs, mißweisende Peilung
magnetic bias Vormagnetisierung
magnetic blow-out magnetische Blaseinrichtung, Blasmagnet (621.316.5)
magnetic blow-out fuse Sicherung mit Blasspule
magnetic brake Magnetbremse, Bremsmagnet (621.317)
magnetic bubble memory Magnetblasenspeicher
magnetic card Magnetkarte
magnetic cathode ray tube Braunsches Rohr mit magnetischer Ablenkung
magnetic character reader Magnetzeichenleser
magnetic chuck Magnetfutter (621.9)
magnetic circuit magnetischer Kreis
magnetic clutch magnetische Kupplung
magnetic coating magnetische Beschichtung
magnetic cohesion magnetische Haftung
magnetic compass magnetischer Kompaß
magnetic contactor elektromagnetisches Schütz
magnetic control magnetische Steuerung
magnetic core Magnetkern
magnetic-core storage Kernspeicher
magnetic couple magnetisches Moment
magnetic creeping magnetische Nachwirkung
magnetic cut-out Magnetschütz
magnetic damping Wirbelstromdämpfung
magnetic declination magnetische Deklination
magnetic deflection magnetische Ablenkung (621.385)
magnetic disc memory Magnetplattenspeicher
magnetic drum Magnettrommel(-Speicher) (681.14)
magnetic elongation Magnetostriktion, magnetostriktive Verlängerung
magnetic fatigue Blechalterung (621.313)
magnetic field magnetisches Feld
magnetic field strength magnetische Feldstärke
magnetic figures (pl.) Feilspanbild
magnetic flux magnetischer Fluß
magnetic flux counter Magnetflußzähler
magnetic flux crowds magnetischer Fluß drängt sich zusammen
magnetic focusing magnetische Strahlkonzentrierung
magnetic force magnetische Feldstärke
magnetic freezing Kleben (eines Relais)
magnetic head Magnetkopf
magnetic hysteresis magnetische Hysteresis
magnetic inclination magnetische Inklination
magnetic induction magnetische Induktion
magnetic ink magnetische Tinte
magnetic intensity magnetische Feldstärke
magnetic interlocking magnetische Verriegelung
magnetic iron ore Magneteisenerz
magnetic joint Stoßstelle im Magnetkreis
magnetic key Magnettaste
magnetic lag magnetische Nachwirkung
magnetic latitude magnetische Breite
magnetic leakage magnetische Streuung

magnetic leakage path magnetischer Streuweg
magnetic lens magnetische Linse
magnetic link record Stahlstäbchenmessung
magnetic loudspeaker magnetischer Lautsprecher
magnetic map magnetische Karte
magnetic memory Magnetspeicher
magnetic meridian magnetischer Meridian
magnetic microphone magnetisches Mikrophon
magnetic mine Magnetmine (623)
magnetic modulator Pungsdrossel
magnetic moment magnetisches Moment
magnetic needle Magnetnadel
magnetic north magnetisch Nord
magnetic ore separator magnetischer Erzscheider, Magnetscheider
magnetic oxide of iron Magneteisenstein
magnetic path magnetischer Weg
magnetic permeability magnetische Permeabilität
magnetic pickup magnetischer Tonabnehmer
magnetic pole Magnetpol, magnetischer Pol
magnetic pole strength magnetische Polstärke
magnetic potential magnetisches Potential
magnetic potential difference magnetische Potentialdifferenz, magnetische Spannung
magnetic proof plane Prüfspule, Spulensonde
magnetic property magnetische Eigenschaft
magnetic pull magnetische Anziehungskraft
magnetic pulley Magnettrommel; Trommel des Magnetscheiders
magnetic quantity magnetische Größe
magnetic quantum number magnetische Quantenzahl
magnetic reactance induktiver Blindwiderstand
magnetic recording magnetische Aufzeichnung, Magnetbandaufnahme
magnetic recording tape Magnetband
magnetic regulator magnetischer Konstanthalter
magnetic release magnetische Auslösung
magnetic reluctance magnetischer Widerstand, Reluktanz
magnetic remanence magnetische Remanenz, Remanenz
magnetic repulsion magnetische Abstoßung
magnetic resistance magnetischer Widerstand
magnetic retentivity Remanenz, Remanenzinduktion
magnetic reversal Ummagnetisierung
magnetic rotation magnetische Drehung (535)
magnetic saturation magnetische Sättigung
magnetic screen magnetische Abschirmung
magnetic sensitivity magnetische Empfindlichkeit (eines Hallgenerators)
magnetic separation Magnetscheidung, magnetische Aufbereitung
magnetic separator Magnetscheider
magnetic shell magnetisches Blatt
magnetic shield magnetische Abschirmung
magnetic shunt magnetischer Nebenschluß
magnetic skin effect Flußverdrängung
magnetic spectrum Beta-Strahlungsanteil (539)
magnetic square Epstein-Apparat
magnetic stability Stabilität (621.318.2)
magnetic storm magnetischer Sturm (538.7)
magnetic strip Magnetstreifen
magnetic susceptibility magnetische Suszeptibilität
magnetic switch Schütz, Magnetschalter
magnetic tape Magnetband, Tonband
magnetic tape device Magnetbandgerät, Tonbandgerät
magnetic-tape reader Magnetbandleser
magnetic-tape recording and reproducing systems in reel-to-reel and cassettes technique Magnetbandsysteme für Schallaufzeichnung in Spulen- und Kassettentechnik
magnetic-tape store Magnetbandspeicher
magnetic tester Magnetometer
magnetic testing magnetische Prüfung
magnetic time relay magnetisches Zeitrelais
magnetic track Magnetspur
magnetic track braking Schienenstrombremse
magnetic transition temperature magnetische Umwandlungstemperatur, Curiepunkt
magnetic trip magnetische Auslösung
magnetic tubes of force magnetische Kraftröhren
magnetic-type microphone magnetisches Mikrophon, Magnetmikrophon
magnetic variation (magnetische) Mißweisung
magnetic viscosity magnetische Nachwirkung
magnetic voltage stabilizer magnetischer Spannungskonstanthalter
magnetic wire store Magnetdrahtspeicher
magnetically-operated switch (berührungsloser) Magnetschalter
magnetics (pl.) Magnetteil, Magnetgestell; magnetische Teile (pl.); Magnetismus, Lehre vom Magnetismus
magnetism Magnetismus

magnetite Magnetit, Magneteisenstein, Magneteisenerz
magnetizability Magnetisierbarkeit
magnetization Magnetisierung
magnetization curve Magnetisierungskurve
magnetization range Magnetisierungsbereich
to magnetize magnetisieren
magnetized watch magnetisierte Uhr
magnetizing Magnetisierungs-...
magnetizing coil Magnetisierungsspule
magnetizing component Magnetisierungsstromanteil (538)
magnetizing current Magnetisierungsstrom
magnetizing-current inrush Einschaltstromstoß (621.313)
magnetizing force magnetomotorische Kraft, MMK
magnetizing winding Magnetisierungswicklung
magneto Zündmagnet, Zündmaschine (621.4)
magneto alternator magnetelektrische Maschine; Rufmaschine (621.394); Fahrraddynamo (629)
magneto bell Induktorwecker
magneto board Klappenschrank (621.395)
magneto call Induktorruf, Induktoranruf (621.395)
magneto compass magnetelektricher Kompaß
magneto-electric generator magnetelektrischer Generator
magneto exploder Zündmaschine (622)
magneto generator Induktor, Kurbelinduktor
magnetohydrodynamics = MHD Magnetohydrodynamik
magnetohydrodynamic generator magnetohydrodynamischer Generator
magneto ignition Magnetzündung (621.4)
magneto inductor Magnetinduktor
magneto-ionic double refraction magnetische Doppelbrechung
magneto-ionic splitting magnetische Aufspaltung
magneto-ohmmeter Kurbelinduktor für Widerstandsmessung
magneto optic magneto-optisch
magneto-optical rotation magneto-optische Drehung
magneto-optics (pl.) Magnetoptik
magnetoresistive element Feldplatte
magneto ringer Induktor, Rufinduktor
magneto ringing Induktoranruf
magneto speed indicator elektrischer Drehzahlmesser
magneto telephone system OB-Telefonie mit Induktorruf
magnetometer Magnetometer
magnetometer method Magnetometer-Verfahren
magnetometry Magnetometrie
magnetomotive force magnetomotorische Kraft, MMK
magneton Magneton
magnetophone Magnetophon
magnetron Magnetron
magnetron oscillations Magnetronschwingungen
magnetron pulling Magnetronfrequenzziehen, Magnetronlastverstimmung
magnetostriction Magnetostriktion
magnetostriction loudspeaker magnetostriktiver Lautsprecher
magnetostriction oscillator magnetostriktiver Schwinger
magnetostrictive magnetostriktiv
magnetostrictive oscillator magnetostriktiver Schwinger
magnetron Magnetron, Magnetfeldröhre
magnification Vergrößerung (535); Gütezahl (621.396)
magnification factor Gütezahl (621.396); Verstärkung (621.396)
magnifier Vergrößerungsapparat; Lupe, Vergrößerungsglas
to magnify vergrößern, verstärken
magnifying glass Vergrößerungsglas, Lupe
magnitude Größe, Ausdehnung; Größe (53), (52)
magnitude of a vector Betrag eines Vektors
main Hauptleitung (621.3); Hauptanschluß (621.395); Netzleitung (621.3); Netzpol (621.3)
main Haupt-..., hauptsächlich; Netz-... (621.3)
main aerial Netzantenne (621.396.62)
main airborne computer Zentraler Bordrechner
main bang Sendeimpuls (621.396.96)
main beam Hauptkeule, Hauptstrahl (621.396.67); Hauptbalken (624)
main-beam killing Hauptstrahlunterdrückung
main cable Hauptkabel, Netzkabel, Speisekabel
main-centre office Knotenamt
main circuit Hauptstromkreis
main circuit connection Netzanschluß
main condition Hauptbedingung, Grundbedingung
main constituent Hauptbestandteil
main contact Hauptkontakt
main current Hauptstrom; Netzstrom
main current-carrying contact Hauptkontakt (621.365.5)

main current relay Hauptstromrelais, Primärrelais, aufgebautes Relais
main distribution centre Hauptverteilung (696)
main distribution frame Hauptverteiler (621.395)
main distributor Hauptverteiler
main exciter Haupterregermaschine
main field Hauptfeld
mainframe Großrechner
main fuse Hauptsicherung
main leakage Luftspaltstreuung (621.313)
main limit switch Haupt-Endausschalter
main line Hauptstrecke, Hauptlinie (625)
main-line current Netzstrom
main lobe Hauptkeule
main lobe of radiation Hauptstrahlzipfel
main pattern Hauptraster (621.396.2)
main pole Hauptpol (621.313)
main power fuse Hauptnetzsicherung
main shaft Hauptschacht (622); Transmissionswelle (621.9)
main spring Uhrfeder, Hauptfeder, Triebfeder (681)
main station Hauptanschluß (621.395); Hauptstation, Zentralstation
main storage Hauptspeicher (681.3)
main switch Hauptschalter
main switchboard Hauptschalttafel
main switchgear connections (pl.) Hauptschaltanlage
main telephone station Teilnehmer Hauptanschluß
main terminal Hauptanschluß
main traffic hour Hauptverkehrsstunde (621.395)
mainmast Hauptmast
mains Netz, Versorgungsnetz; Netz-..., Netzanschluß-..., mit Netzanschluß
mains antenna Lichtantenne, Netzantenne
mains-distribution rack Netzschaltfeld
mains failure Netzausfall
mains-failure generating set Notstromversorgung
mains filter Netzsperre, Störschutz
mains fluctuation Netzspannungsschwankung
mains frequency Netzfrequenz
mains plug Netzstecker
mains power unit Netzanschlußgerät
mains receiver Netzanschlußempfänger
mains set Netzanschlußgerät
mains socket Steckdose
mains switch Netzschalter
mains transformer Netztransformator
mains valve wechselstromgeheizte Röhre
mains variation Netzschwankung
mains voltage Netzspannung
mainspring Uhrfeder, Triebfeder (681)
to maintain unterhalten, aufrechterhalten; unterhalten, pflegen, warten, erhalten; instandhalten; instandsetzen
to maintain Machinery Maschinen warten, Maschinenpark pflegen
to maintain at a potential auf einem Potential halten
maintained-contact switch Stellschalter
maintained-tuning fork Stimmgabelgenerator
maintained vacuum durch Pumpen aufrechterhaltenes Vakuum, gepumpt
maintained-vacuum rectifier an der Pumpe liegender Gleichrichter
maintained vibration Dauerschwingung, erzwungene Schwingung
maintainability Wartbarkeit
maintenance Aufrechterhaltung; Wartung, Pflege, Erhaltung, Instandhaltung
maintenance cost Wartungskosten, Unterhaltungskosten (pl.)
maintenance free wartungsfrei
maintenance instructions Wartungsvorschrift, Wartungsanleitung
maintenance personel Wartungspersonal
maintenance of programs Programmpflege
maintenance schedule Wartungsprogramm
major größer, hauptsächlich, Haupt-...; Dur
major cycle Grundzyklus, Hauptzyklus
major lobe Hauptkeule, größte Keule (621.396.67)
major principal stress größte Hauptspannung (539)
major semi-axis große Halbachse
major repair Hauptinstandsetzung (623)
majority carrier Majoritäts(ladungs)träger (621.382)
make Bauart, Fabrikat, Erzeugnis; Einschalt-...
make-and-break contact Ein- und Auschaltkontakt
make-and-break ignition Abreißzündung
make before break Schließen vor Öffnen, unterbrechungslos schalten
make-before-break contact unterbrechungsloser Umschaltkontakt
make-and-break key Ein- und Ausschalter
make circuit Arbeitsstromkreis
make contact Schließer, Arbeitskontakt
make impulse Schließimpuls
make spark Schließungsfunke
make time Ansprechzeit, Ansprechverzug
to make machen, herstellen, fabrizieren

to make allowance for berücksichtigen, Zuschläge machen für
to make an angle with einen Winkel bilden mit
to make and break the current den Strom ein- und ausschalten
to make a circle einen Kreis schlagen, einen Kreis beschreiben
to make a connection einen Anschluß herstellen
to make contact Kontakt machen, Kontakt herstellen
to make a current einen Strom einschalten
to make a field ein Feld aufbauen
to make a good fit gut passen; gut einpassen, gut passend machen
to be made a good fit gut sitzen, gut passen
hydrogen and chlorine make muriatic acid Wasserstoff und Chlor bilden Salzsäure, H und Cl verbinden sich zu HCl
to make a piece to suit another one einen Teil zu einem anderen passend machen
to make up verarbeiten, fertigmachen, garnieren
to make up with nachfüllen mit, auffüllen mit, ergänzen durch
2 + 2 makes 4 2 + 2 gibt 4
maker Fabrikant, Hersteller
maker's name-plate Firmenschild
makeshift Notbehelf, Behelfsmittel, Behelfs-...
make-up Aufmachung, Ausführungsform
make-up air Zusatzluft (bei Kreislaufkühlung)
making Einschaltung
making capacity Einschaltleistung
making current Einschaltstrom
making-current inrush Einschaltstromstoß
malachite Malachit
maladjustment schlechter Abgleich, schlechte Angleichung
male männlich
male plug Steckerstift, Stecker (621.3)
male radius Außenradius
male screw Schraubenbolzen
male thread Außengewinde
malfunction Störung, fehlerhafte Funktion
malleability Schmiedbarkeit
malleable schmiedbar
malleable cast iron Temperguß
malleable casting schmiedbarer Guß
malleable iron Schmiedeisen
Maltese cross Malteserkreuz
man Mensch; Mann
manhole Mannloch
manhole chimney Arbeitsschacht, Kabelschacht
manhour Arbeitsstunde, Mannstunde
man-made künstlich erzeugt
man-made noise industrielle Funkstörung
manpack radio Tornisterfunkgerät
manpower Arbeitskraft, menschliche Arbeitsleistung
to man bemannen; (mit Personal) besetzen
to manage leiten, verwalten
to manage a factory eine Fabrik betreiben, eine Fabrik leiten
manageable handlich
management Betriebsleitung, Direktion, Verwaltung, Leitung
manager Leiter, Direktor, Betriebsführer
managing Leitungs-..., Direktions-...
managing board Geschäftsleitung, Betriebsführung, Direktion, Vorstand
managing director Generaldirektor, geschäftsführender Vorstand
mandrel Dorn, Drehdorn, Wickeldorn, Mandrill
manganese Mangan
to manipulate behandeln, handhaben, betätigen, manipulieren; umformen, umrechnen (51)
manipulated variable Stellgröße (621.316.7)
manipulation Behandlung, Handhabung; Umformung, Zwischenrechnung (51)
manipulation of Umgehen mit
manipulation of a relation Umformung einer Beziehung (51)
manoevring cam Steuernocken
manometer Manometer, Druckmesser
Mansbridge condenser Wickelkondensator, Kondensatorwickel
mantissa (pl.: **-ae**) Mantisse
manual Handbuch, Beschreibung, Bedienungsanleitung; Hand-..., von Hand
manual central office Handamt, Handvermittlung
manual control Handsteuerung, Handbetätigung, Handbedienung
manual controller handbetätigter Steuerschalter
manual data manuell eingegebene Daten
manual drive Handantrieb
manual exchange Handamt, Handvermittlung (621.395)
manual operation Handbetätigung; Handbetrieb; manueller Betrieb (681.3)
manual ringing Handruf, Induktorruf
manual switchboard Handschrank, Handvermittlung (621.395)
manual telephone system Handvermittlung
manual tuning Handabstimmung, Abstimmung von Hand

manual update service Handbuch-Änderungsdienst
manual weld Handschweißung
manual work Handarbeit
manually von Hand
manufacture Herstellung, Fabrikation, Erzeugung
to manufacture herstellen, erzeugen, fabrizieren
manufacturer Hersteller, Erzeuger, Fabrikant
manufacturing Herstellung, Erzeugung
manufacturing capacity Produktionskapazität
manufacturing process Fabrikationsverfahren, Herstellungsverfahren
manufacturing programme Fabrikationsprogramm
map Karte
map display Kartendarstellungsgerät (623)
to mar verderben, „vermurksen", „versauen"
marble Marmor
marble slab Marmorplatte
Marconi antenna vertikale Eindraht-Antenne
Marconi-Reiss microphone Reiss-Mikrophon
Marconi system Marconi-System
marconigram Funktelegramm
margin Rand, freier Rand; Spielraum, Überschuß; Toleranzgrenze; Verdienstspanne; Sicherheitsabstand
margin of error Fehlergrenze
margin of safety Sicherheitsabstand
marginal ... am Rande; Überschuß-...
marginal check Grenzwertprüfung (681.3)
marine Marine; Handels-Schiffahrt
marine Marine-..., Schiffs-..., See-...
marine galvanometer Spanndrahtgalvanometer
marine motor Schiffsmotor
marine propulsion Schiffsantrieb
marine radio communication Seefunk
marine wiring Schiffsinstallation
maritime maritime
maritime radio navigation service Seenavigationsfunkdienst
maritime service Seefunkdienst
mark Markierung, Marke; Markierungsstrich, Marke; Teilstrich; Zeichen, Kennzeichen; Ziel; Spannmarke (621.9); Bearbeitungsspur (621.9); Mark, D-Mark; Stromschritt (621.394)
below the mark unternormal, schlecht
up to the mark den Vorschriften entsprechend
mark-to-space ratio Impulstastverhältnis
to mark markieren, angeben, kennzeichnen; anzeichnen; festlegen; Spannmarkenverursachen, Bearbeitungsspuren hinterlassen (621.9)
to mark off abtragen, auftragen, abgreifen
to mark off in einteilen in, teilen in
to mark "x" mit „x" bezeichnen; mit einem„x" kennzeichnen
marked erheblich, deutlich ausgeprägt; aufgedruckt, aufgestempelt; gestempelt; laut Aufdruck für ...
marked pole Nordpol (538)
marked ratio Nennübersetzungsverhältnis (621.317)
marker Markierung, Markierungszeichen; Zeitmarkengenerator (621.317); Markierer (621.395.3)
marker beacon Markierungsfunkfeuer (621.396)
marker lamp Schlußlampe (656.25)
marker light Schlußlicht, Schlußlampe (656)
marker pip Markierungsimpuls (621.397)
marker stud Markierungsknopf
marking Markierung; Markierung, Markierungslinie, Markierungszeichen, Kennzeichen
marking-off table Anreißtisch, Anreißplatte (621.9)
marking scraper Reißnadel
marking signal (element) Stromschritt (621.394)
marking tag Bezeichnungsfahne
to marshal ordnen, anordnen; rangieren (625)
marshalling yard Rangierbahnhof
martensite Martensit
Marx rectifier Marx-Gleichrichter
maser (**m**icrowave **a**mplification by **s**timulated **e**mission of **r**adiation) Maser
to mash quetschen, zerquetschen
mask Schablone, Abdeckung, Maske
mask engineering Maskentechnik
mask microphone Maskenmikrophon
mask-programmable read-only memory maskenprogrammierbarer Festwertspeicher
to mask maskieren; verdecken (534), (621.396.62)
to mask out unschädlich machen, kompensieren, unterdrücken
masking Verdeckung (534), (621.396.62); Maskieren (621.382)
masking of the signal by noise Verdeckung der Signale durch Rauschen
mass Masse
mass-absorption coefficient Massenabsorptionskoeffizient (621.386)
mass of body at rest Ruhemasse (539)
mass conservation Massenkonstanz

mass core Massekern, Eisenpulverkern, Pulverkern (621.396)
mass defect Massendefekt (539)
mass difference Massendefekt (539)
mass distortion Massenverschiebung, seitliche Versetzung
mass impregnation Durchimprägnierung
mass moment of inertia Massenträgheitsmoment
mass number Massenzahl (539)
mass production Massenproduktion, Serienfertigung
mass production testing Massenprüfung
mass resistivity Widerstand pro Längeneinheit und Gewichtseinheit
mass separator Isotopentrenner, Isotopentrenneinrichtung (539)
mass spectrograph Massenspektrograph
mass spectrum Massenspektrum
mass storage Massenspeicher, Speichergroßer Kapazität
mass survey Massenuntersuchung, Reihenuntersuchung (61)
mass unit Masseneinheit
massive massiv; klotzig; massebehaftet
massiveness Dichte, Kompaktheit
massless masselos
to mass produce massenfertigen
massy massig, klobig, klotzig
mast Mast (621.396.67), (629)
mast base Mastfuß
mast switch Mastschalter
master Meister (**of:** in); Mutter...
master clock Taktgeber, Zeitgeber; Hauptuhr
master controller Hauptsteuerschalter, Steuerwalze; Fahrschalter (625)
master exchange Muttervermittlungsstelle
master fire-alarm box Hauptfeuermelder
master form Hauptform (621.357)
master frequency meter Frequenzuhr (621.311)
mastergroup Tertiärgruppe (621.395.4)
master key Hauptschlüssel, Passepartout
master motor Leitmotor (621.316.7)
master oscillator Steuersender
master oscillator circuit Steuerkreis, Schwingkreis des Steuersenders
master oscillator with decadic setting dekadische Steuerstufe
master processor Hauptprozessor
master routine Hauptprogramm, ausführendes Programm (621.382)
master service Gemeinschaftsanschluß (621.395)
master-slave configuration Master-Slave Konfiguration
master switch Hauptschalter
mastic Mastix
to masticate kauen (61); mastizieren (67)
mat Matte, Fußmatte, Mattengeflecht
mat matt, mattiert
mat black matt-schwarz
mat-surface glass mattiertes Glas, Mattglas
match Streichholz; passendes Stück, Paßstück; Anpassung
to be a match passen (**for:** zu)
to match anpassen (**to:** an) (627.39); vergleichen (681.3)
to match into einpassen in
matched pair Paßpaar
matcher Anpaßglied
matching Anpassung
matching device Anpassungsglied
matching efficiency Anpassungswirkungsgrad
matching impedance Anpassungsimpedanz
matching line Anpassungsleitung, Leitungstransformator (621.396)
matching section Leitungstransformator, Anpassungsleitung
matching stub Blindspritze, Anpassungsstück
matching transformer Anpassungsübertrager
to mate eingreifen, kämmen
mating gear eingreifendes Rad, kämmendes Rad, zugehöriges Rad
mating surface Paßfläche; ineinandergreifende Fläche, Eingriffsfläche
material Material, Stoff; Werkstoff, Baustoff
material particle Materieteilchen
(materials) testing reactor Prüfreaktor
to materialize materialisieren; (sich) in Materie verwandeln; Wirklichkeit werden
materials Materialien; Material (sing.)
materials testing Materialprüfung
math. = mathematical math. = mathematisch
mathematical mathematisch
mathematical certainty mathematische Gewißheit
mathematical fiction mathematische Annahme
mathematical formulation mathematische Formulierung
mathematical physicist theoretischer Physiker
mathematical physics (pl.) theoretische Physik
mathematical problem mathematisches Problem, mathematische Aufgabe
mathematical proof mathematischer Beweis
mathematical representation mathematische Schreibweise, mathematische Darstellung
mathematician Mathematiker

mathematics (pl.) Mathematik
Mathieu function Mathieusche Funktion
mating contact Gegenschaltstück
matrix Matrix (51); Matrix, Grundmasse (66); Kathodenform (621.357)
matrix character Matrixzeichen
matrix printer Matrixdrucker
matt matt, mattiert (s. auch **mat**)
matte dip Mattierungsbad
matter Stoff, Materie; Material, Masse; Angelegenheit
Matthiessen's standard (Widerstandsnorm für Leitungskupfer)
max. = maximum max. = Maximum
maxigraph Maxigraph, Spitzenschreiber (621.317)
maximum (pl. **-ma**) Maximum, Höchstwert, Scheitelwert
maximum Höchst-..., Größt-..., Maximal-..., maximal
maximum asymetric short-circuit current Stoßkurzschlußstrom
maximum capacity Höchstleistung, höchstmögliche Leistung
maximum circuit breaker Überstromschalter
maximum current Höchststrom, Strommaximum
maximum current circuit breaker Maximalschalter
maximum current relay Maximalrelais, Überstromrelais
maximum cut-out Überstromauslöser, Überstromschütz
maximum deflection Vollausschlag (621.317)
maximum demand Spitzenlast, Höchstbedarf, Lastspitze, Bedarfsspitze
maximum demand indicator Höchstlastanzeiger, Spitzenzähler (621.317)
maximum-demand meter Spitzenzähler (621.317)
maximum equipment line Bahnprofil
maximum limited stress Grenzbeanspruchung
maximum load Höchstlast, Höchstbelastung
maximum output höchste Ausgangsleistung (621.396)
maximum output voltage swing höchste zulässige Schwingungsbreite der Ausgangsspannung
maximum permissible dose Toleranzdosis (539)
maximum permissible exposure Toleranzdosis (539)
maximum permissible level Toleranzdosis
maximum permissible load zulässige Belastung
maximum permissible stress zulässige Beanspruchung
maximum potential höchstes Potential, höchstzulässige Spannung
maximum power Höchstleistung, Größtleistung, Spitzenleistung
maximum power demand Spitzenleistung, Leistungsbedarfsspitze
maximum power transfer höchste Übertragungsleistung, größte Leistungsübertragung
maximum-pressure governor Überdruckwächter
maximum rate of change of the output voltage größte Flankensteilheit der Ausgangsspannung
maximum rated mean forward current Dauergrenzstrom (621.382.2)
maximum rated surge forward current Stoßstrom Grenzwert
maximum relay Maximalrelais
maximum safe pull höchstzulässige Zuglast
maximum scale value Skalenendwert (621.317)
maximum speed Höchstgeschwindigkeit, Höchstdrehzahl
maximum speed potential höchstzulässige Drehzahl, höchstmögliche Drehzahl, Höchstgeschwindigkeit
maximum stress Höchstbeanspruchung, Höchstspannung
maximum compression stress Höchstdruckbeanspruchung
maximum safety stress höchstzulässige Beanspruchung
maximum tensile stress höchste Zugspannung, höchste Zugbeanspruchung
maximum tension stress Höchstzugspannung, höchste Zugbeanspruchung
maximum temperature Höchsttemperatur
maximum safe temperature zulässige Höchsttemperatur
maximum thermometer Maximalthermometer
maximum usable frequency = MUF höchste brauchbare Weitverkehrsfrequenz
maximum value Höchstwert, Größtwert, Maximalwert
maximum voltage Maximalspannung
maximum voltage circuit breaker Überspannungsschalter
maximum voltage relay Überspannungsrelais
maximum weight Höchstgewicht
maxwell Maxwell (538)

Maxwell bridge Maxwellsche Brücke (621.317)
Maxwell field equations Maxwellsche Gleichungen
Maxwell statistics Maxwell-Statistik
MAYDAY MAYDAY (internationales Notzeichen)
MCBF = mean cycles between failure
M. C. W. = modulated continuous wave modulierter Dauerstrich, unmodulierter Betrieb mit Anodenwechselspannung
meadow ore Raseneisenerz
mean Mittelwert, Hauptwert, Häufungswert
mean mittlere(r), Mittel-...; dürftig, schlecht
mean conducting state power loss mittlere Durchlaßverlustleistung (621.314.63)
mean conical (luminous) intensity mittlere Lichtstärke im Nutzkegel
mean deviation mittlerer Hub
mean free path mittlere freie Weglänge
mean hemispherical intensity mittlere hemisphärische Lichtstärke
mean horizontal intensity mittlere horizontale Lichtstärke
mean interconnecting number Mischungsverhältnis (621.395)
mean life mittlere Lebensdauer
mean load mittlere Last, Durchschnittslast
mean lower hemispherical intensity mittlere untere hemisphärische Lichtstärke
mean on-state current Durchlaßstrom-Mittelwert (621.314.63), (621.382.2)
mean output mittlere Ausgangsleistung (621.396); mittlere Dauerleistung (621.313)
mean power mittlere Leistung
mean proportional mittlere Proportionale
mean sea level N. N., Normal Null, mittlere Seehöhe
mean spherical intensity mittlere sphärische Lichtstärke
mean system capacity mittlere Netzleistung
mean temperature mittlere Temperatur
mean-time-between-failures = MTBF mittlerer Ausfallabstand, mittlere Zeit zwischen Ausfällen
mean upper hemispherical intensity mittlere obere hemisphärische Lichtstärke
mean value Mittelwert
mean-value meter Mittelwertmesser
mean value with respect to time zeitlicher Mittelwert
mean zonal intensity mittlere Lichtstärke in der Nutzzone
to meander sich winden, sich schlängeln
means Mittel, Hilfsmittel
means of communication Nachrichtenmittel
means of conveyance Beförderungsmittel, Fördermittel
means of ignition Zündmittel
means of observation Beobachtungsmittel
means of transportation Transportmittel, Beförderungsmittel
measurable meßbar
measurand Regelgröße
measure Maßnahme, Maß; Maßstab, Hohlmaß; Teiler
measure of amplification Maß an Verstärkung, Verstärkungsmaß (621.396)
measure of capacity Hohlmaß
to measure nach Maß
to measure messen, ausmessen
to measure from von ... aus messen
to measure off abmessen, auftragen, abtragen
beam measuring 10 ft in length and 3 × 5 in. in cross-selection Balken mit den Abmessungen 10 Fuß Länge und 3 × 5 Zoll Querschnitt
measured object Meßobjekt
measurement Messung, Abmessung
measurement of angles Winkelmessung
measurement range Meßbereich
measurement of temperature Temperaturmessung
measurements Maßeinheiten
measuring Messen, Messung
measuring messend, Meß-..., mit den Abmessungen
measuring bridge Meßbrücke
measuring bush Meßbuchse
measuring chain Meßkette, (**1 chain = 22 yards** = 20,116 m) (526)
measuring container Meßgefäß, Meßbehälter
measuring device Meßgerät, Meßeinrichtung
measuring equipment Meßeinrichtung, Meßanordnung, Meßgerät
measuring error Meßfehler
measuring head Meßkopf
measuring instrument Meßinstrument
measuring line Meßleitung (621.396); Meßleine (52)
measuring orifice Stauscheibe, Meßrand
measuring pole Meßlatte
measuring range Meßbereich
measuring-range extension Meßbereicherweiterung
measuring section Meßstrecke, Meßabschnitt
measuring table Meßtisch (621.9)
measuring tape Bandmaß

measuring technique Meßtechnik, Meßmethode
measuring transducer Meßwertumformer
measuring tool Meßwerkzeug
measuring voltage Meßspannung
mechanic Schlosser, Mechaniker; Ingenieur; Handwerker
mechanical mechanisch
mechanical advantage Kraftübersetzung, Kraftverstärker
mechanical axis mechanische Achse (537), (54)
mechanical blocking mechanische Verriegelung
mechanical connector mechanischer Verbinder
mechanical drive Stimmgabelsteuerung (621.396.61); mechanischer Antrieb
mechanical drive system mechanisches Antriebssystem
mechanical energy mechanische Energie
mechanical engineer Maschinenbau-Ingenieur, Maschinenbauer
mechanical engineering Maschinenbau
mechanical equivalent of heat mechanisches Wärmeäquivalent
mechanical equivalent of light mechanisches Lichtäquivalent
mechanical filter mechanisches Filter, magneto-restriktives Filter
mechanical image impedance mechanischer Kennwiderstand = $\sqrt{m \cdot s}$
mechanical impedance mechanische Impedanz
mechanical joint Klemmverbindung
mechanical master oscillator Stimmgabelsteuerstufe
mechanical mixture mechanische Mischung, physikalische Mischung, Mischung, Gemenge
mechanical ohm akustisches Ohm, mechanisches Ohm
mechanical operation mechanische Betätigung
mmchanical parts mechanische Teile
mechanical plating Galvanisierung mit bewegten Elektroden
mechanical power mechanische Leistung
the mechanical powers die einfachen Maschinen = Keil, Schraube, schiefe Ebene, Rolle
mechanical property mechanische Eigenschaft
mechanical protection mechanischer Schutz
mechanical quantity mechanische Größe
mechanical reactance mechanischer Blindwiderstand
mechanical rectifier mechanischer Gleichrichter, Kontaktgleichrichter
mechanical resistance mechanischer Wirkwiderstand
mechanical stop mechanischer Anschlag
mechanical strength mechanische Festigkeit
mechanical trip mechanische Auslösung
mechanically operated mechanisch betätigt
mechanically operated circuit breaker automatischer Schalter
mechanically trip free ... mit mechanischer Freiauslösung
mechanician Mechaniker, Monteur, Maschinenschlosser
mechanics (pl.) Mechanik
mechanism Mechanismus, Mechanik
mechanization Mechanisierung
to mechanize mechanisieren
median Mittelhalbierende, Seitenhalbierende; Zentralwert (51)
median Mittel-...
median life zentrale Lebensdauer
median line Mittellinie
median plane Mittelebene
medical ärztlich
medical coil Behandlungsspule (615.84)
medium Mittel, Medium; mittler, Mittel-...
medium frequency Mittelwelle
medium grain mittlere Korngröße
medium grained texture mittelfeines Korn (669)
medium hardness mittlere Härte
medium-hardness mittelhart
medium plate Mittelblech
medium pressure Mitteldruck (621.18); Mittelspannung (621.3)
medium quality mittlere Qualität, mittlere Güte (klasse)
medium-quality mittlerer Güte
medium scale integration MSI mittlerer Integrationsgrad
medium voltage (Spannung zwischen 250 und 650 Volt)
medium wave Mittelwelle (300 kHz ... 3 MHz)
to meet treffen, auftreffen, stoßen gegen, stoßen auf; sich treffen; erfüllen
to meet at a point einem Punkt zufließen, sich in einem Knotenpunkt vereinigen (621.3)
to meet requirements Vorschriften erfüllen, Anforderungen befriedigen
to meet a stop anschlagen, auf einen Anschlag kommen, gegen einen Anschlag laufen
meg..., mega... Meg..., Mega...
megacycle = megacycle per second Megahertz
mega electron volt Mega-Elektronenvolt
megaphone Megaphon, Sprachtrichter

megawatt Megawatt
megger Isolationsprüfer
megohm Megohm
Meissner circuit transformatorische Rückkopplung, Meissner-Schaltung (621.396.61)
melt Schmelze, Schmelzvorgang; Tauen
to melt schmelzen; tauen
to melt down herunterschmelzen
to melt into verschmelzen mit
to melt out ausschmelzen
meltabel schmelzbar
melting Schmelzen, Schmelzvorgang, Schmelz-...
melting crucible Schmelztiegel
melting current Schmelzstrom
melting furnace Schmelzofen
melting point Schmelzpunkt
melting process Schmelzprozeß, Schmelzvorgang
melting time Abschmelzzeit, Ansprechzeit (621.316.9)
membrane Membran
memory Speicher (861.3) (siehe auch **storage**)
memory allocation Speicherzuordnung
memory address register Speicheradressen-Register
memory capacity Speicherkapazität
memory core Speicherkern
memory cycle Speicherzyklus
memory data register Speicherregister
memory mapping [spezielle Adressierungsart]
memory protection Speicherschutz
memory-resident speicherresident
memory-sense amplifier Leseverstärker
mend Flickstelle, Flicken
to mend ausbessern, flicken
menue [Anzeige für den Benutzer über Bedienung oder Bedienungs-Alternativen (681.3)]
meniscus (pl. **-ci**) Meniskus; Konkav-Konvex-Linse (535)
mensurable ausmeßbar
mental arithmetic Kopfrechnen
to mercerize merzerisieren, glänzen
mercerized cotton Glanzgarn, merzerisierte Baumwolle
mercuric Merkuri-...
mercurous Merkuro-...
mercury Quecksilber
mercury arc Quecksilberdampflichtbogen
mercury-arc converter Quecksilberdampfumrichter
mercury-arc inverter Quecksilberdampfwechselrichter
mercury-arc lamp Quecksilberdampflampe
mercury-arc rectifier Quecksilberdampfgleichrichter
mercury break Quecksilber-Unterbrecher
mercury cell Amalgamzelle
mercury column Quecksilbersäule
mercury-discharge lamp Quecksilberdampflampe
mercury electrode Quecksilberelektrode
mercury fulminate Knallquecksilber
mercury gauge Quecksilbermanometer
mercury-in-glass switch Quecksilberschaltröhre
mercury-interrupter Quecksilber-Unterbrecher
mercury-jet Quecksilberstrahl
mercury-jet rectifier Quecksilberstrahlgleichrichter
mercury meter Quecksilberzähler, Elektrolytzähler, Stia-Zähler
mercury motor meter Quecksilber-Motorzähler
mercury pool Quecksilberkathode, Kathodensumpf
mercury-pool-type rectifier Quecksilberdampfgleichrichter mit flüssiger Kathode
mercury rectifier Quecksilber(dampf)gleichrichter
mercury relay Quecksilberrelais
mercury switch Quecksilberschalter
mercury thermometer Quecksilberthermometer
mercury thread Quecksilbersäule, Quecksilberfaden
mercury unit Siemens-Einheit (alte Widerstandseinheit; 1 Siemenseinheit = 0,94 Ohm)
mercury vapour Quecksilberdampf
mercury-vapour discharge Quecksilberentladung
mercury-vapour frequency changer Quecksilberdampfumrichter
mercury-vapour with glass bulb mit Glasgefäß
mercury-vapour with iron tank mit Eisengefäß
mercury-vapour lamp Quecksilberdampflampe
mercury-vapour rectifier Quecksilberdampfgleichrichter
mercury-vapour tube Quecksilberdampfröhre
mercury-wetted contact quecksilbergenetzter Kontakt
meridian Meridian

meridian plane Meridional-Ebene
meridian section Meridianschnitt, Meridionalschnitt
meridional meridional = südlich
Merz-Hunter protective system Fehlerstrom-Differentialschutz
Merz-Price protection Merz-Price-Schutz; Längsdifferentialschutz, Stromdifferentialschutz
mesa technique Mesatechnik (621.382)
mesh Masche, Netzmasche; Maschenweite; Netz, Maschennetz (621.3); Netzmasche, Masche (621.3); Gewebe, Gitter; Eingriff (621.8)
mesh circuit Masche
mesh-connected circuit Polygonschaltung
mesh connection Dreieckschaltung, Polygonschaltung
mesh current Maschenstrom (621.3)
mesh network vermaschtes Netz, Maschennetz
mesh voltage Dreieckspannung, verkettete Spannung
to mesh in Eingriff stehen, kämmen (**with:** mit)
meshing gear Gegenrad
Mesny circuit Mesny-Schaltung
meson Meson
mesotron Mesotron, Meson
message Nachricht, Botschaft; Fernspruch, Rundspruch
message format Spruchschema, Spruchformat
message register Gesprächszähler (654.15)
message-storage exchange Speichervermittlung (621.394)
message switching, store and forward Teilstreckentechnik (621.394)
messenger Tragdraht, Tragseil (621.33), (621.39)
messenger wire Aufhängedraht (621.33)
met. service = meteorological service Wetterdienst
metabolons Zerfallsprodukte (539)
metadyne Metadyne (621.313)
metal Metall; Beschotterung, Schotter (625)
metal base Metallfuß, Metalluntersatz
metal braid Metallgeflecht, Metallbeflechtung
metal case Metallgehäuse
metal-cased conductor Leitung in Stahlpanzerrohr
metal ceramic Metallkeramik
metal-clad metallgekapselt, gußgekapselt
metal-clad building Gebäude mit Metallwänden
metal-clad switchgear gußgekapselte Schaltgeräte (pl.), gekapselte Schaltanlage
metal cladding Kapselung, Metallkapselung
metal coating metallischer Belag; Metallmantel
metal conductor metallischer Leiter
metal content Metallgehalt
metal-cored carbon Metalldochtkohle
metal cutting spanabhebende Bearbeitung
metal detector Metallsuchgerät
metal distribution ratio Schichtdickenverhältnis (621.357)
metal electrode Metallelektrode
metal envelope Metallhülle
metal-envelope tube Metallröhre (621.385), Stahlröhre
metal filament Metalldraht
metal-filament lamp Metallfadenlampe
metal-film resistor Metall-Schichtwiderstand
metal fog dispergierter Metallstaub (621.357)
metal foil Metallfolie, Blattmetall
metal insert Metalleinlage
metal lamp Metalldrahtlampe
metal locator Metallsuchgerät
metal mist dispergierter Metallstaub (621.357)
metal molding Bergmannrohr (696)
metal-nitride semiconductor field-effect transistor Feldeffekttransistor mit Metall-Nitrid-Halbleiteraufbau, MNS-FET
metal oxide semiconductor field-effect transistor Feldeffekttransistor mit Metall-Oxid-Halbleiter-Aufbau, MOS-FET
metal-oxide rectifier Kupferoxydul-Gleichrichter
metal panel Metallwand, Blechwand
metal pipe Metallrohr
metal powder Metallpulver
metal rectifier Trockengleichrichter
metal reinforcement Bewehrung, Armierung
metal-removing work Zerspanungsarbeit; spanabhebende Bearbeitung
metal roof Metalldach
metal sheath Metallmantel
metal shell Metallhülle
metal spraying Metallspritzen, Metallspritzverfahren, Schoopisierung
metal stand Metallstativ, Metallgestell
metal strip Metallstreifen
metal strip fuse Streifensicherung
metal-tank mercury-arc rectifier Eisengleichrichter
metal valve Stahlröhre (621.385)
metal vapour Metalldampf
metal-vapour lamp Metalldampflampe
metal-vapour rectifier Metalldampfgleichrichter
metal „V"-ring Kommutatorring

metal X-ray tube Strahlenschutzröhre (621.386)
metallic metallisch
metallic bond Metallbindung (539)
metallic circuit Leitung; galvanische Verbindung
metallic filament Metalldraht (621.32)
metallic-filament lamp Metalldrahtlampe
metallic-film resistor Metallschichtwiderstand
metallic insulator Metallisolator, Lambda-Viertel-Stütze (621.396)
metallic iron metallisches Eisen
metallic lustre metallischer Glanz
metallic ore Metallerz
metallic oxide Metalloxid
metallic-oxide rectifier Trockengleichrichter
metallic rectifier Trockengleichrichter
metallic return circuit Kabelrückleitung (621.33)
metallic state metallischer Zustand
metalline metallisch, metallähnlich
metallization Metallisierung
to metallize metallisieren
metallized filament metallisierter Leuchtdraht
metallized-paper capacitor MP-Kondensator, Metall-Papier-Kondensator
metallized-paper capacitor with plastic-film dielectric MPF-Kondensator
metallized valve metallisierte Röhre
metallizing Metallisierung
metallographic metallographisch
metallography Metallographie
metalloid Metalloid
metastable metastabil
metastable equilibrium metastabiles Gleichgewicht (54)
metastable output configuration metastabile Ausgangsfiguration (621.374)
meteorological meteorologisch
meteorology Meteorologie
meter Messer, Meßwerk, Meßgerät, Meßinstrument; Zähler
meter calibration Zählerabgleich, Zählerprüfung
meter change-over clock Zählerschaltuhr
meter deflection Zeigerausschlag, Instrumentenanzeige
meter dial Zählerskala
meter disc Zählerscheibe (621.317)
meter display Instrumentenanzeige
meter error Zählerfehler (621.317.5)
meter indication Instrumentenanzeige; Zähleranzeige, Zählerstand
meter jewel Zählerlagerstein
meter key Zähltaste, Gesprächszähltaste (621.395)
meter loss Zählereigenverbrauch
meter mounting board Zählerplatz
meter panel Zählertafel
meter rack Zählergestell (621.396)
meter reader Zählerableser
meter switch Meßbereichumschalter, Meßbereichwähler
meter tester Zählerprüfer, Zählereicher
meter testing Zählerprüfung
meter-testing station Zählerprüfstelle, Zählerprüfamt
meter train Rollenzählwerk (621.317)
meter watts (pl.) aus der Zähleranzeige errechnete mittlere Leistung
meter watthours angezeigter Verbrauch (sing.) (621.317)
meter winding Zählerwicklung
meter Meter
meter ampere Strommoment (621.396.67)
meter bridge Schleifdraht-Meßbrücke
meter candle (obs.) Lux = Lumen/qm
meter cube Kubikmeter; Würfel von 1 m Kantenlänge
meter wire Schleifdraht (621.317)
to meter messen, ausmessen, zumessen
metering Meßwesen, Meß-...
metering device Meßgerät
metering pulse Zählimpuls (621.395)
metering rate Zähltakt (621.395)
metering zone Gebührenzone (621.395)
methane Methan
method Methode, Verfahren
method of analysis Rechenverfahren (51); Analysenverfahren (54)
method of application Auftragsverfahren, Anbringungsweise
method of attack Behandlungsweise, Lösungsweg (51)
method of calculation Rechenverfahren
method of direct deflections Ausschlagverfahren (621.317)
method of extraction Herstellungsverfahren, Darstellung (54)
method of least squares Methode der kleinsten Quadrate
method of measurement Meßverfahren, Meßmethode
method of observation Beobachtungsverfahren
method of operation Arbeitsweise, Betriebsart
method of procedure Behandlungsweise, Art des Vorgehens

method of production Herstellungsverfahren, Fabrikationsart
method of sections Schnittmethode, Schnittverfahren (531)
method of test Prüfverfahren
methylalcohol Methylalkohol, Holzspiritus
metric metrisch
in metric in metrischem System; in metrischem Maß, metrisch
to metric measurement nach metrischem Maß
metric system metrisches System
metric thread metrisches Gewinde
metric waves Meterwellen, UKW
metrological maßkundlich
metrology Meßwesen, Maß- und Gewichtswesen
metropolitan railway Stadtbahn
MF = medium frequency
MFC-signalling (multi frequency code) Mehrfachfrequenzcode-Wählverfahren
mev = mega electron-volt MeV = Mega-e-Volt
mg = milligram (me) mg = Milligramm
mh = millihenry mH = milli-Henry
mho = 1/ohm Siemens
mica Glimmer
mica condenser Glimmerkondensator
mica dielectric Glimmer-Dielektrikum
mica dielectric condenser Glimmerkondensator
mica lamination Glimmerscheibe, Glimmerplättchen
mica sheet Glimmerblättchen
mica slab Glimmerplatte
mica „V" ring Kommutator-Ringisolierung
micanite Mikanit
micarta Mikanit, Mikartit
Michelson-Morley experiment Michelson-Versuch (530.12)
micro-... mikro-...; 10^{-6}
microammeter Mikroamperemeter
micro assembly zusammengesetzte Mikroschaltung
micro-balance Mikrowaage
microcircuit Mikroschaltung, miniaturisierte Schaltung
microclimate Mikroklima
microcomputer Mikrocomputer
microelectronics Mikroelektronik
microfarad Mikrofarad
microfilm Mikrofilm
microgalvanometer Galvanometer mit photoelektrischer Verstärkung
microgram(me) Mikrogramm
micrograph Mikrofotografie, Schliffbild (69)
microgroove Mikrorille
microhenry Mikrohenry
microinstruction Mikrobefehl
micromanipulator Mikromanipulator
microhm Mikro-Ohm
micrometer Mikrometer
micrometer caliper Schraublehre, Mikrometer
micrometer dial Feinmeßuhr
micrometer gap Feinmeßfunkenstrecke
micrometer graduated mit Mikrometerteilung
micrometer screw Mikrometerschraube, Feinstellschraube
micromicrofarad Pikofarad
micron Mikron
microphone Mikrophon, Mikrofon
microphone amplifier Mikrophonverstärker
microphone battery Mikrophon-Batterie, Sprechbatterie
microphone button Sprechtaste (621.395); Mikrophonkapsel
microphone cable Mikrophonkabel, Mikrophonleitung
microphone capsule Mikrophonkapsel
microphone circuit Mikrophonschaltung, Mikrophonkreis
microphone effect Mikrophonie
microphone mount Mikrophonfassung
microphone relay Mikrophonverstärker, mikrophonischer Verstärker
microphone stand Mikrophonstativ
microphone switch Sprechtaste
microphone transmitter Mikrophon
microphonic Mikrophon-...
microphonic action Mikrophon-Effekt (621.385)
microphonic noise Mikrophon-Effekt (621.385)
microphonic valve Röhre mit Mikrophoneffekt, klingende Röhre
microphonics (pl.) Mikrophon-Effekt (621.385)
microphonism Mikrophoneffekt
microporosity Feinporigkeit, Mikroporosität
microprocessor Mikroprozessor
microprocessor slices [siehe bit-slice microprocessor]
microprogram Mikroprogramm
microscope Mikroskop
microscopic lens Mikroskop-Linse
microscopic section Mikroskopschnitt, Schnitt
microsecond Mikrosekunde
micro-section Schliff
micro-strip Streifenleiter, Bandleiter, Bandleitung

microstructure Feinstruktur
microswitch Mikroschalter
microtelephone Handapparat
microtome Mikrotom, Feinschnitt
microwave Mikrowelle, cm-Welle (621.396)
microwave integrated cicuits = MICs integrierte Mikrowellenschaltkreise
microwave radio relay Richtfunk
microwave repeater Richtfunk-Relaisstelle
microwave set Mikrowellengerät (621.396)
to micronize feinmahlen, feinstmahlen
to microscope mikroskopieren
mid mitten, in der Mitte befindlich
in mid air mitten in der Luft, im Fluge
midband Bandmitte (621.396)
midband frequency Bandmittenfrequenz
mid-depth in der Wandmitte
mid-height halbe Höhe
at mid-height auf halber Höhe
mid point Mittelpunkt (51); Mittelpunkt, Sternpunkt (621.3); Mittelanzapfung (621.3)
mid-point protective system Windungsschlußschutz-System
mid position Mittelstellung
midrange Bereichsmitte
midrange speaker Mittelton-Lautsprecher
mid-way halbweges, mitten zwischen
mid-way between mitten zwischen, in der Mitte zwischen
mid-way in potential auf Mittelpotential
midwidth Mittellinie
in midwidth auf der Mittellinie
middle Mittelteil, Mitte
middle mittler, Mittel-
middle conductor Mittelleiter (621.3)
middle distance Mittelgrund (621.397), (77)
middle height mittlere Größe, mittlere Höhe
middle limb Mittelschenkel
middle marker (beacon) Haupteinflugzeichen (621.396.13)
middle point Mittelpunkt
middle range Mittelbereich
middle section Mittelabschnitt, mittlerer Abschnitt
middle size mittlere Größe
middle-size mittelgroß
middle third mittleres Drittel
middle wire Mittelleiter, Mitteldraht
midget Zwerg...
midget base Mignonfassung, Zwergfassung
midget relay Kleinrelais
midget set Zwergempfänger
midget valve Zwergröhre
Mie oscillator Knallfunkensender
migration Wanderung
migration area Wanderfläche (539)
migration constant relative Wanderungsgeschwindigkeit, wahre Überführungszahl (621.35)
migration speed Wanderungsgeschwindigkeit (539)
"mike" Mikrophon
mil Tausendstel Zoll (Längenmaß = 25,4 Mikron)
mild mild, weich
mild steel weicher Stahl, Flußstahl, Schmiedeeisen
mild steel wire Weicheisendraht, Flußstahldraht, Stahldraht
mile Meile
mile per hour per second (Beschleunigungsmaß) (625)
1 mile per hour per second = 0,447 m/s^2
milker Melkmaschine
milkiness Milchigkeit, Trübung
milking Gasen (621.355)
milking generator Einzellenlademaschine (621.355)
mill Mühle; Fabrik; Hütte, Hüttenwerk
mill boarding Faserplatte
mill fittings Werkstattleuchten
to mill mahlen; fräsen (621.9)
to mill out ausfräsen (621.9)
milled-out portion Ausfräsung, ausgefräster Teil
milled slot gefräste Nut (621.313)
millboard Graupappe (67)
miller Fräsmaschine (621.9)
Miller bridge Durchgriffs-Meßbrücke
millesimal tausendstel
milli-... milli-...; 10^{-3}
milliammeter Milli-Amperemeter
milliampere Milli-Ampere
milliamperemeter Milli-Amperemeter
milliampere-second meter Dosismesser (621.386)
milliard (Br.) Milliarde, 10^9
milligram(me) Milligramm
millihenry milli-Henry
milli lambert Millilambert
millilitre Milliliter
millimetre Millimeter
millimetric waves EHF, Millimeterwellen
milli micro-... nano-..., millimicro... = 10^{-9}
milling Fräsen, Fräsvorgang; Mahlen
milling machine Fräsmaschine
milling plant Mahlanlage
million Million, 10^6

millivolt millivolt
millivoltmeter Millivoltmeter
milometer Kilometerzähler (629)
mimic bus diagram Blindschaltbild
mimic-diagram control Anlagenbildsteuerung
mimic-system diagram Netzbild
mine Grube, Bergwerk, Schacht; Mine (623)
mine exploder Zündmaschine (662)
mine-gas indicator Schlagwetteranzeiger
mine-railway Grubenbahn
mine ventilation Grubenbewetterung
mine ventilator Grubenventilator
miner Bergarbeiter
miner's lamp Grubenlampe
miner's safety lamp Sicherheitsgrubenlampe
mineral Mineral
mineral mineralisch
mineral wool Schlackenwolle, Steinwolle
miniature Miniatur-..., Klein..., Kleinst
miniature base Mignonfassung
miniature capacitor Kleinkondensator
miniature motor Kleinstmotor
miniature switch Kleinstschalter, Mikroschalter
miniature triode Zwergtriode
miniature valve Miniaturröhre
miniaturization extreme Verkleinerung, Miniaturisierung
minicartridge Minikassette
minicomputer Kleinrechner, Minicomputer
minifuse Feinsicherung
minim Flüssigkeitsmaß (1 mimim = 1/60 fl. dr. = 0,590 cm^3)
minimization Reduktion auf ein Minimum
to minimize aufs Mindestmaß herunterdrücken
minimum (pl. **-ima**) Minimum, Kleinstwert
minimum Mindest-..., Kleinst-...
minimum audible field Hörschwelle im freien Schallfeld
minimum audible pressure Hörschwelle des Schalldrucks im Ohr
minimum bearing Minimumpeilung
minimum circuit breaker Minimalschalter, Schalter mit Minimalauslösung
minimum clearance Mindestabstand, Mindestschlagweite (621.3)
minimum clearance to ground Mindestabstand nach Erde, Mindestschlagweite gegen Erde
minimum clearing Enttrübung (621.396.96)
minimum cut-out Minimalauslöser, Unterspannungsschütz
minimum demand Mindestbedarf
minimum diameter Kleinstdurchmesser
minimum direction finder Minimumpeiler
minimum discernible signal Kleinstes wahrnehmbares Signal
minimum distance Mindestabstand
minimum enrichment Minimalanreicherung (539)
minimum flashover voltage Mindestüberschlagspannung, unterer Isolationspegel
minimum flow Mindestwasser (626)
minimum fusing current Dauerschmelzstrom, großer Prüfstrom (621.316.9)
minimum-oil circuit breaker ölarmer Schalter
minimum output Mindestleistung
minimum potential Mindestpotential, Potentialschwelle; Zersetzungsspannung (621.385)
minimum power Mindestleistung, Mindestlast, Minimallast, Lasttal (621.2)
minimum-pressure governor Unterdruckwächter
minimum receiver input Empfängerschwelle, Eingangsschwellwert
minimum relay Minimalrelais
minimum triggering level Auslöseschwelle
minimum thermometer Minimal-Thermometer
minimum value Kleinstwert
minimum voltage Mindestspannung
minimum voltage relay Unterspannungsrelais
minimum wage Mindestlohn
minimum wavelength kurzwellige Grenze (621.386)
mining Bergbau, Bergwesen; Abbau, Förderung; Bergwerksbetrieb
mining locomotive Grubenlokomotive
minor kleiner, geringfügig; Moll-...
minor determinant Unterdeterminante, Minor
minor lobe Nebenzipfel (621.396.67)
minor semi axis kleine Halbachse
minority carrier Minoritätsträger (621.382)
minus minus, weniger; erleichtert um
minus pole Minuspol
minus sign Minuszeichen
minus terminal Minusklemme
minute Minute, Zeitminute; Bogenminute; Protokoll
minute book Protokollbuch, Kladde
minute hand Minutenzeiger (681)
to minute notieren, aufzeichnen, Protokoll führen
minute winzig, sehr klein
minutely minutlich, minütlich
mirage Fata morgana

mirror Spiegel
mirror finish Hochglanz, Politur
mirror galvanometer Spiegelgalvanometer
mirror image Spiegelbild
mirror magnetometer Spiegel-Magnetometer
mirror reflection spiegelnde Reflexion
mirror-reflection echoes Mehrfachechos, Umwegechos
to mirror spiegeln, widerspiegeln; wiedergeben, abbilden
misadjustment Fehleinstellung
to misalign verstimmen, aus dem Gleichgewicht bringen
misalignment Fehlausrichtung, schlechte Ausrichtung (621); mangelnder Gleichlauf, Verstimmung (621.396)
to miscalculate (sich) verrechnen, falsch berechnen
miscalculation Rechenfehler, falsche Rechnung
to miscall falsch bezeichnen, falsch benennen
miscibility Mischbarkeit
miscible mischbar
misconstruction Fehlkonstruktion
miscount falsche Zählung, Fehlzählung
to miscount falsch zählen
misfire Fehlzündung (621.4); Versager (623.4)
to misfire fehlzünden (621.4); versagen (623)
misfiring Fehlzünden (621.4); Versagen (623)
to mishear (sich) verhören, falsch hören
misinterpretation falsche Deutung, Fehldeutung
mismatch Fehlanpassung (621.396)
mismatch factor Reflexionsfaktor
mismatch loss Stoßdämpfung (621.395)
mismatching Fehlanpassung, Anpassungsfehler (621.396)
to misname falsch bezeichnen, falsch nennen
misnomer falsche Bezeichnung
misoperation Fehlansprechen
misplaced falsch eingebaut; unangebracht; verlagert, verrutscht, falsch eingelegt (621.7)
missed approach procedure Fehlanflug (581.814.33)
missile Flugkörper, Wurfgeschoß, Geschoß
missile guidance Flugkörperlenkung
missile tracking Flugkörperverfolgung
mist Nebel, Nebeldunst, Dunst
to mistime zeitlich falsch anbringen, zur falschen Zeit tun
to mistune verstimmen (621.396)
mistuning Falschabstimmung, Fehlabstimmung, Verstimmung
M. I. T. = Massachusetts Institute of Technology
mitre Mitra
mitre gear Kegelrad
Mitre valve Kegelventil
to mitre unter 45° zusammenfügen
mix Mischung
to mix mischen; sich mischen (**with:** mit)
mixed mathematics angewandte Mathematik
mixed network Verbundnetz
mixed retroaction magnetische und elektrische Rückkopplung
mixed winding kombinierte Wellen-Schleifen-Wicklung (21.313)
mixer Mischer (66); Mischkammer, Mischbatterie (621.6); Mischkopf (621.396); Mischstufe, Überlagerungsstufe (621.396)
mixer amplifier Mischverstärker
Mixer diode Mischdiode
mixer gain Mischverstärkung
mixer stage Mischstufe
mixer tube Mischröhre, Modulatorröhre, Überlagererröhre
mixer valve Mischröhre, Modulatorröhre, Überlagererröhre
mixing Mischen, Mischvorgang (66), (621.396); Überlagerung (621.396)
mixing blade Rührschaufel, Rührflügel
mixing chamber Mischkammer (621.1)
mixing console Mischpult (654.19)
mixing probe Mischmeßkopf (621.317)
mixing stage Mischstufe (621.396)
mixing unit Mischpult (621.396.7)
mixing valve Mischröhre (621.396); Mischventil (621.646)
mixture Mischung, Gemisch, Gemenge (53), (621.396), (66), (621.43)
m. k. s. units M. K. S.-System (sing.), M. K. S.-Einheiten, technische Einheiten
ml. = millilitre Milliliter
mm = millimetre mm = Millimeter
mmf = magneto-motive force MMK = magnetomotorische Kraft
MMU = memory management unit
mnemonic operation code mnemotechnischer Code
MNOS memory MNOS-Speicher, Metall-Nitrid-Oxid-Silizium-Speicher
mobile beweglich, fahrbar; dünnflüssig
mobile hole Defektelektron
mobile radio Fahrzeugfunk
mobile service beweglicher Funkdienst
mobile station bewegliche Funkstelle
mobile substance dünnflüssige Substanz
mobility Beweglichkeit
mock-... Schein-..., Blind-...
mockup Nachbildung (621.392); Modell, Attrappe

mock window blindes Fenster
mode Typus, Art, Art und Weise; Schwingungsordnung, Schwingungstyp, Wellentyp; Modul; Betriebsart (621.39); Feldtyp (621.372)
mode of connection Anschlußart, Schaltart
mode of motion Schwingungsform, Schwingungstyp (534)
mode of operation Betriebsart
mode of oscillation Schwingungstyp
mode of progression Art der Fortbewegung
mode of propagation Ausbreitungsweg, Art der Ausbreitung
mode of reaction chemisches Verhalten, Reaktionsart
mode of representation Darstellungsart, Schreibweise
mode of working Arbeitsweise
model Modell
model plant Modellanlage, Versuchsanlage, Musteranlage
modem Modem, Modulations-Demodulationsgerät
to moderate mäßigen; bremsen (54)
moderated reactor moderierter Reaktor
moderation Mäßigung; Moderierung (539); Bremsung (539); (54)
moderator Bremssubstanz, Verlangsamer, Moderator, Verzögerer, Moderator, Bremsmittel (54), (539)
moderator control Moderatortrimmung (539)
modification Abart, Modifikation; Änderung, Abänderung, Abwandlung; Vergütung (66)
to modify modifizieren, ändern; abändern, abwandeln; vergüten (66)
modified refractive index modifizierte Brechzahl
moding Frequenzspringen (621.373)
modular system Bausteinsystem, Baukastensystem, modulares System
to modulate modulieren, modeln, einstellen
to modulate upon aufmodulieren auf
to modulate with modulieren mit
modulated carrier modulierter Träger (621.396)
modulated keyed continuous wave A 2-Wellen, mit Tonfrequenztelegraphie modulierte ungedämpfte Wellen
modulated quantity modulierte Größe
modulating voltage Modulationsspannung
modulating wave Modulationsfrequenz
modulation Modulation, Modelung
modulation capability höchstzulässiger Modulationsgrad
modulation characteristic Modulationskennlinie
modulation coil Modulationsspule
modulation depth Modulationsgrad; Modulationstiefe
modulation distortion Modulationsklirrfaktor
modulation envelope Modulationshüllkurve
modulation factor Modulationsgrad
modulation index Phasenhub
modulation meter Modulations-Meßgerät, Modulationsmesser
modulation monitor Modulationskontrollgerät
modulation noise Modulationsrauschen
modulation percentage Modulationsgrad
modulation rate Schrittgeschwindigkeit
modulation suppression Modulationsunterdrückung
modulation-voltage amplifier Modulationsverstärker
modulator Modulator; Modulatorstufe; Modulatorröhre; Impulszentrale (621.397); Mischstufe; Mischröhre
modulator driver Modulations-Treiberstufe
modulator electrode Wehneltelektrode, Steuerelektrode
modulator stage Modulationsstufe, Modulatorstufe, Mischstufe
modulator tube Modulationsröhre, Modulatorröhre, Mischröhre
modulator voltage Modulationsspannung
module Modul (621.833); Baustein, Baugruppe (621.39)
modulus Modulus, Modul (621.8); Amplitude, Betrag, Modulus (51)
modulus of elasticity Elastizitätsmodul
modulus of elasticity in shear Schubmodul
modulus of rigidity Schubmodell
modulus of rupture Bruchfestigkeit (674)
modus Modus, Art und Weise
Mohs scale of hardness Mohssche Härteskala
moiré Moiré (621.397)
moist feucht
moisted angefeuchtet
to moisten feucht machen, anfeuchten, durchfeuchten, benetzen
moistening agent Befeuchtungsmittel
moisture Feuchtigkeit
moisture content Feuchtigkeitsgehalt
moisture gradient Feuchtegradient, Feuchtigkeitsgradient (621.396), (55)
moisture-proof feuchtigkeitsfest; wasserfest; feuchtigkeitsgeschützt, vor Feuchtigkeit geschützt

moisture-proof motor feuchtigkeitsgeschützter Motor
moisture-proof switchgear feuchtigkeitsgeschütztes Schaltgerät
moisture proofing Feuchtigkeitsschutz, Imprägnierung
moisture-repellent feuchtigkeitsabstoßend, nicht benetzbar
moisture-resistant feuchtigkeitsfest, feuchtigkeitsbeständig
moisture-saturated wassergesättigt
moisture-tight feuchtigkeitsdicht
moisture trap Wasserfalle
molar Massen-...; als Ganzes
molar conductivity molare Leitfähigkeit
molar resistivity molarer spezifischer Widerstand
mold (US) Preßform (s. auch **mould**)
mold cavity Formraum
mold closing time Formschließzeit
to mold (US) pressen, formen (s. auch **to mould**)
to mold under pressure verpressen, unter Druck formen
molded bakelite jacket Bakelitgehäuse, Preßstoffgehäuse
molding Preßvorgang; Preß-...
molding compound Preßmasse, Formmasse
molding pressure Preßdruck
molding technique Preßtechnik
mole Mol, Grammolekül (54); Hafenmole, Mole (62); Maulwurf
molecular molekular, Molekular-...
molecular bond Molekülbindung
molecular crystal Molekularkristall
molecular formula chemische Formel
molecular friction molekulare Reibung, innere Reibung
molecular pump Molekularpumpe
molecular size Molekülgröße
molecular theory of magnetization Molekulartheorie (des Magnetismus)
molecular weight Molekulargewicht
molecule Molekül, Molekel
one molecule deep monomolekular, eine Moleküllage dick
one molecule thick monomolekular, von monomolekularer Dicke
molten geschmolzen (nur von Metallen)
molten bath Schmelzbad
molten iron geschmolzenes Eisen
molten pool flüssige Schweiße (621.791)
molybdenum Molybdän
moment Moment, kurze Zeitspanne, Augenblick, Zeitpunkt, Moment; Drehmoment; Bedeutung
moments equation Momentengleichung
moment of a force Moment einer Kraft
moment of inertia Trägkeitsmoment
moment of momentum Drehimpuls
moment about a point Drehmoment um einen Punkt
moment of resistance widerstehendes Moment, Gegenmoment
moment of torsion Drillmoment, Torsionsmoment
momentary momentan, kurzzeitig
momentary centre of motion Wälzpunkt, Drehmittelpunkt
momentary contact Wischkontakt
momentary contact-switch Tastschalter
momentary value Augenblickswert
momentum Impuls, Bewegungsgröße
monatomic einatomig
monaural monaural
monazite sand Monazitsand
Monel metal Monelmetall
monitor Prüf..., Kontroll-..., Überwachungsgerät, Monitor
monitor aerial Kontrollantenne, Überwachungsantenne
monitor console Überwachungstisch, Monitorkonsole
monitor headphone Kontrollhörer
monitor lamp Kontrollampe
monitor program Monitorprogramm (681.3)
to monitor überwachen, steuern, kontrollieren; vermitteln (621.395)
monitoring key Mithörtaste
monitoring radio receiver Kontrollempfänger
monitoring routine Überwachungsprogramm (681.3)
monitoring telephone Tonüberwachung
monkey Rammbär (624); Laufkatze (621.8)
monkey wrench Patentschlüssel, „Engländer", verstellbarer Schraubenschlüssel
monochord Monochord
monochromatic monochromatisch, einfarbig
monochromatic pyrometer Glühfadenpyrometer
monoclinic monoklin
monocrystal Monokristall
monocular monokular
monocyclic alternator (US) Hilfsphasen-Generator
monocyclic system Einphasen-Dreiphasen-Verteilung
monolith Monolith

monolithic monolithisch
monolithic block Monolith, Einkristall
monolithic circuit Festkörperschaltung, monolithische Schaltung, Einkristallschaltung
monolithic conduit muffenloses Betonrohr (621.315.2)
monomode fiber Monomode-Faser
monomolecular monomolekular
monophase einphasig, Einphasen-...
monopole aerial Vertikalantenne
monorail Einschienenbahn
monoscope Monoskop
monostable monostabil
monostable circuit monostabile Schaltung
monostable multivibrator Monostabiler Multivibrator
monostable trigger circuit Monostabiler Multivibrator
monotelephone Resonanz-Telefon
monotonic quantity monotone Größe (51)
monotonous monoton (51)
monotooth alternator Wechselstromgenerator mit konzentrierter Wicklung (621.313)
monotropic monotrop
montegal Al-Ca-Legierung (Handelsname)
monthly monatlich
monthly average Monatsmittel
Moore lamp Moorelichtlampe
mooring Haltetau, Verankerung (629.12), (629.13), (624)
mooring mast Ankermast (629.13)
m. o. p. a. = master oscillator power amplifier Steuersender-Kraftverstärker
Morse Morse
Morse alphabet Morsealphabet
Morse code Morsealphabet, Morseschrift
Morse cone Morsekegel (621.9)
Morse key Morsetaste (621.396)
Morse signal Morsezeichen
Morse sounder Morseticker
Morse telegraph Morsetelegraph
MOS = metal oxide semiconductor Metall-Oxide-Halbleiter
MOS-FET = metal oxide semiconductor field-effect transistor
MOS transistor enhancement mode MOS-Transistor vom Anreicherungstyp
mosaic Mosaikstein, Mosaiksteinchen
mosaic-diagram display Mosaikbetriebsschaubild
mosaic printer Mosaikdrucker
mosaic surface Mosaikfläche
mosaic telegraphy Mosaiktelegraphie
mosaic tube Bildspeicherröhre
mosaics (pl.) Mosaik
Moscicky condenser Minosflasche
most significant bit höchstwertiges Bit
most significant digit Ziffer mit höchstem Stellenwert
mote Sonnenstäubchen
motherboard Hauptplatine, Mutterplatine
mother lye Mutterlauge
mother-of-pearl lustre Perlmutterglanz
mothertape Mutterband (681.84)
motion Bewegung
in motion in Bewegung, in Betrieb
motion picture Kinobild, Kinoaufnahme
motion-picture engineering Kinotechnik
motion work Triebwerk; bewegliche Teile (pl.)
motional aus Bewegung entstehend
motional impedance Rückwirkungsimpedanz, Ersatzwiderstand für die mechanische Rückwirkung
motional resistance Ersatzwiderstand für mechanische Verluste, Rückwirkungswiderstand
motionless bewegungslos
motive treibend, antreibend, Antriebs-...
motive power Antriebskraft, Antriebsleistung
motive-power unit Motorwagen (621.33)
motograph Edison-Lautsprecher
motor Motor (621.313), (621.4)
motor armature Motoranker
motor axle Motorachse (625)
motor base Motorfuß
motor bogie Triebgestell (625)
motor-branch circuit Motorstromkreis
motor break Motorunterbrecher
motor bus Autobus
motor car Kraftwagen, Auto
motor casing Motorgehäuse
motor clutch Motorkupplung
motor coach Motorwagen (625)
motor-coach train Triebwagenzug
motor coil Motorspule
motor with combinded ventilation Motor mit Eigen- und Fremdbelüftung
motor with compound characteristic Motor mit Doppelschlußverhalten
motor converter Motorgenerator, Kaskadenumformer
motor current Motorstrom
motor drive Motorantrieb
motor-driven motorisch angetrieben, mit Motorantrieb, mit Motorbetrieb
motor-driven interrupter Motor-Unterbrecher
motor element Triebsystem (621.395.623)

motor-field control Feldregelung
motor frame Motorgestell, Motorrahmen
motor gear Motorgetriebe
motor-generator set Motorgenerator
motor-generator locomotive Umformerlokomotive
motor for hazardous locations explosionsgeschützter Motor
motor isolator Motortrennschalter
Motor meter Motorzähler
motor mounting Motorblock; Motoreneinbau
motor für operation in explosive gas atmosphere explosionsgeschützter Motor
motor operator Stellmotor
motor with outer rotor Außenläufermotor
motor output Motorleistungsabgabe, Motorleistung
motor protection Motorschutz
motor pump Motorpumpe
motor-reduction unit Getriebemotor
motor with series characteristic Motor mit Reihenschlußverhalten
motor shaft Motorwelle
motor speed Motordrehzahl
motor-speed variation Schwankung der Motordrehzahl
motor with squirrel cage armature Motor mit Kurzschlußläufer
motor starter Motoranlasser
motor starting Anlassen des Motors
motor-starting switch Motoranlaßschalter
motor switch Motorschalter
motor terminal Motorklemme, Motoranschlußklemme
motor-terminal board Motorklemmbrett
motor torque Motordrehmoment
motor transformer Gleichstrom-Umformer
motor-type watthour meter Wattstunden-Motorzähler
motor uniselector Motorwähler (621.395)
motor vehicle Motorfahrzeug, Kraftfahrzeug
motor voltage Motorspannung
motor winding Motorwicklung
to motor als Motor laufen (621.313)
motoring action motorischer Zustand, motorischer Betrieb
to motorboat tuckern (621.396.62)
motory motorisch, bewegend
mould Form, Gußform, Preßform (621); Humus (63); Schimmel; Schimmelpilze (pl.) (s. auch: **mold**)
mould-filling capacity Formfüllungsvermögen (621.7)
to mould formen, bilden (**upon:** nach) (s. auch: **to mold**)
to mould plastics Kunststoff pressen
moulded insulator Preßstoffisolator
moulded plastic Isolierpreßstoff
moulding Formen, Formvorgang; Pressen, Preßvorgang; Preßling; Hohlkehle, Profilleiste (69)
moulding box Formkasten, Gußkasten
moulding compount Preßmasse
moulding powder Preßpulver
moulding shop Gießerei, Formerei, Presserei
moulding time Preßzeit
mouldy schimmelig
mount Fassung, Sockel (621.396)
to mount montieren, einbauen, anbringen; sich montieren lassen; sich anbringen lassen; hochsteigen, aufsteigen
to mount an engine einen Motor einbauen
to mount a specimen eine Probe einspannen
mounting Fassung, Halterung; Montage; Einbau, Aufbau, Aufhängung, Befestigung; Chassis (621.396)
mounting bolt Montageschraube, Befestigungsschraube
mounting bracket Tragkonsole
mounting clip Befestigungsklemme, Befestigungsschelle
mounting device Befestigungsvorrichtung
mounting height Aufhängehöhe, Einbauhöhe
mounting hole Befestigungsloch
mounting lug Befestigungsöse, Aufhängeöse
mounting material Kleinteile
mounting position Einbaustellung, Einbaulage
mounting receptacle Steckdose auf einer Grundplatte
mouse Maus (681.3)
mouth Mund (61); Mündung (91); Mund, Mundstück, Mundloch, Öffnung (621)
mouth area Mündungsquerschnitt (621.396)
mouth-blow pipe Lötrohr
mouth of the horn Trichteröffnung (621.395)
mouthpiece Mundloch, Mundstück; Einsprechöffnung (621.395)
movable beweglich; verschiebbar, verstellbar
movable carriage Schlitten (621.396)
movable crane Laufkran
movable load Verkehrslast, bewegliche Last (624)
movable transformer Wandertransformator
to move bewegen; sich bewegen; transferieren, übertragen (681.3)
to move about (sich) hin- und herbewegen

to move through an angle um einen Winkel drehen
to move back and forth (sich) hin- und herbewegen, vor- und zurückschwingen
to move into line with vor... bringen, in eine Linie bringen mit
to move on (sich) weiterbewegen, weiterlaufen
move instruction Transferbefehl
movement Bewegung; Meßwerk, System (621.317); Werk, Uhrwerk (681)
movement across a field Bewegung in einem Feld, Bewegung durch ein Feld hindurch
mover Triebkraft, Antrieb
movie film Film
movies (pl.) Kino, Lichtspieltheater
moving beweglich; drehbar, Dreh-...
moving arm of the potentiometer Abgriff des Potentiometers, Potentiometerarm
moving coil Drehspule
moving-coil galvanometer Drehspulgalvanometer
moving-coil headphone dynamisches Telefon
moving-coil instrument Drehspulinstrument
moving-coil loudspeaker dynamischer Lautsprecher
moving-coil microphone Tauchspulmikrophon
moving-coil oscillograph Schleifen-Oszillograph
moving-coil pick-up dynamischer Tonabnehmer
moving-coil relay Drehspulrelais
moving-coil speaker dynamischer Lautsprecher
moving-coil support Drehspultragkonstruktion, Lagerung der Drehspule, Drehspulaufhängung
moving-coil telephone Tauchspultelefon, dynamisches Telefon
moving-coil transformer Transformator mit beweglicher Sekundärwicklung
moving-coil voltmeter Drehspulspannungsmesser
moving-coil wattmeter dynamometrisches Wattmeter
moving-conductor instrument dynamometrisches Instrument
moving-conductor loudspeaker dynamischer Lautsprecher
moving-conductor microphone dynamisches Mikrophon
moving contact beweglicher Kontakt
moving-contact member bewegliches Kontaktstück
moving-contact rod Traverse (621.316.5)
moving-dielectric voltmeter Strohhalm-Elektrometer
moving element (bewegliches) System (621.317)
moving force treibende Kraft, Antriebskraft
moving ground nachgiebiger Boden, nachgiebiger Baugrund
moving-iron instrument Dreheiseninstrument
moving-iron loudspeaker magnetischer Lautsprecher
moving-iron voltmeter Dreheisenspannungsmesser, Weicheisenvoltmeter
moving load bewegte Last
moving machinery umlaufende Maschinen (pl.)
moving magnet Drehmagnet; Dreheisen-...
moving-magnet galvanometer Nadelgalvanometer
moving-magnet instrument Dreheiseninstrument
moving-magnet pickup Magnet(-System)-Tonabnehmer
moving mean gleitendes Mittel
moving pictures (pl.) Film, Kino
moving plates bewegliche Platten; beweglicher Plattensatz (sing.), Rotorplatten (621.319.4)
moving staircase Rolltreppe
moving system (bewegliches) System (621.317)
moving-target indication = MTI Anzeige beweglicher Ziele, Festzeichenunterdrückung (621.396.96)
moving-target radar Radar zur Erfassung bewegter Ziele
moving vanes bewegliche Platten, Rotorplatten (621.319.4)
m. p. e. = multiple protective earthing Nullung
m. p. h. = miles per hour Meilen pro Stunde (**1 m. p. h.** = 1,6 km/h)
MSB = most significant bit höchstwertiges Bit
M. S. I. = medium scale integration Integrierte Halbleiterschaltung mittlerer Dichte
M. S. L. = mean sea level N. N. = Meeresspiegel = Normal-Nullniveau
MTBF = mean-time-between failures ausfallfreie Zeit
MTTF = meantime-to-failures mittlere Zeit bis zum Ausfall
MTTFF = meantime-to-first-failure mittlere Zeit bis zum ersten Ausfall
mu my (griech. Buchstabe)
mu factor (Kehrwert des Durchgriffs) (621.385)
Mumetal Mumetall

mud Schmutz, Dreck, Schlamm; Schlamm (621.356)
M. U. F. = maximum usable frequency höchste brauchbare Frequenz, MUF
muffle Muffel (66)
muffle furnace Muffelofen
to muffle dämpfen
muffler Schalldämpfer, Dämpfer (681); Schalldämpfer, Auspufftopf (621.4)
multiaddress Mehradreß...
multianode rectifier Mehranodengleichrichter
multibeam oscillograph Mehrstrahloszillograph
multibreak circuit-breaker Schalter mit Vielfachunterbrechung
multibreak switch Schalter mit Vielfachunterbrechung
multiburst signal Multiburst
multicavity magnetron Vielkammermagnetron
multicavity v. m. tube Mehrkreis-Triftröhre
multicellular voltmeter Multizellular Voltmeter
multichange gear Vielganggetriebe
multichannel Mehrkanal..., Vielkanal...
multichannel carrier telephone system Mehrfach-Trägerfrequenz-Fernsprechsystem
multichannel circuit Trägerfrequenz-Grundleitung, Mehrkanalverbindung
multichannel data recorder Mehrkanalschreiber, Mehrkanalregistrierinstrument
multichannel radio system Mehrkanal-Funkverbindung
multichannel telegraphy Mehrfachtelegraphie
multichannel telephonie Vielkanaltelephonie
multichannel television Mehrkanalfernsehen
multichannel voice-frequency telegraph system Mehrkanal-Tonfrequenztelegraphie
multichip integrated circuit Multichip
multicircuit winding Schleifenwicklung (621.313)
multiconductor cable Mehrleiterkabel, mehradriges Kabel
multicontact gang switch Stufen-Vielfachschalter
multicontact jack Vielfachklinke, vielpolige Kupplung, mehrpolige Kupplung
multicontact plug vielpoliger Stecker, Mehrfachstecker
multicontact socket Vielfachsteckdose, Mehrfachanschlußbuchse
multicore mehradrig, vieladrig
multicore cable mehradriges Kabel, vieladriges Kabel
multicore line mehradrige Leitung
multidestination (circuit) Sammelschaltung
multidirectional illumination Allgemeinbeleuchtung
multifilament lamp Mehrdrahtlampe
multifrequency alternator polumschaltbarer Generator
multifrequency generator polumschaltbarer Generator
multigap arrester Vielfachfunkenstrecke; Rollenblitzableiter (621.395); Kathodenfallableiter (621.316.9)
multigap discharger Vielfachfunkenstrecke
multigrid tube Mehrgitterröhre
multigun tube Mehrstrahl-Kathodenstrahlröhre
multihop transmission Übertragung durch Mehrfachreflexion
multilayer Mehrschicht
multilayer circuit Mehrlagenschaltung
multilayer coil mehrlagige Spule
multilayer printed board Mehrschichtleiterplatte
multilayered mehrlagig
multiloop control system vermaschter Regelkreis
multimeter Vielfachinstrument, Vielfachmeßgerät
multioffice exchange Mehramtsverbindung
multipart insulator mehrteiliger Isolator
multipath propagation Mehrfachwegausbreitung (621.396)
multipath reflection Mehrwegereflexion
multiphase Mehrphasen-..., Drehstrom-...
multiphase alternating current Mehrphasenwechselstrom, Drehstrom
multiphase circuit Drehstromkreis, Mehrphasenschaltung
multiphase current Mehrphasenstrom, Drehstrom
multiphase generator Drehstromgenerator
multiphase meter Drehstromzähler
multiphase motor Drehstrommotor, Mehrphasenmotor
multiphase transformer Mehrphasen-Transformator
multiphase winding mehrphasige Wicklung, Mehrphasenwicklung
multiple (ganzzahliges) Vielfaches
multiple vielfach, Vielfach-..., Mehrfach-...
multiple access Mehrfach-Zugriff, Mehrfach-Zugang (681.3)
multiple-arch dam Gewölbestaumauer
multiple-beam direction finder Mehrstrahlpeilung
multiple call Sammelgespräch
multiple-changeover switch Mehrfachumschalter

multiple circuit Mehrfachleitung (621.315); Schleifenwicklung (621.313)
multiple coil Mehrfachspule (621.313)
multiple common-frame circuitbreaker gekuppelter mehrpoliger Schalter
multiple-conductor system Mehrleitersystem
multiple connection Vielfachverbindung, Mehrfachanschluß
multiple contact switch Stufenschalter (621.316.7); Zellenschalter (621.356)
multiple-current generator Mehrstromgenerator
multiple-earthed system genulltes Netz
multiple feeder Mehrfachspeiseleitung
multiple-folded dipole antenna Mehrfach-Schleifendipolantenne
multiple-groove sheave Mehrrillenscheibe
multiple image Echobild (621.397)
multiple jack Vielfachdose, Vielfachklinke
multiple joint Verzweigungsmuffe, Abzweigungsmuffe (621.315.2)
multiple line Mehrfachleitung
multiple-multipole circuit breaker mehrpoliger Vielfachumschalter
multiple-operator motor-generator set Mehrstellenschweißumformer
multiple-outlet television system Fernseh-Gemeinschaftsempfangsanlage
multiple-pin plug mehrpoliger Stecker
multiple-pole switch mehrpoliger Schalter
multiple protective earthing Nullung
multiple reception Mehrfachempfang
multiple-reflection echo Umwegecho
multiple recorder Mehrfachschreiber
multiple-resistance welding Mehrstellenwiderstandsschweißung
multiple-series connection Serien-Parallelschaltung
multiple signal Mehrfachzeichen (621.396)
multiple-stage mehrstufig
  multiple-stage amplifier mehrstufiger Verstärker
multiple-start mehrgängig
multiple-step path Mehrfachreflexionsweg (621.396)
multiple switchboard Vielfach, Vielfachfeld (621.395)
multiple system Vielfachsystem (621.395); Mehrzellenanlage (621.357)
multiple tariff Doppeltarif, Mehrfachtarif
  multiple-tariff meter Doppeltarifzähler
multiple thread mehrgängiges Gewinde
multiple-tuned aerial Mehrfachantenne
multiple-twin cable mehrpaariges Kabel
multiple-twin quad Dieselhorst-Martin-Vierer
multiple-unit control Allwagensteuerung, Vielfachsteuerung
multiple-unit controller Zugsteuerschalter (625)
multiple-unit tube Mehrfachröhre
multiple-valued function mehrwertige Funktion (51)
multiple-voltage system Mehrspannungsanlage
multiple-way (telegraph) system Mehrfachtelegraphie-System
multiple winding Schleifenwicklung (621.313)
multiple-wire aerial Vieldrahtantenne
multiplex Multiplex-..., Mehrfach-...
  multiplex mode Multiplexbetrieb
  multiplex operation Multiplex-Betrieb
  multiplex system Mehrfachbetrieb, Mehrfachausnutzung, Multiplexbetrieb (621.394)
  multiplex transmission Mehrkanalübertragung, Multiplexübertragung
  multiplex wave winding Serienparallelwicklung, Reihenparallelwicklung (621.313)
  multiplex winding Wellenwicklung (621.313)
to multiplex in Mehrkanalordnung schalten
multiplexer Multiplexer
multiplexing Mehrkanalausnützung, Mehrfachausnutzung (von Leitungen) (621.39); multiplexen (681.3)
multiplicand Multiplikand
multiplication Multiplikation (51), (539), (54)
  multiplication factor Multiplikationsfaktor (539), (54)
  multiplication table Multiplikationstafel
multiplier Vervielfacher (621.385); Vervielfältiger; Multiplikator (51); Vorwiderstand (621.317)
to multiply multiplizieren
  to multiply across überkreuz multiplizieren
  to multiply out ausmultiplizieren
  to multiply together miteinander multiplizieren
  to multiply throughout ganz durchmultiplizieren; erweitern
multiplying power Meßbereichserweiterung
multipolar mehrpolig
  multipolar dynamo mehrpolige Maschine, mehrpoliger Generator
  multipolar generator mehrpoliger Generator
  multipolar motor mehrpoliger Motor
  multipolar winding mehrpolige Wicklung
multipole mehrpolig

multipole-bounce technique Vielfach-Pralltechnik
multipole-circuit breaker vielpoliger Schalter, mehrpoliger Schalter
multiposition switchboard Mehrplatzvermittlung
multiprogramming Mehrprogrammbetrieb
multirange mit mehreren Bereichen, Mehrbereich-...
multirange instrument Vielfachinstrument
multirate meter Mehrfachtarifzähler
multirate tariff switch Tarifschaltuhr
multisection tube Mehrfachröhre, Verbundröhre
multisegment magnetron Vielschlitz-Magnetron
multispeed motor drehzahlumschaltbarer Motor
multistage mehrstufig
multistage amplifier mehrstufiger Verstärker
multistart mehrgängig
multistart thread mehrgängiges Gewinde
multistorey switch mehretagiger Schalter
multistrand wire feindrähtige Litze
multisystem valve Verbundröhre (621.385)
multitrack recording system Mehrspur-Aufnahmeverfahren
multiturn current transformer Mehrwindungswandler (621.317)
multiuser system Mehrplatzsystem, Mehrbenützersystem
multivane capacitor Mehrplattenkondensator
multivibrator Multivibrator, Kippschwinger (621.396)
multiway mehrwegig
multiway cable vielpaariges Kabel
multiway connector Steckerleiste
multiway switch Mehrfachumschalter
multiwire aerial Mehrleiterantenne
multiwire triatic aerial Dreiecksflächenantenne
municipal städtisch, gemeindlich
municipal works städtische Werke, Stadtwerke
muriate of ...... -chlorid
muriatic acid (technische) Salzsäure
Murray loop test Fehlerortsbestimmung durch Schleifenmessung
musa aerial Musa-Antenne
muscovite Muskowit
mush area Verwirrungsgebiet, Interferenzgebiet
mush winding wilde Wicklung
mush-wound coil Träufelspule (621.313)
mushroom Pilz
mushroom-shaped pilzförmig
musical musikalisch
musical frequency Tonfrequenz
musical spark tönender Funke
mutator Umrichter
mute stumm
mute aerial künstliche Antenne
mute (lightning) discharge stumme Entladung
muting stumm, abstimmen
muting Rauschunterdrückung, Rauschsperre
mutual gegenseitig, wechselseitig, Gegen-...
mutual A. C. conductance Steilheit (621.385)
mutual admittance verknüpfter Scheinleitwert
mutual attraction gegenseitige Anziehung
mutual branch gemeinsamer Zweig (621.3)
mutual capacitance Betriebskapazität zwischen zwei Leitern bei Erdung aller übrigen; gegenseitige Kapazität
mutual conductance Steilheit (621.385)
mutual impedance Kopplungs-(Wechselstrom-)Widerstand, verknüpfter Scheinwiderstand
mutual inductance Gegeninduktivität
mutual-inductance coupling induktive Kopplung
mutual induction gegenseitige Induktion
mutual inductometer regelbare Gegeninduktivität, Variometer
mutual inductor veränderliche Gegeninduktivität
mutual interference gegenseitige Beeinflussung
mutual relation Wechselbeziehung
mutual repulsion gegenseitige Abstoßung
mutually perpendicular aufeinander senkrecht, lotrecht zueinander
mutually screened gegeneinander abgeschirmt
M. V. A. = megavoltamperes MVA = Megavoltampere
mycalex Mycalex
myria-... 10^4
myriahertz 10 kHz
myriametric waves Längstwellen, Myriameterwellen

N

nabla Nabla-Operator
nail Nagel (61), (621), (69)
naked nackt
naked eye unbewaffnetes Auge, bloßes Auge
naked light offenes Licht
name plate Firmenschild, Leistungsschild, Bezeichnungsschild (621); Türschild
NAND circuit NAND-Schaltung, NAND-Glied
NAND operation NAND-Verknüpfung
Naperian logarithm natürlicher Logarithmus

Naperian system of logarithms natürliches Logarithmensystem
naphtha Naphtha, Erdöl
naphthaline Naphthalin
narrow eng, schmal, beschränkt
narrow-angle diffusion Schmalwinkelstreuung (635)
narrow band schmales Band, schmales Frequenzband (621.396)
narrow-band reception Schmalbandempfang
narrow-band television Schmalbandfernsehen
narrow-band transmission system Schmalbandübertragungssystem
narrow beam schmaler Strahl, dünner Strahl (535); enges Bündel, scharfe Keule (621.396.67)
narrow bounds enge Grenzen
narrow edge scharfe Kante
narrow flame Stichflamme
narrow frequency range schmales Frequenzband
narrow gauge Schmalspur
narrow-gauge film Schmalfilm
narrow-gauge motor Schmalspur-Bahnmotor
narrow-gauge tram Schmalspurstraßenbahn
narrow pencil schmales Bündel (535)
narrow resonance curve schmale Resonanzkurve, scharfe Resonanzkurve
to narrow schmal machen, verengen; bündeln (621.396)
narrowing Verengung; Bündelung (621.396)
nascent naszierend, in statu nascendi (54)
national national, Reichs-...
National Electric Code (US-Vorschriften, entsprechend den VDE-Bestimmungen)
native land Herkunftsland
natural natürlich, Eigen-...
natural cooling natürliche Kühlung, Eigenkühlung, Selbstkühlung
natural draught natürlicher Zug
natural frequency Eigenfrequenz
natural gas Erdgas
natural law Naturgesetz
natural leak Eigenverluste (621.396.69)
natural logarithm natürlicher Logarithmus
natural magnet natürlicher Magnet
natural mica Naturglimmer
natural oscillation Eigenschwingung
natural period of vibration Eigenschwingungszeit, Eigenschwingungsdauer, Periode der Eigenschwingung
natural philosophy Naturwissenschaft
natural resonance Eigenschwingung, Eigenresonanz
natural science Naturwissenschaft
natural uranium reactor Natururanreaktor
natural ventilation natürliche Belüftung
natural wavelength Eigenwellenlänge
nature Natur; Art
naught Nichts, Null
nautical nautisch
nautical instrument nautisches Instrument
nautical mile Seemeile (6080 ft. = 1,853 km)
naval Marine-..., Flotten-...
nave Kirchenschiff, Schiff (72); Nabe (621)
navigation Navigation
navigation lights Befeuerung
navigation satellite Navigationssatellit
navigational Navigations-...
navigational aid Navigationshilfe
navigational set Navigationsgerät, Navigationseinrichtung
navigator Navigator, Navigationsoffizier
NC = numerical control numerische Steuerung
n-channel field-effect transistor N-Kanal-Feldeffekttransistor, N-Kanal FET
n-conducting n-leitend
n-conductor cable n-Leiter-Kabel
N. D. B. = nondirectional beacon Ungerichtetes Funkfeuer, ZFB, Zielflugbake, Mittelwellenfunkfeuer
n-doped n-dotiert (621.382)
NDRO = non-destructive read-out nichtlöschendes Lesen
near nahe, nahe gelegen, benachbart, dicht bei
near-channel selection Nachbarkanalselektion
near-end cross talk Nahenebensprechen, Mitnebensprechen
near guess gute Schätzung, fast richtige Schätzung
neat unverdünnt, rein (54)
nebula (pl.: **lae**) Nebel (52)
nebular Nebel-... (52)
nebular matter Nebelsubstanz, Materie im Nebel
neck Hals (61), (621)
neck of a roll Zapfen, Wellenzapfen; Walzenzapfen
to neck absetzen, einen Hals geben
to neck down a shaft eine Welle absetzen
to neck down the thread das Gewinde hinterdrehen, Gewinde in eine Rille auslaufen lassen
necking Einschnürung (539); Säulenhals (72)
need for repair Reparaturbedürftigkeit

needle Nadel; Graviernadel, Radiernadel, Gravierstift (621.9); Zeiger (621.317); Kompaßnadel (53); Grammophonnadel (534); Schneidstift (534); Schlagbolzen (623)
needle bearing Nadellager, Walzenlager
needle galvanometer Nadelgalvanometer
needle-like nadelähnlich
needle point Punktierung (621.315.6); Nadelspitze
needle-point gap Nadelfunkenstrecke
needle-point voltage Spitzenspannung, Überschlagspannung zwischen Spitzen
needle scratch Nadelgeräusch
needle-shaped nadelförmig
needle valve Nadelventil
needle vertical force (Nadel-)Auflagekraft
needle wear Nadelabnutzung
negation Negation
negative Negativum; Negativ (77)
negative negativ
negative-amplitude modulation Negativ-Amplitudenmodulation
negative brush negative Bürste (621.313)
negative carbon negative Kohle (621.32)
negative charge negative Ladung
negative conductor negativer Leiter, Minusleiter (621.315)
negative current negativer Strom; Schienenstrom (625)
negative differential conductance region Bereich des negativen differentiellen Leitwertes (621.382)
negative electricity negative Elektrizität
negative electrode negative Elektrode, Kathode
negative element negativer Pol, negative Platte (621.355)
negative feedback Gegenkopplung; negative Rückführung (621.316.7)
negative feeder negative Speiseleitung, Schienenkabel (621.33)
negative-sequence-field impedance Impedanz des gegenläufigen Feldes
negative glow negatives Glimmlicht
negativ-going negativ gerichtet, negativ
negative lead Minusleiter
negative light modulation umgekehrte Lichtmodulation
negative main Minusleiter des Netzes
negative matrix Negativform (621.357)
negative off-state current negativer Sperrstrom (621.314.63)
negative phase-sequence relay Unsymmetrierelais
negative plate negative Platte, Negative (621.355)
negative pole negativer Pol
negative quantity negative Größe
negative repeater NLT-Verstärker
negative resistance negativer Widerstand
negative-resistance characteristic negative Charakteristik, fallende Charakteristik
negative-resistance oscillator Oszillator mit negativem Widerstand
negative retroaction negative Rückkopplung, Gegenkopplung
negative-sequence gegenläufig, linksläufig (621.3)
negative-sequence active power Wirkleistung der gegenläufigen Komponente
negativ-sequence field gegenläufiges Feld
negative-sequence reactance Blindwiderstand für die gegenläufige Komponente
negative-sequence reactive power Blindleistung der gegenläufigen Komponente
negative-sequence resistance Widerstand für die gegenläufige Komponente
negative-sequence symmetrical component gegenläufige symmetrische Komponente (538.55)
negative sign negatives Vorzeichen, Minus-Zeichen
negative swing negative Halbwelle
negative-temperature coefficient = NTC Heißleiter, NTC-Widerstand
negative terminal negative Klemme, Minus-Klemme
negative transconductance negative Steilheit (621.385)
negative-transconductance oscillator Dynatronsender, Dynatron
negative value negativer Wert
negative wire negative Leitung
negatively charged negativ geladen
negator Negationsglied, Negator, NICHT-Glied
negatron Elektron; Negatron, Vierelektrodenröhre mit negativer Charakteristik
negligence Nachlässigkeit, Fahrlässigkeit
negligible vernachlässigbar
negligible by comparison vergleichsweise vernachlässigbar, relativ vernachlässigbar
neighbourhood Nachbarschaft
in the neighbourhood of 20 amps in der Gegend von 20 A
neighbouring benachbart

neon Neon
neon bulb Neonröhre, Neonlampe, Glimmröhre
neon detector Glimmgleichrichter
neon-electric stroboscope Glimmlicht-Stroboskop
neon-glow lamp Glimmlampe
neon-glow tube h. f. voltage indicator Glimmröhrenanzeiger
neon indicator Glimmlichtanzeigeröhre
neon-induction lamp elektrodenlose Glimmlampe
neon lamp Glimmlampe
neon sign Neon-Reklame
neon tube Glimmröhre, Leuchtröhre
neon-tube crest voltmeter Glimmröhren-Scheitelspannungsmesser
neon-tube peak voltmeter Glimmlicht-Scheitelspannungsvoltmeter
neon-tube rectifier Glimmlichtgleichrichter, Glimmgleichrichter
neoprene Neopren
neper Neper, Np. (621.39)
Nernst lamp Nernstlampe
nest Nest
nest of curves Kurvenschar
nest of tubes Röhrenbündel
to nest einpassen, einfügen, unterbringen; verschachteln (681.3)
nested program verschachteltes Programm
net Netz, Punktnetz; Netzwerk
net Netto-..., Gesamt-..., resultierend, Rein-...
net calorific value unterer Heizwert
net effect Endeffekt, Gesamtwirkung
net heat value unterer Heizwert
net loss Restdämpfung
net resultant force Resultierende, Gesamtkraft
net total Gesamtergebnis, Gesamtsumme
net value Nettowert
netting Netz, Netzwerk
network Schaltung, Netz, Netzwerk; Fernmeldenetz
π-network π-Schaltung, Vierpol in π-Schaltung
network analogue Netzwerknachbildung
network with earth-connected neutral Netz mit unmittelbar starr geerdetem Strom
network feeder Netzspeiseleitung
network with insulated neutral Netz mit isoliertem Sternpunkt
network-junction point Netzknotenpunkt
network level Netzebene (621.395)
network master relay Hauptanschlußrelais
network node Netzwerkknoten
network operation Betrieb eines Netzes
network primary distribution system Hochspannungsverteilungsnetz
network protector Netzschutzschalter
network relay Schutzrelais, Netzrelais
network secondary distribution system Niederspannungsverteilungsnetz
network structure Netzstruktur
neuroelectricity Aktionsspannungen (pl.)
neutral Nulleiter, Mittelleiter, Sternpunktleiter (621.3)
neutral neutral
neutral atom neutrales Atom
neutral auto-transformer Erdungsdrossel (621.316.9)
neutral axis neutrale Achse (621.313)
neutral in charge ungeladen
neutral compensator Stützdrossel
neutral conductor Nulleiter, Sternpunktleiter, Mittelleiter, Neutralleiter
neutral conductor current Nulleiterstrom, Mittelleiterstrom
neutral d. c. telegraph system Morse-Einstromsystem
neutral distributor Nulleiterzuführung
neutral gear Leergang
neutral ground Sternpunktserdung, Nullpunktserdung
neutral line neutrale Achse (621.313); Nullinie
neutral main Nulleiter, Mittelleiter
neutral operation Einfachstrom-Betrieb
neutral point Nullpunkt, Sternpunkt, Mittelpunkt
neutral position Nullstellung, Ausgangsstellung
neutral relay polarisiertes Relais
neutral temperature Neutralpunkt (621.362)
neutral terminal Sternpunktklemme
neutral wedge Graukeil
neutral wire Nulleiter, Sternpunktleiter, Mittelleiter, Nulleiterdraht
neutral zone neutrale Zone (621.313)
neutralator Transformator zur Erzeugung eines künstlichen Sternpunktes
neutralization Nullung (621.316); Neutralisierung, Neutralisation (621.396), (54)
to neutralize nullen (621.316); neutralisieren (621.396), (54); (sich) aufheben, (sich) kompensieren, sich auslöschen (51), (53)
neutralized series motor Einphasen-Kommutatormotor mit Kompensationswicklung
neutralizing Nullung (621.316); Neutralisierung (621.396), (54)

neutralizing condenser Neutralisierungskondensator
neutralizing transformer Kompensations-Transformator
neutrino Neutrino
neutrodon Neutralisierungs-Kondensator
neutrodyne Neutrodyne
neutrodyne receiver Neutrodynempfänger
neutron Neutron
neutron absorption Neutroneneinfang, Neutronenabsorption
neutron bombardment Beschießung mit Neutronen
neutron capture Neutroneneinfang
neutron counter Neutronenzählrohr, Neutronenzähler
neutron diffusion Neutronendiffusion
neutron excess Neutronenüberschuß
neutron flux Neutronenstrom, Neutronenfluß
neutron multiplication Neutronenvervielfachung
neutron radiation Neutronenstrahlung
neutron source Neutronenquelle
neutron yield Neutronenausbeute
new neu
new-built neuerbaut, neuerrichtet
new plant Neuanlage, neue Anlage
Newtonian mechanics klassische Mechanik, Newtonsche Mechanik
Newton's rings Newtonsche Ringe
next nächster; dicht bei
next door to nothing unendlich klein (51)
next door to zero beinahe Null (51)
n-gate thyristor anodenseitig steuerbarer Thyristor
n-gon n-Eck
nibble Halbbyte
nichrome Chromnickel
nickel Nickel
nickel bronze Nickelbronze
nickel-cadmium accumulator NC-Sammler, Nickel-Kadmium Batterie
nickel-clad nickelplattiert
nickel iron Nickeleisen
nickel-iron battery NC-Batterie (621.356)
nickel-iron cell NC-Akkumulator, Edison-Akkumulator
nickel-plated vernickelt
nickel plating Vernickelung
nickel silver Neusilber
nickel steel Nickelstahl
nickeline Nickelin
Nicol Nicol, Nicolsches Prisma
Nicol prism Nicolsches Prisma, Nicol
nicoloy (Handelsname für HF-Nickeleisen)
nicore Nickelstaubkern
night Nacht
night effect Nachteffekt (621.396.67)
night error Nachteffekt (621.396.9)
night fighter Nachtjäger (623)
night frequency Nachtfrequenz (621.396)
night glass Nachtglas (538)
night light Nachtlicht; Nachtbeleuchtung
night service Nachtschaltung
night-service number Nachtrufnummer
night vision Nachtsicht
Ni-hard harter Nickelstahl
nil Nichts; Fehlanzeige
niobium Niobium, Niob
to nip kneifen, (kneifend) erfassen
to nip between erfassen mit, einklemmen zwischen
to nip off abzwicken
Nipkow disc Nipkow-(Loch-)Scheibe
nippers (pl.) Kneifzange, Beißzange
nipple Nippel, Schmiernippel, Schmierloch
Ni-resist wärme- und korrosionsfester Ni-Stahl
Ni-Tensyl hochfester Ni-Stahl
niter Salpeter
neton Niton, Emanation
to nitrard nitrierhärten
nitric acid Salpetersäure
nitric oxide Stickoxid
nitrocellulose laquer Nitrolack
nitro compounds Sticksoffverbindungen, Nitro-Verbindungen
nitrogen Stickstoff
nitrogen-filled lamp Lampe mit Stickstoffüllung
nitrogen hardening Nitrierhärtung (621)
nitrogen peroxide Stickstoffperoxid
nitrogen trioxide Stickstofftrioxid
nitrogenous mit Stickstoff, stickstoffhaltig
nitro group Nitrogruppe
nitrous nitros, Stickstoff-..., salpetrig
nitrous acid salpetrige Säure
nitrous fume nitroses Gas
nitrous gas nitroses Gas
nitrous oxide Stickoxydul, Lachgas
nitrous vapour nitroser Dampf
Nixie tube Nixie-Anzeigeröhre
n-level signal wertdiskretes Signal
N-MOS N-Kanal MOS
no kein
no-load Leerlauf, unbelastet
no-load balanced telephone interface factor ausgeglichener Leerlauf-Telefonstörfaktor

no-load characteristic Leerlaufcharakteristik (621.313)
no-load condition Leerlaufzustand
no-load current Leerlaufstrom
no-load isolator Leerschalter
no-load losses Leerlaufverluste
no-load magnetizing current Leerlaufmagnetisierungsstrom
no-load measurement Leerlaufmessung, Leerlaufversuch
no-load saturation curve Leerlaufkennlinie, Magnetisierungskennlinie (621.313)
no-load speed Leerlaufdrehzahl
no-load starting Leeranlauf
no-load state Leerlaufzustand
no-load switching Leerlaufschaltung; stromloses Schalten
no-load voltage Leerlaufspannung
no-loss verlustlos
no-loss condenser verlustloser Kondensator
no-loss dielectric verlustloses Dielektrikum
no-loss insulator verlustloser Isolator
no-signal bias Gittergleichspannung, Gittervorspannung
no-torque position Nullmomentenstellung
no-transmission interval Sendepause
no-volt coil Unterspannungsspule
no-voltage Nullspannungs-...
no-voltage circuit breaker Nullspannungsschalter
no-voltage relay Nullspannungsrelais
no-voltage release Nullspannungsauslöser
noble gas Edelgas
noble metal Edelmetall
noble-metal uniselector switch Edelmetall-Motor-Drehwähler, EMD
NOC = normally open contact Schließer, Arbeitskontakt
noctovision Infrarot-Sehen, Restlichtsehen
noctovisor Infrarot-Bildwandler
nocturnal nächtlich
nodal Knoten-...
nodal exchange Knotenvermittlungsstelle
nodal line Knotenlinie
nodal plane Knotenfläche
nodal point Knotenpunkt; Schwingungsknoten
node Knoten, Wellenknoten, Schwingungsknoten; Verzweigungspunkt eines Netzwerks
Nodon valve Elektrolyt-Gleichrichter
Noeggerath dynamo Noeggerath-Maschine, Unipolarmaschine
noise Rauschen; Störung, Geräusch, Lärm (621.396)
noise absorber Geräuschabsorber, Geräuschdämpfung
noise analysis Geräuschanalyse
noise factor Rauschwert
noise-field intensity meter Störfeldstärkemesser
noise figure Rauschzahl, Rauschfaktor, Grenzempfindlichkeit
noise filter Störfilter, Störsperre
noise generator Rauschgenerator, Rauschsender
noise immunity Störsicherheit
noise killer Krachtöter (621.396.62)
noise level Störpegel, Rauschpegel
noise margin Störspannungsabstand
noise measurement Rauschmessung
noise-measuring set Rauschmeßgerät, Rauschspannungsmesser
noise meter Rauschmesser, Geräuschmeßgerät
noise pickup Aufnahme von Rauschstörungen (621.396)
noise power Rauschleistung
noise-power ratio Grundgeräuschabstand
noise quieting Rauschunterdrückung
noise ratio Rauschzahl
noise suppression Rauschunterdrückung, Krachtötung (621.39); Geräuschunterdrückung
noise suppressor Krachtöter, Rauschunterdrücker
noise temperature Rauschtemperatur
noise voltage Geräuschspannung, Störspannung, Rauschspannung
noise-weighting filter Rauschbewertungsfilter
noiseless geräuschlos
noisy lärmend, geräuschvoll, laut, lärmerfüllt
noisy room lauter Raum, geräuscherfüllter Raum
noisy valve stark rauschende Röhre
nomenclature Nomenklatur
nominal nominell, Nenn-..., Nominal-...
nominal collector-ring voltage Nennerregerspannung (621.313)
nominal current Nennstrom
nominal cut-off frequency Nenngrenzfrequenz (621.392)
nominal maximum circuit hypothetischer Bezugskreis
nominal output Nennleistung
nominal pull-in torque Nenn-Synchronisiermoment (asynchrones Moment bei 95% Nenndrehzahl)

nominal range of use Nennbereich
nominal value Nennwert
nominal value of mean forward current Nennwert des Durchlaßstromes
nominal voltage rating Nennspannung
nomograph Nomogramm
non-... Nicht-...
nonabsorbent nicht saugfähig, nichtabsorbierend
nonapproved circuit nicht zugelassene Leitung
nonassociation cable nicht normgerechtes Kabel
nonautomatic opening reine Handauslösung
nonbounce contacts prellfreie Kontakte
nonbridging contacts nichtüberbrückende Kontakte
noncondensing steam engine Auspuffdampfmaschine
nonconducting nichtleitend
nonconductor Nichtleiter
noncontinuous electrode Nippelelektrode (621.357)
noncorrodible nicht korrodierbar, korrosionsfest
noncorrosive korrosionsfest, nichtkorrodierbar, nicht korrodierend
noncurrent carrying nichtstromführend
noncutting shaping spanlose Formung
noncyclical field wirbelfreies Feld
nondescript unbestimmt, schwer zu definieren, undefinierbar
nondestructive nicht zerstörend, zerstörungsfrei
nondestructive testing of materials zerstörungsfreie Werkstoffprüfung
nondestructive readout memory Speicher ohne Zerstörung des Speicherinhaltes beim Herauslesen der Information
nondirectional ungerichtet
nondirectional antenna ungerichtete Antenne
nondirectional beacon ungerichtetes Funkfeuer, NDB
nondriving wheel Laufrad (625)
nonearthed ungeerdet, nicht geerdet
nonearthed neutral nicht geerdeter Mittelleiter
nonelectric nichtelektrisch
nonequivalence exklusives ODER
nonerasable storage Festspeicher, nicht löschbarer Speicher
nonreactive load Belastung durch Wirkwiderstand
nonexistent nicht vorhanden, fehlend, nicht existierend
nonferrous Nichteisen-...
non-ferrous metal Nichteisenmetall, Buntmetall
nonfusing arcing electrode nicht abschmelzende Elektrode
nongraded unklassifiziert
non-glare reflexfrei
non-impact printer anschlagfreier Drucker
noninductive induktionsfrei
noninductive cable induktionsfreies Kabel
noninductive capacitor induktionsfreier Kondensator (621.396)
noninductive circuit induktionsfreier Kreis
noninductive load induktionsfreie Belastung, ohmsche Belastung
noninductive resistance winkelfreier Widerstand, induktionsfreier Widerstand
noninductively wound bifilar gewickelt, induktionsfrei gewickelt
noninflammability Unbrennbarkeit, Nichtentzündlichkeit
noninflammable nicht brennbar, unbrennbar
noninjurious unschädlich, harmlos
nonintegral nicht ganzzahlig
noninteracting rückwirkungsfrei
noninterchangeable unverwechselbar
nonkinkable flex Gummischlauchleitung
nonlinear nichtlinear
nonlinear circuit nichtlinearer Stromkreis
nonlinear circuit element nichtlineares Schaltelement
nonlinear coil nichtlineare Drossel, Eisenspule
nonlinear crosstalk nichtlineares Nebensprechen
nonlinear distortion nichtlineare Verzerrung
nonlinear distortion factor Klirrfaktor
nonlinear resistor = NLR nichtlinearer Widerstand
nonlinear response nichtlineares Verhalten
nonlinearity Nichtlinearität, Linearitätsmaß
nonluminous nicht leuchtend
nonmachinable unbearbeitbar, nichtzerspanbar
nonmagnetic unmagnetisch
nonmagnetic material unmagnetisches Material
nonmagnetic watch unmagnetische Uhr
nonmetal Nichtmetall
nonmetallic nichtmetallisch
nonmetallic sheathed cable Isoliermantelkabel

nonmicrophonic klingarm (621.385)
nonoperative außer Betrieb, stillstehend
nonoscillatory nicht schwingfähig
nonoxidizing atmosphere Schutzgas, nicht oxidierende Atmosphäre
nonpaged jump [Sprung innerhalb eines Segments]
nonperiodic nicht periodisch
nonphysical primary Primärfarbe
nonpinking klopffest (621.4)
nonpoisonous ungiftig, nicht giftig
nonpolar generator Unipolar-Generator, Unipolarmaschine
nonpolar link homöopolare Bindung (54)
nonpolarized relay unpolarisiertes Relais
nonporous blasenfrei (621.7)
nonpower-fed repeater station eigengespeiste Verstärkerstelle
nonprecedence call gewöhnliches Gespräch, Gespräch ohne Vorrang
nonprinting control character nichtdruckendes Steuerzeichen
nonpropagated nicht ausbreitungsfähig (621.396)
nonquadded cable mehrpaariges Kabel ohne Vierer
nonradioactive nicht radioaktiv, stabil (539)
nonreactive induktionsfrei
nonreactive load induktionsfreie Belastung
nonreactive power Wirkleistung
non-re-entrant mit Freiauslösung
non-re-entrant current limiting Überstromabschaltung
nonrenewable fuse unit Sicherungspatrone, Patronensicherung
nonreflecting reflexionsfrei
nonrepetitive peak reverse voltage Stoßspannung, Stoßspitzenspannung
nonretractive switch Stellschalter
nonresonant line mit Wellenwiderstand abgeschlossene Leitung
nonresonating transformer schwingungsfreier Umspanner
nonreturn valve Rückschlagventil
nonrigid unstarr, nachgiebig
nonsalient nicht vorstehend
nonsalient pole rotor Trommelläufer, Trommelinduktor (621.313)
nonscreen film Laue-Film (621.386)
nonsealed relay Relais mit offenen Kontakten
nonsine waves (pl.) nicht sinusförmige Kurvenform
nonsinusoidal nicht sinusförmig
nonstop ohne Halt, durchfahrend
nonstop switch Durchfahrschalter (621.876)
nonsynchronous asynchron
nonsynchronous motor Asynchronmotor
nontechnical person Nichttechniker, technischer Laie
nontension joint zugentlastete Verbindung, zugentlasteter Verbinder
nontracking kriechstromfest
nontransparent undurchsichtig
nonvolatile nicht flüchtig
nonvolatile storage nichtflüchtiger Speicher, leistungsloser Speicher, energieunabhängiger Speicher
nonagon Neuneck
nonagonal neuneckig
NOR-circuit NOR-Glied, NOR-Schaltung
NOR-operation NOR-Verknüpfung
Nordhausen acid rauchende Schwefelsäure
norm Norm
normal Normale, Lot; Normalpegel; Normaltemperatur; Normaldruck; Normalzustand
normal normal, senkrecht, Normal-...
normal bend 90°-Bogen, Bogen (621.315.6)
normal case Normalfall
normal charging current echter Ladestrom
normal conditions normale Bedingungen
normal conduction current normaler Leitungsstrom
normal current Normalstrom
normal discharge current Entladestrom
normal electrode Normal-Elektrode (621.35), (54)
normale frequency Normalfrequenz, normale Frequenz
normal frequency recovery voltage wiederkehrende Spannung der Betriebsfrequenz
normal impedance Scheinwiderstand bei Normalbetrieb (534.86)
normal incidence senkrechter Einfall
normal induction Höchstinduktion bei reiner Wechselmagnetisierung
normal output Normalleistung, Normalabgabe
normal position Ruhestellung, Normalstellung
normal pressure senkrechter Druck
normal rating normale Nennlast; übliche Bemessung
normal reaction senkrechter Auflagerdruck
normal solution Normallösung (54)
normal speed Normalgeschwindigkeit; normale Drehzahl
normal state Normalzustand

normal stress Normalspannung, senkrechter Druck
normal temperature Normaltemperatur, normale Temperatur
normal value Normalwert
normal voltage Normalspannung, normale Spannung
normal zero Normal-Null, Meeresspiegel
normally-closed contact = NC contact Öffnungskontakt, Öffner, Ruhekontakt
normally-open contact = NO contact Schließer, Arbeitskontakt
normally polarized waves normal polarisierte Wellen
normalization Normalisierung (51), (66)
to normalize normalisieren, normieren (51), (66); weichglühen, normal glühen (66)
normalizing temperature Glühtemperatur
north-light roof Sheddach
north seeking nordweisend
north-seeking pole Nordpol (638)
Northrup furnace HF-Induktionsofen, kernloser Induktionsofen
nose-suspended motor Tatzlagermotor
notation Schreibweise
NOT-circuit Negationsglied, NICHT-Schaltung
NOT-operation Negation
not standard(ized), not to standard nicht normgerecht
notch Kerbe, V-Kerbe, Einschnitt; Schaltstufe, „Knopf"; „Zahn" (621.3)
notch diplexer Filterweiche (621.396)
notch effect Kerbwirkung
notch-impact strength Kerbzähigkeit
notch toughness Kerbfestigkeit
10-notch bar Zehnteile, Zehnerbarren (669.71)
to notch kerben, mit Einschnitten versehen
notched bar impact strength Kerbschlagfestigkeit
notched bar impact test Kerbschlagversuch
note Ton, Klang; Notiz; Note
note amplifier WT-Verstärker (621.394)
notice Ankündigung, Warnung, Notiz; Kündigung
notice board Anschlagbrett; Warnungsschild
notifiable anmeldepflichtig, meldepflichtig
nought Null
nought state Nullzustand
noxious schädlich, gesundheitsschädlich
nozzle Düse, Sprühdüse, Düsenrohr
NP junction NP-Übergang
NPN transistor NPN-Transistor
NPNP device Vierschichtgleichrichter
NRZ = non return to zero Richtungsschrift
NTC = Negative Temperature Coefficient Heißleiter, NTC-Widerstand
N. T. P. = Normal Temperature and Pressure Normalzustand (760 mm Hg und 0 °C)
N-type conduction N-Leitung
N-type semiconductor N-Halbleiter
nuclear Kern-...
nuclear binding energy Kernbindungsenergie
nuclear chain reaction Kernkettenreaktion
nuclear charge Kernladung
nuclear charge number Kernladungszahl
nuclear chemistry Kernchemie
nuclear constant Kernkonstante
nuclear decay Kernzerfall
nuclear disintegration Kernzerfall
nuclear energy Kernenergie
nuclear energy level Kernenergieniveau
nuclear fission Kernspaltung
nuclear force Kernkraft
nuclear fuel Spaltstoff, Kernbrennstoff
nuclear fusion Kernverschmelzung, Kernfusion
nuclear magnetic moment magnetisches Moment des Kerns (539)
nuclear mass Kernmasse
nuclear physicist Kernphysiker
nuclear physics (pl.) Kernphysik
nuclear pile Kernreaktor, Atommeiler
nuclear poison Reaktorgift
nuclear power Kernenergie, Atomenergie
nuclear power plant Kernkraftwerk
nuclear reaction Kernreaktion
nuclear reactor Kernreaktor, Atommeiler
nuclear spin Kernspin
nuclear structure Kernstruktur, Kernbau
nuclear theory Kernphysik
nuclear transformation Kernumwandlung
nuclear transmutation Kernumwandlung
nucleon Nukleon, Kernbaustein
nucleonics (pl.) Kernphysik
nucleonic power plant Kernkraftwerk
nucleus (pl.: **lei**) Kern (539); Kondensationskern; Keim; Kristallisationskern (54)
nuclei number Ordnungszahl, Kernladungszahl
nuclide Nuklid
null Null
null detector Nullanzeige
null drift Abweichung von Null
null indicator Nullanzeiger, Nullinstrument
null method Nullmethode (621.317)
null position Nullstellung (621.3)
nullification Nullabgleich, Auf-Null-Bringen

to nullify auf Null bringen, zu Null machen; zunichte machen
number Zahl, Anzahl; Zahl, Ziffer, (reine) Zahl; Nummer, Ausgabe; Numerus (51)
number of cycles Periodenzahl
number of digits Stellenzahl (51)
number identification Identifizierung des Rufes (621.395)
number of lines of force Kraftlinienzahl
number of neutrons Neutronenzahl
number of pairs of poles Polpaarzahl
number of phases Phasenzahl
number plate Nummernschild
number of poles Polzahl
number representation Zahlendarstellung
number of revolutions per minute Umdrehungszahl, Drehzahl
number triplet Zahlentripel
number of turns Windungszahl
number of windings Zahl der Wicklungen
to number numerieren, beziffern; zählen
numbering box Numerierwerk
numerable zählbar, abzählbar (51)
numeral Zahlwort; Ziffer
numerator Zähler (51)
numeric numerisch
numeric word numerisches Wort
numerical numerisch, zahlenmäßig
numerical aperture numerische Apertur (535)
numerical constant Zahlenwert, Zahlenkonstante
numerical control numerische Steuerung
numerical data Zahlenangabe(n)
numerical display numerische Anzeige
numerical distance numerischer Abstand
numerical example Zahlenbeispiel
numerical factor Zahlenfaktor
numerical-readout tube Ziffernanzeigeröhre
numerical selector Nummernwähler
numerical value zahlenmäßiger Wert, Zahlenwert
numerical working Zahlenrechnung, numerisches Rechnen
numicator Ziffernröhre
nut Mutter (621); Nuß (58)
nut tapping Mutternschneiden
nutation Nutation
nutscrew Schraubenmutter
nuvistor Nuvistor
nylon Nylon
Nyquist slope Nyquist-Charakteristik

O

oakum Werg
object Objekt, Gegenstand; Ziel, Absicht
object code Objektcode [vorübersetztes Programm ohne Adreßbezug]
object carrier Objektträger (535)
object glass Objektiv, Objektivlinse (535)
object language Maschinensprache, Objektsprache
object lens Objektivlinse, Objektiv
object program Objekt-Programm, Maschinenprogramm
objective lens Objektivlinse
oblate abgeplattet, abgeflacht
oblate spheroid abgeplattetes Rotationsellipsoid
obligatory keying speed Pflichttempo (621.394)
oblique geneigt, schief, schräg
oblique angeled schiefwinklig
oblique cone schiefer Kegel
oblique front view Ansicht schräg von vorn
oblique-incidence Schrägeinfall
oblique-incidence transmission Schrägwinkelübertragung
oblong rechteckig, länglich, gestreckt, im Querformat
oblong hole Langloch
oblong spheroid verlängertes Rotationsellipsoid
OBOE Diskus (Navigationsverfahren) (621.396)
obscuration Abschattung, Verdeckung, Überstrahlung
to obscure verdecken, überdecken (auch 621.396); überstrahlen
observation Beobachtung
observation point Beobachtungspunkt, Beobachtungsstelle
observatory Observatorium, Sternwarte (52)
to observe beobachten; beachten
observed fact Beobachtungstatsache
observer Beobachter
obsolete veraltet, überholt, überaltert
obstruction Behinderung, Hindernis, Hemmung
obstruction effect Abschattung (621.396)
obstruction light Hindernisfeuer, Flugwarnfeuer (656.7)
obstruction lighting Hindernisbefeuerung
obstruction of line-of-sight path Sichtbehinderung
obstruction loss Zusatzdämpfung
obtuse stumpf (51)
obtuse angle stumpfer Winkel

obverse Ober-...
obverse side Oberseite
occulting beacon unterbrochenes Feuer (656.7)
occulting light aussetzendes Feuer
occupation Belegung (621.395)
occupation level Belegungszustand (621.395)
occupied bandwidth belegte Bandbreite
ochre Ocker
OCR = optical character recognition optische Zeichenerkennung
octal notation Oktalschreibung (681.3)
octane Oktan
octave Oktave (534); Oktav, Oktavformat (655)
octave filter Oktavsieb
octode Oktode, Achtpolröhre
ocular okular, Augen-...
ocular micrometer Okularmikrometer
odd ungewöhnlich; ungerade (51); übrig
odd legs (pl.) Zirkel mit ungleichen Enden
odd number ungerade Zahl
odd numbered ungeradzahlig
odd-numbered harmonic ungeradzahlige Oberwelle
odd order ungerade Ordnungszahl, von ungerader Ordnungszahl, ungerade
odd thousandths überschießende Tausendstel
20 odd einige zwanzig, zwanzig und etwas
oddly shaped sonderbar geformt, unregelmäßig
oddside falsche Form, Vorform (621.74)
OEM = original equipment manufacturer
oersted Oersted
off weg, ab, weg von; aus... heraus; ausgeschaltet, abgestellt
off the axis aus der Achse verschoben, nicht auf der Achse liegend
off-centre außermittig, exzentrisch
off-centre plan display exzentrische Rundsichtdarstellung, exzentrisches Schirmbild
off-circuit spannungslos
off emergency Notausschalter
off-hook signal Beginnzeichen (621.395)
off-line von der Hauptanlage getrennt, nicht verbunden; off-line (681.3)
off-line operation Off-Line-Betrieb
off-load interruption Leerausschaltung
off-load switch Ohnelastschalter
off peak außerhalb der Spitze
off-peak load Belastung außerhalb der Spitzenzeit
"off"-position „Aus"-Stellung, Abschaltstellung, Ausschaltstellung; Ruhestellung (621.3)
off resonance außerhalb der Resonanz
off-state Sperrzustand
off-state characteristic Sperrkennlinie
off-state power loss Sperrverlustleistung
off-state resistance Sperrwiderstand
off-voltage... spannungslos...
offcut Verschnitt, Abfall
to offer anbieten (621.395)
to offer resistance Widerstand leisten, sich widersetzen, Widerstand entgegensetzen
offering Aufschalten (621.395)
office Büro; Dienst; Aufgabe, Dienstbereich
office selector Amtswähler
office test amtliche Zählernachprüfung
offset Versetzung, Verschiebung; Abbiegung, Umgehung; Doppelbogen; Ableger (63); Offset (655); bleibende Regelabweichung (621.316.7); Versatzbetrieb (621.396)
offset feed dezentrierte Speisung (621.396.67)
offset voltage Offset-Spannung
to offset versetzen, abbiegen, aus der Normalrichtung herausbringen, aus der Normallage herausbringen; ausgleichen; aufheben
to offset a tendency einer Tendenz entgegenwirken, eine Tendenz aufheben
offset abgebogen, versetzt, verschoben
offset bevel gears windschiefe Kegelräder
to be offset in phase phasenverschoben sein; in der Phase verschoben sein
off-the-shelf ab Lager
ogee S-förmige Schweifung
ohm Ohm
ohm's law Ohmsches Gesetz
ohmer Widerstandsmesser
ohmic ohmsch, ohmisch
ohmic drop ohmscher Spannungsabfall
ohmic loss ohmscher Verlust
ohmic resistance ohmscher Widerstand, phasenreiner Widerstand
ohmmeter Ohmmeter, Widerstandsmeßgerät
oil Öl
oil-base type compound Vergußmasse auf Ölbasis
oil-blast (circuit-) breaker Löschkammerschalter
oil-blast explosion and expansion breaker Löschkammerschalter
oil-break circuit breaker automatischer Ölschalter
oil-break fuse Ölsicherung (621.3)
oil-break switch Ölschalter
oil breather Ölbelüfter, Ölentlüfter
oil burner Ölbrenner

oil circuit breaker Ölschalter
oil condenser Ölkondensator
oil-cooled ölgekühlt
oil-cooled starter Ölanlasser
oil-cooled transformer Öltransformator
oil cooling Ölkühlung
oil-electric drive diesel-elektrischer Antrieb
oil-expansion tank Ölausgleichsgefäß (621.314)
oil feed Ölzuführung
oil-feeding reservoir Ölausgleichsgefäß (621.315.2)
oil-filled cable Ölkabel
oil film Ölschicht
oil filter Ölfilter
oil firing Ölfeuerung
oil-fuse cut-out Ölsicherung
oil gauge Ölschauglas
oil-gun nipple Schmiernippel, Abschmiernippel
oil-immersed unter Öl
oil-immersed contactor Ölschütz
oil-immersed forced aircooled transformer Öl-Transformator mit Fremdlüftung (OF)
oil-immersed pump unter Öl laufende Pumpe
oil-immersed starter Ölanlasser
oil-immersed transformer Öltransformator
oil-impregnated ölimprägniert; ölgetränkt
oil-impregnated cable Ölkabel
oil-insulated transformer Öltransformator
oil insulation Ölisolation
oil insulator Ölrinnenisolator
oil-less öllos, ölfrei
oil level Ölspiegel, Ölstand
oil paper Ölpapier
oil pressure Öldruck
oil-proof ölfest, ölbeständig
oil pump Ölpumpe
oil-quenched in Öl abgeschreckt
oil-quenched cut-out Ölsicherung
oil-quenched fuse Ölsicherung
oil quenching Ölhärtung
oil retainer Spritzring (621.8)
oil ring Schmierring
oil-ring lubrication Ölringschmierung
oil-soluble öllöslich
oil starter Ölanlasser
oil strainer Ölsieb
oil switch Ölschalter
oil tank Ölbehälter, Öltank
oil-tight öldicht
oil transformer Öltransformator
to oil ölen, mit Öl schmieren
oiled paper Ölpapier
oiled paper condenser Ölpapierkondensator, Ölkondensator
oilcloth Öltuch; Wachstuch
oiler Ölkanne, Schmierkanne
oiling Ölung, Schmierung
oiling arrangement Schmiervorrichtung
oiling device Schmiereinrichtung
oiling point Schmierstelle
oiless breaker ölfreier Schalter
oillike ölähnlich
oilostatic cable Öldruckkabel
oilstone Abziehstein (621.9)
old alt
old hand alter Fachmann
old material Altmaterial
of old standing traditionell, „schon immer so“
oleic acid Ölsäure, Oleinsäure
oleo-... Öl-...
oleo-resin Ölharz
omnibearing-distance navigation Navigation nach Polarkoordinaten
omnibearing selector Kurswähler
omnibus Omnibus
omnibus bar = bus bar Sammelschiene
omnibus configuration Liniensystem, Reihensystem
omnibus train Bummelzug
omnidirectional ungerichtet
omnidirectional antenna Rundstrahlantenne
omni(-directional)-radio beacon Rundstrahlbake; Drehfunkfeuer
omnirange Rundstrahlfunkfeuer, Drehfunkfeuer
OMR = optical mark reader
on an average im Mittel
on-call channel Reservekanal
on-course direction richtiger Kurs
on edge hochkant
on fire in Brand
on-hook signal Schlußzeichen
on-line on-line, direkt verbunden, schritthaltend (681.3)
on-line operation On-Line-Betrieb
on load belastet, unter Last
on-load Belastungs-..., unter Last
on-load speed Drehzahl bei Belastung
on-load switching Unterlastschaltung
on-load tap-changing equipment Lastschalter, Umschalter für Spannungsregelung unter Last
on-load voltage Belastungsspannung, Spannung bei Belastung, Spannung unter Last, Arbeitsspannung
on-off control Auf-Zu-Regler,

Schwarz-Weiß-Steuerung, Zweipunkt-Regler, Ein-Aus-Regler
on-off keying Ein-Aus-Tastung
on open circuit bei Leerlauf, leerlaufend
on-period Stromführungszeit (621.314)
"on" position „Ein"-Stellung, Einschaltstellung
on-state Durchlaßzustand
on-state characteristic Durchlaßkennlinie
on-state current Durchlaßstrom
on-state resistance Durchlaßwiderstand
on-state threshold voltage Schleusenspannung (621.314.63)
once through cooling Durchlaufkühlung
one ein; einzig, alleinig
one-address instruction Einadressenbefehl
one-at-a-time principle sequenzielles Verfahren, Nacheinanderverfahren
one-brick wall 1 Stein starke Wand
one-chip device Einchip-Element
one-condition Eins-Zustand (681.3)
one-directional einsinnig
one-fourth cycle Viertelperiode
one-hole fixing Einlochbefestigung
one-hour rating Stundenlast (621.3)
one-line-to-earth short circuit einpoliger Kurzschluß
one-man Einmann-...
one-man operation Einmannbedienung, Einmannbetrieb
one-to-one correspondence natürlicher Maßstab (1 : 1)
one-to-one transformer Isoliertransformator, Transformator 1 : 1
one-phase = single phase Einphasen-..., einphasig
one-piece einteilig, ungeteilt
one-piece inductor Massivinduktor (621.313)
one-range winding einetagiger Wickelkopf
one-shot multivibrator monostabiler Multivibrator
one-shot operation Einzelschrittbetrieb
one-turn einwindungs-...
one-turn loop Einwindungsrahmen
one-wave rectifier Halbwellengleichrichter, Einweggleichrichter
one-way Einbahn-... (625); Einweg-... (621.3)
one-way communication Einwegfernmeldeverkehr, Richtungsverkehr
one-way rectifier Einweggleichrichter
one-way reversible communication wechselweiser Richtungsverkehr, Wechselverkehrsverbindung
one-way switch (einfacher) Schalter
one-way trunk gerichtet betriebene Leitung (621.395)
O-network Viereckschaltung, Vierpol in Polygonschaltung
ONIX ONIX [16-Bit-Betriebssystem mit spezifischen Eigenschaften]
onrush Ansturm, Anprall
ooze Schmiere
to ooze herausquellen, herausschwitzen, träufeln, sickern
oozing basin Sickerbecken
opacity Undurchsichtigkeit
opal Opal
opal glas Opalglas, Milchglas
opal lamp Opallampe
opal pane Milchglasscheibe
opal shade Opalschirm
opalescence Opaleszenz; Trübung; Milchigkeit
opalescent glass Opalglas
op amp = operational amplifier Operationsverstärker
opaque trübe, undurchsichtig, undurchlässig
opaque paper Verdunkelungspapier
opaque screen lichtundurchlässiger Schirm
open Freies, Freiluft
in the open im Freien
to open öffnen; sich öffnen, auftrennen
to open after bending nach dem Biegen zurückfedern
to open a block einen Streckenabschnitt freiblocken, zurückblocken
to open a circuit einen Stromkreis öffnen, ausschalten
opened up in ausgebautem Zustand, auseinandergebaut
open öffen, geöffnet; offen, leerlaufend (621.3); licht, offen; gedehnt (621.317)
open-air freie Luft, Freiluft
open-air Freiluft-...
open-air climates Freiluftklima
open-air plant Freiluftanlage
open-air storage Lagerung unter freiem Himmel
open amortisseur winding unterteilte Dämpferwicklung (621.313)
open arc lamp offene Bogenlampe
open belt drive offener Riemenantrieb
open block entblockter Streckenabschnitt
open centre display Schirmbild mit Nullkreisdarstellung
open circuit Leerlaufzustand, offener Stromkreis

open-circuit armature Anker mit offener Wicklung
open-circuit characteristic Leerlaufkennlinie
open-circuit condition Leerlaufzustand
open-circuit control Umschaltung mit Unterbrechung
open-circuit impedance Leerlaufwiderstand
open-circuit loss Leerlaufverlust
on open circuit bei Leerlauf
to be on open circuit leerlaufen
open-circuit output impedance Leerlauf-Ausgangsimpedanz
open-circuit system Arbeitsstromsystem
open-circuit transimpedance Kernwiderstand (621.372.5)
open-circuit transition Reihen-Parallelumschaltung mit Unterbrechung (621.33)
open-circuit voltage Leerlaufspannung
open-circuit winding offene Wicklung (621.313)
to open-circuit auf Leerlauf schalten, sekundär abtrennen, ausschalten
open-circuited leerlaufend
open-coil armature Anker mit nicht in sich geschlossener Wicklung
open-coil armature winding offene Wicklung (621.313)
open contact Arbeitsstromkontakt
open-core transformer Eisentransformator mit Luftspalt
open-ended am Ende offen, leerlaufend
open-ended line offene Leitung, leerlaufende Leitung
open fuse (cut-out) offene Sicherung, Streifensicherung
open-hearth process (Siemens-)Herdverfahren; Siemens-Martin-Prozeß
open-hearth steel Siemens-Martin-Stahl, S. M.-Stahl
open jet tunnel Freistrahlwindkanal
open-loop control Steuerkette (621.316.7)
open-loop control system Steuersystem (ohne Rückführung)
open-loop voltage gain Leerlaufspannungsverstärkung
open machine offene Maschine (621.313)
open-mesh weitmaschig
open motor offener Motor
open neutral aufgelöster Sternpunkt
open order Ausschaltbefehl
open-phase circuit breaker Motorschutzschalter
open-phase protection Phasenbruchschutz
open-phase relay Phasenunterbrechungsrelais
open-primary current transformer Schienenstromwandler, Stabstromwandler
open scale großer Maßstab; offene Skala, gedehnte Skaleneinstellung (621.317)
open-shut Auf-Zu; Schwarz-Weiß (621.317)
open slot offene Nut
open sparking offene Funkenbildung
open star point aufgelöster Sternpunkt
open starter offener Anlasser
open switchgear offene Schaltgeräte (pl.)
open to offen für; anfällig für; unterworfen; zugänglich
open-type hot plate Glühdrahtkochplatte
open-type switchgear offene Schaltanlage
open wagon offener Güterwagen (625.2)
open-wire line Freileitung (621.394), (621.5)
open-wire (pole) line Freileitung auf Holzmasten (621.395)
open-wire twin unabgeschirmte Doppelleitung
opening Öffnung; Ausschaltung (621.3)
opening bridge Klappbrücke
opening of a circuit breaker Öffnung eines Schalters, Ausschaltung, Abschaltung
opening delay Ausschaltverzögerung
opening speed Öffnungsgeschwindigkeit (621.316.5)
opening time Eigenzeit (621.316.5); Ausschaltzeit
operability Betriebsfähigkeit
operable betriebsfähig
operand Rechengröße, Operand (681.3)
to operate betreiben, betätigen; wirken, arbeiten; operieren (61), (35); anziehen (621.318.5)
to operate on steam mit Dampf arbeiten, mit Dampf betrieben werden; mit Dampf betreiben
operate lag Anzugverzögerung
operate time Ansprechzeit
operate winding Anzugswicklung
operating Betriebs-...
Operating aisle Bedienungsgang
Operating ampere-turns aktive Amperewindung
operating characteristic Betriebseigenschaft, Betriebskennlinie
operating code = opcode OP-Code, Befehlscode
operating conditions Betriebsbedingungen
operating crank Stellkurbel, Betätigungskurbel
operating crew Bedienungsmannschaft, Bedienung
operating current Betriebsstrom, Arbeitsstrom
operating data computer Betriebswerterechner

operating device Betätigungseinrichtung, Bedienungsgerät
operating differential Regelabweichung (621.316.7)
operating duty Betriebsart, Betriebsbeanspruchung
operating economy Wirtschaftlichkeit
operating engineer Betriebsingenieur
operating error Bedienungsfehler
operating factor Einschaltdauer = ED
operating force Betätigungskraft, Stellkraft
operating frequency Betriebsfrequenz
operating handle Antriebshebel, Bedienungsgriff
operating hours Betriebsstunden
  operating-hour counter Betriebsstundenzähler
operating instructions Bedienungsanweisung, Betriebsanleitung
operating level Ansprechpegel
operating lever Betätigungshebel, Schalthebel
operating limit Arbeitsbereichsgrenze, zulässige Nutzleistung, Aussteuerungsgrenze
operating manual Bedienungsanleitung, Handbuch
operating mechanism Betätigungsmechanismus
operating overload betriebsmäßige Überlastung
operating point Arbeitspunkt (621.385)
operating position Arbeitsstellung, Betriebsstellung; Ansprechstellung; Betriebslage
operating range Arbeitsbereich
operating region Arbeitsbereich, Umgebung des Arbeitspunktes (621.385)
operating rod Betätigungsstange (621.316.5)
operating room Fernsprechsaal, Vermittlungssaal; Operationssaal, Operationszimmer (61)
operating sequence Arbeitsablauf
operating speed Betriebsdrehzahl; Arbeitsgeschwindigkeit; Zeichengeschwindigkeit (621.396)
operating system Betriebssystem (681.3)
operating temperature Betriebstemperatur
operating temperature range Betriebstemperaturbereich
operating time Betriebsdauer, Betriebszeit; Eigenzeit, Ausschaltzeit (621.316.5)
operating torque Stellmoment
operating voltage Betriebsspannung (621.316.5); Betätigungsspannung
operating wavelength Betriebswellenlänge

operation Betrieb; Betätigung; Arbeitsgang, Operation (621.9); Operation (35), (61)
  operation code Befehlscode
  operation cycle Operationszyklus
  operation factor prozentuale Benutzungsdauer
  in operation in Betrieb, im Betrieb
  operation instruction Bedienungsanweisung
  operation mode Betriebsart
  operation path Operationspfad
  operation period Benutzungsdauer, Betriebsdauer
  operation procedure Betriebsverfahren
  operation scheduling Arbeitsvorbereitung
  operation system Betriebssystem (681.3)
  operation time Betriebszeit
operational betriebsmäßig, betrieblich, Betriebs-...
  operational amplifier Operationsverstärker
  operational calculus Operatorenrechnung
  operational characteristics Betriebsmerkmale
  operational conditions Betriebsbedingungen, betriebliche Bedingungen
  operational data collection Betriebsdatenerfassung
  operational key Betätigungstaste, Auslösetaste
  operational life Lebensdauer
  operational phase Betriebsabschnitt
  operational quality Betriebsgüte
operative wirksam; betriebsbereit
  operative condition betriebsfähiger Zustand
  operative speed Betriebsdrehzahl
operator Operator; Bediener, Bedienungsmann; Maschinenwärter, Bedienungsperson; Dreher (621.94); Vermittlungsbeamtin, „Fräulein vom Amt" (621.395)
  operator's board Abfrageplatz (621.395)
  operator control panel Bedienungsfeld
  operator interception Bescheidansage durch Bedienungspersonal (621.395)
  operator recall Schrankherbeiruf
  operator's telephone set Vermittlungsapparat
opposed brush gegenläufige Bürste
opposing widerstreitend, entgegengesetzt gerichtet
  opposing currents entgegengesetzte Ströme
  opposing field Gegenfeld, entgegengesetztes Feld
  opposing magnetic field of armature Ankergegenfeld
  opposing voltage Gegenspannung
opposite Gegenseite, Gegenkathete (51); Gegenteil
opposite gegenüber; entgegengesetzt,

entgegengesetzt gerichtet, entgegengesetzt liegend; gegenüberliegend
opposite angle gegenüberliegender Winkel
opposite direction entgegengesetzte Richtung
opposite in direction entgegengesetzt gerichtet
to be opposite in direction entgegengesetzte Richtung haben
opposite in phase in Gegenphase
opposite sign entgegengesetztes Vorzeichen
opposite in sign mit entgegengesetztem Vorzeichen
to be opposite in sign entgegengesetztes Vorzeichen haben
optic optisch
optic angle Sehwinkel
optic axis optische Achse (54)
optical optisch
optical axis optische Achse (54)
optical bench optische Bank
optical character recognition optische Zeichenerkennung
optical compensation optischer Bildausgleich
optical coupler Optokoppler
optical fibre Glasfaser
optical flat planparallele Glasplatte
optical grating optisches Gitter
optical horizon optische Sichtweite
optical illusion optische Täuschung
optical image optisches Bild
optical lever Lichtzeiger (621.317)
optical mark reader optischer Markierungsleser
optical pyrometer optisches Pyrometer
optical range optische Sicht (621.396)
optical rotations (pl.) optische Drehung
optical sign Schauzeichen
optical square Winkelspiegel
optical telephony Lichttelefonie
optical waveguides Lichtwellenleiter
optically flat optisch eben
optics (pl.) Optik
option Zusatzeinrichtung
optimization Optimierung
optimum (pl.: **-ma**) Optimum, Bestwert
optimum optimal, Best-..., Optimal-...
optimum coupling kritische Kopplung
optimum load angepaßte Last, günstigste Last
optimum temperature Optimaltemperatur
optimum value Bestwert, Optimalwert
optional wunschgemäß, nach Wunsch
opto electronic Optronik, Optoelektronik
oral mündlich
oral exposition mündliche Darlegung
orbit Kreis, Kreisbahn, Umlaufbahn, gekrümmte Bahn
orbital Kreis-..., kreisend
orbital angular momentum Bahndrehimpuls
orbital electron Hüllenelektron (539)
orbital eccentricity Bahnexzentrizität
orbital motion Bahnbewegung (53)
orbital path Umlaufbahn
orbital plane Bahnebene
OR-circuit ODER-Schaltung
OR-operation ODER-Verknüpfung, Disjunktiv
order Ordnung; Größenordnung; Anweisung, Befehl; Order, Anweisung; Bestellung, Auftrag; Kommando (681.3)
order of accuracy Größenordnung der Genauigkeit
order channel Dienstkanal (621.396)
order of connection Schaltfolge
in order in Ordnung, in der richtigen Reihenfolge, der Reihe nach
in order of length der Länge nach
order of magnitude Größenordnung; Zehnerpotenz
order of merit Qualitätsreihenfolge, Reihenfolge nach der Güte
order of weight dem Gewicht nach geordnet
order wire Dienstleitung (621.395)
order wire circuit Dienstleitung (621.395)
to order je nach Auftrag, nach Bestellung
ordinal (number) Ordnungszahl
ordinary gewöhnlich, üblich; ordentlich
ordinary ray ordentlicher Strahl (53)
ordinary temperature Raumtemperatur, gewöhnliche Temperatur
ordinary wave ordentliche Welle
ordinate Ordinate
ore Erz
organic organisch
organic acid organische Säure
organic chemistry organische Chemie
organic compound organische Verbindung
organic dye organischer Farbstoff
organic moderated reactor Reaktor mit organischer Bremssubstanz
to orient orientieren, ausrichten; einrichten, einordnen
oriented sheet kristallorientiertes Blech (621.314.2)
to orientate ausrichten; sich ausrichten, sich orientieren
orientation Orientierung (548); Winkellage; räumliche Anordnung
orifice Öffnung, Austrittsöffnung, Mündung, Mundloch

origin Nullpunkt, Ursprung (51); Herkunft, Ursprung; Quellpunkt (535)
original Original, Urschrift
original Original-..., Ausgangs-...ursprünglich; originell
original elongation Ausgangsdehnung
original function Ausgangsfunktion
original packing Originalverpackung
to originate hervorbringen; entstehen
originating exchange Ursprungsvermittlungsstelle
originating register Leitregister (621.395)
O-ring Hohlgummidichtung, Schlauchdichtung
ornamental ornamental, Schmuck-...
ornamental lantern Schmuckleuchte
ornamental lighting Schmuckbeleuchtung
orthocentre Schnittpunkt der Lote im Dreieck
orthochromatic orthochromatisch
orthoclase Orthoklas, Feldspat
orthogonal orthogonal
orthogonal curvilinear co-ordinates krummlinige orthogonale Koordination
orthojector circuit breaker Ölstrahlschalter, Querstrahlschalter
to oscillate schwingen; schwanken
oscillating Schwing-..., schwingend
oscillating circuit Schwingkreis, schwingender Kreis
oscillating component Wechselanteil
oscillating crystal Schwingkristall
oscillating current Schwingstrom, oszillierender Strom
oscillating detector Schwingaudion
oscillating electrode Gitter der Bremsfeldröhre
oscillating electromotive force Wechsel-EMK
oscillating field schwingendes Feld
oscillating meter Pendelzähler (621.317)
oscillating period Schwingungsdauer
oscillating potential difference Wechselpotentialdifferenz
oscillating quantity Wechselgröße, Schwinggröße
oscillating voltage Wechselspannung
oscillation Schwingung
oscillation constant (Produkt aus Induktivität und Kapazität eines Schwingungskreises oder die Wurzel daraus)
oscillation frequency Schwingfrequenz
oscillation squeal Rückkopplungspfeifen (621.396)
oscillation transformer HF-Übertrager
oscillator Oszillator; Summer, Schwinger; Schwingungsgenerator
oscillator coil Schwingspule, Schwingkreisspule
oscillator crystal Schwingquarz, Oszillatorquarz
oscillator noise Oszillatorrauschen
oscillator tube Senderöhre
oscillator wavemeter Überlagerungswellenmesser
oscillatory schwingend, periodisch
oscillatory circuit Schwingkreis, schwingfähiger Kreis
oscillatory discharge periodische Entladung, oszillatorische Entladung
oscillatory impedance Hochfrequenzwiderstand
oscillatory power Schwingleistung
oscillatory surge HF-Stoß (621.316.9)
oscillion Schwingaudion
oscillogram Oszillogramm
oscillograph (schreibender) Oszillograph
oscilloscope Oszilloskop, Oszillograph (ohne Registrierung)
oscilloscope tube Oszillographenröhre, Kathodenstrahlröhre, Elektronenstrahlröhre
oscilloscope screen Betrachtungsschirm, Oszillographenschirm
to osculate schmiegen (51)
osculation Schmiegungspunkt (51)
osglim lamp Neon-Helium-Glimmlampe
osmium Osmium
osmium lamp Osmiumlampe
osmose, osmosis Osmose
osmotic osmotisch
osmotic pressure osmotischer Druck
osophone Knochentelefon
Ossana's circle diagram Ossanakreis
ossiphone Knochentelefon
ought Null
ounce Unze (**1 oz** = 28,3495 g)
out aus
out of action außer Betrieb
out and back hin und zurück, hin und her
out of balance aus dem Gleichgewicht, nicht ausbalanciert; Unsymmetrie-...
out-of-balance current Unsymmetriestrom
out-of-balance load Unsymmetrielast, unsymmetrische Belastung
out-of-band radiation Randaussendung
out-of-band signalling systemeigene Wahl, Signalisierung außerhalb des Sprachbandes (621.395)
to be out in a calculation sich verrechnet haben
out of centre exzentrisch, außermittig
out-of-date veraltet

out-of-date machinery veraltete Maschinenanlage
out of operation außer Betrieb
out of order nicht in Ordnung, nicht betriebsfähig
out and out durch und durch, ganz und gar
out-of-phase in der Phase verschoben, phasenverschoben
90° out of phase um 90° verschoben
out-of-phase component falschphasige Komponente
out of plumb vom Lot abweichend
out of position in falscher Lage, in falscher Stellung
out of proportion in keinem Verhältnis stehend, unverhältnismäßig
out of repair nicht mehr zu reparieren
out-of-round Schlag (621.9)
out of service nicht in Betrieb, außer Betrieb
out of sight außer Sicht, außer Sichtweite
out of step außer Tritt
outage Ausfallen, Außerbetriebkommen
outage time Ausfallzeit
to outbalance ausbalancieren
outband Außerband..., außerhalb des Bandesliegend
outband signalling Signalisierung außerhalb des Sprachbands (621.395)
outbreak Ausbruch
outbreak of fire Brandausbruch, Ausbruch eines Feuers
outdoor Freiluft-..., im Freien, Außen-...
outdoor aerial Hochantenne, Außenantenne
outdoor circuit breaker Freiluftschalter
outdoor equipment Freiluftmaterial
outdoor fitting Außenarmatur
outdoor lighting Außenbeleuchtung
outdoor plant Freiluftanlage
outdoor station Freiluftstation
outdoor substation Freiluftstation
outdoor transformer Freiluftumspanner, Freilufttransformator
outdoor type Freilufttype
outdoor-type switchgear Freiluftschaltgeräte (pl.)
outdoors im Freien, draußen
outer außer, Außen-...
outer cable Außenleiter (621.3)
outer conductor Außenleiter, Phasenleiter
outer conductor voltage Spannung zwischen den Außenleitern; Leiterspannung
outer corona protection Außenglimmschutz
outer diameter Außendurchmesser
outer end äußeres Ende, Außenende; äußerer Anschluß (621.3)
outer fringe äußere Linie (535)
outer layer Außenschicht, äußere Lage
outer marker beacon Voreinflugzeichen
outer but one zweitäußerst
outer product vektorielles Produkt, äußeres Produkt
outer sheath Außenmantel
outer-shell electron Valenzelektron, Elektron der äußeren Schale
outer wall Außenwand
outfit Ausrüstung; Satz, Werkzeugsatz
outfitter Ausrüster (629)
outflow Ausfluß, ausfließende Menge
outgassing test Ausgasprüfung
outgoing ausgehend; abgestrahlt, ausgestrahlt (621.396); abgehend
outgoing current wegfließender Strom, abfließender Strom, abgehender Strom
outgoing feeder abgehende Speiseleitung
outgoing junction line abgehende Amtsverbindung
outgoing wave abgehende Welle, Senderwelle
outhouse Anbau, Nebengebäude
to outlast überdauern
outlet Auslaß, Ausgang, Auslaßöffnung; Abzug, Abzugsrohr; Steckdose
outlet air Abluft
outlet socket Ausgangsbuchse
outlet temperature Austrittstemperatur
outlet terminal Ausgangsklemme
outline Umriß, Profil (51); Schema, Entwurf
outline drawing Umrißzeichnung
outline lighting Konturen-Beleuchtung
to outnumber an Zahl übertreffen, zahlenmäßig überlegen sein
to outphase aus der Phase bringen, phasenverschieben
outphasing modulation system Chireix-Modulation
outpouring Sickermenge, Sickerwasser; austretende Flüssigkeit
output Ausstoß; Ausbringen; Fertigungsleistung; Förderleistung; Fördermenge (622); Ausgangsleistung (621.3); Ausgabe (681.3)
output admittance Ausgangsscheinleitwert
output amplifier Ausgangsverstärker, Endverstärker
output block Ausgabeblock
output capacitance Ausgangskapazität
output capacity Speichervermögen, Kapazität (621.355)

output circuit Ausgangskreis
output configuration Ausgangskonfiguration (621.374)
output cord Anschlußleitung, „Ausgangsstrippe"
output-coupling capacitor Anodenkopplungsblock
output-current range Arbeitsbereich des Ausgangsstromes (621.375)
output device Ausgabegerät
output end Empfangsseite, Ausgangsseite, Verbraucherseite, Ausgang, Ende
output equipment Ausgabegerät (681.3)
output factor Belastungsfaktor
output impedance Ausgangsscheinwiderstand, Ausgangsimpedanz
output instruction Ausgabebefehl
output level Ausgangspegel
output-loading capability statische Ausgangsbelastbarkeit (621.375)
output meter Ausgangsspannungsmesser, Ausgangsleistungsmesser; Outputmeter
output quantity Ausgangsgröße (621.316.7)
output rate Ausgabegeschwindigkeit
output referred to a potential potentialbezogener Ausgang
output register Ausgabespeicher
output regulator Leistungsregler
output size Leistungsgröße, Type
output signal Ausgangssignal; Ausgangsbefehl (621.316.7)
output of steam Dampfleistung (621.1)
output terminal Ausgangsklemme
output transfer function Ausgangsübertragungsfunktion
output transformer Ausgangstransformator, Ausgangsübertrager
output unit Ausgabe-Einheit
output value Ausgangswert (621.316.7)
output valve Ausgangsröhre, Endröhre
output variable Ausgangsgröße, Regelimpuls
output voltage Ausgangsspannung
output-voltage swing Ausgangsspannungshub
output volume Lautstärke
to outrange weiter reichen als, an Reichweite übertreffen
outrigger clamp zugentlastete Anschlußklemme
to outshine überstrahlen
outside Außen-...
outside broadcast transmission Zubringerübertragung
outside calipers (pl.) Außentaster
outside coating Außenmantel
outside dimension Außenmaß
outside diameter Außendurchmesser
outside wire Außenleiter, Hauptleiter
outsize Übergröße
outskirt Außenbezirk, Randgebiet
outstation Unterstation (621.398)
outthrust Druck nach außen, Seitenschub
outward auswärts, nach außen zu gerichtet
outward normal äußere Flächennormale
to outweigh aufwiegen; überwiegen
outwork Außenwerk, Vorwerk (623.1)
oval oval
oval cable Kabel mit ovalem Mantel
oval tube ovales Rohr, ovale Röhre
oven Ofen; Heizkammer, Thermostat (621.396)
oven drying Ofentrocknung
overall Gesamt-...
overall amplification Gesamtverstärkung, Betriebsverstärkung
overall attenuation Betriebsdämpfung
overall depth Gesamthöhe
overall dimension Gesamtmaß, Abmessung über alles
overall efficiency Gesamtwirkungsgrad
overall gain Gesamtverstärkung
overall length Gesamtlänge
overall loss Restdämpfung, Betriebsdämpfung
overall-loss plan Dämpfungsplan
overall size Bauhöhe; Baulänge
overall thickness Gesamtdicke
overall voltage Gesamtspannung
overamplification Übersteuerung (621.396.64)
overbalance Übergewicht
to overbalance Übergewicht geben, überbalancieren
overburden überlasten, überbeanspruchen
overcharge Überladung (621.356)
to overcharge überladen (621.356)
overcommutation Überkommutierung
overcompounded überkompoundiert (621.313)
overcompounding Überkompoundierung (621.313)
to overcouple Überkoppeln, überkritisch koppeln
to overcure überbehandeln, überhärten
overcurrent Überstrom, Überlaststrom
overcurrent circuit breaker Überstromschalter, Maximalschalter
overcurrent circuit breaking Überstromauslösung, Maximalauslösung
overcurrent coil Überstromspule
overcurrent protection Überstromschutz
overcurrent relay Überstromrelais, Maximalrelais

overcurrent and reverse current circuit breaker Überstrom-Rückstromschalter
overcurrent trip Überstromauslösung
over-and-undercurrent relay Maximal-Minimalrelais
overcurrent undervoltage circuit breaker Maximalschalter mit Nullspannungsauslösung
overdamped gedämpft, überaperiodisch gedämpft
overdamping überaperiodische Dämpfung, Überdämpfung, überkritische Dämpfung
to overdischarge zu weit entladen (621.356)
to overdose überdosieren
to overdrive übersteuern (621.396.64)
over excitation Übererregung
overexposure Überbelichtung, Überexposition (77)
overexposure to radiation Überbestrahlung, Dosisüberschreitung (61)
overflow Überlauf, Überlaufmenge
overflow data Überlaufdaten
overflow traffic Überlaufverkehr
to overflow überfließen, überlaufen
overflowing water Überlaufwasser
to overfuse übersichern, zu hoch absichern (621.3)
overground oberirdisch
overhang Überhang, Ausladung; Spulenkopf; Stirnverbindungen (pl.) (621.313)
overhang of coils Wickelkopf
to overhang überhängen, überkragen
overhaul Überholung, Instandsetzung
to overhaul überholen, instandsetzen, nachsehen; überholen, vorbeifahren an
overhauling Überholung, Instandsetzung
overhead obenliegend, überkopf (liegend); Decken-...; Gesamt-..., General-...
overhead bit informationsloses Bit
overhead conductor Oberleitung, Fahrleitung (621.33)
overhead contact rail obenliegende Stromschiene
overhead contact system Fahrdrahtsystem, Oberleitungssystem
overhead crane Laufkran
overhead electric hoist Elektrozug (621.387)
overhead electric traveller Elektrolaufkran
overhead line Freileitung (621.3), Fahrleitung, Fahrdraht, Oberleitung (621.33);
overhead line system Freileitungsanlage, Freileitungsnetz
overhead motion Transmission (621.9)
overhead rail collector Fahrdraht-Stromabnehmer
overhead railway Seilschwebebahn
overheads Gemeinkosten
overhead system Fahrdrahtanlage (621.33)
overhead transmission line Freileitung (621.315)
overhead welding Überkopfschweißung
overhead wire Freileitungsdraht (621.315); Oberleitungsdraht (621.88)
to overhear abhören, mithören (621.395)
to overheat überhitzen; sich zu stark erhitzen, zu heiß werden
overhung portion überstehender Teil, Überhang; Ausladung
overhung-type alternator Außenpolgenerator (621.313)
overlap Überlappung, Überdeckung; Profilüberdeckung (626.833); [Betriebsart, bei der mehrere Tätigkeiten gleichzeitig ablaufen (681.3)]
overlap test Spannungsabfallmessung (621.315.2)
to overlap sich überlappen, sich überdecken, sich überschneiden, sich überlagern; überlappen, überlappt anordnen
overlay Transparentauflage; Overlay, Überlagerung (681.3)
overlay network überlagertes Netz (621.395)
to overlay drüberschichten, darüberlagern; sich darüberlegen
overload Überlast, Überbelastung; Überstrom (621.3)
overload capacity Überlastbarkeit, Überlastbarkeitsgrenze
overload circuit breaker Überlastschalter, Maximalschalter, Überstromschalter
overload circuit breaking Überstromauslösung
overload coil Überstromspule
overload control Überlastregelung
overload current Überstrom
overload level Grenzpegel (621.395); Arbeitsbereichsgrenze, zulässige Nutzleistung, Aussteuerungsgrenze
overload protection Überlastschutz, Überstromschutz
overload relay Überstromrelais, Maximalrelais
overload release Überstromauslöser, Maximalauslöser
overload and surge current capability Überstrom- und Stoßstromverhalten
overload switch Überstromschalter, Maximalschalter

overload temperature relay thermisches Überlastrelais
overload trip Überstromautomat, Überstromauslöser
overload tripping Überstromauslösung, Maximalauslösung
overload valve Überdruckventil
to overload überlasten; überbelasten (621), (621.3); übersteuern (621.396)
overmeasure Übermaß
to overmodulate übermodulieren
overmodulation Übermodulation
overneutralization Überneutralisierung
overpotential = overvoltage
overshoot factor Überschwingfaktor
oversize über das Normalmaß, Übermaß
overpower protection Überlastschutz
to overpower überwiegen gegen; überwältigen
overpressure Überdruck, zu hoher Druck
to overrev „überdrehen", Drehzahl zu hoch treiben
to override überdecken, zudecken, übertönen
to overshadow überschatten, überdecken; überstrahlen, verdecken
overshoot Regelungsüberschreitung, Überschwingen
to overshoot überschießen, überschwingen
overshooting Überregelung
overshooting of a lamp Einschaltstoß einer Lampe
to oversize überdimensionieren
overspeed Überdrehzahl, Durchgehen
overspeed device Schnellschluß
overspeed protection Durchgehschutz, Schnellschlußschutz
overspeed trip Schnellschlußauslösung
to overspeed überdrehen, zu schnell laufen lassen
overspeeding Durchgehen
to overstate übertreiben, zu stark darstellen; zuviel anzeigen (621.317)
to overstretch überdehnen
overstrung zu stark angespannt; kreuzsaitig (681)
over-the-horizon radio system Überhorizontverbindung
overtone Oberton (534), (621.3); Oberschwingung (534)
overtone crystal Oberschwingungsquartz
to overturn umkippen, umstürzen, stürzen
overturning moment Kippmoment
overvoltage Überspannung
overvoltage circuit breaker Überspannungsschalter
overvoltage protection Überspannungsschutz (621.313)
overvoltage protective device Überspannungsschutzeinrichtung (621.313)
overvoltage relay Überspannungsrelais (621.313)
overvoltage suppressor Überspannungsableiter
overvoltage tripping Überspannungsauslösung
overwater propagation Ausbreitung über Wasser
overweight Übergewicht
to overwind überdrehen, zu weit aufziehen (681)
to overwrite überschreiben
oviform eiförmig
ovoid eiförmiger Gegenstand, Ei; Ovoid (51); Eierbrikett
ovonic amorpher Halbleiter
O-wave Grundwelle
owner's manual Bedienungsanweisung, Gebrauchsanweisung
oxalic acid Oxalsäure
oxidation Oxidation
oxidation inhibitor Oxidationsschutzmittel
oxidation resistance Oxidationsbeständigkeit
oxidation state Oxidationsstufe
oxidative oxidierend, oxidbildend
oxide Oxid
oxide-coated mit Oxid bedeckt, Oxid-...
oxide-coated cathode Oxidkathode
oxide-coated emitter Oxidkathode
oxide-coated filament Oxidkathode
oxide film Oxidfilm, Oxidhaut, Oxidschicht
oxides of nitrogen nitrose Gase, Stickoxide
oxidization Oxidation
to oxidize oxidieren
oxy-acethylene Azetylen-Sauerstoff
oxy-acethylene cutting torch Azetylen-Schneidbrenner
oxy-acethylene welding Azetylenschweißung
oxy-arc cutting Sauerstoffschneiden
oxygen Sauerstoff
oxygen compound Sauerstoffverbindung
oxygen cutter Schneidbrenner
oxygen cylinder Sauerstoffflasche
oxygen gas Sauerstoffgas
oxygenous sauerstoffhaltig
oxy-hydrogen flame Knallgasflamme
oz. = ounce Unze (**1 oz** = 28,3495 g)
ozone Ozon
ozonizer Ozonisator

P

PABX = private automatic branch exchange automatische Nebenstellanlage
pace Schritt, Schrittlänge (etwa 30 Zoll =75... 80 cm); Tempo
pack Packen, Stapel, Stoß; Packeis; Menge; Satz
to pack packen, verpacken, einpacken; sich verpacken lassen; stopfen, Gleis stopfen (625.1); verdichten (681.3)
package Paket, Gerätebaugruppe; Gehäuse
package density Packungsdichte
packet Paket, Päckchen
packet of energy Energiepaket
packet switching Teilstreckentechnik mit paketweiser Übertragung (621.394)
packet-type switch Pacco-Schalter
packing Packung, Dichtung; Backen (pl.) (621.395.61); Packmaterial, Verpackung
packing block Beilage, Unterlegstück
packing density Packungsdichte
packing of the line Bettung, Schotterbett
packing material Verpackungsmaterial, Packmaterial; Dichtungsmaterial
packing piece Packung, Zwischenlage, Beilage, Futter, Futterstück
packing ring Packungsring, Liederungsring
pad Dämpfungsglied
padder Trimmer, Paddingkondensator (621.396)
padder capacitor Trimmerkondensator, Paddingkondensator, Gleichlaufkondensator (621.396)
padding Fütterung, Wattierung; Padding (621.396)
padding condenser Paddingkondensator, Gleichlaufabstimmkondensator (621.396)
pagefax Pressefax
page [Speicherblock innerhalb des internen Speichers]
page printer Blattschreiber
page-at-a time printer Seitendrucker
paint Farbe, Farbanstrich, Anstrich; Bildspur (621.397)
paint brush Malerpinsel
paint coat Farbanstrich
paint-spray gun Farbspritzpistole
pair Paar, Doppelleitung (621.3); Adernpaar, Doppelader
pair of compasses Zirkel
pair of forceps Pinzette
pair gearing Paarverzahnung
pair generation Paarbildung (539)
pair of leads Leitungspaar, Doppelleitung
pair of poles Polpaar
pair of scales Waage
pair of scissors Schere
pair of tongs Zange
pair of trams Stangenzirkel
pair production Paarbildung (539)
pair of wires Aderpaar
to pair paaren, paarig abteilen, zu zweien gruppieren
paired cable paarverseiltes Kabel, mehrpaariges Kabel
pairing Paarigkeit (der Zeilen) (621.397)
panhead Flachkopf (621.884)
pancake coil Flachspule, Scheibenspule (621.3)
panchromatic panchromatisch
to pane over umnieten
panel Tafel (einer Täfelung), Feld, Brett, Wand, Verkleidung, Fach, Rahmen; Schalttafel, Frontplatte
panel brace Schalttafelstütze
panel dial system Flachschienenwählersystem
panel-frame mounting Schalttafeleinbau (621.316.5)
panel framework Schalttafelrahmen
panel heating Wandheizung, Strahlungsheizung
panel instrument Schalttafelinstrument
panel loads (pl.) Dachlast, Bodenlast
panel meter Schalttafelinstrument
panel mounting Schalttafeleinbau, Schalttafelaufbau
panel-mounting meter Schalttafelinstrument
panel switch Schalttafelschalter
panel switchboard Schalttafel
panic bolt Panikriegel
panic lighting Panikbeleuchtung
panoramic radar Weitwinkelradar
panoramic receiver Panoramaempfänger (621.396)
pantograph Pantograph, Storchschnabel; Stromabnehmer
pantograph collector Scherenstromabnehmer, Pantographstromabnehmer (621.33)
pantograph trolley Scherenstromabnehmer
paper Papier
paper board Pappe
paper capacitor Papierkondensator, Wickelkondensator
paper dielectric Papierisolation
paper-dielectric condenser Papierkondensator
paper feed Papiervorschub
paper-insulated papierisoliert

paper-insulated cable papierisoliertes Kabel, Papierkabel
paper-insulated lead covered cable = PILC cable Papierbleikabel
paper insulation Papierisolation
paper interleaver Papiereinlegevorrichtung
paper lining Papierauskleidung, Papierfutter
paper-tape punch Lochstreifenstanzer
paper-tape reader Lochstreifenleser
paper-tape reperforator Streifenlocher (621.394)
paper-tape store Lochstreifenspeicher
paper throw Papiervorschub
paper winder Papieraufspulvorrichtung
papier mâché (Ausspr. franz.) Papiermaché
PAR = precision approach radar
parabola Parabel
parabolic parabolisch
parabolic aerial Parabolantenne
parabolic mirror Parabolspiegel
parabolic range parabelförmiger Bereich
parabolic reflector Parabolreflektor, Parabolspiegel
paraboloid Paraboloid
paraboloid of revolution Rotationsparaboloid, Umdrehungsparaboloid
paraffin Paraffin
paraffin paper paraffiniertes Papier
paraffined paraffiniert
paraffined paper paraffiniertes Papier
paraffined wire Wachsdraht
parallax Parallaxe
parallel Parallele (51); Breitengrad (91)
parallel of latitude Breitenkreis, Breitengrad (91)
parallel parallel
parallel addition Paralleladdition
parallel access Parallelzugriff
parallel capacitor Parallelkondensator, Überbrückungskondensator
parallel by character zeichenparallel
parallel circuit Parallelstromkreis, Parallelresonanzkreis
parallel code Parallelcode
parallel combination Parallelschaltung
parallel conductor line Paralleldrahtleitung
parallel connection Parallelschaltung
parallel distributor Fahrdraht-Speiseleitung
parallel end gauge Endmaß
parallel faced planparallel
parallel feeder Mehrfachspeiseleitung
parallel to fibre parallel zur Faser
parallel-jaw vice Parallelschraubstock
parallel line protection Doppelleitungsschutz
parallel notch Parallelschaltstufe (621.33)
parallel operation Parallelbetrieb
parallel phase resonance Parallelresonanz
parallel printer Paralleldrucker, Zeilendrucker
parallel resistance Parallelwiderstand, Nebenwiderstand
parallel resonance Parallelresonanz, Stromresonanz
parallel resonant circuit Parallelschwingkreis, Sperrkreis
parallel ruler Parallelen-Lineal
parallel running of alternators Parallelbetrieb von Generatoren (621.311)
parallel-search storage Assoziativspeicher
parallel-serial converter Parallel-Serien-Umsetzer
parallel series Reihenparallel-...
parallel-slide valve Parallelschieber
parallel slots Parallelnuten (621.313)
parallel thread Flachgewinde
parallel wire line Paralleldrahtleitung, Lecherleitung
to parallel parallelschalten; in Parallele setzen (**with:** zu); parallel liegen zu
paralleling reactor Ausgleichsdrossel (621.314.2)
paralleling switch Synchronisierungsschalter
parallelepiped Parallelepiped
parallelism Parallelismus, Analogie
parallelogram Parallelogramm
parallelogram of forces Parallelogramm der Kräfte
paramagnetic paramagnetisch
paramagnetic material paramagnetisches Material
paramagnetism Paramagnetismus
parameter Parameter, Einflußgröße
parametric amplifier parametrischer Verstärker
parametric form Parameterdarstellung
paranthesis runde Klammern (51)
paraphase amplifier Symmetrieverstärker, Phasenumkehrverstärker
parasites wilde Schwingungen
parasitic parasitär, Stör-...
parasitic aerial strahlungsgespeiste Antenne, strahlungserregte Antenne; Sekundärstrahler
parasitic capacitance parasitäre Kapazität
parasitic capture störender Einfang, Wegfang (539)
parasitic current Fremdstrom
parasitic element strahlungserregter Teil, strahlungsgekoppelter Teil
parasitic emission parasitäre Strahlung

parasitic lobe Nebenzipfel, Störzipfel (621.396.677)
parasitic noise Rauschen
parasitic oscillation wilde Schwingung, Störschwingung, parasitäre Schwingung
parasitic reflector Störreflektor
parasitic stopper Schaltung zur Unterdrückung parasitärer Schwingungen
parasitically excited strahlungserregt, strahlungsgekoppelt
paravane Otter (623)
to pare beschneiden
to pare leather Leder anschärfen
parent Ausgangs-..., Grund-...
parent atom Mutteratom, Ausgangsatom
parent compound Ausgangsverbindung
parent field ursprüngliches Feld, Ausgangsfeld
parent metal Grundmetall (621.791)
parent nucleus Ausgangskern
parent section Hauptteil
parent substance Muttersubstanz, Ausgangssubstanz
paring Abschnitt, abgeschnittenes Stück
parity Parität (681.3)
parity bit Paritätsbit
parity check Vollständigkeitsprüfung, Paritätskontrolle
to park on busy warten auf Freiwerden
part Teil, Abschnitt; Teil, Anteil
part-failure rate Teilausfallrate
one part in a million ein Millionstel
three parts of alcohol 3 Teile Alkohol
part by volume Raumanteil, Raumteil
part by weight Gewichtsteil
to part abtrennen, abstechen (621.94); sich trennen, auseinandergehen
partial teilweise, anteilig, partiell, Teil-..., Partial-...
partial barring Teilsperre (621.395)
partial derivate partielle Ableitung
partial differential partielles Differential
partial differential equation partielle Differentialgleichung
partial diffusion unvollständige Streuung (535)
partial disconnection Teilabschaltung; zeitweise Abschaltung
partial earth Erdschluß über Widerstand, unvollkommener Erdschluß
partial eclipse partielle Finsternis (52)
partial failure Teilausfall
partial field Teilfeld
partial fraction Partialbruch
partial fraction expansion Partialbruchzerlegung
partial plating Teilgalvanisierung
partial power Teilleistung
partial pressure Partialdruck, Teildruck (533)
partial short Teilkurzschluß
partial vacuum Unterdruck
partial view Teilansicht
partial winding Teilwicklung
partially restricted extension halbamtsberechtigte Nebenstelle
particle Teilchen, Partikel, Korpuskel
particle accelerator Teilchenbeschleuniger
particle capture Teilcheneinfang
particle counter Teilchenzähler
particle-flux density Teilchenflußdichte (539)
particle velocity Teilchengeschwindigkeit
particoloured vielfarbig, mehrfarbig, bunt
particular besonders, eigen, kennzeichnend; Sonder-..., Spezial-...
particular integral partikuläres Integral
particular solution partikuläre Lösung
partition Trennwand, Zwischenwand; Verschlag
partition wall Trennwand, Zwischenwand
to partition abteilen, abtrennen
partly teilweise, Teil-...
partly closed slot halbgeschlossene Nut, halboffene Nut
parts list Stückliste, Schaltteilliste
party Teilnehmer, Gesprächsteilnehmer (621.396)
party called angerufener Teilnehmer
party calling anrufender Teilnehmer
party line Leitung für Gemeinschaftsanschluß, Gesellschaftsanschluß (621.395)
party-line ringing Morseruf
Paschen's law Paschensches Gesetz
pass Stich, Zug, Arbeitsgang, Durchgang (621.9)
password Kennwort (681.3)
to pass between passen zwischen, hindurchgehen zwischen, sich durchschieben lassen zwischen; zwischen ... durchlaufen
to pass current Strom durchlassen
to pass current freely Strom glatt durchlassen
to pass a current through einen Strom durchschicken durch
to pass into solution in Lösung gehen
to pass through a minimum ein Minimum durchlaufen
to pass through a point durch einen Punkt gehen
passage Durchgang; Durchlaß, Kanal
passage of current Stromdurchgang
passage of electricity through gases Elektrizitätsdurchgang durch Gase, Gasentladung

passage of heat Wärmedurchgang, Wärmeübergang
passband Durchlaßbereich (621.329)
passband characteristic Durchlaßcharakteristik
passenger carrying aircraft Personenflugzeug, Verkehrsflugzeug
passenger lift Personenaufzug
passing contact Wischkontakt
passing loop Überholungsgleis (625.1)
passing through zero Nulldurchgang
passivation layer Passivierungsschicht
passive passiv (669), (621.3)
passive aerial Sekundärstrahler, passiver Strahler
passive electric network passive Schaltung
passive quadripole passiver Vierpol
passive reflector passiver Reflektor, passiver Strahler
passive reflector antenna Rückstrahlantenne
passive repeater passive Relaisstelle
passive state passiver Zustand, Ruhezustand
passive transducer passiver Vierpol
past cut-off ins Sperrgebiet hinein (621.385)
paste Paste, Teig, Brei
pasteboard Pappe
to paste bekleben, kleben; pastieren, schmieren (621.355)
pasted plate Pasteplatte, pastierte Platte (621.355)
patch Korrekturbefehl, Korrekturroutine (681.3)
to patch stöpseln, rangieren
patchboard Schalttafel (621.3)
patch cord Rangierschnur
patch panel Stecktafel
patchy fleckig
patent Patent
patent application Patentanmeldung
patent evasion Patentumgehung
patent infringement Patentverletzung
patent interference (US) Übergreifen der Patentansprüche
patent office Patentamt
patent-right claim Patentanspruch
patent specification Patentbeschreibung, Patentschrift
path Pfad, Weg
path attenuation Übertragungsdämpfung, Ausbreitungsdämpfung
path control Bahnsteuerung
path of current Strompfad, Strombahn
path difference Wegunterschied, Wegdifferenz (621.396)
path finding Wegesuchen (621.395)
path-finding section Wegesuchabschnitt
path of integration Integrationsweg
path length Weglänge, Funkfeldlänge
path loss Ausbreitungsdämpfung, Funkfelddämpfung
path profile Geländeschnitt
path of propagation Ausbreitungsweg
path spin Bahndrehimpuls (53)
pattern Muster; Muster, Modell; Bauart; Modell, Gußmodell; Charakteristik, Diagramm; Schirmbild (621.385)
pattern detection Zeichenerkennung
pattern generator Bildmustergenerator (621.397)
pattern making Modelltischlerei, Modellbau, Modellanfertigung
pattern miller Kopierfräsmaschine
to pattern mit einem Muster versehen
Pauling furnace Pauling-Stickstoffofen
pawl Klinke, Sperrklinke, Sperrhaken
Pawsey stub Symmetrierleitung
PAX = private automatic exchange teilnehmereigene Fernsprechanlage mit Wählbetrieb
pay-off reel Ablaufspule, Ablaufhaspel
pay station Münzfernsprechstelle
PBX = private branch exchange Nebenstellenanlage
p. c. = per cent v. H. = vom Hundert = Prozent
PC board (gedruckte) Leiterplatte
P-channel field effect transistor P-Kanal-Feldeffekttransistor, P-Kanal-FET
P-channel enhancement mode MOSFET P-Kanal MOSFET vom Anreicherungstyp
P. C. D = pitch circle diameter Lochkreisdurchmesser; Teilkreisdurchmesser
PCM = pulse code modulation Pulscodemodulation
p. d. = potential difference Potentialdifferenz, Spannung
PDM = pulse duration modulation Pulsdauermodulation
peak Spitzenwert, Scheitelwert, Amplitude, Höchstwert; Spitze
peak amplitude größte Amplitude, Spitzenwert
peak clipper Spitzenbegrenzer
peak current Spitzenstrom, Stromspitze
peak of a cycle Scheitelwert einer Periode
peak-data transfer rate maximale Datenübertragungsgeschwindigkeit
peak detector Spitzengleichrichter
peak envelope power Spitzenleistung (621.369.61)
peak factor Scheitelfaktor

peak forward gate current Vorwärts-Spitzenstrom, Spitzenwert des Stromes in Durchlaßrichtung
peak forward voltage Spitzenwert der Spannung in Durchlaßrichtung
peak inverse voltage Scheitelwert der Sperrspannung
peak limiter Amplitudenbegrenzer, Begrenzer
peak load Lastspitze, Spitzenlast
peak-making current Scheitelwert des Einschaltstroms
peak meter Spitzenzähler (621.317)
peak output Spitzenleistung
peak-to-peak größte Spannungsschwankung (621.396)
peak plate current Anodenspitzenstrom
peak point Gipfelpunkt, Höckerpunkt
peak power Spitzenleistung (621.397)
peak programm meter Aussteuerungsmesser (654.19)
peak reverse recovery current Sperrverzögerungsstromspitze, Richtstromspitze (621.314.63)
peak transient reverse voltage Stoßspitzensperrspannung
peak to valley point current ratio Gipfel-Tal-Stromverhältnis
peak value Spitzenwert, Scheitelwert, Sprechspitze, Sprachspitze
peak value of turn-off loss Ausschaltverlustleistungsspitze (621.314.63)
peak value of turn-on loss Einschaltverlustleistungsspitze
peak voltage Spitzenspannung, Spannungsspitze
peak voltmeter Scheitelspannungsmesser, Spitzenspannungsmesser
peak white Weißspitze
peak white level Weißwert (621.397)
to peak anspitzen, mit einer Spitze versehen, zuspitzen
peaked spitz, angespitzt, zugespitzt
peaked wave spitze Kurvenform
peaker Anspitzschaltung, Verteilerstufe
peaking Verstärkungsüberhöhung
peaking circuit Impulsverteilerschaltung
peaking coil Spule für Anhebung der oberen (Video-)Frequenzen
peal Klang, Glockenklang
to peal klingen
pear Birne
peardrop Reflektorballon
pear push Hängedruckknopf, Durchknopfbirne
pear switch Schalterbirne, Hängeschalter, Schnurschalter
pearl Perle
pearl lamp mattierte Lampe, Opallampe
PEARL = processor and experiment automation realtime language
pearlite Perlit
pearlitic perlitisch
peck = 2 gal (Hohlmaß, 1 pK = 9,092 l)
pedal Pedal
pedal brake Fußbremse
pedal circuit Schienenstromkreis
pedestal Untersatz, Fuß, Gestell; Schwarzabhebung (621.397); Sockelimpuls
pedestal-type grinding machine Schleifbock
to peen over umnieten, umbördeln
peep hole Schauloch, Guckloch
peg Pflock, Zapfen; Holzhaken; Anschlagstift, Endanschlag; Knebel
pegboard Stecktafel
to peg pflocken
pellette Schicht, Häutchen
pellicle Häutchen
Peltier coefficient Peltier-Koeffizient
Peltier cooling block Peltier-Kühlblock
Peltier effect Peltier-Effekt
Pelton runner Peltonrad
Pelton wheel Peltonrad
pencil Bleistift; Strahlenbündel (535)
pencil beam scharfgebündelter Strahl, Nadelstrahl
to pend hängen, anhängig sein
pendant hängend, Hänge-...; Hängelampe, Pendel, Pendelleuchte
pendant cord Pendelschnur
pendant light Hängeleuchte, Pendelleuchte
pendant push Hängedruckknopf, Druckknopfbirne
pendant socket outlet Hängedose, Hängesteckdose
pendant switch Hängeschalter, Schnurschalter
pendulous pendelnd, Pendel-...
pendulum Pendel
pendulum meter Aronscher Pendelzähler
to pene behämmern, hämmern
to pene over umhämmern, umbördeln
to penetrate durchdringen, eindringen; eintauchen; durchgehen
to penetrate to a depth bis zu einer Tiefe eindringen
to penetrate through durchdringen, durchgehen
penetrating durchdringend

penetrating cosmic radiation durchdringende kosmische Strahlung, (harte) Höhenstrahlung
penetrating power Durchdringungsfähigkeit, Eindringvermögen
penetration Durchdringen, Eindringen, Eindringtiefe; Durchdringung (51); Durchspülen, Durchspülung; Einschneiden (621.9); Einbrand (621.791)
penetration of current Eindringtiefe des Stromes
penetration of matter Durchdringung von Materie
penetration of moisture Durchschlagen von Feuchtigkeit (693)
penetration below optical horizon Reichweite über die optische Sicht
penetration voltage Durchgreifspannung, Sperrschichtberührungsspannung
penetrometer Härtemeßgerät
penlight battery Mignonzelle, Mignonbatterie
penny-in-the-slot Automat, Pennyautomat, Groschenautomat
pennyweight (Gewichtseinheit = 1/20 Unze)
pentagon Fünfeck, Pentagon
pentagonal fünfeckig
pentagram Pentagramm, Drudenfuß
pentagrid Fünfgitterröhre, Heptode
pentagrid converter Mischheptode
pentane Pentan
pentatron Doppeldreipolröhre
pentode Pentode, Fünfpolröhre
pentode-mixer tube Fünfpolmischröhre
percent-modulation indicator Aussteuerungsanzeiger (621.396)
percent ripple prozentuale Kräuselspannung
percentage Prozentsatz, Hundertsatz
percentage of accuracy prozentualer Fehler (621.317.8)
percentage differential relay Prozentdifferentialrelais
percentage error prozentualer Fehler (621.318)
percentage of modulation Modulationsgrad, Aussteuerungsgrad
percentage of moisture Feuchtigkeitsgehalt inv. H., prozentuale Feuchtigkeit, relative Feuchte
perch Stange (Längenmaß = 5 1/2 yd. = 5,029 m)
to percolate durchtropfen, durchtropfen lassen
percolation Durchtropfen
percussion Schlag, Erschütterung, Aufschlag; Aufschlagzündung
percussion drilling machine Schlagbohrmaschine
percussive welding Stoßschweißen
perfect vollkommen, ideal; vollständig, gut, vollendet
perfect alignment saubere Ausrichtung (621.1); guter Gleichlauf (621.396)
perfect circle idealer Kreis, vollständiger Kreis, sauberer Kreis
perfect combustion vollständige Verbrennung
perfect condenser verlustloser Kondensator, idealer Kondensator, winkelfreier Kondensator
perfect crystal Idealkristall
perfect dielectric ideales Dielektrikum, verlustfreies Dielektrikum
perfect diffusion vollständige Streuung (535)
perfect gas ideales Gas
perfect metal ideales Metall
perfect rectifier idealer Gleichrichter
perfect square sauberes Quadrat; vollständiges Quadrat
perfectly inductive rein induktiv
perfectly screened völlig abgeschirmt
to perforate perforieren, lochen
perforated metal screen Metallsieb, Lochblech (621); gelochte Metallfolien-Abschirmung (621.315.2)
perforated plate Lochblech, Siebblech
perforation Perforation, Lochung, Durchlochung; Lochfraß (621.315.2), (696)
to perform ausführen, durchführen
to perform a cycle einen Kreisprozeß durchlaufen, eine Periode durchlaufen
to perform an integration eine Integration ausführen
to perform work Arbeit verrichten
performance Verhalten, Vorführung, Darbietung, Aufführung, Vorstellung; Leistung (621); betriebliches Verhalten
performance capabilities Leistungsfähigkeit, Leistungsumfang
performance characteristic Arbeitskenngröße, Leistungsmerkmal
performance curve Alterungsdiagramm (621.32)
perimeter Umfang, Umfangslänge
period Periode, Zeitabschnitt; Satz; Punkt
1/2 period Halbperiode (621.3); Halbwertszeit (539.16)
period of contact Kontaktdauer (621.3); Eingriffszeit (621.835)
period of half-life Halbwertszeit (539.16)
period of an instrument Schwingungsdauer (621.317)
period resonance Resonanz mit der Eigenschwingung

period of rest Betriebspause, Stillstandszeit
period rise Anlaufzeit (539)
period of rotation Umlaufzeit (52)
period of vibration Schwingungsdauer, Schwingungszeit
periodic(al) periodisch
periodic current periodischer Strom, periodischer Wechselstrom
periodic damping periodische Dämpfung, unteraperiodische Dämpfung
periodic duty Aussetzbetrieb, AB
periodic electromotive force periodische EMK
periodic function periodische Funktion, Kreisfunktion
periodic line Kettenleiter
periodic potential difference periodische Spannung, periodische Potentialdifferenz
periodic quantity periodische Größe
periodic rating Nennleistung für Aussetzbetrieb, AB-Leistung
periodic ring Weiterruf
periodic table Periodensystem (54)
periodic test periodische Überprüfung (621.317)
periodic time Periodendauer, Schwingungsdauer (621.317)
periodic voltage periodische Spannung
periodic wave periodische Welle
periodical line displacement Zeilenversatz
periodicity Periodizität; Frequenz, Periodenzahl; Eigenfrequenz
peripheral Umfangs-...; Peripherie...; peripher
peripheral device Peripheriegerät
peripheral electron Elektron der äußeren Schale
peripheral leakage Luftspaltstreuung (621.313)
peripheral speed Umfangsgeschwindigkeit
peripheral unit Ein-Ausgabegerät, Peripheriegerät
peripheric peripher
periphery Peripherie, Umfang; [Ein-Ausgabegeräte, Speicher und andere Geräte außerhalb des Rechners (681.3)]
periscope Periskop, Sehrohr
perlon Perlon
permalloy Permalloy
permanence Konstanz, Gleichmäßigkeit, gleichmäßiges Andauern
permanence of value zeitliche Konstanz (621.317)
permanency Permanenz; Haltbarkeit; zeitliche Konstanz
permanency of setting bleibende Einstellung, konstante Einstellung
permanent dauernd, permanent, Dauer-...
permanent circuit Standleitung
permanent deformation bleibende Verformung
permanent discharge dauernde Stromentnahme, ununterbrochene Entladung (621.353), (621.356)
permanent echo Festzeichen (621.397)
permanent-echo cancellation Festzeichenunterdrückung, Festzeichenaustastung
permanent extension bleibende Dehnung (539)
permanent ground Dauererdschluß
permanent load Dauerbelastung
permanent magnet Dauermagnet, Permanentmagnet
permanent-magnet moving-coil instrument Drehspulinstrument
permanent-magnet moving-coil loudspeaker permanent-dynamischer Lautsprecher
permanent-magnet moving-iron instrument polarisiertes Dreheiseninstrument
permanent-magnet speaker permanent-dynamischer Lautsprecher
permanent magnetic field permanentes magnetisches Feld
permanent magnetism permanenter Magnetismus
permanent memory permanenter Speicher
permanent noise Dauergeräusch
permanent service Dauerbetrieb, durchgehender Betrieb
permanent set bleibende Verformung, dauernde Dehnung, bleibende Dehnung
permanent store permanenter Speicher
permanent wiring festverlegte Installation
permanently tuned fest abgestimmt
permeability Permeabilität, magnetische Durchlässigkeit (538); Durchlässigkeit (**to:** für)
permeability tuning Permeabilitätsabstimmung, induktive Abstimmung
permeable permeabel; durchlässig (**to:** für)
permeameter Magnetprüfgerät
permeance magnetischer Leitwert, A_L
perminvar Isoperm, Perminvar
permissible zulässig, zugelassen
permissible contact duty zulässige Kontaktbelastung
permissible deviation zulässige Abweichung
permissible limit zulässige Grenze
permissible stress zugelassene Beanspruchung, zulässige Beanspruchung

permissive block permissiver Block, bedingtes Halt (625)
permittance Dielektrizitätskonstante, D. K.
permittivity Verschiebungszahl, Dielektrizitätskonstante, D. K.
permutation Permutation (51)
permutator Kommutatorgleichrichter
permutite Permutit
peroxide Superoxid, Wasserstoffsuperoxid
to per-oxidize höher oxidieren
perpendicular Lot; Senkrechte, Höhe (51)
out of the perpendicular aus dem Lot, nicht senkrecht
perpendicular senkrecht, lotrecht
perpendicular line Senkrechte, Lot (51)
perpendicular magnetization magnetische Queraufzeichnung (681.84)
perpendicular projection Parallelprojektion, Parallelperspektive
perpendicularity Senkrechtstehen, Lotrechtsein
perpetual dauernd, andauernd
perpetual motion machine Perpetuum mobile
persistence Anhalten, Andauern; Nachleuchten, Nachleuchtdauer (621.385); Zeitkonstante (Induktivität/Widerstand)
persistent dauernd, andauernd, Dauer-...; nachleuchtend (621.385)
persistent alarm Dauermeldung
persistent resonator schwach gedämpfter Resonator
persistent screen Nachleuchtschirm
personal alerter Alarmempfänger
Personal Computer [persönlicher Computer, Tisch-Computer]
personal equation Reaktionszeit, Schrecksekunde
person-to-person call Voranmeldegespräch
perspective Perspektive
perspex cover Transparentabdeckung
PERT = programm evaluation and revue technique PERT-Planungstechnik, PERT-Netzplantechnik
pertinax Pertinax
pertinent einschlägig; zusammenhängend (**to:** mit)
pertinent data (einschlägige) Unterlagen
perturbation method Methode der kleinen Störungen
perveance Raumladungskonstante
pervious durchlässig (**to:** für)
pervious stratum wasserdurchlässige Schicht
Petersen coil Petersenspule, Erdschlußspule
petrographic petrographisch
petrography Petrographie
petrol Treibstoff, Kraftstoff, Benzin
petrol-electric car benzin-elektrisches Fahrzeug
petrol-electric generating set Benzingenerator
petroleum Petroleum; Steinöl, Mineralöl, Erdöl
petticoat (Isolator-)Schirm (621.315.6)
petticoat insulator Schirmisolator (621.315.6)
pewter Weißmetall (621); Zinnwaren (pl.)
P. F. = power factor Leistungsfaktor
p-gate thyristor kathodenseitig steuerbarer Thyristor
pH pH(-Wert)
phanotron Glühkathoden-Quecksilberdampfgleichrichter
phantastron Phantastron
phantom Phantom
phantom aerial Phantomantenne
phantom circuit Phantomkreis, Phantomschaltung, Phantomleitung
phantom-circuit loading coil Phantomspule
phantom-circuit repeating coil Phantomkreis-Übertrager
phantom-repeating coil Phantomkreisübertrager, Viererübertrager
pharosage Beleuchtungsstärke
phase Phase, Zustand; Mondphase (52); Phase, Phasenlage (621.3); Phase, Phasenwicklung, Strangwicklung (621.3)
phase advancer Phasenschieber, Blindleistungsmaschine (zur Verbesserung des Leistungsfaktors) (621.313)
phase angle Phasenwinkel; Fehlwinkel, Phasenfehler, Winkelfehler (621.317)
phase-angle control (Phasen-)Anschnittsteuerung
phase-angle correction factor Phasenfehlerkorrektur (621.317)
phase-angle error Winkelfehler (621.317)
phase-angle meter Phasenwinkelmesser
phase balance Phasengleichheit
phase-balance relay Phasenrelais; Unsymmetrierelais
phase break Phasentrennung, Phasenstoß
phase comparison Phasenvergleich
phase-comparison circuit Phasenvergleichschaltung
phase conductor Phasenleiter, Außenleiter
phase constant Phasenbelag, Kettenwinkelmaß (621.392)
phase contrast Phasenkontrast (535)
phase-controlled rectifier phasengesteuerter Gleichrichter
phase converter Phasenschieber (621.313)
phase current Außenleiterstrom

phase delay Phasenlaufzeit, Phasenverzögerung
phase-delay distortion Phasenverzerrung, Laufzeitfehler
phase deviation Phasenhub
phase-deviation meter Phasenhubmesser
phase diagram Zustandsschaubild, Zustandsdiagramm (66)
phase difference Phasenunterschied, Phasenverschiebung
phase discriminator Phasendiskriminator
Phase displacement Phasenverschiebung, Phasenabweichung
phase distortion Laufzeitverzerrung (621.392)
phase error Phasenfehler
phase equalization Laufzeitentzerrung
phase factor Phasenverzerrung (621.375), Leistungsfaktor
phase fading Schwund, Interferenzschwund
phase failure Spannungsrückgang; Leitungsbruch
phase-failure protection Phasenbruchrelais
phase-frequency distortion Phasenverzerrung
in phase in Phase
phase indicator Phasenanzeiger
phase inversion Phasenumkehr
phase-inverter circuit Phasenumkehrschaltung
phase inverter (tube) Phasenumkehrröhre
phase inverting Phasenumkehrung
phase jitter Phasenzittern
phase lag nacheilende Phasenverschiebung, Phasennacheilung
phase lamp Phasenlampe
phase lead Phasenvoreilung
phase locked phasenstarr
phase-locked loop Phasenregelschleife, PLL-Schaltkreis
phase margin Phasensicherheit (gegen Selbsterregung) (621.375)
phase meter Phasenmesser, Leistungsfaktormeßinstrument
phase modifier (rotierender) Phasenschieber, Blindleistungsmaschine
phase modulated phasenmoduliert
phase modulation Phasenmodulation
in phase opposition in Gegenphase, in Phasenopposition
phase-to-phase voltage Dreieckspannung
phase position Phasenlage
phase quadrature 90°-Verschiebung, Quadratur
phase relation Phasenbeziehung
phase relationship Phasenbeziehung, Phasenlage
phase resonance Phasenresonanz
phase response Phasengang
phase reversal Phasenumkehr
phase-reversal protection Unsymmetrieschutz
phase-reversal relay Unsymmetrierelais
phase reversing Phasenumkehr-...
phase-reversing stage Phasenumkehrstufe, Umkehrstufe
phase rotation Phasenfolge
phase-rotation relay Phasenfolgerelais
phase-section break Phasenstoß
phase-selective rectifier Phasengleichrichter
phase sequence Phasenfolge
phase-sequence indicator Phasenfolgezeiger, Drehfeldrichtungsanzeiger
phase-sequence relay Phasenfolgerelais
phase shift Phasenverschiebung
phase-shift control Phasenanschnittsteuerung
phase-shift keying = PSK Phasenumtastung
phase-shift oscillator R-C Summer, R-C Generator
phase shifter Phasenschieber, Phasendreher, Phasenstellglied
phase-shifting transformer Drehtransformator, Phasenschiebertransformator, Schwenktransformator
phase spacing Phasenabstand
phase splitter Phasenspalter
phase-splitting device Phasenspalter
phase swinging Pendelung (621.311)
phase timing Phasenzeit; Phaseneinstellung
phase transformation Phasenumwandlung; Phasentransformierung
phase transformer Phasentransformator
phase turner Phasenschieber (621.3)
phase unbalance Phasenungleichheit; Phasenunsymmetrie
phase undervoltage Unterspannung
phase-undervoltage protection Unterspannungsschutz
phase velocity Phasengeschwindigkeit (621.329)
phase voltage Phasenspannung, Sternspannung, Strangspannung
phase voltmeter Phasenvoltmeter
phase winding Phasenwicklung, Strangwicklung
to phase in Phasenbeziehung setzen; in der Phase einstellen, einphasen
phasing Synchronisierung, Phaseneinstellung
phasing arrangement Phasenschieberschaltung

phasing coil Franklin-Antennenspule, Phasenumkehrspule
phasing-in Synchronisierung
phasing section Anpaßstück
phasor (rare) Zeiger (621.3), (538)
phenol Phenol
phenol formaldehyde resin Bakelitharz, Bakelitlack, Phenol-Formaldehyd-Kunstharz
phenolic paper Hartpapier
phenolphthalein Phenolphthalein
phon Phon
phone Telefon, Telefonhörer
on phone auf Telefonie (621.396)
on the phone telefonisch, am Telefon
phonic drum Tonrad
phonic wheel Tonrad
phono cartridge Tonabnehmersystem
phonogram Tonaufzeichnung, Tonspur (77)
phonograph Phonograph, Plattenspieler
phonograph pick-up Tonabnehmer
phonograph record Schallplatte
phonographic phonographisch
phonographic recorder Diktiergerät
phonography Phonographie
phonometer Lautstärkemesser
phonotelemetry Echolotung, akustische Entfernungsmessung
phoresis Elektrophorese
phosphor Leuchtmasse, Leuchtsubstanz, Phosphor (53)
phosphor bronze Phosphorbronze
phosphor chemical Leuchtmasse
phosphor dots Leuchtstoffpunkte
phosphor triple Farbtripel
to phosphorate phosphorieren (54)
to phosphoresce phosphoreszieren (53)
phosphorescence Phosphoreszenz
phosphorescent phosphoreszierend
phosphoric acid Phosphorsäure
phosphorous phosphorig
phosphorus Phosphor (54)
photic radiation sichtbare Strahlung
photocathode Photokathode
photochemical photochemisch
photochemical cell elektrolytisches Photoelement
photoconductive camera tube Bildaufnahmeröhre mit Photohalbleiterschicht
photoconductive cell Photowiderstand
photoconductive effect Photoleitung
photoconductivity Photoleitfähigkeit
photocoupler Optokoppler
photodiode Photodiode, Foto diode
photoelectric photoelektrisch, lichtelektrisch
photoelectric cell lichtelektrische Zelle, Photozelle
photoelectric conductivity photoelektrische Leitfähigkeit
photoelectric cryptometer photoelektrischer Deckkraftmesser
photoelectric effect Photoeffekt, photoelektrischer Effekt, lichtelektrischer Effekt
photoelectric element lichtempfindliches Teil
photoelectric emission photoelektrische Emission
photoelectric lighting control Dämmerungsschalter
photoelectric photometer elektrisches Photometer
photoelectric reader photoelektrischer Leser
photoelectric relay Lichtrelais
photoelectric tube Photozelle
photoelectric voltage Photospannung
photoelectric yield photoelektrische Ausbeute
photoelectrons Photoelektronen
photoemissive camera tube Bildaufnahmeröhre mit Photokathode
photoemissive cell Photozelle, Zelle mit Photokathode
photoemissivity Photoemission
photoetched photogeätzt
photoflash (Elektronen-) Blitzgerät
photographic photographisch
photographic copy Photokopie
photolitography Photolitografie
photoluminescence Photolumineszenz
photomasking Photomaskierung (621.382)
photometer Photometer, Luxmeter
photometer screen Photometerschirm
photometric photometrisch
photometric integrator Ulbrichtsche Kugel
photometry Photometrie
photomicrogram Mikrophotographie; mikrophotographische Aufnahme
photomultiplier (tube) Elektronenvervielfacher, Sekundärelektronenverstärker
photomultiplier cell Vervielfacher-Photozelle
photon Photon, Lichtquant, Lichtkorpuskel
photoneutron Photoneutron
photonuclear reaction Kern-Photoreaktion
photophone Lichttelefoniegerät
photophoresis Photophorese
photoprint Photokopie
photoprinting machinery Photokopiereinrichtung
photoradiogram Funkbild
photoresist Fotolack, Abdeckschutzlack
photosensitive lichtempfindlich, photoelektrisch

phototelegraphy Bildtelegraphie
phototelephone Lichttelefoniegerät
phototelephony Lichttelefonie
phototransistor Phototransistor
phototube Photozelle
photovoltaic cell Sperrschicht-Photozelle
photovoltaic effect Sperrschicht-Photoeffekt
photronic cell Photoelement
pH-value pH-Wert
physical physikalisch; praktisch, wirklich; körperlich (61); körperlich, geometrisch
physical characteristic physikalische Eigenschaft
physical circuit Stammkreis (621.394/5)
physical configuration räumliche Anordnung, geometrische Anordnung
physical constant physikalische Konstante
physical dimension geometrische Abmessung
physical explanation physikalische Erklärung
physical force Muskelkraft, körperliche Gewalt
physical impossibility physikalische Unmöglichkeit
physical law physikalisches Gesetz
physical length geometrische Länge
physical magnitude physikalische Größe
physical position räumliche Lage
physical property physikalische Eigenschaft
physical quantity physikalische Größe
physical sciences Naturwissenschaften
physical size räumliche Abmessung
physical state physikalischer Zustand
physicist Physiker
physics (pl.) Physik
physiological physiologisch
physiological effect physiologische Wirkung
physiotherapy Elektrotherapie
piano Piano, Klavier
piano steel wire Klaviersaitendraht
pick-off Aufnehmer, Fühler
to pick off abgreifen (621.317)
pickup Abgriff; Meßfühler; Abnehmer, Tonabnehmer; Einstreuung
pick-up by Einstreuung auf; Aufnahme durch
pickup arm Tonarm
pickup cartridge Tonabnehmereinsatz
pickup coil Goniometerspule (621.396); Abnehmerspule, Abgriffspule (621.317); Pickup-Spule
pickup current Ansprechstrom, Anzugstrom (621.318)
pickup head Tonabnehmerkopf
pickup magnet Magnetkopf
pickup range Auffaßbereich (621.397)
pickup spectral characteristic spektrale Aufnahmecharakteristik
pickup value Ansprechwert (621.318)
pickup voltage Ansprechspannung (621.318)
to pick-up abgreifen, abnehmen (621.317); aufnehmen, auffangen (621.396); auffassen (621.397)
picofarad Pikofarad
pictorial bildmäßig
pictorial quality Bildgüte (621.397)
picture Bild, Abbildung
picture black Bildschwarz (621.397)
picture brightness Bildhelligkeit
picture carrier Bildträger
picture channel Bildkanal (621.397)
picture element Bildelement (621.397)
picture frequency Bildwechselzahl, Bildfolgefrequenz
picture instability Bildstandschwankung (621.397)
picture monitor Bildmonitor
picture phone Fernsehtelefon
picture pulse Bildimpuls (621.397)
picture resolution Bildauflösung
picture signal Bildsignal
picture slip Bilddurchlauf (621.397)
picture strip Bildzeile (621.397)
picture-synch ratio Synchronanteil, Verhältnis Weißwert/Synchronwert
picture telegraphy Bildtelegraphie
picture tone Bildträgerfrequenz
picture transmission Bildübertragung, Faksimile-Übertragung
picture-transmission period Bildübertragungsdauer
picture transmitter Bildsender
picture tube Bildröhre, Bildwiedergaberöhre
picture and waveform monitor Bildkontrollgerät
picture white Bildweiß (621.397)
to picture bildlich darstellen
piece Stück; Abschnitt, Stück; Werkstück, Stück
all of a piece ganz aus einem Stück
to piece stückeln, anstückeln, aus Stücken zusammensetzen
to piece together stückweise aneinandersetzen, anstückeln
to piece up zusammenstückeln
to pierce durchstoßen, durchdringen; durchlöchern, lochen (621.9); stanzen (621.96)
pierced solder tag Lötöse
piercing saw Metallaubsäge, Stichsäge
piezoelectric piezoelektrisch
piezoelectric ceramics Piezo-Keramik

piezoelectric coupler Koppelquarz, Dreielektrodenquarz
piezoelectric crystal piezoelektrischer Kristall, Piezoquarz
piezoelectric effect piezoelektrischer Effekt
piezoelectric frequency stabilizer Steuerquarz (621.396)
piezoelectric frequency standard Normalquarz
piezoelectric indicator piezoelektrischer Indikator, piezoelektrischer Schwinger
piezoelectric loudspeaker piezoelektrischer Lautsprecher
piezoelectric microphone Kristallmikrophon, piezoelektrisches Mikrophon
piezoelectric oscillator Schwingquarz
piezoelectric pick-up piezoelektrischer Tonabnehmer
piezoelectric pressure gauge Druckmeßquarz, Kristalldruckabnehmer
piezoelectric relay piezoelektrisches Relais
piezoelectric resonator Quarzschwinger, Schwingquarz
piezoelectricity Piezoelektrizität
piezoquartz Piezoquarz
pig Massel, Masselrinne (66)
pig iron Roheisen, Masseleisen
pigment Pigment, Farbkörper, Farbstoff
pigtail Zopf, Anschlußlitze (621.313)
pigtail lead Lötschwanz, Lötdrähtchen
pile Pfahl, Rammpfahl, Rostpfahl; Stapel, Haufen; Halde (622); Säule (621.3); Atommeiler, Reaktor, Brenner (539)
pile driver Ramme, Pfahlramme
pile poisoning Reaktorvergiftung
piler Stapler
pillbox antenna Segmentantenne
pilot Lotse (629.12); Pilot, Flugzeugführer (629.13)
pilot brush Prüfbürste (621.313); Hilfsbürste (621.313)
pilot bulb Anzeigelampe (621.396.64)
pilot bush Führungsbuchse
pilot carrier Pilotfrequenz, Indexträger
pilot cell Prüfzelle (621.355)
pilot control Pilotüberwachung (621.395)
pilot controller Steuerschaltwalze
pilot core Hilfsader
pilot exciter Hilfserregermaschine
pilot frequency Steuerfrequenz; Pilotfrequenz (621.395)
pilot lamp Meldelampe, Kennlampe, Überwachungslampe; Notlampe
pilot light Kontrollampe
pilot lighting Notbeleuchtung
pilot motor Leitmotor, Servomotor
pilot-operated controller indirekter Regler
pilot pin Führungsstift
pilot plant Versuchsanlage
pilot production Nullserie
pilot protection Streckenschutz (621.315)
pilot reference Pilotton
pilot relay Rufrelais (621.395); Melderelais
pilot signal Meldelampe, Merklampe (621.395)
pilot spark Zündfunke
pilot-spark arc rectifier Zündfunkengleichrichter, Lichtbogengleichrichter mit Funkensteuerung
pilot synchronizing Trägererzeugung mit Hilfsfrequenz (621.396)
pilot valve Steuerventil
pilot voltmeter Netzspannungskontroll-voltmeter, Fernvoltmeter
pilot wire Hilfsdraht, Hilfsader (621.315.2)
pilot-wire controlled network Netz mit drahtferngesteuerten Schalteinrichtungen, Netz mit Hilfsdraht-Rundsteuerung
pilot-wire protection system Hilfsdrahtschutzsystem
pilot-wire regulator Pegelregler mit Hilfsader
to pilot lotsen
pin Nadel, Stecknadel; Stift, Zapfen, Führungsstift; Kegel (79); Stütze, Isolatorstütze (621.315.6); Reißzwecke (74); Sockelstift, Stift (621.385)
pin bearing Zapfenlager (621.317), (681)
pin compatible pinkompatibel, austauschbar
pin configuration Anordnung (Belegung) der Stifte (621.382)
pin contact Stiftkontakt
pin-for-pin replacement (völlige) Austauschbarkeit
pin hinge Fitsche, Fitschenband
pin hole Nadelstich(loch); kleines Loch; Feinlunker (621.7)
pin-hole camera Lochkamera
pin-hole perforation Perforation, Perforierung
pin jack Stöpselloch, Steckerloch
pin joint Bolzenverbindung, Gelenkverbindung
pin-jointed frei aufgelagert
pin plug Bananensteckerstift, Steckerstift
pin punch Durchschlag (621.9)
pin-type insulator Stützisolator, Stützer (621.315.6)

pin-type socket Stiftsockel (621.385)
pin winding gefädelte Wicklung
pin writer Nadeldrucker
to pin anheften, anzwecken, annageln, sich zusetzen
pincers (pl.) Zwickzange
pinch Quetschfuß (621.9), (621.385); Kniff
pinch effect Einschnüreffekt (621.382)
pinch-off Abschnürung
pinch tongs (pl.) Quetschzange
pinch-type lead-in Quetschfußdurchführung
to pinch quetschen, zusammenquetschen; Quetschfuß drücken
to pinch off abkneifen, abquetschen
pincushion correction Kissenentzerrung
pincushion distortion Kissenverzerrung
pine Fichte, Tanne
pinetree aerial Tannenbaumantenne
pinion Ritzel
pinion plays in the wheel Ritzel greift ins Rad
to pink klopfen (621.43)
pink noise rosa Rauschen
pint (Hohlmaß = 1/8 gal = 568 cm^3)
pioneer installation Erstausführung
pip Pips, kleiner Vorsprung (621); Ausschlag, Zacken, Impulsbild, Spannungsstoß (621.397); Echozeichen (621.396)
pipe Röhre, Rohr (621.64); Pfeife (681.8); (als Hohlmaß = 105 gal)
pipe bend Rohrkrümmer, Rohrbiegung
pipe clip Rohrschelle
pipe connection Rohranschluß
pipe fitter Rohrleitungsmonteur, Rohrschlosser
pipe fitting Rohrleitungsmontage; Rohrfitting, Rohrleitungsarmatur, Rohrmuffe
pipe-ground potential Rohr-Boden-Spannung (621.6)
pipe inlet Ansaugrohr, Saugstutzen
pipe joint Rohrmuffe, Rohrverbindung
pipe laying Rohrverlegung
pipeline Rohrleitung, Zufuhrleitung; Versorgungslinie
pipe seam Rohrnaht
pipe-to-soil potential Rohr-Boden-Potential
pipe system Rohrleitungsnetz, Rohrnetz
pipe tee T-Stück; Abzweigstück
pipe thread Rohrgewinde, Gasgewinde
pipe tongs (pl.) Rohrzange
pipe-type internal-gas pressure cable Gasinnendruck-Rohrkabel
pipe ventilated fremdbelüftet
pipe-ventilated motor fremdbelüfteter Motor
pipe welding Rohrschweißung
pipe work Rohrleitungsanlage, Rohrnetz
pipe wrench Rohrschlüssel, Rohrzange
to pipe pfeifen; (Rohre) installieren
to pipe to mit Rohren zuführen zu, mit Rohren anschließen an
pipette Pipette
piping Rohrnetz, Rohrleitungsanlage, Leitungsanordnung, Rohrleitungssystem
pipless bulb spitzenloser Kolben (621.32)
pistol light Handsignalscheinwerfer (656.7)
piston Kolben (621)
piston diaphragm Kolbenmembran (534)
piston displacement Hubraum (621.4)
piston ring Kolbenring
piston trimmer Glasrohrtrimmer
pit Loch, Vertiefung, Grube; Grube, Bergwerk, Zeche (622); Parkett (72)
pits Löcher, Grübchen (621.357)
pit head gear Fördergerüst, Förderanlage
pitch Neigung, Steigung (621.8), (629.1); Teilung (621.8); Höhe, Tonhöhe (534); Länge über alles, Gesamtlänge; Stampfbewegung (629.12); Pech (66)
pitch-black pechschwarz
pitch circle Teilkreis (621.8); Lochkreis
pitch control Längsneigungsregler, Längsneigungssteuerung (629.13)
pitch diameter Teilkreisdurchmesser (621.8); Lochkreisdurchmesser
pitch factor Sehnungsfaktor
pitch of a note Höhe eines Tones
pitch of the roof Dachneigung
pitch of a screw Ganghöhe einer Schraube, Abstand der Gewindegänge
pitch of teeth Zahnteilung (621.8)
to pitch räumlich unterteilen; anordnen (621.8); stampfen (629.12); um die Querachse schwingen (629.13)
pitchblende Pechblende, Uranpecherz
pitched (in der Tonlage) abgestimmt (534)
pith Mark
pith ball Holundermarkkügelchen
pithead Fördergerüst (622)
pithead frame Fördergerüst
pithead gear Förderanlage
pitman Pleuel (621.9)
Pitot static Pitotrohr, Staurohr
Pitot tube Pitotrohr, Staurohr
pitted (punktförmig) angefressen; angeschmort, mit Schmorperlen besetzt (621.3)
pitted contact angefressener Kontakt
pitting Lochfraß (621.8); punktförmige Anfressung; Schmorperle (621.3)
pivot Lagerstift, Drehzapfen, Pinne

pivot bearing Spitzenlager
pivot jaw Drehkontaktwangen (pl.) (621.316.5)
pivot screw Zapfenschraube
to pivot (drehbar) lagern
to pivot around sich drehen um
to pivot on (drehbar) lagern auf; (drehbar) gelagert sein auf
pivoted about drehbar um
pivotal Dreh-...
pivotal centre Drehpunkt
pixel Bildelement
place Platz, Ort, Stelle
places of decimals Dezimalstellen
to four places of decimals auf 4 Stellen (genau)
to place placieren; plazieren; anbringen, hinlegen; einsetzen, eintragen, hinschreiben; einstellen; stellen (621.316.7)
to place (a grid) negative (ein Gitter) negativ machen, negativ vorspannen
to place in parallel parallel schalten (**to:** zu); in Parallele setzen
to place a voltage across eine Spannunganlegen an
plain einfach, gewöhnlich, ohne Besonderheiten; offenkundig; unlegiert
plain bearing (gewöhnliches) Gleitlager
plain cast iron gewöhnliches Gußeisen
plain condenser kapazitätsgerader Kondensator
plain conductor blanker Leiter, unverzinnter Leiter
plain conduit I-Rohr, Installationsrohr
plain cut-off einfacher Ausschalter
plain dipole gestreckter Dipol
plain fuse offene Sicherung, gewöhnliche Sicherung
plain language Klartext
plain lead-covered cable blankes Bleikabel
plain shed insulator Isolierglocke
plain steel conduit Stahlrohrleitung
plain tuning error Abstimmpeilfehler (621.396.9)
plait Zopf, Flechte
to plait verflechten, flechten
plan Plan, Absicht, Projekt; Plan, Grundrißplan, Grundriß, Draufsicht
plan copying Photokopieren
in plan im Grundriß
plan outline Grundrißfigur
plan position indicator PPI, Radarsichtgerät mit drehender Zeitachse, Rundsichtgerät, PPI-Sichtgerät
plan view Grundriß, Lageplan, Grundrißzeichnung
to plan planen, entwerfen, projektieren; vorhaben
planar eben
planar IC integrierte Schaltung in Planartechnik
planar technique Planartechnik
planar transistor Planartransistor
Planck radiation law Plancksches Strahlungsgesetz
Planckian locus Ortskurve des Schwarzen Strahlers
plane Ebene, Fläche, ebene Fläche (51); Flügel, Tragflügel, Flugzeugflügel (629.13); Flugzeug (629.13); Hobel (68)
in the same plane in der gleichen Ebene (51); auf der gleichen Ebene
plane of the paper Zeichenebene
plane of parting Spaltebene
plane of polarization Polarisationsebene
plane of rupture Bruchfläche; Rutschfläche
plane of translation Translationsfläche
plane vector Zeiger, Zeitvektor (538)
plane of weakness schwache Fläche, schwache Stelle
plane plan, eben
plane-parallel planparallel
plane polarized eben polarisiert, eben
plane polarized wave eben polarisierte Welle
plane surface ebene Oberfläche
plane wave ebene Welle
to plane hobeln, glätten
planetary Planeten-..., planetenähnlich
planetary electron Hüllenelektron
planetary gearing Planetengetriebe
planimeter Planimeter
plan position indicator = PPI Rundsichtgerät
plant Anlage, Betrieb, Werk; Pflanze (58)
plant capacity installierte Leistung (621.3)
plant (capacity) factor Ausnutzungsgrad, Belastungsfaktor
plant efficiency Kraftwerkswirkungsgrad
plant manager Betriebsleiter
plant shortage Leistungsmangel (621.311)
Planté accumulator Planté-Akkumulator, formierter Sammler
plasm(a) Plasma
plasma beam Plasmastrahl
plaster Pflaster (61); Gips, Stuck, Verputz, Putz (693)
plastic Kunststoff, Preßstoff
plastic compound Kunststoff, Preßstoff
plastic encapsulation Kunststoffkapselung

plastic-foil capacitor Kunststoffolien-Kondensator
plastic insulation Kunststoffisolation
plastic material plastisches Material, Kunststoff, Preßstoff
plastic mouldings Kunstpreßstoffteile, Preßstoffteile
plastic package Kunststoffgehäuse
plasticity Plastizität, Bildsamkeit
to plasticize plastifizieren, plastisch machen, plastizieren
plasticizer Weichmacher
plate Platte; Blech, Grobblech; Platte (621.356); Platte, Photoplatte (77); Anode, Anodenblech (621.385); Mauerlatte (69); Teller (64); Schild
plate area Plattenfläche (621.356); Anodenfläche (621.385)
plate battery Anodenbatterie (621.396)
plate-bending rolls (pl.) Blechbiegemaschine
plate circuit Anodenkreis
plate condenser Plattenkondensator (621.319.4)
plate conductance (Kehrwert des inneren Widerstandes) (621.385)
plate current Anodenstrom
plate-current modulation Anodenstrommodulation, Heisingmodulation
plate-current saturation Anodensättigung, Anodenstromsättigung; Pseudosättigung (Pentode)
plate detection Anodengleichrichtung
plate dissipation Anodenverlustleistung (621.385)
plate efficiency Wirkungsgrad (der Umsetzung von Gleichstrom auf Hochfrequenz) (621.396.61)
plate fuse Streifensicherung
plate glass Spiegelglas, Flachglas
plate impedance Innenwiderstand, innerer Widerstand (einer Röhre) (621.385)
plate load Anodenbelastungswiderstand, Anodenwiderstand
plate-load resistor Anodenwiderstand (621.396)
plate metal Metallblech
plate-modulated anodenmoduliert
plate modulation Anodenmodulation
plate modulator Anodenmodulator
plate-power detector Anodengleichrichter
plate-protected motor Motor mit Plattenschutz (622.81)
plate resistance innerer Widerstand (einer Röhre) (621.385)
plate saturation Anodensättigung
plate-shearing machine Blechschere (621.9)
plate straightening rolls (pl.) Blechrichtmachine (621.9)
plate supply Anodenspannung, Anodenbetriebsspannung, Anodenstromquelle, Anodenspannungsquelle
plate surface Anodenoberfläche (621.385)
plate tank Anodenschwingkreis
plate-tuned circuit Schaltung mit abgestimmtem Anodenkreis; Anodenabstimmkreis, abgestimmter Anodenkreis
plate voltage Anodenspannung
to plate plattieren, mit Überzug versehen, galvanisieren
plated (galvanisch) vernickelt; galvanisiert; mit einem Blechflansch versehen
plated carbon plattierte Kohle
platinating Platinieren, Platinierung
plating Plattierung; Galvanisierung; Deckschicht
plating dynamo Galvanisiermaschine
plating rack Galvanisiergehänge
plating vat Galvanisierbottich
platinocyanide Platincyanür
platinum Platin
platinum black Platinschwarz, Platinmohr
platinum contact Platinkontakt
platinum iridium Platin-Iridium
platinum rhodium Platinrhodium
platinum silver Platinsilber
platinum sponge Platinschwamm, Platinmohr
platinum wire Platindraht
play Spiel; Spiel (621)
one thousandth of an inch play eintausendstel Zoll Spiel
playback Kontrollwiedergabe (654.19); Wiedergabe
playback amplifier Wiedergabeverstärker
to playback abspielen
plenary Plenar-...
plenary capacitance Betriebskapazität zwischen zwei Leitern bei freiem Potential aller übrigen
plenum heating Luftheizung, Umwälzheizung
pliability Schmiegsamkeit, Faltbarkeit
pliable faltbar; schmiegsam, biegsam
plied fabric mehrlagiges Gewebe
pliers (pl.) Zange, Flachzange, Biegezange
plate impedance Innenwiderstand, innerer Widerstand (einer Röhre) (621.385)
plate load Anodenbelastungswiderstand, Anodenwiderstand

plate-load resistor Anodenwiderstand (621.396)
plate metal Metallblech
plate-modulated anodenmoduliert
plate modulation Anodenmodulation
plate modulator Anodenmodulator
plate-power detector Anodengleichrichter
plate-power supply Anodenspannungsquelle
plate-protected motor Motor mit Plattenschutz (622.81)
plate resistance innerer Widerstand (einer Röhre) (621.385)
plate saturation Anodensättigung
plate-shearing machine Blechschere (621.9)
plate straightening rolls (pl.) Blechrichtmachine (621.9)
plate supply Anodenspannung, Anodenbetriebsspannung, Anodenstromquelle, Anodenspannungsquelle
plate-supply bypass capacitor Anodenspannungssiebkondensator
plate-supply ripple (kleine) Schwankungen (pl.) der Anodenspannung, Anodenbrumm
plate-supply voltage Anodenspannung, Anodenbatteriespannung
plate surface Anodenoberfläche (621.385)
plate tank Anodenschwingkreis
plate terminal Anodenklemme, Anodenanschluß
plate-tuned circuit Schaltung mit abgestimmtem Anodenkreis; Anodenabstimmkreis, abgestimmter Anodenkreis
plate voltage Anodenspannung
plate winding Anodenwicklung (am Transformator)
plate work Blecharbeit
plate working Blechbearbeitung
to plate plattieren, mit Überzug versehen, galvanisieren
plated (galvanisch) vernickelt; galvanisiert; mit einem Blechflansch versehen
plated carbon plattierte Kohle
platen Preßstempel (621.9); Walze (651); Preßplatte, Preßtisch, Druckplatte; Meßteller, Lehrscheibe (621.9)
platform Bahnstein; Plattform, Bühne (625)
platform roof begehbares flaches Dach, Dachterrasse
platinating Platinieren, Platinierung
plating Plattierung; Galvanisierung; Deckschicht
plating dynamo Galvanisiermaschine
plating rack Galvanisiergehänge
plating vat Galvanisierbottich
to platinize platinieren
platinocyanide Platincyanür
platinum Platin
platinum black Platinschwarz, Platinmohr
platinum contact Platinkontakt
platinum iridium Platin-Iridium
platinum rhodium Platinrhodium
platinum silver Platinsilber
platinum sponge Platinschwamm, Platinmohr
platinum wire Platindraht
play Spiel; Spiel (621)
one thousandth of an inch play eintausendstel Zoll Spiel
playback Kontrollwiedergabe (654.19); Wiedergabe
playback amplifier Wiedergabeverstärker
to playback abspielen
plenary Plenar-...
plenary capacitance Betriebskapazität zwischen zwei Leitern bei freiem Potential aller übrigen
plenum heating Luftheizung, Umwälzheizung
pliability Schmiegsamkeit, Faltbarkeit
pliable faltbar; schmiegsam, biegsam
plied fabric mehrlagiges Gewebe
pliers (pl.) Zange, Flachzange, Biegezange
plot Parzelle, Landstück (63); Ort, Platz; Aufzeichnung, Auftragung, Diagramm
to plot auftragen, aufzeichnen, punktweise eintragen; abtragen (51); zu einem Diagramm verarbeiten (621)
to plot against auftragen über... (als Abszisse)
to plot upon a co-ordinate system in ein Koordinatensystem eintragen
to plot a diagram ein Diagramm auftragen
plot out aufzeichnen
to plot to scale maßstäblich auftragen
plotter Zeichner, Kurvenzeichengerät, Kurvenschreiber, Plotter
plough Schlitzschienenstromabnehmer (621.33)
plug Stecker (621.3); Stopfen, Stöpsel (621); Stöpsel (621.317)
plug adapter Schraubstecker, Lampensockelstecker; Übergangsstecker
plug board Schalttafel, Stecktafel, Rangierfeld
plug cartridge Stöpselpatrone
plug commutator Stöpselumschalter
plug-connected mit Steckeranschluß
plug connection Steckverbindung, Steckeranschluß, Stöpselverbindung
plug connector Steckdose
plug cut-out Stöpselschalter
plug fuse Steckersicherung, Stöpselsicherung

plug gauge Lehrdorn, Meßdornlehre, Meßdorn
plug hole Stöpselloch, Steckerloch
plug-in Steck-..., Einsteck-...; Einschub
plug in coil Steckspule
plug-in device Steckvorrichtung
plug-in module steckbarer Baustein, Steckmodul
plug-in printed-circuit board Steckkarte
plug-in program Programmstecker
plug-in relay Stecksockelrelais
plug-in transformer Stecktransformator
plug-in tray steckbarer Einschub
plug-in-type subassembly Steckbaugruppe
plug-in unit Einschub, Steckeinheit
plug in position gesteckter Stöpsel (621.317)
plug out of box gezogener Stöpsel (621.317)
plug pin Steckerstift
plug receptacle Steckerleiste
plug socket Steckerbuchse
plug and socket Steckverbindung, Steckvorrichtung
plug-and-socket connector Steckverbinder, Steckverbindung, Kupplung (621.3)
plug switch Steckerschalter
plug tap Nachschneider, Endbohrer (621.9)
plug-terminated cord Steckerstrippe, Leitung mit Stecker
plug-type base Stiftsockel (621.385)
plug weld Lochschweißung
to plug stecken, stöpseln; Stöpsel stecken (621.317); sich zusetzen, sich verstopfen (69)
to plug a hole ein Loch zustöpseln; ein Loch zustopfen
to plug into einstecken in, mit Stecker anschließen an
to plug into connection an die Steckdose anschließen
pluggable steckbar
plugging Stöpselstecken; Verstopfung (69); Gegenstrombremsung
plugging contactor Gegenstromschütz
plumb Lot, Lotpendel, Senkblei
out of plumb aus dem Lot, nicht senkrecht, schief
plumb lotrecht, senkrecht; genau
plumbago Graphit
plumber Installateur, Klempner
plumbic Blei-..., Plumbi-...
plumbicon Plumbicon (621.397)
plummet Lot, Lotleine (629.12)
plunger Kolben, Tauchkolben (621.6), (66), (621.396); Pufferstempel (625)
plunger electromagnet Tauchspulmagnet
plunger switch Schalter mit Kontakttulpe
plunger-type relay Tauchkolbenrelais, Tauchankerrelais
plural service Doppelversorgung, zweiseitiger Anschluß
plurality Mehrzahl, Pluralität
plus plus; +, und; (zusammen) mit; positiv
plus electricity positive Elektrizität
plus pole Pluspol (621.3)
plus quantity positive Größe (51)
plus side Plusseite
plus sign Pluszeichen
plutonium Plutonium
plutonium pile, plutonium reactor Plutoniumbrenner, Plutoniumreaktor
pluviometer Regenmesser (55)
ply Lage, Schicht, Furnier (67)
plywood Sperrholz, Furnierholz
p. m. = permanent magnet Permanentmagnet, Dauermagnet
p. m. loudspeaker permanentdynamischer Lautsprecher
P. M. G. = Post Master General Generalpostmeister, Postminister
PMOS = p-channel metal oxide semiconductor
PN boundary PN-Grenze, PN-Übergang
PN junction PN-Sperrschicht
PN-junction diode PN-Sperrschichtgleichrichter
pneumatic pneumatisch, Druckluft-..., Preßluft-...
pneumatic control Druckluftsteuerung
pneumatic controller pneumatischer Regler
pneumatic drill Preßluftbohrer
pneumatic house tube Hausrohrpost
pneumatic loudspeaker Druckkammerlautsprecher
pneumatic motor Druckluftmotor, Preßluftmotor
pneumatic operator pneumatisches Stellglied
pneumatic press Druckluftpresse, pneumatische Presse
pneumatic rammer Preßluftramme, Preßluftstampfer
pneumatic tool Preßluftwerkzeug
pneumatic tube system Rohrpostanlage
pneumatically operated switch Schalter mit Druckluftbetätigung
PNP (junction) transistor PNP-(Flächen)Transistor
pocket Tasche; Kammer; Sack, Wärmesack
pocket chamber Taschendosimeter
pocket lamp Taschenlampe
pocket-lamp battery Taschenlampenbatterie

pocket-lamp bulb Taschenlampenbirne
pocket-lamp case Taschenlampengehäuse
pocket meter Taschenmeßgerät
pocketphone Taschenfunkgerät
pocket torch Taschenlampe
pocket voltmeter Taschenvoltmeter
poignant odour stechender Geruch
point Punkt (51); Spitze; Dezimalpunkt = Komma (51); Unterbrecherkontakt (629.113)
point angle spitzer Winkel, Winkel an der Spitze
point of application Anschlußpunkt (621.3); Angriffspunkt (531)
point of balance Schwerpunkt
point of concurrency Pol (eines Strahlenbüschels)
point of congealment Erstarrungspunkt
point contact Punktberührung; Spitzenkontakt
point-contact diode Spitzendiode
point-contact transistor Spitzentransistor
point of contact Berührungspunkt; Kontaktpunkt (621.3); Eingriffspunkt (621.8)
point electrode Spitzenelektrode (621.3)
point of impact Treffpunkt, Aufschlagpunkt, Einschlag
point of incidence Einfallspunkt
point of inflection Umkehrpunkt
point of intersection Schnittpunkt
point lattice Punktgitter
point load Punktbelastung (621.3), (624)
point machine Weichenantrieb
point of a needle Nadelspitze
point nought ones Hundertstel
point 0 ones Hundertstel
point 00 ones Tausendstel
point one = .1 Null Komma Eins = 0,1
point ones Zehntel
point of origin Ursprungspunkt; Ursprung, Anfangspunkt (51)
point to point connection Standverbindung, Punkt-zu-Punkt-Verbindung
point to point control Koordinatensteuerung, Streckensteuerung
point to point transmission Standverbindung, Direktverbindung
point to point radio communication feste Funkverbindung
point of a pole Spitze eines Mastes, Mastspitze
point of rest Nullstellung; Ruheausschlag, stationärer Ausschlag (621.317)
point of sales terminal Kassenterminal
point is a sharp angle Spitze ist scharf, Spitze hat einen scharfen Winkel
point source Punktquelle (621.3); punktförmige Lichtquelle (535)
point of support Auflagepunkt
point of tangency Berührungspunkt
point tongue Weichenzunge
point of voltage application Spannungszuführungspunkt
to point anspitzen, spitzen; zeigen, weisen; ausfugen, verstreichen, Fugen verstreichen (69)
to point at zeigen auf, zeigen nach, hinweisen auf; gerichtet sein nach
to point towards gerichtet auf, ausrichten auf; gerichtet sein auf, gerichtet sein nach, zeigen nach
pointed spitz, zugespitzt, angespitzt, spitzig
pointed fracture zugespitzter Bruch
pointed pole zugespitzter Pfahl
pointed resonance curve spitze Resonanzkurve
pointer Zeiger (621.317) (681.3); Zeigestock; Hinweis
pointer deflection Zeigerausschlag
pointer instrument Zeigerinstrument
pointer knob Zeigerdrehknopf, Einstellknopf
pointolite lamp Punktlichtlampe, Wolframbogenlampe
points (pl.) Weiche (625)
points indicator Weichensignal
points lock Weichenverriegelung, Weichenschloß
poise Gleichgewichtslage; Poise (53)
to poise auswägen, abwägen
poised gewichtsbehaftet, massebehaftet
poisoning Vergiftung (61), (539), (54)
poisoning of an atomic pile Vergiftung eines Atombrenners
poisoning of a catalyst Vergiftung eines Katalysators
Poisson's equation Poissonsche Gleichung
Poisson's ratio Poissonsche Zahl
Poisson's theory of magnetization Poissonsche Magnetisierungstheorie
poke welding Hauptpunktschweißung
polar Polare
polar polar (51), (91)
polar angle Zentriwinkel
polar arc Polarbogen (621.313)
polar capacitor gepolter Kondensator
polar circle Polarkreis
polar contact polarisierter Kontakt
polar co-ordinates Polarkoordinaten
polar d. c. telegraph system Morsedoppelstromsystem
polar diagram Polardiagramm

polar equation Polargleichung, Gleichung in Polarkoordinaten
polar indication Polaranzeige (621.397)
polar link polare Bindung (54)
polar moment of inertia polares Trägheitsmoment
polar operation Doppelstrombetrieb
polar paper Polarkoordinatenpapier, Polarpapier
polar pitch Polteilung
polar region Polargegend
polarity Polarität
to be the same polarity gleiches Vorzeichenhaben
polarity indicator Polprüfer
polarity inversion Vorzeichenumkehr
polarity reversal Vorzeichenumkehr
polarity test Polaritätsprüfung
polarity-test paper Polprüfpapier
polarization Polarisation
polarization error Polarisationsfehler (621.396)
to polarize polarisieren; Nulleiter markieren
polarized ammeter polarisiertes Amperemeter
polarized armature polarisierter Anker (621.318)
polarized electric drainage polarisierte Streustromableitung
polarized induction Vormagnetisierungsinduktion
polarized light polarisiertes Licht
polarized moving-iron instrument polarisiertes Dreheiseninstrument
polarized plug gepolter Stecker
polarized relay polarisiertes Relais
polarized sounder polarisierter Ticker
polarized vane instrument Dreheiseninstrument
polarized wavc polarisierte Welle
polarizing field Vormagnetisierungsfeld
polarizing magnetization Gleichstromvormagnetisierung
polarizing magnetizing force Vormagnetisierungsfeldstärke
polarizing slot Indexnut, Polarisierungsschlitz
polarizer Polarisator (621.35)
polaroid Polarisations-... (535)
polaroid screen Polarisationsfilter
pole Pfahl, Mast, Holzpfahl, Holzmast; Pol (91); Pol, Magnetpol (538); Pol, Polklemme (621.3); Pol (51)
pole arc Polbogen (621.313)
pole butt Mastfuß
pole-change motor polumschaltbarer Motor
pole changer Polwechsler (621.394)
pole changing Polumschaltung
pole-changing control Drehzahlveränderung durch Polumschaltung
pole-changing motor polumschaltbarer Motor
pole-changing switch Polumschalter, Polwechsler
pole core Feldkern
pole face Polfläche (538)
pole-finding paper Polprüfpapier
pole guy Mastabspannung, Spanndraht, Spannanker
pole horn Polschuh, Polspitze
pole indicator Polprüfer
pole line Freileitung, Holzmastleitung
pole mounting Mastaufhängung
pole-mounted circuit breaker Mastschalter
pole piece Polschuh; Pol (621.313)
pole pitch Polteilung (621.313)
pole shank Polkern
pole shim Polabgleichblech, Beilage
pole shoe Polschuh
pole span Polbogen
pole speed ratio synchrone Drehzahl
pole spider Polstern, Polrad (621.313)
pole step Klettereisen (621.315.1)
pole strength Polstärke
pole switch Mastschalter
pole tester Polprüfer
pole tip Polspitze, Polende, Polschuhspitze
pole transformer Masttransformator
pole trolley Rollenstromabnehmer, Stangenstromabnehmer
pole-type condenser Mastkondensator
pole-type isolator Masttrennschalter
pole-type transformer Masttransformator
pole unit Einzelpol (621.316.5)
police Polizei
police radio Polizeifunk
polish Poliermittel; Polierung, Politur; Oberflächengüte, Glätte; polnisch
polish notation polnische Notation
to polish polieren
polished plate glass Spiegelglas
polishing wheel Polierscheibe
polithion Oppanol
to poll abfragen
polling Abfragetechnik (681.3)
polling call Serienverbindung (621.395)
polonium Polonium
polychromatic vielfarbig, polychromatisch
polycyclic mehrphasig, Mehrphasen-...
polyester cast resin Polyestergießharz
poly-ethylene Polyäthylen

polygon Polygon, Vieleck
polygon of forces Krafteck
polygonal vieleckig, polygonal, Vieleck-...
polygonal connection Polygonschaltung (621.3)
polyhedral vielflächig
polyhedron Polyeder, Vielflächner
polymer Polymer; polymer
polymerization Polymerisation
to polymerize polymerisieren, aushärten
polymorphic Mehrstrom-... (621.3)
polymorphic generator Mehrstromgenerator
polynomial Polynom (51) (**in:** von)
polynomial polynom, polynomial
polyphase mehrphasig, Mehrphasen-...; vielphasig
polyphase aggregate mehrphasiges Aggregat
polyphase alternating current mehrphasiger Wechselstrom, Mehrphasenstrom
polyphase alternator Mehrphasengenerator
polyphase circuit Mehrphasenschaltung, Drehstromschaltung
polyphase commutator motor Drehstromkommutatormotor, Mehrphasenkollektormotor, Mehrphasenkommutatormotor
polyphase earth mehrphasiger Erdschluß, Doppelerdschluß
polyphase generator Mehrphasengenerator, Drehstromgenerator
polyphase induction motor Drehstrom-Induktionsmotor, Drehstrom-Asynchronmotor
polyphase motor Drehstrommotor
polyphase set of alternating currents Mehrphasenstromsystem
polyphase set of potential differences Mehrphasenspannungssystem
polyphase shunt commutator motor Mehrphasennebenschlußmotor
polyphase synchronous generator Mehrphasensynchrongenerator
polyphase system Drehstromanlage, Drehstromsystem
polyphase transformer Drehstromumspanner, Drehstromtransformator
polyphase winding Drehstromwicklung, mehrphasige Wicklung
polypropylene capacitor Kunststoffolienkondensator
polyrod aerial dielektrische (Mehrfach-)Antenne, Harke (621.396)
polystage mehrstufig
polystage amplifier mehrstufiger Verstärker
polystyrene Polystyrol
polystyrol Polystyrol
polytechnic school Ingenieurschule, Polytechnikum
polyvinyle chloride Polyvinylchlorid, PVC
pondage Fassungsvermögen, Inhalt
ponderability Wägbarkeit
ponderable wägbar
ponderous schwer, massig
pony motor Andrehmotor, Anwurfmotor
pool Teich (91); Schmelzbad, Schmelze (66); flüssige Kathode (621.314.6)
pool-type rectifier Quecksilberdampfgleichrichter mit flüssiger Kathode
poor arm, minderwertig
poor coal minderwertige Kohle
poor conductor schlechter Leiter
poor design minderwertige Konstruktion
poor insulation schlechte Isolation
poor insulator schlechter Isolator
poor quality schlechte Qualität, minderwertige Güte
poor result schlechtes Ergebnis, mageres Ergebnis
poorly screened schlecht abgeschirmt
pop Körner (621.9); (Pfropfen-)Knall, Puffen, Puff
to pop vom Kellerstapel nehmen (681.3)
poppet valve Schnüffelventil
porcelain Porzellan; Emaille (US)
porcelain cup Porzellanteller, Isolatorenteller, Isolator
porcelain former Porzellankörper
porcelain insulator Porzellanisolator
porcelain insulated porzellanisoliert
porcelain-insulated transformer Porzellanwandler
porcelain interior Porzellaneinlage (621.315.6); Porzellaneinsatz
porcelain-like porzellanähnlich
porch Austastschulter, Schwarzschulter (621.397)
porous porös
porous barrier poröse Wand, Diaphragma
porous cell Diaphragma (621.35)
porous substance poröse Substanz, poröser Stoff
port Hafen (629.12); Pforte (629.12); Öffnung, Auslaßöffnung; Brennermund; Backbord; Anschlußport (621.3.049.76)
portability Tragbarkeit, Fahrbarkeit
portable tragbar, fahrbar
portable electro drill elektrische Handbohrmaschine

portable equipment tragbare Geräte (pl.), fahrbares Gerät
portable instrument tragbares Instrument
portable lamp Handlampe
portable receiver Kofferempfänger
portable set nicht ortsfestes Gerät, tragbares Gerät, fahrbarer Maschinensatz; Kofferradio
portal Portal (72)
portal design Portalbauweise
portative force Tragkraft (621.318)
portion Teil, Abschnitt; Portion
to portion portionsweise abteilen, einteilen, zuteilen
position Position, Stellung, Lage; Angriffspunkt
position control Einstellung der Nullinienhöhe (621.385)
position of equilibrium Gleichgewichtslage
position finding Ortung
position fix Positionsmessung, Positionsfeststellung
position-fixed only nur längsgeführt (539)
position of a force Angriffspunkt einer Kraft (531)
in position in der richtigen Lage, richtig liegend, richtig stehend, in der richtigen Stellung
position indication Stellungsanzeige
position indicator Stockwerksanzeiger; Stellungsanzeige, Stellungsmelder (621.316.7), (681.876)
position light Positionslampe (629.13)
position-light signal Lichtsignal
position line Funkstandlinie, Peilstandlinie (621.396.93)
position of rest Ruhelage, Ruhestellung
position revertive signal Stellungsrückmeldung (621.316.7)
position sensor Wegmeßgeber
position transducer Wegmeßgeber
position-type telemeter Vergleichsgrößen-Fernmeßgerät
to position anordnen, hinlegen; einstellen, einregeln; stellen (621.316.7)
positioning device Einstelleinrichtung
positioning element Regelglied
positional checking Stellungsmelder
positional error Positionsfehler
positional equilibrium Gleichgewicht gegen Verschiebung
positioner Stellantrieb, Stellglied, Steller
positional notation Stellenwertschreibung (51), (681.3)
positional signal Standortidentifizierungssignal (621.396.93)
positioning control (system) Positionierungssteuerung
positioning device Einstelleinrichtung
positioning element Regelglied, Regelorgan
positive Positivum; Positiv (77); Positive, positive Platte (621.355)
positive positiv; zuverlässig; zwangsschlüssig, zwangsläufig
positive air pressure Luftüberdruck
positive-amplitude modulation Positiv-Amplitudenmodulation
positive brush positive Bürste (621.313)
positive carbon positive Kohle
positive charge positive Ladung (621.3)
positive column positive Säule (621.385)
positive conductor positiver Leiter (621.315)
positive current positiver Strom
positive-direction mitläufig (621.392)
positive drive Zahntrieb
positive electricity positive Elektrizität
positive electrode positive Elektrode, Anode
positive electron Positron
positive element positiver Pol
positive feedback positive Rückführung, Rückkopplung (621.316.7); Mittkopplung
positive feeder positive Speiseleitung, Fahrdrahtspeiseleitung
positive-going positiv gerichtet, positiv
positive lead Plusleiter
positive light modulation lichtproportionale Lichtmodulation
positive lock washer zwangsschlüssige Sicherungsscheibe
positive logic positive Logik
positive main Plusleiter
positive matrix Positivform (621.357)
positive number positive Zahl
positive photograph (photographisches) Positiv
positive plate Positive, positive Platte (621.355)
positive pole positiver Pol, Pluspol, Plusklemme
positive quantity positive Größe
positive rays Kanalstrahlen
positive ray analysis Massenspektrographie
positive ray current Kanalstrom, Ionenstrahlstrom (621.385)
positive relative to positiv gegen, positiv gegenüber
positive-sequence mitläufig, rechtsläufig, Mit-... (621.3)
positive-sequence active power Wirkleistung des mitläufigen Systems
positive-sequence reactance Längsreaktanz

positive-sequence reactive power Blindleistung des mitläufigen Systems
positive-sequence symmetrical component mitläufige symmetrische Komponente (538)
positive sign positives Vorzeichen, Pluszeichen
positive terminal positive Klemme, Anode
positive wire positive Leitung, a-Ader (621.395)
positively charged tatsächlich aufgeladen; positiv aufgeladen; positiv geladen
positron Positron
post acceleration Nachbeschleunigung
post-accelerating electrode Nachbeschleunigungselektrode
post-dialling delay Rufverzug
post head Mastaufführung (621.395)
post-indexed addressing nachindizierte Adressierung
post-injection voltage Injektionsfolgespannung (621.382)
post insulator Stützisolator (ohne Stütze)
post office bridge Telegraphenmeßbrücke
pot Topf, Schmelzhafen
pot core Topfkern (621.396)
pot head Kabelendverschluß
pot-type breaker Löschkammerschalter
pot-type coil Topfkernspule
pot-type iron dust core Eisenspulvertopfkern
to pot bechern (621.3)
potash Pottasche, Kali-...
potash lye Kalilauge
potassium Kalium, Kali
potential Potential; Spannung
potential circuit Spannungspfad (621.317)
potential coil Spannungsspule (621.317)
potential coupler Spannungsteiler
potential difference Potentialdifferenz, Spannung
potential divider Spannungsteiler
potential drop Spannungsabfall
potential element Spannungseisen (621.317)
potential equalization Potentialausgleich
potential field Potentialfeld
potential function Potentialfunktion
potential fuse Sicherung des Spannungspfades
potential galvanometer spannungsempfindliches Galvanometer
potential gradient Gradient, Potentialgradient, Potentialgefälle (621.3), (55)
potential indicator Spannungsanzeiger
potential to puncture Durchschlagspannung
potential regulator Spannungsregler
potential slope Potentialgradient, Feldstärke
potential transformer Spannungswandler
potential möglich, potentiell
potential energy potentielle Energie
potentiograph Keinathschreiber
potentiometer Potentiometer, Spannungsteiler
potentiometer arm Abgriffarm des Potentiometers
potentiometer circuit Spannungsteilerschaltung
potentiometer control Spannungsteilerregelung, Potentiometerregelung
in potentiometer fashion in Spannungsteilerschaltung, nach Art eines Spannungsteilers, wie ein Spannungsteiler
potentiometer method Kompensationsmethode
potentiometer pick-up Potentiometerabgriff
potentiometer-type field rheostat Feldspannungsteiler
potentiometer wire Schleifdraht (621.317)
potentiometric potentiometrisch
potentiometric measurement potentiometrische Messung
potentiometric recorder selbstabgleichender Widerstandsgeber, Potentiometerschreiber
potentiometric transducer potentiometrischer Meßwertwandler
Potier diagram Potier-Diagramm, Potier-Dreieck
Potier reactance Statorstreublindwiderstand, Statorstreureaktanz
Potier reactive voltage Statorstreuspannung
Potter-Bucky grid Bleigitter (621.386)
potting Vergießen
potting compound Vergußmasse
Poulsen arc Poulsenbogen, Poulsenlichtbogen
pound Pfund (Währungseinheit = **100 newpence**); Pfund (Masseneinheit = **12 ounces** = 453.6 g)
pound-weight Pfundgewicht
pound weight Gewichtspfund
to pound zermalmen, zerquetschen, zerstampfen, bombardieren
poundage Gewicht, Menge, (gewichtsmäßiges) Fassungsvermögen
poundal Krafteinheit im technischen Maßsystem (**1 poundal = 1 lb. ft/sec^2** = 13825 dyn = 0,13825 N)
pour point Stockpunkt (66)
poured concrete Gußbeton, Gießbeton
pouring Gießen, Gieß-...
pouring lip Tülle
pouring temperature Gießtemperatur
pouring test Auslaufversuch (669)
powder Puder, Pulver; Preßpulver (66)
powder core Massekern

powder spectrogram
Debye-Scherrer-Aufnahme (621.386)
to powder pulvern, pulverisieren
powdered coal Kohlenstaub, Staubkohle
powdered iron Eisenpulver
powdered iron core Eisenpulverkern
powdery pulverförmig, staubförmig
power Leistung; Gewalt, Kraft (621), (621.3); Potenz (51); Vollmacht
power absorption Leistungsverbrauch, Kraftverbrauch, Leistungsaufnahme
power amplification Leistungsverstärkung (621.396.64)
power amplifier Leistungsverstärker, Kraftverstärker, Endverstärker (621.396.64)
power bandwith Leistungsbandbreite
power cable Starkstromkabel, Netzkabel, Energieversorgungskabel
power capacitor Starkstromkondensator
power circuit Starkstromkreis (621.3)
power component Wirkkomponente, Wirkanteil
power consumed aufgenommene Leistung, Leistungsaufnahme
power consumer Verbraucher, Stromabnehmer
power consumption Leistungsaufnahme, Stromverbrauch, Kraftverbrauch
power control Leistungsregelung
power converter Umformer, Leistungswandler (621)
power cord Starkstromanschlußleitung, „Starkstromstrippe"
power cost Stromkosten (pl.)
power current (nachfolgender) Betriebsstrom (621.316.9)
power cut Stromabschaltung
power demand Leistungsbedarf
power derating Leistungsverminderung, Herabsetzen der Leistung
power directional relay Rückleistungsrelais, Stromrichtungsrelais
power dissipation Verlustleistung
power distribution Leistungsbilanz (621.396.67); Stromverteilung (621.315)
power distribution box Stromverteilerkasten
power divider Leistungsteiler
power drain Leistungsaufnahme
power-down mode Bereitschaftsbetrieb [Energie-Sparbetrieb]
power-driven motorisch angetrieben, mit Motorantrieb
power economy Energiewirtschaft
power engineering Starkstromtechnik
power factor Leistungsfaktor (621.3)
power-factor adjusting device Vorrichtungzum 90°-Abgleich (621.317)
power-factor capacitor Phasenschieberkondensator
power-factor compensation Leistungsfaktorverbesserung, cos φ-Kompensation, Blindstromkompensation
power-factor compensator 90°-Abgleich (621.317)
power-factor correction Phasenausgleich, Blindleistungsausgleich, Blindstromausgleich, cos φ-Ausgleich
power-factor indicator Phasenmesser, cos φ-Messer
power-factor meter cos φ-Messer
power-factor regulator cos φ-Regler
power-fail protection unterbrechungsfreie Stromversorgung
power failure Netzausfall
power feeder Hochspannungsspeiseleitung; Speiseleitung
power filter Netzdrossel
power flow Leistungsfluß
power-flux density Strahlungsdichte
power frequency Netzfrequenz, Betriebsfrequenz
power-frequency flash-over voltage Überschlagwechselspannung
power-frequency withstand voltage Wechselstromstehspannung
power fuse Hochleistungssicherung
power gain Leistungsverstärkung, Leistungsgewinn (621.375)
power gas Kraftgas
power-generating plant Erzeugungsanlage, Kraftwerk
power generation Krafterzeugung
power generator Generator, Stromerzeuger
power hacksaw Kaltsäge
power-handling capacity Leistungsvermögen, Leistungsfähigkeit; verarbeitbare Leistung (621.396.64); Übertragungsgrenzleistung (621.315)
power house Krafthaus, Kraftwerk
power increase Leistungsanstieg (621.396.667); Leistungszunahme
power indicator Leistungsmesser
power induction (induziertes) Starkstromgeräusch (621.396)
power input Leistungsaufnahme, aufgenommene Leistung, aufgewandte Leistung, Eingangsleistung
power interchange Leistungsaustausch

power law Potenzgesetz (51)
power level Pegel; Betriebspegel, Leistungspegel (621.395)
power-level recorder Leistungsschreiber
power lighting Starkstrombeleuchtung
power limit Stabilitätsgrenzleistung (621.311)
power line Starkstromleitung
power-line carrier telephony Trägerfrequenztelefonie auf Hochspannungsleitungen
power-line hum Netzbrumm (621.396.62), (621.396.64)
power-line ripple Netzflimmern (621.397)
power load Strombelastung, Kraftbelastung, Last (621.3)
power loss Verlustleistung
power mains Kraftnetz, Starkstromnetz, Netz
power meter Leistungsmesser
power minimum kleinste Last, Lastminimum
power is off Anlage ist spannungslos, es ist abgeschaltet
power operation Kraftbetätigung
power oscillator Leistungssender
power outlet Netzsteckdose
power output Leistungsabgabe, Ausgangsleistung, abgegebene Leistung
power pack Netzanschlußgerät, Netzteil, Stromversorgung (621.396)
power peak Leistungsspitze
power pentode Leistungspentode
power plant Kraftanlage, Kraftwerk, Stromversorgung
power plug Netzstecker, Kraftstromstecker
power rating Nennleistung, Leistungsbemessung
power ratio Leistungsverhältnis (691.396.64)
power ratio of a directional aerial Richtverhältnis einer Richtantenne (Leistung in Strahlrichtung / Leistung senkrecht dazu)
power reactor Leistungsreaktor
power rectifier Leistungsgleichrichter, Netzgleichrichter
power relay Leistungsrelais, Starkstromrelais
power resistor Hochlastwiderstand
power reversal Umkehr der Leistungsrichtung, Leistungsumkehr
power-reversal relay Rückwattrelais, Rückstromrelais
power ringing Maschinenruf
power semiconductor Leistungshalbleiter
power series Potenzreihe
power source Kraftquelle, Stromquelle, Spannungsquelle, Energiequelle
power stage Kraftverstärkerstufe, Endstufe, Leistungsstufe (621.375)
power standing wave ratio Leistungswelligkeitsfaktor
power station Kraftwerk, Zentrale, Kraftzentrale
power-station machinery Kraftwerksmaschinen (pl.)
power-station personnel Kraftwerkspersonal
power-station transformer Kraftwerkstransformator
power supply Stromversorgung, Stromversorgungsgerät, Energieversorgung, Netzanschluß; Netzanschlußgerät (621.396)
power-supply cable Starkstromkabel, Versorgungskabel, Netzanschlußkabel, Zuführungskabel
power-supply line Anschlußleitung, Zuführungsleitung
power-supply regulations Anschlußbedingungen (696)
power-supply set Netzanschlußgerät (621.396)
power-supply switch Netzschalter, Hauptschalter
power-supply system Versorgungssystem
power switch Leistungsschalter, Netzschalter
power switchboard Hauptschalttafel (621.311); Kraftstromschalttafel (696.6)
power switchgroup Fahrschalter (625)
power take-off Abtrieb, Abtriebsleistung (621.8)
power transfer Leistungsübertragung, Energieübertragung (621.315), (621.396)
power-transfer relay Lastumschaltrelais
power transformer Netztransformator, Leistungsumspanner, Leistungstransformator
power transistor Leistungstransistor
power transmission Kraftübertragung (621.315)
power-transmission line Hochspannungsleitung
power tube Leistungsröhre, Endröhre (621.396.62), (621.385)
to power potenzieren (51); mit Leistung versehen, antreiben
to power-up einschalten, mit voller Leistung betreiben
powered Leistungs-..., angetrieben
to be powered by angetrieben werden von
powerful leistungsfähig, stark, kräftig
powerful spring kräftige Feder
Poynting's vector Poyntingscher Vektor, Vektor der Leistungsdichte
PPI = plan position indicator Rundsichtgerät

PPI-scan Rundsichtdarstellung
ppm = parts per million millionstel
PPM = pulse position modulation Pulsphasenmodulation, Pulslagemodulation
PSWR = power standing wave ratio Leistungsreflexionsfaktor
practicability Ausführbarkeit, Durchführbarkeit, Brauchbarkeit, Anwendbarkeit
practical praktisch
practical electric units praktische elektrische Einheiten
practical training praktische Arbeit
practical working Ausübung (34)
practically constant praktisch konstant, merklich konstant
practice Praxis, Brauch, Übung; Praxis (61)
in practice praktisch, in praxi, in der Praxis
practice-type für die Praxis; für Übungszwecke
practice-type set Übungsgerät (621.396)
practice unit Übungsgerät
to practice praktisch anwenden, ausführen, ausüben; üben, trainieren
practicing engineer Betriebsingenieur, Ingenieur der Praxis
practitioner Praktiker
pragauss (Einheit der magnetischen Induktion $= 10^{10}$ G)
pragilbert (Einheit der magnetomotorischen Kraft = 0,1 Gilbert)
pramaxwell (Einheit des magnetischen Flusses $= 10^{8}$ Maxwell)
praoersted (Einheit der magnetischen Feldstärke $= 10^{10}$ Oe)
to pre-adjust voreinstellen, vorher einstellen
pre-adjustment Voreinstellung, vorhergehende Einstellung
preamplifier Vorverstärker
to pre-arrange vorher festlegen, im voraus festlegen
to pre-assign zuteilen (621.396)
to pre-assign frequencies Frequenzen zuteilen
prebeading Voreinschmelzen, Aufbringen einer Einschmelzperle
precast pile Gußbetonpfahl, Betonmast
precautionary measure Vorsichtsmaßnahme, Unfallverhütungsmaßnahme
to precede vorangehen; vorgeschaltet sein, davor liegen (621.3)
preceding stage vorhergehende Stufe, vorangehende Stufe (621.396.62), (621.64); Vorstufe; Eingangsstufe (621.396.62), (621.64)
precedence call Vorranggespräch
precedence designation Vorrangbezeichnung
precious metal Edelmetall
precious stone Edelstein, Juwel
to precipitate niederschlagen, fällen, ausfällen, abscheiden; sich abscheiden, sich niederschlagen
precipitation Niederschlag, Ausscheidung, Fällung (54); Niederschlag, Niederschläge (pl.) (55)
precipitation equipment Entstaubungsanlage
precipitation hardening Ausscheidungshärtung (66)
precipitation noise Entladungsstörung
precipitation static Zischstörung, Entladungsstörung, Sprühstörung (621.396)
precipitation tank Scheidebehälter
precision Präzision, Genauigkeit
precision approach radar PAR, Präzisions-Anflugradar
precision-ground feingeschliffen
precision instrument Präzisionsinstrument, Feinmeßinstrument
precision machine Präzisionsmaschine
precision mechanic Feinmechaniker
precision mechanics (pl.) Feinmechanik
precision mechanism feinmechanisches Gerät
precision-momentary contact-limit switch Feinendtaster, Präzisionsendtaster
precision-range unit Entfernungsfeinmeßstufe (621.396.96)
precision-type Präzisions-..., Fein-...
precision-type instrument Feinmeßinstrument
precision wavemeter Präzisionswellenmesser
precombustion Vorverbrennung
precursor of a nuclide Mutternuklid
to predetermine vorherbestimmen, vorher festlegen
predetermined boundary conditions vorgegebene Randbedingungen
predialing delay Wähltonverzug
to predict vorhersagen
prediction Voraussage, Vorhersage
predictor Warngerät
predictor control voraussehende Regelung
predischarge Vorentladung (621.3)
predryer Vortrockner
pre-emphasis Vorverzerrung, Preemphasis
pre-equalization Aufnahmeentzerrung
preferred direction Vorzugsrichtung (54)
prefix Vorsatz; Vorsilbe; vorderer Index; vorgesetzte Kennung, Kennzahl (621.395)
to prefix davorsetzen
prefocus vor dem Brennpunkt liegend
preform Vorform (66)

to preform vorformen (66)
preformed winding Schablonenwicklung
preheat Vorhitze, Vorerwärmung, Vorglühen (66)
to preheat vorwärmen, vorerwärmen, vorerhitzen, vorglühen (669)
preheater Vorwärmer, Anwärmofen
preheating Vorheizung, Vorwärmung
preheating cabinet Vorwärmschrank (66)
preheating time Vorheizzeit
preheating torch Vorwärmbrenner, Vorwärmflamme (621.791)
pre-ignition Frühzündung
preimpregnation Vorimprägnierung
pre-indexed addressing vorindizierte Adressierung
preknockpulse Vorimpuls
prelim. = preliminary examination Ausleseprüfung
preliminary Vorarbeit, Vorbereitung, Vorstadium
preliminary vorbereitend, Vor-...
preliminary examination Ausleseprüfung, Auswahlprüfung
preliminary research Voruntersuchung (53)
preliminary selector Vorwähler (621.395)
preliminary signal Vorsignal, Ankündigungssignal (625)
preliminary signal-frequency stage HF-Vorstufe (621.396)
preliminary stage Vorstufe (621.396)
preload Vorlast, Vorbelastung (624)
to preload vorbelasten
pre-oscillation current Anschwingstrom
preparation Herstellung, Zubereitung (54); Vorbereitung, Vorbehandlung, Aufbereitung (66); Behandlung, Imprägnierung (621.3); Arznei (61); Präparat (535)
preparation of the mixture Gemischbildung (621.4)
preparatory vorbereitend (**to:** für)
preparatory operation Vorarbeit, vorbereitender Arbeitsgang
preparatory work Vorarbeit
to prepare vorbereiten, aufbereiten, ausarbeiten, herstellen
prepared chalk Schlämmkreide
prepayment meter Münzzähler (621.317)
preplumbed system starrer Hohlleiterzug (621.372)
preread disturb pulse Lese-Vorimpuls
prerequisite Vorbedingung, Vorerfordernis
to preselect vorauswählen, vorauslesen
preselective gearing Getriebe mit Gangvorwahl
preselectivity Vorselektion (621.396.62)
preselector Vorwähler (621.395); HF-Vorkreis
presentation Darstellung
preservation Konservierung (66); Erhaltung, Erhaltungszustand (621); Imprägnierung (674)
preservative Konservierungsmittel
to preset voreinstellen, vorher einstellen, vorbereitend einstellen, vorbereiten
presetting Voreinstellung, vorbereitende Einstellung
press Presse (621.9), (07); Druckpresse (655); Quetschfuß (621.385); Bücherbord
press button Druckknopf (621.3.)
press-button control Druckknopfsteuerung
press-button switch Druckknopfschalter
press die Preßstempel
press glass Preßglas
press-glas base Preßglasteller (621.385)
press load Pressenfüllung, Presseneinsatz
press shop Presserei, Preßwerkstatt
press work Preßarbeit; Preßstück
to press pressen; drücken (621.9); drängen; quetschen, ausquetschen
to press the button auf den Knopf drücken
to press a lever einen Hebel herunterdrücken, einen Hebel betätigen
to press-on aufpressen
to press a pedal ein Pedal betätigen, auf ein Pedal treten
to press a switch einen Schalter einlegen, auf einen Knopf drücken
pressed paper Hartpapier, Preßpapier
pressed stem Quetschfuß
pressboard Preßpappe, Preßspan
pressed water proof druckwassergeschützt
Pressel switch Schalterbirne, Druckknopfschalterbirne
presser plate Druckplatte (621.313)
pressfax Pressefax
pressfit Preßsitz (621.9)
presspahn Preßspan
pressure Druck
pressure-actuated druckbetätigt
pressure-actuated microphone auf Schalldruck ansprechendes Mikrophon, Druckmikrophon
pressure balance Druckausgleich
pressure calibration Druckkammereichung (621.395.61)
pressure-chamber loudspeaker Druckkammerlautsprecher
pressure contact Druckkontakt
pressure die-casting Spritzguß; Druckguß
pressure difference Druckdifferenz

pressure drop Druckabfall, Druckgefälle
pressure fluctuation Druckschwankung
pressure gauge Manometer, Druckmesser
pressure governor Druckregler
pressure gradient Druckgefälle
pressure head Druckhöhe
pressure pipe Druckrohr, Druckrohrleitung
pressure proof druckfest
pressure pump Druckpumpe
pressure reduction Druckreduzierung
pressure-reduction valve Druckreduzierventil, Reduzierventil
pressure regulator Druckregler
pressure relay Druckrelais
pressure relief Sicherheitsventil
pressure-relief valve Drosselventil, Druckminderventil
pressure-responsive druckabhängig
pressure-sensitive druckempfindlich, auf Druck ansprechend
pressure-sensitive bulb Druckdose (621.317), (621.9)
pressure sensor Druckaufnehmer
pressure side Druckseite
pressure spring Druckfeder
pressure transducer Druck-Meßwertwandler
pressure transmitter Druckgeber
pressure wave Druckwelle
pressure welding Preßschweißung
pressurized component Bauteil unter Überdruck
pressurized-water reactor Druckwasserreaktor
prestage Vorstufe
to prestore vorspeichern, zwischenspeichern
pretrigger Vorimpuls
prevention Verhinderung, Verhütung
prevention of passage of noise Schalldämmung
preventive vorbeugend, verhütend
preventive coil Schaltspule, Überbrückungsspule, Überschaltspule, Schaltdrossel (621.316.5)
preventive maintenance Wartung; vorbeugende Unterhaltung
preventive resistance Überschaltwiderstand (621.316.5)
previous vorübergehend; früher
previous magnetic history magnetische Vorgeschichte
PRF = pulse repetition frequency Impulsfolgefrequenz, Impulsfrequenz (621.397)
to prick stechen, durchstechen
to prick a hole ein Loch stechen
primary Primärwicklung (621.314.2); Ausgangsstoff, Ausgangssubstanz (54)
primary primär, erstrangig; Erst-..., Primär-...; Grundfarbe (621.397)
primary air Primärluft (621.1)
primary alcohol primärer Alkohol
primary battery Primärbatterie (621.35)
primary cell Primärelement (621.35)
primary circuit Primärstromkreis
primary clock Hauptuhr
primary coating Grundanstrich (69)
primary coil Primärspule
primary colour Grundfarbe (535)
primary colour picture Primär-Farbauszug (621.397)
primary constants (of lines) bezogene Leitungseinheiten
primary current Primärstrom
primary distribution feeder Hochspannungsanschlußleitung
primary distribution main Hochspannungsverteilungsschiene
primary distribution network Hochspannungsversorgungsnetz
primary distribution system Hochspannungsverteilungsanlage
primary distribution trunk line Hochspannungsanschlußleitung
primary electron Primärelektron (621.385)
primary failure Primärausfall
primary fault Punktierung, Isolationsdurchbruch
primary frequency Hochfrequenz, Arbeitsfrequenz (621.39)
primary inductance Hauptinduktivität, Querinduktivität
primary luminous standard Normalkerze, Normallichtquelle
primary network Hochspannungsnetz
primary radiator Primärstrahler, aktive Antenne, Erreger
primary relay Hauptstromrelais
primary route Leitweg (621.395)
primary service area Hauptversorgungsgebiet; schwundfreie Nahzone (621.396)
primary signal Farbwertsignal (621.397)
primary standard Hauptnormal
primary switchgear connections (pl.) Hauptschaltanlage
primary terminal Primärklemme, Primäranschluß
primary transmission feeder Hochspannungsübertragungsleitung
primary voltage Primärspannung
primary winding Primärwicklung
prime Prime (796.8); Strich (51)

x prime = x′ x Strich = x′
to prime zünden, initiieren (66); grundieren (69.1)
prime erstrangig, auserlesen; Premier-...; Grund-...
prime coat Grundanstrich (68)
prime conductor Leiter erster Klasse
prime mover Antriebsmaschine, Triebkraft
prime number Primzahl
prime power installierte Leistung
primer Zündkapsel (66); Grundanstrich (69); Anfängerlehrbuch
priming Zündung (66); Grundierung (69)
priming colour Grundierungsfarbe
primitive primitiv
primitive period Periode der Grundwelle, Grundperiode
principal hauptsächlich, Haupt-..., Erst-...
principal axis Hauptachse (535), (54)
principal current Hauptstrom (621.314.63)
principal focus Brennpunkt; Brennweite
principal mode Grundtypus (534), (621.396)
principal stress Hauptspannung (539)
principle Prinzip, Grundsatz; Grundbestandteil, Essenz (54)
principle of action Arbeitsprinzip, Wirkungsweise
principle of Archimedes Archimedisches Prinzip
principle of design Bauprinzip, Konstruktionsprinzip
principle of the lever Hebelgesetz
principle of relativity Relativitätsprinzip
principle of superposition Superpositionsprinzip, Überlagerungsprinzip
to print drucken; pausen, Abzüge machen (621.9); kopieren, Abzüge machen (77); mit Druckbuchstaben schreiben
to print out drucken, ausdrucken (681.3)
printed board gedruckte Leiterplatte, Karte, geätzte Platine
printed circuit (board) gedruckte Schaltung
printed wiring gedruckte Verdrahtung
print speed Druckgeschwindigkeit (381.3)
printer Drucker
printing Drucken, Druckvorgang (655); Kopieren (77)
printing reperforator druckender Empfangslocher
printing telegraph Typendrucktelegraph, Fernschreiber
printing telegraph system Fernschreibsystem
printometer Zählerstandsdrucker
printout Ausdruck (681.3); Abzug
priority Vorrang
priority processing Vorrangverarbeitung (682.3), (621.395)
prism Prisma (51), (53); Spektrum (535)
prismatic prismatisch, stäbchenförmig, Prismen-..., Spektral-...
prismatic colour Spektralfarbe
prismatic powder Stäbchenpulver
privacy Geheimhaltung (65); Abhörsicherheit (621.395), (621.396)
private privat
private automatic branch exchange, PABX automatische Nebenstellenanlage
private branch exchange, PBX Nebenstellenanlage
private exchange Hausamt, Hausvermittlung
private extension privater Nebenanschluß (621.395)
private railway Privatbahn, Privateisenbahn
to prize open aufbiegen
probable error wahrscheinlicher Fehler
probability Wahrscheinlichkeit (**of:** für)
probability of delay Wartewahrscheinlichkeit (621.395)
probability laws Gesetze der Wahrscheinlichkeit
probe Sonde (61); Prüfsonde, Prüfabgriff, Prüfstift, Prüfanschluß, Meßfühler, Abtaststift; (elektrostatischer) Abgriff, Sonde, Koppelschleife, Drahtsonde (621.396)
to probe sondieren, prüfen
problem oriented language problemorientierte Programmiersprache
procedure Verfahren, Prozedur
to proceed fortschreiten, vorgehen; weitergehen, weiterlaufen; vorwärtsgehen, vorwärts treiben
proceed-to-send-signal Abrufzeichen (621.395)
proceed-to-send tone Wählton
proceed signal Vorrücksignal (625)
proceeding lines of force austretende Kraftlinien
process Verfahren, Arbeitsverfahren, Prozeß, Arbeitsprozeß; Vorgang
process automation Prozeßautomatisierung
process computer Prozeßrechner
in process of construction im Bau
process control Prozeßsteuerung, Steuerung von Arbeitsabläufen
process-control computer Prozeßrechenanlage
process-control language Prozeßsteuersprache
process-cycle controller Programmregler

process becomes cumulative Prozeß wird zur Kettenreaktion
process of decay Zerfallsvorgang, Zerfallsprozeß
process of design Konstruktionsverfahren
process of growth Wachstumsvorgang
process industries (pl.) Verfahrenstechnik (66)
process of trial and error schrittweise Annäherung, Iterationsverfahren, Probiermethode
to process behandeln, bearbeiten, verarbeiten, entwickeln und fixieren (77); durchsetzen (66)
processing Verfahren, Arbeitsablauf, Aufbereitung
processing instruction Bearbeitungsanweisung, Behandlungsanweisung
processing plant Aufbereitungsanlage
processing time Bearbeitungszeit
processor Rechner, Prozeßrechner; Prozessor
processor-interrupt facility Prozessorunterbrechungsmöglichkeit
to prod anstoßen, antippen
to produce herstellen, machen, erzeugen, ausbringen; hervorrufen, erzeugen; verlängern, weiterführen
to produce a line eine Linie verlängern (51)
to produce machinery Maschinen (pl.) fabrizieren
producer Fabrikant, Hersteller, Erzeuger
producer gas Generatorgas
product Produkt (51); Erzeugnis, Produkt
product of combustion Verbrennungsprodukt
product of decomposition Zersetzungsprodukt
products of mutual intermodulations senderfremde Modulationsprodukte
production Produktion, Fabrikation, Herstellung, Fertigung
production control Fertigungssteuerung, Überwachung der Fabrikation
production planning Arbeitsvorbereitung
production process Arbeitsverfahren, Herstellungsverfahren
production range Fertigungsbreite, Fertigungsprogramm
production reactor Produktionsreaktor
production schedule Fabrikationsprogramm, Terminplan
production stage Serienreife
productive produktiv; ergiebig, fündig; Produktions-...
productive time produktive Zeit
productivity Produktivität, Arbeitsergiebigkeit
profile Profil, Streckenprofil (625.1)
in profile im Profil
profile projector Schattenprojektor
to profile profilieren
profiled wire Profildraht, profilierter Draht
profiling machine Kopierfräsmaschine
to program programmieren
program Programm
program advance Programmfortschaltung
program amplifier Modulationsverstärker (eines Rundfunksenders)
program card Programmkarte
program checking Programmprüfung
program control Programmsteuerung, Zeitplanregelung
program counter Befehlszähler, Programmzähler, Programmschrittzähler
program debugging Fehlersuche in einem Programm
program fault Programmierfehler
program flow Programmablauf
program flowchart Programmablaufplan
program jump Programmverzweigung
program library Programmbibliothek
program loop Programmschleife
program run Programmlauf
program step Programmschritt
program status word Programmzustandswort
program switch Programmschalter
program transmitter Programmgeber, Zeitplangeber
program unit Programmbaustein
programmable read only memory programmierbarer Festspeicher
programmed instruction programmierter Unterricht
programmer Programmierer
programming language Programmiersprache
progressing fortschreitend, verlaufend
progressing wave fortschreitende Welle, hinlaufende Welle
progression Fortschreiten, Weiterlaufen (621.396); Reihe (51)
progressive fortschreitend, anhaltend, weiterlaufend; fortschrittlich
progressive grading Staffel (621.396)
progressive induction seam welding fortlaufende Nahtschweißung
progressive motion fortschreitende Bewegung
progressive scanning Abtastung mit Zeilensprung
progressive spot welding Punktnahtschweißung
progressive stress testing Lebensdauerprüfung mit steigender Belastung
progressive wave aerial Wanderwellenantenne

prohibitive size undiskutierbare Abmessungen (pl.), unmögliche Größe
project Projekt, Plan, Entwurf
project engineering Anlagentechnik
to project vorstehen, hervorstehen, herausragen, herausstechen; projizieren (535), (51)
to project forward nach vorn schleudern, vorwärtsschleudern
to project on projizieren auf (51), (535)
to project out into hinausschießen in
projected area Projektionsfläche
projected scale instrument Projektionsinstrument, Lichtfleckinstrument, Projektionsskaleninstrument
projected surface Projektionsfläche
projecting vorspringend, vorstehend
projecting abutment vorragendes Widerlager, vorspringendes Widerlager
projecting dog vorstehende Nase
projection Vorsprung, vorstehender Teil, Ansatz; Projektion (51), (535); Auskragung
projection lantern Projektionsapparat, Projektionslampe
projection ratio Abbildungsmaßstab
projection weld Warzenschweißung, Buckelschweißung
projector Projektor, Projektionsapparat
projector lamp Projektionslampe
projector-type filament lamp Projektionsglühlampe
prolate verlängert (51)
prolate spheroid verlängertes Rotationsellipsoid
PROM = programmable read only memory programmierbarer Festspeicher, programmierbarer Lesespeicher
promoter Förderer; Beschleuniger, Katalysator (66)
proneness Neigung
proneness to ignition Zündneigung, Zündwilligkeit
prong Spitze, Zinke, Zacke; Pratze; Steckerstift (621.3)
prong grip Stiftfassung, Stiftklammer (621.385)
Prony brake Pronyscher Zaum
proof Beweis; Beleg, Belegstück, Nachweis; Probeabzug, Probeabdruck, Bürstenabzug, Korrekturbogen, Korrekturabzug (655)
proof load Prüflast, Probebelastung
proof plane Prüfspule (621.3), (538)
proof test Abnahmeprüfung, Kontrolle (621.317)
proof-test voltage Prüfspannung
to proof imprägnieren
proofed tape Isolierband
proof sicher, geschützt; imprägniert
proof against sicher vor, geschützt gegen
...-proof ...fest
proofing Imprägnierung
to propagate fortpflanzen, weiterleiten, leiten; sich fortpflanzen, sich ausbreiten
to propagate fire weiterbrennen; Feuerunterhalten (621.315.6)
propagation Ausbreitung, Fortpflanzung, Verbreitung
propagation characteristic Ausbreitungscharakteristik
propagation coefficient Kettenübertragungsmaß
propagation constant Übertragungsbelag, Fortpflanzungskonstante, Übertragungskonstante (621.395)
propagation delay Laufzeitverzögerung
propagation error Ausbreitungsfehler
propagation loss Ausbreitungsdämpfung, Ausbreitungsverlust
propagation pattern Ausbreitungscharakteristik
propagation time Laufzeit
propagation of waves Wellenausbreitung (621.396)
propane Propan
to propel antreiben (629); vorwärtsbewegen
propeller Propeller, Schraube (629.12), (629.13)
propeller fan Schraubenventilator, Schraubenlüfter
propeller pitch Propellersteigung
propeller shaft Propellerwelle (629.13); Antriebswelle, Kardanwelle (629.113); Schraubenwelle (629.12)
propellershaft tunnel Wellentunnel (629.12)
propeller thrust Propellerschub, Schraubenschub
propelling treibend, Trieb-...
propelling equipment Triebwerk (629.13); Antriebsvorrichtung
proper eigentlich, richtig
proper fit guter Sitz
proper fraction echter Bruch
proper operation richtiges Ansprechen
properly fused richtig eingebrannt, guteingebrannt (66)
property Eigenschaft, Fähigkeit; Besitz, Eigentum
property table Eigenschaftstabelle, Tabelle mit Materialkonstanten
proportion Proportion; Anteil; Verhältnis, Teil, Anteil; Ausmaß

in proportion to im Verhältnis zu
to proportion bemessen, dimensionieren, proportionieren; aufteilen, gleichmäßig aufteilen, harmonisch gliedern
to proportion uniformly gleichmäßig verteilen, gleichmäßig aufteilen
proportioning pump Dosierpumpe
proportional proportional, verhältnisgleich (**to:** zu, mit)
proportional action P-Verhalten, Proportionalverhalten
proportional arms Vergleichszweige (621.3)
proportional control P-Regelung
proportional control with disturbance variable feed forward system PZ-Regelung
proportional counter Proportionalzähler (539)
proportional plus derivate action controller PD-Regler = Proportional-Differential-Regelung
proportional plus integral controller PI-Regler, Proportional-Integral-Regler
proportional plus integral plus derivative controller PID-Regler, Proportional-Integral-Differential-Regler
proportional position action controller Regler mit Stellungszuordnung, statischer Regler
proportional reset controller Isodromregler
proportional response P-Übergangsfunktion (621.316.7); Proportionalverhalten
proportional speed controller Regler mit Stellgeschwindigkeitszuordnung
proportionality Proportionalität
propulsion Vortrieb, Antrieb
propulsion equipment Triebwerk (621.34), (625)
propulsion power Antriebsleistung
propulsion unit Triebwerk (629.13)
propulsive Antriebs-..., Vortriebs-..., antreibend
proscenium light Rampenlicht
to protect beschützen, schützen (**against:** gegen), (**from:** vor)
to protect with fuse absichern
protected geschützt; mit Berührungsschutz (621.3)
protected from accidental contact geschützt gegen zufällige Berührung
protected cut-out Handschutzsicherung
protected data transmission gesicherte Datenübertragung
protected frequency geschützte Frequenz
protected fuse geschützte Sicherung
protected machine Maschine mit Berührungsschutz, geschützte Maschine (621.3)
protected motor geschützter Motor
protected slide way verdeckte Führung (621.9)
protected switch Schalter mit Schutzkappe
protected switchgear gekapselte Schaltgeräte (pl.)
protecting schützend, Schutz-...
protecting plate Schutzplatte, Schutzblech (621)
protecting shell Schutzhaube, Schutzhülle
protecting shield Schutzschild, Schutzplatte
protection Schutz
protection against Schutz gegen
accidental contact zufällige Berührung
contact Berührung
contact with tools Berührung mit Werkzeugen
dripwater Tropfwasser
electric-shock hazarding Berührung; (Berührungsschutz)
finger touch Berührung mit den Fingern
harmful dust deposits schädliche Staubablagerungen
intentional touch absichtliche Berührung
internal pressure inneren Überdruck
jet water Strahlwasser
rain Regen
splash water Spritzwasser
touch with the flat of the hand Berührung mit der Hand
water surges Schwallwasser
water under pressure Druckwasser
protection by balanced relays Achterschutz
protection cap Schutzkappe
protection channel Schutzkanal (621.396)
protection cover Schutzdeckel
protection range Schutzbereich
protective Schutz-...
protective apparatus Schutzapparat
protective arrangement Schutzeinrichtung
protective box Schutzkasten
protective casing Schutzgehäuse
protective coating Schutzanstrich, Schutzumhüllung, Schutzschicht, (Korrosions-)Schutzüberzug
protective contact Schutzkontakt
protective covering Schutzanzug, Schutzabdeckung
protective device Schutzvorrichtung, Schutzeinrichtung, Schutzgerät
protective earthing Schutzerdung
protective gap Schutzfunkenstrecke
protective gauze Schutzgaze, Schutznetz

protective gear Schutzvorrichtung
protective ground CCIT-Schutzerde
protective housing Schutzgehäuse, Schutzkappe; Gehäuse
protective jacket Schutzmantel, Schutzhülle, Schutzumschlag (655)
protective layer Schutzschicht
protective lead Schutzleitung, Schutzdraht
protective measure Schutzmaßnahme, Unfallschutzmaßnahme
protective reactance coil Strombegrenzungsdrossel
protective relay Schutzrelais
protective resistor Schutzwiderstand
protective sheath Schutzmantel, Schutzschicht
protective spark gap Schutzfunkenstrecke
protective switch Schutzschalter
protective system Schutzsystem, Schutzart
protector Schutzgerät, Sicherung (621.395)
protector springs Schutzsockel, Sicherungssockel (621.395)
protein Protein, Zelleiweiß
Protodur insulation Protodurisolation
Protoflex tube Protoflexrohr
proton Proton
prototype Prototyp, Baumuster
prototype unit Baumuster
to protract in die Länge ziehen; aufzeichnen, aufreißen
protractor Transporteur, Winkelmesser; Schmiege (621.9)
to protrude herausragen, hervorstehen, hervorragen
protruding grip vorstehender Griff
protrusion Vorsprung
to prove beweisen, prüfen, nachprüfen; sich erweisen als
to prove false sich als falsch erweisen, sich als unrichtig herausstellen
to prove true sich als richtig erweisen, sich bestätigen
proven bewährt
proven application bewährte Anwendung
proven device bewährte Einrichtung
proven measure bewährte Maßnahme
proving Prüfung, Prüf-..., Erprobung
proving ground Erprobungsstelle, Prüffeld; Schießplatz (623)
proving trial Erprobungsversuch
provision Beschaffung, Besorgung; Vorkehrung; Verfügung, Vorschrift, Erlaß, Bestimmung, Festsetzung
to make provision for etwas vorsehen, Vorkehrungen treffen für, vorsorgen für
provisional vorläufig
provisional design vorläufige Konstruktion
provisional wiring provisorische Installation
proximity Nachbarschaft
proximity effect Rückwirkung (durch Anwesenheit in der Nähe); Stromverdrängungseffekt (621.382)
proximity fuse Abstandszünder (623)
proximity switch Annäherungsschalter, berührungsloser Schalter
proximity zone Nahfeld
pseudo-... Pseudo-...
pseudo-gravitational schwerkraftartig
pseudo instruction symbolischer Befehl
pseudo-stable output configuration pseudostabile Ausgangskonfiguration (621.374)
psi = pounds per square inch (US) Pfund pro Quadratzoll (1 psi = 0, 6894 N/cm^2)
psophometer Geräuschspannungsmesser
psophometric curve Psophometerkurve
psophometric noise Geräuschspannung
psophometric power bewertete Geräuschleistung
psophometric weight Geräuschbewertungsfaktor
psophometric weighting curve Geräuschbewertungskurve
PSWR = power standing wave ratio Leistungswelligkeitsfaktor
P-type conduction Defektleitung, Löcherleitung, P-Leitung
P-type impurity P-Verunreinigung (621.382)
P-type semiconductor P-Halbleiter
public öffentlich
public address system Lautsprecheranlage
public communication network öffentliches Fernmeldenetz
public exchange call Amtsgespräch
public exchange connection Amtsanschluß
public mobile land radio service öffentlich beweglicher Landfunk
public power supply öffentliche Energieversorgung
public radio service öffentlicher Funkdienst
public supply mains öffentliches Versorgungsnetz
public telephone station öffentliche Sprechstelle
public utility öffentliche Versorgung; Versorgungsbetrieb; EVU (621.3)
public utility works öffentliche Versorgungsbetriebe
to puddle puddeln, umrühren (69)
puddled iron Puddeleisen
puff Puff, Verpuffung

to puff puffen
puissancegraph Leistungsozillographenschleife
pull Zug, Zugkraft, Zugbeanspruchung, Kraft; Strang, Zug
pull box Zwischendose (621.315.6)
pull-in torque synchronisierendes Moment
pull-in voltage Anzugsspannung, Ansprechspannung
pull-off Fahrdrahtabspannung
pull-off spring Abziehfeder
pull-out unit ausziehbares Teil, Einschub
pull-out torque Kippmoment (621.313)
pull switch Zugschalter
pull-up torque niedrigstes Moment während des Anlaufs
to pull ziehen, eine Zugkraft ausüben; ziehen, mitziehen, mitnehmen (621.396.61); abziehen (66); pullen, rudern (629.12)
to pull from herausziehen aus
to pull into synchronism in Tritt ziehen, in Tritt fallen (621.313), (621.396.61)
to pull off the course vom Wege ablenken
to pull on ziehen an, Zugkraft ausüben auf
to pull out herausziehen; außer Tritt fallen, kippen (621.313)
to pull through hindurchziehen
to pull towards ziehen nach
to pull towards oneself anziehen
pulling crystals Kristallziehen
pulling effect Mitnahmeeffekt
pulling range Mitnahmebereich, Fangbereich
puller Abziehvorrichtung, Ausziehvorrichtung
pulley Rolle, Riemenscheibe, Schnurscheibe, Schnurrolle; Flaschenzug
pulley block Flaschenzug
pulley end Kupplungsseite, Riemenseite, Antriebseite (621.313)
pulley frame Fördergerüst, Seilscheibengerüst
pulper Zerfaserer
to pulsate wechseln, schwanken, pulsieren
pulsating pulsierend, Schwell-...
pulsating component Wechselanteil, Wechselkomponente
pulsating current pulsierender Strom
pulsating direct current pulsierender Gleichstrom, welliger Gleichstrom
pulsating electromotive force pulsierende EMK
pulsating field Wechselfeld, pulsierendes Feld
pulsating flux schwankender Gleichfluß
pulsating potential difference Wechselspannung (mit überlagerter Gleichspannung)
pulsating quantity Wechselgröße (mit überlagertem Gleichanteil)
pulsating voltage Wechselspannung (mit überlagerter Gleichspannung)
pulsation Pulsation, Schwankung; Stoßvorgang
pulse Impuls (621.3), (621.397)
pulse amplifier Impulsverstärker
pulse amplitude Impulsamplitude, Impulshöhe
pulse amplitude analyzer Impulsamplituden-Analysator
pulse amplitude modulation = PAM Pulsamplitudenmodulation
pulse bandwidth Impulsbandbreite
pulse clipper Impulsabtrennstufe
pulse code modulation Pulscodemodulation
pulse command Impulsbefehl
pulse counter Impulszähler
pulse of current Stromstoß, Stromimpuls
pulse-decay time Impulsabfallzeit
pulse delay time Impuls-Verzögerungszeit
pulse-digit spacing Impulsabstand
pulse discriminator Impuls-Diskriminator, Impuls-Verteiler
pulse droop Dachschräge (eines Pulses)
pulse duration Impulsdauer
pulse-duration modulation Pulsdauermodulation
pulse duty factor Verhältnis von Puls- zu Pausenzeit
pulse envelope Impulsform
pulse former Pulsformer
pulse-forming network Impulsformer, Impulsformer-Schaltung
pulse-frequency modulation Pulsfrequenzmodulation
pulse generator Impulsgenerator
pulse group Impulsgruppe, Impulszug, Impulsseite
pulse-height analyzer Pulshöhenanalysator
pulse interval Impulsperiodendauer
pulse-interval modulation Pulsperiodendauermodulation
pulse length Impulsdauer, Impulslänge
pulse-length modulation Pulslängenmodulation
pulse meter Impulszähler
pulse-operated relay Stromstoßrelais
pulse-phase modulation Pulsphasenmodulation
pulse-position modulation Pulsphasenmodulation
pulse recorder aufzeichnender Impulszähler
pulse-recurrence frequency Pulsfolgefrequenz
pulse regeneration Pulsregenerierung

pulse repeater Wählimpulsverstärker (621.395); Impulswiederholer
pulse repetition Impulsfolge
pulse-repetition frequency Impulsfrequenz, Impulsfolgefrequenz (621.397)
pulse-repetition rate Impulsfolgefrequenz
pulse rise time Impulsanstiegszeit
pulse separator Impulstrennstufe
pulse shape Pulsform
pulse shaper Pulsformer, Impulsformer
pulse spacing Impulsabstand
pulse-spacing modulation Pulsperiodendauermodulation
pulse spike Impulsspitze
pulse spring Nummernschalterkontaktfeder
pulse-time modulation Pulszeitmodulation
pulse tilt Dachschräge (eines Pulses)
pulse train Impulsfolge, Impulsserie (621.394)
pulse transformer Impulstransformator (621.397)
pulse voltmeter Impulsspannungsmesser
pulse width Impulsbreite, Impulsdauer
pulse-width modulation Pulsdauermodulation
pulse-width recording Impulsbreitenaufzeichnung, Wechseltaktschrift
pulser Impulszentrale (621.397)
pulsing signal Wählzeichen (621.395)
pulverized pulverisiert, Staub-...
pulverized iron Eisenstaub, Eisenpulver (621.396)
pumice(-stone) Bimsstein
pump Pumpe
pump-fed lubrication Druckschmierung
pump shop Pumpraum (621.386)
to pump pumpen, fördern; pumpen, pendeln (621.313.3)
to pump to high pressure gegen hohen Druckfördern, auf hohen Druck bringen
pumped-storage hydroelectric station Pumpspeicherwerk
pumping Pumpen, Pumpvorgang
pumping arrangement Pumpanlage (621.385)
pumping oven Pumpstand (621.385)
pumping process Pumpverfahren
pumping stem Pumpstempel (621.385), (621.326)
pumpless pumpenlos
pumpless rectifier pumpenloser Gleichrichter
punch Körner (621.9); Durchschlag, Locheisen (621.9); Preßstempel, Stanze (621.9)
punch mark Körnermarke, Körnerschlag, Körner
punch operator Datentypist
punch press Stanze
punch-through voltage Durchgreifspannung, Sperrschichtberührungsspannung (321.682)
to punch stanzen, lochen
to punch-mark ankörnen
punched gelocht, Loch-...
punched card Lochkarte
punched-card interpreter Lochkartenleser
punched-card reader Lochkartenleser
punched-card verifier Lochkartenprüfer
punched plate Lochblech, gelochtes Blech
punched tape Lochstreifen
punched-tape equipment Lochstreifengerät
punched-tape printer Lochstreifendrucker
punched-tape reader Lochstreifenleser
punched-tape recorder Lochstreifenempfänger
puncher Locheisen
punching Stanzteil (621.9)
punching machine Stanze, Lochstanze, Stanzmaschine
punching position Lochstelle (681.3)
punching strain Stanzspannung
puncture Durchschlag, Punktierung (621.3); Reifenfehler, Reifenpanne, Durchschlag, „Plattfuß“ (629.113); Glasdurchschlag (621.385)
puncture of insulation Isolationspunktierung
puncture-proof durchschlagfest
puncture-proof insulator durchschlagfester Isolator
puncture strength Durchschlagfestigkeit (621.3)
puncture voltage Durchschlagsspannung (621.3)
puncture-voltage test Durchschlagprüfung
to puncture durchschlagen, punktieren (621.3)
puncturing voltage Durchschlagsspannung
pungency stechender Geruch, beißender Geschmack
pungent stechend, beißend
pungent smoke beißender Rauch
Pupin coil Pupinspule
to pupinize pupinisieren, bespulen
pupinized cable pupinisiertes Kabel, bespultes Kabel
pure rein
pure capacitance reine Kapazität, verlustlose Kapazität
pure continuous waves reiner Dauerstrich (sing.), ungedämpfte Wellen
pure inductance reine Induktivität
pure mathematics (pl.) reine Mathematik

pure metal Reinmetall, reines Metall
pure resistance reiner Widerstand, winkelfreier Widerstand
pure rotating magnetic field reines magnetisches Drehfeld
pure sine rein sinusförmige Kurve
pure solid carbon Massivkohle
purely inductive rein induktiv
purely resistive rein ohmisch, winkelfrei (621.317)
purification Reinigung, Raffinierung (66); Läuterung (66); Aufbereitung
purification plant Reinigungsanlage, Aufbereitungsanlage
to purify reinigen; läutern (66); aufbereiten
to purify steam Dampf reinigen
purifying magnet Reinheitsmagnet
to purl kluckern
purple boundary Purpurlinie
to purple purpurfarben anlaufen, (sich) purpurn färben
purr Summen, Surren, Schnurren
to purr summen, surren, schnurren
push Druck, Stoß, Schub
push button Druckknopf (621.3)
push-button control Druckknopfsteuerung
push-botton dialling Tastenwahl (621.395)
push-button lift control Druckknopf-Fahrstuhlsteuerung
push-button radio Drucktastenradio, Rundfunkapparat mit Stationswählertasten
push-button selector Drucktastenwähler (621.396.62)
push-button switch Druckknopfschalter, Taster
push-button-type radio set Drucktastenempfänger
push fit Schiebesitz (621.9)
push-pull Gegentakt
push-pull amplifier Gegentaktverstärker
push-pull circuit Gegentaktschaltung
push-pull connection Gegentaktschaltung
push-pull deflection Gegentaktablenkung
push-pull input Gegentakteingang, symmetrischer Eingang
push-pull microphone Differential-Mikrophon, Gegentaktmikrophon
push-pull output Gegentaktausgang
push-pull power amplifier Gegentaktkraftverstärker
push-pull transformer Gegentakttransformator, symmetrischer Übertrager
push rod Schaltstange, Schubstange
push-to-talk button Sprechtaste (621.39)
push-to-talk key Sprechtaste
to push stoßen, drücken, schieben; vorantreiben; auf den Kellerstapel legen (681.3)
to push down herunterdrücken, hinunterdrücken
push-down storage Kellerspeicher
to push-fit sauber einpassen
to push off abstoßen (629.12), (625)
to push out of gear außer Eingriff bringen, auskuppeln
pusher-puller Auf- und Abziehvorrichtung
pushing Frequenzverwerfung
pushing-off device Abdrückvorrichtung
to put setzen, stellen, legen
to put across übersetzen über; anlegen an (621.3); parallel schalten zu (621.3)
to put aside beiseite lassen, beiseite legen
put aside unter Weglassung von
to put the brakes on die Bremsen anziehen
to put a charge on a capacitor eine Ladung auf einen Kondensator bringen
to put down the brushes die Bürsten auflegen, die Bürsten aufsetzen
to put down instructions Anweisungen formulieren
to put energy into Energie hineinstecken in
put-in energy hineingesteckte Energie, aufgewandte Energie
to put into hineinstecken in
to put into a cell in eine Zelle hineinladen (621.35)
to put into operation in Betrieb setzen
to put into practice in die Praxis umsetzen, praktisch ausführen
to put mechanical stresses on mechanisch beanspruchen
to put in motion in Bewegung setzen
to put in parallel with parallelschalten mit
to put a plant out of service eine Anlage außer Betrieb setzen
to put on anlegen an; anziehen
to put on speed schneller fahren, schneller laufen, zulegen
to put out ausbringen, ausstoßen (621.3)
to put out of operation außer Betrieb setzen
to put to test zur Prüfung bringen
to put through durchverbinden (621.395)
to put through to verbinden mit (621.395)
to put a transmitter on the air einen Sender ausstrahlen lassen, senden
to put a voltage across a resistance Spannung an einen Widerstand anlegen

to put a voltage on a winding Spannung an eine Wicklung legen
putty Kitt, Glaserkitt
"p" wire c-Ader
pycnometer Pyknometer
pylon Mast (621.315.1)
pyr (Lichtstärkeeinheit = 0,1 Neue Kerze)
pyramidal pyramidenförmig
pyramidal horn Pyramidenhorn
pyrite Pyrit
pyroconductivity Heißleitfähigkeit
pyroelectric crystal pyroelektrischer Kristall
pyroelectricity Pyro-Elektrizität
pyrometer Pyrometer
pyrometer probe Temperaturfühler
pyrometric pyrometrisch
pyrometric cone Seegerkegel
pyrometric measurement pyrometrische Messung
pyrometry Pyrometrie
pyron detector Pyrit-Detektor
pyrophoric pyrophor

Q

Q-band Q-Band (ca. 39 ... 50 GHz)
Q-circuit = quieting circuit Krachtöterschaltung
Q-factor Gütefaktor, Kreisgüte
"Q" meter Güte-Meßgerät, Q-Meter
quad Vierer, Adervierer (621.395)
quad cable, quadded cable Viererkabel
quad op amp Vierfach-Operationsverstärker
quad pair Doppelsternvierer
quadrangle Viereck
quadrangular viereckig
quadrant Quadrant, Viertelkreis (51); Quadrant (629.12); Schneckensegment, Segmentschnecke (621.8)
quadrant electrometer Quadrantenelektrometer
quadrant-steering gear Quadrantruderanlage
quadrantal viertelkreisig, vierzählig
quadrantal deviation vierzähliger Peilfehler
quadrantal error vierzähliger Fehler (621.396.9)
quadrophony Quadrophonie, Vierkanal-Übertragung
quadratic quadratische Gleichung
quadratic quadratisch
quadratic equation quadratische Gleichung
quadratic scale quadratische Skala (621.317)
quadratic term quadratischer Ausdruck
quadrature Quadratur (51); 90°-Verschiebung (621.3)
quadrature adjustment 90°-Abgleich
quadrature axis Querachse, magnetische Querachse, Querfeldachse (621.313)
quadrature-axis component Querfeldkomponente (621.313)
quadrature-axis component of armature current Querfeldkomponente des Statorstromes
quadrature-axis component of the m. m. f. Querdurchflutung
quadrature-axis subtransient open-circuit time constant Subtransient-Zeitkonstante des Querfeldes bei Leerlauf
quadrature-axis subtransient reactance Subtransient-Reaktanz des Querfeldes
quadrature-axis subtransient short-circuit time constant Subtransient-Zeitkonstante des Querfeldes bei Kurzschluß
quadrature-axis synchronous reactance synchrone Querfeldreaktanz
quadrature-axis transient open-circuit time constant Transientzeitkonstante des Querfeldes bei Leerlauf
quadrature-axis transient reactance Transient-Reaktanz des Querfeldes
quadrature-axis transient short-circuit time constant Transientzeitkonstante des Querfeldes bei Kurzschluß
quadrature-axis voltage Querfeldspannung, vom Querfeld induzierte Spannung, Querfeld-EMK
quadrature of the circle Quadratur des Kreises
quadrature component Blindkomponente
quadrature distortion Quadraturverzerrung
in quadrature um 90° verschoben, senkrecht in der Phase
quadrature modulation Quadraturmodulation
quadrillion (US) Billiarde, 10^{15}
quadripolar vierpolig
quadripole Quadrupol
quadrode Tetrode, Vierpolröhre
quadrophonics Vierkanalübertragung
to quadruple vervierfachen
quadruple bundle Viererbündel
quadrupler Vervierfacher
quadruplex recording Querspurverfahren (681.84)
quadruplex (telegraph) system Quadruplex System
quadruplex bundle (conductor) Viererbündel
to quake beben, zittern

qualimeter Härtemesser (621.386)
qualitative qualitativ
qualitative analysis qualitative Analyse
qualitative investigation qualitative Untersuchung
quality Qualität, Güte, Beschaffenheit; Sorte, Farbe, Färbung (534), Klangfarbe
quality assurance Qualitätssicherung, Gütesicherung
quality control Qualitätsprüfung, Qualitätskontrolle
quality factor Gütefaktor
quality-factor meter Gütefaktormeßgerät
quality inspection Güteprüfung
quality of a sound Klangfarbe eines Tones, Farbe (534)
quality standard Gütegrad
quantitative quantitativ
quantitative analysis quantitative Analyse, Maßanalyse
quantitative result quantitatives Ergebnis
quantity Größe, Menge, Quantität
quantity efficiency Gütegrad (621.355)
quantity of electricity Elektrizitätsmenge
quantity of heat Wärmemenge
quantity of light Lichtmenge (Lichtstrom × Zeit)
quantity of magnetism Magnetismusmenge, Polstärke
quantity to be measured zu messende Größe, Meßgröße
quantity meter Ampere-Stundenzähler, Ah-Zähler
quantity of motion Bewegungsgröße, Impuls
quantity of X-rays Röntgendosis
quantization Quantisierung; Analog-Digitalumsetzung
quantization noise Quantisierungsgeräusch (621.395.4)
to quantize quanteln
quantizer Quantisiereinrichtung, Quantisierer
quantometer Kriechgalvanometer
quantum (pl. **-ta**) Quantum, Menge; Quant (53)
quantum jump Quantensprung
quantum of light Lichtquant
quantum limit Grenzwellenlänge, kurzwellige Grenze (621.386)
quantum-mechanical quantenmechanisch
quantum mechanics (pl.) Quantenmechanik
quantum theory Quantentheorie
quart Quart (796.8)
quart (Hohlmaß = 1/4 gal, etwa 1 l)
quarter Viertel; Mondviertel (52); Quartier (35); Stadtviertel; (Mengenmaß; **1 quarter = 8 bushel** oder **1 quarter = 28 lbs.**); Viertelzentner; Vierteljahr
quarter bend Viertelbogen (66)
quarter circle Viertelkreis
quarter mile Viertelmeile (ca. 400 m)
quarter-phase Zweiphasen-..., zweiphasig
quarter-phase circuit Zweiphasenschaltung
quarter plate Plattenformat 3 1/2 × 4 1/2 Zoll (etwa 9 cm × 12 cm) (77)
quarter size in viertel Größe, Maßstab 1 : 4
quarter turn Vierteldrehung, 90°
quarter wave Viertelwelle
quarter-wave Lambda-Viertel-...
quarter-wavelength Viertelwellenlänge
quarter-wavelength aerial Viertelwellenantenne, Lambda-Viertel-Antenne, Lambda-Viertel-Dipol
quarter-wave line Lambda-Viertel-Leitung
quarter-wave shield Lambda-Viertel-Sperrtopf
quarter-wave skirt dipole/sleeve dipole Sperrtopfantenne
quarter-wave sleeve Lambda-Viertel-Glied
quarter-wave stub support Lambda-Viertel-Stütze
quarter-wave termination Lambda-Viertel-Abschluß
quarter-wave transformer Lambda-Viertel-Transformationsglied
quarter-wave transmission Viertelwellen-Übertragung
to quarter vierteln, durch 4 teilen (51)
quarto Quart-..., Quartformat
quartz Quarz
quartz clock Quarzuhr
quartz crystal Quarzkristall
quartz fibre Quarzfaden
quartz glass Quarzglas
quartz lamp Quarzlampe
quartz oscillator Quarzsender, Quarzoszillator
quartz powder Quarzmehl, Quarzpulver
quartz resonator Schwingquarz
quartz tube Quarzröhre, Quarzkolben
quartz vein Quarzgang
quartz vibrator Schwingquarz
quartzolite heating elements Quarzrohrheizelemente
quasi-optical quasi-optisch
quasi-optical propagation quasi-optische Ausbreitung
quasi-regular fast gesetzmäßig, quasi-regulär
quasi-stationary quasi-stationär

quasi-stationary circuit quasi-stationärer Stromkreis
quaternary quaternär (66)
quaternion Quaternion
quaternion calculus Quaternionenrechnung
quaternionics (pl.) Quaternionenrechnung
to quench ersticken, dämpfen, löschen; abschrecken, plötzlich abkühlen
to quench an arc einen Lichtbogen löschen
to quench coke Koks ablöschen, Kokslöschen
to quench a discharge eine Entladung löschen, eine Entladung abdrosseln
to quench a fire ein Feuer löschen
quench circuit Löschkreis
quenched Lösch... (621.3)
quenched-arc circuit breaker Deion-Schalter
quenched spark Löschfunken
quenched spark gap Löschfunkenstrecke
quenched spark-gap transmitter Löschfunkensender
quenched spark system Löschfunkensystem
quencher Lichtbogenlöscheinrichtung
quenching Löschung; Abschreckung
quenching coil Sperrdrossel
quenching condenser Löschkondensator
quenching device Lichtbogenlöschvorrichtung
quenching frequency Pendelfrequenz (621.396.61)
quenching medium Löschmittel
quenching oscillation Pendelschwingung (621.396.61)
to query abfragen
queue Warteschlange
queue status Zustand der Warteschlange
queuing-in Einfädeln (351.814)
queuing theory Wartezeit-Theorie
quibinary code Quibinärcode
quick schnell
quick access store Schnellspeicher
quick-acting schnellwirkend, Schnell-...
quick-acting regulation Schnellregelung
quick-acting regulator Schnellregler
quick-acting relay Schnellrelais
quick-acting vice Schnellschraubstock, Schnellspann-Schraubstock
quick break Momentunterbrecher
quick-break fuse flinke Sicherung
quick-break switch Schnellschalter
quick-closing valve Schnellschlußventil
quick-disconnect plug Abreißstecker, Schnelltrennstecker
quick-make Schnelleinschalt-...
quick-make-and break switch Momentschalter
quick nut Auslösemutter, Schnellspannmutter
quick-response fuse flinke Sicherung
quick-response relay Schnellrelais
quick-return motion Schnellrücklauf (621.91)
quick-reverse motion Schnellrücklauf (621.91)
quick shut-off device Schnellschlußeinrichtung
quick start Schnellanlauf
quick test Schnellversuch, abgekürzter Versuch
quicklime ungelöschte Kalk
quicksilver Quecksilber
quiescent ruhig
quiescent carrier modulation Modulation mit Trägerunterdrückung in den Pausen
quiescent current Ruhestrom (621.38)
quiescent input current Eingangsruhestrom (621.375)
quiescent-operation Ruhe-..., Arbeitspunkt-... (621.385)
quiescent output voltage Ausgangsruhespannung
quiescent plate voltage Anodengleichspannung, Anodenruhespannung
quiescent point Ruhepunkt, Ruhestand
quiescent push-pull Gegentakt-C-Betrieb
quiescent supply current Ruhestrom
to quiet beschwichtigen; krachtöten (621.396.62)
quiet leise, still, ruhig
quiet automatic volume control verzögerte Schwungregelung
quiet interval Modulationspause (621.396)
quiet operation geräuschloser Betrieb (621)
quiet tuning Stummabstimmung
quieting Krachtötung, Geräuschunterdrückung (621.396.62)
quinhydrone electrode Chinhydronelektrode
quinhydrone half cell Chinhydronelektrode
quinol Hydrochinon (77)
quintal Doppelzentner ≈ 100 kg
quintillion (US) Trillion, 10^{18}
quintode Fünfpolröhre, Pentode
to quiver bibbern
quotient Quotient
quotient meter Quotientenmeßgerät
QPP = quiescent push-pull Gegentakt-C-Betrieb (621.396.64)
qr = quarter Viertelzentner (28 lbs)
Q-value Gütezahl (621.396)

R

rabbit Rohrpostbüchse
R. A. C. = rectified alternating current gleichgerichteter Wechselstrom
race Rennen, Wettrennen, Wettfahrt; Strom, Strömung, Bahn; Rasse
raceway Leitungskanal, Leitungsrohr, Kabelkanal, Kabelrost
racing Rennbetrieb, Rennen, Renn-...; Durchgehen (621.313)
rack Gestell, Gestellrahmen
rack pinion Zahnstangenritzel
rack and pinion device Zahnstangenbetrieb
rack railway Zahnradbahn
rack teeth (pl.) Zahnstange
rack wheel Zahnrad
to rack mit einer Zahnstange antreiben
racon = radar beacon
radar = radio detecting and ranging Radar, Funkmeßgerät
radar-aid to navigation Radarnavigationshilfe
radar air-traffic service Radaranflugkontrolldienst
radar attenuation Radarsignaldämpfung
radar band Radarband
radar beacon Radarabfragefunkfeuer, Navigationsradarbake
radar bomb scoring Radartreffpunktermittlung
radar camouflage Radartarnung
radar contact Radarkontakt
radar control area Radarkontrollbereich
radar control and direction Radarkontrolle und Führung
radar coverage Radarerfassungsbereich
radar deception Radartäuschung
radar detector Radarwarnempfänger
radar display Radarschirmbild
radar duct Wellenleiter (im Radarbereich)
radar early warning system Radarfrühwarnsystem
radar equation Radargleichung
radar equipment Radargerät
radar gap tote Zone
radar height finder Radar-Höhenmeßgerät
radar-homing missile radargesteuerter Flugkörper
radar indicator Radarsichtgerät
radar illumination Radarbeleuchtung
radar-interscan technic Radareinblendverfahren
radar link Radarbildübertragung
radar locator Radarortungsempfänger
radar map Radarbild
radar marker Radarsendebake
radar reflection buoy Radarreflektorboje
radar-reflection interval Radarimpulslaufzeit
radar repeater Radar-Tochtergerät
radar responder Radar-Impulswiederholer
radar response Radarecho
radarscope Radarbildschirm
radar speed meter Radargeschwindigkeitsmesser
radar trace Radarbildspur
radar tracking Radar-Zielverfolgung
radar-traffic beam Radarleitweg
radial radial, radial verlaufend, radial gerichtet
radial commutator Scheibenkollektor, Flachkommutator
radial configuration Sternsystem, sternförmige Anordnung
radial distribution feeder Speiseleitung im Radialnetz
radial drilling machine Radialbohrmaschine
radial duct radialer Kühlschlitz
radial feeder Speiseleitung (eines Strahlennetzes)
radial projection Zentralprojektion
radial slot Radialnut (621.313)
radial speed Radialgeschwindigkeit
radial system of ventilation Radialkühlung (621.313)
radial transmission feeder Radialleitung
radial turbine Radialturbine, Ljungströmturbine
radial velocity Radialgeschwindigkeit
radial ventilating duct radialer Kühlschlitz
radian Radiant (Winkelmaß)
radian frequency Kreisfrequenz, Winkelfrequenz
radiance Strahlung
radiant Strahlungsquelle, Strahler
radiant strahlend
radiant body strahlender Körper, Strahlkörper
radiant efficiency Strahlungsausbeute
radiant energy strahlende Energie
radiant energy density Strahlungsdichte
radiant flux Strahlungsfluß
radiant flux density Strahlungs(fluß)dichte
radiant heat Wärmestrahlung, Strahlungswärme
radiant heat fire Strahlofen
radiant heat therapy Infrarot-Therapie
radiant heater Strahlofen, Strahlungsheizkörper (621.365)
radiant intensity Strahlungsdichte
radiant lamp heater Trockenstrahler

radiant sensitivity of a camera tube Empfindlichkeit einer Bildaufnahmeröhre
to radiate strahlen, ausstrahlen, abstrahlen; strahlenförmig verlaufen
to radiate heat Wärme ausstrahlen
radiated beam Richtstrahl (621.396)
radiated energy ausgestrahlte Energie, abgestrahlte Energie
radiated field Strahlungsfeld (621.396)
radiating guide Hohlleiterantenne
radiating surface Strahlungsdichte
radiation Strahlung, Ausstrahlung (621.396)
radiation coefficient Strahlungsfaktor (536)
radiation constant Strommoment (621.396.67)
radiation counter Strahlungsmesser
radiation damage Strahlenschaden
radiation detector Strahlungsdetektor
radiation dose Strahlungsdosis (539), (621.386)
radiation efficiency Strahlungswirkungsgrad, Antennenwirkungsgrad (621.396.67)
radiation factor Strommoment (621.396.67)
radiation frequency Strahlungsfrequenz
radiation-frequency changer Wellenlängentransformator, Bildwandler (535)
radiation furnace Strahlungsofen
radiation hardness Durchdringungsfähigkeit (539)
radiation hazard Strahlengefährdung
radiation height Strahlungshöhe (621.396.67)
radiation loss Strahlungsverlust
radiation measurement Strahlungsmessung
radiation meter Strahlungsmesser
radiation pattern Strahlungsdiagramm (621.396.67)
radiation pyrometer Strahlungspyrometer
radiation quantum Strahlungsquant
radiation rate meter Dosismesser
radiation resistance Strahlungswiderstand (621.396.67)
radiation shield radioaktive Abschirmung
radiation sickness Strahlungskrankheit (539)
radiation source Strahlungsquelle
radiation stability Strahlungsstabilität (539)
radiation-tight strahlungsdicht
radiative strahlend, mit Strahlung verbunden
radiative capture Teilcheneinfang mit Ausstrahlung, Strahlungseinfang (539)
radiative equilibirum Strahlungsgleichgewicht
radiator Strahlkörper (53); Strahler (621.396); Radiator (69); Kühler (629)
radiator tube Radiatorröhre, Coolidgeröhre (621.385)
radical Radikal (54); Wurzelzeichen
radio Radio, Rundfunk, Tonfunk, Funk..., Hochfrequenztechnik (621.396) (siehe auch **wireless**); radio-... (539)
radioactive radioaktiv
radioactive contamination radioaktive Verseuchung, radioaktive Kontamination
radioactive decay radioaktiver Zerfall
radioactive emanation radioaktive Emanation
radioactive half-life radioaktive Halbwertszeit
radioactive material radioaktiver Stoff
radioactive nucleus radioaktiver Kern
radioactive series radioaktive Zerfallsreihe
radioactive waste radioaktiver Abfall
radioactivity Radioaktivität
radio aids to navigation Navigationsfunkhilfsmittel
radio alert Funkbereitschaft, Funkwache
radio altimeter HF-Höhenmesser, Funkhöhenmesser
radio amateur Funkamateur
radio approach aid Funkanflughilfe
radio astronomy Radioastronomie
radio beacon Funkbake, Navigationsfunkfeuer
radio beam Funkleitstrahl
radio bearing Funkpeilung, Peilwinkel
radio brodcasting Rundfunk
radio channel Funknachrichtenkanal
radio circuit Hochfrequenzkanal; Funkverbindung
radio coil Rundfunkspule, Hochfrequenzspule
radio communication HF-Nachrichtenverkehr, Funkverkehr, Funkverbindung
radio communication guard Funkbereitschaft mit sendebereitem Sender
radio communication service Funkdienst
radio compass Funkkompaß, Funkpeiler, Radiokompaß
radio control Funksteuerung
radio control receiver Kommando-Empfänger
radio deception Funktäuschung
radio detection Funkerfassung
radio detection and ranging Radar
radio detector HF-Gleichrichter
radio direction finding Funkpeilung
radio discipline Funkdisziplin
radio doppler Funk-Doppler
radio engineering Hochfrequenztechnik, Radiotechnik
radio equipment Funkgerät
radio-examination of materials Röntgen-Materialprüfung

radio fadeout Schwund, Mögel-Dellinger-Effekt
radio field intensity Hochfrequenzfeldstärke
radio field-to-noise ratio Störfeldabstand, Verhältnis Nutzfeld/Störfeldstärke
radiofixing aid Funkhilfe zur Standortbestimmung
radio frequency HF = Hochfrequenz
  radio-frequency alternator Hochfrequenzgenerator
  radio-frequency ammeter HF-Strommesser
  radio-frequency amplification Hochfrequenzverstärkung
  radio-frequency bridge Hochfrequenzmeßbrücke
  radio-frequency components HF-Bauteile
  radio-frequency filter HF-Sperre, HF-Filter
  radio-frequency furnace Hochfrequenzofen
  radio-frequency generator Hochfrequenzgenerator
  radio-frequency interference Funkstörung
  radio-frequency oscillator Hochfrequenzoszillator, Hochfrequenzgenerator
  radio-frequency pentode HF-Pentode
  radio-frequency section Hochfrequenzteil (621.396.62)
  radio-frequency pattern Radiofrequenzraster
  radio-frequency resistance HF-Widerstand
  radio-frequency suppressor Entstöreinrichtung
  radio-frequency stage Hochfrequenzstufe
  radio-frequency surges Funkstörungen
  radio-frequency transformer HF-Übertrager
  radio-frequency voltage HF-Spannung
radio fuse Rundfunk-Apparatesicherung
radiogoniometer HF-Goniometer
radiogram Funktelegramm
radiography Röntgenphotographie, Röntgenprüfung
radio guidance system Funkleitverfahren, Fernlenksystem
radio homing Zielflug
radio horizon Radiohorizont, Funkhorizont
radio-inductive interference Starkstromfunkstörung
radio installation Funkanlage
radio interference Funkstörung
  radio-interference level HF-Störpegel
  radio interference suppression Funkentstörung
radio intercept Funkerfassung
radioisotope Radioisotop
radio jamming (beabsichtigte) Funkstörung
radio landing aid Funklandehilfe
radio link Funkverbindung, Funkübertragungsweg, Richtfunkverbindung
radio location Funkortung
radioluminescence Radiolumineszenz
radiomateriology Röntgen-Materialprüfung
radio-mesh navigation Funkgitter-Navigationssystem
radio message Funkspruch
radiometallography Röntgen-Feinstrukturuntersuchung
radiometer Strahlungsmesser
radio navigation Funknavigation
  radio navigation satellite Navigationssatellit
  radio navigation service Navigationsfunkdienst, Ortungsfunkdienst
radio net Funknetz
radionuclide Radionuklid
radio noise Funkstörpegel
  radio noise field intensity Rauschfeldstärke, Störpegel
  radio noise suppression Funkentstörung, Rauschunterdrückung
radio operator Funker
radiopaque undurchlässig für Röntgenstrahlen
radioparent durchlässig für Röntgenstrahlen
radiophare nachteffektfreies Funkfeuer
radiophone Funkfernsprecher, Funksprechgerät
radio phototelegraphy Funkbildübertragung
radio propagation Funkausbreitung
  radio propagation prediction Funkwettervorhersage
radio range beacon Leitstrahlfunkfeuer
  radio range finding Funk-Entfernungsmessung
  radio range leg Leitstrahl eines Kursfunkfeuers
radio receiver Rundfunkempfänger, Radiogerät, Radio
radio receiving station Funkempfangsstelle
radio reception Funkempfang
radio reconnaissance Funkaufklärung
radio relay Richtverbindung
  radio relay station Richtfunkstelle, Relaisfunkstelle
  radio relay transmission system Richtfunksystem
radioscopy Röntgendurchleuchtung
radio search Funksuchempfang
radio set Funkgerät; Rundfunkgerät, Radio, Radiogerät
radio shielding Hochfrequenz-Abschirmung

radio silence Sendeverbot, Funksendeverbot, Funkstille
radiosonde Radio-Sonde
radiosonde balloon Radio-Ballonsonde (55)
radio station Funkstelle, Rundfunksender
radio technology Rundfunktechnik
radio telecommunication Funkverkehr
radio telecontrol Funkfernsteuerung
radio telegraphy drahtlose Telegraphie
radio telemetry Funkfernmessen
radiotelephone Funksprechgerät
radiotelephony Funksprechen, drahtlose Telefonie
radio teleprinter Funkfernschreiber
radio therapy Röntgentherapie
radio transmission drahtlose Nachrichtenübermittlung
radio transmitter Hochfrequenzsender, Sender, Funksender
radio transmission Hörfunk, Funkübertragung
radio warning Funkwarnung
radio waves Funkwellen
to radio funken, durch Hochfrequenz übertragen, im Rundfunk darbieten
to radiotelegraph drahtlos telegraphieren
radium Radium
radium-bearing radiumhalting
radium beryllium neutron source Radium-Beryllium Neutronenquelle
radium compound Radiumverbindung
radium emanation Radium-Emanation, Niton, Radon
radium paint Leuchtfarbe, Radiumfarbe
radium rays (pl.) Radiumstrahlung
radiumtherapy radioaktive Bestrahlung, Radiumtherapie
radius (pl.: **-dii**) Radius, Halbmesser (51); Speiche (61)
radius of a bend Krümmerradius
radius of curvature Krümmungsradius
radius gauge Abrundungslehre
radius of gyration Trägheitsradius
radius vector (pl.: **radii-vectors**) Radiusvektor
to radius abrunden, verrunden
radix Radix, Wurzel, Basis
radix point Komma, Dezimalpunkt (681.3)
radix-two-counter Binärzähler
radix representation Radixschreibweise
radon Radon (539), (54)
rail Schiene
rail bond Schienenverbinder (621.33)
railcar Triebwagen
rail contact Gleiskontakt
rail current Schienenstrom
rail current switch Schienenkontakt
rail effect Schienenbeeinflussung (65)
rail joint Schienenstoß
rail-less system Obus-System
rail motor Triebwagen
railway Eisenbahn
railway call system Bahnsteigrufanlage
railway motor Bahnmotor, Fahrmotor
railway signalling Eisenbahnsignalwesen
railway telegraph Bahntelegraph
rain Regen
rainbow pattern Regenbogenmuster
rain clutter Trübung, Regenstörung (621.396.96)
rain-clutter suppression (Radar-)Enttrübung
rainfall Regen; Niederschlagshöhe (55)
rainfall outage Übertragungsunterbrechung durch Regen (621.371)
rainfall recorder Regenschreiber
rain gauge Regenmesser
rain-proof regenfest, regensicher, regendicht
rain-proof lighting fitting tropfwassergeschützte Leuchte
rain shield Regenschutz
rain test Beregnungsversuch
rain-tight regendicht
rain water Regenwasser
to raise erheben; erhöhen; hochheben, hochziehen, hochfördern, hochbringen
to raise from ... to ... von ... auf ... erhöhen
to raise a selector einen Wähler heben (621.395)
to raise steam Dampf erzeugen, Dampf aufmachen
to raise the temperature through 10 degrees C die Temperatur um 10 °C erhöhen
raised temperature Übertemperatur, erhöhte Temperatur (61)
to raise the voltage die Spannung erhöhen
raising to a power Potenzierung
rake Rechen, Harke; (nach hinten geneigter) Anstellwinkel; Neigung
ram Ramme; Rammbär; Rammsporn (629); Stoßkopf, Preßstempel, Preßbär, Stößel, Preßbalken, Stoßbalken (621.9)
RAM = random access memory Speicher mit wahlfreiem Zugriff, Schreib-Lesespeicher
RAM-Disk [simuliertes Laufwerk im Arbeitsspeicher]
RAM refresh operation RAM-Auffrischoperation
to ram rammen, stampfen
rammed concrete Stampfbeton
Raman spectrum Raman-Spektrum

ramark Radarbake, Radarsendebake
ramp Rampe; Vorfeld (531.814.33); Sägezahn (64.37)
ramp generator Sägezahngenerator
random zufällig, willkürlich, beliebig
random access wahlfreier Zugriff; Zugriff in stochastischer Reihenfolge
random-access storage Speicher mit wahlfreiem Zugriff
at random wahllos, unregelmäßig
random chosen beliebig herausgegriffen
random error zufälliger Fehler, statistischer Fehler
random failures Zufallsausfälle
in random fashion wahllos, regellos
random generator Zufallsgenerator
random interval Zufallsstreubereich (620.1)
random noise weißes Rauschen (621.396)
random number Zufallszahl
random order Zufallsanordnung
random quantity Zufallsgröße
random sample wahllos herausgegriffenes Probestück, Zufallsstichprobe
random test Stichprobenprüfung
random twiggering Fehlauslösung
random voltage Rauschspannung
random-walk methode Monte-Carlo-Methode
range Bereich, Aktionsradius, Reichweite; Streubereich = Differenz zwischen Kleinst- und Größtwert (51); Meßbereich (621.317); Umfang; Schießstand (623); Küchenherd
range of application Anwendungsbereich
range discrimination Entfernungsauflösung
range of error Fehlerbereich
range finder Entfernungsmesser (623)
range gate Entfernungstorschaltung (621.396.96)
range height indicator Radaranzeigegerät (für Entfernung und Höhe)
range marker Entfernungsmarkengeber
range marking Zeitmarkierungs-…, Entfernungsmarkierungs… (621.397)
range of modulation Aussteuerbereich
range of nominal tension Nennspannungsbereich
range-only radar Radarentfernungsmesser
range of optical vision optische Sicht, optische Sichtweite
range preselection Bereichsvorwahl
range of products Umfang des Fertigungsprogramms, Fertigungsprogramm
range of regulation Regelbereich (621.316.7)
range resolution Entfernungsauflösung
range selector (switch) Meßbereichwähler, Bereichswähler
range of speed control Geschwindigkeitsregelbereich, Drehzahlregelbereich
range switch Bereichsumschalter (621.317); Wellenschalter
range of temperature Temperaturbereich
range of a transmitter Senderreichweite
to range rangieren; ordnen, anordnen; reichen, messen der Entfernung
to range from … to … reichen von … bis …
to range up to hinaufreichen bis, reichen bis
rapid schnell, rasch, geschwind
rapid autoreclosing Störungswegschaltung
rapid checking Schnellprüfung
rapid joint Schnellverbindung (621.6)
rapid make-and break mechanism Momentausschaltung
rapid memory Schnellspeicher, Speicher mit kurzer Zugriffszeit
rapid scan Schnellabtastung (621.397)
rapid slope steiles Gefälle
rapid starting schneller Anlauf, Schnellanlauf (621.313); guter Schwingeinsatz (621.396.61)
rapidly moving electrons schnelle Elektronen
rapidity Geschwindigkeit (nicht in räumlichem Sinne)
rare selten
rare earths seltene Erden
rare gases Edelgase
rarefaction Verdünnung
to rarefy verdünnen, selten machen
rarefied gas verdünntes Gas
raster Zeilenraster
rat race Ringhybride
rat tail Antennenzuführungsleitung
ratchet Ratsche, Knarre; Sperrklinke
ratchet bar Zahnklinkenstange
ratchet brace Bohrknarre
ratchet feed Sperrklinkenvorschub
ratchet lever Sperrklinkenhebel
ratchet pawl Sperrklinke
ratchet stop Ratschenklinke
ratchet wheel Sperrad; Schaltrad
rate Maß, Zahl (bezogen auf eine andere), Verhältnis; Steilheit; Geschwindigkeit; … ziffer (in Zusammensetzungen)
at a rate of 10 per hour mit einem Tempo von 10 je Stunde
rate action (control) proportional-differenzierende Regelung, Regelung mit Vorhalt
rate of amplitude change Steilheit (621.385)

rate of change Änderungsgeschwindigkeit
rate of change protection Stromanstiegsschutz
rate of change relay Steilheitsrelais, Stromanstiegsrelais
rate of combustion Verbrennungsgeschwindigkeit
rate of creep Kriechgeschwindigkeit (539)
rate of current change Stromänderungsgeschwindigkeit
rate of evaporation Verdampfungsgeschwindigkeit
rate of flow Fließgeschwindigkeit, Durchflußgeschwindigkeit
rate of formation Bildungsgeschwindigkeit
rate growth Stufenziehen (621.382)
rate gyro Dämpfungskreisel
rate of motion Geschwindigkeit (531)
rate of registering Drehzahl (eines Zählers)
rate of rise Anstiegsgeschwindigkeit, Steilheit
rate signal Geschwindigkeits-Steuergröße (621.5)
rate of stretch Reckgeschwindigkeit, Kriechdehnung
rate of travel of sound Schallausbreitungsgeschwindigkeit
rate of voltage change Spannungsänderungsgeschwindigkeit
rate of voltage rise Steilheit des Spannungsanstiegs
rate of working Arbeitsgeschwindigkeit, Leistung
to rate bemessen, auslegen für; Nennwerte festlegen; abschätzen, bewerten
rated Nenn-...
to be rated ausgelegt sein, bemessen sein; bewertet werden
rated burden Nennbürde (621.317)
rated circuit voltage Nennspannung
rated continuous current (Dauer-)Nennstrom (621.316.5)
rated cross section Nennquerschnitt
rated current Nennstrom
rated excitation Nennerregung
rated frequency Nennfrequenz
rated insulation voltage Isolationsnennspannung
rated interrupting capacity Nennabschaltleistung
rated interrupting current Nennabschaltstrom (621.316.5)
rated kva tap Nennleistungsklemme, Klemme für Nennübersetzung
rated kva of a transformer Nennleistung eines Umspanners
rated load Nennlast, Nennbelastung, Nennleistung
rated (maximum) voltage zulässige Höchstspannung
rated operating voltage Arbeitsspannung
rated output Nennleistung, Nennabgabe
rated overload characteristic Grenzstromkennlinie (621.314.5)
rated power Nennleistung
rated power factor Nenn-Leistungsfaktor
rated primary current Nennprimärstrom
rated primary voltage primäre Nennspannung
rated secondary current sekundärer Nennstrom
rated secondary voltage Nennsekundärspannung
rated service Nennbetrieb
rated short-time current Kurzzeitnennstrom, Kurzzeitstrom (612.316.5)
rated speed Nenndrehzahl, Nenngeschwindigkeit
rated torque Nennmoment
rated value Nennwert, Sollwert
rated voltage Nennspannung
rated voltage drop Nennspannungsabfall
rated working condition Nennwertbelastung
rating Bemessung, Belastbarkeit; Nennleistung; Klassifizierung; Bewertung, Bezugswert
rating of an instrument Nennmeßbereich eines Instrumentes
rating plate Typenschild
ratio Verhältnis; Quotient, Übersetzungsverhältnis; Windungsübersetzungsverhältnis (621.314.2)
ratio of actual steady-state controller output to maximum possible control output Stellgrad (621.316.7)
ratio-arm unit Brückenzweigwiderstand (621.317)
ratio-balance protective system Reaktanzschutz, Netzschutz
ratio box Brückenzweigwiderstand (621.317)
ratio of components Mischungsverhältnis
ratio control Stufenregelung
ratio correction factor Korrekturfaktor für das Übersetzungsverhältnis (621.317)
ratio error Übersetzungsfehler (621.317)
ratio meter Quotientenmesser
ratio-test set Wandlerprüfgerät
ratio of transformation Übersetzungsverhältnis (621.314.2)
ratio by volume Volumverhältnis

ratio of width to thickness Verhältnis Länge zu Breite
rational rational (51); vernünftig
rational electrical units rationale elektrische Einheiten
rationalization Betragbildung; Rationalisierung
to rationalize rationalisieren; reell machen, Betrag bilden (bei komplexen Größen) (51)
rationalized units rationale Einheiten
rattle Klappern, Rasseln, Rattern; Klapper, Rassel
to rattle klappern, rasseln, rattern
rattles (pl.) Rattern, Rattergeräusch (621.396.62)
raw roh, Roh-..., unbearbeitet
raw data unbearbeitete Daten
raw material Rohmaterial, Rohstoff
raw state roher Zustand, unbearbeiteter Zustand
raw steel Rohstahl
raw storage Basisinformation (621.316.7)
rawlbolt Steinschraube
ray Strahl
ray control Strahlsteuerung (621.385)
ray of electrons Elektronenstrahl
ray of ions Ionenstrahl
ray-locking device Strahlfalle (621.385)
raymark = radar beacon
ray of protons Protonenstrahl
ray-protected X-ray tube Strahlenschutzröhre (621.386)
raying Bestrahlung (61)
Rayleigh cycle Rayleigh-Schleife
Rayleigh disk Rayleigh-Scheibe
rayon Rayon, Glanzgarn
rayon yarn Rayongarn, Glanzgarn
RC circuit = resistance-capacitance circuit RC oscillator RC-Generator, RC-Summer
reach Reichweite; Flußabschnitt; Ausladung
reach forward Reichweite, Arbeitstiefe (621.9)
reacquisition time Wiedererfassungszeit
to react rückwirken (**with:** auf); reagieren (**with:** mit) (54)
reactance (induktiver) Blindwiderstand, Reaktanz
reactance bond Drosselverbinder
reactance coil Drosselspule
reactance control method Drosselregelung (621.314.5)
reactance dimmer Verdunkelungsdrossel
reactance(-distance) protection Reaktanzschutz
reactance drop induktiver Spannungsabfall
reactance grounded induktiv geerdet
reactance relay Reaktanzrelais
reactance tie bars (pl.) Drosselkupplung (621.311)
reactance valve Reaktanz-Röhre
reactance voltage Reaktanzspannung, Streuspannung, induktiver Spannungsabfall
reaction Rückwirkung; Auflagerdruck; Reaktion (54); Rückkopplung (621.396)
reaction alternator Reluktanzgenerator
reaction brush holder schräger Bürstenhalter
reaction circuit Rückkopplungskreis
reaction coil Rückkopplungsspule
reaction capacitor Rückkopplungskondensator
reaction generator Drehfeldgenerator
reaction point Auflager, Auflagerpunkt
reaction value Anlaufwert (621.316.7)
reactive reaktionsfähig (54); induktiv, Blind-..., um 90° nacheilend (621.3)
reactive amperage Blindstrom, nacheilender Blindstrom
reactive ampere hour meter Blindstromzähler, Blindstrom-Ampere-Stundenzähler
reactive component Blindkomponente, induktive Komponente (621.3); chemisch reagierender Anteil (54)
reactive current Blindstrom
reactive current compensation Blindstromkompensation
reactive current meter Blindstrommesser, Blindstrom-Amperemeter
reactive drop induktiver Spannungsabfall
reactive factor Blindleistung, Scheinleistung
reactive load Blindlast, induktive Last, nacheilende Last
reactive metering Blindlastzählung
reactive power Blindleistung
reactive power factor nacheilender Leistungsfaktor
reactive power meter Blindleistungszähler
reactive voltage induktive Blindspannung, induktiver Spannungsabfall
reactive voltamps (pl.) induktive Scheinleistung, Blindleistung
reactive volt-ampere(-hour)-meter Blindleistungszähler
reactivity Reaktionsverlust (54); Reaktivität (539)
reactor Drosselspule (621.316); Reaktor-Ofen, Reaktor, Atommeiler (539)
reactor charging Reaktorbeschickung
reactor cell Reaktorzelle
reactor containment Reaktorsicherheitshülle
reactor control Reaktorsteuerung und -regelung
reactor core Reaktorkern, Spaltzone

reactor instrumentation Reaktorüberwachung
reactor output Reaktorleistung
reactor pit Reaktorgrube
reactor service Reaktorbetrieb
reactor sphere Reaktorkuppel
reactor-start motor Motor mit Drosselanlasser
reactor supervision Reaktorüberwachung
reactor vault Reaktorgewölbe
reactor vessel Reaktorbehälter
to read lesen, ablesen, anzeigen (621.317); abfühlen (681.3)
to read (by, from, off) ablesen (an, auf, von)
to read in einlesen, eingeben
to read in a microscope an einem Mikroskop ablesen
to read off an instrument an einem Instrument ablesen
to read off from ablesen an
to read out auslesen, ausgeben
thermometer reads 70 °C Thermometer zeigt 70 °C
to read values Werte anzeigen
read after write Hinterbandkontrolle
read-around number mittlere Angriffszahl
read head Lesekopf
read only memory = ROM Festwertspeicher, (Nur-)Lesespeicher
read out circuit Leseschaltung
read out device Anzeigeeinrichtung
read punch Leselocher
read out tube Anzeigeröhre
reading Ablesung, Anzeige
reading accuracy Ablesegenauigkeit
reading head Lesekopf
to readjust abgleichen, erneut abgleichen, wieder abgleichen, nachtrimmen
readjustment Wiederherstellung; Nachgleich, Neuabgleich, Nachregulierung
ready fertig
ready character Bereitschaftszeichen (681.3)
ready-for-use Fertig-..., betriebsbereit
ready reserve generating set Sofortbereitschaftsaggregat
ready solubility leichte Löslichkeit
ready status Bereit-Zustand
reagent Reagens (54)
reagent bottle Reagensflasche
real Realteil (51)
real reell (51), reell, wirklich, echt
real axis reelle Achse
real coal Steinkohle
real image reelles Bild, echtes Bild
real number reelle Zahl
real power Wirkleistung (621.3)
real term reeller Anteil (51)
real time Echtzeit, Realzeit
real-time operation Real-time Betrieb, Echtzeitbetrieb, schritthaltende Datenverarbeitung
real-time processing Echtzeitverarbeitung
real value Istwert
to realign ausrichten (**to:** nach), sich ausrichten (**to:** auf)
realm Reich, Bereich
realm of experience Reich der Erfahrung
realm of possibility Bereich des Möglichen
realm of pure science Reich der reinen Wissenschaft
to ream reiben, aufreiben (621.9)
reamer Reibahle
reamplification Wiederverstärkung
rear Hinter-..., hinter, rückseitig; Acker-..., achter (629.12)
rear axle Hinterachse
rear connection rückseitiger Anschluß, Anschluß von hinten (621.3)
rear curtain hintere Antennenwand, Reflektorwand (621.396.677)
rear feed Rückwärtsspeisung (621.396.677)
rear light Rücklicht
rear panel Rückwand
rear terminal rückseitige Klemme
rear terminal block rückwärtige Anschlußleiste
rear view Rückansicht, Ansicht von hinten
rear view mirror Rückspiegel (629.113)
to rearrange umschreiben (51); neuordnen; umordnen, umgruppieren; eine Verbindung verlagern (621.395)
rearrangement Umordnung
rearward nach hinten zu gelegen
to reassemble wieder zusammenbauen
Réaumur Reaumur
to rebore nachbohren, ausbohren
rebound Rücksprung, Rückfall, Abprallen
to rebound abprallen, zurückspringen
rebroadcasting transmitter Umsetzer
rebuild Grundüberholung
to rebuild wieder aufbauen
to recalibrate nacheichen
recalibration Nacheichung, Neueichung
to recede rückschreiten, zurückweichen; rückwärts durchlaufen; sich zurückziehen, sich entfernen (**from:** von); rücklaufen (621.9); weglaufen
receipt signal Quittungszeichen (621.39)
to receive empfangen, aufnehmen, bekommen, erhalten

to receive motion from angetrieben werden durch
to receive a signal Zeichen aufnehmen, Signale empfangen (621.396)
received band Empfangsband, Empfangsbandbreite (621.396)
received pressure Spannung am Leitungsende (621.3)
received voltage Spannung am Leitungsende (621.3)
receiver Empfänger; Leitungsende (621.3); Empfänger (621.396); Druckspeicher, Windkessel (621.6); Handapparat, Hörer (621.395), Hörkapsel
receiver cabinet Empfängergehäuse
receiver capsule Hörkapsel
receiver circuitry Empfängerschaltung
receiver chassis Empfänger-Chassis
receiver earth Empfänger-Erdanschluß
receiver-earth terminal Empfänger-Erdklemme
receiver loudspeaker Empfangslautsprecher
receiver output volume Empfängerlautstärke
receiver rest Apparategabel, Gabel (621.395)
receiver valve Empfängerröhre
receiving Empfangs-...
receiving aerial Empfangsantenne, Empfängerantenne
receiving apparatus Empfangsapparat (621.394)
receiving circuit Empfangskreis
receiving coil Empfängerspule
receiving end Empfangsseite, Ausgangsseite, Verbraucherseite, Ausgang, Ende
receiving-end voltage Spannung am Leitungsende
receiving installation Empfangsanlage (621.396)
receiving loop Empfangsrahmen(-Antenne)
receiving perforator Empfangslocher (621.394)
receiving reference equivalent Empfangsbezugsdämpfung (621.39)
receiving station Empfangsstation
receiving tube Empfängerröhre, Empfangsröhre
receptacle Behälter; Stecker; Steckerloch, Steckerleiste
receptacle outlet Steckeranschlußstelle, Steckdose
reception Annahme, Aufnahme, Anmeldung, Empfang; Empfang (621.394), (621.396)
reception via demodulator repeater Ballempfang
receptivity Aufnahmevermögen
recess Einbuchtung, einspringende Vertiefung, Nische; Einschnürung; Loch, Höhlung
to recess einspringen, zurücktreten; einsenken (621.8); mit einer Rille versehen, einschnüren
recessed collar-head screw Halsschraube
recessed switch Einbauschalter, Unterputzschalter
to recharge neu laden, wieder laden, wieder aufladen (621.3)
rechargeable battery aufladbare Batterie
reciprocal reziprok; umgekehrt; wechselseitig, gegenseitig
reciprocal attraction gegenseitige Anziehung
reciprocal lattice reziprokes Gitter (54)
reciprocal ohm Siemens (Leitwerteinheit)
reciprocal process umgekehrter Prozeß
reciprocal relationship Wechselbeziehung
reciprocal replacement gegenseitige Austauschbarkeit
reciprocal (value) Kehrwert
to reciprocate hin- und hergehen, sich hin- und herbewegen
reciprocating hin- und hergehend; Kolben-... (621)
reciprocating steam engine Kolbendampfmaschine
reciprocity Reziprozität
reciprocity theorem Reziprozitätssatz, Reziprozitätstheorem (621.396)
to recirculate umlaufen lassen, im Kreislauf führen
recirculated air Umluft
recirculation Umlauf, Kreislauf
to reckon rechnen, berechnen, auswerten, zählen
to reckon among zählen zu, als zugehörig zu ... bewerten
to reckon from ... to ... von ... bis ... zählen
reckoner Rechentafel, Multiplikationstafel
reckoning Rechnen, Berechnung
reclaimable copper Altkupfer
to reclose wieder einschalten
reclosing relay Wiedereinschaltrelais
reclosing time Wiedereinschaltzeit
recognition Erkennung
recognized zulässig, anerkannt
recognized overload zulässige Überlast
recoil Rückstoß, Rückschlag (539), (623)
to recoil rücklaufen, rückstoßen; neu wickeln, umspulen
recombination Rekombination
recombination rate Rekombinationskoeffizient
to recombine rekombinieren, sich rekombinieren; wieder zusammentreten; wieder zusammenfügen

recommended crest working reverse voltage Nennsperrspannung (621.382.2), (621.314.63)
to recondition überholen, instand setzen, aufarbeiten
reconditioned carrier aufgefrischter Träger, angehobene Trägerwelle (621.396)
reconditioning Überholung, Aufarbeitung, Instandsetzung
to reconnect wieder anschließen, wieder einschalten; umklemmen
reconnection Umschaltung, Wiedereinschaltung
to reconstruct wieder aufbauen, neu aufbauen
reconstructed mica Mikanit; Glimmer-Erzeugnisse (pl.)
reconvergence Wiedervereinigung, Zusammenlaufen
record Registrierung, „Schrieb" (621.317); Schallplatte (681.84); Aufzeichnung, Protokoll, Niederschrift, Liste, Tätigkeitsnachweis
record changer Plattenwechsler
record density Aufzeichnungsdichte
record of measured values Meßprotokoll
record noise Plattengeräusch
record player Plattenspieler
record position Meldeplatz (621.395)
record reader Belegleser
record turntable Plattenteller
to record aufschreiben, aufzeichnen, registrieren; anzeigen; zu Protokoll nehmen
to record faithfully getreu aufzeichnen, sauber wiedergeben
recorded call automatische Wählwiederholung
recorded information service Ansagedienst
recorder Schreiber, Registriergerät, schreibendes Meßinstrument; Rekorder, Aufzeichnungsgerät, Wiedergabegerät
recording Registriervorgang; Registrierung, Aufzeichnung
recording device Aufnahmeeinrichtung, Aufnahmegerät (681.84)
recording Registrier-…, Schreib-…
recording completing trunk Überweisungsleitung, ULMV
recording cutter Schneidnadel, Schneidstichel (681.84)
recording density Schreibdichte, Bitdichte
recording element Registrierteil
recording head Schreibkopf, Aufnahmekopf
recording instrument Registrierinstrument, schreibendes Meßinstrument
recording-level band Aussteuerbereich (681.8)
recording-level indicator Aussteuerungsanzeiger
recording magnet Magnetkopf
recording mode Schreibverfahren
recording pen Tintenschreiber (621.317)
recording potentiometer registrierendes Potentiometer, Keinathschreiber
recording speed Schreibgeschwindigkeit
recording tape Tonband, Magnetband
recording track Aufzeichnungsspur
recording train Rollenzählwerk
recording transmission measuring set Pegelschreiber
recording trunk Überweisungsmeldeleitung, ÜLM
recovered charge Sperrverzugsladung (621.382.2)
recovery Gewinnung, Wiedergewinnung, Rückgewinnung; Erholung, Rückkehr in den Ausgangszustand
recovery effect Sperrträgheit (621.382.2)
recovery time Abklingzeit (621.396.62); Erholungszeit; Verzögerungszeit (621.314.63); Ausgleichszeit, Beruhigungszeit
recovery voltage wiederkehrende Spannung (621.316.5)
recrystallization Rekristallisation
to recrystallize rekristallisieren
rectangle Rechteck
rectangular rechtwinklig, rechteckig
rectangular bar Vierkantstange
rectangular block Quader
rectangular co-ordinates rechtwinklige Koordination
rectangular loop Rechteckschleife
rectangular scan Rechteckabtastung
rectangular timber Kantholz
rectangular tube rechteckiges Rohr
rectangular wave Rechteckwelle
rectification Gleichrichtung (621.3); Rektifikation (54), (66)
rectification curve Gleichrichterkennlinie, dynamische Kennlinie (621.385)
rectifier Gleichrichter (621.314)
rectifier anode Gleichrichteranode
rectifier apparatus Gleichrichtergerät
rectifier cathode Gleichrichterkathode
rectifier cell Gleichrichter(-Element)
rectifier characteristic Gleichrichterkennlinie
rectifier circuit Gleichrichterschaltung, Gleichrichterkreis
rectifier connection Gleichrichterschaltung
rectifier cubicle Gleichrichtergerät
rectifier disc Gleichrichterscheibe
rectifier instrument Gleichrichterinstrument
rectifier modulator Ringmodulator

rectifier photoelectric cell Gleichrichterphotozelle
rectifier stack Gleichrichtersäule
rectifier stage Gleichrichterstufe
rectifier substation Gleichrichterunterwerk, Gleichrichterstation
rectifier transformer Gleichrichterumspanner, Gleichrichtertransformator
rectifier-type instrument Gleichrichter(meß)instrument
rectifier unit Gleichrichtersatz
rectifier valve Gleichrichterröhre
to rectify gleichrichten (621.3); rektifizieren (54); berichtigen, richtigstellen
rectified alternating current gleichgerichteter Wechselstrom
rectified current gleichgerichteter Strom
rectified value elektrolytischer Mittelwert
rectifying Gleichrichter-...
rectifying effect Gleichrichterwirkung, Richtwirkung
rectifying stack Gleichrichtersäule
rectilineal geradlinig, gerade
rectilinear geradlinig, gerade
rectilinear motion geradlinige Bewegung
rectilinear scanning Zeilenabtastung
recto rechte Seite, rechtsseitig
recuperative rekuperativ
recuperator Rekuperator
to recur wiederkehren, sich wiederholen
recurrence Wiederkehr, Wiederholung; Rekursion (51)
recurrence formula Rekursionsformel (51)
recurrence frequency Impulsfrequenz (621.397)
recurrence rate Wiederholungsfolge; Impulsfrequenz (621.397)
recurrence relation Rekursionsformel (51)
recurrent wiederkehrend, sich wiederholend
recurrent circuit Kettenleiter
recurrent network Kettenleiter
recurring wiederkehrend
to recycle rückführen, im Kreislauf führen, rücklaufen lassen (54), (66); wieder aufbereiten
red rot
red brass Rotguß
red-hot rotglühend, rot
red lead Mennige
red magnetism Nordmagnetismus
red metal Rotguß
red shift Rotverschiebung (52)
red shortness Rotbrüchigkeit, Warmbrüchigkeit
red-yellow-blue rot-grün-veil (Phasenfarben)
to redden röten, rot färben
reddish rötlich
reddish-brown rötlich-braun, rotbraun
to redesign umkonstruieren, neu konstruieren
rediffusion Rückdiffusion (54); Wiederausstrahlung (621.396)
redirecting media Rückwurfmittel, Rückstrahlmittel (621.32)
redirecting surfaces Rückstrahlflächen, Rückwurfflächen (621.32)
redness Rotglut
to redouble nochmals verdoppeln, vervielfachen
to redress ausbessern
to redry nachtrocknen (67)
redryer Nachtrockner (67)
to reduce reduzieren, vermindern, verkleinern, herunterdrücken; reduzieren (54); abschwächen (77)
to reduce by vermindern um
reduced carrier transmission Übertragung mit vermindertem Träger
reduced kva tap Anschluß für verminderte Leistungsaufnahme
to reduce to meal zu Mehl mahlen, zu Mehl zerkleinern
to reduce to normal aufs Normale zurückführen
to reduce the output die abgegebene Leistung vermindern
to reduce the pressure den Druck vermindern, Druck reduzieren
to reduce to zerkleinern zu; sichvereinfachen auf, vereinfachen zu, sich zurückführen lassen auf (51)
reduced scale verkleinerter Maßstab
reduced-voltage starting Anlassen mit verminderter Spannung
reducing Reduzier-..., reduzierend, Reduktions-...
reducing agent Reduktionsmittel
reducing fitting Reduktionsstück
reducing joint Reduktionsmuffe (621.315.2)
reducing valve Reduktionsventil
reduction Reduktion, Verminderung, Verkleinerung, Senkung, Abnahme; Reduktion (54); Abschwächung (77); Einschränkung
reduction in amplitude Amplitudenabnahme
reduction factor Reduktionsfaktor, Umrechnungsfaktor
reduction gear Untersetzungsgetriebe
reduction in size Größenverminderung
reduction of speed Abwärtsregelung

reduction unit Untersetzung, Untersetzungsgetriebe
reduction in weight Gewichtsverminderung
reduction work Verkleinerungsarbeit
redundancy Redundanz
redundancy check Redundanzkontrolle
redundant überschüssig, überzählig
reed Rohr, Ried (58); Zunge (681); Schwingzunge (534)
reed contact Schutzgaskontakt, Schutzrohrkontakt
reed frequency meter Zungenfrequenzmesser
reed loudspeaker Zungenlautsprecher
reed relay Reed-Relais, Relais mit Schutzrohrkontakt
reed switch Reed-Schalter
reel (nicht tragende) Rolle; Spule
reel cutting machine Rollenschneidmaschine (67)
reel drum Wickelrolle
to reel spulen, wickeln, rollen
to reel in aufhaspeln; „hineinleiern“ (629.13)
to reel out herausspulen, abspulen; „rausleiern“ (629.13)
to reel up aufspulen, aufhaspeln, auftrommeln, aufwickeln
re-enterable jederzeit wiederverwendbar (681.3)
re-entrant einspringend
re-entrant horn gefalteter Trichter
re-entrant series winding Reihenwicklung (621.313)
reference to Bezugnahme auf, Heranziehung von
reference diode Referenzdiode
reference circuit Eichleitung; Bezugskreis, Bezugsverbindung (621.372)
reference equivalent Bezugsdämpfung
reference frequency Bezugsfrequenz
reference input Führungsgröße
reference letter Bezugszeichen, Buchstabenbezeichnung
reference level Bezugspegel (621.395)
reference pilot Pilotton
reference plane Bezugsebene
reference point Bezugspunkt
reference potential Bezugspotential
reference signal Vergleichssignal
reference temperature Bezugstemperatur
reference time Bezugszeit
reference tone Normalton
reference value Sollwert, Bezugsgröße
reference voltage Vergleichsspannung, Bezugsspannung, Referenzspannung
reference white Bezugsweiß, Referenzweiß
to refine raffinieren, reinigen, verfeinern, feinen
to refine the grain das Korn verfeinern
to refine metal Metall raffinieren, Metall läutern
refined Rein-..., Raffinade-...
refined copper Raffinadekupfer
refinery Raffinerie
to refinish aufpolieren
refit Neuausrüstung (62)
to refit neu ausrüsten, umrüsten
to reflect reflektieren, widerspiegeln, zurückwerfen, zurückspiegeln
reflected capacity Rückwirkungskapazität
reflected glare Glanzreflex (535)
reflected power reflektierte Leistung
reflected resistance übertragener Widerstand, Rückwirkungswiderstand (621.396)
reflected wave rücklaufende Welle, reflektierte Wanderwelle
reflectance Reflexion (535)
reflectance reduction Reflexminderung, Vergütung (535)
reflecting reflektierend; Rück-..., Spiegel-...
reflecting electrode Bremsfeldanode
reflecting galvanometer Spiegelgalvanometer
reflecting instrument Spiegelinstrument
reflecting power Rückstrahlvermögen, Reflexionsvermögen
reflection Reflexion (535), Rückstrahlung (621.396)
reflection coefficient Reflexionsfaktor, Reflexionsvermögen
reflection factor Reflexionsfaktor
reflection loss Stoßdämpfung, Reflexionsverlust
reflection-measuring set Fehlerdämpfungsgerät
reflection ratio Reflexionsfaktor
reflection sounding Echolotung
reflectivity Reflexionsfähigkeit, Reflexionsvermögen
reflectometer Reflektometer, Fehlanpassungsanzeiger
reflector Reflektor (535), (621.32), Richtstrahler, Spiegel (621.396)
reflector antenna Reflektorantenne, Spiegelantenne
reflector floodlight Spiegelscheinwerfer
reflector lamp Scheinwerferlampe, Reflektorlampe
reflectoscope Episkop
reflex Reflex
reflex amplifier Reflexverstärker
reflex circuit Reflexschaltung
reflex-circuit arrangement Reflexschaltung

reflex klystron Reflexklystron
refluent rückfließend, zurückfließend
reflux Rückfluß, Rücklauf (66)
to reflux rückkochen (54), (66)
to refract brechen (535)
refracting brechend, Brech-...
refracting telescope Refraktor (52)
refraction Brechung (535)
refractive Brechungs-..., brechend
refractive gradient Gradient des Brechungsexponenten
refractive index Brechungsindex, Brechungszahl, Brechzahl
refractive modulus modifizierter Brechwert
refresh operation Auffrischoperation (681.3)
refrigerant Kühlmittel
to refrigerate kühlen, gefrieren, gefrieren lassen
refrigerating Kühl-...
refrigerating machine Kältemaschine
refrigeration Kühlung
refrigeration cabinet Kühlschrank
refrigeration locker Kühlschrank
refrigerator Kühlschrank, Kühlanlage, Kühleinrichtung
refrigerator ship Kühlschiff
refusal of acceptance Abnahmeverweigerung
refuse Abfall, Rückstand, Müll; Ausschuß
refuse-destructor furnace Müllverbrennungsofen
to refuse sich weigern, verweigern, ablehnen; wieder schmelzen
to refuse a frequency eine Frequenz nicht durchlassen
to regain wiedererlangen, zurückerhalten
to regain control wieder steuerfähig werden (621.385)
to regelate zusammenfrieren, wieder frieren
regelation Wiedergefrieren, Zusammenfrieren
to regenerate regenerieren, auffrischen; rückkoppeln
regenerated leach liquor aufgefrischte Mutterlauge
regeneration Regenerierung, Auffrischung; Rückkopplung (621.396)
regenerative Regenerativ-...; Rückkopplungs-...
to be regenerative als Rückkopplung wirken
regenerative braking Nutzbremsung, Generatorbremsung (621.33)
regenerative circuit Rückkopplungskreis, Rückkopplungsschaltung
regenerative control Nutzbremsschaltung (621.33)
regenerative detection Rückkopplungsgleichrichtung
regenerative feedback Mitkopplung; positive Rückführung (621.316.7); Rückkopplung (621.396)
regenerative repeater rückgekoppelter Verstärker (621.395)
regenerator rückgekoppelter Verstärker (621.395)
region Gebiet, Bereich, Raum; Versorgungsgebiet; Zone (621.382)
region of convergence Konvergenzbereich
regional gebietlich, Gebiets-..., regional
regional broadcasting station Regionalsender, Bezirkssender
register Zählwerk (621.317); Register (681.3); Stimmumfang, Stimmlage; Rauchschieber, Rauchklappe (66); Eingriff (621.8)
register constant Zählerkonstante
register face Zählwerksskala (621.317)
register key Zähltaste, Gesprächszähltaste (621.395)
register mark Paßmarke, Paßzeichen
register ratio Zählwerkskonstante (621.317)
register reading Zählwerksanzeige, Zählwerksablesung
register translator Nummern-Umsetzer (621.395)
to register zählen, registrieren; aufzeichnen; in Eingriff stehen (621.833); einpassen (621); einschreiben (65)
to register with eingreifen in, in Eingriff stehen mit
to register zero auf Null zeigen
registering instrument zählendes Meßinstrument
registering mechanism Zählwerk
registering Zählung
registration Registrierung, Eintragung; Zählwerksanzeige
registration number Zulassungsnummer (629.113)
regressive rückschreitend, regressiv
to regrind nachschleifen
to regroup umgruppieren, neu anordnen, neu gruppieren
regular regelmäßig, regulär, gesetzmäßig, normal
regular reflection reguläre Reflexion, echte Reflexion (535)
regular reflection factor Koeffizient der regulären Reflexion (535)
regular transmission reguläre Lichtdurchlässigkeit

regular transmission factor Koeffizient der regulären Durchlässigkeit
regularity Regelmäßigkeit
regularity return loss Rückflußdämpfung
to regulate regulieren, einregulieren (621), (627), (681.1)
regulated reguliert; stabilisiert, geregelt (621.3)
regulated frequency regulierte Frequenz
regulated line section Regelungsabschnitt (621.395.4)
regulated power supply stabilisiertes Netzanschlußgerät
regulated quantity Stellgröße
regulated voltage geregelte Spannung
regulated voltage source stabilisierte Stromquelle
regulated voltage supply unit spannungsstabilisierte Stromversorgung
regulating Regulier-..., Regel-...
regulating cell Regelzelle (621.355)
regulating equipment Regeleinrichtung
regulating pilot Regelpilot (621.395.4)
regulating relay Regelrelais
regulating resistance Regulierwiderstand, Regelwiderstand
regulating resistor Regulierwiderstand, Regelwiderstand
regulating switch Regelschalter
regulating transformer Regeltransformator, Regelumspanner, Stelltransformator
regulating unit Stellglied (621.316.7)
regulating valve Regulierventil, Regelventil
regulating winding Regulierwicklung, Regelwicklung
regulation Vorschrift, Bestimmung; Spannungsänderung (bei Abwurf der Nennlast) (621.313); Regelung, Regulierung (siehe auch **control**)
regulation "up" Spannungserhöhung bei Lastabwurf
regulation characteristic Regelcharakteristik
regulator Regler
reguline einwandfreier galvanischer Überzug
reheating Nacherhitzung
to reinforce verstärken, stärken; anfachen (534); armieren, bewehren
reinforced concrete Eisenbeton, Stahlbeton
reinforced-concrete pole Eisenbetonmast
reinforced seam Wulstnaht (621.791)
reinforcing ply Verstärkungseinlage, Kordeinlage (67)
reinforcing rib Verstärkungsrippe
reinforcement Verstärkung, Verstärkungseinlage
Reis microphone Reismikrophon
reject Ausschußstück, beanstandetes Stück
reject pulse Abwurfimpuls (621.3)
to reject zurückweisen, ablehnen, verwerfen; Abnahme verweigern; zurückgeben; nicht durchlassen, sperren (621.392)
to reject a frequency eine Frequenz nicht durchlassen, eine Frequenz sperren
rejection Ausschuß (621); Fehlguß (621.7); Abnahmeverweigerung, Zurückweisung bei der Abnahme; Dämpfung, Unterdrückungsfaktor
rejection band Dämpfungsbereich, Sperrbereich (621.372.5)
rejector circuit Sperrkreis (621.3)
rejector impedance Resonanzwiderstand
rejuvenation Auffrischung, Regenerierung (621.385)
relative relativ, vergleichbar
relative abundance relative Häufigkeit
relative equivalent relative Bezugsdämpfung
relative humidity relative Luftfeuchtigkeit
relative level relativer Pegel
relative permittivity Dielektrizitätszahl, relative Dielektrizitätskonstante
relative position relative Lage
relative rate of ... to Verhältnis von ... zu ...
relative speed Relativgeschwindigkeit, Relativdrehzahl
relative velocity Relativgeschwindigkeit
relativity Relativität
relativity physics Relativitätsphysik
to relax nachlassen, lockern, entspannen; vorentwarnen (35)
relaxation Entspannung, Lockerung; Relaxation (53); Vorentwarnung (35)
relaxation circuit Kippschaltung
relaxation frequency Kippfrequenz
relaxation inverter Wechselrichter
relaxation oscillator Kippgenerator (621.396)
relaxation time Relaxationzeit
relaxation-type oscillator Kippschwingungsgenerator
relay Relais (621.3)
relay bouncing Relaisprellen
relay broadcasting station Rundfunktochtersender, Relaissender
relay chatter Relaisflattern
relay coil Relaiswicklung
relay contact Relaiskontakt
relay control Relaissteuerung
relay device Relaiseinrichtung
relay drops out Relais fällt ab
relay magnet Relaismagnet
relay operates, relay responds Relais zieht an

relay pickup voltage Relais-Ansprechspannung
relay protection Relaisschutz
relay receiver Ballempfänger
relay recorder Relaisschreiber
relay station Tochtersender, Nebensender (621.396); Relaisstelle
relay sticks Relais klebt
relay switch Schaltschutz
relay terminal Relaisklemme
relay-type recording instrument Fallbügelinstrument
relay-type sensor Relaisfühler
relay winding Relaiswicklung
to relay durch Relais weiter übertragen
relaying sounder Morseticker mit Relaissatz
to re-lay wieder verlegen, wieder legen, wieder hinlegen
release Klinke (621); Auslösung, Ausklinkung, Freigabe (621.3)
release of the armature Ankerabfall
release current Auslösestrom, Abfallstrom
release gear Auslösegestände, Verklinkung (621.316.5)
release-guard signal Auslösequittungszeichen (621.395.5)
release key Löschtaste
release lag Abfallverzögerung
release mechanism Ausklinkvorrichtung
release signal Auslösezeichen (621.395.5)
release time Abfallzeit, Auslösezeit
release winding Abwurfwicklung
release wire Auslöseleitung (621.395.34)
to release auslösen (621.395); ausklinken; fallen lassen; entfesseln, loslassen; abfallen, abwerfen (621.318.5)
to release the brake die Bremse lösen
to release a circuit breaker einen Schalter ausklinken, einen Schalter auslösen
to release gases Gase austreten lassen, Gaseabgeben
releasing cam Auslösedaumen
reliable zuverlässig, betriebssicher
reliability Zuverlässigkeit
reliability test Lebensdauerprüfung, Zuverlässigkeitsprüfung
relief Erleichterung; Abhilfe, Nothilfe, Hilfe; Relief (73)
relief angle Freiwinkel (621.9)
relief etching Hochätzung (655)
to relieve hinterschneiden (621.9); erleichtern, entspannen; unterstützen, helfen
relieving Schwabbeln (621.357)
to reload wieder belasten; neu beschicken; umladen, neu beladen
relocatable verschiebbar (681.3)
relocatable program structures verschiebbare Programmstrukturen
relocating loader Lader für verschiebbare Programme (681.3)
reluctance Widerspenstigkeit, Widerstreben; magnetischer Widerstand, Reluktanz (621.3)
reluctance motor Reluktanzmotor
reluctivity spezifischer magnetischer Widerstand = Kehrwert der Permeabilität
to remagnetize wieder aufmagnetisieren, aufmagnetisieren
remainder Rest (61); Rest, Rückstand
remaining restlich, übrig, Rest-...
remains Rückstände (64), (66)
remanence Remanenz, zurückbleibende Magnetisierung
remanent remanent
remanent flux remanenter Fluß
remanent magnetism remanenter Magnetismus
remanent magnetization remanente Magnetisierung
remanent relay Haftrelais
R. E. M. E. = Royal Electrical and Mechanical Engineers Königliches Pionierkorps (Technische Truppe)
remedy Heilmittel, Abhilfsmittel, Abhilfe
to remedy heilen, beseitigen, gutmachen, abhelfen
to remelt umschmelzen (s. **to melt**)
remelt metal Umschmelzmetall
to remigrate rückwandern
remigration Rückwanderung
remnant Rest, Überbleibsel
remodulation Rückmodclung
remote fernliegend, abgelegen; abgesetzt, Fern-...
remote batch processing Stapelfernverarbeitung
remote control Fernsteuerung, Fernbetätigung, Fernwirken
remote-control equipment Fernsteueranlage, Fernwirkanlage
remote controlled ferngesteuert, fernbetätigt
remote cut-off tube Exponentialröhre, Regelröhre (621.385)
remote indicator Fernanzeigegerät, abgesetztes Anzeigegerät
remote-indicating device Fernanzeigevorrichtung
remote manual control Handfernbetätigung
remote metering Fernmessung, Telemetrie

remote operation Fernbetätigung (621.316.5)
remote position indicator Stellungsgeber (621.316.7)
remote power indicator Fernleistungsmesser
remote side abgewandte Seite, Rückseite
remote setting Ferneinstellung
remote station abgesetzte Station
remote tuning Fernabstimmung
remotely piloted vehicle unbemannter Flugkörper
to remount wieder einbauen; wieder aufsteigen
removable abnehmbar, lose
removable change wheel loses Wechselrad
removable element ausfahrbarer Teil (621.316.5)
removable truck Schaltwagen (621.316.5)
removable truck-type switchgear Schaltwagenanlage
to remove wegnehmen, wegführen, abholen; fortnehmen, beseitigen, entfernen; ausbauen, herausnehmen; beheben
to remove the burr abgraten, entgraten
to remove energy Energie entziehen, Energie entnehmen
to remove all pull vollständig von Zug entlasten
to remove rust entrosten, den Rost entfernen
to remove the stress entlasten, entspannen
to remove the waste material den Abfallbeseitigen
to rend raufen, reißen, spalten, zerreißen, zerspalten
to rend apart abspalten, abreißen
to rend asunder losreißen, lossplittern
to render machen; darbieten, vorführen
to render audible hörbar machen
to render visible on a screen auf einem Schirm sichtbar zeigen, vorführen
rendering Verputz (69)
renewable auswechselbar, ersetzbar
renewable fuse unit Streifensicherung, offene Sicherung
renewal Erneuerung, Ersatz
renewal parts Ersatzteile (621.316.5)
rent Riß, Spalte; Miete, Pacht
to renumber umnumerieren, neunumerieren
to repack nachstopfen, nachpacken (625)
repair Reparatur, Instandsetzung, Ausbesserung
repair account Reparaturkonto
in bad repair in schlechtem Erhaltungszustand, schlecht repariert
in good repair in gutem Erhaltungszustand, gut gepflegt, gut erhalten
repair shop Reparaturwerkstatt, Ausbesserungswerkstatt, Instandsetzungswerkstatt
repair test Reparaturprüfung
repair time Instandsetzungszeit
repair work Reparaturarbeit, Instandsetzungsarbeit, Ausbesserungsarbeit
to repair reparieren, ausbessern, instandsetzen
reparability Reparierbarkeit, Wiederherstellbarkeit
reparable reparierbar, wiederherstellbar, ausbesserungsfähig
repeated wiederholt, häufig, mehrfach
repeated impact test Dauerschlagversuch
repeated root Mehrfachwurzel (51)
repeated stress Dauerwechselbeanspruchung
repeater Verstärker, Leitungsverstärker (621.39); Repetieruhr (681.1); Tochtergerät, Tochterinstrument (621.317); Tochterkompaß, Tochter (629.12)
repeater bay Verstärkergestell
repeater coil Fernleitungsübertrager (621.39)
repeater section Verstärkerfeld (621.39)
repeater spacing Verstärkerabstand
repeater station Verstärkeramt (621.395)
repeating Repetier-...
repeating coil Übertrager, Ringübertrager
repeating signal Rückmeldesignal
repeating station Zwischenstation (621.394)
to repel abstoßen, sich abstoßen
to repel back into in ... zurücktreiben
repellent abstoßend
repeller Reflexionselektrode (Klystron)
repeller plate Reflexionsanode (Klystron)
repelling force Abstoßungskraft
repercussion Rückwirkung; Echo, Widerhall
reperforator Empfangslocher (621.394)
repetition Wiederholung, Wiederkehr
repetition frequency Impulsfrequenz (621.397)
repetition instruction Wiederholbefehl (681.3)
repetition job Serienarbeit, gleichmäßig wiederkehrender Arbeitsvorgang
repetition rate Wiederholungsfrequenz; Impulsfrequenz (621.397)
repetition work Serienherstellung
repetitious sich wiederholend, wiederkehrend (51)
repetitive fortlaufend, wiederholt
repetitive peak forward current periodischer Vorwärts-Spitzenstrom
repetitive peak reverse voltage periodische Rückwärts-Spitzensperrspannung
replace parts Ersatzteile

to replace ersetzen; auswechseln, austauschen (**by:** durch), (**with:** gegen); (an die alte Stelle) zurücklegen
to replace a fuse eine Sicherung auswechseln
to replace the handset (den Handapparat) einhängen
to replace the receiver den Hörer auflegen (621.395)
replaceable ersetzbar, auswechselbar
replacement Auswechselung, Ersatz, Ersatzteil
replacement part Ersatzteil
to replenish nachfüllen; Nachladen (621.3); nachliefern
replenisher Nachfüllmasse
replenishment Nachfüllung, Nachladung; Nachlieferung
to replicate nachbilden, kopieren, „abkupfern“
replication Nachbildung
to replot umzeichnen, umtragen, neu auftragen
to repolish neu aufpolieren, nachpolieren
to repound abprallen
to repose ruhen, ruhend liegen
repoussé getrieben, gehämmert (73)
representable darstellbar (**by:** durch)
representation Darstellung
representative beispielhaft; charakteristisch; kennzeichnend (**of:** für)
representative specimen Stichprobe
reprocessing Aufarbeitung (54)
reprocessing plant Aufbereitungsanlage (539)
reproducer Kartendoppler
to reproduce doppeln (681.3)
reproducible reproduzierbar
reproduction Reproduktion; Wiedergabe; Doppeln (681.3)
reproduction factor Multiplikationsfaktor (539)
reproduction of sound Tonwiedergabe
reprogrammable ROM reprogrammierbares ROM
repulsion Abstoßung
repulsion induction motor Repulsionsmotor mit Käfiganker, Repulsionsmotor mit Zusatzkäfig
repulsion motor Repulsionsmotor (621.313)
repulsion-start induction motor Asynchronmotor mit Repulsionsanlauf
repulsion-type ammeter Rundspul-Amperemeter
repulsive abstoßend
repulsive force Abstoßungskraft
request Abrufen (621.39); Anforderung (681.3)
to request anfordern (621.395), (621.398)
request repeat system Fehlererkennungssystem mit Signalwiederholung (621.394)
to require bedürfen, brauchen, nötig haben; fordern, verlangen, beanspruchen
to require power Leistung benötigen, Leistung verlangen
required power Leistungsaufnahme, Leistungsbedarf
to reradiate wieder ausstrahlen, zurückstrahlen
reradiation Rückstrahlung
reradiation error Rückstrahlfehler
rerouting Verbindungsaufbau mit Rücksprung (621.395)
rering Nachrufen (621.395)
to rerun wiederholen (681.3)
reseal voltage Löschspannung (621.316.9)
research Forschung
research activity Forschungstätigkeit
research and development, R and D Forschung und Entwicklung
research institute Forschungsinstitut
research reactor Forschungsreaktor
reserve Reserve, Zurückhaltung; Reserve (621)
reserve bus bar Reserve-Sammelschiene
reserve equipment Reserve; Reservegeräte, Reservemaschinen (pl.)
reserve machinery Reservemaschinenanlage
reserve plant Reserveanlage
reserve protection Reserveschutz
reservoir Reservoir, Speicher, Vorratsbehälter
reservoir capacitor Ladekondensator, Speicherkondensator (621.3)
reset Bremslüftschutz (629); Rückfall (621.318); Rückstellung, Rücksetzen (681.3)
reset input Löscheingang (631.374)
reset contactor Bremslüftschütz (629)
reset magnet Rückstellmagnet
reset pulse Rückstell-Impuls, Resetsignal, Rückstellungssignal
reset switch Bremslüftschalter (629)
reset winding Rückstellwicklung
to reset wieder einstellen, zurückstellen, nullstellen; zurückfallen (621.318.5); löschen (681.3)
resetting Rückstellung, Nachgleich
resetting magnet Rückholmagnet
resetting spring Rückstellfehler
resident innewohnend, vorhanden; permanent verfügbar (681.3)
resident engineer örtlicher Bauleiter
residual Winkelfehler (621.317)
residual bleibend, nachbleibend, Rest-...
residual charge Restladung

residual compensation Reststromschutz (621.316.9)
residual current Reststrom; Fehlerstrom
residual-current operating system Fehlerstromschutzeinrichtung
residual-current region Anlaufstromgebiet (621.385)
residual error Restfehler, Eigenfehler
residual excitation remanente Erregung
residual field Restfeld, Remanenzfeld (621.313)
residual gas Restgas
residual image Nachleuchtbild
residual induction Remanenzinduktion, Remanenz
residual magnetism Restmagnetisierung, remanenter Magnetismus
residual magnetization remanente Magnetisierung
residual octantal error elevationsabhängiger Fehler (621.371)
residual relay Erdstromrelais, Erdschlußrelais
residual-time constant Eigenzeitkonstante
residual voltage Restspannung
residue Rest, Bodensatz, Rückstand (66)
residue curve Oberwellenstromkurve, Kurvenform des Reststromes
residuum (pl. **-dua**) Residuum (51)
resilience Rückfederung
resiliency Rückfederung
resilient rückfedernd, federnd
resilient suspension elastische Aufhängung
resin Harz
resin-bonded paper Hartpapier
resin-cored solder Tinoldraht, Kolophoniumlötdraht
resin dipped kunstharzgetaucht
resin molded kunstharzumpreßt
resin scaled Kunstharz vergossen
resinoid harzähnlich
resinous harzig
resinous electricity Harzelektrizität, negative Elektrizität
resist Abdecker, Abdeckung (621.357), (621.382)
resist etchant Abdeckmittel
resistance Widerstand; Widerstandsfähigkeit (**to**: gegen)
resistance to ageing Alterungsbeständigkeit
resistance alloy Widerstandslegierung
resistance amplifier widerstandsgekoppelter Verstärker
resistance box Stöpselkasten, Widerstandskasten, Stöpselwiderstand (621.317)
resistance braking Widerstandsbremsung
resistance branch Widerstandszweig (621.3)
resistance brazing Widerstandshartlötung
resistance bridge Widerstandsmeßbrücke, Wheatstonesche Brücke
resistance to burning Abbrandfestigkeit
resistance butt welding Widerstandsstumpfschweißung
resistance-capacitance circuit RC-Schaltung
resistance-capacitance coupled amplifier Widerstandsverstärker, R-C-Verstärker, widerstandsgekoppelter Verstärker
resistance-capacitance coupling R-C-Kopplung
resistance-capacitance oscillator R-C-Generator, R-C-Summer
resistance-capacitor combination R-C-Schaltung
resistance coil Widerstandsspule, Widerstandsrolle
resistance to be compared Vergleichswiderstand
resistance to corrosion Korrosionsfestigkeit, Korrosionsbeständigkeit
resistance coupled widerstandsgekoppelt
resistance coupled amplifier widerstandsgekoppelter Verstärker
resistance coupling Widerstandskopplung
resistance drop ohmscher Spannungsabfall
resistance of earthed conductor Erdungswiderstand
resistance to earth Erdungswiderstand
resistance element Widerstandselement, Heizelement
resistance to fatigue Dauerbruchsicherheit
resistance frame Widerstandssatz
resistance of friction Reibungswiderstand
resistance furnace Widerstandsofen
resistance to fusion hoher Schmelzpunkt, Schwerschmelzbarkeit
resistance grounded über einen Widerstandgeerdet
resistance to heat Hitzebeständigkeit
resistance heating Widerstandsheizung; Widerstandserwärmung
resistance to impact Stoßfestigkeit, Schlagbiegefestigkeit
resistance-inductance coupling induktive Kopplung, Drosselkopplung (621.396.64)
resistance loss ohmscher Verlust, joulesche Wärme
resistance mat Widerstandsgewebe, Schniewindtband
resistance noise thermisches Rauschen, Widerstandsrauschen, Wärmerauschen

resistance notch Widerstandsstufe
resistance percussive welding Widerstandshammerschweißung
resistance pyrometer Widerstandspyrometer
resistance relay Wirkwiderstands-Distanzrelais
resistance to scratching Kratzfestigkeit
resistance spot welder Widerstandspunktschweißmaschine
resistance-start motor Motor mit Widerstandsanlasser
resistance temperature coefficient Temperaturkoeffizient des Widerstandes
resistance thermometer Widerstandsthermometer
resistance to train movement Fahrwiderstand
resistance-tuned oscillator RS-Summer, RS-Generator
resistance voltage ohmscher Spannungsabfall
resistance voltage divider Widerstands-Spannungsteiler, ohmscher Spannungsteiler
resistance to wear Verschleißfestigkeit, Abriebfestigkeit
resistance welder Widerstandsschweißgerät
resistance welding Widerstandsschweißung
resistance wire Widerstandsdraht
resistanceless widerstandslos
resistant widerstandsfähig (**to**: gegen)
resistant to acids säurefest, säurebeständig
resistant to cold kältebeständig
resistant to moisture feuchtigkeitsfest
resistant to wear verschleißfest, abriebfest
resistive widerstandsbehaftet, ohmisch, Widerstands-...
resistive branch ohmscher Zweig, Widerstandszweig
resistive cut-off frequency Entdämpfungsgrenzfrequenz
resistive coil Spule mit Widerstand, widerstandsbehaftete Spule, Spule mit Verlusten
resistive component Wirkkomponente (621.3)
resistive conductor widerstandsbehafteter Leiter, Widerstandsschnur
resistive coupling Widerstandskopplung
resistive disk Widerstandsscheibe
resistive film Widerstandsschicht, Widerstandsfolie
resistive inductor Spule mit Widerstand, widerstandsbehaftete Spule, Spule mit Verlusten
resistive load ohmsche Last, Widerstandsbelastung
resistive property Widerstandseigenschaft
resistive reverse current stationärer Sperrstrom
resistive wire widerstandsbehafteter Draht, Widerstandsdraht
resistivity Widerstandseigenschaft; spezifischer Widerstand, Widerstandsbeiwert, Eigenwiderstand (621.3)
resistor Widerstand (= Widerstandskörper)
resistor capacitor transistor logic = RCTL Widerstands-Kondensator-Transistor-Logik
resistor, composition Massewiderstand
resistor core Widerstandskörper
resistor element Widerstandselement, Widerstandskörper
resistor, film type Schichtwiderstand
resistor furnace Widerstandsofen
resistor housing Widerstandsgehäuse
resistor, photoconductive Photowiderstand
resistor wire Widerstandsdraht
resistor, wire wound Drahtwiderstand
to resoften wieder weichmachen
to resolder nachlöten
to resolidify wieder erstarren, wieder festwerden
resoluble zerlegbar, unterteilbar
resolution Auflösung, Auflösungsvermögen; Zerlegung, Unterteilung; Komponentenzerlegung
resolution error Auflösungsfehler
resolution triangle Zerlegungsdreieck
to resolve lösen, auflösen (51), (53); (in Komponenten) zerlegen
to resolve into zerlegen in; zusammenfassen zu (51)
resolved parts rechtwinklige Komponenten
resolver funktionsbestimmendes Element, Lösungselement (681.142); Funktionsdrehmelder, Koordinatenwandler (621.316.7)
resolver differential transmitter Funktionsdifferentialgeber
resolver receiver Funktionsempfänger
resolver synchro Funktionsempfänger, Funktionsdrehmelder (621.316.7)
resolver transformer Funktionsempfänger
resolver transmitter Funktionsgeber (621.316.7)
resolving power Auflösungsvermögen (535)
resolving time Auflösungszeit
resonance Resonanz; Spannungsresonanz (621.3)
above resonance oberhalb der Resonanz
resonance absorption Resonanzabsorption (53); Resonanzeinfang (539.17)
resonance amplifier abgestimmter Verstärker
below resonance unterhalb der Resonanz

resonance chamber Hohlraumresonator
resonance constant s. **oscillation constant**
resonance curve Resonanzkurve
resonance escape s. **to escape resonance**
resonance escape probability Wahrscheinlichkeit, dem Resonanzeinfang zu entgehen (539.17)
resonance frequency Resonanzfrequenz
resonance-frequency meter Resonanzfrequenzmesser
resonance hump Resonanzbuckel
resonance indicator Resonanzanzeiger
off resonance vom Resonanzpunkt entfernt
resonance point Resonanzpunkt
resonant schwingfähig, schwingend; Schwing...
resonant at an applied frequency für eine angelegte Frequenz in Resonanz
resonant cavity Hohlraumresonator
resonant circuit Schwingkreis, Saugkreis, Spannungsresonanzkreis
resonant frequency Resonanzfrequenz, Eigenfrequenz
resonant line Leitungsschwingkreis, schwingende Leitung
resonant method Resonanzverfahren
resonant rise Resonanzanstieg, Resonanzüberhöhung
resonant shunt Überbrückungsschwingkreis
resonant vibration Resonanzschwingung
to resonate in Resonanz schwingen, in Resonanz sein; in Resonanz kommen
to resonate with abstimmen auf, auf Resonanz mit... bringen; in Resonanz mit... kommen, in Resonanz sein mit...
resonating circuit Resonanzkreis, Schwingkreis
resonator Schwinger; Schwingkreis; Hohlraumresonator (534), (621.396)
resonator crystal Schwingquarz, Filterquarz
to resound widerhallen
to respond ansprechen, reagieren
to respond positively positiv reagieren
responder Responder, Antwortsender
response Ansprechen, Verhalten (gegenüber äußeren Einflüssen); Empfindlichkeit (621.317); Ausbeute (54); (Amplituden-)Gang (621.375); Antwort (621.316.7)
response band Ansprechband
response characteristic Frequenzgang (621.396.64)
response curve Frequenzgang, Kennlinie (621.375)
response sensitivity Ansprechempfindlichkeit
response synchro Rückmelder (621.316.7)
response time Beruhigungszeit (621.317); Anlaufzeit (621.316.7); Einschwingzeit, Ansprechzeit; Antwortzeit (681.3)
response value Ansprechwert (621.318)
responsive ansprechfähig, reaktionsfähig
to be responsive to ansprechen auf
responsivity Ansprechvermögen, Ansprechen
responsor Responder, Antwortsender
rest Stillstand, Ruhe; Rest; Auflage (621.9)
at rest in Ruhe, stillstehend, im Ruhezustand
rest energy Ruhe-Energie (539)
rest gas Restgas
to rest ruhen
to rest in ruhen in, aufliegen auf
to rest on, upon aufliegen auf, ruhen auf, sich abstützen auf; beruhen auf
to restart wieder in Gang setzen, wieder anlaufen lassen
to restate neu formulieren, umschreiben (51)
restatement Neuformulierung (51)
resting Ruhe-...
resting carrier frequency feste Trägerfrequenz
resting condition Ruhezustand
resting frequency unmodulierte Trägerfrequenz
restoration Wiederherstellung, Restaurierung
restoration of flux Wiederaufbau des Flusses
to restore wiederherstellen, restaurieren; rückführen, zurückbringen; wiedereinspeichern (681.3)
to restore to zero wieder auf Null bringen, auf Null zurückbringen, auf Null abgleichen
restoring force Rückstellkraft, rückführende Kraft, Richtkraft
restoring force gradient Rückstellsteilheit, Richtkraftänderung
restoring moment Rückstellmoment, Richtmoment
restoring spring Rückstellfeder
restoring torque gradient Rückstellkraft, Richtkraft, Richtmoment (621.317)
restricted gedrängt, beschränkt; begrenzt; begrenzt zugänglich; N. f. D. = Nur für den Dienstgebrauch
restricted extension nicht amtsberechtigte Nebenstelle
restricted range begrenzter Bereich
restricted scale gedrängte Skala (621.317)
restrictive einschränkend, beschränkend, Spar...
restrictive circuit Sparschaltung
to restrike wieder zünden, neuzünden (621.387); rückzünden (621.387)
restriking voltage Rückzündspannung

to result sich ergeben, resultieren (**from:** aus); herauskommen; beruhen (**from:** auf)
resultant Resultierende
resultant resultierend
resultant fault scheinbarer Fehlerort
resultant flux resultierender Fluß
resultant net force resultierende Kraft, Resultierende
resultant nucleus Endkern (539)
resultant voltage resultierende Spannung, verkettete Spannung
to resume wieder einnehmen, wieder aufnehmen
to resume control Steuerfunktion wiedererhalten
to resume position Stellung wieder einnehmen
to resupply wieder zusetzen
to resupply the carrier den Träger wieder zusetzen (621.396)
retail Kleinhandel, Einzelhandel, Detailverkauf
retail dealer, retailer Kleinhändler, Wiederverkäufer, Einzelhändler
to retain beibehalten, festhalten, noch aufweisen; abstützen, umschließen
retained charge zurückbleibende Ladung
to retain the continuity Durchgang behalten, noch Durchgang aufweisen
to retain an image ein Bild festhalten
retained image Nachwirkungsbild (621.397)
retaining vessel Behälter
retaining wall Stützmauer, Behältermauer
retainer Halterung
to retard retardieren, verzögern, bremsen
retard position Bremsstellung, Verzögerungsstellung
retardation Verzögerung, Bremsung, Abbremsung; Verspätung, Nacheilung
retardation coil Phasenabgleichspule (621.317); Siebspule (621.394/5)
retardation network Laufzeitkette, Verzögerungsschaltung
retardation valve Bremsfeldröhre
retarded verspätet, Spät-...
retarded ignition Spätzündung
retarded potential retardiertes Potential
retarder 90°-Spule (621.317)
retarding verzögernd, bremsend, Brems-...
retarding field Bremsfeld
retarding field generator Bremsfeldgenerator, Bremsfeldsender
retarding force Bremskraft
retarding torque Bremsmoment
retention time Speicherzeit (621.396.96)
retentive festhaltend
retentiveness Remanenz
retentivity Remanenzeigenschaft
reticulated netzartig, netzförmig
to retighten nachziehen, noch einmal anziehen
to retin neuverzinnen, nachverzinnen; wieder einlöten
retort Retorte
retort furnace Retortenofen
retrace Rücklauf
to retrace wieder durchlaufen, noch einmal überschreiben (621.385); rückwärts durchlaufen; zurückverfolgen
retracted antenna eingefahrene Antenne
to retransform rückverwandeln, rücktransformieren (51)
retransformation Rückverwandlung, Rücktransformation (51)
retrieval Wiederauffinden (681.3)
to retrim nachtrimmen
retroaction Rückkopplung
retroactive rückwirkend
retrofit Nachrüstung
retrograde rückwärts gerichtet
retrograde motion Rückwärtsbewegung
to retrograde sich nach rückwärts bewegen, rückläufig sein (52)
retrogressive-wave winding gekreuzte Wellenwicklung
to retune nachstimmen
return Rückkehr, Umkehr; Rückleiter, Rückleitung (621.3); Unterprogrammrücksprung (681.3)
return bend Umkehrbogen, Kopfbogen, Doppelkrümmer
return circuit Rückleitung
return current Rückstrom, rückfließender Strom
return-current coefficient Anpassungskoeffizient, Reflexionsfaktor
return feeder Rückleitung
return idler Rücklauftragrolle, Leerbandtragrolle
return loss Anpassungsdämpfung, Reflektionsdämpfung, Fehlerdämpfung
return path Rückleitung, Rückweg (621.3)
return stroke Rückhub (621.4)
return time Rücklaufzeit, Fleckrücklaufzeit (621.385)
return trace Rücklaufspur, Rücklauf, Fleckrücklauf (621.385)
return wire Rückleitung, Rückleitungsdraht, Nulleiter
return to zero Rückkehr nach Null
to return umkehren, rückkehren, zurückkehren;

wiederkehren; zurückziehen, zurückholen (621.9); zurücklaufen lassen (621.89)
to return to reflektieren nach, zurücksenden nach; zurückschalten auf, rückverbinden mit (621.395)
rev = revolution Umdr. = Umdrehung
to rev = to revolve drehen, laufen lassen
reverberate widerhallen
reverberation Nachhall
reverberation time Nachhallzeit
reverberator Reflektor
reverberatory reflektierend, hallend
reverberatory furnace Flammofen, Glühofen
reversal Umkehr, Umkehrung, Umsteuerung; Vorzeichenwechsel; Drehrichtungsumkehr
reversal of the current Stromwendung, Stromrichtungsumkehr
reversal of phase Phasenumkehr
reversal of polarity Umpolung
reversal of sign Vorzeichenwechsel, Vorzeichenumkehr
reversal of stress Lastwechsel, Lastrichtungswechsel
to reverse umkehren, sich umkehren; umsteuern, die Drehrichtung umkehren; zurücksetzen, zurückstoßen, rückwärtsfahren (629.113)
to reverse an engine einen Motor umsteuern
to reverse a motor einen Motor umsteuern
reverse umgekehrt, Rück-...
reverse attenuation Rückwärtsdämpfung
reverse-base bias current Basissperrstrom
reverse bias in Sperrichtung betrieben
reverse-blocking diode thyristor rückwärts sperrende Thyristordiode
reverse-blocking state Rückwärts-Sperrzustand
reverse-compound winding Gegenkompoundwicklung
reverse-conducting state Durchlaßzustand in Rückwärtsrichtung
reverse current Rückstrom; Rückwärtsstrom, Sperrstrom (621.382.2)
reverse-current circuit breaker Rückstromschalter
reverse-current cleaning anodische Reinigung
reverse-current cut-out Rückstromschütz
reverse-current relay Richtungsrelais, Stromrichtungsrelais, Rückstromrelais
reverse-current release Rückstromauslösung
reverse direction Rückwärtsrichtung (621.382)
reverse-frequency position Kehrlage (621.394.6)
reverse flow Rückströmung
reverse-flow check Rückschlagklappe, Rückschlagventil
reverse gear Umkehrgetriebe, Rückwärtsgang
reverse grid current Gitterrückstrom
in reverse umgekehrt
in the reverse order in umgekehrter Reihenfolge
reverse leakage Sperrstrom
reverse off-state Rückwärts-Sperrzustand (621.314.63)
reverse on-state current Rückwärts-Durchlaßstrom
reverse power Rückleistung
reverse power Rückwatt-...
reverse-power circuit breaker Rückwattschalter
reverse-power loss Rückwärtsverlustleistung; Sperrverlust (621.314.5)
reverse-power relay Rückwattrelais, Rückleistungsrelais
reverse-power tripping Rückwattauslösung, Rückstromauslösung
reverse recovery time Sperrverzögerungszeit, Sperrverzug (621.314.63)
reverse side Rückseite
reverse voltage Sperrspannung; Rückwärtsspannung, Spannung in Sperrichtung (621.382)
reverse voltage-current characteristic Rückwärtskennlinie
reversed umgekehrt
reversed bending test Wechselbiegeversuch
reversed biased vorgespannt in Sperrichtung
reversed motion Rückwärtsgang, Rücklauf
reversed-phase sequence gegenläufiger Drehfeldsinn
reversed taper umgekehrter Konus
reverser Fahrtrichtungsschalter (621.33)
reversibility Umkehrbarkeit, Umsteuerbarkeit
reversibility of function Umkehrbarkeit der Wirkungsweise
reversible umkehrbar, reversibel; umzudrehen; umsetzbar; umstellbar; umsteuerbar
reversible action umkehrbarer Vorgang, umkehrbarer Prozeß
reversible counter Zweirichtungszähler (621.374)
reversible current Nachwirkungsstrom
reversible motor Umkehrmotor, umsteuerbarer Motor
reversible permeability reversible Permeabilität
reversible process umkehrbarer Prozeß
reversible reaction umkehrbare Reaktion, umkehrbarer chemischer Vorgang

reversing Umsteuerung, Drehzahlumkehr, Umsteuerungsvorgang
reversing colour sequence wechselnde Farbfolge (621.397)
reversing commutator Umkehrschalter, Umpoler
reversing controller Drehrichtungsumschalter
reversing field Kommutierungsfeld
reversing gear Umkehrgetriebe
reversing key Stromwender, Umschalter, Polwender
reversing motor Umkehrmotor
reversing operation Reversierbetrieb
reversing pole Wendepol
reversing process Umkehrverfahren (655)
reversing starter Umkehranlasser
reversing switch Umpoler, Umschalter, Polumschalter, Umkehrschütz
revertive call Umkehrverbindung (621.395)
revertive control Wähler mit Rückimpulsgabe
to revet verkleiden, verblenden (693)
revetment Verkleidung, Verblendung (693)
revolution Umdrehung
revolution counter Umdrehungszähler
revolution indicator Umdrehungszähler, Drehzahlanzeiger, Tachometer
revolutions per minute U/min = Umdrehungen pro Minute
to revolve umlaufen, sich drehen, rotieren
to revolve through the arc of a circle sich auf einem Kreisbogen drehen
revolving Dreh-..., umlaufend, rotierend
revolving arc furnace Drehrohrlichtbogenofen
revolving armature alternator Generator mit feststehendem Feld, Außenpolgenerator
revolving commutator rotierender Umschalter (621.3)
revolving contact maker umlaufende Kontaktscheibe, Joubertscheibe
revolving counter Umdrehungszähler
revolving drum umlaufende Trommel (621.317)
revolving field Drehfeld
revolving filter Trommelfilter
revolving furnace Drehofen
revolving grate Drehrost
revolving lathe-centre mitlaufende Reitstockspitze
revolving stage Drehbühne
revolving table Drehtisch, Karussell (621.9)
revolving vector Drehstrecke, Zeiger, Zeitvektor (538.55)
revs per min U/min = Umdrehungen pro Minute
rewind Neuwicklung; Ersatzspule
to rewind neuwickeln, umrollen, umspulen, zurückspulen
rewinder Umspuler, Umroller
rewinding Rücklauf (681.84); Umspulen
to rewire neuinstallieren, Leitungen neu legen; zurückkabeln (621.394)
to rewrite umschreiben, umformen (51); noch einmal anschreiben; wiedereinschreiben (681.3)
Reynold's number Reynoldssche Zahl
r. f. = radio frequency (siehe dort) HF = Hochfrequenz
rhenium Rhenium
rheochord Schleifdraht
rheodyne circuit widerstandsgekoppelter Sender
rheomicrophone Widerstandsmikrophon
rheostat Widerstand, Schiebewiderstand
rheostat dimmer Verdunkelungswiderstand, Widerstandsverdunkler
rheostat slider Widerstandsabgriff, Schleifer
rheostatic Widerstands-...
rheostatic braking Widerstandsbremsung
rheostatic-braking controller Widerstandsbremsschalter
rheostatic control Hauptstrom-(Drehzahl-)Regelung
rheostatic controller Widerstandsregler
rheostatic loss Widerstandsverlust
rheostatic starter Widerstandsanlasser
rhodium Rhodium
rhomb Rhombus, Raute
rhombic rhombisch, rautenförmig
rhombic aerial Rhombusantenne
rhomboid Parallelogramm
rhomboid aerial Rhombusantenne
rhombus (pl. **-bi**) Rhombus, Raute
rhythm Rhythmus
rhythmic(al) rhythmisch
rhythmic light Taktfeuer (35)
rib Rippe (61); Rippe, Spant; Auflageleiste
rib stiffener Versteifungsrippe
to rib verrippen, mit Rippen versehen
ribbed radiator Rippenheizkörper
ribbon Band; Farbband (621.394.6)
ribbon cable Bandkabel, Flachbandkabel
ribbon coil Flachspule, Bandspule
ribbon conductor Bandleitung
ribbon microphone Bändchenmikrophon
ribbon resistance Bandwiderstand
ribbon suspension Bandaufhängung
ribbon wire Aufhängeband (621.317); Formdraht
rich in reich an, ...-reich
rich in energy energiereich

rich in harmonics oberwellenreich
Richardson effect Richardson-Effekt
to rid reinigen, befreien (**of:** von)
riddle Grobsieb
to riddle sieben, rütteln
ridge Kamm, Grat, First, Erhöhung
ridge turret Dachreiter
ridge waveguid Steghohlleiter
right richtig; rechts; gerade
right angle rechter Winkel, Rechter
at right angles im rechten Winkel
right-angled rechtwinklig
right-angled triangle rechtwinkliges Dreieck
right ascension Rektaszension (52)
right cone gerader Kegel
right cylinder senkrechter Zylinder (51)
right-hand rechts, rechts gelegen
right-hand rule Rechte-Hand-Regel
right-hand winding rechtsläufige Wicklung
right-handed rechtsgängig, Rechts-...
right-handed machine rechtsläufige Maschine
right-handed rotation Rechtsdrehung (53), (45); Rechtslauf (621.313)
right-handed screw rechtsgängige Schraube
right-handed thread Rechtsgewinde
rightward nach rechts gerichtet
rightward welding Rechtsschweißung
rigid streng, scharf; starr, steif
rigid body starrer Körper
rigid coupling starre Kupplung
rigid regulation starre Vorschrift
rigid rule strenge Regel
rigid steel conduit Stahlpanzerrohr
rigid suspension starre Aufhängung (621.33)
rigid tower Abspannmast
rigid truck Wagen ohne Drehgestell (625)
rigidly coupled starr gekuppelt
rigidity Festigkeit, Steifigkeit, Starrheit, Unnachgiebigkeit
rigorous streng
rigorous analysis strenge Rechnung, genaue Rechnung
rigorous mathematical proof strenger mathematischer Beweis
rim Rand, Umrandung; Radkranz, Felge, Reifen; Zarge (68)
rim drive Reibradantrieb
rim gasket Flanschdichtung
rim speed Umfangsgeschwindigkeit
rim of a wheel Radkranz
ring Ring; Klang, Ton; Klingeln; Telefonanruf
ring anode Ringanode
ring armature Ringanker (621.313)
ring-back Rückruf (621.395)
ring bolt Ösenschraube, Ringschraube, Tragöse
ring bus-bar Ringsammelschiene
ring counter Ringzähler (621.374)
ring filament ringförmig angeordneter Leuchtdraht
ring gear Zahnkranz
ring lubrication Ringschmierung
ring lubricator Schmierring
ring main Ringspeiseleitung (621.3)
ring molecule Ringmolekül
ring-operated system ringförmig betriebenes Netz
ring seal ringförmige Abdichtung, Dichtungsring, Ringwulst
ring segment Ringsegment, Ringabschnitt
ring-shaped ringförmig
ring side b-Ader (621.395)
ring tone Freiton (621.395)
ring-type modulator Ringmodulator
ring winding Ringwicklung, in sich geschlossene Wicklung
ring wire b-Ader (621.395)
to ring klingeln, läuten; rufen (621.395)
to ring back zurückrufen
to ring the bell die Glocke läuten, klingeln
to ring for klingeln nach, läuten nach
to ring off abrufen, abläuten (621.395)
to ring the ringer anrufen, anläuten (621.395)
to ring up herausklingeln; anrufen (621.395)
ringer Wecker, Anrufklingel (621.395)
ringing Ruf-..., Rufstrom-... (621.395); Überschwingen (621.375)
ringing battery Rufstrombatterie, Rufbatterie
ringing converter Rufumformer
ringing current Rufstrom
ringing dynamo Rufmaschine, Rufstromdynamo
ringing frequency Rufstromfrequenz
ringing generator Rufmaschine, Rufstromgenerator
ringing key Ruftaste
ringing oscillator Blockschwinger
ringing power supply Rufstromversorgung
ringing repeater Niederfrequenz-Rufumgehungsschaltung (621.395)
ringing and signalling generator Ruf- und Signalmaschine
ringing time Abklingzeit (621.397); Nachschwingzeit
ringing tone Freiton
ringing wire Klingeldraht

rinse Wässerung, Spülung
ripple kleine Schwankung, Brumm, (leichte) Welligkeit; Kräuselspannung, Wellenspannung (621.3)
ripple blanking Ausblenden (überflüssiger Nullen) (681.3)
ripple content Welligkeitsanteil
ripple current Kräuselstrom
ripple factor Welligkeit
ripple filter Brummfilter, Brummsperre
ripple frequency Frequenz der Welligkeit
ripple and noise Brumm und Rauschen
ripple quantity Kräuselgröße
ripple ratio Kräuselanteil
ripple rejection Glättungsfaktor, Brummspannungsunterdrückung (621.314)
ripple time Abklingzeit (621.375)
ripple tolerance Abklingresttoleranz (621.375)
ripple voltage Brummspannung, Kräuselspannung
to ripple kräuseln, leicht wellen
rise Aufstieg, Anstieg; Ganghöhe (621.8)
rise of current Steigen der Stromstärke, Stromanstieg
rise-in-current control Stromanstiegssteuerung
rise-and-fall pendant Zugpendel (621.32)
rise in temperature Temperaturanstieg
rise time Anstiegszeit, Einschwingzeit
to rise steigen, aufsteigen, ansteigen
current rises Strom nimmt zu
to rise steeply steil ansteigen
temperature rises Temperatur steigt, Temperatur nimmt zu
riser Steiger (621.7); Steigerung (69); Fahne, Kollektorfahne (621.313)
rising steigend
rising characteristic steigende Charakteristik
rising gradient Steigung
rising main Steigleitung
rising power steigende Potenz (51)
to rive reißen, zerreißen
rivet Niet
rivet head Nietkopf
rivet hole Nietloch
rivet joint Nietverbindung
rivet punch Nietkörner, Setzer
rivet set Setzeisen, Nietkopfsetzer
rivet shank Nietschaft
to rivet nieten
riveted work Nietkonstruktion, genietete Stahlkonstruktion
riveter Nietmaschine; Nieter
riveting Nieten; Nietvorgang
riveting hammer Niethammer
riveting machine Nietmaschine
RMM = read mostly memory
r. m. s. = root mean square quadratischer Mittelwert; Effektivwert (621.3)
r. m. s. current Effektivstrom, Effektiv-Wert des Stromes
r. m. s. voltage Effektivspannung, Effektivwert der Spannung
Rn = Radon Radon
road Straße
road-illumination Straßenbeleuchtung
road traffic Straßenverkehr
to roast rösten (54)
roasting Röst-...
roasting kiln Röstofen
to rob energy Energie entziehen
Robinson direction finder umlaufender Leitstrahlpeiler
robot Roboter, Maschinenmensch, Automat
robot pilot automatische Kurssteuerung, Selbststeuergerät
Rochelle salt Rochellesalz
rock Felsen
rock crystal Bergkristall
rock drill Steinbohrer
rock gas Erdgas
rock salt Steinsalz
to rock wiegen, schaukeln, wippen, kippen, wackeln, hin- und herwackeln
rocker Kipphebel, Wiegebalken; Bürstengestell, Bürstenhalterring (621.313)
rocker-dolly switch Wippenschalter
rocker ring Bürstenring
rocking lever Schwinghebel
Rockwell hardness Rockwell-Härte
rod Stange, Stab, Zugstange; Rute (Längenmaß: 1 rod = 5,5 yd = 5,029 m); Schweißdraht, Schweißstab (621.791)
rod aerial Stabantenne
rod blank Stangenmaterial (621.8)
rod collector Stangenstromabnehmer (621.33)
rod driven mit Stangenantrieb
rod lattice Liniengitter, Stabgitter
rod and lever mechanism Schwinghebelanordnung
rod-type carbon resistor Karbowid-Widerstand
rod-type insulator Langstabisolator
rod-type rectifier Stabgleichrichter
rodding duct Stangeneinzug (621.315.2)
Roentgen ... = Röntgen ...
Roget's dancing spiral Roget's Spirale
roll Rolle, Walze (621); rollendes Geräusch, Rollen, Grollen; Rollbewegung (629.13)

roll stabilizer Schlingerdämpfungseinrichtung
roll turner's lathe Walzendrehbank
to roll rollen, walzen; sich abwälzen; walzvergüten (66); kalandrieren (676); rollen (629.12), (629.13); wickeln (621.319.4)
to roll back wiederholen (681.3)
to roll off abrollen
to roll sheets Bleche walzen
to roll together aufeinander abrollen, sich aufeinander abwälzen
rolled gewalzt, Walz-...
rolled glass Rohglas, Walzglas
rolled plate Walzblech
rolled steel Walzstahl
roller Roller, Tragrolle, Laufrolle, Walze
roller ball Rollkugel
roller bearing Rollenlager
roller race Rollenlaufring
roller-spot welder Rollenpunktschweißmaschine
roller switch Rollentrenner
roller-type chain Rollenkette
roller-type contact Rollenschaltstück
rolling Rollen, Walzen; Walzvorgang; Abwälzvorgang, Wälz-...
rolling circle Wälzkreis
rolling-contact bearing Wälzlager
rolling curve Wälzbahn (531)
rolling cut Rollschnitt (621.397)
rolling friction Rollreibung, rollende Reibung
rolling gearing Wälzgetriebe
rolling gears Wälzräder
rolling motion Rollbewegung, Schlingerbewegung; Wälzbewegung
rolling operation Walzvorgang, Walzstich (621.7)
rolling stock rollendes Material, Wagenpark
rollover [gleichzeitige Betätigung mehrerer Tasten]
ROM = read only memory Festwertspeicher, Lesespeicher
Röntgen meter Dosismesser, Röntgendosismesser
Röntgen rays Röntgenstrahlen
Röntgen tube Röntgenröhre
röntgenogram Röntgenaufnahme
röntgenography Röntgenaufnahmetechnik
röntgenology Röntgenologie
röntgenometer Röntgendosismesser
röntgenoscope Röntgenschirm, Fluoreszenzschirm
röntgenoscopy Röntgenschirmtechnik
röntgentherapy Röntgentherapie
roof Dach
roof aerial Dachantenne
roof conductor Dachleitung (69)
room Raum, Platz; Zimmer
room climates Innenraumklima
room lighting Raumbeleuchtung
room noise Raumgeräusch (621.395)
room temperature Raumtemperatur, Zimmertemperatur
root Wurzel (58); Wurzel (51); Zahnwurzel, Zahnfuß (621.8); Fußhöhe (621.9); Zahnfuß (621.313)
root mean square quadratischer Mittelwert (s. auch: **r. m. s.**)
root mean square voltage effektive Spannung, Effektivwert der Spannung
to root Wurzel ziehen (51)
rope Seil (meist textil, mehr als 10 mm Durchm.); Düppelband (623)
rope bracing Verspannung
rope brake Seilbremse
rope connector Seilverbinder
rope core Seilkern, Seele
rope drive Seiltrieb
rope-driven mit Seilantrieb
rope drum Seiltrommel
rope groove Seilnut, Seilrille
rope guide Seilführung
rope haulage Seilförderung
rope-lay cable verseiltes Kabel
rope-lay conductor mehrfach verseiltes Kabel
rope-lift control Seilsteuerung für Fahrstühle
rope railway Seilbahn
rope shackle Seilgeschirr, Flaschenzug
ropeway Seilbahn
ropy fadenziehend
rosin Harz; Kolophonium
rosin connection schlechte Lötstelle, alte Lötstelle
rotary umlaufend, sich drehend, rotierend
rotary actuator Antriebseinrichtung für Drehbewegung
rotary amplifier Verstärkermaschine
rotary arm Dreharm
rotary automatic telephone system Drehwählersystem
rotary beam Drehrichtstrahl, umlaufender Richtstrahl
rotary compressor Turbokompressor, Kreiselverdichter
rotary condenser Synchronphasenschieber
rotary converter rotierender Umformer
rotary current Mehrphasenstrom, Drehstrom
rotary dial system Drehwählersystem

rotary exchange Vermittlung mit Drehwähler (621.395)
rotary felt Filzschwabbel, Filzscheibe, Polierfilz
rotary field Drehfeld
rotary hysteresis Drehfeldummagnetisierung, Drehfeldhysteresis
rotary kiln Drehrohrofen
rotary lift Paternoster (621.34), (621.8)
rotary magnetism Drehfeldmagnetisierung
rotary motion Drehbewegung
rotary movement Drehbewegung
rotary multi position switch Drehstufenschalter
rotary phase advancer rotierender Phasenschieber
rotary phase converter Phasenschiebermaschine, rotierender Phasenschieber
rotary polishing mop Schwabbelscheibe
rotary pump Kreiselpumpe, Turbopumpe
rotary rheostat Kurbelwiderstand
rotary selector Drehwähler
rotary spark gap rotierende Funkenstrecke
rotary stepping switch Drehstufenschalter
rotary switch Drehschalter; Drehwähler (621.395); Drehstufenschalter
rotary switch deck Schaltersegment
rotary synchronizer Synchronoskop
rotary synchroscope Synchronoskop
rotary system Drehwählersystem
rotary transformer Gleichstrom-Gleichstromumformer
rotary voltmeter rotierendes Voltmeter
rotary wire brush Drahtschwabbel, rotierende Drahtbürste
to rotate rotieren, umlaufen, sich drehen; drehen, drehend antreiben
rotating drehend, umlaufend, rotierend; Dreh-...
rotating aerial direction finder Umlaufpeiler
rotating beacon Drehfeuer, Drehfunkfeuer (621.396.93)
rotating element Anker (621.317)
rotating field Drehfeld; rotierendes Feld
rotating field aerial Drehfeldantenne
rotating field instrument Drehfeldinstrument
rotating field magnet Polstern, umlaufendes Polsystem
rotating-field-type alternator Innenpolgenerator (621.313)
rotating head Schraubkopf (531), (62)
rotating insulator switch Schalter mit Drehstützer
rotating joint Drehkupplung
rotating loop direction finder umlaufender Rahmenpeiler
rotating-loop radio transmitter Drehrahmensender
rotating magnetic field magnetisches Drehfeld
rotating plates (pl.) drehbarer Plattensatz (621.319.4)
rotating screen Trommelsieb
rotating table Drehtisch, Karussell (621.9)
rotating vector umlaufender Vektor; Zeiger (621.3)
rotating vector diagram Zeigerdiagramm
rotating wireless beacon Drehfunkbake, Drehfunkfeuer
rotation Drehung, Rotation
rotation through an angle Drehung um einen Winkel
rotation rate Rotationsgeschwindigkeit
rotational rotatorisch, drehend, durch Drehung bewirkt
rotational electromotive force EMK der Drehung (621.313)
rotational equilibrium Gleichgewicht gegen Verdrehung
rotational field Wirbelfeld
rotor Rotor; Läufer, Laufrad; Rotor, Läufer, Induktor (621.313)
rotor bearing Lauflager (681); Traglager
rotor circuit Läuferkreis, Induktorkreis
rotor core Ankerkern, Anker (621.313)
rotor current Läuferstrom, Rotorstrom, Induktorstrom
rotor e. m. f. Läufer-EMK, Rotor-EMK
rotor fed läufergespeist
rotor-fed self-compensated motor läufergespeister kompensierter Induktionsmotor
rotor interconnection Rotorverbindung (621.34), (621.8)
rotor lamination Läuferblech
rotor rim Läuferkranz
rotor section Rotorplattensatz (621.319.4)
rotor starter Läuferanlasser
rotor voltage Läuferspannung
rotor winding Rotorwicklung, Läuferwicklung, Induktorwicklung
rough roh, rauh, grob
rough adjustment Grobabgleich, Grobeinstellung
rough circle roher Kreis, ungefährer Kreis
rough indication rohe Anzeige, rohes Anzeichen

rough measurement rohe Messung, ungefähre Messung
rough sketch rohe Skizze, grobe Skizze
rough state roher Zustand, unbearbeiteter Zustand
rough tuning Grobabstimmung
rough usage grobe Benutzung, rauher Betrieb
to rough schruppen, vordrehen (621.9)
to rough-turn überdrehen, schruppen (621.9)
to roughen aufrauhen, rauhen
roughish rauh, grob
roughish use rohe Behandlung, rauhe Behandlung
round rund; rund, um ... herum, etwa
round about in der Gegend von, so etwa
round conductor Rundleiter
round earth gekrümmte Erdoberfläche (621.396)
round hole Rundloch, rundes Loch
round a mesh in einer Masche, auf dem Umlaufweg um eine Masche, auf einem Umlauf
round-nose tool Rundstahl, Schruppstahl (621.9)
round number runde Zahl
round-off error Rundungsfehler
round stock Rundmaterial (621.9)
round-trip echo Mehrfachecho
round wave guide Rundhohlleiter
round wire Runddraht
round the world echo Erdumlaufecho
to round runden, verrunden, abrunden, rundmachen
to round off abrunden, runden (51)
to round up aufrunden (51); zusammentreiben
rounding of corners Kantenverrundung, Verrundung der Ecken
rounding error Rundungsfehler (681.3)
roundness Rundheit; Unrundheit, Schlag (621.9)
route Weg, Ausbreitungsweg (621.396); Leitweg (621.395)
route beacon Kursfeuer, Flugstreckenfeuer
route data Leitweginformation (621.395)
route layout Trassierung
route of flight Flugstrecke
route translator Zuordner (621.395)
routine Routine, Gewohnheit, täglicher Dienst; Programm, Routine (681.3)
routine geläufig, üblich, normal
routine inspection laufende Prüfung, Stückprüfung
routine test laufende Prüfung, übliche Prüfung
routine-test circuit automatische Prüfeinrichtung
routing Leitweg, Leitwegzuteilung (621.395); Leitungsführung
routing line Leitzeile
routing-line segregation Leitwegaufteilung
row Zeile, Reihe, reihenförmige Anordnung
row of figures Zahlenreihe
row of radiators Strahlerzeile (621.396.67)
row of rivets Nietreihe
row of switches Schalterreihe
row of a table Zeile einer Tabelle
Rowland's method Ringkernprüfung (621.318)
r. p. m. = revolutions per minute Umdr./min = U/min = Umdrehungen pro Minute
R. P. V. = remotely piloted vehicle
RR-alloy Rolls-Royce-Legierung
R-scan waagerechte (verzögerte) Zeitablenkung (621.385)
RS flipflop = reset flipflop RS-Flipflop
RTL = resistor transistor logic Widerstands-Transistor-Logik
royality Lizenzgebühr
rub Reiben, Reibung
to rub reiben
to rub away wegreiben
to rub bare blank reiben, blank scheuern
to rub down verreiben, zerreiben
to rub off abreiben, sich abnutzen (621.94)
to rub out ausreiben, auslöschen, ausradieren
to rub through a sieve durch ein Sieb reiben
rubber Gummi, Kautschuk, Radiergummi; Reiber; Scheuerleiste (629.12)
rubber band Gummiband
rubber-bonded-to-metal Gi-Metall
rubber bumper Gummipuffer
rubber cable gummi-isolierte Leitung, Gummiaderleitung
rubber cushion Gummipuffer
rubber foot Gummifuß, Gummifüßchen, Fußpuffer
rubber gasket Gummidichtung
rubber insulated gummiisoliert
rubber-insulated cord gummiisolierte Litze
rubber-insulated lead-covered cable Gummibleikabel
rubber-insulated steel-wire braided cable Panzerader
rubber-insulated wire gummiisolierter Draht, Gummiaderleitung
rubber insulation Gummiisolation
rubber-metal vibration damper Schwingmetall
rubber packing Gummidichtung, Gummiliderung
rubber pad Gummipolster, Gummipuffer

rubber sheathing Gummi-Umhüllung, Gummimantel
rubber spring Gummipuffer, Gummifeder
rubber tape Isolierband
rubber tubing Gummischlauch
rubber tyre Gummireifen
rubber-vacuum tubing Vakuumgummischlauch
rubber washer Gummischeibe, Gummidichtung
rubbery gummiartig, gummiähnlich
rubbing Reib-...
rubbing action Reibwirkung
rubbing contact Reibkontakt
rubbing speed Gleitgeschwindigkeit
rubidium Rubidium
ruby Rubin
ruby laser Rubinlaser
ruby red rubinrot
rudder telegraph Rudertelegraph
rude roh
rude drawing grobe Zeichnung
rude shock heftiger Schlag
ruggedness Robustheit, Widerstandsfähigkeit
Ruhmkorff coil Ruhmkorff-Induktor
rule Regel, Lehrsatz; Vorschrift; Lineal
rule of three Dreisatzrechnung
rule of thumb Faustregel
by the rule of thumb über den Daumen gepeilt
to rule liniieren, Linien ziehen; Vorschriften machen, verfügen
to rule a line eine Linie ziehen
to rule out ausscheiden, ausschließen, unmöglich machen
ruled paper liniiertes Papier
ruler Lineal, Maßstab, Reißschiene
ruling Gitterstrich (535); Liniierung, Lineatur
ruling machine Teilmaschine (535); Liniiermaschine
ruling pen Reißfeder, Ziehfeder
to rumble rumpeln, grummeln
rumbles (pl.), **rumbling** Gerumpel, Grummeln (621.396)
Rumford photometer Schattenphotometer
run Auslauf, Landeweg (629.13); Betriebszeit, Betriebsabschnitt, Betrieb; Lauf; Gang, Lauf; Gruppe, Klasse, Lieferung, Serie; Lauf, Durchlauf (681.3)
on the run „im Gang"
run-of-river plant Laufkraftwerk, Laufwasserkraftwerk
runtime Laufzeit (681.3)
run-out time Nachlaufzeit
to run rennen, laufen; betreiben, unterhalten; verlaufen, sich abspielen, ablaufen
to run a cable eine Leitung verlegen, eine Leitung legen, ein Kabel legen
a cell runs down ein Element bricht zusammen (651.352)
to run dry auslaufen, leer werden
to run a factory eine Fabrik betreiben
to run freely on sich frei drehen auf
to run from ... to ... zwischen ... und ... liegen (621.3)
to run at high speed auf vollen Tourenlaufen, schnell laufen
to run hot warm werden, auf Temperatur kommen
to run in einlaufen, einlaufen lassen, einfahren
to run into the grid-current ins Gitterstromgebiet kommen
to run loosely freilaufen, leerlaufen
to run at low speed langsam laufen, sich langsam drehen
to run a machine light eine Maschine leerlaufen lassen, Maschine unbelastet laufen lassen
to run off abziehen, ablaufen lassen, abstechen, ablassen; abspulen
to run out ausrollen (621.315.2)
to run out of truth unrund laufen, „eiern"
to run in parallel parallel betrieben, parallel fahren
to run in synchronism synchron laufen
to run a thread through a hole einen Faden durch ein Loch fädeln
to run up sich auferregen (621.313.2)
to run up to speed auf Touren bringen
to run to waste in den Ausguß fließen, ablaufen
runaway Durchstoßen, Abrutschen, Umkippen, Instabilwerden, Weglaufen
runner Läufer (621.2); Einguß, Einlauf (621.7)
runner cup Gießtrichter, Einlaufkopf
running Gang, Lauf; laufend, fortlaufend
running balance dynamische Auswuchtung
running current Laststrom, Belastungsstrom
running down Auslauf
running edge Laufkante
running fit Laufsitz
running position „Betriebs"-Stellung
running speed Betriebsdrehzahl
running test Laufprobe
running torque synchrones Moment
running voltage Betriebsspannung (eines Motors, der mit einer Anlaßspannung angelassen wird)

runway Startbahn, Rollbahn, Landebahn (629.13)
runway approach Landebahnanflug
runway floodlight Landebahnbeleuchtung
runway flush-marker light Startbahnunterflurfeuer (656.7)
runway lights Landebahnfeuer
runway localizing beacon Landefunkfeuer
runway threshold marking Schwellenkennzeichnung
rupture Bruch, Durchbruch, Riß, Zerreißen
to rupture reißen, zerreißen, zerbrechen
rupturing capacity Abschaltleistung (621.3)
rupturing-capacity test Schalterprüfung
rupturing strength Bruchfestigkeit, Reißfestigkeit
rural exchange Landzentrale
rush Stoß, Schwall
rush of current Stromstoß, Stoßstrom, Anlaßstoß
rush hour Hauptverkehrsstunde (621.395); (625)
rush of water Wasserschwall, Wassereinbruch
to rush einbrechen, stürzen
rushing noise Rauschen (621.396)
rust Rost
rust formation Rostbildung, Rostansatz
rust-free nichtrostend
rustless rostfrei
rustless steel rostfreier Stahl
rust preventive Rostschutzmittel; rostverhindernd
rust-preventive agent Rostschutzmittel
rust-preventive compound Rostschutzmasse
rust-proof rostsicher, rostfrei, nichtrostend
rust proofing Rostschutzanstrich, Rostschutzmaßnahme
rust protection Rostschutz
rust resisting schwer rostend, nichtrostend
rust-resisting paint Rostschutzanstrich
to rust rosten, Rost ansetzen
to rust fast festrosten
rusty rostig
ruthenium Ruthenium
Rutherford-Bohr model of the atom Rutherford-Bohrsches Atommodell
rutile Rutil
"r" wire b-Ader (621.394)
Rydberg's constant Rydberg-Konstante

S

saccharimeter Saccharimeter, Polarisationsapparat
sack Sack
sack cell Beutelelement, Lechlanché-Element
sacrificial anode Schutzanode (621.35)
safe sicher; bestimmt; zuverlässig
safe carrying capacity zulässige Belastung; zulässige Strombelastung (621.3)
safeguard Sicherheitsmaßnahme, Schutzmaßnahme, Schutz, Sicherung
safe light Dunkelkammerlampe, Dunkelkammerlicht
safe limit Sicherheitsgrenze
safe load zulässige Belastung, zulässige Last, Tragfähigkeit
safe maximum zulässiger Höchstwert, zugelassenes Maximum
safe working stress zulässige Spannung, zulässige Beanspruchung
to safeguard sichern, schützen; warnen
to safeguard against schützen vor, sichern gegen; warnen vor
safety Sicherheit
safety appliance Sicherheitsvorrichtung
safety base Sicherheitsfilm, unbrennbarer Film
safety belt Sicherheitsgurt
safety breaker Sicherheitsschalter
safety cellulose acetate base Celluloseacetat-Sicherheitsfilm, unbrennbarer Film
safety clearance Sicherheitsabstand
safety code (sing.) Sicherheitsvorschriften
safety curtain eiserner Vorhang
safety device Schutzvorrichtung, Sicherheitsvorrichtung, Unfallschutzvorrichtung
safety factor Sicherheitsfaktor
safety fuse Schmelzsicherung
safety gap Sicherheitsfunkenstrecke
safety glass Sicherheitsglas, Sekurit, nichtsplitterndes Glas
safety lamp Grubenlampe
safety lock Sicherungsschloß
safety pin Sicherheitsstift
safety precaution Sicherheitsvorkehrung, Unfallschutzmaßnahme
safety regulations Sicherheitsvorschriften
safety rule Unfallverhütungsvorschrift, Sicherheitsvorschrift

safety slipping clutch Sicherheitsrutschkupplung
safety valve Sicherheitsventil
safety wire Plombendraht (621.317)
sag Durchhang (67), (621.315.1); Durchbiegung (539)
to sag durchhängen (67), (621.315.1); sich durchbiegen (539)
Saint Elmo's fire Elmsfeuer
sal Salz (54)
sal ammoniac Salmiak
salicylic acid Salicylsäure
salient Vorsprung
salient vorspringend, ausspringend, hervorstehend; hervorstechend, auffällig
salient point vorstehender Punkt
salient pole ausgeprägter Pol (621.613)
salient-pole machine Maschine mit ausgeprägten Polen, Schenkelpolmaschine
salt Salz
salt bath Salzbad
salt-bath furnace Salzbadofen
salt intensifying screen Verstärkerfolie (621.386)
saltspray Salznebel, Salzsprühung
saltspray test Salzsprühprobe (620.1)
same gleich
same-channel operation Gleichkanalbetrieb
same-pitched note gleich hoher Ton
by the same scale im gleichen Maßstab
sample Stichprobe
to sample Proben entnehmen; abtasten
sample and hold circuit Abtast- und Halteschaltung
sampled-data control system Abtastregelsystem
sampling Probeentnahme; Auslese; Beispiel, Muster, Probestück; Abtasten; Sampling (621.317)
sampling control Abtaststeuerung
sampling frequency Abtastfrequenz
sampling inspection Stichprobenkontrolle
sampling oscilloscope Sampling Oszillograph; Abtastoszillograph
sampling period Abtastintervall
sampling test Stichprobenprüfung
sand Sand; Sandbank (629)
sand blast Sandstrahl, Sandstrahlgebläse
sand-blast machine Sandstrahlgebläse
sandblasting Sandstrahlung
sand casting Sandguß (621.7)
sand castings Sandgußteile, Sandgußstücke; Sandguß (sing.) (621.7)
sand paper Sandpapier
sandstone wheel Schleifstein, Sandsteinschleifscheibe
to sand sandstrahlen, sanden; mit Sandpapier abziehen
to sandblast sandstrahlen, sanden
to sandwich zwischenschichten, in die Mitte legen, zwischen zwei Platten legen
sandwich winding Scheibenwicklung
sapphire Saphir
satellite Satellit, Trabant
satellite communication Übertragung über Satelliten, Satellitenfunk
satellite computer Satellitenrechner
satellite earth station Satelliten-Erdefunkstelle
satellite electron Hüllenelektron
satellite exchange Teilamt, Teilvermittlungsstelle (621.395)
satellitic satellitenartig, Trabant-...
to satisfy conditions Bedingungen erfüllen, Bedingungen genügen
saturable sättigungsfähig; mit Sättigungscharakter (621.3)
saturable inductor Sättigungsspule, Eisenkernspule
saturable reactor Eisenkernspule, Eisendrossel
to saturate sättigen (54), (538), (621.1), (621.385)
saturated gesättigt, Satt-...
saturated compound gesättigte Verbindung
saturated iron-core gesättigter Eisenkern
saturated iron-core modulation Drosselmodulation
saturated state gesättigter Zustand
saturated steam Sattdampf
saturated vapour Sattdampf, gesättigter Dampf
saturation Sättigung
saturation amplitude Sättigungsamplitude (621.396.61)
saturation current Sättigungsstrom (621.385)
saturation curve Magnetisierungskurve, Sättigungscharakteristik
saturation density Sättigungsinduktion (538); Sättigungsdichte; Sättigungsfeuchtigkeit
saturation-flux density Sättigungsinduktion (538)
saturation induction Sättigungsinduktion
saturation limiting Sättigungsbegrenzung (621.396.61)
saturation point Sättigungspunkt (536)
saturation temperature Sättigungstemperatur (536), (621.385)
saturation value Sättigungswert
saturation voltage Sättigungsspannung; Restspannung (621.382.3)
sausage aerial Reusenantenne (621.396.67)

to save erretten (**from:** aus, von); aufheben, sparen; sparen, einsparen
to save energy Energie einsparen
to save labour Arbeitskräfte einsparen, Handarbeit sparen
to save space Raum sparen
saving Ersparnis, Einsparung
saving in time Zeitersparnis
saw Säge
saw blade Sägeblatt
sawtooth (pl. **-teeth**) Sägezahn
sawtooth generator Sägezahn-Generator
sawtooth oscillation Sägezahnschwingung
sawtooth signal Sägezahnsignal
sawtooth voltage Sägezahnspannung (621.3)
sawtooth wave Sägezahnkurvenform (621.3)
to saw sägen
to saw cut absägen
to saw to shape nach Maß aussägen
Saybolt universal second (Viskositätsmaß)
3000 S. u. s. = 6 poise
S. B. = simultaneous broadcasting Gleichwellenfunk, Gemeinschaftswellen-Rundfunk
SBA = standard beam approach SBA-Lande-Verfahren, Lorenz-Verfahren
s. b. c. lamp = small bayonet cap lamp Lampe mit kleiner Bajonettfassung, Swankleinsockellampe
scalar Skalar, skalare Größe (51)
scalar skalar
scalar field skalares Feld, Skalarfeld
scalar function skalare Funktion (51)
scalar potential field Potentialfeld, skalares Potentialfeld
scalar product skalares Produkt, inneres Produkt
scalar quantity skalare Größe
scale Skala, Teilung; Maßstab; Abbildungsmaßstab; Waagschale (681)
scale correction Skalenkorrektur
scale division Teilstrich, Teilstrichbreite, Skalenteil
scale drawing maßstäbliche Zeichnung
scale of equal parts gleichmäßig geteilte Skala, proportionale Skala
scale factor Normierungsfaktor (51)
scale mark Skalenteil, Skalenstrich
scale of notation Zahlensystem
scale plate Skalenscheibe (621.396.6)
scale reading Skalenablesung
scale rule Zeichenmaßstab, Dreikantmaßstab
scale-of-two circuit Halbierschaltung (621.374)
scale-of-two counter binare Zählstufe (621.374)
to scale maßstäblich
to scale schuppen, abschuppen; sich schuppen; abblättern; blättern; eine Oxidhaut bilden, Hammerschlag bilden (66); Kesselstein ansetzen (621.1); Kesselstein entfernen (621.1); mit der Skala versehen; maßstäblich abändern; sich (zahlenmäßig) belaufen auf; normieren
to scale from a diagram aus einem Diagramm abgreifen, in einem Diagramm ablesen
to scale down maßstäblich verkleinern
to scale off maßstäblich abgreifen; abblättern (66)
to scale up maßstäblich vergrößern
scalene ungleichseitig, schiefwinklig
scalene triangle ungleichseitiges Dreieck (51)
scaler Teiler, Teilerstufe
scales (pl.) Waage (681)
scaling Zahlenbereichsänderung
scaling factor Normierungsfaktor
scaling down Maßstabverkleinerung
scan Abtastung, Auflösung in Bildelemente (621.397); Schirmbilddarstellung (621.396.96)
scan-current generator Kippstrom-Generator
scan speed Abtastgeschwindigkeit, Strahlgeschwindigkeit, Fleckgeschwindigkeit (621.397)
to scan abtasten, in Bildelemente auflösen; in der Zeilenrichtung ablenken; strichweise überstreichen, in der Zeilenrichtung laufen (621.397); abfragen (621.398); absuchen
scanner Abtaster, Abtastgerät; Bildabtastgerät; Lesestift (681.3)
scandium Skandium
scanning Zeilenablenkung, Abtastung, Bildzerlegung, Bildabtastung (621.397); Lesen (681.3)
scanning agents Taktgebersignale (621.397)
scanning beam abtastender Elektronenstrahl
scanning circuit Ablenkschaltung, Abtastschaltung
scanning coil Zeilenspule, Zeilenablenkspule, Ablenkspule (621.397)
scanning density Rasterfeinheit
scanning generator Zeitablenkgenerator; Kippgenerator (621.397)
scanning head Abtastkopf
scanning line Zeile, Abtastzeile (621.397)
scanning rate Abtastgeschwindigkeit
scanning receiver Suchempfänger
scanning sequence Abtastfolge
scanning speed Abtastgeschwindigkeit
scanning spot Abtastfleck; Schreibfleck
scanning tube Bildabtaströhre

scarf (pl. **-rfs**) Blatt (69)
scarf joint Anblattung, Überblattung (69)
scatter Streuung, Streuausbreitung
scatter link Scatterverbindung
scatter transmission Übertragung durch Streustrahlung, Scatter-Übertragung
scatter propagation Streuausbreitung, Streustrahlungsübertragung
scattered Streu-..., gestreut
scattered light gestreutes Licht, Streulicht
scattered radiation Streustrahlung, gestreute Strahlung
scattered wave Streuwelle
scattered X-rays (pl.) Röntgenstreustrahlung
scattering Streuung
scattering angle Streuwinkel
scattering effect Streuwirkung
scattering medium Lichtstreustoff
scattering screen Streuscheibe (535)
scattering surface Streufläche (621.32)
to scavenge reinigen, spülen (621.4)
scavenging Spül-...
scavenging agent Spülmittel
scavenging air Spülluft
s. c. c. cap = small central contact cap kleiner konzentrischer Sockel
s. c. e. = standard cable equivalent Leitungsdämpfung in **standard miles** (s. dort)
schedule Schema, Plan; Zeitplan; Tabelle, Liste, Verzeichnis
schedule time fahrplanmäßige Zeit, Planzeit
schematic schematisch
schematic diagram Schaltbild, Schemaschaltbild, Schaltschema
schematic representation schematische Darstellung
scheme Schema, Plan; Entwurf
scheme of connections Schaltanordnung
scheme drawing Entwurfszeichnung, Hauptentwurf; Zusammenstellung
to scheme planen, entwerfen
Scherbius control Scherbius-Drehzahlregelung
Scherbius phase advancer Scherbius-Maschine
Schering bridge Schering-Brücke
Schmitt trigger (circuit) Schmitt-Trigger (-Schaltung)
Schottky (barrier) diode Schottky-Diode
Schottky effect Schottky-Effekt
Schottky TTL Schottky TTL
science Wissenschaft
science of electricity Elektrizitätslehre
scientific notation wissenschaftliche Schreibweise
scientist Wissenschaftler
to scintillate funkeln; szintillieren (53)
scintillation Sprühfunken (pl.); Szintillation (53)
scleroscope hardness Skleroskophärte
scope = oscilloscope Kathodenstrahloszillograph, optisches Anzeigegerät, Sichtgerät (621.397)
scope Bereich; Schirmbild (621.396.6)
to be beyond the scope of... den Umfang von... überschreiten
scope of application Anwendungsbereich
scotch (block) Hemmschuh, Radkeil
to scotch verkeilen, Rad feststellen, Keil vorlegen
Scott (connected) transformer Transformator in Scott-Schaltung, Scott-Umspanner
Scott connection Scott-Schaltung
SCR = short-circuit ratio Kurzschlußverhältnis (621.313)
SCR = silicon controlled rectifier Thyristor, gesteuerter Siliziumgleichrichter
scram rod Abstellstab (639)
to scramble verwürfeln, verschlüsseln
scrambeled speech verschlüsselte Sprache
scrambler Mischgerät, Verwürflungsgerät
scrambler circuit Bandspalter, Verwürflungsschaltung (621.395)
scrap Schrott, Abfall
scrap copper Altkupfer
scrap iron Eisenschrott, Alteisen
to scrap verschrotten
to scrape kratzen, schaben, ziehklingen; abwaschen (69)
scraper Ziehklinge, Schaber, Zieheisen
scraper conveyor Transportkratze
scraping Schnitzel, Stückchen; Ziehspan
scratch Kratzer, Riß; Handskizze; Kontaktgeräusch
scratch filter Nadelgeräuschfilter
scratchpad memory Notizbuchspeicher
to scratch kratzen, ankratzen, ritzen, verkratzen
to scratch glass Glas anritzen, Glas verkratzen
scratching Kratzspur, Kratzmarke, Kratzer
screen Schirm, Abschirmung (621.396); Bildschirm, Leuchtschirm (53), (621.385); Filter (53); Schirmgitter (621.385); Windschutzscheibe (629.113); Sieb (66), (622); Schirm, Projektionsschirm, Projektionswand (77); Raster (65)
screen by-pass capacitor Schirmgitterblock (621.385)
screen defroster Frostschutzscheibe
screen demisting Scheibenheizung (629.113)
screen dot Rasterpunkt

screen-dropping resistor Schirmgitterwiderstand, Schirmgittervorwiderstand
screen grid Schirmgitter (621.385)
screen-grid dissipation Schirmgitterverlustleistung
screen-grid valve Schirmgitterröhre (621.385)
screen hole Siebloch (622)
screen luminous persistence characteristic Nachtleuchtdauercharakteristik
screen persistance Nachleuchteffekt
screen print Siebdruck
screen-protected motor halbgeschlossener Motor
screen wire Maschendraht, Siebgewebe
to screen abschirmen (621.396), sieben (66), (622); projizieren (77); durchleuchten (621.386)
screened aerial abgeschirmte Antenne
screened cabin Abschirmkäfig
screened cable abgeschirmtes Kabel, abgeschirmte Leitung
screened conductor abgeschirmter Leiter
screened down lead abgeschirmte Antennenzuleitung
screened loop aerial abgeschirmte Rahmenantenne
screened valve Schirmgitterröhre
screening Abschirmung (621.396); Siebung (622); Durchleuchtung (621.386)
screening box Abschirmkasten
screening cabinet Abschirmgehäuse
screening can Abschirmbecher
screening conductor Schirmleiter
screening effect Abschirmwirkung
screening factor Reduktionsfaktor der Schirmung, Abschirmfaktor
screening plate Abschirmplatte, Abschirmblech
screening sheath Abschirmmantel
screening wire Schutzdraht (621.531.1)
screenlike gitterartig, siebähnlich
screw Schraube (621.8); Spindel, Preßspindel, Hubspindel
screw and wheel Schneckenradgetriebe
screw adjustment Schraubeneinstellung
screw binds Schraube frißt sich fest, Schraube klemmt
screw bolt Schraubenbolzen
screw caliper Schraubmikrometer, Mikrometerschraube
screw cap Verschraubung; Überwurfmutter; Schraubdeckel; Schraubkappe
screw connection Verschraubung
screw conveyor Schraubenförderer, Transportschnecke
screw cutting Gewindeschneiden
screw-cutting machine Gewindeschneidmaschine
screw-down motor Anstellmotor
screw driver Schraubendreher
screw eye Schrauböse, Ösenschraube
screw gearing Schraubverzahnung; Schraubtrieb
screw joint Verschraubung, Schraubverbindung
screw pitch Schraubensteigung, Schraubengang
screw plate stock Gewindeschneidkluppe
screw plug Schraubstopfen
screw-plug cartridge fuse Schraubsicherung
screw press Spindelpresse
screw socket Schraubfassung
screw spindle Gewindespindel
screw stock Schneidkluppe, Windeisen
screw thread Schraubengewinde
screw-thread micrometer Schraubmikrometer
screw-type lamp Lampe mit Edison-Sockel
to screw schrauben, verschrauben, einschrauben; mit Gewinde versehen; sich einschrauben; sich einschrauben lassen
to screw in eine Schraube anziehen
to screw off abschrauben
to screw up zuschrauben
screwed mit Gewinde, mit Gewinde versehen; verschraubt
screwed conduit Stahlpanzerrohr (621.3)
screwed contact Schraubkontakt
screwed coupler Muffenverbinder, Schraubmuffe (696)
screwed fitting Schraubmuffe
screwed holder Gewindesockel, Edison-Sockel
screwed joint Verschraubung
screwed-lamp socket Schraubfassung
screwed socket Schraubmuffe (696)
screwed tubing Stahlpanzerrohr (696)
screwed-type socket Edisonfassung
screwind machine Gewindeschneidmaschine, Gewinderollmaschine
to scribe reißen, anreißen, anzeichnen (621.9)
to scribe arcs of a circle Kreisbögen schlagen (51)
scriber Reißnadel (621.9)
scribing block Parallelreißer
scroll Schnecke (621.8); Papierrolle
scrolling Verschieben (der Bildschirmanzeige)

scruple (kleine Gewichtseinheit im ap. weight:
1 scruple = 20 grains = 1,296 g)
scrutineer Gleichlaufkontrolleinrichtung (621.394)
sculptured keyboard konkave Tastatur
SDLC = synchronous data link control
sea See, Meer
sea clutter Seegangsreflexionen
sea level Meeresspiegel, Meereshöhe
sea return Seegangsecho
seal Dichtung; Abschluß, Verlötung, Verguß; Einschmelzung (621.385)
to seal abdichten; abschließen, verlöten, vergießen, versiegeln; zuschmelzen, abschmelzen
to seal through glass in eine Glaswanne einschmelzen (621.385)
to seal in einschmelzen (621.326) (621.385)
to seal a meter einen Zähler plombieren
to seal against moisture gegen Feuchtigkeit abschließen, gegen Feuchtigkeit abdichten
to seal off abziehen, abschmelzen (621.385)
to seal up abdichten
sealable equipment Geräte (pl.) in geschlossener Ausführung
sealed dicht; verlötet; abgeschmolzen (621.385)
sealed against abgedichtet gegen
sealed assembly rolling Walzschweißung
sealed cable end überkapptes Kabelende, zugebleites Kabelende
sealed-in battery vergossene Batterie
sealed rectifier pumpenloser Gleichrichter
sealed source umschlossener radioaktiver Strahler (539)
sealed system abgeschlossenes System
sealed tube abgeschlossenes Rohr; abgeschmolzene Röhre (621.385)
sealed valve abgeschmolzene Röhre
sealing Abschluß, Verlötung, Verguß; Dichtung, Abdichtung
sealing box Kabelmuffe
sealing chamber Endverschluß (621.315.2)
sealing compound Vergußmasse, Kabelmasse
sealing current Ansprechstrom (621.318)
sealing end Endverschluß (621.315.2)
sealing-in Einschmelzen (72), (621.385)
sealing-in joint Einschmelzung (621.326)
sealing off Abschmelzen, Abziehen (621.32), (621.385)
sealing substance Dichtungssubstanz, Dichtungsmittel
sealing voltage Ansprechspannung (621.318)
sealing weld Dichtschweißung
seam Naht; Lötnaht, Schweißnaht (621.791); Flöz (622)
seam welding Nahtschweißung
seamless nahtlos
seamless pipe nahtloses Rohr
search Suche, Durchsuchung, Nachforschung
search capability Reichweite, Entfernungsbereich (621.396.6)
search coil Prüfspule (621.3); Suchspule
search cycle Suchzyklus
searchlight Scheinwerfer
searchlight control radar Scheinwerferleitradar
searchlight reflector Scheinwerferspiegel
search radar Suchradar
search-type reflector Parabolspiegel (621.396.67)
search winding Prüfspule, Meßspule (621.314)
seasonal jahreszeitlich, Saison-...
seasonal load jahreszeitliche Belastung
seasoned wood gealtertes Holz, abgelagertes Holz
seat Sitz; Passung
seat of a failure Fehlerort, Sitz eines Fehlers
to seat the brushes die Bürsten auflegen (621.313)
seating Sitz, Auflage, Auflagefläche
Sec. = Section Abt. = Abteilung, Abschn. = Abschnitt
sech Kehrwert des Hyperbelkosinus
secohm Ohmsekunde, Henry
secohmmeter Induktivitätsmesser
second Sekunde; Zweiter
second-class zweiter Güteklasse, zweitklassig
second-class conductor Leiter zweiter Klasse (621.3)
second detector zweiter Gleichrichter, ZF-Gleichrichter (621.396.62)
second grade Klasse 2 (621.317)
second-hand material Gebrauchtmaterial
second law of thermodynamics zweiter thermodynamischer Hauptsatz
second mark Sekundenzeichen
second mode zweiter Art, vom Modul 2
second-order zweiter Ordnung
second-order effect Effekt zweiter Ordnung
second-order quantity Größe zweiter Ordnung
second-order small klein zweiter Ordnung (51)
second-order term Ausdruck zweiter Ordnung, Größe zweiter Ordnung
seconds pendulum Sekundenpendel
second voltage range zweiter Spannungsbereich (30... 175 V)

secondary Sekundärwicklung, Zweitwicklung (621.314.2)
secondary sekundär
secondary air Zweitluft, Sekundärluft (621.1)
secondary alcohol sekundärer Alkohol
secondary battery Sekundärbatterie, Akkumulatorenbatterie
secondary cell Sekundärelement, Akkumulator
secondary circuit Sekundärkreis
secondary clock Tochteruhr, Nebenuhr
secondary coil Sekundärspule
secondary contact Hilfskontakt
secondary coverage Fernempfangszone (621.396)
secondary current Sekundärstrom
secondary defect Folgefehler
secondary distribution feeder Niederspannungs-Anschlußleitung
secondary distribution main Niederspannungs-Verteilungsschiene
secondary distribution network Niederspannungs-Versorgungsnetz
secondary distribution system Niederspannungs-Verteilungsanlage
secondary distribution trunk line Niederspannungs-Anschlußleitung
secondary electrode Zwischenelektrode, Mittelelektrode (621.35)
secondary electron Sekundärelektron
secondary electron emission Sekundärelektronen-Emission
secondary electron multiplier Sekundärelektronen-Verstärker
secondary element Sekundärelement, Akkumulator
secondary emission Sekundäremission (621.385)
secondary failure Folgeausfall
secondary fault Fehler durch nachfolgenden Betriebsstrom
secondary leakage sekundäre Streuung
secondary line Nebenbahn, Kleinbahn (625)
secondary lobe Nebenzipfel (621.396.67)
secondary luminous standard Lampennormal (621.32)
secondary metal Umschmelzmetall
secondary neutral grid Nulleiternetz, Vierleiternetz
secondary radar Sekundärradar
secondary radiation Sekundärstrahlung, Streustrahlung
secondary radiator Sekundärstrahler
secondary service area Fernempfangszone (621.396)
secondary standard Unternormal, abgeleitetes Normal
secondary terminal Sekundärklemme (621.3)
secondary voltage Sekundärspannung
secondary winding Sekundärwicklung, Zweitwicklung
secondary wiring Signal- und Steuerleitungen (pl.), Hilfskabel (621.311)
secondary X-rays (pl.) sekundäre Röntgenstrahlung
secrecy Geheimhaltung (621.396)
secret geheim
secret switch Schalter unter Verschluß
to secrete absondern, sekretieren, ausscheiden
secretion Sekret, Sekretion (61); Absonderung, Ausscheidung
section Sektion, Abschnitt; Abschnitt (51); Abschnitt, Querschnitt, Schnitt, Teil, Wicklungsabschnitt, Wicklungsteil (621.313); Glied, Leitungsglied, Kettenglied (621.392); Schaltsegment
section blocking Streckenblock
section box Trennmuffe (621.315.2)
section clearing Streckenentblockung
section drawing Schnittzeichnung
section-fuse board Sicherungsverteilungstafel
section insulator Streckenisolator (621.33)
section iron Profileisen, Profilstahl
section modulus Widerstandsmoment (539.384)
section of a parabola Parabelabschnitt
section plane Schnittebene
section switch Verteilungsschalter
section view Schnittzeichnung, Ansicht im Schnitt
to section unterteilen, in Abschnitte zerlegen; schraffieren
sectional Querschnitts-..., Schnitt-...; Abteilungs-..., Profil-...
sectional area Querschnittsfläche, Schnittfläche
sectional bar Profilstange, Profilstab
sectional coil Scheibenspule
sectional elevation Schnittzeichnung, Aufriß im Schnitt
sectional form Formstück; Querschnittsform
sectional panel unterteilte Tafel
sectional plane Querschnittsebene
sectional shrinkage Querschnittsschrumpfung
sectional steel Profilstahl
sectional vertical antenna mehrfach unterteilte Vertikalantenne
sectional view Ansicht im Schnitt, Schnittzeichnung

to sectionalize unterteilen, in Abschnitte zerlegen
sectionalized bus bars unterteilte Sammelschienen
sectionalizing station Schaltstation
sectionalizing switch Streckenschalter (621.33)
sector Sektor (manchmal auch: Segment)
sector cable Sektorkabel
sector-pattern instrument Sektorinstrument
sector-scanning beacon Sektorfunkfeuer
sectoral horn eindimensionaler Hornstrahler, flacher Hornstrahler, Sektorhorn (621.396.67)
secure sicher, fest, festsitzend, sicher befestigt
secure bonding sichere Abbindung, zuverlässige Abbindung (621.396)
securely screwed festverschraubt
securely wedged fest verkeilt, sicher festgekeilt
to secure sichern, sicherstellen; sicher befestigen, befestigen, feststellen
to secure under terminals unterklemmen (621.3), anklemmen, festklemmen, festspannen
securing Befestigung, Halterung
securing bolt Haltebolzen, Befestigungsschraube
securing screw Halteschraube, Befestigungsschraube
security Sicherheit
security clearance Sicherheitsunbedenklichkeitsbescheinigung
sedimentation Absetzen, Sedimentation, Sedimentierung, Absitzenlassen
Seebeck effect Seebeck-Effekt, thermoelektrischer Effekt
seed (crystal) Impfkristall, Keim (für Kristallbildung)
seesaw Wippe
seesaw motion Wippbewegung, Kippbewegung
to seesaw wippen, hin- und herbewegen
Seger cone Seger-Kegel
segment Segment, Abschnitt (manchmal auch Sektor); Lamelle, Scheibe; Programmabschnitt (681.3)
segment of circle Kreisabschnitt
segment pitch Segmentteilung
segment-shaped segmentförmig
segmental segmentförmig
segmental arch preußische Kappe, Flachbogen, Stichbogen
segmental conductor Sektorleiter
segmental core disc Ankerblechsegment, Segmentankerscheibe (621.313)
to segregate sich entmischen (66); absondern, abscheiden; seigern (66)
segregation Unterteilung, Absonderung; Entmischung (66); Seigerung, Blockseigerung (66); Belegungsvorgang (621.395)
to seize ergreifen, erfassen; festfressen, fressen (681); belegen (621.395)
to seize fast against sich festfressen auf
seizing signal Belegungszeichen (621.395)
seizure Besitzergreifung; Festfressen, Fressen (681); Belegungsvorgang (621.395)
seizure signal Belegungssignal
to select wählen, auswählen, aussuchen
selected point Aufpunkt (51), ausgewählter Punkt
selectance Selektivität (621.396.62)
selecting Wahl-...
selecting switch Wahlschalter
selection Auswahl, Wahl (621.395); Anwahl (621.316.7), (621.398)
selection stage Wahlstufe
selective selektiv; wählerisch; Selektiv-...
selective action Selektivität (621.396)
selective calling Selektivruf
selective command Anwahlbefehl (621.316.7), (621.398)
selective control Anwahlsteuerung (621.316.7), (621.398)
selective digit emitter Impulsgeber (681.3)
selective fading selektiver Schwund (621.396)
selective impulse Wählimpuls (621.395)
selective level meter selektiver Pegelmesser
selective network Siebschaltung
selective paging Personenrufanlage
selective property selektive Eigenschaft
selective protection Selektivschutz (621.3)
selective radiation selektive Strahlung, Linienspektrum
selective reflection selektive Reflexion
selective resonance Einwelligkeit
selective ringing selektiver Anruf, Selektivruf
selectivity Trennschärfe, Selektivität
selectivity characteristic Selektionskurve (621.396.62)
selector Wähler
selector room Wählersaal (621.395)
selector switch Wählerschalter, Wahlschalter
selenium Selen
selenium cell Selenzelle
selenium magnifier Photozellen-Galvanometer-Verstärker, photoelektrischer Kompensator
selenium rectifier Selengleichrichter
selenium resistor (cell) Selenwiderstand, Selenzelle
self-... selbst-..., automatisch

self-acting selbsttätig, automatisch wirkend
self-acting brake automatische Bremse
self-acting controller Regler ohne Hilfsenergie
self-acting cutout automatisches Schaltschütz
self-acting discharger Selbstentlader
self-acting valve Schnüffelventil, Klapperventil
self-adjusting sich selbst abgleichend, selbsteinstellend
self-adjustment Selbstabgleich, automatische Einstellung
self-aligning ball-bearing Pendelkugellager
self-baking electrode Söderberg-Elektrode, selbstbackende Elektrode
self-balancing selbstabgleichend
self-balancing bridge selbstabgleichende Brücke
self-balancing instrument Kompensationsinstrument
self-balancing potentiometer selbstabgleichender Spannungsteiler, Nachlaufwiderstand
self-balancing protective system Summenstromschutz
self-bias automatische Gittervorspannung (621.375)
self-capacitance Eigenkapazität
self-check eingebaute Prüfeinrichtung, Eigenprüfung
self-checking selbstprüfend
self-checking code fehlererkennender Code
self-cleaning selbstreinigend
self-cleaning contact Schleifkontakt, Kontakt mit Selbstreinigung
self-coloured einfarbig
self-contained in sich abgeschlossen; für sich zusammengebaut, selbständig, Einbau-...
self-contained lathe Drehbank mit Einzelantrieb
self-contained instrument Multi-Instrument, Vielfach-Meßgerät
self-contained radio set Kofferempfänger
self-contained repeater station eigengespeiste Verstärkerstelle
self-contained set selbständiges Gerät, vollständiges Gerät
self-controlled automatisch gesteuert
self-cooled selbstgekühlt
self-cooled machine selbstgekühlte Maschine
self-cooled motor selbstgekühlter Motor
self-cooled transformer selbstgekühlter Transformator, selbstgekühlter Umspanner
self-cooling Selbstkühlung
self-correcting code fehlerkorrigierender Code
self-demagnetization Selbst-Entmagnetisierung
self-discharge time constant Selbstentlade-Zeitkonstante
self-excitation Selbsterregung
self-excited selbsterregt
self-excited generator selbsterregter Generator
self-excited transmitter selbsterregter Sender
self-exciting selbsterregt, sich selbst erregend
self-exciting oscillation selbsterregte Schwingung
self-feeding mit automatischem Vorschub
self-field Eigenfeld
self-healing selbstheilend
self-holding circuit Selbsthalteschaltung
self-holding contact Selbsthaltekontakt
self-ignition Selbstentzündung
self-impedance Eingangswiderstand bei Leerlauf an allen anderen Klemmen
self-indicating mit automatischer Anzeige
self-inductance Selbstinduktion, Selbstinduktivität, Eigeninduktivität
self-induction Selbstinduktion
self-inductor Drosselspule, Selbstinduktivität
self-limiting selbst begrenzend, automatisch begrenzt
self-limiting chain reaction selbstbremsende Kettenreaktion
self-linkage Eigenverkettung
self-locking selbstsperrend
self-lubrication Selbstschmierung
self-luminous selbstleuchtend
self-luminous paint Leuchtfarbe
self-maintaining nuclear chain reaction selbständige Kettenreaktion
self-oiling ring lubrication Ringschmierung
self-operated controller Regler ohne Hilfsenergie
self-oscillation selbsterregte Schwingung, Eigenschwingung
self-oscillator selbsterregter Sender, eigenerregter Schwinger
self-oscillator transmitter eigenerregter Sender
self-powered batteriebetrieben, unabhängig vom Netz betrieben
self-propagating allein weiterlaufend
self-propagating reaction Kettenreaktion
self-propelled mit Eigenantrieb
self-propelled carriage Triebwagen
self-protected X-ray tube Strahlenschutzröhre

self-quenched counter tube selbstlöschendes Zählrohr
self-quenching oscillator Pendeloszillator, Pendler
self-recording selbstschreibend
self-rectifying operation Transformatorbetrieb, Wechselstrombetrieb (621.396), (621.396.61)
self-registering selbstaufzeichnend
self-regulating selbstregelnd
self-reset rückfallendes Relais
self-resonance Eigenschwingung
self-resonance point Resonanzpunkt
self-resonant frequency Eigenresonanzfrequenz
self-righting sich selbst einstellend
self-sealing capacitor selbstheilender Kondensator
self-service lift Selbstfahreraufzug
self-stabilizing dynamo Querfeldgenerator, Querfelddynamo
self-starter automatischer Anlasser
self-starting selbstanlaufend
self-starting synchronous motor selbstanlaufender Synchronmotor, Synchronmotor mit asynchronem Anlauf
self-starting oscillator selbst einsetzender Sender, Sender mit weichem Schwingungseinsatz
self-supporting selbsttragend, freitragend
self-supporting steel tower freitragender Stahlmast
self-sustaining selbsthaltend
self-sustaining discharge selbständige Entladung (621.385)
self-synchronizing selbstanlaufend
self-ventilated eigenbelüftet
self-verification Selbstkontrolle
self-weight Eigengewicht
self-whistle Eigenerregung (durch wilde Kopplung), Überlagerungspfeifen (621.396)
self-winding clock Uhr mit Selbstaufzug
selsyn Selsyn-Drehmelder
selsyn system Selsynverfahren
selsyn transformer Selsynmotor; Selsyngeber; Selsynempfänger
semaphore Semaphor; Signalmast (625)
semaphore signal Semaphor, Flügelsignal (625)
semaphoric traffic control system Verkehrssignalanlage
semi-... halb-...
semiabsolute volt (Spannungseinheit = 1 int. Ohm × 1 abs. Amp.)
semi-automatic Halbautomat (621.9); halbautomatisch
semi-automatic lift control halbautomatische Fahrstuhlsteuerung
semi-automatic plating Durchlaufgalvanisierung
semi-automatic substation halbautomatisches Unterwerk
semi-axis Halbachse
semicircle Halbkreis
semicircular halbkreisförmig
semicircular deviation zweizähliger Peilfehler
semicircular error zweizähliger Fehler (621.396.9)
semi-closed halbgeschlossen
semi-closed slot halbgeschlossene Nut (621.313)
semiconducting halbleitend
semiconductor Halbleiter
semiconductor cooling element Halbleiterkühlelement
semiconductor device Halbleiterbauelement
semiconductor diode Halbleiterdiode
semiconductor integrated circuit integrierter Halbleiterkreis
semiconductor junction Halbleiterübergang, Sperrschicht
semiconductor monolithic integrated circuit monolithische integrierte Halbleiterschaltung
semiconductor rectifier diode Halbleitergleichrichter, Einkristallgleichrichterzelle
semiconductor region Halbleiterzone
semiconductor wafer Halbleiterscheibe
semi-distributed winding verteilte Spulenwicklung
semi-dry feuchter Elektrolyt (-Kondensator) (621.396)
semi-dry electrolytic condenser feuchter Elektrolytkondensator (621.396)
semiduplex operation bedingter Gegenverkehr, Semiduplexbetrieb
semi-enclosed halbgeschlossen
semi-enclosed cut-out halbgeschlossene Sicherung
semi-enclosed motor halbgeschlossener Motor, geschützter Motor
semi-enclosed switchgear geschirmtes Schaltgerät
semi-enclosed type geschützte Bauart
semi-finished Halb-..., halbfertig
semi-finished product Halbfabrikat

semi-flush switch halbversenkter Schalter
semi-hollow schalenförmig, gehöhlt
semi-indirect halbindirekt
semi-indirect fittings halbindirekte Leuchten
semi-indirect lighting halbindirekte Beleuchtung
semi-infinite einseitig unbegrenzt (52)
semi-logarithmic(al) halblogarithmisch
semi-logarithmical scale halblogarithmisches Koordinatensystem
semi-mat(t) halbmatt (77), (67)
semi-permeable halbdurchlässig
semi-porous halbdurchlässig
semi-protected machine halbgeschlossene Maschine
semi-recessed switch halbversenkter Schalter
semi-rigid halbstarr, nachgiebig
semi-rigid suspension halbstarre Aufhängung
semi-solid halbfest
semi-spherical halbkugelförmig
semi-strain insulator V-Doppelkette (621.315.7)
semi-sunk halbversenkt (621.8), (621.8); halbunterirdisch (624)
semitone Halbton, halber Ton
semi-transparent halbdurchsichtig, durchscheinend
semi-work halbtechnisches Verfahren
to send senden
to send a current Strom abgeben (**to:** an)
sending Sendung, Ausstrahlung (621.396)
sending end Sendeseite; Eingang, Anfang, Generatorseite, Leitungsanfang (621.392)
sending-end impedance Eingangswiderstand
sending-end voltage Eingangsspannung, Spannung am Leitungsanfang
sending reference equivalent Sendebezugsdämpfung (621.39)
sender Sender (35)
senior engineer Oberingenieur
sensation Empfindung, Wahrnehmung, Eindruck, Sinneseindruck
sensation of light Lichtempfindung, Lichteindruck
sense Sinn, Richtungssinn; Lese... (681.3)
sense adjustment Seitenbestimmung (621.396.9)
sense amplifier Leseverstärker
sense of a bearing Peilsinn (621.396.9)
sense finder Seitenschalter, Schalter für Seitenbestimmung (621.396.9)
sense finding Seitenbestimmung (621.396)
sense of rotation Drehsinn, Umlaufsinn
sense signal Lesesignal
to sense abtasten, fühlen
sensing Abtastung; Lesen (681.3)
sensing device Abtasteinrichtung
sensitiveness Ansprechwert (621.316.7)
sensitivity Empfindlichkeit (in US = Eingangsspannung für 50 mW, Ausgangsleistung bei 450 Hz und 30 % Modulation) (621.396.62)
sensitivity control Empfindlichkeitsregelung (621.396.62)
sensitivity limits Ansprechgrenzen
sensitization Sensibilisierung (77)
to sensitize sensibilisieren (77)
sensitized paper sensibilisiertes Papier; lichtempfindliches Papier (77)
sensor Aufnehmer, Messfühler, Fühler, Sensor (621.316.7)
sensor control Sensortastensteuerung (621.396)
sensory equipment Meßfühlereinrichtung, Sensoreinrichtung (621.316.7)
sentence articulation Satzverständlichkeit
sentinel Marke, Markierung (681.3)
separable trennbar, trennfähig
to separate trennen, scheiden, separieren; sich trennen, auseinandergehen, sich öffnen (621.316.5); separieren, schleudern, zentrifugieren; auseinanderreißen; abschneiden, auseinanderschneiden (621.7)
to separate into zerfallen in, sich einteilen lassen in; trennen in, aufteilen in
separated phase switchgear Schaltanlage mit Phasentrennung
separate getrennt, besonders, einzeln, Einzel-...
separate excitation Fremderregung
separate exciter getrennte Erregermaschine
separate installation Einzeleinbau, Einzelinstallation
separate motor drive Einzelantrieb
separately air-cooled motor Motor mit Fremdbelüftung
separately excited fremderregt
separately excited generator fremderregter Generator
separately lead-covered cable gürtelloses Kabel, Dreibleimantelkabel
separately ventilated fremdbelüftet
separately ventilated machine fremdbelüftete Maschine
separating Trenn-..., auflösend, trennend
separating chamber Ölfalle, Ölfangraum (621.316.5)
separation Separation, Scheidung, Trennung; Abstand, Zwischenraum; Nichtberührung

separation of the variables Trennung der Variablen (51)
separative Trenn-..., trennend, Separier-...
separative process Trennprozeß, Separierverfahren
separator Separator, Scheider; Scheider (621.356); Schleuder; Zentrifuge, Separator
separator plate Trennblech, Blechzwischenlage
septate cable radial unterteiltes konzentrisches Kabel (621.396)
sequel Folge, Folgeerscheinung, Folgende; nachstehende Ausführungen (pl.)
sequence Reihenfolge, Folge
sequence access sequentieller Zugriff
sequence contact Folgekontakt
sequence control Ablaufsteuerung, Folgesteuerung (621.316.7)
sequence counter Befehlszähler (681.3)
in sequence in Stufenschaltung, in Kettenschaltung
sequence of region Zonenfolge (621.382)
sequence switch Programmschalter
sequential aufeinanderfolgend
sequential-access storage (digitaler) Speicher mit sequentiellem Zugriff
sequential circuit Folgeschaltung, sequentielle Schaltung (621.374)
sequential colour television Sequenz-Farbfernseh-Verfahren
sequential hunting geordneter Absuchvorgang (621.395)
sequential scanning zeilenweise Abtastung ohne Zeilensprung
serial fortlaufend, eine Reihe bildend, seriell
serial access serieller Zugriff
serial adder serieller Addierer
serial call Kettengespräch
serial memory serieller Speicher
serial number laufende Nummer
serial-parallel converter Serie-Parallel-Wandler
serial printer Zeichendrucker
serial scanning serielle Abtastung
serial storage Speicher mit seriellem Zugriff
serial and teletype interface Schnittstelle für Fernschreibanschluß und serielle Datenübertragung
serial transmission Serienübertragung
series Reihe (51), (621.3); Serie
series-aiding connection gleichsinnige Hintereinanderschaltung (621.3)
series-arc furnace Doppellichtbogenofen
series-arc lamp Hauptschlußbogenlampe, Reihenschaltungsbogenlampe
series capacitance Querkapazität (621.314.2)
series capacitor reihengeschalteter Kondensator, Vorschaltkondensator
series characteristic Reihenschlußverhalten (621.313)
series circuit Reihenresonanzkreis (621.396)
serious code Reihencode
series coil Hauptschlußfeldspule, Reihenschlußwicklung
series combination Reihenschaltung, Hintereinanderschaltung
series-commutator motor Reihenschlußkommutatormotor
series-compensated motor Einphasen-Kommutatormotor mit Kompensationswicklung
series-conduction motor Wechselstrom-Reihenschlußmotor
series connection Reihenschaltung, Hintereinanderschaltung, Serienschaltung
series construction Reihenbau (69)
series diode reihengeschaltete Diode, Vorschaltdiode
series-distribution system Thury-System
series excitation Hauptschlußerregung, Reihenschlußerregung
series field Reihenfeld, Hauptstromfeld
series-field regulator Hauptstromregler
series-grid resistor Gittervorwiderstand, Gitterdämpfungswiderstand
in series in Reihe, hintereinandergeschaltet, hintereinander
series inductance Serieninduktivität; Vordrossel, vorgeschaltete Induktivität
series-inductance input filter Tiefpaß in T-Schaltung
series induction-conduction motor Déri-Motor
series lamp Reihenschlußlampe, Kettenlampe
series loading Pupinisierung, Bespulung, Reihenbelastung
series of measurements Meßreihe
series modulation Anodenmodulation
series motor Hauptstrommotor, Reihenschlußmotor
series notch Reihenschaltstufe (621.33)
series opposing gegensinnige Hintereinanderschaltung
series-overcurrent tripping direkte Auslösung, aufgebaute Auslösung (621.316.5)
series-parallel Reihenparallel-...
series-parallel battery control Fahrschaltung mit Reihenparallelschaltung der Batterie

series-parallel connection Reihenparallelschaltung
series-parallel switchgear starter Reihenparallelanlasser
series-parallel winding mehrgängige Wellenwicklung, Reihenparallelschaltung
series-phase resonance Reihenresonanz, Phasenresonanz
series production Serienfertigung, Reihenfertigung, Serienfabrikation
series reactance Serien-Blindwiderstand
series representation Reihenersatzschaltbild (621.3)
series resistance Vorwiderstand = Wert des Vorwiderstandes, Eigenwiderstand, Serienwiderstand
series resistor Vorwiderstand, vorgeschalteter Widerstand (621.317)
series resonance Reihenresonanz, Phasenresonanz
series resonant circuit Reihenresonanzkreis
series stator Stromkern, Stromeisen (621.317)
series system Reihenschaltungssystem; Thury-Schaltung
series transformer Stromwandler (621.317), Zusatzumspanner (621.314.2)
series undercurrent tripping direkte Nullstromauslösung
series voltage Reihenspannung, Zusatzspannung
series welding Mehrstellenreihenschweißung
series winding Hauptschlußwicklung, Reihenschlußwicklung
series wound ... mit Reihenschlußwicklung, Reihenschluß-..., Hauptschluß-...
series-wound arc lamp reihenregulierte Bogenlampe
series-wound d. c. motor Gleichstromhauptschlußmotor
series-wound dynamo Hauptschlußgenerator
series-wound generator Hauptschlußgenerator
series-wound motor Hauptschlußmotor, Reihenschlußmotor
serpentine Serpentin (55); Serpentine
serpentine cooler Kühlschlange
to serpentine sich schlängeln, sich winden
serrate, serrated gezähnt, gezackt
serration Zähnung
to serve dienen; bedienen
to serve a district ein Gebiet versorgen
to serve for dienen für, dienen zu
to serve a function einen Zweck erfüllen, einer Aufgabe dienen, eine Funktion haben
to serve a purpose einem Zweck dienen, einen Verwendungszweck erfüllen
to serve the purpose abgearbeitet werden (621.385, Magnetron)
served umbandelt (621.315.2)
served cable umbandeltes Kabel
served lead-covered cable Bleikabel mit Korrosionsschutz
service Betrieb; Betriebsart (621.3); Versorgung; Betrieb, Versorgungsanlage; Wartung, Dienst, Kundendienst
service-aged im Betriebe gealtert
service aerea Versorgungsbereich, Versorgungsgebiet (654.19)
service band Funkdienstband, Betriebsbereich (621.396)
service box Hausanschlußkasten
service cable Versorgungskabel, Zuleitungskabel, Kabelanschluß
service call Dienstgespräch
service capacity Netzkapazität, Netzleistungsfähigkeit (621.35)
service circuit Betriebsstromkreis
service condition Betriebsbedingung
under service conditions unter Betriebsbedingungen
service conductor Hausanschlußleitung
service denial Anschlußsperre (621.395)
service derangement Betriebsstörung
service drop Hausanschlußleitung
service entrance Hausanschluß
service-entrance conductor (äußere) Hausanschlußleitung
service equipment Betriebsgeräte (pl.)
service ground Betriebserde
service head Dachständerkopf
service instruction Dienstanweisung, Wartungsanweisung
service interruption Betriebsunterbrechung, Betriebsstörung
service kit Ersatzteilausstattung
service life Betriebslebensdauer, Lebensdauer, Nutzungsdauer; Entladezeit (621.35)
service limit Lebensdauer
service line Hausanschlußleitung
on service im Dienst, in Betrieb
service program Betriebsprogramm; Wartungsprogramm (681.3)
service quality Dienstgüte (621.395)
service reliability Betriebssicherheit, Zuverlässigkeit im Betrieb
service routine Betriebsprogramm (681.3)
service set Eigenbedarfsmaschine
service test Kapazitätsprüfung (621.35)

service test of a meter Zählerüberprüfung beim Abnehmer
service-time register Betriebsstundenzähler
service voltage Betriebsspannung
serviceable betriebsfähig, dienstfähig
serviceable time Benützerzeit
serving Umbandelung, Umwickelung; Wickellage, Lage; Schutzanstrich (621.315.2)
servo-... Servo-... Steuer-...
servo actuator Stellglied
servo amplifier Servo-Verstärker, Regelverstärker (621.316.7)
servo components Bauteile der Regel und Steuertechnik
servo controller Nachlaufregelung
servo logic Vergleichslogik (621.316.7)
servomechanism Servomechanismus, Folgeregelung
servomotor Servomotor, Stellmotor, Stellantrieb
servo-operated instrument Nachlaufinstrument
servo system automatisches Regelungssystem, Servosystem
s. E. s. cap lamp = smal Edison screw cap lamp Kleinedisonlampe, Mignonlampe
sesqui-side-band system Anderthalbseitenband-Verfahren
set Satz; Gerät; Satz, Maschinensatz; Schränkung (621.9); Setzwerkzeug, Setzhammer (621.8); Gerät, Apparat (621.3)
set analysis Geräte-Untersuchung (621.396)
set (of apparatus) Gerätesatz, Apparatesatz
set-back Rückschlag (55)
set-collar Stellring (621.316.7)
set-head Setzkopf (621.8)
set of change wheels Satz Wechselräder
set of conditions Anzahl von Bedingungen
set constructor Apparatebauer (621.396)
set of co-ordinates Koordinatensystem
set of equations Gleichungssystem
set of forces Kräfteverteilung, Kräftesystem, Kräftegruppe
set of lines Linienbüschel
set of machinery Maschinensatz, Maschinengruppe
set of mathematical instruments Reißzeug
set of plates Plattensatz (621.356), (621.319)
set of readings Ablesungsreihe
set of tables Tabellensammlung
set of teeth Zahnkranz
set of terminals Klemmenpaar (621.3)
set of valves Röhrensatz
set point Einstellwert, Sollwert
set-point adjustment Sollwert-Einsteller (621.316.7)
setscrew Einstellschraube, Stellschraube
setscrewing Stellring, Spannring, Stellmutter
set up (interval) Schwarzabhebung (621.397); Aufbau, Meßanordnung
set-up scale Skala mit unterdrücktem Nullpunkt
set-up time Aufbauzeit (621.374)
to set setzen; einstellen, stellen; einsetzen (66); zurichten, setzen (621.7); setzen (65); abbinden (69)
to set into action in Gang setzen
to set alight in Brand setzen
to set aside außer acht lassen, vernachlässigen
to set at einstellen auf, stellen auf
to set bounds to Grenzen setzen für
to set up a connection eine Verbindung aufbauen (621.39)
to set cores Kerne einlegen (621.7)
to set on edge hochkant stellen
to set in a field einem Feld aussetzen, der Einwirkung eines Feldes unterwerfen
to set on fire in Brand setzen, anzünden
to set free freimachen, herauslösen
to set electrons free Elektronen auslösen
to set limits to Grenzen setzen für, begrenzen
to set in the meridian in den Meridian einstellen
to set in motion in Bewegung versetzen, in Bewegung setzen
to set off zünden, loslassen; aufzeichnen, anreißen
to set off a discharge eine Entladung zünden
to set off a line eine Linie anreißen
to set out auftragen, anreißen, zeichnerisch darstellen; abstecken (52)
to set plumb lotrecht einrichten; ins Lot bringen
to set right in Ordnung bringen, zurecht bringen
to set up aufstellen; hervorrufen, erzeugen; abgleichen
to set up a battery eine Batterie aufstellen
to set up an engine einen Motor aufstellen
to set in vibration in Schwingungen (pl.) versetzen
to set at zero auf Null einstellen
set cement abgebundener Zement
setter Setzer, Einsetzer (66)
setting Einstellung, Einstellwert; Einsatz (66); Satz (78); Fassung (67); Abbinden, Vorgang des Abbindens (69); Aufspannung, Spannvorgang (621.9); Setzen, Sacken (69)

setting accuracy Einstellgenauigkeit
setting of a field Aufbau eines Feldes
setting gauge Einstellehre, Paßlehre
setting knob Einstellknopf
setting mechanism Einstellvorrichtung, Einstellmechanismus
setting range Einstellbereich
setting ring Stellring
setting scale Einstellskala
setting time Abbindezeit; Beruhigungszeit
setting of zero reference Nullpunkteinstellung
to settle sich absetzen, sich setzen; absitzen; sich niederlassen
to settle down to abklingen auf, absinken auf
settling Sedimentierung; Ausgleich
settling basin Klärbecken, Klärteich
settling time Einstellzeit, Einschwingzeit (eines Systems)
seven-wire three-phase system Drehstromsiebenleiternetz
to sever in Stücke teilen, trennen, aufteilen, auftrennen, zerschneiden, ablängen
severance Abschneiden, Ablängen
severe heftig, streng, schwer
severe strain harte Beanspruchung, schwere Beanspruchung
severe test strenge Prüfung
sexagesimal system Sechziger-Teilung
sextant Sextant
s. f. p. m. = surface feet per minute (Umfangsgeschwindigkeit; 1000 s. f. p. m. = 5,09 m/s)
shackle Fessel; Bindung, Schäkel; Gehänge, Traglasche; Einspannbacke; Hebezeug
shackle insulator Abspannisolator, Schnallenisolator, Schäkelisolator
shade Schatten, Schattierung, Tönung, Nuance; Schirm, Lampenschirm, Schirmwand
shade-carrier ring Reflektortragring (621.32)
to shade schattieren, schraffieren, schummern
to shade off abschattieren
shaded daylight gedämpftes Tageslicht
shaded pole Spaltpol
shaded-pole instrument Spaltpolinstrument, Ferrarisinstrument
shaded-pole motor Spaltpolmotor
shading Bildabschattung (621.397)
shading ring Kurzschlußring (621.317), Abschirmring
shadow Schatten
shadow-column instrument Schattenzeigerinstrument
shadow factor Abschattungsfaktor
shadow mask tube (Loch-) Maskenröhre (621.397)
shadow photometer Schattenphotometer
shaft Welle (621); Schaft (621.9); Stiel; Strahl (53); Deichsel (63); Schacht, Grubenschacht (621); Schacht, Kanal (624)
shaft basis system Passungssystem-Einheitswelle
shaft bearing Wellenlager
shaft-bearing bracket Wellenlagerbock
shaft butt Wellenstumpf
shaft collar Wellenring, Anschlagring
shaft current Wellenstrom, Lagerstrom
shaft derrick Förderturm (622)
shaft extension Wellenstumpf
shaft load Last an der Welle, mechanische Belastung (621.34)
shaft of a tool Werkzeugstiel
shaft slot Wellennut
to shaft mit Wellen kuppeln
shafting Wellen-...; Transmission
shaker conveyor Schüttelrutsche, Schüttelförderband (622)
shaking device Rütteleinrichtung
shallow flach, seicht
shallow cup flache Schale
shallow cut flacher Schnitt, leichter Span (621.9)
sham Bold, Vorspiegelung, Täuschung
sham Schein-..., vorgetäuscht
sham target Scheinziel
shank Schaft (621.8), (621.8), (621.9); Stiel
shank of a bolt Schraubenschaft
shank contour Schaftform
shape Gestalt, Form
to be the shape of a Tee T-förmig sein, T-Form haben
to shape formen, gestalten, Form geben
to shape to formen nach
shaped beam antenna Parabolzylinder-Antenne
shaped conductor Sektorleiter (621.315.2)
shaped handle handgerechter Griff
shaping circuit Impulsformerschaltung
shaping machine Shapingmaschine, (Metall-) Hobelmaschine
to share teilen, teilhaben an
to share a load gemeinsam eine Last speisen
shared line Gemeinschaftsleitung
shared-line equipment Gemeinschaftseinrichtung (621.395)
sharp scharf, genau; scharf, spitz
sharp bend 90 °-Bogen, Bogen (621.315.6)

sharp cut-off scharfer unterer Knick der Kennlinie
sharp cut-off frequency scharfe Grenzfrequenz
sharp definition hohe Zeichenschärfe; gute Bildschärfe (621.397)
sharp edge scharfe Kante, scharfe Schneide
sharp fracture scharfkantiger Bruch
sharp image scharfes Bild
sharp needle spitze Nadel
sharp outline scharfer Umriß
sharp report scharfer Knall
sharp tone zu hoher Ton
sharp tool scharfes Werkzeug
sharp transition scharfer Übergang
sharp tuning scharfe Abstimmung, genaue Abstimmung
sharp turn scharfer Knick
sharp wave front steile Wellenfront
sharply defined scharf begrenzt, scharf (535)
sharply peaked mit scharfer Spitze, spitzig
sharply resonant resonanzscharf (621.396)
to sharpen schärfen, verschärfen, schärfer machen
sharpening Verschärfung (621.396); Schärfung
sharpness Schärfe; Scharfkantigkeit
sharpness of directivity Richtschärfe, Bündelungsschärfe
sharpness of resonance Resonanzschärfe
sharpness of the spot Fleckschärfe (621.385)
sharpness of tuning Abstimmschärfe
to shave schaben (621.9)
to shave past vorbeirutschen an
shaving Span, Schnitzel; Hobelspan, Drehspan
shavings Hobelspäne
shavings bin Spankasten, Spänekasten (621.9)
shear Scherung, Scherbeanspruchung; Wange (621.94)
shear breakdown Molekülverkleinerung durch mechanische Beanspruchung (66)
shear force Scherkraft, Schubkraft, Schub
shear legs Bockgerüst, Schere (621.8)
shear modulus Gleitmodul, Schubmodul
shear pin Scherbolzen
shear strain Schubspannung, Scherspannung
shear-strain energy Torsionsenergie
shear strength Schubfestigkeit, Scherfestigkeit
shear stress Schubbeanspruchung, Schubspannung, Scherbeanspruchung
to shear scheren, abscheren; auf Schub beanspruchen
shearing Scherung, Scher-...
shearing effect Scherwirkung
shearing machine Schere, Hebelschere (621.96)
shearing modulus Schubmodul, Gleitmodul
shearing test Scherversuch
shears (pl.) Schere
sheath Scheide, Hülle; Überzug; Mantel (621.315.2)
sheath current Mantelstrom
sheath loss Mantelverluste (pl.)
sheath unbalance Außenerdkopplung
to sheathe umkleiden, ummanteln
sheathed electrode umhüllte Elektrode (621.791)
sheathed wire umhüllte Leitung (621.3)
sheathing Umhüllung, Ummantelung
sheave Rillenscheibe, Seilscheibe
shed Schuppen; Schirm (621.315.6); Lochstreifenabfall (621.394.6)
shed roof Pultdach, Sheddach
shed tape durchgelochter Streifen
shedless tape Lochstreifen ohne Abfall, angelochter Lochstreifen, Schuppenlochstreifen
to shed vergießen, verschütten
to shed light on Licht werfen auf
sheet Blech; Blatt, Bogen (65); Abschnitt (67)
sheet aluminium Aluminiumblech
sheet brass Messingblech
sheet copper Kupferblech
sheet glass Fensterglas, Flachglas
sheet iron Eisenblech
sheet-iron casing Eisenblechgehäuse
sheet laminations Dynamobleche
sheet lightning Flächenblitz
sheet metal Metallblech
sheet mica Spaltglimmer; Mikanit
sheet resistivity Quadratwiderstand
sheet steel Stahlblech
sheet-steel clad stahlblechgekapselt
sheet zinc Zinkblech
sheeting Blech, Blechbelag
shelf (pl. **-ves**) Sims (69); Regal; Bücherbrett, Bücherbord (64); Zwischenbrett, Fach (69)
shelf corrosion Selbstentladung (621.352)
shelf depreciation Wertverlust durch Lagerung
shelf life Lagerfähigkeit, Lebensdauer in unbenutztem Zustand
shelf test Lagerprüfung, Prüfung auf Lagerfähigkeit
shell Muschel, Muschelschale (59); Haube, Hülse; Außenhaut, Mantel, Schale (629.13); Glocke, Schirm (621.315.6); Granate, Geschoß (623); Schale (539); Bedieneroberfläche (681.3)
shell antenna Muschelantenne

shell-like schalenförmig
shell-type motor Anbaumotor, Flanschmotor
shell tracking Geschoßverfolgung
shell-type transformer Manteltransformator
to shell off abblättern
shellac Schellack
shellac solution Schellack(lösung)
shellac varnish Schellack, Schellackanstrich
to sherardize scherardisieren
Schering bridge Schering-Brücke
SHF = Super High Frequency Gigahertzbereich, Zentimeterwellen(bereich)
shield Schild, Schutzschild; Schirm, Abschirmung; Schutzblech
shield can Abschirmbecher (621.396)
shield layer Abschirmlage (621.396)
shield terminal Abschirmanschluß, Abschirmungsklemme
shield winding Abschirmwicklung
to shield abschirmen, schirmen, abschatten (**from:** gegen)
shielded aerial connection Abschirmkabel
shielded arc electrode Arkatom-Elektrode, Elektrode mit gasumhülltem Lichtbogen
shielded arc welding Schutzgasschweißung, Arkatomschweißung
shielded cable geschirmtes Kabel
shielded compartment abgeschirmter Raum
shielded conductor cable H-Kabel, Höchstätter-Kabel
shielded enclosure abgeschirmtes Gehäuse
shielded grid valve Schirmgitterröhre
shielded ignition wiring abgeschirmte Zündanlage
shielded line abgeschirmte Leitung
shielded loop aerial abgeschirmte Rahmenantenne
shielded metal arc welding Schutzgas-Elektrodenschweißung
shielded pair abgeschirmte Doppelleitung
shielded pole instrument Ferrarisinstrument
shielded wattmeter eisengeschirmtes Wattmeter
shielded X-ray tube Strahlenschutzröhre (621.386)
shielding Abschirmung, Schirmung
shielding can Abschirmbecher (621.396)
shielding grid Schirmgitter (621.385)
shielding harness Abschirmgeflecht (621.396)
shift Verschiebung, Versetzung; Schicht, Arbeitszeit; Notbehelf, Behelf
shift in frequency Frequenzverwerfung (621.396)
shift operation [Verschiebung der Binärstellen um je eine Stelle]
shift register Schieberegister (621.374)
to shift verschieben, umschalten (621); schieben, schwenken (621.3)
to shift the belt den Riemen umlegen, den Riemen rüberschieben
to shift bodily als Ganzes verschieben
to shift the frequency die Frequenz umtasten (621.394.1)
to shift the gear das Getriebe umschalten, Gang schalten (629.113)
to shift the lobe die Keule schwenken (621.397)
to shift the phase die Phase drehen, die Phase schieben
shifting Umschaltung, Verschiebung
shifting device Umschalteinrichtung
shim Beilage, Beilageblech, Ausgleichsscheibe
shim member Trimmelement
shim plate Zwischenlage, Beilageblech
shim rod Anpassungsstab (539)
to shim durch Beilagen ausrichten, ausgleichen
shimmer Schimmer, schwacher Lichtschein, Flimmern
to shine scheinen, leuchten; polieren, putzen, saubermachen
to shine up aufpolieren, nachpolieren
ship Schiff
shipborne Schiffs-...
shipborne equipment Bordgerät für Schiffe
shipborne transmitter Schiffssender
by ship per Schiff, auf dem Wasserwege
shipyard Schiffswerft
shock Schlag, Stoß, Ruck; Schlag, Schock (621.3)
shock absorber Stoßdämpfer (629.113)
shock-absorbing mounting stoßdämpfende Aufhängung
shock excitation Stoßerregung (621.313), (621.396); Stoßanregung (534)
shock-excitation-type transmitter stoßerregter Sender
shock excited stoßerregt
shock-excited oscillator stoßerregter Generator, Blockschwinger (621.396)
shock hazard Berührungsgefahr (621.3)
shock load Stoßlast, Stoßbeanspruchung, Stoßbelastung, Laststoß
shock-proof berührungsschutzsicher, mit Berührungsschutz (621.3)
shock-proof switch berührungsschutzsicherer Schalter
shock strength Stoßfestigkeit

shock wave Druckwelle, Druckstoß, Stoßwelle
to shock schockieren, erschüttern
to shock into oscillation zum Schwingen anstoßen
shocking coil Elektrisierungsspule, Induktorium
shoe Schuh, Bremsschuh, Bremsbacke; Grundplatte, Sohle, Sohlplatte; Polschuh
shoot-in bolt Schießdübel
to shoot the sun die Sonne schießen (527)
to shoot out ausstoßen, auswerfen, herausstoßen
shop Werkstatt, Arbeitsraum; Laden; Fachsimpelei
shop operation Arbeitsgang, Bearbeitungsgang
shop parlance Werkstattsprache
shore Strebebalken, Abstützung (624); Ufer, Küste, Strand
shore-based equipment Küstenstation
shore-end cable Küstenkabel
shore-line effect Küsteneffekt (621.371)
to shore abstützen, verstreben
short Kurzschluß (621.3)
short kurz, kurzzeitig
short addressing direkte Adressierung
short chord winding Bruchlochwicklung
short circuit Kurzschluß
short circuit across Kurzschloß von... nach...
short-circuit analysis Kurzschlußberechnung
short-circuit armature Kurzschlußanker
short-circuit beyond Kurzschluß hinter
short-circuit calculation Kurzschlußberechnung
short-circuit characteristic Kurzschlußkennlinie
short-circuit current Kurzschlußstrom
short-circuit forward current transfer ratio Kurzschlußstromverstärkung, Stromverstärkungsfaktor (621.382.3)
short-circuit forward transfer admittance Kurzschluß-Übertragungsadmittanz vorwärts, Transmittanz (621.382.3)
short-circuit impedance Kurzschlußwiderstand
short-circuit input admittance Kurzschluß-Eingangsadmittanz
short-circuit interruption Kurzschlußabschaltung
short-circuit key Kurzschlußtaste
short-circuit limiting reactor Kurzschlußdrossel
short-circuit period Kurzschlußdauer
short-circuit proof kurzschlußfest, kurzschlußsicher
short-circuit ratio Kurzschlußverhältnis (621.313)
short-circuit reverse transfer admittance Remittanz, Kurzschluß-Übertragungsadmittanz rückwärts (621.382.3)
short-circuit test Kurzschlußversuch
short-circuit transmittance Kernleitwert (621.372.5)
short-circuit trip Kurzschlußauslöser
short-circuit voltage Kurzschlußspannung
short-circuiting Kurzschließen, Kurzschlußvorgang
short-circuiting bar Kurzschlußbügel
short-circuiting device Kurzschließer, Kurzschlußvorrichtung
short-circuiting power Kurzschlußleistung
short-circuiting winding Kurzschlußwicklung
short distance geringe Entfernung, kleiner Abstand
short division Division durch Zahlen bis 12
short life kurze Lebensdauer
short-lived kurzlebig
short-living kurzlebig
short of dicht unterhalb; weit entfernt von
short of the frequency dicht unterhalb der Frequenz
short of satisfactory alles andere als zufriedenstellend
short-pitch winding Sehnenwicklung
short range klein Reichweite, kleiner Abstand, Nahbereich
short-range fading Nahschwund
short-range force Nahkraft, Kraft mit kleiner Reichweite (539.17)
short-range radar Nahbereichsradar
short-rated für kurze Belastung ausgelegt
short stroke kurzer Hub
short-stroke Kurzhub-, kurzhubig, ... mit kurzem Hub
in short supply schwer lieferbar
short-term stability Konstanz über kurze Zeit; kurzzeitige Frequenzkonstanz (621.396.61)
short-time Kurzzeit-...
short-time duty Kurzzeitbetrieb, KB (621.313)
short-time rated für kurze Belastung ausgelegt
short-time rating Kurzzeitleistung
short-time stability Frequenzkonstanz über kurze Zeit, Kurzzeitkonstanz (621.396.61)

short-time service Kurzzeitbetrieb (621.313)
short-type coil Kurzspule (621.313)
short waves kurze Wellen (10 bis 100 m Wellenlänge)
short-wave adapter Kurzwellenvorsatz (621.396)
short-wave connection Kurzwellenverbindung
short-wave set Kurzwellengerät
short-wave therapy Kurzwellentherapie
short weight Untergewicht, Mindergewicht
to short kurzschließen
to short-circuit kurzschließen
short-circuited armature Kurzschlußanker
short-circuited line kurzgeschlossene Leitung
short-circuited loop Kurzschlußschleife (621.317)
short-circuited rotor Anker mit kurzgeschlossener Wicklung, Kurzschlußläufer
short-circuited turn Kurzschlußwindung
short-circuited turns test Prüfung auf Windungsschluß
short-circuited winding Kurzschlußwicklung, kurzgeschlossene Wicklung
to short ground gegen Erde kurzschließen, erden
to short out überbrücken
shorted turn Kurzschlußwindung
to be shorted kurzgeschlossen sein
to shorten verkürzen, kürzer machen; sich verkürzen
to shorten the exposition Belichtungszeit verkürzen
shortening condenser Verkürzungskondensator
shorter side Kathete (51); kürzere Seite
shorting Kurzschließen
shorting bar Kurzschlußbügel (621.396)
shorting bridge Kurzschlußbrücke
shorting link Kurzschlußleitung
shorting plug Kurzschlußstecker
shorting plunger Kurzschlußkolben, Kurzschlußschieber (621.396)
shorting strip Kurzschlußlasche
shot Schuß (662); Schrot, Schrotkugel (623.4)
shot effect Schroteffekt (621.385)
shot-firing cable Zündkabel, Schießkabel (622)
shot-firing generator Zündmaschine (662.2)
shot hole Bohrloch, Sprengloch
shot noise Schrotrauschen
shoulder Schulter, Ansatz, Widerlager
shoulder screw Ansatzschraube
to shoulder Abstützen, sich abstützen (**against:** auf)
to shoulder down absetzen, einen Absatz drehen (621)
show cable Bildschirmtext
to shred zerfetzen, zerfasern
shriek Kreischen, Quietschen
to shriek kreischen, quietschen, schrillen
shrink... Schrumpf...
to shrink schrumpfen, schwinden (621); eingehen (67); aufschrumpfen (621.9)
to shrink into place einschrumpfen
to shrink on aufschrumpfen, warm aufziehen
shrinkage Schwindung, Schrumpfung; Schwund, Lunkerbildung
schrinkage cavity Schwindlunker (66)
shrinkage crack Schwindriß, Schwundriß
shrinkage dimension Schwindmaß
shrinkage stress Schrumpfspannung, Schwindspannung
shrink fit Schrumpfpassung
shrinkhole Schwindlunker, Lunker
shrink ring Schrumpfring
shrivel varnish Schrumpflack
shroud Verkleidung
shroud tube Verkleidungsrohr
to shroud verkleiden; einhüllen
shunt Nebenschluß, Parallelschaltung; Nebenwiderstand (621.3); Nebengleis (625); Querglied, Quer-... (621.3)
shunt-arc lamp Nebenschlußbogenlampe
shunt box Nebenwiderstandskasten (621.317)
shunt capacitor Querkondensator (621.392); Überbrückungskondensator, Parallelkapazität (621.375)
shunt capacitor input filter Tiefpaß in Pi-Schaltung
shunt capacity Parallelkapazität, Überbrückungskondensator
shunt characteristic Nebenschlußverhalten, Nebenschlußcharakteristik
shunt circuit Feldstromkreis (621.313); Parallelstromkreis, parallelgeschalteter Zweig, Nebenschluß
shunt coil Nebenschlußwicklung, Nebenschlußfeld (621.313)
shunt conductance per unit length Ableitungsbelag
shunt-conduction motor Wechselstrom-Nebenschluß-Kollektormotor
shunt connection Parallelschaltung
shunt current Querstrom
shunt dynamo Nebenschlußgenerator
shunt excitation Nebenschlußerregung

shunt excited nebenschlußerregt
shunt-fed antenna parallelgespeiste Antenne
shunt field Nebenschlußwicklung
shunt-field relay Relais mit magnetischem Nebenschluß
shunt-field rheostat Nebenschlußfeldregler
in shunt in Parallelschaltung, nebeneinander geschaltet, parallel geschaltet, parallel
in shunt across parallel zu, parallel geschaltet zu
shunt-induction motor läufergespeister Nebenschluß-Kollektormotor
shunt inductor Paralleldrossel, parallelgeschaltete Induktivität
shunt loading Beschaltung mit Querdrosseln
shunt motor Nebenschlußmotor
shunt ratio Meßbereichserweiterung (621.317)
shunt regulating resistance Nebenschlußregelwiderstand
shunt regulation Nebenschlußregelung
shunt regulator Feldregler, Nebenschlußregler
shunt resistance Nebenwiderstand, Parallelwiderstand
shunt resistor parallelgeschalteter Widerstand, Überbrückungswiderstand
shunt rheostat Nebenschlußregler
shunt running Leerlaufvortrieb (621.317)
shunt stator Spannungskern, Spannungseisen (621.317)
shunt terminal Nebenwiderstandsklemme
shunt trip Spannungsauslösung
shunt-trip coil Sekundärauslöser
shunt winding Nebenschlußwicklung
shunt-wound ... mit Nebenschlußwicklung, Nebenschluß-...
shunt-wound dynamo Nebenschlußdynamo, Nebenschlußgenerator
shunt-wound generator Nebenschlußgenerator
shunt-wound machine Nebenschlußmaschine
shunt-wound motor Nebenschlußmotor
to shunt parallelschalten (**across:** zu); überbrücken (**by:** mit); rangieren (625)
to shunt across parallel zu ... anschließen
shunted ammeter Amperemeter mit Nebenwiderstand
shunted condenser überbrückter Kondensator
shunted field control Feldschwächungsregelung (621.33)
shunted meter Strommesser mit Nebenwiderstand
shunting Parallel-... (621.3); Rangieren, Rangier-...
shunting capacity Überbrückungskondensator, Parallelkapazität
shunting effect Nebenschlußwirkung
shunting tower Rangierstellwerk
to shut schließen, zumachen
to shut down stillsetzen, stillegen, schließen, Betrieb einstellen, abschalten, außer Betrieb setzen
to shut in einschließen
to shut off absperren
to shut out ausschließen, aussperren; ausschließen, unterbinden
shutdown Betriebsunterbrechung, Betriebseinstellung
shut-off rod Schnellschlußstab (539)
shutter Verschluß (77); Hohlleitersperre (621.372); Jalousie
shuttle Weberschiff; Rohrpostbüchse
shuttle armature Doppel-T-Anker
shuttle train Pendelzug (626)
side Seite; Seite (51); Flanke
side anchor Querabspannung
side approach Auflaufschuh (621.33)
sideband Seitenband (621.396)
sideband component Seitenbandanteil
sideband energy Seitenbandleistung
sideband frequency Seitenbandfrequenz
sideband noise Seitenbandrauschen
sideband telephony Seitenbandtelefonie
sideband suppression Seitenbandunterdrückung
side bending seitliches Ausweichen
side chain Seitenkette (54); Seitenlinie, Abzweig
side channel Seitenkanal
side circuit Stammkreis (621.395)
side-circuit loading coil Stammspule
side-circuit repeating coil Stammübertrager
side connection seitlicher Anschluß
side crust seitliche Verkrustung (621.355)
sidecutting pliers Seitenschneider
side echo Nebenzipfelecho
side elevation Seitenriß
side flashing Abspringen (69)
side lead seitlicher Anschluß, seitliche Zuführung
side line Nebenlinie
side lobe Nebenzipfel
side-lobe attenuation Nebenzipfeldämpfung
side-lobe suppression Nebenzipfelunterdrückung
side-looking radar Seitensichtradar, Schrägsichtradar
side panel Seitenwand (64)

side pressure Seitendruck, seitlicher Druck
side rod Triebstange (621.33)
side-rod drive Stangenantrieb (621.33)
side-running trolley seitlicher Stromabnehmer
side thrust Seitenschub, Schub
sidetone Mithörton
sidetone attenuation Rückhördämpfung, Mitsprechdämpfung
sidetone level Mithörlautstärke, Kontrollautstärke
sidetone reduction wiring mitsprechfreie Schaltung (621.395)
sidetone-reference equivalent Rückhörbezugsdämpfung
sidetone telephone set Fernsprechapparat ohne Rückhördämpfung, Mithörtelefon
side tool Seitenstahl, Seitenschneider (621.9)
side view Seitenansicht
side waves Seitenbänder (621.396)
to side track umleiten
Siemens electro-dynamometer elektrodynamometrisches Meßinstrument
Siemens-H-armature Siemens-Doppel-T-Anker
Siemens-Hell writing telegraph Siemens-Hell-Schreiber
Siemens-Martins steel Siemens-Martin-Stahl
Siemens moving-coil-type telewriter Siemens-Drehspul-Fernschreiber
Siemens-multiplex printing telegraph Siemens-Schnellschreiber, Siemens-Rekorder
Siemens unit Siemens-Einheit (des Widerstandes) = 0,9407 Ohm
Siemens unit of resistance Siemens-Widerstandseinheit
sieve Sieb
sieve spacing Sieböffnung, Maschenweite
to sieve sieben
to sift sieben, sichten
to sift out aussortieren (621.385)
sifting device Sichter, Siebeinrichtung
sight Sicht, Sichtweite (55); Gesichtssinn; Visier (623)
sight glass Schauglas
sight hole Schauloch
in sight of each other in optischer Sicht (621.396.67)
to sight visieren, anvisieren; sichten, in Sicht von ... kommen
sighting device Visiereinrichtung
sign Zeichen, Vorzeichen; Zeichen, Anzeichen, Merkmal; Schild, Aushängeschild, Firmenschild
sign lights Leuchtreklame
sign logic Vorzeichenlogik, Richtungslogik
sign of addition Pluszeichen, Additionszeichen
sign of current Stromrichtung
sign of differentiation Differentialzeichen
sign of integration Integralzeichen, Integrationszeichen
sign of potential difference Vorzeichen der Spannung
signal Signal; Warnung; Zeichen, Zeichenspannung, Signalspannung (621.396), (621.375); Schaltkennzeichen (621.395.3)
signal aspect Signalbild (656.25)
signal book Code, Schlüsselverzeichnis
signal box Signal-Telefonhäuschen, Stellwerk (625)
signal-circuit controller Signalflügelkontakt
signal comparison Kurvenvergleich, Vergleich des Zeichenverlaufs
signal component Zeichenelement, Signalkomponente
signal-to-cross-talk ratio Grundwert des Nebensprechens (621.395)
signal diode Signaldiode
signal element Schritt (621.394)
signal fidelity Verzerrungsfreiheit eines Signals
signal generator Meßsender, Prüfsender, Pegelsender, Summer (621.396)
signal grid Steuergitter (621.385)
signal ground CCITT-Betriebserde
signal-to-hum ratio Brummabstand
signal-to-image ratio Spiegelfrequenzselektion
signal indication Signalstellung (656.25)
signal inversion Signalinversion
signal lamp Signallampe
signal level Signalpegel, Zeichenpegel (621.396)
signal lever Signalhebel
signal light Blinkscheinwerfer, Signalscheinwerfer
signal motor Stellmotor, Signal-Stellmotor (65)
signal-to-noise ratio Signal-Rauschverhältnis, Signal-Geräuschabstand
signal path Signalweg, Signalverlauf
signal post Signalmast (625)
signal proving Signalmelder (625)
signal received Empfangszeichen
signal is received Zeichen fallen ein (621.396)
signal-repeater light Rangierlichtsignal
signal rope Signalseil
signal-shaping network Zeichen-Entzerrer
signal source Signalquelle (621.39);

Tonfrequenz-Meßsender, Normalgenerator (621.3)
signal strength Feldstärke, Empfangsfeldstärke (621.396)
signal-strength meter Feldstärkemeßgerät
signal tracer Signalverfolger
signal transmitter Signalgeber
signal velocity Signalgeschwindigkeit
signal voltage Steuerspannung, Signalspannung, Eingangsspannung, Empfangsspannung
signal wave Zeichen (621.394)
to signal signalisieren, melden, Signal geben
signalling Signal-..., Warn-...
signalling apparatus Warnapparat
signalling bell Warnglocke, Signalglocke
signalling cable Signalkabel
signalling channel Signalkanal, Signalisierungskanal
signalling device Signaleinrichtung, Warneinrichtung
signalling key Taste (621.394)
signalling lead Signalleitung
signalling line Signalleitung
signalling wire Signalleitung [einpolig]
to signalize kennzeichnen
signatory Unterzeichner; Stationskennzeichen (621.396)
signature Unterschrift; Röntgenbeugungsbild, Debye-Scherrer-Bild (621.386)
signature analyzer Signatur-Analysator
signature reader Unterschriftenleser
signature tune Stationszeichen, Erkennungsmelodie (621.396)
significant kenn...
significant condition of modulation Modulationskennzustand (621.394)
significant interval Kennabschnitt (621.394.6)
significant digit signifikante Ziffer
silence in operation geräuschloser Lauf
to silence stumm machen, stillsetzen; zum Schweigen bringen
silencing device Schalldämpfer (621.4); Krachtöter (621.396)
silent still, stumm, geräuschlos
silent balance Abgleich auf Tonlosigkeit
silent discharge stille Entladung, dunkle Entladung
silent film Stummfilm
silent tuning Stummabstimmung
silent zone Sprungzone, tote Zone (621.371)
silica Kieselsäure
silica lamp Quarzlampe
silica valve Quarzröhre
silicic acid Kieselsäure
silicon Silizium
silicon alloy Silumin
silicon carbide Siliziumkarbid
silicon controlled rectifier, SCR gesteuerter Siliziumgleichrichter, Thyristor
silicon detector Silizium-Detektor
silicon fiberglas laminate Silikon-Glasseiden-Hartfasergewebe
silicon-gate MOS MOS-Schaltung mit Silizium-Gate
silicon-nitride passivation Siliziumnitrid-Passivierung
silicon planar transistor Siliziumplanartransistor
silicon rectifier Siliziumgleichrichter
silicon-on-sapphire Silizium auf Saphir (621.382)
silicon steel legiertes Blech (621.3)
silicon transistor Silizium-Transistor
silicone Silikone
silicone oil Silikone-Öl
silicone rubber Silikone-Gummi
silk Seide
silk-covered seidenumsponnen
silk-covered wire seideumsponnener Draht
silver Silber
silver amalgam Silberamalgam
silver-chloride cell Silber-Chlorid-Element
silver cladding Silberplattierung
silver contact Silberkontakt
silver foil Blattsilber, Silberfolie
silver leaf Blattsilber
silver litharge Silberglätte
silver paper Silberpapier = Zinnfolie
silver-plated versilbert, mit Silberauflage
silversteel Silberstahl
silver storage battery Silberzinkakkumulator
silver voltameter Silber-Voltameter
silver-zinc accumulator Silber-Zink-Akkumulator
to silver versilbern
silvering Versilberung
Simmance-Abady photometer Flimmer-Photometer
to simmer sieden
Simms-Edison torpedo elektrischer Torpedo
simple einfach
simple body Element (54)
simple control einfache Ausschaltung
simple equation lineare Gleichung
simple harmonic current sinusförmiger Strom, einwelliger Strom
simple harmonic law reines Sinusgesetz

simple harmonic quantity sinusförmige Größe
simple rectification Gleichrichtung im Anodenknick
simple sinusoidal current einwelliger Strom, sinusförmiger Strom
simple sinusoidal electromotive force einwellige EMK
simple sinusoidal potential difference rein sinusförmige Spannung
simple sinusoidal voltage rein sinusförmige Spannung
simply mesh-connected circuit Polygonschaltung
simply supported einfach aufgelagert, frei aufgelagert
simplex einfach, Einfach...
simplex circuit Wechselverkehrsverbindung, Simplexverbindung
simplex coil Differentialübertrager
simplex operation Simplex-Betrieb (621.39)
simplex (telegraph) system Simplex-Telegraphie, Wechselverkehr, Richtungsverkehr (621.394)
simplex voice communication Wechselsprechen
simplex winding eingängige Wicklung
to simulate vortäuschen, nachahmen; simulieren, (im Ersatzschaltbild) nachbilden
simulator Simulator; Simulationsprogramm
simultaneity Gleichzeitigkeit
simultaneous gleichzeitig, simultan
simultaneous colour television Simultan-Farbfernseh-Verfahren
simultaneous equations Gleichungen mit mehreren Unbekannten
simultaneous equations in 6 unknowns Gleichungen mit 6 Unbekannten
simultaneous two-wire conversation Zweidraht-Gegensprechverkehr
sin = sine sin = sinus
sin^{-1} arc sin
sine sinus, Sinus
sine current Sinusstrom, sinusförmiger Strom
sine curve Sinuskurve
sine galvanometer Sinus-Bussole
sine law Sinusgesetz
sine law variation sinusförmige Änderung, sinusförmiger Verlauf
sine meter Blindleistungszähler
sine rule Sinussatz
sine-squared pulse generator Rechteck-Pulsgenerator
sine wave Sinuswelle
sine-wave sinusförmig, einwellig, Sinus-...
sine-wave generator Oszillator, Sinuswellengenerator, Sinusgeber
sine-wave voltage Sinusspannung, einwellige Spannung
to sing singen; summen
singing Pfeifen, Rückkopplungspfeifen (621.375)
singing arc tönender Lichtbogen, singender Lichtbogen
singing margin Pfeifpunktabstand
singing point Pfeifpunkt
singing point equivalent Restdämpfung am Pfeifpunkt
singing spark tönender Funken, tönender Lichtbogen
singing suppressor Rückkopplungssperre
to singe sengen, versengen
single einzeln
single address Einzeladressen (681.3)
single-address instruction Einadreß-Befehl
single-acting einfach wirkend
single-acting engine einfach wirkende Maschine
single-anode rectifier Einanodengleichrichter
single armature converter Einanker-Umformer
single-beam direction finding Einstrahlpeilung
single bevel einfache Abschrägung
single-board computer Einplatinencomputer
single-break switch Schalter mit Einfachunterbrechung
single cable Einleiterkabel, konzentrisches Kabel
single catenary suspension einfache Tragdrahtaufhängung (621.33)
single-cavity velocity-modulated tube Einkreis-Triftröhre
single-coil filament Einfachwendel
single conductor Einzelleitung, einzelne Leitung
single-contact einpolig
single-contact relay einpoliges Relais
single-contact switch einpoliger Schalter
single-core einadrig
single-core cable einadriges Kabel
single crystal Einkristall
single current Einfachstrom (621.394)
single-current key Einstrom-Taste (621.394)
single-current (telegraph) system Einfachstromsystem
single-current working Einfachstrom-Betrieb
single-cylinder engine Einzylindermotor
single-dial control Einknopfbedienung (621.396.62)
single-electrode potential Elektrodenpotential
single-element relay Einwicklungsrelais

single-end control Steuerung von einem Führerstand aus (621.33)
single-ended amplifier unsymetrischer Verstärker
single-ended input Eintakteingang
single feeder Einzelspeiseleitung
single-groove sheave Einrillenscheibe
single-hop transmission Übertragung durch Einfachreflexion
single-lamp transformer Kleinspannungstransformator
single layer einzelne Lage
single-layer einlagig
single-layer coil einlagige Spule
single-layer winding einlagige Wicklung, einschichtige Wicklung, einlagige Zylinderspule (621.396)
single line Einzelanschluß (621.395)
single metering Einfachzählung (621.395.1)
single-office exchange Hauptamtvermittlung
single operation Simplex-Betrieb
single-operator motor-generator set Einstellen-Schweißumformer
single-phase einphasig, Einphasen-...
single-phase alternating current Einphasen-Wechselstrom, einphasiger Wechselstrom
single-phase alternator Einphasengenerator
single-phase circuit Einphasenschaltung, Einphasenkreis
single-phase commutator motor Wechselstrom-Kollektormotor
single-phase current Einphasenstrom
single-phase current locomotive Einphasenlokomotive
single-phase generator Einphasengenerator
single-phase induction motor Einphasen-Induktionsmotor
single-phase meter Einphasenzähler
single-phase motor Einphasenmotor, Wechselstrommotor
single-phase series commutator motor Reihenschluß-Kommutatormotor
single-phase system Einphasensystem
single-phase traction Einphasenbahnbetrieb
single-phase traction motor Einphasen-Bahnmotor
single-phase synchronous generator Einphasen-Synchrongenerator
single-phase transformer Einphasentransformator
single-phase winding Einphasenwicklung
single-picture Einzelbild
single-pivot instrument Meßinstrument mit Hängelager
single-point load Einzellast, punktförmige Last
single-polarity pulse unipolarer Impuls
single-pole einpolig
single-pole circuit breaker einpoliger Schalter
single-pole cut-out einpoliges Schütz
single-pole double-throw switch einpoliger Umschalter
single-pole fuse einpolige Sicherung
single-pole switch einpoliger Schalter
single-post pantograph-type isolating switch Einsäulentrenner
single potential Elektrodenpotential
single-race ball bearing einreihiges Kugellager
single ring coil Kreisringspule, Toroidspule
single-shot multivibrator monostabiler Multivibrator
single-shot operation Einzelschrittbetrieb (681.3)
single-sideband Einseitenband-...
single-sideband transmission Einseitenbandübertragung
single-side band transmitter Einseitenbandsender
single sided einseitig
single sided diskette einseitig beschreibbare Diskette
single-signal receiver Einwellenempfänger
single-stage einstufig, Einstufen-...
single-step operation Einzelschrittbetrieb
single-stroke bell Einschlagwecker
single supply operation Betrieb mit nur einer Versorgungsspannung
single-throw circuit breaker gewöhnlicher Schalter, Ein-Aus-Schalter
single-throw switch Ein- und Ausschalter
single-track eingleisig
single-track recording Einspurtechnik (681.84)
single-trolley system Eindraht-Oberleitungssystem
single turn einzelne Windung
single-turn coil Einstabspule (621.313); Einwindungsrahmen (621.396)
single-turn current transformer Schienenstromwandler, Stabstromwandler
single-turn loop Einwindungsrahmen (621.396)
single-turn transformer Einleiterwandler (621.317)
single-valued einwertig, eindeutig (51)
single-valued function eindeutige Funktion

single-wave rectification Einweg-Gleichrichtung
single-wavelength working Einwellenbetrieb
single-way switch einfacher Einschalter (621.3)
single-winding multispeed motor drehzahlumschaltbarer Motor, polumschaltbarer Motor
single wire einzelner Draht; Eindraht-...
single-wire aerial Eindrahtantenne
single-wire cable einadrige Kabel, Einleiterkabel
single-wire conductor einpolige Leitung
single-wire line eindrähtige Leitung, einpolige Leitung
single-wound armature Anker mit eingängiger Wicklung, eingängig gewickelter Anker
singly re-entrant winding eingängige Wicklung, Einschleifenwicklung
singular singulär, einzigartig, einmalig; singulär (51)
singularity Singularität (51)
sinh. = hyperbolic sine sinh = Hyperbelsinus
sink Senke (621.39), (51); Ausgang, Ende, Verbraucherseite (621.392), Ausguß (64)
sink current [Ausgangsstrom bei Ausgang „L"]
to sink sinken, untergehen; herabsinken, versinken, sich senken; versickern; versenken
to sink a pole einen Mast setzen
to sink into the wall in die Wand einbauen
to sink to zero auf Null absinken
sinking Versenkung; Vertiefung; Abnehmen, Abnahme, Fallen
sinking in a workpiece Vertiefung in einem Werkstück
sinoidal (US) sinusförmig (**s. sinusoidal**)
to sinter sintern, fritten
sintered bearing Sinterlager
sintered carbide Sinterkarbid, Widia
sintered coal Sinterkohle
sintered magnet Sintermagnet
sintered metal Sintermetall
sintered-metallic Sintermetall-...
sinuosity Geschlängel, Windung
sinuous sich windend, geschlängelt
sinusoidal sinusförmig
sinusoidal current sinusförmiger Strom, einwelliger Strom
sinusoidal field sinusförmig veränderliches Feld
sinusoidal quantity sinusförmige Größe
sinusoidal sweep sinusförmige Zeitablenkung (621.385)
sinusoidal therapy Wechselstromtherapie
sinusoidal voltage sinusförmige Spannung
sinusoidal wave Sinuswelle
siren Sirene
sisal Sisal, Sisalhanf
site Bauplatz, Aufstellungsort; Standort, Lage; Baustelle
site characteristics örtliche Gegebenheiten
site error durch Aufstellungsort bedingte Peilfehler
site-error correction Funkbeschickung
site plan Lageplan
site work Arbeit auf der Baustelle
in situ an Ort und Stelle
six sechs
six-digits figure sechsstellige Zahl
six-electrode valve Sechspolröhre
six-phase sechsphasig
six-phase circuit Sechsphasenschaltung
six-phase current Sechsphasenstrom
six-phase transformer Sechsphasenumspanner
six-position meter switch Sechsfachumschalter
six-sided sechsseitig
six-wire three-phase system offenes Dreiphasensystem
sixfold sechsfach
size Größe (54); Größe, Größenklasse (52); Nummer; Leim (676)
below size unter Nennmaß, mit Untermaß
size of current Größe des Stromes
size of wire Drahtstärke
to size leimen (67); klassieren, sortieren
sizeable merklich, ansehnlich
sizing Sortierung (nach der Größe), Klassierung; Leimung (67)
sizing screen Klassiersieb
sizzle Prasseln, Spratzen
to sizzle prasseln, spratzen
sizzling noise prasselndes Geräusch
skate Schlittschuh
skate machine Schienenbremse
skeleton Skelett, Gerüst; Skelett, Gerippe
in skeleton form schematisch, in schematischer Form
skeleton-type switchboard Schaltpult (621.3)
sketch Skizze, Zeichnung
sketching board Reißbrett
skew schief, windschief; Schräglauf (681.84)
skew-coil winding zweietagige Wicklung mit gleichen Spulen
skew error Schrägfehler (621.396.6)
to skew schräg stellen, (Nuten) verschränken (621.313)
skewed schief gequetscht, verquetscht
skewed slots schräge Nuten

skiagram Röntgenaufnahme
skiatron Blauschriftröhre, Dunkelschriftröhre (621.385)
skid Rutsche, Schleifschuh, Kufe, Schlitten
skid bar Gleitkufe, Gleitschiene
to skid rutschen, ausrutschen; schliddern, bremsen, hemmen
skimmings (pl.) Schaum (669)
skin Haut; Haut, Außenhaut, Hülle (629.12), (629.13); Haut, Gußhaut (621.7)
skin affection Hautschädigung, Hautschaden (61)
skin-deep oberflächlich
skin depth Eindringtiefe
skin effect Hautwirkung, Stromverdrängung, Skineffekt (621.3)
skin stresses Oberflächenspannungen
skin tracking Verfolgen nach Zielrückstrahlung (621.396.96)
to skin abisolieren
skip Lore, Kippwagen; Hüpfer; Überspringen, Leerbefehl (681.3)
skip distance Sprungentfernung (621.396)
skip keying Impulsfolge-Frequenzteilung (621.396.96)
skip zone tote Zone, Sprungzone (621.396)
to skip hüpfen, springen; überspringen, überhüpfen
skipping übergreifen (621.395.3)
skirt Randgebiet
sky Himmel
skylight Oberlicht, Dachluke, Bodenfenster
sky wave Raumwelle (621.396)
slab Platte, Tafel; Schicht, Leiste, Sims (69); Barren (66)
slab coil Flachspule
slab covering Plattenbelag
slack lose, wacklig, schlackrig, locker, nicht angezogen (621.8)
slack-cable switch Schlaffseilschalter (621.876)
slack end loses Ende
slack hours (pl.) verkehrsschwache Zeit
slack joint Wackelkontakt, lose Verbindung
slack-rope switch Schlaffseilschalter (621.876)
slack side lose Seite
slack side of a belt Riemenrücklauf
slack water Wasseranstauung
to slack locker machen, lösen, lose machen
to slack off a nut eine Mutter lockern
to slacken lockern, lösen; sich lockern, sich lösen, locker werden
slag Schlacke
to slake löschen (69)
slaked lime gelöschter Kalk
slant schräge
slant height schief gemessene Höhe
slant range schief gemessener Abstand, Schrägentfernung
to slant abschrägen, schief legen; schief liegen
slanting schief, geneigt
to slash schlitzen, schleißen
slat Leistchen
S. lat. = South latitude südl. Br. = südlicher Breite
slate Schiefer; Dachschiefer, Schieferziegel
slaty schiefrig
slave Folge..., Tochter...
slave clock Nebenuhr
slave exchange Tochtervermittlungsstelle, Untervermittlung
slave flip-flop Tochter-Flip-Flop, Slave Flip-Flop
slave motor Folgemotor
slave station Tochtersender (621.397); Unterstation
slave system Untersystem, Tochtersystem
slave transmitter Tochtersender (621.397)
to slave parallelführen
"SL" cable = separately lead covered cable Dreimantel-Kabel
sledge Schlitten; Vorschlaghammer (621.7)
sledge coil Variometerkopplungsspule
sleet Graupeln; Rauhreifansatz, Eislast (621.3)
sleet hood Eisschutzhaube, Rauhreifschutzhaube
sleet-proof rauhreifsicher
sleeve Ärmel (64); Muffe, Ring, Futter (621); Gleitmuffe, Futterring, Buchse (621)
sleeve coupling Muffenkupplung
sleeve dipole Rohrdipol, Hülsendipol, Koaxialdipol mit Sperrtopf
sleeve joint Muffenverbindung, Vermuffung (621.315.2)
sleeve-stab antenna Koaxialantenne mit Gegengewicht
sleeve valve Zylinderschieber
sleeve wire Stöpselringanschluß (621.395)
sleeving Schutzschlauch
slenderness ratio Schlankheitsgrad
slew rate Anstiegsgeschwindigkeit (621.375)
slice Slice, Scheibe
slicer Doppelbegrenzer, Amplitudensiebschaltung
slide Gleitbahn; Schlitten; Schieber; Objektträger (535); Lichtbild, Diapositiv, Glasbild (77)
slideback method Kompensationsverfahren

slideback voltage Gegenspannung (bei Kompensation)
slide bar Gleitschiene
slide-in chassis Einschub
slide rail Spannschiene (621.313)
slide resistance Schiebewiderstand
slide rule Rechenschieber
slide switch Schiebeschalter
slide transformer Stelltransformator (mit Schiebekontakt)
slide wire Schleifdraht (621.317)
slide-wire bridge Schleifdrahtmeßbrücke
to slide gleiten, rutschen; schieben; längsdrehen, runddrehen (621.9)
slider Schleifer, Schleifkontakt, Abgriff (621.3); Zunge (68)
sliding Gleit-...; Gleiten, Rutschen
sliding action Gleiten (621.8)
sliding bar Gleitbügel (621.3)
sliding caliper Schublehre, Schieblehre
sliding clip Abgriffschelle (621.3)
sliding contact Gleitkontakt, Schleifkontakt, Kontaktschieber (621.3)
sliding-contact maker Schleifkontakt (621.317)
sliding door Schiebetür
sliding fit Gleitpassung, Gleitsitz
sliding key Gleitkeil, Schiebekeil
sliding resistance Schiebewiderstand
sliding scale Gleitskala
sliding-surface bearing Gleitlager
sliding switch Schubschalter
sliding voltage gleitende Spannung
slight leicht, geringfügig
slight displacement geringfügige Verschiebung
slight sparking Perlfeuer
slightly curved schwach gekrümmt
slightly low etwas zu klein, etwas zu niedrig
slightly taper leicht konisch
slim-rod pulling Dünnziehen (621.382)
slinging wire Einhängedraht (621.357)
slip Schlupf, Schlupffrequenz, Schlupfdrehzahl; Zettel; Gleiten, Rutschen; Slip (629.12)
slip casting Schlickerguß (66)
slip conduit Falzrohr, Bergmannrohr (621.3)
slip fitter Ansatzstutzen
slip frequency Schlupffrequenz
slip gauge Endmaß, Paßblock
slip-joint tubing Falzrohr (621.3)
slip of metal Metallstreifen
slip-on current transformer Aufschiebestromwandler
slip-over current transformer überschiebbarer Kabelstromwandler
slip plane Gleitfläche (66)
slip power Schleifringleistung (621.313)
slip regulator Schlupfregler
slip resistance of grains Gleitwiderstand an den Korngrenzen
slip ring Schleifring (621.313)
slip-ring bush Schleifringisolation
slip-ring current Schleifringstrom
slip-ring induction motor Schleifringmotor, Asynchronmotor mit Schleifringläufer
slip-ring phase advancer Phasenschieber-Hintermaschine
slip-ring rotor Schleifringläufer
slip-ring spider Schleifringspeichen (pl.)
slip-ring starter Schleifringanlasser
slip-ring voltage Schleifringspannung
to slip schlüpfen, gleiten, rutschen
a belt slips ein Riemen schlüpft
to slip into einstecken in
to slip on aufstecken
to slip out herausschlüpfen; herausziehen
to slip a plug into the jack den Stecker in die Klinke stecken
to slip through durchrutschen
slippage Schlupf
slipper Hemmschuh (625.2)
slippery schlüpfrig, rutschig
slipping Rutschen, Gleiten; Verschränken (621.395)
slipping clutch Rutschkupplung
slipping friction gleitende Reibung, Gleitreibung
slit Schlitz, Spalt; Längsschlitz
to slit schlitzen, längsschlitzen
slitting saw Metallkreissäge
sliver Span, Splitter, Spänchen
slope Neigung, Gefälle; Anstieg, Steilheit; Böschung; Flanke, Abfall (621.373)
slope resistance Ersatzwiderstand (621.382.2)
slope time Flankenzeit
to slope schräg abfallen, abschrägen, abflachen
to slope back nach hinten abfallen
sloping abfallend, schräg, schief
sloping tube boiler Schrägrohrkessel
slot Schlitz; Nut, Nute (621.313); Spalt (629.13); Spalt, Schlitz (621.396); Einschnitt
slot aerial Schlitzantenne
slot bridge Nutbrücke (621.313)
slot factor Wicklungsfaktor (621.313)
slot-fed aerial schlitzgespeister Dipol
slot insulator Nutisolation
slot leakage Nutenstreuung (621.313)
slot liner Isolierhülse, Nutenisolation, Nuthülse (621.313)

slot lining Nutauskleidung
slot meter Münzzähler
slot permeance Nutzleitwert (621.313)
slot pitch Nutteilung (621.313)
slot ripples Nutenobertöne
slot-space factor Nutfüllfaktor
slot system unterirdische Stromzuführung, Schlitzschienensystem (621.33)
slot-type magnetron Spaltmagnetron
slot wedge Nutenkeil; Deckleiste (621.313)
slot weld Schlitzschweißung (621.791)
slot width Spaltbreite (629.13); Schlitzbreite; Nutbreite (621.313)
to slot schlitzen; nuten, mit Nuten versehen (621.313); einschneiden
slotted conduit Schlitzstromzuführung (621.313)
slotted core genuteter Anker, Nutenanker (621.313)
slottet core armature genuteter Anker (621.313)
slotted cylinder antenna Rohrschlitzantenne
slotted disk Schlitzscheibe
slotted-head screw Schlitzkopfschraube
slotted line geschlitzte Hohlrohrleitung, Meßleitung
slotted sieve Schlitzsieb
slotted washer Schlitzscheibe, geschlitzte Unterlegscheibe
slotting schlitzend, Nuten-...
slow langsam; unempfindlich (77)
slow acting relay Relais mit Ansprechverzögerung
slow clock nachgehende Uhr
slow film unempfindlicher Film
slow fuse träge Sicherung
slow-motion dial Feintriebskala (621.396.62)
slow-motion sequence Zeitlupe
slow-motion starter Langsamanlasser
slow-neutron fission Spaltung mit langsamen Neutronen, Spaltung mit thermischen Neutronen
slow reactor langsamer Reaktor, Reaktor mit langsamen Neutronen
slow-release relay abfallverzögertes Relais
slow scan TV Fernsehübertragung mit langsamer Abtastung, Schmalband-Fernsehanlage
slow speed geringe Geschwindigkeit, Langsamkeit
slow-speed jet Leerlaufdüse (621.4), (629.113)
slow-speed machine langsam laufende Maschine
to slow verlangsamen, langsam fahren
to slow down verlangsamen, langsamer werden; verzögern, abbremsen
to slow up langsamer machen, verzögern
slowing-down area Bremsfläche (539)
slowing switch Langsamschalter, Einfahrschalter (621.876)
sludging Schlammbildung, Verschlammung
slug Wurmschraube, Madenschraube (621); Gußzeile (65); Barren (539); Metallklotz, Metallklümpchen; Stumpf, Pfropfen; Spulenkern (zur induktiven Abstimmung); Kurzschlußwindung, Kupfermantel (621.318)
sluggish träge, kriechend (621.313); zäh, zähflüssig; schwergängig
sluggish liquid zähe Flüssigkeit
sluggish measuring instrument träges Meßinstrument
sluggishness Zähflüssigkeit, Trägheit
slum Ölschlamm, Ölharz
to slur verschleifen, ineinanderziehen; ineinander übergehen
slurry Schleim, schleimige Masse
slurry reactor Schlammreaktor
small klein
small-area contact Kleinflächenkontakt
small bayonet cap (s. b. c.) Klein-Swansockel
small bore kleine Bohrung, enge Bohrung
small-bore engine Motor mit kleinem Zylinderdurchmesser
small central contact (s. c. c.) cap kleiner konzentrischer Sockel
small deviations from equilibrium kleine Störungen vom Gleichgewichtszustand, kleine Abweichungen vom Gleichgewichtszustand
small diameter coaxial cable carrier system TF-System für Kleinkoaxialpaare
small Edison screw (s. E. s.) cap Mignon-Gewindesockel (621.32)
small end dünnes Ende
small rain leichter Regen, Sprühregen
small in relations to klein gegen...
small scale kleiner Maßstab
small-scale integration niedriger Integrationsgrad
small signal Kleinsignal
small-signal capacitance differentielle Kapazität, Kleinsignalkapazität
small-signal short-circuit forward current transfer ratio Kleinsignal-Kurzschlußstromverstärkung
small tools Handwerkszeug (sing.)
small-type Klein-...

small wiring Signal- und Steuerleitungen (pl.) (621.311)
smalls (pl.) Klein, Grus (621.1), (621.622)
SMD = surface mounted device oberflächenmontiertes Bauelement
smear Schmiere
to smear schmieren; beschmieren (**with:** mit)
to smear on aufschmieren
to smear out ausschmieren, breitschmieren
smeared-out verschmiert, unscharf (535)
smearing Nachzieheffekt (621.397)
to smelt schmelzen, verschmelzen; ausschmelzen, erschmelzen, verhütten
smelting Schmelzvorgang, Verhüttung, Schmelz-...
smelting furnace Schmelzofen
smelting loss Schmelzverlust
smelting works Hüttenwerk (sing.)
smith Schmied
smith chart Smith Diagramm
smoke Rauch
smoke consumer Rauchverzehrer
smoke detector Rauchmelder
smoke indicator Rauchmelder
to smoke rauchen; rauchen, Rauch abgeben; räuchern; rußen, qualmen; ausräuchern
smokeless rauchlos
smoking Rauchen; Rußen; Räuchern
to smooth glätten (621.3), (621.9); abflachen (621.3)
to smooth a curve eine Kurve ausgleichen
to smooth down glattstreichen, ausbügeln
to smooth out glätten, ausglätten, einebnen; glätten, ausgleichen (51)
smooth glatt; stetig, stufenlos (621.3); sanft
smooth core glatter Anker, ungenuteter Anker (621.313)
smooth core armature glatter Anker (621.313)
smooth file Feinfeile, Schlichtfeile
smooth running leichter Lauf, reibungsloser Lauf
smooth surface glatte Oberfläche
smoothing Glättung (621.3), (621.9)
smoothing capacitor Glättungskondensator
smoothing choke Glättungsdrossel
smoothing circuit Glättungsschaltung
smoothing effect Glättungswirkung, ausgleichende Wirkung
smoothing iron Bügeleisen
smoothing resistance Glättungswiderstand, Beruhigungswiderstand
smother Qualm
to smother ersticken, unterdrücken; schwelen; erstickt werden
smothered-arc furnace Lichtbogenwiderstandsofen
to smoulder schwelen, kokeln, glimmen, schmoren
smudge Schmutzfleck
snake Einziehdraht (696)
snap Schnappen, Klappen; kurze Zeit
snap-action contact Sprungkontakt, Schnappkontakt
snap-action switch Schnappschalter, Mikroschalter
snap gauge Rachenlehre
snap lock Schnappverschluß
snap ring Sprengring
snap switch Schnappschalter, Drehschalter, „Knipser"
to snap abschnappen; schnappen; reißen; abreißen; knipsen (621.3)
to snap in einschnappen, einrasten; anknipsen
to snap off wegspringen, abplatzen, abschnappen; abzwacken, abschlagen; ausknipsen (621.3)
to snap on einschalten, anknipsen
to snap the light on das Licht anknipsen
to snap out herausschnappen; ausknipsen
to snap the lamps out die Lampen ausknipsen
snappy schnappend, federnd, einschnappend
snappy blow federnder Schlag
sneak current Kriechstrom (aus Starkstromanlagen) (621.39)
sneak path unerwünschter Nebenschluß
to sniff schnüffeln; Wind bekommen
sniffing valve Schnüffelventil
snip Schnippel, Schnippelchen
snook rectifier Nadelgleichrichter
snow Schnee, Grieß; Hintergrundrauschen (621.396.96)
snubber Anschlagpuffer
snug bequem, gemütlich
snug fit bequemer Sitz; Feinsitz, Festsitz
soak Einweichen (64); Dauerglühung (66); Regenguß (55)
to soak tränken, durchfeuchten, einweichen (64); dauerglühen (66)
to soak in einziehen
to soak through durchweichen
to soak up aufsaugen
soaking in Nachwirkungsladung
soaking out Nachentladung
soaring zone Aufwindzone, Aufwindschlauch
"SO" cable SO-Kabel, Sonnemann-Kabel

socket Sockel, Fuß, Fußhülse; Fassung; Buchse, Gegenstecker; Fußlager; Röhrensockel (621.385)
socket head Innensechskant
socket-head cap screw Inbusschraube
socket joint Kugelgelenk (621); Steckverbindung (621.3)
socket outlet Steckdose, Wandsteckdose
socket-outlet adapter Doppelstecker, Mehrfachstecker (621.385)
socket panel Buchsenfeld
socket receptacle Sockelfeder, Sockelkontakt (621.385)
socket switch Sockelschalter, Fassungsschalter
socket terminal Röhrensockel; Sockelanschluß
socket wrench Steckschlüssel
socle Sockel (72)
soda Soda, Natron
soda lye Natronlauge
sodium Natrium
sodium-cooled reactor natriumgekühlter Reaktor
sodium discharge Natriumdampfentladung
sodium-discharge lamp Natriumdampflampe
sodium(-graphit) reactor Natriumgraphitreaktor
sodium hydroxide Natriumhydroxid, Ätznatron
sodium-vapour lamp Natriumdampflampe
Soederberg electrode Soederberg-Elektrode, selbstbackende Elektrode
soffit Soffitte, Gewölbeleibung
soft weich; lose; locker; gashaltig (621.385)
soft annealed weichgeglüht
soft annealed wire weichgeglühter Draht
soft developer weicher Entwickler
soft flame kühle Flamme
soft iron Weicheisen
soft iron core Weicheisenkern
soft iron instrument Weicheiseninstrument
soft iron mass Weicheisenstück
soft keying Weichtasten
soft magnetic weich magnetisch
soft photoelectric cell gasgefüllte Photozelle
soft soap Schmierseife
soft solder Weichlot
soft soldering Weichlöten
soft steel weicher Stahl, ungehärteter Stahl, weiches Eisen
soft tube (US) gashaltige Röhre, gasgefüllte Röhre (621.385), (621.6)
soft tube rectifier gasgefüllter Gleichrichter
soft valve gasgefüllte Röhre (621.385), (621.6)
soft water weiches Wasser (621.1), (64)
soft wax Klebwachs (53)
soft X-rays weiche Röntgenstrahlung (sing.)
to soften weich machen (621.1), (77); weich werden
softening Erweichung, Weichwerden
softening point Erweichungspunkt (67)
to soft-solder weichlöten
software Dienstleistung; Software, nichtmaterielle Mittel, Programmausstattung (681.3)
software compatible softwarekompatibel
software design Softwareentwurf, Softwareentwicklung
software engineering Softwareentwicklung
soil Boden, Erdboden, Baugrund
soil conductivity Leitfähigkeit des Erdbodens
sol Sol (54)
solar solar, Solar-..., Sonnen-...
solar activity Sonnentätigkeit
solar battery Solarbatterie, Sonnenbatterie
solar cell Solarzelle, Sonnenzelle
solar constant Solarkonstante
solar generator Solargenerator
solar noise Solarstörung (621.396)
solar power Sonnenenergie
solar power station Sonnenkraftwerk
solar power supply solare Stromversorgung
solar rays (pl.) Sonnenstrahlung
Solari coherer Quecksilber-Kohärer
solder Lot, Lötmittel, Lötzinn
solder cream Lötpaste
solder-cream dispenser Lötpastendosierer
solder iron Lötkolben
solder metal Lötmetall, Lot, Lötzinn
solder tag Lötfahne
solder wire Lötdraht
to solder löten
to solder on auflöten
to solder a seam eine Naht löten
soldered joint Lötverbindung, Lötstelle
soldering Löten, Lötvorgang, Lötung
soldering appliances Lötgeräte; Lötwerkzeug (sing.)
soldering blow pipe Lötlampe
soldering flux Lötflußmittel
soldering iron Lötkolben
soldering lug Lötschuh
soldering paste Lötfett
soldering seam Lötnaht
soldering solution Lötwasser
soldering tag Lötfahne, Lötanschluß
soldering-tag terminal strip Lötösenstreifen
soldering tin Lötzinn

solderless lötfrei
solderless wrap Würgeverbindung, lötfreie Wickelverbindung
solenoid Spule, Zylinderspule, Solenoid
solenoid blow-out elektrodynamische Blaseinrichtung (621.316.5)
solenoid brake Magnetbremse (621.33)
solenoid-operated magnetisch betätigt, Magnet-...
solenoid operation Magnetbetätigung (621.316.5)
solenoid relay Spulenrelais, Tauchkernrelais
solenoid starter Magnetanlasser (621.386)
solenoid switch magnetisch betätigter Schalter
solenoid valve Magnetventil, magnetisch gesteuertes Ventil
solenoidal Spulen-...
solenoidal field quellenfreies Feld
solenoidal magnetism Längsmagnetismus
solenoidal magnetization Längsmagnetisierung
solid Festkörper, Körper
solid dielectricum festes Dielektrikum
solid of revolution Rotationskörper
solid fest; massiv; solide
solid angle Raumwinkel, Ecke
solid-back microphone Kohleblock-Mikrophon
solid carbon Massivkohle
solid circular vollkreisförmig, massiv rund
solid conductor massiver Leiter
solid drawn nahtlos gezogen
solid electrolyte fester Elektrolyt
solid enclosure dichter Verschluß, dichter Abschluß
solid line ausgezogene Linie
solid make solide Bauart
solid matter fester Stoff
solid phase feste Phase (54)
solid-phase reaction Reaktion in der festen Phase
solid pole Massiveisenpol
solid rotor einteiliger Läufer
solid silver Massivsilber, massives Silber
solid solution feste Lösung, Mischkristall
solid state Festkörper, fester Zustand
solid-state circuit Festkörperschaltung, Halbleiterschaltung
solid-state diffusion Diffusion im festen Zustand (621.382)
solid-state image sensor Halbleiter-Bildsensor
solid-state physics Festkörperphysik
solid substance feste Masse, fester Körper
solid system of switchgear insultation gekapselte Schaltanlage mit Massefüllung
solid-type cable Massekabel
solid wire Massivdraht
solidly earthed system starre Erdung, starr geerdetes Netz
solidly grounded starr geerdet
solidification Erstarrung, Festwerden, Erstarren
solidification point Erstarrungspunkt
to solidify erstarren, fest werden
solidifying point Erstarrungspunkt, Stockpunkt
solubility Löslichkeit (54)
soluble löslich (54)
soluble glass Wasserglas
soluble solder leichtschmelzendes Lot
solute gelöste Substanz
solution Lösung (51), (54)
solution potential Lösungspotential (**against:** gegen)
solution tension Lösungsdruck
solution voltage Lösungsspannung
solvability Lösbarkeit, Löslichkeit (51)
solvable lösbar (51)
to solve lösen, auflösen (51)
to solve equations Gleichungen lösen
to solve for auflösen nach (51)
solvent Lösungsmittel
solvent lösend, Lösungs-...
solvent agent Lösungsmittel
solvent extraction Extraktion, Auszug
solvent liquid Lösungsflüssigkeit
sonar = sound navigation and ranging, Sonar
sonde Sonde; elektrischer Abgriff (621.396)
sonic Schall-..., akustisch
sonic barrier Grenzgeschwindigkeit, Schallgeschwindigkeit, Schallgrenze (629.13)
sonic delay line akustische Verzögerungsleitung
sonic depth sounder Echolot
sonic frequency Tonfrequenz
sonic speed Schallgeschwindigkeit
sonic sounding Echolotung
sono buyo Sonar-Boje, Sonoboje
soot Ruß
to sorb sorbieren
to sort sortieren
to sort out aussortieren
to sort with passen zu
sorting Sortierung, Sortiervorgang
sound Schall, Ton, Klang (534); Sonde (61)
sound absorbant Schalldämmstoff, Schluckstoff; schallabsorbierend, schalldämmend

sound absorption Schallabsorption, Schallschluckung
sound-absorption coefficient Schall-Absorptionskoeffizient
sound absorptive schallschluckend
sound analyser Klanganalysator, Oktavsieb
sound articulation Lautverständlichkeit
sound bandwith (NF-)Bandbreite; Tonbandbreite (681.84)
sound-carrier frequency Tonträger
sound-carrier trap Tonträgersperre
sound channel Tonkanal (621.39)
sound discriminator Tonfrequenzgleichrichter (621.397)
sound dosimeter Schalldosimeter
sound echo Schallecho, akustisches Echo
sound film Tonfilm
sound-film projector Tonfilmvorführapparat, Tonfilmprojektor
sound generator Schallsender, Tongenerator
sound head Tonkopf, Sprechkopf
sound-image tape recorder Ton-Bild-Bandgerät
sound insulation Schalldämmung
sound intensity Schallstärke
sound level Lautstärke, Schallpegel
sound-level meter Lautstärkemesser, Schalldruckmesser
sound minimum Tonminimum (621.317)
sound mixing Tonmischung
sound modulated tonmoduliert
sound-particle velocity Schallschnelle
sound-powered telephone batterieloser Fernsprecher
sound pressure Schalldruck
sound-pressure level Schalldruckpegel
sound-program circuit Tonanschlußleitung (621.396.9)
sound-program control equipment Tonkontrolleinrichtung
sound-program exchange circuit Tonaustauschleitung (621.396.9)
sound propagation medium Schall(-Ausbreitungs-)Medium
sound-proof schalldicht, schallgeschützt
sound proofing Schallschutz, Schalldämmung
sound-radiation impedance Schallstrahlungsimpedanz
sound-radiation reactance Schallstrahlungsreaktanz
sound-radiation resistance Schallstrahlungswiderstand
sound radio Tonfunk
sound ranging Schallmeß-...
sound receiver Schallaufnehmer, Schallempfänger
sound reception Hörempfang
sound recording Tonaufnahme, Tonaufzeichnung
sound-recording medium Tonträger
sound-reflecting schallreflektierend
sound reproduction Tonwiedergabe
sound signal akustisches Signal
sound source Schallquelle
sound system Lautsprecheranlage, Übertragungsanlage, elektroakustische Anlage
sound transmission Schallübertragung
sound-transmission factor Schalltransmissionsgrad
sound trap Tonfalle; Saugkreis (621.397)
sound velocity Schallgeschwindigkeit
sound vibration Schallschwingung
sound wave Schallwelle
sound work Akustik
to sound ertönen; ertönen lassen; sondieren; loten
to sound an alarm Warnung betätigen; akustisches Signal auslösen
sounder Klopfer (621.394)
sounding Sondierung, Lotung
sounding device Loteinrichtung, Lotvorrichtung
source Quelle, Herkunft; Quelle (51); Energiequelle, Spannungsquelle, Stromquelle (621.3); Eingang, Anfang (621.392); source (621.382.3)
source coding Quellenkodierung
source current Source-Strom (621.382); [Ausgangstrom bei Ausgang „H"]
source data Ursprungsdaten
source of defects Schadenquelle, Fehlerquelle
source of energy Energiequelle
source of errors Fehlerquelle (51), (621.317)
source of heat Wärmequelle
source impedance Quellenwiderstand, Generator-Innenwiderstand
source of interference Störungsquelle
source language Quellsprache, Ursprungssprache
source of light Lichtquelle
source material Ausgangsmaterial (539)
source of neutrons Neutronenquelle
source of power Leistungsquelle, Energiequelle
source program Quellprogramm
a source sends a current via ... eine Stromquelle treibt einen Strom durch ...
source of sound Schallquelle

source of supply Nachschubquelle; Stromquelle, Energiequelle
source of trouble Störungsquelle, Störungsursache
sourdene Schwingungsdämpfer (621.31)
south Süden
south-seeking südweisend
south-seeking pole Südpol (538)
s. p. = single phase Einphasen...
s. p. = single pole einpolig
space Raum, leerer Raum, Weltraum; Zwischenraum; Abstand, Lücke, Lückenbreite; Wickelraum (621.31); offene Rubrik, Spatium (655)
space charge Raumladung (621.3)
space-charge effect Raumladungswirkung
space-charge grid Raumladegitter (621.385)
space-charge limitation Raumladungsbegrenzung
space-charge limited raumladungsbegrenzt
space-charge region Raumladungsgebiet
space cloth Absorptionsmaterial (621.396)
space conditions Raumverhältnisse
space craft Flugkörper, Raumfahrzeug
space current Raumladungsstrom, Raumladestrom, raumladungsbegrenzter Strom
space diagonal Raumdiagonale
space diagram Lageplan
space diversity Raum-Diversity
space-division system Raumvielfachsystem (621.395)
space factor Füllfaktor (621.313)
space heater Raumheizkörper (621.365)
in space räumlich, im Raume
in space and time räumlich und zeitlich, in Raum und Zeit
space lattice Raumgitter
space permeability Permeabilität des leeren Raumes, Induktionskonstante
space radio Weltraumfunk, Satellitenrundfunk
space rays (pl.) Raumstrahlung (621.396)
space requirements (pl.) Raumbedarf
space-saving raumsparend, platzsparend
space technique Raumfahrttechnik
space variation räumliche Änderung
space wave Freiraumwelle
space winding unterbrochene Wendel (621.32)
space-wound mit Abstand gewickelt, auf Abstand gewickelt
to space räumlich anordnen, mit Abstand voneinander hinlegen
to space apart getrennt anordnen
to space 180° apart um 180 Grad getrennt anordnen
to space at intervals of in Abständen von... anordnen
spaced getrennt, mit Abstand angeordnet
spaced aerial Antenne für Mehrfachempfang
spaced-aerial reception Empfang mit Mehrfachantenne, Raum-Diversity-Empfang
spaced... apart im Abstand von... voneinander
spaced-loop direction finder Doppelrahmenpeiler
spaced winding auf Abstand gewickelte Wicklung
spacer Abstandshalter, Abstandsstück
spacer ring Distanzring, Abstandsring
spacing Abstand, Zwischenraum
spacing bar Abstandsstück
spacing between transmitters Senderabstand
spacing bush Abstandsbuchse
spacing current Pausenstrom (621.394)
spacing of the humps Abstand der Resonanzspitzen, Abstand der Buckel
spacing key Leertaste (621.394.6)
spacing ratio Lampenabstand/Lampenhöhe
spacing wave Zeichenpause, Pausenpegel (621.394)
to spaddle spachteln
spade end offene Lötöse; offener Kabelschuh (621.3)
span Spannweite (629.13), (624), (72); Flügelspannung (629.13); Spannweite, Stützweite, Mastabstand (621.315.1)
span length Spannweite, Stützweite
span wire Abspanndraht, Tragdraht (621.332)
span-wire suspension Spanndrahtaufhängung
to span spannen, überspannen, überbrücken; führen über, aufliegen auf
Spanish windlass Seilknebel
spanner Schraubenschlüssel
spar Strebe, Holm (629.13); Spat (54); Sparren (69)
spare Reserveteil, Ersatzteil; Reserve-..., Ersatz-...
spare battery Reservebatterie, Ersatzbatterie
spare conductor Ersatzader, Reserveader
spare engine Reservemotor
spare equipment Reserveteile (pl.)
spare fuse Ersatzsicherung
spare lamp Ersatzlampe
spare machinery Ersatzmaschinen, Reservemaschinen (pl.)

spare part Ersatzteil
spare-parts list Ersatzteilliste
spare-part number Ersatzteilnummer
to spare sparen, einsparen; schonen
sparingly soluble schwer löslich
spark Funke
spark absorbing Funkenlöschung
spark arrester Funkenfang (625)
spark ball Kugelelektrode (621.38)
spark coil Zündspule
spark capacitor Störschutzkondensator, Überbrückungskondensator (621.318)
spark discharge Funkenentladung
spark erosion Elektroerosion, Funkenabtragung
spark frequency Funkenfrequenz
spark gap Funkenstrecke
spark-gap protector Funkenstrecken-Überspannungsschutz
spark killer Störschutzkondensator, Überbrückungskondensator (621.318)
spark length Funkenschlagweite, Schlagweite, Überschlagsweg
spark machining Funkenabtragung, Elektroerosion
spark at make Schließungsfunke
spark-over Überschlag
spark-over test Überschlagsprüfung
spark-over voltage Überschlagsspannung
spark photography Funkenphotographie
spark plug Zündkerze (621.43)
spark quenching Funkenlöschung (621.318.5)
spark-quenching device Funkenlöscheinrichtung
spark recorder Funkenschreiber
spark resistance Funkenwiderstand
spark spectroscopy Funkenspektroskopie
spark spectrum Funkenspektrum
spark transmitter Funkensender
to spark funken, Funken bilden
to spark-over überschlagen
to spark through durchschlagen
sparking commutator funkender Kommutator
sparking Funkenbildung, Funken (621.3); Funkenauswurf (625.2)
sparking coil Zündspule
sparking contact Abbrennkontakt
sparking distance Schlagweite
sparking limit Grenzlast für funkenfreie Kommutierung
sparking plug Zündkerze
sparking potential Funkenspannung, Überschlagsspannung
sparking voltage Funkenspannung, Überschlagsspannung
sparkle Funkeln, Gefunkel
to sparkle funkeln
sparkless funkenfrei
sparkless collection funkenfreie Kommutierung
sparkless commutation funkenfreie Kommutierung
sparklet Fünkchen
sparse knapp, spärlich
spatial räumlich, im Raum
spatial distribution räumliche Verteilung
to spatter (mit Schmutz) bespritzen
to speak sprechen
to speak at besprechen, auf etwas sprechen
to speak into a microphone ein Mikrophon besprechen
speaker Lautsprecher (621.396.62)
speaker cone Lautsprechermembran, Konus, Konusmembran
speakerfield Lautsprechererregung
speaker housing Lautsprechergehäuse
speaking Sprech-..., Sprach-...
speaking clock Zeitansage (621.395)
speaking current Sprechstrom
speaking key Sprechtaste, Abfragetaste
speaking position Sprechstellung (621.395)
speaking trumpet Sprachrohr, Megaphon
special speziell, Spezial-..., Sonder-...
special brick Formstein, Sonderstein
special cap Spezialsockel (621.32)
special case Sonderfall, Spezialfall
special character Sonderzeichen
special design Sonderkonstruktion, Spezialausführung
special equipment Sonderausrüstung
special information tone Hinweiston (621.395)
special knowledge Spezialkenntnisse (pl.)
special purpose motor Spezialmotor
special-tariff meter Tarifzähler
special type Spezialtype, Sonderbauart
special-type machinery Sondermaschinen (pl.)
specialist Spezialist, Fachmann
specialistic speziell, spezialisiert
specific spezifisch, Art-...; besonder, speziell
specific acoustical impedance spezifischer Schallwiderstand
to be specific ins Detail gehen
specific conductance (conductivity) spezifischer Leitwert, Leitfähigkeit
specific consumption spezifischer Verbrauch
specific damping Dämpfungsmaß (621.392)

specific density spezifische Dichte, Wichte, Artgewicht
specific dielectric strength Durchschlagsfestigkeit
specific gravity spezifisches Gewicht, Artgewicht
specific heat spezifische Wärme
specific inductive capacitance (capacity) Dielektrizitätskonstante, DK
specific ionic mobility Ionenbeweglichkeit
specific magnetic resistance spezifischer magnetischer Widerstand (Kehrwert der Permeabilität)
specific power loss spezifische Verlustleistung
specific pressure spezifischer Druck; Bodenpressung (624)
specific reluctance spezifischer magnetischer Widerstand
specific resistance spezifischer Widerstand, Artwiderstand
specific rotation spezifische Drehung (535.56)
specific thermal resistance (Kehrwert der Wärmeleitfähigkeit)
specific utilization coefficient Ausnutzungszahl (621.313)
specific utilization coefficients (pl.) spezifische Belastung (621.313)
specific volume spezifisches Volumen
specific weight spezifisches Gewicht, Artgewicht
specification Beschreibung; Vorschrift, Spezifikation
specifications technische Bedingungen, Spezifikation
to specify vorschreiben, verlangen, fordern
specified limit vorgeschriebener Grenzwert
specified path vorgeschriebener Weg (51)
specimen Probestück, Prüfobjekt, Probe; Muster, Beispiel; Prüfstab, Probestab, Probe (620)
speck Klecks
speck of dust Staubkörnchen, Sonnenstäubchen
spectral spektral, Spektral-...
spectral analysis Spektralanalyse
spectral distribution coefficient Spektralwert
spectral emissivity spektrale Verteilung der Ausstrahlung
spectral radiant intensity Spektralanteil der Strahlungsdichte
spectral resolution spektrale Auflösung
spectral sensitivity spektrale Empfindlichkeit, spektrale Empfindlichkeitsverteilung
spectral shift reactor Reaktor mit Spektralsteuerung
spectrometer Spektrometer
spectrometric spektrometrisch
spectrophotometer Spektral-Photometer
spectroscope Spektroskop
spectroscopic(al) spektroskopisch
spectroscopic analysis Spektralanalyse
spectrum (pl. **-tra**) Spektrum (pl. -tren)
spectrum analyzer Spektrum-Analysator
spectrum of the n-th order Spektrum n-ter Order
specular reflection Glanzwinkelreflexion
speech Sprache, Sprechsinn; Ansprache, Vortrag
speech amplifier Sprechverstärker, Sprachverstärker
speech audiometer Sprachaudiometer
speech band Sprechfrequenzband
speech circuit Sprechkreis
speech coil Lautsprecherspule, Triebspule (621.395.623)
speech current Sprechstrom (621.395)
speech decoding Sprachentschlüsselung
speech energy Sprechleistung
speech-excited von der Sprache angeregt, durch die Sprache ausgelöst
speech frequency Sprachfrequenz
speech intelligibility Sprachverständlichkeit
speech inverter Sprachinverter (621.395.1)
speech level Sprachpegel
speech-level meter Aussteuerungsmeßgerät
speech-modulated sprachmoduliert
speech-modulated continuous waves ungedämpfte sprachmodulierte Wellen
speech path Sprechweg
speech power Sprechleistung
speech protection Sprachschutz
speech range Sprachband (621.395)
speech restorer Trägerzusatzeinrichtung (621.396)
speech scrambler Sprachverwürfler
speech security Sprachsicherheit
speech signal Sprechstrom, Modulationsspannung, Modulation
speech synthesizer Sprach-Synthesizer, Sprach-IC
speech transformer Tonfrequenzübertrager
speed Geschwindigkeit, Schnelligkeit; Drehzahl, Umdrehungszahl; Gang, Geschwindigkeitsstufe, Drehzahlstufe, Getriebestufe
speed-adjusting rheostat Drehzahlregelwiderstand
speed adjustment Drehzahleinstellung (621.9)
speed of adjustment Stellgeschwindigkeit (621.316.7)

speed box Schaltgetriebe, Getriebekasten (621.9)
speed characteristic Drehzahlkennlinie, Drehzahlverhalten
speed of circulation Umlaufgeschwindigkeit, Umwälzgeschwindigkeit
speed control Drehzahlsteuerung, Drehzahlregelung
speed controller Drehzahlregler, Geschwindigkeitsregler
speed above the critical überkritische Drehzahl
speed-distance curve Geschwindigkeits-Weg-Diagramm
speed of driving end Abtriebsdrehzahl
speed of ejection Austrittsgeschwindigkeit
speed governing Drehzahlregelung
speed indicator Drehzahlanzeiger, Geschwindigkeitsmesser
speed limit Geschwindigkeitsbeschränkung, (zulässige) Höchstgeschwindigkeit
speed-limit indicator Höchstgeschwindigkeitsanzeiger (656.25)
speed meter Geschwindigkeitsmesser
speed of phase Phasengeschwindigkeit
speed of propagation Ausbreitungsgeschwindigkeit, Fortpflanzungsgeschwindigkeit
speed range Drehzahlbereich
speed ratio Übersetzung (621.83)
speed-regulating rheostat Drehzahlregelwiderstand
speed regulation Drehzahlregelung
speed regulator Drehzahlregler
speed of response Ansprechgeschwindigkeit
speed restriction Fahrtbeschränkung, Geschwindigkeitsbeschränkung
speed-torque characteristic Drehzahlkennlinie (621.313)
speed-torque curve Drehzahlkennlinie (621.313)
speed transformer Übersetzungsgetriebe (621.8)
speed of transmission Übertragungsgeschwindigkeit, Sendegeschwindigkeit
speed variation Geschwindigkeitsschwankung
to speed schnell fahren
to speed up beschleunigen, schneller fahren, zulegen; sich beschleunigen, schneller werden
spell Zeitraum, Periode
spelter Hartlot; Zink, Rohzink, unreines Zink
to spend ausgeben, verbrauchen
to spend power Leistung verbrauchen
to spend voltage Spannung verbrauchen
spent ausgebraucht, abgearbeitet, abgenutzt, matt
spent acid verbrauchte Säure, matt gewordene Säure
spent electron abgearbeitetes Elektron (621.385)
spent-fuel element abgebranntes Brennelement
spent solution ausgebrauchtes Bad, verbrauchte Lösung (77)
spermaceti Walöl
sp. gr. = specific gravity spez. Gew. = spezifisches Gewicht, Wichte
sphere Kugel (51); Bereich, Gebiet, Sphäre
sphere of cohesion Anziehungsbereich
sphere gap Kugelfunkenstrecke
sphere-gap voltmeter Kugelfunkenstrecken-Voltmeter
sphere photometer Kugelphotometer, Ulbrichtsche Kugel
spherical kugelförmig, sphärisch, Kugel-...
spherical aberration sphärische Aberration
spherical cavity kugelförmiger Hohlraum, Hohlkugel
spherical geometry sphärische Geometrie
spherical lens sphärische Linse
spherical polars Kugelkoordinaten
spherical-reduction factor (Umrechnungszahl von mittlerer horizontaler auf mittlere sphärische Lichtstärke)
spherical resonator Hohlkugelresonator
spherical surface Kugelfläche
spherical triangle sphärisches Dreieck, Kugeldreieck
spherical trigonometry sphärische Trigonometrie
spherically seated kugelförmig gelagert
spheric wave Kugelwelle
spheroid Ellipsoid, Rotationsellipsoid
spheroid of revolution Rotationsellipsoid
spheroidal ellipsoidähnlich, kugelig
sphygmograph Pulsschreiber
sphygmophone Pulsmikrophon
spider Spinne (59); Kardangelenk; Spinne (534); Polrad, Rotorstern (621.313)
spider lines (pl.) Fadenkreuz (535)
spiegeleisen Spiegeleisen
spigot Pimpel; (Faß-)Zapfen
to spigot fügen
to spigot into one another ineinanderfügen (628); fugen (69)
spike Spitze, Eisenspitze; Spike (79); Schienennagel (625)
spill Lambdahalbe-Düppel
spill current Fehlerstrom (621.316.9)

spill-over Überlauf (621.394)
to spill vergießen, verschütten, ausgießen, ausschütten, überlaufen lassen
spin Spin (53); Drall, Drehbewegung
spin drier Trockenschleuder, Wäscheschleuder (643)
spin of the electrons Elektronenspin
to spin spinnen (677); sich drehen, herumdrehen, herumlaufen, herumwirbeln; drehend antreiben (621); drücken (621.9); trudeln (629.13)
spinning reserve (leer) mitlaufende Reserve (621.311)
spindle Spindel (67), (621); Antriebswelle, Drehachse (621); Drehachse (621.317)
spindle bar Drehachse, senkrechter Stift
spindle oil Spindelöl
spinel Spinell
spinning Drücken, Drückvorgang (621.9); Spinnen, Spinnvorgang (67)
spinning lathe Drückbank
spinthariscope Szintillationsschirm (539)
spintharometer Funkenmikrometer
spiral Spirale, Schneckenlinie, Wendel; Locke (621.9)
to spiral sich spiralig winden
to spiral out sich herauswinden; herausleiern
spiralled key Spiralkeil
spiral spiralig, Spiral-...
spiral antenna Spiralantenne
spiral-bevel gear Spiralkegelrad
spiral coil Flachspule
spiral connection Spiralstirnverbindung
spiral conveyor Förderschnecke
spiral four cable Feldfernkabel
spiral gear Spiralzahnrad
spiral groove Spiralnut
spiral line Spirallinie, Spirale, Schneckenlinie
spiral scan Spiralabtastung
spiral spring Spiralfeder
spiral teeth Spiralzähne
spiral turn Spiralwindung
spiral winding Wellenwicklung
spiral-wound electrode Netzmantelelektrode
to spiralize spiralig aufwinden, aufspulen
spirit Alkohol, Spiritus
spirit level Wasserwaage
splash Spritzer, Schwall
splash buffle Ablenkplatte (621.314.6)
splash-proof schwallwasserfest
splash-proof instrument schwallwassergeschütztes Instrument
splash-proof machine schwallwassergeschützte Maschine
splash-proof motor schwallwassergeschützter Motor
splash-proof switchgear schwallwassergeschützte Schaltgeräte (pl.)
splash water Schwallwasser
to splash spritzen, bespritzen
to splay aufweiten, sich erweitern
to splay out aufspleißen; verbreitern, sich verbreitern
splice Spleißstelle
to splice spleißen
spliced joint Spleißverbindung, gespleißte Verbindung
splicing ear Fahrdrahtverbinder, Verbinder
splicing sleeve Lötseilverbinder, Lötverbinder
spline Wellenkeil, Wellennut
to spline mit axialen Nuten versehen, nuten
splined shaft genutete Welle
splinter Splitter
to split spalten, aufspalten; schlitzen, aufschlitzen; sich spalten, aufspalten (535)
to split apart trennen
to split the difference sich in die Differenz teilen
to split into teilen in, aufspalten in
to split off (sich) abspalten
to split up aufspalten, zerlegen
split gespalten, geschlitzt, Schlitz-...
split anode Schlitzanode
split-anode magnetron Schlitzanodenmagnetron, Zweischlitzmagnetron
split case geteiltes Gehäuse
split-conductor cable Mehrfachleiterkabel, Teilleiterkabel
split-conductor protective system Parallelzweigschutz, Querdifferentialschutz
split-core type transformer instrument Zangenstromwandler, Anlegewandler
split-flow reactor Reaktor mit geteiltem Kühlmittelfluß
split frame geteiltes Gehäuse (621.313)
split lug Schlitzring
split order wire circuit Mehrfachdienstleitung
split phase Hilfsphase
split-phase motor Motor mit Hilfsphase
split pin Splint
split pole Spaltpol
split-pole motor Spaltpolmotor
split-pole rotary converter Spaltpolumformer
split ring Schlitzring, Sprengring
split screen geteilter Bildschirm
split tape-wound core Schnittbandkern

split-tube balun Schlitztransformator
split waveguide Quetschhohlleiter
splitting Aufspaltung, Spaltung
spoke Speiche, Radspeiche; Sprosse, Leitersprosse
spoke shave Zieheisen
sponge Schwamm
sponge rubber Schwammgummi
spongy schwammig
spontaneous spontan
spontaneous combustion Selbstentzündung
spontaneous ignition Selbstentzündung
spontaneous nuclear reaction spontane Kettenreaktion
spontaneous radiation charakteristische Strahlung (621.385)
spool Spule, Spulenkasten, Spulentrommel, Bandspule
spool end Spulenwange
spool head Spulenwange
spool winding Kastenwicklung
spoon-type contact Löffelkontakt
sporadic sporadisch
spot Punkt (51); Fleck, Ort, Stelle, Punkt; Fleck, Punkt, Leuchtfleck (621.385); Punkt (621.791); Spurelement (681.84)
spot of arcing Schmorstelle
spot lamp Punktlichtlampe
spot light Lichtmarke (621.317); Suchlicht (629.113)
spot-light scanning Lichtpunktabtastung
spot lighting Scheinwerferbeleuchtung, Punktanstrahlung
spot of light Leuchtfleck (621.385)
spot shape Fleckform, Gestalt des Fleckes (621.385)
spot size Fleckgröße (621.385)
spot speed Abtastgeschwindigkeit
spot welder Punktschweißmaschine
spot welding Punktschweißung
spot-welding machine Punktschweißmaschine
spot wobble Zeilenwobbelung (621.397)
to spot orten, auffinden, ausmachen
to spot out ausblühen
to spot-solder punktlöten
to spot-weld punktschweißen, punkten
spotting out Ausfleckung (621.357)
spout Tülle, Ausguß
to spout ausgießen, gießen
spray Sprühschleier, Nebel, Sprühnebel; Spritzauftrag, Spritzlackierung, Sprühvorgang, Besprühung
spray arrester Abdeckscheibe (621.355)
spray cooler Sprühkühler, Rieselkühler
spray etcher Sprühätzvorrichtung
spray gun Spritzpistole
spray-lacquer painting Spritzlackierung
to spray on aufspritzen, aufsprühen
sprayer Zerstäuber
spraying Spritzvorgang, Zersprühen; Spritzanlage, Farbspritzanlage
spraying process Sprühprozeß, Sprühverfahren
spraying varnish Spritzlack
spread Ausbreitung, Spielraum; Keulenbreite, Strahlbreite (621.396.67); Spannweite (624)
spread factor Sehnungsfaktor (621.313)
to spread ausbreiten, sich ausbreiten; bestreichen, beschmieren (**with:** mit)
to spread apart mit Abstand anordnen
spreader Abstandshalter, Spreize
sprig Stift, Drahtstift; Kernstift (621.74)
to sprig anheften, anstiften (leicht) annageln
spring Feder; Federung, Federkraft (621)
spring-adjusting plier Justierzange, Kontaktfeder-Justierzange
spring axle gefederte Achse (625.2)
spring balance Federwaage
spring blade Blattfeder
spring bolt Federbolzen
spring bow Nullenzirkel
spring bush Federbuchse
spring calipers (pl.) Federtaster
spring catch Schnappverschluß
spring clip Federklammer, Federpakethalter
spring-closing valve federbelastetes Ventil
spring contact Federkontakt
spring control Federrückstellung, Richtfeder
spring control instrument Instrument mit Federrichtkraft
spring-controlled federbelastet; mit Federrückstellung
spring hook Federhaken
spring leaf Blattfeder, Federblatt
spring-loaded federbelastet
spring lock Schnappschloß
spring motor Federantrieb, Federmotor
spring-mounted federgelagert
spring-operated federangetrieben, federbetätigt
spring-operated clock Federuhr
spring-operated valve Federventil
spring plate Federteller
spring plunger Federkolben, federbelasteter Kolben
spring pressure Federdruck
spring pressure gauge Federmanometer
spring reserve Uhrwerksreserve (621.317)

spring ring Federring, Springring
spring seat Federteller
spring steel Federstahl
spring suspension Federsockel, federnde Aufhängung
spring washer Federscheibe, federnde Unterlegscheibe
spring winder Federwickelmaschine
to spring federn; auseinanderklaffen
to spring out nach außen federn
to spring up entspringen, entstehen, sich erheben
springing Federung, Abfederung (629.113)
to sprinkle sprengeln, besprengen, versprengen, fein verteilen, aufstreuen, streuen; sprinkeln
sprinkler Sprinkler
sprinkler nozzle Sprinklerdüse
sprinkler plant Sprinkleranlage
sprinkling bottle Spritzflasche
sprocket Sprosse; Stiftzahn, Zahnstift, Kettenradzahn
sprocket wheel Kettenrad
spun glas Glasseide, Glaswolle
spun glass wool felt Glaswollenlage
spur Sporn, Dorn
spur gear Zahnrad, Stirnrad
spur gearing Stirnradverzahnung, Stirntrieb
spur wheel Zahnkranz
spurious unecht, ungewollt, unerwünscht, störend
spurious bearing Fehlpeilung, fehlerhafte Peilung
spurious effect Störwirkung
spurious emission Nebenwellenausstrahlung, unerwünschte Ausstrahlung
spurious-emission suppression Nebenwellendämpfung
spurious modulation Störmodulation
spurious oscillation wilde Schwingung (621.396)
spurious radiation Störausstrahlung, unerwünschte Strahlung (621.396.61)
spurious resistance Zusatzwiderstand
spurious resonance (unerwünschte) Nebenresonanz
spurious response Ansprechen auf Nebenfrequenzen (621.395)
spurious signal Störsignal
spur wheel Stirnrad, Zahnrad
to sputter zerstäuben, verstäuben; spritzen (621.791); stottern (621.4); aufstäuben
sputtering Zerstäubung, Plasma- oder Kathodenzerstäubung
sq = square q = Quadrat-...
sq. ft. = square foot Quadratfuß = 930 cm^2
sq. in. = square inch Quadratzoll = 6,45 cm^2
sq. m. = square meter m^2 = Quadratmeter
sq. mile = square mile Quadratmeile = 2,59 km^2
sq. yd. = square yard Quadratyard = 0,836 m^2
square Quadrat; Quadrat, Quadratzahl; Platz; Winkel, Anreißwinkel, Meßwinkel
square on a line Quadrat über einer Linie
square of a number Quadrat einer Zahl
to square quadrieren (51); begleichen, bereinigen; senkrecht abschneiden (621.9)
to square off rechtwinklig abschneiden
to square out a corner eine Ecke scharf ausdrehen
squared zum Quadrat, ins Quadrat erhoben
square viereckig, quadratisch; senkrecht
square bar Vierkantstange, Vierkanteisen
square base quadratische Grundfläche
square brackets eckige Klammern
square corner rechtwinklige Ecke
square-edge joint Bördelschweißung
square headed mit Vierkantkopf
square-headed bolt Schraube mit Vierkantkopf
square law quadratisches Gesetz (51)
square-law capacitor wellengerader Kondensator
square-law detector quadratischer Detektor, quadratischer Gleichrichter (621.396)
square measure Flächenmaß
square pulse Rechteckimpuls
square root Quadratwurzel
square-root extraction device Radiziereinrichtung (621.316.7)
square section viereckiger Querschnitt, quadratischer Querschnitt
square-section Vierkant-...
square-section pole Vierkantpfahl
square-section spring Vierkantfeder
square-section wire Vierkantdraht, Formdraht
square shaft Vierkantwelle
square stone Quader, Quaderstein
square thread Flachgewinde
square to senkrecht auf, senkrecht zu
square-topped oben flach geschnitten
square-topped pulse Rechteckimpuls
square unit Flächeneinheit
square wave Rechteckwelle (621.3)
square-wave generator Rechteckgenerator
square with senkrecht zu, senkrecht auf
square yard Quadratfuß (**1 sq. yd.** = 0,836 m^2)
squaring the circle Quadratur des Kreises

squaveness ratio Seitenverhältnis, Rechteckigkeitsverhältnis
to squeak quieken, quietschen, piepen
to squeal pfeifen, heulen
squeeze Quetsche, Presse; Quetschung, Quetschvorgang; Quetschfuß (621.385)
  squeeze head Quetschkopf, Preßstempel
  squeeze piston Quetschkolben
to squeeze quetschen, ausquetschen, pressen
  to squeeze in dazwischenquetschen, hineinquetschen
  to squeeze out herausquetschen (**of:** aus)
    to squeeze out of concentricity unrund quetschen, verquetschen, verdrücken
  to squeeze through durchquetschen
squeezable waveguide Quetschhohlleiter
squeezing Quetschvorgang, Quetschen, Ausquetschen
  squeezing device Quetschvorrichtung
Squegger oscillator Blockschwinger, Pendeloszillator
squelch Krachtöterschaltung, Rauschsperre (621.396.62)
  squelch circuit Krachtöterschaltung, automatische Geräuschsperre
to squelch im Schlamm ersticken; krachtöten, unterdrücken (621.396.62)
squint Schiebwinkel (621.396.67)
to squint schielen
  squinting lobe schielende Charakteristik (621.396.677)
squirrel cage Käfiganker, Käfigläufer (621.313)
  squirrel-cage induction motor Asynchronmotor mit Kurzschlußanker
  squirrel-cage motor Kurzschlußmotor
  squirrel-cage repulsion motor Repulsionsmotor mit Kurzschlußanker
  squirrel-cage rotor Kurzschlußanker, Anker mit kurzgeschlossener Wicklung
  squirrel-cage winding Käfigwicklung, Kurzschlußkäfig
to squirt spritzen, sprudeln
  squirted filament gespritzter Leuchtdraht (621.32)
squitter unerwünschte Transponderauslösung
  squitterpulse Füllimpuls (621.376)
S. R. E. = surveillance radar equipment Rundsichtradargerät
S/S = step by step (s. dort)
SSB-transmission = single-sideband transmission Einseitenband-Übertragung
SSR = secondary surveillance radar Sekundärrundsichtradar
SSTV = slow scan TV Schmalbandfernsehanlage
ST = short time Kurzzeit(-spannungsprüfung)
st. = stone Stein = 14 lbs.
stability Stabilität; Festigkeit, Widerstandsfähigkeit
  stability condition Stabilitätsbedingung
  stability factor Kippleistung/Nennleistung (621.31)
  stability limit Stabilitätsgrenze; Kippleistung (621.311)
stabilivolt Stabilovolt, Glimmstabilisator
stabilization Stabilisierung, Konstanthaltung
to stabilize stabilisieren; glätten
  stabilized feed-back amplifier durch Gegenkopplung stabilisierter Verstärker, gegengekoppelter Verstärker
  stabilized power supply geregeltes Stromversorgungsgerät
  stabilized shunt-wound motor kompoundierter Nebenschlußmotor
stabilizing resistance Stabilisierungswiderstand
stabilizing winding Tertiärwicklung
stabilizer Stabilisator (621.385); Glättungsschaltung (621.3)
  stabilizer circuit Stabilisierungsschaltung
stable stabil
  stable equilibrium stabiles Gleichgewicht
    stable equilibrium point stabiler Gleichgewichtszustand, stabiler Gleichgewichtspunkt
  stable foundation stabiles Fundament
  stable in frequency frequenzkonstant, frequenzstabil
  stable position stabile Lage
  stable trigger circuit bistabiler Multivibrator
  stable vertical gyro Horizontierungskreisel
stack Miete, Schober (63); Meiler (66); Schornstein (66); Stapel, Schichtung; Säule (621.3); Stapelspeicher, Kellerspeicher (681.3)
  stack gas Rauchgas
  stack pointer Stapelzeiger, Kellerzeiger
to stack stapeln, aufstapeln, schichten, aufreihen
  stacked array (vertikale) Antennenzeile
stacker Stapeleinrichtung; Höhenförderer, Stapler (63)
staff (pl. **-aves**) Stab, Belegschaft, Personal; Notenlinien (pl.) (78)
  staff-finding system Personensuchanlage
  staff locator Personensuchanlage, Rufanlage
stage Stadium, Stufe, Zustand; Stufe, Abschnitt; Objekttisch (535); Tisch (621.9); Bühne (79)
  stage-by-stage switching schritthaltender Verbindungsaufbau

stage lighting Bühnenbeleuchtung
stage-lighting control board Stellwarte
stage pocket Anschlußkästchen
stage switchboard Bühnenschalttafel
stagger Staffelung, Versetzung
to stagger staffeln, versetzen, versetzt anordnen
staggered brushes versetzte Bürsten (621.313)
staggered pulse repetition frequency versetzte Impulsfolgefrequenz
staggered spot welding Zickzack-Punktschweißung
stagnation Stagnation, Stillstand
stain Fleck, Farbfleck; Beize (67)
stain spots Farbflecke, Flecke (621.357); Ausblühung (621.1)
to stain beflecken, fleckig machen; beizen
stained fleckig, bunt
stainless fleckenlos, makellos; korrosionsfest, rostfrei
stainless steel rostfreier Stahl, nichtrostender Stahl, korrosionsfester Stahl
stair Stufe, Treppenstufe
staircase Treppenhaus
staircase voltage Treppenspannung
stair lighting Treppenbeleuchtung
stake Pfahl, Markierungspfahl
to stake (off) abstecken, mit Pfählen markieren
to stall überziehen (629.13); abreißen (533)
at stall im festgebremsten Zustand
to stall an engine einen Motor zum Stillstand bringen (Überlast), einen Motor abwürgen
stalling torque abgebremstes Drehmoment
stalloy siliziumlegiertes Blech, Transformatorenblech
stamina Kraft, Ausdauer, Stehvermögen
stamp Stampfen, Aufstampfen; Pochwerk, Pochstempel (622); Stampfer (624); Stempel (651); Stempelaufdruck (651); Briefmarke
to stamp drücken, stanzen (621.9); warmpressen, pressen (621.9); stempeln (651); frankieren, freimachen, Briefmarke aufkleben (656); stampfen, pochen, zerkleinern
to stamp letters on Buchstaben einschlagen auf
stamped poles lamellierte Pole
stamping Blechschnitt
stanchion Pfosten, Stütze, Stützpfeiler
stand Untersatz, Unterbau; Gestell, Ständer, Stativ; Tribüne, Stand (796)
stand-by Reserve-..., Bereitschafts..., Ersatz...
stand-by battery Reservebatterie
stand-by equipment Reservegerät
stand-by generating set Notstromversorgung
stand-by machinery Reservemaschinen (pl.)
stand-by period Wartezeit, Stillstandszeit in Reserve
stand-by set Notstrom-Aggregat
stand-by switch Bereitschaftsschalter
standpipe Standrohr
standstill Stillstand
at standstill im Stillstand
to stand stehen; stellen, aufstellen; aushalten, gewachsen sein
to stand by daneben stehen, in Reserve stehen (siehe auch **stand-by**)
to stand on edge hochkant stehen
to stand for für... stehen, bezeichnen, bedeuten, entsprechen
to stand rough work rauhen Betrieb aushalten, rauhem Betrieb gewachsen sein
to stand still stillstehen, stilliegen
to stand a test eine Prüfung aushalten, eine Prüfung bestehen
standard Standard, Normalstück, Normalmaß; Musterstück, Vorbild; Durchschnittswert, Norm; Standarte, Flagge (35); Stehlampe (621.31); Ständer, Ständerrahmen, Walzenständer, Preßständer (621.9)
standards Normen, Normalien
standard of accuracy Güteklasse, Klasse (621.317)
standard atmosphere Normalklima
standard atmospheric conditions atmosphärische Normalbedingungen
standard bar Normalstange, Normalstab
standard book Normvorschriften (pl.), Standardbuch
standard cable Norm-Kabel
standard cable equivalent Dämpfung in Standard Cable Miles
Standard Cable Mile (Dämpfungsmaß) (621.395)
standard candle (obs.) internationale Kerze, Normalkerze
standard capacitor Normalkondensator
standard cell Normalelement (621.3)
standard code Normvorschriften (pl.) (656.25)
standard of comparison Vergleichsnormal
standard component Normbauteil
standard conversion of taped programs Magnetbandnormwandlung
standard copper Normalkupfer, Elektrolytkupfer
standard design normale Bauart, Normalkonstruktion, Einheitsbauweise, Regelausführung, Serienausführung
standard deviation Standard-Abweichung, Streuung (519.2)

standard electrode potential Normalpotential
standard equipment practice Einheitsbauweise
standard frame Normalkern, Normalpaket (621.313)
standard frequency Normalfrequenz
standard frequency generator Normalfrequenzgenerator, Eichsender (621.395)
standard frequency transmission Normalfrequenz-Aussendung
standard gauge Normalspur (625)
standard gauge railway normalspurige Eisenbahn, Vollbahn
standard gauge track normalspuriges Gleis, Vollbahngleis
standard inductance Normalinduktivität
standard lamp Normallampe, Lampennormal (535); Stehlampe (643)
standard level Normalpegel (621.392) (6 mW entsprechend 0 dB)
standard meter Normalmeter, Urmeter
standard musical pitch Normalton, Kammerton
standard noise figure Standard-Rauschzahl
standard noise generator Rauschgenerator
standard observer Normalbetrachter
standard ohm Normal-Ohm
standard part Normteil
standard pattern Normalmuster, Normalausführung, Regelbauart
standard product Markenartikel
standard radio atmosphere Standard-Atmosphäre
standard refraction Standard-Brechung
standard register Vierstellenzählwerk (621.317)
standard resistance Normalwiderstand
standard section Normalprofil
standard shape Normalform, übliche Form
standard size Normalgröße, Normaltype; gängige Größe; Normalformat
standard socket Normalsockel (621.32), (621.385)
standard specifications Pflichtenheft
standard tape Bezugsband (681.3)
standard of temperature Normaltemperatur
standard test method Norm-Prüfverfahren
standard test vessel Normal-Prüfgefäß
standard thread Normgewinde
standard time Normalzeit
standard transmission line Eichleitung
standard tuning frequency Norm-Stimmton
standard type Normtype, Normalbauart
standard voltage Normspannung
standard weight Normalgewicht
standard yard Normalyard, Uryard
standardization Normung; Normierung
to standardize normen
standardizing Eichung, Normung
standardizing laboratory Eichamt; Prüfamt, Zählerprüfamt
standardizing room Eichraum, Zählereichwerkstatt
standards Normalien, Normen
standards committee Normenausschuß
standards institution Eichanstalt, Eichamt; Bundesanstalt
standards room Eichraum, Zählerwerkstatt
standby Bereitschaft
standing laufend, dauernd, dauernd gültig; stehend
standing current Dauerstrom, Ruhestrom (621.385)
standing d. c. component überlagerter Gleichstromanteil (621.397)
standing lamp Tischlampe
on standing beim Stehenlassen
standing order Geschäftsordnung, Dienstanweisung
standing wave stehende Welle (621.392)
standing-wave meter Welligkeitsmesser
standing-wave pattern Spannungsverteilung (621.396), (621.317)
standing-wave ratio Stehwellenverhältnis, Welligkeit (621.396), (621.317)
stannic Zinn-...
stannic chloride Zinnchlorid
stannous Zinn-...
stannous chloride Zinnchlorür
staple Stapel (67); Klammer, Heftklammer (651), (655); Krampe (69); Bein, Stützbein (621.8)
staple fiber Stapelfaser
to staple heften (651), (655)
star Stern (52); Sternschaltung (621.3); Leuchtsatz, Leuchtstern, Leuchtkugel (66)
star-connected sterngeschaltet, in Stern geschaltet
star-connected circuit Sternschaltung
star connection Sternschaltung
star-delta Sterndreieck
star-delta control Sterndreieck-Schaltung
star-delta starter Sterndreieck-Anlasser
star-delta starting Sterndreieck-Anlauf
star-delta switch Sterndreieck-Schalter
star-delta switching starter Sterndreieck-Anlasser
star-delta throw-over switch Sterndreieck-Umschalter

star-delta transformation Sterndreieck-Transformation
star handle Kreuzgriff, Speichengriff
star point Sternpunkt
star-quad cable Sternviererkabel
star-shaped sternförmig
star-shaped network Sternnetz
star-tracking system Sternnachführungssystem
star-type network Sternnetz, sternförmiges Netz
star voltage Sternspannung, Phasenspannung, Strangspannung
starch Stärke, Dextrin
to starch stärken
start Anfang, Start; Anlauf, Anlaufvorgang; Vorsprung
start-of-block signal Blockbeginnzeichen
start-dialling signal Abrufzeichen (621.395.3)
start distance Anlauflänge (681.3), (681.84)
start order Anlaßbefehl
start-pulsing signal Abrufzeichen (621.395.3)
start signal Beginnzeichen
start-stop Start-Stop, Geh-Steh
start-stop distortion Start-Stop-Verzerrungsgrad
start-stop printer Start-Stop-Schreiber, Geh-Steh-Schreiber
start-stop telegraphy Start-Stop-Telegraphie, Geh-Steh-Telegraphie; Springschreibertelegraphie
start of a tap Gewindeanschnitt
start-up Anlauf(-Vorgang) (621.165), (621.35); (621.374)
start winding Anlaufwicklung
to start anfangen, beginnen; anlassen
to start into action anstoßen
to start at beginnen bei, anfangen um
to start at infinity unmerklich anfangen
to start an engine einen Motor anwerfen, einen Motor starten (621.4)
to start under load unter Last anlassen; unter Last anlaufen
to start a motor einen Motor anlassen
to start registering anlaufen (eines Zählers) (621.317)
to start up anfahren; sich ausbilden, sich entwickeln
to start up cold kalt anfahren, kalt einschalten (621.314.6)
starter Anlasser (621.3), (621.4); Starter (621.4)
starter battery Anlaßbatterie, Starterbatterie
starter contact Anlasserstufe, Anlasserkontakt
starter pinion Anlasserritzel
starter pushbutton Startknopf (629.113)
starter resistor Anlaßwiderstand, Anlasser, Widerstandsanlasser
starting Anlauf, Anlaßvorgang, Anlaufvorgang; Einschaltung (621.313); Inbetriebnahme
starting anode Zündanode (621.314.5)
starting box Anlasser (621.313)
starting capacitor Anlaufkondensator
starting capacity Anlaßleistung
starting coil Anlaufspule, Reibungskompensationsspule (621.317)
starting condition Anlaufbedingung
starting connection Anlaßschaltung
starting counter torque Anlaufgegenmoment
starting current Anlaufstrom, Anlaßstrom; Anschwingstrom (621.37); Anzugsstrom (621.318)
starting current rush Anlaßstromstoß, Einschaltstromstoß
starting device Anwurfvorrichtung
starting handle Anlasserkurbel, Anwurfkurbel (621.4)
starting lever Anlaßhebel
starting of a machine Inbetriebsetzung einer Maschine, Ingangsetzen einer Maschine
starting motor Anwurfmotor, Anlasser (621.4), (629.113)
starting point Ausgangspunkt, Anfangspunkt
the starting point chosen der gewählte Ausgangspunkt
starting position Ausgangsstellung
starting power Anlaufleistung
starting process Anlaufvorgang, Anlaßvorgang
starting reactor Anlaßdrossel
starting resistance Anlaßwiderstand
starting resistor Anlaßwiderstand
starting rheostat Anlaßwiderstand
starting and safety clutch Anlauf- und Sicherheitskupplung
starting sequence Einschaltfolge
starting sheet Niederschlagselektrode (621.357)
starting switch Anlaßschalter
starting time Anlaßzeit
starting torque Anlaufmoment, Anzugsmoment, Anfahrdrehmoment
starting tractive force Anfahrzugkraft
starting tractive power Anfahrzugleistung
starting transformer Anlaßtransformator
starting value Ausgangswert
starting winding Anlaßwicklung, Anlaufwicklung, Anlaufkäfig

state Zustand, Stand; Zustand, Aggregatzustand; logischer Zustand (621.374)
state of aggregation Aggregatzustand
state of alert Alarmzustand
state of the art Stand der Technik
state of the atmosphere Luftzustand
state of disrepair nicht mehr reparaturfähiger Zustand
state of equilibrium Gleichgewichtszustand
state of fusion Schmelzzustand
state of strain Spannungszustand
in a state of flux im Fluß
statement Feststellung, Darstellung; Anweisung (681.3)
statement of particulars Spezifikation
static atmosphärische Störung (621.396)
static eliminator Störbefreiungsschaltung, Krachtöter (621.396.62)
static noise Störgeräusch, Störpegel, atmosphärischer Störpegel, Rauschen (621.396.62)
static elimination Störbefreiung
static statisch
static balancer Ausgleichstransformator (621.316.1)
static breeze elektrischer Wind (537)
static characteristic statische Kennlinie (621.385)
static charge statische Ladung
static condenser statischer Kondensator, Starkstromkondensator
static converter statischer Umformer, Stromrichter
static current Ruhestrom (621.385)
static discharge statische Entladung
static electricity statische Elektrizität
static electrification statische Aufladung
static electrode potential statisches Elektrodenpotential (ohne Strombelastung)
static end ring Schirmring (621.314.2)
static frequency changer statischer Frequenzwandler, Umrichter
static friction statische Reibung, ruhende Reibung
static harmonic absorber Oberwellensieb, Saugkreisschaltung
static hysteresis Wechselfeldhysteresis
static induction motor Drehregler (621.313)
static inverter statischer Wechselrichter
static RAM statisches RAM
static sensitivity Gleichlichtempfindlichkeit (621.385)
static state statischer Zustand, Ruhezustand
static store statischer Speicher (681.3)
static torque Drehmoment bei festgebremstem Rotor
static value of the forward current transfer ratio Gleichstromverhältnis (621.382.3)
staticizer Serien-Parallel-Wandler
station Bahnhof, Station (625); Station, Sendestation, Sender (621.396); Stelle, Station
station on the air sendende Station, Station im Sendebetrieb
station line Teilnehmeranschlußleitung (621.395)
station ringer Wecker, Teilnehmerwecker
station-service machine Eigenbedarfsmaschine
station-service switchgear Eigenschaltanlage
to station stationieren, aufstellen, placieren
stationary stillstehend, ruhend, stationär
stationary armature Statoranker, Ständeanker
stationary battery ortsfeste Batterie
stationary blade Leitschaufel
stationary coil feststehende Spule, Feldspule, Statorspule
stationary contact member fester Kontakt, feststehender Kontakt
stationary element ruhender Teil, feststehender Teil
stationary field stationäres Feld
stationary lamp ortsfeste Lampe
stationary pattern ruhendes Bild, stillstehendes Bild (621.385)
stationary structure feste Anlage (621.316.5)
stationary transformer ortsfester Umspanner, ortsfester Transformator
stationary wave stehende Welle
stationary winding stillstehende Wicklung
statistical statistisch
statistical distribution statistische Verteilung
statistical quality control statistische Qualitätskontrolle
statistical research Großzahlforschung
statistician Statistiker
statistics (pl.) Statistik
stator Ständer, Stator (621.313), (621.319.4)
stator coil Ständerspule, Statorspule
stator conductor Ständerstab
stator core Statorkern, Ständerblechpaket
stator current Ständerstrom, Statorstrom
stator e. m. f. Stator-EMK, induzierte Ständerspannung
stator fed ständergespeist
stator frame Statorgehäuse
stator lamination Statorblech, Ständerblech
stator leakage Ständerstreuung
stator plate Statorplatte, feststehende Platte (621.319.4)

stator slot Ständernut
stator vane Statorplatte, feststehende Platte (621.319.4)
stator winding Ständerwicklung, Statorwicklung
stator yoke Ständerjoch
status Zustand
status bit Zustandsbit
status word register Statuswortregister
statute mile gesetzliche Meile, Britische Landmeile (**1 stat. mile = 1760 yard** = 1,60934 km)
stay Abspannseil, Verspannungstau; Strebe; Mastanker; Auflage (621.9); Lünette (621.9); Ständer (621.9)
stay bolt Stehbolzen
stay rod Stehbolzen, Zugstange
stay wire Abspanndraht
to stay absteifen, abspannen, verstreben; bleiben
steadiness Konstanz, zeitliche Konstanz
steady konstant, zeitlich konstant, stetig, gleichmäßig; Ruhe-..., Gleichstrom-...
steady anode current Anodenruhestrom, Anodengleichstrom
steady brace Fahrleitungsstütze (621.3)
steady component Gleichstromkomponente, Gleichstromanteil
steady condition stationärer Zustand
steady light Gleichlicht; konstante Lichtquelle
steady load gleichmäßige Last, ruhende Belastung
steady-operation Ruhe-..., Arbeitspunkt-... (621.385)
steady pace gleichmäßiges Tempo
steady pressure konstanter Druck, ruhender Druck
steady span Seitenhalterung (621.3)
steady state stationärer Endzustand, Gleichgewichtszustand, eingeschwungener Zustand
steady-state behaviour Beharrungsverhalten
steady-state condition stationärer Endzustand
steady-state error bleibende Regelabweichung
steady-state gain Übertragungsfaktor
steady-state power limit statische Kippleistung (621.311)
steady-state stability statische Stabilität
steady-state stability factor statische Kippleistung/Nennleistung (621.311)
steady-state stability limit statische Kippleistung (621.311)
steady supply Gleichstromzuführung, Zufuhr der Betriebsspannung (621.385)
steady values Ruhewerte, Werte am Arbeitspunkt (621.385)
steady voltage Gleichspannung, Betriebsspannung (621.385)
steam Dampf, Wasserdampf; Brüden
steam coil Dampfschlange, Heizschlange
steam-electric generating set Dampfkraftgenerator
steam engine Dampfmaschine
steam gauge Dampfdruckmesser, Manometer
steam heating Dampfheizung
steam jet Dampfstrahl
steam-jet air ejector Dampfstrahlpumpe
steam power Dampfkraft
steam pressure Dampfdruck
steam-pressure gauge Dampfdruckmanometer
steam-proof dampfdicht
steam pump Dampfpumpe
steam purifier Dampfreiniger
steam superheater Überhitzer
steam tension Dampfspannung
steam-tight dampfdicht
steam trap Kondenstopf
steam turbine Dampfturbine
steam-turbine electric locomotive dampfturboelektrische Lokomotive
to steam dampfen, Dampf abgeben; dampfen, fahren; dämpfen, mit Dampf behandeln
steaming-hot kochend heiß
steaming 20 knots mit 20 Knoten Fahrt (629.12)
steaming Dämpf-..., Dampf-...
steaming pit Dämpfgrube (674)
steamy dunstig, dampferfüllt, dunsterfüllt
stearic acid Stearinsäure
stearin Stearin
steatite Steatit
steel Stahl
steel-alkaline battery NC-Sammler, alkalischer Sammler, Edisonbatterie
steel band Stahlband
steel-band armouring Stahlbandbewehrung
steel casting Stahlguß, Stahlgußstück
steel castings (pl.) Stahlguß
steel conductor Eisenleiter
steel conduit Stahlrohr, Stahlpanzerrohr (621.3)
steel core Stahlkern, Stahlseele
steel-cored aluminium Stahlaluminium
steel cylinder Stahlflasche

steel-cylinder mercury-arc rectifier Eisengleichrichter
steel disc Stahlscheibe
steel facing galvanische Verstählung
steel flat Flachstahl
steel forging Stahlschmiedestück
steel pipe Panzerrohr
steel punch Stahlstempel, stählerner Durchschlag
steel reinforcement Stahleinlage
steel rope Drahtseil, Stahlseil
steel section Profilstahl
steel sheet Stahlblech
steel shot Stahlschrot, Stahlkies
steel spring Stahlfeder
steel strip Stahlband, Bandstahl
steel tank Stahlgefäß
steel-tank rectifier Eisengleichrichter (621.3)
steel tape Stahlbandmaß
steel-tape armoured stahlbandbewehrt
steel tower Eisenmast, Stahlmast
steel wire Stahldraht
steel-wire brush Stahlbürste, Stahldrahtbürste
steel-wire armoured cable stahldrahtbewehrtes Kabel
steel wool Stahlwolle; Stahlspäne (pl.)
steel yard Laufgewichtswaage
steep steil
steep front of the wave steile Wellenfront
steep gradient steiles Gefälle
steep taper steiler Konus
steeply falling side steile Flanke (621.396)
to steepen versteilen, steiler machen
steeping Tränkung, Imprägnierung
steepness of the wave front Steilheit der Wellenfront
to steer steuern
steerable antenna Antenne mit schwenkbarer Charakteristik
Stefan-Boltzmann law of radiation Stefan-Boltzmannsches Strahlungsgesetz
Steinmetz coefficient Hysteresiskoeffizient der Steinmetz-Formel
Steinmetz connection Drehstrom-Zweiphasentransformator
stellar Stellar-..., Stern-...
stellit Stellit
stem Lampenfuß (621.326); Stiel
stem-type thermostat Stabausdehnungsthermostat
stencil Schablone
step Schritt; Stufe (621.3), (629.12), (69)
step back Pilgerschritt
step-back relay Rückschaltrelais
step-back welding Pilgerschrittschweißung
step bearing Fußlager, Spurlager
step box Spurlager
step-by-step schrittweise, stufenweise; schritthaltend (621.39)
step-by-step automatic telephone system Schrittwählersystem
step-by-step dial system Schrittwählersystem
step-by-step integration stufenweise Integration, schrittweise Integration
step-by-step motor Schrittschaltwerk
step-by-step movement schrittweise Bewegung
step-by-step operation Einzelschrittbetrieb; schritthaltender Verbindungsaufbau (621.395)
step-by-step switch Schrittschaltwähler, Schrittwähler
step-by-step switching Verbindungsaufbau mit direkter Wählereinstellung
step-cone pulley Stufenscheibe
step counter Schrittzähler
step-down Herunterübersetzung
step-down gear Untersetzungsgetriebe
step-down ratio Abwärtsübersetzung, Untersetzung
step-down transformer Abwärtstransformator, Abspanner
step function Stufenfunktion (51)
step-function response Pulsamplitudengang
in step in Tritt, synchron in Phase
in step with synchron mit
stepless regulation stufenlose Regelung, kontinuierliche Regelung
stepless speed control stufenlose Drehzahlregelung
step method of testing Spannungsprüfung mit stufenweiser Spannungssteigerung
out of step außer Tritt
step of a resistance Widerstandsstufe
step response Sprungantwort
step of the starter Anlaßstufe
step stress stufenweise Belastungssteigerung
step-strobe marker Stufenmarkengeber (621.396.6)
step switch Stufenschalter
step-up Aufwärtsübersetzung
step-up gear Übersetzungsgetriebe
step-up ratio Aufwärtsübersetzung
step-up transformer Aufwärtstransformator, Aufspanner
step voltage regulator Spannungsstufenregler
to step schreiten, Schritte machen; abstufen

to step back zurückschreiten, rückwärtsgehen
to step down herunterübersetzen (621.3), (621.83); heruntertransformieren (621.3)
to step the voltage down die Spannung heruntertransformieren, abspannen
to step forward voranschreiten
to step off (mit dem Zirkel) abstecken
to step on to fortschreiten zu
to step up hinauftransformieren, aufspannen (621.3); hinaufübersetzen (621.3), (621.83)
stepping motor Schrittmotor
stepping rate Schrittgeschwindigkeit
stepping relay Wählerrelais (621.395), Schrittrelais, Zählrelais
stepping switch Schaltwähler; Schrittschaltwerk
Stephenson reversing gear Kulissensteuerung (621.1)
stepped treppenförmig, gestuft, Stufen-...
stepped cone Stufenscheibe
stepped-curve distance-time protection Distanzschutz mit Stufenkennlinie
stepped gear Stufenrad, Stufenzahnrad
stepped grate Treppenrost (621.18)
stepped grate producer Treppenrostgenerator
stepped leader stufenweise vorwachsende Blitzentladung (551)
stepped pulley Stufenscheibe
stepped winding Treppenwicklung
stepper motor Schrittmotor
steradiancy Richtstrahlungsdichte (621.32)
stereo... Stereo..., räumlich
stereo amplifier Stereoverstärker
stereo console Stereogerät, Stereotruhe
stereo mixer Stereomischpult
stereo recorder Stereo-Aufnahme- und Wiedergabegerät, Stereorekorder
stereo sound Raumklang, Stereowiedergabe
stereophonic stereophonisch
stereophonic effect stereophonischer Effekt, Raumwirkung
stereophonic reception Stereo-Empfang
stereophonic reproduction stereophonische Wiedergabe
stereophonic sound system Stereophonie
stereophonic TV sound stereophoner Fernsehton
stibium Antimon
stick Stange, Knüppel, Stock; Stück, Stange
stick operated für Schaltstangenbetätigung (621.316.5)
stick of sealing wax Stange Siegellack
stick of solder Lotstab, Lotstreifen
to stick stecken; kleben, ankleben, aufkleben; hängenbleiben, klebenbleiben; sich festsetzen; festhängen
stick circuit Haltestromkreis (621.318)
to stick fast festhaften
to stick on aufstecken, aufnadeln; aufkleben
to stick to anhaften
to stick together zusammenkleben; zusammenhängen; zusammenstecken
stickiness Klebrigkeit
sticking haftend, Haft-..., klebend, festhängend
sticking friction Haftreibung
sticky klebrig, schmierig, haftend
sticky liquid zähe Flüssigkeit, klebrige Flüssigkeit
sticky stuff Klebzeug, Klebstoff
sticky surface klebrige Oberfläche, schmierige Oberfläche
stiff steif, starr, biegungssteif, biegungsstarr
stiff cardboard steifer Karton, steife Pappe
stiff paste steife Paste, steifer Brei
to stiffen versteifen, steifer werden, sich verfestigen; steifer machen
stiffener Versteifung, Versteifungsrippe, Versteifungsleiste
stiffening Versteifung
stiffness Steifigkeit, Steife
stilb Stilb (veraltet)
still ruhig, still
still air ruhende Luft
still picture feststehendes Bild (77)
still water stehendes Wasser, stilles Wasser
stillage Stellage, Gestell
to stimulate anregen (534)
stimulation Anregung (534)
sting ray Nadelstrahl, scharf gebündelter Strahl
stir Umrühren, Rührvorgang
to stir rühren, umrühren
to stir into einrühren in, darunterrühren
stirrer Rührer, Quirl
stirring Rühr-...
stirring rods (pl.) Rührer
stirring spider Rührkreuz, Rührstern
stirrup Steigbügel; Gehänge (681)
to stitch nähen, sticken; heften (655)
stitching machine Heftmaschine (655)
stitching needle Heftnadel (655)
stochastic noise stochastisches Rauschen
stock Lager, Vorrat; Stock, Grundstock; Schaft, stumpfes Ende; Klappe (621.9); Unterlage (58), (63)
from stock vom Lager, ab Lager
in stock am Lager, vorrätig

out of stock nicht (mehr) am Lager, ausverkauft
stock taking Bestandsaufnahme, Inventur
to stock stapeln, horten; auf Lager nehmen, auf Lager halten; einlagern; vorrätig haben
stocks (pl.) Stapel (629.12)
STOL = short take-off and landing Kurzstart (629.13)
stop Halt, Haltepunkt, Haltestelle; Anschlag, Endanschlag (621); Ende, Schluß; Punkt (41); Klappe (681)
stop bar Anschlagstange, Anschlagstift
stop button Abstellknopf
stop collar Anschlagring
stop gap Notbehelf, Notmaßnahme
stop joint Sperrmuffe (621.315.2)
stop lever Arretierhebel
stop peg Anschlag, Anschlagstift, Endanschlag
stop pin Anschlagstift
stop screw Anschlagschraube, Arretierschraube
to stop anhalten, bremsen, halten; stoppen, stehenbleiben; stopfen, zustopfen, verstopfen
to stop hunt Freiwahl unterbinden (621.395)
stoppage Aufhören, Anhalten, Halt, Stillstand, Betriebsunterbrechung, Betriebsstörung
stopper Stopfen, Verschlußpfropfen, Spund
to stopper zustöpseln, zukorken, zupfropfen
stopping off Abdecken (621.357)
stopwatch Stoppuhr
storage Speicher (681.3); Aufbewahrung, Lagerung, Speicherung (siehe auch **memory**)
storage action Speicherwirkung
storage address Speicheradresse
storage battery Sammlerbatterie, Akkumulatorenbatterie
storage-battery electric locomotive Akkumulatorenlokomotive
storage bin Vorratsbehälter, Lagergefäß, Silo
storage bottle Vorratsflasche, Speicherflasche
storage bottle for compressed gases Gasflasche, Stahlflasche
storage capacity Speichervermögen, Kapazität (621.355); Speicherkapazität (681.3)
storage cell Sammler, Akkumulator (621.355); Speicherzelle (681.3)
storage circuit Speicherkreis (681.3)
storage-cycle time Speicherzykluszeit
storage element Speicherelement
storage ground Vorratsplatz, Speicherplatz, Lagerplatz
storage life Lagerungsfähigkeit
storage location Speicherplatz
storage medium Datenträger
storage oscilloscope Speicheroszilloskop
storage-power station Speicherkraftwerk
storage tank Vorratsbehälter, Speicherbehälter
storage target Speicherplatte
storage temperature Lagerungstemperatur
storage tube Speicherröhre
store Lager, Laden, Speicher
storehouse Lagerhaus, Speicher, Magazin
storehouse of energy Energiespeicher
store instruction Speicherbefehl (681.3)
in store auf Lager, auf Vorrat
to store speichern (681.3), lagern, stapeln; aufheben, aufhäufen
to store heat Wärme speichern
stored energy gespeicherte Energie; Speichervermögen (538)
stored-program controlled speicherprogrammiert
storing Lagerung, Speicherung
storing capacity Speichervermögen
stout steif, stabil, kräftig; kurz und dick
stout card(board) steifer Karton
stout construction solide Bauweise
stout wire steifer Draht, dicker Draht
stove Ofen, Zimmerofen
to stove darren (63); einbrennen, im Ofen härten, ofenhärten (66)
straddle Spreizsitz; Spreize, Spreizstütze
to straddle eingabeln (623); spreizen
straggling cuttings lockere Späne
straight gerade
straight axle Laufachse (625)
straight course gerader Kurs
straight current transformer Schienenstromwandler, Stabstromwandler
straight-cut control Streckensteuerung (621.9)
straight-draw moulding machine Aufrecht-Formmaschine
straight edge gerade Kante; Stahllineal, Lineal; Richtscheit (69)
straight-forward network gestrecktes Koppelnetz
straight gears (pl.) geradverzahnte Räder; Geradverzahnung
straight joint Muffe, Durchgangsmuffe (621.315.2)
straight line gerade Linie, Gerade (51)
straight-line capacitor kapazitätsgerader Kondensator
straight-line frequency capacitor frequenzgerader Kondensator
straight-line motion geradlinige Bewegung
straight-line path geradliniger Weg; Großkreisausbreitung (621.396)

straight-line portion geradliniger Teil
straight-line wavelength capacitor wellengerader Kondensator
out of straight krumm, verbogen
straight path-control Bahnsteuerung
straight polarity welding Schweißung mit negativem Schweißstab
straight receiver Geradeaus-Empfänger
straight-run gerades Stück, gerades Leitungsstück
straight-sided geradflankig
straight storage system reines Batteriesystem (621.32), (621.33)
straight through direkt durchgehend
straight-through current transformer Schienenstromwandler, Stabstromwandler
straight-through joint Längsverbindung, glatte Verbindung
straight-through ray direkter Strahl (621.396)
straight-up-and-down filament Langdraht-Leuchtdraht (621.32)
straight wire gerader Draht
to straighten richten, geraderichten, ausrichten, begradigen
to straighten out glätten, begradigen
strain Anstrengung, Beanspruchung, Formänderung, Dehnung, Zerrung
strain car Abspannöse (621.32)
strain clamp Abspannklemme
strain gauge Dehnungsmesser, Dehnungsmeßstreifen
strain hardening Spannungsvergütung
strain insulator Abspannisolator (621.315.6)
strain portal structure Abspannportal
strain structure Abspannmast
strain-type insulator Abspannisolator (621.315.6)
strain variation diagram Dehnungsdiagramm
strain viewer optischer Spannungsprüfer (66)
to strain verformen, dehnen; beanspruchen; seihen, durchsieben, sieben
to strain a solution eine Lösung durchseihen
strainer Filter, Sieb; Seihtuch; Saugkorb; Ankerspanner (621)
strainer rack Filterrechen, Einlaufrechen (621.2)
straining Beanspruchung; Dehnung, Zerrung; Siebung, Seihung, Nutschen; Abspannung
strand Strand (91); Litzendrähtchen, Litze, Ader; Litze, Litzenseil
x-strand cable x-drähtiges Seil
to strand stranden, auf Strand laufen, scheitern (629.12); verseilen, verlitzen
stranded verseilt, aus Litze, verlitzt
stranded cable Litzendraht, verseiltes Kabel
stranded conductor Litzenleiter, Seil, Leiterseil
stranded rope Drahtseil, Litzenseil
stranded wire verseilter Draht, Litzendraht, Litze
strap Band, Bandage; Riemen, Gurt; Lasche (621.791), (621.9)
strap brake Bandbremse (629.113)
strap coil Bandspule
strap connection Laschenverbindung (621.3)
strap hinge Scharnierband
strap iron Bandeisen
strap key Momentdruckknopfschalter
to strap gurten, schnallen, festschnallen; zusammenschalten
strapping diagram Schaltskizze
to stratify schichten
stratum (pl. **-ta**) Schicht (55); Schichtung, Lage
stray Streukapazität (621.396); Streu-...
to stray herumirren, vagabundieren, streunen
strays (pl.) Schaltkapazität; atmosphärische Störungen (pl.) (621.396)
stray capacitance Streukapazität
stray capacity Streukapazität, Nebenkapazität
stray coupling Streukopplung, wilde Kopplung, ungewollte Kopplung
stray current Streustrom, vagabundierender Strom, Erdstrom
stray earth current vagabundierender Strom, Erdstrom
stray-electrostatic field elektrostatisches Streufeld
stray field Streufeld
stray flux Streufluß
stray light Streulicht
stray losses Zusatzverluste (621.313)
stray magnetic field magnetisches Streufeld
stray magnetic flux magnetischer Streufluß
stray pick-up Einstreuung, Einwirkung von Streufeldern (621.396)
stray power Zusatzverluste (pl.) (621.313)
stray-wiring capacity Schaltkapazität
straw Düppel (623)
streak Streif, Streifen
to streak streifen, streifig machen
streaking Fahneneffekt (621.396.96)
streaky streifig
stream Strom, Strömung, Flut
stream of electrons Strom von Elektronen
stream of energy Energieströmung, Energiestrom
stream of light Strom von Licht
streamline Stromlinie

streamline shape Stromlinienform
stream lubrication laufende Schmierung
to stream strömen, wehen, flattern
streamer Fangentladung, Leuchtfaden; Düppelstreifen (623)
to streamline stromlinienförmig machen
streamlined stromlinienförmig
streamlined production Fließfertigung
street Straße
street illumination Straßenbeleuchtung
street lighting Straßenbeleuchtung
street noise Straßenlärm
strength Stärke, Festigkeit, Widerstandsfähigkeit; Stärke, Mannschaftsstärke
strength of a current Stärke eines Stromes
strength of draught Zugstärke, Zug (621.1)
strength factor Beanspruchungsfaktor, Ausnutzungsfaktor
strength of the field Stärke des Feldes
strength of magnetic pole magnetische Polstärke
strength of a rope Festigkeit eines Seiles
strength testing Festigkeitsprüfung
to strengthen verstärken
strengthener Verstärkung, Einlage
strengthening Verstärkung
strengthening boss Verstärkungswarze
stress Zug, Zugbeanspruchung, Zugspannung; Spannung, Spannungszustand; Feldstärke; Betonung
stress coat Reißlack (62)
stress concentration Kerbwirkung, Spannungsüberhöhung
stress cracking Spannungsrißbildung
stress diagram Kräfteplan, Cremonaplan
stress relief Zugentlastung (621.3), Entspannung, Spannungsfreimachen
stress-relief heat-treatment Spannungsfreiglühen
stress-strain diagram Spannungs-Dehnungs-Schaubild, Zerreißdiagramm, Feinmeßdiagramm
stress-variation diagram Spannungsverteilungsdiagramm
to stress beanspruchen (621), (621.3); auf Zug beanspruchen; betonen
stressing Beanspruchung, Belastung
stretch Strich, Gebiet, Strecke; Landstrich; Streckung, Zerrung, Reckung, bleibende Dehnung
to stretch sich ausdehnen, sich erstrecken; strecken, recken, spannen; sich verlängern, sich dehnen, sich recken
to stretch tight strammen, strammziehen
to stretch a belt tight einen Riemen straff anziehen, einen Riemen straff spannen
stretched cord gespannte Saite
stretchless unnachgiebig, nicht dehnbar
to strew in einstreuen
stria (pl. **-iae**) Schliere
striated schlierig, streifig
striation Schlierenbildung
strict streng, genau
strict program store reiner Programmspeicher
strike Streik, Ausstand
strike measure Streichmaß
to strike schlagen, treffen auf, auftreffen auf, stoßen auf; auffallen auf, stoßen gegen, aufprallen auf, anschlagen, anstreichen; streiken
to strike against anprallen an
to strike an arc einen Lichtbogen zünden, einen Lichtbogen einleiten
to strike back zurückschlagen, rückzünden
to strike a balance eine Bilanz ziehen
to strike a blow einen Schlag führen
to strike a circle einen Kreis schlagen
to strike a contact einen Kontakt berühren, auf einen Kontakt auftreffen
to strike a discharge eine Entladung zünden
to strike a match ein Streichholz anreißen, ein Streichholz anzünden
to strike off terminal figures Endziffernstreichen, letzte Stellen weglassen
to strike oil auf Öl fündig werden; „einen guten Griff machen"
to strike a spark einen Funken ziehen
striker Pendelschlaghammer (62), Schlagbolzen (623)
striker pin Auslösestift
striking auffallend, auffällig; Streich-..., Zünd-...
striking by bulb tilting Kippzündung (621.314.6)
striking current Mindestzündstrom
striking by dipping electrode Tauchzündung (621.314.6)
striking distance Schlagweite
striking particle stoßendes Teilchen, stoßende Partikel
striking potential Zündspannung (621.385)
striking surface Reibfläche, Streichfläche
string Saite; Faden, Bindfaden, Seil, „Strippe"; Isolatorenkette, Kette (621.315.6); Zeichenfolge, Datenfolge (621.394), (681.3)
string efficiency Kettenwirkungsgrad (621.315.6)

string galvanometer Saitengalvanometer
to string spannen, ausspannen, anspannen
to string from ... to ... ausspannen zwischen ... und ..., von ... nach ... spannen
stringer Zugstrebe, Verspannung
stringy saitenähnlich
strip Streifen, Band
strip aluminium Aluminiumband
strip-chart recorder Bandschreiber (621.317)
strip conductor Bandleiter, Leiterband
strip frequency Zeilenfrequenz (621.397)
strip lighting Soffittenbeleuchtung
strip-line technique Bandleitertechnik, Streifenleitertechnik
strip resistor Bandwiderstand
strip steel Bandstahl
strip-wound armature Flachkupferwicklung (621.313)
strip-wound toroidal core Bandringkern
to strip abstreifen; abisolieren, abmanteln
to strip a screw thread Gewinde ausreißen
to strip a wire einen Draht abisolieren (621.315.3)
stripped atom abgestreiftes Atom
stripe Streifen, Farbstreifen
stripper Abstreifer
stripping Extraktion (54)
stripping compound Trennschicht (621.35)
stripping pliers Abziehzange
stripping stage Extraktionsstufe (54)
strobe Meßmarke, Marke; Taktimpuls (681.3)
strobe marker Meßmarkengeber (621.396.96)
strobe pulse Markierungsimpuls (621.396.96)
strobing pulse Auswerteimpuls
stroboglow Glimmlichtstroboskop
stroboscope Stroboskop
stroboscopic stroboskopisch
stroboscopic device Stroboskop
stroboscopic effect stroboskopische Wirkung
stroboscopic method stroboskopisches Verfahren
stroboscopic meter disc stroposkopische Zählerscheibe
stroke Schlag, Schlaganfall (61); Schlag, Takt; Hub, Takt; Schlag, Blitzschlag; Schrägstrich (/); Strich, Federstrich; Zug; Anschlag
stroke of the saw Sägenzug
stroke of a typewriter Anschlag einer Schreibmaschine
strong stark, fest, kräftig
strong coupling feste Kopplung
strong in compression druckfest
strong current starker Strom, kräftiger Strom
strong light starkes Licht
strong line kräftige Linie, dicke Linie
strong mixture reiches Gemisch (621.4)
strong signal kräftiger Empfang (621.396)
strongly ionized hochionisiert
strontium Strontium
Strowger automatic telephone system Strowger-Wählsystem
Strowger selector Strowgerwähler, Hebdrehwähler, Viereckwähler
struck particle getroffenes Teilchen, gestoßene Partikel
structural Bau-..., Struktur-..., baulich
structural element Bauteil, tragender Bauteil
structural formula Strukturformel (54)
structural joint mechanisch beanspruchte Verbindung
structural part Bauteil, Konstruktionselement
structural return loss Rückflußdämpfung, Restdämpfung
structural section Profil, Profileisen, Walzträger
structural steel Baustahl
structure Struktur, Gefüge (53); Gefüge (66); Gebilde; Baulichkeit, Konstruktion, Gerüst (69); Mast
structure conflict [Leitungsnäherung, bei der ein Umbruch der einen die andere in Mitleidenschaft ziehen kann]
structured strukturiert
structured array strukturierte logische Schaltung
structured programming strukturierte Programmierung
structured statement strukturierte Anweisung (681.3)
structureless strukturfrei, strukturlos
structures Bauten, Baukonstruktionen
strut Strebe, Druckstab, Stütze
strut frame Sprengwerk
to strut verstreben, versteifen, abstützen, stützen
st.-t. a. = steel-tape armoured stahlbandbewehrt
stub Abzweig, Stichleitung (621.3); Stutzen, Stumpf (621); Blindspritze, Blindleitung (621.396)
stub attachment Blindspritzenanschluß (621.396)
stub axle Achsstummel
stub line Leitungsansatz, Leitungsabzweig, Blindspritze (621.396)
stub location Lage der Stichleitung (621.315), Lage der Blindspritze (621.396)
stub shaft Wellenstumpf
stub-supported line Leitung auf Metallisolatoren (621.396)

Stub's wire gauge Stub's Drahtlehre
stuck klebend, hängengeblieben (s. **to stick**)
stuck relay hängengebliebenes Relais, klebendes Relais
stud Zapfen, Lagerzapfen, Stift; Stiftschraube (621.8); Meßzapfen; Amboß (681)
stud bolt Stiftschraube
stud-current transformer Schienenstromwandler, Durchsteckstromwandler
studio Studio, Aufnahmeraum, Senderaum (621.396), (621.397), Atelier (7)
studio mixer Trickmischgerät
stuff Stoff, Material; „Zeug", „Zeugs"
stuffing Füllsel, Füllmasse; Polstermaterial; Füllzeichen (621.376)
stuffing box Stopfbuchse, Stopfbüchse
stuffing-box packing Stopfbüchspackung
stuffing gland Stopfbüchse, Stopfbuchse
stump Stumpf, Stummel
sturdy kräftig, widerstandsfähig
style Art, Form, Bauart, Machart, Stil
stylo Schreibstift
stylos Nadel, Abspielnadel
stylos printer Matrixdrucker
subaqueous Unterwasser-...
subassembly Untergruppe, Baugruppe
subatomic subatomar
subatomic energy Atomkernenergie
subatomic particle Kernbauteil
subaudible frequency Infraschallfrequenz
subaudio frequency Infraschallfrequenz, unhörbar tiefe Frequenz
subaudio frequency telegraph system Unterlagerungstelegraphie
subcadmium neutrons Subkadmiumneutronen
subcarrier Gruppenträgerfrequenz (621.395); Farbträger (621.397); Hilfsträger (621.394)
subcarrier frequency modulation Hilfsträgermodulation
subcarrier modulation Unterlagerungsmodelung
subcentre Verteilungspunkt, Hauptverteilungspunkt
subcircuit Abzweig; Nebenstromkreis (621.3)
subcircular fast kreisförmig
subcontrol system Modulationskontrolle (621.396)
subcritical unterkritisch
subcritical reactor unterkritischer Reaktor
to subdivide into unterteilen in
subdivided capacitor Kondensatorkasten
subdivision Unterteilung, Einteilung; Unterabteilung, Untergruppe
subdued light gedämpftes Licht
subexchange Untervermittlung
subfeeder Speiseleitung, Verteilungsleitung
subharmonic Unterfrequenz, harmonische Unterwelle; ganzzahliger Bruchteil; subharmonisch
subjacent (unmittelbar) darunter liegend
subject Subjekt (41); Gegenstand, Thema
to subject unterwerfen, unterziehen
to subject to heat-treatment einer Wärmebehandlung unterziehen
subjective subjektiv
subjective S/N-ratio visueller Störabstand
subjective tone subjektiv wahrgenommener Ton
sublimate Sublimat
sublimation Sublimierung, Sublimation
submains Verteilersammelschienen
submarine U-Boot
submarine detection U-Boot-Anzeige, U-Boot-Erfassung (623)
submarine Unterwasser-..., See-...
submarine bell Unterwasserschallsender
submarine cable Seekabel, Unterwasserkabel
submarine mine Seemine
submarine power cable Starkstromseekabel
submarine repeater Seekabelverstärker, Unterwasserverstärker
submarine signal Unterwasserschallsignal
submarine telegraph system Seekabeltelegraphie
submarine telephony Seekabeltelefonie
to submerge überschwemmen, überdecken; eintauchen, tauchen
submerged aerial Unterwasserantenne
submerged repeater Unterwasserverstärker
submersibility Tauchfähigkeit
submersible tauchfähig
submersible motor Unterwassermotor, Tauchmotor
submersion Überschwemmung, Tauchvorgang
submodulator Modulationsverstärker
submultiple ganzzahliger Bruchteil, Unterfrequenz
subrack Baugruppenträger
subprogram Unterprogramm
subroutine Unterprogramm (681.3)
subscriber Abonnent (65); Teilnehmer (621.395)
subscriber's apparatus Endstelleneinrichtung
subscriber's extension station Nebenstelle
subscriber's jack Teilnehmerklinke (621.395)
subscriber's line Amtsleitung; Telefonanschluß, Teilnehmerleitung, Anschlußleitung

subscriber's loop Teilnehmerschleife
subscriber's main station Hauptanschluß, Hauptstelle (621.39)
subscriber's multiple Vielfachschaltung
subscriber's number Teilnehmerrufnummer
subscriber ringing signal Teilnehmerruf
subscriber's set Fernsprechapparat
subscriber's rotary line switch Vorwähler
subscriber's stage Teilnehmerwahlstufe
subscriber's station Teilnehmeranschluß
subscriber verification Teilnehmerkennung
subscript unterer Index
subscript number Indexzahl
to subside sich setzen, sich senken, sich absetzen; fallen, abklingen
subsidence Abklingen; Senkung, Setzung
subsidiary Hilfs-..., Unter-..., Tochter-...
subsidiary conduit Anschlußrohrkanal
subsidiary vibration Hilfsschwingung, Nebenresonanz
subsoil Untergrund, Baugrund
subsonic frequency Infraschallfrequenz, Unterschallfrequenz
substance Stoff, Substanz
in substance im ganzen, in der Hauptsache
substandard abgeleitetes Normal, Unternormal
substandard grade Klasse 0,5 (621.317)
substandard meter Vergleichszähler (621.317)
substandard quality minderwertige Qualität
substantial wesentlich, bedeutend, kräftig
substantially independent merklich unabhängig
substantially linear praktisch linear, merklich linear
substantially made stabil gebaut
substation Unterwerk (621.3)
substation switchboard Unterwerks-Schalttafel
substitute Ersatz, Ersatzmittel, Ersatzstoff, Neustoff, Austauschstoff; Vertreter
to substitute austauschen (**for:** gegen); ersetzen (**by:** durch), einsetzen, substituieren (51)
to substitute limits Grenzen einsetzen (51)
substitutional Ersatz-..., Substitutions-...
substitutional resistance Ersatzwiderstand
substitutional method Substitutionsmethode, Substitutionsverfahren
substrate Trägerplättchen, Substrat, Träger, Trägermaterial
substruction Unterbau (69), (625)
subsurface work unterirdisches Werk
subsynchronous untersynchron
subsynchronous reluctance motor Reluktanz-Langsamläufer
to subtend gegenüberliegen zu, überspannen (51)
subterranean unterirdisch
subterranean water Grundwasser
subtlety Subtilität
subtotal Zwischensumme
to subtract abziehen, subtrahieren (**from:** von)
subtraction Subtraktion (51)
subtractive polarity gegenpolige Schaltung, entgegengesetzte Polung
subtrahend Subtrahend
subtransient reactance (of generator) Generatorstoßstreu-Reaktanz
suburban Vorort-...
suburban line Vorortstrecke (625)
subway Untergrundbahn (US); Unterführung
subway-type transformer unterirdischer Transformator
subwoofer Tiefbaß-Lautsprecher, Baßlautsprecher
sub-zero unter Null
sub-zero temperature Temperatur unter Null, Frost
succession Folge
in succession hintereinander, aufeinanderfolgend
succession of phases Phasenfolge
successive folgend, aufeinanderfolgend, aneinander anschließend, nacheinander eintretend
successive approximation measuring method Annäherungsmeßverfahren
successive pulses aufeinanderfolgende Impulse (621.374)
to suck saugen, ansaugen
sucking inductor Saugdrossel
suction Ansaugung, Absaugung; Saug-...
sudden short-wave fade out Mögel-Dellinger-Effekt
suds (pl.) Schaum; Löschschaum
suffix Suffix, Anhängsel; unterer Index (51); Anhängesilbe, Nachsilbe; nachgesetzte Kennung (621.395)
to suffix unten anhängen, als Index zufügen (51)
sulphate Sulfat
sulphate paper Natronpapier
to sulphate sulfatieren (621.356)
sulphating Sulfatierung (621.356)
sulphite Sulfit
sulphur Schwefel
sulphuretted hydrogen Schwefelwasserstoff
sulphuric acid Schwefelsäure
sulphurous acid schweflige Säure
sum Summe; Summe, Geldsumme; Rechenaufgabe
sum check Summenprobe
sum of digits Quersumme

sum frequency Summenfrequenz
sum total Gesamtsumme
to sum summieren, addieren, zusammenzählen
to sum up to sich addieren zu, zusammen ausmachen, zusammen kosten
summable addierbar
summary Zusammenfassung, Kurzdarstellung
summation Summation, Addierung, Addition
summation instrument summierendes Meßgerät
summation term Summenausdruck (51)
summation transformer Summenstromwandler
summator Summiereinrichtung
summatory attachment Summenzählerzusatz
summatory meter Summenzähler
summing amplifier Summierverstärker
summing network Summierglied
sump Sumpf; Ablaufgrube; Ölwanne (629.113), (621.9); Ölsumpf (621.3)
sun Sonne
against the sun gegen den Uhrzeigersinn
sun engine Sonnenkraftmaschine
sunlight lamp Tageslichtlampe
sunlight power Sonnenkraft, Sonnenenergie
with the sun mit dem Uhrzeigersinn
sunk key Nutenkeil, Wellenkeil, Feder
sunk switch versenkter Schalter, Unterputzschalter (696)
super Flächen-...
super ft. Quadratfuß
super über-..., höchst-...
superaudio frequency Ultraschallfrequenz
superaudio frequency modulated waves ultraschallmodulierte Wellen
superaudio telegraph system Überlagerungstelegraphie
superconductive supraleitend
superconductivity Supraleitfähigkeit
superconductor Supraleiter
supercontrol tube Exponentialröhre (621.385)
to supercool unterkühlen
supercritical überkritisch
supercritical reactor überkritischer Reaktor
superexcitation Stoßerregung
superexcited condenser Stoßerregungs-Phasenschieber
superfatted überfettet
superficial oberflächlich
superficial layer Deckschicht, Oberflächenschicht
superficial measure Flächenmaß
superficies Oberfläche (flächenmäßig)
supergroup Sekundärgruppe (621.395.4)
supergroup allocation Übertragungslage der Sekundärgruppen (621.395.4)
supergroup modulator Sekundärgruppenumsetzer
superheating Überhitzung
superhet Super, Superhet, Überlagerungsempfänger
superheterodyne Superhet, Überlagerungs-...
superheterodyne receiver Überlagerungsempfänger, Superhet-Empfänger, Superhet, Super
superheterodyne reception Überlagerungsempfang
to superheterodyne sich überlagern (621.396)
superhigh besonders hoch
superhigh frequency range Zentimeterwellenbereich (621.396)
supericonoscope Superikonoskop
to superimpose aufeinanderlagern, überlagern
to superimpose on überlagern über
superimposed überlagert, Zusatz-...
superimposed circuit Phantomkreis
superimposed current überlagerter Strom
superimposed field überlagertes Feld
superimposed load überlagerte Belastung, Zusatzlast (624)
superimposed telegraphy Überlagerungstelegraphie
to superintend überwachen, Aufsicht führen über
superintendent Aufsichtsbeamter, Überwachungsbeamter; Wärter
superior überlegen, den Durchschnitt überschreitend, erstklassig
superior in überlegen hinsichtlich
superior limit obere Grenze
superior material besseres Material, hochwertiges Material
superior quality bessere Qualität, bessere Sorte
superior to größer als, besser als
superior workmanship erstklassige Wertarbeit
superload Überlast, Zusatzlast
to superload überlasten, überbelasten
supermaster group Quartärgruppe (621.395.4)
supernatant überschüssig, oben stehend, obenschwimmend (54)
superphantom circuit Achterkreis
to superpose überlagern; übertragen, abbilden (51); übereinander anordnen
superposed circuit Phantomkreis, Kunstschaltung
superposed current überlagerter Strom
superposed ringing überlagerter Gleichstromruf

superposed ringing current Rufstromüberlagerung, überlagerter Rufgleichstrom
superposed telegraphy Überlagerungstelegraphie
superposition Überlagerung, Superposition
superpower valve Leistungsröhre, Kraftverstärkerröhre
superpressure Höchstdruck
superpressure plant Höchstdruckanlage
superrefraction Überstandard-Brechung
superregenerative receiver Pendelrückkopplungsempfänger
supersaturation Übersättigung
superscript oberer Index (51)
supersonic Überschall-, Ultraschall-...
supersonic delay line Ultraschall-Laufzeitkette
supersonic frequency Überschallfrequenz, Ultraschallfrequenz
supersonic gauge Ultraschall-Meßlehre, Ultraschall-Wandstärkenmesser
supersonic sounding Ultraschallot
supersynchronous übersynchron
supertension cable Höchstspannungskabel
supertonic (obs.) = **supersonic**
to supervise überwachen
supervisor Überwachungsbeamter; Aufsicht (621.395)
supervisory Überwachungs-...
supervisory board Aufsichtsbehörde
supervisory channel Überwachungskanal
supervisory control halbautomatische Steuerung
supervisory lamp Überwachungslampe; Schlußlampe (625)
supervisory relay Überwachungsrelais
supervisory remote control and telemetering system Fernwirksystem
supervisory signal Besetztklappe (621.395)
supervisory system Überwachungssystem
supervisory work Überwachungsarbeit, Aufsichtstätigkeit
supple nachgiebig, geschmeidig
supplement Supplement (51)
supplementary Ergänzungs-..., Zusatz-..., Nachtrags-...
supplementary anode Hilfsanode
supplementary data ergänzende Unterlagen, Zusatzunterlagen
supply Versorgung, Lieferung, Versorgungsnetz, Netz
supply authority Versorgungsunternehmen
supply current Netzstrom
supply of energy Energiezufuhr, Energieversorgung
supply of heat Wärmeversorgung, Wärmezufuhr
supply line Zuleitung, Versorgungsleitung; Zufuhrstrecke
supply line wire Zuleitungsdraht
supply mains Versorgungsnetz, Netzanschlußleitung, Netz
supply mains to a house Hausanschlußleitung, Hausanschluß
supply meter Netzzähler
supply of power Leistungszufuhr, Energiezufuhr, Stromversorgung
supply station Kraftwerk
supply voltage Versorgungsspannung, Speisespannung, Netzspannung
supply-voltage rejection ratio Empfindlichkeit gegen Versorgungsspannungsänderungen
to supply liefern, beliefern, versorgen, zuführen
to supply current Strom liefern
to supply energy Energie zuführen
support Unterstützung, Halterung; Auflager, Unterstützungspunkt, Einspannung; Aushalten, Vertragen, Ertragen; Stütze, Mast
support arm Tragarm
support box Befestigungsdose (621.315.6)
support point Unterstützungspunkt, Auflagerpunkt
to support unterstützen, tragen, haltern; erhalten, aufrechterhalten
to support combustion die Verbrennung unterhalten
supported on a knife edge schneidengelagert (681)
supporting Trag-..., Stütz-..., Unterstützungs-...
supporting beam Tragbalken, Unterzug
supporting cable Tragseil, Tragkabel
supporting frame Gestell, Traggerüst
supporting insulator Stützisolator, Stützer
supporting strang Tragseil (621.39)
supporting structure Tragkonstruktion (621.33)
supporting tower Abspannturm, Abspannmast, Tragturm, Brückenturm (624)
supporting wire Tragdraht, Stützdraht, Halterung
supports (pl.) Tragkonstruktion
to suppress unterdrücken, weglassen
to suppress the carrier den Träger unterdrücken (621.396)
to suppress digits Stellen weglassen, runden (51)

to suppress oscillations Schwingungen unterdrücken
suppressed automatic volume control verzögerte Schwundregelung
suppressed-carrier transmission Nachrichtenübertragung mit unterdrücktem Träger
suppressed scale unterdrückte Skala, Skala mit unterdrücktem Nullpunkt (621.317)
suppressed-zero instrument Instrument mit unterdrücktem Nullpunkt
suppression Unterdrückung, Verminderung
suppression of carrier Trägerunterdrückung
suppression choke Sperrdrossel
suppression of cross modulation Unterdrückung der Kreuzmodulationsprodukte
suppression of harmonics Oberwellenverminderung
suppression of the sky-wave Unterdrückung der Raumwelle
suppression of statics Funkentstörung
suppressor Begrenzerschaltung; Drosselschaltung; Bremsgitter (621.385)
suppressor circuit Begrenzerschaltung (621.396.62)
suppressor grid Bremsgitter (621.385)
suppressor-grid modulation Bremsgittermodulation
suppressor modulation Bremsgittermodulation
suppressor resistance Entstörungswiderstand (621.396), (629.113)
supra-acoustic frequency Ultraschallfrequenz
supra-acoustic telegraph systems (pl.) Überlagerungstelegraphie
supraconductivity Supra-Leitfähigkeit
supraconductor Supraleiter
supradyne reception Überlagerungsempfang
to surcharge überladen, überlasten
surd irrationale Zahl
surd irrational
surd number irrationale Zahl
surface Oberfläche, Fläche; Fläche, Tragfläche, Flügel (629.13)
surface abrasion Abrieb
surface-to-air missile Boden-Luft-Flugkörper
surface area Oberfläche, Flächeninhalt der Oberfläche
surface-barrier transistor Oberflächensperrschichttransistor
surface brightness Oberflächenhelligkeit, Leuchtdichte, Flächenhelligkeit
surface charge Oberflächenladung
surface conductivity Oberflächenleitfähigkeit
surface of constant phase Phasenfläche (535), (621.396)
surface of contact Berührungsfläche
surface cooling Oberflächenkühlung
surface cracking Oberflächenrisse (pl.)
surface defect Außenlunker (66); Oberflächenfehler
surface of discontinuity Unstetigkeitsfläche
surface duct Oberflächenleitschicht, Oberflächendukt (621.396)
surface earthing Oberflächenerdung, Banderdung
surface effect Oberflächenwirkung, Oberflächeneffekt
surface feet per minute Umfangsgeschwindigkeit
surface finish Oberflächenbeschaffenheit, Oberflächengüte
surface force Oberflächenkraft
surface friction Oberflächenreibung
surface gauge Parallelreißer, Höhenreißer
surface grinding Planschleifen
surface ground Oberflächenerder, Banderder
surface hardened oberflächengehärtet
surface hardening Oberflächenhärtung
surface installations Anlagen über Tage (622)
surface of integration Integrationsfläche
surface leakage (Oberflächen-)Kriechstrom (621.3)
surface metal raceway blechabgedeckter Kabelkanal; Bergmannrohr, J-Rohr (696)
surface mounted device oberflächenmontiertes Bauelement
surface mounting Aufputzinstallation
surface plane Oberflächenebene
surface plant Übertagebetrieb (622), oberirdische Anlage
surface protection Oberflächenschutz
surface quality Oberflächengüte, Bearbeitungsgüte
surface recombination Oberflächenrekombination
surface resistance Oberflächenwiderstand
surface resistivity Oberflächenwiderstand
surface shear stress Randschubspannung
surface skin Oberflächenschicht
surface soil Mutterboden
surface speed Umfangsgeschwindigkeit
surface-surface missile Boden-Boden-Flugkörper
surface switch Wandschalter (696.6)
surface tension Oberflächenspannung
surface treatment Oberflächenbehandlung
surface voltage gradient Schrittspannung

surface wave Bodenwelle, Oberflächenwelle
surface-wave guide Oberflächenwellenleitung
surface-wave transmission line Drahtwellenleitung
surface wiring Verlegung über Putz, Leitungsverlegung auf Putz (621.3)
to surface Straßendecke herstellen (625); plandrehen (621.9); an die Oberfläche bringen; auftauchen (629.12)
surfaced submarine aufgetauchtes U-Boot
surge Welle, Wanderwelle, Stoß, Stromstoß, Überspannungswelle
surge-absorbing capacitors Trägerstaueffekt-Beschaltung (621.314.6)
surge arrester Überspannungsableiter
surge-crest ammeter Stahlstäbchen-Amperemeter
surge current Wellenstrom, Stoßstrom
surge of current Stromstoß, Stromwelle
surge dissipator Überspannungsableiter
surge diverter Überspannungsableiter
surge-electrode current Überlaststrom bei Rückzündung oder Kurzschluß (621.382)
surge frequency Stoßfrequenz (621.3)
surge front Wellenfront, Wellenstirn (621.3)
surge generator Stoßgenerator
surge impedance Wellenwiderstand (621.3)
surge load Stoßbelastung (621.3)
surge-power generator Stoßleistungsgenerator
surge-pressure test Sprungwellenprobe
surge protection Wanderwellenschutz, Überspannungsschutz, Stoßspannungsschutz
surge-protection capacitor Überspannungsschutzkondensator
surge recorder Klydonograph, Überspannungszählwerk
surge relay Stromanstiegrelais
surge reverse power dissipation Stoß-Rückwärtsverlustleistung (621.382.2)
surge voltage Stoßspannung, Wellenspannung
surge-voltage generator Stoßspannungsanlage, Stoßspannungsgenerator
surge-voltage protector Überspannungsbegrenzer
surge-voltage stress Stoßspannungsbeanspruchung
surgical chirurgisch
surgical diathermy Hochfrequenzchirurgie
surgical instruments ärztliche Instrumente, chirurgische Instrumente
surgical lamp Arztlampe
surplus Mehraufwand; Überschuß, Restbestand
surplus air Luftüberschuß
surplus length Übermaß, Überlänge
surplus load Mehrbelastung
surrogate Surrogat, Ersatzgegenstand, Ersatzstoff
surrounding Umgebungs-...
surrounding noise Umgebungslärm, Raumgeräusch
surveillance radar Rundsicht-Radar, Überwachungsradar
survivale craft station Rettungsgerätfunkstelle
susceptance Blindleitwert
susceptibility Suszeptibilität, Aufnahmevermögen
susceptible leicht geneigt (**to:** zu); anfällig (**to:** gegen)
susceptible to trouble störanfällig
to suspend aufhängen; suspendieren; sperren (621.395)
suspended schwebend
suspended ash Flugasche
suspended coil type galvanometer Drehspulgalvanometer mit Fadenaufhängung
suspended insulator Hängeisolator
suspended matter Schwebeteilchen (pl.), Plankton
suspended pump Bohrlochpumpe
suspended railway Schwebebahn
suspension Suspension, Aufschwemmung (54); Aufhängung, Aufhängedraht, Gehänge; Hänge-..., federnde Aufhängung, Federung (629.113)
suspension attachment Aufhängung, Aufhängevorrichtung
suspension band Aufhängeband
suspension bracket Ausleger, Traverse (621.315)
suspension bridge Hängebrücke
suspension chain Tragkette (621.315.6)
suspension clamp Hängeklemme (621.315.1)
suspension distance Deckenabstand (621.32)
suspension eye Aufhängeöse
suspension insulator Hängekette, Hängeisolator
suspension insulator weight Tragkettenzusatzgewicht
suspension mounting Aufhängung
suspension push Hängeschalter, Schaltbirne
suspension rail Hängeschiene
suspension railway Schwebebahn
suspension rod Aufhängestange
suspension spring Aufhängungsfeder, Aufhängefeder
suspension strand Tragseil (621.39)

suspension switch Hängeschalter, Schnurschalter
suspension tower Tragmast
suspension type Hänge-...
suspension-type insulator Hängeisolator, Hängekette
suspension wire Aufhängekabel, Tragkabel
to sustain unterhalten, aufrechterhalten, stützen
sustained oscillation ungedämpfte Schwingung, erzwungene Schwingung
sustained short-circuit Dauerkurzschluß
sustained short-circuit current Dauerkurzschlußstrom
sustained vibration erzwungene Schwingung
to swab aufwischen, aufputzen
to swage hämmern, verdichten; quetschen, gesenkschmieden
swage block Schmiedegesenk, Gesenk; Lochplatte (621.7)
swage forging Gesenkschmiedestück
swaged wire gehämmerter Draht, Coolidge-Draht
swamp ore Raseneisenerz
to swamp verdecken, überdecken (621.396)
swamping resistance Spulenvorwiderstand (621.317)
swan-necked insulator Isolator mit Schwanenhalsstütze
Swan socket Swansockel, Bajonettsockel
to swap austauschen (681.3)
swarf Späne (pl.)
to sway sich hin- und herbewegen, hin- und herbewegt werden; schwojen (629.12)
to sweat schwitzen; löten, anschmelzen
to sweat on anlöten, anschmelzen
sweating thimble Kabelschuh, Lötschuh
sweep Zeitablenkung, Ablenkung, Zeitachse, Wobbelhub, Hub (621.385); Bewegung; Krümmung, Strich
sweep amplifier Ablenkverstärker
sweep circuit Ablenkschaltung, Zeitkreis
sweep displacement Zeitbasisverschiebung
sweep frequency Wobbelfrequenz
sweep generator Sägezahngenerator, (Zeit-)Ablenkgenerator; Wobbelsender
sweep repetition rate Kippfrequenz, Ablenkfrequenz (621.397)
sweep voltage Ablenkspannung, Zeitspannung
sweep width Wobbelhub
to sweep streichen, wischen; fahren, gleiten (629.11)
to sweep around the periphery den Umfang überstreichen, umfahren
to sweep the horizon den Horizont absuchen, den Horizont überblicken
to sweep out an area ein Gebiet absuchen (621.397)
to sweep over a range einen Bereich überstreichen, einen Bereich überdecken, einen Bereich durchwandern
swell Quellung, Schwellung, Wulst; Dünung, Seegang; Rollschweller (681)
swelling Wulst, Durchmesservergrößerung; Schwellung, Beule (61)
S. W. G. = British Standard Wire Gauge Britische Normal-Drahtlehre
swift schnell
swift electron schnelles Elektron
swimming-pool reactor Bassin-Reaktor, Schwimmbadreaktor
swing Schwingung; Ausschwingpendel; Schwankung, Ausschlag, Hub; (621.396)
swing on the dial Ausschlag auf der Skala; Verstellung der Skala
swing-out panel ausschwenkbares Schaltfeld
to swing schwingen (sich) schaukeln; auslenken; schwenken
to swing along entlang schwingen an
to swing aside ausschwingen, beiseite drehen
to swing back zurückschwingen
to swing back and forth hin- und herschwingen
to swing clear of von ... wegschwenken; frei an ... vorbeischwingen
to swing out ausschwingen
to swing out of mesh außer Eingriff bringen, auskuppeln (621.8)
to swing outward vorklappen, nach außenklappen
to swing past vorbeischwingen, vorbeischwenken
to swing through an angle um einen Winkel drehen, um einen Winkel schwenken
to swing up herausklappen
swinging bracket Schwenkarm
swinging choke Sättigungssiebdrossel
swinging magnet type galvanometer Nadelgalvanometer
S-wire Meßdraht (621.315.2)
swirl Wirbel
to swirl wirbeln, schwenken
to swirl around herumwirbeln
switch Schalter, Taster, Wähler (621.3); Weiche (625)
switch base Schaltergrundplatte
switchboard Vermittlungsschrank; Schalttafel
switchboard cell Schaltkasten

switchboard connection Schalttafelanschluß
switchboard cord Vermittlungsschnur, Schnurpaar
switchboard cubicle Schaltzelle
switchboard drop Schaltkreise einer Handvermittlung
switchboard exchange Schrankvermittlung
switchboard frame Schalttafelgerüst, Schalttafelrahmen
switchboard instrument Schalttafelinstrument
switchboard lamp Schranklampe (621.395)
switchboard meter Schalttafelzähler
switchboard mounting Schalttafelaufbau, Montage auf der Schalttafel
switchboard panel Schalttafelfeld, Schalttafelwand
switchboard section Vermittlungsschrank (621.395); Schalttafelfeld, Schaltfeld (621.316)
switchboard skeleton Schalttafelgerüst
switchboard-type Schalttafelbauart, Schalttafel-...
switchboard-type instrument Schalttafelinstrument
switch box Schaltkasten (621.395); Stellwerk (625)
switch cabin Bahnunterstation, Unterwerk, Schaltstelle (621.332)
switch clicks Schaltknacken
switch connection Schalteranschluß
switch contact Schaltkontakt
switch cover Schalterkappe, Schalterdeckel, Schalterhaube
switch cradle Gabel, Apparategabel, Gabelschalter (621.395)
switch deck Schalterebene
switch dissipation Schaltverlust
switch enclosure Schaltkasten
switchgear Schaltgeräte (pl.); Schaltanlage
switchgear cubicle Schalterzelle
switchgear pedestal Schaltergestell, Schaltgerüst
switchgear pillar Schaltsäule (621.316)
switchgear truck Schaltwagen
switch fuse Einsatzsicherung
switch hook Gabelumschalter, Schaltstange, Schalthaken
switch house Schalthaus
switch-interlocking device Schaltsperre
switch knife Schaltmesser
switch knob Schalterknopf, Schaltknebel
switch lampholder Hahnfassung (621.32)
switch lever Schalthebel
switch machine Weichenstellvorrichtung
switch-machine lever Weichenstellwerk
switch-machine point detector Weichenstellungs-Kontrollschalter
switch-machine pole changer Umkehrschalter (625)
switch pedal Schaltpedal, Fußschalter
switch pillar Schaltsäule (621.316)
switch plate Schalterabdeckblech
switch plug Schaltstecker
switch point Schalterstelle, Schaltereinbaupunkt
switch position Schalterstellung
switch-position indicator Schaltstellungsanzeiger
switch room Wählersaal (621.395)
switch setting Schalterstellung
switch signal Weichenstellungssignal
switch socket Schaltfassung
switch spring (obs.) Klinke (621.395)
switch starter Anlaßschalter
switch station Schaltwarte
switch tank Schalterkessel
switchyard Freiluftschaltanlage (621.3); Rangierbahnhof (625)
switch wafer Schaltscheibe (621.396)
to switch schalten (621.3); rangieren (625)
to switch back zurückschalten (**to:** auf)
to switch back and forth hin- und herschwenken (621.396.67)
to switch in einschalten
to switch in parallel with parallel zu... schalten, parallel schalten zu
to switch off ausschalten, abschalten
to switch on einschalten, zuschalten
to switch on to anschalten an, anlegen an
to switch out ausschalten
switched beam direction finder Umtastpeiler, AN-Peiler, Leitstrahlpeiler
switched-mode power supply Schaltnetzteil
switched power Schaltleistung
switched public telephone network öffentliches Fernsprechnetz
switching Schaltung, Schalten, Schaltvorgang (621.3); Rangieren (625)
switching angle Schaltphase
switching algebra Schaltalgebra
switching apparatus Schaltapparat
switching arrangement Schaltanordnung, Schaltvorrichtung
switching capacity Schaltleistung
switching center Schalt- oder Vermittlungsstelle
switching command Schaltbefehl

switching control pilot Schaltpilot (621.395.4)
switching device Schaltvorrichtung, Schalteinrichtung, Schaltgerät
switching diode Schaltdiode
switching element Koppelelement (621.395); Schaltelement (621.316.5)
switching engine Rangierlokomotive
switching logic Schaltalgebra
switching matrix Koppelvielfach (621.395)
switching mechanism Schaltwerk
switching network Koppelnetz (621.395)
switching point Schaltzelle
switching power loss Schaltverlustleistung
switching quality Vermittlungsgüte
switching row Koppelreihe (621.395)
switching stage Koppelstufe (621.395)
switching station Schaltstation
switching structure Schaltgerüst
switching surge Schaltüberspannung
switching time Schaltzeit
switching torque Umschaltmoment (621.313)
switching transient Schaltüberspannung
switching tube Schaltröhre
swivel Drehlager; Schwenkvorrichtung
swivel arm Lenkachshebel (629.113)
swivel slide Winkelsupport (621.9)
swivel table drehbarer Tisch, Drehtisch; schwenkbarer Tisch, Schwenktisch
swivel truck Drehgestell
swivel vice schwenkbarer Schraubstock
to swivel schwenken, herumdrehen
swivelling trolley head Stromabnehmerdrehstuhl
syenite Syenit
syllable articulation Silbenverständlichkeit
sylphon bellows (pl.) Wellrohr
symbol Symbol; Schaltzeichen (621.3); Buchstabe (51); Sinnbild
in symbols symbolisch, in Buchstaben
symbolic symbolisch
symbolic language symbolische Sprache, Symbolsprache
symbolic notation symbolische Schreibweise (621.3)
symbolic representation symbolische Darstellung
symbolism Symbolismus
symmetrical symmetrisch, mit gleichlaufenden Halbwellen (621.3)
symmetrical alternating current Wechselstrom mit gleichverlaufenden Halbwellen
symmetrical components symmetrische Komponenten (538)
symmetrical heterostatic circuit Quadrantenschaltung; Nadelschaltung
symmetrical light distribution symmetrische Lichtverteilung
symmetrical load symmetrische Last, symmetrische Belastung
symmetrical polyphase set of potential differences symmetrisches Mehrphasenspannungssystem
symmetrical short-circuit current symmetrischer Kurzschlußstrom
symmetrical three-phase system symmetrisches Drehstromsystem
symmetrical transducer symmetrischer Vierpol
symmetrical voltage symmetrisches Drehstromsystem
symmetrically cyclicly magnetized condition eingeschwungener Zustand der symmetrischen Ummagnetisierung
to symmetrize symmetrieren
symmetry Symmetrie
sympathetic sympathisch; mitlaufend, mitbewegt, nachlaufend
symmetrization Symmetrierung
sympathy Sympathie
in sympathy with gleichlaufend mit, in gleichem Maße wie
to move in sympathy with ... nachlaufen, sich mit ... mitbewegen
symptom Symptom, Anzeichen, Kennzeichen (**of:** von, für)
Synaut motor synchronisierter Asynchronmotor
sync = synchron
sync character Synchronisationszeichen
sync pulse Synchronsignal, Synchronimpuls (621.397)
sync regeneration Impulserneuerung
synchro Drehfeldsystem, Drehmelder
synchro control differential transmitter Steuerdifferentialgeber
synchro control receiver Steuerempfänger, Drehmelder
synchro control transformer Steuerempfänger, Drehmelder
synchro mesh Synchronschaltung (629.113)
synchro resolver Koordinatenwandler
synchro system Drehfeldsystem
synchro torque receiver Momentempfänger
synchro torque transmitter Momentgeber
synchronism Synchronismus, Gleichlauf
synchronism indicator Synchronoskop
synchronization Synchronisierung, Parallelschaltung
synchronization-clock system Impulsgeber

synchronization control Synchronsteuerung
synchronization of oscillators Mitziehen von Sendern (621.396)
synchronization-signal generator Impulsgeber
synchronization-signal-generator timing unit Taktgeber
to synchronize synchronisieren; parallelschalten (621.311)
synchronized induction motor synchronisierter Asynchronmotor
synchronized rotation Synchronlauf
synchronizer Synchronisierwandler, Synchronisator
synchronizing Synchronisier-...
synchronizing bus bar Synchronisierungs-Schiene
synchronizing circuit Synchronisierschaltung, Synchronisierleitung
synchronizing current synchronisierender Strom
synchronizing dark method Dunkelschaltung (621.311)
synchronizing equipment Synchronisiereinrichtung
synchronizing force synchronisierende Kraft
synchronizing of images Bildsynchronisierung (621.397)
synchronizing lamp Synchronisierlampe
synchronizing moment synchronisierendes Moment
synchronizing pilot Gleichlaufpilot (621.395.4)
synchronizing pulse Synchronisierimpuls
synchronizing pulse distribution amplifier Impulsverteilerschaltung
synchronizing reactor Synchronisierungsdrossel
synchronizing relay Synchronisierungsrelais
synchronizing signal Synchronsignal (621.397)
synchronizing-signal level Synchronwert
synchronizing torque synchronisierendes Moment
synchronizing transformer Synchronisierwandler
synchronizing voltmeter Synchronisiervoltmeter
synchronoscope Synchronoskop
synchronous synchron
synchronous alternator Synchrongenerator
synchronous booster converter Einankerumformer-Zusatzmaschine
synchronous capacitor synchroner Phasenschieber
synchronous clock Synchronuhr
synchronous compensator Blindleistungsmaschine
synchronous converter Einankerumformer
synchronous generator Synchrongenerator
synchronous impedance synchrone Impedanz (Verhältnis des Erregerstromes für Nennstrom als Dauerkurzschlußstrom zum fiktiven Erregerstrom bei Nennspannung auf der verlängerten Anfangsleerlaufkennlinie)
synchronous induction motor synchronisierter Asynchronmotor
synchronous induction motor generator Synchron-Asynchron-Umformer
synchronous machine Synchronmaschine
synchronous motor Synchronmotor
synchronous modifier synchroner Phasenschieber
synchronous operation Synchronbetrieb
synchronous reactance synchrone Reaktanz (621.313.3)
synchronous reactor Blindleistungsmaschine
synchronous spark gap Synchronfunkenstrecke
synchronous speed synchrone Drehzahl
synchronous telegraph systems Synchrontelegraphie-Systeme
synchronous tie elektrische Welle (621.87)
synchronous time motor Synchronuhr-Motor
synchronous torque synchrones Moment
synchronous-type vibrator Zerhacker mit Wiedergleichrichtung
synchron separator Amplitudensieb (621.397)
synduct motor Motor mit senkrechter Welle
synopsis Zusammenstellung, Übersicht; Synopse (55)
synthesis (pl. **-eses**) Synthese
to synthesize synthetisieren (54); zusammensetzen (621.397)
synthesizer Frequenzgenerator (mit Frequenzsynthese); Synthesizer
synthetic synthetisch, künstlich
synthetic insulating material synthetisches Isoliermaterial, Isolierkunststoff
synthetic material Kunststoff
synthetic plastics Kunstpreßstoffe
synthetic process Synthese-Verfahren, Synthese
synthetic resin Kunstharz
synthetic resin glue Kunstharzleim
synthetic resin paper board Hartpapierplatte
synthetic resin paper rod Hartpapierstange
synthetic resin varnish Kunstharzlack
synthetic rubber Kunstgummi, synthetischer Gummi
synthetical synthetisch

syntonization (obs.) Abstimmung
syntony (obs.) Abstimmzustand, Resonanz
system System, Anordnung; Netz (621.3)
system capacity Netzleistung; Leistung des Systems
system of co-ordinates Koordinatensystem
system-dispatching center Netzleitstelle
system earth Betriebserde
system-earthing conductor Erdsammelleitung
system efficiency Netzwirkungsgrad (621.315); Systemwirkungsgrad
system engineering Anlagentechnik, Systementwicklung
system fault Netzstörung
system ground Betriebserde
system interconnection Netzkupplung, Netz-Parallelbetrieb
system linking Netzkupplung (621.3)
system loss Systemdämpfung
system of plates Plattensatz (621.356), (621.319.4)
system-protective equipment Netzschutz
system of pulleys Riementrieb
system of reference Bezugssystem (51), (53)
system reserve Netzreserve (621.311)
system of units Maßsystem (53)
systematic systematisch
systematic error systematischer Fehler, systembedingter Fehler
systematization Systematisierung
to systematize systematisieren
systoflex Rüschschlauch
systoflex insulating tube Excelsior-Isolierrohr

T

tab Klappe, Brettchen, Tafel; Kennschild, Bezeichnungsschild, Beschriftungsschild; Drucktastenauslöser; Landeklappe (629.13)
table Tafel, Tabelle; Tisch, Tafel
table of correction Eichtabelle
table lamp Tischlampe
table of logarithms Logarithmentafel
table lookup capability [Tabellen-Durchlauf-Möglichkeit]
table of measures Meßtafel
table standard Tischlampe
table of truth Wahrheitstabelle
table vice Maschinenschraubstock
table of weights Gewichtstafel, Gewichtstabelle
table work Tabellenwerk
tablet Täfelchen, Gedenktafel; Tablette (61)
tabloid Tablette
tabular tabellarisch
tabular difference Tabellendifferenz, Tafeldifferenz
tabular value Tabellenwert, Tafelwert
to tabulate tabulieren, tabellieren
tabulation Tabulierung, Tabellierung
tabulator Tabelliermaschine
TACAN = tactical air navigation system Tacan
tachogenerator Tachodynamo
tachograph Tachograph
tachometer Tachometer, Geschwindigkeitsmesser
tachometric tachometrisch
tachometric course Kollisionskurs (623)
tachometric device Geschwindigkeitsmeßeinrichtung
tacho-pulse generator Tachosignalgeber
tack Zwecke, (kleiner) Nagel; Heftzwecke; Tacks, Tecks (68); Segelschlag, Kreuzschlag (629.12)
tack rivet Heftniet
tack weld Heftschweißung
to tack heften, anheften; heftschweißen; kreuzen (629.12)
to tack weld heftschweißen, heften
tackle Geschirr; Segelgeschirr (629.12); Hebezeug, Flaschenzug, Seilgeschirr (621.8)
to tackle anpacken, in die Hand nehmen, in Angriff nehmen, sich abgeben mit; takeln (629.12)
tacky zähklebrig
T-aerial T-Antenne
tag Anschlußfahne (621.313); Anhänger, Etikett, Aufhänger, Anhängeschild; Marke, Ziel (681.3)
tag block Lötösenstreifen
tag-end connector plug Tannenbaumverteiler
tail Schwanz, Schwanzende, Hinterteil
tail eating Aufspulen einer Verbindung (621.395)
tail lamp Rücklicht (629.113)
tail light Rücklicht (629.113); Schlußlicht (625)
tail race Unterwasserkanal, Unterwassergraben
tail of surge wave Rücken der Überspannungswelle
tailstock Reitstock (621.9)
tail-time constant Rückenzeitkonstante
to tail off schwanzförmig auslaufen, allmählich abklingen (621.385)

take-off Start, Startvorgang; Startpunkt, Absprung
take-off power Startleistung (629.13)
to take nehmen, annehmen, aufnehmen, fortnehmen; bringen
to take along mitnehmen (621.8)
to take a bearing, to take bearings on ... anpeilen
to take the constant of a galvanometer die Konstante eines Galvanometers bestimmen
to take current Strom aufnehmen
to take a curve eine Kurve aufnehmen, eine Kurve ermitteln
to take fire Feuer fangen
to take the gauge of abmessen, Maß nehmen an; schätzen
to take an inclined position sich schief stellen
to take off the load die Last abschalten
to take measure Maß nehmen
to take moments Momente berechnen, Momente ansetzen
to take observations Beobachtungen anstellen, Ablesungen machen
to take out of herausnehmen aus, herausholen aus
to take out of a face aus einer Fläche austreten lassen
to take out a patent ein Patent nehmen
to take out through herausführen durch, ausführen durch (621.385)
to take p at a time p gleichzeitig nehmen, Gruppen von p Stück bilden (51)
taken p at a time zu p, zu p Stück (51), in Gruppen zu p
to take to pieces auseinandernehmen, in Stücke zerlegen
to take a pressure einen Druck aufnehmen
to take readings Ablesungen machen
to take in relation in Beziehung setzen
to take samples Proben entnehmen
to take in service in Betrieb nehmen
to take the swings Spitzen abfahren, Frequenz fahren (621.311)
to take two paths sich verzweigen (621.3)
to take up aufnehmen; demontieren
to take up a battery eine Batterie auseinandernehmen
to take up an engine einen Motor demontieren, einen Motor aufnehmen
to take up moisture Feuchtigkeitaufnehmen
to take up wear Abnutzung ausgleichen
to take watts Leistung aufnehmen
talc Talk
talk Gespräch, Unterhaltung
talk on the telephone Telefongespräch, telefonische Unterhaltung
to talk sprechen, sich unterhalten
to talk shop fachsimpeln
talker echo Gegenecho, gegenläufiges Echo (621.395)
talkie Tonfilm
talking Ton-..., Sprech-..., Sprach-...
talking battery Sprechbatterie (621.395)
talking film Tonfilm, Sprechfilm
talking key Vermittlungstaste, Abfragetaste
talking picture Tonfilm, Sprechfilm
talking test Sprechprobe, Verständigungsprobe, Leitungsprobe (621.395)
talking tone Sprachfrequenz
tall hoch, groß
tall pylon hoher Mast
tall tower hoher Turm, hoher Mast
to tally zueinanderpassen, zusammenpassen
tallygraph schreibendes Fernmeßinstrument
tallymeter Fernmeßinstrument
to tamp tupfen; zusammendrücken, stampfen; verdämmen (66); stopfen, zustopfen (66)
tamper proof geschützt gegen unbefugten Eingriff
tamping Verdämmungsmaterial (62); Lehmpfropf (66)
tan = tangent tan = tangens (51)
$\mathbf{tan^{-1}}$ arctg
tandem Tandem
tandem central office Knotenamt
tandem connection Kaskadenschaltung (621.313)
tandem exchange Knotenvermittlung
in tandem hintereinandergeschaltet, in Kaskade geschaltet; aneinandergereiht
tandem-knife switch mehrpoliger Messerschalter
tandem office Knotenvermittlungsamt (621.395)
tandem selector Anrufsucher
tandem trunk Durchgangsleitung (621.394)
tang Ahle (621.9); Heft; Hall, Klang (534)
to tang hallen, klingen
tangent Tangente (51); Tangens (51)
tangent galvanometer Tangentenbussole, Nadelgalvanometer
tangent ray Tangentialstrahl
tangent berührend
to be tangent to berühren, tangieren
tangential tangential (**to:** zu, an)
tangential force Schubspannung; Tangentialkraft

tangential pressure Tangentialdruck
tangential stress tangentiale Feldstärke
tangential thrust tangentialer Schub
to tangle verwirren, sich verschlingen, sich verheddern
tank Tank, Behälter, Trog; Tank, Panzer, Panzerwagen (623); Schwingkreis (621.396); Schleppkanal (629.12)
tank circuit Topfkreis; Oszillatorschwingkreis, abgestimmter Anodenkreis, Schwungradkreis (621.396.61, 62); Parallelschwingkreis, Tankkreis (621.396)
tank-circuit (circulating) current Anodenschwingkreisstrom, Schwungradstrom (621.396)
tank-circuit oscillator Hauptschwingkreis, Anodenschwingkreis, Sender mit Anodenschwingkreis (621.396.61)
tank fittings Kesselarmaturen (621.314.2), (621.316.5)
tank lifter Kesselabsenkvorrichtung (621.316.5)
tank lining Kesselauskleidung (621.316.5)
tank voltage Badspannung (621.357)
tankage Behälterinhalt
tanker Tanker, Tankschiff
tantalum Tantal
tantalum electrolytic capacitor Tantal-Elektrolyt-Kondensator
tantalum lamp Tantallampe
tantalytic capacitor Tantal-Elektrolyt-Kondensator
tap Klopfen, (leichter) Schlag; Anzapfung, Zapfstelle, Zapfhahn (621); Anzapfung, Abgriff (621.3); Gewindebohrer (621.9)
tap borer Zapfenbohrer
tap changing Anzapfungsumschaltung
tap-changing gear Anzapfungsschalter
tap-changing switch Regelschalter
tap-changing transformer Regeltransformator
tap-field control [Drehzahlregelung durch Anzapfung der Feldwicklung]
tap winding Stufenwicklung (621.314.2)
tap wrench Windeisen, Kluppe (621.99)
to tap antippen, anklopfen, klopfen, leicht anschlagen; anzapfen, zapfen, anstechen; Gewinde schneiden
to tap into zwischenschalten, einschalten in (621.3)
to tap off abstechen (669)
to tap off from abgreifen an, abgreifen von (621.3)
to tap on to anlegen an, anzapfen bei (621.3)
to tap a telephone line eine Telefonleitung anzapfen, sich an eine Telefonleitung anschalten
tapping point Anzapfstelle, Abgriffpunkt
tape Band; Isolierband (621.3); Bandmaß, Lochstreifen (621.394); Magnetband (681.3, 621.397)
tape cartridge Bandkassette
tape controlled lochstreifengesteuert
tape-copying system Lochstreifenvervielfältigungseinrichtung
tape core Bandkern
tape drive Bandantrieb, Magnetbandlaufwerk
tape drum Bandführungstrommel (681.84), Bandtrommel
tape feed Bandvorschub, Bandzuführung
tape galvanometer Bändchengalvanometer
tape lead-in Bandeinziehvorrichtung
tape mark Bandmarke
tape measure Maßband
tape perforator Streifenlocher
tape-punch set Streifenlocher
tape reader (Loch-)Streifenleser
tape recorder Morseschreiber (621.394); Bandschreiber (621.317); Magnetbandgerät, Tonbandgerät (681.84)
tape recording Morseschrift; (Magnet-)Bandaufnahme (681.84)
tape reel Bandrolle
tape reproducer Streifendoppler
tape skew Schräglauf des Bandes
tape speed Bandgeschwindigkeit
tape store Bandspeicher
tape teleprinter Streifenschreiber
tape transmitter Lochstreifensender
tape winder Streifenaufwickler
to tape umwickeln, umbandeln, bewickeln; auf Band aufzeichnen (621.397)
taper Konus, Kegel; Konizität
taper bush kegelige Lagerschale, Kegelschale
taper gauge Bohrungslehre
taper hole konisches Loch, konische Bohrung
taper-loaded cable Kabel mit längsveränderlicher Bespulung
taper piece Verjüngung, Reduktion, Reduktionsstück (66)
taper pin Kegelstift
taper-roller bearing Kegelrollenlager
taper slot Parallelzahnnut
taper tap konischer Gewindebohrer, Vorschneider (621.9)
taper kegelig, konisch
to taper sich verjüngen, spitz zulaufen; verjüngen, spitz machen, kegelig machen

to taper off spitz auslaufen; abklingen
tapered konisch, kegelig; spitz zulaufend, keilförmig
tapered line Transformationsleitung
tapered pin Kegelstift
tapered reamer Reibahle, konische Reibahle
tapered roller Kegelrolle
tapered thread konisches Gewinde
taping Bewicklung, Umbandelung (621.3)
tapped mit einem Abgriff versehen angezapft (621.3); Gewinde-...
tapped hole Gewindeloch
tapped voltage transformer Anzapftransformator
tapped Anschlag (621)
tapped switch Kippschalter
tapping Anzapfung (621.3); Gewindeschneiden (621.9)
tapping clip Abgriffschelle
tapping drill Gewindebohrer, Bohrer für Gewindeloch
tapping off Anzapfung an
tapping point Abgriffpunkt
tapping size Vorbohrdurchmesser (621.99)
tar Teer
to tar teeren
target Antikathode, Auftreffektrode (621.386); Ziel, Sollziffer, Planziffer (621); Speicherplatte, Target (621.396.96)
target acquisition Zielerfassung (623)
target nucleus beschossener Kern (539)
target number Sollzahl, Planzahl
target seeking equipment Zielsuchgerät
target tracking radar Zielverfolgungsradar (623)
to tarnish trüben, mattmachen; anlaufen; trübe werden
tarry teerig
taut straff, stramm, angezogen
taut band Spannband
taut-band suspension Spannbandlagerung
taut-tape attachment Spannbügelschalter (621.394.6)
to tauten strammen, anspannen
tax Abgabe, Gebühr
tax indication Gebührenanzeige
tax metering Gebührenzählung
taxiphone Münzfernsprecher
taxiway Rollfeld, Rollbahn
taxiway marker light Rollbahnfeuer
Taylor's series Taylorsche Reihe
T. C. = total carbon Gesamt-C-Gehalt, gesamter Kohlenstoffgehalt
team Mannschaft, Arbeitsgemeinschaft, Gruppe
teamwork Gemeinschaftsarbeit
tearing Zeilenreißen (621.397)
to tease strähnen, zerfasern
teasing machine Strähnmaschine, Kämmaschine
teaser winding Höhenwicklung
technical technisch
technical college Ingenieurschule
technical data technische Angaben, technische Daten
technical knowledge technische Wissenschaften (pl.); technisches Wissen
technical magazine technische Zeitschrift
technical manual technische Anleitung, Bedienungsanleitung, technische Beschreibung, Handbuch
technical parlance Fachsprache (des Ingenieurs)
technical term Fachausdruck
technical university Technische Universität, Technische Hochschule
technicality technische Einzelheit
technician Techniker
technics (pl.) Technik
technique Technik, Verfahren
technological technologisch, technisch
technology Technologie, Technik
tectonic tektonisch
Tee T; T-Profil
Tee-bar T-Profil, T-Eisen
Tee-joint T-Stoß, Abzweigklemme
Tee-section T-Querschnitt
Tee-slot T-Schlitz, Rostschlitz
Tee-weld T-Schweißung
to tee in Stich schalten, abzweigen (621.3)
teed feeder Stichspeiseleitung, Gabelleitung
teleadjusting Ferneinstellung
teleautography Fernbildübertragung, Bildübertragung
telecine equipment Filmabtaster
to telecommand fernsteuern
telecommunication Nachrichtentechnik, Fernmeldetechnik
telecommunication line Fernmeldeverbindung
telecommunication network Fernmeldenetz
telecontrol Fernwirk...
telecontrol station Fernwirkstation
to telecontrol fernsteuern, fernwirken
telectal (Al-Li-Legierung)
telefilm Fernsehfilm
telegram Telegramm
telegraph Telegraph, Telegraphen-...
telegraph broadcast unit Telegraphie-Rundschreibgerät

telegraph cable Telegraphenkabel
telegraph channel Telegraphenkanal
telegraph character Telegraphiezeichen
telegraph circuit Telegraphiekreis, Telegraphenleitung
telegraph code Telegraphenalphabet
telegraph converter Telegraphie-Umsetzer
telegraph distortion Telegraphieverzerrung, Zeichenverzerrung, Schrittverzerrung
telegraph distributor Kanalumschalter, Verteilerschalter (621.394)
telegraph instrument Telegraphenapparat
telegraph key Telegraphentaste, Morsetaster, Taste
telegraph line Telegraphenleitung
telegraph line current Fernschreiblinienstrom
telegraph message file and retrieval system Telegraphie-Speichersystem
telegraph noise Telegraphiegeräusch (621.39)
telegraph pole Telegraphenmast
telegraph receiver Telegraphenapparat, Empfänger
telegraph relay Telegraphenrelais
telegraph repeater Telegraphieverstärker
telegraph retransmission system Fernschreibteilstreckensystem
telegraph signal Telegraphiezeichen
telegraph-signal distortion Zeichenverzerrung, Schrittverzerrung (621.394)
telegraph-signal element Telegraphierschritt
telegraph sounder Ticker
telegraph and telephone line Fernmeldeleitung
telegraph transmission Telegraphieverkehr
telegraph transmission speed Telegraphiergeschwindigkeit
telegraph transmitter Telegraphenapparat, Geber; Sendeteil eines Telegraphenapparates
telegraph wire Telegraphendraht
to telegraph telegraphieren, drahten
telegraphic telegraphisch
telegraphic automatic relay system automatische Telegraphie-Teilstreckenvermittlung
telegraphic connection telegraphische Verbindung
telegraphist Telegraphist, Telegraphenbeamter
telegraphone (obs.) Diktaphon, Magnetophon
telegraphy Telegraphie
telegraphy code Telegraphen-Alphabet
teleindication Fernanzeige
telemeter Fernmeßinstrument
telemetering Fernmessen
telemetering device Fernmessung, Fernmeßeinrichtung
telemetry Telemetrie, Fernmessen
telephone Telephon, Telefon; Fernsprecher, Fernsprech-...
telephone apparatus Telefonapparat
telephone bell Telefonklingel
telephone booth Telefonkabine, Fernsprechzelle
telephone box Telefonzelle
telephone cable Telefonkabel, Fernsprechkabel
telephone call Telefonanruf, Telefongespräch
telephone channel Telefonkanal, Sprechkanal
telephone charges Telefongebühren
telephone circuit Sprechstromkreis, Fernsprechleitung
telephone connection Fernsprechanschluß, Fernsprechverbindung
telephone conversation Telefongespräch
telephone current Sprechstrom, Telefoniestrom, Modulation
telephone directory Telefonbuch, Telefonverzeichnis
telephone drop Telefon-Hausanschlußleitung
telephone engineering Telefontechnik, Fernsprechtechnik
telephone equipment Telefoniegerät, Fernsprechgerät
telephone exchange Fernsprechamt, Fernsprechvermittlung
telephone fees Telefongebühren
telephone flex Telefonschnur
telephone frequency Sprachfrequenz
telephone headset Doppelkopfhörer, Kopfhörer
telephone installation Telefonanlage, Fernsprechanlage
telephone interference factor Kopplung (bei Schwachstrombeeinflussung); Fernsprechstörfaktor
telephone line Telefonleitung
telephone meter Gesprächszähler (621.395)
telephone network Fernsprechnetz
telephone operator Telefonist, Fernsprechbeamter; Fernsprechbeamtin, Vermittlungsbeamtin
telephone pair Aderpaar
telephone receiver Hörer, Telefonhörer
telephone relay Fernsprechrelais, Telefonrelais
telephone repeater Verstärker, Leitungsverstärker, Fernsprechverstärker
telephone service Telefondienst
telephone set Telefonapparat, Apparat

telephone station Fernsprechanlage, Teilnehmeranlage, Teilnehmersprechstelle
telephone subscriber Fernsprechteilnehmer
telephone switchboard Vermittlungsschrank
telephone switching Fernsprechvermittlung
telephone tandem exchange Fernsprech-Knotenvermittlung
telephone terminal Leitungsklemme; Fernsprech-Endeinrichtung
telephone terminal exchange Fernsprech-Endvermittlung
telephone toll office Fernsprech-Fernvermittlung
telephone transformer Übertrager, Ausgangsübertrager
telephone transmission Telefonverkehr, Fernsprechübertragung
telephone transmitter Mikrophon
telephone trunk exchange Fernsprech-Fernvermittlung
telephone-type arrester Glimmstrecke, Überspannungsableiter
telephone wire Telefondraht, Fernsprechleitung
to telephone telefonieren, telephonieren, fernsprechen (**to:** mit)
telephonic telefonisch
telephonic frequency Tonfrequenz, Sprachfrequenz
telephonic message telefonische Nachricht
telephonic repeater Leitungsverstärker
telephonist Telefonist, Vermittlungsbeamter, Vermittlungsbeamtin
telephonist release Amtstrennung
telephonometry Telefon-Meßtechnik
telephony Telefonie, Fernsprechwesen
telephony and telegraphy Telegraphie und Telefonie, Fernmeldewesen, Drahtnachrichtentechnik
telephotograph Fernschreibbild
to telephotograph bildübertragen
telephotography Fernbildübertragung
to teleprint fernschreiben
teleprinter Fernschreiber, Ferndrucker, Fernschreibmaschine
teleprinter exchange Fernschreibvermittlung
teleprinter network Telexnetz
teleprocessing Datenfernverarbeitung
Teleran = television radar air navigation Funkbildnavigation
telerecording Fernseh-Aufzeichnung
telescope Teleskop, Fernrohr
to telescope sich einschieben, sich zusammenschieben, sich einschieben lassen
telescoping section ausziehbares Leitungsstück, Posaune (621.396)
telescopic Teleskop-..., Einschieb-...
telescopic leg Teleskopbein, Stativbein
telescopic section ausziehbares Rohrstück
telescopic tube Teleskoprohr, ausziehbares Rohr
teleswitching Fernschalten
telesyn Telesyn-Drehmelder (621.316.7)
telethermometer Fernthermometer
teletransmission Fernübertragung
teletransmitter Ferngeber
teletype Ferndruck-..., Fernschreib-...
teletype equipment Fernschreibgerät
teletype network Fernschreibnetz
teletypewriter Fernschreibmaschine
teletypewriter exchange Fernschreibvermittlung
to televise fernsehen, durch Fernsehen übertragen
television Fernsehen, Fernsehfunk
television aerial Fernsehantenne
television broadcasting Fernsehrundfunk, Fernsehen
television program center Fernsehschaltstelle
television channel Fernsehkanal
television-channel spacing Fernsehkanalabstand
television line Fernsehleitung
television picture Fernsehbild
television receiver Fernsehempfänger
television reception Fernsehempfang
television recorder Video-Recorder
television set Fernsehgerät, Fernsehempfänger
television standard Fernsehnorm
television standards converter Fernsehnormwandler
television studio Fernsehstudio
television system Fernsehanlage
television system converter Fernseh-Normwandler
television test chart Testbild
television test signals Fernsehprüfsignale
television translator Fernsehumsetzer
television transmitter Fernsehsender
television transposer Fernsehumsetzer
television tube Bildröhre
telewriter Fernschreiber
to tell erzählen, berichten; erkennen (**by:** an)
telling circuit Meldeleitung
telltale Merk-..., Schau-...
telltale board Anzeigetafel
telltale lamp Merklampe, Schaulampe, Kontrollampe, Überwachungslampe

telltale picture Anschauungsbild
telltale transmitter Kontrollsender
tellurium Tellur
telpher Elektrohängebahn
temper Härte, Härtegrad
temper colour Anlauffarbe
temper hardening Anlaßhärtung, Glühhärtung
to temper tempern, anlassen
temperature Temperatur
temperature-actuated switch Temperaturschalter, Thermostat
temperature balance Temperaturgleichgewicht; Temperaturausgleich
temperature on Cent scale Temperatur nach Grad Celsius
temperature coefficient Temperaturkoeffizient
temperature control Temperatursteuerung, Temperaturregelung
temperature-cycle test Temperaturwechselprüfung
temperature-dependent attenuation balance Temperaturdämpfungsausgleich
temperature detector Temperaturfühler
temperature drop Temperaturabfall
temperature entropy diagram t, s-Diagramm
temperature of fusion Schmelztemperatur
temperature gauge Temperaturmeßgerät
temperature of ignition Flammpunkt, Entzündungstemperatur
temperature inversion Temperaturinversion
temperature measurement Temperaturmessung
temperature radiation Temperaturstrahlung
temperature radiator Temperaturstrahler
temperature range Temperaturbereich
temperature of reaction Reaktionstemperatur
temperature recorder Temperaturschreiber, Thermograph
temperature relay Temperaturrelais, thermisches Relais
temperature ripple Temperaturwelligkeit
temperature rise Übertemperatur, Erwärmung (621.313); Temperaturzunahme
temperature sensor Temperaturfühler
temperature stability Temperaturkonstanz
temperature switch Thermoschalter
temperature transmitter Temperaturgeber
temperature variation Temperaturschwankung
tempering Temperung, Anlaßvorgang
tempering colour Anlauffarbe
template Schablone, Bohrschablone, Bohrlehre; Modell
templet = template
temporary zeitweilig, temporär, vorübergehend, provisorisch
temporary service kurzzeitiger Betrieb
temporary set-up provisorischer Aufbau
temporary stoppage kurzzeitige Betriebsunterbrechung
temporary storage Arbeitsspeicher, Zwischenspeicher
temporary wiring vorläufige Leitungsverlegung, vorläufige Installation
to tend pflegen, warten (621); dahin wirken, dahin neigen; bestrebt sein, bemüht sein
to tend to streben nach, gehen nach, hinziehen auf, neigen zu
to tend to a limit einer Grenze zustreben, nach einer Grenze gehen
tendance Wartung, Pflege (621)
tendency Tendenz, Neigung (**to:** zu) (**towards:** nach)
tendency to oscillation Schwingneigung
tender Tender, Kohlenwagen (625); Maschinenwärter (621); Angebot, Anschlag
tenon Zapfen
tenon jointing Zapfverbindung (69)
tens counting tube Zehner-Zählröhre
tense straff, gespannt
tense cord gespanntes Seil, straffe Schnur
tense string gespannte Saite
tensile dehnbar; Zug- ...
tensile force Zugkraft
tensile load Zuglast, Zugbeanspruchung
tensile member Zugglied, Zugstab
tensile spring Zugfeder
tensile strain Zugdehnung, Zugbeanspruchung
tensile strength Zugfestigkeit
tensile stress Zugspannung, Zugbeanspruchung
tensile test Zugversuch
tensility Dehnbarkeit
tension Spannung (539), (621.3); Zug, Zugspannung
tension on the belt Riemenzug, Riemenspannung
tension boom Zuggurt
tension-compression fatigue test Zug-Druck-Dauerprüfung
in tension unter Zug stehend, auf Zugbeansprucht
tension indicator Spannungsanzeiger
tension insulator Abspannisolator
tension joint Verbinder, Seilverbinder
tension member Zugglied, Zugstab, auf Zug beanspruchtes Bauelement
tension pulley Spannrolle

tension side Zugseite (539.3)
tension test Zugversuch, Zerreißversuch
tension-test bar Zerreißstab, Prüfstab
unter tension unter Zugspannung stehend
to tension spannen
tensioning device Spannvorrichtung
tensor Tensor (51)
tenter Spannrahmen, Spanner
tenter hook Spannhaken
tenth zehnter; Zehntel
tenth-second timer Zehntelsekunden-Schalter
tenuous air dünne Luft (551.5)
to tepefy lau machen
tephigram Temp (551.5)
teredo tape Kabelschutzbinde (621.395.5)
term Klammerausdruck (51); Semester (Br.); Dritteljahr, Schulzeitabschnitt; Bedingung; Bezeichnung, Ausdruck, Fachausdruck; Energieterm, Energie-Niveau (539)
term in brackets Klammerausdruck (51)
in terms of gemessen in, ausgedrückt durch; in der Ausdrucksweise von ...
term in parenthesis Klammerausdruck (51)
term of a sequence Glied einer Reihe (51)
to term bezeichnen, benennen
terminal Klemme, Anschlußklemme (621.3); Endstück, letztes Ende, Endeinrichtung; Empfangsgebäude, Abfertigungsgebäude (33); Datenstation, Terminal (681.3)
terminal aera surveillance radar Nahverkehrsbereichsradar
terminal block Klemmbrett, Klemmleiste (621.313)
terminal board Klemmbrett, Klemmleiste (621.313)
terminal box Klemmenkasten; Endverteiler; Anschlußkasten; Verteilerkasten; Endgerät; Bediengerät, Endverschluß (621.315.2)
terminal condition Abschlußbedingung; Bedingung am Leitungsende
terminal connector Klemmenverbinder
terminal contact Endkontakt, Begrenzungskontakt
terminal exchange Endvermittlung
terminal equipment Endschaltung (621.9); Endgerät, Endeinrichtung (621.39); Ein-Ausgabegerät (681.3)
terminal head Anschlußkopf, Klemmenkopf
terminal impedance Klemmenscheinwiderstand
terminal insulator Klemme (621.314.2); Abspannisolator (621.315.1)
terminal interface Terminal-Interface
terminal lead äußere Zuführung, Klemmenzuleitung (621.385)
terminal limit switch Grenzstockwerkschalter, Endschalter (621.876)
terminal per line Anschluß je Anschlußleitung (621.395)
terminal lug Anschlußöse, Endbundklemme
terminal part Endteil
terminal pole Abspannmast, Endmast (621.315.1)
terminal posts Batterieklemmen (621.355)
terminal pressure Enddruck (624)
terminal repeater Endverstärker (621.395)
terminal resistance Abschlußwiderstand
terminal room Verteilerraum (621.395)
terminal screw Anschlußschraube, Klemmschraube
terminal station Endstation (625); Datenstation (621.394)
terminal per station Anschluß je Sprechstelle (621.395)
terminal strip Anschlußleiste, Klemmenleiste
terminal stud Anschlußbolzen
terminal switch Endschalter, Endausschalter
terminal voltage Klemmenspannung
terminal wire Anschlußdraht, Zuleitung
to terminate beenden, beendigen, abschließen; mit einem Abschluß versehen, mit einem Abschlußwiderstand versehen, abschließen (621.3)
to terminate in abschließen mit; abgeschlossen sein mit (621.3)
terminated line abgeschlossene Leitung, reflexionsfrei abgeschlossene Leitung
terminating Abschluß-...
terminating impedance Abschlußwiderstand (allgemein); Ausgangsklemmenimpedanz
terminating resistance (ohmscher) Abschlußwiderstand
terminating set Gabelschaltung (621.395)
terminating transformer Abschlußübertrager
termination Beendigung, Abschluß, Ende; Abschluß, Abschlußwiderstand (621.3)
termination amplifier Gabelverstärker (621.395)
termination cable Abschlußkabel
termination equipment Endstelleneinrichtung
terminology Terminologie, Ausdrucksweise, Bezeichnungsweise
ternary ternär; Dreistoff-... (669)
ternary system Dreistoffsystem
terrain Terrain, Gelände; Geländeform
terrain-avoidance radar Hindernis-Warn-Radar

terrain-clearance indicator Funk-Feinhöhenanzeiger
terrain-clearance warning indicator Hindernis-Warnanzeiger
terrestrial terrestrisch, Erd-...
terrestrial heat Erdwärme, terrestrische Wärme
terrestrial magnetism Erdmagnetismus
tertiary Tertiärwicklung
tertiary tertiär, Dritt-, Tertiär-...
tertiary winding Tertiärwicklung
Tesla coil Tesla-Spule
Tesla current Tesla-Strom
Tesla transformer Tesla-Transformator
tesselation mosaikartige Unterteilung, schachbrettartige Unterteilung
tessera (pl. **-rae**) Mosaiksteinchen
test Versuch, Probe, Prüfung, Untersuchung
test of accuracy Maßhaltigkeitsprüfung, Genauigkeitsprüfung
test atmosphere Prüfklima
test bar Prüfstab
test bench Prüfstand, Prüfraum, Prüftisch
test board Meßplatz, Prüfplatz
test box Prüftafel (621.316)
test cage geschirmter Meßraum
test call Probeanruf, Leitungsprobe
test cap Zubleiung, Endverbleiung (621.315.2)
test certificate Prüfschein, Abnahmeprotokoll
test connector Prüfstecker
test cord Prüfdraht, Prüfleitung, „Prüfstrippe“, Prüfschnur
test department Prüffeld
test desk Meßplatz, Prüfplatz
test to destruction Zerreißprüfung, Prüfung bis zur Zerstörung
test equipment Prüfgeräte (pl.); Prüfeinrichtung
test floor Prüffeld
test frequency Prüffrequenz, Meßfrequenz
test glass Reagenzglas
test indicator Prüfzeiger
test jack Prüfklinke (621.395)
test lamp Prüflampe, Prüfling (621.32)
test-lead point Prüfspitze
test of life Lebensdauerprüfung
test line Prüfzeile (621.397)
test load Prüflast, Prüfbelastung
test loading Probebelastung; Prüfbelastung
test log Prüfprotokoll
test method Prüfverfahren
test oscillator Prüfsender
test paper Reagenzpapier (54); Polpapier (621.3)
test pattern Testbild
test piece Probestück, Prüfobjekt
test rack Prüfgestell
test report Prüfbericht, Abnahmeprotokoll, Prüfprotokoll
test result Prüfungsergebnis, Versuchsergebnis
test set Prüfgerät
test shield Erdungsmantel (621.315.2)
test specimen Prüfstab, Probestück
test stand Prüfstand, Prüfeinrichtung
test-signal generator Prüfsignalgeber
test socket Prüfbuchse
test team Meßtrupp; Versuchsgruppe
test terminal Prüfklemme
test tone Prüfton, tonfrequente Prüfspannung
test tube Reagenzglas
test vessel Prüfgefäß
test voltage Prüfspannung
to test untersuchen, prüfen, erproben
to test by bending biegeprüfen
to test under brake load abbremsen (621.1)
to test for continuity auf Durchgang prüfen
to test to destruction bis zum Bruch prüfen, bis zur Zerstörung prüfen
to test for parallel auf Parallelität prüfen
to test through auf Durchgang prüfen
to test a tube Röhre (auf Emission) prüfen
tester Prüfer, Probenehmer; Prüfgerät
testing Prüfung, Prüfvorgang
testing accuracy Prüfgenauigkeit
testing apparatus Prüfapparat
testing bush Prüfring (621.395)
testing centre Prüffeld
testing current Prüfstrom
testing instrument Prüfinstrument
testing laboratory Prüfamt, Prüflabor
testing machine Prüfmaschine, Zerreißmaschine
testing of materials Materialprüfung, Materialuntersuchung
testing potential Prüfspannung
testing pressure Prüfdruck (621.1); Prüfspannung (621.3)
testing purpose Prüfzweck
testing reactor Prüfreaktor
testing room Prüfraum, Prüfzimmer
testing set Prüfsatz, Prüfeinrichtung
testing station Prüfstation, Prüfanstalt, Prüfstelle
testing terminal Prüfklemme
testing transformer Prüftransformator
testing unit Prüfsatz, Prüfgerät
testing wire Prüfdraht

to tether anbinden, festmachen (63); fesseln, elastisch binden (539)
tetragon Viereck
tetragonal tetragonal, viereckig
tetragonal thread Trapezgewinde
tetravalent vierwertig (54)
tetrode Tetrode, Vierpolröhre
tetrode transistor Transistor-Tetrode
text editor Text-editor
textile Textilstoff, Faserstoff
textile fabric Textilgewebe
texture Textur, Gefüge
thallium Thallium
thalofide cell Thalliumsulfidzelle
to thaw tauen, auftauen
T-head bolt Hammerkopfschraube
theodolite Theodolit
theorem Theorem, Lehrsatz
theory Theorie
theory of electrostatics Elektrostatik
theory of relativity Relativitätstheorie
therm Einheit der Wärmemenge = 10^5 **B. Th. U.**
thermal thermisch, Wärme-...
thermal agitation Wärmerauschen, thermische Unruhe
thermal agitation noise Wärmerauschen
thermal agitation voltage Rauschspannung
thermal breeder langsamer Brutreaktor, thermischer Brüter
thermal breakdown thermische Instabilität, thermischer Durchbruch (621.382)
thermal capacitance Wärmekapazität
thermal capacity Wärmekapazität
thermal circuit breaker Schalter mit thermischer Auslösung, Motorschutzschalter
thermal condition Wärmezustand, Temperaturzustand
thermal conduction Wärmeleitung
thermal conductivity Wärmeleitfähigkeit
thermal conductor Wärmeleiter
thermal contact Wärmekontakt, thermischer Kontakt
thermal converter Thermoumformer
thermal cross Thermokreuz, Thermoumformer
thermal cross section thermischer Wirkungsquerschnitt (539)
thermal cut-out thermischer Auslöser, thermisches Schütz
thermal delay switch Thermoschalter
thermal derating factor thermischer Reduzierkoeffizient (621.392)
thermal diffusion Thermodiffusion
thermal dissipation Wärmeabfuhr, Wärmeabgabe
thermal drift thermisch bedingte Abweichung
thermal effect Wärmewirkung
thermal efficiency thermischer Wirkungsgrad
thermal energy Wärmeenergie
thermal equivalent of work mechanisches Wärmeequivalent
thermal expansion Wärmedehnung
thermal fission thermische Spaltung (539)
thermal flasher thermische Blinkschalteinrichtung
thermal flashing relay Thermoblinkrelais, Bimetall-Blinkrelais
thermal imaging (night sight) device Wärmebildgerät (623)
thermal impedance Wärmewiderstand (621.382)
thermal impedance für one single pulse Impulswärmewiderstand
thermal impedance under pulse conditions Pulswärmewiderstand
thermal instrument Thermo-Meßinstrument
thermal insulating wärmeisolierend
thermal insulation Wärmeisolierung, Wärmeisolation
thermal motion thermische Bewegung, Wärmebewegung
thermal neutron thermisches Neutron, langsames Neutron
thermal-neutron fission Spaltung mit thermischen Elektronen
thermal noise thermisches Rauschen, Widerstandsrauschen, Wärmerauschen, Johnson-Rauschen
thermal ohm thermisches Ohm = GradCelsius/Watt
thermal overload capacity thermische Überlastbarkeit
thermal printer Thermodrucker
thermal power Wärmekraft, Wärmeleistung
thermal power plant Wärmekraftanlage, Wärmekraftwerk
thermal property Wärmeeigenschaft, thermische Eigenschaft
thermal radiator Wärmestrahler
thermal reactor thermischer Reaktor, Reaktor mit langsamen Neutronen
thermal relay Thermoverstärker (621.317); Thermorelais, Bimetallrelais
thermal release thermische Auslösung
thermal resistance Wärmewiderstand, Wärmeleitungswiderstand

thermal resistivity spezifischer Wärmeleitungswiderstand; Wärmefestigkeit, Wärmebeständigkeit
thermal runaway thermisches Weglaufen, thermische Instabilität
thermal sensor Temperaturfühler
thermal shock plötzliche Temperaturänderung, Temperatursturz, Abschreckung; Wärmeschock
thermal stability Wärmebeständigkeit
thermal station Wärmekraftwerk, thermisches Kraftwerk
thermal storage heating Speicherheizung
thermal-storage water heater Heißwasserspeicher
thermal stress Wärmespannung
thermal telephone Hitzdrahttelefon
thermal test Wärmeprüfung
thermal time constant Wärmezeitkonstante
thermal treatment Wärmebehandlung, thermische Behandlung
thermal trip thermischer Auslöser
thermal tripping thermische Auslösung
thermal unit Wärmeeinheit, Einheit der Wärmemenge
thermal velocity neutron thermisches Neutron
thermal writing thermische Aufzeichnung
thermel thermo-elektrisches Thermometer
thermic thermisch
thermionic Elektronen-..., Glühkathoden-...
thermionic amplifier Röhrenverstärker
thermionic cathode Glühkathode, thermische Kathode
thermionic converter Glühkathodenstromrichter (621.316.7); thermischer Wandler (621.313)
thermionic current thermischer Elektronenstrom
thermionic detector Gleichrichteröhre (621.396)
thermionic emission thermische Elektronenemission
thermionic energy conversion thermionische Energieumwandlung
thermionic rectifier Glühkathodengleichrichter, Röhrengleichrichter
thermionic valve Glühkathodenröhre, Elektronenröhre
thermions thermische Ionen, emittierte Ionen
thermistor Thermistor, Heißleiter, temperaturabhängiger Widerstand
thermite Thermit
thermite welding Thermitschweißung
thermo-bimetallic instrument Bimetallmeßinstrument
thermochemical thermochemisch
thermochemical equation thermische Umsatzgleichung
thermochemical generator Brennstoffelement
thermochemistry Thermochemie
thermo converter Thermo-Umformer (621.317)
thermocouple Thermoelement
thermocouple instrument Thermoumformer-Instrument
thermocouple junction Lötstelle des Thermoelements
thermocouple pyrometer thermoelektrisches Pyrometer
thermocouple thermometer thermoelektrisches Thermometer
thermodetector Barretter
thermodynamic thermodynamisch
thermodynamic efficiency thermodynamischer Wirkungsgrad
thermodynamics (pl.) Thermodynamik
thermoelectric thermoelektrisch
thermoelectric characteristic Kennlinie eines Thermoelements
thermoelectric cooler thermoelektrischer Kühlblock
thermoelectric couple Thermoelement
thermoelectric current thermoelektrischer Strom
thermoelectric effect thermoelektrischer Effekt, Seebeck-Effekt
thermoelectric force Thermokraft (621.36)
thermoelectric generating set Wärme-Kraftmaschinensatz
thermoelectric generator Thermosäule
thermoelectric inversion Polaritätsumschlag der Thermospannung
thermoelectric junction Thermoelement, Kontaktstelle
thermoelectric pile Thermosäule
thermoelectric power Thermokraft
thermoelectric pyrometer thermoelektrisches Pyrometer
thermoelectric relay Thermorelais
thermoelectric series thermoelektrische Spannungsreihe
thermoelectric thermometer thermoelektrisches Thermometer
thermoelectric voltage Thermospannung
thermoelectricity Thermoelektrizität
thermoelectromotive force Thermo-EMK
thermoelement Thermoumformer (621.317)
thermo E. M. F. Thermokraft, Thermospannung

thermoexpansion Wärmedehnung
thermoexpansion instrument Hitzdrahtinstrument
thermogalvanometer Thermo-Galvanometer, Duddell-Galvanometer
thermogenerator Thermosäule
thermojunction Thermoelement, Lötstelle
thermometer Thermometer
thermometer probe Temperaturfühler
thermometer scale Thermometerskala
thermometer screen Thermometerhütte (55)
thermometric thermometrisch
thermometry Temperaturbestimmung, Thermometrie
thermonuclear reaction thermonukleare Reaktion
thermophile auf Wärme ansprechend
thermophone Thermophon
thermopile Thermosäule
thermoplastic Thermoplast, thermoplastisches Material
thermoplastic thermoplastisch (67)
thermorelay Thermorelais, thermoelektrischer Verstärker
thermoscope Thermoskop, Wärmeanzeiger
thermosetting warmhärtend
thermosetting glue warmhärtender Leim
thermosetting plastic warmhärtender Kunststoff
thermostat Thermostat, Wärmeschrank; Temperaturschalter
thermostat regulator Temperaturregler, Regelthermostat
thermostat relay Thermostatrelais, thermisches Relais
thermostatic thermostatisch
thermostatic control Temperaturregler, Thermostat
thermostatic controller Thermostatregler
thermostatic regulator Thermostat, Temperaturregler
thermovoltmeter Thermo-Umformer-Voltmeter
Thevenin's theorem Satz von der Ersatzstromquelle
thick dick
10 in. thick 10 Zoll dick, 10 Zoll stark
thick-film circuit Dickschichtschaltung
thick line dicker Strich, starke Linie
thick with dicht bedeckt mit, gefüllt mit
thick wire dicker Draht, starker Draht; dickdrähtig
thick-wire coil dickdrähtige Spule
to thicken dick werden, dicker werden, sich verdichten (551.5); dicker machen
thickener Verdickungsmittel (665)
thickness Dicke, Stärke, Wandstärke; Lage (621.3)
thickness gauge Dickenlehre
two thickness zwei Lagen, zwei Schichten
thickness vibration Dickenschwingung
thickness vibration of a crystal Kristalldickenschwingung
thickness of the wall Wandstärke
thiefproof diebessicher
thimble Fingerhut, Nähring; Meßring (681.2), (531.7)
thimbleful fingerhutvoll
to thin verdünnen, dünnflüssig machen, dünner machen; anspitzen, zuspitzen
thin dünn
thin air dünne Luft
thin film Dünnfilm
thin-film circuit Dünnfilmschaltung
thin-film rectifier Sperrschichtgleichrichter, Trockengleichrichter
thin-film resistor Dünnfilmwiderstand
thin flame Stichflamme
thin-line transmission system Schmalbandübertragungssystem
thin-rod pulling Dünnziehen (621.382)
thin section Dünnschliff (535.8)
thin sheet Feinblech
thin slab dünne Platte
thin wall geringe Wandstärke, dünne Wand
thin-walled dünnwandig
thin wire dünner Draht
thin-wire dünndrähtig
thin-wire coil dünndrähtige Spule
thin-wire winding dünndrähtige Wicklung
third dritter; Drittel
third class dritte Klasse, drittklassig
third class conductor negativer Widerstand
third harmonic dritte Harmonische, dritte Oberwelle
third power dritte Potenz
third rail dritte Schiene, Stromschiene (621.33)
third-rail clearance line Stromschienenprofil
third-rail collector Stromschienenabnehmer
third-rail electric locomotive Stromschienenlokomotive
third-rail insulator Stromschienenisolator
third-rail system Stromschienensystem
third root dritte Wurzel, Kubikwurzel
third voltage range Spannungsbereich von 176 bis 250 V
third wire Nulleiter, Mittelleiter (im Gleichstromnetz)
Thomson balance Stromwaage

Thomson bridge Thomsonsche Brücke, Doppelbrücke (621.317)
Thomson effect Thomson-Effekt
Thomson meter Thomson-Zähler
Thomson-Varley slide Doppelkurbel (621.317)
thoria Thoroxid, Thoriumoxid
to thoriate thorieren
thoriated filament thorierte Kathode (621.385)
thoriated tungsten thoriertes Wolfram
thoriated tungsten cathode thorierte Wolframkathode
thorium Thorium
thousand tausend
one in a thousand ein Promille, ein v. T.
thousands selector Tausender-Wähler
thread Faden, Zwirn (66); Draht; Gewinde (621.8)
thread cutting Gewindeschneiden
thread-cutting die Schneideisen
thread flank Gewindeflanke
thread galvanometer Fadengalvanometer
thread gauge Gewindelehre
thread-milling cutter Gewindefräser
thread recorder Fallbügelschreiber
thread tap Gewindebohrer, Gewindeschneider
thread worm Schneckenrad
to thread einfädeln; aufziehen, auffädeln; Gewinde schneiden, mit Gewinde versehen
to thread beads Perlen aufziehen, Perlen auffädeln
to thread through a circuit durch einen Stromkreis hindurchtreten (lassen), mit einem Stromkreis verketten, mit einem Stromkreis verkettet sein
to thread a winding eine Wicklung einfädeln, eine Wicklung fädeln
threaded mit Gewinde, Gewinde-...
threaded collar Gewindemuffe
threaded-in winding eingefädelte Wicklung, gefädelte Wicklung
threaded part Gewindeteil
threading Gewindeschneiden; Einfädeln, Durchfädeln
threading die Schneidbacke, Schneideisen
threading head Schneidkopf (621.99)
threading pin Gewindestift
three drei
three-address code Drei-Adreß-Code
three-ammeter method Drei-Amperemeter-Leistungsmessung (621.317)
three-aspect signal dreibegriffiges Signal
three-centre arch Korbbogen
three-channel pregroup Vorgruppe (621.395.4)
three-colour process Dreifarbenverfahren (77)
three-component alloy Dreistofflegierung
three-contact dreipolig
three-contact plug dreipoliger Stecker
three-core dreiadrig
three-core cable dreiadriges Kabel
three-cornered dreieckig
three-db-bandwidth Halbwertsbreite
three-db coupler Drei-db-Koppler
three-dial decade condenser Drei-Dekaden-Kurbelkondensator
three-dimensional dreidimensional
three-dimensional problem dreidimensionales Problem
three-dimensional sound Raumklang
three-electrode Dreielektroden-..., dreipolig, Dreipol-...
three-electrode switch dreipoliger Schalter
three-electrode valve Dreipolröhre, Triode, Dreielektrodenröhre
three-foot lead „Meterstrippe“
three-gun color-picture tube Dreistrahlfarbbildröhre
three-gun shadow-mask tube Dreistrahlfarbbildröhre, Lochmaskenfarbbildröhre
three halves power dreihalbte Potenz
three-heat switch Dreistufenschalter, Kochplattenumschalter
three-in-one-valve Dreifachröhre
three-jaw chuck Dreibackenfutter
three-level modulation Drei-Pegel-Modulation
three-phase Dreiphasen-..., Drehstrom-...
three-phase alternating current Dreiphasenwechselstrom, Drehstrom
three-phase alternator Drehstromgenerator
three-phase bridge connection Drehstrom-Brückenschaltung
three-phase circuit Drehstromkreis
three-phase commutator motor Drehstromkommutatormotor, Drehstromkollektormotor
three-phase current Drehstrom
three-phase four-wire system Drehstrom-Vierleiteranlage, Drehstromnetz mit Nulleiter
three-phase generator Drehstromgenerator
three-phase induction motor Drehstrom-Asynchronmotor
three-phase locomotive Drehstromlokomotive
three-phase meter Drehstromzähler
three-phase motor Drehstrommotor
three-phase railway Drehstrombahn

three-phase rotor winding dreiphasige Rotorwicklung, dreiphasige Läuferwicklung
three-phase six-wire system Drehstromsystem in offener Schaltung
three-phase squirrel cage motor Drehstrom-Kurzschlußanker-Motor
three-phase supply network Drehstromnetz
three-phase starter Drehstromanlasser
three-phase switch Drehstromschalter
three-phase three-wire circuit Drehstrom-Dreileiternetz, Drehstromnetz ohne Nulleiter
three-phase three-wire system Drehstromanlage ohne Nulleiter
three-phase traction Drehstrombahnbetrieb
three-phase transformer Drehstromtransformator, Drehstromumspanner
three-phase winding Drehstromwicklung, dreiphasige Wicklung, dreisträngige Wicklung
three-pin Dreistift-..., dreipolig
three-pin plug dreipoliger Stecker; Dreistiftstecker
three-point switch Dreifachumschalter
three-point system Zieldeckungsverfahren, Dreipunktverfahren
three-pole dreipolig
three-pole box dreipolige Abzweigdose
three-pole circuit breaker dreipoliger Schalter, dreipoliger Ausschalter
three-pole jack dreipolige Klinke, dreipolige Steckdose
three-pole plug dreipoliger Stecker
three-pole switch dreipoliger Schalter
three-position relay Dreistellungsrelais, Schaltstellungszeiger
three quarter dreiviertel
three-quarter wavelength dreiviertel Wellenlänge
three-range winding dreietagige Wicklung
three-share bed Dreiwangenbett (621.9)
three-square Dreikant-...
three-square file Dreikantfeile, Sägefeile
three-stage dreistufig, Dreistufen-...
three-stage amplifier dreistufiger Verstärker
three-state output Drei-Zustands-Ausgang (621.374)
three-storeyed dreigeschossig, dreietagig
three-storeyed switch dreietagiger Schalter
three-tier winding dreietagige Wicklung
three-valve Dreiröhren-...
three-valve receiver Dreiröhrenempfänger
three-voltmeter method Dreivoltmeter-Methode, Dreivoltmeter-Leistungsmessung
three-wattmeter method Dreiwattmeter-Schaltung
three-way Dreiweg-...
three-way-speaker Drei-Weg-Box
three-way switch Dreifachumschalter
three-winding transformer Dreiwicklungsumspanner, Dreiwickler, Dreiwicklungstransformator
three-wire Dreileiter-...
three-wire compensator Wechselstromkompensator
three-wire distribution Dreileiter-Verteilung, Dreileiternetz
three-wire dynamo Dreileitungsgenerator
three-wire system Dreileiteranlage, Dreileiternetz
three-wire transformer Dreileiter-Transformator, Ausgleichstransformator
threshing machine Dreschmaschine
threshold Schwelle (69); Schwellwert, Schwelle (53), Ansprechwort
threshold of audibility Hörschwelle, Schwelleder Hörempfindung
threshold of hearing Hörschwelle
threshold howl Einsatzpfeifen (621.392)
threshold kinetic energy Schwellenenergie
threshold level Schwellwert
threshold marker light Schwellenfeuer (351.814)
threshold logic Schwellwertlogik
threshold value Ansprechwert, Ansprechschwelle, Einsatzwert
throat Kehle; Gurgel, Schlund (61); Bruchdicke, Bruchquerschnitt (539); Halsansatz (621.396.67)
throat of a horn Trichterhals (534)
throat area Halsquerschnitt (621.396.67), (534)
throat microphone Kehlkopfmikrophon
throat-type microphone Kehlkopfmikrophon
throttle Drosselventil, Absperrventil
throttle switch Strombegrenzungsschalter
throttle valve Drosselventil; Drosselklappe (629.113)
to throttle drosseln (621.1)
throttling Drosselung (621.1)
throttling-type governor Drosselregler
through durchgehend
troughgoing durchgehend
through-group filter Primärgruppendurchschaltefilter

through hardened durchgehärtet
through hardening Durchhärtung
through hardness durchgehende Härte
through level Meßpegel (621.395)
through-mastergroup filter Tertiärgruppendurchschaltefilter (621.395.4)
throughput Durchsatz, Umschlag; Durchgangsleistung (621.3)
through-supergroup filter Sekundärgruppendurchschaltefilter
through-supermaster group filter Quartärgruppendurchschaltefilter (621.395.4)
through-type current transformer Aufsteckstromwandler
through wiring durchgehende Leitung
throw Wurf; Ausladung (624); ballistischer Ausschlag, Stoßausschlag (621.317)
throw-off force Schalteröffnungskraft, Kurzschlußauslösekraft
throw-over contact Umschaltkontakt
throw-over relay Kipprelais
throw-over switch Umschalter, Wechselschalter
to throw werfen (**at**: nach)
to throw a beam einen Strahl aussenden (621.396)
to throw in einwerfen, einkuppeln
to throw in and out of operation in und außer Betrieb setzen
to throw in resistance Widerstand einschalten
to throw ... into circuit ... in den Stromkreis einschalten
to throw light on Licht werfen auf
to throw off abwerfen
to throw off the belt den Riemen abwerfen
to throw on to (plötzlich) einschalten auf, anschalten an
to throw out of alignment aus der Achse herausdrücken
to throw out of gear auskuppeln
to throw over umschalten, überschalten
to throw over a point eine Weiche umlegen, eine Weiche stellen
to throw a switch to einen Schalter umlegen auf
to throw to idle auf Leerlauf schalten (621.4)
throwing-off Abwurf
throwing-on the load Lastzuschaltung
throwing power Eindringtiefe (621.357)
thrust Schub, Druck
thrust bearing Drucklager
thrust flange Druckflansch, Längsdruckflansch
thrust washer Druckring, Druckscheibe (621.8)
to thrust schieben, drücken, Schub ausüben
thumb nut Flügelmutter
thumb screw Flügelschraube, Flügelmutter
thumb tack Reißpinne, Reißbrettstift
thumbwheel Rändelrad
thump Klopfen, Kopfton; Bums, Knacks; Telegraphiegeräusch
thunder Donner
thunderbolt Blitzschlag, Donnerschlag
thunderstorm Gewitter
thunderstorm location Gewitterortung
Thury regulator Thuryregler
Thury transmission system Thury-Anlage, Thury-System
thyratron Thyratron, Stromtor
thyristor Thyristor
thyrite Thyrit (spannungsabhängiges Widerstandsmaterial)
tick Ticken; Haken, Häkchen
to tick ticken
ticker Ticker (621.396)
ticket-stamping machine Fahrkartendrucker
tickler coil Rückkopplungsspule
tidal Gezeiten-...
tidal-power plant Flutkraftwerk, Gezeitenkraftwerk
tide Flut, Gezeitenströmung
tide gauge Pegel
tie Zugstab, Zugstrebe, Strebe, Zuganker
tie bar Zugstab, Zugstange, Zuganker
tie beam Zugbalken, Ankerbalken
tie bolt Spannbolzen
tie feeder Kuppelleitung
tie line Querverbindung (621.395)
tie trunk Querverbindung (621.395)
tie wire Seilschlinge (621.315.6); Zugdraht, Ankerdraht
to tie spannen, abspannen, verspannen, abbinden
to tie back zurückbinden
to tie down herunterbinden
to tie off a cable ein Kabel abbinden
to tie together zusammenbinden, verbinden, vereinigen; zusammenschalten (621.3)
to tie up binden, als Einsatz verwenden, als Füllung verwenden (54); zubinden
tied by elastic elastisch gebunden
tier Lage, Schicht; Strahlerebene (321.696.67); (of winding) Etage
tight dicht; straff, gespannt, angespannt
tight cord gespannter Strick, straffer Strick
tight coupling feste Kopplung
tight fit strammer Sitz
to be a tight fit stramm sitzen
tight-fitting dicht passend, stramm sitzend
tight knot fester Knoten

tight push fit fester Schiebesitz
tight rope straffes Seil
tight side of a belt Zugseite eines Riemens
tight texture dichtes Gefüge
to tighten anziehen, spannen; abdichten, dichten
tightening Dichtungs-...; Spann-...
tightening belt Spannriemen
tightening compound Dichtungsmasse
tightening key Spannschlüssel
tightening pulley Spannrolle
tightness Dichtigkeit
tile (Mosaik-)Baustein (621.316.5)
tile with electrical function funktioneller Baustein
tilt Schieflage, Schräglage, Neigung; Abstrahlwinkel (621.396.67)
tilt angle Neigungswinkel
tilt gauge Neigungslehre
to tilt kippen, schrägstellen, neigen, kanten
to tilt over überkippen, umkippen, kanten
to tilt with respect to schrägstellen gegen
tiltable schrägstellbar
tiltable press Kipp-Presse
tilting Kippung, Neigung, Kippvorgang; Kipp-...
tilting furnace Kippofen
tilting-insulator switch Schalter mit Kippstützer
tilting moment Kippmoment
tilting movement Kippbewegung
tilting movement of the bulb Kippbewegung des Kolbens (621.314.6)
tilting tendency Kippneigung, Neigung zum Kippen
timber Nutzholz, Bauholz; Holzbalken
timbre Klangfarbe, Timbre (534)
time Zeit, Uhrzeit
time amplitude Zeiger, Zeitvektor (621.3)
time announcement Zeitansage
time axis Zeitachse (51)
time base Zeitablenkung, Zeitnullinie (621.385); Zeitbasis (621.317)
time-base generator Zeitbasis-Generator
time-clock Stechuhr, Kontrolluhr
time constant Zeitkonstante
time-continuous signal zeitkontinuierliches Signal
time-current characteristic of a fuse Kennlinie einer Sicherung
time cycle zeitlicher Zyklus
time decrease of permeability Permeabilitätsabnahme durch Alterung
time delay zeitlicher Verzögerung, Verzögerungszeit; Laufzeit (621.392), (621.396.62)
time-delay circuit breaker Schalter mit verzögerter Auslösung
time-delay connection Verzögerungsschaltung
time-delay relay Zeitrelais, Verzögerungsrelais
time derivative zeitliche Ableitung
time-detined signal zeitdiskretes Signal
time discriminator Zeitdiskriminator
time-distribution system Zentraluhrenanlage
time-division multiplex Zeitmultiplex, Zeitvielfach
time element Zeitglied, Zeitrelais (621.3)
time-equalization circuit Zeitfehlerausgleich
time factor Zeitkonstante
time function Funktion der Zeit, zeitlich veränderliche Funktion (51)
time fuse Zeitzünder (623)
time gate Zeitgatter
time grading Zeitstaffelung (621.316)
time to half-crest on the wave tail Rückenhalbwertszeit
time history zeitlicher Ablauf, zeitliche Folge
time to impulse flashover Stoßüberschlagsverzögerung
time interval Pause; Zeitabschnitt, Zeitraum
time-interval meter Zeitmesser, Zeitintervallmesser
time litter zeitliche Verwackelung (621.397)
time keeper Zeitnehmer (621.9)
time-keeping quality Ganggenauigkeit (681)
time lag Zeitverzögerung, Verzug; Nacheilung (621.3)
time-lag attachment Zeitelement (621.318)
time-lag circuit Verzögerungsschaltung
time-lag circuit breaking verzögerte Abschaltung
time-lag tripping verzögerte Auslösung
time limit zeitliche Begrenzung, zeitliches Limit
time-limit circuit breaker Zeitschalter
time-limit relay Zeitrelais
time-limit start relay Anlaßzeitrelais
time-limit switch Zeitschalter
time-loading test Zeitlastprüfung
time meter Zeitzähler
time metering Zeitmessung
time of persistence Dauer des Vorhandenseins, Dauer, Dauer des Bestehens; Nachleuchtdauer (621.396.96)
time-pulse metering Zeitimpulszählung
time-pulse sender Zeittaktgebermaschine

time-rate of change zeitliche Änderungsgeschwindigkeit
time rating Versuchsdauer (621.313); Prüfdauer
time release Zeitauslösung
time resolution Zeitauflösung, Zeitauflösungsvermögen
time scale Zeitskala, Zeitmaßstab
time schedule Programm, zeitliches Programm, Terminplan
time-sharing operation Zeitmultiplexbetrieb; Teilnehmerverfahren, Timesharing (681.3)
time signal Zeitsignal, Zeitzeichen
time slice Zeitscheibe (681.3)
timeslot Zeitscheibe
time stability zeitliche Konstanz
time standard Zeitnormal
time of swing Schwingungsdauer
time switch Zeitschalter, Schaltuhr
time table Fahrplan, Zeitplan
time temperature program Temperaturprogramm
time-tested dauererprobt, lange bewährt
time trace Zeitmarke
time-travel control Zeit-Weg-Steuerung
time undervoltage protection verzögerter Unterspannungsschutz
time-variable zeitlich veränderlich
time vector Zeiger, Zeitvektor (538)
time yield zeitliche Nachwirkung
to time Zeitnehmen, Zeit abstoppen, Zeitmessen; Zeit festlegen, Zeit bestimmen, zeitlich bemessen
timed zeitlich festgelegt, zeitlich bemessen, zeitlich gesteuert
timed pulse zeitlich definierter Impuls (621.373)
timed recall Wiederanruf
timed spark transmitter gesteuerter Funkensender
timer Zeitgeber (681.3); Zeitschalter, Impulsgeber, Impulszentrale (621.373); Taktgeber (621.396.96); Zeitglied, Zeitelement
timing Zeitbemessung, zeitliche Festlegung
timing circuit Zeitgeber
timing of a clock Gangregulierung einer Uhr
timing diagram Impulsdiagramm
timing element Zeitgeber
timing generator Taktsignalgenerator
timing pulse Synchronisierungsimpuls, Synchronisierimpuls
timing range (zeitlicher) Einstellbereich
timing relay Zeitrelais, Verzögerungsrelais
timing shaft Synchronisierwelle
timing stroke Synchronisierimpuls, Taktimpuls
tin Zinn; Büchse, Weißblechbüchse, Konservendose
tin-coated verzinnt
tin-coated wire verzinnter Draht
tin coating Verzinnung
tin foil Stanniol, Silberpapier
tin-foil electrode Stanniol-Elektrode, Stanniolbelag
tin plate Weißblech
tin smith Klempner, Spengler
tin-smith's shears (pl.) Blechschere
tin tack verzinnter Nagel
to tin verzinnen; in Konservendosen einmachen, konservieren
to tin-plate verzinnen
tinned conductor verzinnter Leiter
tinned copper wire verzinnter Kupferdraht
tinned iron Weißblech
tinned wire verzinnter Draht
tincal Borax
tindery zunderig, verzundert
tinge Farbton, Färbung, Tönung, Hauch
to tinge färben, tönen, farbig anhauchen
tinning Verzinnung
tinsel cord Litzenleitung (Litzen bestehen aus Band)
tint Tönung, Färbung, Farbton
to tint färben, tönen
tip Spitze; Scheitelwert (einer Kurve); Neigung, Schiefstellung; Schneidblättchen (621.9); Brennermundstück, Brennermund (621.791); Elektrode (621.791)
tip clearance Kopfspiel (621.8)
tip of the hysteresis loop Spitze der Hysteresisschleife
tip of the plug pin Spitze des Steckerstiftes
tip of the tool Stahlspitze, Schneidkante
tip wire a-Ader (621.395)
to tip mit einer Schneide versehen, mit einer Spitze versehen; neigen, schiefstellen; ausschütten, auskippen, herunterkippen, neigen, kippen, abkippen; an der Schneide belegen, mit einer Auflage versehen (621.9)
to tip up hochkippen, schiefstellen
tipped an der Spitze beschlagen
tipped tool Hartmetallwerkzeug, Werkzeug mit Widiaschneide, Stahl mit Schneideauflage
tipping Kippen, Ausschütten, Entleeren
tipping stage Kippbühne
tire (US) = **tyre** (Br.) Reifen, Gummireifen; Radkranz

to tire ermüden, müde machen; müde werden, ermüden; bereifen (US)
T-iron T-Eisen
Tirrill regulator Tirrill-Regler
tissue Gewebe; Mullgewebe, Gaze
tissue paper Seidenpapier
titania Titanoxid, Rutil
titanium Titan
to titrate titrieren
titration Titrierung, Titration
T-joint Abzweigmuffe, T-Muffe (621.315.2)
T-network T-Schaltung, Vierpol in T-Schaltung
to and fro hin und her
Toepler-Holtz machine Toepler-Holtz-Influenzmaschine
Toepler machine Toepler-Influenzmaschine
toggle Winkelhebel, Kipphebel, Kniehebel, Wippe; bistabile Kippschaltung; Knebel (629.12)
toggle action Kniehebelwirkung
toggle circuit Flipflop-Schaltung
toggle lever Kniehebel
toggle switch Kellog-Schalter, Kippschalter
toggle trigger bistabiler Multivibrator
token [Kurzbefehl in BASIC]
tolerance Toleranz (erlaubtes) Abmaß
tolerance indicator Toleranzanzeiger
tolerance on resistance Widerstandstoleranz
tolerance standard Toleranzdosis (621.396), (539)
tolerance on thickness Dickentoleranz
to tolerate tolerieren, zulassen
toll Schnellverkehrs-... (Br.); Fernverkehrs-... (US)
toll board Schnellverkehrsschrank
toll cable Fernkabel
toll call Ferngespräch
toll centre Knotenvermittlungsstelle, Überweisungsvermittlung
toll exchange Fernvermittlung
toll line Schnellverkehrsleitung (Br.); Fernleitung (US) (621.395)
toll office Schnellamt (Br.); Fernamt (US) (621.395)
toll-switching trunk Überweisungsleitung, ÜL
toluene Toluol
tommy bar Spanneisen, Windstift (621.8)
ton Tonne (Gewichtseinheit)
tons ft. Tonnenfuß (1 t. ft. = 1 t. × 1 ft. = 0,305 mt)
tons sw. in. Tonnen pro Quadratzoll (1 t./sq. in. = 156 kg/cm^2)
tons/sw. yd. Tonnen pro Quadratyard (1 t./sw. yd. = 1,2 t/m^2)
tonalities Hörtöne
tone Ton (534); Ton, Tönung (77)
tone arm Tonarm
tone cadence Tonsignal-Rhythmus
tone-compensated tonrichtig, gehörrichtig (621.396.62)
tone-compensated volume control gehörrichtiger Lautstärkenregler
tone control Tonblende, Klangregler (621.396.62)
tone-control aperture Tonwertblende
tone corrector Klangfarbenregler
tone frequency Tonfrequenz
tone-frequency telegraphy Tonfrequenztelegraphie
tone generator Summer, Tonfrequenzgenerator
tone-modulated waves tonfrequent modulierte Wellen
tone ring Tonruf
tone oscillator Tongenerator (621.396.61)
tone source Tonfrequenzspannungsquelle
tone wedge Tonwertskala
tongs (pl.) Zange
tongue Zunge (61); Lasche (68); Feder (69)
to tongue mit einer Feder versehen (69)
tongued flooring gespundeter Fußboden
tonic train sinusförmig modulierter Wellenzug
tonic-train signalling Tonfrequenz-Zeichengebung
tonnage Tonnage; Menge in t
tool Werkzeug; Drehstahl (621.9); Hilfsprogramm (681.3)
tool bit Schneidplättchen (621.9)
tool box Werkzeugkasten, Werkzeugkiste; Sammlung von Hilfsprogrammen (681.3)
tool chest Werkzeugkasten
tool kit Werkzeugsatz, Werkzeugkiste, Werkzeugkasten, Werkzeugbesteck
tool maker Werkzeugmacher, Werkzeugschlosser
tool mark Bearbeitungsspur
tool setter Maschineneinrichter
tool steel Werkzeugstahl
tool tip Schneidplättchen, Hartmetallauflage (621.9)
to tool mit Werkzeugen bestücken (621.9); behauen (69); prägen (655)
tooling accuracy Bearbeitungsgenauigkeit
tooling sequence Bearbeitungsfolge
toot Tuten
tooth (pl. **teeth**) Zahn (61); (621)
tooth breadth Zahnbreite
tooth breakage Zahnbruch (621.8)

tooth clutch Zahnkupplung
tooth contour Zahnform, Zahnprofil
tooth-pulsation frequency Zahnfrequenz, Nutfrequenz (621.313)
tooth ratio Zahnbreite, Nutbreite
tooth ripples Nutenobertöne
tooth saturation Zahnsättigung (621.313)
tooth tip Zahnspitze (621.313)
tooth-top leakage Zahnkopfstreuung (621.313)
to tooth zähnen, mit Zähnen versehen
toothed bar Zahnstange
toothed rim Zahnkranz
toothed wheel Zahnrad
toothed wheel gear Zahnradgetriebe
top Spitze, Scheitel, oberster Punkt; Gewindespitze (621.8); Oberkasten (621.74); Topp (629.12)
top cap Anodenkappe, Gitterkappe, Röhrenkappe (621.385)
top coating Deckanstrich
top crust Deckkruste (621.355)
top edge Oberkante
top face Oberseite, obere Stirnfläche
top-fed antenna obengespeiste Antenne
top frame Schalterdeckel (621.315.6)
top limit obere Grenze
top-loaded aerial dachbelastete Antenne
top pressure Druck von oben
top of section Oberkante Profil
top speed Höchstgeschwindigkeit
top of the stroke oberer Totpunkt
top vane oberste Platte (621.319.4)
to top maximal erreichen, sich höchstens belaufen auf; die Spitze erreichen; auflaufen zu; die Spitze abschneiden; kappen
to top 120 lbs auf 120 Pfund auflaufen; 120 Pfund erreichen
topping up Nachfüllen (621.355)
topaz Topas
topographic(al) topographisch
topographic map (US) Generalstabskarte
topography Topographie
torch Fackel; Brenner (621.791)
torch case Taschenlampe, Taschenlampenhülse
torch light Taschenlampe; Fackel, Pechfackel
torch orifice Brennermund (621.791)
tornado Tornado, Wirbelsturm
toroid Toroid, ringförmiger Körper
toroid core Ringkern
toroidal toroidförmig, ringförmig
toroidal coil Toroidspule, Ringspule
toroidal transformer Ringkerntransformator
toroidal winding Ringwicklung
torque Drehmoment, Triebmoment (621.317)
torque amplifier Momentenverstärker, Drehmomentenverstärker
torque arm Hebelarm
torque differential receiver Momentdifferentialempfänger, Drehmelder, Differentialempfänger
torque differential transmitter Momentdifferentialgeber
torque meter Momentenwaage, Drehmomentmesser
torque moment Drehmoment
torque motor Drehmomentantrieb, Motor mit konstantem Drehmoment
torque output Ausgangsdrehmoment, Antriebsmoment (621.34)
torque on a shaft Moment an einer Welle
torque at starting Moment beim Anlauf, Anlaufmoment
torque synchro Momentdrehmelder
torque transmitter Momentgeber
torque wrench Momentenschlüssel
torrid zone heiße Zone
torsiometer Momentenwaage (621.313)
torsion Torsion, Drehung, Drillung, Verdrehung
torsion balance Torsionswaage
torsion bar Torsionsfeder
torsion galvanometer Torsionsgalvanometer
torsion meter Torsionsmeßgerät, Verdrehungsmeßgerät
torsion-string galvanometer Spannbandgalvanometer, Schleifengalvanometer
torsion test Torsionsversuch, Torsionsprüfung, Verdrehungsprüfung
torsional Verdrehungs-..., Drill-..., Torsions-...
torsional force Torsionskraft, Drehkraft
torsional modulus Schubmodul
torsional moment Torsionsmoment, Drehmoment
torsional strain Verdrehungsbeanspruchung, Torsionsbeanspruchung
torsional strength Schubfestigkeit, Verdrehungsfestigkeit, Torsionsfestigkeit
torsional stress Torsionsspannung
torsional vibration Torsionsschwingung, Drillschwingung
torsional vibration of a crystal Kristalldrehungsschwingung
torus Torus, Ringkörper, Ring
tot = total
to tot = to total
total Gesamtzahl, Gesamtsumme, Summe

total Gesamt-..., total, vollständig
total barring Vollsperre (621.395)
total break time Ausschaltdauer
total capacitance Totalkapazität
total carbon Gesamtgehalt an Kohlenstoff
total characteristic Steuerspannungskennlinie (621.385)
total current regulation Generatorstromregelung (621.32), (621.33)
total cyanide Gesamtgehalt an Cyankali (621.357)
total eclipse totale Finsternis
total efficiency Gesamtwirkungsgrad
total emission Kathodenergiebigkeit, Kathodensättigungsstrom
total emissivity Gesamtstrahlung
total failure Gesamtausfall
total flux Gesamtfluß
total harmonic content Gesamtgehalt an Oberwellen
total heat Wärmeinhalt
total heating surface gesamte Heizfläche
total load Gesamtlast, Gesamtbelastung
total loss Gesamtverlust, Gesamtdämpfung
total loss reference equivalent Gesamtbezugsdämpfung
total modulation vollkommene Modulation
total output Gesamtleistung
total output panel Gesamtleistungsfeld (621.311)
total power loss Gesamtverlustleistung
total range of controller output Regelbereich (621.316.7)
total reflection totale Reflexion
total reflection factor Koeffizient der Gesamtreflexion
total resistance Gesamtwiderstand (621.3)
total result Gesamtergebnis
total stock Gesamtbestand
total sum Gesamtsumme
total transmission loss Systemdämpfung
total voltage Gesamtspannung
totally closed slot (ganz) geschlossene Nut
totally enclosed vollständig gekapselt
totally enclosed drive Innenantrieb (621.9)
totally enclosed fan-cooled machine ganz geschlossene Maschine mit Mantelkühlung
totally enclosed machine ganz geschlossene Maschine, Maschine mit Luftkreislauf
totally enclosed motor gekapselter Motor
totally enclosed switchgear gekapseltes Schaltgerät
to total zusammen ergeben, sich addieren zu, sich belaufen auf, zusammenzählen, summieren
to totalize zusammenzählen, addieren
totalizing anemometer Windwegmesser
totalizing relay Summenrelais
totalizer Summenzählwerk
tottery wackelig
touch Berührung; Anschlag; Anflug (**of:** von)
touch button Sensortaste
touch-down point light Aufsetzpunktfeuer
touch of the piano Anschlag des Klaviers
touchpanel keyboard Tipp-Tastenfeld, Sensor-Tastenfeld
touch-sensitive switch berührungsempfindlicher Schalter, Sensorschalter
touch terminal berührungsempfindliches Terminal
touch tuning Druckknopfabstimmung (621.396.62)
to touch berühren, anfassen; berühren, tangieren (51)
to touch up schärfen, nachschärfen (621.9); schönen, retuschieren (77)
tough zäh, widerstandsfähig, schwer bearbeitbar
tough copper Raffinadekupfer
tough rubber sheathed vollgummi-isoliert, Vollgummi-... (621.3)
tough steel zäher Stahl, schwer bearbeitbarer Stahl
tough temper zähes Gefüge
tough wood zähes Holz
to toughen zäher machen
toughness Zähigkeit
tourmaline Turmalin
tow hook Zughaken, Schlepphaken
to tow schleppen; treideln (626)
towage Schleppen; Treideln
tower Turm; Mast, Eisenmast (621.315); Tower, Flughafenkontrolle (357.814)
tower base Mastfuß
tower-earthing resistance Masterdungswiderstand
tower footing Mastfuß
tower loading Mastbelastung, Mastbeanspruchung
tower radiator Maststrahler, selbstschwingender Mast
tower winding Turmförderung
towing Schleppen; Treideln
towing cable Schleppseil
towing engine Treidellokomotive
towing line Schleppseil

township (US) (Flächenmaß = **6** × **6 sq.miles** ≈ 10 × 10 km²)
T. R. = transmitter receiver
T. R. box Sende-Empfangsverriegelung; Nullode (621.396.96)
T. R. tube Empfängersperröhre
trace Spur, Schreibspur, Leuchtspur (621.385); Spur, winzige Menge (54); Trasse (621.315); Ablaufverfolgung (681.3); Leiterbahn (621.374)
to trace aufspüren, verfolgen; nachweisen (54); einer Spur folgen; schreiben, zeichnen; trassieren; abtasten, abfühlen
to trace out herausfinden, aufspüren; aufschreiben, aufzeichnen (621.385)
to trace out a curve eine Kurve beschreiben
traceable zurückführbar (**to:** auf); auffindbar, nachweisbar
tracer Leuchtspurgeschoß; Träger; radioaktives Spurenelement; Kennfaden, Beilauffaden (621.315.2); Fühler (621.316.7)
tracer control Fühlersteuerung
tracer element radioaktives Spurenelement, Tracer-Element
tracing Pause, Lichtpause; Durchpausen, Durchzeichnen; Trassierung; Ablaufverfolgung (681.3)
tracing cloth Pausleinwand
tracing paper Pauspapier, Zeichenpapier
tracing program Überwachungsprogramm (681.3)
track Pfad, Spur, Bahn; Gleis, Fahrbahn, Geleise, Schienenstrang; Spurweite; Kette, Fahrkette, Raupenkette; Flugweg, Zielweg, Kurs über Grund (351.814)
track brake Gleisbremse (621.33)
track circuit Schienenstromkreis, Gleisstromkreis
track-diagram control-panel Gleisbildstelltisch
track-diagram switchbox Gleisbildstellwerk
track display Flugwegdarstellung, Zielwegdarstellung
track element schienenseitiger Geräteteil (625); Spurelement (681.84)
track guidance system Leitliniensystem (623)
track indicator panel Gleisbildtafel (625)
track instrument Gleiskontakt (65)
track-interlock system Spurplanstellwerk
track magnet Gleismagnet
track rail bond Schienenverbinder, Schienenstoßverbinder
track relay Gleisrelais
track return circuit Gleisrückleitung
track switch Schienenschalter, Streckenschalter
track wheel Laufrad
track width Spurweite, Spurbreite (681.84)
to track verfolgen, in der gleichen Bahn nachlaufen, spuren; nachführen; gleichlaufen (621.396.62)
tracker wire Betätigungsseil, Zugseil
tracking Trimmen, Gleichlaufabstimmung (621.396.62); Zielverfolgung (623); Kriechspurbildung (621.315.6); Spureinstellung (681.84)
tracking capacitor Padding-Kondensator, Gleichlauftrimmer
tracking radar Verfolgungsradar
trackless nicht schienengebunden
trackless trolley bus Obus, Oberleitungsbus
trackway element schienenseitiger Geräteteil (625)
traction Zug, Zugkraft
traction battery Fahrzeugbatterie
traction circuit Speisestromkreis
traction engine Zugmaschine
traction generator Bahnstromgenerator, Fahrzeuggenerator (621.33)
traction lamp Zugbeleuchtungslampe
traction motor Bahnmotor, Fahrmotor (621.33)
traction rectifier Bahngleichrichter
traction switch Fahrschalter (621.33)
traction transformer Bahntransformator, Bahnumspanner
traction supply Bahnstromversorgung
traction-supply line Bahnstromleitung
tractive Zug-...
tractive effort Radumfangskraft, Zugkraft
tractive force Zugkraft (621.33); (538)
tractive power Zugleistung (621.33)
tractive power of a magnet Zugkraft (eines Magneten), Tragkraft (621.318.2)
tractive power unit Motorwagen (621.33)
tractor Traktor, Schlepper, Trecker, Zugmaschine
trade mark Fabrikmarke, Warenzeichen
trade name Handelsbezeichnung
traffic Verkehr; Handel; Fernmeldeverkehr
traffic balancing Verkehrsausgleich (621.39)
traffic capacity Leistung, Verkehrsleistung eines Leitungsbündels (621.395)
traffic carried Belastung (621.395)
traffic control Verkehrsregelung
traffic-control system Verkehrsregelung, Verkehrssignalanlage
traffic density Verkehrsdichte

traffic distributor Anrufverteiler (621.395)
traffic flow Verkehrswert (621.395)
traffic intensity Verkehrswert (621.395)
traffic interruption Verkehrsunterbrechung, Verkehrsstörung
traffic light Verkehrsampel
traffic load Verkehrsbelastung (621.395)
traffic offered Angebot (621.395)
traffic signalling equipment Verkehrssignalanlage
traffic volume Verkehrsmenge (621.395)
T-rail T-Schiene
to rail anhängen, schleppen lassen, schleppen, nachschleppen, dranhängen
trailer Anhänger
trailer car Anhänger, Anhängerwagen
trailer control Anhängerfahrschaltung, Fahrschaltung vom Anhänger aus (621.33)
trailing Schlepp-..., Hinter-..., nachgeschleppt
trailing aerial Schleppantenne
trailing axle Hinterachse
trailing bogie Laufgestell (625)
trailing brush mitläufige Bürste
trailing brush holder schräger Bürstenhalter
trailing cable Schleppseil; Schleppkabel, loses Zuführungskabel
trailing car Anhänger, Anhängewagen
trailing contact Schleppkontakt
trailing edge Hinterkante, ablaufende Kante, abfallende Flanke (621.3), (629.113)
trailing pole horn ablaufende Polkante
trailing tip ablaufende Polkante
trailing wire Schleppdraht, Schleppantenne
train Zug (625); Zug, Kette, Folge; Rädersatz (621.8); Zündschnur (66)
train control Zugsteuerung
train-control relay Geschwindigkeitsüberwachungsrelais (625)
train deceleration Zugbremsung
train describer Zugfolgeanzeiger (625)
train of gears Rädersatz (621.8)
train lighting Zugbeleuchtung
train-lighting dynamo Zugbeleuchtungs-Dynamo
train shunting radio (service) Rangierfunk
train speed Zuggeschwindigkeit
train tonnage Zuggewicht, Zuglast
train of waves Wellenzug
train of wheels Rädergetriebe, Rädersatz, Räderfolge (621.8)
training Ausbildung, Fortbildung
training reactor Ausbildungsreaktor
trajectory Trajektorie, Flugbahn
tram Straßenbahn, Straßenbahnwagen; „Hund“, Förderwagen (622)
tramcar Straßenbahnwagen
tram mains Straßenbahnoberleitung, Straßenbahnleitung
tram motor Straßenbahnmotor
tram rail Straßenbahnschiene, Rillenschiene
tram standard Straßenbahnmast, Fahrleitungsmast
tramway Straßenbahn
tramway motor Straßenbahnmotor
tramway pole Straßenbahnmast
trammel head Stangenzirkel
trams (pl.) Stangenzirkel
transadmittance Transadmittenz (621.372.5); Steilheit (621.385)
transatlantic transatlantisch
transatlantic cable transatlantisches Kabel
transceiver Sende-Empfangsgerät
transcendental transzendent (51)
transcendental function transzendente Funktion
transcendental number transzendente Zahl
transconductance Steilheit (621.385); Gegenwirkleitwert
transcontinental transkontinental
to transcribe umschreiben, übertragen (65)
transducer Energiewandler, Meßwandler, Meßwertaufnehmer, Meßwertumformer, Wandler; Umformer (621.316.7)
transducer loss Wirkdämpfung
transducer power gain Betriebs-Leistungsverstärkung (621.375)
transductor Transduktor, Magnetverstärker
transfer Transfer, Transferierung, Übertragung; Transport
transfer box Hausanschlußkasten (621.315.6)
transfer characteristic Steuerkennlinie (621.385); Übertragungskennlinie (621.375), (621.312.5)
transfer circuit Dienstleitung (621.395)
transfer coefficient Übertragungsmaß
transfer command Transferbefehl
transfer constant Betriebsübertragungsmaß
transfer curve Übertragungskurve
transfer current Transferstrom
transfer of energy Energieübertragung
transfer impedance eingekoppelter Widerstand (621.396)
transfer instruction Transferbefehl
transfer lag Übertragungsverzug (621.5)
transfer operator Verschiebungsoperator (51)
transfer rate Übertragungsgeschwindigkeit (621.394)

transfer switch unterbrechungsfreier Umschalter
transfer value Transferwert
to transfer transferieren, übertragen; überführen, transportieren; übertragen (51)
to transfer back rückübertragen; rückwirken (621.396)
transfiguration Transfiguration, Umwandlung
to transfigure transfigurieren, umwandeln
transform Transformierte (51)
to transform transformieren, umwandeln (51), (54); transformieren, umspannen (621.3)
to transform to umwandeln in; sich umwandeln in, übergehen in
transformation Transformation, Umwandlung, Umsetzung (54); Transformation, Umwandlung (51); Transformation, Umspannung (621.3)
transformation constant Zerfallskonstante (539)
transformation period Halbwertzeit (539)
transformation ratio Übersetzungsverhältnis
transformer Transformator, Umspanner (621.3); Übertrager (621.39)
transformer bank Transformatorensatz
transformer bar Stabtransformator
transformer bay Transformatorenfeld
transformer bobbin Transformatorspule, Umspannerwicklung
transformer cell Transformatorenzelle, Umspannerzelle
transformer circuit Transformatorenkreis
transformer-connected meter Wandlerzähler
transformer core Transformatorenkern, Umspannerkern
transformer-coupled übertragergekoppelt
transformer coupling Übertragerkopplung, transformatorische Kopplung, induktive Kopplung
transformer current Transformatorenstrom, Umspannerstrom
transformer dissipation Transformatorverluste, Umspannerverluste (pl.)
transformer effect Transformatoreneffekt
transformer failure Transformatorenausfall
transformer-fed transformatorisch gespeist
transformer generator Drehfeldumformer
transformer input vom Transformator aufgenommene Leistung
transformer lamination Transformatorenblech
transformer losses Transformatorverluste
transformer oil Transformatorenöl
transformer output vom Transformator abgegebene Leistung
transformer-overcurrent tripping indirekte Auslösung, Stromwandlerauslösung
transformer plant Umspannwerk
transformer plate Transformatorenblech
transformer primary winding primäre Transformatorwicklung
transformer protection Transformatorenschutz
transformer-protective device Transformatoren-Schutzapparat
transformer rating Transformatorenleistung
transformer relay Umspannerrelais
transformer secondary winding sekundäre Transformatorwicklung
transformer sheet Transformatorenblech
transformer stampings (pl.) Transformatorblechschnitt
transformer station Umspannwerk, Transformatorenstation
transformer sub-station Umspannwerk
transformer switch Transformatorenschalter
transformer tank Transformatorenkessel
transformer terminal Transformatorenklemme, Übertragerklemme
transformer type Transformatorentype, Umspannerbauart
transformer-undercurrent tripping indirekte Nullstromauslösung
transformer vault Transformatorenzelle, Umspannerzelle
transformer voltage Transformatorspannung
transformer winding Umspannerwicklung, Transformatorenwicklung
transformer yoke Transformatorenjoch
transhybrid loss Gabelübergangsdämpfung
transhorizon link Überhorizontverbindung
transhorizon propagation Ausbreitung über den Horizont
transient Ausgleichsvorgang, Überspannung
transient vorübergehend, flüchtig, einmalig, zeitweilig
transient current freier Strom, Ausgleichsstrom
transient current ratio in saturation Umschaltstromverhältnis bei Sättigung, Umschaltstromverhältnis bei Übersteuerung (621.382.3)
transient distortion Sprungverzerrung
transient effect Laufzeitverzerrung
transient fault selbstlöschender Fehler, vorübergehender Fehler
transient noise Geräusch durch Einschwingvorgänge (621.372)
transient oscillation Ausgleichschwingung

transient phenomenon Ausgleichsvorgang; Einschwingvorgang
transient power limit dynamische Kippleistung (621.311)
transient protection Überspannungsschutz
transient protective device Wanderwellenschutz
transient reactance Stoßkurzschlußreaktanz
transient reaction Ausgleichsvorgang, freier Vorgang
transient recovery voltage beim Ausgleichsvorgang wiederkehrende Spannung (621.316.5)
transient recovery voltage rate Steilheit der beim Ausgleichsvorgang wiederkehrenden Spannung (621.316.5)
transient response Einschwingverhalten, Sprungantwort
transient stability dynamische Stabilität (621.313)
transient stability factor dynamische Kippleistung/Nennleistung (621.31)
transient stability limit Grenze der dynamischen Stabilität
transient state Einschwingzustand (eines Systems)
transient thermal resistance transienter Wärmewiderstand (621.382)
transient voltage Überspannung, freie Spannung, Ausgleichsspannung; Stoßspannung
transient wave Wanderwelle
transiently stable dynamisch stabil
transimpedance Transimpedanz
transistor Transistor
transistor chopper Transistor-Zerhackerschaltung
transistor current gain Stromverstärkung (eines Transistors)
transistor parameter test set Transistorprüfgerät, Transistor-Parameter-Meßgerät
transistor (radio) Transistorradio
transistor switch Transistorschalter, Schalttransistor
transistorized transistorisiert
transit Transit, Durchgang; Transport, Durchgangstransport; Durchgang (52)
transit exchange Durchgangsvermittlungsstelle
in transit to auf dem Transport nach
transit(-phase) angle Laufzeitwinkel
transit register Durchgangsregister (621.395)
transit time Laufzeit, Übergangszeit; Durchgangszeit (621.382)
transit-time tube Laufzeitröhre
transit transmission Transitübertragung (621.395)
transite [Handelsname für ein verlustarmes Isoliermaterial]
transition Übergang, Umwandlung; Einschwingvorgang, Übergangsvorgang, Stoßvorgang
transition anode Kommutierungsanode
transition curve Übergangsbogen (625.1)
transition frequency Übergangsfrequenz, Dämmerungsfrequenz (621.396.67); Transitfrequenz (621.382.2)
transition interval Übergangsstadium; Laufzeit (621.395); Umwandlungszeit
transition period Übergangszeit
transition point Übergangspunkt, Umwandlungspunkt (669)
transition range Übergangsgebiet
transition region Übergang, Zonenübergang
transition resistance Überschaltwiderstand
transition time Übergangszeit
transition stop Zwischenschaltstellung
transition temperature Umwandlungstemperatur, Umwandlungspunkt
transition zone Übergangszone (69), (621.791)
transitional Übergangs-...
transitory vorübergehend
transitory line-to-ground fault Erdschlußwischer
transitron Transitron, Dynatron (621.396.61)
to translate übersetzen; verschieben, translatorisch bewegen (53)
to translate bodily als Ganzes verschieben
to translate from ... to ... vom ... bis ... übersetzen; von ... nach verschieben
translating relay Zeichenumsetzrelais
translation Übersetzung; Translation, Verschiebung (53)
translator Übersetzer; Zeichenumsetzer (621.394); [Programm zur Übersetzung einer Programmsprache in eine andere (681.3)]
translay protective system Längsdifferentialschutz
translucent durchscheinend
transmissibility Durchlässigkeit (535)
transmissible durchlässig (535); übertragbar (621.396)
transmission Übertragung; Kraftübertragung, Getriebe (621.3); Sendung (621.396)
transmission admittance Übertragungsadmittanz
transmission band Sendeband; Durchlaßbereich (621.396)

transmission box Getriebekasten (621.9), (629.113)
transmission capsule Sprechkapsel
transmission coefficient Übertragungsfaktor
transmission deception Täuschung durch Ausstrahlung (623)
transmission dynamometer Durchgangsleistungsmesser
transmission efficiency Übertragungswirkungsgrad
transmission of energy Energieübertragung
transmission factor Durchlässigkeit (535); Übertragungsfaktor
transmission feeder Übertragungsleitung
transmission gain Übertragungsgewinn, Verstärkung
transmission gear Übersetzungsgetriebe, Vorgelege (621)
transmission of heat Wärmeübergang
transmission jamming Störung durch Ausstrahlung (623)
transmission lag Übertragungsverzug, Übertragungszeit
transmission level Pegel, Sendepegel (621.395)
transmission line Fernleitung, Übertragungsleitung, Antennenleitung; Hochspannungsleitung (621.3); Energieleitung (621.396)
transmission-line amplifier Leitungsverstärker
transmission-line ground Leitungserde
transmission-line theory Leitungstheorie
transmission loss Übertragungsdämpfung, Leitungsdämpfung, Durchgangsdämpfung (621.395); Leitungsverluste (pl.) (621.315)
transmission measuring set Pegelmeßgerät
transmission mode Übertragungsverfahren, Übertragungsart
transmission network Übertragungsnetz
transmission path Übertragungsweg; Nachrichtenpfad
transmission performance Übertragungsgüte
transmission plan Dämpfungsplan (621.395)
transmission of power Kraftübertragung
transmission range Übertragungsbereich, Durchlaßbereich (621.395); Reichweite (621.396)
transmission rate Übertragungsgeschwindigkeit
transmission ratio Übersetzung (621.8)
transmission reference equivalent Sendebezugsdämpfung
transmission reference system Ureichkreis (621.395)
transmission region Durchlaßbereich (621.392), (621.396)
transmission route Trasse, Leitungstrasse, Leitungsweg (621.315.1)
transmission shaft Getriebewelle
transmission of sound Schallübertragung; Schalldurchgang
transmission speed Übertragungsgeschwindigkeit
transmssion system Hochspannungsnetz, Übertragungsnetz, Fernleitungsnetz
transmission test set Übertragungsmeßgerät
transmission tower Hochspannungsmast, Leitungsmast
transmission unit (T. U.) Bel
to transmit senden (621.396); durchlassen (621.392), (535); übermitteln, weiterleiten
transmit distortion Sendeverzerrung
transmit-receive switch Empfängersperrschalter, Sende-Empfangsumschalter
transmitted energie Sendeenergie
transmitted frequency Sendefrequenz
transmitted wave Sendewelle
transmittance Durchlässigkeit (535)
transmitter Sender (621.396); Sprechkapsel, Mikrofon (621.395)
transmitter amplifier Sendeverstärker
transmitter-combining filter Senderweiche
transmitter control station Sendezentrale
transmitter distributor Lochstreifensender
transmitter finder Anrufsucher (621.395)
transmitter frequency Sendefrequenz
transmitter gyro Mutterkreisel
transmitter noise (US) Mikrofongeräusch
transmitter output valve Senderendröhre
transmitter-receiver Sender-Empfänger, Sende-Empfangsgerät
transmitter valve Senderöhre
transmitting Sende-...
transmitting aerial Sendeantenne
transmitting apparatus Sendeeinrichtung (621.394)
transmitting channel Sendekanal (621.395), (621.396)
transmitting circuit Sendekreis
transmitting earth Erdung auf der Sendeseite (621.394)
transmitting power Sendeleistung
transmitting relay Senderelais
transmitting set Sender, Sendegerät (621.396)
transmitting station Sender, Sendestation (621.395.396)
transmitting valve Senderöhre

transmutation Umwandlung (539)
transoceanic transozeanisch, Übersee-...
transparence Durchsichtigkeit, Transparenz
transparency Transparenz
transparent transparent, durchsichtig
transparent colour Lasur, Lasurfarbe
transparent paper Transparentpapier, Pauspapier
transparent region Durchlaßbereich (535); „Fenster" (535)
transponder Antworteinrichtung, Transponder
transport Transport; Transportgelegenheit, Transportmittel
transport cross section Transportquerschnitt (539)
transport number Überführungszahl (621.35)
transport by rail Schienentransport
transport ratio Feinwanderungskonstante (621.316.5)
to transport transportieren, befördern
transported winding Gitterstabwicklung
transportable transportabel, transportfähig, tragbar
transportation Beförderung, Beförderungsgelegenheit
transportation of contact material Materialwanderung (621.316.5)
transportation crane Laufkran
to transpose übertragen; verdrillen (621.315.1), (621.392); auf die andere Seite der Gleichung bringen; verschränken
transposed conductor verdrillter Stab, Röbl-Stab
transposed winding ausgekreuzte Wicklung, geröbelte Wicklung
transposer Umsetzer
transposition Verdrillung (621.315.1), (621.392); Leitungsverklinkung, Teilleiterweiche (621.313); Auf-die-andere-Seite-Bringen, Hinüberschaffen (51); Kreuzung (von Leitungen)
transposition error Transpositionsfehler
transposition insulator Verdrillungsisolator
transrector Transformator mit Gleichrichter
transuranic element Transuran
transversal recording Querspurverfahren (681.84)
transverse Transversal-..., Quer-..., quergerichtet
transverse conductance Querleitwert
transverse current Querstrom
transverse current microphone Querstrommikrophon
transverse differential protection Querdifferentialschutz
transverse dimension Querabmessung
transverse electric wave TE-Welle, H-Welle
transverse field Querfeld
transverse magnetic wave TM-Welle, E-Welle
transverse magnetization Transversalaufzeichnung (681.84)
transverse motion Querbewegung
transverse pressure Querdruck
transverse voltage Beeinflussungsspannung; Querspannung
transverse wave Transversalwelle
transverter Hochspannungs-Kommutatorgleichrichter
trap Falle; Syphon, Geruchsverschluß (69); Fangstelle, Zeithaftstelle (621.382); Sperrkreis
10 kc trap 10-kHz-Sperre (621.396.62)
trap circuit Haltestromkreis (621.318), (625)
to trap fangen, festhalten, abfangen
trapped electron angelagertes Elektron, eingefangenes Elektron
trapped gases Gaseinschlüsse (621.7), (66)
trapped mode Wellenleiter-Ausbreitungstyp
trapezium Trapez (51)
trapezoid Viereck (51)
trapezoidal viereckig
trapezoidal pulse Trapezimpuls
trautonium Trautonium
travel Weg, Weglänge; Kolbenweg (621.1), (621.4); Reise
travel control Wegregelung (621.316.7)
travel of radio waves Wellenausbreitung
travel-reversing switch Endumschalter (621.9)
to travel reisen; laufen; sich ausbreiten (621.396)
to travel in a circle auf einem Kreise laufen, sich auf einem Kreise bewegen
to travel down entlanglaufen an
to travel forwards and backwards hin- und zurücklaufen (621.392)
to travel through space sich räumlich ausbreiten
to travel with the velocity of sich mit einer Geschwindigkeit von ... ausbreiten
traveller Reisender, Fahrgast; Laufkran (621.8)
travelling Wander-..., Lauf-...
travelling bridge Schiebebühne, Verladebrücke
travelling crab Laufkatze
travelling crane Laufkran
travelling overvoltage Überspannungswanderwelle
travelling substation fahrbares Unterwerk
travelling wave Wanderwelle (621.3), laufende Welle (534), (621.3)
travelling-wave aerial Wanderwellenantenne
travelling-wave tube Wanderfeldröhre

travelling weight Laufgewicht
traverse Durchlauf, Umlauf
traverse limiting switch Endausschalter (621.9)
traverse movement Längsbewegung; Vorschub (621.9)
traverse of the piston Kolbenweg, Kolbenhub
to traverse durchlaufen, überstreichen; durchsetzen; vorschieben (621.9)
to traverse a cycle eine Periode durchlaufen
tray Tablett, (flache) Schale, Untersatz; flacher Vorratsbehälter; Schubkasten; Einschub; Etage, Tragbrett (621.356)
treadle Pedal, Fußtritt
treadle operation Fußantrieb, Tretantrieb
treble Sopran; Dreifaches (51)
treble control Hochtonblende (621.396.62)
treble equalization Höhenentzerrung
treble loudspeaker Hochtonlautsprecher, Hochtöner
to treble verdreifachen
treble dreifach
tree Baum
tree-like baumförmig, verästelt
tree system of wiring Steigeleitungssystem (696)
treeing Bleibaumbildung (621.355)
trefoil Kleeblatt
trega- Tera- (10^{12})
tregohm Teraohm
tremble zittern
trembler Kontakthammer, Kontaktklöppel, Wagnerscher Hammer
trembler bell Klingel mit Unterbrecher
trembler coil Klingelspule
trench Graben; Kabelgraben, Kabelschacht (621.315.2); Schützengraben (623)
to trench Gräben ziehen; Kabelgraben ausheben
trend Entwicklungsrichtung, Gang
to trend eine Richtung haben, geneigt sein, hinneigen (**to:** zu, nach)
trestle Bock, Tischbock, Gestellblock
t. r. f. = tuned radio frequency Abstimmfrequenz
triac Triac, bidirektionaler Thyristor
trial Prozeß, Verfahren, Gerichtsverfahren; Erprobung, Untersuchung, Probe, Prüfung, Abnahmeversuch
trial cut Probeschnitt
trial and error Probieren
trial and error method Probierverfahren, Näherungsverfahren, Iterationsmethode
trial function Ausgangsfunktion (51)
trial run Probelauf, Erprobung
triangle Dreieck (51); Zeichendreieck, Dreibock, Bockgerüst; Triangel (78)
triangular dreieckig, Dreiecks-...
triangular file Dreikantfeile, Sägenfeile
triangular hole Dreiecksloch
triangular impuls Dreieckimpuls
triangular scale Dreikantmaßstab
triangular thread Dreiecksgewinde, scharfgängiges Gewinde
triatomic dreiatomig
triboelectricity Reibungselektrizität
trichlorethylene Trichloräthylen
trichromatic dreifarbig
trickle Rieseln, Tröpfeln
trickle charge Dauerladung (621.355)
trickle charger Dauerladegerät
trickle pool Sickergrube
to trickle rieseln, tröpfeln
trickling cooler Rieselkühler
triclinic triklin
trick sequence Trickaufnahme
trifoliate dreiblättrig
trifurcating box (Dreifach-)Verzweigungsmuffe (621.315.2)
trig = trigonometry Trigo = Trigonometrie
trigatron Triggerröhre, Trigatron
trigger Abzug (623); Auslöser (77); Auslöseimpuls, Trigger (621.374)
trigger-action relay Kipprelais
trigger circuit Auslösekreis, Kippschaltung, Impulsgeberschaltung, Triggerkreis (621.396.96)
trigger device Auslöseeinrichtung
trigger oscillator (Auslöse-)Impulsgenerator
trigger pulse Ziehimpuls, Auslöseimpuls (621.374)
trigger relay Fallklappenrelais
trigger signal Auslösesignal
trigger spark Zündfunke, Auslösefunke, Hilfsfunke, Steuerfunke
to trigger abziehen, abdrücken (623.4); auslösen, triggern (621.374), (77)
triggering Auslösung
triggering input Setzeingang (631.374)
triggering level Auslösepegel
triggering voltage Kippspannung
trigonometrical trigonometrisch (51)
trigonometria function Kreisfunktion, Winkelfunktion
trigonometry Trigonometrie
trilateral dreiseitig
trillion (Br) Trillion, 10^{18}
trillion (US) Billion, 10^{12}
to trim trimmen (621.396), (622); beschneiden, zurechtschneiden; besäumen (67)
to trim away wegschneiden

to trim of burr entgraten, vom Grat befreien
to trim the edges Kanten bestoßen, Kanten beschneiden, Kanten behobeln
to trim off abschneiden, wegschneiden
to trim to size auf Maß schneiden, auf Format schneiden, auf Größe schneiden
trimmer Trimmer, Kohlenschipper (622); Trimmer, Trimmkondensator, Abgleichkondensator (621.396); Trimmklappe (629.13)
trimming Trimmen (621.396) (622), (629.12/13); Besäumen, Beschneiden (67)
trimming condenser Trimmer, Trimmkondensator (621.396)
trinitrotoluene Trinitrotoluol
trinomial Trinom, Trinomialausdruck (51)
triode Triode, Dreipolröhre
triode circuit, triode connection Triodenschaltung
triode thyristor Thyristortriode
triograph Kathodenstrahl-Kardiograph
trip Auslöser (621.3)
trip circuit Auslösestromkreis
trip coil Auslösespule, Ausschaltspule
trip-free relay Freiauslösungsrelais
trip-free release Freiauslösung
trip gear Auslösegestänge
trip latch Auslöseklinke
trip point Auslösepunkt
trip relay Auslöserelais
trip switch Betätigungsschalter, Auslöseschalter
to trip auslösen, ausklinken (621.3)
to trip a gas tube ein Gasentladungsrohr zünden
to trip a switch einen Schalter auslösen
to trip a valve ein Ventil auslösen
triphase = three-phase Drehstrom
triplane Tripelspiegel
triple dreifach, Dreifach-...
triple bond Dreifachbindung
triple concentric dreifach konzentrisch
triple concentric cable dreifach konzentrisches Kabel
triple-frequency harmonic dritte Harmonische, dritte Oberwelle
triple-grid Dreigitter-...
triple-grid tube Dreigitterröhre, Fünfpolröhre, Pentode
triple-pole dreipolig
triple pole circuit-breaker dreipoliger Schalter
triple-pole plug dreipoliger Stecker
triple-pole switch dreipoliger Umschalter
triple reduction Dreifachuntersetzung
triple-throw switch Dreifach-Umschalter
triple reentrant winding dreigängige Wicklung
tripler Verdreifacher, Frequenzverdreifacher
triplet Tripel (51); Triplett (535)
triplex Dreifach-...
triplex cable Dreileiterkabel, dreipoliges Kabel
triplex glass Dreischichtglas, Sicherheitsglas
triplex winding Dreischleifenwicklung
tripod Dreibein, Dreifußstativ, Stativ
tripolar dreipolig
tripping Auslösung, Abschaltung
tripping action Auslösewirkung, Auslösung
tripping circuit Auslösekreis, Betätigungsstromkreis
tripping current Auslösestrom
tripping device Auslöser, Auslöseeinrichtung, Ausklinkvorrichtung (621.3)
tripping effect Auslösewirkung
tripping gear Auslösegestänge, Auslösegetriebe
tripping magnet Auslösemagnet
tripping mechanism Auslösegestänge, Auslösegetriebe
tripping piece Auslöser, Auslöseglied
tripping speed Auslösegeschwindigkeit, Ausklinkdrehzahl
tripping time Auslösezeit, Ausschaltzeit (621.316.5)
tripping voltage Auslösespannung (621.316.5)
to trisect dreiteilen (51)
trisection Dreiteilung
tristimulus value Farbwert (621.397)
trivalent dreiwertig
trivector (meter) Scheinleistungszähler
trivet Dreifuß
trolitul Trolitul
trolley Draisine (625); Rollbahre (61); Obus (621.33); Abnehmerrolle (621.33); Laufkatze (621.8)
trolley base Stromabnehmerfuß
trolley boom Stromabnehmerstange
trolley bus Oberleitungs-Omnibus, Obus, Trolleybus
trolley bush Stromabnehmerrollenlager
trolley car Triebwagen (625)
trolley-electric locomotive Fahrdrahtlokomotive
trolley frog Fahrdrahtweiche
trolley harp Stromabnehmergabel
trolley head Stromabnehmerkopf
trolley line Obus-Linie; Fahrleitung

trolley locomotive Lokomotive mit Rollenstromabnehmer
trolley omnibus Obus, Oberleitungs-Omnibus
trolley pole Straßenbahnmast
trolley retriever Rückholseil (621.33)
trolley reverser Stromabnehmerkippvorrichtung
trolley shoe Stromabnehmer
trolley standard Stromabnehmerachse
trolley system Fahrdrahtsystem, Oberleitungssystem
trolley track Kranlaufbahn, Krangleis; Kranschienen (pl.)
trolley voltage Fahrdrahtspannung
trolley wheel Stromabnehmerrolle
trolley wire Fahrdraht
trombone Posaune (681)
tropical tropisch
tropical design Tropenausführung, tropenfeste Ausführung
tropical-proof tropenfest
tropical switch Schalter auf Isolierfüßen
tropicalized tropensicher
trop-in winding geträufelte Wicklung
tropopause Tropopause
troposcatter transmission troposphärische Überhorizontverbindung
troposphere Troposphäre
tropospheric radio duct troposphärischer Wellenleiter
tropospheric reflection troposphärische Reflexion
tropospheric scattering troposphärische Streuung, troposphärische Streuausbreitung
trotyl = trinitrotoluene Trinitrotoluol
trouble Schwierigkeit; Störung; Unruhe
trouble desk Störannahmestelle
trouble shooter Fehlersucher, Störungssucher (621.3), (621.396)
trouble shooting Störungssuche, Fehlersuche
troublefree störungsfrei
trough Trog, Mulde, Rinne; Wellental (621.3)
trough meter Minimalspannungsmesser (621.317)
trough-shaped muldenförmig
troughing Kabelkanalformstein
troughing idler Lastbandtragrolle
troughing roller Förderbandtragrolle
Troy Troy (Gewichtssystem)
t. r. s. = tough rubber sheathed Gummischlauch-... (621.3)
t. r. switch = transmitter receiver switch Sende-Empfangs-Umschalter; Nullode (621.397)

truck Lastwagen, Lastkraftwagen, Lkw.; Güterwagen, Waggon (625.2); Förderwagen, Hund (622)
truck balance Waggonwaage, Gleiswaage
truck engine Lastwagenmotor
truck-generator suspension Aufhängung der Lichtmaschine am Wagen (625)
truck-type circuit breaker Ausziehschalter
truck-type switchgear Schaltwagen
to true auf richtiges Maß bringen; zuschneiden, auf Maß schneiden, beschneiden
to true up wieder auf Maß bringen, wieder in die richtige Form bringen
to true up a commutator einen Kommutator nachdrehen
true wahr, echt; richtig, genau; maßhaltig, genau, nach Maß (621)
true a. c. rms Wechselspannungs-Effektivwert
true arc voltage Gesamtlichtbogenspannung (637)
true bearing rechtweisender Kurs, rechtweisende Peilung
true dielectric echter Nichtleiter
true inductance wahre Induktivität
true ohm absolutes Ohm
true power wirkliche Leistung, Wirkleistung
true ratio wirkliches Übersetzungsverhältnis (621.317)
true remanence wahre Remanenz
true resistance Echtwiderstand, ohmscher Widerstand
true to genau nach ..., entsprechend
true to shape formgetreu
true to size maßgerecht
true to type typenrecht
true value wahrer Wert, richtiger Wert
true watts (pl.) Wirkleistung
to truncate abstumpfen, stumpf abschneiden; runden (681); abbrechen (681.3)
truncated cone Kegelstumpf
truncated pyramid Pyramidenstumpf
trunk Verbindungsleitung, Abnehmerleitung, VL (621.395); Sammelschiene; Rumpf (61); Stamm (58)
trunk-busy tone Wegebesetztton
trunk cable Fernkabel
trunk call (Br.) Ferngespräch
trunk circuit Fernleitung (621.395)
trunk code Ortskennzahl
trunk connection Fernverbindung
trunk dialling Fernwahl
trunk exchange Fernamt, Fernvermittlung
trunk feeder Hauptspeiseleitung (621.315)
trunk hunting Leitungsuchen (621.395)

trunk line Fernleitung, Amtsleitung (621.395); Hauptlinie, Hauptstrecke (625)
trunk-line conduit Hauptkabelkanal
trunk operator Fernamtsbeamtin
trunk position Fernplatz
trunk switchboard Fernschrank (621.395)
trunking Hauptrohrleitung, Hauptrohr, Hauptleitung
truth Wahrheit; maßgerechter Zustand; genaue Lage
out of truth ungenau; falsch sitzend
truth table Funktionstabelle, Wertetabelle, Wahrheitstabelle (621.374)
to try versuchen; erproben, ausprobieren; beanspruchen, mitnehmen
try-square Winkel, Anschlagwinkel (621.9)
T-section T-Profil, T-Eisen (624), T-Glied (621.392)
T-shape T-Form
T-shaped T-förmig
T-square Reißschiene
TTL = transistor-transistor logic
T. U. = transmission unit Bel
tub Wanne, Kübel; Förderwagen, Hund (622)
tube Röhre, Rohr (621.6); Röhre (US) (siehe auch **valve**) (621.385); Schutzrohr (621.36); U-Bahn, Untergrundbahn; Untergrundbahntunnel, U-Bahntunnel; Tubus (535); Schlauch (629.113/118)
tube base Röhrensockel
tube characteristic Röhrenkennlinie
tube checker unit Röhrenprüfgerät
tube clamp Röhrenhalterung, Röhrenklemme
tube complement Röhrensatz, Röhrenbestückung
tube data manual Röhrendatenbuch
tube drop Spannungsabfall in der Röhre (621.314.6)
tube electrometer Röhrenelektrometer
tube extractor Röhrenauszieher
tube face Röhrenboden, Leuchtschirm (621.385)
tube of flux Flußröhre
tube of force Kraftröhre
tube fuse Röhrensicherung
tube hardness Röhrenhärte, Vakuumgüte
tube heat dissipation Röhrenverlustleistung
tube heating time Röhrenanheizzeit
tube hum Röhrenbrumm
tube lamp Röhrenlampe
tube life Röhrenlebensdauer
tube neck Röhrenhals (621.385)
tube plate resistance innerer Röhrenwiderstand (621.385)
tube railway Untergrundbahn
tube replacement Röhrenersatz (621.396)
tube system Rohrsystem
tube test Röhrenprüfung
tube test set Röhrenprüfgerät
tube voltmeter Röhrenvoltmeter
tube voltage drop Brennspannung (621.385)
tubing Rohrleitungsanlage, Rohr; Schlauch
tubular rund, röhrenförmig, rohrförmig
tubular boiler Röhrenkessel
tubular condenser Röhrenkondensator (621.1); Röhrchenkondensator (621.396)
tubular cover runde Hülse, Hülse
tubular heater Rohrheizkörper, Heizrohrelement
tubular lamp Röhrenlampe, Soffittenlampe
tubular lamp holder Fassung für Leuchtröhren, Soffittenfassung
tubular mast Rohrmast
tubular pole Rohrmast
tubular railway Untergrundbahn
tubular rivet Rohrniet, Hohlniet
tubular steel Stahlrohr, Stahlröhre
tubular steel frame Stahlrohrrahmen
tubular-type resistor Rohrwiderstand (621.316.8)
tubular-type tank Röhrenkessel (621.314.2)
tubular vector field quellenfreies Feld
Tudsbury machine Druckluft-Influenzmaschine
tue iron Gebläserohr (669)
tuft Büschel, Bausch, Haufen
tuft discharge Büschelentladung
tumbler Zuhaltung (68); Kipp...
tumbler device Schwenkvorrichtung, Kippvorrichtung
tumbler drum Schaltwalze
tumbler gear Schwenkrad (621.8)
tumbler shaft Nockenwelle, Kippwelle
tumbler switch Kippschalter, Kellogschalter
tunable abstimmbar
tunable range Abstimmbereich
tune Melodie, Weise
to tune abstimmen
to tune for resonance auf Resonanz abstimmen
to tune in for abstimmen auf, einstellen auf (621.396.62)
to tune out wegstimmen
tuned abgestimmt
tuned amplification abgestimmte Verstärkung
tuned amplifier Resonanzverstärker, abgestimmter Verstärker
tuned anode circuit abgestimmter Anodenkreis
tuned anode coupling Schwingkreiskopplung (621.375)

tuned circuit abgestimmter Kreis, Schwingkreis
tuned circuit element Abstimmkreisglied
tuned filter abgestimmtes Filter, abgestimmter Siebkreis
tuned grid Gitterabstimmkreis, abgestimmter Gitterkreis
tuned grid circuit abgestimmter Gitterkreis, Gitterschwingkreis
tuned plate abgestimmter Anodenschwingkreis
tuned radio-frequency amplifier abgestimmter Hochfrequenzverstärker
tuned radio-frequency circuit abgestimmter HF-Kreis
tuned r. f. receiver Geradeausempfänger
tuned-reed frequency meter Zungenfrequenzmesser
tuned-reed rectifier Schwinggleichrichter
tuned relay abgestimmtes Relais, Resonanzrelais
tuned resonating cavity abgestimmter Hohlraumresonator
tuned rope abgestimmtes Düppelband
tuner Klavierstimmer (681); Anpassungsglied (621.396); Tuner, Abstimmeinheit, Kanalwähler (621.396); UKW-Empfangsteil
tuning Abstimmung, Abstimm ...
tuning coil Abstimmspule
tuning scale Abstimmskala
tuning variable-capacitance diode Abstimmdiode
tungstate screen Kalziumwolframatfolie (621.386)
tungsten Wolfram
tungsten arc lamp Wolframbogenlampe, Punktlichtlampe
tungsten carbide Wolframkarbid, Hartmetall
tungsten carbide bit Hartmetall-Schneidplättchen
tungsten carbide tool Hartmetall-Werkzeug
tungsten filament Wolframkathode, Wolframheizfaden (621.385); Wolframleuchtfaden (621.32)
tungsten filament lamp Wolframdrahtlampe
tungsten filament valve Wolframdrahtröhre, Röhre mit Wolframdrahtkathode
tungsten lamp Wolframlampe, Metalldrahtlampe
tungsten sensitivity Empfindlichkeit bei Glühlampenlicht (621.385)
tuning Abstimmung; Abstimm- ...
tuning capacitor Abstimmkondensator
tuning-capacitor control Abstimmknopf (621.396.62)
tuning coil Abstimmspule
tuning condenser Abstimmkondensator
tuning control Abstimmeinrichtung, Abstimmknopf
tuning curve Resonanzkurve, Durchstimmkurve
tuning equipment Abstimmittel (pl.)
tuning error Abstimmfehler
tuning fork Stimmgabel
tuning-fork clock stimmgabelgesteuerte Uhr
tuning-fork oscillator Stimmgabeloszillator
tuning indicator Abstimmanzeiger (621.396.62)
tuning inductor Abstimmspule
tuning-in for Abstimmung auf
tuning knob Abstimmknopf
tuning pitch Normstimmton
tuning point Abstimmpunkt
tuning position Abstimmstellung
tuning precision Abstimmschärfe
tuning range Abstimmbereich, Durchstimmbereich, Ziehbereich (621.396.61)
tuning stub Abstimm-Blindspritze
tuning switch Wellenschalter
tunnel Tunnel; Schacht (622); Windkanal (533)
tunnel action Tunnelung, Tunnelvorgang (621.382)
tunnel diode Tunneldiode
tunnel effect Tunneleffekt
tunnel furnace Tunnelofen
tunnel kiln Tunnelofen
tunnel winding Fädelwicklung
tunnel-wound armature gefädelter Anker (621.313)
to tunnel tunneln (621.382)
turbid trübe
turbidity Trübung, trüber Niederschlag
turbine Turbine
turbine-electric locomotive turboelektrische Lokomotive
turbine generator Turbogenerator
turbine vane Turbinenschaufel (621.2)
turbine wheel Turbinenrad
turbo- ... Turbo- ...
turbo alternator Turbogenerator
turbo generator Turbogenerator
turbo-super charger Turbolader
turbulator Querstromlöschkammer
turbulence Turbulenz
turbulent turbulent, Wirbel- ...
turbulent flow turbulente Strömung
turn Biegung, Wendung, Kurve; Windung (621.3); Umdrehung, Drehung; Schicht
turn and bank indicator Wende- und Lagezeiger (629.1)

turns count Windungszählung
turn heading towards Kurve nach, Biegung nach
in turn der Reihe nach; andererseits
turn indicator Wendezeiger
turn-over moulding machine Umklappformmaschine (621.74)
turn ratio Windungsverhältnis (621.3)
turns ratio Windungsverhältnis, Übersetzungsverhältnis (621.3)
turn of a screw Schraubenwindung
turn switch Drehschalter
turn table Drehscheibe; Plattenteller (681.84)
to turn drehen (621.94); drechseln (688); sich drehen, sich wenden; wenden (687)
to turn through an angle um einen Winkel drehen
to turn a commutator einen Kollektor abdrehen
to turn down abstellen, kleinstellen, herunterdrehen; ablehnen
to turn an edge over einen Rand umbiegen
to turn in einliefern, zurückgeben
to turn into umsetzen in, verwandeln in, umformen in, umwandeln in, sich verwandeln in
to turn to the extreme left ganz nach links drehen
to turn off abstellen, abdrehen, abschalten
turn-off delay time Abschaltverzögerungszeit
turn-off loss Ausschaltverlustleistung
turn-off time Abschaltzeit
turn-off thyristor Abschaltthyristor
to turn on aufdrehen (621.396.62); anstellen, einschalten
turn-on time Einschaltzeit
turn-on loss Einschaltverlustleistung
to turn on and off an- und abstellen (621.1); ein- und ausschalten, zu- und abschalten (621.3)
to turn on the power supply Stromeinschalten, Stromversorgung einschalten
to turn out ausliefern, ausstoßen; sich herausstellen als; ausdrehen
to turn out false sich als falsch herausstellen
to turn out insufficient sich als unzureichend erweisen
to turn round umdrehen; (sich) drehen um; rotieren um
to turn round through ... drehen um einen Winkel von ...
to turn the scale den Ausschlag geben
to turn to the stop bis zum Anschlag drehen
to turn a switch einen Schalter betätigen, einen Schalter herumdrehen
to turn to umschlagen in, sich verwandeln in
to turn to red nach Rot umschlagen
to turn to yellow vergilben, gelb werden
to turn up nach oben drehen, herumdrehen; in Erscheinung treten
to turn all the way to ... ganz nach ... drehen, bis zum Anschlag nach ... drehen
turned bolt gedrehte Schraube, Maschinenschraube
turned shaft gedrehte Welle
turnbuckle Ankerspannschoß (621)
turning Drehung, Krümmung; Drehvorgang, Drehen (621.9); Straßenabzweig, Dreh-...
turning and boring mill Karussell-Drehbank
turning chisel Drechselstahl, Stichel
turning circle Drehkreis
turning couple Kräftepaar
turning effect Drehmoment, Drehwirkung
turning lathe Drechselbank, Drehbank
turning lever Drehhebel
turning moment Drehmoment
turning point Wendepunkt; Scheitelpunkt, Gipfelpunkt
turnkey project schlüsselfertiges Projekt
turnover Umsatz; Umpolung (621.3); Umpoleffekt (621.317)
turnover characteristic Kennlinienknick
turnscrew Schraubenzieher
turnstile Drehkreuz
turnstile aerial Drehkreuzantenne, Quirlantenne, Kreuzdipol, Turnstile-Antenne
turntable Drehscheibe (625); Drehtisch (621.9); Plattenteller (681.84)
turpentine Terpentin
"turps" Terpentin
turquoise Türkis
tuyere Windrohr, Windform (66); Gebläserohr (621)
TV = television (siehe dort) Fernsehen; Fernsehfunk
TV home receiver Fernsehrundfunkempfänger, Fernsehgerät, Fernseher
TV satellite Fernsehsatellit
tweeter Hochtonlautsprecher, Hochtöner
tweezers (pl.) Pinzette
twilight Zwielicht
twin Doppel-..., Zwei-..., Zwillings-...
twin cable Zweileiterkabel
twin concentric cable doppelt konzentrisches Kabel
twin contact Zwillingskontakt, Doppelkontakt

twin-contact wires (pl.) Doppelfahrdraht, Zwillingsfahrdraht
twin conductor Doppelleiter, Doppelleitung
twin-engined zweimotorig
twin flex doppeladrige Litze
twin flexible cord doppeladrige Litze
twin formation Zwillingsbildung (548)
twin-induction regulator Doppeldrehregler
twin lead symmetrische Doppelleitung (621.395)
twin triode Doppeltriode, Doppel-Dreipolröhre
twin wire Doppelader
twin-wire cable zweiadriges Kabel, Doppelleitungskabel
to twin Zwillinge bilden, verzwillingen (548)
to twin up together verdrillen (621.3)
twinned crystals Zwillingskristalle, Kristallzwillinge
twinning Zwillingsbildung (54); Paarigkeit (der Zeilen) (621.397)
to twinkle glitzern, funkeln
"T" wire a-Ader
twist Verdrehung, Drillung, Torsion
twist drill Spiralbohrer
twist flex verdrillte Litze
twist system einfache Viererverdrillung
to twist verdrehen, verdrillen, drillen, drehen, auf Torsion beanspruchen; zusammendrehen, verdrillen
to twist off abwürgen, abdrehen
to twist up zusammendrehen, verdrillen
twisted conductor verdrillter Stab (621.313)
twisted drill Schlangenbohrer
twisted flex verdrillte Litze
twisted flexible cord verdrillte Litze
twisted gear schrägverzahntes Rad
twisted pair verdrillte Doppelleitung (621.3)
twisted sleeve joint Würgeverbindung mit Hülse
twisted-strip galvanometer Bändchengalvanometer
twisted winding Gitterstabwicklung
twister Rohrschlüssel, Rohrzange
twisting Drehung, Torsion, Drillung
twisting moment Drillmoment, Torsionsmoment, Drehmoment
two zwei
two-address code Zweiadreßbefehlscode
two-band telephony Zweiband-Telefonie
two-circuit method (for meter testing) Zweikreisschaltung (zur Zählerprüfung), Sparschaltung
two-circuit single-interruption switch Serienschalter (696)
two-circuit tuner Gleichlaufabstimmelement von zwei Kreisen
two-circuit winding zweizweigige Wellenwicklung (621.313)
two-conductor wiring zweipolige Installation
two-core cable zweiadriges Kabel
two-digit zweistellig
two-dimensional zweidimensional
two-edged zweischneidig
two-electrode rectification Diodengleichrichtung
two-electrode valve Zweielektrodenröhre, Diode, Zweipolröhre
two-element relay Zweiwicklungsrelais
two-fold zwiefältig, zwiefach
two-fold increase of Verdoppelung von, Vergrößerung von ... auf das Doppelte
two-layer zweilagig
two-layer winding zweilagige Wicklung
two-limits regulator Zweipunktregler
two-motion selector Koordinatenwähler (621.395)
two-out-of-five code Zwei-aus-fünf-Code
two-part commutator zweiteiliger Kollektor
two-party line Zweieranschluß
two-phase zweiphasig, Zweiphasen-...
two-phase alternator Zweiphasengenerator
two-phase circuit Zweiphasenschaltung
two-phase current Zweiphasenstrom
two-phase four-wire system Zweiphasen-Vierleiter-System
two-phase generator Zweiphasen-Generator
two-phase meter Zweiphasenzähler
two-phase motor Zweiphasenmotor
two-phase three-wire system Zweiphasen-Dreileiter-System
two-phase winding Zweiphasenwicklung
two-pin plug zweipoliger Stecker
two-ply zweilagig, doppellagig, zweischichtig
two-point Zweipunkt
two-point stabilizer Zweipunktregler
two-pole Zweipol-..., zweipolig
two-pole armature zweipoliger Anker, Doppel-T-Anker
two-pole plug zweipoliger Stecker
two-pole socket zweipolige Steckdose
two-pole switch zweipoliger Schalter
two-port network Vierpol
two-position action-controler Auf-Zu-Regler, Zweipunktregler, Ein-Aus-Regler
two-position range selector Zweibereichschalter

two-purpose valve Zweizweckröhre, Mehrfachröhre
two-range winding zweietagige Wicklung
two-rate meter Doppeltarifzähler
two-resonator clystron Zweikammerklystron
two-speed Zweigang-...
two-speed gear Zweiganggetriebe
two-speed pole-changing motor zweifach polumschaltbarer Motor
two-stage zweistufig, Zweistufen-...
two-step controller Zweipunktregler
two-step controller action Zweipunktverhalten
two-step relay zweistufiges Relais
two-stroke Zweitakt-...
two-terminal zweipolig, Zweipol-...
two-terminal impedance passiver Zweipol (621.3)
two-terminal network Zweipol
two-terminal source aktiver Zweipol (621.3)
two-tier winding zweietagige Wicklung
two-track commutator Zweibahn-Kommutator
two-tube Zweiröhren-... (US)
two-tube amplifier Zweiröhrenverstärker
two-valve Zweiröhren-... (Br.)
two-valve receiver Zweiröhrenempfänger
two-wattmeter method Aronschaltung, Zweiwattmeterschaltung
two-wave rectification Vollwellengleichrichtung
two-way break-before-make contact Umschaltkontakt mit Unterbrechung
two-way repeater Zweidrahtverstärker
two-way switch Umschalter, Wechselschalter
two-way telephone system Gegensprechanlage
two-way wiring Wechselschaltung
two-wire Zweileiter-... (621.3); Zweidraht-... (621.395)
two-wire circuit Zweidrahtkreis, Zweidrahtschaltung, Zweidrahtleitung
two-wire line Doppelleitung, Zweidrahtleitung
two-wire meter Zweileiterzähler
two-wire repeater Zweidrahtverstärker
two-wire system Zweileiteranlage, Zweileiternetz (621.3); Zweidrahtsystem (621.395)
two-wire system amplifier Zweidrahtverstärker
two-wire telephone circuit Zweidrahtkreis, Zweidrahtschaltung
two-wire telephone line Zweidrahtfernsprechleitung

type Typ, Type, Typus; Druck, Druckschrift, Satz, Schrift (655); Type, Druckbuchstabe, Buchstabe, Letter (655); Type (681.61)
type "A 1" waves A 1-Wellen
type acceptance test Typenprüfung
type approval Typenzulassung, Legalisierung
type lever Typenhebel (681.61)
type of construction Bauart
type of modulation Modulationsart
type of operation Verkehrsart, Betriebsart
type-printing telegraph Typendrucktelegraph
type of protection Schutzart
21 type repeater Zweidrahtbrückenverstärker, Einrohr-Leitungsverstärker, Einröhren-Leitungsverstärker (621.395)
22 type repeater Zweidrahtgabelverstärker, Zweirohr-Leitungsverstärker, Zweidraht-Verstärker (621.395)
type of service Betriebsart
type of transmission Übertragungsart
to type tippen, maschineschreiben
type wheel Typenrad
to typewrite maschineschreiben, tippen
typewriter Schreibmaschine
typification Typisierung
to typify typisieren
tyre Reifen, Decke (629.113); Radkranz (625)
to tyre bereifen, mit Reifen versehen

U

UART = universal asynchronous receiver-transmitter
u. h. f. = ultrahigh frequency dm-Wellen, dm-Gebiet
u. h. f. range Dezimeterbereich, dm-Gebiet
Ulbricht globe Ulbrichtsche Kugel
ultimate schließlich, letztlich, letztmöglich, höchstmöglich, höchstzulässig, letzt
ultimate capacity Endleistung, Leistung im Endausbau
ultimate constituents letzte Bestandteile
ultimate damping Grenzdämpfung (539)
ultimate limit äußerste Grenze
ultimate limit switch Endausschalter
ultimate particle Elementarteilchen
ultimate sensitivity Grenzempfindlichkeit (621.396.62)
ultimate strength Bruchfestigkeit, Zerreißfestigkeit
ultimate tensile strength Zugfestigkeit

ultimate vacuum Grenzvakuum
ultra besonders, besonders hoch, besonders groß, über-...
ultra-acoustic teleprinter system Überlagerungs-Fernschreiber-System
ultra-audible frequency Ultraschallfrequenz
ultra-audion Rückkopplungsaudion
ultra-gamma radiation Ultrastrahlung, Höhenstrahlung
ultrahigh frequency dm-Wellen (pl.)
ultra-incandescent lamp Oxidschichtglühlampe
ultramicrometer Kondensator-Mikrometer
ultramicroscope Übermikroskop
ultraphotic radiation UV-Strahlung
ultrashort wave ultrakurze Welle, UKW
ultrasonic Ultraschall
ultrasonic brazing Ultraschallschweißen
ultrasonic cleaning tank Ultraschall-Reinigungsgerät
ultrasonic generator Ultraschallschwinger
ultrasonic irradiation Beschallung, Ultraschallbestrahlung
ultrasound Ultraschall
ultraviolet ultraviolett, UV
ultraviolet erasable PROM = EPROM UV-löschbares PROM
ultraviolet lamp UV-Lampe, Ultraviolettlampe
ultraviolet light UV-Licht, Ultraviolettlicht
ultraviolet rays (pl.) Ultraviolettstrahlung
ultraviolet therapy UV-Therapie
ultra-X-rays (pl.) Höhenstrahlung, durchdringende Strahlung, kosmische Strahlung
umbrella aerial Schirmantenne
umbrella insulator Schirmisolator
umbrella-type alternator Schirmgenerator (621.313)
unabsorbed field strength ungedämpfte Feldstärke
unaided eye unbewaffnetes Auge
unalloyed unlegiert
unarmoured unbewehrt, nicht bewehrt (621.315.2); nicht gepanzert (623)
unarmoured cable unbewehrtes Kabel
unassigned answer offene Amtsabfrage
unattached nicht gebunden, frei
unattached electron freies Elektron
unattended ungewartet, unbedient, bedienungslos, unbemannt
unattended substation bedienungsloses Unterwerk, unbemanntes Unterwerk, automatisches Unterwerk
unattenuated (räumlich) ungedämpft
unbalance Unbalance, Abweichung vom Gleichgewicht (621); Unsymmetrie, Abgleichfehler (621.3)
unbalance attenuation Unsymmetriedämpfung
to unbalance aus dem Gleichgewicht bringen, labil machen
unbalanced nicht im Gleichgewicht; nicht ausbalanciert, nicht ausgewuchtet, unausgewuchtet (621); unsymmetrisch (621.3)
unbalanced deflection erdunsymmetrische Ablenkung (621.385)
unbalanced to ground unsymmetrisch gegen Masse
unbalanced line unsymmetrische Leitung
unbalanced load unsymmetrische Belastung (621.3); Schieflast
unbalanced magnetic pull einseitiger magnetischer Zug
unbalanced mass unausgewuchtete Masse, Unwucht
unbalanced three-phase system unbalanciertes Drehstromsystem
unbalanced voltages (pl.) unsymmetrisches Spannungsdreieck
to unbend (wieder) gerade biegen
unbiased ohne Vorspannung
to unbind freimachen, entbinden
unblanking pulse Austastimpuls, Hellsteuerungsimpuls
unbleached ungebleicht
to unblock entblocken, freigeben, entsperren
to unbolt losschrauben
unbounded unbegrenzt
unburnt unverbrannt
unbypassed unüberbrückt, ohne Nebenschluß (621.3)
uncertainty Unbestimmtheit
uncharged ungeladen (537), (621.355)
uncharged state ungeladener Zustand (537), (621.355)
unclean unsauber, unrein, schmutzig
to uncoil abwickeln, abspulen, abrollen, abhaspeln
uncoiler Ablaufspule, Ablaufhaspel, Abrollhaspel
uncombined frei (54), (66)
uncombined carbon freier Kohlenstoff
un-come-at-able unzugänglich
uncomplete dialling unvollständige Wahl (621.395)
unconditional jump unbedingter Sprung (681.3)
unconfined nicht abgeschlossen; unverdämmt (66)
uncontrollable nicht steuerbar, unkontrollierbar

uncontrolled ungesteuert, nicht gesteuert
to uncouple auskuppeln, entkuppeln; loskuppeln, abkuppeln, abhängen (625)
to uncover abdecken, freigeben, aufdecken
uncritical kritiklos, unkritisch; unkritisch (621.396)
unctuous fettig, schmierig
to uncurl aufwickeln, abwickeln; richten
undamped ungedämpft
  undamped oscillation ungedämpfte Schwingung
  undamped wave ungedämpfte Welle
under unter
  under foot unter den Füßen, am Erdboden
  under investigation in der Untersuchung, zu untersuchend
  under load belastet, unter Last
  under low pressure unter niedrigem Druck, bei niedrigem Druck
undercoat Grundanstrich, Grundierung
undercommutation Unterkommutierung
undercurrent zu schwacher Strom
  untercurrent circuit breaker Minimalschalter
  undercurrent circuit breaking Minimalauslösung
  undercurrent relay Minimalrelais, Stromrückgangsrelais
undercut unterschnitten; unterätzt
underdamped unterkritisch gedämpft, unteraperiodisch
unterdamping unterkritische Dämpfung, unteraperiodische Dämpfung
to underexpose unterexponieren, unterbelichten
underexposure Unterbelichtung
underfrequency protection Unterfrequenzschutz
underground Untergrund, Untergrundbahn
unterground unterirdisch, unter Tage, Erd-...
  underground cable Erdkabel
  underground conduit Kabelrohr (621.315.2)
  underground construction Tiefbau (624)
  underground contact rail unterirdische Stromschiene
  underground conveyance unterirdischer Transport, Beförderung unter Tage
  unterground installation Anlage unter Tage (622)
  underground laying Erdverlegung
  underground pipe unterirdisches Rohr
  underground power station Kavernenkraftwerk
underhung mountain Hängeanordnung (621.316.5)
to underline unterstreichen
underload Unterlast, zu geringe Last
  underload circuit breaker Minimalschalter
  underload circuit breaking Minimal-Auslösung
  underload relay Minimalrelais
to underpass unterschreiten
to underrun unterbelasten, zu schwach belasten
to underheat unterheizen (621.385)
undershot unterschlächtig (621.2)
underside Unterseite
undersize Untergröße, Untermaß
undersized zu klein, mit Untermaß
understandable verständlich (auch 621.395)
undervoltage circuit breaking Nullspannungsauslösung, Unterspannungsauslösung
  undervoltage coil Unterspannungsspule (621.316)
  undervoltage protection Unterspannungs-Schutzsystem
  undervoltage relay Unterspannungsrelais, Nullspannungsrelais
  undervoltage tripping Unterspannungsauslösung, Nullspannungsauslösung
underwater ... Unterwasser ...
  underwater cable Seekabel
  underwater floodlighting Unterwasserbeleuchtung
  underwater sound detector Unterwasserhorchgerät
  underwater sound projector Unterwasser-Schallsender
  underwater sound ranging Unterwasser-Schallortung
underweight Untergewicht, Mindergewicht
underweight zu leicht
undetectable nicht feststellbar
undetected error rate Restfehlerquote
undetermined unbestimmt
undiluted unverdünnt
undissociated nicht dissoziiert
undistorted unverzerrt
  undistorted power output unverzerrte Ausgangsleistung (621.396.62), (621.375)
  undistorted signal power unverzerrte Sprechleistung (621.396.62)
  undistorted wave unverzerrte Welle
to undo lösen (621.882); aufheben, rückgängig machen
to undulate schwanken
  undulating light Flackerfeuer (65)
undulation Schwankung, Wellung, Wellenbewegung
undulator Undulator (621.394)
undulatory schwankend, wellenförmig
  undulatory current Wellenstrom

unechoic room schalltoter Raum
unending endlos, unendlich
unending series unendliche Reihe (51)
unequal ungleich, ungleichmäßig, verschieden
unequal angle ungleichschenkliges Winkeleisen
uneven uneben, ungleichmäßig
uneven commutator unrunder Kollektor
uneven running unruhiger Lauf
to unfasten abschrauben, lösen
unfinished unbearbeitet, nicht fertig bearbeitet
unflawed fehlerlos, makellos
unformatted unformatiert
to ungear auskuppeln
unglazed unglasiert
ungrounded unbegründet; ungeerdet
ungrounded system ungeerdetes System, ungeerdete Anlage
unguent Schmiermittel, „Schmiere“
to unhook abhaken, loshaken; abheben (621.395)
uniaxial einachsig
unidirectional einsinnig, einseitig gerichtet, einsinnig verknüpft; gleichgerichtet (621.3); gleichsinnig (621.3); rückwirkungsfrei (621.53)
unidirectional aerial Richtantenne mit einseitiger Charakteristik
unidirectional characteristic einseitige Richtcharakteristik
unidirectional current gleichgerichteter Strom, einsinniger Strom, Strom gleichbleibender Richtung
unidirectional flow einsinnige Strömung
unidirectional illumination Anleuchtung, Anstrahlung
unidirectional receiver Richtempfänger (621.396)
unidirectional transmitter Richtstrahler
unified field theory allgemeine Feldtheorie
unifilar einfädig, Unifilar-..., Einfaden-...
unifilar elastic control Einfadenaufhängung, Fadenaufhängung
unifilar electrometer Einfaden-Elektrometer
unifilar suspension Fadenaufhängung (621.317)
uniform einheitlich, gleichförmig, gleichmäßig
uniform field homogenes Feld
uniform line homogene Leitung
uniform motion gleichförmige Bewegung
uniform-spectrum random noise weißes Rauschen
uniform velocity gleichförmige Geschwindigkeit
uniformly accelerated gleichförmig beschleunigt
uniformly retarded gleichförmig verzögert
unijunction transistor Unijunction-Transistor
unilateral einseitig
unilateral conductivity unipolare Leitfähigkeit
unilateral drive einseitiger Antrieb
uninflammable unbrennbar, unentzündlich
uninsulated unisoliert, blank
uninsulated cable blankes Kabel
uninsulated wire blanker Draht
unintelligible unverständlich
unintelligible crosstalk unverständliches Nebensprechen
uninverted crosstalk verständliches Nebensprechen
Unionmelt welding Ellira-Schweißung
uniphase (obs.) Einphasen-..., einphasig
unipivot instrument Meßinstrument mit Hängelager
unipolar unipolar
unipolar generator Unipolargenerator
unipolar transistor Unipolartransistor
unipotential cathode Äquipotentialkathode
uniqueness of a solution Eindeutigkeit einer Lösung (51)
uniselector Drehwähler (621.395)
unison Einklang, Gleichklang
to be unison with abgestimmt sein auf
unit Satz, Einheit, Maschinensatz, Bauteil, Bauelement; Einheit (53)
unit area Flächeneinheit
unit-area capacitance Kapazitätsbelag (des Elektrolytkondensators), Kapazität pro Flächeneinheit
unit of brightness Einheit der Leuchtdichte
unit call Gesprächseinheit
unit charge Einheit der Ladung, Ladungseinheit (53), Gebühreneinheit (621.395)
unit circle Einheitskreis
units-counting tube Einer-Zählröhre
unit current Einheitsstrom, Strom Eins
unit of current Stromeinheit
unit distance apart im Abstand Eins
unit electric flux Einheit des elektrischen Flusses, Einheitsfluß
unit of equipment Geräteteil
unit of erythemal flux Erythemdosis
unit fee Gebühreneinheit
unit feeder Einfachspeiseleitung
unit force Einheit der Kraft, Krafteinheit, Kraft Eins
unit of force Krafteinheit
unit discontinuous function Einheitssprung, Sprungfunktion (51)

unit of heat Wärmeeinheit
unit impulse Einheitsstoß
unit length Längeneinheit, Länge Eins
unit load spezifische Belastung
unit magnetic pole magnetischer Einheitspol
unit operation Schaltzyklus (621.316.5)
per unit auf die Einheit bezogen
"per unit" admittance bezogener Leitwert, Leitwertbelag
"per unit" susceptance bezogener Blindleitwert, Blindleitwertbelag
unit pole Einheitspol
unit quantity of primary colour Grundfarbeneinheit
unit resistance Widerstandseinheit, Widerstand Eins
unit of resistance Widerstandseinheit
unit step function Einheitssprung, Sprungfunktion
unit-system operation Blockbetrieb (621.311)
unit (of) time Zeiteinheit, Einheit der Zeit, Zeit Eins
unit tube of force Kraftröhre, Einheitsröhre
unit vector Einheitsvektor
unit velocity Geschwindigkeit Eins
unit (of) voltage Spannungseinheit, Spannung Eins
unitary die Einheiten betreffend
unitary ratio Umrechnungszahl (vom el.magn. zum el.stat. c. g. s.-System) = Lichtgeschwindigkeit
to unite vereinigen, verbinden, zusammentun; sich vereinigen; sich verbinden, eine Verbindung eingehen (54)
unity Einheit, der Wert Eins, die Zahl Eins
unity current transfer ratio Stromverstärkung 1
unity gain Verstärkung Eins
unity power factor Phasenwinkel Null, Leistungsfaktor Eins
unity power factor impedance phasenreiner Widerstand, rein ohmscher Widerstand
unity turn ratio Übersetzungsverhältnis Eins, Windungsverhältnis Eins
univalent einwertig
universal Universal-..., universell
universal battery Zentralbatterie = ZB
universal battery system Z. B.-System
universal coupling Universalkupplung
universal decimal classification Dezimalklassifikation
universal joint Kardangelenk, Universalgelenk, Kreuzgelenk
universal lighting Allgemeinbeleuchtung
universal machine Universalmaschine
universal-mains Allstrom-...
universal-mains radio set Allstromradiogerät
universal motor Universalmotor, Allstrommotor
universal valve Universalröhre
universality Universalität, universelle Verwendbarkeit
unknown Unbekannte (51)
unknown unbekannt
unknown property unbekannte Eigenschaft
unknown quantity unbekannte Größe
to unlatch ausklinken
unlatching mechanism Ausklinkvorrichtung
unlicensed listener Schwarzhörer
unlicensed transmitter Schwarzsender
unlike ungleich, anders als, abweichend von; entgegengesetzt gerichtet; ungleichnamig
unlike currents entgegengesetzt gerichtete Ströme
unlike pair Doppelleitung mit entgegengesetzt gerichteten Strömen, Hin- und Rückleitung
unlike poles ungleichnamige Pole
unlike signs ungleiche Vorzeichen
unlimited unbegrenzt, unbeschränkt
to unload entladen, abladen, löschen; abziehen (66); aussetzen (66); entlasten
unloaded aerial unbelastete Antenne
unloaded cable unbespultes Kabel
unloaded wavelength freie Wellenlänge, Eigenwelle der Antenne ohne Abstimmelemente
unloading Entladung, Löschung; Entlastung
to unlock entriegeln, aufriegeln; entfesseln, lösen
unmagnetized unmagnetisch
to unmake (eine Verbindung) lösen, trennen
unmarked pole Südpol (538)
unmatched nicht angepaßt
unmeasurable unmeßbar
unmodulated unmoduliert
unmodulated carrier unmodulierter Träger
unmodulated interval Sendepause
unmodulated key-controlled continuous waves A 1-Wellen
unmodulated signal unmoduliertes Zeichen, Dauerstrich (621.396)
unmodulated wave unmodulierte Welle, Dauerstrich
to unpack auspacken
to unplug Stöpsel ziehen (621.317)
unpurified ungereinigt
unrationalized units nicht-rationale Einheiten
to unravel entwirren, entknäueln
unreactive nicht reaktionsfähig, passiv (54)
unreadable unleserlich, nicht zu lesen

unreal nicht reell, irreal
to unreel abspulen, abwickeln
unreliable unzuverlässig
unrestricted extension voll-amtsberechtigte Nebenstelle
to unrivet losnieten, abnieten
to unroll aufrollen, aufwickeln, entrollen
unsafe unzulässig, gefährlich
  unsafe load Überlast, unzulässige Überlastung
  unsafely loaded überlastet
unsaturated ungesättigt
  unsaturated compound ungesättigte Verbindung
  unsaturated standard cell ungesättigtes Normalelement
unscreened ungeschirmt, unabgeschirmt (621.396); ungesiebt
to unscrew abschrauben, losschrauben, aufschrauben
  to unscrew out of herausschrauben aus, herauswinden aus
  to unscrew right off völlig abschrauben
unselective nicht selektiv
unseparated ungetrennt, nicht separiert
unserviceable unbrauchbar, unbenutzbar, nicht betriebsfähig
unshaded nicht schraffiert, unschraffiert
unsharp unscharf
to unsheathe abmanteln (621.315.2)
unshielded unabgeschirmt
unshrinkable nicht einlaufend (67)
unshrinking nicht einlaufend, nicht eingehend (67)
unshunted unüberbrückt, ohne Nebenwiderstand
to unsolder loslöten, ablöten, auslöten
unsolvable unlösbar
unsprung ungefedert
  unsprung motor weight ungefedertes Motorgewicht (621.33)
  unsprung weight ungefederte Massen (pl.) (629.113), (625)
unstable unstabil, nicht stabil, labil
  unstable equilibrium labiles Gleichgewicht
unsteady unruhig, unbeständig
  unsteady image unruhiges Bild (621.397)
unstrained nicht beansprucht
unstressed nicht beansprucht, unbelastet
to unsweat auflöten, ablöten, loslöten
unsymmetrical unsymmetrisch
unsymmetry Unsymmetrie
to untangle entwirren
to unthread losbinden, abbinden, aufbinden; ausfädeln
to untie losbinden
untight undicht
untimely unzeitgemäß, zur unpassenden Zeit
untuned nicht abgestimmt, unabgestimmt
  untuned aerial aperiodische Antenne
  untuned circuit aperiodischer Kreis
  untuned secondary circuit aperiodischer Sekundärkreis
to untwine aufdrehen, aufdrillen
to untwist aufdrehen, aufdrillen; sich aufdrehen
untwisted unverdrillt
unused time Ruhezeit
unvarnished unlackiert
unvarying unveränderlich
unverified unbestätigt, unbewiesen
unwanted responses Störresonanzen
unweighted noise Fremdspannung (621.395)
unweighted signal-to-noise ratio Fremdspannungsabstand
unwinding spool Ablaufspule
to unwrap auswickeln
up auf, nach oben zu
  6 up 6 nach oben (51)
  to be up to gewachsen sein, leisten können
  the key is up die Taste ist offen (621.394)
  up to the mark vollbefriedigend; den Vorschriften entsprechend
  up-and-down motion Auf-und-ab-Bewegung
upcoiler Aufwickelhaspel
to upgrade verbessern, Qualität verbessern
upkeep Unterhaltung, Wartung, Pflege, Erhaltung
  upkeep work Unterhaltungsarbeiten (pl.)
uplift Abheben (629.13)
upper ober, höher gelegen, Ober-...
  upper band Oberband
  upper boom Obergurt
  upper change-point oberer Umwandlungspunkt
  upper flange Obergurt, oberer Flansch
  upper-out-off frequency obere Grenzfrequenz
  upper half obere Hälfte, Oberteil
  upper hemispherical flux oberer hemisphärischer Lichtstrom
  upper lubrication Oberschmierung
  upper reaches (pl.) Oberlauf (626)
  upper side band oberes Seitenband
upright Pfosten, Stempel, Türstock; Ständer, Säule (621.9)
upright aufrecht, aufrechtstehend, senkrecht
  upright drilling machine Säulenbohrmaschine
  upright side senkrechte Seite (51)
upset Verdickung, Wulst, Anstauchung (621.791)
  upset gear Stauchgetriebe (621.791)

to upset stauchen, anstauchen; quetschen, pressen; umwerfen, über den Haufen werfen, umkippen
upside down auf den Kopf gestellt
upsweep column Peitschenmast
uptake Aufnahme
uptake of energy Energieaufnahme
uptilt Diagramm-Anhebung, Hauptabstrahlrichtung in der Elevationsebene (621.396.67)
uptime Betriebszeit, Verfügbarkeit
upturned umgekippt, umgestürzt
upward aufwärts, nach aufwärts, nach oben zu
upward flux nach oben gerichteter Lichtstrom
upward motion Aufwärtsbewegung
upward stream Fangentladung, Aufwärtsblitz
upward streamer Fangentladung, Aufwärtsblitz
upward trend steigende Tendenz
upward welding senkrechte Schweißung nach oben
uranium Uran
uranium atom Uran-Atom
uranium bar Uranstab
uranium bomb Uranbombe
uranium compound Uranverbindung
uranium deposit Uranlagerstätte
uranium fission Uranspaltung
uranium hexafluoride Uranhexafluorid
uranium isotope Uranisotop
uranium ore Uranerz
uranium oxide Uranoxid
uranium pile Uranbrenner, Uranreaktor (539)
uranium power plant Urankraftwerk
uranium rays (pl.) Uranstrahlung
uranium reactor Uranreaktor
uranium rod Uranstab
urban städtisch, Stadt-...
urban railway Stadtbahn
urea Harnstoff
urea resin Harnstoffharz
urea resin glue Harnstoffkunstharzleim
to urge drängen, dringen, zwängen
to urge an electron ein Elektron treiben
usability Brauchbarkeit, Verwendbarkeit
usage Gebrauch, Beanspruchung, Betriebsbeanspruchung
USART = universal synchronous-asynchronous receiver-transmitter
U-section U-Profil
used gebraucht; verbraucht, ausgebraucht
useful nützlich, nutzbar, Nutz-...
useful cross-section Nutzquerschnitt (621.313)
useful current Nutzstrom, Wirkstrom
useful effect Nutzleistung
useful field intensity Nutzfeldstärke
useful flux Nutzfluß
useful level Nutzpegel
useful life Nutzungsdauer, Lebensdauer, Benutzungsdauer
useful load Nutzlast
useful screen area Nutzbildfläche (621.397)
useful signal Nutzsignal
user Benutzer, Verbraucher, Kunde, Anwender
user of electric power Stromverbraucher, Stromabnehmer
user program Anwender-Programm
user software Anwender-Software
user's manual Handbuch, Beschreibung
user terminal Benutzerstation, Teilnehmer-Ein-Ausgabegerät
U-shape U-Form
U-shaped U-förmig
USW = ultra short waves UKW = Ultra-Kurzwellen
utility program Dienstprogramm (681.3)
utility service Versorgungsbetrieb
utilization Benutzung, Nutzbarmachung, Verwendung
utilization equipment Verbraucheranlage
utilization factor Ausnutzungsfaktor (621.314)
utilization of material Materialausnutzung
U. T. S. = ultimate tensile strength Zugfestigkeit
U. T. S. elongation Bruchdehnung
U-Tube U-Rohr
uviol lamp Uviol-Lampe

V

VA = volt-ampere VA = Voltampere
vacancy Gitterlücke, Fehlstelle (591)
vacant leer, unbenutzt, vakant
vacant line freie Leitung, nicht besetzte Leitung (621.395)
to vacate leer machen, frei machen, räumen
vacating Ausfädeln (351.814.33)
to vacillate hin- und herschwanken
in vacuo im Vakuum, unter Vakuum
vacuo junction Vakuum-Thermoelement
vacuous space Hohlraum
vacuum Vakuum, Unterdruck
vacuum brake Vakuum-Bremse
vacuum capacitor Vakuumkondensator
vacuum cleaner Staubsauger

vacuum diffusion pump Dampfstrahldiffusionspumpe, Diffusionspumpe
vacuum flask Thermosflasche
vacuum gauge Vakuummesser
vacuum guide Bandführungssegment (681.84)
vacuum impregnation Vakuum-Imprägnierung
vacuum lamp Vakuumlampe
vacuum oven Vakuumofen
vacuum pack Vakuumpackung
vacuum pan Vakuumtopf
vacuum pump Vakuumpumpe
vacuum seal Vakuumdichtung
vacuum servo brake Unterdruckbremse
vacuum switch Vakuumschalter
vacuum tube (US) Vakuumröhre, Hochvakuumröhre (621.385)
vacuum-tube amplifier Röhrenverstärker
vacuum-tube circuit Röhrenschaltung
vacuum-tube detector Gleichrichterröhre, Hochvakuumgleichrichterröhre
vacuum-tube generator Röhrensender, Röhrensummer
vacuum-tube lamp Gasentladungslampe
vacuum tube oscillator Röhrensummer, Röhrensender
vacuum-tube rectifier Röhrengleichrichter
vacuum-tube transmitter Röhrensender
vacuum-tube voltmeter Röhrenvoltmeter
vacuum tubing Vakuumschlauch
vacuum valve (Br.) Hochvakuumröhre (621.385)
V-aerial V-Antenne
vagabond current Erdstrom, vagabundierender Strom
valence Valenz, Wertigkeit
valence band Valenzband (539)
valence electron Valenzelektron
valency Valenz, Wertigkeit
valency bond Valenzbindung
valency link Valenzbindung
valley point Talpunkt
valley-point current Talstrom
value Wert
value engineering analysis Wertanalyse
value of the open-circuit reverse voltage transfer ratio Leerlaufspannungsrückwirkung (621.382.3)
valve Röhre, Rohr (621.385); Ventil, Schieber, Absperrorgan (621.646); Gleichrichterventil (621.396) (siehe auch **tube** und **vacuum tube**)
valve adapter Zwischensockel (621.385)
valve amplifier Röhrenverstärker
valve base Röhrenfuß, Röhrensockel
valve detector Gleichrichterröhre (621.396)
valve effect Ventilwirkung
valve envelope Röhrenkolben
valve generator Röhrengenerator, Röhrensummer
valve hiss Röhrenrauschen
valve holder Röhrensockel, Röhrenfassung (621.396)
valve noise Röhrenrauschen, Schrotrauschen
valve-operated induction furnace Induktionsofen mit Röhrensender
valve oscillator Röhrensummer, Röhrensender
valve ratio Gleichrichteverhältnis, Sperrwirkung
valve receiver Röhrenempfänger
valve rectifier Röhrengleichrichter
valve tester Röhrenprüfgerät
valve transmitter Röhrensender
valve-type lightning arrester Kathodenfallableiter
valve voltmeter Röhrenvoltmeter
van Vorhut, Vorläufer; Lieferwagen, Güterwagen
van den Graaff machine van-de-Graaff-Generator, Bandgenerator
van der Waals's law van der Waalssches Gesetz
vanadium Vanadium
vane Leitblech, Schaufel (621.2); Fahne, Dämpfungsflügel (621.317); Platte (621.319.4); Flügel, Mühlenflügel, Propellerblatt, Propellerflügel (533); Windfahne, Wetterfahne (55)
vane-type relay Drehfahnenrelais, Stellungszeiger
to vanish verschwinden; gegen Null gehen (51)
V-antenna (US) V-Antenne
vapor (US) Dampf (s. auch **vapour**)
to vapor deposit aufdampfen
vaporization Verdampfung, Verdampfen, Verdunstung
to vaporize verdampfen, verdunsten
vaporizer Verdampfer; Verdunster
vaporizing Verdampfungs-...
vaporizing plant Verdampferanlage
vaporizing point Verdampfungspunkt
vaporizing temperature Verdampfungstemperatur
vaporous dampfartig, dampfförmig
vapour (Br.) Dampf
vapour absorption Dampfabsorption
vapour absorption type refrigerator Absorptionskühlschrank
vapour cooling Verdampfungskühlung

vapour diffusion pump Dampfstrahlpumpe, Diffusionspumpe
vapour-proof dampfdicht
vapour-proof fitting Feuchtraumleuchte (621.32)
vapour-proof wiring Feuchtrauminstallation
vapour tension Dampfdruck
vapourizing equipment Aufdampfanlage
varactor diode Varactor Diode, Kapazitätsdiode
varh Blind-Wh
varhour meter Blindstromzähler, Blindleistungszähler
variable veränderlich, variabel, regelbar
variable air condenser Luftdrehkondensator
variable amount variable Summe, variabler Betrag
variable-area (sound) track Zackenschrift, Amplitudenschrift
variable attenuator Dämpfungsregler
variable-capacitance diode Kapazitätsdiode
variable-capacitance standard Normaldrehkondensator
variable capacitor Drehkondensator, veränderbarer Kondensator
variable coupling veränderliche Kopplung
variable drive regelbarer Antrieb
variable field veränderliches Feld
variable film resistor Schichtdrehwiderstand
variable frequency veränderliche Frequenz
variable-frequency oscillator durchstimmbarer Sender, Summer mit veränderlicher Frequenz
variable head welding torch Wechselbrenner (621.791)
variable inductance Variometer, veränderliche Induktivität
variable inductor veränderliche Selbstinduktivität, Variometer
variable of integration Integrationsvariable
variable-length line Posaune (621.317), (621.396)
variable load wechselnde Belastung
variable mu veränderliche Steilheit (621.385)
variable mu action Exponentialwirkung (621.385)
variable mu characteristic Exponentialkennlinie, Regelcharakteristik (621.385)
variable mu tube (US) Exponentialröhre, Regelröhre
variable mu valve (Br.) Exponentialröhre, Regelröhre
variable oil capacitor Öldrehkondensator
variable ratio veränderliches Übersetzungsverhältnis
variable ratio transformer Regeltransformator, Regelumspanner
variable-reactance amplifier parametrischer Verstärker
variable resistance Regelwiderstand, veränderlicher Widerstand
variable resistor veränderlicher Widerstand, Drehwiderstand
variable service conditions wechselnde Betriebsbedingungen
variable slider Abgriffschelle (621.3)
variable-slope valve Exponentialröhre (621.385)
variable speed veränderliche Geschwindigkeit; veränderliche Drehzahl
variable-speed transmission Regelgetriebe
variable-speed motor drehzahlregelbarer Motor, Regelmotor; steuerbarer Motor
variable voltage veränderbare Spannung
variable-voltage control Drehzahlregelung durch Spannungsregelung, Ankerregelung; Steuerung durch Spannungsänderung
variable-voltage generator regelbarer Generator; Steuergenerator
variable-voltage source steuerbare Spannungsquelle
variable-voltage speed control Ankerspannungsregelung
variable-voltage transformer Regeltransformator, Regelumspanner
variable wire-wound resistor Drahtdrehwiderstand
variation Variante, Spielart, Abwandlung; Schwankung, Veränderung, Variation
variation in capacity Kapazitätsschwankung
variation in load Lastschwankung
variation range Schwankungsbereich
variation in signal strength Lautstärkeschwankung im Empfang
variation from standard Abweichung von der Norm
variation in temperature Temperaturschwankung
variation in voltage Spannungsschwankung
variational Variations-...
variational calculus Variationsrechnung
variational method Variationsverfahren, Variationsrechnung
variegated bunt, buntfleckig
variocoupler Kopplungsvariometer
variolosser regelbares Dämpfungsglied
variometer Variometer

varistor Varistor, spannungsabhängiger Widerstand
Varley's test [Fehlerortsbestimmung für Kabel]
varnish Lack, Lackanstrich, Lackierung
varnish impregnation Lackimprägnierung
to varnish lackieren
varnished lackiert, Lack-...
varnished cambric Lackleinen
varnished cambric cable Kabel mit Lackleinenisolation
varnished flex tubing Lackschlauch, Rüschschlauch
varnished manilla paper Manila-Lackpapier
varnished paper Lackpapier
varnished silk Lackseide
varnished sleeving Lackschlauch, Rüschschlauch (621.396)
varnished wire Lackdraht
to vary (sich) ändern, schwanken, voneinander abweichen, auseinandergehen
to vary about... um... als Mittelwert schwanken
to vary with the time sich mit der Zeit ändern, zeitlich veränderlich sein
varying wechselnd, schwankend; Wechsel-...
varying component Wechselstromanteil, Wechselkomponente
varying duty wechselnde Betriebsart
varying speed wechselnde Drehzahl
varying-speed motor Regelmotor
varying-voltage control Hauptstromregelung (621.34)
vaseline Vaseline
vat Bütte, Bottich
vault Volte (79); Gewölbe (72)
vault-type transformer Kellerstationstransformator
V-belt Keilriemen
V-belt drive Keilriemenantrieb
V-belting Keilriemenantrieb
V-block Kreuzschiene; Wellenbock, Dreikantauflage (621.9)
VCO = voltage-controlled oscillator spannungsgesteuerter Oszillator
VCR = vertical redundancy check
VDR = voltage-dependent resistor spannungsabhängiger Widerstand
VDU = video display unit
vector Vektor
vector admittance Leitwertoperator
vector amplification komplexer Verstärkungsfaktor
vector analysis Vektoranalysis
vector diagram Zeigerdiagramm
vector field Vektorfeld
vector function Vektorfunktion
vector impedance komplexer Widerstand, Widerstandsoperator
vector line gerichtete Strecke, Vektor
vector notation Vektorschreibweise, Zeigerschreibweise
vector potential field Vektorpotentialfeld
vector power Zeiger der Scheinleistung
vector power factor Leistungsfaktor
vector product Vektorprodukt, äußeres Produkt; vektorielles Produkt
vector quantity Vektorgröße, vektorielle Größe
vector reprensation Vektordarstellung, Zeigerdarstellung
vector sum Vektorsumme, vektorielle Summe, Zeigersumme
vector volt-amperes (pl.) Scheinleistung
vectoral vektoriell, Vektor-...
vectoral difference vektorielle Differenz; Zeigerdifferenz
vectoral sum Vektorsumme, Zeigersumme
vectored interrupt gerichteter interrupt
vectorial vektoriell
vectorscop Vektorskop
Vee Vau (Zusammensetzungen s. unter **V-belt, V-block** usw.)
to veer sich drehen (55)
veering winds drehende Winde
vegetable pflanzlich, vegetabil
vegetable oil pflanzliches Öl
vehemence Vehemenz, Wucht
vehement vehement, wuchtig
vehicle Fahrzeug
vehicle battery Fahrzeugbatterie, Autobatterie
vehicle cell Fahrzeugakkumulator
velocity Geschwindigkeit
velocity of detonation Detonationsgeschwindigkeit
velocity distribution Geschwindigkeitsverteilung
velocity of efflux Ausflußgeschwindigkeit, Ausströmungsgeschwindigkeit
velocity energy kinetische Energie, lebendige Kraft, Wucht
velocity of light Lichtgeschwindigkeit
velocity microphone auf Schallschnelle ansprechendes Mikrophon, Druckgradientenmikrophon
velocity-modulated geschwindigkeitsmoduliert, laufzeitmoduliert (621.396)
velocity-modulated amplifier Laufzeitverstärker

velocity-modulated oscillator Laufzeitoszillator
velocity-modulated tube Laufzeitröhre
velocity modulation Geschwindigkeitsmodulation
velocity of progression Fortschreitgeschwindigkeit
velocity of propagation Fortpflanzungsgeschwindigkeit, Ausbreitungsgeschwindigkeit
velocity resonance Phasenresonanz
vent Luftloch, Lüftungsloch, Belüftungsloch; Steiger (621.7)
venthood Abzugshaube
to vent lüften, belüften; entlüften (621.7); atmen lassen (621.3)
to ventilate lüften, belüften, ventilieren
ventilated gelüftet, belüftet, ventiliert; luftgekühlt (621.313)
ventilated motor luftgekühlter Motor
ventilated totally enclosed motor Motor mit Mantelkühlung
ventilated transformer luftgekühlter Transformator
ventilating Lüftungs-..., Kühl..., Ventilations-...
ventilating air channel Lüftungskanal
ventilating aperture Belüftungsöffnung
ventilating current Wetterstrom (622)
ventilating duct Ventilationskanal, Luftkanal; Kühlschlitz (621.313)
ventilating fan Lüfter; Lüfterrad (621.313); Kühlgebläse (621.314); Ventilator
ventilating hole Lüftungsloch, Luftloch
ventilating plant Belüftungsanlage
ventilating shaft Luftschacht, Belüftungsschacht
ventilating space Kühlschlitzbreite (621.313)
ventilation Lüftung, Belüftung, Ventilation; Kühlung (621.313); Bewetterung (622)
ventilation system Belüftungsanlage
ventilator Ventilator, Lüfter
venting system Belüftungssystem
Venturi tube Venturi-Rohr
Verdet's constant Verdetsche Konstante (535)
verdigris Grünspan
verge Rand, Kante
verge-punched card randgelochte Karte
to verge grenzen an, berühren
verification Verifikation, anschaulicher Beweis; Bestätigung, Nachweis
verificatory bestätigend
verificatory test Kontrollversuch
to verify verifizieren, experimentell nachweisen, wahrscheinlich machen
vermillion Zinnober
vernier Nonius
vernier caliper rule Schieblehre, Schublehre
vernier condenser Feinabstimmkondensator
vernier scale Noniusskala, Feineinstellskala
versorial force Richtkraft
versorium Elektrometernadel
vertex (pl. **-ices**) Scheitel, Spitze; Eckpunkt, Spitze, Ecke (51)
vertex of the parabola Parabelscheitel
vertex feed Scheitelspeisung (621.396.67)
vertex plate Erregerreflektor
vertical Vertikale, Lot, Senkrechte
vertical vertikal, senkrecht, lotrecht
vertical angle Winkel an der Spitze (51)
vertical antenna Vertikalantenne
vertical break switch Schalter mit senkrechter Schaltstrecke
vertical control Vertikalbildlageregler
vertical coverage Höhenbedeckung
vertical current luftelektrischer Leitungsstrom, luftelektrischer Vertikalstrom
vertical deflection Vertikalablenkung
vertical directivity vertikale Bündelung
vertical displacement vertikale Verschiebung
vertical effect Hochantenneneffekt (621.396)
vertical force Auflagekraft
vertical height Höhe (im Dreieck)
vertical hold control vertikaler Bildfang
vertical incidence Senkrecht-Einfall (621.396)
vertical interval reference signal Prüfzeilenreferenzsignal (621.397)
vertical interval test signal Prüfzeilenmeßsignal
vertical polar diagram Vertikaldiagramm (621.396.67)
vertical polarization vertikale Polarisation
vertical projection vertikale Projektion
vertical redundancy check Querprüfung (681.3)
vertical scan transformer Vertikal-Ablenktransformator
vertical section Vertikalschnitt, senkrechter Schnitt
vertical shaft senkrechter Schacht; senkrechte Welle
vertical shaft motor Motor mit senkrechter Welle
vertical sweep vertikale Abtastung, vertikale Ablenkung (621.396.96)
vertical switchboard Schalttafel
vertical synchronizing pulses Vertikalsynchronsignal

vertically polarized senkrecht polarisiert
verticity Richtkraft
very high höchst
very high frequency UKW-Frequenz, Meterwellen, Ultrakurzwellen
very high frequency reception UKW-Empfang = Ultra-Kurzwellen-Empfang
very high tension Höchstspannung
very high tension line Höchstspannungsleitung
very low frequency Längswellen, Myriameterwellen
very short range Nahbereich
vestige Spur, Rest
vestigial restlich, Rest-...
vestigial sideband Restseitenband, Seitenbandrest (621.396)
vestigial-sideband bandwidth Breite der Nyquistflanke (621.397)
vestigial-sideband filter Restseitenbandfilter
vestigial-sideband modulation Restseitenbandmodulation
VF = voice frequency
VF-telegraphy WT, Wechselstromtelegraphie
V-groove V-Rille, Keilrille
to V-groove auskreuzen, ausvauen (621.791)
V-grooved mit Keilrille
V-grooved wheel Keilriemenscheibe
v. h. f. = very high frequency UKW-Frequenzgebiet
v. h. f.-omnidirectional range = VOR UKW-Drehfunkfeuer
via net loss Durchgangsdämpfung
vibrant schwingend, vibrierend
to vibrate schwingen
to vibrate about schwingen um
vibrating Schwing-...
vibrating circuit Vibrator-Schaltung (621.394)
vibrating grate Schüttelrost
vibrating power Schwingleistung
vibrating reed Schwingzunge (621.317)
vibrating-reed electrometer Schwingkondensator-Elektrometer
vibrating-reed instrument Zungenfrequenzmesser
vibrating-reed rectifier Schwinggleichrichter
vibrating screen Schwingsieb, Schüttelsieb
vibrating sieve Schwingsieb, Rüttelsieb
vibration Schwingung, Erschütterung, Vibration
vibration absorbing schwingungsdämpfend, schwingungsschluckend
vibration damper Schwingungsdämpfer (621.315.1)
vibration-free schwingungsfest, schwingungsfrei
vibration galvanometer Vibrationsgalvanometer
vibration insulation Schwingungsisolation
vibration isolator Schwingungsdämpfer, Schwingungspuffer
vibration meter Schwingungsmesser
vibration microphonism Vibrationsmikrophonie
vibration-proof schwingungsfest, schüttelfest, erschütterungsfrei
vibration table Schütteltisch
vibration test Schüttelprüfung
vibrational Schwingungs-..., Schwing-...
vibrational characteristic Schwingungseigenschaft
vibrator Schwinger (534); Schütteleinrichtung (621); Zerhacker, Wechselrichter (621.314)
vibrator unit Zerhackerpatrone (621.3)
vibratory Schwing-...
vibratory impulse Schwingimpuls
vibratory sieve Schwingsieb
vibratory stresses (pl.) Schwingungsbeanspruchung
vibrograph Schwingungsmesser
vice Schraubstock
vice bench Werkbank
vice jaw Schraubstockbacke
video Fernseh-... Video-...
video amplifier Videoverstärker
video carrier Bildträger
video control and regenerating amplifier Video-Regel- und Regenerierverstärker
video crossbar system Video-Koppelsystem, Video-Kreuzschienenverteiler
video crosspoint connector Video-Koppelpunkt
video display unit Videosichtgerät
video frequency Videofrequenz
video-frequency amplifier Videoverstärker
video-frequency generator Videomeßgenerator
video head Videokopf
video image Fernsehbild
video mapping Kartenvideo-Einblendung (623)
video mixer Bildmischgerät
videophone Bildfernsprecher
video recorder Bildaufzeichnungsgerät, Video Recorder
video signal Video-Signal, Fernseh-Frequenzgemisch
video tape recorder Fernsehaufzeichnungsgerät, Bildaufzeichnungsgerät, Videobandgerät

video tape recording magnetische Videosignal-Aufzeichnung, MAZ, Bildaufzeichnung
video technique Fernsehtechnik, Videotechnik
video test-signal generator Video-Prüfsignalgenerator
video transmission Fernsehübertragung, Fernsehsendung
video transmitter Fernsehsender
vidicon camera tube Vidikon
vidicon telecine Vidikon-Filmabtaster
view Sicht, Aussicht; Ansicht, Meinung
view finder Sucher (77)
viewphone Fernsehtelefon
viewing Gesichts-..., Betrachtungs-...
viewing angle Bildwinkel, Betrachtungswinkel
viewing distance Betrachtungsabstand
viewing screen Bildschirm, Betrachtungsschirm
viewing storage tube Sichtspeicherröhre
vinylchloride Vinylchlorid
virgin „jungfräulich", Original-...
virgin metal Neumetall, Frischmetall
virtual virtuell, gedacht
virtual address virtuelle Adresse
virtual cathode virtuelle Kathode
virtual displacement virtuelle Verschiebung
virtual height of ionospheric layers scheinbare Höhe der Ionosphäre
virtual image virtuelles Bild
virtual junction temperature Ersatzsperrschichttemperatur
virtual mass mitschwingende Masse, mitbewegte Wassermasse (534)
virtual memory virtueller Speicher
virtual short circuit ideeller Kurzschluß
virtual temperature innere Ersatztemperatur (621.382)
virtual value Effektivwert (621.3)
virtual voltage Effektivspannung (621.3)
viscose Viskose
viscose silk Viskoseseide
viscosity Viskosität
viscous viskos, zähflüssig, zäh
viscous damping Flüssigkeitsdämpfung
viscous flow laminare Strömung, Laminarströmung
viscous fluid zähe Flüssigkeit
viscous impingement-type filter ölbenetztes Filter
viscous rayon Viskoseglanzgarn, Viskosegarn
visibility Sichtbarkeit; Sicht, Sichtweite (551.5)
visibility factor Anteil des sichtbaren Lichts
visible sichtbar
visible indication Sichtanzeige, sichtbare Anzeige
visible paging optische Personensuchanlage
visible phenomenon sichtbare Erscheinung
visible radiation sichtbare Strahlung, optische Strahlung
vision Gesicht, Gesichtssinn, Sehen
vision carrier Bildträger (621.397)
vision frequency Fernsehfrequenz, Bildträger
vision-frequency transmitter Fernsehbildsender
vision modulation Bildmodulation
vision signal Fernsehsignal
visional Seh-...
visual sichtbar, Sicht-..., mit Sichtanzeige
visual acuity Sehschärfe
visual aid chart Anschauungsbild
visual-aural radio range Vierkursfunkfeuer mit Sicht- und Höranzeige
visual call sign optisches Rufzeichen
visual contact approach Sichtanflug
visual detector Sichtanzeiger, objektive Anzeigeeinrichtung (621.317)
visual display unit Sichtgerät, Bildschirmgerät; Anzeigeeinheit
visual distance Sehweite (535)
visual field Gesichtsfeld
visual image sichtbares Bild
visual indication optische Anzeige, sichtbare Anzeige
visual indicator Schauzeichen
visual means Sichtgerät
visual observation Sichtbeobachtung
visual signal optische Meldung
visual tuning optische Abstimmung, magisches Auge
visual-type ... mit Sichtanzeige
vita glass UV-durchlässiges Glas
vita rays (pl.) physiologische UV-Strahlung
vital lebenswichtig, entscheidend
vital error entscheidender Fehler
to vitiate verderben, zunichte machen
vitiated air verdorbene Luft, verbrauchte Luft
vitreous glasähnlich, glasig
vitreous electricity Glaselektrizität, positive Elektrizität
vitreous enamel Emailleglas, Emaille
vitreous enamelling Emaillierung, Feueremaillierung
vitreous humour Glaskörper (611.84)
vitrification Brennvorgang (66)
to vitrify verglasen
vitriol Vitriol
vitriolate of... ...-sulfat

v. l. f. = very low frequency Längstwellen
VLSI = very large scale integration Höchstintegration
vocal Stimm-..., Sprech-...
vocal chords Stimmbänder
vocoder = voice codes to recreate Vocoder
vodas = voice-operated device, anti-singing Sprachsperre, Sprechsperre (621.395)
vogad = voice operated gain adjusting device sprachgesteuerter Pegelregler
voice Stimme, Sprache
voice band Sprachband
voice call sign Sprechrufzeichen
voice channel Sprechkanal, Sprachkanal
voice coil Lautsprecherspule
voice communication Sprachverständigung
voice current Sprechstrom
voice frequency Sprachfrequenz, Tonfrequenz
voice-frequency carrier telegraphy Wechselstromtelegraphie
voice-frequency range Sprachfrequenzbereich, Tonfrequenzbereich
voice-frequency ringing Tonfrequenzruf
voice-frequency signalling relay set Tonfrequenzrufumsetzer
voice-frequency signalling transmitter Tonfrequenzrufumsetzer, Tonfrequenzrufgenerator
voice-frequency telegraph system WT-Telegraphie
voice-frequency telephony Tonfrequenztelefonie
voice grade circuit Sprechkreis
voice-modulated sprachmoduliert
voice modulation Sprachmodulation
voice-operated gain adjusting device sprachgesteuerter Pegelregler, Vogad
voice-operated relay sprachgesteuertes Relais, Sprechrelais (621.395)
voice operation demonstrator Sprachsimulator
voice peak Sprechspitze
voice power Sprechleistung
void Hohlraum, Blase, Lunker (66); Zwischenraum; Leere
volatile flüchtig, leicht verdampfbar
volatile constituent flüchtiger Bestandteil
volatile matter flüchtiger Stoff
volatile storage energieabhängiger Speicher, flüchtiger Speicher
volatile substance flüchtige Substanz
volatility Flüchtigkeit
volatilization Verflüchtigung
to volatilize verflüchtigen, abdampfen
volt Volt
volt-amperes (pl.) VA, Scheinleistung
volt-ampere-hour meter Volt-Ampere-Stunden-Zähler
volt-ampere meter Scheinleistungsmesser
volt-ampere reactive hour Blindwattstunde
volt box Spannungsteiler (für Kompensationsapparat) (621.317)
volt gauge Spannungszeiger
volt line Einheit des magnetischen Flusses = 1 Voltsekunde = 10^8 M = 1 Wb
volts lost Spannungsabfall
12-volt operation Zwölf-Volt-Betrieb, 12-V-Betrieb
volt vector Spannungszeiger
Volta effect Volta-Effekt, Kontaktspannung
voltage Spannung; Stromspannung
voltage amplification Spannungsverstärkung
voltage-amplification factor Spannungsverstärkungsfaktor
voltage amplifier Spannungsverstärker, Vorverstärker
voltage antinode Spannungsbauch
voltage applied angelegte Spannung
voltage attenuation Spannungsdämpfung
voltage balance Spannungsgleichgewicht
voltage between lines Leiterspannung, verkettete Spannung
voltage between phases Leiterspannung, verkettete Spannung
voltage breakdown Spannungsausfall
voltage change Spannungsveränderung
voltage circuit Spannungspfad (621.317)
voltage coil Spannungsspule
voltage commutation Spannungsumschaltung
voltage comparator Spannungskomparator
voltage control Spannungsregelung
voltage-controlled spannungsbetätigt, spannungsgeregelt
voltage-current characteristic Strom-Spannungs-Kennlinie
voltage dependent spannungsabhängig
voltage detector Spannungsprüfer, Spannungsfühler
voltage divider Spannungsteiler
voltage-dividing resistor Potentiometer, Spannungsteilerwiderstand
voltage division Spannungsteilung
voltage doubler Spannungsverdoppler
voltage-doubler circuit Spannungsverdopplungsschaltung
voltage doubling Spannungsverdopplung
voltage-doubling circuit Verdopplerschaltung

voltage drop Spannungsabfall (**across:** an)
voltage-dropping resistance Vorwiderstand
voltage equivalent of thermal energy Temperaturspannung (621.382)
voltage-fed spannungsgespeist
voltage feedback Spannungsrückkopplung
voltage fluctuation Spannungsschwankung
voltage follower Impedanzwandler
voltage from ... to ... Spannung zwischen ... und ...
voltage gain Spannungsverstärkung
voltage gradient Potentialgefälle, Spannungsgradient
voltage to ground Spannung gegen Erde
voltage impulse Spannungsstoß
voltage indicator Spannungsanzeiger, Spannungssucher
voltage injection Spannungseinführung
voltage level Spannungspegel, absoluter Spannungspegel
voltage lost in ... Spannungsabfall an ...
voltage magnet Spannungseisen (621.317)
voltage magnification Spannungsüberhöhung (621.317)
voltage multiplier Vorwiderstand (621.317)
voltage to neutral Strangspannung, Phasenspannung, Spannung gegen den Sternpunkt
voltage node Spannungsknoten
voltage-operated spannungsbetätigt
voltage peak Spannungsspitze
voltage phase-balance protection Spannungs-Unsymmetrieschutz
voltage probe Spannungssonde (621.396)
voltage range Spannungsmeßbereich
voltage-range switch Spannungs-Meßbereichumschalter
voltage rating Spannungsauslegung; Spannungssicherheit; Nennspannung
voltage ratio Leerlaufübersetzungsverhältnis, Spannungsübersetzung
voltage recovery Spannungsrückkehr
voltage reference Bezugsspannung
voltage-reference diode Referenzdiode, Spannungsreferenzdiode, Vergleichsspannungsdiode
voltage regulation Spannungsstabilisierung, Spannungsregelung; Spannungsanstieg bei Lastabwurf (621.313)
voltage regulator Spannungsregler, Spannungsstabilisator
voltage-regulator diode Referenzdiode, Z-Diode
voltage relay Spannungsrelais
voltage responsive spannungsabhängig
voltage restraint ... mit Spannungsanwurf (621.318)
voltage ripple Welligkeit, Kräuselspannung
voltage saturation Spannungssättigung
voltage of self-induction Selbstinduktionsspannung
voltage-sensitive resistor spannungsabhängiger Widerstand
voltage stabilizer Spannungskonstanthalter, Spannungsregler
voltage standard Spannungsnormal
voltage standing-wave ratio Welligkeitsfaktor, Stehwellenverhältnis
voltage stress elektrische Feldstärke, elektrische Beanspruchung
voltage surge Sprungwelle, Spannungsstoß, Spannungswelle, Überspannung
voltage swing Spannungsschwankung
voltage of the system verkettete Spannung, Leiterspannung
voltage test Spannungsprüfung
voltage tester Spannungsprüfer
voltage to Spannung gegen
voltage transfer Spannungsübersetzung (621.396)
voltage transformer Spannungswandler
voltage triangle Spannungsdreieck
voltage tripler Spannungsverdreifacher
voltage-tripler circuit Verdreifacherschaltung
voltage-tuned spannungsabgestimmt (621.396.61) [Klystron]
voltage-type telemeter Fernmeßgerät mit Spannungsübertragung
voltage variation Spannungsschwankung
voltage vector Spannungszeiger
voltage zero Nulldurchgang der Spannung
voltaic galvanisch
voltaic battery Voltasche Säule
voltaic cell galvanisches Element
voltaic electricity Volta-Spannung, Kontaktelektrizität
voltaic pile Voltasche Säule
voltameter Voltameter
voltmeter Voltmeter, Spannungsmesser
voltmeter switch Voltmeterumschalter
volume Volumen, Rauminhalt; Lautstärke (621.396.62); Pegel (621.395)
volume compressor Dynamikbegrenzer (621.396.61)
volume control Lautstärkenreglung
volume distribution räumliche Verteilung
volume equivalent Bezugsdämpfung

volume expander Dynamikdehner (621.396.62)
volume indicator Pegelzeiger (621.395); Ausgangsleistungsmesser (621.936.62)
volume moment Volummoment (53)
volume range Dynamik (621.396.61/62)
volume resistivity spezifischer Widerstand, Würfelwiderstand (621.3)
volume for volume zu gleichen Teilen, Liter auf Liter; unter Annahme gleicher Volumina
volumetric(al) volumetrisch, raummäßig
volumetric coverage Raumbedeckung
volumetric extension Volumenausdehnung
volumetric measurement Hohlmaße, Raummaße (pl.); Inhaltsvermessung, Volumbestimmung
volute Spirale
volute chamber Spiralgehäuse, Einlaufkammer (621.2)
VOR = V. H. F. omni-range UKW-Drehfunkfeuer
vortex (pl. **-tices**) Wirbel (533), (532)
vortex shedding Wirbelablösung
vortical Wirbel-... (532), (533)
voucher Quittung, Bestätigung
V. P. N. = Vickers pyramid number Vickers-Härtezahl
"V" ring Kollektorspannung
V-rope Keilriemen
VR tube = voltage regulating tube Stabilisationsröhre, Stabilovoltröhre
V. S. B. = vestigal sideband Restseitenband
V-shape V-Form
V-shaped V-förmig
VSWR = voltage standig wave ratio Welligkeitsfaktor
V-thread Dreiecksgewinde
V-threaded scharfgängig
VTOL = vertical take-off and landing Senkrechtstart
VTR = video tape recorder Videorecorder, Fernsehaufzeichnungsgerät
V-type contact Messerkontakt (621.316.5)
VU-meter VU-Meter (621.375)
vulcanite Hartgummi
vulcanization Vulkanisation, Vulkanisierung
to vulcanize vulkanisieren
vulcanized fibre Vulkanfiber
vulcanizer Vulkanisierapparat
vulgar fraction gemeiner Bruch

W

wad Pfropfen, Stopfen, Bausch
wad of air Luftpfropfen, Luftschwaden
to wad zupfropfen, festpfropfen
wadding Bausch; Pfropfmaterial
wafer Schalterebene (621.396); Waffel, Oblate, Scheibe; Kristallscheibe (621.382)
to wafer mit einer Oblate zukleben; in Scheiben schneiden, aus Scheiben aufbauen
wafered switch Segmentschalter (621.396)
wag(g)on Waggon, Lastwagen; (offener) Güterwagen
Wagner earth Wagnersche Hilfserde (621.317)
Wagner earthing device Wagnersche Hilfsbrücke
wainst Einschnürung (539)
waiting Warten, Wartezeit
waiting-on time Wartezeit (621.9)
waiting-passenger indicator Stockwerksruf (621.876)
waiting program wartendes Programm
waiting traffic Wartebelastung (621.395)
wake-up service Weckdienst
walkie-talkie Handfunksprecher, (tragbares) Sende-Empfangsgerät
walkman Walkman
wall Wand, Mauer
wall-board Wandplatte, Hartfaserplatte; Wandtafel
wall bracket Wandkonsole, Konsole
wall bushing Wanddurchführung, Außendurchführung
wall-entrance bushing Einführung, Einführungsisolator, Wanddurchführung (621.315.6)
wall-entrance insulator Einführungsisolator, Einführung, Außendurchführung
wall fitting Wandleuchte
wall mounting Wandbefestigung; Wandmontage (621.316.5)
wall plug Wandstecker (621.3)
wall socket Anschlußdose (621.395); Steckdose (an der Wand) (621.3)
wall-type Wand-...
wall-type telephone Wandtelefon, Wandapparat (621.395)
to wander wandern; auswandern (621.317); verlaufen (621.9)
wander plug Wanderstecker (621.317)
wandering Strahlauswanderung (621.396)
wane Abnehmen, Zustand des Abnehmens
to wane abnehmen, dahinschwinden

want Bedarf, Mangel, Not
want of balance Nichtausbalancierung, Unbalance; Unsymmetrie (621.3)
for want of fuel wegen Brennstoffmangels, mangels Brennstoff
to want benötigen, brauchen, nötig haben
to want a gentle heating vorsichtig erhitzt werden müssen
wanted carrier Nutzträger
wanted emission geforderte Sendeleistung
wanted signal Nutzsignal
war Krieg
war head Gefechtskopf (623)
war plane Militärflugzeug
war ship Kriegsschiff
to warble wobbeln
ward Pförtnerloge; Station (61); Zuhaltung (683.33)
warehouse Warenlager, Speicher
warehouse of energy Energiespeicher
to warm wärmen, anwärmen, hochheizen; sich erwärmen, warm werden, sich aufheizen
to warm an engine einen Motor anwärmen, einen Motor warmlaufen lassen
to warm throughout durchwärmen
to warm up anwärmen, aufheizen, hochheizen; sich erwärmen
warm warm
warm junction warme Lötstelle (621.36)
warm-up time Vorwärmzeit, Einlaufzeit, Aufwärmzeit
warming Anwärmen, Aufheizung
on warming beim Anwärmen, bei Erhitzung
warming-up Anheizung, Anwärmung
warming-up period Anwärmzeit (621.1), Anheizzeit (621.385)
warmth Wärme, Warmsein
warmth of the body Körperwärme
warning Warnung
warning light Warnlampe, Meldelampe
warning panel Warntableau
warning receiver Warnempfänger
warp Kette, Aufzug (67)
to warp aufziehen (67); sich verwerfen, sich verziehen; sich wellen, sich werfen; verholen (629.12)
warped krumm, wellig, verzogen
warping Verwerfen, Verziehen, Werfen
warranty Gewährleistung, Garantie
wash Waschprozeß, Waschen (64); Waschen, Spülen, Unterwaschen, Spülwirkung (627); Anstrich, Leimfarbenanstrich (69)
wash boiler Waschkessel, Wäschekocher
to wash waschen, sich waschen; waschen, sich waschen lassen (64); wässern, auswaschen (54), (77); spülen, waschen (627); anstreichen, mit Leimfarbe anstreichen (69)
to wash in hineinspülen
to wash over überstreichen, überpinseln
washable abwaschbar, waschbar
washer Unterlegscheibe, Lochscheibe, Dichtungsscheibe, Ring, Dichtungsring
washer dryer Wäschetrockner, Wäscheschleuder
washing Wäsche; Anstrich (69)
washing flask Waschflasche (54)
washing machine Waschmaschine
wastage Ausschuß (621.9); Verschwendung, Vergeudung
waste Abfall, Abfallmaterial, Abfallstoff; Verschwendung
waste heat Abwärme
waste heat recovery Abwärmeverwertung
waste instruction Leerbefehl (681.3)
waste material Abfallmaterial
waste oil Abfallöl, Altöl
waste piece Abfallstück, Verschnittrest, Ausschnitt (621.9)
waste product Abfallprodukt
waste steam Abdampf
waste-steam turbine Abdampfturbine
waste time Verlustzeit, Zeitverlust; Leerlauf, Leerlaufzeit; Leerzeit (621.9)
waste traffic Blindverkehr
waste water Abwasser
to waste verschwenden, vergeuden, nutzlos vertun
to waste watts elektrische Energie (nutzlos) in Wärme umsetzen
wasted energy Verlustenergie, Leerlaufarbeit
wasteful verlustreich, verlustbehaftet
waster Ausschlußstück, Ausschuß; Fehlguß (621.7)
wastrel Ausschußstück
watch Uhr, Taschenuhr; Wache, Wachzeit
watchmaker Uhrmacher
watchman Wachmann, Wächter
watchman's clock Wächterkontrolluhr, Stechuhr
watchman-control system Wächterkontrollanlage
watch rate Gang (61)
watch-rate recorder Zeitwaage
water Wasser
water absorption Wasseraufnahme
water bath Wasserbad
water-boiler reactor Siedewasserreaktor

water bottle Waschflasche, Vorlage (54); Wasserflasche
water brake Wasserbremse, Wirbelbremse (621)
water circulation Wasserumlauf, Wasserkreislauf
water cistern Wassertank, Wasserbehälter
water-cleaning plant Wasserreinigungsanlage
water column Wassersäule
  water-column arrester Wasserwiderstandsableiter
water conduit Wasserleitung
water content Wassergehalt
water-cooled wassergekühlt
  water-cooled engine wassergekühlter Motor (621.4)
  water-cooled motor wassergekühlter Motor
  water-cooled reactor wassergekühlter Reaktor
  water-cooled resistance wassergekühlter Widerstand
  water-cooled transformer wassergekühlter Transformator, Umspanner mit Wasserkühlung
  water-cooled valve wassergekühlte Röhre
water cooler Wasserkühler
water cooling Wasserkühlung
water of crystallization Kristallwasser
water equivalent Wasserwert
waterfall Wasserfall; Gefälle, Gefällhöhe (621.2)
water feeder Wasserzuleitung
water gas Wassergas
  water-gas generation Wassergaserzeugung
  water-gas generator Wassergasgenerator
water gauge Wasserstand, Wasserstandsanzeiger
  water-gauge glass Wasserstandsglas
water-glass Wasserglas (54)
water hammer Wasserschlag
water head Wassersäule
water heating Warmwasserbereitung
water jacket Wassermantel
  water-jacketed mit einem Wassermantel
water jet Wasserstrahl
  water-jet arrester Wasserstrahlerder
  water-jet pump Wasserstrahlpumpe
water-level Wasserstand, Wasserspiegel, Wasserhöhe; Wasserwaage
  water-level gauge Wasserstandsrohr
  water-level indicator Wasserstandsanzeiger
water main Wasserleitung
water mains Wasserleitungsnetz
water pipe Wasserrohr
water power Wasserkraft
  water-power plant Wasserkraftwerk, Wasserkraftanlage
  water-power station Wasserkraftwerk
water pressure Wasserdruck
waterproof wasserdicht
  waterproof luminaire wasserdichte Leuchte
  water proof serving wasserdichte Imprägnierung
water pump Wasserpumpe
water-pumping station Wasserwerk
water-raising pump Wasserhaltungspumpe (622)
water rate Wassertarif, Wassergeld
water-saving Wassereinsparung; wassersparend
water screen Wasserrechen, Einlaufrechen (621.2)
water supply Wasserzufuhr, Wasserleitung, Wasserversorgung
  water-supply pipe Wasserleitungsrohr
  water-supply pump Wasserpumpe, Wasserleitungspumpe
water tank Wasserbehälter
water thrust Wasserdruck
water-tight wasserdicht
  watertight fittings wasserdichte Leuchten
  watertight gland wasserdichte Kabelstopfbuchse
  watertight machine schwallwassergeschützte Maschine
  watertight motor schwallwassergeschützter Motor
  watertight under pressure druckwasserdicht
  watertight switchgear schwallwassergeschützte Schaltgeräte (pl.)
water tower Wasserturm
water tube Wasserrohr
  water-tube boiler Wasserrohrkessel
water turbine Wasserturbine
water vapour Wasserdampf
water work(s) Wasserwerk
to water wässern, bewässern; Wasser aufnehmen; mit Wasser versorgen; Wasser abgeben, Wasser absondern; gießen (63)
to waterproof wasserdicht machen, imprägnieren
watertightness Wasserdichtigkeit
watery wässerig, wasserhaltig
watt Watt
  watt-hour Wattstunde
    watt-hour capacity Wattstundenkapazität (621.35)
    watt-hour constant Kehrwert der Zählerkonstante in kWh/Umdr.

watt-hour efficiency Wirkungsgrad (621.355)
watt-hour meter Wattstundenzähler
wattman Wagenführer, Straßenbahnfahrer
wattmeter Wirkleistungsmesser
wattmeter method Eisen-Verlustmessung, Verlustziffermessung (621.318)
watt-second constant Kehrwert der Zählerkonstante in kWs/Umdr.
wattage Leistung (621.3)
wattage rating Nennleistung
wattful Wirk-...
wattful current Wirkstrom
wattful power Wirkleistung
wattful voltage Wirkspannung
wattless wattlos, Blind-...; verlustfrei
wattless component Blindanteil, Blindkomponente
wattless component meter Blindstromzähler
wattless current Blindstrom, wattloser Strom
wattless power Blindleistung
wattman Wagenführer, Straßenbahnfahrer
wave Welle; Kurvenform
wave analysis Kurvenformuntersuchung
wave analyzer harmonischer Analysator, Frequenzanalysator
wave angle Strahlungswinkel; Einfallswinkel
wave antenna Langdrahtantenne, Beverage-Antenne
wave band Wellenbereich, Frequenzband
wave-change switch Wellenschalter, Wellenumschalter
wave clutter Seegangsechos
wave connection Wellenschaltung, Wellenwicklung (621.313)
wave crest Wellenberg
wave detector Detektor
wave diffraction Wellenbrechung
wave distortion Verzerrung der Kurvenform
wave duct Dukt, Leitschicht, Wellenleiter
wave filter Kettenleiter, Siebkette, Wellensieb (621.392)
waveform Kurvenform
waveform generator Funktionsgenerator
waveform instrument Kurvenform-Meßgerät
waveform monitor Bildkontrolloszillograf (621.397)
wavefront Wellenstirn, Wellenfront (621.3)
wave guide Wellenleiter; Hohlrohrleitung, Hohlkabel, Hohlleiter (621.396)
wave-guide attenuator Hohlleiterdämpfungsglied
wave-guide cutoff frequency Hohlleitergrenzfrequenz
wave-guide dispersion Hohlkabeldispersion
wave-guide elbow Hohlleiterknie
wave-guide filter Wellenfilter
wave-guide lens Hohlleiterlinse
wave-guide plunger Hohlleiterabstimmkolben
wave-guide shim Hohlleiterkontaktblech
wave-guide test bench Hohlrohrmeßleitung, Meßleitung
wave-guide twist Hohlleiterverdrehung
wave head Wellenstirn, Wellenkopf
wave impedance komplexer Feldwiderstand, Wellenwiderstand (des Raumes)
wave-interference error Peilfehler durch Mehrwegeausbreitung
wavelength Wellenlänge
wavelength constant Phasenbelag (621.392)
wave lens Hochfrequenzlinse
wave-like wellenähnlich, wellenartig
wave-mechanical wellenmechanisch
wave mechanics (pl.) Wellenmechanik (53)
wavemeter Wellenmesser
wave-mode conversion Wellenumwandlungen in Hohlkabeln
wave motion Wellenbewegung
wave-motion equation Wellengleichung (51)
wave propagation Wellenausbreitung
wave radiation Wellenstrahlung
wave recorder Seegangschreiber (629.12)
wave shape Wellenform; Kurvenform
wave tail Wellenrücken
wave telephony Trägerfrequenz-Telefonie
wave tilt Neigung der Wellenfront
wave train Wellenzug
wave trap Sperrkreis, Saugkreis (621.396.62)
wave velocity Wellengeschwindigkeit, Phasengeschwindigkeit
wave winding Wellenwicklung (621.313)
wave-wound wellengewickelt (621.613)
wave-wound armature Anker mit Wellenwicklung
to wave wellen
to wave-shape Impuls formen (621.397)
waved rule gewellte Linie, Wellenlinie
waving Fließfiguren (pl.) (539.3)
wavy wellig, gewellt
wax Wachs
wax-coated gewachst
wax-coated string Wachsdraht, gewachster Faden
wax formation Wachsbildung (621.315.2)
wax mould Wachsmatrize, Wachsform (621.357)
wax string Wachsfaden, gewachste „Strippe"
to wax wachsen

waxed wire Wachsdraht
waxy wachsähnlich
way Weg; Art, Weise
by way of (auf dem Wege) über ..., vermittels
by way of expedient behelfsmäßig
by way of the ground über Erde (621.3)
n-way switch n-facher Umschalter, n-fach-Umschalter
wayside Wegseite, Straßenseite, Schienenseite
wayside element schienenseitiger Geräteteil (65)
weak schwach, wenig fest
weak acid schwache Säure
weak coupling lose Kupplung
weak current schwacher Strom
weak current engineering Schwachstromtechnik
weak design schwache Konstruktion
weak echo schwaches Echo, leises Echo
weak field schwaches Feld
weak note leiser Ton
weak signal schwacher Empfang
weak spot Schwachstelle
weakly excited schwach erregt
to weaken schwächen, schwächer machen
to weaken the field das Feld schwächen (621.313)
to weaken the strength die Festigkeit verringern
weakening Schwächung, Verminderung
weakening of the field Feldschwächung
weakening of a permanent magnet Nachlassen eines Permanentmagneten, Alterung eines Dauermagneten
weakness Schwäche, Mangel an Widerstandsfähigkeit; wunder Punkt, schwacher Punkt
weakness in design wunder Punkt in der Konstruktion, Konstruktionsschwäche
wear Abnutzung, Verschleiß
wear-resisting verschleißfest, abriebfest
wear resistivity Verschleißfestigkeit, Abriebfestigkeit
wear and tear Abnutzung
to wear abnutzen, verschleißen, abreiben; tragen, anziehen; sich halten, sich tragenlassen
to wear away wegreiben, abnutzen
to wear down abnutzen
to wear off abnutzen, wegscheuern, sich abschleifen
to wear out sich ausleiern; sich ausreiben, sich ausschleifen
wearability Abnutzungsfestigkeit
wearable abnutzbar, abnutzfest
wearing depth Verschleißdicke, Verschleißtiefe
weather Wetter
weather-beaten dem Wetter ausgesetzt
weather chart Wetterkarte
weather forecast Wetter-Vorhersage
weatherproof wetterfest; spritzwassergeschützt (621.3)
weatherproof fittings Freiluftleuchten
weatherproof insulated wire wetterfeste Leitung
weatherproof machine wettergeschützte Maschine
weatherproof motor wettersicherer Motor
weatherproof switchgear Freiluftschaltgeräte (pl.)
to weather verwittern; der Witterung aussetzen; der Witterung ausgesetzt sein
weathering test Verwitterungsprüfung
to weave weben
weaver's knot Weberknoten
web Band, Streifen, Abschnitt (67); Bahn, Streifen (67); Kurbelwange (621.1), (621.4); Steg (621.9)
web cleat Stegwinkel (621.9), (624)
web plate Stegblech
webbing Kantenverstärkung (67)
Weber electro-dynamometer dynamometrisches Galvanometer
Weber = Wb (internat. Einheit des magnetischen Flusses = 10^8 M = 1 Voltsekunde)
wedge Keil; Keil, Feder, Wellenkeil
wedge contact Keilkontakt
wedge cotter Befestigungskeil, Beilagekeil
wedge coupling Keilkupplung
wedge lock Keilschloß
wedge photometer Graukeilphotometer
wedge-shaped keilförmig
to wedge keilen, aufkeilen, unterkeilen, verkeilen; sich verkeilen
to wedge between zwischenkeilen
to wedge in between dazwischenkeilen
to wedge into festkeilen in
to wedge open aufkeilen, mit Keilen öffnen
to wedge tightly festkeilen; sich festkeilen
wedging Keilverbindung, Verkeilung
Wehnelt interrupter Wehnelt-Unterbrecher
to weigh wiegen, abwägen, bewerten
to weigh out auswiegen, abwiegen
to weigh up aufwiegen, ausgleichen
to weigh upon schwer lasten auf
weighbridge Brückenwaage

weight Gewicht; Gewicht, Gewichtsstück, Gewichtsstein
weight per axle Ausdruck, Achslast
weight coefficient spezifische Leistung [auf Gewichtseinheit bezogen] (621.313)
weight of core-and-coil assembly aktives Gewicht (621.314.2)
weights and measures Maß und Gewicht (sing.)
weight rod Lastarm, Lasthebel
weight of volume Volumgewicht
to weight belasten, Gewichte auflegen, beschweren
weighted average gewogenes Mittel
weighted current bewerteter Strom
weighted noise Geräuschspannung (621.39)
weighting factor Bewertungsfaktor
weighting network Bewertungsfilter
weightless gewichtslos, masselos
Weiß magneton Weißsches Magneton
weld Schweiße, Schweißmaterial; Schweißung
weld layer Schweißlage
weld metal Schweiße (621.791)
weld puddle Schweiße
to weld schweißen
to weld on anschweißen
weldability Schweißbarkeit
weldable schweißbar, schweißfähig
welded geschweißt, Schweiß-...
welded bridge geschweißte Brücke
welded-contact rectifier Sperrschichtgleichrichter
welded frame Schweißrahmen
welded joint Schweißverbindung
welded seam geschweißte Naht, Schweißnaht
welded steel section Schweißprofil
welded tubular steel frame geschweißter Stahlrohrrahmen
welded work Schweißkonstruktion
welder Schweißer; Schweißmaschine
welder throat Schweißmaschinenmaul
welding Schweißen, Schweißvorgang
welding arc Schweißlichtbogen
welding-arc voltage Schweißspannung (621.791)
welding burner Schweißbrenner
welding circuit Schweißstromkreis
welding converter Schweißumformer
welding current Schweißstrom
welding electrode Schweißelektrode
welding fillet Schweißraupe
welding flame Schweißflamme
welding flux Schweißpaste, Flußmittel, Schweißmittel
welding furnace Schweißofen
welding generator Schweißgenerator
welding head Schweißkopf
welding jig Schweißvorrichtung
welding leads Schweißkabel, Schweißzuleitungen
welding machine Schweißmaschine
welding operator Schweißer
welding rod Schweißstab, Schweißdraht
welding seam Schweißnaht
welding set Schweißgerät
welding shop Schweißwerkstatt
welding source Schweißstromquelle
welding test Schweißversuch, Schweißprüfung
welding time Schweißzeit, Schweißdauer
welding timer Schweißtakter
welding tongs (pl.) Schweißzange
welding torch Schweißbrenner
welding transformer Schweißtransformator, Schweißumspanner
welding unit Schweißaggregat
to well quellen, fließen
to well out ausfließen, austreten
well gut, wohl
well-constructed gut gebaut
well-cooled gut gekühlt
well-defined wohldefiniert, sauber festgelegt
well-established gut definiert, feststehend, sicher feststehend
well-founded wohlbegründet
well-insulated gut isoliert
well-proportioned wohlproportioniert
well-stoppered gut verschlossen, gut zugekorkt
well-timed zeitlich gut gewählt, im rechten Augenblick angebracht
well-tried erprobt, wohlerprobt, bewährt
well-ventilated gut gelüftet
Weston normal cell Weston-Element, Normalelement
wet naß, feucht
wet arcing Regenüberschlag (621.3)
wet arcing distance Regenüberschlagsweg
wet battery nasse Batterie (621.35)
wet beaten schmierig (67)
wet cell nasses Element (621.35)
wet condenser nasser Elektrolytkondensator
wet flashover test Regenüberschlagsprüfung, Spannungsprüfung bei Regen
wet flashover voltage Regenüberschlagspannung
wet spark-over test Regenüberschlagprüfung
wet spell feuchte Zeit, Regenzeit, Regenperiode
wet steam Naßdampf
wet test Beregnungsprüfung (621.3)

wetness Nässe, nasser Zustand; Schmierigkeit (676)
to wet anfeuchten, benetzen, feucht machen, naß machen
wh = watthour Wh = Wattstunde
wharf (pl. **-ves**) Kai, Ladekai, Verladestraße, Anlegeplatz
Wheatstone automatic telegraph system Wheatstone-Telegraphie-System
Wheatstone bridge Wheatstone-Brücke
Wheatstone's rheostat Spiralschleifdraht
Wheatstone's telegraph Wheatstonescher Schnelltelegraph
wheel Rad, Scheibe; Riemenscheibe
wheel base Radstand, Achsabstand, Radgestell
wheel and disc drive Reibradgetriebe
wheel drag Bremsschuh (625)
wheel flange Spurkranz (625)
wheel gear Zahnradgetriebe
wheel hub Radnabe
wheel key Wellenkeil
wheel load Raddruck
wheel printer Typenraddrucker
wheel rim Radkranz, Spurkranz (625)
wheel stud Achsstummel
wheel tyre Radreifen; Radkranz (625.2)
to wheel rollen, fahren, auf Rädern laufen
wheeled ... -rädrig
wheelless räderlos
to whet wetzen
whetstone Wetzstein
whiffletree switch Schaltpyramide
to whine winseln, heulen, singen
to whip peitschen
to whip sidewards abrutschen, seitlich abschnappen
whip aerial Peitschenantenne
whirl Wirbel
to whirl wirbeln, herumwirbeln
whistle Pfeife; Pfeifton (621.396)
to whistle pfeifen
whistling note Pfeifton
Whit. = Whitworth Whitworth-Gewinde
white weiß
white coal weiße Kohle = Wasserkraft
white frost Rauhreif
white heat Weißglut
white hot weißglühend
white level Weißpegel (621.397)
white metal Weißmetall, Lagermetall
white noise weißes Rauschen, statistisch verteiltes Rauschen
white-noise meter Weißrauschmesser
whitesmith Klempner
whiter-than-white Weißspitze (621.397)
white wire Nulleiter
Whitworth thread Whitworth-Gewinde
whity weißlich
whiz Zischen, Gezisch
to whiz zischen
whole ganz, vollständig
whole-coiled winding Wicklung mit einer Spule je Pol
whole number ganze Zahl
W. I. = wrought iron Schmiedeeisen
wide breit, groß, weitgehend; weitab
wide-angle diffusion Weitwinkelstreuung (535)
wide-angle lighting fitting Breitstrahler
wide-angle luminaire Breitstrahler
wide-aperture direction-finder Großbasispeiler
wide band breites Frequenzband, Breitband (621.396)
wide-band amplification Breitbandverstärkung
wide-band antenna Breitbandantenne
wide band-pass filter Breitbandfilter
wide-band radio relay system Breitband-Richtfunksystem
wide-band receiver Breitbandempfänger
wide-band reception Breitbandempfang
wide-base Großbasis-...
wide field Weitwinkel (535), (77)
wide frequency range breites Frequenzband
wide interval großer Zeitraum, lange Pause
wide mesh weitmaschig
wide-open receiver Breitbandempfänger
wide range breiter Bereich, großer Bereich
wide tolerance große Toleranz
widely spaced weitabständig, licht
to widen erweitern, aufweiten; sich erweitern, breiter werden
to widen out aufweiten
Widia Widia
width Breite, Weite; Fabrikationsbreite
width of band Lichtbandbreite (621.317)
width coding Pulslängenkodierung
width control Bildbreiteregler
width of the gap Breite des Luftspaltes, Luftspalt
width-to-height ratio Seitenverhältnis, Verhältnis Breite zur Höhe
width of meshes Maschenweite
width of the passband Durchlaßbreite
width of the resonance curve Resonanzbreite, Breite der Resonanzkurve
width of span Spannweite
width of tooth Zahnbreite
width of tooth at top Zahnkopfbreite

width of the track Spurweite
width vibration Dickenschwingung
Wiedemann-Frantz constant Wiedemann-Frantzsche Zahl
Wien bridge Wiensche Brücke
Wien displacement law Wiensches Verschiebungsgesetz
Wien radiation law Wiensches Strahlungsgesetz
wig-wag signal Winksignal
Wigner release Ausheizung von Wigner-Energie
willemite Willemit, Zinksilikat
Wilson cloud chamber Wilsonsche Nebelkammer
Wimshurst machine Wimshurst-Maschine
winch Spill, Seilwinde, Seiltrommel; Windkurbel
winch drum Seiltrommel
Winchester (disk) Winchesterplatte (harte Magnetplatte)
wind Wind (55); Blasinstrumente (pl.) (681)
windage Winddruck; Belüftung, Ventilation (621.313)
windage losses Luftreibungsverluste, Ventilationsverluste
wind-driven generator Windkraftgenerator
wind force Windkraft, Windstärke
wind gauge Windmesser, Windgeschwindigkeitsmesser, Anemometer
wind indicator Windrichtungsanzeiger
wind load Windlast
windmill Windmühle, Flügelrad
windmill sock Windsack
windmill-type anemometer Flügelradanemometer
wind power Windkraft
wind-power plant Windkraftanlage
wind-power station Windkraftwerk
wind pressure Winddruck (621.315.1)
windscreen Windschutzscheibe (629.113)
windshield Windschutz; Windschutzscheibe (629.113)
windshield wiper Scheibenwischer (629.113)
wind suction Sog, Windsog
wind thrust Winddruck
wind tunnel Windkanal (533)
to wind wickeln, winden, aufwickeln, aufziehen (681); fördern (622)
to wind a coil eine Spule wickeln
to wind off abwickeln
to wind on aufwickeln auf
to wind on reels auf Spulen wickeln, aufspulen
to wind a spring eine Feder wickeln
to wind up aufziehen; fördern; abwickeln
to wind up a clock eine Uhr aufziehen
to wind with bewickeln mit
windage Winddruck; Belüftung, Ventilation (621.313)
windage losses Luftreibungsverluste, Ventilationsverluste
winder Fördermaschine
winding Wicklung (621.3); Förderung (622); Vorlauf (681.84)
winding breakdown Wicklungsdurchschlag
winding capacitance Wicklungskapazität
winding coefficient Sehnungsfaktor (621.313)
winding controller Fahrtregler (622)
winding cross section Wickelquerschnitt, Wicklungsquerschnitt
winding data Wickeldaten
winding device Wickeleinrichtung
winding diagram Wicklungsschaltbild
winding engine Fördermaschine
winding factor Sehnungsfaktor; Windungsfaktor
winding gear Fördergetriebe
winding inductance Wicklungsinduktivität
winding machine Wickelmaschine (621.3); Fördermaschine (622)
winding pitch Wicklungsteilung, Wicklungsschritt
winding plant Förderanlage
winding puncture Wicklungsdurchschlag
winding resistance Wicklungswiderstand
winding rope Förderseil
winding shield Wicklungsabdeckung, Wicklungsschild (621.313)
winding space Wickelraum
winding specifications Wickeldaten
winding speed Fördergeschwindigkeit
winding stairs (pl.) Wendeltreppe
winding temperature rise Windungserwärmung
winding type current transformer Durchsteckwandler
winding tower Förderturm (22)
winding turns Windungszahl
to windmill im Luftstrom leer mitlaufen
window Düppel (623); Fenster
window burst Düppelstoß (623)
window fan Fensterventilator
window-type current transformer Querlochstromwandler
windy windig
wing Flügel (59); Flügel, Tragflügel (629.13); Flügel, Gebäudeflügel
wing of a building Gebäudeflügel
wing nut Flügelmutter
wing pump Flügelpumpe
wing tip Flügelspitze

wing-tip tight Positionslampe (629.13)
winged nut Flügelmutter
to wink zwinkern, blinken
winking light Flackerlampe
winking signal Flackerzeichen
Winter-Eichberg motor Winter-Eichberg-Motor
to wipe wischen, abwischen
to wipe dry trockenwischen
to wipe out auswischen; auslöschen; sich aufheben, sich kompensieren
wipe contact Schleifkontakt, Reibkontakt; Wischkontakt; rollender Schnitt (621.397)
wipe-out Gitterblockierung, Zustopfung (621.396.62)
wipe-out area Durchschlagsgebiet (621.396)
wipe spark Abschaltfunke, Abreißfunke
wiping gland Lötmuffe, Lötkragen
wiping head Löschkopf
wiper Wählerkontaktarm, Schleifer
wire Draht; Telegramm
wire armour Drahtbewehrung (621.315.2)
wire-armoured cable drahtbewehrtes Kabel
wire armouring Drahtarmierung, Drahtbewehrung (621.315.2)
wire bar Drahtbarren
wire bond Drahtbondung
wire bow Drahtbügel
wire bracing Drahtverspannung
wire break Drahtbruch
wire broadcasting Drahtfunk
wire brush Drahtbürste
wire coil Drahtspule
wire communication circuit Drahtverbindung
wire cutter Drahtschere, Drahtschneider
wire edge Grat (621)
wire fence Drahtzaun
wire gauge Drahtlehre
wire gauze Drahtgaze, Drahtgewebe
wire grid Drahtgitter
wire guard Drahtkorb, Schutzkorb
wire guide Leitungshalterung
wire insulation Drahtisolation
wire lamp Drahtlampe
wire line Drahtleitung
wire mesh Maschendraht, Drahtgeflecht
wire nail Drahtstift
wire netting Drahtnetz, Drahtgewebe, Drahtgeflecht, Drahtgaze
wire nippers (pl.) Drahtzange, Drahtschneider
wire pin Drahtnadel
wire printer Nadeldrucker
wire recorder Drahtmagnetophon
wire resistance Drahtwiderstand
wire rope Drahtseil
wire screen Maschendraht, Gaze
wire-spring relay Drahtfederrelais
wire staple Drahtheftklammer
wire stripper Abisoliergerät
wire-tack Drahtstift
wire telephony and telegraphy Drahtnachrichtentechnik
wire-wound drahtgewickelt
wire-wound armature drahtgewickelter Anker, Drahtanker
wire-wound heating element Drahtheizkörper
wire-wound potentiometer Drahtpotentiometer
wire-wound resistor Drahtwiderstand (621.396)
wire-wound rotor drahtgewickelter Läufer
wire-wrap connection Wire-Wrap-Verbindung, gewickelte Verbindung, lötfreie Wickelverbindung
to wire Leitungen ziehen, Drähte verlegen, installieren; schalten, beschalten, verdrahten (621.3); mit eingerolltem Rand versehen (621.396); einrollen, randrollen (621.9); telegraphieren, drahten
to wire-brush mit der Drahtbürste reinigen, drahtbürsten
wired Draht-...
wired broadcasting Drahtfunk
wired glass Drahtglas
wired or Phantom-Oder (681.374)
wired-program control verbindungsprogrammierte Regelung
wired wireless Drahtfunk
to wireless drahtlos telegraphieren, funken
wireless drahtlos, Radio-..., Funk-... (siehe auch) radio
wireless aerial Rundfunkantenne, Radioantenne
wireless beacon Funkbake (656.7)
wireless beam Funkrichtstrahl
wireless call signal drahtloser Ruf
wireless communication drahtlose Nachrichtenübermittlung, Radioverbindung, Funkverbindung
wireless compass Funkpeiler, Funkkompaß
wireless direction finder Funkpeileinrichtung, Funkpeiler
wireless echo Funkechozeichen, Erdumlaufzeichen
wireless engineering Funktechnik, Radiotechnik
wireless fix Funk-Ortsbestimmung, Funkortung
wireless message Funktelegramm

wireless range finder Funkmeßgerät, Radar
wireless receiver Radioempfänger, Funkempfänger
wireless receiving aerial Radioempfangsantenne
wireless receiving equipment Funkempfangseinrichtung
wireless receiving set Funkempfangsgerät
wireless section Radio-Abteilung
wireless station Funkstation
wireless telegraphy Funktelegraphie, drahtlose Telegraphie
wireless telephony drahtlose Telefonie
wireless telephony and telegraphy Funknachrichtenwesen, drahtlose Nachrichtentechnik
wireless television Fernsehen, Funkfernsehen
wireless transmission Funkübertragung
wireless transmitter Funksender, drahtloser Sender
wiring Leitungsverlegung, Installation (621.3); Verdrahtung (621.396); Schaltung
wiring arrangement Installation, Leitungsanordnung
wiring capacity Schaltkapazität, Kapazität der Leitungen
wiring diagram Kabelplan, Leitungsplan; Schaltbild, Stromlaufplan
wiring harness Kabelbaum
wiring installation Leitungsinstallation
wiring point Anschlußstelle
wiring regulation Installationsvorschrift
wiring resistance Verdrahtungswiderstand, Widerstand der Zuleitungen
wiring scheme Schaltschema, Schaltplan, Leitungsplan
to withdraw zurückziehen, herausziehen, herausholen; entfernen, entnehmen; sich zurückziehen, sich entfernen
withdrawable zurückziehbar, ausziehbar
withdrawal handle Ausziehgriff
within innerhalb, in
within earshot in Hörweite
within easy reach leicht zu erreichen
within hearing in Hörweite
within limits in gewissen Grenzen
within a margin of mit einem Spielraum von, mit einer Genauigkeit von
within a mile innerhalb einer Meile
within reach in Reichweite
within sight in Sichtweite
to withstand aushalten, widerstehen
W. long. = Western longitude westl. Lg. = westliche Länge
to wobble wobbeln (621.396); eiern, unrund laufen, schlagen (621.9)
wobbler Wobbler
wobbulator Wobbelsender, Wobbler
Wollaston wire Wollastondraht
Wommelsdorf machine Wommelsdorf-Maschine
wood Holz
wood fibre slab Holzfaserplatte
wood screw Holzschraube
wooden hölzern
wooden pole Holzmast
Woodruff key Halbrundkeil
woof Kette, Aufzug (67)
woofer Tieftonlautsprecher
wool Wolle
woolly wollig
woolly reproduction dumpfe Wiedergabe (621.396.62)
word Wort; Buchstabengruppe, Gruppe, Fünfergruppe (621.396)
word articulation Wortverständlichkeit
word intelligibility Wortverständlichkeit
word length Wortlänge
words per minute Gruppen pro Minute (s. **w. p. m.**)
word-organized storage wortorganisierter Speicher
word processing Textverarbeitung
work Arbeit, Werk, Arbeitsstelle; Werkstück; Werk, Triebwerk, Uhrwerk
workbench Werkbank
workday Arbeitstag, Werktag
work done in deformation Verformungsarbeit
work function Austrittsarbeit, Austrittspotential, Austrittsspannung (621.385)
work-hardened verfestigt, spannungsvergütet
work-hardening Verfestigung
work-holding power Spannkraft (621.8)
work law Energieprinzip, Energiesatz
workmanlike fachmännisch, werkgerecht
work material Werkstoff, Arbeitsmaterial
workpiece Werkstück
work room Arbeitsraum, Werkstatt, Betriebsraum
workshop Werkstatt
workshop calculation Werkstattberechnung, Werkstattkalkulation
workshop equipment Werkstatteinrichtung
workshop installation Werkstattinstallation
workshop for repair work Reparaturwerkstatt
workshop wiring Werkstattinstallation (621.3)

work simplification Rationalisierung
workstation Arbeitsplatz
work table Aufspanntisch (621.9)
to work arbeiten; betätigen, in Gang setzen; betreiben; bearbeiten, verarbeiten
to work backward zurückrechnen, rückwärtsrechnen, rückschließen (51)
to work by betreiben mit
worked by steam mit Dampf betrieben
to work examples Beispiele ausrechnen
to work from the bar von der Stange arbeiten (621.9)
to work from rod von der Stange arbeiten
to work from scratch nach einer Skizze arbeiten
to work into a load auf eine Belastung arbeiten
to work loose losschütteln, sich losarbeiten (621.3)
to work a machine hard eine Maschine stark beanspruchen
to work metal Metall bearbeiten, Metall verarbeiten
to work out herausarbeiten; ausarbeiten
to work out as sich ergeben als, herauskommen als; ausrechnen zu (51)
to work out at sich ergeben als, herauskommen als; ausrechnen zu, ausrechnen mit (51)
to work out correctly richtig herauskommen, richtig ausrechnen
to work to n places of decimals auf n Stellen genau (be-)rechnen
to work through a high resistance über einen hohen Widerstand arbeiten, auf einen hohen Außenwiderstand arbeiten (621.385)
to work up into verarbeiten zu
workability Bearbeitbarkeit, Verwendbarkeit
workable bearbeitbar, verwendbar
worker Arbeiter
working Arbeit, Betrieb, Arbeiten
working allowance Bearbeitungszugabe (621.7)
working capacity Arbeitsvermögen, Leistungsfähigkeit
working condition betriebsfähiger Zustand
working conditions Betriebsbedingungen; Betriebszustand (sing.)
working current Betriebsstrom
working cycle Arbeitszyklus, Arbeitsspiel
working depth Arbeitstiefe, Eingriffstiefe (621.833)
working drawing Werkstattzeichnung
working face Arbeitsfläche, Bearbeitungsfläche
working flux Hauptfluß, Nutzfluß (621.313)
working frequency Arbeitsfrequenz, Betriebsfrequenz
working instructions Betriebsvorschriften
working level Betriebspegel (621.3)
working matrix Arbeitsmatrix (621.374)
working method Bearbeitungsverfahren
working operation Bearbeitungsgang
in working order in betriebsmäßigem Zustand
working platform Arbeitsbühne
working point Arbeitspunkt
working pool Arbeitsgemeinschaft
working power Betriebskraft
working pressure Betriebsdruck, Arbeitsdruck
working print Werkstattpause
working process Arbeitsprozeß, Arbeitsgang, Arbeitsverfahren
working register Arbeitsregister
working rule Arbeitsregel
working space Arbeitsraum, Raum zum Arbeiten
working speed Betriebsdrehzahl, Arbeitsgeschwindigkeit
working standard Werkstattnormal, Unternormal (621.32)
working storage Arbeitsspeicher
working stress betriebsmäßige Beanspruchung
working stroke Arbeitshub
working substance Arbeitsmedium
working table Arbeitstabelle (621.374)
working temperature Betriebstemperatur, Arbeitstemperatur
working-temperature range Betriebstemperaturbereich
working time Arbeitszeit, Betriebszeit
working tools (pl.) Handwerkzeug
working value Mindestbetätigungsspannung (621.318); Steuergröße (621.316.7)
working voltage Betriebsspannung, Arbeitsspannung, Klemmspannung bei Belastung (621.352)
works Werk (sing.)
works manager Werkdirektor
works railway Werkbahn, Werksbahn
worm Wurm (59); Schnecke (621.83)
worm conveyor Förderschnecke
worm drive Schneckenantrieb
worm-driven mit Schneckenantrieb
worm gear Schneckengetriebe; Schneckenrad
worm gearing Schneckengetriebe
worm pinion Schneckenrad, Schneckenritzel
worm reducing gear Schneckentrieb, Schneckenuntersetzung
worm wheel Schneckenrad

worm-wheel transmission Schneckenradübersetzung
worst case ungünstigster Fall, Grenzfall
wound gewickelt (s. auch **to wind**)
wound rotor Schleifringläufer, gewickelter Läufer, bewickelter Rotor
wound-rotor induction motor Asynchronmotor mit Schleifringläufer
woven gewebt (s. auch **to weave**)
woven glass Glasgewebe
woven sheet Widerstandsgewebe, Schniewindtband (621.36)
woven wire Drahtgewebe
wove Rumpelspannung
w. p. m. = words per minute Wörter pro Minute (621.394)
20 w. p. m. 100 Buchstaben pro Minute, Tempo 100
to wrap einwickeln, umhüllen, umwickeln, einhüllen
to wrap around herumwickeln, herumlegen
to wrap in einwickeln, einhüllen
to wrap round herumwickeln, herumlegen
wrapped connection gewickelte Verbindung, Wire-Wrap-Verbindung
wrapped joint Wickellötverbindung
wrapper Hülle, Umschlag, Streifband
wrapping Verpackung, Umhüllung
wrapping paper Einwickelpapier, Umschlagpapier, Einschlagpapier, Packpapier
wrapping test Wickelprobe (620)
wrench Schraubenschlüssel; Windeisen, Kluppe
to wring wringen, auswringen
to wring gauge blocks together Endmaßeansprengen
to wring out auswringen (648)
wringing fit Haftsitz, Würgesitz
wrinkle Falte, Runzel
to wrinkle Falten werfen, faltig werden, runzlig werden, sich runzeln
wrinkly runzlig
to write schreiben
to write down niederschreiben; anschreiben (51)
to write down an equation eine Gleichung anschreiben, eine Gleichung ansetzen
to write in einschreiben (681.3)
to write out ausgeben, schreiben (681.3)
write head Schreibkopf
write-read head Lese-Schreibkopf
writing Schreiben, Urkunde
writing beam Schreibstrahl
writing head Schreibkopf
writing magnet Schreibmagnet
writing materials Schreibmaterialien; Schreibmaterial (sing.)
writing signal Einschreibsignal (681.3)
writing telegraph Drucktelegraph
writing telegraph system Drucktelegraphiersystem
wrought bearbeitet, geschmiedet, geknetet
wrought aluminium alloy Aluminiumknetlegierung
wrought goods Schmiedestücke
wrought iron Schmiedeeisen, Puddeleisen
wt = weight Gew. = Gewicht
wye connection Sternschaltung

X

X-intercept Schnittpunkt mit der X-Achse
X-radiation Röntgenstrahlung
X-ray Röntgenstrahl
X-ray apparatus Röntgenapparat
X-ray crystal analysis Röntgen-Feinstrukturprüfung
X-ray crystallography Röntgen-Feinstrukturuntersuchung
X-ray diffraction Röntgenstrahlenbeugung
X-ray diffraction pattern Röntgenbeugungsbild
X-ray dosimeter Röntgendosismesser
X-ray emitter Röntgenstrahlquelle
X-ray equipment Röntgeneinrichtung, Röntgenanlage
X-ray examination Röntgenprüfung (621.7)
X-ray film Röntgenfilm
X-ray photograph Röntgenaufnahme
X-ray picture Röntgenbild
X-ray spectrograph Röntgenspektrogramm
X-ray spectrometer Röntgenspektrometer
X-ray spectroscope Röntgenspektroskop
X-ray spectrum Röntgenspektrum
X-ray transformer Röntgentransformator
X-ray tube Röntgenröhre
X-ray video amplifier tube Röntgenbildverstärker
to X-ray durchleuchten, röntgenprüfen
X-rays (pl.) Röntgenstrahlung
"X" stopper Krachtöter (621.396.62)
X-tube Röntgenröhre
XY plotter Plotter, X-Y-Kurvenschreiber; Koordinatenschreiber

XY recorder Koordinatenschreiber, X-Y-Schreiber
xenon Xenon
xerographic printer xerographischer Drucker
xylene Xylol
xylole Xylol
xylonite Zelluloid

Y

Y Y, Stern-... (621.3)
Y-connected circuit Sternschaltung
Y-connection Sternschaltung
"Y"-delta starter Stern-Dreieck-Anlasser, Stern-Dreieck-Schalter
Y-joint Abzweigmuffe, T-Muffe (621.315.2)
"Y" voltage Strangspannung, Phasenspannung
Y-winding sterngeschaltete Wicklung
Yagi array Yagi-Antenne
yard Yard (**1 yard = 3 feet** = 0,924 m); Hof
yardmeasure Metermaß
yardstick Zollstock, Maßstab
Yale lock Yaleschloß, Sicherheitsschloß
to yaw schieben (629.13)
yellow metal Gelbmetall, Buntmetall
yield Ausbeute, Ausbringen, Ertrag, Nutzertrag
yield curve Ausbeutekurve
yield of neutrons Neutronenausbeute
yield point Streckgrenze, Fließgrenze
to yield ergeben; liefern, hergeben; ergeben (51); elastisch nachgeben, federn (539)
to yield neutrons Neutronen liefern
to yield a result ein Ergebnis liefern
to yield to nachgeben
YIG = yttrium iron garnet Yttrium-Eisen-Granat
yoke Joch, Magnetjoch
yoke lamination Jochblech
Young's modulus Elastizitätsmodul

Z

Zamboni's dry pile Zambonische Säule
zapping Zerstörung (eines Bauelementes)
Zener breakdown Zener-Durchbruch
Zener diode Zenerdiode, Z-Diode
Zener voltage Zenerspannung
zenith Zenit
zenithal zenital
Zeppelin antenna Zeppelin-Antenne
zero Null
zero-access storage zugriffszeitfreie Speicherung
zero adjuster Nullstelleinrichtung (621.317)
zero adjusting Nullpunkteinstellung, Nullpunkteinstell-...
zero adjustment Nulleinstellung, Nullstellung
zero beat Schwebungsnull
zero burden (of current transformer) äußerer Kurzschluß (beim Stromwandler) (621.317)
zero burden (of potential transformer) Leerlauf (beim Spannungswandler)
zero-center galvanometer Galvanometer mit Nullpunkt in der Mitte
zero-center instrument Meßinstrument mit Nullpunkt in der Skalenmitte
zero compression Nullunterdrückung (beim Speichern) (681.3)
zero condition Nullzustand (681.3)
zero cross Nulldurchgang
zero crossing detector Nulldetektor
zero current Stromnulldurchgang
zero error Abweichung vom Nullpunkt, Nullpunktfehler
zero-field resistive residual voltage Widerstands-Restspannung (eines Halleffekt-Bauelementes)
zero flag Nullkennzeichnung
zero frequency Nullfrequenz (621.395.4)
zero level Nullniveau, Nullpegel
zero mark Nullstrich
zero marker Z-Markierungsfeuer
zero method Nullmethode
zero of operating range Einsatzpunkt (621.316.7)
zero phase-sequence relay Erdschlußrelais
zero point Nullpunkt
zero-point anode Sternpunktsanode
zero potential Nullpotential
zero-power reactor Nulleistungsreaktor, Forschungsreaktor
zero-reference level Bezugspegel Null
zero response Empfangsnullstelle (621.396.67)
zero root Nullösung, Nullwurzel (51)
zero r. p. m. Stillstand
zero-sequence Null-... (621.3)
zero-sequence reactance Blindwiderstand für das Nullsystem
zero-sequence reactive power Blindleistungdes Nullsystems

zero-sequence symmetrical component symmetrische Nullkomponente
zero setting Einstellung auf Null, Nulleinstellung
zero shift Nullpunktswanderung (621.317)
zero-speed monitor Stillstandswächter
zero-speed resonance plugging switch Resonanzbremswächter
zero suppression Nullunterdrückung
zero subcarrier chromaticity Farbart bei Farbträgernull
zero-time reference Nullbezugszeit
zero'th nullter
zero-type dynamometer Nullwattmeter
zero value Nullwert, Nullstelle
zero voltage Nullspannung, Spannung Null, Spannungsnulldurchgang
to zero auf Null setzen
zig-zag Zickzack
zig-zag air gap leakage Zickzack-Streuung (621.313)
zig-zag connection Zickzackschaltung
zig-zag flux Zickzackstreufluß
zig-zag path Zickzackweg
zig-zag riveting Zickzacknietung
to zigzag im Zickzack fahren; zacken (629.12)
zinc Zink
zinc alloy die casting Zinkspritzguß
zinc blende Zinkblende
zinc electrode Zinkelektrode
zinc-lead accumulator Zink-Blei-Sammler
zinc ore Zinkerz
Zinc sulphate Zinksulphat
Zinc sulphide Zinksulphid
zinc white Zinkweiß, Zinkoxid
zinc wire Zinkdraht
to zinc verzinken
zip-fastener Reißverschluß
zirkonium Zirkonium
zirkonium lamp Zirkonlampe
zirkonium window Zirkonfenster (621.386)
Z-iron Z-Eisen
Z-shape Z-Form
Z-shaped Z-förmig
zone Zone, Gebiet, Bereich
zone blanking Sektoraustastung (621.396.96)
zone of luminosity Leuchtzone
zone of silence tote Zone (621.371)
zoner Zoner; Verzoner (621.395.34)
zoning verzonen (621.395)
zoom lense Variooptik, Gummilinse
to zoom hochziehen, hochreißen (629.13)

Anhang

Umrechnungszahlen veralteter Meßeinheiten technischer und physikalischer Größen in internationale Einheiten

Sowohl Großbritannien als auch die USA haben 1975/1976 durch Gesetz das internationale Si-Einheitensystem (Système International d'Unité) eingeführt, das auf dem metrischen System beruht. Im täglichen Gebrauch werden jedoch die veralteten, nicht gesetzlichen Einheiten auch weiterhin verwendet (z. B. *Barrel* als Maßeinheit für Erdöl), und natürlich findet sich in der älteren Literatur eine Fülle überholter Einheiten. Sie können mit Hilfe der nachfolgenden Tabellen in internationale Si-Einheiten umgerechnet werden.

Bei Angaben der englischen und US-Einheiten ist die englische Schreibweise – Dezimalpunkt – benutzt; das Komma dient zur Stellenunterteilung. Bei den deutschen Einheiten dient das Komma zur Dezimaltrennung.

Übersicht

I. Längenmaße

1 A.U. = 1 Angstroem Unit: 1 A.E. = 1 Ångström = 10^{-8} cm

1 mil = .001 inch = 25,4 µm = 0,00254 cm

1 point = 1/6 line = 1/72 inch = .01389 inch = 0,3528 mm (veraltet, nur noch in der Drucktechnik benutzt, hier aber vom deutschen typographischen Punkt verschieden)

1 line = 1/40 inch = 0,635 mm

1 inch = 1 in. = 1000 mil (meist unterteilt in 1/2, 1/4, 1/8, 1/16, 1/32, 1/64, 1/128; in US ist die dezimale Unterteilung gebräuchlicher)

= 2,540 051 cm (US)
= 2,539 978 cm (Br.)
= 2,5400 (für alle technischen Zwecke)

1 link = 1 li. = 7.29 in. = .66 ft. = 20,1168 cm

1 span = 9 in. = 22,86 cm

1 hand = 4 inches = 10,16 cm (US)

1 foot = 1 ft. = 12 in.

= 30,479 921 cm (Br.)
= 30,480 061 cm (US)
= 30,480 cm (für alle technischen Zwecke)

1 yard = 1 yd. = 3 ft. = 36 in.
= 0,9144 m (für alle technischen Zwecke)

1 fathom = 6 ft. = 2 yd. = 1,8288 m

1 pole = 1 rod = 1 lug = (1 perch) = 5.5 yd.
= 25 link = 5,0292 m

1 chain = 4 rd. = 100 link = 66 ft. = 20,1168 m
(1 Gunter's chain = 1 surveyor's chain = 1 ch.) (Vermessung)

1 chain = 100 ft. = 30,48 m (= 1 Ramsden's chain)
= 1 engineer's chain (Maschinenbau)

1 furlong = 1 fur. = 220 yd. = 40 rd. = 1/8 mile = 201,168 m

1 mile = 1 statute mile = 1 st.m. = 8 fur. = 1760 yd.
= 1,6093472 km (US)
= 1,6093426 km (Br.)

1 nautical mile = 1 geographical mile = (1 knot)
= 6,080 ft. (rd. eine Bogenminute des Erdmeridians)
= 1,85318 km (Br.)
= 6,080.27 ft. (mittlere Bogenminute des Erdmeridians)
= 1,85325 km (US)
= 1,852 km (international nautical mile)

1 Admiralty mile = 6,086.5 ft. (Bogenminute des Erdäquators
= 1,855 162 km (Br.)

1 league = 3 miles = 3 st.m. = 4,828 km

II. Flächenmaße

1 barn = 10^{-24} cm²

1 circular mil = Fläche eines Kreises von 1 mil Durchmesser
= $\pi/4$ mil² = 0,000506 700 mm²

1 square mil = 10^{-6} sq.in. = 0,000645 mm² = 645 µm²

1 circular inch = Fläche eines Kreises von 1 Zoll Durchmesser
= $\pi/4$ inch² = 5,067 cm²

1 square inch = 1 sq.in. = 1 Quadratzoll = 6,4516 cm²

1 square link = 1 sq.li. = 62.76 sq.in. = 404,7 cm²

1 square foot = 1 sq.ft. = 144 sq.in. = 0,0929 m²

1 square yard = 1 sq.yd. = 9 sq.ft. = 1 Quadratyard = 0,8361 m²

1 square = 100 sq.ft. = 9,29 m² (im Bauwesen)

1 square pole = 1 square rod = 1 sq.rd. = 1 perch = 30.25 sq.yd. = 625 sq.li. = 25,293 m²

1 square chain = 1 sq.ch. = 16 sq.rd. = 404,7 m²

1 rood = 40 sq.rd. = 40 perches = 1011,7 m² = 10,117 a (1 Rute)

1 acre = 1 A = 4 rood = 10 sq.ch. = 4840 sq.yd. = 43,560 sq.ft. = 0,4047 ha
= 1,6 (preußische) Morgen

1 square mile = 1sq.mi. = 1 section (US) = 640 A = 3,097,600 sq.yd. = 2,59 km²

1 township (US) = 6 · 6 square miles = 93,24 km²

III. Volumeneinheiten

A. Raummaße

1 cubic mil = 10^{-9} cu.in. = $16{,}387 \cdot 10^{-6}$ mm³

1 cubic inch = 1 cu.in.
= 16,387 021 cm³ (Br.)
= 16,387 162 cm³ (US)
= 16,387 cm³ (für alle technischen Zwecke)

1 cubic foot = 1 cu.ft. = 1728 cu.in.
= 28,31677 dm³ (Br.)
= 28,31702 dm³ (US.)
= 28,317 dm³ (für alle technischen Zwecke)

1 cubic yard = 1 cu.yd. = 27 cu.ft. = 46,656 cu.in.
= 0,76455287 m³ (Br.)
= 0,76455945 m³ (US.)
= 0,7646 m³ (für alle technischen Zwecke)

1 shipping ton = 40 cu.ft. = 1,133 m³ (Schiffsfracht)

1 chaldron = 1,309 m³(Br.)

1 register ton = 100 cu.ft. = 2,832 m³ (Schiffbau)

1 cord = 128 cu.ft. = 3,625 m³ (US)

1 cubic fathom = 216 cu.ft. = 6,116 m³

1 acrefoot = 1 A · 1 ft. = 43,560 cu.ft. = 1233,5 m³

1 cubic rod = 1 cu.rod = 127,199 m³

B. Hohlmaße für Flüssigkeiten

	Br.	US
1 minim = 1 min.	59,19 mm³	61,61 mm³
1 fluid drachm = 1 fl.dr. = 60 min.	3,552 cm³	3,697 cm³
1 fluid ounce = 1fl.oz. = 8 fl.dr. = 480 min	28,4131 cm³	29,574 cm³
1 gallon = 1 gal = 8 pt.	1 Imp. gallon 160 fl.oz. 4,545 963 l	1 US. gallon 128 fl.oz. 3,785 435 l
1 gill	142 cm³	118,3 cm³
1 pint = 1 pt. = 4 gills	568,2 cm³	473,2 cm³
1 quart = 2 pt.	1,137 l	0,946 l
1 pottle = 2 quarts	2,273 l	–

1 barrel (Erdöl) = 35 Imp.petr.gal.
= 42 US.petr.gal. = 158,7581

C. Schüttmaße (Hohlmaße für trockene Stoffe)

	Br.	US
1 dry pint = 1 pt.	–	0,551 l
1 dry quart = 1 qt.	–	1,101 l
1 peck = 1 pk.	2 Imp.gal. 9,092 l	8 qt 8,810 l
1 bushel = 1 bu. = 4 pk.	36,37 l	35,24 l
1 quarter = 8 bu.	290,94 l	–

IV. Einheiten für Massen

1 grain = 1 gr. = 64,798919 mg
1 dram = 1 dr.av. = 27,334 gr. = 1,772 g
1 ounce = 1 oz.av. = 16 dr.av. = 473,5 gr. = 28,349 g
1 pound = 1 lb. = (1 lb.av.) = 16 oz.av. = 7000 gr. = 0,453 592 kg
1 stone = 14 lbs. = 6,35 kg
1 quarter (US) = 25 lbs. = 11,34 kg
1 quarter (Br.) = 28 lbs. = 12,70 kg
1 central = 1 short hundredweight = 1 cwt.sh. = 100 lbs. = 45,36 kg
1 hundredweight (Br.) = 1 cwt. = 112 lbs. = 50,80 kg
1 short ton (US und Schiffahrt) = 1 sh.ton = 20 cwt.sh. = 2000 lbs. = 907,18 kg
1 long ton = 1 l.ton = 20 cwt. = 2240 lbs. = 1016 kg = 1,016 t

V. Temperatureinheiten

Festpunkte	Celsius C	Kelvin K	Fahrenheit F	Rankine Rank	
Absoluter Nullpunkt	–273,15	0	–459,69	0	Grad
Eispunkt	0	273,15	32	491,682	Grad
Normale Körpertemperatur	37	–	98,5	–	Grad
Dampfpunkt	100	373,15	212	671,67	Grad

Bei Temperatur differenzen gilt:

1 °C = 1 K = 9/5 °F = 9/5 °Rank

1 °F = 1 °Rank = 5/9 °C = 5/9 K

Bei Temperatur angaben gilt:

x °C = (x + 273,15) K = (1,8 · x + 32) °F = (1,8 · x + 491,7) °Rank

y °F = 5/9 (y – 32) °C = (y + 459,7) °Rank = 5/9 (y + 459,7) K

VI. Einheiten der Wärmemenge

1 British Thermal Unit ist die Wärmemenge, die erforderlich ist, um 1lb. Wasser um 1 °F zu erhitzen.

1 British Thermal Unit = 1 B. Th. U = 1 BTU = 1,055056 kJ (J = Joule)

1 Therm = 10^5 BTU = 1055056 kJ

1 Centrigrade heat unit = 1 CHU ist die Wärmemenge, die erforderlich ist, um 1 lb. Wasser um 1 °C zu erhitzen. 1 CHU = 9/5 B. Th. U. = 1,899 kJ

VII. Zusammengesetzte Einheiten

A. Geschwindigkeiten

1 ft./min. = 0,508m/s

1 mile/hour = 1,609 km/h

Beschleunigungen

1 ft./sec^2 = 0,305 m/s^2

Erdbeschleunigung: g = 32.2 ft./s^2
= 9,81 m/s^2

B. Gewichte

1 lb./cu.ft. = 0,01602 g/cm^3

1 lb./cu.in. = 27,68 g/cm^3

1 lb./gal. (Br.) = 0,0998 kg/l

1 lb./gal. (US) = 0,1198 kg/l

1 long ton/cu.yd. = 1329 kg/m^3

1 grain/gallon = 14,26 mg/l

C. Kräfte

1 poundal = 1 lb.ft/s^2 = 13 825 dyn = 0,13825 N (= Newton)
1 gr.wt. = 0,6355 Nm
1 lb.wt. = 4,4483 N

D. Drücke und Spannungen

1 lb./sq.in. = 1 psi. = 0,68948 N/cm^2
1 lb./sq.ft. = 47,876 N/m^2
1 long ton/sq.in. = 1544,5 N/cm^2
1 long ton/sq.ft. = 10,728 N/cm^2
1 long ton/sq.yd. = 1,196 N/cm^2
1 inch of water = 1 Zoll Wassersäule = 2,54 cm WS = 2,49 mbar
1 foot of water = 1 Fuß Wassersäule = 30,5 cm WS = 29,891 mbar
1 inch of mercury = 1 Zoll Quecksilbersäule = 25,4 Torr = 33,864 mbar

E. Drehmomente

1 in.lb. = 112,8 Nm (Newton-Meter)
1 ft.lb. = 1356,3 Nm
1 in.ton = 253 · 10^3 Nm
1 ft.ton = 0,300 m t

F. Energie und Arbeit

1 footpoundal = 1 ft · 1 poundal = 1 lb.ft./s^2 = 0,04214 J
1 ft.lb. = 1 ft. · 1 lb. (wt.) = 1,33805 J
1 British Thermal Unit = 1 B.Th.U. = 1 BTU = 0,293 Wh = 1,055056 kJ
1 HP hour = 0,746 kWh = 1,0139 PSh = 2,6841 · 10^6 J

G. Leistung

1 ft.lb./s = 0,00182 HP = 1,353 W
1 HP = 550 lb.ft./s = 33,000 ft.lb./min = 1,0139 PS = 0,746 kW
1 BTU/s = 1,05506 kW
1 Therm/hour = 29,31 kW

H. Elektrische Einheiten

In der Elektrotechnik wird überwiegend mit den Einheiten des metrischen Maßsystems gerechnet. Einige Umrechnungszahlen für gelegentlich verwendete Einheiten sind:

1 millivolt/ft. = 3,28 mV/m
1 amp/sq.in. = 0,155 A/cm^2 = 1,55 mA/mm^2
1 Ohm · circ.mil/ft. = 1,68 · 10^{-3} Ohm · mm^2/m
1 volt/in. = 0,394 V/cm
1 amp/in. = 0,394 A/cm
1 Maxwell/sq. in. = 0,155 · 10^{-4} T
1 mho = 1/Ohm = 1 S = 1 Siemens
1 bel = 1,15 N = 1,15 Neper
1 db = 1 decibel = 0,115 N
1 db/mile = 0,0717 N/km

VIII. Lichttechnische Einheiten

Lichtstärke luminous intensity

1 candle-power = 1 candela = 1 Candela = 1 cd
1 Hefner candle = 0,903 cd

Lichtstrom luminous flux

1 Lumen = 1 lm = 1 cd sr

Beleuchtungsstärke illumination

1 Lux = 1 lx = 1 lm/m^2
1 footcandle = 1 lm/ft^2 = 10,7639 lx

Leuchtdichte brightness

1 Candela/Meter^2 = 1 cd/m^2
1 lambert = 1 la = 3183,10 cd/m^2
1 foot-lambert = 1 ft la = 3,426 cd/m^2
1 candela per square inch = 1 cd/sq in = 1550,0 cd/m^2
1 candela per square foot = 1 cd/sq ft = 10,76 cd/m^2
1 stilb = 1 sb = 1 cd/m^2
1 apostilb = 1 asb = $(1/\pi) \cdot (cd/m^2)$

IX. Wasserhärten

1 D.G. = 1 deutscher Härtegrad = 10 mg CaO in 1 l Wasser
1 English Degree of Hardness = 1 engl. Härtegrad = 10 mg $CaCO_3$ in 0,7 l Wasser
1 p.p.m. = 1 part per million = 1 US-Härtegrad = 1 mg Ca in 1 l Wasser
Es entsprechen: 1 D.G. = 1,25 engl. Härte = 17,8 p.p.m.